The New Public Health

Second Edition

The New Public Health

Second Edition

Theodore H. Tulchinsky, MD, MPH
Braun School of Public Health and Community Medicine
Hebrew University-Hadassah, Ein Karem
Jerusalem, Israel

Elena A. Varavikova, MD, MPH, PhD
I. M. Sechenov Moscow Medical Academy
Moscow, Russian Federation

With Foreword by

John Last, MD, DPH, FRACP, FFPH, FRCPC, FACPM, FACE
MD (Hon) Edinburgh and Uppsala
Professor Emeritus, University of Ottawa
Ottawa, Ontario, Canada

AMSTERDAM • BOSTON • HEIDELBERG • LONDON • NEW YORK • OXFORD • PARIS
SAN DIEGO • SAN FRANCISCO • SINGAPORE • SYDNEY • TOKYO
Academic Press is an Imprint of Elsevier

Elsevier Academic Press
30 Corporate Drive, Suite 400, Burlington, MA 01803, USA
525 B Street, Suite 1900, San Diego, California 92101-4495, USA
84 Theobald's Road, London WC1X 8RR, UK

This book is printed on acid-free paper. ∞

Library of Congress Cataloging-in-Publication Data
Tulchinsky, Theodore H.
  The new public health / authors, Theodore H. Tulchinsky, Elena A. Varavikova ; with
foreword by John Last. – 2nd ed.
      p. ; cm.
  Includes bibliographical references and index.
  ISBN 978-0-12-370890-8 (hardcover : alk. paper) 1. Public health. I. Varavikova,
Elena. II. Title.
  [DNLM: 1. Public Health–trends. WA 100 T917n 2009]
  RA425.T77 2009
  362.1–dc22

                         2008027586

British Library Cataloguing in Publication Data
A catalogue record for this book is available from the British Library

ISBN 13: 978-0-12-370890-8

For all information on all Elsevier Academic Press publications
visit our Web site at www.elsevierdirect.com

Printed in the United States of America
08   09   10   9   8   7   6   5   4   3   2   1

Working together to grow
libraries in developing countries

www.elsevier.com | www.bookaid.org | www.sabre.org

ELSEVIER   BOOK AID
            International   Sabre Foundation

Dedication

We wish to dedicate this book to our families, professional colleagues, and friends who sustained and supported us in the hard labor for five years on the first edition and two years on the second edition through the ups and downs of such an endeavor.

Theodore Tulchinsky: I would like to dedicate this book to my family; to my wife, Joan, and our children Daniel, Joel, and Karen, and their children; to my late parents Ann and Harry Tulchinsky; and to my sisters Norma and Ruth and brother Gerald and their families.

Elena Varavikova: I wish to especially thank Natalya, Tatyana, Inna, and Harvey for their support and inspiration.

We jointly dedicate this book to all those who sustained us and contributed to its development and its translations into Russian and many languages of Eastern Europe and Central Asia, including Albanian, Bulgarian, Georgian, Moldovan, Macedonian, Mongolian, Romanian, and Uzbek. We make a special dedication to Milton and Ruth Roemer, whose contributions to public health in the United States and globally were enormous, and whose friendship and encouragement with this book were very special to its preparation.

Milton and Ruth Roemer

This special dedication to Ruth and Milton Roemer is in recognition of their leadership in public health and health care organization, and also for their warm support and friendship while the first edition of this textbook was still in its early stages. Milton wrote the Foreword for the first edition, and Ruth helped us to connect with Academic Press of San Diego, our publisher.

Milton Roemer was a world-renowned scholar in many areas of public health, including international health, primary care, rural health, and health care organization. Roemer's many notable achievements included studies showing that in an insured population, a hospital bed built is a hospital bed filled—a finding that contributed to the enactment of certificate of need legislation and comprehensive health planning. This finding was so robust that it bears his name: Roemer's Law, as discussed in this book.

Ruth Roemer was a role model and leader in many public health issues such as tobacco control, mental health, fluoridation of community water, and abortion rights. Her work from 1993 onward contributed to the WHO Framework Convention on Tobacco Control of 2003. Milton died in Los Angeles in 2001 at age 84, and Ruth in 2005, at age 89.

MILTON I. ROEMER

Milton I. Roemer, MD, MPH, was one of the most influential public health figures of the twentieth century; his work over more than six decades benefited the lives of millions of people all over the world. He worked in 71 countries and published 32 books and 430 articles on the social aspects of health services. During his career he consistently proved prescient in foreseeing health trends and possibilities, including development of HMOs, promoting the role of ambulatory care, and documenting the need for national health insurance for the entire U.S. population. He advocated development of doctoral training in health administration to prepare students for leadership in public health practice, and established an endowed fellowship to support students in the program.

Roemer received his master's degree in sociology from Cornell University in 1939, his MD from New York University in 1940, and a public health degree from the University of Michigan in 1943. He served at all levels of health administration—county, state, national, and international. As a medical officer of the New Jersey State Health Department, he supervised 92 venereal disease clinics. During World War II, he served as a member of the commissioned corps of the U.S. Public Health Service.

His 1948 book *Rural Health and Medical Care*, with F. D. Mott, was the first to systematically analyze rural health care needs and services in the United States. As county health officer of Monongalia County, West Virginia, he introduced public health innovations, including pioneering a cancer detection clinic for the mining community. Roemer was among the first to advocate integration of public health and medical care. In 1953, he was appointed director of medical and hospital services of the Saskatchewan, Canada, Department of Public Health, North America's first social insurance program for hospital care. After teaching at Yale and Cornell, Roemer came to the University of California–Los Angeles (UCLA)

School of Public Health in 1962, where he taught health administration, conducted research and continued to publish for 38 years. The capstone of Roemer's many publications was his two-volume work, *National Health Systems of the World*, a monumental comparative analysis of national health systems of countries of the world.

RUTH ROEMER

Ruth Roemer, JD, was a pioneer in public health law and advocacy whose career spanned more than 50 years. Roemer made lasting contributions in areas that included reproductive health services, environmental health, tobacco control, and health services organization. She remained an influential figure in public health well into her 80s, initiating what later became the International Framework Convention on Tobacco Control, adopted by the World Health Organization in 2003.

Born Ruth Joy Rosenbaum in Hartford, Connecticut, Roemer graduated from Cornell Law School and began her career as a labor lawyer. She switched to health law in the 1960s after participating in a landmark study of the laws governing admission to mental hospitals in the State of New York. She joined the UCLA School of Public Health faculty in 1962, and promptly became principal organizer and vice president of the California Committee on Therapeutic Abortion, to provide public education and leadership in reform of California's century-old abortion law. The group spearheaded abortion law reform in California in 1967, 6 years before the U.S. Supreme Court's decision in *Roe v. Wade*. In ensuing years she made her mark with analyses of the functions, education, and regulation of health personnel and as an ardent campaigner for fluoridation of public water supplies in California and worldwide. In an early 1970s study she examined the gap between the advanced technology to roll back pollution and its application to protect people's health and living conditions. She pioneered with a seminal work on tobacco control with a world review of tobacco control legislation for the World Health Organization, first published in 1982. She was active in international tobacco control conferences, and helped prepare a document that led to WHO's first international convention on tobacco control and the treaty adopted in 2003. Ruth Roemer taught health law, ethics, and policy at UCLA.

Contents

Foreword   xvii
Preface to the Second Edition   xix
Acknowledgments   xxi
Introduction to the First Edition   xxiii

1. A History of Public Health   1

Introduction   1
Prehistoric Societies   1
The Ancient World   2
The Early Medieval Period (Fifth to
    Tenth Centuries CE)   4
The Late Medieval Period (Eleventh to
    Fifteenth Centuries)   4
The Renaissance (1500–1750)   6
Enlightenment, Science, and Revolution (1750–1830)   7
    Eighteenth-Century Reforms   8
    Applied Epidemiology   8
    Jenner and Vaccination   9
Foundations of Health Statistics and Epidemiology   9
Social Reform and the Sanitary
    Movement (1830–1875)   11
    Snow on Cholera   13
    Germ Versus Miasma Theories   14
Hospital Reform   15
The Bacteriologic Revolution   16
    Pasteur, Cohn, Koch, and Lister   16
    Vector-Borne Disease   17
Microbiology and Immunology   18
    Poliomyelitis   18
    Advances in Treatment of Infectious Diseases   19
Maternal and Child Health   19
Nutrition in Public Health   20
Military Medicine   21
Internationalization of Health   23
The Epidemiologic Transition   23
Achievements of Public Health in the
    Twentieth Century   25
Creating and Managing Health Systems   26
Summary   26
Historical Markers   28
Electronic Resources   31

Recommended Readings   31
Bibliography   32

2. Expanding the Concept of
Public Health   33

Introduction   33
Concepts of Public Health   33
Evolution of Public Health   35
Health and Disease   36
    Host–Agent–Environment Paradigm   37
The Natural History of Disease   38
Society and Health   39
Modes of Prevention   41
    Health Promotion   41
    Primary Prevention   42
    Secondary Prevention   43
    Tertiary Prevention   43
Demographic and Epidemiologic Transition   43
Interdependence of Health Services   44
Defining Public Health   44
    Social Medicine and Community Health   45
    Social Hygiene Eugenics, and Corruption
        of Public Health Concepts   46
    Medical Ecology   46
    Community-Oriented Primary Care   46
World Health Organization's Definition of Health   47
    Alma-Ata: Health for All   47
Selective Primary Care   49
The Risk Approach   50
The Case for Action   50
Political Economy and Health   52
Health and Development   53
Health Systems: The Case for Reform   54
Advocacy and Consumerism   55
    Professional Advocacy and Resistance   55
    Consumerism   57
The Health Field Concept   58
The Value of Medical Care in Public Health   58
Health Targets   59
    United States Health Targets   59
International Health Targets   61
    European Health Targets   61

United Kingdom Health Targets	61
Individual and Community Participation in Health	62
Ottawa Charter for Health Promotion	62
State and Community Models of Health Promotion	63
Healthy Cities/Towns/Municipalities	63
Human Ecology and Health Promotion	65
Defining Public Health Standards	65
Integrative Approaches to Public Health	66
The Future of Public Health	67
The New Public Health	68
Summary	68
Electronic Resources	70
Recommended Readings	71
Bibliography	71

3. Measuring and Evaluating the Health of a Population — 73

Introduction	73
Demography	74
Fertility	75
Population Pyramid	76
Life Expectancy	76
Epidemiology	77
Social Epidemiology	80
Epidemiology in Building Health Policy	82
Definitions and Methods of Epidemiology	82
Rates and Ratios	82
Measures of Morbidity	84
Prevalence Rates	85
Measures of Mortality	85
Social Classification	87
Sentinel Events	87
The Burden of Disease	88
Years of Potential Life Lost	88
Qualitative Measures of Morbidity and Mortality	88
Measurement	89
Research and Survey Methods	89
Variables	90
The Null Hypothesis	90
Confounders	91
Sampling	91
Randomization	91
Normal Distribution	92
Standardization of Rates	92
Direct Method of Standardization	93
Indirect Method of Standardization	93
Potential Errors in Measurement	94
Reliability	94
Validity	95
Screening for Disease	95
Epidemiologic Studies	96

Observational Studies	96
Experimental Epidemiology	99
Establishing Causal Relationships	100
Notification of Diseases	100
Special Registries and Reporting Systems	101
Disease Classification	103
Hospital Discharge Information	103
Health Information Systems (Informatics)	105
WHO European Region Health for All Database	106
Surveillance, Reporting, and Publication	107
Assessing the Health of the Individual	107
Assessment of Population Health	108
Defining the Population	109
Socioeconomic Status	109
Nutrition	111
Environment and Occupation	111
Health Care Financing and Organization	111
Health Care Resources	111
Utilization of Services	112
Health Care Outcomes	112
Quality of Care	113
Self-Assessment of Health	114
Costs and Benefits	114
Effects of Intervention	114
Qualitative Methods	114
Summary — From Information to Knowledge to Policy	116
Electronic Resources	117
Recommended Readings	117
Bibliography	118

4. Communicable Diseases — 121

Introduction	121
Public Health and the Control of Communicable Disease	122
The Nature of Communicable Disease	122
Host–Agent–Environment Triad	123
Classifications of Communicable Diseases	124
Modes of Transmission of Disease	125
Immunity	125
Surveillance	126
Health Care–Associated Infections	128
Endemic and Epidemic Disease	128
Epidemic Investigation	129
Control of Communicable Diseases	130
Treatment	130
Methods of Prevention	130
Vaccine-Preventable Diseases	131
Immunization Coverage	133
Vaccine-Preventable Diseases	144

Essentials of an Immunization Program 144
  Regulation of Vaccines 146
  Vaccine Development 146
Control/Eradication of Infectious
    Diseases 147
  Smallpox 147
  Eradication of Poliomyelitis 148
  Other Candidates for Eradication 149
  Future Candidates for Eradication 149
Vector-Borne Diseases 156
  Malaria 156
  Rickettsial Infections 157
  Arboviruses (Arthropod-Borne
    Viral Diseases) 157
  Lyme Disease 160
Parasitic Diseases 161
  Echinococcosis 161
  Tapeworm 162
  Onchocerciasis 162
  Dracunculiasis 162
  Schistosomiasis 163
  Leishmaniasis 163
  Trypanosomiasis 164
  Other Parasitic Diseases 164
Legionnaire's Disease 165
Leprosy 165
Trachoma 165
Sexually Transmitted Infections 166
  Syphilis 166
  Gonorrhea 166
  Other Sexually Transmitted Infections 166
  Control of Sexually Transmitted Infections 167
  HIV/AIDS 168
Diarrheal Diseases 170
  Salmonella 170
  Shigella 170
  Escherichia coli 171
  Cholera 171
  Viral Gastroenteritis 171
  Parasitic Gastroenteritis 172
  A Program Approach to
    Diarrheal Disease Control 172
Acute Respiratory Infections 173
Inequalities in Control of
    Communicable Diseases 174
Communicable Disease Control
    in the New Public Health 175
Summary 176
Electronic Resources 176
Recommended Readings 177
Bibliography 177

5. Noncommunicable Conditions 181

Introduction 181

The Rise of Chronic Disease 182
The Burden of Chronic Conditions 183
Risk Factors and Causation of Chronic Conditions 184
Chronic Manifestations of Infectious Diseases 187
  Cardiovascular Diseases 188
Chronic Lung Disease 195
  Asthma 195
  Chronic Obstructive Pulmonary Disease 196
  Restrictive Lung Diseases 196
  Occupational Lung Diseases 196
Diabetes Mellitus 197
  Prevention of Diabetes and Its Complications 199
End-Stage Renal Disease 200
Cancer 200
  Prevention of Cancer 203
Chronic Liver Disease 204
Disabling Conditions 205
  Arthritis and Musculoskeletal Disorders 205
  Osteoporosis 205
  Degenerative Osteoarthritis 206
  Rheumatoid Arthritis and Gout 206
  Low Back Syndromes 206
Neurologic Disorders 206
  Alzheimer's Disease 206
  Parkinson's Disease 206
  Multiple Sclerosis 206
  Epilepsy or Seizures 207
  Brain and Spinal Cord Injury 207
Visual Disorders 207
Hearing Disorders 208
Trauma, Violence, and Injury 208
  Motor Vehicle Accidents 210
  Domestic Violence 211
  Suicide and Suicide Attempts 212
  Homicide 212
  Prevention of Violence 212
Chronic Conditions and the New Public Health 213
Summary 213
Electronic Resources 213
Recommended Readings 214
Bibliography 215

6. Family Health 217

Introduction 217
The Family Unit 218
Maternal Health 219
  Fertility 219
  Public Health Concerns of Fertility 220
  Family Planning 220
  Maternal Mortality and Morbidity 221
Pregnancy Care 223
  High-Risk Pregnancy 224
Labor and Delivery 226
  Safe Motherhood Initiatives 227

Care of the Newborn   227
Care in the Puerperium   229
Genetic and Birth Disorders   230
  Rhesus Hemolytic Disease of
    the Newborn   231
  Neural Tube Defects   231
  Cerebral Palsy   232
  Intellectual Disability   232
  Down Syndrome   233
  Cystic Fibrosis   233
  Sickle-Cell Disease   233
  Thalassemia   233
  Phenylketonuria   234
  Congenital Hypothyroidism   235
  Fetal Alcohol Syndrome   235
  Tay-Sachs Disease   235
  G6PD   235
  Familial Mediterranean Fever   235
Infant and Child Health   235
  Fetal and Infant Mortality   236
Infancy Care and Feeding   237
Anticipatory Counseling   238
Documentation, Records,
    and Monitoring   238
The Preschooler (Ages 1–5 Years)   240
School and Adolescent Health   241
  Smoking   242
  Alcohol Abuse   243
  Drug Abuse   243
  Sexual Risk Behavior   244
  Dietary Risk Behavior   244
  Physical Activity   245
  Violence and Gang Behavior   245
Adult Health   245
Women's Health   246
Men's Health   248
Health of Older Adults   249
  Health Maintenance for Older Adults   250
Summary   254
Electronic Resources   255
Recommended Readings   256
Bibliography   256

7. Special Community Health Needs   259

Introduction   259
Mental Health   260
  Historical Changes in Methods of Treatment   262
  Mental Health Epidemiology   263
  Mental Disorder Syndromes   265
  Controversies in Mental Health Policies   269
  Community-Oriented Mental Health   270
  Prevention and Health Promotion   271
Mental Disability   271

Oral Health   272
  Fluoridation   273
  Periodontal Disease   273
  Dental Care   274
  Oral Cancer   274
Physical Disability and Rehabilitation   274
Special Group Health Needs   276
Gay and Lesbian Health   276
Native Peoples' Health   277
Prisoners' Health   278
Migrant Population Health   280
Homeless Population Health   280
Refugee Health   281
Military Medicine   282
Health in Disasters   283
Summary   286
Electronic Resources   287
Recommended Readings   287
Bibliography   288

8. Nutrition and Food Safety   291

Introduction   291
Development of Nutrition in
    Public Health   292
Nutrition in a Global Context   292
Nutrition and Infection   293
Functions of Food   294
  Composition of the Human Body   294
Human Nutritional Requirements   294
  Carbohydrates   295
  Proteins   295
  Fats and Oils   295
  Vitamins   296
  Minerals and Trace Elements   296
Growth   296
Measuring Body Mass   298
Recommended Dietary Intakes   299
Disorders of Undernutrition   300
  Underweight: Protein-Energy Malnutrition (PEM)   303
  Failure to Thrive   304
  Marasmus   304
  Kwashiorkor   304
  Vitamin A Deficiency   304
  Vitamin D Deficiency (Rickets and Osteomalacia)   306
  Vitamin C Deficiency   308
  Vitamin K Deficiency (Hemorrhagic
    Disease of the Newborn)   308
  Vitamin B Deficiencies   308
  Iron-Deficiency Anemia   310
  Iodine-Deficiency Diseases   310
  Osteoporosis   311
  Eating Disorders   311
Diseases of Overnutrition   312

Overweight/Obesity 312
Cardiovascular Diseases 313
Cancer 313
Nutrition in Pregnancy and Lactation 314
Promoting Healthy Diets and Lifestyles 315
Dietary Guidelines 315
Vitamin and Mineral Enrichment of Basic
    Foods 315
    Controversy in Food Enrichment 316
Food and Nutrition Policy 317
    The Evolution of a Federal Role 317
    Nutrition Issues in Development Policies 318
    The Role of the Private Sector and NGOs 319
    The Role of Health Providers 319
Nutrition Monitoring and Evaluation 320
    Standard Reference Populations 320
    Measuring Deviation from the
        Reference Population 322
Food Quality and Safety 322
Nutrition and the New Public Health 326
Summary 327
Electronic Resources 328
Recommended Readings 328
Bibliography 329

9. Environmental and Occupational
   Health 333

ENVIRONMENTAL HEALTH 333
Introduction 333
Environmental Issues 334
Geographic and Environmental Epidemiology 335
Environmental Targets 335
Global Environmental Change 335
    Climate Change 337
    Environmental Impact on Health Burden
        of Disease 337
Community Water Supplies 338
    Waterborne Diseases 339
Sewage Collection and Treatment 340
Toxins 343
    Toxic Effects on Fertility 343
    Toxic Effects of Lead in the Environment 344
Agricultural and Environmental Hazards 346
Air Pollution 346
    The External Environment 346
    Methyl Tertiary Butyl Ether 348
Indoor Pollution 349
    Radon Gas 349
    Outdoor–Indoor Pollutants 349
    Biological Pollutants 350
    Sick Building Syndrome 350
Hazardous or Toxic Wastes 350
    Minimata Disease 351
    Toxic Waste Management 351

Radiation 351
    Ionizing Radiation 351
    Non-Ionizing Radiation 352
Environmental Impact 352
    Emergency Events Involving Hazardous
        Substances 352
    Man-Made Disasters, War, Terrorism 354
    Preventing and Managing Environmental
        Emergencies 354
Environmental Health Organization 355
OCCUPATIONAL HEALTH 356
Introduction 356
Development of Occupational Health 358
The Health of Workers 359
The Burden of Occupational Morbidity
    and Mortality 359
    Occupational Health Priorities in the
        United States 360
International Issues in Occupational Health 360
National and Management
        Responsibilities 361
    Standards and Monitoring 362
Occupational Health Targets 363
Toxicity at the Workplace and in the Environment 363
    Lead 364
    Asbestos 364
    Silica 365
    Cotton Dust (Byssinosis) 366
    Vinyl Chloride 366
    Agent Orange 366
Workplace Violence 366
Occupational Health in Clinical Practice 367
Inspecting the Place of Work 368
Risk Assessment 368
Preventing Disasters in the Workplace 368
Occupation and the New Public Health 369
Summary 369
Electronic Resources 370
Recommended Readings 371
Bibliography — Water Quality and Waterborne
    Disease 372
Bibliography — Occupational and Environmental
    Health 373

10. Organization of Public Health
    Systems 375

Introduction 375
Government and Health of the Nation 376
    Federal and Unitary States 377
    Checks and Balances in Health Authority 377
    Government and the Individual 378
Functions of Public Health 378
    Regulatory Functions of Public Health
        Agencies 379

Methods of Providing or Assuring
  Services — Direct or Indirect?   379
Nongovernmental Roles in Health   380
Disasters and Public Health Preparedness   381
Medical Practice and Public Health   382
Incentives and Regulation   383
  Promotion of Research and Teaching   385
  Accreditation and Quality Regulation   386
National Government Public
  Health Services   386
State Government Public Health Services   388
Local Health Authorities   389
Monitoring Health Status   392
National Health Targets   393
Universal Health Coverage and the
  New Public Health   393
Hospitals in the New Public Health   395
  Hospital Classification   396
  Supply of Hospital Beds   396
  The Changing Role of the Hospital   398
  Regulation of Hospitals   399
The Uninsured as a Public Health
  Challenge   399
Summary   400
Electronic Resources   401
Recommended Readings   401
Bibliography   402

11. Measuring Costs: The Economics
    of Health   405

Introduction   405
Economic Issues of Health Systems   406
  Investing in Health   407
  National Health Care Spending   408
Basic Concepts in Health Economics   408
Supply, Need, Demand, and
    Utilization of Health Services   408
  Normative Needs   409
  Felt Need   409
  Expressed Need   410
  Comparative Need   410
  Demand   410
  Supply   411
  Grossman's Demand Model   411
Competition in Health Care   412
Elasticities of Demand   413
Measuring Costs   415
Economic Measures of Health Status   415
Cost–Effectiveness Analysis   416
Cost–Benefit Analysis   417
Basic Assessment Scheme for
    Intervention Costs and Consequences   419
The Value of Human Life   419

Health Financing — The Macroeconomic Level   421
Costs of Illness   424
  Costs and Variations in Medical Practice   424
  Cost Containment   425
Medical and Hospital
    Care — Microeconomics   426
  Payment for Doctor's Services   426
  Payment for Comprehensive Care   426
Health Maintenance and Managed Care
    Organizations   427
District Health Systems   428
Paying for Hospital Care   429
Capital Costs   430
Hospital Supply, Utilization, and Costs   430
Modified Market Forces   432
Economics and the New Public Health   432
Summary   434
Electronic Resources   434
Recommended Readings   435
Bibliography   435

12. Planning and Managing
    Health Systems   439

Introduction   439
Health Policy and Planning as Context   440
The Elements of Organizations   442
Scientific Management   442
Bureaucratic Pyramidal Organizations   442
Organizations as Energy Systems   443
  Cybernetics and Management   444
Target-Oriented Management   445
  Operations Research   445
  Management by Objectives   445
Human Relations Management   445
  The Hawthorne Effect   446
  Maslow's Hierarchy of Needs   446
  Theory X–Theory Y   447
Network Organization   448
Total Quality Management   449
Changing Human Behavior   450
Empowerment   451
Strategic Management of Health Systems   451
Health System Organization Models   452
  Functional Model   453
  Corporate Model   453
  Matrix Model   454
Skills for Management   454
The Chief Executive Officer of
    Health Organizations   455
Community Participation   456
Integration — Lateral and Vertical   456
Norms and Performance Indicators   457
Health Promotion and Advocacy   457

Philanthropy and Volunteerism   457
New Organizational Models   458
New Projects and Their Evaluation   458
Systems Approach and
     the New Public Health   460
Summary   461
Electronic Resources   461
Recommended Readings   461
Bibliography   462

13. National Health Systems   465

Introduction   466
Health Systems in Developed Countries   467
Evolution of Health Systems   467
The United States   469
Federal Health Initiatives   469
Medicare and Medicaid   471
The Changing Health Care Environment   472
Health Information   474
Health Targets   475
Social Inequities   476
The Dilemma of the Uninsured   477
Summary   478
Canada   478
Reform Pressures and Initiatives   479
Provincial Health Reforms   480
Health Status   480
Summary   481
The United Kingdom   482
The National Health Service   482
Structural Reforms of the
      National Health Service   482
Reforms Since 1990   482
Social Inequities   484
Health Promotion   484
Health Reforms   484
Primary Care Trusts   485
Summary   486
The Nordic Countries   486
Sweden   487
Denmark   489
Norway   489
Finland   489
Western Europe   491
Germany   491
The Netherlands   494
Russia   495
The Soviet Model   496
Epidemiologic Transition   497
Post-Soviet Reform   499
Future Prospects   501
Summary   501
Israel   502

Origins of the Israeli Health System   502
Health Resources and Expenditures   503
Health Reforms   503
Mental Health   504
Healthy Israel 2020   504
Summary   505
Health Systems in Developing
     Countries   506
Federal Republic of Nigeria   507
Summary   511
Latin America and the Caribbean   512
Colombia   512
Asia   514
China   515
Japan   519
Comparing National Health Systems   520
Economic Issues in National
      Health Systems   521
Reforming National Health Systems   523
Summary   525
Electronic Resources   526
Recommended Readings   528
Bibliography   529

14. Human Resources for Health Care   535

Introduction   535
Overview of Human Resources   536
Human Resources Planning   538
Supply and Demand   539
Basic Medical Education   541
Postgraduate Medical Training   543
Specialization and Family Practice   544
Training in Preventive Medicine   545
Nursing Education   545
In-Service and Continuing Education   547
Accreditation of Medical Educational or
      Training Facilities   547
The Range of Health Disciplines   548
Licensure and Supervision   548
Constraints on the Health Care Provider   549
New Health Professions   550
Nurse Practitioners   550
Physician Assistants   551
Feldshers   552
Community Health Workers   552
Alternative Medicine   553
Changing the Balance   554
Education for Public Health and
      Health Management   554
Health Policy and Management of
      Human Resources   560
Summary   561
Electronic Resources   561

Recommended Readings   562
Bibliography   562

15. Health Technology, Quality, Law, and Ethics   565

Introduction   565
Innovation, Regulation, and Quality
    Control   566
Appropriate Health Technology   568
Health Technology Assessment   570
    Technology Assessment in Hospitals   571
    Technology Assessment in Prevention and
        Health Promotion   572
    Technology Assessment in National
        Health Systems   573
Dissemination of Technology   574
Diffusion of Technology   576
Quality Assurance   577
    Adverse Events and Negligence   577
    Licensure and Certification   578
    Health Facility Accreditation   578
    Peer Review   580
    Algorithms and Clinical Guidelines   581
Organization of Care   584
    Diagnosis-Related Groups   584
    Managed Care   585
Performance Indicators   585
Consumerism and Quality   585
The Public Interest   586
Total Quality Management   587
Public Health Law   588
    Environmental Health   591
    Public Health Law Reform   591
Ethical Issues in Public Health   591
    Individual and Community Rights   593
    Ethics in Public Health Research   595
    Ethics in Patient Care   596
    Ethics in Public Health   596
    Human Experimentation   597
    Sanctity of Life Versus Euthanasia   598
    The Imperative to Act or Not Act in
        Public Health   598

Summary   599
Electronic Resources   600
Recommended Readings   601
Bibliography   602

16. Globalization of Health   605

Introduction   605
The Global Health Situation   607
Priorities in Global Health   608
    Poverty–Illness–Population–Environment 610
    Child Health   613
    Maternal Health   613
    Population Growth   614
    Malnutrition   616
    The Fight Against HIV/AIDS and Other
        Communicable Diseases   617
    Chronic Disease   618
    Disaster Management   619
    Environment   620
    Global Partnership for Development   620
Development and Health   620
Organization for International Health   621
The World Health Organization   621
    The United Nations Children's
        Fund (UNICEF)   624
Nongovernmental Organizations   624
The World Bank   625
Trends in Global Health   626
Emerging Infectious Disease Threats   627
Expanding National Health Capacity   630
Global Health and the New Public Health   631
Summary   632
Electronic Resources   634
Recommended Readings   634
Bibliography   635
Publications and Journals   636

Glossary of Terms   637
Index   659

Foreword

The foreword to the first edition of this book (2000) was written by the eminent authority on public health, the late Milton Roemer. It is fitting that the authors have dedicated this edition to him and to his late wife Ruth Roemer, also a distinguished contributor to public health. It is an honor to be invited to follow where Milton Roemer led, and to contribute the Foreword to this second edition of the book.

I read the first edition with admiration and great interest. Admiration because the two authors had done such a splendid job in covering all the important bases so thoroughly; and with great interest because I had by then edited several editions of a large, comprehensive textbook of public health and preventive medicine; had written one of my own; and had been teaching, professing, and mentoring on many aspects of the subject for several decades.

This book is based on rich experience in public health, and on courses offered at the Hadassah-Hebrew University School of Public Health, and is published in both English and Russian, ready for use in graduate programs in Russia and elsewhere in the Slavic world as well as throughout the English-speaking world. The new edition again justifies the book's title, *The New Public Health*. There is up-to-date information on emerging public health problems, including SARS and avian influenza and its human variant, as well as the latest advice on old problems of hygiene and sanitation. It is an excellent text for learners with some uniquely valuable chapters and discussions of topics rarely covered in textbooks and monographs on public health sciences, the practice of public health in the field, and the organization and administration of public health services. It is gratifying to read a book that includes accounts of the history of public health, and that has chapters on economic, legal, and ethical aspects of public health, as well as very comprehensive and current accounts of all the widely recognized domains of the important set of disciplines and skills that make up this essential component of all organized human communities. The links to relevant websites that are scattered throughout the text and in the reference lists are another valuable feature. It is useful to have a website for the book which includes a guide for teachers of general survey overview courses in public health, especially as that website provides links to case studies of classics of public health with a discussion of their current relevance.

As mentioned, the first edition of this book has been translated into Russian, Bulgarian, Romanian, Moldovan, Uzbek, and Mongolian and will soon be translated into Albanian and Georgian. It is used as a standard text throughout the countries of the former Soviet Union, in newly developing schools of public health, and in traditional hygiene training centers. This is important because of the great challenge of very high mortality rates from preventable chronic diseases. Change in the epidemiology and demography of these countries necessitates a reorientation of public health thinking toward health promotion and changes in management of the health system to address chronic diseases as well as communicable disease. This book serves this purpose very well indeed.

Many schools of public health focus on specialized and compartmentalized courses each with their own textbooks and monographs, but students often do not get a general overview of the field such as this book presents.

Therefore it is understandable that this book has been widely accepted and has become a standard course text. Graduate students in master's and similar courses in all the public health sciences, and their teachers and mentors have very good reason to be grateful to Drs. Tulchinsky and Varavikova.

John Last, MD, DPH, FRACP, FFPH, FRCPC,
FACPM, FACE
MD (Hon) Edinburgh and Uppsala
Emeritus Professor of Epidemiology and
Community Medicine
University of Ottawa
Ottawa, Canada

Preface to the Second Edition

The first edition of this book was published in Russian in 1999, followed by publication in 2000 in English. Since then it has been translated and published in Bulgarian, Uzbek, Macedonian, Mongolian, Moldovan, and Romanian and currently is being translated to the Albanian and Georgian languages. It has been used widely in schools of public health in the United States and Europe and in the translations in new schools of public health which have developed in those countries.

The phrase *The New Public Health*, not original to our first edition, was described in WHO publications and several texts in the United Kingdom and in Australia, but in more limited contexts than our intent. This book was originally prepared for publication in Russia where the idea of a New Public Health was and is sorely needed for the health of the population during a critical period of transition with high rates of preventable mortality. This book is primarily about ideas and experience of public health, both classical and new in recognizing the centrality of policy and management of health systems in the New Public Health.

A Teaching Guide based on this textbook was prepared and circulated, along with a model curriculum to all the new schools of public health using this textbook. We believe this has been helpful to the faculties and students of the new schools in Russia (Moscow, Chelyabinsk, and Tver), Kazakhstan, Uzbekistan, Albania, Macedonia, and Moldova, Mongolia, and others.

We are gratified that this book has had wide impact in many former socialist countries. It provided many veteran teachers with fresh material in their own language from the wide world of science and practice which were largely inaccessible in former days. It provided students in new schools of public health in those countries with access to modern material in public health when they lacked even the most basic of libraries and Internet connection. Working with best public health professionals of many countries we have learned a lot, through the teaching, work on the book translation, discussions of major concepts, current concerns, and country-specific issues.

The English edition is widely used in the United States and elsewhere as a general introductory text in schools of public health, which was our intent and hope. The second edition is meant to follow in this path.

Our reviewers from the United States and Europe, provided by our publisher, gave us extremely valuable feedback and professional opinion that inspired our work on the second edition.

A general textbook is needed for the many students coming into public health with undergraduate education insufficient to generally orient to public health studies; and that includes those coming from medicine and nursing as well as those from the social sciences.

As an introductory text, its merit, we hope, is in its wall-to-wall coverage, with clear recognition of the need of specialized textbooks in the individual fields of study making up public health such as epidemiology and research methods, qualitative research, and the social sciences, economics, and others. Our experience suggests that a standard textbook is especially important in development of a new school of public health and will provide overview of the main concepts and core subject fields anywhere.

Our intent was to emphasize "the broad view," or "the big picture" with sufficient real-life examples to make it realistic and sufficient content to provide the student to the holistic picture and to introduce the many specialized aspects of public health. We have tried to maintain that approach in the second edition with old and new examples and references, and a stress on the historical process, as a basis for thinking of the future.

We hope this book helps teach students ideas and the concepts of public health along with the evolving technologies, experience, and ethical professional standards of this profession. There are so many controversies in public health that the new entry-level person may be confused without a sense of direction and methods to form judgments on these issues. We hope that we achieve a forward-looking approach as well as a review of past and present successes.

A quote from the *New England Journal of Medicine* of March 28, 2008, perhaps illustrates this objective:

Many important lifesaving advances have been made by taking one crucial step forward at a time. Consider the worldwide effort to eradicate polio ... an essentially transformative idea — that had occurred centuries earlier in the mind of Edward Jenner, who observed that milkmaids who had been exposed to cowpox became immune to smallpox, a far more deadly disease.

How can we capture such transformative innovation in order to address the problems in global health? First, it is clear that innovation does not take place only in the United States or Western Europe ... new ideas can be fleeting.... Innovation frequently arises from the lessons of repeated failure.... opportunities to capture novel approaches that can transform a field ... can come from anywhere.

Each year, 9.7 million children die before 5 years of age, 4 million of them within the first month of life and the vast majority of them in the poorest countries in the world.... Most of these deaths can be averted with the application of existing tools, but in some cases only new ideas will provide practical and effective solutions.[1]

Finally, we endorse the statement attributed to former U.S. Surgeon General C. Everett Koop that public health is needed by everyone all the time, while medical care is needed by many people some of the time. Of course both are needed and a world facing preventable morbidity and premature death of millions annually from preventable disease means that training people in the ideas and ways of public health are essential to civil societies both local and global. We hope this book will help graduate and undergraduate students and practitioners of public health to understand this, to be better able to contribute to reducing this sad and unnecessary loss of health and of life.

Public health has proven its effectiveness in saving lives and improving quality of life. The context associated with this process is continuously expanding but needs a conceptual basis, which we call the New Public Health.

T. H. Tulchinsky, MD, MPH
E. A. Varavikova, MD, MPH, PhD
May 16, 2008

---

[1]Yamada, T. 2008. In Search of New Ideas for Global Health. *New England Journal of Medicine*, 358:1324–1325.

Acknowledgments

The second edition of this textbook owes its genesis to the many people who contributed to the first edition published in 2000, including friends and colleagues: the Open Society Institute of New York, the American Joint Distribution Committee, the State University of New York at Albany School of Public Health, the University of California–Los Angeles (UCLA) School of Public Health, the Braun School of Public Health, Hebrew University-Hadassah, Jerusalem, the I. M. Sechenov Moscow Medical Academy, to those who translated this book into many languages of South East Europe and the former Soviet Union, to those who reviewed the first edition and provided many ideas for inclusion in the second edition, and to many supporters mentioned in the acknowledgments of that edition.

We wish to especially thank Joan Bickford (Winnipeg, Canada) whose support and assistance in important stages of preparation of this edition were of very great importance for us. Many others whose contributions to updating and revising this book have been enormously helpful include Igor M. Denisov (Moscow Medical Academy), Leon Epstein (Braun SPH), Becca Feldman (Braun SPH), Gary Ginsberg (Braun SPH), Soumik Kalita (Braun SPH), Maria Koleilat (UCLA SPH), Natalia Koroleva (London School of Hygiene and Tropical Medicine), Meira Weinstein (Braun SPH), Denitsa Radeva (Braun SPH), Thaddeus Davíd Ritch de Herrera (Braun SPH), Edward Salakhov (Semashko Institute, Moscow), Suzanne Wilnerminc (Braun SPH), and Yael Wolfe (Braun SPH).

Our warmest thanks go to Mara Conner, Renske Van Dijk, Julie Ochs, and the editorial staff of Life and Biomedical Sciences, Elsevier/Academic Press, who have worked with us closely and supportively through the process of preparing this book.

Of course, we could not have developed this second edition of our book without the encouragement and very constructive input of family, friends, colleagues, and students. We are deeply grateful for their support and contributions to the international flavor of the book. The common goal is to improve access to public health knowledge and to contribute to the development of public health in many countries globally. The final responsibility is, of course, with the authors.

T. H. Tulchinsky, MD, MPH
E. A. Varavikova, MD, MPH, PhD

Introduction to the First Edition

The idea for this book evolved from many years of teaching the principles of health organization to students of public health from Africa, Latin America, the Caribbean, Asia, the United States, eastern Europe, and Russia. It also emerged from the practice of public health in a wide variety of settings, including Canada, Israel, the West Bank and Gaza, the United States, Colombia, Azerbaijan, and Russia. While working together on a review of the health situation in Russia during 1992–1995 for the World Bank, we concluded that there was a need for a new textbook of public health bringing current thinking in the broad field to new students and veteran practitioners with an international orientation.

This book was originally prepared for publication in Russia where the idea that a New Public Health is sorely needed for the health of the population during a critical period of transition. This publication in the United States, we hope, will be relevant worldwide. Health systems everywhere are undergoing reform toward a population-based approach in addition to traditional individual care. Both the Russian and English editions are intended to reach people entering the field of public health as undergraduate students or entry-level graduate students who want or need an overview of the broad scope of the field. We also hope to reach physicians, nurses, managers, policymakers, and many others in the health field to define a new approach, linking the fields of public health and clinical services.

The term *New* in no way deprecates the known and trusted elements of the *Old* in public health. On the contrary, the New is a rediscovery and stands on the shoulders of the Old. The Greeks had gods of therapy and prevention in health, Aesculapius and Hygeia. The Mosaic law in health is based on the principles of *Pikuah Nefesh*, or sanctity of life, and *Tikkun Olam*, literally repairing the world in the sense of correcting the faults in human society. The ancient Greek and Mosaic traditions together with subsequent social organizational philosophies of health as a right and scientific advances provide the basis for the scientific and ethical approaches of the New Public Health.

Traditionally, *public health* has been defined as health of populations and communities. However, the NPH focuses on health of the individual, just as for any other medical practitioner or health care worker, working both directly with individuals and indirectly through communities and populations. The indirect approach is to reduce the risk factors in the environment, whether physical or social, while the direct approach works with the individual patient or client as does a doctor treating the patient.

As medical science evolved, especially since the mid-twentieth century, the effectiveness of prevention and treatment services organized separately, sometimes with conflicting values, and mostly with separate financing and administrative modalities. The New Public Health is a synthesis of classical public health based on the experience of the past several centuries of applied public health with the biomedical, clinical, and social sciences; technology; management; and experience of health issues as they have developed and continue to develop.

The NPH includes all possible activities known to be useful and effective in promoting health and in the prevention, treatment, and rehabilitation of diseases for the individual, the community, and the population as a whole. It includes preventive medicine, environmental and community health, as well as personal health services. It provides standards relevant to any country whether developed or developing, but application of the specifics depends on the particular health problems and economic status of each country, or region within a country.

The NPH links traditional hallmarks of public health, such as sanitation, communicable disease control, maternal and child health, and epidemiology, with clinical services, health systems management, and health promotion. It recognizes that the health of the individual and the community is directly and indirectly affected by social and economic factors. An understanding of these concepts is essential to the design of effective health care interventions to prevent the occurrence of diseases or their complications. The NPH takes into account the realities of resource allocation and economic issues and priorities in health. It recognizes that resources for health care are limited even in the wealthiest societies, so that choices must be made as to the balance of programs and services provided, often made under the imperative of cost constraint and substitution of one type of service for another. Above all, it stresses that society and the individual have rights and responsibilities in promoting and maintaining health through direct services and through healthy environmental and community health services.

Improving the health of the individual requires both the direct and indirect approaches. Examples cover a wide range of issues from control of infectious diseases, especially vaccine-preventable diseases, to nutrition, such as fortification of food and prevention of iron-deficiency anemia. Many issues in the New Public Health deal with both the individual and the community, including prevention of waterborne disease, assurance of access to medical care through health insurance, organization of home and chronic care, district health systems, prevention of asbestos-related disease, birth defects or thalassemia, development of new health professions such as community health workers, and many others.

The major components of public health each have records of great achievements and failures, but they are part of modern civilization and the desires of all other societies to emulate. While no one is exempt from death ultimately, no mother or father anywhere wants to lose a child, or a parent, especially from a preventable disease or condition. Yet, while we have learned, through a combination of science, political philosophy, and experience, that well-planned interventions can greatly reduce the risk of that happening, we have also failed to implement that knowledge as widely as we have implemented treatment services.

The New Public Health is not so much a concept as it is a philosophy which endeavors to broaden the older understanding of public health so that, for example, it includes the health of the individual in addition to the health of populations, and seeks to address such contemporary health issues as are concerned with equitable access to health services, the environment, political governance and social and economic development. It seeks to put health in the development framework to ensure that health is protected in public policy. Above all, the New Public Health is concerned with action. It is concerned with finding a blueprint to address many of the burning issues of our time, but also with identifying implementable strategies in the endeavor to solve these problems.[1]

The New Public Health incorporates a wide range of interventions in the physical and social environment, health behavior, and biomedical methods, along with health care organization and financing. The possibility to change not only individual lives but also the quality of life in communities draws health professionals from all walks of life who are dedicated to improving the public's health through a combination of their various disciplines. The social advocacy role with the application of up-to-date biomedical and social sciences makes this a challenging and rewarding field. Yet, public health is also the art of the possible. We cannot solve all problems of poverty and injustice, but we can improve survival and quality of life, step by step, one acre at a time, to achieve wondrous miracles as we enter the new century.

The New Public Health defines measurable targets of improved health of the individual and the community. It addresses both the social and physical environment as well as the personal services that address individual health needs. The NPH brings together aspects of public health that are community-oriented and personal care that is individual-oriented. One can no longer be separated from the other if we are to address the health needs of society in the twenty-first century.

T. H. Tulchinsky, MD, MPH
E. A. Varavikova, MD, MPH, PhD
June 2000

[1]Ncayiyana, D., Goldstein, G., Goon, E., Yach, D. 1995. *New Public Health and WHO's Ninth General Program of Work: A Discussion Paper.* Geneva: World Health Organization.

Chapter 1

A History of Public Health

Introduction
Prehistoric Societies
The Ancient World
The Early Medieval Period (Fifth to Tenth Centuries CE)
The Late Medieval Period (Eleventh to Fifteenth Centuries)
The Renaissance (1500–1750)
Enlightenment, Science, and Revolution (1750–1830)
    Eighteenth-Century Reforms
    Applied Epidemiology
    Jenner and Vaccination
Foundations of Health Statistics and Epidemiology
Social Reform and the Sanitary Movement (1830–1875)
    Snow on Cholera
    Germ Versus Miasma Theories
Hospital Reform
The Bacteriologic Revolution
    Pasteur, Cohn, Koch, and Lister
    Vector-Borne Disease
Microbiology and Immunology
    Poliomyelitis
    Advances in Treatment of Infectious Diseases
Maternal and Child Health
Nutrition in Public Health
Military Medicine
Internationalization of Health
The Epidemiologic Transition
Achievements of Public Health in the Twentieth Century
Creating and Managing Health Systems
Summary
Historical Markers
Electronic Resources
Recommended Readings
Bibliography

INTRODUCTION

History provides a perspective to develop an understanding of health problems of communities and how to cope with them. We see through the eyes of the past how societies conceptualized and dealt with disease. All societies must face the realities of disease and death, and develop concepts and methods to manage them. These coping strategies form part of a worldview associated with a set of cultural or scientific beliefs, which in turn help to determine the curative and preventive approaches to health.

The history of public health is a story of the search for effective means of securing health and preventing disease in the population. Epidemic and endemic infectious disease stimulated thought and innovation in disease prevention on a pragmatic basis, often before the causation was established scientifically. The prevention of disease in populations revolves around defining diseases, measuring their occurrence, and seeking effective interventions.

Public health evolved with trial and error and with expanding scientific medical knowledge, often stimulated by war and natural disasters. The need for organized health services grew as part of the development of community life, and in particular, urbanization. Religious and societal beliefs influenced approaches to explaining and attempting to control communicable disease by sanitation, town planning, and provision of medical care. Where religious and social systems repressed scientific investigation and spread of knowledge, they were capable of inhibiting development of public health.

Modern society still faces the ancient scourges of malaria, cholera, and plague, as well as the more prominent killers: obesity, cardiovascular disease, mental depression, trauma, and cancer. The advent of AIDS, SARS, avian influenza, and emerging drug-resistant microorganisms forces us to seek new ways of preventing their potentially serious consequences to society. Diseases, natural disasters, and man-made catastrophes including war, terrorism, and genocide are always threats to human civilization. The evolution of public health continues; pathogens change, as do the environment and the host. In order to face challenges ahead, it is important to have an understanding of the past.

PREHISTORIC SOCIETIES

Earth is considered to be 4.5 billion years old, with the earliest stone tools dating from 2.5 million years BCE representing the presence of antecedents of man. *Homo erectus* lived

1

from 1.5 million to 500,000 years ago and *Homo sapiens Neanderthalensis* at about 110,000 BCE. The Paleolithic Age is the earliest stage of man's development where organized societal structures are known to have existed. These social structures consisted of people living in bands which survived by hunting and gathering food. There is evidence of use of fire going back some 230,000 years, and increasing sophistication of stone tools, jewelry, cave paintings, and religious symbols during this period. Modern man evolved from *Homo sapiens,* probably originating in Africa and the Middle East about 90,000 ye~~ in Europe during the Ice Age perio~ BCE. During this time, man spread masses following the retreating glacie~ at 11,000–8000 BCE.

A Mesolithic Age or transitional phase of evolution from hunter-gatherer societies into the Neolithic Age of food-raising societies occurred at different periods in various parts of the world, first in the Middle East from 9000 to 8000 BCE onward, reaching Europe about 3000 BCE. The change from hunting, fishing, and gathering modes of survival to agriculture was first evidenced by domestication of animals and then growing of wheat, barley, corn, root crops, and vegetables. Associated skills of food storage and cooking, pottery, basket weaving, ovens, smelting, trade, and other skills led to improved survival techniques and population growth gradually spread throughout the world.

Communal habitation became essential to adaptation to changing environmental conditions and hazards allowing population growth and geographic expansion. At each stage of human biological, technological, and social evolution, man coexisted with diseases associated with the environment and living patterns, seeking herbal and mystical treatments for the maladies. Man called on the supernatural and magic to appease these forces and prevent plagues, famines, and disasters. Shamans or witch doctors attempted to remove harm by magical or religious practices along with herbal treatments acquired through trial and error.

Nutrition and exposure to communicable disease changed as mankind evolved. Social organization included tools and skills for hunting, clothing, shelter, fire for warmth and cooking of food for use and storage, burial of the dead, and removal of waste products from living areas. Adaptation of human society to the environment has been and remains a central issue in health to the present time. This is a recurrent theme in the development of public health, facing daunting new challenges of adaptation and balance with the environment.

THE ANCIENT WORLD

Development of agriculture served growing populations unable to exist solely from hunting, stimulating the organization of more complex societies able to share in production and in irrigation systems. Division of labor, trade, commerce, and government were associated with development of urban societies. Growth of population and communal living led to improved standards of living but also created new health hazards including spread of diseases. As in our time, these challenges required community action to prevent disease and promote survival.

Eastern societies were the birthplace of world civilization. Empirical and religious traditions were mixed. Superstition and shamanism coexisted with practical knowledge of herbal medicines, midwifery, management of wounds or broken bones, and trepanation to remove "evil spirits" that resulted from blood clots inside the skull. All were part of communal life with variations in historical and cultural development. The advent of writing led to medical documentation. Requirements of medical conduct were spelled out as part of the general legal Code of Hammurabi in Mesopotamia (circa 1700 BCE). This included regulation of physician fees and punishment for failure and set a legal base for the secular practice of medicine. Many of the main traditions of medicine were those based on magic or that derived from religion. Often medical practice was based on belief in the supernatural, and healers were believed to have a religious calling. Training of medical practitioners, regulation of their practice, and ethical standards evolved in a number of ancient societies.

Some cultures equated cleanliness with godliness and associated hygiene with religious beliefs and practices. Chinese, Egyptian, Hebrew, Indian, and Incan societies all provided sanitary amenities as part of the religious belief system and took measures to provide water, sewage, and drainage systems. This allowed for successful urban settlement and reinforced the beliefs upon which such practices were based. Personal hygiene was part of religious practice. Technical achievements in providing hygiene at the community level slowly evolved as part of urban society.

Chinese practice in the twenty-first to eleventh centuries BCE included digging of wells for drinking water; from the eleventh to the seventh centuries BCE this included use of protective measures for drinking water and destruction of rats and rabid animals. In the second century BCE, Chinese communities were using sewers and latrines. The basic concept of health was that of countervailing forces between the principles of yin (female) and yang (male), with emphasis on a balanced lifestyle. Medical care emphasized diet, herbal medicine, hygiene, massage, and acupuncture.

Ancient cities in India were planned with building codes, street paving, and covered sewer drains built of bricks and mortar. Indian medicine originated in herbalism associated with the mythical gods. Between 800 and 200 BCE, Ayurvedic medicine developed and with it, medical schools and public hospitals. Between 800 BCE and 400

CE, major texts of medicine and surgery were written. Primarily focused in the Indus Valley, the golden age of ancient Indian medicine began in approximately 800 BCE. Personal hygiene, sanitation, and water supply engineering were emphasized in the laws of Manu. Pioneering physicians, supported by Buddhist kings, developed the use of drugs and surgery, and established schools of medicine and public hospitals as part of state medicine. Indian medicine played a leading role throughout Asia, as did Greek medicine in Europe and the Arab countries. With the Mogul invasion of 600 CE, state support declined, and with it, Indian medicine.

Ancient Egyptian intensive agriculture and irrigation practices were associated with widespread parasitic disease. The cities had stone masonry gutters for drainage, and personal hygiene was highly emphasized. Egyptian medicine developed surgical skills and organization of medical care, including specialization and training that greatly influenced the development of Greek medicine. The Eberus Papyrus, written 3400 years ago, gives an extensive description of Egyptian medical science, including isolation of infected surgical patients.

The Hebrew Mosaic Law of the five Books of Moses (circa 1000 BCE) stressed prevention of disease through regulation of personal and community hygiene, reproductive and maternal health, isolation of lepers and other "unclean conditions," and family and personal sexual conduct as part of religious practice. It also laid a basis for medical and public health jurisprudence. Personal and community responsibility for health included a mandatory day of rest, limits on slavery and guarantees of the rights of slaves and workers, protection of water supplies, sanitation of communities and camps, waste disposal, and food protection, all codified in detailed religious obligations. Food regulation prevented use of diseased or unclean animals, and prescribed methods of slaughter improved the possibility of preservation of the meat. While there was an element of viewing illness as a punishment for sin, there was also an ethical and social stress on the value of human life with an obligation to seek and provide care. The concepts of sanctity of human life (*Pikuah Nefesh*) and improving the quality of life on Earth (*Tikun Olam*) were given overriding religious and social roles in community life. In this tradition, the saving of a single human life was considered "as if one saved the whole world," with an ethical imperative to achieve a better earthly life for all. The Mosaic Law, which forms the basis for Judaism, Christianity, and Islam, codified health behaviors for the individual and for society, all of which have continued into the modern era as basic concepts in environmental and social hygiene.

In Cretan and Minoan societies, climate and environment were recognized as playing a role in disease causation. Malaria was related to swampy and lowland areas, and prevention involved planning the location of settlements.

Ancient Greece placed high emphasis on healthful living habits in terms of personal hygiene, nutrition, physical fitness, and community sanitation. Hippocrates articulated the clinical methods of observation and documentation and a code of ethics of medical practice. He articulated the relationship between disease patterns and the natural environment (Air, Water, and Places) which dominated epidemiologic thinking until the nineteenth century. Preservation of health was seen as a balance of forces: exercise and rest, nutrition and excretion, and recognizing the importance of age and sex variables in health needs. Disease was seen as having natural causation, and medical care was valued, with the city-state providing free medical services for the poor and for slaves. City officials were appointed to look after public drains and water supply, providing organized sanitary and public health services. Hippocrates gave medicine both a scientific and ethical spirit lasting to the present time.

Ancient Rome adopted much of the Greek philosophy and experience concerning health matters with high levels of achievement and new innovations in the development of public health. The Romans were extremely skilled in engineering of water supply, sewage and drainage systems, public baths and latrines, town planning, sanitation of military encampments, and medical care. Roman law also regulated businesses and medical practice. The influence of the Roman Empire resulted in the transfer of these ideas throughout much of Europe and the Middle East. Rome itself had access to clean water via 10 aqueducts supplying ample water for the citizens. Rome also built public drains. By the early first century the aqueducts allowed people to have 600–900 liters per person per day of household water from mountains. Marshlands were drained to reduce the malarial threat. Public baths were built to serve the poor, and fountains were built in private homes for the wealthy. Streets were paved, and organized garbage disposal served the cities.

Roman military medicine included well-designed sanitation systems, food supplies, and surgical services. Roman medicine, based on superstition and religious rites, with slaves as physicians, developed from Greek physicians who brought their skills and knowledge to Rome after the destruction of Corinth in 146 BCE. Training as apprentices, Roman physicians achieved a highly respected role in society. Hospitals and municipal doctors were employed by Roman cities to provide free care to the poor and the slaves, but physicians also engaged in private practice, mostly on retainers to families. Occupational health was described with measures to reduce known risks such as lead exposure, particularly in mining. Weights and measures were standardized and supervised. Rome made important contributions to the public health tradition of sanitation, urban planning, and organized medical care. Galen, Rome's leading physician, perpetuated the fame of Hippocrates through his medical writings, basing medical assessment on the four humors of man (sanguine, phlegmatic, choleric, and melancholic). These ideas dominated European medical thought for nearly 1500 years until the advent of modern science.

THE EARLY MEDIEVAL PERIOD (FIFTH TO TENTH CENTURIES CE)

The Roman Empire disappeared as an organized entity following the sacking of Rome in the fifth century CE. The eastern empire survived in Constantinople, with a highly centralized government. Later conquered by the Muslims, it provided continuity for Greek and Roman teachings in health. The western empire integrated Christian and pagan cultures, looking at disease as punishment for sin. Possession by the devil and witchcraft were accepted as causes of disease. Prayer, penitence, and exorcising witches were accepted means of dealing with health problems. The ensuing period of history was dominated in health, as in all other spheres of human life, by the Christian doctrine institutionalized by the Church. The secular political structure was dominated by feudalism and serfdom, associated with a strong military landowning class in Europe.

Church interpretation of disease was related to original or acquired sin. Man's destiny was to suffer on Earth and hope for a better life in heaven. The appropriate intervention in this philosophy was to provide comfort and care through the charity of church institutions. The idea of prevention was seen as interfering with the will of God. Monasteries with well-developed sanitary facilities were located on major travel routes and provided hospices for travelers. The monasteries were the sole centers of learning and for medical care. They emphasized the tradition of care of the sick and the poor as a charitable duty of the righteous and initiated hospitals. These institutions provided care and support for the poor, as well as efforts to cope with epidemic and endemic disease.

Most physicians were monks guided by Church doctrine and ethics. Medical scholarship was based primarily on the teachings of Galen. Women practicing herbal medicine were branded as witches. Education and knowledge were under clerical dominance. Scholasticism, or the study of what was already written, stultified the development of descriptive or experimental science. The largely rural population of the European medieval world lived with poor nutrition, education, housing, sanitary, and hygienic conditions. Endemic and epidemic diseases resulted in high infant, child, and adult mortality. Commonly, 75 percent of newborns died before the age of five. Maternal mortality was high. Leprosy, malaria, measles, and smallpox were established endemic diseases with many other less well-documented infectious diseases.

Between the seventh and tenth centuries, outside the area of Church domination, Muslim medicine flourished under Islamic rule primarily in Persia and later Baghdad and Cairo; Rhazes and Ibn Sinna (Avicenna) translated and adapted ancient Greek and Mosaic teachings, adding clinical skills developed in medical academies and hospitals. Piped water supplies were documented in Cairo in the ninth century. Great medical academies were established, including one in conquered Spain at Cordova. The Cordova Medical Academy was a principal center for medical knowledge and scholarship prior to the expulsion of Muslims and Jews from Spain and the Inquisition. The Academy helped stimulate European medical thinking and the beginnings of western medical science in anatomy, physiology, and descriptive clinical medicine.

THE LATE MEDIEVAL PERIOD (ELEVENTH TO FIFTEENTH CENTURIES)

In the later feudal period, ancient Hebraic and Greco-Roman concepts of health were preserved and flourished in the Muslim Empire. The twelfth-century Jewish philosopher-physician Moses Maimonides, trained in Cordova and expelled to Cairo, helped synthesize Roman, Greek, and Arabic medicine with Mosaic concepts of communicable disease isolation and sanitation.

Monastery hospitals were established between the eighth and twelfth centuries to provide charity and care to ease the suffering of the sick and dying. Monastery hospitals were described in the eleventh century in Russia. Monasteries provided centers of literacy, medical care, and the ethic of caring for the sick patient as an act of charity. The monastery hospitals were gradually supplanted by municipal, voluntary, and guild hospitals developed in the twelfth to sixteenth centuries. By the fifteenth century, Britain had 750 hospitals. Medical care insurance was provided by guilds to its members and their families. Hospitals employed doctors, and the wealthy had access to private doctors.

In the early middle ages, most physicians in Europe were monks, and the medical literature was compiled from ancient sources. In 1131 and 1215, Papal rulings increasingly restricted clerics from doing medical work, thus promoting secular medical practice. In 1224, Emperor Frederick II of Sicily published decrees regulating medical practice, establishing licensing requirements: medical training (3 years of philosophy, 5 years of medicine), 1 year of supervised practice, then examination followed by licensure. Similar ordinances were published in Spain in 1238 and in Germany in 1347.

The Crusades (1096–1270 CE) exposed Europe to Arabic medical concepts, as well as leprosy. The Hospitallers, a religious order of knights, developed hospitals in Rhodes, Malta, and London to serve returning pilgrims and crusaders. The Muslim world had hospitals, such as Al Mansour in Cairo, available to all as a service provided by the government. Growing contact between the Crusaders and the

Muslims through war, conquest, cohabitation, and trade introduced Arabic culture and diseases, and revised ancient knowledge of medicine and hygiene.

Leprosy became a widespread disease in Europe, particularly among the poor, during the early Middle Ages, but the problem was severely accentuated during and following the Crusades, reaching a peak during the thirteenth to fourteenth centuries. Isolation in leprosaria was common. In France alone, there were 2000 leprosaria in the fourteenth century. This disease has caused massive suffering and lingers until now. The development of modern antimicrobials has cured millions of leprosy (Hansen's disease) cases and with early case finding and multidrug therapy, this disease and its disabling and deadly effects are now largely a matter of history.

As rural serfdom and feudalism were declining in Western Europe, cities developed with crowded and unsanitary conditions. Towns and cities developed in Europe with royal charters for self-government, primarily located at the sites of former Roman settlements and at river crossings related to trade routes. The Church provided stability in society, but repressed new ideas and imposed its authority particularly via the Inquisition. Established by Pope Gregory in 1231, the Inquisition was renewed and intensified, especially in Spain in 1478 by Pope Sixtus IV, to exterminate heretics, Jews, and anyone seen as a challenge to the accepted Papal dogmas.

Universities established under royal charters in Paris, Bologna, Padua, Naples, Oxford, Cambridge, and others set the base for scholarship outside the realm of the Church. In the twelfth and thirteenth centuries there was a burst of creativity in Europe, with inventions including the compass, the mechanical clock, the waterwheel, the windmill, and the loom. Physical and intellectual exploration opened up with the travels of Marco Polo and the writings of Thomas Aquinas, Roger Bacon, and Dante. Trade, commerce, and travel flourished.

Medical schools were established in medieval universities in Salerno, Italy, in the tenth century and in universities throughout Europe in the twelfth to fifteenth centuries: in Paris (1110), Bologna (1158), Oxford (1167), Montpellier (1181), Cambridge (1209), Padua (1222), Toulouse (1233), Seville (1254), Prague (1348), Krakow (1364), Vienna (1365), Heidelberg (1386), Glasgow (1451), Basel (1460), and Copenhagen (1478). Physicians, recruited from the new middle class, were trained in scholastic traditions based on translations of Arabic literature and the ancient Roman and Greek texts, mainly Aristotle, Hippocrates, and Galen, but with some more current texts, mainly written by Arab and Jewish physicians.

Growth exacerbated public health problems in the newly walled commercial and industrial towns leading to eventual emergencies, which demanded solution. Crowding, poor nutrition and sanitation, lack of adequate water sources and drainage, unpaved streets, keeping of animals in towns, and lack of organized waste disposal created conditions for widespread infectious diseases. Municipalities developed protected water sites (cisterns, wells, and springs) and public fountains with municipal regulation and supervision. Piped community water supplies were developed in Dublin, Basel, and Bruges (Belgium) in the thirteenth century. Between the eleventh and fifteenth centuries, Novgorod in Russia used clay and wooden pipes for water supplies. Municipal bath houses were available.

Medical care was still largely oriented to symptom relief with few resources to draw upon. Traditional folk medicine survived especially in rural areas, but was suppressed by the Church as witchcraft. Physicians provided services for those able to pay, but medical knowledge was a mix of pragmatism, mysticism, and sheer lack of scientific knowledge. Conditions were ripe for vast epidemics of smallpox, cholera, measles, and other epidemic diseases fanned by the debased conditions of life and vastly destructive warfare raging throughout Europe.

The Black Death (mainly pneumonic and bubonic plague), due to *Yersinia pestis* infection transmitted by fleas on rodents, was brought from the steppes of central Asia to Europe with the Mongol invasions, and then transmitted via extensive trade routes throughout Europe by sea and overland. The Black Death was also introduced to China with Mongol invasions, bringing tremendous slaughter, halving of the population of China between 1200 and 1400 CE. Between the eleventh and thirteenth centuries, during the Mongol–Tatar conquests, many widespread epidemics, including plague, were recorded in Rus (now Russia). The plagues traveled rapidly with armies, caravan traders, and later by shipping as world trade expanded in the fourteenth to fifteenth centuries (see Box 1.1). The plague ravaged most of Europe between 1346 and 1350, killing between 24 and 50 million people; approximately one-third of the population, leaving vast areas of Europe underpopulated. Despite local efforts to prevent disease by quarantine and isolation of the sick, the disease devastated whole communities.

Fear of a new and deadly disease, lack of knowledge, speculation, and rumor led to countermeasures which often exacerbated the spread of epidemics (as occurred in the late twentieth century with the AIDS epidemic). In Western Europe, public and religious ceremonies and burials were promoted, which increased contact with infected persons. The misconception that cats were the cause of plague led to their slaughter, when they could have helped to stem the tide of disease brought by rats and by their fleas to humans. Hygienic practices limited the spread of plague in Jewish ghettoes, leading to the blaming of the plague's spread on the Jews, and widespread massacres, especially in Germany and central Europe.

Box 1.1 "This is the End of the World": The Black Death

"Rumors of a terrible plague supposedly arising in China and spreading through Tartary (Central Asia) to India and Persia, Mesopotamia, Syria, Egypt and all of Asia Minor had reached Europe in 1346. They told of a death toll so devastating that all of India was said to be depopulated, whole territories covered by dead bodies, other areas with no one left alive. As added up by Pope Clement VI at Avignon, the total of reported dead reached 23,840,000. In the absence of a concept of contagion, no serious alarm was felt in Europe until the trading ships brought their black burden of pestilence into Messina while other infected ships from the Levant carried it to Genoa and Venice.

By January 1348 it penetrated France via Marseille, and North Africa via Tunis. Ship-borne along coasts and navigable rivers, it spread westward from Marseille through the ports of Languedoc to Spain and northward up the Rhone to Avignon, where it arrived in March. It reached Narbonne, Montpellier, Carcassone, and Toulouse between February and May, and at the same time in Italy spread to Rome and Florence and their hinterlands. Between June and August it reached Bordeaux, Lyon, and Paris, spread to Burgundy and Normandy into southern England. From Italy during the summer it crossed the Alps into Switzerland and reached eastward to Hungary.

In a given area the plague accomplished its kill within four to six months and then faded, except in the larger cities, where, rooting into the close-quartered population, it abated during the winter, only to appear in spring and rage for another six months."

Source: Tuchman, B. W. A *Distant Mirror: The Calamitous 14th Century*, op cit. page 93.

Seaport cities in the fourteenth century began to apply the biblical injunction to separate lepers by keeping ships coming from places with the plague waiting in remote parts of the harbor, initially for 30 days (*treutina*), then for 40 days (*quarantina*) (Ragusa in 1465 and Venice in 1485), establishing the public health act of quarantine, which on a pragmatic basis was found to reduce the chance of entry of the plague. Towns along major overland trading routes in Russia took measures to restrict movement in homes, streets, and entire towns during epidemics. All over Europe, municipal efforts to enforce isolation broke down as crowds gathered and were uncontrolled by inadequate police forces. In 1630, all officers of the Board of Health of Florence, Italy, were excommunicated because of efforts to prevent spread of the contagion by isolation of cases, thereby interfering with religious ceremonies to assuage God's wrath through appeals to divine providence.

The plague continued to strike with epidemics in London in 1665, Marseille in 1720, Moscow in 1771, and Russia, India, and the Middle East through the nineteenth century. In sixteenth-century Russia, Novgorod banned public funerals during plague epidemics, and in the seventeenth century, Czar Boris Godunov banned trade, prohibited religious and other ceremonies, and instituted quarantine-type measures. Plague continued into the twentieth century (see *The Plague* by Albert Camus) with epidemics in Australia (1900), China (1911), Egypt (1940), and India (1995). The disease is endemic in rodents in many parts of the world, including the United States; however, modern sanitation, pest control, and treatment have greatly reduced the potential for large-scale plague epidemics.

Guilds organized to protect economic interests of traders and skilled craftsmen developed mutual benefit funds to provide financial assistance and other benefits for illness, death, widows and orphans, and medical care, as well as burial benefits for members and their families. The guilds wielded strong political powers during the late middle ages. These brotherhoods provided a tradition later expressed in the mutual benefit or Friendly Societies, sick funds, and insurance for health care based on employment groups. This tradition has continued in western countries, where labor unions are among the leading advocates for the health of workers and their families.

The fourteenth century saw a devastation of the population of Europe by plague, wars, and the breakdown of feudal society. It also set the stage for the agricultural revolution and later the industrial revolution. The period following the Black Death was innovative and dynamic. Lack of farm labor led to innovations in agriculture. Enclosures of common grazing land reduced spread of disease among animals, increased field crop productivity, and improved sheep farming, leading to development of the wool and textile industries and the search for energy sources, industrialization, and international markets.

THE RENAISSANCE (1500–1750)

Commerce, industry, trade, merchant fleets, and voyages of discovery to seek new markets led to the development of a moneyed middle class and wealthy cities. In this period, mines, foundries, and industrial plants flourished, creating new goods and wealth. Partly as a result of the trade generated and the increased movement of goods and people, vast epidemics of syphilis, typhus, smallpox, measles, and the plague continued to spread across Europe. Malaria was still widespread throughout Europe. Rickets, scarlet fever, and scurvy, particularly among sailors, were rampant. Pollution and crowding in industrial areas resulted in centuries-long epidemics of environmental disease, particularly among the urban working class.

A virulent form of syphilis, allegedly brought back from America by the crews of Columbus, spread rapidly throughout Europe between 1495 and 1503, when it was first described by Fracastorus. Control measures tried in various cities included examination and registration of prostitutes, closure of communal bath houses, isolation in special hospitals, reporting of disease, and expulsion of sick prostitutes or strangers. The disease gradually decreased in virulence, but it remains a major public health problem.

In Russia, Czar Ivan IV (Ivan the Terrible) in the sixteenth century arranged to hire the court physician of Queen Elizabeth I, who brought with him to Moscow a group of physicians and pharmacists to serve the court. The Russian army had a tradition of regimental doctors. In the mid-seventeenth century, the czarist administration developed pharmacies in major centers throughout the country for military and civilian needs, and established a State Pharmacy Department to control pharmacies and medications, education of doctors, military medicine, quarantine, forensic medicine, and medical libraries. Government revenues from manufacturing, sale, and encouragement of vodka provided for these services. Preparation of military doctors (*Lekars*) with 5–7 years of training was instituted in 1654. Hospitals were mainly provided by monasteries, serving both civilian and military needs. In 1682, the first civic hospital was opened in Moscow, and in the same year, two hospitals were opened also in Moscow by the central government for care of patients and training of *Lekars*.

In European countries, growth of cities with industrialization and massive influx of the rural poor brought the focus of public health needs to the doorsteps of municipal governments. The breakdown of feudalism, the decline of the monasteries, and the land enclosures dispossessed the rural poor. Municipal and voluntary organizations increasingly developed hospitals, replacing those previously run by monastic orders. In 1601, the British Elizabethan Poor Laws defined the local parish government as being responsible for the health and social well-being of the poor, a system later brought to the New World by British colonists. Municipal control of sanitation was weak. Each citizen was in theory held responsible for cleaning his part of the street, but hygienic standards were low with animal and human wastes freely accumulating.

During the Renaissance, the sciences of anatomy, physiology, chemistry, microscopy, and clinical medicine opened medicine to a scientific base. Medical schools in universities developed affiliations with hospitals, promoting clinical observation with increasing precision in description of disease. The contagion theory of disease, described in 1546 by Fracastorus and later Paracelsus, including the terms *infection* and *disinfection*, was contrary to the until-then sacrosanct miasma teachings of Galen.

From 1538, parish registers of christenings and burials were published in England as weekly and annual abstracts, known as the *Bills of Mortality*. Beginning in 1629, national annual Bills of Mortality included tabulation of death by cause. On the basis of the Bills of Mortality, novelist Daniel Defoe described the plague epidemic of London of 1665 100 years later.

In 1662, John Graunt in England published *Natural and Political Observations Upon the Bills of Mortality*. He compiled and interpreted mortality figures by inductive reasoning, demonstrating the regularity of certain social and vital phenomena. He showed statistical relationships between mortality and living conditions. Graunt's work was important because it was the first instance of statistical analysis of mortality data, providing a foundation for use of health statistics in the planning of health services. This established the sciences of demography and vital statistics and methods of analysis, providing basic measurements for health status evaluation with mortality rates by age, sex, and location. Also in 1662, William Petty took the first census in Ireland. In addition, he studied statistics on the supply of physicians and hospitals.

Microscopy, developed by Antony van Leeuwenhoek in 1676, provided a method of study of microorganisms. In the seventeenth century, the great medical centers were located in Leyden, Paris, and Montpelier. Bernardino Ramazzini published the first modern comprehensive treatise on occupational diseases in 1700.

In Russia, Peter the Great (1682–1725) initiated political, cultural, and health reforms. He sent young aristocrats to study sciences and technology in Western Europe, including medicine. He established the first hospital-based medical school in St. Petersburg and then in other centers, mainly to train military doctors. He established the Anatomical Museum of the Imperial Academy of Sciences in St. Petersburg in 1717, and initiated a census of males for military service in 1722. In 1724, V. N. Tateshev carried out a survey by questionnaire of all regions of the Russian empire regarding epidemic disease and methods of treatment.

ENLIGHTENMENT, SCIENCE, AND REVOLUTION (1750–1830)

The Enlightenment, a dynamic period of social, economic, and political thought, provided great impetus for emancipation and rapid advancement of science and agriculture, technology, and industrial power. Changes in many spheres of life were exemplified by the American and French revolutions, along with the economic theory of Adam Smith (author of *The Wealth of Nations*), which developed political and economic rights of the individual. Improvements in agriculture created greater productivity and better nutrition. These were associated with higher birth rates and falling death rates, leading to rapid population growth.

The agricultural revolution during the sixteenth and seventeenth centuries, based on mechanization and in larger land units of production with less manpower, led to rural depopulation, provided excess workers to staff the factories, mines, ships, homes, and shops of the industrial revolution, expanding commerce, and a growing middle class. Exploration and colonization provided expansion of markets that fueled the industrial revolution, growth of science, technology, and wealth.

An agricultural revolution improved methods of production of grain, milk, and meat through better land use, animal husbandry, and farm machinery resulting in greater agricultural productivity and food supply. Later, introduction of new crops from the Americas, including the potato, the tomato, peppers, and maize, contributed to a general improvement in nutrition. This was supplemented by increasing availability of cod from the Grand Banks of the Atlantic, adding protein to the common diet.

Industrialized urban centers grew rapidly. Crowded cities were ill-equipped to house and provide services for the growing working class. Urban areas suffered from crowding, poor housing, sanitation, and nutrition; harsh working conditions which produced appalling health conditions. During this period, documentation and statistical analysis developed in various forms, becoming the basis for social sciences including demography and epidemiology. Intellectual movements of the eighteenth century defined the rights of man and gave rise to revolutionary movements to promote liberty and release from tyrannical rule, as in the American and French revolutions of 1775 and 1789. Following the final defeat of Napoleon at Waterloo in 1815, conservative governments were faced with strong middle-class movements for reform of social conditions, with important implications for health.

Eighteenth-Century Reforms

The period of enlightenment and reason was led by philosophers Locke, Diderot, Voltaire, Rousseau, and others. These men produced a new approach to science and knowledge derived from observations and systematic testing of ideas as opposed to instinctive or innate knowledge as the basis for human progress. The idea of the rights of man contributed to the American and French revolutions, but also to a widening belief that society was obliged to serve all rather than just the privileged. This had a profound impact on approaches to health and societal issues.

The late eighteenth century was a period of growth and development of clinical medicine, surgery, and therapeutics, as well as of the sciences of chemistry, physics, physiology, and anatomy. From the 1750s onward, voluntary hospitals were established in major urban centers in Britain, America, and Eurasia. Medical–social reform involving hospitals, prisons, and lazarettos (leprosy hospitals) in Britain, led by John

Howard (who published *On the State of Prisons* in 1777), produced substantive improvements in these institutions. During the French Revolution, Philippe Pinel removed the chains from patients at the Bicetre Mental Hospital near Paris and fostered reform of insane asylums. Reforms were further carried out in Britain by the Society of Friends (the Quakers), who built the York Retreat, providing humane care as an alternative to the inhuman conditions of the York Asylum.

Although Ramazzini's monumental work on occupational diseases was published in 1700, little progress was made in applying epidemiologic principles to this field. In the latter part of the century though, interest in the health of sailors and soldiers led to important developments in military and naval medicine. Studies of prevalent diseases were carried out by pioneering physicians among workers in various trades, such as metalworkers, bakers, shoemakers, and hat makers, and identified causative agents and thus methods of prevention. Observational studies of Percival Potts on scrotal cancer in chimney sweeps (1775), and Baker on the Devonshire colic (lead poisoning) in 1767, helped to lay the basis for development of investigative epidemiology.

Pioneers and supporting movements successfully agitated for reform in Britain through the parliamentary system. The anti-gin movement, aided by the popular newspapers (the "penny press") and the brilliant engravings of Hogarth, helped produce legal, social, and police reforms in English townships. Conditions for sailors in the British navy were improved following the explorations of Captain James Cook during the period 1766–1779, and the Spithead mutiny of 1797 and adoption of health recommendations of Captain James Lind. The United States developed the Marine Hospitals Service for treatment and quarantine of sailors in 1798, which later became the U.S. Public Health Service. The antislavery movement led by protestant Christian churches goaded the British government to ban slavery in 1797 and the slave trade in 1807, using the Royal Navy to sweep the slave trade from the seas during the early part of the nineteenth century.

Applied Epidemiology

Scurvy (the Black Death of the Sea) was a major health problem among sailors during long voyages. In 1498, Vasco da Gama lost 55 crewmen to scurvy during his voyages, and in 1535, Jacques Cartier's crew suffered severely from scurvy on his voyage of discovery to Canada. During the sixteenth century, Dutch sailors knew of the value of fresh vegetables and citrus fruit in preventing scurvy.

Purchas (1601) and John Woodall (a British naval doctor) (1617) recommended use of lemons and oranges in treatment of scurvy, but this was not widely practiced. During the seventeenth to eighteenth centuries, Russian military

End Tues

Box 1.2 James Lind and Scurvy, 1747

Captain James Lind, a physician serving Britain's Royal Navy, developed a hypothesis regarding the cause of scurvy based on clinical observations. In May 1747, on HMS *Salisbury*, Lind conducted the first controlled clinical epidemiologic trial by treating 12 sailors sick with scurvy with six different dietary regimens. The two sailors who were fed oranges and lemons became well and fit for duty within 6 days, while the others remained sick. He concluded that citrus fruits would treat and prevent scurvy. In 1757, he published his *Treatise on the Scurvy: An Inquiry on the Nature, Causes and Cure of that Disease.*

This discovery was adopted by progressive sea captains and aided Captain Cook in his voyages of discovery in the South Pacific in 1768–1771. By 1795, the Royal Navy adopted routine issuance of lime juice to sailors to prevent scurvy. This extended the time naval ships could remain at sea, which was crucial in their blockade of Napoleonic Europe.

Lind also instigated reforms in living conditions for sailors, thus contributing to improvement in their health and fitness and the functioning of the fleet.

---

practice included antiscorbutic preparations, and use of sauerkraut for this purpose became common in European armies. Scurvy was a major cause of sickness and death among sailors when supplies of fruit and vegetables ran out; it caused disease and death which seriously limited long voyages and contributed to frequent mutinies at sea.

In 1740–1744, a British naval squadron of seven ships and nearly 2000 men led by Commodore George Anson left Plymouth, circumnavigating the globe, returned to England with one ship and 145 men after losing most of the crews to scurvy. In 1747, James Lind carried out a pioneering epidemiologic investigation on scurvy among sailors on long voyages, leading some 50 years later to adoption of lemon or lime juice as a routine nutrition supplement for British sailors. Vitamins were not understood or isolated until almost 150 years later, but Lind's scientific technique of careful observation, hypothesis formulation and testing, followed by documentation established clinical epidemiologic investigation of nutrition in public health (see Box 1.2).

Jenner and Vaccination

Smallpox, a devastating and disfiguring epidemic disease, ravaged all parts of the world and was known since the third century BCE. Described first by Rhazes in the tenth century, the disease was confused with measles and was widespread in Asia, the Middle East, and Europe during the Middle Ages. It was a designated cause of death in the Bills of Mortality in 1629 in London. Epidemics of smallpox occurred throughout the seventeenth to eighteenth and into the nineteenth centuries primarily as a disease of childhood, with mortality rates of 25 to 40 percent or more and disfiguring sequelae.

Smallpox was a key factor in the near elimination of the Aztec and other societies in Central and South America following the Spanish invasion. Traditions of prevention of this disease by inoculation or transmission of the disease to healthy persons to prevent them from a more virulent form during epidemics were reported in ancient China. This practice called variolation was first brought to England in 1721 by Lady Mary Montagu, wife of the British ambassador to Constantinople, where it was common practice. It was widely adopted in England in the mid-eighteenth century, when the disease affected millions of people in Europe. Catherine the Great in Russia had her son inoculated by variolation by a leading English practitioner.

Edward Jenner was the first to use vaccination with cowpox to prevent smallpox in 1796 (Box 1.3), initiating one of the most dramatically successful endeavors of public health, culminating in the eventual eradication of this dreaded killing and disfiguring disease. In 1800, vaccination was adopted by the British armed forces, and the practice spread to Europe, the Americas, and the British Empire. Denmark made vaccination mandatory in the early nineteenth century and soon eradicated smallpox locally. Despite some professional opposition, the practice spread rapidly from the upper classes and voluntary groups to the common people because of the fear of smallpox. Vaccination later became compulsory in many countries, leading to the ultimate public health achievement: global eradication of smallpox in the late twentieth century.

FOUNDATIONS OF HEALTH STATISTICS AND EPIDEMIOLOGY

Registration of births and deaths forms the basis of demography. Epidemiology as a discipline borrows from demography, sociology, and statistics. The basis of scientific reasoning in these fields emerged in the early seventeenth century with inductive reasoning enunciated by Francis Bacon and applied by Robert Boyle in chemistry, Isaac Newton in physics, William Petty in economics, and John Graunt in demography. Bacon's writing inspired a whole generation of scientists in different fields and led to the founding of the Royal Society.

In 1722, Peter the Great began Russia's system of registration of births of male infants for military purposes.

Box 1.3 Jenner and Smallpox

In 1796, Edward Jenner (1749–1823), a country physician in Gloustershire, England, investigated local folklore that milkmaids were immune to smallpox because of their exposure to cowpox. He took matter from a cowpox pustule on a milkmaid, Sarah Nelmes, and applied it with scratches to the skin of a youngster named James Phipps, who was then inoculated with smallpox. He did not develop the disease. Jenner's 1798 publication, An Enquiry into the Causes and Effects of the Variolae Vaccina, described his wide-scale vaccination and its successful protection against smallpox, and prophesied that "the annihilation of the smallpox, the most dreadful scourge of the human species, must be the result of this practice."

He developed vaccination as a method to replace variolation, which was exposure of persons to the pustular matter of cases of smallpox, originally documented in ancient China in 320 CE. Variolation was practiced widely in the eighteenth century and constituted a very lucrative medical business. Opposition to vaccination was intense, and Jenner's contribution was ignored by the scientific and medical establishment of the day, but rewarded by Parliament. Vaccination was adopted as a universal practice by the British military in 1800 and by Denmark in 1803. Vaccination became an increasingly wide practice during the nineteenth century. In 1977, the last case of smallpox was identified and smallpox eradication was declared by the World Health Organization in 1980.

Remaining stocks of the virus in the United States and Russia were to be destroyed in 1999, but this was delayed and following the 9/11 attack on the Twin Towers in New York City, the threat of bioterrorism was taken seriously including the possibility of use of smallpox. As a result, vaccination was reinstated for "first responders" including fire, police, and hospital staff in the United States and other countries.

Source: WHO. Smallpox, http://www.bt.cdc.gov/agent/smallpox/ [accessed April 27, 2008]

In 1755, M. V. Lomonosov led initiatives in establishing the study of demography in Russia. He carried out surveys and studies of birth statistics, infant mortality, and quality of medical care, alcoholism, and worker's health. He brought the results of these studies to the attention of the government, which led to improved training of doctors and midwives, and epidemic control measures. Lomonosov also helped initiate the medical faculty of Moscow University (1765).

Daniel Bernoulli, a member of a European family of mathematicians, constructed life tables based on available data showing that variolation against smallpox conferred lifelong immunity and vaccination at birth increased life expectancy. Following the French Revolution, health statistics flourished in the mid-nineteenth century in the work of Pierre Louis, who is considered the founder of modern epidemiology. Louis conducted several important observational studies, including one showing that bloodletting, then a common form of therapy, was not effective, leading to a decline in this harmful practice. His students included Marc D'Epigne in France, William Farr in Britain, and others in the United States who became the pioneers in spreading "la methode numerique" in medicine.

Health statistics for social and public health reform took an important place in the work of Edwin Chadwick, Lemuel Shattuck, and Florence Nightingale. Recognizing the critical importance of accurate statistical information in health planning and disease prevention, Edwin Chadwick's work led to legislation establishing the Registrar-General's Office of Britain in 1836. William Farr became its director-general and placed the focus of this office on public health. Farr's analysis of mortality in Liverpool, for example, showed that barely half of its native-born lived to their sixth birthday, whereas in England overall the median age at death was 45 years. As a result, Parliament passed the Liverpool Sanitary Act of 1846, creating a legislated sanitary code, a Medical Officer of Health position, and a local health authority.

The London Epidemiological Society, founded in 1850, became an active investigative and lobbying group for public health action. Its work on smallpox led to passage of the Vaccination Act of 1853, establishing compulsory vaccination in the United Kingdom. William Budd, a student of Louis and founding member of the London Epidemiological Society, investigated outbreaks of typhoid fever in his home village in 1839, establishing it as a contagious self-propagating disease spread by microorganisms, discrediting the miasma theory.

In 1842 in Boston, Massachusetts, Lemuel Shattuck initiated a statewide registration of vital statistics, which became a model elsewhere in the United States. His report was a landmark in the evolution of public health administration and planning. This provided a detailed account of data collection by age, sex, race, occupation, and uniform nomenclature for causes of diseases and death. He emphasized the importance of a routine system for exchanging data and information.

In the later part of the nineteenth century, Florence Nightingale highlighted the value of a hospital discharge information system. She promoted collection and use of statistics that could be derived from the records of patients treated in hospitals. Her work led to improved management and design of hospitals, military medicine, and nursing as a profession.

SOCIAL REFORM AND THE SANITARY MOVEMENT (1830–1875)

Following the English civil war in 1646, veterans of the Parliamentary Army called on the government to provide free schools and free medical care throughout the country as part of democratic reform. However, they failed to sustain interest or gain support for their revolutionary ideas amidst postwar religious conflicts and restoration of the monarchy.

In Russia, the role of the state in health was promoted following initiatives of Peter the Great to introduce western medicine to the country. During the rule of Catherine the Great, under the supervision of Count Orlov, an epidemic of plague in Moscow (1771–1772) was suppressed by incentive payments to bring the sick for care. In 1784, a Russian physician, I. L. Danilevsky, defended a doctoral dissertation on "Government power — the best doctor." In the eighteenth and nineteenth centuries, reform movements promoted health initiatives by government. While these movements were suppressed (the Decembrists, 1825–1830) and liberal reform steps reversed, their ideas influenced later reforms in Russia.

Following the revolution in France, the Constituent Assembly established a Health Commission. A national assistance program for indigents was established. Steps were taken to strengthen the *Bureaux de Sante* (Offices of Health) of municipalities which had previously dealt primarily with epidemics. In 1802, the Paris Bureau addressed a wide range of public health concerns, including sanitation, food control, health statistics, occupational health, first aid, and medical care issues. The other major cities of France followed with similar programs over the next 20 years, and in 1848 a central national health authority was established. Child welfare services were also developed in France in the middle part of the nineteenth century. The reporting of vital statistics became reliable in the German states and even more so in France, fostering the development of epidemiologic analysis of causes of death.

The governmental approach to public health was articulated by Johann Peter Franck for the Germanic states in his monumental series of books, *A Complete System of Medical Police* (1779–1817). This text explained the government's role in states with strong central governments and how to achieve health reform through administrative action. State regulations were to govern public health and personal health practices including marriage, procreation, and pregnancy. He promoted dental care, rest following delivery and maternity benefits, school health, food hygiene, housing standards, sanitation, sewage disposal, and clean water supplies. In this system, municipal authorities were responsible for keeping cities and towns clean and for monitoring vital statistics, military medicine, venereal diseases, hospitals, and communicable disease.

This system emphasized a strong, even authoritarian role of the state in promoting public health including provision of prepaid medical care. It was a comprehensive and coherent approach to public health, emphasizing the key roles of municipal and higher levels of government. This work was influential in Russia where Franck spent the years 1805–1807 as director of the St. Petersburg Medical Academy. Because of its primary reliance on authoritarian governmental roles, however, this approach was resisted in most western countries, especially following the collapse of absolutist government ideas following the Napoleonic period.

Municipal (voluntary) boards of health were established in some British and American cities in the late eighteenth and early nineteenth centuries. A Central Board of Health was established in Britain in 1805, primarily to govern quarantine regulations to prevent entry of yellow fever and cholera into the country. Town life improved as sanitation, paving, lighting, sewers, iron water pipes, and water filtration were introduced, although organization for development of such services was inadequate. Multiple agencies and private water companies provided unsupervised and overlapping services. London City Corporation had nearly 100 paving, lighting, and cleansing boards, 172 welfare boards, and numerous other health-related authorities in 1830. These were later consolidated into the London Board of Works in 1855.

In Great Britain, early nineteenth-century reforms were stimulated by the Philosophic Radicals led by Jeremy Bentham, who advocated dealing with public problems in a rational and scientific way, initiating a reform movement utilizing parliamentary, legal, and educational means. Economic and social philosophers in Britain, including Adam Smith and Jeremy Bentham, argued for liberalism, rationalism, free trade, political rights, and social reform, all contributing to "the greatest good for the greatest number." Labor law reforms (the Mines and the Factory acts) banning children and women from underground work in the mines and regulating reduction in the workday to 10 hours were adopted by the British Parliament in the 1830s to 1840s. The spread of railroads and steamships, the penny post (1840), and telegraphs (1846), combined with growing literacy and compulsory primary education introduced in Britain in 1876, dramatically altered local and world communication.

The British Poor Law Amendment Act of 1834 replaced the old Elizabethan Poor Laws, shifting responsibility for welfare of the poor from the local parish to the central government's Poor Law Commission. The parishes were unable to cope with the needs of the rural poor whose condition was deteriorating with loss of land rights due to agricultural innovations and enclosures. The old system was breaking down, and the new industrialization needed workers, miners, sailors, and soldiers. The new conditions forced the poor to move from rural areas to the growing industrial towns. The urban poor suffered or were forced into workhouses while resistance to reform led to more radicalization, unsuccessful revolution, followed by deep political conservatism.

Deteriorating housing, sanitation, and work conditions in Britain in the 1830s resulted in rising mortality rates recorded in the Bills of Mortality. Industrial cities like Manchester (1795) had established voluntary boards of health, but they lacked the authority to alter fundamental conditions to control epidemics and urban decay. The boards of health were unable to deal with sewage, garbage, animal control, crowded slum housing, privies, adulterated foods and medicines, industrial polluters, or other social or environmental risk sources. Legislation in the 1830s in Britain and Canada improved the ability of municipalities and boards of health to cope with oversight of community water supplies and sanitation.

Under pressure from reformists and the Health of Towns Association, the British government commissioned Edwin Chadwick to undertake a study which led to the *Report on the Sanitary Conditions of the Labouring Population of Great Britain* (1842), which led to a further series of reforms through the Poor Law Commission. The British Parliament passed the Health of Towns Act and the Public Health Act of 1848. This established the General Board of Health, mainly to ensure safety of community water supplies and drainage, establishing municipal boards of health in the major cities and rural local authorities, along with housing legislation, and other reforms. Despite setbacks due to reaction to these developments, the basis was laid for the "Sanitary Revolution," dealing with urban sanitation and health conditions, as well as cholera, typhoid, and tuberculosis control.

In 1850, the Massachusetts Sanitary Commission, chaired by Lemuel Shattuck, was established to look into similar conditions in that state. Boards of health established earlier in the century became efficiently organized and effective in sanitary reform in the United States. The report of that committee has become a classic public health document. Reissued in the 1970s, it remains a useful model for a comprehensive approach to public health.

The Chadwick (1842) and Shattuck (1850) reports developed the concept of municipal boards of health based on public health law with a public mandate to supervise and regulate community sanitation. This included urban planning, zoning, restriction of animals and industry in residential areas, regulation of working conditions, and other aspects of community infrastructure, setting the basis of public health infrastructure in the English-speaking world and beyond for the next century.

The interaction between sanitation and social hygiene was a theme promoted by Rudolph Virchow, the founder of cellular pathology and a social–medical philosopher. He was a leading German physician in the mid-nineteenth century, despite being an anticontagionist (i.e., a "miasmic"). He promoted the ideas of observation, hypothesis, and experimentation, helping to establish the scientific method and dispel philosophic approaches to medical issues. He was a social activist and linked health of the people to social and economic conditions, emphasizing the need for political solutions. Virchow played an important part in the 1848 revolutions in Central and Western Europe, the same year as the publication of the *Communist Manifesto* by Karl Marx. These all contributed to growing pressure on governments by workers' groups to promote better living, working, and health conditions in the 1870s.

In 1869, the Massachusetts State Board of Health was established and in the same year a Royal Sanitary Commission was appointed in the United Kingdom. The American Public Health Association (APHA), established in 1872, served as a professional educational and lobbying group to promote the interests of public health in the United States, often successfully prodding federal, state, and local governments to act in the public interests in this field. The APHA definition of appropriate services at each level of government continues to set standards and guidelines for local health authorities. The organization of local, state, and national public health activities over the twentieth century in the United States owes much to the professional leadership and lobbying skills of the APHA.

Max von Pettenkoffer in 1873 studied the high mortality rates of Munich, comparing them to rapidly declining rates in London. His public lectures on the value of health to a city led to sanitary reforms, as were being achieved in Berlin at the same time under Virchow's leadership. Pettenkoffer introduced laboratory analysis to public health practice and established the first academic chair in hygiene and public health, emphasizing the scientific basis for public health. He is considered to be the first professor of experimental hygiene. Pettenkoffer promoted the concept of the value of a healthy city, stressing that health is the result of a number of factors and public health is a community concern, and that measures taken to help those in need benefit the entire community.

In 1861, Russia freed the serfs and returned independence to universities. Departments of hygiene were established in the university medical schools in the 1860s and 1870s to train future hygienists, and to carry out studies of sanitary and health conditions in manufacturing industries. F. F. Erisman, a pioneer in sanitary research in Russia, promoted the connection between experimental science, social hygiene, and medicine, and he established a school of hygiene in 1890, later closed by the czarist government. In 1864, the government initiated the Zemstvos system of providing medical care in rural areas as a governmental program. These health reforms were implemented in 34 of 78 regions of Russia, before the Revolution. Prior to these reforms, medical services in rural areas were practically nonexistent. Epidemics and the high mortality of the working population induced the nobility and new manufacturers in rural towns to promote Zemstvos' public medical services. In rural areas previously served by doctors based in the towns traveling to the villages, local hospitals and delivery homes were established. The Russian medical profession largely supported free public medical care as a fundamental right.

In 1881, Otto von Bismarck, Chancellor of Germany, introduced legislation providing mandatory insurance for injury and illness, and survivor benefits for workers in industrial plants, and then in 1883, introduced social insurance for health care of workers and their families, based on mandatory payments from workers' salaries and employer contributions. In the United Kingdom in 1911, Prime Minister Lloyd George established compulsory insurance for workers and their families for medical care based on capitation payment for general practitioner services. This was followed by similar programs in Russia in 1912 and in virtually all central and western European countries by the 1930s. In 1918, Vladimir Lenin established the state-operated health program, named for its founder Nikolai Semashko, bringing health care to the wide reaches of the Soviet Union. These programs led to the wide recognition of the principle of social solidarity with governmental responsibility for health of the population in virtually all developed countries by the 1960s (see Chapter 13). In the United States, pensions were established for Civil War veterans, widows and orphans, and were made a national Social Security system only in 1935. Health care insurance was developed through trade unions, and only extended to governmental medical care insurance for the elderly and the poor in 1965.

The harsh conditions in the urban industrial and mining centers of Europe during the industrial revolution led to efforts in social reform preceding and contributing to sanitary reform even before the germ theory of disease causation was proven and the science of microbiology established. Pioneering breakthroughs, based on trial and error, challenged the established dogmas of the time, produced the sanitary revolution, still one of the important foundations of public health.

Snow on Cholera

The great cholera pandemics originated in India between 1825 and 1854 and spread via increasingly rapid transportation to Europe and North America. Moscow lost some 33,000 people in the cholera epidemic of 1829, which recurred in 1830–1831. In Paris, the 1832 cholera epidemic killed over 18,000 people (just over 2 percent of the population) in 6 months.

Between 1848 and 1854, a series of outbreaks of cholera occurred in London with large-scale loss of life. The highest rates were in areas of the city where two water companies supplied homes with overlapping water mains. One of these (the Lambeth Company) then moved its water intake to a less polluted part of the Thames River, while the Southwark and Vauxhall company left its intake in a part of the river heavily polluted with sewage. John Snow, a founding member of the London Epidemiological Society and anesthetist to Queen Victoria, investigated an outbreak of cholera in Soho from August to September, 1854, in the area adjacent to Broad Street. He traced some 500 cholera deaths occurring in a 10-day period. Cases either lived close to or used the Broad Street pump for drinking water. He determined that brewery workers and poorhouse residents in the area, using uncontaminated wells, escaped the epidemic. Snow concluded that the Broad Street pump was probably contaminated. He persuaded the authorities to remove the handle from the pump, and the already subsiding epidemic disappeared within a few days.

During September to October, 1854, Snow investigated another outbreak, again suspecting water transmission. He identified cases of mortality from cholera by their place of residence and which water company supplied the home (Table 1.1). Snow calculated the cholera rates in a 4-week period in homes supplied by each of the two companies. Homes supplied by the Southwark and Vauxhall Water Company were affected by high cholera death rates while adjacent homes supplied by the Lambeth Company had rates lower than the rest of London. This provided overwhelming epidemiologic support for Snow's hypothesis that the cholera epidemic source was the contaminated water from the Thames River, distributed to homes in a large area of south London.

This investigation, with a study and control group occurring in an actual disease outbreak, strengthened the

TABLE 1.1 Deaths from Cholera Epidemic in Districts of London Supplied by Two Water Companies, 7 Weeks, 1854[a]

Water supply company	Number of houses	Deaths from cholera	Cholera deaths 10,000 houses
Southwark and Vauxhall	40,046	1,263	315
Lambeth	26,107	98	37
Rest of London	256,423	1,422	59

[a]Source: Snow J. On the mode of transmission of cholera. In: *Snow on Cholera: A Reprint of Two Papers*. New York: The Commonwealth Fund, 1936.

germ theory supporters who were still opposed by powerful forces. It also led to legislation mandating filtration of water companies' supplies in 1857. *Vibrio cholerae* was not isolated until 1883 during an investigation of waterborne cholera outbreaks in Egypt by Robert Koch. Snow's work on cholera has become one of the classic epidemiologic investigations, studied to this day for its scientific imagination and thoroughness, this despite preceding the discovery of the causative organism nearly 30 years later.

Snow's work on cholera stimulated more research into causes of enteric diseases. William Budd, physician at the Bristol Royal Infirmary, was a pioneering exponent of the germ theory of disease. He carried out a number of epidemiologic investigations of typhoid fever in the 1850s, finding waterborne episodes of the disease. He investigated an outbreak in 1853 in Cowbridge, a small Welsh village, where a ball attracted 140 participants from surrounding Counties. Almost immediately afterward, many of those attending the ball became sick with typhoid fever. He found that a person with typhoid had been at the location some days before and that his excreta had been disposed of near a well from which water was drawn for the ball. Budd then concluded that water was the vehicle of transmission of the disease. He investigated other outbreaks and summarized his reports in *Typhoid Fever: Its Nature, Mode of Transmission and Prevention,* published in 1873, which is a classic work on waterborne transmission of enteric disease. These investigations contributed to the movement to disinfect public water systems on a preventive basis.

The brilliant epidemiologic studies of Snow and Budd set a new direction in epidemiology and public health practice, not only with waterborne disease. They established a standard for investigation of the distribution of disease in populations with the object of finding a way to interrupt the transmission of disease. Improved sanitation and water safety developed in urban and rural population centers contributed greatly to improved survival and decrease in cholera and typhoid epidemics. However, globally, waterborne disease remains a major cause of morbidity and mortality especially among children to the present day.

Germ Versus Miasma Theories

Until the early and middle parts of the nineteenth century, the causation of disease was hotly debated. The miasma theory, holding that disease was the result of environmental emanations or miasmas, went back to Greek and Roman medicine, and Hippocrates (Air, Water, and Places). Miasmists believed that disease was caused by infectious mists or noxious vapors emanating from filth in the towns and that the method of prevention of infectious diseases was to clean the streets of garbage, sewage, animal carcasses, and wastes that were features of urban

living. This provided the basis for the Sanitary Movement, with great benefit to improving health conditions. The miasma theory had strong proponents well into the later part of the nineteenth century.

The contagion or germ theory gained ground, despite the lack of scientific proof, on the basis of biblical and Middle Ages experience with isolation of lepers and quarantine of other infectious conditions. In 1546, Fracastorus published *De Contagione,* a treatise on microbiological organisms as the case of specific diseases. The germ theory was strengthened by the work of Antony van Leeuwenhoek, who invented the microscope in 1676. The invention of this apparatus is considered to be a watershed in the history of science. His research showing small microorganisms led to his recognition as a Fellow of the Royal Society of England in 1680. The germ theorists believed that microbes, such as those described by van Leeuwenhoek, were the cause of diseases which could be transmitted from person to person or by contact with sewage or contaminated water.

Major contributions to resolving this issue came from the epidemiologic studies of Snow and Budd in the 1850s, proving waterborne transmission of cholera and typhoid. The classic study of a measles epidemic in the remote Faroe Islands by Peter Panum in 1846 clearly showed person-to-person transmission of this disease, its incubation period, and the lifelong natural immunity exposure gives (Box 1.4). The dispute continued, however, with miasmists or sanitationists arguing with equal vehemence.

While the science of the issue was debated until the end of the nineteenth century, the practical application of sanitary reform was promoted by both theories. Increasing attention to sewage, water safety, and removal of waste products by organized municipal activities was adopted in European and North American cities. The sanitary revolution proceeded while the debates raged and solid scientific proof of the germ theory accumulated, primarily in the 1880s. Fear of cholera stimulated New York City to establish a Board of Health in 1866. In the city of Hamburg, Germany, a Board of Health was established in 1892 only after a cholera epidemic attacked the city, while neighboring Altona remained cholera-free because it had established a water filtration plant.

The specific causation of disease (the germ theory) has been a vital part of the development of public health. The bacteriologic revolution (see later section entitled "Bacteriologic Revolution"), led by the work of Louis Pasteur and Robert Koch, provided enormous benefit to medicine and public health. But those who argued that disease is environmental in origin (the miasma theory) also contributed to public health because of their recognition of the importance of social or other environmental factors, such as poor sanitation and housing conditions or nutritional status, all of which increase susceptibility to specific agents of disease, or the severity of disease.

Box 1.4 Panum on Measles in the Faroe Islands, 1846

Peter Ludwig Panum, a 26-year-old newly graduated medical doctor from the University of Copenhagen, was sent to the Faroe Islands by the Danish government to investigate an outbreak of measles in 1846. There had been no measles since 1781 in the islands, located in the far reaches of the North Atlantic. During the 1846 epidemic, about 6000 of the 7782 islanders were stricken with measles and 102 of them died of the disease or its sequelae. Panum visited all isolated corners of the islands, tracing the chain of transmission of the disease from location to location, and the immunity of those exposed during the 1781 epidemic.

From his well-documented observations he concluded, contrary to prevailing opinion, that measles is a contagious disease spread from person to person, and that one attack gives lifelong immunity. His superb report clearly demonstrated the contagious nature of the disease, its incubation period, and that it is not a disease of "spontaneous generation," nor is it generally dispersed in the atmosphere and spread as a "miasma," proving that isolation of cases was an effective intervention.

Despite availability of an inexpensive, highly effective, and safe vaccine since the 1960s, and elimination of domestic circulation of the virus in many countries, measles remains a serious global health problem in 2008. An estimated 250 thousand children died of measles in 2006, and outbreaks of measles are occurring in countries that were thought to be measles free as a result of imported cases and local spread. Measles elimination is possible with the two dose policy with current vaccines if immunization is given high priority and with determined national and international priority and efforts.

Source: Panum, P. L. *Observations Made During the Epidemic of Measles on the Faroe Islands in the Year 1846.* From: Roueche, B. (ed.) op. cit. WHO. Measles, http://www.who.int/mediacentre/factsheets/fs286/en/ [accessed April 27, 2008]

HOSPITAL REFORM

Hospitals developed by monasteries as charitable services were supplanted by voluntary or municipal hospitals mainly for the poor during and after the Renaissance. Reforms in hospital care evolved along with the sanitary revolution. In eighteenth-century Europe, hospitals operated by religious orders of nuns and by municipal or charitable organizations were dangerous cesspools of pestilence because of lack of knowledge about and practice of infection control, concentration of patients with highly communicable diseases, and transmission of disease by medical and other staff.

Reforms in hospitals in England were stimulated by the reports of John Howard in the late eighteenth century, becoming part of wider social reform in the early part of the nineteenth century. Professional reform in hospital organization and care started in the latter half of the nineteenth century under the influence of Florence Nightingale, Oliver Wendel Holmes, and Ignaz Semmelweiss. Clinical–epidemiologic studies of "antiseptic principles" provided a new, scientific approach to improvement in health care.

In the 1840s, puerperal fever was a major cause of death in childbirth, and was the subject of investigation by Holmes in the United States, who argued that this was due to a contagion. In 1846, Semmelweiss, a Hungarian obstetrician at the Vienna Lying-In Hospital, suspected that deaths from puerperal fever were the result of contamination on the hands of physicians transmitted from autopsy material to living patients. He showed that death rates among women attended by medical personnel were two to five times the rates among those attended by midwives. By requiring doctors and medical students to soak their hands in chlorinated lime after autopsies, he reduced the mortality rates among the medically attended women to the rate of the midwife-attended group.

Semmelweiss's work, although carefully documented, was slow to be accepted by the medical community, taking some 40 years for general adoption. His pioneering investigation of childbed fever (streptococcal infection in childbirth) in Vienna contributed to improvement in obstetrics and a reduction in maternal mortality. In the 1850s, prevention of blindness in newborns by prophylactic use of silver nitrate eyedrops, developed by Karl Crede in Leipzig, spread rapidly through the medical world (Box 1.5). This practice continues to be a standard in prevention of ophthalmia neonatorum.

Florence Nightingale's momentous work in nursing and hospital administration in the Crimean War (1854–1856) established the professions of nursing and modern hospital administration. In the 1860s, she emphasized the importance of poor law and workhouse reform and training special district nurses for care of the sick poor at home. Nightingale's subsequent long and successful campaigns to raise standards of military medicine, hospital planning, supply services and management, hospital statistics, and community health nursing were outstanding contributions to the development of modern, organized health care and antisepsis.

Despite all the cumulative progress over the past 150 years, such as the advent of sterile techniques and antibiotics, hospital-acquired infection remains a serious public health problem to the present time with multidrug-resistant organisms and a persistent failure of regular use of simple hand washing between patient care by doctors and nurses, and in antiseptic measures need for invasive procedures such as central venous and bladder catheters.

Box 1.5 Crede and Prevention of Gonococcal Ophthalmia Neonatorum

Gonorrhea was common in all levels of society in nineteenth-century Europe and ophthalmic infection of newborns was a widespread cause of infection, scarring, and blindness. Carl Franz Crede, professor of obstetrics at the University of Leipzig, attempted to treat neonatal gonococcal ophthalmic infection with many medications. Crede discovered the use of silver nitrate as a treatment and introduced its use as a preventive measure during the period 1854–1860 with astonishing success.

The prophylactic use of silver nitrate spread rapidly hospital by hospital, but it was decades before it was mandated widely because of widespread medical opposition to this innovation. It was only in 1879 that the gonococcus organism was discovered by Neisser. Estimates of children saved from blindness by this procedure in Europe during the nineteenth century are as high as one million.

Source: Schmidt A. 2007. Gonorrheal ophthalmia neonatorum. In *Pediatric Infectious Diseases Revisited*. Birkhäuser Basel. http://www.springerlink.com/content/xtu8475716207264/ [accessed April 27, 2008]

THE BACTERIOLOGIC REVOLUTION

In the third quarter of the nineteenth century, the sanitary movement rapidly spread through the cities of Europe with demonstrable success in reducing disease in areas served by sewage drains, improved water supplies, street paving, and waste removal. At the same time, innovations occurred in hospitals, stressing hygiene and professionalization of nursing and administration. These were accompanied by breakthroughs in establishing scientific and practical applications of bacteriology and immunology.

Pasteur, Cohn, Koch, and Lister

In the 1850s to 1870s, Louis Pasteur, a French professor of chemistry, brilliantly developed the basis for modern bacteriology as a cornerstone of public health. He established a scientific, experimental proof for the germ theory with his demonstration in 1854 of anaerobic microbial fermentation. Between 1856 and 1860 he showed how to prevent wine from spoilage due to contamination from foreign organisms by heating the wine to a certain temperature before bottling it to kill the undesired ferments. This led to the process of "pasteurization." Asked to investigate the threatened destruction of the French silk industry by epidemics destroying the silkworms, he discovered microorganisms (1865) causing the disease and devised new growing conditions which eliminated the problem, raising scientific and industrial interest in the germ theory. This was followed by similar work in the beer industry (1871).

Pasteur went on to develop the science of immunology by working with vaccines. He produced vaccines from attenuation or weakening an organism's strength by passing it successively through animals, recovering it and retransmitting it to other animals. In 1881, he inoculated hens with attenuated cultures of chicken cholera and then in an inspired experiment challenged them with virulent organisms and found them to be immune. In 1883, he produced a similar protective vaccine for swine erysipelas, and then in 1884–1885, a vaccine for rabies.

Rabies was widely feared as a disease transmitted to humans through bites of infected animals and was universally fatal. Pasteur reasoned that the disease affected the nervous system and was transmitted in saliva. He injected material from infected animals, attenuated to produce protective antibodies but not the disease. In 1885, a 14-year-old boy from Alsace was severely bitten by a rabid dog. Local physicians agreed that because death was certain, Pasteur, a chemist and not a physician, be allowed to treat the boy with a course of immunization. The boy, Joseph Meister, survived, and similar cases were brought to Pasteur and the person successfully immunized. Pasteur was criticized in medical circles, but both the general public and scientific circles soon recognized his enormous contribution to public health.

Ferdinand Julius Cohn (1828–1898), professor of Botany at Breslau University, developed and systematized the science of bacteriology using morphology, staining, and media characteristics of microorganisms, and trained a key generation of microbiological investigators. One student, Robert Koch (1843–1910), a German rural district medical officer, investigated anthrax using mice inoculated with blood from sick cattle, transmitting the disease for more than 20 generations. He developed basic bacteriologic techniques including methods of culturing and staining bacteria. He demonstrated the organism causing anthrax, recovered it from sick animals, and passed it through several generations of animals, proving the transmission of specific disease by specific microorganisms.

In 1882, Koch demonstrated and cultured the tubercle bacillus. He then headed the German Cholera Commission visiting Egypt and India in 1883, isolating and identifying *Vibrio cholerae* (Nobel Prize, 1905).

He demonstrated the efficacy of water filtration in preventing transmission of enteric disease including cholera. In 1883, Koch, adapting postulates on causation of disease from clinician–pathologist Jacob Henle (1809–1885), established criteria for attribution of causation of a disease to a particular parasite or agent (Box 1.6). These were fundamental to establishment of the science of bacteriology and the relationship of microorganisms to disease causation.

The Koch-Henle postulates in their pure form were too rigid, and would limit identification of causes of many diseases, but they were important in establishing germ theory and the scientific basis of bacteriology, dispelling the many other theories of disease still widespread in the late nineteenth century. These postulates served as guidelines for evidence of causation, but had limitations in that not all microbiologic agents can be grown in pure culture, some organisms undergo antigenic drift or change in antigenicity, and there is no animal host for some organisms. Koch's postulates were later adapted by Evans (1976) to include noninfectious disease-causing agents, such as cholesterol, following the changing emphasis in epidemiology of noninfectious diseases.

In the mid-1860s, Joseph Lister in Edinburgh, under the influence of Pasteur's work and with students of Semmelweiss, developed a theory of "antisepsis." His 1865 publication *On the Antiseptic Principle in the Practice of Surgery* described the use of carbolic acid to spray operating theaters and to cleanse surgical wounds, applying the germ theory with great benefit to surgical outcomes. Lister's work on chemical disinfection for surgery in 1865 was a pragmatic development and a major advance in surgical practice; an important contribution to establishing the germ theory in nineteenth-century medicine.

Box 1.6 The Koch-Henle Postulates on Microorganisms as the Cause of Disease

1. The organism (agent) must be shown to be present in every case of the disease by isolation in pure culture;
2. The agent should not be found in cases of any other disease;
3. Once isolated, the agent should be grown in a series of cultures, and then must be capable of reproducing the disease in experimental animals;
4. The agent must then be recovered from the disease produced in experimental animals.

Source: Last, J. M. 2001. A *Dictionary of Epidemiology.* 4th ed. New York: Oxford University Press.
Last, J. M. 2006. A *Dictionary of Public Health* and http://www.medterms.com/script/main/art.asp?articlekey=7105 [accessed December 24, 2007]

Vector-Borne Disease

Studies of disease transmission defined the importance of carriers (i.e., those who can transmit a disease without showing clinical symptoms) in transmission of diphtheria, typhoid, and meningitis. This promoted studies of diseases borne by intermediate hosts or vectors. Parasitic diseases of animals and man were investigated in many centers during the nineteenth century, including Guinea worm disease, tapeworms, filariasis, and veterinary parasitic diseases such as Texas cattle fever. David Bruce demonstrated transmission of nagana (animal African trypanosomiasis), a disease of cattle and horses in Zululand, South Africa, in 1894–1895, caused by a trypanosome parasite transmitted by the tsetse fly, leading to environmental methods of control of disease transmission. Alexandre Yersin and Shibasabro Kitasato discovered the plague bacillus in 1894, and in 1898 French epidemiologist P. L. Simmond demonstrated the plague was a disease of rats spread by fleas to humans.

Malarial parasites were identified by French army surgeon Alphonse Laveran (Nobel Prize, 1907) in Algeria in 1880. Mosquitoes were suspected as the method of transmission by many nineteenth-century investigators, and in 1897 Ronald Ross (Nobel Prize, 1902), a British army doctor in India, Patrick Manson in England, and Benvenuto Grassi in Rome demonstrated transmission of malaria by the *Anopheles* mosquito. Yellow fever, probably imported by the slave trade from Africa, was endemic in the southern United States but spread to northern cities in the late eighteenth century. An outbreak in Philadelphia in 1798 killed nearly 8 percent of the population. Outbreaks in New York killed 732 people in 1795, 2086 in 1798, and 606 in 1803. The Caribbean and Central America were endemic with both yellow fever and malaria.

The conquest of yellow fever also contributed to establishing the germ or contagion theory versus the miasma theory when the work of Cuban physician Carlos Finlay was confirmed by Walter Reed in 1901. His studies in Cuba proved the mosquito-borne nature of the disease as a transmissible disease via an intermediate host (vector) but not contagious between humans. William Gorgas applied this to vector control activities and protection of sick persons from contact with mosquitoes, resulting in an eradication of yellow fever in Havana within 8 months, and in the Panama Canal Zone within 16 months (Box 1.7). This work showed a potential for control of vector-borne disease that has had important success in control of many tropical diseases, including yellow fever and, currently, Guinea worm disease and onchocerciasis. Malaria, although it has come under control in many parts of the world, has resurged in many tropical countries since the 1960s.

Box 1.7 Havana and Panama: Control of Yellow Fever and Malaria, 1901–1906

The United States Army Commission on Yellow Fever led by Walter Reed, an Army doctor, worked with Cuban physicians Carlos Finlay and Jesse Lazear to experiment with yellow fever transmission in Cuba in 1901. Working with volunteers, he demonstrated transmission of the disease from person to person by the specific mosquito, *Stegomyia fasciata*. The Commission accepted that "the mosquito acts as the intermediate host for the parasite of Yellow Fever."

Another U.S. army doctor, William Gorgas, applied the new knowledge of transmission of yellow fever and the life cycle of the vector mosquito. He organized a campaign to control the transmission of yellow fever in Havana, isolating clinical cases from mosquitoes and eliminating the breeding places for the *Stegomyia* with Mosquito Brigades. Yellow fever was eradicated in Havana within 8 months. This showed the potential for control of other mosquito-borne diseases, principally malaria with its specific vector, *Anopheles*. Gorgas then successfully applied mosquito control to prevent both yellow fever and malaria between 1904 and 1906, permitting construction of the Panama Canal.

MICROBIOLOGY AND IMMUNOLOGY

Ilya Ilyich Mechnikov in Russia in 1883 described phagocytosis, a process in which white cells in the blood surround and destroy bacteria, and his elaboration of the processes of inflammation and humoral and cellular response led to a joint Nobel Prize in 1908 with Paul Ehrlich. Other investigators searched for the bactericidal or immunological properties of blood that enabled cell-free blood or serum to destroy bacteria. This work greatly strengthened the scientific bases for bacteriology and immunology.

Pasteur's co-workers, Emile Roux and Alexandre Yersin, isolated and grew the causative organism for diphtheria and suggested that the organism produced a poison or toxin which caused the lethal effects of the disease. In 1890, Karl Fraenkel in Berlin published his work showing that inoculating guinea pigs with attenuated diphtheria organism could produce immunity. At the same time, Emile Behring in Germany with Japanese co-worker Shibasaburo Kitasato produced evidence of immunity in rabbits and mice to tetanus bacilli. Behring also developed a protective immunization against diphtheria in humans with active immunization as well as an antitoxin for passive immunization of an already infected person (Nobel Prize, 1901). By 1894, diphtheria antitoxin was ready for general use. The isolation and identification of new disease-causing organisms proceeded rapidly in the last decades of the nineteenth century. The diphtheria organism

was discovered in 1885 by Edwin Klebs and Friedrich Loeffler (students of Koch), and a vaccine for it was developed in 1912, leading to the control of this disease in many parts of the world. Between 1876 and 1898, many pathogenic organisms were identified, providing a basis for advances in vaccine development.

During the last quarter of the nineteenth century, it was clear that inoculation of attenuated microorganisms could produce protection through active immunization of a host by generating antibodies to that organism, which would protect the individual when exposed to the virulent (wild) organism. Passive immunization could be achieved in an already infected person by injecting the serum of animals infected with attenuated organisms. The serum from that animal helps to counter effects of the toxins produced by an invading organism. Pasteur's vaccines were followed by those of Waldemar Haffkine, a bacteriologist working in India; the first microbiologist to develop and use vaccines against cholera and bubonic plague, after testing on himself. Other pioneering achievements include those of Richard Pfeiffer and Carroll Wright for typhoid, Albert Calmette and Alphonse Guerin for tuberculosis, and Arnold Theiler and Theobald Smith for yellow fever.

The twentieth century has seen the flowering of immunology in the prevention of important diseases in animals and in man based on the pioneering work of Jenner, Pasteur, Koch, and those who followed. Many major childhood infectious diseases have come under control by immunization in one of the outstanding achievements of twentieth-century public health.

Poliomyelitis

Poliomyelitis was endemic in most parts of the world prior to World War II, causing widespread crippling of infants and children, hence its common name of "infantile paralysis." The most famous polio patient was Franklin Delano Roosevelt, crippled by polio in his early 30s, who went on to become president of the United States. During the 1940s and 1950s, poliomyelitis occurred in massive epidemics affecting thousands of North American children and young adults, with national hysteria of fear of this disease because of its crippling and killing power. In 1952, 52,000 cases of poliomyelitis were reported in the United States, bringing a national response and support for the "March of Dimes" Infantile Paralysis Association for research and field vaccine trials.

Based on the development of methods for isolating and growing the virus by John Enders and colleagues, Jonas Salk developed an inactivated vaccine in 1955 and Albert Sabin a live attenuated vaccine in 1961. Salk's field trial proved the safety and efficacy of his vaccine in preventing poliomyelitis. Sabin's vaccine proved to be cheaper and easier to use on a mass basis and is still the mainstay of

---

Box 1.8 Enders, Salk, Sabin, and Eradicating Poliomyelitis

In the early 1950s, John Enders and colleagues developed methods of growing polio virus in laboratory conditions, for which they were awarded a Nobel Prize. At the University of Pittsburgh, Jonas Salk (1914–1995) developed the first inactivated (killed) vaccine under sponsorship of a large voluntary organization which mobilized the resources to fight this dreaded disease. Salk conducted the largest field trial ever involving 1.8 million children in 1954. The vaccine was rapidly licensed and quickly developed and distributed in North America and Europe, interrupting the epidemic cycle and rapidly reducing polio incidence to low levels.

Albert Sabin (1906–1994) at the University of Cincinnati developed a live, attenuated vaccine given orally (OPV), which was approved for use in 1961. This vaccine has many advantages: it is given easily, spreads its benefits to nonimmunized persons, and is inexpensive. It became the vaccine of choice and was used widely reducing polio to a negligible disease in most developed countries within a few years. Sabin also pioneered application of OPV through national immunization days in South America which contributed to control of polio there, and more recently, in other countries such as China and India.

In 1987, the World Health Organization declared the target of eradication of poliomyelitis by the year 2000. With the help of international and national commitm_____ ricas were declared polio-free in 1990. Poli_____ _____tinues in only several countries _____ _____ _____ _____s-mission of the disease _____ _____cine is being adopted by _____ continues to be used in most p_____

SARS
seen Acut Respartory Syndrom

---

polio eradication worldwide (Box 1.8). Conquest of this dreaded, disabling disease has provided one of the most dramatic achievements of public health in the twentieth and early twenty-first centuries with good prospects for elimination of poliomyelitis by 2010 (see Box 1.8).

Advances in Treatment of Infectious Diseases

Since World War II, _____ s in immunology as applied to public health led t_____ l and in some cases potential eradication of _____ rtussis, tetanus, poliomyelitis, measles, _____ and more recently hepatitis B an_____ _____zae type b. The future in this field _____ lay a central role in public health _____ st century.

Important

Treatment o_____ has also played a vital part in reducing _____ e and limiting its spread. In 1909, Paul Ehr_____ the Nobel Prize in 1908 jointly with Meth_____ seeking a "magic bullet," discovered an effect_____ imicrobial agent for syphilis (Salvarsan). Later more_____ portant antimicrobial antibiotics were discovered in the 1920s, followed by the sulfa drugs in the 1930s, and the antibiotics and penicillin and streptomycin in the 1940s by Alexander Fleming and Selman Waksman (Nobel Prizes, 1945 and 1952). These and later generations of antibiotics have proven powerful tools in the treatment of infectious diseases.

Antibiotics and vaccines, along with improved nutrition, general health, and social welfare, led to dramatic reductions in infectious disease morbidity and mortality. As a result, optimistic forecasts of the conquest of communicable disease led to widespread complacency in the medical and research communities by the late twentieth century. In the 1990s, organisms resistant to available antibiotics constituted a major problem for public health and health care systems. Resistant organisms are now evolving as quickly as newer generation antimicrobials can be developed, threatening a return of diseases once thought to be under control. The pandemic of AIDS and other emerging and re-emerging diseases like SARS will require new strategies in treatment and prevention including new vaccines, antibiotics, chemotherapeutic agents, and risk reduction through community education.

MATERNAL AND CHILD HEALTH

Preventive care for the special health needs of women and children developed as public concerns in the late nineteenth century. Public awareness of severe conditions of women's and children's labor grew to include the effects on health of poverty, poor living conditions and general hygiene, home deliveries, lack of prenatal care, and poor nutrition.

Preventive care as a service separate from curative medical services for women and children was initiated in the unsanitary urban slums of industrial cities in nineteenth-century France in the form of milk stations (*gouttes de lait*). One village in France instituted an incentive payment to mothers whose babies lived to 1 year; this resulted in a decline in infant mortality from 300 per 1000 to 200 per 1000 within a few years. The plan was later expanded to a complete child welfare effort, especially promoting breast-feeding and a clean supply of milk to children, which had dramatic effects in reducing infant deaths.

The concept of child health spread to other parts of Europe and the United States with the development of pediatrics as a specialty and an emphasis on appropriate child nutrition. Henry Koplik in 1889 and Nathan Strauss in 1893 promoted centers to provide safe milk to pregnant women and children in the slums of New York City in order to combat summer diarrhea. The Henry Street Mission, serving poor immigrant areas, developed the model of visiting nurses and public milk stations. The concept of the "milk station," combined with home visits, was pioneered by Lillian Wald, who coined the term *district nurse* or *public health nurse*. This became the basis for public prenatal, postnatal, and well-child care as well as school health supervision. Visiting Nursing Associations (VNAs) gradually developed throughout the United States to provide such services. Physicians' services in the United States were mainly provided on a fee-for-service basis for those able to pay, with charitable services in large city hospitals. The concept of direct provision of care to those in need by local authorities and by voluntary charitable associations, with separation between preventive and curative services, is still a model of health care in many countries.

In Jerusalem, from 1902, Shaare Zedek Hospital kept cows to provide safe milk for infants and pregnant women. In 1911, two public health nurses came from New York to Jerusalem to establish milk stations (*Tipat Halav*, "drop of milk") for poor pregnant women and children. This model became the standard method of Maternal and Child Health (MCH) provision throughout Israel, operating parallel to the Sick Funds which provided medical care. The separation between preventive and curative services persists to the present, and is sustained by the Israeli national government's obligation to assure basic preventive care to all regardless of insurance or ability to pay.

In the Soviet Union, institution of the state health plan in 1918 by Nikolai Aleksandrovich Semashko gave emphasis to maternal and child health, along with epidemic and communicable disease control. All services were provided free as a state responsibility through an expanding network of polyclinics and other services, and prenatal and child care centers, including preventive checkups, home visits, and vaccinations. Infant mortality declined rapidly even in the Asian republics with previously poor health conditions.

During the 1990s, the United States was having difficulty immunizing children in high poverty areas of urban centers and adopted immunization as part of their Women, Infants, and Children (WIC) food support program for poor pregnant women and children, helping to achieve much higher coverage levels in years since.

The emphasis placed on maternal and child health continues to be a keystone of public health. Care of children and women in relation to fertility is the application of what later came to be called the "risk approach," where attention is focused on designing health programs for the most vulnerable groups in the population.

NUTRITION IN PUBLIC HEALTH

As infectious disease control and later maternal and child health became public health issues in the eighteenth to nineteenth centuries, nutrition gained recognition from the work of pioneers such as James Lind (see preceding section entitled "Applied Epidemiology"). In 1882, Kanehiro Takaki, surgeon-general of the Japanese navy, reduced incidence of beriberi among naval crews by adding meat and vegetables to their diet of rice. In 1900, Christiaan Eijkman, a Dutch medical officer in the East Indies, found that inmates of prison camps who ate polished rice developed beriberi, while those eating whole rice did not. He also produced beriberi experimentally in fowls on a diet of polished rice, thus establishing the etiology of the disease as a deficiency condition and fulfilling a nutritional epidemiologic hypothesis. Eijkman was awarded the Nobel Prize in physiology or medicine in 1929.

In the United States, the pioneering Pure Food and Drug Act was passed in 1906, stimulated by journalistic exposures of conditions in the food industry and Upton Sinclair's famous 1906 novel *The Jungle*. The legislation established federal authority in food and labeling standards, originally for interstate commerce, but later for the entire country. This provided for a federal regulatory agency and regulations for food standards. The Food and Drug Administration (FDA) has pioneered nutritional standards now used throughout the world.

In the early part of the twentieth century, the U.S. Department of Agriculture (USDA) supported "land grant colleges" and rural counties to establish an extension service to promote agricultural improvement and good nutrition in poor agricultural areas of the country. These services, along with local women's organizations, helped create a mass movement to improve good nutrition, canning surplus foods, house gardening, home poultry production, home nursing, furniture refinishing, and other skills that helped farm families survive the years of economic depression and drought, promoting better nutrition through education and community participation.

In 1911, the chemical nature of vitamin D was discovered, and a year later, Kasimir Funk coined the term vitamin ("vital amine"). In 1914, Joseph Goldberger of the U.S. Public Health Service established the dietary causes of pellagra and in 1928 he discovered the pellagra-preventing factor in yeast (Box 1.9). In 1916, U.S. investigators defined fat-soluble vitamin A and water-soluble vitamin B, which was later shown to be more than one factor. In 1922, Elmer McCollum identified vitamin D in cod liver oil, which became a staple in child care for many decades. In the period 1931–1937, fluoride in drinking water was found to prevent tooth decay, and in 1932 vitamin C was isolated from lemon juice.

Iodization of salt to prevent iodine deficiency disorders (IDD) has been one of the greatest successes and

---

Box 1.9 Goldberger on Pellagra

"Mal de la rosa," first described in Spain by Casal in 1735, was common in northern Italy, when, in 1771, Frappolli described "pelle agra" or farmers' skin, common among poor farmers whose diet was mainly corn flour. In 1818, Hameau described a widespread skin disease among poor farmers in southern France. Roussel investigated and concluded that pellagra was endemic and due to poverty rather than a diet heavy in corn. His recommended reforms were implemented by the Department of Agriculture and raised standards of living among the poor farmers, including growing wheat and potatoes instead of corn, and the disease disappeared by the beginning of the twentieth century. Similar measures in Italy reduced growth of corn, and here too the disease disappeared.

The disorder was thought to be due to a toxin in raw corn or produced by digestion in the intestine. Lambrozo, in Verona, Italy, reported many cases of pellagra among mental hospital patients, concluding that it was due to toxic material in corn. At the beginning of the twentieth century, the corn theory was less accepted and the common view was that pellagra was an infectious disease. British investigator L. V. Sambon, discoverer of the role of the tsetse fly in trypanosomiasis in 1910 took the view that the disease was mosquito-borne.

Pellagra was first reported in the United States in 1906 as an epidemic in a mental hospital in Alabama. In the first decades of the twentieth century, pellagra was considered the leading public health problem in the southern United States, where poverty was rampant. The medical community had no ideas as to the cause or prevention of this widespread disease, generally believed to be infectious in origin.

In 1913, Joseph Goldberger was appointed by the U.S. Public Health Service to investigate pellagra. He had previously worked on yellow fever, dengue, and typhus. He visited psychiatric hospitals and orphanages with endemic pellagra and was struck by the observation that the staff was not affected, suggesting that the disease was not infectious but may have been due to the diet. In one mental hospital, he eliminated pellagra by adding milk and eggs to the diet and concluded that the disease was due to a lack of vitamins and preventable by a change in diet alone. Goldberger, trained in infectious disease, was able to recognize non-transmission from patients to staff and went on to establish the nutritional basis of this disease and, along with Lind, established nutritional epidemiology in public health.

---

failures of twentieth-century public health. From studies in Zurich and in the United States, the efficacy of iodine supplements in preventing goiter was demonstrated. Morton's iodized salt became a national standard in the United States; an early and noble example of voluntary public health action by private industry. In Canada in 1979, iodized salt became mandatory along with other vitamin and mineral fortification of bread and milk (see Chapter 8).

Rickets, still common in industrialized countries prior to World War II and into the 1950s, virtually disappeared following fortification of milk with vitamin D. Prevention of IDD by salt iodization has become an important goal in international health, and progress is being made toward universal iodization of salt in many countries where goiter, cretinism, and iodine deficiency are still endemic.

The international movement to promote proper nutrition is vital to reduce the toll of the malnutrition–infection cycle in developing countries. No less important is prevention of noncommunicable diseases associated with over-nutrition, including cardiovascular diseases, diabetes, and some cancers in industrialized nations. Nutrition is a key issue in the New Public Health, with international movements to eradicate vitamin and mineral (micronutrient) deficiency conditions, all of which are important, widespread, and preventable.

MILITARY MEDICINE

Professional armies evolved with urban civilizations and developed in the ancient world from about 4000 BCE. Since organized conflict began, armies have had to deal with the health of soldiers as well as treatment of the wounded. Injunctions on military and civilian camp siting and sanitation were clearly spelled out in the Bible (Old Testament). Roman armies excelled at construction of camps with care and concern for hygienic conditions, food, and medical services for the soldiers. Throughout history, examples of defeat of armies by disease and lack of support services prove the need for serious attention to the health and care of the soldier. Studies of casualties of war in major conflicts contribute not only to military medicine but to knowledge of the care of civilian populations in natural or man-made disasters.

As the armies and weapons become increasingly powerful, the care of the sick and wounded became more complex. Military medicine perfected knowledge and skills in taking care of wounded at the battlefield and preventing loss of life. Epidemics in armies have killed more troops than have weapons, so treating and preventing disease are a part of military medicine. Many medical discoveries have been implemented in the army and later in civil society, including surgery, vaccination, antibiotics, nutrition, and others.

The Roman Empire developed military medicine as professional armies spread across the known world. The Roman army included physicians to provide medical care for the legions, beginning with ensuring that only the fittest (and most intelligent) candidates were recruited. Once in service, the military medical corps strove to ensure the general health of the soldier by a continuous stress of hygiene. The design of legion fortifications and encampments ensured a healthy environment for the troops. Following the destruction of Rome and later the Eastern Empire, the Roman military medical tradition disappeared. Military medicine during the Middle Ages was relatively primitive.

Jean Henri Dunant, a young Swiss businessman, arrived in Solferino, Italy, on the evening of the battle fought on June 24, 1859, between the French-Sardinian allies against the Austrian army. Some 38,000 injured, dying, and dead soldiers remained on the battlefield, with little attempt to provide care. Dunant took the initiative to organize volunteers from the local civilian population, especially the women and girls, to provide assistance to the injured and sick soldiers. He organized the purchase of needed materials and helped erect makeshift hospitals, providing care for all without regard to their affiliation in the conflict, along with volunteer doctors and Austrian doctors captured by the French.

After returning to Geneva, Dunant published a book about his experience, *A Memory of Solferino*, describing the battle, its costs, and the chaos afterwards. He proposed that in the future a neutral organization should be established to provide care to wounded soldiers. His work led to the First Geneva Convention on the treatment of non-combatants and prisoners of war, followed by the foundation of the International Committee of the Red Cross in 1863, now ratified by 194 countries. He was awarded the first Nobel Prize for Peace in 1901.

The Crimean War was a medical disaster for the British Army with higher mortality from disease than from battle largely as a result of poorly organized sanitation, supply, and medical services. Mortality rates among British amputees in the Scutari Hospital averaged nearly 30 percent. Of every 100 men in the French forces admitted to military hospitals, 42 percent died — a hospital mortality rate equivalent to that of the Middle Ages. Florence Nightingale with her 18 trained nurses in November 1854 after the battle of Balaklava introduced basic standards of hygiene, nutrition and sanitation, and administration in the British military hospitals. Upon her arrival, Nightingale reported a hospital mortality rate at Scutari of 44 percent. As a result of her efforts, the rate dropped to 2 percent by the end of the war.

Nightingale's work made an enormous contribution to knowledge and practice of hospital organization and management. On the opposing side of the same Crimean War, Nikolai Perogov, a military surgeon in the Russian czarist

[handwritten note: Beriberi — disease from lack of vitamin B-1 (thiamin)]

army, developed rectal anesthesia for field surgery, triage by degree of severity, and hygiene of wounds. He defined improved systems for management used in war theaters, which had applicability to hospitals. The French army in World War I developed the triage system of casualty clearance used worldwide in military and disaster situations.

Naval sailors on long sea voyages and the epidemiologic study of scurvy, followed a century later by the work on beriberi, were important steps in identifying nutrition and its importance to public health. Bismarck's establishment of national health insurance and other benefits for workers was partly based on the need to improve the health of the general population in order to build mass armies of healthy conscripts (see Chapter 13). During conscription to the U.S. Army in World War I, high rates of rejection of draftees as medically unfit for military service raised concerns for national health standards. Finding high rates of goiter in the draftees led to efforts to identify high-risk areas and to reduce iodine deficiency in the civilian population by iodization of salt.

In the wake of World War I, a massive pandemic of influenza killed some 20 million people. The Spanish flu pandemic lasted from 1918 to 1919. Older estimates say it killed 40 to 50 million people, while current estimates say 50 to 100 million people worldwide died in this pandemic, described as "the greatest medical holocaust in history" and may have killed as many people as the Black Death. The Spanish flu, closely following the huge losses of World War I, was to a large degree spread in the close quarters of army camps and mass movement of troops, with a high percentage of the deaths occurring among young men, the group most affected by the war itself.

Epidemics of louse-borne typhus in Russia following the war and the Russian Revolution contributed to the chaos of the period. This prompted Lenin's statement "Either socialism will conquer the louse, or the louse will conquer socialism." In World War II, sulfa drugs, antimalarials, and antibiotics made enormous contributions to the Allied war effort and later to general health care and preserving the health of the population. Lessons learned in war for protection of soldiers from disease and treatment of burns, crash injuries, amputations, battle fatigue, and many other forms of trauma were brought back to civilian health systems. Much of modern medical technology was first developed for or tested by the military. As an example, sonar radio wave mechanisms developed to detect submarines were adapted after World War II as ultrasound, now a standard noninvasive instrument in medical care.

In the twentieth century, the destructiveness of war increased enormously with chemical, biological, and nuclear weapons of mass destruction. The Nuremberg Trials addressed the Holocaust and unethical medical experimentation of the Nazi military on civilian and military prisoners. The International Declaration of Human Rights (1948) and the Helsinki Declaration (1964) set new standards for medical and research ethics, with important implications for public health.

The brutalities of wars against civilian populations have tragically recurred even near the end of the twentieth century in genocidal warfare in Iraq, Rwanda, and the former Yugoslavia. Those tragedies produced massive casualties and numerous refugees, with resultant public health crises requiring intervention by local and international health agencies. International and national public health agencies have major responsibility for prevention of extension of some of the mass tragedies of the twentieth century recurring, perhaps on a bigger scale in the twenty-first century with the spread of the potential for chemical, biological, and potentially nuclear terrorism.

INTERNATIONALIZATION OF HEALTH

Cooperation in health has been a part of international diplomacy from the first international conference on cholera in 1851 in Cairo to the health organization of the League of Nations after World War I, and into modern times. Following World War II, international health began to promote widespread application of public health technology, such as immunization, to developing countries. The World Health Organization (WHO) was founded in 1946 with a charter defining health as "the complete state of physical, social and mental well-being, and not merely the absence of disease."

The tradition of international cooperation is continued by organizations such as WHO, the International Red Cross/Red Crescent (IRC), United Nations Children's Fund (UNICEF), and many others. Under the leadership of WHO, eradication of smallpox by 1977 was achieved through united action, showing that major threats to health could be controlled through international cooperation. The potential for eradication of polio further demonstrates this principle.

The global spread of disease has taken enormous tolls of human life with global proportions and the threat continues in the twenty-first century. Globalization of public health threats can emerge and spread rapidly, as seen with the HIV pandemic since the 1980s and SARS in 2003. More recently, concerns have grown for potentially devastating pandemic influenza, such as the H5N1 virus strain known as avian influenza. Chronic diseases, the commonest causes of mortality and disability in the industrialized countries in the latter half of the twentieth century, are now also predominant in most developing countries with growing middle-class communities. Tobacco use, obesity, diabetes, heart disease, and cancer are among the leading causes of morbidity and mortality in the contemporary world. The toll of violence, often overlooked as a public health problem, cannot be overstated. Many other factors affect health globally, including environmental degradation with global warming, accumulating ozone and toxic wastes, acid rain, nuclear accidents, loss of nature reserves such as the Amazon basin rain forests, and the human tragedy of chronic poverty of many developing countries. Global health issues are by their very nature beyond the capacity of individual or even groups of countries to solve. They require organized common efforts of governments, international agencies, and nongovernmental organizations to cooperate with each other, with industry, and with the media to bring about change and reduce the common hazards that abuse of the environment and social gaps cause.

Tobacco is the leading preventable risk factor for premature mortality worldwide. Estimates of 5 million annual deaths attributable to smoking do not adequately measure the impact of the tobacco pandemic, as tobacco use contributes and exists as a co-morbidity factor to a very wide spectrum of disease and has justly become a major issue of public health. Over half of the estimated 650 million people currently smoking will die of effects of this addiction and if current smoking patterns continue, it will cause more than 10 million deaths yearly by 2020. Recognition of tobacco control is gaining support of researchers, medical professionals, politicians, and communities around the world. The CDC sees recognition of tobacco as a public health hazard as one of the greatest public health achievements of the twentieth century in the United States. The WHO's Framework Convention on Tobacco Control adopted by the 56th World Health Assembly in 2003 placed elimination of tobacco use as one of the greatest public health challenges, but one far from fruition.

Bringing health care to all the people is as great a challenge as feeding a rapidly growing global population. Successes in eradication of smallpox and control of many other diseases by public health measures show the potential for concerted international cooperation and action targeted to specific objectives that reduce disease and suffering.

THE EPIDEMIOLOGIC TRANSITION

As societies evolve, so do patterns of disease. These changes are partly the result of public health and medical care but just as surely are due to improved standards of living, nutrition, housing, and economic security, as well as changes in fertility and other family and social factors. As disease patterns change, so do appropriate strategies for intervention.

During the first half of the twentieth century, infectious diseases predominated as causes of death even in the developed countries. Since World War II, a major shift in epidemiologic patterns has taken place in the industrialized countries, with the decline in infectious diseases and an increase in the noninfectious diseases as causes of death. Increases in longevity have occurred primarily from declining infant and child mortality, improved nutrition, control of vaccine-preventable diseases, and the advent of antibiotics for treatment of acute infectious diseases. The rising incidence of cardiovascular diseases and cancer affects primarily older people, leading to a growing emphasis in epidemiologic investigations on causative risk factors for these noninfectious diseases.

Studies of the distribution of noninfectious diseases in specific groups go back many centuries when the Romans reported excess death rates among specific occupational groups. These studies were updated by Ramazzini in the early eighteenth century. As noted earlier, in eighteenth-century London, Percival Potts documented that cancer of the scrotum was more common among chimney sweeps than in the general population. Nutritional epidemiologic studies, from Lind on scurvy among sailors in 1747 to Goldberger on pellagra in the southern United States in 1914, focused on nutritional causes of noninfectious diseases in public health.

Observational epidemiologic studies of "natural experiments" produced enormously important data in the early 1950s, when pioneering investigators in the United Kingdom, Richard Doll, Austin Bradford Hill, and James Peto, demonstrated a relationship between tobacco use and lung cancer. They followed the mortality patterns of British physicians from different causes, especially lung cancer. They found that mortality rates from lung cancer were 10 times higher in smokers than in nonsmokers. Epidemiologic studies pointing to the relationship of diet and hypertension with cardiovascular diseases also provided critically important material for public health policy, and raised public concern and consciousness in western countries of the impact of lifestyle on public health. In this new era of public health, the complementary relationship between miasma and germ theories is recognized (Box 1.10). These issues are discussed subsequently throughout this book.

In the mid-twentieth century, while communicable diseases were coming under control, risks related to modern living developed; cardiovascular disease, trauma, cancer, and other chronic diseases have become the predominant causes of premature death. These are more complex than the infectious diseases, both in causation and the means of prevention. Still, public health interventions have shown surprising success in combating this set of mortality patterns, with a combination of improved medical care and activities under the general title of health promotion.

At the beginning of the twenty-first century, the need to link public health with clinical medical care and organization of services is increasingly apparent. The decline in coronary heart disease mortality is accompanied by a slow increase in morbidity, and recent epidemiologic evidence shows new risk factors not directly related to

---

Box 1.10 Complementary Contributions of the Miasma and Germ Theories

The miasma theory (i.e., the concept that airborne vapors or "miasmata" caused most diseases) in the mid-nineteenth century competed with the germ theory (i.e., specific microorganisms cause specific diseases). The latter gained pre-eminence among scientists and biological sciences, yet the miasma theory was the basis for action by sanitary reformers. Miasma explained why cholera and other diseases were epidemic in places where the undrained sewage water was foul-smelling. Their endeavors led to improved sanitation systems, which led to decreased episodes of cholera. The connection between dirtiness and diseases led to public health reforms and encouraged cleanliness. The miasma theory was consistent with the observations that disease was associated with poor sanitation and foul odors, and that sanitary improvements reduced disease.

These two theories continued to compete until today. The wider version of the miasma theory is that environmental and social conditions are the main factors in disease, in contrast to the more biologically oriented approach of germ theory of infectious diseases and as applied to chronic disease related to toxins (e.g., smoking) or nutritional indicators (e.g., blood lipids and micronutrient deficiency conditions). Clearly, both are operative, with improved infectious disease control and environmental and social conditions all contributing to improved longevity and reduced burden of many diseases. However, large gaps remain between rich and poor as a result of differential social, economic, and cultural differences.

In 2007, the British Medical Journal conducted an opinion survey of the most important medical innovations of all time. The clear winner was the "sanitary revolution as greatest medical advance since 1840." The problem remains today that millions of people die annually from lack of modern sanitation of safe drinking water and poor sewage and solid waste disposal. Other large-scale killers are infectious diseases for which excellent vaccines or other management tools exist.

Both the germ and miasma ideas are important elements of global health. Even in developed countries, management of infectious diseases is still a major public health issue, and includes the annual loss of life from influenza, pneumonia, medically related diseases such as multidrug-resistant tuberculosis, and the rise of drug-resistant organisms easily controlled by antibiotics a short generation ago.

Sources: Ferriman, A. 2007. BMJ readers choose the "sanitary revolution" as greatest medical advance since 1840. British Medical Journal, 334:111.
Mackenbach, J. P. 2007. Sanitation: pragmatism works. British Medical Journal, 334(suppl 1):s17.

lifestyle, but requiring longitudinal preventive care to avert early recurrence and premature death. Progress continues into the twenty-first century as new challenges arise.

ACHIEVEMENTS OF PUBLIC HEALTH IN THE TWENTIETH CENTURY

The foundations of public health organization were laid in the second half of the nineteenth and first half of the twentieth centuries. Water sanitation, waste removal, and food control developed at municipal and higher levels of government, establishment of organized local public health offices with state and federal grants, and improved vaccination technology all contributed to the control of communicable diseases. Organized public health services implemented the regulatory and service components of public health in developed countries, with national standards for food and drug safety, state licensing, and discipline in the health professions.

At the beginning of the twentieth century, there were few effective medical treatments for disease, but improved public health standards resulted in reduced mortality and increased longevity. As medical technology improved following World War II with antibiotics, antihypertensives, and antipsychotic therapeutic agents, the focus was on curative medical care, with a widening chasm between public health and medicine. In our time, a new interest in the commonality between the two is emerging as new methods of organizing and financing health care develop, to contain the rising costs of health care and increase utilization of preventive medicine.

National and state efforts to promote public health during the twentieth century widened in scope of activities and financing programs. This required linkage between governmental and nongovernmental activities for effective public health services. Dramatic scientific innovations brought vaccines and antibiotics which along with improved nutrition and living standards, helped to control infectious disease as the major cause of death. In the developed countries, the advent of national or voluntary health insurance on a wide scale opened access to health care to high percentages of the population.

The modern era of public health from the 1960s to today has brought a new focus on noninfectious disease epidemiology and prevention. Studies of the impact of diet and smoking on cardiovascular diseases and smoking on lung cancer isolated preventable risks for chronic disease. As a result of these and similar studies of disease and injury related to the environment, modern public health has, through health promotion and consumer advocacy, played a significant role in mortality and morbidity

reduction for a spectrum of diseases. For prevention of premature disease and death, more comprehensive approaches will be needed by public health and health care providers than have been developed to date.

The twentieth century saw great achievements in public health in the industrialized countries, indeed throughout the world. The Centers for Disease Control reviewed these achievements in a series of publications in 1999 which represent the potential for public health and if not a universal "gold standard," at least a well-documented set of achievements and potential for public health everywhere (see Box 1.11).

The dream of international and national health agencies to achieve *Health for All* faces serious obstacles of inequities, lack of resources, distortions with overdevelopment of some services at the expense of others, and competing priorities. Managing health care to use resources more effectively is now a concern of every health professional. At the same time, public expectations are high for unlimited access to care, including the specialized and highly technical services that can overwhelm budgetary and personnel resources available. All nations, wealthy or poor, face the problem of managing limited resources. How that will be achieved is part of the challenge we discuss as the New Public Health.

End

Box 1.11 Ten Great Achievements of Public Health in the United States in the Twentieth Century

During the twentieth century, the health and life expectancy in the United States improved dramatically. Since 1900, average lifespan lengthened by >30 years; 25 years of this gain is attributable to advances in public health. *Morbidity and Mortality Weekly Report* (MMWR) profiled 10 public health achievements in a series of reports published in 1999. This reflects similar public health achievements in many industrialized countries.

1. Control of infectious disease
2. Vaccination
3. Motor vehicle safety
4. Safer workplaces
5. Decline in deaths from coronary heart disease, strokes
6. Safer and healthier foods
7. Healthier mothers and babies
8. Family planning
9. Fluoridation of drinking water
10. Recognition of tobacco as a health hazard

Source: CDC. Ten Great Public Health Achievements — United States, 1900–1999. *Morbidity and Mortality Weekly Reports*, 48(12):241–243. www.cdc.gov/mmwr

CREATING AND MANAGING HEALTH SYSTEMS

Provision of medical care to the entire population is one of the great challenges of public health. Governments of all political stripes are active in the field of health policy, as insurers, providers, or regulators of health care. As will be discussed in subsequent chapters, nations have many reasons to ensure health for all, just as they promote universal education and literacy. National interests in the late nineteenth and early twentieth centuries were defined to include having healthy populations, especially for workers and soldiers, and for national prestige. Responsibility for the health of a nation included measures for prevention of disease, but also financing and prepayment for medical and hospital care. National policies gradually took on measures to promote health, structures to evaluate health of the nation, and modification of policies to keep up with changing needs.

The health of a population requires access to medical and hospital services as well as preventive care, a healthy environment, and a health promotion and policy orientation. Greek and Roman cities appointed doctors to provide free care for the poor and the slaves. Medieval guilds provided free medical services to their members. In 1883, Germany introduced compulsory national health insurance to ensure healthy workers and army recruits, which would provide a political advantage. In 1911, Britain's Chancellor of the Exchequer, Lloyd George, instituted the National Insurance Act, providing compulsory health insurance for workers and their families. In 1918, following the October Revolution, the Soviet Union created a comprehensive state-operated health system with an emphasis on prevention, providing free comprehensive care in all parts of the country.

During the 1920s, national health insurance was expanded in many countries in Europe. Following the Great Depression of the 1930s and hopes raised by the victory in World War II, important social and health legislation was enacted to provide health care to the populations of Britain, Canada, and the United States. In Britain the welfare state including the National Health Service (NHS) was developed by the Labour Government. In Canada, a more gradual development took place in the period from 1940–1970, including the establishment of national pensions and a national health insurance program. In the United States, social legislation has been slow in coming following the defeat of national health insurance legislation in Congress in 1946 and long-standing ideological opposition to "socialized medicine," but in 1965, universal coverage of the population over age 65 (Medicare) was instituted and coverage for the poor under Medicaid soon followed. Inadequate coverage of workers and low-income American families is still a serious problem. In the latter part of the twentieth century, virtually every country recognizes the importance of health for the social and economic well-being of its population.

The term *health systems* may imply a formalized structure or a network of functions that work together to meet the needs of a population through health insurance or health service systems. Private health insurance is still the dominant mode in the United States, but the elderly and the poor are covered by government health insurance (see Chapter 13). The American public health community is currently seeking means to achieve universal health coverage. Prepayment for health is financed through general tax revenues in many countries, and in others through payment by workers and employers to social security systems. Both developed and developing countries are involved in financing health care as well as research and training of health professionals.

Industrialized countries share increasing concerns of cost escalation, with health expenditure costs hindering general economic growth. While health care is a large-scale employer in all developed countries, high and rising expenditures for health, reaching 16 percent of GDP in the United States and around 10 percent in many other western countries, is a major factor in stimulating health care reform. Many countries are struggling to keep up with rising costs and competition from other social needs, such as education, employment, and social welfare, all of which are important for national health and well-being. Some economic theories allocate no economic value to a person except as an employee and a consumer. Liberal and social democratic political philosophies advocate an ethical concern and societal responsibility for health. Both approaches now concur that health has social and economic value. The very success of public health has produced a large increase in the percentage of the elderly in the population, raising ethical and economic questions regarding health care consumption, allocation of services, and social support systems.

For developing countries, providing health care for the entire population is a distant dream. Limited resources and overspending on high-technology facilities in larger cities leave little funding for primary care for the rural and urban poor. Despite this, there has been real progress in implementing fundamental services such as immunization and prenatal care. Still, millions of preventable deaths occur annually because of lack of basic primary care programs.

SUMMARY

The history of public health is directly related to the evolution of thinking about health. Ancient societies in one way or another realized the connection between sanitation and health and the role of personal hygiene, nutrition, and fitness. The sanctity of human life (*Pikuah Nefesh*) established an overriding human responsibility to save life derived from Mosaic Law from 1500 BCE. The scientific and ethical basis of medicine was also based on the teachings of Hippocrates in the fourth century BCE. Sanitation, hygiene, good nutrition, and physical fitness all had roots in ancient societies including obligations of the society to provide care for the poor. These ethical foundations support efforts to preserve life even at the expense of other religious or civil ordinances.

Social and religious systems linked disease to sin and punishment by higher powers, viewed investigation or intervention by society (except for relief of pain and suffering) as interference with God's will. Childbirth was associated with pain, disease, and frequent death as a general concept of "in sorrow shall you bring forth children." Health care was seen as a religious charitable responsibility to ease the suffering of sinners.

The clear need and responsibility of society to protect itself by preventing entry or transmission of infectious diseases was driven home by pandemics of leprosy, plague, syphilis, smallpox, measles, and other communicable diseases in the Middle Ages. The diseases themselves evolved, and pragmatic measures were gradually found to control their spread, including isolation of lepers, quarantine of ships, and closure of public bath houses. Epidemiologic investigations of cholera, typhoid, occupational diseases, and nutritional deficiency disorders in the eighteenth and nineteenth centuries began to show causal relationships and effective methods of intervention before scientific proof of causation was established. Even in contemporary times, public health practice continues on a pragmatic basis, often before full scientific basis of the causation of many diseases has been worked out. Public health organizations to ensure basic community sanitation and other modalities of prevention evolved through the development of local health authorities, fostered, financed, and supervised by civic, state, or provincial and national health authorities as governments became increasingly involved in health issues.

Freeing human thought from restrictive dogmas which limited scientific exploration of health and disease fostered the search for the natural causation of disease. This was of paramount importance in seeking interventions and preventive activities. This concept, first articulated in ancient Greek medicine, provided the basis for clinical and scientific observations leading to the successes of public health over the past two centuries. The epidemiologic method led to public health interventions before the biological basis of disease was determined. Sanitation to prevent disease was accepted in many ancient societies, and codified in some as part of civil and religious obligations. Lind's investigation of scurvy, Jenner's discovery of vaccination to prevent smallpox, and Snow's investigation of cholera in London demonstrated disease causation in modern scientific epidemiologic terms, and were accepted despite lack of contemporary biochemical or bacteriologic proof. They helped to formulate the core methodology of public health.

Public health has developed through pioneering epidemiologic studies, devising forms of preventive medicine, and community health promotion. Reforms pioneered in many areas, from abolition of slavery and serfdom to provision of state-legislated health insurance, have all improved the health and well-being of the general population. In the last years of the twentieth century, the relationship between health and social and economic development gained recognition internationally. The twentieth century has seen a dramatic expansion of the scientific basis for medicine and public health. Immunology, microbiology, pharmacology, toxicology, and epidemiology have provided powerful tools and resulted in improved health status of populations. New medical knowledge and technology have come to be available to the general public in many countries in the industrialized world through the advent of health insurance. In this century, virtually all industrialized countries established systems of assuring access to care for all the population as essential for the health of the individual and the population.

Major historical concepts have had profound effects on the development of public health. Sickness as punishment for sin prevented attempts to control disease over many centuries. This mentality persists in modern times by "blaming the victim"; AIDS patients are seen as deserving their fate because of their behavior, workers are believed to become injured because of their own negligence, and the obese person and the smoker are believed to deserve their illnesses because of weakness in the way they conduct their lives. The sanctity of human life, improving the world, and human rights are fundamental to the ethics and values of public health, as is charity in care in which there is a societal and professional responsibility for kindness and relief of suffering. Ethical controversies are still important in many diverse areas such as universal health insurance, food fortification, fluoridation of water supplies, managed care, reproductive health, cost–benefit analysis, euthanasia, and care of sex workers and prisoners and many others.

Acceptance of the right to health for all by the founders of the United Nations and the WHO added a universal element to the mission of public health. This concept was embodied in the constitution of the WHO and given more concrete form in the *Health for All* concept of Alma-Ata, which emphasized the right of health care for everyone and the responsibility of governments to ensure that right. This concept also articulates the primary importance of prevention and primary care, which became a vital issue in competition for resources between public health and hospital-oriented health care.

The lessons of history are important in public health. Basic issues of public health need to be revived because new challenges for health appear and old ones re-emerge. The philosophical and ethical basis of modern public health is a belief in the inherent worth of the individual and his or her human right to a safe and healthful environment.

The health and well-being of the individual and the community are interdependent. Investment in health, as in education, is a contributor to economic growth, as healthy and educated individuals contribute to a creative and economically productive society.

The New Public Health is derived from the experience of history. Organized activity to prevent disease and promote health had to be relearned from the ancient and post-industrial revolution worlds. As the twenty-first century begins, we must learn from a wider framework how to use all health modalities, including clinical and prevention-oriented services, to effectively and economically preserve, protect, and promote the health of individuals and of greater society. The New Public Health, as public health did in the past, faces ethical issues which relate to health expenditures, priorities, and social philosophy. Throughout the course of this book, we discuss these issues and attempt to illustrate a balanced, modern approach toward the New Public Health.

HISTORICAL MARKERS

3000 BCE	Dawn of Sumerian, Egyptian, and Minoan cultures — drains, flush toilets
2000 BCE	Indus valley — urban society with sanitation facilities
1700 BCE	The Code of Hammurabi — rules governing medical practice
1500 BCE	Mosaic Law — personal, food, and camp hygiene, segregating lepers, overriding duty of sanctity of human life (Pikuah Nefesh) and improving the world (Tikun Olam) as religious imperatives
400 BCE	Greece — personal hygiene, fitness, nutrition, sanitation, municipal doctors, occupational health; Hippocrates — clinical and epidemic observation and environmental health
500 BCE to 500 CE	Rome — aqueducts, baths, sanitation, municipal planning, and sanitation services, public baths, municipal doctors, military, and occupational health
170 CE	Galen — physiology, anatomy, humors dominated western medicine until 1500 CE
500–1000	Europe — destruction of Roman society and the rise of Christianity; sickness as punishment for sin; mortification of the flesh, prayer, fasting, and faith as therapy; poor nutrition and hygiene, pandemics; anti-science; care of the sick as religious duty
700–1200	Islam — preservation of ancient health knowledge, schools of medicine, Arab–Jewish medical advances (Ibn Sinna and Maimonides)
1000+	Universities and hospitals in Middle East and Europe
1000+	Rise of cities, trade, and commerce, craft guilds, municipal hospitals
1096–1272	Crusades — contact with Arabic medicine, hospital orders of knights, leprosy
1268	Roger Bacon publishes treatise on use of eyeglasses to improve vision
1348	Venice — board of health and quarantine established

1348–1350	Black Death — origins in Asia, spread by armies of Genghis Khan, world pandemic kills 60 million in fourteenth century, one-third to one-half of the population of Europe
1300	Pandemics — bubonic plague, smallpox, leprosy, diphtheria, typhoid, measles, influenza, tuberculosis, anthrax, trachoma, scabies, and others until eighteenth century
1400–1600	Renaissance and enlightenment, decline of feudalism, rise of urban middle class, trade, commerce, exploration, new technology, arts, science, anatomy, microscopy, physiology, surgery, clinical medicine, hospitals (religious, municipal, voluntary)
1518	Royal College of Physicians founded in London
1532	Bills of Mortality published
1546	Girolamo Fracastorus publishes De Contagione — the germ theory
1562–1601	Elizabethan Poor Laws — responsibility for the poor on local government
1628	William Harvey publishes findings on circulation of the blood
1629	London Bills of Mortality specify causes of death
1639	Massachusetts law requires recording of births and deaths
1660s	Leyden University strengthens anatomical education
1661	John Graunt founds medical statistics
1661	Rene Descartes publishes first treatise on physiology
1662	Royal Society of London founded by Francis Bacon
1665	Great Plague of London
1673	Antony van Leeuwenhoek — microscope, observes sperm and bacteria
1667	Pandemics of smallpox in London; pandemic of malaria in Europe
1687	William Petty publishes Essays in Political Arithmetic
1700	Bernardino Ramazzini publishes compendium of occupational diseases
1701	London — 75% of newborns die before fifth birthday
1701	Variolation against smallpox practiced in Constantinople, isolation practiced in Massachusetts
1710	English Quarantine Act
1720+	London — voluntary teaching in hospitals; Guy's, Westminster
1721	Lady Mary Montagu introduces inoculation for smallpox to England
1730	Science and scientific medicine; Rights of Man, encyclopedias, agricultural and industrial revolutions, population growth — high birth rates, falling death rates
1733	Obstetrical forceps invented
1733	Stephen Hales measures blood pressure
1747	James Lind — case control study of scurvy in sailors
1750	British naval hospitals established
1750	John Hunter establishes modern surgical practice and teaching
1752	William Smellie publishes textbook of midwifery
1762	Jean Jacques Rousseau publishes Social Contract
1775	Percival Pott investigates scrotal cancer in chimney sweeps
1777	John Howard promotes prison and hospital reform in England

1779	Johann Frank promotes Medical Police in Germany
1785	William Withering — discovers foxglove (*Digitalis*) treatment of dropsy
1788	Legislation to protect boys employed as chimney sweepers
1796	Edward Jenner — vaccinates 24 children against smallpox from milkmaid's cowpox pustules
1796	British Admiralty adopts daily issue of lime juice for sailors at sea to prevent scurvy
1797	Massachusetts legislation permitting local boards of health
1798	Philippe Pinel removes chains from insane in Bicetre Asylum
1798	President John Adams signs law for care of sick and injured seamen, establishing marine hospital service, later becoming U.S. Public Health Service (1912)
1800	Britain and U.S. — Municipal Boards of Health
1800	Vaccination adopted by British army and navy
1800	Adam Smith, Jeremy Bentham — economic, social philosophers
1801	Vaccination mandatory in Denmark, local eradication of smallpox
1801	First national census, United Kingdom
1804	Modern chemistry established — Humphrey Davey, John Dalton
1807	Abolition Act — mandates eradication of international slave trade by the Royal Navy
1827	Carl von Baer in St. Petersburg establishes science of embryology
1834	Poor Law Amendment Act documents harsh state of urban working class in United States
1837	United Kingdom National Vaccination
1830s–1840s	Sanitary and social reform, growth of science; voluntary societies for reform, boards of health, mines and factory acts — improving work conditions
1842	Edwin Chadwick — Sanitary Commission links poverty and disease
1844	Horace Wells — anesthesia in dentistry, then surgery
1848	U.K. Parliament passes Public Health Act establishing the General Board of Health
1850	Massachusetts — Shattuck Report of Sanitary Commission
1852	Adolph Chatin uses iodine for prophylaxis of goiter
1854	John Snow — waterborne cholera in London: the Broad Street pump
1854	Florence Nightingale, modern nursing and hospital reform — Crimean War
1855	London — mandatory filtration of water supplies and consolidation of sanitation authorities
1858	Louis Pasteur proves no spontaneous generation of life
1858	Rudolph Virchow publishes *Cellular Pathology*; pioneer in political–social health context
1858	Public Health and Local Government Act and Medical Act in United Kingdom — local health authorities and national licensing of physicians
1859	Charles Darwin publishes *On the Origin of Species*
1861	Emancipation of the serfs in Russia
1861	Ignaz Semmelweiss publishes *The Cause, Concept and Prophylaxis of Puerperal Fever*
1862	Louis Pasteur publishes findings on microbial causes of disease

1862	Florence Nightingale founds St. Thomas' Hospital School of Nursing
1862	Sanitary Commission during U.S. Civil War
1862	Emancipation of slaves in United States
1864	Boston bans use of milk from diseased cows
1864	Russia — rural health as tax-supported local service through Zemstvos
1864	First International Geneva Convention and founding of International Committee of the Red Cross
1866	Gregor Johann Mandel, Czech monk, basic laws of heredity, basis of genetics
1867	Joseph Lister describes use of carbolic spray for antisepsis
1869	Dimitri Ivanovitch Mendeleev — periodic tables
1872	American Public Health Association founded
1872	Milk stations established in New York immigrant slums
1876	Robert Koch discovers anthrax bacillus
1876	Neisser discovers *Gonococcus* organism
1879	U.S. National Board of Health established
1879	U.S. Food and Drug Administration established
1880	Typhoid bacillus discovered (Laveran); leprosy organism (Hansen); malaria organism (Laveran)
1882	Robert Koch discovers the *Tuberculosis* organism, tubercle bacillus
1883	Otto von Bismarck introduces social security with workmen's compensation, national health insurance for workers and their families in Germany
1883	Robert Koch discovers bacillus of cholera
1883	Louis Pasteur vaccinates against anthrax
1885	Takaki in Japanese navy describes beriberi and recommends dietary change
1884	*Diphtheria*, *Staphylococcus*, *Streptococcus*, *Tetanus* organisms identified
1885	Pasteur develops rabies vaccine; Escherich discovers *coli* bacillus
1886	Karl Fraenkel discovers *Pneumococcus* organism
1887	Malta fever or brucellosis (Bruce) and chancroid (Ducrey) organisms identified
1887	U.S. National Institutes of Health founded
1890	Anti-tetanus serum (ATS)
1892	Gas gangrene organism discovered by Welch and Nuttal
1894	Plague organism discovered (Yersin, Kitasato); botulism organism (Van Ermengem)
1895	Louis Pasteur develops vaccine for rabies
1895	Wilhelm Roentgen — discovers electromagnetic waves (x rays) for diagnostic imaging
1895	Emil von Behring develops diphtheria vaccine (Nobel Prize, 1901)
1897	London School of Hygiene and Tropical Medicine founded
1897	Felix Hoffman — synthesizes acetylsalicylic acid (aspirin)
1905	Abraham Flexner — major report on medical education in United States
1905	Workman's Compensation Acts in Canada
1906	U.S. Pure Food and Drug Act passed by Congress
1910	Paul Ehrlich — chemotherapy use of arsenical salvarsan for treatment of syphilis
1911	Lloyd George, United Kingdom compulsory health insurance for workers

1911	Kasimir Funk investigates "vital amines" and names them vitamins	1954	Richard Doll reports on link of smoking and lung cancer
1912	Health insurance for industrial workers in Russia	1954	Jonas Salk's inactivated poliomyelitis vaccine licensed
1912	U.S. Children's Bureau and U.S. Public Health Service established	1955	Michael Buonocore discovers dental sealants
1914	Joseph Goldberger investigates cause and prevention of pellagra	1956	Gregory Pincus reports first successful trials of birth control pills
1915	Johns Hopkins and Harvard Schools of Public Health founded	1960	Albert Sabin — live poliomyelitis vaccine licensed
1915	Tetanus prophylaxis and antitoxin for gas gangrene	1961	American Academy of Pediatrics recommends routine vitamin K for all newborns
1918–1919	Pandemic of Spanish flu (influenza) kills some 20 million people	1963	Measles vaccine licensed
1904	Ivan Petrovitch wins Nobel Prize for work in conditioned reflexes, neurophysiology	1964	U.S. Surgeon General's Report on Smoking (Luther Terry)
1918	Nikolai Semashko introduces U.S.S.R. national health plan	1965	United States passes Medicare for the elderly, Medicaid for the poor
1921	Frederick Banting and Charles Best discover insulin in Toronto	1966	U.S. National Traffic and Motor Vehicle Safety Act
1923	Health Organization of League of Nations established	1967	Mumps vaccine licensed
		1970	Rubella vaccine licensed
1924	David Cowie promotes widespread ionization of salt in the U.S.; Morton's iodized salt popularized in North America	1971	Canada has universal health insurance in all provinces
		1974	LaLonde Report — New Perspectives on the Health of Canadians
1926	Pertussis vaccine developed	1977	WHO adopts Health for All by the Year 2000
1928	Alexander Fleming discovers penicillin	1977	Last known outbreak of smallpox reported in Somalia
1928	George Papanicolaou develops Pap smear for early detection of cancer of cervix	1978	Alma-Ata Conference on Primary Health Care
1929–1936	The Great Depression — widespread economic collapse, unemployment, poverty, and social distress in industrialized countries	1978	Hepatitis B vaccine licensed
		1979	Canada adopts mandatory vitamin/mineral enrichment of foods
1930	U.S. Food and Drug Administration established	1979	WHO declares eradication of smallpox achieved
1935	President Roosevelt — Social Security Act and the New Deal in the United States	1981	First recognition of cases of acquired immune deficiency syndrome (AIDS)
1940	Charles Drew describes storage and use of blood plasma for transfusion	1985	WHO European Region Health Targets
		1985	*Haemophilus influenzae* b (Hib) vaccine licensed by FDA
1941	Norman Gregg reports rubella in pregnancy causing congenital anomalies	1985	Luc Montaignier publishes genetic sequence of HIV
1939–1945	World War II, food fortification in United States, Canada, and Britain; U.K. National Hospital Service — wartime nationalization of hospitals; (William) Beveridge Report in the United Kingdom — the "Welfare State" (1942); U.S. National Centers for Disease Control established; U.S. Emergency Maternity and Infant Care for families of servicemen; U.S.S.R. wartime emergency medical structure; Nazi Holocaust of 6 million Jews and many others	1988	American College of Obstetricians, Gynecologists recommends annual pap smears for all women
		1989	WHO targets eradication of polio by the year 2000
		1989	Warren and Marshall — *Helicobacter pylori* as treatable cause of peptic ulcers
		1989	International Convention on the Rights of the Child
		1990	World Summit on Children, New York
		1990	World Conference on Education for All, Jomtien
		1990	W. F. Anderson performs first successful gene therapy
		1990	Newly emerging and reemerging diseases (HIV, Marburg, Ebola, cholera, mad cow disease, tuberculosis) and multidrug-resistant organisms
1945	Grand Rapids MI; Newburgh, NY; and Brantford, Ontario — first cities to fluoridate water supplies		
1946	World Health Organization founded	1991	Folic acid proven to prevent neural tube defects
1946	National health insurance defeated in U.S. Congress	1992	United Nations Conference on Environment and Development, Rio de Janeiro
1946	U.S. Congress Hill–Burton Act supports local hospital construction up to 4.5 beds/1000 population	1992	International Conference on Nutrition
		1993	World Conference on Human Rights, Vienna
1946	Tommy Douglas — Saskatchewan provincial hospital insurance plan	1993	*World Development Report: Investing in Health* published by World Bank
1947	Nuremberg Doctors Trial of Nazi crimes against humanity	1993	Russian Federation approves compulsory national health insurance
1948	International Declaration of Human Rights	1994	International Conference on Population and Development, Cairo
1948	United Kingdom establishes National Health Service	1994	Clinton National Health Insurance plan defeated in U.S. Congress
1953	James Watson and Francis Crick discover the double helix structure of DNA (Nobel Prize 1962)	1995	World Summit for Social Development, Copenhagen
1954	Framingham study of heart disease risk factors		

1995	United Nations Fourth World Conference on Women, Beijing
1996	Second United Nations Conference on Human Settlement (Habitat II), Istanbul
1996	Explosive growth of managed care plan coverage in the United States
1997	Legal suits for damages against tobacco companies for costs of health effects of smoking, 33 states in the United States and ot[...] countries
1998	Clinton proposed legislation on patien[...] managed care
1998	FDA approves rotavirus vaccine
1998	WHO *Health for All in the Twenty-First Century* adopted
1998	U.S. National Academy of Sciences recommends routine vitamin supplements for adults
1998	Bologna Declaration on post-graduate education in Europe adopts BA, MA, and PhD levels
1998	United States and Canada adopt mandatory fortification of flour with folic acid to prevent birth defects
1999	U.S. Congress passes legislation regulating patients' rights in managed care
1999	Master Settlement Agreement between U.S. states and tobacco companies for $206 billion for Medicaid damages
2001	9/11 Terrorism and mass casualties in United States
2001	Anthrax and bioterrorism
2003	SARS epidemic in China reaches Toronto; 8098 total cases with 774 deaths
2004	Tsunami and mass casualties in southeast Asia
2005	Hurricanes Katrina and Rita with widespread devastation and mass casualties
2006	Bird flu of H5N1 virus threatens world pandemic
2006	Human papillomavirus (HPV) vaccine approved by FDA for prevention of cervical cancer
2006	Medicare Part D prescription drug plan for seniors instituted in United States
2007	HPV vaccine in wide use for preteen girls in industrialized countries

ELECTRONIC RESOURCES

Centers for Disease Control. 1999. Ten Great Public Health Achievements in the 20th Century, http://www.cdc.gov/od/oc/media/tengpha.htm [accessed April 25, 2008]

Center for History in Public Health. March, 2008. London School of Hygiene and Tropical Medicine, http://www.lshtm.ac.uk/history/ [accessed April 20, 2008]

Columbia University. 2008. Program in the History of Public Health and Medicine, http://cpmcnet.columbia.edu/dept/hphm/ [accessed April 20, 2008]

Google newstimeline of public health. 2008, http://news.google.com/archivesearch?hl=en&q=of+public+health&um=1&ie=UTF-8&scoring=t&sa=X&oi=archive&ct=title

House of Commons, UK. 2001, http://www.publications.parliament.uk/pa/cm200001/cmselect/cmhealth/30/3008.htm [accessed April 20, 2008]

Kondratas R. Images from the History of the U.S. Public Health Service. 1998. Department of Health and Human Services, http://www.nlm.nih.gov/exhibition/phs_history/contents.html [accessed April 20, 2008]

Nobel Prize Medicine. 2008, http://nobelprize.org/nobel_prizes/medicine/laureates/ [accessed April 21, 2008]

http://nobelprize.org/nobel_prizes/medicine/ [accessed April 23, 2008]

Rollins School Public Health Emory University Atlanta. 2008. History of Public Health Infolinks, http://www.sph.emory.edu/PHIL/PHILhistory.php [accessed April 20, 2008]

RECOMMENDED READINGS

End Fri

[...] well child care. *Pediat-*[...]

Centers for Disease Con[...] Public Health Achievements, United States, 1900-1999. *Morbidity and Mortality Weekly Reports*, 48:241–243.

Centers for Disease Control. 1999. Achievements in Public Health, 1900-1999: Impact of Vaccines Universally Recommended for Children — United States, 1990-1998. *Morbidity and Mortality Weekly Reports*, 48:243–248.

Centers for Disease Control. 1999. Achievements in Public Health, 1900-1999: Motor-Vehicle Safety: A 20th Century Public Health Achievement. *Morbidity and Mortality Weekly Reports*, 48:369–374.

Centers for Disease Control. 1999. Achievements in Public Health, 1900-1999: Improvements in Workplace Safety — United States, 1900-1999. *Morbidity and Mortality Weekly Reports*, 48:461–469.

Centers for Disease Control. 1999. Achievements in Public Health, 1900-1999: Control of Infectious Diseases. *Morbidity and Mortality Weekly Reports*, 48:621–629.

Centers for Disease Control. 1999. Achievements in Public Health, 1900-1999: Decline in Deaths from Heart Disease and Stroke — United States, 1900-1999. *Morbidity and Mortality Weekly Reports*, 48:649–656.

Centers for Disease Control. 1999. Achievements in Public Health, 1900-1999: Healthier Mothers and Babies. *Morbidity and Mortality Weekly Reports*, 48:849–858.

Centers for Disease Control. 1999. Achievements in Public Health, 1900-1999: Safer and Healthier Foods. *Morbidity and Mortality Weekly Reports*, 48:905–913.

Centers for Disease Control. 1999. Achievements in Public Health, 1900-1999: Fluoridation of Drinking Water to Prevent Dental Caries. *Morbidity and Mortality Weekly Reports*, 48:933–940.

Centers for Disease Control. 1999. Achievements in Public Health, 1900-1999: Tobacco Use — United States, 1900-1999. *Morbidity and Mortality Weekly Reports*, 48:986–993.

Centers for Disease Control. 1999. Achievements in Public Health, 1900-1999: Family Planning. *Morbidity and Mortality Weekly Reports*, 48:1073–1080.

DeBuono, B. A. 2005. *Milestones in Public Health: Accomplishments in Public Health Over the Last 100 Years.* New York: Pfizer Inc.

Johnson, N. P., Mueller, J. 2002. Updating the accounts: Global mortality of the 1918–1920 "Spanish" influenza pandemic. *Bulletin of the History of Medicine*, 76:105–115.

Larson, E. 1989. Innovations in health care: Antisepsis as a case study. *American Journal of Public Health*, 79:92–99.

Markel, H. 1987. When it rains it pours: Endemic goiter, iodized salt, and David Murray Cowie, MD. *American Journal of Public Health*, 77:219–229.

Monteiro, L. A. 1985. Florence Nightingale on public health nursing. *American Journal of Public Health*, 75:181–186.

Ottaviani, R., Vanni, P., Baccolo, G. M., Guerin, E., Vanni, D. 2003. The First Nobel Peace Prize, Henry Dunant (founder of the International Red Cross) and His "Mémoirs." *Vesalius*, 9:20–27.

Rosen, G. 1958. *A History of Public Health.* New York: MD Publications: Republished as Expanded Edition. Baltimore, MD: Johns Hopkins University Press, 1993.

Roueche, B. (ed.). 1963. *Curiosities of Medicine: An Assembly of Medical Diversions 1552–1962.* London: Victor Gollancz Ltd.

Taubenberger, J. K., Morens, D. M. 2006. 1918 influenza: The mother of all pandemics. *Emerging Infectious Diseases* [serial on the Internet]. 2006 January [accessed January 2008]. Available at http://www.cdc.gov/ncidod/EID/vol12no01/05-0979.htm.

BIBLIOGRAPHY

Barkan, I. D. 1985. Industry invites regulation: The passage of the Pure Food and Drug Act of 1906. *American Journal of Public Health*, 75:18–26.

Buehler-Wilkerson, K. 1993. Bringing care to the people: Lillian Wald's legacy to public health nursing. *American Journal of Public Health*, 83:1778–1786.

Camus, A. 1947. *The Plague.* Middlesex, England: Penguin Modern Classics.

Carter, K. D. 1991. The development of Pasteur's concept of disease causation and the emergence of specific causes in nineteenth-century medicine. *Bulletin of the History of Medicine*, 65:528–548.

Centers for Disease Control. 1999. Ten Great Public Health Achievements — United States, 1900-1999. *Morbidity and Mortality Weekly Reports*, 48:241–243. www.cdc.gov/mmwr.

Crosby, W. H. Book Review 1993: of Gabriel, R. A., Metz, K. S. *A History of Military Medicine.* New York: Greenwood Press, 1992. *New England Journal of Medicine,* 328:1427–1428.

Diamond, J. 1997. *Guns, Germs and Steel: The Fates of Human Societies.* New York: W. W. Norton Co.

Dunn P. M. 2002. Dr William Farr of Shropshire (1807-1883): Obstetric mortality and training. *Archives of Disease in Childhood — Fetal and Neonatal Edition Online,* 8:F67–69.

Dunn, P. M. 2005. Ignacz Semmelweis (1818-1865) of Budapest and the prevention of puerperal fever. *Archives of Disease in Childhood — Fetal and Neonatal Edition Online,* 90:F345–348.

Dunn, P. M. 2007. Perinatal lessons from the past: Sir Norman Gregg, ChM, MC, of Sydney (1892-1966) and rubella embryopathy. *Archives of Disease in Childhood — Fetal and Neonatal Edition Online,* 92:F513–514.

Garrison, F. H. 1929. *An Introduction to the History of Medicine,* Fourth Edition. Republished by WB Saunders Co., Philadelphia, 1966.

Grob, G. N. 1985. The origins of American psychiatric epidemiology. *American Journal of Public Health*, 75:229–236.

Hollingshead, A. B., Redlich, F. C. 2007. [excerpts from] *Social Class and Mental Illness: A Community Study.* New York: John Wiley, 1958. Reprinted in *American Journal of Public Health*, 97:1756–1757.

Hughes, J. G. 1993. Conception and creation of the American Academy of Pediatrics. *Pediatrics*, 92:469–470.

Knobler, S., Mack, A., Mahmoud, A., Lemon, S. 2005. *The Threat of Pandemic Influenza: Are We Ready? Workshop Summary.* Washington, DC: The National Academies Press.

Mack, A. (ed.). 1991. *The Time of the Plague: The History and Social Consequences of Lethal Epidemic Disease.* New York: New York University Press.

Mackenbach, J. P. 2007. Sanitation: pragmatism works. *British Medical Journal*, 334(suppl_1):s17.

Marti-Ibanez, F. (ed.). 1960. *Henry E. Sigerist on the History of Medicine.* New York: MD Publications.

Massachusetts Sanitary Commission. 1850. *Report of a General Plan for the Promotion of Public and Personal Health, Sanitary Survey of the State.* Reprinted by Arno Press & *The New York Times,* New York, 1972.

McCullough, D. 1977. *The Path Between the Seas: The Creation of the Panama Canal 1870–1914.* New York: Touchstone.

McNeill, W. H. 1989. *Plagues and Peoples.* New York: Anchor Books.

Monteiro, L. A. 1985. Florence Nightingale on public health nursing. *American Journal of Public Health*, 75:181–186.

Plotkin, S. L., Plotkin, S. A. 1994. A short history of vaccination. *In* S. A. Plotkin, E. A. Mortimer (eds.). *Vaccines,* Second Edition. Philadelphia: WB Saunders.

Rajakumar, K., Greenspan, S. L., Thomas, S. B., Holick, M. F. 2007. Solar ultraviolet radiation and vitamin D: A historical perspective. *American Journal of Public Health*, 97:1746–1754.

Rather, L. J. (ed.). 1958. *Disease, Life, and Man: Selected Essays by Rudolf Virchow.* Stanford: Stanford University Press.

Roberts, D. E., Heinrich, J. 1985. Public health nursing comes of age. *American Journal of Public Health*, 75:1162–1172.

Roemer, M. I. (ed.). 1960. *Sigerist on the Sociology of Medicine.* New York: MD Publications.

Roemer, M. I. 1988. Resistance to innovation: The case of the community health center. *American Journal of Public Health*, 78:1234–1239.

Rosenberg, C. E. 1992. *Explaining Epidemics and Other Studies in the History of Medicine.* Cambridge, UK: Cambridge University Press.

Rosenberg, C. E. 2007. Erwin H. Ackerknecht, social medicine, and the history of medicine. *Bulletin of the History of Medicine*, 8:511–532.

Scrimshaw, N. S. 2007. Fifty-five-year personal experience with human nutrition worldwide. *Annual Review of Nutrition*, 27:1–18.

Sinclair, U. 1906. *The Jungle.* New York: Doubleday, Jabber & Company.

Slaughter, F. G. 1950. *Immortal Magyar: Semmelweiss, Conqueror of Childbed Fever.* New York: Henry Schuman.

Smith, I. S. 1990. *Patenting the Sun: Polio and the Salk Vaccine.* New York: Wm. Morrow & Co.

Snow, J. 1936. *Snow on Cholera: A Reprint of Two Papers.* New York: Commonwealth Fund.

Snow, S. J. [Book review] 2004. *Cholera, Chloroform, and the Science of Medicine: A Life of John Snow.* By Peter Vinten-Johansen, Howard Brody, Nigel Paneth, Stephen Rachman, Michael Rip, and David Zuck. *New England Journal of Medicine,* 350:90–91.

Sorokina, T. S. 1994. *History of Medicine.* Moscow: PAIMS [in Russian].

Starr, C. G. 1991. *A History of the Ancient World.* New York: Oxford University Press.

Tuchman, B. W. 1978. *A Distant Mirror: The Calamitous Fourteenth Century.* New York: Alfred A. Knopf Inc.

Wills, C. 1978. *Plagues: Their Origin, History, and Future.* London: Flamingo Press.

Chapter 2

Expanding the Concept of Public Health

Introduction
Concepts of Public Health
Evolution of Public Health
Health and Disease
    Host–Agent–Environment Paradigm
The Natural History of Disease
Society and Health
Modes of Prevention
    Health Promotion
    Primary Prevention
    Secondary Prevention
    Tertiary Prevention
Demographic and Epidemiologic Transition
Interdependence of Health Services
Defining Public Health
    Social Medicine and Community Health
    Social Hygiene, Eugenics, and Corruption of Public
        Health Concepts
    Medical Ecology
    Community-Oriented Primary Care
World Health Organization's Definition of Health
    Alma-Ata: Health for All
Selective Primary Care
The Risk Approach
The Case for Action
Political Economy and Health
Health and Development
Health Systems: The Case for Reform
Advocacy and Consumerism
    Professional Advocacy and Resistance
    Consumerism
The Health Field Concept
The Value of Medical Care in Public Health
Health Targets
    United States Health Targets
International Health Targets
    European Health Targets
    United Kingdom Health Targets
Individual and Community Participation in Health
    Ottawa Charter for Health Promotion
    State and Community Models of Health Promotion
    Healthy Cities/Towns/Municipalities

Human Ecology and Health Promotion
Defining Public Health Standards
Integrative Approaches to Public Health
The Future of Public Health
The New Public Health
Summary
Electronic Resources
Recommended Readings
Bibliography

INTRODUCTION

The evolution of public health from its ancient and recent roots, in the past two centuries especially, has been a continuing process, with revolutionary leaps forward with important continuing and new challenges. Everything in the New Public Health is about preventing disease, injuries, disabilities, and death while promoting a healthy environment and conditions for current and future generations. But in addition, the New Public Health addresses overall health policy, resource allocation, as well as the organization, management, and provision of medical care and of health systems.

The study of history (see Chapter 1) helps the student and the practitioner to understand the process of change, to define where we came from, and to try to understand where we are going. It is vital to recognize and understand change in order to be able to deal with radical changes of direction that occur and will continue to develop in health needs in the context of environmental demographic and societal changes with knowledge gained from social and physical sciences, practice, and economics. For the coming generations, this is not only about the quality of life, but survival of society itself.

CONCEPTS OF PUBLIC HEALTH

Concepts of public health continue to evolve. As a professional field, public health requires specialists trained with deep knowledge of its evolution, scientific advances, and best practices, old and modern. It demands sophisticated professional and managerial skills, the ability to address a problem, reasoning to define the issues, and to advocate, initiate, develop, and implement new and revised programs. It calls for profoundly humanistic values and a sense of responsibility toward protecting and improving the health of communities and every individual.

Health of mind and body is so fundamental to the good life that if we believe that men have any personal rights at all as human beings, then they have an absolute moral right to such a measure of good health as society and society alone is able to give them.

Aristotle, circa 320 BCE (As quoted by Sargent Shriver, Dedication Ceremonies, Health Services Center, Watts, Los Angeles, California, September 16, 1967)

In the past, public health was seen as a discipline which studies and implements measures for control of communicable diseases, primarily by sanitation and vaccination. The sanitary revolution, which came before the development of modern bacteriology, made an enormous contribution to improved health, but many other societal factors including improved nutrition, education, and housing were no less important for population health. Maternal and child health, occupational health, and many other aspects of a growing public health network of activities played important roles, as have the physical and social environment and personal habits of living in determining health status.

The scope of public health has changed along with growth of the medical, social, and public health sciences, public expectations, and practical experience. Taken together, these have all contributed to changes in concepts of disease and their causes. Health systems that fail to adjust to changes in fundamental concepts of public health suffer from immense inequity and burdens of preventable disease, disability, and death. In this chapter, we examine expanding concepts of public health, leading to the development of a New Public Health.

Public health has evolved as a multidisciplinary field that includes the use of basic and applied science, education, social sciences, economics, management, and communication skills to promote the welfare of the individual and the community. It is greater than the sum of its component elements and includes the art and politics of the funding and coordination of the wide diversity of community and individual health services.

The concept of health in body and mind has ancient origins. They continue to be fundamental to individuals and societies, and part of the fundamental rights of all humans to have knowledge of healthful lifestyles and to have access to those measures of good health that society alone is able to provide, such as immunization programs, food and drug safety and quality standards, environmental and occupational health, and universal access to high-quality primary and specialty medical and other vital health services. This holistic view of balance and equilibrium

may be a renaissance of classical Greek and biblical traditions, applied with the broad new knowledge and experience of public health and medical care of the nineteenth, twentieth, and the early decade of the twenty-first centuries as change continues to challenge our capacity to adapt.

The competing nineteenth-century germ and miasma theories of biological and environmental causation of illness each contributed to the development of sanitation, hygiene, immunization, and understanding of the biological and social determinants of disease and health. They come together in the twenty-first century encompassed in a holistic New Public Health addressing individual and population health needs. Medicine and public health professionals both engage in organization and in direct caregiving. Both need a thorough understanding of the issues that are included in the New Public Health, how they evolved, interact, are put together in organizations, and how they are financed and operated in various parts of the world in order to understand changes going on before their eyes.

The New Public Health is comprehensive in scope. It relates to or encompasses all community and individual activities directed toward improving the environment for health, reducing factors that contribute to the burden of disease, and fostering those factors that relate directly to improved health. Its programs range broadly from immunization, health promotion, and child care, to food labeling and fortification, as well as to the assurance of well-managed, accessible health care services. A strong public health system should have adequate preparedness for natural and man-made disasters, as seen in the recent tsunami, hurricanes, biological or other attacks by terrorists, wars, conflicts, and genocidal terrorism (Box 2.1).

Profound changes are taking place in the world population — and public health is crucial to respond accordingly to mass migration to the cities, fewer children, extended life expectancy, and the increase in the population of older people who are subject to more chronic diseases and disabilities in a changing physical, social, and economic climate. Health systems are challenged with continuing reforms, while experiencing strong influence of pharmaceuticals and medicalization of health care.

The concepts of health promotion and disease prevention have become part of the foundation of public health. Parallel scientific advances in molecular biology, genetics, pharmacogenomics, imaging, information technology, computerization, and bio- and nanotechnologies hold great promise for improving the productivity of the health care system. Advances in technology with more rapid and less expensive drug and vaccine development, with improved safety and effectiveness of drugs, fewer adverse reactions, will over time greatly increase efficiency in prevention and treatment modalities.

The New Public Health is important in that it links classical topics of public health with adaptation in the organization and financing of personal health services.

Box 2.1 The New Public Health

The mission of the NPH is to maximize human health and well-being for individuals and communities, nationally and globally.

To achieve this, the NPH works with:

1. Total societal efforts to maximize quality of life and health, economic growth and equity for all societies;
2. Prevention and treatment of illness and disability;
3. Environmental, biological, occupational, social and economic factors that endanger health and human life addressing:
   a. Diseases and infirmity, trauma and injuries;
   b. Local and global environment and ecology;
   c. Healthful nutrition, food security, availability, quality, safety and affordability of food products;
   d. Disaster, natural and man-made including war, terrorism, genocide
   e. Population groups at special risk and with specific health needs
4. Health policy and management of health systems recognizing economic and quality standards of medical, hospital and other professional care to health of individuals and populations;
5. Research and promotion of wide application of current international best practices and standards;
6. Training of professional public health workforce and education of all health workers in the principles and practices of public health;
7. Mobilize the best available international evidence from scientific and epidemiologic studies and best practices recognized as contributing to the overall goal;
8. Maintain and promote individual and community rights to health with equity and high professional and ethical standards.

It involves a changed paradigm of public health to incorporate new advances in political, economic, and social sciences. Failure at the political level to appreciate the role of public health in disease control holds back many societies in economic and social development. At the same time, organized public health systems need to work to reduce inequities between and inside countries to ensure equal access to care. It also demands special attention through health promotion activities of all kinds at national and local societal levels to provide access for groups with special risks and needs to medical and community health care with the currently available and newly developing knowledge and technologies.

The great gap between available capabilities to prevent and treat disease with actually reaching all those in need is still the source of great international and internal inequities. These inequities exist not only between developed and developing countries, but also within transition countries, mid-level developing countries, and those newly emerging with rapid economic development. The historical experience of public health will help to develop the applications of existing and new knowledge and the importance of social solidarity in implementation of the new discoveries for every member of the society, despite socioeconomic, ethnic, or other differences.

The vitally important political will and leadership, adequate financing, and organization systems in the health setting must be supported by well-trained staff for planning, management, and monitoring functions of a health system. The political will and professional support are indispensable in a world of limited resources, with high public expectations and the growing possibilities of effectiveness of public health programs. This requires well-developed information and knowledge management systems to provide the feedback and control data needed for good management. It includes responsibilities and coordination at all levels of government and by nongovernment organizations (NGOs) and participation of a well-informed media and strong professional and consumer organizations. No less important are clear designations of responsibilities of the individual for his or her own health, and of the provider of care for humane, high-quality professional care.

EVOLUTION OF PUBLIC HEALTH

Many changes have signaled a need for transformation toward the New Public Health. Religion, although still a major political and policymaking force in many countries, is no longer the central organizing power in most societies. Organized societies have evolved from large extended families and tribes to rural societies, cities, and national governments. With the growth of industrialized urban communities, rapid transport, and extensive trade and commerce in multinational economic systems, the health of individuals and communities has became more than just a personal, family, and local problem. An individual is not only a citizen of the village, city, or country in which he or she lives, but of a "global village."

The agricultural revolutions and international explorations of the fifteenth to seventeenth centuries that increased food supply and diversity were followed only much later by knowledge of nutrition as a public health issue. The scientific revolution of the seventeenth to nineteenth centuries provided the basics to describe and analyze the spread of disease and the poisonous effects of the industrial revolution, including crowded living conditions and pollution of the environment with serious ecological damage. In the latter part of the twentieth century, a new agricultural "green revolution" had a great impact in reducing human deprivation internationally, yet the full benefits of healthier societies are yet to be realized in the large populations living in abject poverty of sub-Saharan Africa, Southeast Asia, and other parts of the world.

These and other societal changes discussed in Chapter 1 have enabled public health to expand its potential and horizons, while developing its pragmatic and scientific base. Organized public health of the twentieth century proved effective in reducing the burden of infectious diseases and has contributed to improved quality of life and longevity by many years. In the last half-century, chronic diseases have become the primary causes of morbidity and mortality in the developed countries and increasingly in developing countries. Growing scientific and epidemiologic knowledge increases the capacity to deal with these diseases. Many aspects of public health can only be influenced by the behavior of and risks to the health of individuals. These require interventions that are more complex and relate to societal environmental and community standards and expectations as much as personal lifestyle. The dividing line between communicable and noncommunicable diseases changes over time as scientific advances showed the causation of chronic conditions by infectious agents and their prevention by curing the infection, as in *Helicobacter pylori* and peptic ulcers, and in prevention of cancer of the liver and cervix by immunization for hepatitis B and human papilloma virus (HPV) vaccines, respectively.

Chronic diseases have come to center stage in what came to be called "the epidemiologic transition," as infectious diseases came under increasing control. This in part has created a need for reform in the economics and management of health systems due to rapidly rising costs, aging of the population, the rise of obesity and diabetes and other chronic conditions, and expanding capacity to deal with public health emergencies. Reform is also needed in international assistance to help less-developed nations build the essential infrastructure to sustain public health in the struggle to combat AIDS, malaria, TB, and the major causes of preventable infant-, childhood-, and motherhood- related deaths.

The nearly universal recognition of the rights of people to have access to health care of acceptable quality by international standards is a challenge of political will and leadership backed up by adequate staffing with public health–trained people and organizations. The interconnectedness of managing health systems is part of the New Public Health. Setting the priorities and resource allocation to address these challenges requires public health training and orientation of the professionals and institutions participating in the policy, management, and economics of health systems. Conversely those who manage such institutions are recognizing the need for a wide background in public health training in order to fulfill their tasks effectively. Concepts such as objectives, targets, priorities, cost-effectiveness, and evaluation have become part of the New Public Health agenda. An understanding of how these concepts evolved helps the future health provider or manager cope with the complexities of mixing science, humanity, and effective management of resources

Box 2.2 Breslow: The Continuing Epidemiologic Transition

First Public Health Era — the control of communicable diseases
Second Public Health Era — the rise and fall of chronic diseases
Third Public Health Era — the development of long and high-quality life

Source: Breslow L. 2006. Health measurement in the third era of health. *American Journal of Public Health*, 96:17–19.

to achieve higher standards of health and to cope with new issues as they develop in the broad scope of the New Public Health for the twenty-first century, in what Breslow calls the "Third Public Health Era" of long and healthy quality of life (Box 2.2).

HEALTH AND DISEASE

Health can be defined from many perspectives, ranging from statistics of mortality, life expectancy, and morbidity rates to idealized versions of human and societal perfection, as in the World Health Organization (WHO) founding charter: "Health is a state of complete physical, mental and social well being, and not merely the absence of disease or infirmity."

A more operational definition of health is a state of equilibrium of the person with the biological, physical, and social environment, with the object of maximum functional capability. Health is thus seen as a state characterized by anatomic, physiologic, and psychological integrity, and an optimal functional capability in the family, work, and societal roles (including coping with associated stresses), a feeling of well-being, and freedom from risk of disease and premature death.

There are many interrelated factors in disease and in their management. In 1878, Claude Bernard described the phenomenon of adaptation and adjustment of the internal milieu of the living organism to physiologic processes. This concept is fundamental to medicine. It is also central to public health because understanding the spectrum of events and factors between health and disease is basic to the identification of contributory factors affecting the balance toward health, and to seeking the points of potential intervention to reverse the imbalance.

As described in Chapter 1, from the time of Hippocrates and Galen, diseases were thought to be due to humors and miasma or emanations from the environment. The miasma theory, while without basis in fact, was acted on in the early to mid-nineteenth century with practical measures to improve sanitation, housing, and social conditions with successful results. The competing germ theory developed by pioneering epidemiologists (Panum, Snow, and Budd), scientists (Pasteur, Cohn, and Koch), and practitioners (Lister and Semmelweiss) led to the science of bacteriology and a revolution in practical public health measures. The combined application of these two theories has been the basis of classic public health, with enormous benefits coming in the control of infectious disease (Box 2.3).

Host–Agent–Environment Paradigm

In the host–agent–environment paradigm, a harmful agent comes through a sympathetic environment into contact with a susceptible host, causing a specific disease. This idea dominated public health thinking until the mid-twentieth century. The host is the person who has or is at risk for a specific disease. The agent is the organism or direct cause of the disease. The environment includes the external factors which influence the host, his or her susceptibility to the agent, and the vector which transmits or carries the agent to the host from the environment. This explains the causation and transmission of many diseases. This paradigm (Figure 2.1), in effect, joins together the contagion and miasma theories of disease causation. A specific agent, a method of transmission, and a susceptible host are involved in an interaction, which are central to the infectivity or severity of the disease. The environment can provide the carrier or vector of an infective (or toxic) agent, and it also contributes factors to host susceptibility; for example, unemployment, poverty, or low education level.

The Expanded Host–Agent–Environment Paradigm

The expanded host–agent–environment paradigm widens the definition of each of the three components (Figure 2.2), in relation to both acute infectious and chronic noninfectious disease epidemiology. In the latter half of the twentieth century, this expanded host–agent–environment paradigm took on added importance in dealing with the complex of factors related to chronic diseases, now the leading causes of disease and premature mortality in the developed world, and increasingly in developing countries.

Interventions to change host, environmental, or agent factors are the essence of public health. In infectious disease control, the biological agent may be removed by pasteurization of food products or filtration and disinfection (chlorination) of water supplies to prevent transmission of waterborne disease. The host may be altered by immunization to provide immunity to the infective organism. The environment may be changed to prevent transmission by destroying the vector or its reservoir of the disease. A combination of these interventions can be used against a specific risk factor, toxic or nutritional deficiency, infectious organism, or disease process.

Vaccine-preventable diseases may require both routine and special activities to boost herd immunity to protect the individual and the community. For other infectious diseases for which there is no vaccine (for example, malaria), control involves a broad range of activities including case finding and treatment to improve the individual's health and to reduce the reservoir of the disease in the population, as well as vector control to reduce the mosquito population. Tuberculosis control requires not only case finding and treatment, but understanding the contributing

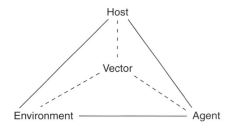

FIGURE 2.1 The host–agent–environment paradigm.

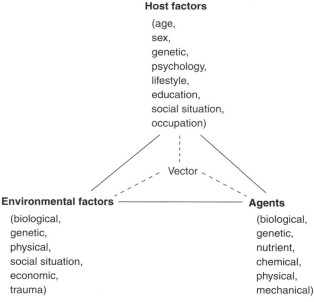

FIGURE 2.2 The expanded host–agent–environment paradigm.

factors of social conditions, diseases with tuberculosis as a secondary condition (drug abuse and AIDS), agent resistance to treatment, and the inability of patients or carriers to complete treatment without supervision. Sexually transmitted infections (STIs), not controllable by vaccines, require a combination of personal behavior change, health education, medical care, and skilled epidemiology.

With noninfectious diseases, intervention is even more complex, involving human behavior factors and a wide range of legal, administrative, and educational issues. There may be multiple risk factors, which have a compounding effect in disease causation, and they may be harder to alter than infectious diseases factors. For example, smoking in and of itself is a risk factor for lung cancer, but exposure to asbestos fibers has a compounding effect. Preventing exposure may be easier than smoking cessation. Reducing trauma morbidity and mortality is equally problematic.

The identification of a single specific cause of a disease is scientifically and practically of great value in modern public health, enabling such direct interventions as use of vaccines or antibiotics to protect or treat individuals from infection by a causative organism, toxin, deficiency condition, or social factor. The cumulative effects of several contributing or risk factors in disease causation are also of great significance in many disease processes, in relation to the infectious diseases such as tuberculosis, or chronic diseases such as the cardiovascular group.

The health of an individual is affected by risk factors intrinsic to that person as well as by external factors. Intrinsic factors include the biological ones that the individual inherits and those life habits he or she acquires, such as smoking, overeating, or engaging in other high-risk behaviors. External factors affecting individual health include the environment, the socioeconomic and psychological state of the person, the family, and the society in which he or she lives. Education, culture, and religion are also contributing factors to individual and community health.

There are factors that relate to health of the individual in which the society or the community can play a direct role. One of these is provision of medical care. Another is to ensure that the environment and community services include safety factors that reduce the chance of injury and disease, or include protective measures; for example, fluoridation of a community water supply to improve dental health and seat belt or helmet laws to reduce motor vehicle injury and death. These modifying factors may affect the response of the individual or the spread of an epidemic (see Chapter 3). An epidemic may also include chronic disease, because common risk factors may cause an excess of cases in a susceptible population group, in comparison to the situation before the risk factor appeared, or in comparison to a group not exposed to the risk factor. This includes rapid changes or "epidemics" in such conditions as type II diabetes, asthma, cardiovascular diseases, trauma, and other noninfectious disorders.

THE NATURAL HISTORY OF DISEASE

Disease is a dynamic process, not only of causation, but also of incubation or gradual development, severity, and the effects of interventions intended to modify outcome. Knowledge of the natural history of disease is fundamental to understanding where and with what means intervention can have the greatest chance for successful interruption or change in the disease process for the patient, family, or community.

The natural history of a disease is the course of that disease from beginning to end. This includes the factors that relate to its initiation; its clinical course leading up to resolution, cure, continuation, or long-term sequelae (further stages or complications of a disease); and environmental or intrinsic (genetic or lifestyle) factors and their effects at all stages of the disease. The effects of intervention at any stage of the disease are part of the disease process (Figure 2.3).

As discussed above, disease occurs in an individual when agent, host, and environment interact to create adverse conditions of health. The agent may be an infectious organism, a chemical exposure, a genetic defect, or a deficiency condition. A form of individual or social behavior may lead to injury or disease, such as reckless driving or risky sexual behavior. The host may be immune or susceptible as a result of many contributing social and environmental factors. The environment includes the vector, which may be a malaria-bearing mosquito, a contaminated needle shared by drug users, lead-contaminated paint, or an abusive family situation.

Assuming a natural state of "wellness" — that is, optimal health or a sense of well-being, function, and absence

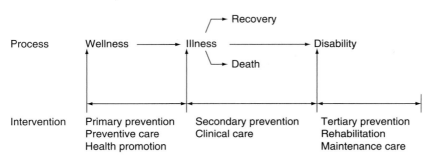

FIGURE 2.3 The process of disease and intervention.

of disease — a disease process may begin with a course of a disease, infectious or noninfectious, following a somewhat characteristic pattern described by clinicians and epidemiologists. Preclinical or predisposing events may be detected by a clinical history, with determination of risk including possible exposure or presence of other risk factors. Interventions, before and during the process, are intended to affect the later course of the disease.

The clinical course of a disease, or its laboratory or radiologic findings, may be altered by medical or public health intervention, leading to the resolution or continuation of the disease with fewer or less severe secondary sequelae. Thus the intervention becomes part of the natural history of the disease. The natural history of an infectious disease in a population will be affected by the extent of prior vaccination or previous exposure in the community. Diseases particular to children are often so because the adult population is immune from previous exposure or vaccinations. Measles and diphtheria, primarily childhood diseases, now affect adults to a large extent because they are less protected by naturally acquired immunity or are vulnerable when their immunity wanes due to inadequate vaccination in childhood.

In chronic disease management, high costs to the patient and the health system accrue where preventive services or management are inadequate or there is a failure to apply the necessary interventions. The progress of a diabetic to severe complications such as cardiovascular, renal, and ocular disease is delayed or reduced by good management of the condition, with a combination of smoking cessation, diet, exercise, and medications with good medical supervision. The patient with advanced chronic obstructive pulmonary disease or congestive heart failure may be managed well and remain stable with smoking avoidance, careful management of medications, immunizations against influenza and pneumonia, and other care needs. Where these are not applied or if they fail, the patient might well require long and expensive medical and hospital care. Failure to provide adequate supportive care will show up in ways more costly to the health system and will prove more life-threatening to the patient.

As in an individual, the phenomenon of a disease in a population may follow a course in which many factors interplay, and where interventions affect the natural course of the disease. The epidemiologic patterns of an infectious disease can be assessed in their patterns in the population, just as they can for individual cases. The classic mid-nineteenth century description of measles in the Faroe Islands by Panum showed transmission and the epidemic nature of the disease as well as the protective effect of acquired immunity (see Chapter 1). Similar more contemporary breakthroughs in medical, epidemiologic, and social sciences have produced enormous benefit to mankind as discussed throughout this text, with some examples

including eradication of smallpox and in the coming years poliomyelitis, measles, leprosy, and other dreaded diseases known for millennia; the near-elimination of rheumatic heart disease and peptic ulcers in the industrialized countries; vast reduction in mortality from stroke and coronary heart disease; and vaccines for prevention of cancers (hepatitis B and human papillomavirus). These and other great achievements of the twentieth and early part of the twenty-first centuries hold great promise for mankind in the coming decades, but great challenges lie ahead as well. The biggest challenge is to bring the benefits of known public health capacity to the poorest population of each country and the poorest populations globally.

In contemporary public health, fears of a pandemic of avian influenza are based on transmission of avian or other zoonotic viruses to humans and then their adaptation permitting human-to-human spread. With large numbers of people living in close contact with more animals (wild and domestic fowl), such as in China and Southeast Asia, and rapid transportation to the far ends of the earth, the potential for global spread of disease is almost beyond historical precedent. Indeed, many human infectious diseases are zoonotic in origin, and transferred from natural wildlife reservoirs to humans either directly or via domestic or peridomestic animals. Monitoring or immunization of domestic animals requires a combination of multidisciplinary zoonotic disease management strategies, public education and awareness, with veterinary public health monitoring and control issues. Rift Valley fever, equine encephalitis, and more recently SARS and avian influenza associated with bird-borne viral disease which can affect man, each show the terrible dangers of pandemic diseases.

SOCIETY AND HEALTH

The health of populations, like the health of individuals, depends on societal factors no less than on genetics, personal risk factors, and medical services. Social inequalities in health have been understood and documented in public health over the centuries. The Chadwick and Shattuck reports of 1840–1850 documented the relationship of poverty and bad sanitation, housing, and working conditions with high mortality, and ushered in the idea of social epidemiology. Political and social ideologies thought that the welfare state, including universal health care systems of one type or another, would eliminate social and geographic differences in health status.

From the introduction of compulsory health insurance in Germany in the 1880s to the failed attempt in the United States at national health insurance in 1995 (see Chapters 1 and 13), social reforms to deal with inequalities in health have focused on improving access to medical and hospital care. Almost all industrialized countries developed such systems, and their contribution to improve health status was an important part of social progress, especially since World War II.

But even in societies with universal access to health care, persons of lower socioeconomic status suffer higher rates of morbidity and mortality from a wide variety of diseases. The Black Report (Douglas Black) in the United Kingdom in the early 1980s pointed out that the Class V population (unskilled laborers) had twice the total and specific mortality rates of the Class I population (professional and business) for virtually all disease categories, from infant mortality to death from cancer. The report was shocking because all Britons have had access to the comprehensive National Health Service since its inception in 1948, with access to a complete range of free services, close relations to their general practitioners, and good access to specialty services. These findings initiated reappraisals of the social factors that had previously been regarded as the academic interests of medical sociologists and anthropologists and marginal to medical care. More recent studies and reviews of regional, ethnic, and socioeconomic differentials in health care access, morbidity, mortality, and patterns indicate that health inequities are present in all societies including the United Kingdom, the United States, and others even with universal health insurance or services.

Although the epidemiology of cardiovascular disease shows direct relationship of the now classic risk factors of stress, smoking, poor diet, and physical inactivity, differences in mortality from cardiovascular disease between different classes among British civil servants are not entirely explainable by these factors. The differences are also affected by social and economic issues that may relate to the psychological needs of the individual, such as the degree of control people have over their own lives. Blue-collar workers have less control over their lives than their white-collar counterparts, and have higher rates of coronary heart disease mortality than higher social classes. Other work shows the effects of migration, unemployment, drastic social and political change, and binge drinking, along with protective effects of healthy lifestyle, religiosity, and family support systems in cardiovascular diseases.

Social conditions affect disease distribution in all societies. In the United States and Western Europe, tuberculosis has re-emerged as a significant public health problem in urban areas partly because of high-risk population groups, owing to poverty and alienation from society, as in the cases of homelessness, drug abuse, and HIV infection. In countries of Eastern Europe and the former Soviet Union, the recent rise in TB incidence has resulted from various social and economic factors in the early 1990s, including the large-scale release of prisoners. In both cases, diagnosis and prescription of medication are inadequate, and the community at large becomes at risk because of the development of antibiotic-resistant strains of tubercle bacillus readily spread by inadequately treated carriers, acting as human vectors.

Studies of socioeconomic status (SES) and health are applicable and valuable in many settings. In Alameda County, California, differences in mortality between black and white population groups in terms of survival from cancer became insignificant when controlled for social class. A 30-year follow-up study of the county population reported that low-income families in California are more likely to have physical and mental problems that interfere with daily life, contributing to further impoverishment.

Studies in Finland and other locations showed that lower SES women use less preventive care such as Pap smears for cervical cancer than women of higher SES, despite having greater risk for cervical cancer. Factors leading to SES inequalities relate to differences in risk behavior, social and emotional distress, occupational factors including exposures to toxic substances, a feeling of lack of control over one's own life, and inadequate family or community social support systems.

Marmot et al. restate (2007) that health and disease are multifactorial. The risk factors associated with them require health care systems to take into account the social, physical, and psychological factors that otherwise will limit the effectiveness of even the best medical care. "This applies to interventions by the medical care provider as well as by the wider health system, including prevention and public health. The paradigm of host–agent–environment is also important in the wider context in which the sociopolitical environment and organized efforts of intervention affect the epidemiologic and individual clinical course of disease. The health system is meant to affect the occurrence or outcome of disease, either directly by primary prevention or treatment, or indirectly by reducing community or individual risk factors."

The World Health Organization Commission on Inequities (2007) states: "The gross inequalities in health that we see within and between countries present a challenge to the world. That there should be a spread of life expectancy of 48 years among countries and 20 years or more within countries is not inevitable. A burgeoning volume of research identifies social factors at the root of much of these inequalities in health. Social determinants are relevant to communicable and non-communicable disease alike. Health status, therefore, should be of concern to policy makers in every sector, not solely those involved in health policy. As a response to this global challenge, WHO is launching a Commission on Social Determinants of Health, which will review the evidence, raise societal debate, and recommend policies with the goal of improving health of the world's most vulnerable people. A major thrust of the Commission is turning public health knowledge into political action."

The effects of social conditions on health can be partly offset by interventions intended to promote healthful conditions; for example, improved sanitation, or through good quality primary and secondary health services, used efficiently and effectively made available to all. The approaches to preventing the disease or its complications may require physical changes in the environment, such as removal of the Broad Street pump handle to stop the cholera epidemic in London, or altering diets as in Goldberger's work on pellagra. Some of the great successes of public health have and continue to be low technology; some examples include DDT-impregnated bed nets, oral rehydration solutions, treatment and cure of peptic ulcers, exercise and diet to reduce obesity, hand washing in hospitals, community health workers, condoms and circumcision for prevention of sexually transmitted infections including HIV and cancer of the cervix, and many others.

The societal context in terms of employment, social security, female education, recreation, family income, cost of living, housing, and homelessness is relevant to the health status of a population. Income distribution in a wealthy country may leave a wide gap between the upper and lower socioeconomic groups, which affects health status. The media have great power to sway public perception of health issues by choosing what to publish and the context in which to present information to society. Modern media may influence an individual's tendency to overestimate the risk of some health issues while underestimating the risk to others, ultimately influencing health choices. The New Public Health has a responsibility for advocacy of societal conditions supportive of good health.

MODES OF PREVENTION

An ultimate goal of public health is to improve health and to prevent widespread disease occurrence in the population and in an individual. The methods of achieving this are wide and varied. When an objective has been defined in preventing disease, the next step is to identify suitable and feasible methods of achieving it, or a strategy with tactical objectives. This determines the method of operation and the resources needed to carry it out. The methods of public health are categorized as health promotion and primary, secondary, and tertiary prevention (Box 2.4).

Health Promotion

Health promotion is the process of enabling people and communities to increase control over factors that influence their health, and thereby to improve their health (adapted from *Ottawa Charter of Health Promotion*, 1986) (Box 2.5). Health promotion is a guiding concept involving activities intended to enhance individual and community health and well-being. It seeks to increase involvement and control of the individual and the community in their own health. It acts to improve health and social welfare, and to reduce specific determinants of diseases and risk factors that adversely affect the health, well-being, and productive capacities of an individual or society, setting targets based on the size of the problem but also the feasibility of successful intervention, in a cost-effective way. This can be through direct contact with the patient or risk group, or act indirectly through changes in the environment, legislation, or public policy. Control of AIDS relies on an array of interventions that promote change in sexual behavior and other contributory risks such as sharing of needles among drug users, screening of blood supply, safe hygienic practices in health care settings, and education of groups at risk such as teenagers, sex workers, migrant workers, and many others. It is also a clinical problem in that patients need antiretroviral therapy but this becomes a management and policy issue for making these drugs available and reducing their price so that they are affordable by the poor countries most affected. This is an example of the challenge and effectiveness of Health Promotion and the New Public Health.

Health promotion is a key element of the New Public Health and is applicable in the community, the clinic or hospital, and in all other service settings. Some health promotion activities are government legislative and regulatory interventions such as mandating the use of seat belts in cars, requiring that children be immunized to come to school, declaring that certain basic foods must have essential minerals and vitamins added to prevent nutritional deficiency disorders in vulnerable population groups, and mandating that all newborns should be given prophylactic vitamin K to prevent hemorrhagic disease of the newborn. Setting food and drug standards and raising taxes on cigarettes and alcohol to reduce their consumption are also part of

---

Box 2.4 Modes of Prevention

Health promotion Fostering national, community, and individual knowledge, attitudes, practices, policies, and standards conducive to good health; promoting legislative, social, or environmental conditions and individual self-care that reduce individual and community risk; and creating a healthful environment. It is directed toward action on the determinants of health.

Primary prevention Preventing a disease from occurring.

Secondary prevention Making an early diagnosis and giving prompt and effective treatment to stop progress or shorten the duration and prevent complications from an already existing disease process.

Tertiary prevention Preventing long-term impairments or disabilities as sequelae; restoring and maintaining optimal function once the disease process has stabilized; for instance, promoting functional rehabilitation.

Box 2.5 The Elements of Health Promotion

1. Address the population as a whole in health-related issues, in everyday life as well as people at risk for specific diseases;
2. Direct action to risk factors or causes of illness or death;
3. Undertake activist approach to seek out and remedy risk factors in the community that adversely affect health;
4. Promote factors that contribute to a better condition of health of the population;
5. Initiate actions against health hazards, including communication, education, legislation, fiscal measures, organizational change, community development, and spontaneous local activities;
6. Involve public participation in defining problems and deciding on action;
7. Advocate relevant environmental, health, and social policy;
8. Encourage health professional participation in health education and health advocacy;
9. Advocate for health based on human rights and solidarity;
10. Invest in sustainable policies, actions, and infrastructure to address the determinants of health;
11. Build capacity for policy development, leadership, health promotion practice, knowledge transfer and research, and health literacy;
12. Regulate and legislate to ensure a high level of protection from harm and enable equal opportunity for health and well-being for all people;
13. Partner and build alliances with public, private, nongovernmental, and international organizations and civil society to create sustainable actions;
14. Make the promotion of health central to the global development agenda.

International conferences following on the Ottawa Charter have been held in Adelaide in 1988, Sundsvall in 1991, Jakarta in 1999, Mexico in 2000, and Bangkok in 2005. The principles of health promotion have been reiterated and have influenced public policy regarding public health as well as the private sector.

Source: Adapted from World Health Organization. *Ottawa Charter for Health Promotion.* Geneva: World Health Organization, 1986, and *Bangkok Charter for Health Promotion in a Globalized World,* 2005. Other conferences are available at http://www.who.int/healthpromotion/conferences/en/index.html [accessed February 2008]

---

health promotion. Promoting healthy lifestyle is a major known obesity preventive activity. Health promotion is practiced by organizations and persons with many professional backgrounds working toward common goals of improvement in health and quality of individual and community life. Initiative may come from government with dedicated allocation of funds to address specific health issues, from donors, or from advocacy or community groups or individuals to promote a specific or general cause in health. Raising awareness and informing people about health and lifestyle factors that might put them at risk requires teaching young people about the dangers of sexually transmitted infections (STIs), smoking, and alcohol abuse to reduce risks associated with their social behavior. It might include disseminating information on healthy nutrition; for example, the need for folic acid supplements for women of childbearing age, and multiple vitamins for the elderly. Community and peer group attitudes and standards affect individual behavior. Health promotion endeavors to create a climate of knowledge, attitudes, beliefs, and practices that are associated with better health outcomes.

Primary Prevention

Primary prevention refers to those activities that are undertaken to prevent disease and injury from occurring. Primary prevention works with both the individual and

the community. It may be directed at the host to increase resistance to the agent (such as in immunization or cessation of smoking), or may be directed at environmental activities to reduce conditions favorable to the vector for a biological agent, such as mosquito vectors of malaria or dengue fever. Examples of such measures abound. Immunization of children prevents diseases such as tetanus, pertussis, and diphtheria. Chlorination of drinking water prevents transmission of waterborne gastroenteric diseases. Wearing seat belts in motor vehicles prevents much serious injury and death in road crashes. Reducing the availability of firearms reduces injury and death from intentional, accidental, or random violence.

Primary prevention also includes activities within the health system that can lead to better health. This may mean, for example, setting standards and ensuring that doctors not only are informed of appropriate immunization practices and modern prenatal care, but also are aware of their role in preventing cerebrovascular, coronary, and other diseases such as cancer of the lung. In this role, the health care provider serves as a teacher and guide, as well as a diagnostician and therapist. Like health promotion, primary prevention does not depend on doctors alone; both work to raise individual consciousness of self-care, mainly by raising awareness and information levels and empowering the individual and the community to improve self-care, to reduce risk factors, and to live healthier lifestyles.

Secondary Prevention

Secondary prevention is the early diagnosis and management to prevent complications from a disease. Public health interventions to prevent spread of disease include the identification of sources of the disease and the implementation of steps to stop it, as shown in Snow's closure of the Broad Street pump. Secondary prevention includes steps to isolate cases and treat or immunize contacts so as to prevent further cases of meningitis or measles in outbreaks and needle exchange programs for intravenous drug users or distribution of condoms to teenagers, drug users, and sex workers help to prevent the spread of STIs and AIDS in schools or colleges. Promotion of circumcision is shown to be effective in reducing transmission of HIV and HPV (the causative organism for cancer of the cervix).

All health care providers have a role in secondary prevention; for example, in preventing strokes by early and adequate care of hypertension. The child who has an untreated streptococcal infection of the throat may develop complications which are serious and potentially life-threatening, including rheumatic fever, rheumatic valvular heart disease, and glomerulonephritis. When a patient is found to have elevated blood pressure, this should be advised for continuing management by appropriate diet and weight loss if obese, regular physical exercise, and long-term medication with regular follow-up by a health provider in order to reduce the risk of stroke and other complications. In the case of injury, competent emergency care, safe transportation, and good trauma care may reduce the chance of death and/or permanent handicap. Screening and high-quality care in the community prevent complications of diabetes, including heart, kidney, eye, and peripheral vascular disease; they can also prevent hospitalizations, amputations, and strokes, thus lengthening and improving the quality of life. Health care systems need to be actively engaged in secondary prevention, not only as individual doctor's services, but also as organized systems of care.

Public health also has a strong interest in promoting high-quality care in secondary and tertiary care hospital centers in such areas of treatment as acute myocardial infarction, stroke, and injury in order to prevent irreversible damage. This includes quality of care reviews to promote adequate long-term post–myocardial infarction (MI) care with aspirin and beta blockers or other medication to prevent or delay recurrence and second or third MIs. The role of high-quality transportation and care in emergency facilities of hospitals in public health is vital to prevention of long-term damage and disability so that cardiac care systems including catheterization and use of stents and bypass procedures are important elements of health care policy and resource allocation, not only in capital cities but accessible to regional populations as well.

Tertiary Prevention

Tertiary prevention involves activities directed at the host or patient, but also at the social and physical environment in order to promote rehabilitation, restoration, and maintenance of maximum function after the disease and its complications have stabilized. The person who has undergone a cerebrovascular accident or trauma will come to a stage where active rehabilitation can help to restore lost functions and prevent recurrence or further complications. The public health system has a direct role in promotion of disability-friendly legislation and standards of building, housing, and support services for the chronically ill, the handicapped, and the elderly. This also involves working with many governmental social and educational departments, but also with advocacy groups, NGOs, and families. It may also include promotion of disability-friendly workplaces and social service centers.

Treatment for an MI or a fractured hip now includes early rehabilitation in order to promote early and maximum recovery with restoration to full function. Providing a wheelchair, special toilet facilities, doors, ramps, and transportation services for paraplegics are often the most vital factors in rehabilitation. Public health agencies work with groups in the community concerned with promoting help for specific categories of risk group, disease, or disability to reduce discrimination. Community action is often needed to eliminate financial physical or social barriers, promote community awareness, and finance special equipment or other needs of these groups. Close follow-up and management of chronic disease, physical and mental, requires home care and assuring an appropriate medical regimen including drugs, diet, exercise, and support services. The follow-up of chronically ill persons to supervise the taking of medications, monitor changes, and support them in maximizing their independent capacity in activities of daily living is an essential element of the New Public Health.

DEMOGRAPHIC AND EPIDEMIOLOGIC TRANSITION

Public health uses a population approach to reach many of its objectives. This requires defining the population, including trends of change in the age–sex distribution of the population, fertility and birth, spread of disease and disability, mortality, marriage and migration, and socioeconomic factors. The reduction of infectious disease as the major cause of mortality, coupled with declining fertility rates, resulted in changes in the age composition, or a demographic transition. Demographic changes, such as fertility and mortality patterns, are important factors in changing the age distribution of the population, resulting in a greater proportion of people surviving to older ages. Declining infant mortality, increasing educational levels of women, the availability of birth control, and other social and economic factors lead to changes in fertility patterns and the demographic transition—an aging of the population—with important effects on health services needs.

The age and gender distribution of a population affects and is affected by patterns of disease. Change in epidemiologic patterns, or an epidemiologic shift, is a change in predominant patterns of morbidity and mortality. The transition of infectious diseases becoming less prominent as causes of morbidity and mortality and being replaced by chronic and noninfectious diseases has occurred in both developed and developing countries. The decline in mortality from chronic diseases, such as cardiovascular disease, represents a new stage of epidemiologic transition, creating an aging population with higher standards of health but also long-term community support and care needs. Monitoring and responding to these changes are fundamental responsibilities of public health, and readiness to react to new, local, or generalized changes in epidemiologic patterns is vital to the New Public Health.

Societies are not totally homogeneous in ethnic composition, levels of affluence, or other social markers. A society classified as developing may have substantial numbers of persons with incomes which promote overnutrition, so that disease patterns may include diseases of excesses, such as diabetes. On the other hand, affluent societies include population groups with disease patterns of poverty, including poor nutrition and low birth weight babies.

A further stage of epidemiologic transition has been occurring in the industrialized countries since the 1960s, with dramatic reductions in mortality from coronary heart disease, stroke, and, to a lesser extent, trauma. The interpretation of this epidemiologic transition is still not perfectly clear. How it occurred in the industrialized western countries but not in those of the former Soviet Union is a question whose answer is vital to the future of health in Russia and some countries of Eastern Europe. Developing countries must also prepare to cope with epidemics of noninfectious diseases, and all countries face renewed challenges from infectious diseases with antibiotic resistance or newly appearing infectious agents posing major public health threats.

Demographic change in a country may reflect social and political decisions and health system priorities from decades before. Russia's rapid population decline since the 1990s, China's gender imbalance with shortage of millions of young women, Egypt's rapid population growth outstripping economic capacity and many other examples indicate the severity and societal importance of capacity to analyze and formulate public health and social policies to address such fundamental sociopolitical issues.

INTERDEPENDENCE OF HEALTH SERVICES

The challenge of keeping populations and individuals healthy is reflected in modern health services. Each component of a health service may have developed with different historical emphases, operating independently as a separate service under different administrative auspices and funding systems, competing for limited health care resources. In this situation, preventive community care receives less attention and resources than more costly treatment services. Figure 2.4 suggests a set of health services in an interactive relationship to serve a community or defined population, but the emphasis should be on the interdependence of these services one to the other and all to the comprehensive network in order to achieve effective use of resources and a balanced set of services for the patient, the client or patient population, and the community.

Clinical medicine and public health each play major roles in primary, secondary, and tertiary prevention. Each may function separately in their roles in the community, but optimal success lies in their integrated efforts. Allocation of resources should promote management and planning practices to promote this integration. There is a functional interdependence of all elements of health care serving a definable population. The components of health services for a population group interact with the patient or the patient as the central figure. Effectiveness in use of resources means that providing the service most appropriate for meeting the individual's or group's needs at a point in time are those that should be applied. Long stays in an acute care hospital often occur when home care or community support services are inadequate. This is wasteful and destructive to good patient care and costly to the health care economy. Linkage and a balance of services that meet individual and community needs promote effective and efficient use of resources.

Separate organization and financing of services place barriers to appropriate provision of services for both the community and the individual patient. The interdependence of services is a challenge in health care organization for the future. Where there is competition for limited resources, pressures for tertiary services often receive priority over programs to prevent children from dying of preventable diseases. Public health must be seen in the context of all health care and must play an influential role in promoting prevention at all levels. Clinical services need public health in order to provide prevention and community health services that reduce the burden of disease, disability, and dependence on the institutional setting.

DEFINING PUBLIC HEALTH

Health was traditionally thought of as a state of absence of disease, pain, or disability, but has gradually been expanded to include physical, mental, and societal well-being. In 1920, C. E. A. Winslow, professor of public health at Yale University, defined public health as follows:

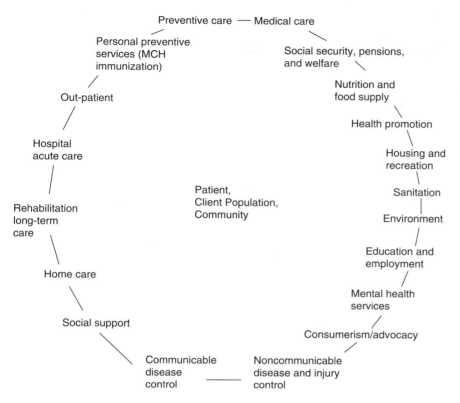

FIGURE 2.4 Community health as a network of services serving a defined population.

Public health is the Science and Art of (1) preventing disease, (2) prolonging life, and (3) promoting health and efficiency through organized community effort for:

    (a) the sanitation of the environment,

    (b) the control of communicable infections,

    (c) the education of the individual in personal hygiene,

    (d) the organization of medical and nursing services for the early diagnosis and preventive treatment of disease, and

    (e) the development of social machinery to ensure everyone a standard of living adequate for the maintenance of health, so organizing these benefits as to enable every citizen to enjoy his birthright of health and longevity. (As quoted in Institute of Medicine. 1988. The Future of Public Health. *Washington, DC: National Academy Press.)*

Winslow's far-reaching definition remains a valid framework but unfulfilled when clinical medicine and public health have financing and management barriers between them. In many countries, isolation from the financing and provision of medical and nursing care services left public health the task of meeting the health needs of the poor and underserved population groups with inadequate resources and recognition. Health insurance for medical and hospital care has in recent years been more open to incorporating evidence-based preventive care, but the organization of public health has lacked the same level of attention. In some countries, the limitations have been conceptual in that public health was defined primarily in terms of control of infectious, environmental, and occupational diseases.

Terms such as *social hygiene, preventive medicine, community medicine, social medicine,* and others have been used to denote public health over the past century. Preventive medicine is a combination of some elements of public health with clinical medicine. Public health deals with the individual just as the clinical health care provider does, as in the case of immunization programs, follow-up of certain illnesses, and other personal clinical services. Clinical medicine also deals in the area of prevention in management of patients with hypertension or diabetes, and in doing so prevents the serious complications of these diseases. Preventive medicine focuses on a medical or clinical function, or what might be called *personal preventive care,* with stress on risk groups in the community and national efforts for health promotion.

Social Medicine and Community Health

Social medicine looks at illness in a social context, but lacks the environmental and regulatory functions of public health. Community health implies a local form of health intervention, whereas public health more clearly implies a global approach, which includes action at the international, national, state, and local levels. There are issues in health that cannot be dealt with at the individual, family, or community levels, requiring global strategies and intervention programs.

The Social Medicine movement primarily developed as an academic discipline and arose from ideas of European physicians during the industrial revolution. It examines statistical data showing, as in various governmental reports in the mid-nineteenth century, that poverty among the working class was associated with short life expectancy and that social conditions were key factors in the health of populations and individuals. This movement became the basis for departments in medical faculties and public health education throughout the world.

Social Hygiene, Eugenics, and Corruption of Public Health Concepts

The ethical base of public health in Europe developed in the context of its successes in the nineteenth and early twentieth centuries. But the twentieth century was also replete with extremism and wide-scale abuse of human rights, with mass executions, deportations, and starvation as official policy in fascist and Stalinist regimes. The "social and racial hygiene" and the eugenics movements led to the medicalization of sterilization in the United States and other countries, and then murder in Nazi Germany first of the mentally and physically handicapped and then "racial inferiors." These eugenics theories were used by Nazi Germany to justify medically supervised killing of hundreds of thousands of helpless incapacitated individuals, and this was linked to wider genocide and the Holocaust, with the killing of 6 million of Jews in industrialized systems of mass murder and corrupt medical experimentation on prisoners. Following World War II, the ethics of medical experimentation (and public health) were codified in the Nuremberg Code and Universal Declaration of Human Rights based on lessons learned from these and other atrocities inflicted on civilian populations (see Chapter 15).

Threats of genocide, ethnic cleansing, and terrorism are still present on the world stage, often justified by current warped versions of racial hygienic theories. Genocidal incitement and actual genocide and terrorism have recurred in the last decades of the twentieth century and into the twenty-first century in the former Yugoslav republics, Africa (Rwanda and Darfur), south Asia, and elsewhere. Terrorism against civilians has become a worldwide phenomenon with threats of biological and chemical agents, and potentially with nuclear capacity. Asymmetrical warfare of insurgents using innocent civilians for cover, as other forms of warfare, carries with it grave dangers to public health, human rights, and international stability.

Medical Ecology

In 1961, Kerr White and colleagues defined *medical ecology* as population-based research providing the foundation for management of health care quality. This concept stresses a population approach, including those not attending and those using health services. This concept was based on previous work on quality of care, randomized clinical trials, medical audit, and structure–process–outcome research. It also addressed health care quality and management.

These themes influenced medical research by stressing the population from which clinical cases emerge as well as public health research with clinical outcome measures, themes that recur in development of health services research and, later, evidence-based medicine. This led to development of the Agency for Health Care Policy and Research and Development in the U.S. Department of Health and Human Services and evidence-based practice centers to synthesize fundamental knowledge for development of information for decision-making tools such as clinical guidelines, algorithms, or pathways. Clinical guidelines and recommended best practices have become part of the New Public Health to promote quality of patient care and public health programming. This can include recommended standards; for example, follow-up care of the post–myocardial infarction patient, an internationally recommended immunization schedule, recommended dietary intake or food fortification standards, and mandatory vitamin K and eye care for all newborns.

Community-Oriented Primary Care

Community-oriented primary care (COPC) is an approach to primary health care that links community epidemiology and appropriate primary care, using proactive responses to the priority needs identified. COPC, originally pioneered in South Africa and Israel by Sidney and Emily Kark and colleagues in the 1950s and 1960s, stresses that medical services in the community need to be molded to the needs of the population, as defined by epidemiologic analysis. COPC involves community outreach and education, as well as clinical preventive and treatment services.

COPC focuses on community epidemiology and an active problem-solving approach. This differs from national or larger-scale planning that sometimes loses sight of the local nature of health problems or risk factors. COPC combines clinical and epidemiologic skills, defines needed interventions, and promotes community involvement and access to health care. It is based on linkages between the different elements of a comprehensive basket of services along with attention to the social and physical environment. A multidisciplinary team and outreach services are important for the program, and community development is part of the process.

In the United States, the COPC concept has influenced health care planning for poor areas, especially provision of federally funded community health centers in attempts to provide health care for the underserved since the 1960s. In more recent years, COPC has gained wider acceptance in the United States, where it is associated with family

---

Box 2.6 Features of Community-Oriented Primary Health Care (COPC)

1. Essential features
   a. Clinical and epidemiologic skills
   b. A defined population
   c. Defined programs to address community health issues
   d. Community involvement
   e. Accessibility to health care—reducing geographic, fiscal, social, and cultural barriers

2. Desirable features
   a. Integration/coordination of curative, rehabilitative, preventive, and promotive care
   b. Comprehensive approach extending to behavioral, social, and environmental factors
   c. Multidisciplinary team
   d. Mobility and outreach
   e. Community development

Sources: Tollman, S. 1991. Community oriented primary care: Origins, evolution, applications. *Social Science and Medicine*, 32:633–642.
Epstein L, Gofin J, Gofin R, Neumark Y. 2002. The Jerusalem experience: three decades of service, research, and training in community-oriented primary care. *American Journal of Public Health*, 92: 1717–1721.

---

physician training and community health planning based on the risk approach and "managed care" systems. Indeed, the three approaches are mutually complementary (Box 2.6). As the emphasis on health care reform in the late 1990s moved toward managed care, the principles of COPC were and will continue to be important in promoting health and primary prevention in all its modalities, as well as tertiary prevention with follow-up and maintenance of the health of the chronically ill.

COPC stresses that all aspects of health care have moved toward prevention based on measurable health issues in the community. Through either formal or informal linkages between health services, the elements of COPC are part of the daily work of health care providers and community services systems. The U.S. Institute of Medicine issued the *Report on Primary Care* in 1995, defining *primary care* as "the provision of integrated, accessible health care services by clinicians who are accountable for addressing the majority of personal health care needs, developing a sustained partnership with patients and practicing in the context of the family and the community." This formulation was criticized by the American Public Health Association as lacking a public health perspective and failing to take into account both the individual and the community health approaches. It is just this gap that COPC tries to bridge.

The community, whether local or national, is the site of action for many public health interventions. Moreover, understanding the characteristics of the community is vital to a successful community-oriented approach. By the 1980s, new patterns of public health began to emerge, including all measures used to improve the health of the community and at the same time working to protect and promote the health of the individual. The range of activities to achieve these general goals is very wide, including individual patient care systems and the community-wide activities that affect the health and well-being of the individual. These include the financing and management of health systems, evaluation of the health status of the population, and steps to improve the quality of health care. They place reliance on health promotion activities to change environmental risk

factors for disease and death. They promote integrative and multisectoral approaches and the international health teamwork required for global progress in health.

WORLD HEALTH ORGANIZATION'S DEFINITION OF HEALTH

The definition of health in the charter of the World Health Organization (WHO) as a complete state of physical, mental, and social well-being had the ring of utopianism and irrelevance to states struggling to provide even minimal care in severely adverse political, economic, social, and environmental conditions (Box 2.7). In 1977, a more modest goal was set for attainment of a level of health compatible with maximum feasible social and economic productivity. One needs to recognize that health and disease are on a dynamic continuum that affects everyone. The mission for public health is to use a wide range of methods to prevent disease and premature death, and improve quality of life for the benefit of individuals and the community.

In the 1960s, most industrialized countries were concentrating energies and financing in health care on providing access to medical services through national insurance schemes. Developing countries were often spending scarce resources trying to emulate this trend. The WHO was concentrating on categorical programs, such as eradication of smallpox and malaria, as well as the Expanded Program of Immunization and similar specific efforts. At the same time, there was a growing concern that developing countries were placing too much emphasis and expenditures on curative services and not enough on prevention and primary care.

Alma-Ata: Health for All

The WHO and the United Nations Children's Fund (UNICEF) a sponsored conference held in Alma-Ata, Kazakhstan, in 1978, which was convened to refocus on

---

Box 2.7 Definitions of Health and Mission of World Health Organization

The World Health Organization defines *health* as "a state of complete physical, mental and social well-being, not merely the absence of disease or infirmity" (WHO Constitution, 1948).

In 1978 at the Alma-Ata Conference on Primary Health Care, the WHO related health to "social and economic productivity in setting as a target the attainment by all the people of the world of a level of health that will permit them to lead a socially and economically productive life." Three general programs of work for the periods 1984–1989, 1990–1995, and 1996–2001 were formulated as the basis of national and international activity to promote health.

In 1995, the WHO, recognizing changing world conditions of demography, epidemiology, environment, and political and economic status, addressed the unmet needs of developing countries and health management needs in the industrialized countries, calling for international commitment to "attain targets that will make significant progress towards improving equity and ensuring sustainable health development."

The 1999 object of the WHO is restated as "the attainment by all peoples of the highest possible level of health," as defined in the WHO constitution, by a wide range of functions in promoting technical cooperation, assisting governments, and providing technical assistance, international cooperation, and standards.

Source: *New Challenges for Public Health: Report of an International Meeting.* 1996. Geneva: World Health Organization, and www://who.org/aboutwho [accessed February 24, 2008]

---

primary care. The Alma-Ata Declaration stated that health is a basic human right, and that governments are responsible to assure that right for their citizens and to develop appropriate strategies to fulfill this promise. This proposition has come to be increasingly accepted in the international community. The conference stressed the right and duty of people to participate in the planning and implementation of their health care. It advocated the use of scientifically, socially, and economically sound technology. Joint action through intersectoral cooperation was also emphasized.

The Alma-Ata Declaration focused on primary health care as the appropriate method of assuming adequate access to health care for all (Box 2.8). This approach was

---

Box 2.8 Declaration of Alma-Ata, 1978; A Summary of Primary Health Care (PHC)

1. Reaffirms that health is a state of complete physical, mental, and social well-being, and not merely the absence of disease or infirmity, and is a fundamental human right.
2. Existing gross inequalities in the health status of the people, particularly between developed and developing countries as well as within countries, is of common concern to all countries.
3. Governments have a responsibility for the health of their people. The people have the right and duty to participate in planning and implementation of their health care.
4. A main social target is the attainment, by all peoples of the world by the year 2000, of a level of health that will permit them to lead a socially and economically productive life.
5. PHC is essential health care based on practical, scientifically sound, and socially acceptable methods and technology.
6. It is the first level of contact of individuals, the family, and the national health system bringing health care as close as possible to where people live and work, as the first element of a continuing health care process.
7. PHC evolves from the conditions and characteristics of the country and its communities, based on the application of social, biomedical, and health services research and public health experience.
8. PHC addresses the main health problems in the community, providing promotive, preventive, curative, and rehabilitative services accordingly.
9. PHC includes the following:
   a. Education concerning prevailing health problems and methods of preventing and controlling them;
   b. Promotion of food supply and proper nutrition;
   c. Adequate supply of safe water and basic sanitation;
   d. Maternal and child health care, including family planning;
   e. Immunization against the major infectious diseases;
   f. Prevention of locally endemic diseases;
   g. Appropriate treatment of common diseases and injuries;
   h. The provision of essential drugs;
   i. Relies on all health workers . . . to work as a health team.
10. All governments should formulate national health policies, strategies and plans, mobilize political will and resources, used rationally, to ensure PHC for all people.

Source: http://www.euro.who.int/AboutWHO/Policy/20010827_1 [accessed February 24, 2008]

endorsed in the World Health Assembly (WHA) in 1977 under the banner "Health for All by the Year 2000" (HFA 2000). This was a landmark decision and has had important practical results. Many countries have gradually come to accept the notion of placing priority on primary care, resisting the temptation to spend high percentages of health care resources on high-tech and costly medicine. Spreading these same resources into highly cost-effective primary care, such as immunization and nutrition programs, provides greater benefit to individuals and to society as a whole.

Alma-Ata provided a new sense of direction for health policy, applicable to developing countries and in a different way to the developed countries. During the 1980s, the Health for All concept influenced national health policies in the developing countries with signs of progress in immunization coverage, for example, but the initiative was diluted as an unintended consequence by more categorical programs such as eradication of poliomyelitis. For example, developing countries have accepted immunization and diarrheal disease control as high-priority issues and achieved remarkable success in raising immunization coverage from some 10 percent to over 75 percent in just a decade.

Developed countries addressed these principles in different ways. In these countries, the concept of primary health care led directly to important conceptual developments in health. National health targets and guidelines are now common in many countries and are integral parts of national health planning. Reforms of the British National Health Service — for example, as discussed in Chapter 13, pay increases for family physicians and encouraging group practice with public health nursing support — have become widespread in the United Kingdom. Progressive health maintenance organizations, such as Kaiser Permanente in the United States and district health systems in Canada, have emphasized integrated approaches to health care for registered or geographically defined populations (see Chapters 11 to 13). This systematic approach to individual and community health is part of the New Public Health.

The interactions among community public health, personal health services, and health-related behavior, including their management, are the essence of the New Public Health. How the health system is organized and managed affects the health of the individual and the population, as does the quality of providers. Health information systems with epidemiologic, economic, and sociodemographic analysis are vital to monitor health status and allow for changing priorities and management. Well-qualified personnel are essential to provide services, manage the system, and carry out relevant research and health policy analysis. Diffusion of data, health information, and responsibility help to provide a responsive and comprehensive approach to meet the health needs of the individual and community. The physical, social, economic, and even the political environments are important determinants of health status of the population and the individual. Joint action (intersectoral cooperation) between public and nongovernmental or community-based organizations is needed to achieve the well-being of the individual in a healthy society.

In the 1980s and 1990s, these ideas became part of an evolving New Public Health, spurred by epidemiologic changes, health economics, the development of managed care linking health systems, and prepayment. Knowledge and self-care skills, as well as community action to reduce health risks, are no less important in this than the roles of medical practitioners and institutional care. All are parts of a coherent holistic approach to health.

SELECTIVE PRIMARY CARE

The concept of selective primary care, articulated in the 1960s by Walsh and Warren, addresses the needs of developing countries to select those interventions on a broad scale that would have the greatest positive impact on health, taking into account limited resources such as money, facilities, and manpower.

The term *selective primary care* is meant to define national priorities that are based not on the greatest causes of morbidity or mortality, but on common conditions of epidemiologic importance for which there are effective and simple preventive measures. Throughout health planning, there is an implicit or explicit selection of priorities for allocation of resources. Even in primary care, selection of targets is a part of the process of resource allocation. In modern public health, this process is more explicit. A country with limited resources and a high birth rate will emphasize maternal and child health before investing in geriatric care.

This concept has become part of the microeconomics of health care and technology assessment, discussed in Chapters 11 and 15, respectively, and is used widely in setting priorities and resource allocation. In primary care in developing countries, cost-effective interventions have been articulated by many international organizations, including iodization of salt, use of oral rehydration therapy (ORT) for diarrheal diseases, vitamin A supplementation for all children, expanded programs of immunization, and others that have the potential for saving hundreds of thousands of lives yearly at low cost. In developed countries, health promotion targeted to reduce accidents and risk factors, such as smoking, high-fat diets, and lack of exercise for cardiovascular diseases are low-cost public health interventions that save lives and reduce use of hospital care.

Targeting specific diseases is essential for efforts to control tuberculosis or eradicate polio, but at the same time, development of a comprehensive primary care infrastructure may be equally or even more important than the single disease approach. Some disease entities such as HIV/AIDS attract donor funding more readily than basic infrastructure services such as immunization and this can sometimes be detrimental to addressing overall health needs of the population and neglected but also important diseases.

THE RISK APPROACH

The risk approach selects population groups on the basis of risk and helps determine interventions' priorities to reduce morbidity and mortality. The measure of health risk is taken as a proxy for need, so that the risk approach provides something for all, but more for those in need — in proportion to that need. In epidemiologic terms, these are persons with higher relative risk or attributed risk.

Some groups in the general population are at higher risk than others for specific conditions. The Expanded Programme on Immunization (EPI), Control of Diarrhoeal Diseases (CD), and Acute Respiratory Disease (ARD) programs of the WHO are risk approaches to tackling fundamental public health problems of children in developing countries.

Public health places considerable emphasis on maternal and child health because these are vulnerable periods in life for specific health problems. Pregnancy care is based on a basic level of care for all, with continuous assessment of risk factors that require a higher intensity of follow-up. Prenatal care helps identify factors that increase the risk for the pregnant woman or her fetus/newborn. Efforts directed toward these special risk groups have the potential to reduce morbidity and mortality. High-risk identification, assessment, and management are vital to a successful maternal care program.

Similarly, routine infant care is designed not only to promote the health of infants, but also to find the earliest possible indications of deviation and the need for further assessment and intervention to prevent a worsening of the condition. Low birth weight babies are at greater risk for many hazards and should be given special treatment. Screening of all babies is done on a routine basis for birth defects or congenital conditions such as hypothyroidism (CH), phenylketonuria (PKU), and other metabolic and hematologic diseases. Screening must be followed by investigating and treating those found to have a clinical deficiency. This is an important element of infant care because infancy itself is a risk factor.

As will be discussed in Chapter 6 and others, epidemiology has come to focus on the risk approach with screening based on known genetic, social, nutritional, environmental, occupational, behavioral, or other factors contributing to the risk for disease. The risk approach has the advantage of specificity and is often used to initiate new programs directed at special categories of need. This approach can lead to narrow and somewhat rigid programs that may be difficult to integrate into a more general or comprehensive approach, but until universal programs can be achieved, selective targeted approaches are justifiable. Indeed, even when universal health coverage is established, it is still important to address health needs or issues of groups at special risk.

Work to achieve defined targets means making difficult choices. The supply and utilization of some services will limit availability for other services. There is an interaction, sometimes positive, sometimes negative, between competing needs and the health status of a population.

THE CASE FOR ACTION

Public health identifies needs by measuring and comparing the incidence or prevalence of the condition in a defined population with that in other comparable population groups and defines targets to reduce or eliminate the risk of disease. It determines ways of intervening in the natural epidemiology of the disease, and develops a program to reduce or even eliminate the disease.

Because of the interdependence of health services, as well as the total financial burden of health care, it is essential to look at the costs of providing health care, and how resources should be allocated to achieve the best results possible. Health economics has become a fundamental methodology in policy determination. The costs of health care, the supply of services, the needs for health care or other health-promoting intervention, and effective means of using resources to meet goals are fundamental in the New Public Health. It is possible to err widely in health planning if one set of factors is over- or underemphasized. Excessive supply of one service diminishes availability of resources for other needed investments in health. If diseases are not prevented or their sequelae not well managed, patients must use costly health care services and are unable to perform their normal social functions such as learning at school or work performance. Lack of investment in health promotion and primary prevention creates a larger reliance on institutional care, driving health costs upward. This restricts flexibility in meeting patients' needs. The interaction of supply and demand for health services is an important determinant of the political economy of health care. Health and its place in national priorities are determined by the social and political philosophy of a government.

The case for action, or the justification for a public health intervention, is a complex of epidemiologic, economic, and public policy factors (Table 2.1). Each disease or group of diseases requires its own case for action. The justification for public health intervention requires sufficient evidence of the incidence/prevalence of the disease

TABLE 2.1 The Case for Action: Factors in Justifying Public Health Interventions

Ethics and potential	Issues
The right to health	Public expectation and social norms
Public advocacy	Concerned groups, the media, an individual
Need — epidemiologic and clinical	Morbidity, mortality, functional disability, physiologic indicators
Available technology	Documented effectiveness, safety, experience, acceptability, affordability
Precedent — "state of the art"	Good public health practice; standards from leading centers of excellence, not necessarily consensus
Research and cumulative evidence	New evidence from research and practice should be published in peer-reviewed journals and made accessible to all policymakers and practitioners
Teaching public health	All health professionals should have broad introductory courses in public health
Legal constraints and liability	Law and court decisions Providers, managed care, and governments
Costs and benefits	Direct cost to health system; indirect cost to the individual, family, and society
Acceptability	Media and public opinion
Leadership	Political and professional
Quality of life	Optimizing human potential in healthy communities

(see Chapter 3); the effectiveness and safety of an intervention; risk factors; safe means at hand to intervene; the human, social, and economic cost of the disease; political factors; and a policy decision as to the priority of the problem. This often depends on subjective factors, such as the guiding philosophy of the health system and the way it allocates resources.

Some interventions are so well established that no new justification is required to make the case, and the only question is how to do it most effectively. For example, infant vaccination is a cost-effective program for the protection of the individual child and the population as a whole. Whether provided as a public service or as a clinical preventive measure by a private medical practitioner, it is in the interest of public health that all children be immunized.

An outbreak of diarrheal disease in a kindergarten presents an obvious case for action, and a public health system must respond on an emergency basis, with selection of the most suitable mode of intervention. The considerations in developing a case for action are outlined above. Need is based on clinical and epidemiologic evidence, but also on the importance of an intervention in the eyes of the public. The technology available, its effectiveness and safety, and accumulated experience are important in the equation, as are acceptability and affordability of appropriate interventions. The precedents for use of an intervention are also important. On epidemiologic evidence, if the preventive practice has been seen to provide reduction in risk for the individual and for the population, then there is good reason to implement it. The costs and benefits must be examined as part of the justification to help in the selection of health priorities.

Health systems research examines the efficiency of health care and promotes improved efficiency and effective use of resources. This is a vital function in determining how best to use resources and meet current health needs. Past emphasis on hospital care at the expense of less development of primary care and prevention is still a common issue particularly in former Soviet and developing countries where high percentages of total health expenditures go to acute hospital care with long length of stay with allocation to community health care. The cost of this imbalance is high mortality from preventable diseases.

New drugs, vaccines, and medical equipment are constantly becoming available, and each new addition needs to be examined among the national health priorities. Sometimes, due to cost, a country cannot afford to add a new vaccine to the routine. However, when there is good medical evidence for the vaccine, it can be applied for those at greatest risk. Although there are ethical issues involved, it may be necessary to advise parents or family members to independently purchase the vaccine. Clearly, recommending individual purchase of a vaccine is counter to the principle of equity and solidarity. On the other hand, failure to advise parents of potential benefits to their children creates other ethical problems.

Mass screening programs involving complete physical examinations have not been found to be cost-effective or to significantly reduce disease. In the 1950s to 1960s, routine general health examinations were promoted as an effective method of finding disease early. Since the late 1970s, a selective and specific approach to screening has become widely accepted. This involves defining risk categories for specific diseases and bearing in mind the potential for remedial action. Early case finding of breast cancer by routine mammography has been found to be effective after age 35, and Papanicolaou smear testing to discover cancer of the cervix is timed according to risk category.

The factor of contribution to quality of life should be considered. A vaccine for varicella may be justified partly for the prevention of deaths or illness from chickenpox.

A stronger argument is often based on the fact that this is a disease that causes children up to 2 weeks of moderate illness and may require parents to stay home with the child, resulting in economic loss to the parent and society. The fact that this vaccination prevents the occurrence of herpes zoster or shingles later in life may also be a justification. Widespread adoption of hepatitis B vaccine is justified on the grounds that it prevents cancer of the liver, liver cirrhosis, and hepatic failure in a small percentage of the population affected.

How many cases of a disease are enough to justify an intervention? One or several cases of some diseases, such as poliomyelitis, may be considered an epidemic in that each case constitutes or is an indicator of a wider threat. A single case of polio suggests that another 1000 persons are infected but have not developed a recognized clinical condition. Such a case constitutes a public health emergency, and forceful organization to meet a crisis is needed. Current standards are such that even one case of measles imported into a population free of the disease may cause a large outbreak as occurred in Britain, France, and Israel during 2007–2008, by contacts on an aircraft, at family gatherings, or even in medical settings. A measles epidemic indicates a failure of public health policy and practice. Screening for some cancers, such as cervix and colon, are cost effective. Screening of all newborns for congenital disorders is important because each case discovered early and treated effectively saves a lifetime of care for serious disability.

Assessing a public health intervention to prevent the disease or reduce its impact requires measurement of the disease in the population and its economic impact. There is no simple formula to justify a particular intervention, but the cost–benefit approach is now commonly required to make such a case for action. Sometimes public opinion and political leadership may oppose the views of the professional community, or may impose limitations of policy or funds that prevent its implementation. Conversely, professional groups may press for additional resources that compete for limited resources available to provide other needed health activities. Both the professionals of the health system and the general public need full access to health-related information to take part in such debates in a constructive way. To maintain progress, a system must examine new technologies and justify their adoption or rejection (Chapter 15).

POLITICAL ECONOMY AND HEALTH

As the concept of public health has evolved, and the value of medical care has improved through scientific and technological advances, societies have identified health as a legitimate area of activity of collective bargaining and government. With this process, the need for managing health care resources became more clearly defined as a public responsibility. In industrialized countries each with very different political makeup, national responsibility for universal access to health became part of the social ethos. With that, the financing and managing of health services became part of a broad concept of public health, and economics, planning, and management came to be part of the New Public Health (discussed in Chapters 10–13).

Social, ethical, and political philosophies have profound effects on policy decisions including allocation of public monies and resources. Investment in public health is an integral part of socioeconomic development. Governments are major suppliers of funds and leadership in health infrastructure development, provision of health services, and health payment systems. They also play a central role in the development of health promotion and regulation of the environment, food, and drugs essential for community health.

In liberal social democracies, the individual is deemed to have a right to health care. The state accepts responsibility to assure availability, accessibility, and quality of care. In many developed countries, government has also taken responsibility to arrange funding and services that are equitably accessible and of high quality. Health care financing may involve taxation, allocation, or special mandatory requirements on employers to pay for health insurance. Services may be provided by a state-financed and -regulated service or through NGOs and/or private service mechanisms. These systems allocate between 6 percent and 14 percent of gross national product (GNP) to health services, with some governments funding over 80 percent of health expenditures; for example, Canada and the United Kingdom.

In Marxist states, the state organizes all aspects of health care with the philosophy that every citizen is entitled to equity in access to health services. The state health system manages research, manpower training, and service delivery, even if operational aspects are decentralized to local health authorities. This model applied primarily to the Soviet model of health services. These systems, except for Cuba, placed financing of health low on the national priority, with funding less than 4 percent of GNP. In the shift to market economies in the 1990s, some former Socialist countries, such as Russia, are struggling with declining health status and a difficult shift from a strongly centralized health system to a decentralized system with diffusion of powers and responsibilities. Promotion of market concepts in former Soviet countries has reduced access to care and created a serious dilemma for their governments.

Former colonial countries, independent since the 1950s and 1960s, largely carried on the governmental health structures established in the colonial times. Most developing countries have given health a relatively low place in budgetary allotment, with expenditures under 3 percent of GNP. During the 1980s, there has been a trend in developing countries toward decentralization of health services and greater roles for NGOs, and the development of health insurance. Some, influenced by medical concepts of their former mother countries, fostered development of specialty medicine in the major centers with little emphasis on the rural majority population. Soviet influence in many ex-colonial countries promoted state-operated systems. The WHO promoted primary care, but the allocations favored city-based specialty care. Israel, as an ex-colony, used British ideas of public health together with central European Sick Funds and maternal and child health as major streams of development until the mid-1990s.

A growing new conservatism in the 1980s and 1990s in the industrialized countries is a restatement of old values in which market economics and individualistic social values are placed above the common good concepts of liberalism and socialism in its various forms. In the more extreme forms of this concept, the individual is responsible for his or her own health, including payment, and has a choice of health care providers that will respond with high-quality personalized care.

Market forces, meaning competition in financing and provision of health services with rationing of services, based on fees or private insurance and willingness and ability to pay, have become part of the ideology of the new conservatism. This assumes that the patient (i.e., the consumer) will select the best service for his or her need, while the provider best able to meet consumer expectations will thrive. In its purest form, the state has no role in providing or financing of health services except those directly related to community protection and promotion of a healthful environment without interfering with individual choices. The state ensures that there are sufficient health care providers and allows market forces to determine prices and distribution of services with minimal regulation. The United States retains this policy in a highly modified form, with 85 percent of the population covered by some form of private or public insurance systems.

Modified market forces in health care are part of health reforms in many countries as they seek not only to ensure quality health care for all, but also to constrain costs. A free market in health care is costly and ultimately inefficient because it encourages inflation of provider incomes or budgets and increasing utilization of highly technical services. Further, even in the most free market societies, the economy of health care is highly influenced by many factors outside the control of the consumer and provider. The total national health expenditures in the United States rose rapidly until reaching 14.6 percent of GDP in 2005, the highest of any country, despite serious deficiencies for those without any or having very inadequate health insurance (in total more than 30 percent of the population). This is compared to some 10 percent of GDP in Canada, which has universal health insurance

under public administration. Following the 1994 defeat of President Clinton's national health program, the conservative Congress and the business community took steps to expand managed care in order to control costs resulting in a revolution in health care in the United States (Chapters 11 and 13).

Market reforms are being implemented in many "socialized" health systems. These may be through incentives to promote achievement of performance indicators, such as full immunization coverage. Others are using control of supply, such as hospital beds or licensed physicians, as methods of reducing overutilization of services that generates increasing costs. Market mechanisms in health are aimed not only at the individual but also at the provider. Incentive payment systems must work to protect the patient's legitimate needs, and conversely incentives that might reduce quality of care should be avoided. Fee for service promotes high rates of services such as surgery. Increasing private practice and user fees can adversely affect middle- and low-income groups, as well as employers, by raising costs of health insurance. Managed care systems, with restraint on fee-for-service medical practice, has emerged as a positive response to the market approach. Incentive systems in payments for services may be altered by government or insurance agencies in order to promote rational use of services, such as reduction of hospital stays. The free market approach is affecting planning of health insurance systems in previously highly centralized health systems in developing countries as well as redevelopment of health systems in former Soviet countries.

Despite political differences, reform of health systems has become a common factor in virtually all health systems since the 1990s, as each government searches for cost-effectiveness, quality of care, and universality of coverage. The new paradigm of health care reform sees the convergence of different systems to common principles. National responsibility for health goals and health promotion leads to national financing of health care with regional and managed care systems. Most developed countries have long since adopted national health insurance or service systems. Some governments may, as in the United States, insure only the highest-risk groups such as the elderly and the poor, leaving the working and middle classes to seek private insurers. The nature and direction of health care reform affecting coverage of the population are of central importance in the New Public Health because of its effects on allocation of resources and on the health of the population.

HEALTH AND DEVELOPMENT

Individuals in good health are better able to study and learn, and be more productive in their work. Improvements in the standard of living have long been known to contribute to improved public health; however, the converse has not always been recognized. Investment in health care was not considered high priority in many countries where economic considerations directed investment to the "productive" sectors such as manufacturing and large-scale infrastructure projects, such as hydroelectric dams.

Whether health is a contributor to economic development or a drain on societies' resources has been a fundamental debate between socially and market-oriented advocates. Classic economic theory, both free enterprise and Marxist, has tended to regard health as a drain on economies, distracting investment needed for economic growth. As a result, in many countries health has been given low priority in budgetary allocation, even when the major source of financing is governmental. This belief among economists and banking institutions prevented loans for health development on the grounds that such funds should focus on creating jobs and better incomes, before investing in health infrastructure. Consequently, development of health care has been hampered.

A socially oriented approach sees investment in health as necessary for the protection and development of "human capital," just as investment in education is needed for the long-term benefit of the economy of a country. In 1993, the World Bank, *World Development Report: Investing in Health,* articulated a new approach to economics in which health, along with education and social development, are seen as essential pre-conditions and contributors for economic development. While many in the health field have long recognized the importance of health for social and economic improvement, its adoption by leading international development banking may mark a turning point for investment in developing nations, so that health may be a contender for increased development loans.

The concept of an essential package of services discussed in that report establishes priorities in low- and middle-income countries for efficient use of resources based on the burden of disease and cost-effectiveness analysis of services. It includes both preventive and curative services targeted to specific health problems. It also recommends support for a comprehensive primary care such as for children with infrastructure development including maternity and hospital care, medical and nursing outreach services, and community action to improve sanitation such as safe water supplies.

Reorientation of government spending on health is increasingly being adopted as in the United Kingdom to improve equity in access for the poor and other neglected sectors or regions of society with added funding for relatively deprived areas to improve primary care services. Differential capitation funding as a form of affirmative action to provide for high-needs populations is a useful concept in public health terms to address the inequities still widely prevalent in many countries.

HEALTH SYSTEMS: THE CASE FOR REFORM

As medical care has gradually become more involved in prevention, and as it has gradually moved into the era of managed care, the gap between public health and clinical medicine has narrowed. As noted above, many countries are engaged in reforms in their health care systems. The motivation is partly derived from the need for cost containment, or to extend health care coverage to underserved parts of the population. Countries without universal health care may still have serious inequities in distribution of or access to services, and may seek reform to reduce those inequities, perhaps under political pressures to improve the provision of services. The incentives, or case for reform, centers on cost constraint, regional equity, and preserving or developing universal access and quality of care.

In some settings, a health system may fail to keep pace with developments in prevention and in clinical medicine. Some countries have overdeveloped medical and hospital care, neglecting important initiatives to reduce risk of disease. The process of reform requires setting standards to measure health status and the balance of services to optimize health. A health service can set a target of immunizing 95 percent of infants with a national immunization schedule, but requires a system to monitor performance and incentives for changes.

A health system may have failed to adapt to changing needs of the population through lack or misuse of a health information program. As a result, the system may err seriously in its allocation of resources, with excessive emphasis on hospital care and insufficient attention to primary and preventive care. All health services should have mechanisms for correctly gathering and analyzing needed data for monitoring the incidence of disease and other health indicators, such as hospital utilization, ambulatory care, and preventive care patterns. For example, the United Kingdom's National Health Service periodically undertakes a restructuring process of parts of the system to improve the efficiency of service. This involves organizational changes and decentralization with regional allocation of resources.

Health systems are under pressures of changing demographic and epidemiologic patterns as well as public expectations, rising costs of new technology, financing, and organizational change. New problems must be continually addressed with selection of priority issues and the most effective methods chosen. Reforms may create unanticipated problems, such as professional or public dissatisfaction, which must be evaluated, monitored, and addressed as part of the evolution of public health.

ADVOCACY AND CONSUMERISM

Literacy, freedom of the press, and increasing public concern for social and health issues have contributed to the development of public health. The British medical community lobbied for restrictions on the sale of gin in the 1780s in order to reduce its damage to the working class. In the late eighteenth and the nineteenth centuries, reforms in society and sanitation were largely the result of strongly organized advocacy groups influencing public opinion through the press. Such pressure stimulated governments to act in regulating working conditions of mines and factories. Abolition of the slave trade and its suppression by the British navy in the early nineteenth century resulted from advocacy groups and their effects on public opinion through the press. Vaccination against smallpox was promoted by privately organized citizen groups, until later taken up by local and national government authorities.

Advocacy is the act of individuals or groups publicly pleading for, supporting, espousing, or recommending a cause or course of action. The advocacy role of reform movements of the nineteenth century was the basis of the development of modern organized public health. This ranged from the reform of mental hospitals, nutrition for sailors, and labor laws to improve working conditions for women and children, to the promotion of universal education and improved living conditions for the working population. Reforms on these and other issues resulted from the stirring of the public consciousness by advocacy groups and the public media, all of which generated political decisions in parliaments (Box 2.9). Such reforms were in large part motivated by fear of revolution throughout Europe in the mid-nineteenth century and the early part of the twentieth century.

Trade unions, and before them medieval guilds, fought to improve hours and conditions of work as well as social and health benefits for their members. In the United States, collective bargaining through trade unions achieved widespread coverage of the working population under voluntary health insurance. Unions and some industries pioneered prepaid group practice, the predecessor of health maintenance organizations and managed care.

Through raising public consciousness on many issues, advocacy groups pressure governments to enact legislation to restrict smoking in public places, prohibit tobacco advertising, and mandate the use of bicycle helmets. Advocacy groups play an important role in advancing health based on disease groups, such as cancer, multiple sclerosis, and thalassemia, or advancing health issues, such as the organizations promoting breastfeeding, environmental improvement, or smoking reduction. Some organizations finance services or facilities not usually provided within insured health programs. Such organizations, which can number in the hundreds in a country, advocate the importance of their special concern and play an important role in innovation and meeting community health needs. Advocacy groups, including trade unions, professional groups, women's groups, self-help groups, and so many others, focus on specific issues and have made major contributions to advancing the New Public Health.

Professional Advocacy and Resistance

The history of public health is replete with pioneers whose discoveries led to strong opposition and sometimes violent rejection by conservative elements and vested interests in medical, public, or political circles. Opposition to Jennerian vaccination, the rejection of Semmelweiss by colleagues in Vienna, and the opposition to the work of Pasteur, Florence Nightingale, and many others may deter other innovators. Opposition to Jenner's vaccination lasted well into the late nineteenth century in some areas, but its supporters gradually gained ascendancy, ultimately leading to global eradication of smallpox. These and other

---

Box 2.9 The Plimsoll Line

Political activism for reform in nineteenth-century Britain led to banning and suppressing the slave trade and terrible working conditions for miners and factory workers and other major political reform.

In keeping with this tradition, Samuel Plimsoll, British Member of Parliament elected for Derby in 1868, conducted a solo campaign for the safety of seamen. His book, *Our Seamen*, described ships sent to sea so heavily laden with coal and iron that their decks were awash. Seriously overloaded ships, deliberately sent to sea by unscrupulous owners, frequently capsized at sea, drowning many crewmembers, with the owners collecting inflated insurance fees.

Overloading was the major cause of wrecks and thousands of deaths in the British shipping industry. Plimsoll pleaded for mandatory load-line markers to prevent any ships putting to sea when the marker was not clearly visible. Powerful shipping interests fought him every inch of the way, but he succeeded in having a Royal Commission established, leading to an act of Parliament mandating the "Plimsoll Line," the safe carrying capacity of cargo ships.

This regulation was adopted by the U.S. Bureau of Shipping as the Load Line Act in 1929 and is now standard practice worldwide.

pioneers led the way to improved health, often after bitter controversy on topics later accepted and which, in retrospect, seem to be obvious.

Advocacy has sometimes had the support of the medical profession but slow response by public authorities. David Marine of the Cleveland Clinic and David Cowie, professor of pediatrics at the University of Michigan, proposed prevention of goiter by iodization of salt. Marine carried out a series of studies in fish, and then in a controlled clinical trial among schoolgirls in 1917–1919, with startlingly positive results in reducing the prevalence of goiter. Cowie campaigned for iodization of salt, with support from the medical profession. In 1924, he convinced a private manufacturer to produce Morton's iodized salt, which rapidly became popular throughout North America. Similarly, iodized salt came to be used in many parts of Europe. This was achieved mostly without governmental support or legislation, and iodine deficiency disorders (IDDs) remain a widespread condition estimated to have affected 1.6 billion people worldwide in 1995. The target of international eradication of IDDs by 2000 was set at the World Summit for Children in 1990, and WHO called for universal iodization of salt in 1994.

Professional organizations have contributed to promoting causes such as child and women's health, and environmental and occupational health. The American Academy of Pediatrics has contributed to establishing and promoting high standards of care for infants and children in the United States, and to child health internationally. Hospital accreditation has been used for decades in the United States, Canada, and more recently in Australia and the United Kingdom. It has helped to raise standards of hospital facilities and care by carrying out systematic peer review of hospitals, nursing homes, primary care facilities, mental hospitals, as well as ambulatory care centers and public health agencies.

Public health needs to be aware of negative advocacy, sometimes based on professional conservatism or economic self-interest. Professional organizations can also serve as advocates of the status quo in the face of change. Opposition by the American Medical Association (AMA) and the health insurance industry to national health insurance in the United States has been strong and successful for many decades. In some cases, the vested interest of one profession may block the legitimate development of others, such as when ophthalmologists lobbied successfully against the development of optometry, now widely accepted as a legitimate profession.

Jenner's discovery of vaccination with cowpox to prevent smallpox was adopted rapidly and widely. However, intense opposition by organized groups of anti-vaccinationists, often led by those opposed to government intervention in health issues and supported by doctors with lucrative variolation practices, delayed adoption of vaccination for many decades. Fluoridation of drinking water is the most effective public health measure for preventing dental caries, but it is still widely opposed, and in some places the legislation has been removed even after implementation, by well-organized anti-fluoridation campaigns. Opposition to legislated restrictions on private ownership of assault weapons and handguns is intense in the United States, led by well-organized, well-funded, and politically powerful lobby groups, despite the amount of morbidity and mortality due to gun-associated violent acts.

Progress may be blocked where all decisions are made in closed discussions, not subject to open scrutiny and debate. Public health personnel working in the civil service of organized systems of government may not be at liberty to promote public health causes. However, professional organizations may then serve as forums for the essential professional and public debate needed for progress in the field. Professional organizations such as the American Public Health Association (APHA) provide effective lobbying for the interests of public health programs and can make an important impact on public policy. In mid-1996, efforts by the secretary of Health and Human Services (HHS) in the United States brought together leaders of public health with representatives of the AMA and academic medical centers to try to find areas of common interest and willingness to promote the health of the population.

Public advocacy has played an especially important role in focusing attention on ecological issues (Box 2.10). In 1995, Greenpeace, an international environmental activist group, struggled to prevent dumping of an oil rig in the North Sea and forced a major oil company to find another solution that would be less damaging to the environment. It also carried on efforts to stop renewal of testing of atomic bombs by France in the South Pacific. International protests led to cessation of almost all testing of nuclear weapons. International concern over global warming has led to

---

Box 2.10 An Enemy of the People

Advocacy is a function in public health that has been important in promoting advances in the field, and one that sometimes places the advocate in conflict with established patterns and organizations. One of the classic descriptions of this function is in Henrik Ibsen's play *An Enemy of the People*, in which the hero, a young doctor, discovers that the water in his community is contaminated. This knowledge is suppressed by the town's leadership, led by his brother the mayor, because it would adversely affect plans to develop a tourist industry in the small Norwegian town in the late nineteenth century. The young doctor is driven from the town, having been declared an "enemy of the people" and a potential risk. The term took on a far more sinister and dangerous meaning in George Orwell's novel *1984* and in totalitarian regimes of the 1930s to the present time.

growing efforts to stem the tide of air pollution from fossil fules, coal-burning electrical production, and other manifestations of CO_2 and toxic contamination of the environment. Progress is far from certain as newly enriched countries such as China and India follow the rising consumption patterns of western countries. Public advocacy and rejection of wanton destruction of the global ecology may be the only way to prod consumers, governments, and corporate entities such as the energy and transportation industries to change direction.

In the latter part of the twentieth century and first decade of the twenty-first century, prominent international personalities and entertainers have taken up causes such as the removal of land mines in war-torn countries, illiteracy in disadvantaged populations, and funding for antiretroviral drugs for African countries to reduce maternal–fetal transmission of HIV and to provide care for the large numbers of cases of AIDS devastating many countries of sub-Saharan Africa. The role of Rotary International in polio eradication efforts and the public/private consortium for promoting immunization has been led in recent years by GAVI (Global Alliance for Vaccines and Immunization) with participation by WHO, UNICEF, the World Bank, the Gates Foundation, vaccine manufacturers, and others. This has had an important impact on extending immunization to protect and save lives of millions of children in deprived countries not yet able to provide fundamental prevention programs such as immunization at adequate levels.

International conferences help to create a worldwide climate of advocacy for health issues. International sanitary conferences in the nineteenth century were convened in response to the cholera epidemics. International conferences continue in the twenty-first century to serve as venues for advocacy on a global scale, bringing forward issues in public health that are beyond the scope of individual nations. WHO, UNICEF, and other international organizations perform this role on a continuing basis (Chapter 16).

Consumerism

Consumerism is a movement that promotes the interests of the purchaser of goods or services. In the 1960s, a new form of consumer advocacy emerged from the civil rights and antiwar movement in the United States. Concern was focused on the environment, occupational health, and the rights of the consumer. Rachel Carson stimulated concern by dramatizing the effects of DDT on wildlife and the environment. This period gave rise to environmental advocacy efforts worldwide, and even a political movement, the Greens, in Western Europe.

Ralph Nader showed the power of the advocate or "whistle-blower" who publicizes health hazards to stimulate active public debate on a host of issues related to the public well-being. Nader, a consumer advocate lawyer, developed a strategy for fighting against business and government activities and products which endangered public health and safety. His 1965 book *Unsafe at Any Speed* took issue with the U.S. automobile industry for emphasizing profit and style over safety. This led to the enactment of the National Traffic and Motor Safety Act of 1966, establishing safety standards for new cars. This was followed by a series of enactments including design and emission standards and seat belt regulations. Nader's work continues to promote consumer interests in a wide variety of fields, including the meat and poultry industries, coal mines, and promotes greater government regulatory powers regarding pesticide usage, food additives, consumer protection laws, rights to knowledge of contents, and safety standards.

Consumerism has become an integral part of free market economies, and the educated consumer does influence the quality, content, and price of products. Greater awareness of nutrition in health has influenced food manufacturers to improve packaging, content labeling, enrichment with vitamins and minerals, and advertisement to promote those values. Low fat dietary products are available because of an increasingly sophisticated public concerned over dietary factors in cardiovascular diseases. The same process occurred in safe toys and clothing for children, automobile safety features such as car seats for infants, and other innovations that quickly became industry standards in the industrialized world.

Consumerism can also be exploited by pharmaceutical companies with negative impacts on the health system, especially in the advertising of health products which leads to unnecessary visits to health providers and pressures for approval to obtain the product. The Internet has provided everyone access to a vast array of information and opinion. This has opened access to current literature otherwise unavailable because of inadequate library resources medical and other health professionals or policymakers have. The very freedom of information the Internet allows, however, also provides a vehicle for extremist and fringe groups to promote disinformation such as "vaccination causes autism, fluoridation causes cancer" which can cause considerable difficulty for basic public health programs.

Advocacy and voluntarism go hand in hand. Voluntarism takes many forms, including raising funds for the development of services or operating services needed in the community. It may be in the form of fund-raising to build clinics or hospitals in the community, or to provide medical equipment to the elderly or handicapped. It may take the form of retirees and teenagers working as hospital volunteers to provide services that are not available through paid staff, and to provide a sense of community caring for the sick in the best traditions of religious or municipal concerns. This can also be extended to prevention as in support for immunization programs, assistance for the handicapped and elderly in transportation,

Meals-on-Wheels, and many other services that may not be included in the "basket of services" provided by the state, health insurance, or public health services.

Community involvement can take many forms, and so can voluntarism. The pioneering role of women's organizations in promoting literacy, health services, and nutrition in North America during the latter part of the nineteenth and the early twentieth centuries profoundly affected the health of the population. The advocacy function is enhanced when an organization mobilizes voluntary activity and funds to promote changes or needed services, sometimes forcing official health agencies or insurance systems to revise their attitudes and programs to meet these needs.

THE HEALTH FIELD CONCEPT

By the early 1970s, Canada's system of federally supported provincial health insurance plans covered all of the country. The federal Minister of Health, Marc LaLonde, initiated a review of the national health situation, in view of concern over rapidly increasing costs of health care. This led to articulation of the Health Field concept in 1974, which defined health as a result of four major factors: human biology, environment, behavior, and health care organization (Box 2.11). Lifestyle and environmental factors were seen as important contributors to the morbidity and mortality in modern societies. This concept gained wide acceptance, promoting new initiatives that placed stress on health promotion in response to environmental and lifestyle factors. Conversely, reliance primarily on medical care to solve all health problems could be counterproductive.

The Health Field concept came at a time when many epidemiologic studies were identifying risk factors for cardiovascular diseases and cancers that related to personal habits, such as diet, exercise, and smoking. The concept advocated that public policy needed to address individual lifestyle as part of the overall effort to improve health status. As a result, the Canadian federal government established health promotion as a new activity. This quickly spread to many other jurisdictions and gained wide acceptance in many industrialized countries.

Concern was expressed that this concept could become a justification for a "blame the victim" approach, in which those ill with a disease related to personal lifestyles, such as smokers or AIDS patients, are seen as having chosen to contract the disease. Such a patient might then be considered not entitled to all benefits of insurance or care that others may receive. The result may be a restrictive approach to care and treatment that would be unethical in the public health tradition and probably illegal in western jurisprudence. This concept was also used to justify withdrawal from federal commitments in cost sharing and escape from facing controversial health reform in the national health insurance program.

THE VALUE OF MEDICAL CARE IN PUBLIC HEALTH

During the 1960s and 1970s, outspoken critics of health care systems, such as Ivan Illytch, questioned the value of medical care for the health of the public. This became a widely discussed, somewhat nihilistic, view toward medical care, and was influential in promoting skepticism regarding the value of the biomedical mode of health care, and antagonism toward the medical profession.

In 1976, Thomas McKeown presented a historical–epidemiologic analysis showing that up to the 1950s, medical care had only limited impact on mortality rates, although improvements in surgery and obstetrics were notable. He showed that crude death rates in England averaged about 30 per 1000 population from 1541 to 1750, declined steeply to 22 per 1000 in 1851, 15 per 1000 in 1901, and 12 per 1000 in 1951 when medical care became truly effective. McKeown concluded that much of the improvement in health status over the past several centuries was due to reduced mortality from infectious diseases. This he related to limitation of family size, increased food supplies, improved nutrition and sanitation, specific preventive and therapeutic measures, and overall gains in quality of life for growing elements of the population. He cautioned against placing excessive reliance for health on medical care, much of which was of unproved effectiveness.

This skepticism of the biomedical model of health care was part of wider anti-establishment feelings of the 1960s and 1970s in North America. In 1984, Milton Roemer pointed out that the advent of vaccines, antibiotics, antihypertensives, and other medications contributed to great

---

Box 2.11 The Health Field Concept—Marc LaLonde

Definition:
Health is a result of factors associated with genetic inheritance, the environment, and personal lifestyle, and of medical care. Promotion of healthy lifestyles can improve health and reduce the need for medical care.
Elements:
1. Genetic and biological factors;
2. Behavioral and attitudinal factors (lifestyle);
3. Environment, including economic, social, cultural, and physical factors;
4. The organization of health care systems.

Source: LaLonde, M. 1974. A New Perspective on the Health of Canadians: A Working Document. Ottawa: Information Canada.

improvements in infant and child care, and in management of infectious diseases, hypertension, diabetes, and other conditions. Therapeutic gains continue to arrive from teaching centers around the world. Vaccine, pharmaceutical, and diagnostic equipment manufacturers continue to provide important innovations that have important benefits, but also raise the cost of health care. The latter issue is one which has stimulated the search for reforms and listing of priorities.

The value of medical care to public health and vice versa has not always been clear, neither to public health personnel nor to clinicians. The achievements of modern public health in controlling infectious diseases, and even more so in reducing the mortality and morbidity associated with chronic diseases such as stroke and coronary heart disease, were in reality a shared achievement between clinical medicine and public health (Chapter 5).

Preventive medicine has become part of all medical practice, with disease prevention through early diagnosis and health promotion through individual and community-focused activities. Risk factor evaluation determines appropriate screening and individual and community-based interventions. Medical care is crucial in controlling hypertension and in reducing the complications and mortality from coronary heart disease. New modalities of treatment are reducing death rates from first-time acute myocardial infarctions. Better management of diabetes prevents early onset of complications. At the same time, the contribution of public health to improving outcomes of medical care is equally important. Control of the vaccine-preventable diseases, improved nutrition, and preparation for motherhood contribute to improved maternal and infant outcomes. Promotions of reduced exposure to risk factors for chronic disease are a task shared by public health and clinical medical services. Both clinical medicine and public health contribute to improved health status and both are interdependent and integral elements of the New Public Health.

HEALTH TARGETS

During the 1950s, many new management concepts emerged in the business community, such as management by objective, coined by Peter Drucker and developed at General Motors, with variants such as zero-based budgeting developed in the U.S. Department of Defense. They focused the activities of an organization and its budget on targets, rather than on previous allocation of resources. These concepts were applied in other spheres, but they influenced thinking in health, whose professionals were seeking new ways to approach health planning. The logical application was to define health targets and to promote the efficient use of resources to achieve those targets. This occurred in the United States and soon after in the WHO European region. In both cases, a wide-scale process of discussion

and consensus building was used before reaching definitive targets. This process contributed to the adoption of the targets by many countries in Europe as well as by states and many professional and consumer organizations. The United States developed national health objectives in 1979 for the year 1990 and subsequently for the year 2000, with monitoring of progress in their achievement and development of further targets for 2010. Beginning in 1987, state health profiles are prepared by the Epidemiology Program Office of the Centers for Disease Control based on 18 health indicators recommended by a consensus panel representing public health associations and organizations.

The eight Millennium Development Goals (MDGs) adopted by the United Nations in 2000 include halving extreme poverty, reducing child mortality by 2/3, improving maternal health, halting the spread of HIV/AIDS, malaria, and other diseases, and providing universal primary education, all by the target date of 2015. This forms a common blueprint agreed to by all the world's countries and the world's leading development institutions. The process has galvanized unprecedented efforts to meet the needs of the world's poorest, yet 2008 reviews of progress indicate that most developing nations will not meet the targets at current rates of progress. This requires sustained efforts to develop the primary care infrastructure: improved reporting and epidemiologic monitoring, consultative mechanisms, and consensus by international agencies, national governments, and nongovernmental agencies. The achievement of the targets will require sustained international support and national commitment. Nevertheless, defining a target is crucial to the process.

There are encouraging signs that national governments are influenced by the general movement to place greater emphasis in resource allocation and planning on primary care to achieve internationally recognized goals and targets. The successful elimination of smallpox, rising immunization coverage in the developing countries, and increasing implementation of salt iodization have shown that such goals are achievable.

United States Health Targets

While the United States has not succeeded in developing universal health care access, it has a strong tradition of public health and health advocacy. Federal, state, and local health authorities have worked out cooperative arrangements for financing and supervising public health and other services. With growing recognition in the 1970s that medical services alone would not achieve better health results, health policy leadership in the federal government formulated a new approach, in the form of developing specific health targets for the nation.

In 1979, the surgeon general of the United States published the Report on Health Promotion and Disease

Prevention (*Healthy People*). This document set five overall health goals for each of the major age groups for the year 1990, accompanied by 226 specific health objectives. New targets for the year 2000 were developed in three broad areas: to increase healthy life spans, to reduce health disparities, and to achieve access to preventive health care for all Americans. These broad goals are supported by 297 specific targets in 22 health priority areas, each one divided into four major categories: health promotion, health protection, preventive services, and surveillance systems. This set the public health agenda on the basis of measurable indicators that can be assessed year by year.

Leading Health Indicators selected for 2010 incorporate the original 467 objectives in *Healthy People 2010* which served as a basis for planning public health activities for many state and community health initiatives. For each of the Leading Health Indicators, specific objectives and sub-objectives derived from *Healthy People 2010*

are used to monitor progress. The specific objectives and sub-objectives used to track progress toward the Leading Health Indicators are listed in Table 2.2.

The process of working toward health targets in the United States has moved down from the federal level of government to the state and local levels. Professional organizations, NGOs, as well as community and fraternal organizations are also involved. The states are encouraged to prepare their own targets and implementation plans as a condition for federal grants, and many states require county health departments to prepare local profiles and targets.

Diffusion of this approach encourages state and local initiatives to meet measurable program targets. It also sets a different agenda for local prestige in competitive terms, with less emphasis on the size of the local hospital or other agencies than on having the lowest infant mortality or the least infectious disease among neighboring local authorities.

TABLE 2.2 *Healthy People 2010* Objectives and Sub-Objectives

Objectives	Sub-objectives
Physical activity	Increase the proportion of adults who engage in moderate physical activity for at least 30 minutes per day 5 or more days per week or vigorous physical activity for at least 20 minutes per day 3 or more days per week. Increase the proportion of adolescents who engage in vigorous physical activity that promotes cardiorespiratory fitness 3 or more days per week for 20 or more minutes per occasion.
Overweight and obesity	Reduce the proportion of children and adolescents aged 6–19 who are overweight or obese. Reduce the proportion of adults who are obese.
Tobacco use	Reduce tobacco use by adults — cigarette smoking. Reduce tobacco use by adolescents — cigarette smoking.
Substance abuse	Increase the proportion of adolescents not using alcohol or any illicit drugs during the past 30 days. Reduce the proportion of adults using any illicit drug during the past 30 days. Reduce the proportion of persons aged 18 years and older engaging in binge drinking of alcoholic beverages.
Responsible sexual behavior	Increase the proportion of sexually active persons who use condoms. Increase the proportion of adolescents who abstain from sexual intercourse or use condoms if currently sexually active.
Mental health	Increase the proportion of adults aged 18 years and older with recognized depression who receive treatment.
Injury and violence	Reduce deaths caused by motor vehicle accidents. Reduce homicides.
Environmental quality	Reduce proportion of persons exposed to air that does not meet the U.S. Environmental Protection Agency's standards for harmful air pollutants, ozone. Reduce proportion of nonsmokers exposed to environmental tobacco smoke.
Immunization	Increase proportion of young children and adolescents who receive all vaccines recommended for universal administration for at least 5 years. Increase proportion of noninstitutionalized adults vaccinated annually against influenza and against pneumococcal disease.
Access to health care	Increase the proportion of persons with health insurance. Increase the proportion of persons of all ages who have a specific source of ongoing care. Increase the proportion of pregnant women who receive early and adequate prenatal care beginning in the first trimester of pregnancy.

Source: U.S. Healthy People, Midcourse Review, http://www.healthypeople.gov/data/midcourse/html/appendix/AppendixE.htm [accessed February 21, 2008]

INTERNATIONAL HEALTH TARGETS

European Health Targets

The WHO European Region document *Health 21 — Health for All in the 21st Century* addresses health in the twenty-first century, with 21 principles and objectives for improving the health of Europeans, within and between countries of Europe. The *Health 21* Targets include:

1. Closing the health gap between countries;
2. Closing the health gap within countries;
3. A healthy start in life (supportive family policies);
4. Health of young people (policies to reduce child abuse, accidents, drug use, unwanted pregnancies);
5. Healthy aging (policies to improve health, self-esteem, and independence before dependence emerges);
6. Improving mental health;
7. Reducing communicable diseases;
8. Reducing noncommunicable diseases;
9. Reducing injury from violence and accidents;
10. A healthy and safe physical environment;
11. Healthier living (fiscal, agricultural, and retail policies that increase the availability of and access to and consumption of vegetables and fruits);
12. Reducing harm from alcohol, drugs, and tobacco;
13. A settings approach to health action (homes should be designed and built in a manner conducive to sustainable health and the environment);
14. Multisectoral responsibility for health;
15. An integrated health sector and much stronger emphasis on primary care;
16. Managing for quality of care using the European health for all indicators to focus on outcomes and compare the effectiveness of different inputs;
17. Equitable and sustainable funding of health services;
18. Developing human resources (educational programs for providers and managers based on the principles of the Health for All policy);
19. Research and knowledge: health programs based on scientific evidence;
20. Mobilizing partners for health (engaging the media/TV/Internet);
21. Policies and strategies for *Health for All* — national, targeted policies based on *Health for All*

United Kingdom Health Targets

There are competing demands in society for expenditure by the government, so making the best use of resources — money and people — is therefore an important objective. Key subjects chosen for action were ischemic heart disease and stroke, cancer, mental illness, HIV and sexual health, and accidents (Box 2.12).

Box 2.12 NHS National Targets, Scotland 2003

Targets for reducing health inequalities

Teenage pregnancy 20 percent reduction in teenage pregnancies among those aged 13–15: target date 2010.

Dental health Children aged 12 should have, on average, no more than 1.5 teeth decayed, missing, or filled: target date 2005.

Smoking Reduce smoking among young people (12–15 age group) to 11 percent: target date 2010; reduce rate of smoking among adults (16–64 age group) in all social classes to 31 percent: target date 2010; reduce the proportion of women who smoke during pregnancy by 9 percent to 20 percent: target date 2010.

Physical activity 50 percent of all adults (aged 16+) accumulating a minimum of 30 minutes per day of moderate physical activity on 5 or more days per week; 80 percent of all children (aged 2–15) accumulating 1 hour per day of physical activity on 5 or more days per week.

Breastfeeding More than 50 percent of women should breastfeed their babies at 6 weeks: target date 2005.

Diet Increase the proportion of the population consuming increased levels of fruits and vegetables, carbohydrates, and fish as defined by the Scottish Dietary Targets: target date 2005. Increase the proportion of the population consuming decreased levels of fat, sugar, and salt as defined by the Scottish Dietary Targets: target date 2005.

Immunization/Vaccination 70 percent of people over age 65 vaccinated against flu: annual target; 95 percent uptake target for all childhood vaccinations (ongoing).

Low birth weight babies To reduce incidence of low birth weight babies by 10 percent: target date 2005.

Eye and dental checks We will invest in health promotion and, as a priority, we will systematically introduce free eye and dental checks for all before 2007.

Screening tests Hearing tests for all newborn babies; breast screening target 70 percent: ongoing; cervical screening target 80 percent: ongoing.

CHD/Stroke 50 percent reduction in the age-standardized mortality rate from CHD and stroke in people aged under 75: target date 2010.

Source: NHS National Targets, Scotland, http://www.scotland.gov.uk/Publications/2003/10/18432/28416 [accessed April 29, 2008]

INDIVIDUAL AND COMMUNITY PARTICIPATION IN HEALTH

National policy in health ultimately relates to health of the individual. The various concepts outlined in the health field concept, community-oriented primary health care, health targets, and effective management of health systems, can only be effective if the individual and his or her community are knowledgeable participants in seeking solutions. Involving the individual in his or her own health status requires raising levels of awareness, knowledge, and action. The methods used to achieve these goals include health counseling, health education, and health promotion (Figure 2.5).

Health counseling has always been a part of health care between the doctor or nurse and the patient. It raises levels of awareness of health issues of the individual patient. Health education has long been part of public health, dealing with promoting consciousness of health issues in selected target population groups. Health promotion incorporates the work of health education but takes health issues to the policy level of government and involves all levels of government and NGOs in a more comprehensive approach to a healthier environment and personal lifestyles.

Health counseling, health education, and health promotion are among the most cost-effective interventions for improving the health of the public. While costs of health care are rising rapidly, demands to control cost increases should lead to greater stress on prevention, and adoption of health education and promotion as an integral part of modern life. This should be carried out in schools, the workplace, the community, commercial locations (e.g., shopping centers), recreation centers, and in the political agenda.

Psychologist Abraham Maslow described a hierarchy of needs of human beings. Every human has basic requirements including physiological needs of safety, water, food, warmth, and shelter. Higher levels of needs include recognition, community, and self-fulfillment. These insights supported observations of efficiency studies such as those of Elton Mayo in the famous Hawthorne effect in the 1920s, showing that workers increased productivity when acknowledged by management in the objectives of the organization (see Chapter 12). In health terms, these translate

into factors that motivate people to positive health activities when all barriers to health care are reduced.

Modern public health faces the problem of motivating people to change behavior; sometimes this requires legislation, enforcement, and penalties for failure to comply, such as in mandating car seat belt use. In others it requires sustained performance by the individual, such as the use of condoms to reduce the risk of STI and/or HIV transmission. Over time, this has been developed into a concept known as knowledge, attitudes, beliefs, and practices (KABP), a measurable complex that cumulatively affects health behavior (see Chapter 3). There is often a divergence between knowledge and practice; for example, the knowledge of the importance of safe driving, yet not putting this into practice. This concept is sometimes referred to as the *KABP gap*.

The health belief model has been a basis for health education programs, whereby a person's readiness to take action for health stems from a perceived threat of disease, a recognition of susceptibility to disease and its potential severity, and the value of health. Action by an individual may be triggered by concern and by knowledge. Barriers to appropriate action may be psychological, financial, or physical, including fear, time loss, and inconvenience. Spurring action to avoid risk to health is one of the fundamental goals in modern health care. The health belief model is important in defining any health intervention in that it addresses the emotional, intellectual, and other barriers to taking steps to prevent or treat disease.

Health awareness at the community and individual levels depends on basic education levels. Mothers in developing countries with primary or secondary school education are more successful in infant and child care than less-educated women. Agricultural and health extension services reaching out to poor and uneducated farm families in North America in the 1920s were able to raise consciousness of safe self-health practices and good nutrition, and when this was supplemented by basic health education in the schools, generational differences could be seen in levels of awareness of the importance of balanced nutrition. Secondary prevention with diabetics and patients with coronary heart disease hinges on education and awareness of nutritional and physical activity patterns needed to prevent or delay a subsequent myocardial infarction.

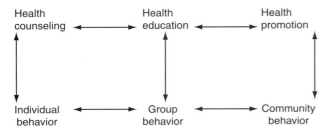

FIGURE 2.5   Health counseling–health education–health promotion.

Ottawa Charter for Health Promotion

The WHO sponsored the First International Conference on Health Promotion held in Ottawa, Canada, in 1986. The resulting *Ottawa Charter* defined health promotion and set out five key areas of action: building healthy public policy, creating supportive environments, strengthening community action, developing personal skills, and reorienting health services. The *Ottawa Charter* called on all countries to:

put health on the agenda of policy-makers in all sectors and at all levels, directing them to be aware of the health consequences of their decisions and to accept responsibility for health. Health promotion policy combines diverse but complementary approaches, including legislation, fiscal measures, taxation, and organizational change. It is a coordinated action that leads to health, income, and social policies that foster greater equity. Joint action contributes to ensuring safer and healthier goods and services, healthier public services, and cleaner, more enjoyable environments. Health promotion policies require the identification of obstacles to the adoption of healthy public policies in non-health sectors, and ways of removing them. The aim must be to make the healthier choice the easier choice for policy makers as well. [Source: Health and Welfare Canada — World Health Organization, 1986.]

State and Community Models of Health Promotion

An effective approach to health promotion was developed in Australia where in the State of Victoria revenue from a cigarette tax has been set aside for health promotion purposes. This has the effect of discouraging smoking, and at the same time finances health promotion activities and provides a focus for health advocacy in terms of promoting cessation of cigarette advertising at sports events or on television. It also allows for assistance to community groups and local authorities to develop health promotion activities at the workplace, in schools, and at places of recreation. Health activity in the workplace involves reduction of work hazards as well as promotion of healthy diet, physical fitness, and avoidance of risk factors such as smoking and alcohol abuse.

In the Australian model, health promotion is not the only persuasion of people to change their life habits; it also involves legislation and enforcement toward environmental changes that promote health. For example, this involves mandatory filtration, chlorination, and fluoridation for community water supplies to reduce waterborne disease and to promote dental health. It also involves vitamin and mineral enrichment of basic foods to prevent micronutrient deficiencies. These are at the level of national or state policy, and are vital to a health promotion program and local community action.

Community-based programs to reduce chronic disease using the concept of community-wide health promotion have developed in a wide variety of settings. Such a program to reduce risk factors for cardiovascular disease was pioneered in the North Karelia Project in Finland. This project was initiated as a result of pressures from the affected population of the province, which was aware of the high incidence of mortality from heart disease. Finland had the highest rates of coronary heart disease in

the world and the rural area of North Karelia was even higher than the national average. The project was a regional effort involving all levels of society, including official and voluntary organizations, to try to reduce risk factors for coronary heart disease. After 15 years of follow-up, there was a substantial decline in mortality with similar decline in a neighboring province taken for comparison, although the decline began earlier in North Karelia.

In many areas where health promotion has been attempted as a strategy, community-wide activity has developed with participation of NGOs or any valid community group as initiators or participants. Healthy Heart programs have developed widely with health fairs, sponsored by charitable or fraternal societies, schools, or church groups, to provide a focus for leadership in program development. A wider approach to addressing health problems in the community has developed into an international movement of "healthy cities."

Healthy Cities/Towns/Municipalities

Following deliberations of the Health of Towns Commission chaired by Edwin Chadwick, the Health of Towns Association was founded in 1844 by Southwood Smith, a prominent reform leader of the Sanitary Movement, to advocate change to reduce the terrible living conditions of much of the population of cities in the United Kingdom. The Association established branches in many cities and promoted sanitary legislation and public awareness of the "Sanitary Idea" that overcrowding, inadequate sanitation, and absence of safe water and food created the conditions under which epidemic disease could thrive. In the 1980s, Ilona Kickbush, Trevor Hancock, and others promoted renewal of the idea that local authorities have a responsibility to build health issues into their planning and development processes.

Healthy Cities is an approach to health promotion that emerged in the 1980s, promoting urban community action on a broad front of health promotion issues (Table 2.3). Activities include environmental projects (such as recycling of waste products), improved recreational facilities for youth to reduce violence and drug abuse, health fairs to promote health awareness, and screening programs for hypertension, breast cancer, and others. It combines health promotion with consumerism and returns to the tradition of local public health action and advocacy.

The municipality, in conjunction with many NGOs, develops a consultative process and program development approach to improving the physical and social life of the urban environment and the health of the population. In 1995, the Healthy Cities movement involved 18 countries with 375 cities in Europe, Canada, the United States, the United Kingdom, South America, Israel, and Australia, an increase from 18 cities in 1986. The Box 2.13 model

TABLE 2.3 The Qualities of Healthy Cities

1. A safe, clean physical environment
2. A stable ecosystem sustainable in the long-term
3. A strong, supportive, nonexploitive community
4. A high level of participation/control by the public over decisions affecting their lives and well-being
5. Meeting basic needs (food, water, shelter, income, safety, and work) for all of the city's people
6. A wide variety of experiences and resources for contact, interactions, communication
7. A diverse, vital, and innovative city
8. Connection with the past — cultural, biological heritage
9. A form compatible with these objectives
10. An optimum level of public health and medical services
11. High health status; positive health, low levels of disease

Source: World Health Organization. 1995. *Twenty Steps for Developing a Healthy Cities Project*, 2nd Edition. Copenhagen: Regional Office for Europe.

now extends to small municipalities, often with populations of fewer than 10,000.

A typical healthy city has a population in the multiple thousands, often multilingual, with an average middle class income. A Healthy Cities project builds a coalition of municipal and voluntary groups working together in a continuing effort to improve quality of service, facilities, and living environment. The city is divided into neighborhoods, engaged in a wide range of activities fostered by the project. Municipalities have traditionally had a leading role in sanitation, safe water supply, building and zoning laws and regulation, and many other responsibilities in public health (see Chapter 10). The Healthy Cities or Communities movement has elevated this to a higher level with policies to promote health in all actions. Some examples are listed below:

1. Traffic circles, crosswalks, and road bumps to slow urban vehicle traffic and improve pedestrian safety
2. Nuisance abatement in local quarries
3. Tree planting and gardens in poor and low income neighborhoods
4. Primary and secondary school physical and security improvements
5. Neighborhood profiles
6. Cooperative housing for low income families
7. Intercultural communication
8. Recreational facilities for youth
9. Restoration of neglected sites — green spaces
10. Extension of public parks
11. Youth and community activities
12. Reduction of drug and crime environment
13. Safe houses for battered women and homeless
14. Community centers for older adults

The importance of working with senior levels of government, other departments in the municipalities, religious organizations, private donors, and the NGO sector to innovate and especially to improve conditions in poverty areas of cities is a vital role for health-oriented local political leadership.

Box 2.13 European Healthy Cities Movement, 2003–2008

Cities continuing in the WHO European Network will be developing and implementing partnership-based, intersectoral plans for developing health documented in up-to-date city health profiles. This includes core developmental themes of healthy urban planning and health impact assessment.

Healthy urban planning Urban planners and policymakers should be encouraged to integrate health considerations in their planning strategies and initiatives with emphasis on equity, well-being, sustainable development, and community safety.

Health impact assessment Health impact assessment processes should be applied within cities to support intersectoral action for promoting health and reducing inequality.

Healthy aging Healthy aging works to address the needs of older people related to health care and the quality of life with special emphasis on active and independent living, creating supportive environments, and ensuring access to sensitive and appropriate services.

Sustained local support Cities must have sustained local government support and support from key decision makers (stakeholders) in other sectors for the Healthy Cities principles and goals.

Coordinator and steering group Cities must have a full-time identified coordinator (or the equivalent) who is fluent in English and administrative and technical support for their Healthy City initiative. Cities must also have a steering group involving political and executive-level decision makers from the key sectors necessary to ensure delivery.

Partnership on core developmental themes Cities must work in partnership with WHO as the testing ground for developing knowledge, tools, and expertise on the core developmental themes.

Source: WHO Healthy Cities Network, http://www.euro.who.int/document/E81924.pdf [accessed February 21, 2008]

HUMAN ECOLOGY AND HEALTH PROMOTION

Human ecology, a term introduced in the 1920s and revived in the 1970s, attempted to apply theory from plant and animal life to human communities. It evolved as a branch of demography, sociology, and anthropology, addressing the social and cultural contexts of disease, health risks, and human behavior. Human ecology addresses the interaction of humans with and adaptation to their social and physical environment.

Parallel subdisciplines of social, community, and environmental psychology, medical sociology, anthropology, and other social sciences contributed to the development of this academic field with wide applications in health-related issues. This led to incorporation of qualitative research methods alongside the quantitative research methods traditionally emphasized in public health, providing crucial insights into many public health issues where human behavior is a key risk factor.

Health education developed as a discipline and function within public health systems in school health, rural nutrition, military medicine, occupational health, and many other aspects of preventive-oriented health care, and is discussed in later chapters of this text. Directed at behavior modification through information and raising awareness of consequences of risk behavior, this has become a long-standing and major element of public health practice in recent times, being almost the only effective tool to fight the epidemic of HIV and the rising epidemic of obesity and diabetes.

Health promotion as an idea evolved, in part, from the LaLonde Health Field Concepts and from growing realization in the 1970s that access to medical care was necessary but not sufficient to improve the health of a population. The integration of the health behavior model, social ecological approach, environmental enhancement, or social engineering formed the basis of the social ecology approach to defining and addressing health issues (Table 2.4).

Individual behavior depends on many surrounding factors, while community health also relies on the individual; the two cannot be isolated from one another. The ecological perspective in health promotion works toward changing people's behavior to enhance health. It takes into account factors not related to individual behavior. These are determined by the political, social, and economic environment. It applies broad community, regional, or national approaches that are needed to address severe public health problems, such as control of HIV infection, tuberculosis, malnutrition, STIs, cardiovascular disorders, violence and trauma, and cancer.

DEFINING PUBLIC HEALTH STANDARDS

The American Public Health Association (APHA) formulation of the public health role in 1995 entitled *The Future of Public Health in America* was presented at the annual meeting in 1996. The APHA periodically revises standards and guidelines for organized public health services provided by federal, state, and local governments (Table 2.5). These reflect the profession of public health as envisioned in the United States where access to medical care is limited for large numbers of the population because of lack of universal health insurance. Public health in the United States has been very innovative in determining risk groups in need of special care and finding direct and indirect methods of meeting those needs.

European countries such as Finland have called for setting public health into all public policy, which is reflective of the

TABLE 2.4 Health Promotion Approaches: Behavior Modification, Environment Enhancement, and Social Ecology

Health behavior model change and lifestyle modification	Social ecological approaches	Environment enhancement: national, municipal, and community based
Behavior modification	Cultural change models of health	Universal access to health care
Social learning theory	Biopsychosocial models of health	Environmental health
Health belief model	Stressful life events	Industrial hygiene
Theory of reasoned action	Ecology of human development	Social security
Theory of planned behavior	Public health psychology	Societal support
Risk perception theory	Medical sociology	Community organization
Fear arousal	Ethnography	Ergonomics/human factors
Protection-motivation theory	Social epidemiology	Health monitoring epidemiology
Health communications	Social ecology of health	Urban planning, architecture
Mass media	Community health promotion	Regulation of housing, zoning
	Public policy initiatives	Injury and disaster control
	Healthy communities	Food and drug control
		Nutrition and food fortification

Source: Modified from: Stokols, D. 1996. Translating social ecological theory into guidelines for community health promotion. *American Journal of Health Promotion*, 10:282–298.

TABLE 2.5 Standards for Public Health Services, American Public Health Association 1995

Vision	Healthy people in healthy communities
Mission	Promote health and prevent disease
Goals	Prevent epidemics and spread of disease Protect against environmental hazards Prevent injuries Promote and encourage healthy behaviors Respond to disasters and assist communities in recovery Assure the quality and accessibility of health services
Essential Services	Monitor health status to identify community problems Diagnose and investigate health problems in the community Inform, educate, and empower people about health issues Develop policies and plans that support individual and community health efforts Enforce laws and regulations that protect health and ensure safety Link people to needed personal health services and assure provision of health care when otherwise unavailable Assure an expert public health workforce Evaluate effectiveness, accessibility, and quality of health services Research for new insights and innovative solutions to health problems

Source: American Public Health Association. *The Nation's Health*, March 1995.

vital role local and county governments can play in developing health-oriented policies. This includes policies in housing, recreation, regulation of industrial pollution, road safety, promotion of smoke-free environments, bicycle paths, health impact assessment, and many other applications of health principles in public policy.

INTEGRATIVE APPROACHES TO PUBLIC HEALTH

Public health involves both direct and indirect approaches. Direct measures in public health include immunization of children, modern birth control, hypertension, and diabetes case finding. Indirect methods used in public health protect the individual by community-wide means, such as raising standards of environmental safety, ensuring a safe water supply, sewage disposal, and improved nutrition (Box 2.14).

In public health practice, the direct and indirect approaches are both relevant. To reduce morbidity and mortality from diarrheal diseases requires an adequate supply of safe water, and also education of the individual in hygiene and the mother in use of oral rehydration therapy (ORT). The targets of public health action therefore include the individual, family, community, region, or nation.

The targets for protection in infectious disease control are both the individual and the total group at risk. For vaccine-preventable diseases, immunization protects the individual but also has an indirect effect by reducing the risk even for nonimmunized persons. In control of some diseases, individual case finding and management reduce risk of the disease in others and the community. For example, tuberculosis requires case finding and adequate care among high-risk groups as a key to community control. In malaria control, case finding and treatment are essential together with environmental action to reduce the vector population, to prevent transmission of the organism by the mosquito to a new host.

Control of noncommunicable diseases, where there is no vaccine for mass application, depends on the knowledge, attitudes, beliefs, and practices of individuals at risk. In this

---

Box 2.14 Why Health Systems Matter to the Social Determinants of Health Inequity

1. **General population benefits** Health systems offer general population benefits that go beyond preventing and treating illness. Appropriately designed and managed, they:
   - provide a vehicle to improve people's lives, protecting them from the vulnerability of sickness, generating a sense of life security, and building common purpose within society;
   - ensure that all population groups are included in the processes and benefits of socioeconomic development; and
   - generate the political support needed to sustain them over time.

2. **Promote health equity** Health systems promote health equity when their design and management specifically consider the circumstances and needs of socially disadvantaged and marginalized populations, including women, the poor, and groups who experience stigma and discrimination, enabling social action by these groups and the civil society organizations supporting them.

3. **Contribute to achieving the Millennium Development Goals** Health systems can, when appropriately designed and managed, contribute to achieving the Millennium Development Goals.

Source: Gilson, L., Doherty, J., Loewenson, R., Francis, V. 2007. *Final Report — Knowledge Network on Health Systems — June 2007*. WHO — Commission on Social Determinants of Health — CSDOH. http://www.who.int/social_determinants/resources/csdh_media/hskn_final_2007_en.pdf [accessed February 21, 2008]

case, the social context is of vital importance as is the quality of care to which the individual has access. Control and prevention of noninfectious diseases involve strategies using individual and population-based methods. Individual or clinical measures include professional advice on how best to reduce the risk of the disease by early diagnosis and implementation of appropriate therapy. Population-based measures involve indirect measures with government action banning cigarette advertising, or direct taxation on cigarettes. Mandating food quality standards, such as limiting the fat content of meat, and requiring food labeling laws are part of control of cardiovascular diseases.

The way individuals act is central to the objective of reducing disease, because many noninfectious diseases are dependent on behavioral risk factors of the individual's choosing. Changing the behavior of the individual means addressing the way one sees his or her own needs. This can be influenced by the provision of information, but how a person sees his or her own needs is more complex than that. An individual may define needs differently than the society or the health system. Reducing smoking among women may be difficult to achieve if smoking is thought to reduce appetite and food intake, given the social message that "slim is beautiful." Reducing smoking among young people is similarly difficult if smoking is seen as fashionable and diseases such as lung cancer seem very remote. Recognizing how individuals define needs helps the health system design programs that influence behavior that is associated with disease.

Public health has become linked to wider issues as health care systems are reformed to take on both individual and population-based approaches. Public health and mainstream medicine found increasingly common ground in addressing the issues of chronic disease, growing attention to health promotion, and economics-driven health care reform. At the same time, the social ecology approaches showed success in slowing major causes of disease, including heart disease and AIDS, and the biomedical sciences provided major new technology for preventing major health problems including cancer, heart disease, genetic disorders, and infectious diseases.

Technological innovations unheard of just a few years ago are now commonplace, in some cases driving up costs of care and in others replacing older and less effective care. At the same time, resistance of important pathogenic microorganisms to antibiotics and pesticides is producing new changes from diseases once thought to be under control, and "newly emerging infectious diseases" challenge the entire health community. New generations of antibiotics, antidepressants, antihypertensive medications, and other treatment methods are changing the way many conditions are treated. Research and development of the biomedical sciences are providing means of prevention and treatment that profoundly affect disease patterns where they are effectively applied.

The technological and organizational revolutions in health care are accompanied by many ethical, economic, and legal dilemmas. The choices in health care include heart transplantation, an expensive life-saving procedure, which may compete with provision of funds and manpower resources for immunizations for poor children or for health promotion to reduce smoking and other risk factors for chronic disease. New means of detecting and treating acute conditions such as myocardial infarction and peptic ulcers are reducing hospital stays, and improving long-term survival and quality of life. Imaging technology has been an important development in medicine since the advent of X-rays in the early twentieth century. Technology has forged ahead with CT, MRI, and others. New technologies enabling lower cost devices, electronic transmission, and distant reading of transmitted imaging will open possibilities for advanced diagnostic capacities to rural and less developed countries and communities. Molecular biology has provided methods of identifying and tracking movement of viruses such as polio and measles from place to place, greatly expanding potential for appropriate intervention.

The choices in resource allocation can be difficult. In part, these add political commitment to improve health, competent professionally trained public health personnel, public's level of health information and legal protection, whether it be through individuals, advocacy, or regulatory approaches for patients' rights. These are factors in a widening methodology of public health.

THE FUTURE OF PUBLIC HEALTH

Public health issues have received a new recognition in recent years because of a number of factors: a growing understanding among the populace at different levels in different countries that health behavior is a factor in health status and that public health is vital for protection against natural or man-made disasters. The challenges are also increasingly understood: preparation for bioterrorism, avian influenza, rising rates of diabetes and obesity; high mortality rates from cancer; and a wish for prevention to be effective.

The Millennium Development Goals selected by the United Nations in 2000 have eight global targets for the year 2015, including four directly related to public health (Box 2.15). This is both recognition and a challenge to the international community and public health as a profession. Formal education in newly developing schools of public health is increasing in Europe and in Capitals. But there is delay in establishing centers of postgraduate education and research in many developing countries which are concentrating their educational resources on training physicians. Many physicians from developing nations are moving to the developed countries, which have become dependent on

---

Box 2.15 Millennium Development Goals by 2015

---

1. Cut poverty by half ($1/day)
2. Universal primary education
3. Gender equality in education
4. Cut child mortality rate and <5 mortality rate by two-thirds
5. Reduce maternal MR 75 percent
6. Reduce HIV/AIDS and malaria
7. Sustainable environment
8. Implement fully sustainable development strategies

---

these countries for a significant part of their supply of medical doctors. Progress in implementation of the MDGs is mixed in sub-Saharan Africa, making some progress in immunization, but falling back on other goals.

A Harvard review of the future of public health in late 2007 recognized these concerns along with the changing population dynamics of demography; economics of prevention versus expensive treatment costs; and the economics of health care. The report included other concerns of the environment and its potential for a disaster of global warming and the potential for the development of basic and medical sciences in genetics, nanotechnology, and molecular biology.

At the same time, the effectiveness of health promotion has shown dramatic successes in reducing the toll of AIDS; reducing smoking; increasing consciousness of nutrition and physical fitness in the population; and public consciousness of the tragic effects of poverty and poor education on health status. The ethics of public health issues are complex and changing with awareness that failure to act on strong evidence-based policies is itself ethically problematic. The future of public health is not as a solo professional sector; it is at the heart of health systems, without which societies are open to chronic and infectious diseases that are preventable, affecting the society as a whole in economic and development matters.

There is an expanding role of private donors with global health efforts such as the Rotary Club and the polio eradication program, GAVI with immunization and bed nets in sub-Saharan Africa, and bilateral donor countries help in reducing the toll of AIDS in sub-Saharan Africa.

THE NEW PUBLIC HEALTH

A WHO meeting in November 1995 on "new challenges for health" reported that the New Public Health was an extension, rather than a substitution, of the traditional public health. It described organized efforts of society to develop healthy public policies: to promote health, to prevent disease, and to foster social equity within a framework of sustainable development (Table 2.6). A new,

revitalized public health must continue to fulfill the traditional functions of sanitation, protection, and related regulatory activities, but in addition to its expanded functions:

The New Public Health is not so much a concept as it is a philosophy which endeavors to broaden the older understanding of public health so that, for example, it includes the health of the individual in addition to the health of populations, and seeks to address such contemporary health issues as are concerned with equitable access to health services, the environment, political governance and social and economic development. It seeks to put health in the development framework to ensure that health is protected in public policy. Above all, the New Public Health is concerned with action. It is concerned with finding a blueprint to address many of the burning issues of our time, but also with identifying implementable strategies in the endeavor to solve these problems. [Source: Ncayiyona, et al., 1995.]

The New Public Health is therefore still evolving as a concept or approach drawing on many ideas and experiences in public health throughout the world. It is influenced by a growing recognition of social inequality in health, even in developed countries with universal health programs, and an acknowledgment of the failure of state-operated health services to cope with dramatic changes in disease patterns affecting their populations. The World Bank evaluation of cost-effective public health and medical interventions to reduce the burden of disease also contributed to the need to seek and apply new approaches to health. The New Public Health synthesizes traditional public health with management of personal services and community action for a holistic approach.

SUMMARY

The object of public health, like that of clinical medicine, is better health for the individual and for society. Public health works to achieve this through indirect methods, such as by improving the environment, or through direct means such as preventive care for mothers and infants or other at-risk groups. Clinical care focuses directly on the individual patient, mostly at the time of illness. But the health of the individual depends on the health promotion and social programs of the society, just as the well-being of a society depends on the health of its citizens. The New Public Health consists of a wide range of programs and activities that link individual and societal health.

The "old" public health was concerned largely with the consequences of unhealthy settlements and with safety of food, air, and water. It also targeted the infectious, toxic, and traumatic causes of death, which predominated among young people and were associated with poverty.

A summary of the great achievements of public health in the twentieth century in the industrialized world is

TABLE 2.6 Origins and Synthesis of the New Public Health

Classical public health	Social ecology	Biomedical care	Organization and financing
To End of Nineteenth Century			

Food and personal hygiene
Settlement health
Quarantine
Nutrition/fitness
Vital statistics
Epidemiology
Sanitation, miasma theory
Municipal organization
Bacteriology, germ theory
Vaccines, immunology
Control of infectious diseases
Maternal and child health
Health education
	Church and
	  serfdom
	Renaissance
	Agricultural
	  revolution
	Improved
	  nutrition
	Rise of cities
	Rights of man
	Industrial
	  revolution
	Labor laws
	Universal
	  education
	Social reform
	Political
	  revolution
	Information
	  revolution
		Basic sciences
		Clinical sciences
		Medical education
		Hospitals: church, municipal,
		  voluntary, university
		Specialization
		Therapeutics
		Antisepsis
		Vaccines
			Private payment for the rich
			Municipal doctors for poor
			Charity, church, voluntary hospital
			  care
			Guilds, mutual benefit, friendly
			  societies for medical, pensions,
			  burial benefits
			National health insurance for workers
			  and families
			Sick funds and voluntary health
			  insurance

To the 1980s			

Epidemiologic transition
Declining mortality and birth
  rates, aging of population
Demographic transition
Decreasing infectious disease
Increase in noninfectious disease
International health
Eradication of smallpox
	Aging of
	  population
	Rising
	  expectations
	Lifestyle and risk
	  factors
	Social inequities
	Social security
	The welfare state
	Governmental
	  responsibility
	  for health
	Advocacy
	Health promotion
		Advancing medical sciences
		Clinical specialization
		Diagnostics, imaging, laboratory
		  technology
		Therapeutics, antibiotics,
		  antihypertensives, cardiac,
		  psychotropic drugs
		Preventive medicine
		Home care
		Long-term care
		Hospital versus community care
		Ambulatory surgery
			Collective bargaining health
			  benefits
			Government responsibility
			National health insurance or national
			  health service
			Rising costs of health care
			Imbalance of hospital and primary
			  care
			Health maintenance organizations
			Cost-benefit evaluation
			Rationalization
			Reforms

2000 and Beyond — The New Public Health			

Policy coordination
Evaluation of health status
Health promotion
Regulation of food, drugs, water,
  worksite, toxic agents, trauma,
  environmental risk factors
Communicable disease control
Control chronic disease
Reduce risk factors
Special needs groups
Mental health
Dental health
Health information systems
Epidemiologic systems
Planning and management
	National health
	  policy
	Resource
	  allocation
	Economic
	  development
	Social context
	Social security
	Ecology and
	  environment
	Nutrition and
	  food policy
	Healthy public
	  policy
	Healthy
	  communities
	Intersectoral
	  cooperation
	Advocacy
	Voluntarism
	Community
	  participation
		University medical schools
		Postgraduate education
		Health management training
		Peer review systems
		Accreditation
		Quality of care (TQM)
		Targeted research
		Balance hospital/community care,
		  long-term care, home care, elderly
		  housing, community services
		Integrated health systems
		Managed care systems
		Ethical issues
			National health targets
			Decentralization/diffusion of
			  implementation
			District health systems
			Managed care systems (HMOs)
			Modified market mechanisms,
			  regulation of supply, incentives, fee
			  control, competition, managed care
			Management accountability
			Economic assessment
			Integrated health systems

Box 2.16 Application of the New Public Health

The New Public Health (NPH) is a comprehensive approach to protecting and promoting the health status of the individual and the society, based on a balance of sanitary, environmental, health promotion, personal, and community-oriented preventive services, coordinated with a wide range of curative, rehabilitative, and long-term care services.

The NPH requires an organized context of national, regional, and local governmental and nongovernmental programs with the object of creating healthful social, nutritional, and physical environmental conditions. The content, quality, organization, and management of component services and programs are all vital to its successful implementation.

Whether managed in a diffused or centralized structure, the NPH requires a systems approach acting toward achievement of defined objectives and specified targets. The NPH works through many channels to promote better health. This includes all levels of government and parallel ministries; groups promoting advocacy, academic, professional, and consumer interests; private and public enterprises; insurance, pharmaceutical, and medical products industries; the farming and food industries; media, entertainment, and sports industries; legislative and law enforcement agencies; and others.

The NPH is based on responsibility and accountability for defined populations in which financial systems promote achievement of these targets through effective and efficient management, and cost-effective use of financial, human, and other resources. It requires continuous monitoring of epidemiologic, economic, and social aspects of health status as an integral part of the process of management, evaluation, and planning for improved health.

The NPH provides a framework for industrialized and developing countries, as well as countries in political–economic transition such as those of the former Soviet system. They are at different stages of economic, epidemiologic, and sociopolitical development, each attempting to assure adequate health for its population with limited resources.

---

included in Chapter 1 and throughout this text. These achievements are reflective of public health gains throughout the industrialized world and are beginning to affect policies in countries in transition from the socialist period. Countries emerging from developing status are also facing the dual burden of infectious and maternal/child health issues along with growing exposure to the chronic diseases of developed nations such as cardiovascular diseases, obesity, and diabetes.

The continuing dilemma of health in the impoverished population of the world is addressed in the Millennium Development Goals. Jeffrey Sachs, Director of the Earth Institute at Columbia University and of the UN Millennium Project, states,

"Sixty years ago, at the launch of the World Health Organization, the world's governments declared health to be a fundamental human right without distinction of race, religion, political belief, economic or social condition. Thirty years ago, in Alma-Ata, the world's governments called for health for all by the year 2000, mainly through the expansion of access to primary health facilities and services. While the world missed that target by a long shot, we can still achieve it, at remarkably low cost. Ten key steps can bring us to health for all in the next few years" (Scientific American, *Dec. 2007*). Sachs goes on to outline a program of international aid to help the developing countries of sub-Saharan Africa to reinforce some of the gains and experience of recent years (Box 2.16).

The New Public Health has emerged as a concept to meet a whole new set of conditions — those associated with increasing longevity and aging of the population, with the growing importance of chronic diseases, with inequalities in health in and between affluent and developing societies, with local and global environmental and ecological damage. Many of the underlying factors are believed to be amenable to prevention through social, environmental, or behavioral change and effective use of medical care.

The New Public Health idea evolved since Alma-Ata, which articulated the concept of Health for All, followed by a trend in the late 1970s to establish health targets as a basis for health planning. During the late 1980s and early 1990s, the debate on the future of public health in the Americas intensified as health professionals looked for new models and approaches to public health research, training, and practice. This helped redefine traditional approaches of social, community, and preventive medicine. The search for the "new" in public health continued with a return to the Health for All concept and a growing realization that health of the individual and of the society involves the management of personal care services and community prevention.

The challenges are many, and affect all countries with differing balances, but there is a common need to seek better survival and health for their citizens.

ELECTRONIC RESOURCES

Alliance for Health Policy and Systems Research, June 2007. What is health policy and systems research and why does it matter? http://www.who.int/alliance-hpsr/resources/AllBriefNote1_5.pdf [accessed February 14, 2008]

Alliance for Health Policy and Systems Research, June 2007. Health system strengthening interventions: Making the case for impact evaluation. http://www.who.int/alliance-hpsr/resources/All BriefNote2_3.pdf [accessed April 29, 2008]

Declaration of Alma-Ata. International Conference on Primary Health Care, Alma-Ata, USSR, September 6–12, 1978. http://www.who.int/hpr/NPH/docs/declaration_almaata.pdf [accessed April 29, 2008]

Department of Health and Human Services. 2008. Healthy People Midcourse Review. http://www.healthypeople.gov/Data/midcourse/ [accessed February 12, 2008]

GAVI (Global Alliance for Vaccine and Immunization). http://www.gavialliance.org/ [accessed April 29, 2008]

Healthy People 2010. http://www.healthypeople.gov/LHI/lhiwhat.htm

Millennium Development Goals. Available at http://www.un.org/millenniumgoals/pdf/mdg2007-progress.pdf [accessed February 12, 2008]

Sachs, J. 2007. Primary Health for All (Extended version). *Scientific American*. http://www.sciam.com/article.cfm?id=primary-health-for-all-extended [accessed April 29, 2008]

United Nations Millennium Development Goals. 2008. http://www.un.org/millenniumgoals/ [accessed April 29, 2008]

World Bank. *World Development Report 1993*. http://www.healthypeople.gov/LHI/lhiwhat.htm [accessed April 29, 2008]

World Health Organization. 2008. Report to the Executive Board, 122nd session, Dr. Margaret Chan, Director-General of the World Health Organization, January 21, 2008. Available at http://www.who.int/dg/speeches/2008/20080121_eb/en/print.html [accessed February 25, 2008]

RECOMMENDED READINGS

Black, D. 1993. Deprivation and health. *British Medical Journal*, 307:1630–1631.

Centers for Disease Control. 1991. Consensus set of health status indicators for the general assessment of community health status. *Morbidity and Mortality Weekly Reports*, 40:449–451.

Declaration of Alma Ata. 1978. Available at http://www.who.int/hpr/NPH/docs/declaration_almaata.pdf [accessed February 14, 2008].

Editorial. 2006. Introducing social medicine. *Social Medicine*, 1:1–4.

Gilson, L., Doherty, J., Loewenson, R., Francis, V. 2007. *Challenging Inequity Through Health Systems: Final Report — Knowledge Network on Health Systems — June 2007.* WHO — Commission on the Social Determinants of Health (CSDOH). Geneva: World Health Organization.

Green, L. W., Richard, L., Potvin, L. 1996. Ecological foundations of health promotion. *American Journal of Health Promotion*, 10:314–328.

Hancock, T. 1993. The evolution, impact and significance of Healthy Cities/Healthy Communities. *Journal of Public Health Policy*, 14:5–18.

Maiese, D. R. 1998. Data challenges and successes with Healthy People. *Healthy People 2000 Statistics and Surveillance,* Centers for Disease Control and Prevention, National Center for Health Statistics, 9:1–8.

Marmot, M. 2005. Social determinants of health inequalities. *Lancet*, 365:1099–1104.

Preamble to the Constitution of the World Health Organization as adopted by the International Health Conference, New York, June 19–22, 1946; signed on July 22, 1946, by the representatives of 61 States (Official Records of the World Health Organization, no. 2, p. 100) and entered into force on April 7, 1948. Available at http://www.who.int/about/definition/en/print.html [accessed February 14, 2008].

Roemer, M. 1984. The value of medical care for health promotion. *American Journal of Public Health*, 74:243–248.

Sachs, J. D. 2008. Primary Care (Extended Version): Ten key actions could globally ensure a basic human right at almost unnoticeable cost. *Scientific American Magazine* (January, 2008).

Schmidd, T. L., Pratt, M., Howze, E. 1995. Policy as intervention: Environmental and policy approaches to the prevention of cardiovascular diseases. *American Journal of Public Health*, 85:1207–1211.

Shea, S. (editorial). 1992. Community health, community risks, community action. *American Journal of Public Health*, 82:785–787.

Smith, G. D., Egger, M. 1992. Socioeconomic differences in mortality in Britain and the United States. *American Journal of Public Health*, 82:1079–1081.

Stokols, D. 1996. Translating social ecology theory into guidelines for community health promotion. *American Journal of Health Promotion*, 10:282–298.

Tollman, S. 1991. Community oriented primary care: Origins, evolution, applications. *Social Science and Medicine*, 32:633–642.

Walsh, J. A., Warren, K. S. 1979. Selective primary health care — an interim strategy for disease control in developing countries. *New England Journal of Medicine*, 301:967–974.

White, K., Williams, T. F., Greenberg, B. G. 1961. The ecology of medical care. *New England Journal of Medicine*, 265:885–892.

BIBLIOGRAPHY

American Public Health Association. 1991. *Health Communities 2000: Model Standards for Community Attainment of the Year 2000 National Health Objectives*, Third Edition. Washington, DC: APHA.

American Public Health Association. 1995. Washington, DC: *The Nation's Health*, March 1995.

Berry, T. R., Wharf-Higgins, J., Naylor, P. J. 2007. SARS wars: An examination of the quantity and construction of health information in the news media. *Health Communication*, 21:35–44.

Bloom, B. R. 2008. The Future of Public Health: Millennial Symposium Series. Harvard School of Public Health. http://www.hsph.harvard.edu/foph/ [accessed February 17, 2007].

Bootery, B., Kickbusch, I. (eds.). 1991. *Health Promotion Research: Towards a New Social Epidemiology*. WHO Regional Publications, European Series, No. 37. Copenhagen: World Health Organization.

Downie, R. S., Fyfe, C., Tannahill, A. 1990. *Health Promotion: Models and Values*. Oxford: Oxford University Press.

Health and Welfare Canada — World Health Organization. 1986. *Ottawa Charter for Health Promotion: An International Conference on Health Promotion*, Ottawa, Canada.

Institute of Medicine. 1988. *The Future of Public Health*. Washington, DC: National Academy Press.

Institute of Medicine. 2003. *The Future of the Public's Health in the 21st Century*. Washington, DC: National Academy Press.

Kark, S. L. 1981. *Epidemiology and Community Medicine*. New York: Appleton-Century-Crofts.

LaLonde, M. 1974. *A New Perspective on the Health of Canadians: A Working Document*. Ottawa: Information Canada.

Lasker, R. D. (ed.). 1997. *Medicine and Public Health: The Power of Collaboration*. New York: The New York Academy of Medicine.

Martin, C., McQueen, C. J. (eds.). 1989. *Readings for a New Public Health*. Edinburgh: Edinburgh University Press.

McKeown, T. 1979. *The Role of Medicine*. Oxford: Blackwell.

Ncayiyana, D., Goldstein, G., Goon, E., Yach, D. 1995. *New Public Health and WHO's Ninth General Program of Work: A Discussion Paper*. Geneva: World Health Organization.

Nutting, P. A. (ed.). 1990. *Community-Oriented Primary Care: From Principles to Practice.* Albuquerque: University of New Mexico Press.

Pan American Health Organization. 1992. *The Crisis in Public Health: Reflections for the Debate.* Washington, DC: PAHO.

Rose, G. 1993. *The Strategy of Preventive Medicine.* Oxford: Oxford University Press.

Rychetnik, L., Hawe, P., Barratt, A., Frommer, M. 2004. A glossary for evidence based public health. *Journal of Epidemiology and Community Health 2004,* 58:538–545.

Secretary of State for Health. 1991. *The Health of the Nation: A Consultative Document for Health in England.* London: Her Majesty's Stationery Office. Reprinted 1995.

Siegel, P. Z., Frazier, E. L., Mariolis, P., Brackbill, R. M., Smith, C. 1993. Behavioral risk factor surveillance, 1991: Monitoring progress toward the nation's year 2000 health objectives. *Morbidity and Mortality Weekly Report,* 42:1–21.

Smith, A., Jacobson, B. 1988. *The Nation's Health: A Strategy for the 1990s.* King Edward's Hospital Fund for London. London: Oxford University Press.

Stahl, T., Wismar, M., Ollila, E., Lahtinen, E., Leppo, K. (eds.). 2006. *Health in All Policies: Prospects and Potentials.* Helsinki, Finland: Ministry of Social Affairs and Health with the European Observatory on Health Systems and Policies.

Suhrcke, M., Rocco, L., McKee, M. 2007. *Health: A Vital Investment for Economic Development in Eastern Europe and Central Asia.* European Observatory on Health Systems and Policies. Copenhagen: World Health Organization, European Region Office.

United Nations Climate Change Conference — Bali, December 3–14, 2007. United Nations Framework Convention on Climate Change.

U.S. Public Health Service. *Health United States* 1992. Hyattsville, MD: U.S. Department of Health and Human Services, Public Health Service.

U.S. Public Health Service. *Health United States* 1998. Hyattsville, MD: U.S. Department of Health and Human Services, Public Health Service.

White, K. L. 1991. *Healing the Schism: Epidemiology, Medicine, and the Public's Health.* New York: Springer-Verlag.

World Bank. 1993. *World Development Report: Investing in Health.* New York: Oxford University Press.

World Health Organization. 1978. *Alma-Ata 1978. Primary Health Care.* Geneva: World Health Organization.

World Health Organization. 1994. *Information Support for New Public Health Action at the District Level.* Report of a WHO Expert Committee. Technical Support Series Number 845. Geneva: World Health Organization.

World Health Organization. 2000. *World Health Report 2000: Health Systems: Improving Performance.* Geneva: World Health Organization.

World Health Organization. 2007. *The World Health Report 2007 — A safer future: global public health security in the 21st century.* Geneva: World Health Organization.

World Health Organization, Regional Office for Europe. 1985. *Targets for Health for All: Targets in Support of the European Strategy for Health for All.* Copenhagen: World Health Organization Regional Office for Europe.

World Health Organization, Regional Office for Europe. 1995. *Twenty Steps for Developing a Healthy Cities Project,* Second Edition. Copenhagen: World Health Organization, European Regional Office.

World Health Organization Europe. 1999. Health 21 — *Health for All in the 21st Century.* Copenhagen: World Health Organization.

Chapter 3

Measuring and Evaluating the Health of a Population

Introduction
Demography
    Fertility
    Population Pyramid
Life Expectancy
Epidemiology
Social Epidemiology
Epidemiology in Building Health Policy
Definitions and Methods of Epidemiology
    Rates and Ratios
    Measures of Morbidity
    Prevalence Rates
    Measures of Mortality
    Social Classification
Sentinel Events
The Burden of Disease
Years of Potential Life Lost
    Qualitative Measures of Morbidity and Mortality
Measurement
    Research and Survey Methods
    Variables
    The Null Hypothesis
    Confounders
    Sampling
    Randomization
Normal Distribution
Standardization of Rates
    Direct Method of Standardization
    Indirect Method of Standardization
Potential Errors in Measurement
    Reliability
    Validity
Screening for Disease
Epidemiologic Studies
    Observational Studies
    Experimental Epidemiology
Establishing Causal Relationships
Notification of Diseases
Special Registries and Reporting Systems
Disease Classification
Hospital Discharge Information

Health Information Systems (Informatics)
    WHO European Region Health for All Database
Surveillance, Reporting, and Publication
Assessing the Health of the Individual
Assessment of Population Health
    Defining the Population
    Socioeconomic Status
    Nutrition
    Environment and Occupation
Health Care Financing and Organization
    Health Care Resources
    Utilization of Services
    Health Care Outcomes
    Quality of Care
    Self-Assessment of Health
    Costs and Benefits
    Effects of Intervention
    Qualitative Methods
Summary — From Information to Knowledge to Policy
Electronic Resources
Recommended Readings
Bibliography

INTRODUCTION

The history of health, its concepts, and scientific development have been discussed in previous chapters. Measuring the health of populations is fundamental to improving their health status. Traditionally public health deals with the health of populations, while the New Public Health deals with the health of both individuals and population groups. This chapter discusses how measurements are used to describe, analyze, prescribe, and justify interventions to protect and improve the health of populations and of individuals and monitor the outcomes of interventions.

The public health professional working with individual and community health needs to acquire the knowledge and skills necessary to measure and interpret the factors that relate to disease and health, both in the individual and in population groups. Demography and epidemiology are the basis of health information systems, but the social and basic medical sciences are also vitally important in understanding public health, providing an expanding array of health status indicators and measures of the impact of interventions.

Demography deals with the recording of the characteristics and trends of a population and its characteristics over time. Epidemiology measures the distribution, causes, control, and outcomes of disease in population groups. It provides the basic tools for quantification of the extent of disease, its patterns of change, and associated risk factors. Epidemiology also provides basic information needed for planning, evaluating, and managing health services. Other disciplines provide additional information and insights needed for community and national health assessment. These include the social sciences (sociology, psychology, anthropology, and economics), as well as clinical fields such as pediatrics and geriatrics, and basic sciences such as microbiology, immunology, and genetics.

This chapter is an introduction to epidemiology and health information systems intended to familiarize the student with basic terms, concepts, and methods. The scope of this text does not lend itself to detailed discussion of biostatistics and epidemiologic methods, but instead focuses on the basic ideas and their relevance to the New Public Health. This chapter is meant to provide a general overview of the role of epidemiology and health information systems in the context of the New Public Health; it cannot serve as an authoritative, detailed text on the subject. Specialized texts are listed in the bibliography at the end of this chapter.

DEMOGRAPHY

Demography is "the study of populations, especially with reference to size and density, fertility, mortality, growth, age distribution, migration, and vital statistics and the interaction of all these with social and economic conditions" (Last, 2001). Demography is based on vital statistics reporting and special surveys of population size and density; it measures trends over time. It includes indices such as fertility, birth, and death rates; rural–urban residential patterns; marriage and divorce rates and migrations; as well as their interaction with social and economic conditions. Since public health deals with disease as it occurs in the population, the definition of populations and their characteristics is fundamental.

Vital statistics include births; deaths; and population by age, sex, location of residence, marital status, socioeconomic status, and migration. Birth data are derived from mandatory reporting of births and mortality data from compulsory death certificates. Other sources of data are population registries including marriage, divorce, adoption, emigration, and immigration, rural–urban residential

patterns as well as economic and labor force statistics compiled by governmental agencies, census data, and data from special household surveys.

A census is an enumeration of the population, recording the identity of all persons in every residence at a specified time. The census provides important information on all members of the household, including age, date of birth, sex, occupation, national origin, marital status, income, relation to head of the household, literacy, education level, and health status (e.g., permanent disabling conditions). Other information on the home and its facilities may be included. A census may assign persons according to their location at the time of the enumeration (*de facto*) or to the usual place of residence (*de jure*). A census tract is the smallest geographic area for which census data are aggregated and published. Data for larger geographic areas (metropolitan statistical areas) are also published. More extensive data may be collected for representative samples of the population. These are carried out over a period of years by a specialized national agency (e.g., Bureau of the Census in the United States and the Central Bureau of Statistics, Office of Population, Censuses and Surveys in the United Kingdom).

Census data are published in multiple-volume series with availability for research on computer disks, CD-ROMs, and on the Internet. Usually inter-census surveys are carried out to determine trends in important economic or demographic data such as individual and family incomes, nutrition, employment, and other social indicators. Accuracy of such a complex and costly process cannot be 100 percent, but great care is taken to maximize response and standardization in interview methods and processing to assure precision. Despite its limitations, the census is accepted as the basis of statistical definition of a population. This is well established in developed countries, but is problematic in developing countries where birth and death registration may be inadequate, requiring community-based registration systems. Life expectancy at birth is a common measure used to compare health status in and between countries (Table 3.1).

Demographic transition is a long-term trend of declining birth and death rates, resulting in substantive change in the age distribution of a population. Population age and gender distribution is mainly affected by birth and death rates, as well as other factors such as migration, economics, war, political and social change, famine, or natural disasters. Changing population patterns also accompany economic development, and demographic transition may be characterized by the following stages:

1. *Traditional:* high and balanced birth and death rates;
2. *Transitional:* falling death rates and sustained birth rates;
3. *Low stationary:* low and balanced birth and death rates;
4. *Graying of the population:* increased proportion of elderly as a result of decreasing birth and death rates, and increasing life expectancy;

5. *Regression:* migration or increasing death rates among young adults due to trauma, AIDS, early cardiovascular disease mortality, or war resulting in steady or declining longevity (i.e., demographic regression).

TABLE 3.1 Life Expectancy at Birth (years), 1960–2005, Selected OECD Countries

	1960	1970	1980	1990	2000	2004/05
Canada			75.3	77.6	79.3	80.2*
Denmark	72.4	73.3	74.3	74.9	76.9	77.9
Finland	69.0	70.8	73.4	74.9	77.6	78.9
France	70.3	72.2	74.3	76.9	79.0	80.3
Germany	69.6	70.4	72.9	75.2	78.0	79.0
Ireland	70.0	71.2	72.9	74.9	76.5	79.5
Japan	67.8	72.0	76.1	78.9	81.2	82.1
Korea	52.4	62.2	65.9	71.4	76.0	78.5
Netherlands	73.5	73.7	75.9	77.0	78.0	79.4
New Zealand	71.3	71.5	73.2	75.4	78.7	79.6
Sweden	73.1	74.7	75.8	77.6	79.7	80.6
United Kingdom	70.8	71.9	73.2	75.7	77.8	79.0
United States	69.9	70.9	73.7	75.3	76.8	77.8*

Source: OECD, 2007.
Note: *2004 data.

Fertility, mortality, disease patterns, and migration are the major influences on this transition within the population. The many factors that affect fertility decline and increasing longevity are outlined in Box 3.1. Education of women, urbanization, improved hygiene and preventive care, economic improvement with better living conditions, and declining mortality of infants and children are the major factors. This is an important issue in developing countries where high fertility rates and declining mortality of children contribute to rapid population growth and poverty.

Birth rates in the industrialized countries have fallen over the past half-century with economic prosperity, efficient and easily available methods of birth control, and greater opportunities for women in the workforce. In some countries, access to prenatal diagnosis of the gender of the fetus has resulted in wide-scale abortion of females because of birth policies, with parental preference for male children in China and India as examples. This is resulting in a major lack of young women in the population with attendant social and political effects.

Fertility

Fertility is bearing of living children and is clearly determined by more than biological potential. Fertility is a complex issue influenced by cultural, social, economic, religious, and even political factors. Although economic prosperity may initially promote higher birth rates, increases in education levels and economic prospects, as well as in survival of those born, are generally related to restraint in birth rates and natural population growth (Box 3.2).

Box 3.1 Factors in Fertility Decline and Increasing Longevity

Factors in fertility decline

1. Education, especially of women;
2. Decreasing infant and child mortality reduces pressure for more children to ensure survivors;
3. Economic development, improved standards of living, expectations, and income levels;
4. Urbanization — changes family needs compared to rural society;
5. Birth control, supply, accessibility, and knowledge;
6. Government policy promoting fertility control as a health measure;
7. Mass media increases awareness of birth control, and aspiration to higher standards of living;
8. Health system development and improved access to medical care;
9. Changing economic status, social role, and self-image of women;
10. Changing social, religious, and political and ideological values.

Factors in increasing longevity

1. Increasing family income and standards of living;
2. Improved nutrition including improved food supply, distribution, quality, and nutritional knowledge;
3. Control of infectious diseases;
4. Reduction in noninfectious disease mortality;
5. Safe water, sewage and garbage disposal, and adequate housing conditions;
6. Disease prevention, reducing risk factors, promoting healthy lifestyle;
7. Clinical care services with improved access and quality;
8. Health promotion and education activities of the society, community, and individual;
9. Social security systems; for example, child allowances, pensions, national health insurance;
10. Improved conditions of employment and recreation, economic and social well-being.

Box 3.2 Commonly Used Fertility Rates

1. *Crude birth rate* (CBR) is the number of live births in a population over a given period, usually one calendar year, divided by the midyear population of the same jurisdiction, multiplied by 1000.
2. *Total fertility rate* (TFR) is the average number of children that a woman would bear if all women live to the end of their child-bearing years, and bear children according to age-specific fertility rates; most accurately answering the question "how many children does a woman have, on average?"

Source: Modified from Last, J. M. (ed.). 2007. A *Dictionary of Public Health*. New York: Oxford University Press.

Population Pyramid

A population pyramid provides a graphic demonstration of the percentage of men and women in each age group in a total population (Figure 3.1). A country or region with a wide population base has a high birth rate and a large percentage of its population under age 15; usually accompanied by limited resources, it is a formula for continued poverty. A population pyramid with a narrow base (i.e., few young people) and a growing elderly population will have a smaller workforce to provide the economic base for the "dependent age" population (i.e., both the young and the old).

With a smaller working-age population to support these social costs of "dependent" subgroups, adverse economic consequences may prejudice costly pension and health services for the population. Other factors may also affect the population pyramid; for example, the loss of a large number of people during wartime. This loss affects a particular age–sex group as well as fertility patterns both during and after the war; for example, a postwar "baby boom." With aging of the population in many countries due to low birth rates and increasing longevity, the concepts of "dependent" population groups of those under age 15 and those over 65 as a percentage of the total population is increasingly relevant to social and economic planning.

LIFE EXPECTANCY

Life expectancy is an important health status indicator based on average number of years a person at a given age may be expected to live given current mortality rates. Life expectancy can be measured at age 0, or any other specific age, representing expected survival time once a person has reached that age; for example, at age 15, 60, or 75 by gender and by ethnic group.

Between 1970 and 2004 (Figure 3.2), the variation by sex and race for those age 65 was similar; at birth, variation by sex and race remains constant. Life expectancy at birth in the United States from 1900–2000 increased dramatically in the first half of the century, reflecting mainly the reduction in infectious diseases and adverse conditions of maternity

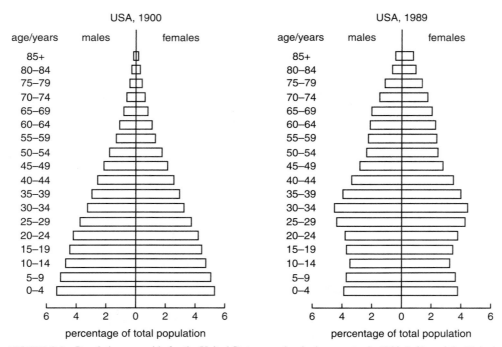

FIGURE 3.1 Population pyramids for the United States as a developing country in 1900 (left), and the United States in 1989 (right). Source: Gray, A. (ed.). 1993. *World Health and Disease*. Milton Keynes: Open University Press, with permission.

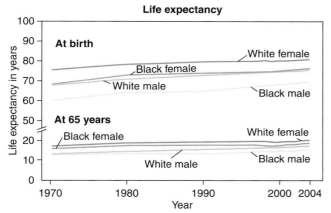

FIGURE 3.2 Life expectancy at birth and age 65, United States, 1970–2004. Source: U.S. Department of Health and Human Services. *Health United States, 2007*, with Chartbook on Health of Americans (Figure 18).

and infancy. The second half of the century was characterized by an increase and then a decrease in cardiovascular diseases as a cause of mortality and an increase in cancer and trauma-related deaths, so that life expectancy increased, but at a lower rate than in the earlier period.

Life expectancy at birth increased dramatically in the United States from 47.3 years in 1900 to 68.2 years in 1950. Since the 1950s, life expectancy increased to 73.7 years in 1980, and to 77.8 years in 2004 (75.2 years for men and 80.4 for women), (Health United States, 2006 and 2007 Figure 3.2). In other parts of the world, life expectancy is even longer, as seen in Figure 3.3 of selected European countries.

Life expectancy is also used in chronic disease epidemiology to summarize patterns of mortality and survival

in a population, such as a population with breast cancer. This is important in clinical epidemiology where studies of effectiveness of specific interventions are assessed. Life expectancy is quite different for male and female genders; thus gender is an important factor in assessment of disease prevalence and also in effectiveness of interventions.

Demography is becoming an major political and social issue in countries where demographic transition is resulting in major shifts in population makeup. Russia is experiencing a major reduction in population with low birth and low life expectancy and less severely in Western European countries. China implemented a "one child per family policy and with male preference now faces a major gender imbalance with excess male and deficit in female population. Developing countries with high birth rates are experiencing population growth exceeding economic growth capacity. The United States has the benefit of steadily improving life expectancy and high immigration rates to offset low birth rates. Japan and many European countries with very high life expectancy and low birth rates face declining and aging population. These population transitions have important political and economic implications in every country and regions within countries (see Chapter 13).

EPIDEMIOLOGY

Health care providers are generally oriented toward individual patient assessment and care. However, even the specialized clinician must have a basic understanding that disease is not an event isolated to an individual, but affects population groups and communities alike. The clinician

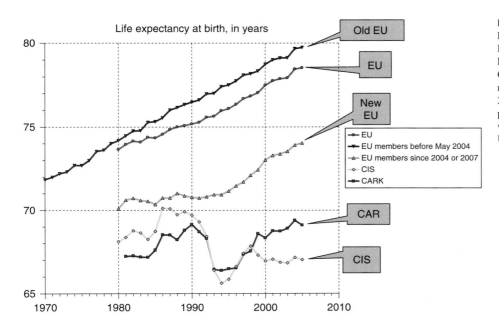

FIGURE 3.3 Life expectancy at birth, European Region, 1970–2004. Source: Health for All database, WHO European Region, November 2007 Note: CARK = Central Asian Republics, Old EU = members of the European Union before 2004, New EU = members of the European Union after 2004, CIS = Commonwealth of Independent States (Russia, Ukraine, Byelorussia).

must be aware of the potential for epidemic disease as well as the statistical risk factors for a noncommunicable disease that affects the individual patient, such as cardiovascular disease, in order to determine management over the long term.

Epidemiology is the study of health events in a population. Its purpose is to help understand disease processes and outcomes, determine factors in causation, and provide direction for medical or public health interventions. The distribution and determinants of health-related states, conditions, or events in defined populations are important to identification of potential interventions and priorities to control health problems. The ultimate purpose of epidemiology is to promote, protect, and restore health. This includes surveillance, observation, and hypothesis generation and testing in analytic research and experiments (Box 3.3). Determinants include physical, biological, social, cultural, and behavioral factors. Distribution includes analysis by time (e.g., month, season, time of day), place, groups of persons affected. Health-related states and events include disease, causes of death, behaviors (e.g., use of tobacco), compliance with public health intervention (e.g., immunizations), and access to and use of health services (Last, 2007).

Epidemiologic studies include descriptive studies of routinely collected and reported data on mortality, morbidity, and related factors. They focus on the distribution of a disease or risk factor by time, place, and person characteristics and form a crucial basis for public health activities and evaluation. Analytic epidemiologic studies are based on hypothesis testing and include observational studies such as cross-sectional, case control, and cohort studies as well as intervention studies, including clinical and program trials. They focus on exposures and outcomes attempting to determine their associations.

Variables are "any attribute, phenomenon, or event that can have different values" (Last, 2001). This includes all the physical, biological, social, cultural, and behavioral factors that influence health. Health-related states and events include diseases, causes of death, behavior such as use of tobacco, compliance with preventive regimens, and provision and use of health services. Health events occur in population groups and the study of epidemiology requires definition of the events and the population studied. Specified populations are those with common, identifiable characteristics that can be quantified, such as gender, age, ethnicity, and region of residence. The goals and methods of epidemiology are outlined in Box 3.4.

Classically, the clinician examines a patient who presents himself or herself for medical care, including preventive action, while the epidemiologist studies a defined population. Both evaluate the effects of preventive or treatment measures and share the need to understand risk factors and the natural process of disease. Epidemiology studies a particular disease in a population, taking into account factors such as age, sex, ethnicity, exposure to known or suspected risk factors, and socioeconomic patterns, as well as the effect of various interventions. This is undertaken in order to understand the natural history of disease, its diagnostic criteria, appropriate methods of prevention or management, outcomes to be expected, and the costs and benefits of the different methods of control.

Clinicians and epidemiologists depend on each other, and need the work of other fields, such as health economics and management, and the documented experience of

---

Box 3.3 State of Health of the Individual and Community: Determinants and Measures

Factors

- Biology — age, gender, genetics
- Geography — urban, rural, climate, nomadic
- Economics — GDP per capita, family income, unemployment, living standards
- Social, cultural, religious, and economic factors
- Education — literacy, gender differences, higher education
- Lifestyle, personal habits — diet, smoking, exercise, sexual habits
- Occupation — accidents, toxic exposures
- Environment — exposure to toxins, carcinogens, infectious agents
- Nutrition — diet, food supply, cost and quality, fortification
- Health services and insurance — accessibility, quality of care, comprehensiveness
- Public health infrastructure and policies

Measures

- Demography — births, deaths, marriages, divorces
- Infrastructure — safe water, food, air, solid waste disposal, transport measures
- Health insurance — coverage, comprehensiveness
- Resources — hospital beds per 1000, medical personnel
- Process — utilization, immunization, hospitalization rates
- Outcomes
  - Mortality — by age, gender, cause
  - Morbidity — by cause, time, place, common exposure
  - Physiologic indicators — growth, BMI, anemia
  - Functions of daily living and disability
- Quality measures — accreditation, peer review, quality improvement
- Knowledge, attitudes, beliefs, practices
- Satisfaction and self-assessment
- Costs and benefits

interventions to improve care and efficient use of resources. This also relies on interaction with the various disciplines within public health, health policy, health systems management, and clinical medicine. Difficult choices in public policy on allocation of resources must be made with many factors in mind, including epidemiology of the condition, cost-effectiveness of intervention, and ethical questions.

In the nineteenth to twentieth centuries, a profound transition occurred in the industrialized countries as the diseases of "pestilence and famine" waned, and chronic diseases became the leading causes of death. Many of these were associated with man-made environmental problems and personal lifestyle. This epidemiologic transition took place, initially primarily because of the cumulative effects of successful public health activities such as environmental sanitation and food safety, later communicable disease control through the success of vaccines and antibiotics in reducing the major diseases of childhood, and improvements in living conditions. A "second era" in public health occurred during the latter half of the twentieth century with the rise and fall of chronic disease in the industrialized countries, but this era is still a great challenge in countries in transition (former socialist countries, especially in the Russian Federation, Ukraine CIS, and

central Asian republics), and increasingly in developing countries as well. Now a "third era" of health has arrived with people living well into their 70s and 80s, often not only free from serious morbidity but leading vibrant and active lives requiring an orientation not only of personal perspectives, but also adjustments in the community and in the health system (Breslow, 2006).

Particularly during the 1950s and 1960s, rising standards of living in the industrialized world were associated with increases in noncommunicable diseases, including cardiovascular diseases, malignancies associated with smoking, other "lifestyle" diseases, and trauma associated with industrialization, violence, self-injury, and motor vehicle accidents. This transition is playing an important role in the disease patterns of developing countries as they urbanize and the middle class grows.

Since the 1960s, a new and equally profound epidemiologic transition has occurred with the decline of heart disease, stroke, and trauma as causes of death, in the industrialized world and also increasingly in developing countries. These have contributed to increasing longevity. Greater health consciousness and self-care, improved social security for the elderly and disabled and vulnerable adults, and advances in medical care have contributed to this phenomenon.

In the 1980s, a dramatic new epidemiologic challenge appeared with the advent of a pandemic of HIV infection and a return of diseases thought to have been under control. Potentially dangerous infectious diseases can be transmitted far from their original habitat with the rapid transportation and movement of populations, including migrants, tourists, and other travelers throughout the globe. Other infectious diseases are becoming resistant to available treatments, with multidrug-resistant (MDR) infectious diseases, especially tuberculosis.

The HIV/AIDS epidemic (see Chapter 4) created a new and deadly situation with a worldwide pandemic and yet control of its spread was achieved in the industrialized countries through scientific achievements and application of public health measures (Figure 3.4). This work has involved partnerships and cooperative activities among governmental, international organizations, bilateral aid, nongovernmental organizations (NGOs), and private agencies to establish screening, education, risk-reduction programs, prophylaxis, and treatment with antiretroviral drugs to improve clinical care and prevent transmission of the virus.

"Newly emerging" diseases are a notable threat to the gains made in the health status of the industrialized world, and an even greater threat to the struggling health systems of developing countries (see Chapters 4 and 16). However, the chief threat to the public's health remains the massive deprivation in developing countries and poverty still present in the industrialized countries. Newly emerging diseases presented a growing challenge to public health

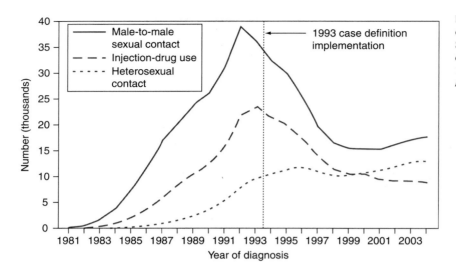

FIGURE 3.4 New cases of AIDS by category of transmission, United States, 1981–2004. Source: Centers for Disease Control. 2006. Epidemiology of HIV/AIDS — United States, 1981–2005. *Morbidity and Mortality Weekly Report,* 55;589–592.

with new disease entities (AIDS, Ebola, SARS, avian influenza) along with renewed threats from diseases present for centuries with MDR cases of tuberculosis, methicillin-resistant *Staphylococcus aureus* (MRSA), and others from abuse of antibiotics and molecular shifts in the organisms. Even diseases thought to have been brought under control by vaccines have reappeared, such as pertussis and measles, in imported and localized epidemic forms.

In the 1990s, there were new breakthroughs in the epidemiology of infections causing highly prevalent chronic diseases. A new infectious agent, the prion, was identified by Stanley Pruziner (Nobel Prize, 1997) as transmitting Creutzfeld-Jakob disease, a serious degenerative and fatal neurologic disorder. A new bacteria first identified in the 1980s, *Helicobacter pylori*, was shown to be the cause of peptic ulcers and cancer of the stomach (B. J. Marshall and J. R. Warren, Nobel Prize, 2005). The previously known relationship of hepatitis B to cancer of the liver and chronic cirrhosis took on new importance as an effective and inexpensive vaccine became available and the finding that nutritional deficiencies are co-factors in a variety of diseases.

In the first years of the twenty-first century, human papillomavirus (HPV), a sexually transmitted virus, was identified as the cause of cancer of the cervix. An effective vaccine was approved in the United States by the FDA in 2006 and is already being used in the industrialized countries, but is too costly for developing countries where it is most needed. It provides the means to control and possibly eliminate one of the leading causes of cancer in women worldwide, but will need to be used along with Pap smear screening for many years to come. Such breakthroughs in medical sciences and public health practice demonstrate the vital importance of combining epidemiologic and clinical investigations to confirm these relationships and to seek out preventive mechanisms.

SOCIAL EPIDEMIOLOGY

Epidemiology has evolved from its origins as a factor in sanitary statistics in the first half of the nineteenth century, as exemplified in the political arithmetic and vital statistics of Farr and the social statistics of Chadwick and Shattuck. It helped foster the sanitary movement and public health benefits through development of drains, sewage systems, and community sanitation. In the late nineteenth century through the first half of the twentieth century, epidemiology was associated with the germ theory of single agents relating to one specific disease, and public health activities focused on interruption of transmission or primary prevention through vaccinations. In the latter half of the twentieth century, chronic disease epidemiology showed associations among multiple risk factors and outcomes, without full understanding of the intervening factors or pathogenesis. Some landmarks of epidemiology are shown in Box 3.5. They are further discussed in Chapters 1, 4, 5, 8, and 13.

Chronic disease epidemiology led to application of risk-control public health measures, affecting lifestyle (diet, exercise, smoking), products (food, guns, cars), and environment (pollution, passive smoking). A new era of epidemiology is emerging in the twenty-first century in which organization, information, and application of biomedical technology are vital in population health. This involves a wider, multidisciplinary approach, where statisticians, economists, social scientists, health systems managers, and epidemiologists bring different skills to a more complex paradigm of public health.

Social inequalities in morbidity and mortality have become a major field of interest in epidemiologic studies for many years. A study of late-stage diagnosis of colorectal cancer in New York State showed women and African-Americans as more likely to have late-stage cancer than

---

Box 3.5 Selected Landmarks in Epidemiology

Vital statistics and social epidemiology

1662	Graunt publishes *Natural and Political Observations Made upon the Bills of Mortality*
1836	Registrar General's Office established by U.K. Parliament
1842	Chadwick — *Report on the Sanitary Condition of the Labouring Population of Great Britain*
1848	Virchow — "medicine is a social science"
1858	Simon maps mortality by district in relation to social and environmental conditions
1974	The LaLonde Report — *New Perspectives on the Health of Canadians* — concept of lifestyle, genetics, environment, and medical care as health status factors
1982	Black Report — showing social class differences in mortality in the United Kingdom
1995	Beijing Conference on Women, empowerment for health of women and children
2001	United Nations Millennium Development Goals (MDGs) and Human Development Index (HDI) monitoring

Infectious disease epidemiology

1796	Jenner uses cowpox to vaccinate against smallpox
1854	Snow identifies and interrupts water transmission of cholera in London
1882	Koch discovers tubercle bacillus and cholera; Koch-Henle postulates on causation
1978	Eradication of smallpox achieved
1980s	HIV and other newly emerging or resurging infectious diseases
1990s	Vaccines for hepatitis B to prevent cancer of liver
2000s	Elimination of yaws, poliomyelitis, leprosy, dracunculiasis, and measles being achieved
2000s	Bioterrorism threat
2000s	SARS, avian flu, multidrug-resistant organisms
2007	H5N1 avian influenza spread and threatened pandemic

Noninfectious disease epidemiology

1747	Lind demonstrates prevention of scurvy by citrus fruits
1775	Pott shows excess cancer of the scrotum in chimney sweeps
1914	Goldberger demonstrates nutritional cause of pellagra
1950	Doll and Hill relate cigarette smoking to lung cancer
1954	Framingham study reports on heart disease risk factors
1960s	Decreasing mortality from cardiovascular diseases, trauma
1990s	Infections as causes of chronic diseases; *Helicobacter pylori* causing peptic ulcers and cancer of stomach
1990s	Vaccines for hepatitis B to prevent cancer of liver
2006	Human papillomavirus vaccines to prevent cancer of cervix

Health policy epidemiology

1883	Bismarck initiates workers' compensation and national health insurance
1917	Semashko establishes Soviet state health system
1948	United Kingdom establishes National Health Service
1964	U.S. Surgeon General Report on Smoking and Health
1965	U.S. Medicare and Medicaid as amendments to Social Security Act of 1935
1978	Declaration on Alma-Ata and Health for All 2000
1979	U.S. Surgeon General — Healthy People and health targets
1990s	Managed care expansion in United States
2000s	Health reforms in central and eastern Europe, Commonwealth of Independent States, central Asia, and emerging developing countries
2008	Universal health insurance issue in U.S. presidential election

Source: Adapted from Susser, M., and Susser, E. 1996. Choosing a future for epidemiology: Eras and paradigms. *American Journal of Public Health*, 86:668–673.

---

men and whites. Individuals living in areas of low socioeconomic status (SES) were significantly more likely to be diagnosed at a later stage than those living in higher SES areas. Similar patterns of socioeconomic disparity in mortality have been shown among men in the state of São Paulo, Brazil, with the poor having three times greater rates of mortality than the wealthy minority. In contrast, a study in Denmark of regional and social class variation in relative risk of death showed little social variation except for persons with no known address. Social inequities in health occur in virtually all societies, even those with "universal" access to health care, including the United Kingdom, Israel, and many others, with differences in physical access to care; differences in lifestyle and risk factors; SES; and knowledge, attitudes, and practices related to health and health care.

Social epidemiology in some senses reflects the nineteenth-century traditions of Virchow, Chadwick, Shattuck, and Farr (see Chapter 1), and a return to the miasma theory of disease, in which health of populations is largely determined by environmental factors of society, and that to understand causation of disease, it is essential to understand its historical and social context. This social epidemiology necessarily incorporates qualitative methodologies based on the social sciences in addition to the quantitative epidemiologic tools of measuring associations between exposure and disease in individuals or groups. The New Public Health integrates the qualitative and quantitative methods with management sciences based on successful applications of all these modalities to public health issues over the past century and more.

EPIDEMIOLOGY IN BUILDING HEALTH POLICY

Epidemiology proved itself in enormously successful interventions for public health in the first half of the nineteenth century. The golden period of infectious disease epidemiology in the late nineteenth and first half of the twentieth centuries established the basis for control of communicable disease, a revolution still in process. During the mid-twentieth century, development of chronic disease epidemiology provided the basis for health promotion and lifestyle changes contributing to reduced morbidity and mortality from cardiovascular disease and the potential for control of cancer, trauma, and other noncommunicable conditions. In the case of HIV/AIDS, health promotion was the only tool available until the antiretroviral drugs became available, but the hoped-for vaccine is still off in the future. There is an important role for controlled trials for preventive modalities and treatment in relation to chronic disease (mammography, hormone therapy, and many others).

The fundamentals of epidemiology are as vital for the student of health sciences as are the study of bacteriology, biochemistry, or surgery. It is equally important that health planners, economists, and others concerned with the policy aspects of health be conversant with epidemiology. This is so they understand the need to adapt health services to changes occurring in the epidemiologic and technological aspects in health and in society, as well as the application of data from studies to the changing needs for health care.

Epidemiology is also essential to formulation of policy and operation of a health system. It is essential for the smooth functioning of a health system, as a method of analysis, and as a monitoring tool. Assessment and monitoring of the health status of a population are, by their very nature, multifactorial. Preliminary and, perhaps, impressionistic reading of the situation makes use of data available from routine sources and serves to generate hypotheses for testing. Evaluation is a more formal and systematic approach in determining the quality of the health of a population as objectively as possible. All evaluations need to look at the input, process, and output of a system. The epidemiologic method is applied to measurements (indicators) of inputs (resources) of a health system, the process (manner) of their utilization, and outcomes of care (indicators of morbidity, mortality, or functional status of a population).

Epidemiology and demography are necessary, but not sufficient alone, for determination of health policy. Other factors include the funds, human resources, and facilities availability, and their utilization, community attitudes, and political will. Epidemiology, health care financing, and resource allocation relate to supply and demand and ultimately to policy. These are all issues of great importance to management of health systems and addressing the changing needs of an aging population with growing obesity, diabetes, and other chronic diseases, while infectious diseases continue to play an important role in population health. They are also of importance in addressing issues of inequalities in health even in countries with universal access through national health insurance. In the United States, without universal health insurance a serious gap in social policy has many downstream effects perpetuating social and ethnic disparities in health.

Analysis of these complex factors provides the intelligence or feedback for managing the broad scope of public health. The New Public Health integrates assessment, evaluation, and epidemiologic analysis with the organization, supply of health care, and other factors relating to health of the community as a whole. These disciplines provide vital material to link population health needs and the use of resources.

Epidemiology is addressing new challenges of social equity in health; this has become an important part of modern epidemiology with the growing understanding that social conditions and cultural background are key factors not only in disease incidence and prevalence, but in access to health care, both preventive and curative. While the epidemiologic identification of health inequality/inequity has been important in identifying the extent and severity of the issues, it is in the utilization of these data in policymaking and action where the real challenge lies. Gender, sexual preference and behavior, ethnicity, place of residence, income, family status, religion and religiosity, social connectivity, occupation, and education are all part of the health–sickness spectrum.

DEFINITIONS AND METHODS OF EPIDEMIOLOGY

Rates and Ratios

Measuring the extent of a disease (or a risk factor) in a population relates known cases to a population base, expressed as a rate or a risk. Comparing the extent of a disease or a risk factor among population groups can be expressed as a ratio (or a relative risk).

Relative risk is the probability that individuals in one group will develop a certain condition (a disease being studied) within a defined time period as compared to people in another group, but this does not necessarily apply to each individual person. As a probability, it therefore ranges from 0 to 1. It is usually calculated as the cumulative incidence of a certain disease condition. Cumulative incidence implies that a defined population group is followed during a certain time period to allow the identification of incidence of new cases. When including new and preexisting cases the term used is *prevalence*. The denominator is a fixed population about which information is available on the condition under study for each individual within that population.

Not all study designs allow such identification of cumulative incidence because the study population (representing the denominator) is often dynamic (with persons entering and leaving the study population). Thus it is often impossible to correctly calculate risk. In order to overcome this problem, rates may be used as the estimation of risk, on condition that:

a. The time of follow-up is relatively short and
b. The rate of disease over the studied time period is fairly constant.

Rates are measures of frequency of a phenomenon, such as occurrence of a health event (A), in a defined population (B), in a given time period. The components of a rate are the numerator (A) defining the number of cases of a specified condition, over the denominator (B) defining the population, in a specified time frame in which the events occurred in a defined population/place/region/country. A multiplier to convert the fraction to a decimal number may be used for convenient comparisons between the frequency of the event in different population groups.

The *risk group* may be the entire population that is defined by a geographic area, an occupational group, a school, a health service or insurance system, or any other specified groups of people as defined by occupation, place of work, or by lifestyle. The population may also be persons who share a risk factor for disease, such as smokers, substance abusers, sex workers, or persons attending a celebration who eat certain foods which may be the common source of a disease outbreak.

Designating the population-at-risk is a crucial aspect of any epidemiologic study and is subject to common errors. Defining the number of cases of a disease or the risk factors being studied is essential to provide the numerator of the rate or ratio. This is also difficult because not all cases of a disease may be reported. So the numerator may be an underestimation of the true value in the population. This can occur with common infectious diseases (e.g., mumps, rubella) or where many cases of disease are not clinically diagnosed and therefore go unreported. The same applies in chronic diseases (e.g., diabetes mellitus) for many reasons, including nonpresentation to the medical system of asymptomatic cases, unclear case definition, and medical error. There may be discrepancies in reporting such as in coronary heart disease where symptoms differ quite significantly between men and women, or when access to care varies between people in different socioeconomic groups.

Crude rates are summary rates based on the actual number of events (e.g., births or deaths) reported in a total population in a given time period. Cause-specific rates measure for specified conditions (e.g., TB) occurring in the total population or in a designated population group (e.g., age–sex groups) in a specified time period. The population used for annual rate calculations is usually estimated at July 1 of that year or may use an average for the entire year (see Box 3.6).

Odds ratios are fractions (A/B), where A and B are two separate and distinct quantities. A ratio may compare mortality rates from a specific disease in two population groups, one group exposed to a risk factor compared to the mortality rate in an unexposed population. This is called a *relative risk* (*RR*), which is described later in this chapter. If rates are determined for several population groups then the rate ratio is the ratio of the rates which allows for a comparison of the rates between the different population groups.

A *proportion* is a ratio where the numerator is included in the denominator population, such as describing the

---

Box 3.6 Commonly Used Mortality Rates and Ratios

1. *Crude Death Rate* (CDR) = number of deaths from all causes per 1000 population in a given year = A/B × 1000 (total deaths/average population × 1000);
2. *Age-Specific Mortality Rate* = number of deaths of persons in the specified age group per 1000 live population in that age group over a period of time, usually a year;
3. *Cause-Specific Mortality Rate* = number of deaths from a specific cause per 100,000 live population (estimated on July 1 of the given year); for example, annual number of deaths from lung cancer in a given year = 400 in a population of 1 million = 400/1,000,000 = 40 lung cancer deaths per 100,000 population;
4. *Case Fatality Rate* (CFR) = number of deaths from a specified cause during a given time period over the number of diagnosed cases of that disease during the same time period × 100; for example, 10 deaths from measles among 5000 cases is a CFR of 10/5000 × 100 = 0.2%;
5. *Proportional Mortality Rate* (PMR) (for a specific cause) = the number of deaths from that cause in a specified time period over the total number of deaths in that population in the same time period × 100; for example, 25 deaths from motor vehicle accidents/1000 total deaths from all causes × 100 = 2.5% (the denominator includes the numerator).
6. *Standardized Mortality Rate or Ratio* (SMR) = the ratio of the number of deaths from a specified condition observed in a study population over the number that would be expected if the study population had the same specific rates as the standard population × 100.

Source: Modified from Last, J. M. (ed.). 2001. A *Dictionary of Epidemiology*, 4th ed. New York: Oxford University Press.

number of cases found in a given population, or the proportion of persons with a certain attribute or risk factor within the defined population; for example, the proportion of smokers within a certain community.

When cases are relatively rare, an approximation can be made using the total population (including both the disease-free and the cases) as the numerator. In such conditions, the odds ratio may serve as a good estimate of the relative risk.

Measures of Morbidity

Morbidity is a departure, subjective or objective, from a state of physiologic and psychological well-being or normal function. It can be measured as the number of persons who are ill, periods or spells of illness, or duration of illnesses (days, weeks, months). Morbidity is also described in terms of frequency or severity. Disability or incapacity rates measure the extent of long-term reduction of a person's capacity to function in society (Box 3.7).

Morbidity data are derived from reported communicable diseases or chronic, genetic, and other conditions for which there are established, recognized reporting systems and registries, usually operated by ministries or departments of health for the population of their jurisdiction. This provides databases for monitoring and providing direction for etiological studies and for priorities and avenues for intervention to control the spread of disease. Morbidity is measured by incidence and prevalence rates, as well as severity and duration, though these are not usually available on routine reporting and may require special investigation. Incidence is more useful for acute conditions, whereas prevalence is more important in measuring chronic disease and assessing the long-term impact of a disease.

Latency is the time period between exposure to a disease-causing agent and the appearance or manifestation of the disease. For an infectious disease, it is called the *incubation period*. A disease may appear clinically days, weeks, months, or even years after exposure to the causative agent, whether it is microbiological, toxic, carcinogenic, or traumatic.

An attack rate is a specific incidence rate expressed as the percentage of the exposed population suffering from the disease. When the population is at risk for a limited period of time, such as during an epidemic, the study period can readily encompass the entire epidemic. The attack rate gives a measure of the extent of the epidemic and may provide information needed to control it. For example, if an epidemic of measles spreads from the initial, or index, cases with an increasing attack rate among the exposed population, a change in vaccination tactics and control measures may be needed in order to avoid rapid spread to other vulnerable groups.

Incidence rates measure the frequency of health-related events in a certain population during a specified time period. The denominator for incidence rates is defined as the "population at risk," in which the studied events may occur. For example, the incidence rate for breast cancer in a certain region will be the number of new cases diagnosed over a 1-year period, divided by the total number of women in that region. An attack rate is the cumulative incidence of infectious cases in a group, observed over a period during an epidemic, either by identification of cases or by sero-epidemiology (Last, 2001).

There are a number of ways to define the denominator for incidence rates:

1. *Ordinary incidence rate* is used when calculating incidence rates in a changing population; for instance, where there is a natural movement in and out of the studied population (due to births, deaths, and migration). In that case, the average size of the population in the specified period is used as the denominator, usually including both the "population at risk" and cases already with the disease (prevalence). Although only the "population at risk" should theoretically be included, such an approximation is often made. The reason for this is that when the condition is relatively

---

Box 3.7 Measures of Frequency of Disease in Population Groups

$$\text{Rate} = \frac{\text{The number of cases in a given time period}}{\text{The population at risk in same time period}} \times N^{th}$$

Where N^{th} = 100, 1000, 10,000, 100,000, or 1,000,000; Period: usually = 1 year; Population: mid-year (July 1) estimate.

Incidence rate defines the rate at which new health-related events occur in a population. The numerator is the number of new events occurring in a defined period (usually 1 year); the denominator is the population at risk of experiencing the event during this period.

Prevalence is the total number of all individuals who have an attribute or a given disease or condition at a point in time or a designated time period. The *prevalence rate* is the number of individuals with the attribute divided by the population at risk, at that point in time (*point prevalence*), or midway through the period (*period prevalence*).

Attack rate defines the cumulative number of cases of a specified disease among the population known to be exposed to that disease over a defined period of time.

Source: Modified from Last, J. M. (ed.). 2001. A *Dictionary of Epidemiology*, 4th ed. New York: Oxford University Press.

rare, the influence that prevalence cases will have over the denominator can be considered negligible. Additionally, the information about prevalence cases is often not available.

2. *Cumulative incidence rate* is usually calculated in longitudinal epidemiologic studies. When a cohort (a group of people), initially free from the disease, is being followed during a certain period of time, incidence cases can be identified as they occur. The sum of those incidence cases is referred to as "cumulative incidence." Here, the denominator only includes the "population at risk"; therefore, cumulative incidence may also be termed *risk* of the condition (Abramson, 2001).

3. *Person-time incidence rate or incidence density* is usually used in follow-up studies in which individuals are "at risk" or may be followed up during different periods of time. In this case, the total number of events is divided by the sum of all subjects' periods at risk, measured, for example, in years, months, or days. In order to calculate the denominator, each individual's "period at risk" must be calculated, measuring the time from the beginning of the follow-up until withdrawal from the study (either due to occurrence of the condition under study, or to "censoring"; i.e., any other reason causing cessation of follow-up).

Prevalence Rates

Unlike incidence (indication of occurrence), prevalence is the measure of the total existing situation of a health-related condition or risk factor, including old and new cases. A prevalence rate measures the proportion of individuals having that condition within a defined population group at or during a specified time. Several measures of prevalence rates exist:

1. *Point prevalence:* the proportion of persons with the condition being studied at a certain point in time is divided by the size of the group or population. Point prevalence is influenced by the incidence rate of the condition, as well as its mean duration up to death or recovery.

2. *Period prevalence:* the proportion of persons who developed the condition before and during the specified period. The denominator includes all the individuals who have or had the condition during the defined period, including those who left, died, or recovered during that period. It allows comparison over time with the same or other population groups. Thus morbidity from a specific condition during 1 year can be compared to previous years, weeks, or months, and between countries or regions in a country.

3. *Lifetime prevalence:* the proportion of persons who have had the condition at any time during their lives; for instance, those who have or had the condition divided by the total population.

The prevalence rate is calculated on the basis of the number of cases and the number of persons exposed, and may be compared to the nonexposed population. Estimation of case prevalence in an exposed population may be underreported if insufficient time has elapsed for a disease with a long latency period. An example of period prevalence is the number of cases of cancer among persons exposed to a carcinogenic agent in the past; for example, mesothelioma cases occurring in a former asbestos worker population over a 30-year latency period following exposure.

Measures of Mortality

A death rate (mortality rate) is the incidence rate that measures the frequency of deaths over a given period of time in a defined population. Mortality rates may be standardized to allow comparability between population groups and may be specific to defined diseases or conditions.

Modern epidemiology originated in studies of mortality derived from the Bills of Mortality (publication of deaths by location and cause) in the United Kingdom by John Graunt in 1662. Mortality data are based on the mandatory reporting of all deaths. A standard national death certificate is vital for public health as it provides basic information needed for demographic and epidemiologic purposes. Box 3.8 presents the data required in a standard death certificate as modified in 2003 in the United States, although the format may vary from country to country.

Death certificates are mandatory in most countries and must be signed by a licensed physician before the body can be buried or cremated and before insurance payment or inheritance can occur. The contents of the death certificate are important because the medically certified cause of death is the basis for mortality statistics. Personal data includes the age, sex, ethnicity, place of residence, and other variables such as occupation and injury. Completeness of reporting, accuracy of diagnosis, and coding of causes of death may limit the conclusions that can be drawn from such data. In practice, however, the data reported in large disease categories are an acceptable guide to actual events.

Analysis of causes of death may take into account more than one diagnosis so as to determine the underlying causes of death such as diabetes. This seems straightforward, but standardization of reporting causes of death is far from simple. Doctors who fill in the form may vary in their perception of diagnosis and the difference between immediate and underlying cause of death. In developing countries, data from death certificates may not be available and determination of leading causes of death may have to be studied by "verbal autopsies" conducted as part of community surveys.

Box 3.8 Data Recorded on United States Standard Certificates of Death, Revised 2003

1. Name
2. Sex
3. Social security number
4. Age at last birthday (<day in hours; <1 year in months)
5. Date of birth
6. Birthplace
7. Residence — state, city, town of residence, street address, apartment no., ZIP code
8. Member of armed forces ever, Y/N
9. Marital status
10. Surviving spouse — maiden name
11. Father's name
12. Mother (maiden name)
13. Informant's name, relationship, mailing address
14–17. Place of death, name of hospital, institution, county, state
18–21. Method and place of disposition, burial/cremation — location
22–23. Funeral director's signature, license number
24–31. Date pronounced dead and time, signature of doctor pronouncing death, license number, date signed

32. Causes of death, and duration
Immediate (final) cause (e.g., pneumonia)
Preceding cause (e.g., chronic ischemic heart disease)
Underlying cause (e.g., diabetes)
Other significant conditions
33–34. Autopsy, Y/N; did findings complete cause of death?
35. Did tobacco use contribute to death?
36. Female — pregnancy-related?
37. Manner of death — natural, accident, suicide, homicide, to be determined
38–41. Date, time, place of injury, work-related
42–44. Describe injury
45–49. Physician's signature, address, license number, date
50. Registrar date filed
51. Education of decedent
52–53. Ethnicity — Hispanic, black, white, American Indian, etc.
54. Decedent's usual occupation
55. Kind of business/industry

Source: http://www.cdc.gov/nchs/data/dvs/DEATH11-03final-acc.pdf [accessed November 23, 2007]

Causes of death recorded on the death certificate include the immediate cause of death (e.g., cardiac arrest); the second and third lines include contributing conditions (e.g., acute myocardial infarction and congestive heart failure); with the fourth line being the underlying cause (e.g., coronary heart disease). The death certificate is filed with a public registry office and forwarded to a vital records office where the causes of death are recorded by a registrar trained to federal standards to interpret and code medical diagnoses, according to the 10th International Classification of Disease (ICD-10), adopted by the World Health Organization (WHO) in 1990.

Overall patterns of mortality are examined by age, sex, and ethnic group, and by cause of death. Mortality trends will be discussed under communicable and noncommunicable disease in Chapters 4 and 5. National mortality trends give vital information on disease and changing epidemiologic patterns, allowing for regional and international comparisons, and help to define health programs and targets (Box 3.9).

Mortality patterns can be studied in a particular year or over time. A cohort is usually a group of persons born in a particular year, but it can be any defined group being followed epidemiologically. Cohorts of persons born in particular years can be followed to observe and compare mortality patterns. With suitable age standardization, the

Box 3.9 Commonly Used Mortality Rates and Ratios

1. *Crude death rate* (CDR) = number of deaths during a specified period divided by the number of persons at risk of dying during the period × 1000 (or other decimal).
2. *Age-specific mortality rate* = number of deaths of persons in the specified age group in an area in a year divided by the average population in that age group in the area in that year × 1000 (or other decimal).
3. *Cause-specific mortality rate* = number of deaths from a specific cause (in an area in a year) divided by the average population (in the area in that year); for example, if the annual number of deaths from lung cancer in a given year is 400 in a population of 1 million, the cause-specific mortality rate is 400/1,000,000 = 40 lung cancer deaths per 100,000 population.
4. *Case fatality rate* (CFR) = number of deaths from a disease during a given period over the number of diagnosed cases of that disease during the same period × 100 (or other decimal); for example, 10 deaths from measles among 5000 cases gives a CFR of 10/5000 × 100 = 0.2%.

Source: Adapted from Last, J. M. (ed.). 2001. A *Dictionary of Epidemiology*, 4th ed. New York: Oxford University Press [with permission].

mortality pattern of men born, for example, in 1900, 1920, 1940, and 1990 can be compared with each other.

Mortality statistics are fundamental to epidemiology and provide some of the most reliable data available. Epidemiologic analysis of mortality data depends on the registration of deaths with basic demographic data and causation of death as recorded by the physician certifying it. Total, age, and sex-specific mortality are usually calculated on an annual basis, with the midyear population as the denominator. This provides crude, age-specific, cause-specific, and proportional mortality rates from which standardized mortality rates are calculated. Case fatality rates (CFRs) relate mortality from a cause to the incidence or prevalence of that disease.

Changes in mortality patterns may occur as a result of a number of factors affecting the outcome of a disease, such as changes in socioeconomic conditions, disease prevention, or methods of treatment. Diagnostic criteria or accuracy of death certificates may also change over time. Thus, a change in mortality may reflect a change in incidence of the disease or CFRs related to treatment methods and access to care, or changes in definition or classification of diseases.

Social Classification

The British Registrar General's Classification of Occupations was established in 1911 and is updated every 10 years (Box 3.10). It is easy to use and provides an excellent demographic and epidemiologic tool that has been used in many studies of disease outcomes. It can help to illustrate the different health experiences of the various social classes, even within the universal National Health Service. It has become part of the database of vital statistics and morbidity patterns in the United Kingdom.

The United States and most other western countries do not have social class data recorded on death certificates and therefore proxy measures of social classification must be used. Ethnicity, national origin, education, or religion are all such proxy measures. In the United States, race is recorded on death certificates and these mortality data can be analyzed by racial groups including American Indian or Alaskan Native, Asian or Pacific Islander, Black, Hispanic, and White. Education level and occupation are also recorded, but mortality data are generally presented by racial group, not social indicators.

The interrelationship between ethnicity and disease or mortality often masks other socioeconomic factors, such as higher levels of poverty or reduced access to medical care among African-American and Hispanic groups in the United States or immigrant groups in European countries. Because there are wide variations in socioeconomic and educational levels within ethnic or racial groups, and many confounding factors in ethnicity or race that might affect disease patterns, analysis of data classified this way should be interpreted carefully.

Social class is increasingly identified as a major variable in health status. It serves as a proxy measure for many health-related issues, such as nutrition, access to care, and dependence on occupations with hazards, with little opportunity for personal development, or lacking security. Social class variations in health status exist even where universal access health systems are well established, even in countries with universal health insurance or services; for example, the United Kingdom, Sweden, and Israel. However, social differences are less pronounced in the Nordic countries where social gaps are generally narrower than in countries with less-developed social welfare systems. This is, however, coming under pressure from immigrant and migrant worker populations which have become significant demographically and in health issues.

SENTINEL EVENTS

Sentinel events are taken as measures of problems in a health care process. They are events, such as avoidable deaths, which should be uncommon if all goes well with preventive and curative care of acceptable standards are in place. Avoidable deaths will vary according to the state of health development of a country, and each country may define its own sentinel events for review. Sentinel events include maternal or surgical deaths, medication errors, or infections occurring in hospital that may jeopardize the health of a patient. Such events occur frequently enough to pose both health risks and an economic burden to the hospital, the insurer, and of course, the patient.

In infectious disease epidemiology, the index case is the first or initial group of cases of a condition that come to attention, providing the first clues in an outbreak or epidemic. In noninfectious conditions, the sentinel event may be a death, where the investigation of the circumstances may help to understand the process of the disease or the

---

Box 3.10 British Registrar General's Classification of Occupation of Head of Household

Class I	Professional and business occupations (e.g., physicians, banker)
Class II	Intermediate occupations (e.g., schoolteacher, storekeeper)
Class III	Non-manual occupations (e.g., clerk)
Class III	Manual skilled occupations (e.g., foreman)
Class IV	Partly skilled occupations (e.g., salesperson, factory worker)
Class V	Unskilled occupations (e.g., porter, waiter)

care that was received. Any case of polio or several epidemiologically linked cases of measles in a country previously free from the disease may be considered sentinel events which should not happen and their investigation may show errors of omission or commission which explain the event and which point to needed remedial action.

Reporting and data systems should be arranged to indicate avoidable deaths from vital records or hospital discharge information systems. Comparison between areas might also include avoidable deaths as a health status indicator. There are selected conditions which are generally preventable or treatable, and therefore warrant investigation when they occur. Maternal deaths (i.e., deaths associated in time and related to pregnancy or the postpartum period), deaths within 24 hours of hospital admission, or deaths following surgery (usually within 48 hours) are examples of sentinel events which are uncommon and should always be investigated. Deaths from appendicitis or appendectomy, tonsillectomy, hysterectomy, tubal ligation, or other elective surgical procedures should be investigated as sentinel avoidable deaths until other explanatory factors are found. Nosocomial (hospital-acquired) infections are a major cause of mortality, increased length of hospital stay, and health care expenditures. This requires an active program of surveillance and prevention within the care setting.

With the advent of newly emerging frightening diseases such as SARS, Ebola, and avian flu, the development of rapid reporting of cases of suspect infectious disease takes on a new urgency. This is even more worrying with the potential for bioterrorism in the twenty-first century, raising specters not seen before in modern public health. Hospital emergency rooms and doctors' offices in the community become front-line monitoring sites for such disease and this depends on current information on possible symptoms or forms of presentation of such disease even in the earliest stages of their development. This is fundamental to detect and localize outbreaks of such highly dangerous infectious diseases.

THE BURDEN OF DISEASE

Burden of disease refers to the combined measurement of mortality and nonfatal health outcomes. The assessment of burden of disease (BOD) serves to design, test, and implement methodologies to aid in setting priorities for effective allocation of health resources. The challenge is to develop valid, reliable, comparable, and comprehensive measures of population health and comparative assessments of the burden of diseases, injuries, and risk factors. This assessment can then be linked with the investigation of costs, efficacy, and effectiveness of major health interventions in order to establish appropriate cost-effectiveness estimations, which should be a major tool in policy design and decision making.

The BOD is an important epidemiologic research instrument. This approach recognizes that social and other factors contribute to diseases which are multifactorial in origin. These estimations, combining economic and epidemiologic data use disability-adjusted life year (DALY) as the unit of measurement of the burden of disease, representing the loss of 1 year of "healthy" life.

YEARS OF POTENTIAL LIFE LOST

Years of potential life lost (YPLL) are calculated based on age-specific rates of mortality or disability. They provide a refinement in epidemiology which has added important new perspectives in the analysis of specific problems. The leading causes of death in the United States, as in most developed countries, are coronary heart disease, cancer, and stroke. However, when the data are examined from the point of view of YPLL, trauma (unintentional injuries, homicides, and suicides) becomes the leading cause of death. YPLL is a better reflection of the impact of diseases on a society than other mortality rates because it is age-related, showing the relative impact of early mortality, which should be taken into account when determining national health priorities. Trends in YPLL for the years 1980–1996 are shown in Table 3.2, showing a large drop in YPLL from total and most specific causes. There was also a substantial decline in YPLL for total and some categories from 1995–1996, especially in HIV, suicide, and homicide deaths.

Qualitative Measures of Morbidity and Mortality

Quality-adjusted life years (QALYs) and DALYs are calculations of morbidity introduced in the international health literature (Box 3.11). They serve as statistical measures of the burden of disease, allowing for international comparisons. Other terms used include disability-free life expectancy (DFLE) and health expectancy — both measures of mortality, morbidity, and impairment or disability. BOD measures are used to assess cost-effectiveness of specific interventions (see Glossary).

The World Bank calculates the variation in BOD between demographic regions, varying from nearly 600 DALYs lost per 1000 population in sub-Saharan African countries compared to approximately 120/1000 in the industrialized countries. These measures are being used in economic analyses of health status, helping to focus on outcome measures to justify resource allocation by comparing benefits in terms of reduced mortality and morbidity.

TABLE 3.2 Years of Potential Life Lost Before Age 75 per 100,000 Population, for Selected Causes of Death, United States, Selected Years, 1980-2004, Age-Adjusted, All Persons

	1980	1990	2000	2004
All causes	10,449	9,086	7,578	7,271
Diseases of heart	2,239	1,618	1,253	1,129
Ischemic heart disease	1,729	1,154	842	721
Cerebrovascular diseases	358	260	223	198
Malignant neoplasms	2,109	2,004	1,674	1,543
Trachea, bronchus, and lung	549	561	443	403
Colorectal	190	165	142	127
Prostate	85	97	64	56
Breast	463	452	333	302
Chronic lower respiratory diseases	169	187	188	174
Influenza and pneumonia	160	142	87	79
Chronic liver disease and cirrhosis	300	197	164	154
Diabetes mellitus	134	156	178	178
Human immunodeficiency virus	—	384	175	143
Unintentional injuries	1,544	1,162	1,027	1,098
Motor vehicle-related injuries	913	716	574	568
Suicide	392	393	335	353
Homicide	426	417	267	265

Source: Health United States, 2007, http://www.cdc.gov/nchs/hus.htm [accessed December 20, 2007]

Note: Data are based on death certificates. Rates rounded.

MEASUREMENT

Epidemiology and public health are dependent on quantitative and qualitative observations to establish relationships and possible points of intervention. Therefore, an appreciation of methods of handling statistics and their interpretation is fundamental. A complete presentation of this field is beyond the scope of this text; however, some general concepts are important to establish.

Research and Survey Methods

The scope and depth of research methods and the many other quantitative and qualitative sciences related to conducting investigations of health and disease in population groups is now an important element of training in public health. It is an area of public health that is basic not only for research but also in reading the literature of a dynamic field such as public health, and in the design of policies, intervention programs, resource allocation, and the management of health systems. Research and surveys are integral parts of public health practice, and especially of academic public health. Familiarity with their basic principles is an important part of the preparation of public health professionals and a responsibility of academic centers training the public health workforce.

A thorough review of the peer-reviewed literature is a prerequisite for development of a study, with skills in use of Internet search engines such as Pubmed and Medline, as well as important sources such as the CDC Atlanta, WHO, and other respected professional bodies. Organized literature reviews are called Cochrane Reviews for leading British epidemiologist Archie Cochrane, using meta-analysis. This is a formal method of review and analysis of multiple studies of a causal relationship of a therapeutic or preventive measure that yields a quantitative aggregate summary of all results. It includes selection of studies of similar design, mostly of randomized control trials, pooling of the data to make a larger sample. This increases the chance that any change and comparison of study and control groups would be statistically significant, but also based on critical analysis and selection of those studies meeting acceptable criteria of methodology as well. A 2007 study reported in *Lancet* on meta-analysis of previous studies showed significant benefit of folic acid supplementation in reducing incidence and severity of stroke, whereas individual studies were equivocal or showed change that was not statistically significant.

The formulation of a study question and its hypothesis includes defining its purposes and objectives. This leads to basic study design, definition and selection of the study population, sample selection, and selection of variables to be measured. A study is dependent on funding and the presentation of the proposal is crucial for success. The study design requires development and testing of survey instruments, organization of the study team, and collection of data. Assessment of reliability and validity of the data is a key part of preparing it for analysis. Training in research methods is thus integral to studies of epidemiology and descriptive and inferential statistics.

Qualitative methods, including quantitative measures used in the social and behavioral sciences based on the social sciences, are also important in public health, with health behavior as a basis for "lifestyle," or personal choices. These are also applied to societal conditions, cultural, socioeconomic and geographic factors, and support systems which are all related to fundamental risk factors for some diseases and their severity, access to health care, and health outcomes. They are also related to organizational systems, management of health systems, economics, and professional interactions.

Box 3.11 Measures of the Burden of Disease

1. **Potential years of life lost (PYLL)** PYLL is a measure of the relative impact of various diseases and lethal forces on society. It highlights the loss to society as a result of youthful or early deaths. The figure for PYLL due to a particular cause is the sum, over all persons dying of that cause, of the years that these persons would have lived had they reached a specified age.

2. **Disability-adjusted life year (DALY)** DALYs are units for measuring the global burden of disease and the effectiveness of health interventions and changes in living conditions. DALYs are calculated as the present value of future years of disability-free life that are lost as a result of premature death or disability occurring in a particular year. DALY is a summary measure of population health, and includes two components: DALY = YLL + YLD, where YLL = years of life lost due to premature mortality; and YLD = years lost due to disability.

The all-inclusive formula for the calculation of DALYs reflects the total amount of healthy life lost incorporating both years lost from premature mortality and healthy life lost from some degree of disability.

3. **Quality-adjusted life year (QALY)** QALYs are an adjustment or reduction of life expectancy reflecting chronic conditions, disability, or handicap, derived from survey, hospital discharge, or other data. Numerical weighting of severity of disability is established on the basis of patient and health professional judgment.

DALYs are calculated by a formula which includes five main components: the duration of time lost due to a death at each age, disability weights, age-weights, time preference (expressed as a discounting function), and the integration of health measures among a population.

For further information on the procedure to calculate DALYs see Human Capital Development and Operations Policy; HCO Working Papers: The DALY Definition, Measurement and Potential Use, http://www.disabilityworld.org/June-July2000/International/DALY.html [accessed April 21, 2008] and http://www.who.int/healthinfo/boddaly/en/ [accessed April 21, 2008]

Sources: Last, J. M. (ed.). 2001. A *Dictionary of Epidemiology*, 4th ed. New York: Oxford University Press [with permission].
Harvard School of Public Health, Burden of Disease Unit, http://www.hsph.harvard.edu/organizations/bdu [accessed April 21, 2008]
WHO, Global Burden Of Disease, http://www.who.int/topics/global_burden_of_disease/en/
Murray, C. J. 1994. Quantifying the burden of disease: The technical basis for disability-adjusted life years. *Bulletin of the World Health Organization*, 72: 429–445.
Murray, C. J., Laakso, T., Shibuya, K., Hill, K., Lopez, A. D. 2007. Can we achieve Millennium Development Goal 4? New analysis of country trends and forecasts of under-5 mortality to 2015. *Lancet*, 370:1040–1054.

In these areas, the applicable social sciences include sociology, psychology, anthropology, political science, organization theory, and information technology. "Social marketing" is based on study of human behavior and how to change it. Public health campaigns against risk factors such as smoking or high-risk sexual behavior depend on such knowledge of awareness, attitudes, behavior, and practices. Qualitative studies are more exploratory and developmental in pursuit of non-numerical aspects of the study question and relate to attitudes, concerns, fears, and social aspects of study questions crucial to success in public health. Examples include studies of teen pregnancies, parental concerns regarding new vaccines, sexual practices such as condom use, interfamilial relationships and their impact on risk behavior and antisocial behavior, and smoking-related issues. Epidemiologic and qualitative studies can be complementary to each other, providing important scientific evidence related to real public health issues of national and international importance.

Interpretation of statistical events requires a familiarity with methods of gathering the basic information and in processing it. Statistics is "the science and art of collecting, summarizing, and analyzing data that are subject to random variation" (Last, 2001). Biostatistics is the application of statistics to biological problems.

Variables

A variable is any factor being studied which is considered to affect health status and which can be measured. It may be an attribute, phenomenon, or event that can have different values, such as age, sex, socioeconomic status (SES), behavior, other disease conditions, characteristics of the health care system, or exposure to a toxic or infectious agent. A dependent variable is the outcome being studied. An independent variable is the characteristic being observed or measured which is hypothesized to cause or contribute to the event or outcome being studied, but is not itself influenced by that event. For instance, in the study of the association between smoking and coronary heart disease, smoking (described as the average number of cigarettes smoked per day, for example) is the independent variable, or the exposure. Coronary heart disease is the dependent variable, or the outcome.

The Null Hypothesis

The null hypothesis is the assumption that one variable has no association with another variable, and that two or more populations being studied do not differ from one another. A statistical test is used to decide whether the null hypothesis may be rejected or accepted; that is, the probability that any differences observed may be due to chance alone and not indicative of a real difference. If the probability of chance alone explaining the observed differences is very low then the null hypothesis may be rejected, suggesting that the studied association or difference may actually exist. The definition of the threshold for "low probability" depends upon the decision of the level of significance required. Statistical testing thus provides the basis for inference or decisions regarding the results of a study as statistically significant and to what degree.

Confounders

A confounding variable (confounder) is a factor other than the one being studied that is associated both with the disease (dependent variable) and with the factor being studied (independent variable). A confounding variable may distort or mask the effects of another variable on the disease in question. For example, a hypothesis that coffee drinkers have more heart disease than non-coffee drinkers may be influenced by another factor (Figure 3.5). Coffee drinkers may smoke more cigarettes than non-coffee drinkers, so smoking is a confounding variable in the study of the association between coffee drinking and heart disease. The increase in heart disease may be due to the smoking and not the coffee.

We are often limited to observational studies for evidence of causal relations. Experimental studies may not be possible for many technical, ethical, financial, or other reasons. The proper causal interpretation of the relations from carefully developed epidemiologic studies is vital to the development of effective measures of prevention.

Sampling

The majority of epidemiologic studies cannot collect information about every individual in the target population (the general population of a country or a region, or a defined group of people). Therefore, a sample, which is chosen from that target population, is defined and used as the study population, for which all the required information is collected. The appropriate choice of the study population is crucial

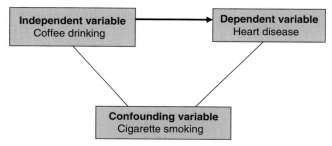

FIGURE 3.5 Independent, dependent, and confounding variables in a study.

to ensure that the results obtained from the study can later be generalized to the general population. Thus a sample must be selected randomly, representative of the general population, and of sufficient size as to increase the likelihood (or probability) that the results obtained from the sample are close enough to the actual situation in the general population (i.e., where the level of significance of statistical test is acceptable) (see Box 3.12).

Randomization

The main sampling methods are described as follows.

Simple random — when all individuals in the population have an equal chance of being selected, the group is known as a *random sample*. This is often done through assigning each person in the group a number and then selecting the sample from a table of random numbers until the desired sample size is reached.

Systematic randomization — every *n*th unit is selected.

Stratified randomization — the population is divided into strata (subgroups) and simple randomization is applied within subgroups. For example, if 20 percent of the population is in the age group 40–59 and 20 percent of the sample comes from this age group, and similarly for other age groups, then all strata are fairly represented with regard to numbers of persons in the sample.

Cluster sampling — The population is non-randomly divided into subgroups (such as households, schools in a city, or classes in a school) and clusters (subgroups) are randomly selected.

Multistep sampling — groups are randomly selected, and then individuals within groups are chosen.

A non-random sample is one in which a form of bias is introduced into the sampling process. For example, a convenience sample is a group of persons who are readily accessible, such as volunteer blood donors or people who appear at a health fair for blood pressure examination. The bias in such samples is that there may be a self-selection process not representative of the total population. A selection of a group at special risk, for example, might entail choosing districts with known low immunization coverage in order to attempt to determine the reasons for this. Such a study would then be applicable to those districts and, although not generalizable to the total population, could provide valuable information affecting the immunization program.

Conclusions based on sample results may be attributable to the population from which the sample is taken. Extrapolation to the total population or a different population is a judgment, justified, but must be qualified by description of the sampling methods used and the potential biases. Despite these limitations, careful sampling is essential for assessing a particular characteristic in a larger population and should give results which are reproducible by other investigators.

Box 3.12 Sample Size

Principles:

1. If samples are drawn from a population, the larger the number of samples and their sizes, the higher the probability that their average value (of the parameter under study) is equal to the value in the population.

2. Because sample sizes are often limited, sampling error must be taken into account (i.e., the probable difference between the value in the sample and in the population).

3. The size of the sampling error is affected by the size of the sample drawn. Increasing the sample size decreases the size of the sampling error.

The principles of sampling rest on the assumption that samples are randomly obtained.

Factors in the calculation of sample size:

Type 1 error The risk of a false positive result (α) (i.e., the chance of detecting a statistically significant difference when there is no real difference).

Type 2 error The risk of a false negative result (β) (i.e., the likelihood of not detecting a significant difference when there really is a difference that is greater than the specified threshold).

The power of a study is its ability to demonstrate an association if one exists. It is determined by several factors, including the frequency of the condition under study, the magnitude of the effect, the study design, and sample size. It is defined as the chance of not getting a false negative result and is equal to 1 − β (type 2 error).

The calculation of sample size is beyond the scope of this text and is found in free computer programs such as Win Pepi (J. H. Abramson, June 24, 2007) available at http://www.brixtonhealth.com/pepi4windows.html and Epi Info (Centers for Disease Control, April 30, 2007) at http://www.cdc.gov/epiinfo/.

Adapted from Last, J. M. (ed.). 2001. A *Dictionary of Epidemiology*, 4th ed. New York: Oxford University Press [with permission].

NORMAL DISTRIBUTION

The evaluation of certain characteristics in a population group is based on the assumption of normal distribution (nutrition, height, weight). A normal distribution is a continuous, symmetrical, bell-shaped frequency distribution of observations (Figure 3.6). A normal distribution has upper and lower values that may extend to infinity, but it has an arithmetic mean, mode, and median from a central point.

Mean, median, and mode are measures of central tendency in a group of numbers (Box 3.13). The symmetrical bell-shaped (Gaussian) curve represents the normal distribution of biological characteristics, such as heart rate, height, weight, or blood pressure in a normal population group. In such a distribution, approximately two-thirds of the observations fall within one standard deviation and approximately 95 percent fall within two standard deviations of the mean.

Normality may be defined in several senses. It is a range of variation in a given population, within two standard deviations below and above the mean, or between specified percentiles, for example, the 10th and 90th of the distribution. *Normally* also refers to the limits of a range of a test or measurement and is an indication of the finding being conducive to good health.

Deciding when a group of observations is "normal" or "abnormal" requires defining cut-off points, both in clinical medicine and in epidemiology. In clinical medicine, deciding what is a normal blood pressure, cholesterol level, or growth of a child is based on norms determined from a large number of observations of what is assumed to be a "normal" population. For example, growth patterns of children used as an international standard are based on data derived from a white, middle-class American population (see Chapter 6).

STANDARDIZATION OF RATES

Standardization of rates is important in comparing data between populations of different age and sex distribution and to remove, as far as possible, the effects of confounders in epidemiologic studies. Comparing mortality

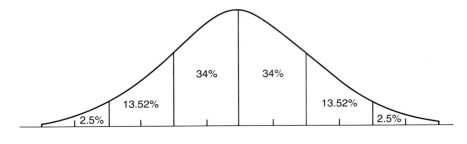

FIGURE 3.6 Normal distribution. Source: Last, J. M. (ed.). 2001. *A Dictionary of Epidemiology*, 4th ed. New York: Oxford University Press, p. 114.

Box 3.13 Summarizing a Group of Numbers

1. **Mean** The average value of the observations (i.e., the sum of values of the observations divided by the number of observations).
2. **Median** The midpoint value where half of the observations are equal or less, and half are equal or greater. It is the middle observation when a set of observation numbers is arranged in increasing values.
3. **Mode** The most frequently occurring value in a set of observations. In a normal distribution, the mean, median, and mode are all equal to one another.
4. **Standard deviation** The common method of summary of how widely spread or dispersed the observed values are from the mean of the observations.
5. **Confidence intervals** The range or interval within which the true value of a variable, such as mean, proportion, or rate, lies at a specified degree of probability (e.g., 95% or 99%). This indicates how precisely the results of an analysis based on a sample approaches the true value of the rate in the population which the sample is meant to represent.

Sources: Adapted from Last, J. M. (ed.). 2001. A *Dictionary of Epidemiology*, 4th ed. New York: Oxford University Press [with permission].
Last, J. M. 2007. A *Dictionary of Public Health*. Oxford University Press.

---

rates in one country, for example, will require using a standard population such as that of the United States in 1940 to compare 1940 mortality with subsequent rates, thus removing the effects of changes in the age and sex composition of the populations. Without standardization of the population, the age–gender changes would act as confounders when describing distributions or comparing mortality or disease incidence between two or more designated groups.

Standardization uses a "standard population" selected to adjust for differences in the distribution of the relevant variables between groups being compared, or between the sample used in a study and the population it was chosen from. The standard population in this procedure is one in which the age and sex composition is known and therefore is used as a benchmark in order to compare rates for a number of different population groups. Comparisons between different states in the United States or countries in Europe would use a U.S., European, or world population distribution. Standardization can be done by direct or indirect methods.

It is important to note that although age and sex distributions frequently act as confounders, many other variables may affect the outcome being measured, depending on the study. For example, smoking and coffee drinking may act as confounders when studying the association between physical activity and coronary heart disease. In case such a confounding effect is present and identified, then the study analysis must control for the confounder in order to correctly assess the main study variables.

Direct Method of Standardization

The direct method of standardization is used when age–gender–specific mortality rates are known for the populations being compared to a standard population. These rates are then applied to the standard population to calculate the *expected numbers of deaths for each group* in the population, as if its composition (with respect to the variable we standardize for) were the same as in the standard population. They are then summed and divided by the total standard population to give a summary adjusted rate. Standardized death rates can be calculated for particular diseases. For example, if one is comparing lung cancer death rates in a number of countries to see if there are differences that might be attributed to external factors such as air pollution patterns, the data for each city can be compared by using standardized (cause-specific) mortality rates.

The direct standardization of rates is an important method of comparing mortality patterns between cities, districts, regions, and countries. Table 3.3 shows the important differences seen in presenting data with age standardization, where a country such as Egypt, with a low crude death rate from circulatory diseases because of the youthfulness of the population but very high age-specific rates. Age-specific rates are also useful for comparative purposes.

Indirect Method of Standardization

When age–gender–specific mortality rates for the study population are not available or if the numbers in some

TABLE 3.3 Crude and Age-Standardized Mortality Rates (per 100,000 Population) for Diseases of the Circulatory System in Selected Countries, 1980

Country	Crude rate	Standardized rate, all ages	Age-specific rate 45–54 years	Age-specific rate 55–64 years
Finland	491	277	204	631
New Zealand	369	254	184	599
France	368	164	97	266
Japan	247	154	95	227
Egypt	192	299	301	790
Venezuela	115	219	177	497
Mexico	95	163	132	327

Source: Beaglehole, R., Bonita, R., Kjellstrom, T. 1993. *Basic Epidemiology*. Geneva: World Health Organization, page 25; calculated from data in World Health Statistics Annual, 1986.

Box 3.14 Standardized Mortality Rates (SMRs) and Standardized Incidence Ratios (SIRs)

$$\text{SMR (SIR)} = O/E = \frac{\text{Observed deaths (cases)}}{\text{Expected deaths (cases)} \times 100}$$

TABLE 3.4 Mesothelioma Deaths Among Former Asbestos Workers in Israel, 1950–1990

Study group (*n*)	4,401
No. of mesothelioma deaths in study group	26
Expected deaths from national population rates	0.12
Standardized mortality rate (SMR)	26/0.12 = 216.7

Note: Expected deaths derived from applying age-specific mesothelioma mortality rates of total population of Israel to the study group.
Source: Tulchinsky, T. H., Ginsberg, G. M., et al. Mesothelioma mortality among former asbestos-cement workers in Israel, 1953-1990. *Israel Journal of Medical Sciences*, 1992;28:542–547.

---

age groups are too small, the indirect method of standardization is used. This method uses known age–gender–specific rates from a standard population to calculate the expected number of the same health event for the population being studied, given that population's distribution (Box 3.14). The expected number of deaths or cases thus calculated is then compared to the actually observed number of deaths or cases. The ratio of observed to expected is then multiplied by 100 (or another decimal multiplier) to give the standardized mortality ratio (SMR), which now shows the comparisons free from confounding factors such as different age distributions.

The SMR thus allows for comparison of one national, regional, or other defined population group where the age–gender–specific rates are not available, to a selected standard population for which these specific rates are known. This same method is also used to calculate morbidity as standardized incidence ratios (SIRs) or other health-related observations.

Comparing mortality or morbidity rates in European countries is made accessible to all by the Health for All Database. It compares rates (mortality, morbidity, health resources, utilization, lifestyle, and other). Data for all countries are standardized to the European population standard, so the reported rates are comparable.

Standardized mortality (incidence) ratios (SMRs or SIRs) are therefore the crude rate or the total number of deaths or cases occurring in the study group, compared to the expected number of deaths if that population had experienced the same death (or incidence) rates as the standard population. The standard population provides a strong base of comparison as it is larger in size, with less likelihood of random variation.

Standardized mortality ratios can be calculated for a specific population group at special risk and compared to a standard population to see if it is vulnerable to higher rates. A group of persons who have been employed in a certain industry and exposed to asbestos may after a long latency period develop mesothelioma. The SMR for a population of former asbestos workers in a 25-year follow-up study is seen in Table 3.4.

In the United Kingdom, the SMR is used as the adjustment factor for allocation of funds to district health authorities. Following a lengthy examination of many alternatives, the SMR was believed to incorporate many variables affecting health, including age, sex, socioeconomic, and environmental factors. Areas with higher than expected

mortality may have more disease or higher case fatality rates resulting from a greater prevalence of risk factors (genetic, environmental, and/or socioeconomic). Excess mortality may also be due to less access to or poorer quality of health care. Extra resources are made available on this basis to deal with the poorer health status of the population. This is a practical method of addressing regional differences in health, providing a high degree of equity in resource allocation. It takes into account greater need in some areas than in others. The SMR applies epidemiologic methods to improve management practice in health.

POTENTIAL ERRORS IN MEASUREMENT

Data must be assessed as to its validity and reliability. It should also be considered for its biological plausibility (Box 3.15). These all affect the degree to which inferences can be made and generalizations drawn from the study sample.

Reliability

Reproducibility or reliability is the degree of stability of the data when the measurement is repeated under similar conditions. If the findings of two researchers carrying out the same test (such as the measurement of blood pressure) are very close, the observations show a high degree of interobserver reproducibility. However, it is common in medicine that standardization of even relatively objective measurements by different observers, such as radiologists reading the same x-rays or cardiologists reading the same cardiograms, show high degrees of variability. Instrument standardization, observer training in common standards, and standardization of recording observations are needed to ensure acceptable standards of reliability in any data set. Measuring the same patient at different times can produce different results (as in measuring blood pressure or blood sugar), such that standardization of

Box 3.15 Observation Measurement Issues in Epidemiology

1. **Validity** degree to which a measure actually measures what it claims to measure.
2. **Accuracy** extent to which a measure conforms to or agrees with the true value.
3. **Precision** quality of being sharply defined.
4. **Reliability, reproducibility** stability seen when a measure is repeated under similar conditions.
5. **Instrumental error** includes all sources of variation inherent in the test itself.
6. **Digit preference** consistent bias by observer rounding of numbers (e.g., to the nearest whole number).
7. **Interobserver variation** differences in observation between different observers of the same phenomenon.
8. **Individual observer variation** when the same observer records the same observation differently due to changes within the observer, not the observed.
9. **Bias** an effect or inference that departs systematically from the true value.
10. **Spurious** an apparent but not genuine epidemiologic relationship.

Adapted from Last, J. M. (ed.). 2001. A *Dictionary of Epidemiology*, 4th ed. New York: Oxford University Press [with permission].

Box 3.16 Sources of Bias

The reliability of a data set may be compromised by systematic biases in the data collection or processing. Such biases include the following:

1. **Assumption bias** errors from faulty logic, premises, or assumptions on which the study is based.
2. **Response bias** systematic error due to differences between those who choose or volunteer for a study as compared to those who do not.
3. **Selection bias** error due to inclusion of those who appear and are included in a study, leaving out those who did not arrive because they had died, were cured without care, were not interested, and so forth.
4. **Sampling bias** error when sampling methodology does not ensure that all members of the reference population have a known and equal chance of being selected for the sample.
5. **Observer bias** error due to differences between observers; may be between observers (interobserver) or by the same observer on different occasions (intraobserver).
6. **Detection bias** systematic error due to faulty methods of diagnosis or verification of cases in a survey.
7. **Design bias** systematic bias due to faulty design of the study.
8. **Information bias** flaws in measuring exposure or outcome resulting in data being not comparable.
9. **Measuring instrument bias** faulty calibration, inaccurate measuring instruments, contaminated reagents, incorrect dilutions/mixing of reagents, flawed questionnaire.
10. **Interviewer bias** conscious or subconscious selection in gathering of data.
11. **Reporting bias** self-report selective reporting, suppressing, or exaggerating of information; for example, history of STIs.
12. **Publication bias** editors prefer positive results so that a distorted perception of an issue may occur.
13. **Bias due to withdrawals** loss of cases from the sample by withdrawal or nonappearance in follow-up.
14. **Ascertainment bias** error due to the type of patients seen by the observer, or in the diagnostic process affected by the culture, customs, or idiosyncrasies of the provider of care.

Source: Adapted from Last, J. M. (ed.). 2001. A *Dictionary of Epidemiology*, 4th ed. New York: Oxford University Press [with permission].

conditions of recording or timing the test are essential to ensure comparable data. Standardization of test requires, as part of quality control, sending samples tested in one laboratory to a reference laboratory to see if the test results are the same. It is important to minimize sources of bias (Box 3.16).

Validity

Validity refers to the degree that a measurement actually measures what it claims to measure. This includes the representativeness of the sample and the nature of the population from which the sample is taken. It includes the nature of the phenomenon being tested and whether the sampling method takes it into account, such as when a condition changes with age, does the sample take that into account, or whether the content of the testing, such as a questionnaire, truly reflects the nature of the phenomenon being studied. The generalizability of a set of findings from a study using white middle-class males or U.S. nurses as subjects may not be applicable to females or males of other ethnic or SES, or populations with different sociocultural environments.

SCREENING FOR DISEASE

Screening for disease may be carried out on a mass basis of a whole population, as was commonly done in the past

for tuberculosis. When done with a number of tests it is called *multiphasic screening*. Screening may target a group at special risk, such as blood lead screening among workers exposed to lead at their place of work or on children living in the vicinity of a plant using lead.

Screening is an essential part of patient care when the caregiver routinely tests, for example, for blood pressure, blood sugar, or cholesterol. Accuracy of a test is usually measured in terms of sensitivity and specificity. Targeted screening may be required by law as in the case of newborn screening for phenylketonuria (PKU), hypothyroidism, and other congenital disorders. The value of the screening test is defined as to its degree of sensitivity and specificity, as well as its costs and benefits for screening or not screening.

Sensitivity is the proportion of truly diseased persons in the screened population who are identified as such by a screening test. This is sometimes called the *true positive rate*. Specificity is the proportion of truly nondiseased persons who are identified as not having the disease; that is, it measures the probability of correctly identifying a nondiseased person with a screening test, or the *true negative rate*. A test which produces too many false positives or false negatives is not valid (Box 3.17).

False negatives occur when a negative laboratory result appears in a person who has the condition for which the test is being conducted. The condition is present but does not show up on the initial screening test or data set. If screening for PKU is done too soon after birth, some cases may be missed and will only appear later. False negatives can compromise the effectiveness of the screening program.

False positive results are those cases in which a positive laboratory result occurs in a person without the condition for which the test is being conducted. Not everyone with an isolated elevated reading of blood pressure has true hypertension. False positive results must be checked because they cannot be excluded without confirmation by more specific testing, such as repeated blood pressure readings. Precision is the quality of sharp definition of the test. If a laboratory test for environmental contamination is accurate to parts per billion as compared to parts per million, then the precision is enhanced.

Screening for disease and risk factors is a common and necessary part of public health. In order to be valuable, screening requires a valid test and a significant condition with a high prevalence in the population. Screening for breast cancer, carcinoma of the cervix, and many other conditions is part of the armamentarium of public health and contributes to lowering mortality and improving survival rates for these diseases. Screening of newborns is important for a condition which is serious and treatable but uncommon as in PKU and congenital hypothyroidism which is more common and treatable (Chapter 6). PKU is a manageable condition with a strict diet to prevent serious consequences of the abnormal biochemical condition. Screening for these and other birth disorders, cancer of the cervix, cancer of the colon, and many other conditions is now accepted in standard clinical guidelines, while screening for breast cancer is recommended but is under review as to its cost-effectiveness.

EPIDEMIOLOGIC STUDIES

Epidemiologic methods of study are important, not only to define the extent of disease in the population, but also to look for specific risk or causal factors for the disease. Epidemiologic studies permit analysis of a risk factor, variable, or an intervention (such as a new vaccine or drug). This permits testing of new hypotheses and innovations in medicine and public health.

Epidemiologic studies are classified as observational or experimental (Figure 3.7). The observational study allows nature to take its course, with no intervention, as opposed to experimental studies, involving interventions.

Observational Studies

Observational studies are those where the population is studied, but nature is allowed to take its course. They may be descriptive or analytical. Descriptive studies are limited to describing the occurrence of a disease in a population, which is often the first step in investigation, as it may provide clues for more in-depth investigation. Analytical studies go further by looking for specific variables which may be causally associated with the disease.

Descriptive Epidemiology

Descriptive epidemiology uses observational studies of the distribution of disease in terms of person, place, and time. The study describes the distribution of a set of variables, without regard to causal or other hypotheses. Personal factors include age, sex, SES, educational level, ethnicity, and occupation. The place of occurrence can be defined by natural or political boundaries, and can also include such variables as location of residence, work, school, or

---

Box 3.17 Screening Tests: Validity, Sensitivity, and Specificity

Screening test	Disease present	Disease absent	Total
Test positive	True positive (A)	False positive (B)	A+B
Test negative	False negative (C)	True negative (D)	C+D
Total	A+B	B+D	A+B+C+D

Note: Sensitivity = TP/TP + FN
Specificity = TN/TN + FP
Positive predictive value = TP/TP + FP
Negative predictive value = TN/TN + FN

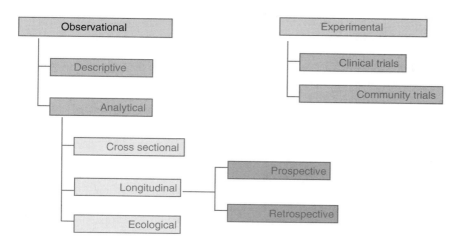

FIGURE 3.7 Classification of epidemiologic studies. Source: Abramson, J. H. Abramson, Z. H. 1999. *Survey Methods in Community Medicine*, 5th ed. Edinburgh: Churchill Livingstone.

recreation. Time factors include time trends, which are generally divided into three types:

1. *Secular trends:* long-term variations;
2. *Cyclic changes:* periodic fluctuations on an annual or other basis;
3. *Short-term fluctuations:* as is seen in epidemic disease outbreaks.

Time trends contribute to understanding the natural history of epidemics of acute infectious diseases such as measles or waterborne disease, as well as noncommunicable diseases such as stroke or cancer. The study of a "natural experiment" when a public health situation is occurring may provide valuable experience and hypothesis for further investigation. Epidemiology also examines the frequency and distribution of potential health indicators and health-related events (such as smoking).

Analytical Studies

Analytical studies are concerned with establishing causes or contributory risk factors to disease, including social, economic, psychological, or political conditions that impinge on health. This helps to define programs to intervene in order to reduce the burden of disease in the population. Analytical epidemiology has made vital contributions to modern medicine through identification of key risk factors, such as higher rates of lung cancer among smokers and higher rates of stroke among people with hypertension. Analytical studies may include cross-sectional (or prevalence studies), as well as retrospective, prospective, and ecological studies.

Analytical studies may be individual or group-based studies. Individual-based epidemiologic studies collect information about individuals (both the exposure and the outcome status should be known for each individual within the study). An ecological study is one in which the units of analysis are populations or groups of a population rather than individuals.

Ecological Studies

Ecological studies, also known as *group-based studies,* compare the mean (or summary) values of exposures and outcomes of different population groups. For example, a study analyzing the association between the gross domestic product of different countries and the prevalence of malnutrition in those countries is an ecological study. However, conclusions from ecological studies should be drawn carefully, as the mean values may not be truly representative of the actual situation, and further, because exposures and outcomes for individuals are not established. If a group has high prevalence of an exposure (e.g., oral contraceptive use among women) and an outcome (e.g., prevalence of heart disease), we still don't know whether the individuals with a positive exposure status are also those with a positive outcome. Drawing a conclusion from this apparent relationship is a bias termed an *ecological fallacy.* The association between aggregated variables based on group characteristics does not necessarily represent the association at the individual level. However, ecological studies are important for population health monitoring and for generation of hypotheses for further study.

Studies showing an apparent correlation between quality of drinking water and mortality rates from heart disease have not been substantiated as indicating a "cause–effect" relationship. It would be an inappropriate conclusion (ecological fallacy) to infer from this finding alone that exposure to water of a particular level of hardness necessarily influences an individual's chances of developing or dying of heart disease.

Ecological studies can be valuable in generating hypotheses for further investigation and intervention. For example, comparison of SMRs for disease categories from routine mortality sets can identify regions with high rates of a specific disease, such as lung cancer or diabetes-related conditions, or motor vehicle accidents, which require follow-up, investigation, and possibly intervention even before more complete epidemiologic studies can be carried out. Studies have shown higher rates of cardiovascular disease mortality for African-Americans compared to whites in the United States. However, further analysis shows that there are gradients for cardiovascular mortality for both whites and African-Americans according to median family income, such that SES emerges as a more important factor than race.

Cross-Sectional Studies or Prevalence Studies

These examine the relationship between specific diseases and health-related factors as they exist in individuals in a population at a particular time. The population may then be divided into subgroups, with and without the disease, and the characteristics of each member of each group analyzed for different variables; for example, age, sex, region of residence, occupation, and social class. Comparisons of these variables may indicate a higher risk for disease in one population group as compared to an otherwise similar comparable population.

This type of study is relatively simple and easy to perform. However, it has some serious drawbacks resulting from the simultaneous measurement of both exposure and outcome. When investigating two variables (a presumed exposure and presumed outcome) it may be impossible to determine which one is the exposure and which is the outcome, as we have no information about a time relationship. For example, a cross-sectional study of body mass index (BMI) and blood pressure may find that high BMI correlates with high blood pressure, but will not be able to indicate whether persons with high BMI had an increase in blood pressure or if persons with high blood pressure became fatter. A cross-sectional study may fail to produce valuable information where the main studied exposure or outcome is only present during a short period of time. If the exposure is short, the recovery from the outcome condition is rapid, or its case-fatality is high, it is unlikely that their assessment at one point in time will actually reflect all the exposures and outcomes.

Case–Control Studies

Case–control studies are observational studies of persons with the disease (or other outcome variable of interest) and a suitable control group of persons without the disease. These studies are retrospective, taking a known outcome status (e.g., disease status) and looking at the exposure status. They compare two similar population groups for their exposure outcomes, one with the disease or condition and the other without. An example is the study of the occurrence of a serious upper limb defect (phocomelia) in children born in Germany in the late 1950s, which showed that of those born with this defect, 41 out of 46 mothers had taken the medication thalidomide as an antinausea pill promoted for use during pregnancy, whereas none of the 300 control mothers with normal children had done so. This study was confirmed by studies in other European countries which had licensed thalidomide, which led to stopping approval for this drug by the FDA in the United States, and later to its banning in countries where it was already in use. Case–control studies are defined as retrospective (defining the outcome status and then looking at the exposure).

The odds ratio is commonly used to summarize findings of case–control studies. It is a ratio of the odds of exposure among cases to the odds of exposure among controls. Case–control studies may be vital to define the differences between the sick and the control groups in an epidemic or outbreak situation.

Case-control studies are ideal for the study of rare disease or conditions which are slow to evolve, as they permit assembling a group of cases of appropriate size for analysis, without requiring an extremely large study population. This presents an important advantage as it reduces the cost and time necessary for the study of such conditions. However, a case–control study is prone to various sources of bias, notably "recall bias," where persons with/without a studied outcome may tend to better remember their exposure status. For example, a study of environmental exposures during pregnancy and fetus malformation may reveal a higher proportion of exposures among women who had an affected fetus because they were more aware and recalled all potential hazards that may have caused the severe adverse outcome of their pregnancies.

Cohort Studies

Cohort studies are also referred to as prospective, longitudinal, or follow-up studies. They examine a population which is initially free from the disease, dividing the population into subgroups according to exposure to a potential risk factor. Such studies can yield the magnitude of risk or incidence rates of the disease under study. The relative risk (a ratio of risk of disease in the two groups; that is, exposed and nonexposed) can then be calculated. Where risk cannot be determined, the rates of disease for each group (exposed and nonexposed) is determined and the rate ratio may be determined for comparison of risk.

The main disadvantages of cohort studies are the follow-up time they require (when persons may leave the study or be lost to follow-up, an important source of bias) as well as relatively large study populations needed in order to assure the appearance of a sufficient number of cases for analysis. Long follow-up and large samples usually imply high costs, and make cohort studies less suitable for the investigation of rare diseases or conditions which develop slowly. On the other hand, cohort studies present many advantages in terms of the reliability of the information collected, as all exposures may be assessed by the investigators at the beginning of the study and outcomes are identified as they appear during the study period so that there is no doubt about time relationships.

Cohort (prospective) studies permit the observation of many outcomes from long-term follow-up of a selected population to ascertain morbidity and mortality data not readily available in the general population reporting systems. The British doctors' smoking habits study initiated by Richard Doll and Bradford Hill was carried on from 1951–2001 showing the harmful effects of smoking in terms of lung cancer, coronary heart disease, and early mortality. The Framingham Study initiated in 1949 has provided a wealth of epidemiologic information of risk factors for cardiovascular disease in the population of Framingham, Massachusetts (see Chapter 5). Many epidemiologic prospective studies followed selected population groups such as the nurses study by Walter Willett at the Harvard School of Public Health, providing a major source of new information on the health of women. This is the largest cohort study of women, tracking over 120,000 nurses since 1976.

Retrospective or historical cohort analysis looks back at earlier records of groups with a specific disease and their earlier life experience. This includes factors such as smoking, birth weight, obesity, hypertension, or exposure to toxic substances such as asbestos in relation to current morbidity and mortality from lung cancer, coronary heart disease, diabetes, and mesothelioma.

In summary, observational studies of particular population groups have provided important public health advances over the past 50 years. A natural experiment is a situation in which naturally occurring circumstances result in two similar population groups, one exposed to a supposed causal factor and one not exposed as a study or control group. The term *natural experiment* is derived from John Snow's 1850 study of Londoners exposed to drinking water supplied by two different water companies, one group having high rates of cholera and the other low rates. This term is currently used in investigating epidemiologic events, regarding each event as a unique situation for which relevant factors need to be defined and, to the extent possible, linked to the disease.

Experimental Epidemiology

Experimental studies are studies of conditions under the direct control of the investigator, conducted as closely as possible to a laboratory experiment. Experimental epidemiology involves changing a variable and measuring the effect in one or more population groups. Clinical epidemiology applies experimental epidemiologic research methods to clinical problems and practice. It includes promoting the use of epidemiologic knowledge in the clinical care of individual patients. Clinical epidemiology also contributes knowledge to the planning and operation of health care systems and clinical and community trials.

Controlled Trials

Controlled trials are epidemiologic experiments designed to study an intervention (preventive or therapeutic). It requires a random method of allocating the cases to the experimental or the control groups, and then both are observed for change over time in relation to the condition being studied. If the people in both the test and control groups do not know which group they are in, the study is called *blind*. If in addition the people judging the outcome are also not aware whether the person tested is in the test or control group, the trial is called *double blind*. Further, if those analyzing the data also do not know who was in each group, the study may be called *triple blind*. This helps to avoid various biases which limit the value of a study. If the difference in outcomes is statistically significant for the control group and the treatment group, then the treatment is deemed to have been effective. Assignment to the treatment or the control group is by random selection.

While randomized controlled trials (RCTs) are considered the gold standard in clinical epidemiology, they are often not available for important policy issues and would be unethical to conduct because denying the benefits of a known positive intervention would be ethically unacceptable. They are also often difficult for policymaking generalization because of inherent limitations in the methodological limitations and resources available for the study.

Field Trials

Field trials follow people who are disease-free in two groups, one with and one without a specific intervention, to determine if the intervention affects the risk of developing the disease. This is often used to test a new vaccine in a susceptible population. The field trial conducted by Jonas Salk of inactivated poliomyelitis vaccine in 1956 demonstrated its protective effect and safety in some 1.5 million American children, and was subsequently adopted throughout the world. Field trials are part of the process of approval for new vaccines and medications.

Community Trials

Community trials are conducted on whole communities measuring the effect of a risk factor or intervention. This cannot easily be randomized because the entire community is selected, and it may not be possible to isolate the community from changes going on in the general population. Community-based heart disease prevention programs have been undertaken in many settings, such as in North Karelia, Finland, and others in the United States such as the Minnesota Heart Health Project, Pawtucket Heart Health Project in Rhode Island, (CHAD project) in Kiryat Yovel, Jerusalem, and many others. These are difficult to evaluate, with a conflict between experimental design and community realities. Regional programs for prevention of heart disease cannot be isolated from time trends in the surrounding communities, limiting the interpretation of outcomes measured. Nevertheless, community trials are necessary in evaluating health interventions to reduce risks or adverse health outcomes. They often rely on performance or utilization indicators as proxies. For example, a village health worker program may increase earlier and more frequent use of prenatal care or immunization coverage, but measurement of outcome variables may be hard to establish in field conditions mainly because of lack of reliable data.

ESTABLISHING CAUSAL RELATIONSHIPS

Classically, the search for causation in medicine and in public health is for the agent–host–vector relationship, with the agent being a specific causative organism. In infectious disease epidemiology, this has provided the scientific basis for immunology and control of vaccine-preventable diseases, and for sanitation to prevent transmission of food- and waterborne diseases. Criteria for attributing causation for communicable disease were established in the nineteenth century by Jacob Henle and Robert Koch (Box 3.18).

Criteria for causation include strength of the association, biological plausibility, consistency with other investigations, and dose–response relationship. Biological plausibility is a test of the plausibility of a causal association based on existing biological or medical knowledge. Consistency with other investigations means that the findings are similar to those of other studies. The dose–response relationship is that in which a change in

amount, intensity, or duration of exposure is associated with a change (increase or decrease) in a specified outcome.

Even in infectious disease control, the public health reality is often more complex than the single causation model. Tuberculosis deaths fell during the nineteenth century, presumably due to improved nutrition and living conditions, and were further reduced in the early part of the twentieth century before the antibiotic era by a combination of improved nutrition and symptomatic treatment. Mortality from measles dropped dramatically despite its endemicity (the continuing presence of a disease in a given geographic area) prior to the successful vaccine introduced in the 1960s. This can be attributed to rising standards of living and improved means of treatment of complications. Even today, the mortality rate from measles is seen to be affected by improving the nutrition of children and by vitamin A supplementation.

For noncommunicable diseases, causation is even more clearly multifactorial, and a risk factor for one disease may also be a contributor to increased risk for another disease. Diet has been established as a major risk factor for coronary heart disease, as well as diabetes and hypertension. Diabetes is a major risk factor for coronary heart disease, stroke, renal, eye, and peripheral vascular disease. Nutrition is an important contributor to certain cancers, so that the multiple-factor causation of disease cannot be ignored.

Risk factors for disease are those aspects of personal behavior or lifestyle, occupational or environmental exposure, social and economic conditions, and inborn or inherited characteristics which, on the basis of epidemiologic evidence, are known to be associated with health-related conditions considered important to prevent. Noninfectious diseases are often related to and exacerbated by a number of risk factors, so that measurement of prevalence of risk factors, or intervening variables, is important to epidemiologic assessment of the future risk of such diseases. The prevalence of smoking, as an example, may serve as an indicator of the future potential of lung cancer and cardiovascular diseases. BMI, blood pressure, and serum cholesterol levels measured in the community serve as indicators of risk for coronary heart disease (Box 3.19). These measurements indicate individual and community risk, and to measure effectiveness of health promotion programs.

NOTIFICATION OF DISEASES

Morbidity data are reported by doctors, usually based on compulsory reporting of specific infectious and noninfectious diseases. Some diseases such as plague, cholera, yellow fever, louse-borne typhus, and louse-borne relapsing fever are notifiable by international convention. Locally

---

Box 3.18 Henle-Koch Postulates on Microorganisms as the Cause of Disease

1. The organism (agent) must be shown to be present in every case of the disease and must be isolated, cultured, and identified;
2. It must produce the disease when a pure culture is given to a susceptible animal;
3. The organism must be recoverable from the animal.

Source: Last, J. M. (ed.). 2001. A *Dictionary of Epidemiology*, 4th ed. New York: Oxford University Press [with permission].

Box 3.19 Criteria for Causation in Chronic Disease — The Evans Postulates

1. Prevalence of the disease should be significantly higher in those exposed to the hypothesized cause than in controls not so exposed.
2. Exposure to the hypothesized cause should be more frequent among those with the disease than in controls without the disease, when all other risk factors are held constant.
3. Incidence of the disease should be significantly higher in those exposed to the hypothesized cause than in controls not so exposed, as shown by prospective studies.
4. The disease should follow exposure to the hypothesized causative agent with a normal or log-normal distribution of incubation periods.
5. A spectrum of host responses should follow exposure to the hypothesized agent along a logical biological gradient from mild to severe.
6. A measurable host response following exposure to the hypothesized cause should have a high probability of appearing in those lacking this before exposure (e.g.,

antibody, cancer cell) or should increase in magnitude if present before exposure. This response pattern should occur infrequently in persons not so exposed.

7. Experimental reproduction of the disease should occur more frequently in animals or humans appropriately exposed to the hypothesized cause than in those not so exposed; this exposure may be deliberate in volunteers, experimentally induced in the laboratory, or may represent a regulation of a natural exposure.
8. Elimination or modification of the hypothesized cause should decrease the incidence of the disease (e.g., attenuation of a virus, removal of tar from cigarettes).
9. Prevention or modification of the host's response on exposure to the hypothesized cause should decrease or eliminate the disease (e.g., immunization, drugs to lower cholesterol, specific lymphocyte transfer factor in cancer).
10. All of the relationships and findings should make biological and epidemiologic sense.

Sources: Evans, A. S. Causation and disease: The Henle-Koch postulates revisited. 1976. *Yale Journal of Biology and Medicine*, 49:175–195.
Last, J. M. (ed.). 2001. A *Dictionary of Epidemiology*, 4th ed. New York: Oxford University Press [with permission].

---

endemic diseases are notifiable under national public health laws in order to monitor their prevalence and the impact of public health measures (see Chapter 4). Additional diseases reported routinely in other countries include water- and food-borne disease, chemical poisonings, botulism, leishmaniasis, septicemia, *Chlamydia trachomatis* (genital), gonococcal ophthalmia, and listeriosis. Other diseases or health events may be added to routine reporting (or to special surveys) according to endemic environmental conditions. Reporting of infectious diseases is one of the most important foundations of public health practice.

SPECIAL REGISTRIES AND REPORTING SYSTEMS

Special registries are important to establish a basis for the epidemiologic study of vital health events pertinent to the population and clinical states important to population health. These include mandatory reporting and special registries and surveys. They are vital for monitoring the health of a population and epidemiologic information to guide health policy, whether it is for an acute infectious disease challenge or a long-term chronic disease problem such as cardiovascular disease and diabetes. The range of such reporting systems is necessarily very wide (Table 3.5), with recent additions to mandatory reporting of child and elder abuse. Priorities may vary from country to country, but the basic registry needs in health care include a range of conditions, including infectious diseases,

cancer, birth defects, and hospital discharge information systems. Data from cancer, birth defect, and low birth weight registries can give valuable clues for environmental exposures of public health importance.

Ideally, disease registries and reporting systems should be coordinated into unified health information systems. The United States has an effective network of such reporting systems, such as the Census Bureau, the Department of Health and Human Services, state health departments, and the CDC with a variety of surveillance systems and a regular weekly publication with periodic special reports on special surveys and routine reports of disease incidence and prevalence. Individual identification numbers, such as Social Security numbers, for each member of the population enables the use of data from related special registries. However, protective measures must be in place to ensure privacy and to prevent the misuse of these data for unethical purposes. Safeguard mechanisms can be built into data systems to protect the privacy of the individual. This is a special problem in the United States with a large unregistered immigrant population, many of whom receive services from public programs, but are also in jeopardy from the threat of possible deportation by federal immigration authorities.

Linkages among data sets allow important epidemiologic correlations to be studied. For example, linking data sets for cancer registries, vital records, pollution indicators, and hospital discharge information systems may enhance investigation of specific medical conditions, such as monitoring longevity and hospital use for childhood

TABLE 3.5 Public Health Mandatory or Voluntary Reporting and Registries

Mandatory	Special registries or surveys
Vital events — birth, death, marriages, and divorces	Cancer registries
Notifiable infectious diseases, including STDs and HIV, tuberculosis	Chronic diseases registries
	Neurologic disorders registries
Birth weights	Diabetes registries
Birth defect registries	Coronary heart disease
Congenital screening for PKU, hypothyroidism	Thalassemia
	Sickle cell disease
Abortions and other pregnancy events	Mental illness — psychiatric conditions
Hospital discharge information systems	Nutritional status indicators surveys
Battered children, wives	Growth and development indicators
Domestic violence	
Motor vehicle accident injuries	Blind and partially sighted persons
Air and water quality monitoring	Deaf and hearing-impaired
	Disability surveys
Environmental hazards and monitoring	At-risk workers' groups
	Behavioral risk factors surveys
Occupational safety and health hazards	Internet and news media obituaries
Animal disease monitoring	Influenza — sentinel reporting centers
Vaccine and drug reactions	Autism registries
Hospital infections and incident reports	Alzheimer's and other dementias
Poison control centers	Toxic substance and poison control centers
Injuries, trauma	
Workers' compensation	Hazardous waste sites
School absence	Psychiatric/mental health
Public health laboratories	Leukemia and lymphoma registries
Social security — Medicare, Medicaid, special categories (e.g., end-stage renal disease patients)	Cystic fibrosis registries
	Self-rated health status surveys
Hospital discharge information systems	Sentinel sites for influenza reporting
Blood bank	Behavioral risk factors surveys (e.g., smoking, teen pregnancies, car seat belt use)
Public health laboratories	
Veterinary public health surveillance	
Animal reservoirs and health	Nutritional surveys (e.g., NHANES)
Vaccine and drug reactions	Growth and development indicators
Hospital (nosocomial) infections	Health insurance systems utilization
Injuries	
Poisonings (e.g., poison control centers)	Performance indicators (e.g., GP immunization and preventive service coverage rates, hospital utilization)
Violence and trauma (i.e., emergency services)	
Battered children	
Domestic and elder abuse	

Source: Adapted from Declich, S., Carter, A. O. 1994. Public health surveillance: Historical origins, methods and evaluation. *Bulletin of the World Health Organization*, 72:285–304.
New York State Department of Health. 1999. Chronic Disease Registries, http://www.health.state.ny.us/diseases/chronic/diseaser.htm [accessed May 1, 2008].

Box 3.20 Identification of Hemorrhagic Disease of Newborns in Review of Vital Records and Follow-Up Study in New York State

Studies from vital statistics registries may raise epidemiologic questions or hypotheses which need further investigation. Special surveys become important as the follow-up to initial findings. Intervention can then be planned on the basis of these investigations. As an example, a review of vital statistics in New York State (1987) showed 32 infant deaths reported during the 1980s attributed to hemorrhagic disease of the newborn (HDN), a disease preventable by prophylactic vitamin K injections of newborns.

A follow-up study of the State Hospital Discharge Information system showed a substantial number of hospital discharges with the diagnosis of HDN (first to fourth diagnosis) during the same time period. A case record review of infant deaths with HDN as a diagnosis (first to fourth diagnosis) showed that two-thirds of the cases did not receive vitamin K at all, or not until after bleeding had already begun.

As a result, the State Department of Health adopted mandatory vitamin K prophylaxis for newborns. Record linkage between hospitalization data and the individual cases would have made such a study more readily achievable.

This study led to adoption of vitamin K injection for all infants in New York State and subsequently in most states in the U.S.

Source: Tulchinsky, T. H., et al. 1993. *American Journal of Public Health*, 83: 1166–1168.

cancer. It may also be used to compare morbidity and mortality patterns for specific conditions by comparing hospitalizations with mortality patterns (Box 3.20). Birth defect registries are very important as there are many interventions which can reduce birth defects, and monitoring incidence of new cases and rates will help to evaluate the effectiveness of interventions such as fortification of flour with folic acid.

The importance of records linkage might also be demonstrated by the following epidemiologic question. Mortality from cardiovascular disease has fallen dramatically in industrialized countries since its peak in the early 1960s. This can be attributed to many factors, including changes in nutrition, smoking, and other risk factors, but also to improved medical care for hypertension and for acute coronary events, as well as long-term cardiac rehabilitation and care. The prevalence of the basic disease process may not have declined, but primary and secondary prevention is much improved. Studies linking hospitalization patterns with preventive action such as smoking education laws and case fatality rates for cardiovascular diseases are helping provide support for prevention and new modalities of care.

DISEASE CLASSIFICATION

Because comparative statistics are vital in monitoring the health status of a population, it has been essential to develop internationally accepted standard nomenclature and a coding system in order to minimize differences in classification. The Bills of Mortality used in the seventeenth century defined 17 categories. Classification of disease by anatomic sites or body system was initiated by William Farr at the Second International Statistical Congress in Paris in 1855.

After World War I, the League of Nations supervised revisions of the *International Classification of Diseases* (ICD), and since the 1948 sixth revision, the ICD has been updated at about 10-year intervals by the WHO. The tenth revision of the *International Classification of Diseases* (ICD-10) came into general use in 1993. The classification is broken down into many subcategories with coding to indicate precise disease and procedure groups (Table 3.6). Similarly, classification of mental health disorders has been developed (Chapter 7).

HOSPITAL DISCHARGE INFORMATION

Admission to a hospital is a major medical event, not less important from an epidemiologic point of view than the reporting of a death or an infectious disease. A hospital discharge data system is an informational, planning, budgeting, epidemiologic, and quality control tool in modern health care. It involves gathering a basic data set on all hospital discharges, input of data into a central file on a regular basis, and processing the data for administrative and epidemiologic purposes. This requires a basic data retrieval form for all hospitalized patients and a system of reporting and analysis with computerized data retrieval preferred.

Hospital statistics were originally promoted by Florence Nightingale in the nineteenth century (Box 3.21). The Uniform Hospital Discharge Information System (UHDIS) evolved due to growing recognition of the importance of hospital utilization in the economics of health care. Introduced in the 1960s by the U.S. National Center for Health Statistics (NCHS), it provided the basis for development of diagnosis-related groups (DRGs), which have become the major mode of payment for hospitals in the United States and in some other countries since the 1980s. Use of the International Classification of Diseases allows for comparisons among data sets, regions, and countries.

A central governmental professional unit is needed at the state level to plan, train, and supervise data retrieval and to process and interpret the output data. Data provided by all hospitals provide a complete picture of the entire population using all hospital services, rather than just those services provided by an individual hospital in the region. This is necessary, as people residing in a hospital catchment area may be hospitalized in another region by referral or for emergency care.

TABLE 3.6 *International Classification of Diseases* (ICD-10)

1.	Certain infectious and parasitic diseases	A00-B99
2.	Neoplasms	C00-D48
3.	Diseases of the blood and blood-forming organs and certain disorders involving the immune mechanism	D50-D89
4.	Endocrine, nutritional, and metabolic diseases	E00-E90
5.	Mental and behavioral disorders	F00-F99
6.	Diseases of the nervous system	G00-G99
7.	Diseases of the eye and adnexa	H00-H59
8.	Diseases of the ear and mastoid process	H60-H95
9.	Diseases of the circulatory system	I00-I99
10.	Diseases of the respiratory system	J00-J99
11.	Diseases of the digestive system	K00-K93
12.	Diseases of skin and subcutaneous tissue	L00-L99
13.	Diseases of musculoskeletal system, connective tissue	M00-M99
14.	Diseases of the genitourinary system	N00-N99
15.	Pregnancy, childbirth, and puerperium	O00-O99
16.	Certain conditions originating in perinatal period	P00-P95
17.	Congenital malformations, chromosomal abnormalities	Q00-Q99
18.	Symptoms, signs, and abnormal clinical or laboratory findings not classified elsewhere	R00-R99
19.	Injury, poisoning, and some other external causes	S00-T98
20.	External causes of morbidity and mortality	V01-Y98
21.	Factors influencing health and contact with health services	Z00-Z99

Source: World Health Organization, http://www.who.int/classifications/icd/en/ [accessed July 16, 2007]

Developing countries need assistance in developing basic registration systems of births, deaths, and other vital events. WHO estimates that tens of millions of such events occur annually without registration or reporting. At the same time, the understaffed primary care services compile daily records with large amounts of indigestible data on ambulatory care utilization. Instead, scarce financial and personnel resources should be focused on more significant and higher-quality data associated with hospitalizations. Fewer centers are involved in hospital care than in ambulatory care, so that data retrieval is easier to control. Most importantly, the less common event of hospitalization is medically and epidemiologically more

---

Box 3.21 Uniform Hospital Discharge Information Systems (UHDIS)

1. **Planning** Organizing based on admission and surgical rates, utilization by age and sex, diagnosis, length of stay, and "small area analysis" which compares practice patterns and use or excess and waste of resources; search for new methods to promote patient flow to alternative care facilities (e.g., minimal supervisory residential care, ambulatory, or home care).

2. **"Case-mix" analysis** Makeup of the hospital case load, looking for common diagnoses or rare events which might be of epidemiologic significance, or may have administrative and quality control importance.

3. **Budgeting** Planning within the hospital and in relation to referral sources based on utilization patterns by diagnosis and department.

4. **Quality of care monitoring** Determination of aberrant practice, complications, or outcomes (e.g., excess surgical rates, infections, mortality). OECD in 2007 includes many measures of hospitalization as quality of care measures, including in-hospital case fatality rates for myocardial infarction, for strokes, cancer of colon, and avoidable hospital admissions for asthma and asthma mortality rates.

5. **Epidemiology** Tracing and mapping epidemics of communicable diseases and identifying localizations and sources; using "tracer conditions" to pick out medically and epidemiologically significant events such as strokes or diabetes mellitus; supplementing international, national, or regional mortality data.

6. **Research** Through case finding of particular clinical events which may then be analyzed for related variables (e.g., incidence of coronary heart disease to compare with mortality patterns, intracranial hemorrhages, and administration of prophylactic vitamin K to newborns, or follow-up of patients with coronary artery bypass procedures).

7. **Linkage with other registries** Linkage with death records, cancer, birth defects, or other special disease registries; relating hospitalization events to special disease registries, such as birth defects, cystic fibrosis, asbestosis, and mesothelioma; supplementing a cancer registry.

8. **Economic analysis** This is an essential aspect of modern health care and the use of hospital care and its alternatives, central to health economics; linked data from various registries and hospitalization data can provide data for important cost-effectiveness and other economic planning models.

Sources: OECD Health at a Glance 2007, available at http://oecd.org/health/healthataglance [accessed November 13, 2007]

---

significant because it consumes 40 percent to 75 percent of health care financing. A UHDIS may be seen as a priority information system after the reporting of infectious diseases, mortality, cancer, and birth defects.

The three primary users of information flow in a hospital information system are clinical medicine, epidemiology, and managerial services. However, much of the development of information systems in recent years has been for managerial purposes. Good data should be easy to interpret for managers and clinicians alike. This requires informatics staff (knowledgeable of modern technology) to tailor the data reporting method so that the manager and others can analyze the data for their needs. The data should be in a manageable format and training should be provided for its users.

Hospital discharge provides a basis for epidemiologic monitoring and control of diseases and simple research information. Analysis of hospital discharge data, especially mortality, surgical complications, and excessive length of stay, are important indicators of efficiency and quality of care. Interregional variations in hospital utilizations provide a clear premise for designing and implementing policies. When so much surgery is done on an outpatient basis or with endoscopic methods, long lengths of stay in hospital are unjustified from the patient welfare point of view as well as from the economic aspect that is important to the health insurance or health service system (Box 3.22).

Hospital discharge data studies permit case-mix studies, show trends in care patterns, patient safety conditions, and provide a basis for peer review within a hospital and between hospitals. They provide material for analysis and

---

Box 3.22 Community Diagnosis

Community Oriented Primary Care (COPC) is a systematic approach to the practice of medicine in the community built on the combination of primary care and epidemiology. The community may be defined geographically, for example, residents of an area served by a polyclinic or community health center. It also applies to members of a health care system, for example, members of a health maintenance organization or a special population community (e.g., elderly, children, teens, HIV-positive populations, migrants, ex-prisoners).

Community diagnosis includes:

- Community definition
- Community characterization
- Problem prioritization
- Detailed assessment
- Intervention
- Evaluation

Source: Mullan, F., Epstein, L. 2002. Community-Oriented Primary Care: New relevance in a changing world. *American Journal of Public Health*, 92:1748–1755.

policy formation at the clinical level, as well as for hospital management and planning; for example, in development of ambulatory care, reducing admissions, and length of stay for services better provided on an outpatient basis.

Hospitalizations are fewer in number, with rates varying by age group. Limitations of the data include factors such as lack of standardization of diagnostic criteria. Some patients do not reach a hospital for economic or other reasons; they may have transportation problems, or may have died prior to admission. Others may be unaware of the existence of some health services or are simply afraid of them. Moreover, the denominator for rates is missing because the hospitals may not have a defined catchment population. Nevertheless, hospital discharge information is an important tool for planning, monitoring, and evaluating health services.

Ambulatory care utilization is generally vast in numbers and too large a data set for effective monitoring. The number of ambulatory care visits may range from 4 to 10 per person per year, depending on the country. Ambulatory care data are of poorer quality because they are usually in broad categories of diagnosis, such as musculoskeletal and respiratory complaints, which comprise the bulk of visits. Ambulatory care can be monitored selectively through sampling or monitoring of representative sentinel centers to provide examples for wider replication. Specific components of ambulatory care should be monitored, such as infants and school-age children receiving immunizations, attendance for prenatal care, birth control services, screening for hypertension and diabetes, or breast cancer screening, as particular health goals. With increasing trends for ambulatory care surgery and medical care, linkage of such data with inpatient care is needed in order to ensure continuity of comparisons with previous patterns of care.

HEALTH INFORMATION SYSTEMS (INFORMATICS)

Information is needed for the management of any health system. It is vital to establishing objectives, developing programs, and managing the use of resources. Modern information technology, or informatics, provides the tools for analysis and policy formation to adjust the service. This is as much a part of health care as the cardiograph or ultrasound machines. It provides the feedback, "imaging," or cybernetics potential for management.

Dissemination of information is no less vital than its collection or interpretation in central offices. Reporting of vital data is meaningless unless the data are processed and fed back to the service system in a regular, timely, and usable fashion or in current computer terminology in a user-friendly manner. Modern health information monitors the operation of a health care system. This includes component parts such as objects (hospital buildings), persons (health personnel), services, policy (equity), finance,

organization, administration, regulation, quality assurance, and health promotion. The component parts interact to support the system as a whole. Interaction is made possible through information and communications technology and driven by financing and organizational imperatives.

Health care services are a source of increasing expense to governments and individuals. As a result, governments throughout the world are recognizing the importance of health information for effective health services management and planning. The requirement for public accountability has led to the design of policies to ensure appropriate quantity, quality, and effectiveness of care with the best use of resources. This has created substantial requirements for information.

Public health informatics is the systematic application of information and computer science to clinical and public health practice, research, and learning. This includes use of computerized medical and hospital records, use of clinical and preventive care guidelines, and disease registry information retrieval.

Each country must develop its own health information system and uniform health information systems, such as that developed by WHO European Region (Box 3.23). That system provides a timely (current or real-time) spectrum of vital statistics, demography, and key outcome measures, as well as data on health care resources and utilization. Each country should provide local, district, community or municipal, and regional health profiles. This information should be widely distributed and available for analysis and discussion to the media, the public, and health professionals. Data are of little value if locked away and unavailable for regular circulation and dissemination to a wide audience, who require this information in order

---

Box 3.23 Functions of Health Information Systems

1. **Comparisons** Using historical, regional, national, or international patterns and standards.
2. **Assessment** An overview of health status of a population based on available data, the professional literature, field visits, and interviews with key health personnel and community representatives.
3. **Evaluation** Monitoring use of resources, performance, and outcomes of programs as part of total quality management.
4. **Prediction** Using current data to predict trends in disease and utilization patterns, costs, potential outcomes, program planning, policy formulation, and priority setting.
5. **Explanation** Data to understand disease patterns, risk factors, and service utilization of a population of a district and determine causal relations, or need for intervention.
6. **Planning** Data are needed for planning responses to public health problems and monitoring the outcomes of interventions.

to make an informed contribution to policy analysis and formation.

Precision is limited by the quality of the data, but even limited data are extremely important in epidemiology and for health planning. Some infectious diseases are reported less stringently than others, partly because of lesser concern by physicians, but also because the clinical presentation may be atypical, or some cases may be entirely subclinical. A clinical case of polio may represent a hundred subclinical cases. Many infectious diseases of public health importance (e.g., measles, rubella) are underreported because nonimmunized, vulnerable children may not be brought to medical care despite mandatory reporting requirements, while some reported cases are unconfirmed by laboratory evidence. Nevertheless, reported cases are the basis for monitoring and policy formation. Awareness of the direction and magnitude of errors will enable the user to determine the validity of the data.

Making health information data available on a routine basis to providers and managers of services helps promote an awareness of the overall operation of the health system in which they are involved. Information provides the basis for accountability, which implies that the provider of care or the manager of a health system is responsible for and must report on the results of his or her work, including unintended outcomes. Any system of service requires a system of accountability in order to maintain standards and to provide the consumer with an assurance of quality care.

In a centrally managed system, reporting of services provided is part of the chain of command. In a decentralized system, such data may be derived from billing patterns from hospitals or physician payments. It is then transferred to the higher levels of the health service administration and used for decision making and planning. Those who provide the data should be informed of the outcome, including resultant operational decisions.

The World Health Organization Technical Committee on Information Systems emphasizes that the more active and innovative a health policy is, the greater the need for information. Data collection and processing require planning, training, and continuing monitoring. While massive data banks are not helpful, well-selected and widely available information systems targeted to vital events in the health process can promote flexibility and relevance in the planning of health services.

WHO European Region Health for All Database

The World Health Organization European Region makes available an outstanding database as a free service. It provides some 500 health indicators for all countries in the European region of WHO, and is updated twice yearly. It can be accessed at http://www.euro.who.int/ under Publications and Data. It can be downloaded to a computer and unzipped to provide continuous access to up-to-date data on demographics, mortality, morbidity, lifestyle, resources, and utilization data, and presented as time trends or single-year comparisons of all countries in the region or as a single-year map. It is excellent for teaching purposes and the graphs and data can be downloaded to Microsoft PowerPoint or Word documents. An example is shown in Figure 3.8 comparing life expectancy at birth in France, U.K., Israel, Russia, European average, EU members before 2004, and EU members since 2004 (i.e., countries of eastern Europe and the Baltic states).

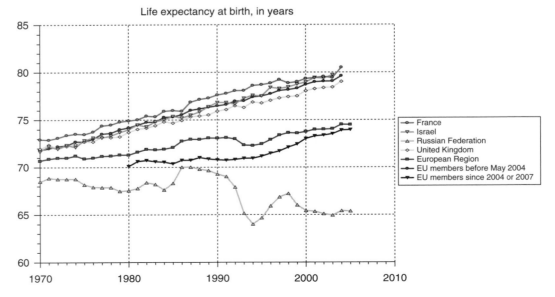

FIGURE 3.8 Life expectancy at birth, European Region, 1970–2005. Source: World Health Organization, European Region Health for All database, November 2007.

SURVEILLANCE, REPORTING, AND PUBLICATION

Publication and wide distribution of weekly summaries of specified reportable diseases are essential to maintain the viability of reporting and promote meaningful use of the data (Box 3.24). The Centers for Disease Control (CDC) of the U.S. Public Health Service publishes and widely distributes the *Morbidity and Mortality Weekly Report* (MMWR), reporting on national and international epidemiologic events through surveys and special reports. The weekly report is supplemented by in-depth special reviews of important public health topics.

WHO publishes the *Weekly Epidemiologic Record* (WER) which reports country epidemiologic events and offers a summary of infectious diseases internationally. Other countries publish their own weekly or monthly epidemiologic summaries of reportable diseases and related laboratory findings, such as the *Canada Communicable Disease Report*, as well as *Chronic Diseases in Canada* for noninfectious diseases and related laboratory findings.

Reporting systems and publication of the data are both vital to epidemiologic monitoring of infectious and noninfectious disease trends. Regular circulation to field personnel increases the sense of awareness and participation in epidemiologic monitoring and shows that the reporting is put to good use. Awareness of the reported data helps local health providers and managers in managing their services more effectively (Box 3.25).

Providing current data as the events unfold promotes a sense of involvement and challenge for achievement of goals, such as high coverage of immunization and rapid control of disease outbreaks.

The Internet has become a vital tool for public health, for reporting and obtaining data, and for access to the world literature. Many resources such as the MMWR, WER, and major journals are available online. Newsgroups enable convenient and immediate discussions by professionals on particular topics, such as Promed for current infectious disease reporting from around the world (Chapter 4). Similarly, the Internet permits literature searches and access to interest groups on virtually any topic in health. This allows people to be in contact with and to obtain support from many others in their field. The WHO home page (http://www.who.int/en/) provides access to its component departments and regional offices.

ASSESSING THE HEALTH OF THE INDIVIDUAL

Physicians and other health professionals are trained to assess the health of the individual patient seeking care (Box 3.26). This involves more than dealing with the chief complaint, requiring a history of the present illness, as well as a wider review of body functions, family and occupational history, physical examination, and laboratory and

---

Box 3.24 Factors Affecting the Value of Data

1. **Relevancy** Are we getting the right data? Are some data collected no longer useful?
2. **Coverage** Do the data help identify high-risk groups?
3. **Quality** How good do data need to be useful? Limitations of data are a factor in decision making.
4. **Acceptability** Are the data collected acceptable in terms of design, cost, and ethical standards?
5. **Timeliness** How recent are the data? How long a time-series is needed to show temporal patterns?
6. **Accessibility** Are the data available to those who need them? Are the data suitable for publication? Are they published and distributed on the Internet and hard copy?
7. **Usability** Are the data in a usable format? Are they presented in a user friendly manner (i.e., easy to access and use for the non-specialists)? Can one generate summaries, graphs, and tabulations from the data?
8. **Cost** What does it cost to collect and process the data? Are the data available to students and researchers without prohibitive cost?
9. **Validity** To what extent do the data relate to the issue of concern?
10. **Specificity and sensitivity** Were the raw data collected using accurate measures (i.e., measures with a high capacity of detecting actual cases and determining non-cases as such)?
11. **Data aggregation and reporting** Are data reported by disease, category of service, social indicators, and region of residence? What is the population at risk?
12. **Biological plausibility** Is the observed or presumed causal association compatible with existing biological and medical science? Can it be explained from a biological perspective?
13. **Equity** Do the data show inter-regional and social class variation and inequity?
14. **Dissemination** Information obtained, collated, and analyzed must be organized and available to those who report the raw data, who need data to monitor health status, and who plan health services and health promotion needs of the population.

Source: Last, J. M. 2001. *Dictionary of Epidemiology*, 4th ed. New York: Oxford University Press [with permission].

Box 3.25 Evidence- and Best Practice-Based Public Health

Evidence-based evaluation of policies to improve health and reduce inequalities, prioritization, and providing resources for these policies requires four basic types of information:

- A detailed assessment of the magnitude and impact of health problems in the population, including information on the causes of loss of health in the population both in terms of diseases and injury, and risk factors or broader determinants;
- Information on health expenditure and health infrastructure (a national system of health accounts) detailing the availability of resources for health improvement and what the resources are currently used for;
- Information on the cost-effectiveness of available technologies and strategies for improving health;
- Information on inequalities in health status, health determinants, and access to and use of health services (including both prevention and treatment services).

Performance-based measures have become essential elements of public health policy and implementation strategies. These are generally based on professional consensus criteria determined by Delphi methods of consultation. These may be translated into "gold standards" and health targets. They may be used for performance monitoring and indeed administrative payment systems to encourage their complete implementation. Examples of performance indicators for payment include immunization coverage, Pap smears, and mammograms for patients registered with British general practitioners.

The concept of health targets has become an essential element of U.S. public health policy with *Healthy People 2010* at the federal level with state compliance with such measures. When reviewing policy issues in public health, currently accepted practices used in other countries with recognized stature in this field should be taken into account, as well as recommendations by respected international health agencies such as WHO, UNICEF, and others.

Sources: Australia's national agency for health and welfare statistics: Mathers, C., Vos, T., Stevenson, C., November 1999, Australian Institute of Health and Welfare, http://www.aihw.gov.au/publications/health/bdia/bdia-c01.pdf [accessed December 3, 2006]
Mays, G. P., McHugh, M. C., Shim, K., Perry, N., Lenoawey, D., Halverson, P. K., Moonsinghe, R., 2006. Institutional and economic determinants of public health system performance. *American Journal of Public Health*, 96:523–531.

---

imaging tests. Defining a differential diagnosis and treatment for a presumptive diagnosis allow for follow-up of a patient to observe the course of the disease, the outcomes of diagnostic tests, and the effects of intervention. Caregivers must take into account the effects of the process on the patient, the family, and the community. Providers must also be concerned about costs of care, alternative methods of looking after the patient to meet changing needs, and promote early and maximum recovery. Continuous monitoring and reevaluation are key parts of the process. There are many parallels in care of the individual and care of the population.

Box 3.26 Assessing the Health Status of the Individual

1. Current chief complaint;
2. Personal data — age, sex, ethnicity, education, marital status, children, living situation;
3. Occupational history;
4. Family history;
5. Personal history;
6. Functional inquiry — systems review;
7. Summary of risk factors — family history, hypertension, diabetes, smoking, sedentary lifestyle, high-fat diet, occupation, alcohol use, stress, other;
8. History of the present illness;
9. Physical examination;
10. Differential diagnosis;
11. Other medical problems;
12. Investigation: laboratory, cardiographic, imaging, other;
13. Presumptive or working diagnosis;
14. Treatment and its effects;
15. Definitive diagnosis;
16. Management of other medical problems;
17. Follow-up management and monitoring;
18. Counseling regarding long-term health needs.

ASSESSMENT OF POPULATION HEALTH

Health service administration is being increasingly decentralized in many countries, and the concept of healthy cities/municipalities is becoming more widespread. These developments have increased the need and value of health profiles at the community, county, and district levels.

This type of health profile provides management with regular monitoring of the health situation, including resources, utilization, morbidity, and mortality. This is the application of modern health informatics at a community level and does not require advanced computer capacity or skills. Annual reports in a standard format using all existing data sources can be brought together in a user-friendly manner to provide valuable health status monitoring.

District or community health information systems increase the potential for local health authorities and communities to have greater power in determining local health policy. National health authorities need to provide the guidance on health targets and resources that may be used flexibly to meet local needs. But supervision and regulation by national health authorities are essential to assure that resources are well used and that targets are being met, as well as to reduce inequalities between regions.

European Region of the World Health Organization has developed a user-friendly computer program for 1000 health indicators, including sociodemographic, mortality, morbidity, health resources, utilization, and lifestyle indicators. These can readily be produced in tabular or graphic form with time trends and mapping capability. This is accessible free of charge to everyone with a personal computer, Internet access, and modest computer skills via http://www.euro.who.int/hfadb.

As with individual health assessment, evaluation of the health status of a population is based on the accumulation of a portfolio of observations and data from a variety of sources and their interpretation, with comparisons to international, national, or regional patterns or standards. Community health assessment (CHA) begins with identification of the main health problems or chief complaints as understood by key health professionals and the community, or from regular community health profiles.

Information should be derived about the community's SES, the resources available for health care, how they are distributed, and how services are utilized, as well as morbidity, mortality, and other "outcome" measures which help to describe or compare health status (Table 3.7). Health measures include how care is provided, how it governs or monitors itself, and how the system is accountable for its component services. The knowledge, attitudes, beliefs, and practices (KABP) of the people and health providers and the way in which society addresses risk factors for ill health may also be important determinants of health status.

Gathering the data necessary to monitor itself should be part of the standard functions of a health system. This provides for accountability in use of public resources and maintains a self-correcting feature of the system. CHAs help point out health risk factors at the population level, and if carried out in a timely and regular fashion, changes can be made without inordinately long waiting periods and without any unnecessary increase in morbidity or mortality.

The CHA is part of the health planning process; it may be designed to monitor the impact of an intervention program meant to deal with a particular health problem, such as coronary heart disease, or a set of risk factors for disease, such as smoking. The CHA is also part of program evaluation, especially in community trials, with an evaluation protocol based on a multiphasic approach and data from many sources.

Defining the Population

The population served by a health system must be defined in terms of age and gender distribution. This is one of the key factors in the planning of health care services, as different age groups have different needs. Women, children, and the elderly utilize more health services and institutional care. The demographic pyramid is an excellent graphic summary of the population distribution. The health status of the elderly is affected by the major chronic diseases and the associated disability and mortality patterns. While increasing longevity is associated with a healthier elderly population, the demand for care still grows with age. The elderly, and increasingly the very elderly (those over 85), are high users of health services, including institutional care in hospitals and long-term facilities.

Socioeconomic Status

Health is affected by standards of living and therefore analysis of income and its distribution is a component of the process of assessing the health status of a population. The national average income is often represented by the GNP (gross national product) or GDP (gross domestic product) per capita; for instance, the average of the total production of goods and services of a nation. Real income may vary by state or district, ethnic group, educational levels, sex, or family size. These and many other factors may affect the distribution of wealth in the population.

Living conditions as reflected in housing standards, density of housing, and crowding (persons per room or per square meter) are dependent on family income. Services, such as electricity, running water, indoor toilet and bathing facilities, as well as other service facilities in the home (e.g., refrigerators, toilets, baths, stoves, central heating and air conditioning) are also important measures of health-related socioeconomic conditions. Adverse economic conditions prejudice health status in measurable ways. In developing countries, the poverty–disease–malnutrition cycle affects children, women, and the elderly predominately, reducing potential for economic growth. Even in industrialized countries, there is unevenness in the patterns of income and of health status; the health status of the upper social class is much better than that of the unskilled workers for many health indicators. Where there are large gaps between the rich and the poor, such as in the United States, there is poorer health status than in countries with smaller social gaps such as Japan and the Scandinavian countries.

Educational level of parents is an important factor in family health. In the case of the father in a family, education level is often a direct determinant of income. In the case of the mother, education relates to income, but even more strongly to successful health care of infants and children. Mothers with higher levels of education, as measured by years of school attendance, are more likely to absorb new knowledge regarding self-care in pregnancy and care of the infant in areas such as nutrition, immunization, and

TABLE 3.7 Evaluation of Population Health of a Community, District, State, or Country

Factors	Topics	Example indicators
Geography	Climate, topography, density, urban/rural	Tropical, temperate, mountainous, desert Distances from medical facilities
Demography	Vital statistics	Population size, age/sex, and urban/rural
Socioeconomic	Ethnic, cultural, religious practices Community, family economic status	Per capita and family income, education, literacy (women), employment, religious affiliation, social attitudes, occupations
Nutrition	Supply, affordability, use of major food groups Food safety and quality Food fortification	Under- and overnutrition Risk group identification Monitoring child growth patterns, anemia
Environment and occupational	Water, air, waste and sewage disposal, toxic wastes, radiologic hazards Industrial or agricultural toxic materials	Ambient air pollutants, bacteriologic and chemical qualities of community and recreational water, radiation and radon levels, lead levels in soil, water
Public health infrastructure	Organization, training and deployment of public health functions and personnel	Legal and regulatory functions Schools of public health Research capacity in epidemiology, public health
Health care system	Organization Prepaid coverage Finance total and internal allocation	Decentralized administration and finances Integration of local services and finances Total resources; % GNP and per capita (US $) spent on health care; % population with full, partial, or no health benefit insurance
Health resources	Expenditures per capita Hospital beds per capita Long-term care facilities Clinics Personnel, doctors, nurses per capita	Expenditure by type of service, preventive, curative, hospital Acute care beds/1,000 Special hospital beds/1,000 Long-term care facilities/1,000 Doctors, nurses/10,000
Health care utilization	Hospitals, general, chronic, and mental Ambulatory care Preventive services	Admissions and days of care/1,000 population Physician visits per person per year Immunization coverage at age 2 years Ambulatory surgery, home care measures
Process (quality) of care	Professional care standards Accreditation by external agency Peer-review mechanisms Records review Mortality case review Clinical guidelines	Criteria for surgery, second opinion Immunization and child health monitoring rates Correction of deficiencies from accreditation Departmental reviews of caesarian, infection rates Maternal and infant mortality case by case reviews Computerized medical records
Health outcomes	Morbidity Mortality Functional/physiologic status "Tracer conditions"	Infectious and chronic disease incidence/prevalence Infant, child, maternal, age-sex–specific mortality rates by cause, cardiovascular disease, trauma Anemia of infancy, pregnancy, blood lead levels Lower limb amputation rates
Costs and benefits	Examine specific diseases, procedures, services or health promotion	Cost-benefit of second dose of measles vaccine, bicycle helmets, air bags in cars, antismoking campaign among high school girls
Knowledge, attitude, beliefs, practices (KABP)	General population Risk groups Patients Patients' families Health providers	Diet, smoking, eating, moderate alcohol use, exercise Diabetes, hypertension, Birth control, rights of women AIDS/STI-related issues

routine baby care. Better educated women tend to have fewer pregnancies, not only because of knowledge of need and methods of birth control, but also because of greater self-awareness and different life goals. Ethnic, cultural, political, and religious beliefs and practices have important implications for health, in such areas as the status of women, mental health, family structure, nutrition, substance use and abuse, as well as birth control and abortion. These beliefs and practices can affect attitudes toward issues such as national health insurance and the funding of health care.

Nutrition

Appropriate nutrition, overnutrition, and undernutrition are fundamental determinants of the health of a population. Overnutrition places a heavy burden of morbidity and mortality on the health system, with such diseases as diabetes, coronary heart disease, hypertension, and stroke and their complications. Undernutrition in the form of gross malnutrition is rare in the industrialized countries, but extremely common in many developing nations. In all societies there are groups at risk for overt or subclinical malnutrition, such as anemia, iodine deficiency, osteoporosis, and others. A society which acts to prevent malnutrition in vulnerable groups is acting on behalf of the vulnerable groups in the population and indicates the well-being of that society. Public health and economic measures to promote good quality of food and its accessibility to the population, fortification of basic foods, school lunch programs, and meal services for the elderly and chronically ill are health promotion programs that show the level of organized community responsibility for its members (Chapter 8).

Special surveys, such as low birth weight or nutritional status conditions, are needed to provide nutrition status data. Monitoring of nutrition status, discussed in detail in Chapter 8, is of fundamental importance to population health evaluation. Periodic large-scale national surveys, such as the National Health and Nutrition Examination Surveys (NHANES), initiated in 1971 in the United States, provide meaningful information on nutrition status in the country. Within the United States, the surveys provide vital information for adjusting recommended dietary allowances and national, state, or local nutrition programs. This is of great importance for the food industry which is obliged to follow federal government standards of labeling and content of packaged and processed foods.

Environment and Occupation

Safety of community water, management of solid and toxic wastes, air and noise pollution, and ambient air standards are all factors in health of the community. Organized public health has traditionally focused on these issues, but they remain public policy issues in virtually all countries and internationally. Healthy societies are dealing with these issues with a very high degree of public awareness, sometimes overcoming strong economic interest groups to force improved attention to the environment by governments, communities, and businesses. Environment includes housing, recreation, schools, parks, urban and rural planning, and many other aspects of community life that are addressed in "healthy community" initiatives. Employment of children and work in hazardous industries are health issues. Societies that tolerate toxic and dangerous work settings create health hazards that are preventable, but costly to treat. Unemployment, job insecurity, loss of health insurance with change of employer, job-related injury or disease, and low income levels for many workers all contribute to poor health (Chapter 9). Where health insurance is related to employment, as in the United States, health protection can be a major factor in relation to losing or changing place of employment.

The development of the New Public Health has moved national agendas and local authorities with major roles in improving health of populations. The idea of community diagnosis and community-oriented primary care (Box 3.22) has played an important part in this process. It is of vital importance in developing countries where the infrastructure for prevention and primary care remain weak. In countries in transition from the Soviet system of health care, reform should address the imbalance of excessive expenditure on hospitalization with inadequate development of primary care, and health promotion. Countries in transition should address high rates of mortality from cardiovascular disease and trauma (see Chapters 11 and 13).

HEALTH CARE FINANCING AND ORGANIZATION

How a nation finances and organizes health care is an important aspect of health status evaluation. Where there is universal coverage of the population, either through health insurance or through a state-operated health care program, the population in principle has equity in access to care. Financial access, however, does not guarantee actual access because the distribution and supply of services are important variables in utilization. Financing and organization of health services are related issues, discussed in Chapters 10–13, that must be recognized as part of the process of assessing the health status of the population of a country or region. Assurance of access to medical and hospital care does not assure that appropriate or effective services are provided.

How services link facilities of different levels of intensity of care and costs is a basic issue in health reform in many countries. How preventive care is provided to special groups in the population (infants, children, adults, the elderly, the chronically ill) and how these fit together as a holistic entity, interacting to serve the community, are important in determining the status of health and health costs of a community or a country.

Health Care Resources

While overall expenditures for health are important determinants of the level of health care available, no less important is how these resources are spent; that is, what the internal financial allocation is within total health. The major resources for health care are in primary care services, hospitals, and long-term care facilities. All countries have limited health financial resources for health expenditures, so that to a great extent one aspect of the health system can only grow at the expense of another.

Hospitals are the largest segment of the health care system in terms of expenditures and may consume more than 50 percent of total expenditures. The supply of hospital beds is, therefore, a central factor in the health care economy. The number of hospital beds per 1000 population is a key indicator for health economics. The hospital bed-to-population ratio varies widely, from 2.5–16 care beds per 1000 in Organisation of Economic Cooperation and Development (OECD) countries, with most countries reducing hospital bed supplies rapidly since the 1980s.

Age distribution of the population affects morbidity and therefore hospitalizations; countries with a high percentage of elderly may need more hospital facilities, as well as alternative care services, such as home care and long-term institutional care services. Innovations in health care organization are influencing health planning, with many developed countries reducing acute care hospital admissions and length of stay by a variety of incentive and management systems (see Chapters 11–13). Health planning requires facing up to political and other pressures to sustain or even increase levels of hospital bed-to-population ratios beyond real need, at the expense of other more appropriate alternative services. The absence of organized home care programs is an indicator of inadequate planning to address the needs of the elderly and chronically ill in a society.

The ratio of medical doctors per 10,000 population also varies widely. A high ratio may indicate an overpopulation of specialists and a lack of primary care services, while a low ratio may indicate a need for training more physicians. Countries in Eastern Europe have high doctor-to-population ratios and low ratings on health status indicators (such as SMRs for trauma) than in countries with fewer doctors. Nurse-to-population ratios are also equally variable, but typically, many countries which have high levels of physician-to-population ratios have relatively low numbers of nursing personnel. The number of nurses registered to practice often overstates the actual supply because many nurses never practice following graduation, work only part-time, or stay in the profession only for a short period.

Excessive supply of medical doctors, inequitable distribution, relative shortages of nurses, inefficient development of community health programs, and inefficient use of community health workers are important issues in many countries (see Chapter 15). These all have economic and health outcome implications, requiring continuous review and reassessment in each country, and application of lessons learned from other countries.

The organization of health services, discussed in Chapters 10–12, is an important factor in the efficiency and quality of care. Community health services are a hallmark of provision of primary care to address population health needs, while many health systems in the past especially emphasized hospital and other institutional care in their norms and financial incentives.

Utilization of Services

Rapid cost increases have fostered a search for efficient ways of organizing and financing health services. In the United States, the development of the diagnosis-related group method of payment for hospital services has reduced hospital length of stay. Health maintenance organizations (HMOs) have been successful in providing comprehensive care with less hospitalization and fewer hospital beds than traditional fee-for-service practice. This has led policymakers and the business community to focus on "managed care" systems to meet the need to extend insurance coverage and to control costs.

While supply of services is important, actual utilization patterns are also a valuable part of the overall evaluation program. Hospital care is a key issue because of its dominance in the economics of health care. Monitoring hospital performance indicators can play an important role in determining the effective functioning of the health care system.

Surgical and other procedure rates are continuing issues in health systems management. Hysterectomy rates vary widely among Canadian provinces, from 639 per 100,000 in Newfoundland to 426 per 100,000 in Alberta, and vary by a factor of 18 within Ontario on a county-to-county basis. A study conducted in Saskatchewan showed that the introduction of mandatory second opinions resulted in dramatic reductions in hysterectomy rates. Appendectomy rates in Germany are as high as three times the rates in other countries, with no epidemiologic explanation.

Studies abound in the United States showing differential utilization of health services by African-American and white populations for coronary heart bypass procedures, for localized as compared to radical surgery for lumps in the breast, and for mammography and other services currently considered to be of benefit to the patient. These differences generally are primarily due to differences in health insurance coverage, but other socioeconomic or ethnic variables may also be responsible. Excess surgical procedures, for example, cesarean sections, are a widespread problem in countries where fee-for-service is the method of payment, but the amount of surgery is also related to the number of surgeons and fee-for-service payments.

Health Care Outcomes

While it is clear that health status is affected by many social and economic factors, the general state of the country's health is often described by epidemiologic indicators, such as mortality and morbidity rates as indicators of health status. Epidemiologic information on communicable and noncommunicable diseases helps to determine a potential for intervention and alteration of the natural history of the disease.

Outcomes can include morbidity, mortality, physiologic, and functional measures (Box 3.27). They may also include measures of self-assessment of health status; risk behavior such as smoking or engaging in unsafe sexual practices; or knowledge, attitude, and beliefs of health-related issues. These measures may be part of evaluation of health status of a population or a program meant to cause change.

Outcome indicators include a variety of measures from routine data sources and special surveys. DALYs and QALYs (described earlier) attempt to quantify mortality and quality of life measures for comparisons and for analysis of specific interventions. In addition, physiologic or functional indicators such as activities of daily living

measure patient performance. Special surveys for clinical signs of undernutrition such as anthropometric measures (growth and body size) should be supplemented by biochemical-level and hematological surveys to establish patterns of undernutrition. Special surveys of nutrition status and disability, school performance, and other indicators of functional status are important aspects of health status evaluation.

Quality of Care

The quality of care (see Chapter 11) is part of evaluation of health in any population. How available funds are spent to address the health problems specific to that population is part of the community health assessment. The findings of such evaluations are meant to affect resource allocation and address unmet needs. Health care is increasingly being evaluated by managers of health insurance programs,

---

Box 3.27 "Outcome Indicators" of Health Status of a Population

Outcome is a variable with a value which varies according to the outcome or the effectiveness of an intervention (Last, 2007), taking into account independent variables, such as more general changes occurring in the same time frame. Examples include:

1. **Mortality-related indicators**
   a. Infant and child mortality rates (IMRs);
   b. Maternal mortality rates (MMRs);
   c. Crude mortality rates (CMRs);
   d. Age-specific mortality rates;
   e. Cause-specific mortality rates — infectious, noninfectious;
   f. Case fatality rates as a measure of the success of medical care;
   g. Life expectancy (LE) at ages 0, 1, 65, and other ages;
   h. Standardized mortality rates (SMRs) — total specific;
   i. Years of potential life lost (YPLL) — a measure of the impact of mortality on different age groups to reflect relative impact of diseases or conditions on the population;
   j. Quality-adjusted life years (QALYs) — an adjustment of life expectancy by inclusion of chronic conditions with impairment, disability, or handicap;
   k. Disability-adjusted life years (DALYs) — a measurement based on adjustment of life expectancy and includes the estimated effect of long-term disability.

2. **Morbidity outcome indicators**
   a. Incidence of vaccine-preventable disease;
   b. Incidence of waterborne disease;
   c. Incidence of food-borne disease;
   d. Incidence/prevalence of tuberculosis;
   e. Incidence/prevalence of STDs/AIDS;
   f. Incidence of malaria, other tropical diseases;
   g. Prevalence of noninfectious diseases — cardiovascular diseases, diabetes, cancer, trauma;
   h. Prevalence of disabling conditions;
   i. Prevalence of risk factors.

3. **Behavioral indicators**
   a. Knowledge, attitudes, beliefs, practices regarding risk factors — smoking; alcohol and drug use; unsafe sexual practices; high-risk behavior regarding motor vehicles, violence, drug usage, suicide;
   b. Compliance with immunization, preventive care, medical treatment and advice, physical fitness, suitable weight.

4. **Physiological indicators**
   a. Nutritional status — growth patterns of infants and children; body mass index of adults; dietary patterns;
   b. Hematologic and biochemical indicators (blood sugar; cholesterol; lipids; vitamins A, B, C, D); anemia among infants, children, and women; iodine status; environment.

5. **Functional indicators**
   a. Work and school absence;
   b. Psychomotor function;
   c. Work capacity;
   d. School performance;
   e. Fitness test performance;
   f. Activities of daily living (ADL);
   g. Cognitive capacity.

whether as health maintenance organizations or veteran's health services and Medicare of the U.S. federal government, or by international organizations (such as WHO, UNICEF, OECD, UNDP), seeing health as an economic investment, and international comparisons, as in the Human Development Index and Health for All database. Data systems for epidemiologic studies and for population health monitoring include the most basic reporting systems of infectious diseases, vital statistics, and also from special disease registries such as a birth defect registry, special surveys such as NHANES on nutrition status (see Chapter 8), and hospitalizations as seminal health events or "tracer conditions" to provide vital material to study and compare effectiveness of health systems, and indeed individual provider performance.

There are other important indicators of quality health systems – health system responsiveness and patient or population satisfaction. Responsiveness is a measure of ease of access and comfort level of clients with "consumer-friendly" and psychologically supportive facilities and staff for the population served.

Practices in prescription drug use may indicate utilization much beyond accepted clinical guidelines, as in the use of proton pump inhibitors in the U.K. NHS. This drug is important but overused and costs the NHS an estimated £100 million annually, whereas less costly methods are just as effective. The use of such analysis of data sets on prescription drug use is of great importance to the economic survival of health systems permitting limited resources to be used to better effect for unmet health needs.

Self-Assessment of Health

Data on self-assessment of health are used along with household expenditure and nutrition surveys to provide information on health-related experience of selected samples of the population, sometimes by household interviews and by telephone surveys. These may yield estimates of poverty, illness, or inequality for small areas for which no or few other data are available. Reliability of recall and reporting are limiting, but this does provide important information which cannot be measured in other ways. Health surveys are vital to monitoring population health and self-assessment is an important component of ongoing monitoring, and to measure inequalities within a health system.

Costs and Benefits

Analysis of costs and benefits, reviewed in more detail in Chapter 11 on economics and health policy, will be mentioned here only briefly. Evaluation of the health status of a population requires examination of the choices made in resource allocation in a particular geographic area. This is of concern not only to the planner, but also to the provider of health care and to the public. If priorities in resource allocation promote highly technological medicine, then primary care may lag behind in resources, and the health status of the population may be compromised. Cost–benefit analyses can contribute to establishing priorities within a health care system.

Effects of Intervention

Adoption of *Haemophilus influenzae* vaccine for infant immunization will result in an almost immediate drop in *H. influenzae* meningitis and pneumonia, as adoption of a two-dose policy for measles vaccination leads to a very rapid reduction in measles morbidity and mortality. Other interventions in public health affect an epidemic curve more slowly, as smoking reduction actions lead to reduced hospitalization and mortality from coronary heart disease.

Many interventions in preventive medicine and public health are complementary so that a doctor's advice to quit smoking and antismoking legislation are mutually reinforcing of the same message. The "natural history of disease" is affected by many sociological and economic factors as well as medical or public health interventions. The dramatic reduction of coronary heart disease mortality, but not necessarily morbidity, is attributable to improved medical care, preventive medical care, and to wider public health activities related to improving knowledge, attitudes, and practices for lifestyle change. These themes were discussed in Chapters 1 and 2, and will recur in coming chapters of this book as part of the continuously changing New Public Health.

Qualitative Methods

In public health it is often necessary to investigate how the social, physical, or policy environment influences people's perceptions and behavior. Quantitative survey research is best at yielding rates, proportions, associations, and correlations, but quantitative methods can only gain a finite understanding of a problem. Quantitative surveys emphasize structure, consistency, precisely worded questions, and analysis methods to quantify experiences and produce measurable outcomes. They are, however, limited in their ability to generate knowledge of social influences and processes by understanding what they mean to people (Box 3.28).

Qualitative research (QR) helps to elucidate patterns of shared understanding and variability in these patterns, and help to understand how the social, physical, or policy environment influences people's perceptions and behavior. The emphasis is on exploration, and relies on synergy between design and discovery, and thus is valuable for program evaluation. It does this by focusing on both verbal and nonverbal language using a more unstructured interview format so participants can answer for as long and as openly as they choose.

Box 3.28 Quantitative Versus Qualitative Research

Quantitative	Qualitative
Methodological approaches	
• Positivism (i.e., uses numbers to define relationships via closed-ended answers); "experience distant," experimental, empirical means • Structured by hypotheses • Sampling representative of population • Data collection: surveys with closed answers • Analysis: turning beliefs, behaviors, or attitudes into numbers to support hypotheses	• Interpretivist; "experience near" • Structured for exploration • Theory or question driven • Sampling — often purposive (i.e., selection of sample of persons with depth of knowledge of research topic) • Data collection, observations use semi-structured or unstructured questions eliciting open answers • Analysis: summarizing interactions, behaviors, or attitudes
Research questions	
• Precisely worded questions • Aim to quantify experiences and produce measurable outcomes • Place emphasis on structure • Reliability, uniformity, objectivity, and freedom from bias are paramount	• Aim to explore and gain insight into behavior and perceptions • How do people interpret and experience their interactions and perceptions and/or attitudes? • Never fixed: discovery and exploration, synergy between design and discovery

Source: Feldman, B. 2007. Unpublished information. Personal communication.

QR is guided by the research problem which fuels research questions. A conceptual framework is often applied to keep the research directed and dictates the combination of questions asked such as ones based on experiences, behaviors, opinions, values, concerns, or knowledge. QR research tends to be dynamic, using questions and approaches which evolve as new insights are gained. Approaches to data collection can take the form of words, images, and observations; observation, in-depth interviews, and focus groups are the fundamental approaches to QR. Other methods such as documentary research and videotaping can play an important role in gaining participants' perspectives as well.

As an example, research focusing on high birth rates among indigenous adolescent women in rural Mexico would require quantitative surveys to provide relevant data such as the percentage of women pregnant in the age group 15–19, the probability that a woman will use a contraceptive method, frequency of abortions, or the risk of her dying from pregnancy. QR would be able to elucidate factors such as misconceptions regarding contraception, parental opinions about adolescent pregnancy, and beliefs and problems regarding accessing pre- and postnatal care. QR methods can operate independently or complement quantitative instruments either by proceeding or preceding them, depending on the study goals.

Entering the field through acknowledging and consulting with "gatekeepers" or leaders of the potential research site population helps in accessing members of the community. It also facilitates follow-up, such as identifying local people to work with, presenting oneself and the research to key stakeholders, and recruiting participants. Researchers often visit common meeting places, chat with potential participants, and subsequently select a sample purposively based on readiness of individuals to participate, as well as their demographic characteristics to represent a defined subgroup. Sampling can be varied and depending on strategy, may select homogeneous, heterogeneous, extreme, or typical participants. Pilot testing often follows in order to assess how well the objectives of the study are fulfilled, and provides the opportunity to circumvent any constraints and obstacles before study initiation.

One-on-one interviewing allows participants to play an active role in determining the direction of the interview (in-depth interview or IDI). Questions follow the flow of conversation and the interview has a conversational quality. The interviews can take the form of unstructured informal conversations, or can be semi-structured or structured. They generate empirical data as participants talk freely about their experiences and beliefs. This is an effective approach when inquiring about sensitive information and when assessing individual's opinions and perceptions rather than understanding community norms and customs. IDIs can highlight the differences between individuals, elicit detailed information, and also provide a forum for follow-up questions.

Internal review boards (IRBs) are research monitoring bodies or committees, sometimes called *Helsinki*

Committees, whose approval is required for research funding and publication purposes. IRBs require that all precautions are taken so that participants are not exposed to harm by the study, and that the project is scientifically sound. They will require that follow-up care be provided with referrals, that a researcher/practitioner is clear about his or her role boundaries, and by ensuring that appropriate information and support are available.

Consent requires that participants be informed that research is not therapeutic. Some situations do not require consent when it is made clear that participants understand the study. Confidentiality must be maintained (e.g., the secure storage of tapes and transcripts), using as few details about participants as possible. This is to prevent anxiety and distress, exploitation, misrepresentation, and identification of participants in published papers. Validation for respondent refers to the process whereby researchers feed back the analysis to the participants before the findings are published.

SUMMARY — FROM INFORMATION TO KNOWLEDGE TO POLICY

Information is the basis for planning, organizing, managing, and providing high-quality care. The process begins with basic vital statistics and the epidemiology of infectious and noninfectious diseases in order to identify and quantify the health needs of the population. It extends into health information systems in order to manage and monitor the functioning of the health care system. Surveillance of health events at national, regional, and community levels depends on building information systems and linking data to provide community health profiles. This is fundamental to monitoring and managing health systems. It requires clear policy to ensure that information systems do not exist to serve only those who process the data at national levels, but are returned to the community level and linked with other data sources in readily usable formats (Box 3.29).

Information is widely available in health statistics and published data of all kinds, today more than ever on the Internet. Health policy formulation requires seeking the appropriate information and making intelligent use of it. Educating health workers in using information coordination and streamlining of data helps them understand the relevance and impact of their actions. Information systems and the flow of properly organized and disseminated data are vital for management. They are as important to the functioning of the system as an intelligence service is to a military operation. The vast and expensive mechanism of a health service operates in the dark without a continuously monitoring information system and applied research methods of epidemiology.

Throughout the world, health care systems are under critical scrutiny because of concerns over costs, accessibility, appropriateness, quality, and outcomes of care.

Box 3.29 Evidence-Based Public Health and the Burden of Proof

The Hippocratic Oath specifies: do good and do no harm. This has found expression in the Precautionary Principle, a contemporary redefinition of Bradford Hill's case for action. According to this principle, when in doubt about the possible presence of a hazard, the burden of proof is shifted from showing presence of risk to showing total absence of risk.

This principle creates a dilemma in public health and in clinical medicine in that it suggests the normal evidence required for action is without validity, implying any possible risk of intervention outweighs the risk of nonintervention.

Great care is warranted when introducing new public health interventions, but the weight of evidence must include not only epidemiologic studies but policies derived from Delphi consultative procedures and successful use of the intervention in large population groups over long periods of time, without substantive evidence of harmful effect.

A balance between the Precautionary Principle, the experience of "good pubic health practice" and epidemiologic evidence is often a delicate judgment, but is nonetheless essential for policy in this field. Last's definition of *evidence-based public* health is wise, with "application of best available experience in setting public health policies and priorities. The evidence comes from official vital and health statistics and from peer reviewed publications in epidemiology, sociology, economics and other relevant disciplines."

Failure to act on best practices and cumulative evidence can be an ethical and indeed a legal problem (see Chapter 15), where inordinate delay in implementing scientific and practical positive experience with public health interventions can allow serious morbidity and mortality to go unchecked when they are preventable.

The time lag between adequate scientific evidence and positive experience with good public health practices can be very long and measures that can save or improve the quality of life for large numbers of people are delayed in implementation due to lack of political motivation, priorities, and active or passive resistance by professional or lobby groups with other agendas.

Delay in adoption of a two-dose policy for measles vaccination and slow implementation in some developing countries cost millions of lives. Banning of DDT in the 1960s due to environmental concerns contributed to the resurgence of malaria, again costing millions of lives. The responsibility to keep up with scientific and best public health practices is an important responsibility of public health in balance with due precaution.

Source: Adapted from Last, J. M. 2007. A *Dictionary of Public Health.* New York: Oxford University Press.

The effectiveness of a health system is frequently on the political agenda. Quality assurance and accountability are critical in the operation of any health system. Health expenditures must be increasingly justified in terms of their need and cost-effectiveness, policy formulation, strategies, and priorities, taking into account economic, sociological, and political factors.

Curbing the soaring costs of health care is a necessity and not a matter of choice for governments and individuals if the WHO policy of Health for All is to be achieved. One means of reaching the goals and objectives of this policy is to develop an efficient health information system. Knowing the population, the epidemiologic patterns of its diseases, and its health care services and utilization, are all part of monitoring and feedback systems essential to allow the health system to evaluate health status, and keep pace with changes. They are therefore essential elements of the New Public Health.

ELECTRONIC RESOURCES

American College of Epidemiology. http://www.acepidemiology2.org/about/history.asp [accessed May 1, 2008]

American Public Health Association. http://www.alpha.org [accessed May 1, 2008]

As part of this drive, the Health Metrics Network is releasing a Monitoring Vital Events resource kit CD-ROM. This kit contains the tools and reference texts that countries can use to guide them in their work toward full civil registration, http://www.who.int/healthmetrics/tools/logbook/en/move/web/index.html

Australia's national agency for health and welfare statistics. Mathers, C., Vos, T., Stevenson, C., November 1999. Australian Institute of Health and Welfare, http://www.aihw.gov.au/publications/health/bdia/bdia-c01.pdf [accessed December 3, 2006]

Burden of Disease, http://www.hsph.harvard.edu/organizations/bdu/gbdmain.htm

Canada, Statistics Canada. http://www.statcan.ca [accessed May 1, 2008]

Canadian Institute for Health Information. http://www.cihi/stats/canhe.htm

Census Bureau — Statistical Abstract of the United States, http://www.census.gov [accessed July 16, 2007]

Dartmouth Atlas of Health Care in the United States. http://www.dartmouthatlas.org/ [accessed July 8, 2007]

Eurosurveillance contents. http://www.eurosurveillance.org/ [accessed May 2, 2008]

Gudnason, T., Briem, H. Euro Surveillance. 2008. An interactive central database of vaccinations in Iceland, http://www. eurosurveillance.org/edition/v13n02/080110_04.asp [accessed May 2, 2008]

Health Metrics Network is a global partnership of the World Health Organization (WHO) to address the lack of reliable health information in developing countries, http://www.who.int/healthmetrics/en/ [accessed May 1, 2008]

Health United States. 2006. http://www.cdc.gov/nchs/hus.htm [accessed February 23, 2007]

Health United States. 2007. http://www.cdc.gov/nchs/hus.htm [accessed December 20, 2007]

International Epidemiologic Association. November 2007. Good Epidemiologic Practice: IEA Guidelines for Epidemiologic Practice (GEP):

IEA Guidelines for Proper Conduct in Epidemiologic Research, http://www.dundee.ac.uk/iea/ [accessed December 22, 2007]

International Epidemiological Association, http://www.dundee.ac.uk/iea/ [accessed May 1, 2008]

National Center for Health Statistics, http://www.cdc.gov/nchs/ [accessed July 16, 2007]

National Center for Health Statistics, http://www.cdc.gov/nchs/

National Health and Nutrition Examination Survey (NHANES). http://www.cdc.gov/nchs/nhanes.htm [accessed May 4, 2008]

National Information Center on Health Service Research and Health Technology, Health Statistics Sources, http://www.nih.gov/nichsr/stats/contents/contents.html

New York State Department of health, through its agency — Statewide Planning Research and Cooperative System (SPARCS). http://www.health.state.ny.us/statistics/sparcs/index.htm [accessed May 2, 2008]

Organization for Economic Development (OECD) in Figures. http://www.oecd.org/home/0,3305,en_2649_201185_1_1_1_1_1,00.html [accessed May 1, 2008]

Pan American Health Organization (PAHO). http://www.paho.org/ [accessed May 2, 2008]

Population Health Intelligence System (PHIS) website for Ireland and Northern Ireland's Population Health Observatory (INIsPHO). 2006. access through the INIsPHO website or directly at http://www.inispho.org/phis, http://www.inispho.ie/phis, and http://inispho.org.uk/phis

United Kingdom Government Statistical Service. http://www.statistics.gov.uk [accessed May 2, 2008]

U.K. Department of Health. http://www.dh.gov.uk/en/index.htm [accessed May 2, 2008]

United States Centers for Disease Control and Prevention (CDC). http://www.cdc.gov/

U.S. Census Bureau. 2008. The Statistical Abstract of the United States. The National Data Book, http://www.census.gov/compendia/statab/; http://www.census.gov/compendia/statab/brief.html [accessed May 1, 2008]

World Health Organization (WHO). http://www.who.int/whosis/ [accessed May 2, 2008]

WHO European Region, Health for All Data Set, WHO Copenhagen. http://www.euro.who.int/ [accessed November 26, 2007]

World Health Report. http://www.who.int/whr/en/ [accessed July 16, 2007]

RECOMMENDED READINGS

Birkmeyer, J. D., Siewers, A. E., Finlayson, E. V., Stukel, T. A., Lucas, F. L., Batista, I., Welch, H. G., Wennberg, D. E. 2002. Hospital volume and surgical mortality in the United States. *New England Journal of Medicine*, 346:1128–1137.

Black, D. 1993. Deprivation and health. *British Medical Journal*, 307:1630–1631.

Boak, M. B., M'ikanatha, N. M., Day, R. S., Harrison, L. H. 2007. Internet death notices as a novel source of mortality surveillance data. *American Journal of Epidemiology*, December 12, 2007 [Epub ahead of print].

Bravata, D. M., McDonald, K. M., Smith, W. M., Rydzak, C., Szeto, H., Buckeridge, D. L., Haberland, C., Owens, D. K. 2004. Systematic review: Surveillance systems for early detection of bioterrorism-related diseases. *Annals of Internal Medicine*, 40:910–922.

Breslow, L. 2006. Health measurement in the Third Era of Health. *American Journal of Public Health*, 96:17–19.

Centers for Disease Control and Prevention. 2007. Carbon monoxide-related deaths — United States, 1999–2004. *Morbidity and Mortality Weekly Report*, 56:1309–1312.

Declich, S., Carter, A. O. 1994. Public health surveillance: Historical origins, methods and evaluation. *Bulletin of the World Health Organization*, 72:285–304.

Evans, A. S. 1976. Causation and disease: The Henle-Koch postulates revisited. *Yale Journal of Biology and Medicine*, 49:175–195.

Evans, A. S., Mueller, N. E. 1990. Viruses and cancer. Causal associations. *Annals of Epidemiology*, 1:71–92.

Forgacs, I., Loganayagam, A. [editorial]. 2008. Overprescribing proton pump inhibitors is expensive and not evidence based. *British Medical Journal*, 336:2–3.

Harnden, A., Grant, C., Harrison, T., et al. 2006. Whooping cough in school age children with persistent cough: Prospective cohort study in primary care. *British Medical Journal*, 333:174–177.

Jemal, A., Thun, M. J., Ward, E. E., Henley, S. J., Cokkinides, V. E., Murray, T. E. 2008. Mortality from leading causes by education and race in the United States, 2001. *American Journal of Preventive Medicine*, 34:1–8.

Kim, S. S., Frimpong, J. A., Rivers, P. A., Kronenfeld, J. J. 2007. Effects of maternal and provider characteristics on up-to-date immunization status of children aged 19 to 35 months. *American Journal of Public Health*, 97:259–266.

Kozak, L. J., DeFrances, C. J., Hall, M. J. 2006. National hospital discharge survey: 2004 annual summary with detailed diagnosis and procedure data. *Vital Health Statistics*, 13:1–209.

Lilienfeld, D. 2007. Celebration: William Farr (1807–1883): An appreciation on the 200th anniversary of his birth. *International Journal of Epidemiology*, 36:985–987.

Lippeveld, T., Sauerborn, R., and Bodart, C. (eds.). 2000. *Design and Implementation of Health Information Systems*. Geneva: World Health Organization.

Liu, J. H., Zingmond, D. S., McGory, M. L., SooHoo, N. F., Ettner, S. L., Brook, R. H., Ko, C. Y. 2006. Disparities in the utilization of high-volume hospitals for complex surgery. *Journal of the American Medical Association*, 296:1973–1980.

Mathers, C. D., Loncar, D. 2006. Projections of global mortality and burden of disease from 2002 to 2030. *Public Library of Science Medicine (PLoS Med)*, 3(11): e442doi:10.1371/journal.pmed.0030442.

Mays, G. P., McHugh, M. C., Shim, K., Perry, N., Lenoawey, D., Halverson, P. K., Moonsinghe, R. 2006. Institutional and economic determinants of public health systems performance. *American Journal of Public Health*, 96:523–531.

Morabia, A. Epidemiologic interactions, complexity, and the lonesome death of Max von Pettenkofer. *American Journal of Epidemiology*, 166:1233–1238.

Murray, C. J. L. 1994. Quantifying the burden of disease: The technical basis for disability-adjusted life years. *Bulletin of the World Health Organization*, 72:429–445.

Murray, C. J., Laakso, T., Shibuya, K., Hill, K., Lopez, A. D. 2007. Can we achieve Millennium Development Goal 4? New analysis of country trends and forecasts of under-5 mortality to 2015. *Lancet*. September 22, 2007, 370(9592):1040–1054.

Pearce, N. 1996. Traditional epidemiology, modern epidemiology, and public health. *American Journal of Public Health*, 86:678–683.

Persell, D. J., Robinson, C. H. 2008. Detection and early identification in bioterrorism events. *Family and Community Health*, 31:4–16.

Singh, G. K., Kogan, M. D. 2007. Persistent socioeconomic disparities in infant, neonatal, and postneonatal mortality rates in the United States, 1969–2001. *Pediatrics*, 119:928–939.

Singh, G. K., Kogan, M. D. 2007. Widening socioeconomic disparities in US childhood mortality, 1969–2000. *American Journal of Public Health*, 97:1658–1665.

Susser, M., Susser, E. 1996. Choosing a future for epidemiology: I. Eras and paradigms; and II. From black box to Chinese boxes and eco-epidemiology. *American Journal of Public Health*, 86:668–673.

Tarozzi, A., Deaton, A. 2007. Using census and survey data to estimate poverty and inequality for small areas. http://www.princeton.edu/~rpds/downloads/tarozzi_deaton_small_areas_final.pdf [accessed January 15, 2008].

Tulchinsky, T. H., Patton, M. M., Randolph, L. A., Myer, M. R., Linden, J. V. 1993. Mandating vitamin K prophylaxis for newborns in New York State. *American Journal of Public Health*, 83:1166–1168.

Tulchinsky, T. H., Ginsberg, G. M., Ishovitz, J., Shihab, S., Fischbein, A., Richter, E. D. 1998. Cancer in ex-asbestos cement workers in Israel, 1953-1992. *American Journal of Industrial Medicine*, 35:1–8.

Weijerman, M. E., van Furth, A. M., Vonk Noordegraaf, A., van Wouwe, J. P., Broers, C. J., Gemke, R. J. 2008. Prevalence, neonatal characteristics, and first-year mortality of Down syndrome: A national study. *Journal of Pediatrics*, 152:15–19.

Yabroff, K. R., Gordis, L. 2003. Assessment of a national health interview survey-based method of measuring community socioeconomic status. *Annals of Epidemiology*, 13:721–726.

BIBLIOGRAPHY

Abramson, J. H. 2001. *Making Sense of Data*, 3rd ed. New York: Oxford University Press.

Abramson, J. H., Abramson, Z. H. 1999. *Survey Methods in Community Medicine*, 5th ed. Edinburgh: Churchill Livingstone.

Abramson, J. H., Abramson, Z. H. 2008. *Research Methods in Public Health*, forthcoming in 2008.

Anand, S., Fabienne, P., Amartya, S. 2006. *Public Health, Ethics, and Equity*. New York: Oxford University Press.

Beaglehole, R., Bonita, R., Kjellstrom, T. 1993. *Basic Epidemiology*. Geneva: World Health Organization.

Bennet, S., Woods, T., Liyanage, W. M., Smith, D. L. 1991. A simplified general method for cluster-sample surveys of health in developing countries. *World Health Statistics Quarterly*, 44:98–106.

Chen, J. T., Rehkopf, D. H., Waterman, P. D., Subramanian, S. V., Coull, B. A., Cohen, B., Elliott, P., Cuzick, J., English, D., and Stern, R. (eds.). 1992. *Geographical and Environmental Epidemiology: Methods for Small Area Studies*. World Health Organization, Regional Office for Europe, Copenhagen: Oxford University Press.

Fleming, S. T., Scuttchfield, F. D., Tucker, T. C. 2000. *Managerial Epidemiology*. Chicago: American College of Healthcare Executives.

Ginsberg, G. M., Tulchinsky, T. H. 1992. Regional differences in cancer incidence and mortality in Israel: Possible leads to occupational causes. *Israel Journal of Medical Sciences*, 28:534–543.

Gordis, L. 1996. *Epidemiology*. Philadelphia: W. B. Saunders Co.

Ibrahim, M. A. 1985. *Epidemiology and Health Policy*. Rockville, MD: Aspen.

International Epidemiologic Association. 1996.

Kuller, L. H. [editorial]. 1995. The use of existing databases in morbidity and mortality studies. *American Journal of Public Health*, 85:1198–1199.

Lang, T., Duceimetiere, P. 1995. Premature cardiovascular mortality in France: Divergent evolution between social categories from 1970–1990. *International Journal of Epidemiology*, 24:331–339.

Last, J. M. (ed.). 2001. *A Dictionary of Epidemiology,* 4th ed. New York: Oxford University Press.

Last, J. M. (ed.). 2007. *A Dictionary of Public Health.* New York: Oxford University Press.

Levi, F., Lucchini, F., Negri, F., La Vecchia, C. 2002. Trends in mortality from cardiovascular and cerebrovascular diseases in Europe and other areas of the world. *Heart*, 88:119–124.

Lilienfeld, D. E., Stolley, P. D. 1994. *Foundations of Epidemiology*, 3rd ed. New York: Oxford University Press.

Mays, G. P., McHugh, M. C., Shim, K., Perry, N., Lenoawey, D., Halverson, P. K., Moonsinghe, R. 2006. Institutional and economic determinants of public health system performance. *American Journal of Public Health*, 96:523–531.

Mullan, F., Epstein, L. 2002. Community-oriented primary care: New relevance in a changing world. *American Journal of Public Health*, 92:1748–1755.

Oleske, D. M. (ed.). 1995. *Epidemiology and the Delivery of Health Care Services.* New York: Plenum Press.

Ostrem, M., Krieger, N. 2006. Mapping and measuring social disparities in premature mortality: The impact of census tract poverty within and across Boston neighborhoods, 1999–2001. *Journal of Urban Health*, 83:1063–1084.

Poikolainen, K., Eskola, J. 1995. Regional and social class variation in the relative risk of death from amenable causes in the city of Helsinki, 1980–1986. *International Journal of Epidemiology*, 24:114–118.

Rice, D. P. 2001. *Health Statistics: Past, Present, Future. Towards a Health Statistics System for the 21st Century: Summary of Workshop.* Washington, DC: The National Academic Press.

Richter, E. D., Laster, R. 2004. The Precautionary Principle, epidemiology and the ethics of delay. *International Journal of Occupational Medicine and Environmental Health*, 17(1):9–16.

Schwartz, R. M., Gagnon, D. E., Muri, J. H., Zhao, Q. R., Kellogg, R. 1999. Administrative data for quality improvement. *Pediatrics*, 103:291–301.

Singh, G. K. 2003. Area deprivation and widening inequalities in US mortality, 1969–1998. *American Journal of Public Health*, 93:1137–1143.

Smith, G. D., Egger, M. [editorial]. 1992. Socioeconomic differences in mortality in Britain and the United States. *American Journal of Public Health*, 82:1079–1081.

Subramanian, S. V., Chen, J. T., Rehkopf, D. H., Waterman, P. D., Krieger, N. 2005. Racial disparities in context: A multilevel analysis of neighborhood variations in poverty and excess mortality among black populations in Massachusetts. *American Journal of Public Health*, 95(2):260–265.

Swedish Hospital Discharge Registry. http://www.sos.se/epc/english/ParEng.htm [accessed April 24, 2005]

Tulchinsky, T. H. 1982. Evaluation of personal health services as a basis for health planning: A review with applications for Israel. *Israel Journal of Medical Sciences*, 18:197–209.

Tulchinsky, T. H., Ginsberg, G. M., Shihab, S., Laster, R. 1992. Mesothelioma mortality among an Israeli ex-asbestos worker population 1953–1990. *Israel Journal of Medical Sciences*, 28:543–547.

Tulchinsky, T. H., Patton, M. M., Randolph, L. A., Meyer, M. R., Linden, J. V. 1993. Mandating vitamin K prophylaxis for newborns in New York State. *American Journal of Public Health*, 1993;83:1166–1168.

Ulin, P. R., Robinson, E. T., Tolley, E. E. 2005. *Qualitative Methods in Public Health. A Field Guide for Applied Research.* San Francisco: Jossey-Bass.

United States Public Health Service. 2007. *Health United States, 2007, With Chartbook on Trends in the Health of Americans.* Hyattsville, MD: U.S. Department of Health and Human Services.

Willett, W. 1998. *Nutritional Epidemiology*, 2nd ed. Monographs in Epidemiology and Biostatistics, Volume 30. New York: Oxford University Press.

Chapter 4

Communicable Diseases

Introduction
Public Health and the Control of Communicable Disease
The Nature of Communicable Disease
Host–Agent–Environment Triad
Classifications of Communicable Diseases
Modes of Transmission of Disease
Immunity
Surveillance
Health Care–Associated Infections
Endemic and Epidemic Disease
    Epidemic Investigation
Control of Communicable Diseases
    Treatment
    Methods of Prevention
Vaccine-Preventable Diseases
    Immunization Coverage
Vaccine-Preventable Diseases
Essentials of an Immunization Program
    Regulation of Vaccines
    Vaccine Development
Control/Eradication of Infectious Diseases
    Smallpox
    Eradication of Poliomyelitis
    Other Candidates for Eradication
    Future Candidates for Eradication
Vector-Borne Diseases
    Malaria
    Rickettsial Infections
    Arboviruses (Arthropod-Borne Viral Diseases)
    Lyme Disease
Parasitic Diseases
    Echinococcosis
    Tapeworm
    Onchocerciasis
    Dracunculiasis
    Schistosomiasis
    Leishmaniasis
    Trypanosomiasis
    Other Parasitic Diseases
Legionnaire's Disease
Leprosy
Trachoma
Sexually Transmitted Infections
    Syphilis
    Gonorrhea
    Other Sexually Transmitted Infections
    Control of Sexually Transmitted Infections
    HIV/AIDS
Diarrheal Diseases
    Salmonella
    Shigella
    Escherichia coli
    Cholera
    Viral Gastroenteritis
    Parasitic Gastroenteritis
    A Program Approach to Diarrheal Disease Control
Acute Respiratory Infections
Inequalities in Control of Communicable Diseases
Communicable Disease Control in the New Public Health
Summary
Electronic Resources
Recommended Readings
Bibliography

INTRODUCTION

Despite advances in medical sciences and public health, infectious diseases remain a central task of public health in the twenty-first century, especially HIV/AIDS, TB, malaria, SARS, avian flu, and others. Globalization has facilitated the spread of many infectious agents to all corners of the globe. Mass travel, economic globalization, and climate changes along with accelerating urbanization of human populations are causing environmental disruption, including global warming. There are and will be more consequences in international transmission of infectious diseases than are now known, in humans and wildlife.

This chapter describes communicable diseases and programs for their prevention, control, elimination, and eradication. Control of communicable disease requires a systems approach using available resources effectively, mobilizing environmental measures, immunization, and clinical and health systems. Rapid transportation and communication make a virus outbreak in any part of the world an international concern, both for professionals and the general public. A basic understanding of infectious diseases is therefore an expectation of any student, just as a general knowledge of family health, chronic disease, nutrition, and economics are part of the modern public health culture.

The material presented in this chapter is intended to give an introduction for the student or review for the public health practitioner, with an emphasis on the applied aspects of communicable disease control. We have relied for the content of this chapter on several standard references, especially Heymann's *Control of Communicable Diseases Manual*, 18th ed. *WHO Vaccine Preventable Diseases Monitoring System: 2007 Global Summary*; *Jawetz, Melnick and Adelberg's Medical Microbiology*, 21st ed., along with *Morbidity and Mortality Weekly Report* of the Centers for Disease Control and Prevention (CDC), and WHO's Weekly Epidemiologic Record (WER). ProMed is a highly recommended Harvard University–based Web site, frequent update source of current infectious disease outbreaks around the world, available with free subscription at Web site listed in electronic resources. We also have relied on electronic sources such as PubMed, the American Academy of Pediatrics, World Health Organization (WHO), and United Nations Children's Fund (UNICEF) Web sites, as well as library access journals. The references listed will augment the limited discussion possible in this text.

PUBLIC HEALTH AND THE CONTROL OF COMMUNICABLE DISEASE

Organized public health grew out of the sanitation movement of the mid-nineteenth century which sought to reduce the environmental and social factors in communicable disease (Box 4.1). Traditionally, the prevention and control of communicable diseases has been accomplished by sanitation, safe water and food supply, isolation, and immunization.

Box 4.1 Communicable Disease

A communicable disease "is an illness due to a specific infectious agent or its toxic products that arises through transmission of that agent or its products from an infected person, animal, or inanimate reservoir to a susceptible host." Transmission may be direct from person to person, or indirect through an intermediate plant or animal host, vector, or the inanimate environment.

Source: Heymann, D. L. (ed.) 2004. *Control of Communicable Diseases Manual*, 18th ed. Washington, DC: American Public Health Association.

The potential for infectious disease to disturb or destroy human life still exists and may increase as infectious diseases evolve and escape current man-made control mechanisms. The spread of plague throughout Europe and Asia in the fourteenth century and subsequent pandemics of smallpox, tuberculosis, syphilis, measles, cholera, and influenza show the explosive potential and epidemic nature of infectious diseases. The spread of AIDS since the 1980s; ongoing cholera epidemics in Asia, Africa, and South America; and diphtheria in the former Soviet Union in the 1990s, remind us why communicable disease control is still one of the major responsibilities of public health.

Both the miasma (environment–host) and bacteriologic (agent–host) theories contributed to great achievements in the control of communicable disease in the first half of the twentieth century. The emergence of the germ theory in the late nineteenth century led to the sciences of bacteriology and immunology, growing out of the work of Jenner, Pasteur, Koch, Lister, and many others (see Chapter 1). The control of the vaccine-preventable diseases has been a boon to humankind, saving countless millions of lives and providing a cornerstone for public health. Despite this, millions of children still die annually from preventable diseases. Infectious diseases of childhood are still tragically undercontrolled internationally. Infectious diseases also undermine the health of other vulnerable groups in the population, such as the elderly and the chronically ill, thereby playing a major role in the economics of health care.

Great strides have been made in the control of communicable diseases through environmental sanitation, safe foods, vaccination, and antibiotics, as seen in Figure 4.1, in the United States and equally in the other industrialized countries. However, the field of infectious disease continues to be dynamic and challenging. Emerging infectious disease threats from new diseases not previously identified, such as HIV and SARS, or new variants of old diseases with resistance to current methods of treatment together provide great challenges to public health. Increasing resistance to therapeutic agents augments the need for new strategies and coordination between public health and clinical services. Understanding the principles and methods of communicable disease control and eradication is important for all health providers and public health personnel.

THE NATURE OF COMMUNICABLE DISEASE

An infectious disease may or may not be clinically manifest so that a person may carry the disease agent without having clinical illness. Acute infectious diseases are intense or short-term, but may have long-term sequelae of great public health importance, such as post-streptococcal glomerulonephritis or rheumatic heart disease. Other infectious diseases are chronic with their own long-term

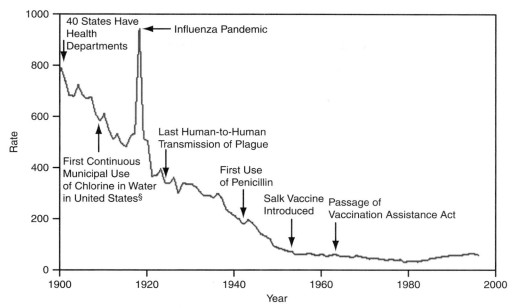

FIGURE 4.1 Crude death rate from infectious diseases – United States 1900–1999. Source: Centers for Disease Control. 1999. Achievements in Public Health, 1900–1999: Control of Infectious Diseases. *Morbidity and Mortality Weekly Report*, 48:621–629.

effects, such as HIV infection or peptic ulcers. Infections may have both short-term and long-term morbidity, as with viral hepatitis infections. The stages of infectious disease include:

1. Exposure and infection;
2. Presymptomatic/prodromal stage;
3. Nonmanifested or subclinical disease;
4. Clinically manifested disease and its progression;
5. Resolution, recovery, remission, relapse, suprainfection, or death; and
6. Long-term sequelae.

Each disease has its own characteristic organism and natural history from onset to resolution. Many infectious diseases may remain at a presymptomatic or subclinical stage without progressing to clinical symptoms and signs, but may be transmissible to other persons. Even a subclinical disease may cause an immunologic effect, producing immunity. The drama of infectious disease is exemplified in the tragic event of the plague in the fourteenth century and its periodic recurrence as in the epidemic of 1665 in London, described by Daniel Defoe (Box 4.2).

HOST–AGENT–ENVIRONMENT TRIAD

The host–agent–environment triad, discussed in Chapter 2, is fundamental to the success of understanding transmission of infectious diseases and their control, including those well known, those changing their patterns, and those

Box 4.2 Daniel Defoe—A Journal of the Plague Year, London, 1665

"It was about the beginning of September 1664, that I, among the rest of my neighbors, heard, in ordinary discourse that the plague had returned again in Holland; for it had been very violent there, and particularly in Amsterdam and Rotterdam, in the year 1663, whither they say, it was brought, some said from Italy, others from the Levant, among some goods, which were brought home by their Turkey fleet; others said it was brought from Candia; others from Cyprus. It mattered not from whence it came; but all agreed it was come into Holland again.

"It was now mid-July and the plague, which had chiefly raged at the other end of town … began to now come eastwards toward the part where I lived. It was to be observed, indeed, that it did not come straight on toward us; for the city, that is to say within the walls, was indifferently healthy still; nor was it got over the water into Southwark; for though there died that week 1,268 of all distempers, whereof it might be supposed above 900 died of the plague, yet there was but 28 in Southwark, Lambeth parish included; whereas in the parishes of St. Giles and St. Martin-in-the-Fields alone there died 421."

Source: Defoe, D. 1723. A *Journal of the Plague Year*. Winnipeg: Meridian Classic, 1984, reprint.

newly emerging or escaping current methods of control. Infection occurs when the organism successfully invades the host's body, where it multiplies and produces an illness.

A host is a person or other living animal, including birds and arthropods, who provides a place for growth and sustenance to an infectious agent under natural conditions. Some organisms, such as protozoa or helminths, may pass successive stages of their life cycle in different hosts, but the definitive host is the one in which the organism passes its sexual stage. The intermediate host is where the parasite passes the larval or asexual stage. A transport host is a carrier in which the organism remains alive, but does not develop.

An agent of an infectious disease is necessary, but not always sufficient to cause a disease or disorder. The infective dose is the quantity of the organism needed to cause clinical disease. A disease may have a single agent as a cause, or it may occur as a result of the agent in company with contributory factors, whose presence is also essential for the development of the disease. A disease may be present in an infected person in a dormant form such as tuberculosis, or a subclinical stage such as poliomyelitis or HIV, without clinical paralytic disease in the case of polio or before clinical AIDS appears in the case of HIV. The virulence or pathogenicity of an infective agent is the capacity of an infectious agent to enter the host, replicate, damage tissue, and cause disease. Virulence describes severity of clinical disease and may vary among serotypes or strains of the same agent.

The environment provides a reservoir for the organism and the mode of transmission by which the organism reaches a new host. The reservoir is the natural habitat where an infectious agent lives and multiplies, from which it can be transmitted directly or indirectly to a new host. Reservoirs may be in people, animals, arthropods, plants, soil, or substances in which an organism normally lives and multiplies, and on which it depends for survival or in which it survives in a dormant form. A fomite is an inanimate object contaminated with infectious material which may transmit disease, such as improperly sanitized medical equipment.

Contacts are persons or animals who have been in association with an infected person, animal, fomite, or environment that might provide a risk for acquiring the infective agent. Persons or animals that harbor a specific infectious agent, often in the absence of discernible clinical disease, and who serve as a source of infection or contamination of food, water, or other materials, are carriers. A carrier may have an unapparent infection (a healthy carrier) or may be in the incubation or convalescent stage of the infection.

CLASSIFICATIONS OF COMMUNICABLE DISEASES

Communicable diseases may be classified by a variety of methods: by clinical syndrome, mode of transmission, methods of prevention (e.g., vaccine-preventable), or by major organism classification; that is, viral, bacterial, fungal, and parasitic disease.

A virus is a nucleic acid molecule (RNA or DNA) encapsulated in a protein coat or capsid. The virus is not a complete cell and can only replicate inside a living cell. The capsid may have a protective lipid-containing envelope. The capsid and envelope facilitate attachment and penetration into host cells, and often contain virulence factors. Inside the host cell, the nucleic acid molecule utilizes cellular proteins and processes for virus replication. Prions — discovered in recent years (Stanley Prusiner, Nobel Prize, 1997) — are proteins, which in a properly folded state, induce disease. As infectious agents, prions cause a number of degenerative central nervous system diseases, including spongiform encephalopathy in livestock (mad cow disease and scrapie) and humans (variant Creutzfeldt-Jakob disease).

Bacteria are unicellular organisms that reproduce sexually or asexually and can exist in an environment with oxygen (aerobic) or in a situation lacking oxygen (anaerobic). Some may enter a dormant state and form spores where they are protected from the environment and may remain viable for years. Bacteria include a nucleus of chromosomal DNA material within a membrane surrounded by cytoplasm, itself usually enclosed by a cellular membrane. Bacteria are classified by morphology and growth conditions, including coloration under Gram stain (gram-negative or gram-positive), microscopic morphology, immunologic (antigen) or molecular (DNA) markers, or by the diseases they may cause. Bacteria include both indigenous flora (normal resident) bacteria and pathogenic (disease-causing) bacteria. Pathogenic bacteria cause disease by invading, overcoming natural or acquired resistance, and multiplying in the body. Bacteria may produce a toxin or poison that can affect a body site distant from where the bacterial replication occurs, such as in tetanus. Bacteria may also initiate an excessive immune response, producing damage to other body tissues away from the site of infection (e.g., acute rheumatic fever and glomerulonephritis).

Mycoses are infections caused by molds and yeasts. Clinical manifestations of fungal disease range from relatively mild superficial infection to systemic, life-threatening conditions. Immunocompromised individuals are at elevated risk. *Cryptococcus, Candida, Aspergillus*, and *Mucor* molds are among the leading causes of morbidity in HIV disease and among immunosuppressed populations. *Pneumocystis jiroveci* (formerly *P. carinii*), once thought to be a protozoan, is now classified a fungus, based on genetic analysis. Common dermatophytic infections, known as tinea, are caused by fungi invading the hair, skin, or nails, and occur in nearly all living organisms.

Parasitology studies protozoa, helminths, and arthropods that live within, on, or at the expense of a host. Protozoa include oxygen-producing, unicellular organisms such as the flagellates *Giardia* and *Trichomonas,* and amoebae such as *Entamoeba*, in enteric and gynecologic disorders. Sporozoa are parasites with complex life cycles in different hosts, such as cryptosporidium or malarial parasites. Helminths are worms that infest humans, especially in poor sanitation and tropical areas. Arthropods, the most numerous of animal species, include lice, fleas, sandflies, blackflies, and ticks, are important disease vectors. They can live at the body's surface (ectoparasites) and transmit bacterial, viral, rickettsial, or other diseases, or by oral–fecal transmission, such as *Shigella* and *E. coli*, in or via biological effects within the host such as in malaria. This group constitutes one of the most important public health threats globally and their control is a continuing challenge for public health.

MODES OF TRANSMISSION OF DISEASE

Transmission of diseases is by the spread of an infectious agent from a source or reservoir to a person (Table 4.1). Direct transmission from one host to another occurs during touching; biting; kissing; sexual intercourse; projection via droplets, as in sneezing, coughing, or spitting; or by entry through the skin. Indirect transmission includes via aerosols of long-lasting suspended particles in air and fecal–oral transmission such as food- and waterborne as well as by poor hygienic conditions with fomites, such as soiled clothes, handkerchiefs, toys, or other objects. Transmission in medical settings is common and preventable by hand washing and sterile techniques.

Vector-borne diseases are transmitted via crawling or flying insects, in some cases with multiplication and development of the organism in the vector, as in malaria.

The subsequent transmission to humans is by injection of salivary gland fluid during biting or by deposition of feces, urine, or other material capable of penetrating the skin through a bite wound or other trauma. Transmission may occur with insects as a transport mechanism, as in *Shigella* on the legs of a housefly.

Airborne transmission occurs indirectly via infective organisms in small aerosols that may remain suspended for long periods of time and which easily enter the respiratory tract. This occurs frequently with viruses such as influenza, the common cold, and measles. Particles of dust may spread organisms from soil, clothing, or bedding.

Vertical transmission occurs from one generation to another or from one stage of the insect life cycle to another stage. Maternal–infant transmission occurs during pregnancy (transplacental), delivery (as in gonorrhea), or breastfeeding (e.g., HIV, with transfer of infectious agents from mother to fetus or newborn).

IMMUNITY

Resistance to infectious diseases is related to many host and environmental factors, including age, sex, pregnancy, nutrition, trauma, fatigue, living and socioeconomic conditions, and emotional status. Good nutritional status has a protective effect and bolsters immunocompetence. Vitamin A supplements reduce complication rates of measles and enteric infections. Tuberculosis may be present in an individual whose resistance is sufficient to prevent clinical disease, but the infected person is a carrier of an organism which can be transmitted to another or cause clinical disease if the person's susceptibility is reduced (Box 4.3).

Immunity is resistance to infection resulting from presence of specific antibodies and complement proteins or cells that act on the microorganism associated with a

TABLE 4.1 Classification of Infectious Diseases by Principal Modes of Transmission

Mode	Method	Examples
Direct	Airborne (droplet and aerosols)	Viral exanthems (measles), streptococcal diseases, various upper and lower respiratory tract diseases, tuberculosis, Legionnaire's disease, influenza
Direct	Physical contact	Leprosy, impetigo, scabies, anthrax
Direct	Sexual contact	HIV, syphilis, gonorrhea, herpes genitalis, hepatitis B, chlamydia, human papillomavirus
Indirect	Blood and blood products	HIV, hepatitis B, hepatitis C
Indirect	Oral-fecal Hygiene Food-borne Water-borne	Cholera, *Shigella*, *Salmonella*, typhoid, botulism, *Campylobacter*, *Staphylococcus aureus*, cryptosporidium, *Listeria*, worms, *Giardia*, hepatitis A, rotavirus, enteroviruses, poliovirus, adenoviruses, *Entamoeba histolytica*
Indirect	Transcutaneous	Vector-borne via insects (arthropod): malaria, viral hemorrhagic fevers, schistosomiasis, plague Animal bite (zoonoses): rabies Health care (iatrogenic): hospital infections, HIV, hepatitis B Self-injected (illicit drug users): HIV, hepatitis B
Vertical	Congenital Maternal-fetal	Congenital rubella syndrome, congenital syphilis, gonorrheal ophthalmia, cytomegalovirus (CMV) HIV, rubella, syphilis, hepatitis B, gonorrhea, chlamydia

---

Box 4.3 Basic Terms in Immunology of Infectious Diseases

Infectious agent a pathogenic organism (e.g., virus, bacteria, rickettsia, fungus, protozoa, or helminth) capable of producing infection or an infectious disease.

Infection the process of entry, development, and proliferation of an infectious agent in the body tissue of a living organism (human, animal, or plant) overcoming body defense mechanisms, resulting in an inapparent or clinically manifest disease.

Antigen a substance (e.g., protein, polysaccharide) capable of inducing specific response mechanisms in the body. An antigen may be introduced into the body by invasion of an infectious agent, by immunization, inhalation, ingestion, or through the skin, wounds, or via transplantation.

Antibody a protein molecule formed by the body in response to a foreign substance (an antigen) or acquired by passive transfer. Antibodies bind to the specific antigen that elicits its production, causing the infective agent to be susceptible to immune defense mechanisms against infections (e.g., humoral and cellular).

Immunoglobulins antibodies that meet different types of antigenic challenges. They are present in blood or other body fluids, and can cross from a mother to fetus *in utero*, providing protection during part of the first year of life. There are five major classes (IgG, IgM, IgA, IgD, and IgE) and subclasses based on molecular weight.

Antisera or antitoxin materials prepared in animals for use in passive immunization against infection or toxins.

Source: Brooks, G. E., Butel, J. S., Morse, S. A. 2004 *Jawetz, Melnick and Adelberg's Medical Microbiology*, 23rd ed. Stamford, CT: Appleton & Lange.

---

specific disease or toxin. Immunity can be acquired by response to an organism or its antigenic components in the body of a person having the infective organism, producing natural immunity, or by immunization. In active immunization, the body responds to introduced antigens by producing antibodies. Passive immunity is temporary, by the passage of preformed antibody from mother to infant in breast milk or injection of preformed immunoglobulins. The body also reacts to infective antigens by cellular responses, including those that directly defend against invading organisms and other cells which produce antibodies.

The immune response is the resistance of a body to specific infectious organisms or their toxins provided by a complex interaction including:

Humoral
  a. B cells (bone marrow and spleen) produce antibodies which circulate in the blood.
  b. Complement proteins, a humoral response which causes lysis of foreign cells.

Cell Mediated
  c. T cell immunity is provided by sensitization of lymphocytes of thymus origin to mature into cytotoxic cells capable of destroying virus-infected or foreign cells.
  d. Phagocytosis, a cellular mechanism which ingests microorganisms (macrophages and neutrophils).

SURVEILLANCE

Surveillance of disease is the continuous scrutiny of all aspects of occurrence and spread of disease pertinent to effective control of that disease. Maintaining ongoing surveillance is one of the basic duties of a public health system, and is vital to the control of communicable disease, providing the essential data for tracking of disease, planning interventions, and responding to future disease challenges. Surveillance of infectious disease incidence relies on reports of notifiable diseases by physicians, supplemented by individual and summary reports of public health laboratories. Such a system must concern itself with the completeness and quality of reporting and potential errors and artifacts. Quality is maintained by seeking clinical and laboratory support to confirm first reports. Completeness, rapidity, and quality of reporting by physicians and laboratories should be emphasized in undergraduate and postgraduate medical education. Enforcement of legal sanctions may be needed where standards are not met. Surveillance of infectious diseases includes the following:

1. Morbidity reports from clinics to public health offices;
2. Mortality reports from attending doctors to vital records;
3. Reports from selected sentinel centers, e.g., emergency rooms, pediatric centers;
4. Special field investigations of epidemics or individual cases;
5. Laboratory monitoring of infectious agents and therapeutic response in population samples;
6. Data on supply, use, and side effects of vaccines, toxoids, immunoglobulins;
7. Data on vector control activities such as insecticide use;
8. Immunity levels in samples of the population at risk;
9. Review of current literature on the disease;
10. Epidemiologic and clinical reports from other jurisdictions.

Epidemiologic monitoring based on individual and aggregated reports of infectious diseases provide data vital to planning interventions at the community level or for individual patients, along with other information sources such as hospital discharge data and monitoring of sentinel centers. These may be specific medical or community sites that are representative of the population and are able to provide good levels of reporting to monitor an area or population group. A sentinel center can be a pediatric practice site, a hospital emergency room, or other location which will provide a "finger on the pulse" to assess suspicious changes occurring in the community. It can also include monitoring in a location previously known for disease transmission, such as Hong Kong in relation to influenza typing for vaccine planning, production, and distribution.

Epidemiologic analysis provided by government public health agencies should be published weekly, monthly, and annually and distributed to a wide audience of public health and health-related professionals throughout the country. Feedback is vital in order to promote involvement and improved quality of data, as well as to allow evaluation of local situations in comparison to other areas. In a federal system of government, national agencies report regularly on all state or provincial health patterns. State or provincial health authorities provide data to the counties and cities in their jurisdictions. Such data should also be readily available to researchers in other government agencies and academic settings for further research and analysis.

Notifiable diseases are those which a physician is legally required to report to state or local public health officials, by reason of their contagiousness, severity, frequency, or other public health importance (Table 4.2). Public health laboratory services provide validation of clinical and epidemiologic reports. They also provide day-to-day supervision of public health conditions, and can monitor communicable disease and vaccine efficacy and coverage. In addition, they support standards of clinical laboratories in biochemistry, microbiology, and genetic screening.

With newly emerging diseases and those spread far from their previously known habitat, and most especially because of the threats of pandemics such as SARS and more worryingly avian influenza, surveillance for human and animal disease is crucial to the societies we live in, including the global society. The first diagnosis of a

TABLE 4.2 Notifiable Infectious Diseases in the United States, 2007

AIDS/ HIV	Mumps
Arboviral disease	Pertussis (whooping cough)
Anthrax	Plague
Botulism	Poliomyelitis, paralytic or non-paralytic
Brucellosis (undulant fever)	Psittacosis
Chancroid	Q-fever
Chlamydia trachomatis, genital infection	Rabies (animal and human)
Cholera	Rocky Mountain spotted fever
Coccidiomycosis	Rubella and rubella congenital syndrome
Cryptosporidiosis	Rubella congenital syndrome
Cyclosporiasis	Salmonellosis
Diphtheria	Severe acute respiratory syndrome–associated coronavirus
Escherichia coli, Shiga toxin-producing (STEC)	Shigellosis
Erlichiosis	Smallpox
Giardiasis	Streptococcal disease, invasive group A
Gonorrhea	Streptococcal pneumonia, pediatric or drug-resistant invasive
Haemophilus influenzae, invasive disease	Streptococcal toxic shock syndrome
Hansen's disease (leprosy)	Syphilis (primary, secondary, latent, late, congenital)
Hantavirus pulmonary syndrome	Tetanus
Hemolytic uremic syndrome (post-diarrhea)	Toxic shock syndrome, streptococcal and non-streptococcal
Hepatitis, viral A, B, C, other	Trichinellosis
Influenza, pediatric mortality or novel influenza A	Tuberculosis
Legionnellosis	Tularemia
Lyme disease	Typhoid fever
Malaria	Vancomycin-resistant or intermediate Staphylococcus aureus
Measles	Varicella
Meningococcal disease	Vibriosis (non-cholera) Yellow fever

Note: Other diseases for which individual state monitoring may be required include: amebiasis, meningitis (aseptic and other bacterial), campylobacteriosis, dengue fever, genital herpes, genital warts, granuloma inguinale, leptospirosis, listeriosis, Lymphogranuloma venereum, mucopurulent cervicitis, nongonococcal urethritis, pelvic inflammatory disease, post-streptococcal disease, and others.
Source: Centers for Disease Control. 2007. www.cdc.gov/epo/mmwr/preview/mmwrhtml/0047449.htm (accessed October 14, 2007)

strange new disease entity may lead to its identification and practical measures to halt its spread. When it comes through anticipated or surprise epidemics and pandemics, and the real threat of bioterrorism, then multisectoral preparation and training are crucial.

HEALTH CARE–ASSOCIATED INFECTIONS

Health care institution–associated infections (HAI) are among the leading communicable and preventable causes of morbidity and mortality throughout the world. Nosocomial infections are those wherein a patient is exposed to and contracts disease while hospitalized. While great strides have been made in hospital sanitation, HAI still occurs in as many as 10 percent of admissions in developed countries. Recent CDC estimates place the number of nosocomial infections in the United States for 2002 at 1.7 million, a higher incidence than any notifiable disease. With a case mortality of nearly 6 percent, HAIs are also among the most deadly. Although progress has been made in HAI prevention, organisms implicated are becoming resistant to conventional therapy. In particular, methicillin-resistant *Staphylococcus aureus* (MRSA) which is among the most virulent and treatment-resistant bacteria, now accounts for over 50 percent of wound infections in many hospitals. Rare reports of vancomycin-resistant *S. aureus* (VRSA) cause alarm, proving antibiotic resistance transferred from other species. Treatment options for VRSA and vancomycin-resistant *Enterococcus* sp. are extremely limited, with concern these organisms could spread or become resistant to the few known effective therapies. The increasing number of immunodeficient patients has increased the importance of prevention of nosocomial infections (Box 4.4).

Where standards of infection control are lacking in both developed and developing countries, hospital staff and patients are vulnerable to serious infection. Of note,

tuberculosis and hepatitis B exposure is common among health care workers, but preventable through airborne precautions and vaccination, respectively. In developing countries, deadly emerging viruses, such as avian influenza H5N1 and Ebola viruses infect nursing, medical, and other staff as secondary cases.

A great obstacle in quantifying the impact of HAI is lack of uniform and clear case definitions, as well as reliance, in most countries, on voluntary reporting by institutions. While many recommendations have been made, notably by the Society for Healthcare Epidemiology of America in 2003, no uniform regulations have been established to mandate reporting of HAIs. Much work has been focused on prevention, though. Standard Precautions (formerly known as Universal Precautions) are a set of basic practices by which health care workers may reduce the spread of nosocomial infection among patients, visitors, and staff, as well as protect health workers from occupationally acquired disease. These include adequate hand washing hygiene and use of protective barriers suited to specific risks. Expanded precautions and mandatory use of organism-specific clinical guidelines are necessary procedures in many health care institutions as protective measures. Organizational policy must be established for each institution by an integrated and authoritative department of infection control and epidemiology.

The cost of nosocomial infections is a major consideration in planning health budgets. Reducing the risk of HAI justifies substantial expenditures for hospital epidemiology and infection control activities. With diagnosis-related group (DRG) payment for hospital care (by diagnosis rather than by days of stay), the effective manager has a major incentive to minimize the risk of nosocomial infections to the improvement of patient care, because infections can greatly prolong hospital stays, raising patient dissatisfaction and health care costs.

ENDEMIC AND EPIDEMIC DISEASE

An endemic disease is the continuous usual presence of a disease or infectious agent in a given geographic area or population group. Hyperendemic is a state of persistence of high levels of incidence of the disease. Holoendemic means that the disease appears early in life and affects most of the population, as in malaria or hepatitis A and B in some regions.

An epidemic is the occurrence in a community or region of a number of cases of an illness in excess of the usual or expected number of cases, or health-related behaviors (e.g., smoking) or events (e.g., motor vehicle accidents). The number of cases constituting an epidemic varies with the disease, and factors such as previous epidemiologic patterns of the disease, time and place of the occurrence, and the population involved must be taken

---

Box 4.4 Health Care Facility Recommendations for Standard Precautions

Key Elements at a Glance
1. Hand hygiene
2. Gloves
3. Facial protection (eyes, nose, mouth)
4. Gown
5. Prevention of needlestick injuries
6. Respiratory hygiene and cough etiquette
7. Environmental cleaning
8. Linens
9. Waste disposal
10. Patient care equipment

Source: Adapted from WHO guidelines on hand hygiene in health care.

into account. A single case of a disease long absent from an area, such as polio, constitutes an epidemic, and therefore is a public health emergency because a clinical case may represent a hundred carriers with nonparalytic or sub-clinical poliomyelitis. Two to three or more cases such as measles or any unusual disease locally that are linked in time and place may be considered sufficient evidence of transmission and presumed to be an epidemic. A pandemic is occurrence of a disease on a wide scale over a very wide area, crossing international boundaries, affecting a large proportion of the world.

Epidemic Investigation

Each epidemic should be regarded as a unique natural experiment. The investigation of an epidemic requires preparation and field investigation in conjunction with local health and other relevant authorities. Verification of cases and the scope of the epidemic will require case definition and laboratory confirmation. Tabulation of known cases according to time, place, and person are important for immediate control measures and formulation of the hypothesis as to the nature of the epidemic. An epidemic curve is a graphic plotting of the distribution of cases by the time of onset or reporting, which gives a picture of the timing, spread, and extent of the disease from the time of the initial index cases and the secondary spread.

Epidemic investigation requires a series of steps. This starts with confirmation of the initial report and preliminary investigation, defining who is affected, determining the nature of the illness and confirming the clinical diagnosis, and recording when and where the first (index) and follow-up (secondary) cases occurred, and how the disease was transmitted. Samples are taken from index case patients (e.g., blood, feces, throat swabs) as well as from possible reservoirs (e.g., food, water, sewage, environment). A working hypothesis is established based on the first findings, taking into account all plausible explanations. The epidemic pattern is studied, establishing common source or risk factors, such as food, water, contact, environment, and drawing a time line of cases to define the epidemic curve.

How many are ill (the numerator) and what is the population at risk (the denominator) establish the attack rate; namely, the percentage of sick among those exposed to the common factor. What is a reasonable explanation of the occurrence: Is there a previous pattern, with the present episode a recurrence or new event? Consultation with colleagues and the literature helps to establish both a biological and epidemiologic plausibility. What steps are needed to prevent spread and recurrence of the disease? Coordination with relevant health and other officials and providers is required to establish surveillance and control systems, document and distribute reports, and respond to the public's right to know.

The first reports of excess cases may come from a medical clinic or hospital. The initial (sentinel or index) cases provide the first clues that may point to a common source. Investigation of an epidemic is designed to quickly elucidate the cause and points of potential intervention to stop its continuation. This requires skilled investigation and interpretation. The term "epidemiologic investigation" means a broad review of all evidence related to a topic, not just one epidemic or outbreak. Epidemiologic investigations have defined many public health problems. Rubella syndrome, Legionnaire's disease, AIDS, Lyme disease, and hantavirus diseases were first identified clinically when unusually large numbers of cases appeared with common features. The suspicions that were raised led to a search for causes and the identification of control methods.

A working hypothesis of the nature of an epidemic is developed based on the initial assessment, the type of presentation, the condition involved, and previous local, regional, national, and international experience. The hypothesis provides the basis for further investigation, control measures, and planning additional clinical and laboratory studies. Surveillance will then monitor the effectiveness of control measures. Communication of findings to local, regional, national, and international health reporting systems is important for sharing the knowledge with other potential support groups or other areas where similar epidemics may occur.

The Centers for Disease Control and Prevention (CDC), originally organized in 1946 as the Office for Malaria Control in War Areas, is part of the U.S. Public Health Service. As of 1993, the CDC had a budget of $1.5 billion, and its 7300 employees include epidemiologists, microbiologists, and many other professionals, By 2007, the CDC budget had reached $9 billion dollars with 8467 employees. The CDC includes national centers for environmental health and injury control, chronic disease prevention and health promotion, infectious diseases, prevention services, health statistics, occupational safety and health, and international health. Recently, however, budget reductions have imposed limits of capacity in such areas as overseas work.

The Epidemic Intelligence Service (EIS) of the CDC in the United States is an excellent model for the organization of the national control of communicable diseases. Clinicians are trained to carry out epidemiologic investigations as part of training to become public health professionals. EIS officers are assigned to state health departments, other public health units, and research centers as part of their training, carrying out epidemic investigation and special tasks in disease control.

The CDC, in cooperation with the WHO, has developed and offers free of charge a personal computer program to support field epidemiology, including epidemic investigations (EPI-INFO), which can be accessed and downloaded from the Internet. This program should be adopted widely in order to improve field investigations, to encourage reporting in real time, and to develop high standards in this discipline.[1]

CDC's *Morbidity and Mortality Weekly Report* (*MMWR*) is a weekly publication of the CDC's epidemiologic data, also available free on the Internet. It includes special summaries of reportable infectious diseases as well as noncommunicable diseases of epidemiologic interest. *MMWR* publishes periodic special reports of important infectious and noninfectious diseases with comprehensive reviews of the literature and recent investigative work by the CDC and other organizations. *MMWR* published a review of Ten Great Achievements of Public Health in the United States in the Twentieth Century, which included control of communicable disease and vaccine-preventable diseases, as well as improvements in occupational health, maternal and child health, motor vehicle accidents, and cardiovascular and other chronic diseases and conditions (see Chapter 1).

CONTROL OF COMMUNICABLE DISEASES

Although an infectious disease is an event affecting an individual, it is transmissible to others, and therefore infection control requires both individual and community measures. Control of a disease is reduction in its incidence, prevalence, morbidity, and mortality. Elimination of a disease in a specified geographic area may be achieved as a result of intervention programs such as individual protection against tetanus; elimination of infections such as measles requires stoppage of circulation of the organism. Eradication is success in reduction to zero of naturally occurring incidence, such as with smallpox. Extinction means that a specific organism no longer exists in nature or in laboratories.

Public health applies a wide variety of tools for the prevention of infectious diseases and their transmission. It includes activities ranging from filtration and disinfection of community drinking water to environmental vector control, pasteurization of milk, and immunization programs (see Table 4.3). No less important are organized programs to promote self-protection, case finding, and effective treatment of infections to stop their spread to other susceptible persons (e.g., HIV, sexually transmitted infections, tuberculosis, malaria). Planning measures to control and eradicate specific communicable diseases is one of the principal activities of public health and remains so for the twenty-first century.

Treatment

Treating an infection once it has occurred is vital to the control of a communicable disease. Each person infected may become a vector and continue the chain of transmission. Successful treatment of the infected person reduces the potential for an uninfected contact person to acquire the infection. Bacteriostatic agents or drugs such as sulfonamides inhibit growth or stop replication of the organism, allowing normal body defenses to overcome the organism. Bactericidal drugs such as penicillin act to kill pathogenic organisms.

Traditional medical emphasis on single antibiotics has changed to use of multiple drug combinations for tuberculosis and more recently for hospital-acquired infections. Antibiotics have made enormous contributions to clinical medicine and public health. However, pathogenic organisms are able to adapt or mutate and develop resistance to antibiotics, resulting in drug resistance. Wide-scale use of antibiotics has led to increasing incidence of resistant organisms. Multidrug resistance constitutes one of the major public health challenges in the twenty-first century. Antiviral agents (e.g., Ribavirin) are important additions to medical treatment potential, as are "cocktails" of antiviral agents for management of HIV infection, known as highly active antiretroviral treatment (HAART). Prudent antibiotic use requires the attention of clinicians and their teachers as well as the public health community and health care managers, representing the interaction of health issues across the entire spectrum of services.

Methods of Prevention

Organized public health services are responsible for advocating legislation and for regulating and monitoring programs to prevent infectious disease occurrence and/or spread. They function to educate the population in measures to reduce or prevent the spread of disease.

Health promotion is one of the most essential instruments of infectious disease control. It promotes compliance and community support of preventive measures. These include personal hygiene and safe handling of water, milk, and food supplies. In sexually transmitted infections, health education is the major method of prevention.

Each of the infectious diseases or groups of infectious diseases has one or more preventive or control approaches (Table 4.3). These may involve the coordinated intervention of different disciplines and modalities, including epidemiologic monitoring, laboratory confirmation, environmental measures, immunization, and health education. This requires teamwork and organized collaboration.

Very great progress has been made in infectious disease control by clinical, public health, and societal means since 1900 in the industrialized countries and since the 1970s in the developing world. This is attributable to

---

[1]Epidemiologic investigation may be arranged at the CDC or WHO by contacting the Epidemiology Program Office, Mailstop G34, Centers for Disease Control, Atlanta, Georgia 30333, or by telephone 404-639-2709 or fax 404-639-3296; or the World Health Organization, 1211 Geneva 27, Switzerland, or by telephone 41-22-791-2111 or fax 41-22-791-0746.

TABLE 4.3 Methods of Prevention or Control of Infectious Diseases by Type of Organism

Control of major infectious diseases	Viruses	Bacteria	Parasites
Vaccination: pre-exposure to protect individuals and the community (herd immunity); post-exposure for individual protection (e.g., for rabies following animal bite, or contact after exposure to measles cases); or immunization of animals to prevent infected meat or milk transfer of disease to humans (e.g., brucellosis)	Rabies, polio, measles, rubella, mumps, hepatitis B, influenza, varicella, hepatitis A, human papillomavirus (HPV)	Diphtheria, pertussis, tetanus, tuberculosis, anthrax, brucella, pneumococcal pneumonia, *Haemophilus influenzae* type b	
Environmental measures: water and sewage control (e.g., chlorination of water to reduce burden of gastroenteric disease), vector control, antimosquito control measures (draining pooled water, larvicides, insecticides, repellants, protective bed nets and clothing)	Hepatitis A, rotaviruses, polio, arboviruses, tick- and mosquito-borne viruses	Salmonella, shigella, cholera, Legionnaire's disease, E. *coli*	Malaria, onchocerciasis, dracunculiasis, schistosomiasis, elephantiasis, worms
Education/social/behavior measures: to promote self-care and self-protection to reduce risk (e.g., safe sexual practices to prevent STIs and HIV), needle exchange, condom distribution among risk groups	HIV, human papillomavirus (HPV), hepatitis B and C	Diarrheal diseases, syphilis, gonorrhea, chancroid	Malaria, scabies, onchocerciasis, dracunculiasis
Animal and food control: to reduce transmission by pasteurization of milk, veterinary supervision of meat production and distribution, food hygiene and safety measures, radiation of food	Rabies	Brucellosis, coliforms, salmonellosis, shigellosis	Tapeworms
Case finding and treatment: to cure or prevent transmission and reduce the carrier population (e.g., blood, sputum screening)	Rabies, herpes, cytomegalovirus (CMV), HIV, hepatitis C	Tuberculosis, STIs, rheumatic fever	Malaria, worms, dracunculiasis, leprosy, onchocerciasis, schistosomiasis
Occupational measures: to protect persons exposed at place of work (e.g., immunization of food handlers, health care and child care workers)	HIV, hepatitis A and B, measles, rubella, arboviruses	Brucellosis, tuberculosis, anthrax	Hydatid cyst, trichinosis

a variety of factors, including organized public health services; the rapid development and wide use of new and improved vaccines and antibiotics; better access to health care; and improved sanitation, living conditions, and nutrition. Triumphs have been achieved in the eradication of smallpox and in the increasing control of other vaccine-preventable diseases. However, there remain serious problems with TB, STIs, malaria, new infections such as HIV, and an increase in multidrug-resistant organisms.

VACCINE-PREVENTABLE DISEASES

Vaccines are one of the most important tools of public health in the control of infectious diseases, particularly for child health. Vaccine-preventable diseases (VPDs) are those diseases preventable by currently available vaccines (Table 4.4). The term *vaccine* is derived from use of cowpox (vaccinia virus) to stimulate immunity to smallpox, first demonstrated by Jenner in 1796, and is generally used for all immunizing agents.

The body responds to invasion of disease-causing organisms by antigen–antibody reactions and cellular responses. Together, these act to restrain or destroy the disease-causing potential. Strengthening this defense mechanism through immunization has become one of the significant achievements of public health, preventing loss of literally millions of lives, and with very great potential for future generations as well (Box 4.5).

TABLE 4.4 Annual Incidence of Selected Vaccine-Preventable Infectious Diseases in Rates per 100,000 Population, Selected Years, United States, 1950–2004

Disease	1950	1960	1970	1980	1990	2000	2004
Diphtheria	3.8	0.5	0.2	0	0	0	0
Pertussis	79.8	8.2	2.1	0.8	1.8	2.9	8.9
Poliomyelitis	22.0	1.8	0	0	0	0	0
Measles	211.0	245.4	23.2	6.0	11.2	0.03	0.04
Mumps	na	na	55.6	3.9	2.2	0.08	0.09
Rubella	na	na	27.8	1.7	0.5	0.06	0
Hepatitis A	na	na	27.8	12.8	12.6	4.9	2.0
Hepatitis B	na	na	4.1	8.4	8.5	3.0	2.1

Note: na = not available.
Source: *Health United States*, 1998, 2006.

Box 4.5 Definitions of Immunizing Agents and Processes

Vaccines: a suspension of live or killed microorganisms or antigenic portion of those agents presented to a potential host to induce immunity to prevent the specific disease caused by that organism. Preparation of vaccines may be from:

a. Live attenuated organisms which have been passed repeatedly in tissue culture or chick embryos so that they have lost their capacity to cause disease but retain an ability to induce antibody response, such as polio-Sabin, measles, rubella, mumps, yellow fever, BCG, typhoid, and plague.

b. Inactivated or killed organisms which have been killed by heat or chemicals but retain an ability to induce antibody response; they are generally safe but less efficacious than live vaccines and require multiple doses, such as polio-Salk, influenza, rabies, and Japanese encephalitis.

c. Cellular fractions, usually of a polysaccharide fraction of the cell wall of a disease-causing organism, such as pneumococcal pneumonia or meningococcal meningitis.

d. Recombinant vaccines produced by recombinant DNA methods in which specific DNA sequences are inserted by molecular engineering techniques, such as DNA sequences spliced to vaccinia virus grown in cell culture to produce influenza and hepatitis B vaccines.

Toxoids or antisera: modified toxins are made nontoxic to stimulate formation of an antitoxin, such as tetanus, diphtheria, botulism, and gas gangrene.

Immune globulin: an antibody-containing solution derived from immunized animals or human blood plasma, used primarily for short-term passive immunization (e.g., rabies, IgG globulin for immunocompromised persons).

Antitoxin: an antibody derived from serum of animals after stimulation with specific antigens and used to provide passive immunity (e.g., tetanus, snake and scorpion venom).

Source: Brooks, G. E., Butel, J. S., Morse, S. A. 2004. *Jawetz, Melnick and Adelberg's Medical Microbiology*, 23rd ed. Stamford, CT: Appleton & Lange; *Harrison's Textbook of Internal Medicine* (2005).

Immunization (vaccination) is a process used to increase host resistance to specific microorganisms to prevent them from causing disease. It induces primary and secondary responses in the human or animal body:

a. Primary response occurs on first exposure to an antigen. After a lag or latent period of 3–14 days (depending on the antigen), specific antibodies appear in the blood. Antibody production ceases after several weeks but memory cells that can recognize the antigen and respond to it remain ready to respond to a further challenge by the same antigen.

b. Secondary (booster) response is the response to a second and subsequent exposure to an antigen. The lag period is shorter than the primary response, with the peak being higher and lasting longer. The antibodies produced have a higher affinity for the antigen, and a much smaller dose of the antigen is required to initiate a response. Booster doses of vaccines are used to activate memory cells to strengthen immunity.

c. Immunologic memory exists even when circulating antibodies are insufficient to protect against the antigen. When the body is exposed to the same antigen again, it responds by rapidly producing high levels of antibody to destroy the antigen before it can replicate and cause disease.

Immunization protects susceptible individuals from communicable disease by administration of a living modified agent, a subunit of the agent, a suspension of killed organisms, or an inactivated toxin (see Table 4.5) to stimulate development of antibodies to that agent. In disease control, individual immunity may also protect another individual.

Herd immunity occurs when sufficient numbers of persons are protected (naturally or by immunization) against a specific infectious disease, reducing circulation of the organism, and thereby lowering the chance of an unprotected person becoming infected. Each pathogen has different characteristics of infectivity, and therefore different levels of herd immunity are required to protect the nonimmune individual.

Immunization Coverage

The critical proportion of a population that must be immunized in order to interrupt local circulation of the organism varies from disease to disease. Eradication of smallpox was achieved with approximately 80 percent world coverage, followed by concentration on new case findings and immunization of contacts and surrounding communities.

TABLE 4.5 Development of Vaccines by Period of Development and Type of Vaccine

Period/Century	Live attenuated	Killed, whole organism	Purified protein or polysaccharide	Genetically engineered
Eighteenth century	Smallpox (1798)	na	na	na
Nineteenth century	Rabies (1885)	Hog cholera (1886)	Diphtheria antitoxin (1888)	na
	Cholera (1896) Plague (1897)	Typhoid (1896)		
Early twentieth century	BCG tuberculosis (1927)	Pertussis (1926)	Diphtheria (1923)	na
	Yellow fever (1935)	Influenza (1936)	Tetanus toxoid (1927)	
		Rickettsia (1936)	Influenza A (1936)	
Post–World War II	Yellow fever (1953)	Influenza (1945)	Diphtheria toxoid (1949)	na
	Polio, Sabin (1963)	Tetanus toxoid (1949)	Pneumococcus (1976, 1983)	
	Measles (1963)	Typhoid (1952)	Meningococcus (1962)	
	Mumps (1967)	Polio, Salk (1955)	Tick-borne encephalitis	
	Rubella (1970) MMR (1971)	Anthrax (1970)		
1980–1999	Adenovirus (1980)	Rabies (1980, human diploid cell)	Haemophilus influenzae type b (1985)	Hepatitis B (1987) recombinant (yeast or mammalian cell derived)
	Typhoid (1992, 1995) (Salmonella Type21a, Vi)	Japanese encephalitis (1993)	Hepatitis B (1981, plasma)	
	Varicella (1995)		Pertussis, acellular (1993)	
	Lyme disease (1998) Rotavirus (1998)	Hepatitis A (1995)		
2000–2010	New vaccines for pneumococcal, meningococcal disease, influenza, parainfluenza, Human papillomavirus (HPV), respiratory syncytial virus			
Anticipated	H. pylori, Streptococcus, HIV, hepatitis C, adenoviruses, malaria			

Note: Years developed or licensed in the United States.
na: Not available.
Centers for Disease Control. Vaccines universally recommended for children — United States, 1990–98. *Morbidity and Mortality Weekly Report*, 48:243–248.
Sources: Modified from Plotkin, S. A., Orentein, W. A. 2003. *Vaccines*, 4th ed. Philadelphia: W. B. Saunders.

For highly infectious diseases such as measles, immunization coverage of over 95 percent is needed to achieve local eradication.

Immunization coverage in a community must be monitored in order to gauge the extent of protection and need for program modification to achieve targets of disease control. Immunization coverage is expressed as a proportion in which the numerator is the number of persons in the target group immunized at a specific age, and the denominator is the number of persons in the target cohort who should have been immunized according to the accepted standard:

Vaccine coverage =

$$\frac{\text{no. persons immunized in specific age group}}{\text{no. persons in the age group during that year}} \times 100$$

Immunization coverage in the United States is regularly monitored by the National Immunization Survey, a telephone-based questionnaire of households from all 50 states, as well as selected areas at high risk for inadequate levels of vaccination. An initial telephone survey is followed by confirmation, where possible, from documentation from the parents or health care providers. The childhood immunization survey for 2006 examined children aged 19–35 months. The results show 85 percent of U.S. children having received four or more (4+) doses of DTaP (diphtheria, tetanus, acellular pertussis), 93 percent with three or more (3+) doses of oral or injected polio vaccine, and 93 percent with three or more (3+) doses of *Haemophilus influenzae* type b (Hib). Hepatitis B coverage (3+) greatly increased to 93 percent, while institution of pneumococcal (3+) and varicella (1+) vaccination policies has rapidly achieved 87 percent and 89 percent, respectively. Despite these gains, only 77 percent of children received all vaccinations at recommended ages.

Present technology allows for control or eradication of important infectious diseases that still cause millions of deaths globally each year. Other important infectious diseases are still not subject to vaccine control because of difficulties in their development. In some cases, a microorganism can mutate with changes. Viruses can undergo antigenic shifts in their molecular structure, producing completely new subtypes of the organism. Hosts previously exposed to other strains may have little or no immunity to the new strains.

Antigenic drift refers to relatively minor antigenic changes which occur in viruses. This is responsible for frequent epidemics. Antigenic shift is believed to explain the occurrence of new strains of influenza virus, necessitating annual reformulation of the influenza vaccine. New variants of poliovirus strains are similar enough to three main types that immunity to one strain is carried over to the new strain. Molecular epidemiology is a powerful genetic technique used to determine geographic origin, permitting tracking of the spread of infectious organisms and epidemics.

Combination of more than one vaccine is now common practice with a trend to enlarging the cocktail of vaccines in order to minimize the number of injections and visits required. This reduces staff time and costs, as well as increasing convenience and compliance by the public. There are virtually no contraindications to use of multiple antigens simultaneously. Examples of vaccine cocktails include DTaP in combination with *Haemophilus influenzae* type b, poliomyelitis, varicella, or MMR (measles, mumps, and rubella) vaccines.

Interventions in the form of effective vaccination save millions of lives each year and contribute to improved health of countless children and adults throughout the world. Vaccination is accepted as one of the most cost-effective health interventions currently available. Continuous policy review is needed regarding allocation of adequate resources, logistical organization, and continued scientific effort to seek effective, safe, and inexpensive vaccines for other important diseases such as malaria and HIV. Molecular technology, producing recombinant vaccines such as those for hepatitis A and B, holds promise for important vaccine breakthroughs in the decades ahead.

Internationally, much progress was made in the 1980s in the control of VPDs. At the end of the 1970s, fewer than 10 percent of the world's children were being immunized. WHO, UNICEF, and other international organizations mobilized to promote an Expanded Programme on Immunization (EPI) with a target of reaching 80 percent coverage by 1990. WHO in 2007 reports that diphtheria, pertussis and tetanus (DPT 3 doses), polio (3 doses), and measles coverages globally reached 80–90 percent in 2006. *Haemophilus influenzae* b vaccination (3 doses) reached 90 percent of the population of the Americas region, 44 percent of the European region, and 24 percent of African region of WHO in 2006. Immunization is preventing some 3 million child deaths annually in developing countries. Bacille Calmette-Guérin (BCG) coverage internationally rose from 31 percent to 89 percent; poliomyelitis with OPV (three doses) from 24 percent to 85 percent, and tetanus toxoid for pregnant women from 14 percent to 57 percent. Recent declines in coverage have occurred in many parts of the world though, notably in Sudan, Burma, and other regions affected by violent conflicts.

The challenge remains to achieve control or eradication of VPDs, thus saving millions more lives. Part of the HFA stresses the EPI approach, which includes immunization against diphtheria, pertussis, tetanus, poliomyelitis, measles, and tuberculosis. An extended form of this is the EPI PLUS program which combines EPI with immunization against hepatitis B and yellow fever and, where appropriate, supplementation with vitamin A and iodine. The success in international eradication of smallpox has been followed with major progress toward eradication of polio-myelitis, measles, and other important infectious diseases.

Diphtheria

Diphtheria is an acute bacterial disease of the tonsils, nasopharynx, and larynx caused by the organism *Coryne-bacterium diphtheriae*. It occurs in colder months in temperate climates where the organism is present in human hosts and is spread by contact with patients or carriers. It has an incubation period of 2–5 days. In the past, this was primarily an infection of children and was a major contributor to child mortality in the pre-vaccine and pre-antibiotic eras. Diphtheria has been virtually eliminated in countries with well-established immunization programs.

In the 1980s, an outbreak of diphtheria occurred in the countries of the former Soviet Union among people over age 15. It reached epidemic proportions in the 1990s, with 140,000 cases (1991–1995) with 1100 deaths in 1994 in Russia alone. This indicates a failure of the vaccination program in several respects: it used only three doses of DPT in infancy, no boosters were given at school age or subsequently, the efficacy of diphtheria vaccine may have been low, and coverage was below 80 percent.

Efforts to control the present epidemic include mass vaccination campaigns for persons over 3 years of age with a single dose of DT (diphtheria and tetanus) and increasing coverage of routine DPT vaccines to four doses by age 2 years. The epidemic and its control measures have led to improved coverage with DT for those over 18 years, and 93 percent coverage among children aged 12–23 months.

WHO recommends three doses of DPT in the first year of life and a booster at primary school entry, as well as at enrollment at college, military, or other organized settings. This is considered by many to be insufficient to produce long-lasting immunity. The United States and other industrialized countries use a four-dose schedule and recommend periodic boosters for adults with DT.

Pertussis

Pertussis is an acute bacterial disease of the respiratory tract caused by the bacillus *Bordetella pertussis*. After an initial coldlike (catarrhal) stage, the patient develops a severe cough which comes in spasms (paroxysms). The disease can last 1–2 months. The paroxysms can become violent and may be followed by a characteristic crowing or high-pitched inspiratory whooping sound, followed by expulsion of tenacious clear sputum, often followed by vomiting. In poorly immunized populations and those with malnutrition, pneumonia often follows, and death is common.

Pertussis declined dramatically in the industrialized countries as a result of widespread coverage with DPT. However, because the pertussis component of early vaccines caused rare reactions, many physicians and parents avoided its use, instead opting for DT alone, leaving children susceptible to infection. During the 1970s in the United Kingdom, many physicians recommended against vaccination with DPT. As a result, pertussis incidence increased with substantial mortality rates. This led to a reappraisal of the immunization program, with institution of incentive payments to general practitioners for completion of vaccination schedules. As a result of these measures, vaccination coverage, with resulting pertussis control, improved dramatically in the United Kingdom. A new acellular vaccine is now in widespread use and will be safer with fewer and less severe reactions in infants, increasing the potential for improved confidence and support for routine vaccination. The new vaccine is used in the United States and other industrialized countries, and forms part of the U.S. recommended vaccination schedule. Although most western European countries are advanced in use of vaccines, there is no Europe-wide equivalent of the CDC-recommended immunization schedule for the region, which will be coming up for discussion in European Union health forums.

The CDC reports estimates of childhood vaccination coverage in the United States with ≥ 3 doses of pertussis-containing vaccine have exceeded 90 percent since 1994. However, reported pertussis cases increased from a historic low of 1010 cases in 1976 to 11,647 in 2003, with a substantial increase in reported cases among adolescents, who become susceptible to pertussis approximately 6–10 years after their childhood vaccination. This is attributed to waning immunity and lack of booster doses, so booster doses in adolescence are now recommended.

Pertussis continues to be a public health threat and recurs wherever there is inadequate immunization in infancy. In addition, recent epidemics have been noted in adults who have lost childhood immunity. While the disease generally follows a milder course in healthy adults, this raised concerns of a reservoir for infection of children and the immunocompromised. To eliminate this risk, pertussis booster vaccination is recommended during adolescence and again in adulthood.

Tetanus

Tetanus is an acute disease caused by an exotoxin of the tetanus bacillus (*Clostridium tetani*) which grows anaerobically at the site of an injury. The bacillus is universally present in the environment and enters the human body via penetrating injuries. Following an incubation period of 3–21 days, it causes an acute condition of painful muscular contractions. Unless there is modern medical care available, patients are at risk of high case fatality rates of 30–90 percent (highest in infants and the elderly).

Antitetanus serum (ATS) was discovered in 1890, and during World War I, ATS contributed to saving the lives of many thousands of wounded soldiers. Tetanus toxoid was developed in 1993. The organism, because of its universal presence in the environment, cannot be eradicated. However, the disease can be controlled by effective immunization of every child during infancy and school age. Adults should receive routine boosters of tetanus toxoid once very decade.

Newborns are infected by tetanus spores (tetanus neonatorum) where unsanitary conditions or practices are present. It can occur when traditional birth attendants at home deliveries use unclean instruments to sever the umbilical cord, or dress the severed cord with contaminated material. Tetanus neonatorum remains a serious public health problem in developing countries. Immunization of pregnant women and women of childbearing age is reducing the problem by conferring passive immunity to the newborn. The training of traditional birth attendants in hygienic practices and the use of medically supervised birth centers for delivery also decrease the incidence of tetanus neonatorum.

Elimination of tetanus neonatorum was made a health target by the World Summit of Children in 1990. In that year, the number of deaths from neonatal tetanus was reported by WHO as 25,293 infants worldwide, declining to 8376 in 2006 (112 countries reporting). Immunization of pregnant women increased from under 20 percent in 1984 to 69 percent in 2006.

Tetanus cases have declined dramatically in the United States, but the disease still occurs mainly among older adults. The CDC reports during 1990–2001, a total of 534 cases of tetanus were reported; 301 (56 percent) cases occurred among adults aged 19–64 years and 201 (38 percent) among adults aged $\geq$65 years. Data from a national population-based serosurvey indicated the prevalence of immunity to tetanus was >80 percent among adults aged 20–39 years but declined with increasing age. This supports current recommendations to give booster doses of tetanus (with diphtheria) vaccine for adolescents and adults every 10 years.

Poliomyelitis

Poliovirus infection may be asymptomatic or cause an acute nonspecific febrile illness. It may reach more severe forms of aseptic meningitis and acute flaccid paralysis with long-term residual paralysis or death during the acute phase. Poliomyelitis is transmitted mainly by direct person-to-person contact, but also via sewage contamination. Large-scale epidemics of disease, with attendant paralysis and death, occurred in industrialized countries in the 1940s and 1950s, engendering widespread fear and panic and thousands of clinical cases of "infantile paralysis."

Growth of the poliovirus by John Enders and colleagues in tissue culture in 1949 led to development and wide-scale testing of the first inactivated (killed) polio vaccine by Jonas Salk in the mid-1950s and great hope and outstanding success in the control of this much feared disease. Development of the live attenuated oral poliomyelitis vaccine (OPV) by Albert Sabin, licensed in 1960, added a major new dimension to poliomyelitis control because of the effectiveness, low cost, and ease of administration of the vaccine. The two vaccines in their more modern forms, enhanced strength inactivated polio vaccine (eIPV), and triple oral polio vaccine (TOPV), have been used in different settings with great success.

Oral polio vaccine (OPV) induces both humoral and cellular, including intestinal, immunity. The presence of OPV in the environment by contact with immunized infants and via excreta of immunized persons in the sewage gives a booster effect in the community. Immunization using OPV, in both routine and National Immunization Days (NIDs) has proven effective in dramatically reducing poliomyelitis and circulation of the wild virus in many parts of the world. Use of the eIPV produces early and high levels of circulating antibodies, as well as protecting against the vaccine-associated disease.

In rare cases, OPV can cause vaccine-associated paralytic poliomyelitis (VAPP), with a risk of 1 case per 520,000 with initial doses, and 1 case per over 12 million with subsequent doses. Approximately 8 to 10 cases of VAPP occurred annually in the United States during the 1990s following the elimination of natural transmission. The CDC changed recommendations to IPV use in 1999, out of concern that VAPP risk would outweigh risk of local wild polio from imported cases. Many developed countries have followed suit. While this eliminates risk for VAPP, concerns have risen that herd immunity may be reduced due to shorter memory and lower intestinal immunity noted with IPV use.

Controversy as to the relative advantages of each vaccine continues. The OPV program of mass repeated vaccination in control of poliomyelitis in the Americas established the primacy of OPV in practical public health, and the momentum to eradicate poliomyelitis is building. OPV requires multiple doses to achieve protective antibody levels. Where there are many enteroviruses in the environment, interference in the uptake of OPV may result in cases of paralytic poliomyelitis among persons who have received 3 or even 4 doses of adequate OPV. Use of IPV as initial protection eliminates this problem. During the 1970s and 1980s, a combined approach bolstering IPV immunity with OPV boosters showed promise in Gaza and Israel, where natural poliovirus was eradicated. Although sequential use of IPV and OPV was adopted as part of the routine infant immunization program in the United States in 1997, current programs use IPV alone. IPV has been adopted as the exclusive polio vaccine in most of the industrialized countries, while developing countries continue relying on the less costly and easier to administer OPV. Mop-up campaigns using monovalent

OPV (Type 1) in still-endemic areas such as specific regions of India and Nigeria are being promoted.

There are concerns that exclusive use of either vaccine alone will not lead to the desired goal of eradication of poliomyelitis. In 1988, the polio eradication initiative was launched. Progress in global eradication of polio has been impressive. Global coverage of infants with three doses of OPV reached 85 percent in 2005 as compared to 83 percent in 1995 (UNICEF). During the same period, OPV coverage in the the African region of WHO increased from 51 percent in 2000 to 75–80 percent in 2006. National immunization days (NIDs) are conducted in many countries throughout the world, achieving coverage of over 400 million children annually. Mop-up operations to reinforce coverage of children in still-endemic areas are proceeding, along with increased emphasis on acute flaccid paralysis (AFP) monitoring. Worldwide clinical cases from wild poliovirus have been reduced to 2000 per year in 2006. By the end of 2006 four countries remained endemic for polio: India (676 cases); Pakistan (40 cases); Afghanistan (31 cases); and Nigeria (1125 cases); another 13 countries in Africa, the Middle East, and Southeast Asia reported clinical cases of poliomyelitis due to active transmission in 2006.

With continued national and international emphasis, and support of WHO, Rotary International, UNICEF, donor countries, there is real prospect of a world without polio.

Measles

Measles is an acute disease caused by a virus of the *Paramyxovirus* family. It is highly infectious with a very high ratio of clinical to subclinical cases (99/1). Measles has a characteristic clinical presentation with fever, rhinorrhea, white spots (Koplik spots) on the membranes of the mouth, and a red blotchy rash appearing on day 3–7 lasting 4–7 days. Mortality rates are high in young children with compromised nutritional status, especially vitamin A deficiency.

The measles virus evolved from a virus disease of cattle (rinderpest) some 3000–5000 years ago, becoming an important disease of humans with high mortality rates in debilitated, poorly nourished children, and significant mortality and morbidity even in industrialized countries. In the pre-vaccine era, measles was endemic worldwide, and remains a major childhood infectious disease.

Single-dose immunization failed to meet control or eradication requirements even in the most developed parts of the world. A live vaccine, licensed in 1963, was later replaced by a more effective and heat-stable vaccine, but still with a primary vaccination failure rate (i.e., fails to produce protective antibodies) of 4–8 percent, and secondary failure rate (i.e., produces antibodies but protection is lost over time) of 4 percent. A two-dose policy incorporates a booster dose, usually at school age, in addition to maximum feasible infant coverage of children in the 9–15 month period (timing varies in different countries).

Catch-up campaigns among school-age children should be carried out until the routine two-dose policy has time to take full effect. Nearly universal primary education in developing countries offers an opportunity for mass coverage of school-age children with a second dose of measles vaccine and a resulting increase of herd immunity to reduce the transmission of the virus. The two-dose policy adopted in many countries should be supplemented with catch-up campaigns in schools to provide the booster effect for those previously immunized and to cover those previously unimmunized, especially in developing countries.

The CDC considers that domestic transmission in the United States has been interrupted and that most localized outbreaks were traceable to imported cases. South America and the Caribbean countries are now considered free from indigenous measles, based on their successful use of NIDs, although a large epidemic occurred in 1999 in Brazil. Eradication of measles is feasible in the second decade of this century, if a two-dose policy is used and sustained with high priority globally, supplemented by catch-up campaigns to older children and young adults, and outbreak control.

Measles eradication is one of the central targets on WHO's agenda, with emphasis placed on reducing mortality and secondary on gradual eradication of the disease. Measles deaths have fallen by 60 percent worldwide since 1999 from an estimated 873,000 deaths in 1999 to 345,000 in 2005. In Africa in this period, measles deaths fell by 75 percent, from an estimated 506,000 to 126,000, with 90 percent of children under the age of five mostly dying from complications such as severe diarrhea, pneumonia, and encephalitis.

International transmission of the virus in carriers has led to importation and subsequent epidemics even in countries thought to have achieved local eradication, with outbreaks in 2006–2008 in the United Kingdom, Switzerland (2250 cases), Austria, France, Italy, and other countries. Israel had an epidemic of over 1200 cases in 2007–2008 following an imported case. The Health Protection Agency in the United Kingdom in July 2008 declared measles to be endemic for the first time in 14 years due to a decade of poor coverage with the measles vaccine. The United States had an annual average of 64 cases during 2000–2007, but an increase in 2008.

The WHO strategy of partnership with national governments and NGOs such as he Measles Initiative, GAVI, and others, includes:

- provision of one dose of measles vaccine for all infants via routine health services;
- a second dose for children through mass vaccination campaigns;
- effective surveillance for measles; and
- enhanced care, including the provision of supplemental vitamin A.

WHO has promoted measles vaccination campaigns along with other life-saving interventions such as bed nets to protect against malaria, deworming medicine, and vitamin A supplements to expand the contact occasion to reduce child death rates in keeping with the Millennium Development Goals between 1990 and 2015. Elimination of measles as a public health problem, and even eradication, are feasible goals in the second decade of the next century and critical to achieving the Millennium Development Goals target of reducing child mortality by the year 2015. This topic deserves to be one of the highest professional and political priorities of international and national donor and public health agencies as well as national governments.

Mumps

Mumps is an acute viral disease characterized by fever, swelling, and tenderness usually of the parotid glands, but also other glands. The incubation period ranges between 12 and 25 days. Orchitis, or inflammation of the testicles, occurs in 20–30 percent of postpubertal males and oophoritis, or inflammation of the ovaries, in 5 percent of postpubertal females. Sterility is an extremely rare result of mumps. Central nervous system involvement can occur in the form of aseptic meningitis, almost always without sequelae. Encephalitis is reported in 1–2 per 10,000 cases with an overall case fatality rate of 0.01 percent. Pancreatitis, neuritis, nerve deafness, mastitis, nephritis, thyroiditis, and pericarditis, although rare, may occur. Most persons born before 1957 are immune to the disease, because of the nearly universal exposure to the disease before that time.

The live attenuated vaccine introduced in the United States in 1967 is available as a single vaccine or in combination with measles and rubella as the measles-mumps-rubella (MMR) vaccine. It provides long-lasting immunity in 95 percent of cases. Mumps vaccine is now recommended in a two-dose policy with the first dose of MMR given between 12 and 15 months of age and a second dose given either at school entry or in early adolescence. MMR in two doses is now standard policy in the United States, Sweden, Canada, Israel, and other countries. The incidence of mumps has consequently declined rapidly. However, it is still a threat.

During 2004–2005, the United Kingdom experienced a nationwide epidemic of mumps, which peaked during 2005 when over 56,000 cases were reported in England and Wales, mostly aged 15–24 years, and most of whom had not been eligible for routine mumps vaccination. Figure 4.2 shows the epidemic curve during the period 2004–2005, as published in *MMWR*. The episode can be traced back to the period of controversy regarding use of MMR vaccine and increased susceptibility in a partly immunized population in the age group that received only one dose of the vaccine, if at all.

Poland also experienced a major outbreak of mumps in 2005–2006, mostly among children aged 5–9 years. The United Kingdom had more than 100,000 mumps cases in 2004 to 2005; the United States had 4000 cases in a Midwest outbreak in 2006. Canada reported more than 450 cases of mumps among university students in the spring of 2007 (WHO; U.S. Centers for Disease Control; Health Canada). In the tourist summer periods of 2004 and 2005, 39 patients with mumps had been hospitalized in Crete and Greece, and almost all were young tourists from Britain. The disease is spreading among the Greek population as well; 6 cases have been reported. Many countries in Europe still do not use MMR or a two-dose policy; thus they are vulnerable to mumps along with rubella outbreaks. MMR vaccination should be adopted as an international standard with two doses for all children and catch-up for school-age children. Local eradication of this disease is worthwhile and should be part of a basic international immunization program.

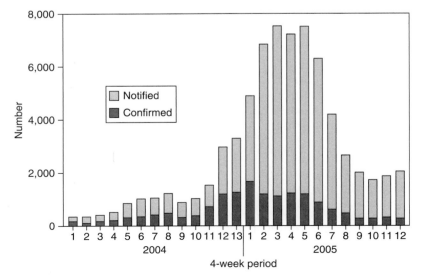

FIGURE 4.2 Curve of mumps epidemic, United Kingdom, 2004–2005. Source: Centers for Disease Control. 2006. Mumps Epidemic — United Kingdom, 2004–2005. *Morbidity and Mortality Weekly Report*, 55:173–175.

Rubella

Rubella (German measles) is generally a mild viral disease with lymphadenopathy and a diffuse, raised red rash. Low-grade fever, malaise, coryza, and lymphadenopathy characterize the prodromal period. The incubation period is usually 16–18 days. Differentiation from scarlet fever, measles, or other febrile diseases with rash may require laboratory testing and recovery of the virus from nasopharyngeal, blood, stool, and urine specimens (Box 4.6).

Congenital rubella syndrome (CRS) occurs with single or multiple congenital anomalies including deafness, cataracts, microophthalmia, congenital glaucoma, microcephaly, meningoencephalitis, congenital heart defects, and others. Moderate and severe cases are recognizable at birth, but mild cases may not be detected for months or years after birth. Insulin-dependent diabetes is suspected as a late sequela of congenital rubella. Each case of CRS is estimated to cost some $250,000 in health care during the patient's lifetime.

Prior to availability of the attenuated live rubella vaccine in 1969, the disease was universally endemic, with epidemics or peak incidence every 6–9 years. In unvaccinated populations, rubella is primarily a disease of childhood. In areas where children are well-vaccinated, adolescent and young adult infection is more apparent, with epidemics in institutions, colleges, and among military personnel.

A sharp reduction of rubella cases was seen in the United States following introduction of the vaccine in 1970, but increased in 1978, following rubella epidemics in 1976–1978. A further reduction in cases was followed by a sharp upswing of rubella and CRS in 1988–1990. An outbreak of rubella among the Amish in the United States, who refuse immunization on religious grounds, resulted in seven cases of CRS in 1991. It is now thought that vaccination of sufficient numbers in the United States reduced circulation of the virus and protected most vulnerable groups in the population. Most industrialized countries adopted MMR in the 1990s and a two-dose policy subsequently. Rubella and CRS incidence dropped dramatically. Controversy in the United Kingdom in the early 2000s led to reduced MMR usage and an increase in cases of measles and rubella. This was subsequently improved by providing incentive payments for general practitioners with 100 percent age-specific immunization coverage.

Some parts of Europe failed to adopt MMR vaccine use and have suffered recurrent outbreaks of these diseases. A number of outbreaks were reported in 2005–2007. Poland reported 7946 cases of rubella in 2005 (20.8 per 100,000 population), an increase of 64 percent compared to 2004. MMR was added to the routine immunization schedule at the end of 2003.

In November 2003, Italy approved a national plan for the elimination of measles and congenital rubella, with the aim of reducing and maintaining the incidence of congenital rubella syndrome (CRS) to less than 1 case per 100,000 live births by 2007.

As of 2007, there is no common recommended childhood immunization program for the European Union or for European Region of WHO. This leaves each country to develop its own and provides no guidelines for countries in transition from the socialist period operating with obsolescent immunization practices and only very slowly adopting western standards. Many have not yet adopted MMR. WHO considers eradication of measles and rubella of higher priority than mumps, but suggests the combination MMR vaccine be used.

In the past, immunization policy for rubella in some countries was to vaccinate schoolgirls aged 12 and women after pregnancy to protect them for the period of fertility. The current approach is to give a routine dose of MMR in early childhood, followed by a second dose in early school age to reduce the pool of susceptible persons. Women of reproductive age should be tested to confirm immunity before pregnancy and immunized if not already immune. Should a woman become infected during pregnancy, termination of pregnancy as previously recommended is now managed with hyperimmune globulin.

The infection of pregnant women during their first trimester of pregnancy is the primary public health implication of rubella. The emotional and financial burden of CRS, including the cost of treatment of its congenital defects, makes this vaccination program cost-effective. Its inclusion in a modern immunization program is fully justified. Elimination of CRS syndrome should be one of the primary goals of a program for prevention of VPDs in developed and developing countries. Adoption of MMR and the two-dose policy will gradually lead to eradication of rubella and rubella syndrome.

Viral Hepatitis

Viral hepatitis is a group of diseases of increasing public health importance due to its large-scale worldwide prevalence, its serious consequences, and our increasing ability to take preventive action. Viral hepatic infectious diseases each have specific etiologic, clinical, epidemiologic,

---

Box 4.6 Discovery of Rubella Syndrome

In 1942, Norman Gregg, an Australian ophthalmologist, noted an epidemic of cases of congenital cataract in newborns associated with a history of rubella in the mother during the first trimester. Subsequent investigation demonstrated that intrauterine death, spontaneous abortion, and congenital anomalies occur commonly when rubella occurs early in pregnancy.

serologic, and pathologic characteristics. They have important short- and long-term sequelae. Vaccine development is of high priority for control and ultimate eradication.

Hepatitis A

Hepatitis A (HAV) is mainly transmitted by the fecal–oral route. Clinical severity varies from a mild illness of 1–2 weeks to a debilitating illness lasting several months. The norm is complete recovery within 9 weeks, but a fulminating or even fatal hepatitis can occur. Severity of the disease worsens with increasing age. HAV is sporadic/endemic worldwide.

Improving sanitation raises the age of exposure, with accompanying complications. It now occurs particularly in persons from industrialized countries who are exposed to situations of poor hygiene, contaminated food products, or among young adults when traveling to areas where the disease is endemic. Common source outbreaks occur in school-age children and young adults from case contact or from food contaminated by infected handlers. Hepatitis A may be a serious public health problem in a disaster situation.

Prevention involves improving personal and community hygiene, with safe chlorinated water and proper food handling. Short-term risk for infection for people exposed to HAV may be reduced with prompt administration of HAV immune globulin. Hepatitis A vaccine is now recommended for all children over 12 months of age, as well as for persons traveling to endemic areas or at increased risk of exposure or morbidity. CDC reports 33 percent of the U.S. population were ever infected with HAV, but there is no chronic carrier state. HAV immunization is being adopted for routine prevention programs in some countries, including the United States, and is used for pre-exposure prevention, but immune globulin is still used for post-exposure protection. As costs of the vaccine come down, its widespread routine use may be recommended.

Hepatitis B

Hepatitis B (HBV) was once thought to be transmitted only by injections of blood or blood products. It is now known to be present in all body fluids and easily transmissible by household and sexual contact, perinatal spread from mother to newborn, and between toddlers. However, it is not usually spread by the oral–fecal route.

Hepatitis B virus is endemic worldwide and is especially prevalent in developing countries. Carrier status with persistent viremia is estimated by CDC to be 1.25 million in the United States, with 4.9 percent of the population ever infected. Carrier rates are 5–8 percent in sub-Saharan Africa but between 8 and 15 percent of babies become infected in some parts of the world, so that routine immunization is recommended. Carriers have detectable levels of HBsAg, the surface antigen (i.e., Australian antigen), in their blood.

Transmission from mother to child and between children by unsafe injections and sexual contact are common. High-risk groups who should be immunized in developed countries include health care workers, intravenous drug users; men who have sex with men; persons with high numbers of sexual partners; those receiving tattoos, body piercing, or acupuncture treatments; and residents or staff of institutions such as group homes and prisons. Immunocompromised and hemodialysis patients are commonly carriers of HBV. HBV may also be spread in a health system by use of inadequately sterilized reusable syringes, as in China and the former Soviet Union. Transmission is reduced by screening blood and blood products for HBsAg and strict technique for handling blood and body fluids in health settings.

HBV is clinically recognizable in less than 10 percent of infected children but is apparent in 30–50 percent of infected adults. Clinically HBV has an insidious onset with anorexia, abdominal discomfort, nausea, vomiting, and jaundice. The disease can vary in severity from subclinical, very mild, to fulminating liver necrosis and death. It is a major cause of primary liver cancer, chronic liver disease, and liver failure, all devastating to health and expensive to treat.

Hepatitis B virus is considered to be the cause of 60 percent of primary cancer of the liver in the world and the most common carcinogen after cigarette smoking. The WHO estimates that more than 2 billion people alive today have been infected with HBV. It is also estimated that 350 million persons are chronic carriers of HBV, with an estimated 1–1.5 million deaths per year from cirrhosis or primary liver cancer. This makes hepatitis B control a vital issue in the revision of health priorities in many countries.

Strict discipline in blood banks and testing of all blood donations for HBV, as well as HIV, and hepatitis C, is mandatory, with destruction of those donations with positive tests. Contacts should be immunized following exposure with HBV immunoglobulin and HBV vaccine. The inexpensive recombinant HBV vaccine should be adopted by all countries and included in routine vaccination of infants. Catch-up immunization for older children is also desirable. Immunization programs should include those exposed at work, such as health, prison, and sex workers and adults in group settings. HBV immunization has been included in WHO's EPI-PLUS expanded program of immunization.

Hepatitis C

First identified in 1989, and previously known as non-A, non-B hepatitis, hepatitis C (HCV) has an insidious onset with jaundice, fatigue, abdominal pain, nausea, and vomiting. It may cause mild to moderate illness, but chronicity is common, progressing to cirrhosis and liver failure. WHO estimates there are 170–180 million persons chronically infected with HCV and 3–4 million newly infected globally each year. The CDC estimates that 3.2 million Americans are chronically infected with HCV, with 8000–10,000 resulting deaths per annum, and HCV is the main cause of liver transplants. HCV is transmitted most commonly in blood products, but also among injecting drug users (90 percent of intravenous drug users were HCV-positive in a Vancouver study in 1998), and is also a risk for health workers. The disease may also occur in dialysis centers and other medical situations. Person-to-person spread is unclear. Prevention of transmission includes routine testing of blood donations, antiviral treatment of blood products, needle exchange programs, and hygiene. The WHO in 1998 declared hepatitis prevention as a major public health crisis, stressing that this "silent epidemic" is being neglected and that screening of blood products is vital to reduce transmission of this disease as for HIV.

It is a major cause of liver cirrhosis, end-stage liver disease, and hepatocellular carcinoma. The virus is primarily transmitted parenterally. No immunization is available at present; however, research is currently directed at vaccine development. Interferon and ribavirin combination therapy is not curative, but may reduce symptoms and prevent HCV-associated cancers. Partly due to the virus's genetic diversity, it evades the host immune response and it has been difficult to develop vaccines.

Significant advances have been made in the treatment of chronic hepatitis C virus infection. Currently, the combination of interferon-alpha and ribavirin is the standard treatment for chronic hepatitis C virus infection, and leads to long-term eradication of the virus in approximately 54 percent of people. Treatment is expensive and has significant adverse effects. Prevention of transmission is primarily addressed to intravenous drug users. Developing countries have high levels of this a disease and very little in the way of resources to deal with it until a vaccine is developed.

Hepatitis D

Hepatitis D virus (HDV), also known as delta hepatitis, may be self-limiting or progress to chronic hepatitis. It is caused by a virus-like particle which requires HBV to reproduce. HDV infects cells along with HBV as a coinfection or in chronic carriers of HBV. HDV occurs worldwide in the same groups at risk for HBV. It also occurs in epidemics and is endemic in South America, Africa, and among drug users. Prevention is by measures similar to those for HBV. Management for HDV is by passive immunity with immunoglobulin for contacts and high-risk groups, and should include HBV vaccination as the diseases often coincide. There is currently no vaccine for HDV.

Hepatitis E

Hepatitis E virus has an epidemiologic and clinical course similar to that of HAV, and is a particle with an incubation period of 15–64 days. There is no evidence of a chronic form of HEV. One striking characteristic of HEV is its high mortality rate among pregnant women. Infection results from waterborne epidemics or as sporadic cases in areas with poor hygiene, spread via the oral–fecal route. It is a hazard in disaster situations with crowding and poor sanitary conditions. Prevention is by safe management of water supplies and sanitation. Treatment is supportive and symptom-directed; passive immunization is not helpful and no vaccine is currently available.

Haemophilus Influenzae Type B

Haemophilus influenzae type b (Hib) is a bacteria which causes meningitis and other serious infections in children. Before the introduction of effective vaccines, as many as 1 in 200 children developed invasive Hib infection. Two-thirds of these had Hib meningitis, with a case fatality rate of 2–5 percent. Long-term sequelae such as hearing impairment and neurologic deficits occurred in 15–30 percent of survivors.

The first Hib vaccine was licensed in 1985, based on capsular material from the bacteria. Extensive clinical trials in Finland demonstrated a high degree of efficacy, but less impressive results were in seen in postmarketing efficacy studies. By 1989, a conjugate vaccine based on an additional protein cell capsular factor capable of enhancing the immunologic response was introduced. Several conjugate vaccines are now available. This vaccine is now widely used in the industrialized countries and the 2006 WHO Advisory on Immunization Programs (SAGE) recommended wide-scale adoption of this important new vaccine.

The conjugate vaccines are now combined with DTaP as their schedule is simultaneous with that of the DTaP. Hib vaccine has been found to be cost-effective, despite being as costly as all the basic vaccines combined (i.e., DPT, OPV, MMR, and HBV). For this reason, its use thus far has been limited to industrialized countries. The vaccine is a valuable addition to the immunologic armamentarium. It showed dramatic results in local eradication of this serious early childhood infection in a number of European countries and a sharp reduction in the United States. The price of the vaccine has also fallen dramatically since the mid-1990s. As a result, in 1997, WHO recommended inclusion of Hib vaccine in routine immunization programs in developing countries.

Influenza

Influenza is an acute viral respiratory illness characterized by fever, headache, myalgia, prostration, and cough. Transmission is rapid by close contact with infected individuals and by airborne particles with an incubation period of 1–5 days. It is generally mild and self-limited with recovery in 2–7 days. However, in certain population groups, such as the elderly and chronically ill, infection can lead to severe sequelae. Gastrointestinal symptoms commonly occur in children. During epidemics, mortality rates from respiratory diseases increase because of the large numbers of persons affected, although the case fatality rates are generally low.

Over the past century, influenza pandemics have occurred in 1889, 1918, 1957, and 1968, while epidemics are annual events. The influenza pandemic of 1918 caused millions of deaths among young adults, by some estimates killing more than had died in World War I. The influenza pandemic of 1918 killed nearly 50 million people worldwide and was characterized by an atypical mortality curve. Influenza usually mostly affects the very old and the very young. The principal group suffering from the 1918 pandemic was young men between the ages of 30 and 60 years, many in army training camps, as well as in the general population. Fear of recurrence of this pandemic led the CDC to launch a massive immunization program in the United States in 1976 to prevent swine flu (the virus was a strain antigenically similar to that of the 1918 pandemic influenza) from spreading from an isolated outbreak in an army camp. The effort was stopped after millions of persons were immunized with an urgently produced vaccine when serious reactions occurred (Guillain-Barré syndrome, a type of paralysis), and when no further cases of swine flu were seen. This demonstrated the difficulty of extrapolating scenarios from a historical experience (Box 4.7).

Box 4.7 High-Risk Groups Recommended for Annual Influenza Vaccination

1. Children between 6 months and 5 years of age.
2. Pregnant women.
3. Adults over 50 years of age.
4. Adults and children with chronic medical conditions.
5. Residents of long-term care facilities, such as nursing homes.
6. Persons in contact with high-risk individuals or populations.
7. Caregivers and contacts of infants and at-risk children.

Notes: Other groups should obtain medical advice regarding influenza risk and vaccination, such as immuno-suppressed patients and those receiving chronic aspirin therapy, among others.

People with allergy to previous flu shot, eggs, or other vaccine components, or with history or risk for Guillain-Barré syndrome may not be candidates for vaccination; obtain medical advice.

Source: Centers for Disease Control. 2007. Prevention & Control of Influenza — Recommendations of the Advisory Committee on Immunization Practices (ACIP). MMWR, Jul 13; 56(RR-06):1–54.

In recent years, concern has again arisen regarding the likelihood of virulent influenza pandemics. Of particular note is the influenza A H5N1 strain, known as avian influenza. WHO reports that from 2003 to October 2007, the number of confirmed human cases was 332 (204 deaths). Although relatively few human-to-human transmissions have been documented, this virus has rapidly spread among wild and domestic bird populations throughout Asia and much of the world. People who contact infected birds or poultry are at risk for severe disease, with over 60 percent case mortality. A minor mutation or genetic conjugation with a known human strain could result in a virus as deadly and contagious as the swine flu of 1918. It is estimated that up to 1.9 million people in the United States could die if such an outbreak occurs. Extensive international plans have been developed for intervention, should a virulent influenza pandemic occur. These include several vaccines with specificity to known virus strains. As many of the most devastating global communicable disease emergencies of recent centuries have been associated with highly pathogenic respiratory viruses, health systems and emergency plans must be prepared in the case of a pandemic. Active surveillance using sentinel chicken flocks now under surveillance for West Nile fever could be used to provide early warning of entry of the bird-borne disease into a specific region and help to trigger activation of response mechanisms.

Each year, epidemiologic services of the WHO and collaborating centers such as the CDC recommend which strains should be used in vaccine preparation for use among susceptible population groups. These vaccines are prepared with the current anticipated epidemic strains. The three main types of influenza (A, B, and C) have different epidemiologic characteristics. Type A and its subtypes, which are subject to antigenic shift, are associated with widespread epidemics and pandemics. Type B undergoes antigenic drift and is associated with less widespread epidemics. Influenza type C is even more localized.

Active immunization against the prevailing wild strain of influenza virus produces a 70–80 percent level of protection in high-risk groups. The benefits of annual immunization outweigh the costs, and it has proven to be effective in reducing cases of influenza and its secondary complications such as pneumonia and death from respiratory complications in high-risk groups.

Avian (H5N1) influenza is a threat to the world's population because of its potential to become a pandemic on the scale of the 1917–1918 flu epidemic. It is a bird-borne zoonotic disease so far affecting fowl such as chickens and turkeys contacted by infected wild fowl. Sensitive and robust surveillance measures are required to detect any evidence that the virus has changed and acquired the ability to transmit between humans. Surveillance is largely passive in relying on reports of infected wild and domestic fowl and most important human cases. The major concern is for detection of human-to-human transmission and thus a threat to transform this disease into a local, regional, and world pandemic in a matter of months.

International efforts to improve national and local capacities in surveillance and response to this threat are vital to review the scale of the threat should the leap from animal-to-human to human-to-human transmission occur. An integral part of the pandemic planning response in the United Kingdom was the creation in 2005 of the U.K. National H5 Laboratory Network, capable of rapidly and accurately identifying potential human H5N1 infections in all regions of the United Kingdom and the Republic of Ireland.

The CDC relies on seven systems for national influenza surveillance, four of which operate year-round: 1) the WHO and the National Respiratory and Enteric Virus Surveillance System (NREVSS) collaborating laboratory systems; 2) the U.S. Influenza Sentinel Provider Surveillance System; 3) the 122 Cities Mortality Reporting System; and 4) a national surveillance system that records pediatric deaths associated with laboratory-confirmed influenza.

Pneumococcal Disease

Pneumococcal diseases, which are caused by *Streptococcus pneumoniae,* include pneumonia, meningitis, and otitis media. Together, these constitute the world's leading cause of vaccine-preventable child mortality; over 1 million children die from pneumococcal diseases each year. The 23 capsular types of pneumococci selected out of 83 known types of the organism for the polysaccharide vaccine (PPV23) are those responsible for 88 percent of pneumococcal pneumonia cases and 10–25 percent of all pneumonia cases in the United States. This vaccine has been found to be cost-effective for high-risk groups, including persons with chronic disease, HIV carriers, patients whose spleens were removed, the elderly, and those with immunosuppressive conditions. It should be included in preventive-oriented health programs, especially for long-term care of the chronically ill. In addition a 7-valent conjugate vaccine (PCV7) is now available for children under 2 years of age, the highest risk age group for pneumococcal disease mortality. The WHO and CDC recommend PCV7 for children under 2 years old and PPV23 for adults over 65 years of age. In addition, others at risk for respiratory disease or pneumococcal infection should be vaccinated.

Varicella (Chickenpox, Shingles, Herpes Zoster)

Varicella is an acute, generalized viral disease caused by the varicella zoster virus (VZV). Despite its reputation as an innocuous disease of childhood, varicella patients can be quite ill. A mild fever and characteristic generalized red rash last for a few hours, followed by vesicles occurring in successive crops over various areas of the body. Affected areas may include the membranes of the eyes, mouth, and respiratory tract. The disease may be so mild as to escape observation or may be quite severe, especially in adults. Death can occur from viral pneumonia in adults and sepsis or encephalitis in children. Neonates whose mothers develop the disease within 2 days of delivery are at increased risk, with a case fatality rate of up to 30 percent.

Long-term sequelae include herpes zoster or shingles with a severely painful, vesicular rash along the distribution of sensory nerves, which can last for months. Its occurrence increases with age and it is primarily seen in the elderly. It can, however, occur in immunocompromised children (especially those on cancer chemotherapy), AIDS patients, and others. Some 15 percent of a population will experience herpes zoster during their lifetimes. Reye syndrome is an increasingly rare but serious complication from varicella or influenza type b. It occurs in children and affects the liver and central nervous system. Congenital varicella syndrome with birth defects similar to congenital rubella syndrome has been identified, emphasizing the importance of effective immunization against VZV. Varicella vaccine is now recommended for routine immunization at age 12–18 months in the United States, with catch-up for nonimmunized children and adults, especially nonpregnant women of childbearing age. To maintain immunity in adolescence and adulthood, booster vaccinations after age 13 and again after age 50 are effective in those who have no history of VZV infection or evidence of immunity. Varicella vaccine is likely to be added to a "cocktail vaccine" containing DPT, polio (IPV), and Hib.

Meningococcal Meningitis

Meningococcal meningitis, caused by the bacterium *Neisseria meningitides,* is characterized by headache, fever, neck stiffness, delirium, coma, and/or convulsions. The incubation period is 2–10 days. It has a case fatality rate of 5–15 percent if treated early and adequately, but rises up to 50 percent in the absence of treatment. There are several important strains (A, B, C, X, Y, and Z). Serogroups A and C are the main causes of epidemics, with B causing sporadic cases and local outbreaks. Transmission is by direct contact and droplet spread.

Meningitis (group A) is common in sub-Saharan African countries, but epidemics have occurred worldwide. During epidemics, children, teenagers, and young adults are the most severely affected. In developed countries, outbreaks occur most frequently in military and college student populations. In 1997, meningococcal meningitis spread widely in the "meningitis belt" in Central Africa.

Epidemic control is achieved by mass chemoprophy-laxis with antibiotics (e.g., rifampin or sulfa drugs) among case contacts, although the emergence of resistant strains is a concern. Vaccines against serotypes A and C (biva-lent) or A, C, Y, and W-135 are available. Their use is effective in epidemic control and prevention institutions and military recruits, especially for A and C serogroups. Recommendations are to immunize using the tetravalent conjugate vaccine (MCV4) during preadolescent years, so immunity is established prior to residential education or military service.

VACCINE-PREVENTABLE DISEASES

VPDs are still among the leading causes of death in developing countries and many mid-level developing or transition countries are not using the full potential of vaccines currently available to protect their children. VPDs are a fundamental aspect of public health not only because of the success achieved in saving millions of lives, but in the enormous potential for future developments that may have equally valuable contribution to length and quality of life. However, the potential of even currently available vaccines is not yet fully realized and traditional practices in many countries are slow to adopt the newer vaccines, and their great life-saving capacity. The following table from WHO (Table 4.6) summarizes the scale of preventable deaths from VPDs.

ESSENTIALS OF AN IMMUNIZATION PROGRAM

Vaccination is one of the key modalities of primary prevention. Immunization is cost-effective and prevents wide-scale disease and death, with high levels of safety. Despite the general consensus in public health regarding the central role of vaccination, there are many areas of controversy and unfulfilled expectations.

A vaccination program should aim at 95 percent or higher coverage at appropriate times, including infants, schoolchildren, and adults. Immunization policy should be adapted from current international standards applying the best available program to national circumstances and financial capacities (Box 4.8). Public health personnel with expertise in VPD control are needed to advise ministries of health and the practicing pediatric community on current issues in vaccination and to monitor implementation and evolution of control programs. Controversies and changing views are common to immunization policy, so that discussions must be conducted on a continuing basis. Policy should be under continuing review by a government-appointed national immunization advisory committee, including professionals from public health, academia, immunology, laboratory sciences, economics, and relevant clinical fields.

Vaccine supply should be adequate and continuous. Supplies should be ordered from known manufacturers meeting international standards of good manufacturing

TABLE 4.6 Estimated Number of Deaths in 2002 from Vaccine-Preventable Diseases (VPDs) Among Children <5 Years in 2002–2004. Diphtheria-Tetanus-Pertussis DTP Vaccine Coverage and Numbers of Unreached Infants and Incompletely Vaccinated Infants by WHO Region

WHO region	No. of deaths	% coverage with 1 dose of DTP	No. of unreached infants*	% coverage with 3 doses of DTP	No. of incompletely vaccinated infants[†]
African	1,113,000	78	5,607,000	66	3,048,000
American	44,000	96	562,000	92	659,000
Eastern Mediterranean	353,000	86	1,948,000	78	1,186,000
European	32,000	96	458,000	94	158,000
South East Asian	757,000	77	8,082,000	69	2,959,000
Western Pacific	251,000	96	1,051,000	90	1,302,000
Total	**2,550,000**	**86**	**17,708,000**	**78**	**9,312,000**

*Number of surviving infants who did not receive 1 dose of DTP, calculated on the basis of WHO/UNICEF estimates of vaccination coverage with 1 dose of DTP and estimates of surviving infants from World Population Prospects: The 2004 Revision.
[†]Number of surviving infants who did not receive 3 doses of DTP; unvaccinated infants were excluded.
World Health Organization Annual Report.

Box 4.8 Recommended Childhood Immunization Schedule, for Persons Aged 0–6, United States, 2006

	Range of recommended ages		Certain high-risk groups

Vaccine	Birth	1 month	2 months	4 months	6 months	12 months	15 months	18 months	19–23 months	2–3 years	4–6 years
Hepatitis B	HepB	HepB			HepB						
Rotavirus			Rota	Rota	Rota						
Diphtheria, tetanus, pertussis			DTaP	DTaP	DTaP		DTaP				DTaP
Haemophilus influenzae type b			Hib	Hib	Hib	Hib					
Pneumococcal			PCV	PCV	PCV	PCV				PPV	
Inactivated poliovirus			IPV	IPV	IPV						IPV
Influenza					Influenza (Yearly)						
Measles, mumps, rubella						MMR					MMR
Varicella						Varicella					Varicella
Hepatitis A						HepA (z doses)				HepA Series	
Meningococcal										MCV4	

This schedule indicates the recommended ages for routine administration of currently licensed childhood vaccines, as of December 1, 2007, for children aged 0 through 6 years. Additional information is available at www.cdc.gov/vaccines/recs/schedules. Any dose not administered at the recommended age should be administered at any subsequent visit, when indicated and feasible. Additional vaccines may be licensed and recommended during the year. Licensed combination vaccines may be used whenever any components of the combination are indicated and other components of the vaccine are not contraindicated and if approved by the Food and Drug Administration for that dose of the series. Providers should consult the respective Advisory Committee on Immunization Practices statement for detailed recommendations, including for high-risk conditions: http://www.cdc.gov/vaccines/pubs/ACIP-list.htm. Clinically significant adverse events that follow immunization should be reported to the Vaccine Adverse Event Reporting System (VAERS). Guidance about how to obtain and complete a VAERS form is available at www.vaers.hhs.gov or by telephone, 800–822–7967.

Source: Centers for Disease Control. 2008. Recommended Childhood and Adolescent Immunization Schedule — United States, Jan 6; 54(52);Q1–Q4.
Note: [a]OPV, Oral polio vaccine; IPV, inactivated polio vaccine; DPT or DPaT, diphtheria, pertussis, tetanus; preferably the acellular preparation (DTaP) and tetanus toxoid (DPT4 can be given at age 12 months if 6 months has elapsed since previous DPT); Td, diphtheria and tetanus; MMR, measles, mumps, rubella; Hep B, hepatitis B; Hib, *Haemophilus influenzae* type b; Var, varicella zoster virus; RV, rotavirus; #, for those who are not immunized in infancy.
[b]During 1999, the recommendation for poliovirus was changed to three doses of IPV in infancy.

practice. All batches should be tested for safety and efficacy prior to release for use. There should be an adequate and continuously monitored cold chain to protect against high temperatures for heat-labile vaccines, sera, and other active biological preparations. The cold chain should include all stages of storage, transport, and maintenance at the site of usage. Only disposable syringes should be used in vaccination programs to prevent any possible transmission of blood-borne infection.

A vaccination program depends on a readily available service with no barriers or unnecessary prerequisites, free to parents or with a minimum fee, to administer vaccines in disposable syringes by properly trained individuals using patient-oriented and community-oriented approaches. Ongoing education and training on current immunization practices are needed. Incentive payments by insuring agency or managed care systems promote complete, on-time coverage. All clinical encounters should be used to screen, immunize, and educate parents/guardians.

Contraindications to vaccination are very few; vaccines may be given even during mild illness with or without fever, during antibiotic therapy, during convalescence from illness, following recent exposure to an infectious disease, and to persons having a history of mild/moderate local reactions, convulsions, or family history of sudden infant death syndrome (SIDS). Simultaneous administration of vaccines and vaccine "cocktails" reduces the number of visits and thereby improves coverage; there are no known interferences between vaccine antigens.

Accurate, complete recording with computerization of records and automatic reminders helps promote compliance, as does co-scheduling of immunization appointments with other services. Adverse events should be reported promptly, accurately, and completely. A tracking system should operate with reminders of upcoming or overdue immunizations; use mail, telephone, and home visits, especially for high-risk families, with semiannual audits to assess coverage and review patient records in the population served to determine the percentage of children covered by their second birthday. Tracking should identify children needing completion of the immunization schedule and assess the quality of documentation. It is important to maintain up-to-date, easily retrievable medical protocols where vaccines are administered, noting vaccine dosage, contraindications, and management of adverse events.

All health care providers and managers should be trained in education, promotion, and management of immunization policy. Health education should target parents as well as the general public. Monitoring of vaccines used and children immunized, individually and by category of vaccination, can be facilitated by computerization of immunization records, or regular manual review of child care records. Where immunization is done by physicians in private practice, as in the United States, determination of coverage is by periodic surveys.

Regulation of Vaccines

Inspection of vaccines for safety, purity, potency, and standards is part of the regulatory function. Vaccines are defined as biological products and are therefore subject to regulation by national health authorities. In the United States, this comes under the legislative authority of the Public Health Service Act, as well as the Food, Drug and Cosmetics Act, with applicable regulations in the Code of Federal Regulations. The federal agency empowered to carry out this regulatory function is the Center for Drugs and Biologics of the Federal Food and Drug Administration.

Litigation regarding adverse effects of vaccines led to inflation of legal costs and efforts to limit court settlements. The U.S. federal government enacted the Child Vaccine Injury Act of 1988, establishing the National Vaccine Injury Compensation Program (VICP). This legislation requires providers to document vaccines given and to report on complications or reactions. It was intended to pay benefits to persons injured by vaccines faster and by means of a less expensive procedure than a civil suit for resolving claims. Using this no-fault system, petitioners do not need to prove that manufacturers or vaccine givers were at fault. They must only prove that the vaccine is related to the injury in order to receive compensation. The vaccines covered by this legislation include Hib, HAV, HBV, HPV, influenza, meningococcal, pneumococcal, rotavirus, VZV, DTaP/TdaP, MMR, OPV, and IPV.

Newly recommended vaccines for children and adolescents have nearly doubled in number since 2000, and the cost of fully vaccinating a child has increased dramatically in the past decade. Funding of the extensive recommended schedule is a problem in all countries where this is provided as a public health service or where it is covered by health insurance. In the United States, with a lack of health insurance for some 15 percent of the population and low levels of coverage for another 15 percent, lack of coverage for immunization is a significant problem. Many of the poorest children are covered under the Women, Infants, and Children nutrition support program, but others in the working poor population may lack access. This is an issue of debate in current political struggles to provide universal coverage for children.

Vaccine Development

Development of vaccines from Jenner in the eighteenth century to the advent of recombinant hepatitis B vaccine in 1987, and of vaccines for acellular pertussis, varicella, hepatitis A, and rotavirus in the 1990s, has provided one of the pillars of public health and led to enormous saving of human life. Vaccines for viral infections in humans for HIV, respiratory syncytial virus, Epstein-Barr virus, dengue fever, and hantavirus are under intense research with genetic approaches using recombinant techniques. The potential for the future of vaccines will be greatly influenced by scientific advances in genetic and molecular technology, with potential for development of vaccines attached to bacteria or protein in plants, which may be given in combination against an increasing range of organisms or toxins.

Recombinant DNA technology has revolutionized basic and biomedical research since the 1970s. The industry of biotechnology has produced important diagnostic tests, such as for HIV, with great potential for vaccine development. Traditional whole organism vaccines, alive or killed, may contain toxic products that may cause mild to severe reactions. Subunit vaccines are prepared from components of a whole organism. This avoids the use of live organisms that can cause the disease or create toxic products which cause reactions. Subunit vaccines traditionally prepared by inactivation of partially purified toxins are costly, difficult to prepare, and weakly immunogenic. Recombinant techniques are an important development for production of new whole cell or subunit vaccines that are safe, inexpensive, and more productive of antibodies than other approaches. Their potential contribution to the future of immunology is enormous.

Molecular biology and genetic engineering have made it feasible to create new, improved, and less costly vaccines. New vaccines should be inexpensive, easily administered, capable of being stored and transported without refrigeration, and given orally. The search for inexpensive and effective vaccines for groups of viruses causing diarrheal diseases led to development of the rotavirus vaccine. Some "edible" research focuses on the genetic programming of plants to produce vaccines and DNA. Vaccine manufacturers, who spend huge sums of money and years of research on new products, tend to work on those which will bring great financial rewards for the company and are critical to the local health care community. This has led to less effort being made in developing vaccines for diseases such as malaria which affect primarily the developing world. Industry plays a crucial role for continued progress in the field; therefore, work must be done to establish incentives for research, development, and application of vaccine technology from a global perspective.

CONTROL/ERADICATION OF INFECTIOUS DISEASES

Since the eradication of smallpox, much attention has focused on the possibility of similarly eradicating other diseases, and a list of potential candidates has emerged. Some of these have been abandoned because of practical difficulties with current technology. Diseases that have been under discussion for eradication have included measles, polio, and some tropical diseases, such as malaria and dracunculiasis. *Eradication* is defined as the achievement of a situation whereby no further cases of a disease

occur anywhere and continued control measures are unnecessary. Reducing epidemics of infectious diseases, through control and eradication in selected areas or target groups, can in certain instances achieve eradication of the disease. Local eradication can be achieved where domestic circulation of an organism is interrupted with cases occurring from importation only. This requires a strong, sustained immunization program with adaptation to meet needs of importation of carriers and changing epidemiologic patterns.

Smallpox

Smallpox was one of the major pandemic diseases of the Middle Ages and its recorded history goes back to antiquity. Prevention of smallpox was discussed in ancient China by Ho Kung (circa 320 CE), and inoculation against the disease was practiced there from the eleventh century CE. Prevention was carried out by nasal inhalation of powdered dried smallpox scabs. Exposure of children to smallpox when the mortality rate was lowest assumed a weakened form of the disease, and it was observed that a person could only have smallpox once in a lifetime. Isolation and quarantine were widely practiced in Europe during the sixteenth and seventeenth centuries.

Variolation was the practice of inoculating youngsters with material from scabs of pustules from mild cases of smallpox in the hope that they would develop a mild form of the disease. Although this practice was associated with substantial mortality, it was widely adopted because mortality from variolation was well below that of smallpox acquired during epidemics. Introduced into England in 1721 (see Chapter 1) it was commonly practiced as a lucrative medical specialty during the eighteenth century. In the 1720s, variolation was also introduced into the American colonies, Russia, and subsequently Sweden and Denmark.

Despite all efforts, in the early eighteenth century, smallpox was a leading cause of death in all age groups. Toward the end of the eighteenth century, an estimated 400,000 persons died annually from smallpox in Europe. Vaccination, or the use of cowpox vaccinia virus to protect against smallpox, was initiated late in the eighteenth century. In 1774, a cattle breeder in Yorkshire, England, inoculated his wife and two children with cowpox to protect them during a smallpox epidemic. In 1796, Edward Jenner, an English rural general practitioner, experimented with inoculation from a milkmaid's cowpox pustule to a healthy youngster, who subsequently proved resistant to smallpox by variolation (see Chapter 1). Vaccination, the deliberate inoculation of cowpox material, was slow to be adopted universally, but by 1801, over 100,000 persons in England were vaccinated. Vaccination gathered support in the nineteenth century in military establishments and in some countries that adopted it universally.

Opposition to vaccination remained strong for nearly a century based on religious grounds, observed failures of vaccination to give lifelong immunity, and because it was seen as an infringement of the state on the rights of the individual. Often the protest was led by medical variolationists whose medical practice and large incomes were threatened by the mass movement to vaccination. Resistance was also offered by "sanitarians" who opposed the germ theory and thought cleanliness was the best method of prevention. Universal vaccination was increasingly adopted in Europe and America in the early nineteenth century and eradication of smallpox in developed countries was achieved by the mid-twentieth century.

In 1958, the Soviet Union proposed to the World Health Assembly a program to eradicate smallpox internationally and subsequently donated 140 million doses of vaccine per year as part of the 250 million needed to promote vaccination of at least 80 percent of the world population. In 1967, WHO adopted a target for the eradication of smallpox. The program included a massive increase in coverage to reduce the circulation of the virus through person-to-person contact. Where smallpox was endemic, with a substantial number of unvaccinated persons, the aim of the mass vaccination phase was 80 percent coverage.

Increasing vaccination coverage in developing countries reduced the disease to periodic and increasingly localized outbreaks. In 1967, 33 countries were considered endemic for smallpox, and another 11 experienced importation of cases. By 1970, the number of endemic countries was down to 17, and by 1973 only 6 countries were still endemic, including India, Pakistan, Bangladesh, and Nepal. In these countries, a new strategy was needed, based on a search for cases and vaccination of all contacts, working with a case incidence below 5 per 100,000. The program then moved into the consolidation phase, with emphasis on vaccination of newborns and new arrivals. Surveillance and case detection were improved with case contact or risk group vaccination. The maintenance phase began when surveillance and reporting were switched to the national or regional health service with intensive follow-up of any suspect case. The mass epidemic era had been controlled by mass vaccination, reducing the total burden of the disease, but eradication required the isolation of individual cases with vaccination of potential contacts.

Technical innovations greatly eased the problems associated with mass vaccination worldwide. During the 1920s, there was wide variation in sources of smallpox vaccine. In the 1930s, efforts to standardize and further attenuate the strains used reduced complication rates from vaccinations. The development of lyophilization (freeze-drying) of the vaccine in England in the 1950s made a heat-stable vaccine that could be effective in tropical field conditions in developing countries. The invention of the bifurcated needle (Rubin, 1961) allowed for easier and more widespread vaccination by lesser trained personnel in remote areas. The net result of these innovations was increased world coverage and a reduction in the spread of the disease. Smallpox became more and more confined by increasing herd immunity, thus allowing transition to the phase of monitoring and isolation of individual cases.

In 1977 the last case of smallpox was identified in Somalia, and in 1980 the WHO declared the disease eradicated. No subsequent cases have been found except for several associated with a laboratory accident in the United Kingdom in 1978. The cost of the smallpox eradication program was $112 million or $8 million per year. Worldwide savings are estimated at $1 billion annually. This monumental public health achievement set the precedent for eradication of other infectious diseases. The World Health Assembly recommended destruction of the last two remaining stocks of the smallpox virus in Atlanta and Moscow in 1999. This was delayed in 1999 due to concern that illegal stocks may be held by some states or terrorists for potential use as weapons of mass destruction, concern regarding the appearance of monkeypox, and a wish to use the virus for further research. Today, virus stocks are handled only in select laboratories with high security. In addition, emergency plans have been developed, including the immunization of key health workers to limit the extent of a bioterror-engendered epidemic.

Eradication of Poliomyelitis

In 1988, the WHO established a target of eradication of poliomyelitis. Although polio epidemics continue, largely in countries with limited access to public health, the burden of disease worldwide has been greatly reduced. At the initiation of the polio eradication campaign, 350,000 cases of childhood paralysis were attributed to polio in 125 countries. By 2006, this number was reduced to 68,000. Only four countries have never achieved wild-poliovirus interruption: Afghanistan, India, Nigeria, and Pakistan. Support from member countries and international agencies such as UNICEF and Rotary International has led to wide-scale increases in immunization coverage throughout the world. The WHO promotes use of OPV as part of routine infant immunization on National Immunization Days (NIDs). This strategy has been successful in the Americas, Europe, and China, but several countries remain problematic.

Eradication of wild poliomyelitis will require flexibility in vaccination strategies and may require the combined approach, using OPV and IPV, as adopted in the United States in 1997 to prevent vaccine-associated clinical cases. Currently, IPV is largely used only in countries where interruption has occurred. Lack of intestinal immunity may be a risk for imported polio though. The combination of OPV and IPV may be needed where enteric disease is common and leads to interference in OPV uptake, especially in tropical areas where endemic poliovirus and diarrheal diseases are still found. In 2004, polio made a resurgence in Nigeria and in 2005 in a number of countries thought to be under control. The use of OPV has been put in doubt by recent decisions in the industrialized countries to follow the U.S. example of IPV only. The developing countries will need to rely on OPV in the coming years because of high cost and limited supply of IPV.

Other Candidates for Eradication

Success in the eradication of smallpox, followed by increasing control and the prospect of eradication of poliomyelitis, as well as other VPDs, has led to optimism in identifying other diseases which could potentially be eradicated or eliminated as public health problems. A list of potential candidates has been identified. Some of these have since been abandoned due to practical and technological difficulties. Diseases that have been under discussion for eradication have included measles, TB, and tropical diseases such as malaria and dracunculiasis.

Eradication of malaria was thought to be possible in the 1950s when major gains were seen in malaria control by aggressive case environmental control, case finding, and management. However, lack of sustained vector control and an effective vaccine has prevented global eradication. Malaria control suffered serious setbacks because of failure in political resolve and capacity to continue support needed for necessary programs. In the 1960s and 1970s, control efforts were not sustained in many countries, and a dreadful comeback of the disease occurred in Africa and Asia in the 1980s. The emergence of mosquitoes resistant to insecticides, and parasite strains of the parasite resistant to antimalarial drugs have made control even more difficult and expensive.

Renewed efforts in malaria control may require new approaches. Use of community health workers (CHWs) in small villages in highly endemic regions of Colombia resulted in a major drop in malaria mortality during the 1990s. The CHWs investigate suspect cases by taking clinical histories and blood smears. A presumptive diagnosis is made clinically or by local examination of blood smears. Therapy is instituted rapidly, and the patient is followed. Quality control monitoring shows high levels of accuracy in reading of slides compared to professional laboratories. Use of DDT-impregnated bed nets has become a major method of prevention. The banning of DDT completely since the 1960s is now seen as an overreaction to legitimate concerns for its widespread use, but vector control remains a key element of malaria control, and DDT has a role in this. Since 2006, WHO has recommended use of DDT-impregnated bed nets and limited uses of DDT for protecting homes to reduce the risk of infection of children especially.

In the late 1970s, there was widespread discussion in the literature of the potential for eradication of measles and TB. Measles eradication was set back as breakthrough epidemics occurred in the United States, Canada, and many other countries during the 1980s and early 1990s, but regional eradication was achieved combining the two-dose policy with catch-up campaigns for older children or in National Immunization Days, as in the Caribbean countries.

Tuberculosis has also increased in the United States and several European countries for the first time in many decades. Unrealistic expectations can lead to inappropriate assessments and policy when confounding factors alter the epidemiologic course of events. Such is the case with TB, where control and eradication have receded from the picture. This deadly disease has returned to developed countries, partly in association with the HIV infection and multidrug-resistant strains, as well as homelessness, rising prison populations, poverty, and other deleterious social conditions. Directly observed therapy is an important recent breakthrough, more effective in use of available technology, and it will play a major role in TB control in the twenty-first century.

Future Candidates for Eradication

A decade after the eradication of smallpox was achieved, the International Task Force for Disease Eradication (ITFDE) was established to systematically evaluate the potential for global eradicability of candidate diseases. Its goals were to identify specific barriers to the eradication of these diseases that might be surmountable and to promote eradication efforts (Box 4.9).

The subject of eradication versus control of infectious diseases is of central public health importance as technology expands the armamentarium of immunization and vector control into the twenty-first century. The control of epidemics, followed by interruption of transmission and ultimately eradication, will save countless lives and prevent serious damage to children throughout the world. The smallpox achievement, momentous in itself, points to the potential for the eradication of other deadly diseases. The skillful use of existing and new technology is an important priority in the New Public Health. Flexibility and adaptability are as vital as resources and personnel.

Selecting diseases for eradication is not purely a professional issue of resources such as vaccines and manpower, organization, and financing. It is also a matter of political will and perception of the burden of disease. There will be many controversies. The CDC published criteria for selection of disease for eradication as shown in Table 4.7.

The WHO, in a 1998 review of health targets in the field of infectious disease control for the twenty-first century, selected the following targets: eradication of Chagas' disease by 2010; eradication of neonatal tetanus by 2010; eradication of leprosy by 2010; eradication of measles by 2020; eradication of trachoma by 2020; and reversing the current trend of increasing tuberculosis and HIV/AIDS. Many of these campaigns have achieved interim goals. Although primary targets for eradication, such as polio, have proven problematic, the coming years appear to be a horizon for

Box 4.9 Criteria for Assessing Eradicability of
Diseases, International Task Force for Disease
Eradication (ITFDE)

1. Scientific feasibility
   a. Epidemiologic vulnerability; lack of nonhuman reservoir, ease of spread, no natural immunity, relapse potential;
   b. Effective practical intervention available; vaccine or other primary preventive or curative treatment, or vectoricide that is safe, inexpensive, long-lasting, and easily used in the field;
   c. Demonstrated feasibility of elimination in specific locations, such as an island or other geographic unit.
2. Political will/popular support
   a. Perceived burden of the disease; morbidity, mortality, disability, and costs of care in developed and developing countries;
   b. Expected cost of eradication;
   c. Synergy of implementation with other programs;
   d. Reasons for eradication versus control.

Source: World Health Organization. 1992. Update International Task Force for Disease Eradication 1991. *Morbidity and Mortality Weekly Report*, 41:40–42.

breakthroughs and elimination of many preventable diseases. A review of onchocerciasis eradication efforts in 2007 concluded that eradication of the disease could be achieved in the Americas but not yet in Africa, but that achievements to date should be preserved by cooperative efforts of WHO, the World Bank, UNDP, and others. The struggle to eliminate and potentially eradicate important diseases, such as was achieved with smallpox, will require many years of strong political and funding support as well as a strong cadre of public health workforce with new scientific breakthroughs (such as malaria and HIV vaccines), but the movement even when partially successful is saving millions of lives and improving quality of life for many more.

Tuberculosis

Tuberculosis (TB) is caused by a group of organisms including *Mycobacterium tuberculosis* in humans and *M. bovis* in cattle. The disease is primarily found in humans, but it is also a disease of cattle and occasionally other primates in certain regions of the world. It is transmitted via airborne droplet nuclei from persons with pulmonary or laryngeal TB during coughing, sneezing, talking, or singing. The initial infection may go unnoticed, but tuberculin sensitivity

TABLE 4.7 Potential Disease Candidates for Control and Eradication, 1998

Organism	Control — Elimination as a public health problem	Eradicable — Regional/global
Bacterial diseases	Pertussis Neonatal tetanus Congenital syphilis Trachoma Tuberculosis Leprosy	Diphtheria *Haemophilus influenzae* type b
Viral disease	Hepatitis B Hepatitis A Yellow fever Rabies Japanese encephalitis	Poliomyelitis Measles Rubella Mumps
Parasitic disease	Malaria Chagas' disease Helminthic infestation Schistosomiasis Leishmaniasis, visceral	Echinococcus Teniasis
Noninfectious disease	Lead poisoning Silicosis Protein energy malnutrition Micronutrient malnutrition 　Iodine deficiency 　Vitamin A deficiency 　Folic acid deficiency 　Iron deficiency	

Source: Goodman, R. A., Foster, K. L., Trowbridge, F. L., Figuero, J. P. (eds.). 1998. Global Disease Elimination and Eradication as Public Health Strategies: Proceedings of a Conference Held in Atlanta, Georgia, USA, 23–25 February 1998. *Bulletin of the World Health Organization*,76 Supplement 2:1–161.

appears within a few weeks. About 95 percent of those infected enter a latent phase with a lifelong risk of reactivation. Approximately 5 percent go from initial infection to pulmonary TB. Less commonly, the infection develops as extrapulmonary TB, involving meninges, lymph nodes, pleura, pericardium, bones, kidneys, or other organs.

Untreated, about half of the patients with active TB will die of the disease within 2 years, but modern chemotherapy almost always results in a cure. Pulmonary TB symptoms include cough and weight loss, with clinical findings on chest examination and confirmation by findings of tubercle bacilli in stained smears of sputum and, if possible, growth of the organism on culture media, and changes in the chest x-ray. Tuberculosis affects people in their adult working years, with 80–90 percent of cases in persons between the ages of 15 and 49. Its devastating effects on the workforce and economic development contribute to a high cost-effectiveness for TB control.

Nearly one-third of the world's population is infected with tuberculosis. In 2005, there were over 8.8 million new cases and nearly 1.6 million deaths. During 2005, new cases of TB included 3.0 million in southeast Asia and 2.5 million in Africa, where HIV disease has become the leading comorbidity and risk for TB mortality. Between 1990 and 1999, WHO estimates there were 88 million new cases of TB, of which 8 million cases were in association with HIV infection. During the 1990s, an estimated 30 million persons died of TB, including 2.9 million with HIV infection. The 2008 Global Tuberculosis Control Report reported 9.2 million new cases in 2006 including 400,000 new cases of multidrug-resistant TB; approximately 1.5 million died of TB in 2006.

A new and dangerous period for TB resurgence has resulted from parallel epidemiologic events: first, the advent of HIV infection and, second, the occurrence of multidrug-resistant TB (MDRTB); that is, organisms resistant at least to both isoniazid (INH) and rifampin, two mainstays of TB treatment. MDRTB can have a case fatality rate as high as 70 percent. HIV reduces cellular immunity so that people with latent TB have a high risk of activation of the disease. It is estimated that HIV-negative persons have a 5–10 percent lifetime risk of TB; HIV-positive people have a risk of 10 percent per year of developing clinical tuberculosis (Box 4.10).

Drug resistance, the long period of treatment, and the socioeconomic profile of most TB patients combine to require a new approach to therapy. Directly observed treatment, short-course (DOTS), has shown itself to be highly effective with patients in poor self-care settings, such as the homeless, drug users, and those with AIDS. The strategy of DOTS uses community health workers to visit the patient and observes him or her taking the various medications, providing both incentive, support, and moral coercion to complete the needed 6–8 month therapy. DOTs has been shown to cure up to 95 percent of cases, at a cost of as little as $11 over the period of treatment per patient. It is one of the few hopes of containing the current TB pandemic.

Box 4.10 Principal Issues of Control of Tuberculosis

- Identifying persons with clinically active TB;
- Diagnostic methods — clinical suspicion, sputum smear for bacteriologic examination, tuberculin skin testing, chest radiograph;
- Case finding and investigation programs in high-risk groups;
- Contact investigation;
- Isolation techniques only during initial therapy;
- Treatment, mainly ambulatory, of persons with clinically active TB;
- Investigation and treatment of contacts;
- Directly observed treatment, short-course (DOTS), where compliance is suspect;
- Environmental control in treatment settings to reduce droplet infection;
- Educate health care providers on suspicion of TB and investigation of suspects.

In 2006, WHO rededicated itself to TB control with the "Stop TB Strategy" for control of tuberculosis over the next decade. The plan calls for new guidelines for control, new aid funds for developing countries, and enlistment of NGOs to assist in the fight. The new guidelines stress short-term chemotherapy in well-managed programs of DOTS, stressing strict compliance with therapy for infectious cases with a goal of an 85 percent cure rate. Even under adverse conditions, DOTS produces excellent results. It is one of the most cost-effective health interventions combining public health and clinical medical approaches. Primary goals of the Stop TB Strategy are to reduce TB incidence and mortality by 50 percent by 2015, compared to 1990, and to eliminate TB as a public health problem by the year 2050 (Figure 4.3).

Tuberculosis incidence in the United States decreased steadily until 1985, increased in 1990, and has declined again since (Figure 4.4). From 1986 to 1992, there was an excess of 51,600 cases over the expected rate if the previous decline in case incidence had continued. This rise was largely due to the HIV/AIDS epidemic and the emergence of MDRTB, but also greater concentration among immigrants from areas of higher TB incidence, drug abusers, the homeless, and those with limited access to health care. This is particularly true in New York City, where MDRTB has appeared in outbreaks among prison inmates and hospital staff.

TB incidence in the United States declined due to stronger TB control programs that promptly identify persons with TB and ensure completion of appropriate therapy. Aggressive staff training, outreach, and case management approaches were vital to this success. Concern over rising rates among recent immigrants and the continued challenge of HIV/AIDS and coincidental transmission of hepatitis A, B, and C among drug users and marginal population groups show that continued support for TB control is needed.

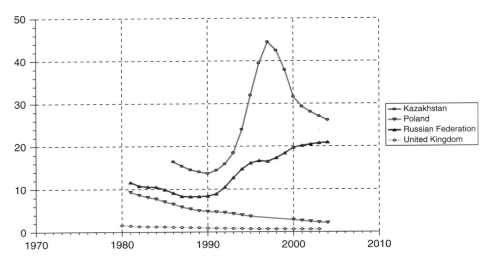

FIGURE 4.3 SDR for Tuberculosis in Poland, Kazakhstan, Russia, and the UK. Note: Standardized Death Rates per 100,000 population. Source: Health for All database WHO European Region, November 2007, http://www. euro.who.int/hfadb [accessed May 24, 2008]

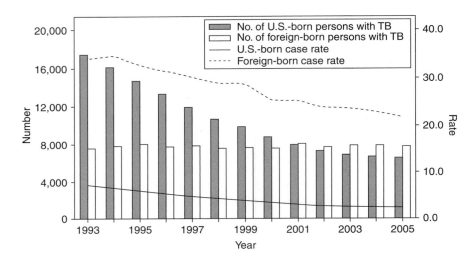

FIGURE 4.4 Tuberculosis numbers and rates, by origin of birth and year, United States, 1993–2005. Source: CDC, Reported Tuberculosis in the United States, 2005, Atlanta, GA: U.S. DHHS, CDC, 2006, http://www.cdc.gov/tb/surv/surv2005/PDF/TBSurv FULLReport.pdf [accessed June 2, 2008]

Bacille Calmette-Guérin (BCG) is an attenuated strain of the tubercle bacillus used widely as a vaccination to prevent TB, especially in high-incidence areas. It induces tuberculin sensitivity or an antigen–antibody reaction in which antibodies produced may be somewhat protective against the tubercle bacillus in 90 percent of vaccinees. Although the support for its general use is contradictory, there is evidence from case–control and contact studies of positive protection against TB meningitis and disseminated TB in children under the age of 5. In some developed, low-incidence countries, it is not used routinely but selectively. It may also be used in asymptomatic HIV-positive persons or other high-risk groups.

The BCG vaccine for tuberculosis remains controversial. While used widely internationally, in the United States and other industrialized countries, it is thought to hinder rather than help in the fight against TB. This concern is based on the usefulness of tuberculin testing for diagnosis of the disease. Where BCG has been administered, the diagnostic value of tuberculin testing is reduced, especially in the period soon after the BCG is used. Studies showing equivocal benefit of BCG in preventing tuberculosis have added to the controversy. While those in the field in the United States continue to oppose the use of BCG, internationally it is still felt to be of benefit in preventing TB, primarily in children. Currently, the WHO recommends use of a single dose of BCG as close to birth as possible as part of the EPI.

A 1994 meta-analysis of the literature of BCG carried out by the Technology Assessment Group at Harvard School of Public Health concluded that on average, BCG vaccine significantly reduces the risk of TB by 50 percent. Protection is observed across many populations, study designs, and forms of TB. Age at vaccination did not enhance BCG efficacy. Protection against tuberculosis death, meningitis, and disseminated disease is higher than for total TB cases, although this result may reflect reduced error in disease classification rather than greater BCG efficacy.

Limitations of current chemotherapy and the only available vaccine, BCG, in the fight against TB make the continued search for new vaccines and therapeutics vital, possibly aided by new methods in design of vaccines and drugs. However, the struggle is now best fought by a combination of DOTS and DOTS plus (for multidrug-resistant strains) along with poverty alleviation and nutritional improvements in vulnerable population groups (Box 4.11).

In 1993, the WHO adopted a national case management strategy, DOTS to reduce the increasing global burden of TB, especially in developing countries. The five elements of the DOTS strategy are sustainable government commitment, quality assurance of sputum microscopy, standardized short-course treatment (including direct observation therapy), regular supply of drugs, and establishment of reporting and recording systems.

The goal of DOTS is to reduce TB morbidity and mortality and the chance of *Mycobacterium tuberculosis* developing resistance to primary treatment drugs. Target goals of TB control adopted in 1991 by the World Health Assembly include ≥70 percent detection rate of the estimated incidence of sputum-smear–positive pulmonary TB (PTB+) and ≥85 percent cure rate for newly detected PTB+ cases. The ≥85 percent cure rate was adopted on the basis of accumulated experience in Africa and certain districts of China.

Performance indicators in the DOTS program use the proportion detected of PTB+, which is the most infectious form of TB. PTB+ is associated with high mortality and is the most effective form of TB to use for bacteriologic monitoring of treatment progress. The proportion of newly detected PTB+ cases among the total number of adults with PTB reflects the proper application of diagnostic criteria. In countries with a medium or high TB burden, when necessary laboratory resources are available and sputum smears for microscopy are taken from TB patients, PTB+ accounts for >50 percent of all TB cases and >65 percent of new PTB cases in adults. Achieving a high (i.e., ≥85 percent) cure rate for PTB+ is a critical priority for TB-control programs. Failure to achieve this rate results in continued infectiousness and possible development of MDR TB, resistant to at least isoniazid and rifampin.

Tuberculosis control remains feasible with current medical and public health methods. Deterioration in its control should not lead to despair and passivity. The recent trend to successful control by DOTS despite the growing problem of MDRTB suggest that control and gradual reduction can be achieved by an activist, community outreach approach. The WHO in 2006 reaffirmed TB control as one of its major priorities, expressing grave concern that the MDR organism, now widely spread in countries of Asia, eastern Europe, and the former Soviet Union, may spread the disease much more widely. The disease constitutes one of the great challenges to public health.

Extremely drug-resistant TB, XMDR-TB, has become a central concern in addressing the current TB epidemic and is part of a WHO-led strategy in this field. The Millennium Development Goals include TB with specific targets endorsed by the Stop TB Partnership:

- 2005: detect at least 70 percent of new sputum smear-positive TB cases and cure at least 85 percent of these cases;
- by 2015: reduce prevalence of and death due to TB by 50 percent relative to 1990;
- by 2050: eliminate TB as a public health problem (<1 case per million population).

Streptococcal Diseases

Acute infectious diseases caused by group A streptococci include streptococcal sore throat, scarlet fever, puerperal fever, septicemia, erysipelas, cellulitis, mastoiditis, otitis media, pneumonia, peritonsillitis (quinsy), wound infections, toxic shock syndrome, and fasciitis, the "flesh eating bacteria." *Streptococcus pyogenes* group A includes some 80 serologically distinct types which vary in geographic location and clinical significance. Transmission is by droplet, person-to-person direct contact, or food infected by carriers. Important complications from a public health point of view include acute rheumatic fever and acute glomerulonephritis, but also skin infections and pneumonia.

Box 4.11 Control of Tuberculosis in Kazakhstan Using DOTS

To manage the increasing burden of TB in the country, in 1998 the Kazakhstan Ministry of Health adopted and implemented a new National Tuberculosis Program (NTP), whose objectives and target goals are in accord with the DOTS strategy. To implement the DOTS strategy in Kazakhstan, primary health care physicians and TB specialists received training in case-detection policy, and laboratories were equipped with binocular microscopes. Unfavorable treatment outcomes for new PTB+ cases were associated with alcohol abuse, homelessness, previous incarceration, unemployment, being male, and urban residence. The epidemic curve peaked in 1998 with a continuous decline since. Treatment of multidrug-resistant TB is more costly and complex, but has become an essential part of international TB work.

Sources: Centers for Disease Control. 2006. Progress toward tuberculosis control and determinants of treatment outcomes — Kazakhstan, 2000–2002. *Morbidity and Mortality Weekly Report*, 55 (SUP01);11–15.
World Health Organization. 2003. *Treatment of Tuberculosis: Guidelines for National Programs*. Geneva: World Health Organization (WHO/CDS/TB/2003.313).

Acute rheumatic fever (ARF) is a complication of strep A infection that has virtually disappeared from industrialized countries as a result of improved standards of living and antibiotic therapy. Mortality rates from rheumatic fever and rheumatic heart disease has declined steadily over the last 3 decades, largely due to increase in availability and use of antibiotics. In developing nations and lower socioeconomic areas where rheumatic fever is more prevalent, ARF is a major cause of death and disability in children and adolescents. However, outbreaks were recorded in the United States in 1985, and an increasing number of cases have been seen since 1990. In developing countries, rheumatic fever remains a serious public health problem affecting school-age children, particularly those in crowded living arrangements. Long-term sequelae include disease of the mitral and aortic heart valves, which require cardiac care and surgery for repair or replacement with artificial valves.

Acute glomerulonephritis is a reaction to toxins of the streptococcal infection in the kidney tissue. This can result in long-term kidney failure and the need for dialysis or kidney transplantation. This disease has become far less common in the industrialized countries, but remains a public health problem in developing countries.

Group B streptococci (GBS) are related organisms. They commonly colonize the reproductive tract of women of reproductive age, and are the leading cause of meningitis in newborn infants. As with other strains of beta-hemolytic streptococci, treatment with penicillin (or appropriate therapy for allergic patients) is effective. Women should be screened for GBS at 35–37 weeks of pregnancy and treated during labor and delivery. If screening tests are unavailable, the risk of infection is high; recommendations are to treat prophylactically.

The streptococcal diseases are controllable by early diagnosis and treatment with antibiotics. This is a major function of primary care systems. Recent increases in rheumatic fever may herald a return of the problem, perhaps due to inadequate access to primary care in the United States for large sectors of the population, along with crowding and possibly poor access to medical care due to lack of or inadequate health insurance.

Where access to primary care services is limited, infections with streptococci can result in a heavy burden of chronic heart and kidney disease with substantial health, emotional, and financial tolls. Measures to improve access to care and public information are needed to assure rapid and effective care to prevent chronic and costly conditions.

Zoonoses

Zoonoses are infectious diseases transmissible from vertebrate animals to humans. Common examples of zoonoses of public health importance in nonindustrialized countries include brucellosis and rabies. In industrialized countries, salmonellosis, mad cow disease, and influenza have reinforced the importance of relationships of animal and human health. Strong cooperation between public health and veterinary public health authorities are required to monitor and to prevent such diseases. Zoonoses have been described and known over many centuries. They involve all types of agents: bacteria, parasites, viruses, and unconventional agents. Bacterial organisms transmitted by animals include salmonellosis and campylobacteriosis, anthrax, brucellosis, *E. coli,* leptospirosis, plague, shigellosis, and tularemia.

Viruses transmitted by animals include *rabies,* which is a disease of carnivores and bats mainly transmissible to humans by bites. Almost all persons severely exposed to rabid animals will die if not treated. An estimated 55,000 persons, mainly children, die of this disease in the world every year mostly from infected dogs. Control measures focus on immunization of domestic animals and household pets. Infected dog-bite transmission is responsible for most human deaths.

Other viral zoonoses are avian influenza, Crimean-Congo hemorrhagic fever, Ebola, and Rift Valley fever. Bovine spongiform encephalopathy is thought to be the cause of variant Creutzfeldt-Jakob disease (vCJD), which is a neurologic disease different from CJD, leading to death in humans.

Other important zoonoses are brucellosis and echinococcosis/hydatidosis, for example. Zoonoses still represent significant and often neglected public health threats. They are affecting hundreds of thousands of people especially in developing countries, although many are preventable, with essential veterinary public health measures.

Brucellosis

Brucellosis is a disease occurring in cattle (*Brucella abortus*), in dogs (*Br. canis*), in goats and sheep (*Br. melitensis*), and in pigs (*Br. suis*). Humans are affected mainly through ingestion of contaminated milk products, by contact, or inhalation. Brucellosis (also known as relapsing, undulant, Malta, or Mediterranean fever) is a systemic bacterial disease of acute or insidious onset characterized by fever, headache, weakness, sweating, chills, arthralgia, depression, weight loss, and generalized malaise. Spread is by contact with tissues, blood, urine, vaginal discharges, but mainly by ingestion of raw milk and dairy products from infected animals. The disease may last from a few days to a year or more. Complications include osteoarthritis and relapses. Case fatality is under 2 percent, but disability is common and can be pronounced.

The disease is primarily seen in Mediterranean countries, the Middle East, India, central Asia, and Central and South America. Brucellosis occurs primarily as an occupational disease of persons working with and in contact with tissues, blood, and urine of infected animals, especially goats and sheep. It is an occupational hazard for veterinarians, packing house workers, butchers, tanners, and laboratory workers. It is also transmitted to consumers of unpasteurized milk from infected animals. Animal vectors include wild animals, so that eradication is virtually impossible. Diagnosis is confirmed by laboratory findings of the organism in blood or other tissue samples, or with rising antibody titers in the blood, with confirmation by blood cultures.

Clinical cases are treated with antibiotics. Epidemiologic investigation may help track down contaminated animal flocks. Routine immunization of animals, monitoring of animals in high-risk areas, quarantining sick animals, destroying infected animals, and pasteurizing milk and milk products prevent spread of the disease. Control measures include educating farmers and the public not to use unpasteurized milk. Individuals who work with animals (cattle, swine, goats, sheep, dogs, coyotes) should take special precautions when handling animal carcasses and materials. Testing animals, destroying carriers, and enforcing mandatory pasteurization will restrict the spread of the disease. This is an economic as well as public health problem, requiring full cooperation between ministries of health and of agriculture.

Rabies

Rabies is primarily a disease of animals, with a variety of wild animals serving as a reservoir for this disease, including foxes, wolves, bats, skunks, and raccoons, who may infect domestic animals such as dogs, cats, and farm animals. Animal bites break the skin or mucous membrane, allowing entry of the virus from the infected saliva into the bloodstream. The incubation period of the virus is 2–8 weeks; it can be as long as several years or as short as 5 days, so that postexposure preventive treatment is a public health emergency.

The clinical disease often begins with a feeling of apprehension, headache, pyrexia, followed by muscle spasms, acute encephalitis, and death. Fear of water ("hydrophobia") or fear of swallowing is a characteristic of the disease. Rabies is almost always fatal within a week of onset of symptoms. There is no effective treatment, so control relies on vaccination of animals, rapid prophylaxis of exposed people, and prevention of contact with biting and scratching animals. The disease is estimated to cause 30,000 deaths annually, primarily in developing countries. It is uncommon in developed countries.

Rabies control focuses on prevention in humans, domestic animals, and wildlife. Prevention in humans is based on preexposure prophylaxis for groups at risk (e.g., veterinarians, zoo workers) and post-exposure immune globulin and vaccine administration for persons bitten by potentially rabid animals. Because reducing exposure of pets to wild animals is difficult, immunization of domestic animals is one of the most important preventive measures. Prevention in domestic animals is by mandatory immunization of household pets. All domestic animals should be immunized at age 3 months and revaccinated according to veterinary instructions.

Prevention in wild animals to reduce the reservoir is successful in achieving local eradication in settings where reentry from neighboring settings is limited. Since 1978, the use of oral rabies immunization has been successful in reducing the population of wild animals infected by the rabies virus. Rabies eradication efforts, using aerial distribution of baits containing fox rabies vaccine in affected areas of Belgium, France, Germany, Italy, and Luxembourg, have been under way since 1989. The number of rabies cases in these affected areas has declined by some 70 percent. Switzerland is now virtually rabies-free because of this vaccination program. The potential exists for focal eradication, especially on islands or in partially restricted areas with limited possibilities of wild animal entry. Livestock need not be routinely immunized against rabies, except in high-risk areas. Where bats are major reservoirs of the disease, as in the United States, eradication is not presently feasible.

Salmonella

Salmonella, discussed later in this chapter under diarrheal diseases, is one of the most common infectious diseases among animals and is easily spread to humans via poultry, meat, eggs, and dairy products. Transmission may also occur from contact with infected animals, particularly reptile pets. Specific antigenic types are associated with food-borne transmission to humans, causing generalized illness and gastroenteritis. Severity of the disease varies widely, but the diseases can be devastating among vulnerable population groups, such as young children, the elderly, and the immunocompromised. Epidemiologic investigation of common food source outbreaks may uncover hazardous food handling practices. Laboratory confirmation or serotypes help in monitoring the disease. Prevention is by maintaining high standards of food hygiene in processing, inspection and regulation, food handling practices, and hygiene education.

Anthrax

Bacillus anthracis causes a bacterial infection in herbivore animals. Its spores contaminate soil worldwide. It affects humans exposed in occupational settings. Transmission is cutaneous by contact, gastrointestinal by ingestion, or respiratory by inhalation. It has gained recent attention (Iraq, 1997) as a highly potent agent for germ warfare or terrorism. In 2001, anthrax was used as a bioterror agent against the United States. Twenty-two people were infected, with a 50 percent case mortality rate.

Although most *B. anthracis* strains are susceptible to common antibiotics, concerns over the existence of resistant "weaponized anthrax" has prompted extensive planning to counter the possibility of repeated attacks. Limited supplies of vaccine are available; however, in the absence of an epidemic, use is only justified for veterinarians, key public health workers, and soldiers or laboratory personnel with higher risk for exposure.

Creutzfeldt-Jakob Disease

Creutzfeldt-Jakob disease is a degenerative disease of the central nervous system linked to consumption of beef from cattle infected with bovine spongiform encephalopathy. It is transmitted by prion proteins in animal feed prepared from contaminated animal material and in transplanted organs. This disease was identified in the United Kingdom linked to infected cattle leading to a 1997 ban on British beef in many parts of the world and slaughter of large numbers of potentially contaminated animals.

Other Major Zoonotic Diseases

The tapeworm causing diphyllobothriasis (*Diphyllobothrium latum*) is widespread in North American freshwater fish, passing from crustacean to fish to humans by eating raw freshwater fish. It is especially common among Inuit peoples and may be asymptomatic or cause severe general and abdominal disorder. Food hygiene (freezing and cooking of meat) is recommended; treatment is by anthelmintics.

Leptospiroses are a group of zoonotic bacterial diseases found worldwide in rats, raccoons, and domestic animals. It affects farmers, sewer workers, dairy and abattoir workers, veterinarians, military personnel, and miners with transmission by exposure to or ingestion of urine-contaminated water or tissues of infected animals. It is often asymptomatic or mild, but may cause generalized illness like influenza, meningitis, or encephalitis. Prevention requires education of the public in self-protection and immunization of workers in hazardous occupations, along with immunization and segregation of domestic animals and control of wild animals.

VECTOR-BORNE DISEASES

Vector-borne diseases are a group of diseases in which the infectious agent is transmitted to humans by crawling or flying insects. The vector is the intermediary between the reservoir and the host. Both the vector and the host may be affected by climatic conditions; mosquitoes thrive in warm, wet weather, and are suppressed by cold weather; humans may wear less protective clothing in warm weather.

Malaria

The only important reservoir of malaria is humans. Its mode of transmission is from person to person via the bite of an infected female *Anopheles* mosquito (Ronald Ross,

Nobel Prize, 1902). The causative organism is a single cell parasite with four species: *Plasmodium vivax, P. malariae, P. falciparum,* and *P. ovale*. Clinical symptoms are produced by the parasite invading and destroying red blood cells. The incubation period is approximately 12–30 days, depending on the specific *Plasmodium* involved. Some strains of *P. vivax* may have a protracted incubation period of 8–10 months and even longer for *P. ovale*. The disease can also be transmitted through infected blood transfusions. Confirmation of diagnosis is by demonstrating malaria parasites on blood smears.

Falciparum malaria, the most serious form, presents with fever, chills, sweats, and headache. It may progress to jaundice, bleeding disorders, shock, renal or liver failure, encephalopathy, coma, and death. Prompt treatment is essential. Case fatality rates in untreated children and adults are above 10 percent. An untreated attack may last 18 months. Other forms of malaria may present as a nonspecific fever. Relapse of the *P. ovale* may occur up to 5 years after initial infection; malaria may persist in chronic form for up to 50 years.

Malaria control advanced during the 1940s to 1960s through improved chloroquine treatment and use of DDT for vector control with optimism for eradication of the disease. However, control regressed in many developing countries as allocations for environmental control and case findings/treatment were reduced. There has also been an increase in drug resistance, so that this disease is now an extremely serious public health problem in many parts of the world. The need for a vaccine for malaria control is now more apparent than ever.

WHO estimates sub-Saharan Africa had 270 million new malaria cases, with 5 percent in children up to age 5. Over 1 million deaths occur annually from malaria, more than two-thirds of them in sub-Saharan Africa, and up to 50 percent of health expenditures go to treatment of malaria patients. Large areas, particularly in forest or savannah regions with high rainfall, are holoendemic. In higher altitudes, endemicity is lower, but epidemics do occur. Chloroquine-resistant *P. falciparum* has spread throughout Africa, accompanied by an increasing incidence of severe clinical forms of the disease. The World Bank estimates that 11 percent of all disability-adjusted life years (DALYs) lost per year in sub-Saharan Africa are from malaria, which places a heavy economic burden on the health systems.

In the Americas, the number of cases detected has risen every year since 1974, and the WHO estimates there to have been 2.2–2.5 million cases in 1991. The nine most endemic countries in the Americas achieved a 60 percent reduction in malaria mortality between 1994 and 1997. In 2002, CDC reports that of 1337 malaria cases in the United States, all but 5 were imported, i.e., acquired in malaria-endemic countries.

Malaria kills over 1 million people annually and infects between 350 and 500 million (WHO, 2008). Sub-Saharan Africa is the hardest hit region with 90 percent of these deaths, especially among children, and it has a serious impact on health and economic development. There is an increase in resistant strains to the major available drugs and of the mosquitoes to insecticides in use.

Vector control, case finding, and treatment remain the mainstay of control. Use of insecticide-impregnated bed nets and curtains, residual house spraying, and strengthened vector control activities are important, as are early diagnosis and carefully monitored treatment with monitoring for resistance. Control of malaria will ultimately depend on a safe, effective, and inexpensive vaccine. Attempts to develop a malaria vaccine have been unsuccessful to date due to the large number of genetic types of *P. falciparum* even in localized areas. Twenty-three prospective *P. falciparum* vaccines are currently in clinical trials, with some reported effectiveness. Research in vaccines for malaria has also been hampered by the fact that it is a relatively low priority for vaccine manufacturers because of the minimal potential for financial benefit. Because of increasing drug resistance, research on malaria has concentrated on the pharmacologic aspects of the disease. Effective control of malaria will require both new drugs for resistant infections, and primary prevention through vector control, with hopes for eventual vaccine development.

In 1998, WHO initiated a campaign to "Roll Back Malaria" and maintain the dream of eradication in the future; Malaria is included in MDG6 and Rollback Malaria for the period 2006–2015. Effective low-technology interventions include community-based case finding, early treatment with good quality insecticides, and vector control. The use of community health workers and widespread provison of insecticide-impregnated bed nets in endemic areas has shown promising results. Local control and even eradication can be achieved with currently available technology. This requires an integration of public health and clinical approaches with strong political commitment internationally and nationally in the affected countries.

Rickettsial Infections

The Rickettsieae are obligate parasites; they can only replicate in living cells, but otherwise they have characteristics of bacteria. This is a group of clinically similar diseases, usually characterized by severe headache, fever, myalgia, rash, and capillary bleeding causing damage to brain, lungs, kidneys, and heart. Identification is by serological testing for antibodies, but the organisms can also be cultured in laboratory animals, embryonic eggs, or in cell cultures. The organisms are transmitted by arthropod vectors such as lice, fleas, ticks, and mites. The diseases

caused millions of deaths during war and famine periods prior to the advent of antibiotics.

These diseases appear in nature in ways that make them impossible to eradicate, but clinical diagnosis, host protection, and vector control can help reduce the burden of disease and deal with outbreaks that may occur. Public education regarding self-protection, appropriate clothing, tick removal, and localized control measures such as spraying and habitat modification are useful.

Epidemic typhus, first identified in 1836, is due to *Rickettsia prowazekii*. Spread primarily by the body louse, typhus was the cause of an estimated 3 million deaths, especially during war and famine, in Poland and the Soviet Union from 1915–1922. Untreated, the fatality rate is 5–40 percent. Typhus responds well to antibiotics. It is currently largely confined to endemic foci in central Africa, central Asia, eastern Europe, and South America. It is preventable by hygiene and pediculicides such as DDT and lindane. A vaccine is available for exposed laboratory personnel.

Murine typhus is a mild form of typhus due to *Rickettsia typhi*, which is found worldwide and spread in rodent reservoirs. Scrub typhus, also known as Tsutsugamushi or Japanese river fever, is located throughout the Far East and the Pacific islands, and was a serious health problem for U.S. armed forces in the Pacific during World War II. It is spread by the *Rickettsia tsutsugamushi* and has a wide variation in case fatality according to region, organism, and age of patient.

Rocky Mountain spotted fever is a well-known and deadly form of tick-borne typhus due to *Rickettsia rickettsii*, occurring in western North America, Europe, and Asia. Q fever is a tick-borne disease caused by *Coxiella burnetii* and is worldwide in distribution, usually associated with farm workers, in both acute and chronic forms. Regular antitick spraying of sheep, cows, and goats helps protect exposed workers. Protective clothing and regular removal of body ticks help protect exposed persons.

Arboviruses (Arthropod-Borne Viral Diseases)

Arthropod-borne viral diseases are caused by a diverse group of viruses which are transmitted between vertebrate animals (often farm animals or small rodents) and people by the bite of blood-feeding vectors such as mosquitoes, ticks, and sandflies and by direct contact with infected animal carcasses. Usually the viruses have the capacity to multiply in the salivary glands of the vector, but some are carried mechanically in their mouth parts.

These viruses cause acute central nervous system infections (meningoencephalitis), myocarditis, or undifferentiated viral illnesses with polyarthritis and rashes, or severe hemorrhagic febrile illnesses. Arbovirus diseases are often asymptomatic in vertebrates but may be severe in humans. Over 250 antigenetically distinct arboviruses are associated with disease in humans, varying from benign fevers of short duration to severe hemorrhagic fevers. Each has a specific geographic location and vector and specific clinical and virologic characteristics, but they can spread globally via travelers and become endemic in new regions. They are of international public health importance because of the potential for spread via natural phenomena and modern rapid transportation of vectors and persons incubating the disease or ill with it, with potential for further spreading at the point of destination.

Encephalitides

Arboviruses are responsible for a large number of encephalitic diseases characterized by mode of transmission and geographic area. Mosquito-borne arboviruses causing encephalitis include Eastern and Western Equine, Venezuelan, Japanese, and Murray Hill encephalitides. Japanese encephalitis is caused by a mosquito-borne arbovirus found in Asia and is associated with rice-growing areas. It is characterized by headache, fever, convulsions, and paralysis, with fatality rates in severe cases as high as 60 percent. A currently available vaccine is used routinely in endemic areas (Japan, Korea, Thailand, India, and Taiwan) and for persons traveling to infected areas. Tick-borne arboviruses causing encephalitis include the Powassan virus, which occurs sporadically in the United States and Canada. Tick-borne encephalitis is endemic in eastern Europe, Scandinavia, and the former Soviet Union.

West Nile Virus (WNV)

This virus identified first in Africa in 1937, with wide distribution in Europe, southern USSR, the Middle East, Africa, and South Asia, appearing in Egypt and Israel in the 1950s, has now become worldwide in scope. An epidemic of mosquito-borne encephalitis in New York City in 1999 included 54 cases and 6 deaths due to the West Nile Fever virus, never before found in the United States. The virus has since spread throughout North America, with animal and human cases in nearly every state of the United States. Use of strategically located sentinel chicken flocks has been very effective in determining the geographical distribution of WNV and predicting local risks for infection. When birds test positive in a new area, health care providers are alerted to the signs and symptoms of WNV, increasing the effectiveness of surveillance, early intervention, and prevention. This highly successful model may potentially be applied to other zoonotic diseases. While only 20 percent of infected individuals develop clinical disease, the consequences of infection can be severe, especially in elderly and immunocompromised people. Treatments for WNV are supportive

and there is no vaccine available. The only effective means of prevention are vector control programs and personal protection against mosquito bites.

In the United States, the CDC reports that during 2006, WNV transmission to humans or animals recurred and expanded into many countries that had not previously reported transmission. Cases of WNV neuroinvasive disease (WNND) increased from 2003–2006. Extrapolations from past serosurveys suggested an estimated 41,750 cases of non-neuroinvasive WNV disease occurred in 2006 as compared to the 2770 reported cases. The spread of WNF indicated need for continuing surveillance, mosquito control, promotion of personal protection from mosquito bites, and research into additional prevention strategies.

Chikungunya

Chikungunya fever is a viral disease spread by the bite of infected mosquitoes. It was mainly located in India and Southeast Asia, causing a severe denguelike illness that is mostly nonfatal. The disease has spread to Europe with outbreaks in France and later Italy following importation from India by a single traveler. Because it was a large outbreak (over 197 cases) in 2007, concerns were raised that it might become endemic.

Rift Valley Fever

Rift Valley fever (RVF) is a virus spread by mosquitoes and other insect vectors. It affects animals and humans who are in direct contact with the meat or blood of affected animals. The virus causes a generalized illness in humans with encephalitis, hemorrhages, retinitis and retinal hemorrhage leading to partial or total blindness, and death (1–2 percent). It also causes universal abortion in ewes and a high percentage of death in lambs.

The normal habitat is in the Rift Valley of eastern Africa (the Great Syrian–African Rift), often spreading to southern Africa, depending on climactic conditions. The primary reservoir and vector is the *Aedes* mosquito, and affected animals serve to multiply the virus which is transmitted by other vectors and direct contact with animal fluids to humans.

An unusual spread of RVF northward to the Sudan and along the Aswan Dam reservoir to Egypt in 1977–1978 caused hundreds of thousands of animal deaths, with 18,000 human cases and 598 deaths. RVF appeared again in Egypt in 1993. This disease is suspected to be one of the ten plagues of Egypt leading to the exodus of the Children of Israel from Egypt during pharaonic-biblical times.

In 1997, an outbreak of RVF in Kenya, initially thought to be anthrax, with hundreds of cases and dozens of deaths, was related to abnormal rainy season and vector conditions. Satellite monitoring of rainfall and vegetation is being used to predict epidemics in Kenya and surrounding countries. Animal immunization, monitoring, vector control, and reduced contact with infected animals can limit the spread of this disease.

RVF has reappeared in the Middle East in Yemen and Saudi Arabia since 2000 and may have become endemic in the region.

Hemorrhagic Fevers

Arboviruses can also cause hemorrhagic fevers. These are acute febrile illnesses, with extensive hemorrhagic phenomena (internal and external), liver damage, shock, and often high mortality rates. The potential for international transmission is high.

Yellow Fever

Yellow fever is an acute viral disease of short duration and varying severity with jaundice. It can progress to liver disease and severe intestinal bleeding. The case fatality rate is 5 percent in endemic areas, but may be as high as 50 percent in nonendemic areas and in epidemics. It caused major epidemics in the Americas in the past, but was controlled by elimination of the vector, *Aedes aegypti*. A live attenuated vaccine is used in routine immunization endemic areas and recommended for travelers to infected areas. Determining the mode of transmission and vector control of yellow fever played a major role in the development of public health (see Chapter 1). In 1997, the WHO reported 200,000 cases and 30,000 deaths from yellow fever globally. Originally imported to the Americas from Africa, yellow fever is endemic in 10 countries in Central and South America, and is now spreading as well as in Asia.

Dengue Hemorrhagic Fever

Dengue hemorrhagic fever is an acute sudden-onset viral disease, with 3–5 days of fever, intense headache, myalgia, arthralgia, gastrointestinal disturbance, and rash. Hemorrhagic phenomena can cause case fatality rates of up to 50 percent. Epidemics can be explosive, but adequate treatment can greatly reduce the number of deaths. Dengue occurs in southeast Asia, the Pacific Islands, Australia, West Africa, the Caribbean, and Central and South America. An epidemic in Cuba in 1981 included more than 500,000 cases, and 158 deaths. Vector control of the *A. aegypti* mosquito resulted in control of the disease during the 1950s–1970s, but reinfestation of mosquitoes led to increased transmission and epidemics in the Pacific Islands, Caribbean, and Central and South America in the 1980s and 1990s (Box 4.12).

Outbreaks in Vietnam included 370,000 cases in 1987, another 116,000 cases in 1990, and a similar-sized outbreak in 1997. Indonesia had over 13,000 cases in 1997 with 240 deaths, and in 1998 over 19,000 cases (January–May) with at least 531 deaths. In 1998, epidemics of dengue were reported in Fiji, the Cook Islands, New Caledonia, and northern Australia. Monkeys are the main reservoir, and the vector is the *A. aegypti* mosquito. No vaccine is currently available, and management is by vector control.

Box 4.12 Dengue and Dengue Haemorrhagic Fever

Dengue fever, a severe influenza-like illness, and dengue hemorrhagic fever are closely related conditions caused by four distinct viruses transmitted by *Aedes aegypti* mosquitoes. Dengue is the world's most important mosquito-borne virus disease. A total of 2,500 million people worldwide are at risk of infection. An estimated 20 million cases occur each year, of whom 500,000 need to be hospitalized. This is a spreading problem, especially in cities in tropical and subtropical areas. Major outbreaks were reported in Colombia, Cuba, and many other locations in 1997.

"The geographical spread of both the mosquito vectors and the viruses has led to the global resurgence of epidemic dengue fever and emergence of dengue hemorrhagic fever (dengue/DHF) in the past 25 years with the development of hyperendemicity in many urban centers of the tropics.

"Recovery from infection by one provides lifelong immunity against that serotype but confers only partial and transient protection against subsequent infection by the other three. There is good evidence that sequential infection increases the risk of more serious disease resulting in DHF.

"DHF was first recognized in the 1950s during the dengue epidemics in the Philippines and Thailand. By 1970, nine countries had experienced epidemic DHF and now, the number has increased more than fourfold and continues to rise. Today emerging DHF cases are causing increased dengue epidemics in the Americas, and in Asia, where all four dengue viruses are endemic, DHF has become a leading cause of hospitalization and death among children in several countries."

Currently vector control is the available method for dengue and DHF prevention and control but research on dengue vaccines for public health use is in process.

Sources: World Health Organization. 1998. *World Health Report* 1998. Geneva: World Health Organization.
WHO and Dengue at WHO Geneva, http://www.who.int/esr/disease/dengue/en/ [accessed October 30, 2007].

Other Hemorrhagic Fevers

Lassa Fever

Lassa fever was first isolated in Lassa, Nigeria, in 1969 and is widely distributed in west Africa, with 200,000–400,000 cases and 5000 deaths annually. It is spread by direct contact with blood, urine, or secretions of infected rodents and by direct person-to-person contact in hospital settings. The disease is characterized by a persistent or spiking fever for 2–4 weeks, and may include severe hypotension, shock, and hemorrhaging. The case fatality rate is 15 percent.

Marburg Disease

Marburg disease is a viral disease with sudden onset of generalized illness, malaise, fever, myalgia, headache, diarrhea, vomiting, rash, and hemorrhages. It was first seen in Marburg, Germany, in 1967, following exposure to green monkeys. Person-to-person spread occurs via blood, secretions, organs, and semen. Case fatality rates can be over 50 percent.

Ebola Fever

Ebola fever is a viral disease with sudden onset of generalized illness, malaise, fever, myalgia, headache, diarrhea, vomiting, rash, and hemorrhages. It was first found in Zaire and Sudan in 1976 in outbreaks which killed more than 400 persons. It is spread from person to person by the blood, vomitus, urine, stools, and other secretions of sick patients, with a short incubation period. The disease has case fatality rates of up to 90 percent. An outbreak of Ebola among laboratory monkeys in a medical laboratory near Washington, DC, was contained with no human cases. The reservoir for the virus is thought to be rodents.

An outbreak of Ebola in May 1995 in the town of Kikwit, Zaire, killed 245 persons out of 316 cases (78 percent case fatality rate). This outbreak caused international concern that the disease could spread, but it remained localized. Another outbreak of Ebola virus occurred in Gabon in early 1996, with 37 cases, 21 of whom had direct exposure to an infected monkey, the remainder by human-to-human contact, or not established; 21 of the cases died (57 percent). Frequent similar epidemics have continued. This disease is considered highly dangerous unless outbreaks are effectively controlled. Once identified, an Ebola epidemic becomes an international emergency; public health workers from across the world are involved in control and intervention through WHO- and CDC-directed projects. In Zaire, lack of basic sanitary supplies, such as surgical gloves for hospitals, almost ensures that this disease will spread when it recurs.

Lyme Disease

Lyme disease is characterized by the presence of a rash, musculoskeletal, neurologic, and cardiovascular symptoms. Confirmation is by laboratory investigation. It is the most common vector-borne disease in the United States, with 64,000 cases reported between 2003 and 2005. It primarily affects children in the 5–14 age group and adults aged 30–49. Lyme disease is preventable by avoiding contact with ticks, by applying insect repellant, wearing long pants and long sleeves in infected areas, and by the early removal of attached ticks. Several U.S. manufacturers have developed vaccines; however, implementation of programs has been difficult due to difficulties in adverse event reporting and tracking (Box 4.13).

Box 4.13 Lyme Disease

In the mid 1970s, a mother of two young boys who were recently diagnosed with arthritis in the town of Lyme, Connecticut, conducted a private investigation among other town residents. She mapped each of the six arthritis cases in the town, cases which had occurred in a short time span among boys living in close proximity. This suggested that this syndrome of "juvenile rheumatoid arthritis" was perhaps connected with the boys playing in the woods. She presented her data to the head of rheumatology at Yale Medical School in New Haven, who investigated this "cluster of a new disease entity." Some parents reported that their sons had experienced tick bites and a rash before onset of the arthritis. A tick-borne, spiral-shaped bacterium, a spirochete, *Borrelia burgdorferi*, was identified as the organism, and *Ixodes* ticks were shown to be the vector. Cases respond well to antibiotic therapy.

Lyme disease infects approximately 20,000 people per year. Risk is highest in the northeast, north central, and mid-Atlantic regions. The disease accounts for over 90% of vector-borne disease in the United States and was the ninth leading reported infection in 1995. Lyme disease has been identified in many parts of North America, Europe, the former Soviet Union, China, and Japan. Personal hygiene for protection from ticks and environmental modification are important to limit spread of the disease.

Sources: Centers for Disease Control. 2007. *MMWR,*, 56:573–576.
Centers for Disease Control. 1996. *MMWR,* 45:481–484.
Centers for Disease Control. 1997. *MMWR,* 46:23.
Lyme disease website, available at http://www.cdc.gov/ncidad/disease/lyme/lyme.htm

PARASITIC DISEASES

Medically important parasites are animals that live, take nourishment, and thrive in the body of a host, which may or may not harm the host, but never bring benefit. They include unicellular organisms such as protozoa (malaria, *Giardia*, amebiasis, and *Cryptosporidium*), and helminths (worms), which are categorized as nematodes, cestodes, and trematodes (Box 4.14).

Public health continues to face the problems of parasitic diseases in the developing world. Increasingly, parasitic diseases are being recognized in industrialized countries. Giardiasis and *Cryptosporidium* infections in waterborne and other outbreaks have occurred in the United States. Parasitic diseases such as malaria are among the most common causes of illness and death in the world. Milder illnesses such as giardiasis and trichomoniasis cause widespread morbidity. Intestinal infestations with worms may cause severe complications, although they commonly cause chronic low-grade symptomatology and iron-deficiency anemia. Deworming every six months has become an effective part of the Expanded Programme of Immunization (EPI plus) along with Vitamin A supplementation and insecticide impregnated bed nets for children.

Echinococcosis

Echinococcosis (hydatid cyst disease) is infection with *Echinococcus granulosus*, a small tapeworm commonly found in dogs. The tapeworm forms unilocular (single, noncompartmental) cysts in the host, primarily in the liver and lungs, but they can also grow in the kidney, spleen, central nervous system, or in bones. Cysts, which may grow up to 10 cm in size, may be asymptomatic or, if untreated, may cause severe symptoms and even death. This parasite is common where dogs are used with herd grazing animals and also have intimate contact with humans.

The Middle East, Greece, Sardinia, north Africa, and South America are endemic areas, as are a few areas in

---

Box 4.14 Neglected Tropical Diseases

"At least 1 billion people suffer from one or more neglected tropical diseases (NTDs), such as Buruli ulcer, cholera, cysticercosis, dracunculiasis (guinea-worm disease), food-borne trematode infections (such as fascioliasis), hydatidosis, leishmaniasis, lymphatic filariasis, onchocerciasis, schistosomiasis, soil-transmitted helminthiasis, trachoma and trypanosomiasis, although there are other estimates that suggest the number could be much higher. Several of these diseases, and others such as dengue, are vector-borne. Often, those populations most affected are also the poorest and most vulnerable and are found mainly in tropical and subtropical areas of the world. Some diseases affect individuals throughout their lives, causing a high degree of morbidity and social stigmatization and abuse.

"For a large group of these diseases — mainly helminthic infections — effective, inexpensive or donated drugs are available for their prevention and control. These tools, when used on a large scale, are able to wipe out the burden caused by these ancient scourges of humanity. For leprosy, treatment with effective antibiotics is leading to the elimination of this ancient disabling disease. There is also a cost-effective approach to treating yaws that could lead to elimination and final eradication of this debilitating disease that may cause gross deformation. In the case of blinding trachoma, the use of the recommended strategy (SAFE) of an effective antibiotic is enhancing the progress toward final elimination. Large-scale, regular treatment plays a central role in the control of many NTDs such as filariasis, onchocerciasis, schistosomiasis, and soil-transmitted nematode infections. For example, regular chemotherapy against intestinal worms reduces mortality and morbidity in preschool children, improves the nutritional status and academic performance of schoolchildren, and improves the health and well-being of pregnant women and their babies.

"There is a second group of NTDs for which the only clinical option currently available is systematic case-finding and management at an early stage. These diseases include Buruli ulcer, Chagas' disease, cholera and other diarrheal diseases, human African trypanosomiasis, and leishmaniasis. Simple diagnostic tools and safe and effective treatment regimens need to be developed urgently for some of these diseases. However, even for these infections, systematic use of the present, imperfect tools at an early stage can dramatically reduce mortality and morbidity. For others, vector control tools are available and present the main method of transmission control, as in the case of Chagas' disease.

There are examples of great successes in the fight against both of these groups of NTDs. Since 1985, 14.5 million patients have been cured of leprosy through multi-drug therapy; today, less than a million people are affected by the disease. Before the start of the Guinea-worm Eradication Programme in the early 1980s, an estimated 3.5 million people in 20 endemic countries were infected with the disease. In 2005, only about 10,000 cases were reported in 9 endemic countries, and the programme is moving towards eradication. Onchocerciasis has freed more than 25 million hectares of previously onchocerciasis-infected land available for resettlement and agricultural cultivation, thereby considerably improving development prospects in Africa and Latin America.

Increased awareness and advocacy are needed to draw attention to the realistic prospect of reducing the negative impact of NTDs on the health and social and economic well-being of affected communities."

Source: World Health Organization. http://www.who.int/neglected_diseases/en/ [accessed October 10, 2007]

the United States and Canada. The human disease has been eliminated in Cyprus and Australia. While the dog is the major host, intermediate hosts include sheep, cattle, pigs, horses, moose, and wolves. Preventive measures include education in food and animal contact hygiene, destroying wild and stray dogs, and keeping dogs from the viscera of slaughtered animals.

A similar, but multilocular, cystic hydatid disease is widely found in wild animal hosts in areas of the northern hemisphere, including central Europe, the former Soviet Union, Japan, Alaska, Canada, and the north-central United States.

Another echinococcal disease (*Echinococcus vogeli*) is found in South America, where its natural host is the bush dog and its intermediate host is the rat. The domestic dog also serves as a source of human infection.

Surgical resection is not always successful, and long-term medical treatment may be required. Control is through awareness and hygiene as well as the control of wild animals that come in contact with humans and domestic animals. Control may require cooperation between neighboring countries.

Tapeworm

Tapeworm infestation (taeniasis) is common in tropical countries where hygienic standards are low. Beef (*Taenia saginata*) and pork (*T. solium*) tapeworms are common where animals are fed with water or food exposed to human feces. *T. solium* is especially deadly; delay in diagnosis and treatment may lead to severe disease, including neurologic cysticercosis. In developing countries, infection is associated with pork consumption, while in the United States, several epidemics have occurred from eating carnivorous game animals such as mountain lions and bears. Freezing or cooking meat, especially that of pigs and carnivorous mammals, is essential to destroy the tapeworm. Fish tapeworm (*Diphyllobothrium latum*) is common in populations living primarily on uncooked fish, such as Inuit, Eastern European, and Scandinavian. These tapeworms are usually associated with northern climates.

Toddlers are especially susceptible to dog tapeworm (*Dipylidium caninum*), which is present worldwide, and domestic pets are often the source of oral–fecal transmission of the eggs. The disease is usually asymptomatic. Similarly, dwarf tapeworm (*Hymenolepis nana*) is transmitted through oral–fecal contamination from person to person, or via contaminated food or water. Rat tapeworm (*Hymenolepis diminuta*) also mostly affects young children.

Onchocerciasis

Onchocerciasis (river blindness) is a disease caused by a parasitic worm, which produces millions of larvae that move through the body causing intense itching, debilitation, and eventually blindness. The disease is spread by a blackfly that transmits the larva from infected to uninfected people. It is primarily located in sub-Saharan Africa and in Latin America, with over 120 million persons at risk. Control is by a combination of activities including environmental control by larvicidal sprays to reduce the vector population, protection of potential hosts by protective clothing and insect repellents, and case treatment.

A WHO-initiated program for onchocerciasis control started in 1974 is sponsored by four international agencies: the Food and Agriculture Organization (FAO), the United Nations Development Program (UNDP), the World Bank, and WHO. It covers 11 countries in sub-Saharan Africa, focusing on control of the blackfly by destroying its larvae, mainly via insecticides sprayed from the air. The Vision 2020 program of the WHO aims for control of river blindness by the year 2020.

The program has been successful in protecting some 30 million persons and helping 1.5 million infected persons to recover from this disease. WHO estimates that the program prevented 500,000 cases of blindness by 2000 and has freed 25 million hectares of land for resettlement and cultivation. The program cost $570 million. This investment is considered by the World Bank to have a return of 16–28 percent in terms of large-scale land reuse and improved output of the population. A WHO program, the African Program for Onchocerciasis Control (APOC), started in 1996, includes Ivermectin and selective vector control efforts by spraying for the blackfly. This involves 30 countries in Africa, and 6 in a similar program in South America (see http://www/who.int/ocp).

Dracunculiasis

Dracunculiasis (Guinea worm disease) is a parasitic disease of great public health importance in India, Pakistan, and central and west Africa. It is an infection of the subcutaneous and deeper tissues caused by a large (60 cm) nematode, usually affecting the lower extremities and causing pain and disability. The nematode causes a burning blister on the skin when it is ready to release its eggs. After the blister ruptures, the worm discharges larvae whenever the extremity is in water. The eggs are ingested in contaminated water and the larvae released migrate through the viscera to locate as adults in the subcutaneous tissue of the leg. Incubation is about 12 months. Larvae released in water are ingested by minute crustaceans and remain infective for as long as a month.

Prevention is based on improving the safety of water supplies and by preventing contamination by infected persons. Education of persons in endemic areas to stay out of water sources and to filter drinking water reduces transmission. Insecticides remove the crustaceans. Chlorine also kills the larvae and the crustaceans which prologue larval infectivity. There is no vaccine. Treatment is helpful, but not definitive.

Dracunculiasis was traditionally endemic in a belt from west Africa through the Middle East to India and central Asia. It was successfully eliminated from central Asia and Iran and has disappeared from the Middle East and from some African countries (Gambia and Guinea).

WHO has promoted the eradication of dracunculiasis. Major progress has been made in this direction. Worldwide prevalence is reported to have been reduced from 12 million cases in 1980 to 3 million in 1990, 152,814 in 1996, and 77,863 cases in 1997. Eradication was anticipated for 2000; however, the Guinea worm remains endemic in several developing African nations. India's reported cases fell from 17,000 in 1987 to 900 in 1992, and the country was free from transmission in 1997. In 1997, formerly high-prevalence countries such as Kenya reported no cases in 1997, while Chad, Senegal, Cameroon, Yemen, and the Central African Republic reported fewer than 30 cases each.

The WHO eradication program was developed successfully as an independent program with its own direction and field staff, but further progress will require the integration of this program with other basic primary care programs in order to be self-sustaining as an integral part of community health. Community-based surveillance systems for this disease are being converted to work for monitoring of other health conditions in the community.

Schistosomiasis

Schistosomiasis is a parasitic infection caused by the trematode (blood fluke) and transmitted from person to person via an intermediate host, the snail. It is endemic in 74 countries in Africa, South America, the Caribbean, and Asia. There are an estimated 200 million persons infected worldwide and more than 600 million at risk for the disease. The clinical symptoms include fever, nausea, vomiting, abdominal pain, diarrhea, and hematuria. The organisms *Schistosoma mansoni* and *S. japonicum* cause intestinal and hepatic symptoms, including diarrhea and abdominal pain. *Schistosoma haematobium* affects the genitourinary tract, causing chronic cystitis, pyelonephritis, with high risk for bladder cancer, the ninth most common cause of cancer deaths globally. A recently identified species, *S. intercalatum,* is genetically unique, but thought to cause both intestinal and genitourinary disease. *S. intercalatum* is largely identified in inhabitants and immigrants from western Africa. Infection by all schistosomes is acquired by skin contact with fresh water containing contaminated snails. The cercariae of the organism penetrate the skin, and in the human host it matures into an adult worm that mates and produces eggs. The eggs are disseminated

to other parts of the body from the worm's location in the veins surrounding the bladder or the intestines, and may result in neurologic symptoms.

Eggs may be detected under microscopic examination of urine and stools. Sensitive serologic tests are also available. Treatment is effective against all three major species of schistosomiasis. Eradication of the disease can be achieved with the use of irrigation canals, prevention of contamination of water sources by urine and feces of infected persons, treatment of infected persons, destruction of snails, and health education in affected areas. Persons exposed to freshwater lakes, streams, and rivers in endemic areas should be warned of the danger of infection. Mass chemotherapy in communities at risk and improved water and sanitation facilities are resulting in improved control of this disease.

Leishmaniasis

Leishmaniasis causes both cutaneous and visceral disease. The cutaneous form is a chronic ulcer of the skin, called by various names (e.g., rose of Jericho, oriental sore, and Aleppo boil). It is caused by *Leishmania tropica, L. brasiliensis, L. mexicana,* or the *L. donovani* complex. This chronic ulcer may last from weeks to more than a year. Diagnosis is by biopsy, culture, and serologic tests. The organism multiplies in the gut of sandflies (*Phlebotomus* and *Lutzomi*) and is transmitted to humans, dogs, and rodents through bites. The parasites may remain in the untreated lesion for 5–24 months, and the lesion does not heal until the parasites are eliminated.

Prevention is through limiting exposure to the phlebotomines and reducing the sandfly population by environmental control measures. Insecticide use near breeding places and homes has been successful in destroying the vector sandflies in their breeding places. Case detection and treatment reduce the incidence of new cases. There is no vaccine, and treatment is with specific antimonials and antibiotics.

Visceral leishmaniasis (kala azar) is a chronic systemic disease in which the parasite multiplies in the cells of the host's visceral organs. The disease is characterized by fever, the enlargement of the liver and spleen, lymphadenopathy, anemia, leukopenia, and progressive weakness and emaciation. Diagnosis is by culture of the organism from biopsy or aspirated material, or by demonstration of intracellular (Leishman–Donovan) bodies in stained smears from bone marrow, spleen, liver, or blood.

Kala azar is a rural disease occurring in the Indian subcontinent, China, the southern republics of the former U.S.S.R., the Middle East, Latin America, and sub-Saharan Africa. It usually occurs as scattered cases among infants, children, and adolescents. Transmission is by the bite of the infected sandfly with an incubation period of 2–4 months.

There is no vaccine, but specific treatment is effective and environmental control measures reduce the disease prevalence. This includes the use of antimalarial insecticides. In localities where the dog population has been reduced, the disease is less prevalent.

Trypanosomiasis

African Trypanosomiasis (Sleeping Sickness)

Sleeping sickness is a fatal degenerative neurologic disease caused by *Trypanosoma brucei*, transmitted by the tsetse fly, primarily in the African savannahs, affecting cattle and humans. Subspecies are known to cause both acute and chronic forms of sleeping sickness. Some 55 million persons are at risk in sub-Saharan Africa. Between 1998 and 2004, renewed surveillance and control reduced the incidence of African trypanosomiasis from 38,000 to approximately 18,000. Prevention depends on vector control, and effective treatment of human cases.

Chagas' Disease (American Trypanosomiasis)

Chagas' disease is a chronic vector- and blood transfusion–borne parasitic disease (*Trypanosoma cruzi*) which causes significant disability and death. It affects some 17 million persons mainly in Central and South America, with some 300,000 new cases and 45,000 deaths occurring annually. About 30 percent of affected persons develop severe heart disease. While vaccine development is not likely due to the ability of trypanosome antigens to cause autoimmunity and rapid immunologic drift of the organism, two drugs have been developed which show effectiveness in limiting early chronic disease. Brazil achieved elimination of transmission in 1998, after Uruguay (1996) and Venezuela (1997), and followed by Argentina (1999). While the initial WHO elimination goal by 2010 now seems unfeasible, efforts continue to dramatically reduce the incidence of *T. cruzi* infection.

Control is difficult, but control measures include reducing the animal host and vector insect population in its habitat by ecological and insecticide measures, education of the population in prevention by clothing, bed nets, and repellents, and with chemotherapy for case management.

Other Parasitic Diseases

Amebiasis

Amebiasis is an infection with a protozoan parasite (*Entamoeba histolytica*) which exists as an infective cyst. Infestation may be asymptomatic or cause acute, severe diarrhea with blood and mucus, alternating with constipation. *E. histolytica* infection sometimes results in invasive abdominal infestation, severe liver disease, and death.

Amebic colitis can be confused with ulcerative colitis. Diagnosis is by microscopic examination of fresh fecal specimens showing trophozoites or cysts. Transmission is generally via ingestion of fecal-contaminated food or water containing cysts, or by oral–anal sexual practices. Amebiasis is found worldwide. Sand filtration of community water supplies removes nearly all cysts. Suspect water should be boiled. Education regarding hygienic practices with safe food and water handling and disposal of human feces is the basis for control.

Ascariasis

Ascariasis is infestation of the small intestine with the roundworm *Ascaris lumbricoides,* which may appear in the stool, occasionally the nose or mouth, or may be coughed up from lung infestation. The roundworm is very common in tropical countries, where infestation may reach or exceed 50 percent of the population. Children aged 3–8 years are especially susceptible. Infestation can cause pulmonary symptoms and frequently contributes to malnutrition, especially iron-deficiency anemia. Transmission is by ingestion of infective eggs, common among children playing in contaminated areas, or via the ingestion of uncooked products of infected soil. Eggs may remain viable in the soil for years. Vermox and other treatments are effective. Prevention is through education, adequate sanitary facilities for excretion, and improved hygienic practices, especially with food. Use of human feces for fertilizer, even after partial treatment, may spread the infestation. Mass treatment is indicated in high prevalence communities.

Pinworm Disease (Enterobiasis)

Pinworm disease (oxyuriasis) is common worldwide in all socioeconomic classes; however, it is more widespread when crowded and unsanitary living conditions exist. The *Enterobius vermicularis* infestation of the intestine may be asymptomatic or may cause severe perianal itching or vulvovaginitis. It primarily affects schoolchildren and preschoolers. More severe complications may occur. Adult worms may be seen visually or identified by microscopic examination of stool specimens or perianal swabs. Transmission is by the oral–fecal ingestion of eggs. The larvae grow in the small intestine and upper colon. Prevention is by educating the public regarding hygiene and adequate sanitary facilities, as well as by treating cases and investigating contacts. Treatment is the same as for ascariasis. Mass treatment is indicated in high prevalence communities.

Ectoparasites

Ectoparasites include scabies (*Sarcoptes scabiei*), the common bed bug (*Cimex lectularius*), fleas, and lice, including the body louse (*Pediculus humanis*), pubic louse

(*Phthirius pubis*), and the head louse (*Pediculus humanus capitis*). Their severity ranges from nuisance value to serious public health hazard. Head lice are common in schoolchildren worldwide and are mainly a distressing nuisance. The body louse serves as a vector for epidemic typhus, trench fever, and louse-borne relapsing fever. In disaster situations, disinfection and hygienic practices may be essential to prevent epidemic typhus. The flea plays an important role in the spread of the plague by transmitting the organism from the rat to humans. Control of rats has reduced the flea population, but during war and disasters, rat and flea populations may thrive. Scabies, which is caused by a mite, is common worldwide and is transmitted from person to person. The mite burrows under the skin and causes intense itching. All of these ectoparasites are preventable by proper hygiene and the treatment of cases. The spread of these diseases is rapid and therefore warrants attention in school health and public health policy.

LEGIONNAIRE'S DISEASE

Legionnaire's disease (Legionnellosis) is an acute bacterial disease caused by *Legionnelae*, a gram-negative group of bacilli, with 35 species and many serological groups. The first documented case was reported in the United States in 1947, and the first disease outbreak was reported in the United States in 1976 among participants of a veteran's convention in Philadelphia. General malaise, anorexia, myalgia, and headache are followed by fever, cough, abdominal pain, and diarrhea. Pneumonia followed by respiratory failure may follow. The case fatality rate can be as high as 40 percent of hospitalized cases. A milder, nonpneumonic form of the disease (Pontiac fever) is associated with virtually no mortality.

The organism is found in water reservoirs and is transmitted through heating, cooling, and air conditioning systems, as well as from tap water, showers, saunas, and jacuzzi baths. The disease has been reported worldwide. Significant epidemics have occurred on cruise ships, where insufficient air conditioning sanitation and an older, more susceptible clientele are a dangerous combination. Prevention requires the cleaning of water towers and cooling systems, including whirlpool spas. Hyperchlorination of water systems and the replacement of filters is required where cases and/or organisms have been identified. Antibiotic treatment with erythromycin is effective.

LEPROSY

Leprosy (Hansen's disease) was widely prevalent in Europe and Mediterranean countries for many centuries, with some 19,000 leprosaria in the year 1300. Leprosy was largely wiped out during the Black Death in the fourteenth century, but continued in endemic form until the twentieth century. Leprosy is a chronic bacterial infection of the skin, peripheral nerves, and upper airway. In the lepromatous form, there is diffuse infiltration of the skin nodules and macules, usually bilateral and extensive. The tuberculoid form of the disease is characterized by clearly demarcated skin lesions with peripheral nerve involvement. Diagnosis is based on clinical examination of the skin and signs of peripheral nerve damage, skin scrapings, and skin biopsy.

Transmission of the *Mycobacterium leprae* organism is by close contact from person to person, with incubation periods of between 9 months and 20 years (average of 4–8 years). Rifampin and other medications make the patient noninfectious in a short time, so that ambulatory treatment is possible. Multidrug therapy (MDT) has been shown to be highly effective in combating the disease, with a very low relapse rate. Treatment with MDT ensures that the bacillus does not develop drug resistance. The increase has been associated with improved case finding. BCG may be useful in reducing tuberculoid leprosy among contacts. Investigation of contacts over 5 years is recommended.

The disease is still highly endemic primarily in five countries: India, Brazil, Indonesia, Myanmar, and Bangladesh, and is still present in some 80 countries in southeast Asia, including the Philippines and Myanmar, sub-Saharan Africa, the Middle East (Sudan, Egypt, Iran), and in some parts of Latin America (Mexico, Colombia) with isolated cases in the United States. World prevalence has declined from 10.5 million cases in 1980, 5.5 million in 1990, to fewer than 300,000 in 2004. WHO aimed to eliminate leprosy as a public health problem by 2000, defined as prevalence of fewer than 1 per 10,000 population, or fewer than 300,000 cases. The achievement of this goal has been a major historical event in public health. WHO reports that "the number of new cases detected globally has fallen by more than 40,019 cases (a 13.4 percent decrease) during 2006 compared with 2005. During the past 5 years, the global number of new cases detected has continued to decrease dramatically, at an average rate of nearly 20 percent per year," and "pockets of high endemicity still remain in some areas of Angola, Brazil, Central African Republic, Democratic Republic of Congo, India, Madagascar, Mozambique, Nepal, and the United Republic of Tanzania. These countries remain highly committed to eliminating the disease, and continue to intensify their leprosy control activities."

TRACHOMA

Trachoma is currently responsible for 6 million blind persons or 15 percent of total blindness in the world. The causative organism, *Chlamydia trachomatis,* is a bacteria which can survive only within a cell. It is spread through contact with eye discharges, usually by flies, or household items (e.g., handkerchiefs, washcloths). Trachoma is common in poor rural areas of Central America, Brazil, Africa, parts of Asia, and some countries in the eastern Mediterranean. The resulting infection leads to conjunctival scarring and if untreated, to blindness. WHO estimates there are 148 million cases of active disease in 46 endemic countries. Hygiene, vector control, and treatment with antibiotic eye ointments or simple surgery for scarring of eyelids and inturned eyelashes prevent the blindness. A new drug, azithromycin, is effective in curing the disease. The WHO is promoting a program for the global elimination of trachoma using azithromycin and hygiene education in endemic areas.

Chlamydia (*Chlamydia pneumonia*) infection is known to be a widespread chronic risk factor for coronary artery disease. Intraarterial infection, among other contributing etiologies, contributes to plaque formation, thromboembolic occlusion of arteries, and myocardial infarction. While antibiotic treatment of chlamydias as a preventive measure for heart disease has not been used, this could potentially reduce the burden of the leading worldwide cause of death at a relatively low cost.

SEXUALLY TRANSMITTED INFECTIONS

Sexually transmitted infections (STIs) are widespread internationally with an estimated 330 million new cases per year, with 5.8 million new cases, over 30 million total cases, and 2.3 million deaths (1997). AIDS has captured world attention over the past decade. The global burden of STIs is enormous (Table 4.8), and the public health and social consequences are devastating in many countries.

Sexually transmitted infections, especially in women, may be asymptomatic, so that severe sequelae may occur before patients seek care. Infection by one STI increases risk of infection by other diseases in this group.

Syphilis

Syphilis is caused by the spirochete *Treponema pallidum*. After an incubation period of 10–90 days (mean is 21 days), primary syphilis develops as a painless ulcer or chancre on the penis, cervix, nose, mouth, or anus, lasting 4–6 weeks. The patient may first present with secondary syphilis 6–8 weeks (up to 12 weeks) after infection with a general rash and malaise, fever, hair loss, arthritis, and jaundice. These symptoms spontaneously disappear within weeks or up to 12 months later. Tertiary syphilis may appear 5–20 years after initial infection. Complications of tertiary syphilis include catastrophic cardiovascular and central nervous system conditions. Early antibiotic treatment is highly effective when given in a large initial dose, but longer-term therapy may be needed if treatment is delayed.

TABLE 4.8 Prevalence of Sexually Transmitted Infections by WHO Region, 1999

Region	Estimated cases 1999 (millions)
Australia, New Zealand	1
Southeast Asia	151
East Asia and Pacific	18
East Europe and Central Asia	22
Latin America and Caribbean	38
North America	14
Sub-Saharan Africa	69
North Africa and Middle East	10
Total All Regions	323

Sources: World Health Organization, http://www.who.int/hiv/pub/sti/who_hiv_aids_2001.02.pdf [accessed October 23, 2007].
World Health Organization, http://www.who.int/reproductive-health/publications/rtis_gep/fpmethods.htm [accessed October 23, 2007].
World Health Organization. 2001. Global prevalence and incidence of selected curable sexually transmitted diseases: Overview and estimates. Geneva: World Health Organization.

Gonorrhea

Gonorrhea (GC) is caused by the bacterium *Neisseria gonorrhoeae*. The incubation period is 1–14 days. Gonorrhea is often associated with concurrent chlamydia infection. In women, GC may be asymptomatic or it may cause vaginal discharge, pain on urination, bleeding on intercourse, or lower abdominal pain. Untreated, it can lead to sterility. In men, GC causes urethral discharge and painful urination. Treatment with antibiotics ends infectivity, but untreated cases can be infectious for months. Drug resistance to penicillin, tetracycline, and quinolones has emerged in many countries so that more expensive and often unavailable drugs are necessary for treatment. Prevention of gonococcal eye infection in newborns is based on routine use of antibiotic ointments in the eyes of newborns.

Other Sexually Transmitted Infections

Chancroid

Chancroid is caused by *Haemophilus ducreyi*. In women, chancroids may cause a painful, irregular ulcer near the vagina, resulting in pain on intercourse, urination, and defecation, but it may be asymptomatic. In men it causes a painful, irregular ulcer on the penis. The incubation period is usually 3–5 days, but may be up to 14 days. An individual is infectious as long as there are ulcers, usually 1–3 months. Treatment is by erythromycin or azithromycin.

Herpes Simplex

Herpes simplex is caused by herpes simplex virus types 1 and 2 and has an incubation period of 2–12 days. Genital herpes causes painful blisters around the mouth, vagina, penis, or anus. The genital lesions are infectious for 7–12 days. Herpes may lead to central nervous system meningoencephalitis infection. It can be transmitted to newborns during vaginal delivery, causing infection, encephalitis, and death. Cesarean delivery is therefore necessary when a mother is infected. Antiviral drugs are used in treatment, orally, topically, or intravenously.

Chlamydia

Chlamydia is caused by *Chlamydia trachomatis*. It is the most common sexually transmitted infection in the United States, where reported incidence has increased to nearly 1 million in 2005–2006. Underreporting is a major problem; actual incidence is estimated at more than twice that reported. In women, it is usually asymptomatic but may cause vaginal discharge, spotting, pain on urination, lower abdominal pain, and pelvic inflammatory disease (PID). In newborns, chlamydia may cause eye and respiratory infections. In men, chlamydia causes urethral discharge and pain on urination. The incubation period is 7–21 days and the infectious period is unknown. Treatment for chlamydia is doxycycline, azithromycin, or erythromycin. Because co-transmission with gonorrhea is extremely common, CDC recommends treatment for both diseases when either is confirmed. Chlamydia infection, not necessarily venereal in transmission, may be transmitted to newborns of infected mothers. *Chlamydia pneumoniae* is suspected and under investigation as a possible cause or contributor in coronary heart disease.

Trichomoniasis

Trichomoniasis is caused by *Trichomonas vaginalis*. The incubation period is 4–20 days (mean is 7 days). In women, trichomoniasis may be asymptomatic or may cause a frothy vaginal discharge with foul odor, and painful urination and intercourse. In men, the disease is usually mild, causing pain on urination. Treatment is by metronidazole taken orally. Without treatment, the disease may persist and remain infectious for years.

Human Papillom Virus (HPV)

HPV is endemic throughout the world and the leading cause of cervical neoplasia and cancer of the cervix. HPV includes many types associated with venereal warts (condylomas). An effective vaccine against the most common carcinogenic strains is now available and recommended for young women to prevent cervical cancer, a breakthrough of enormous importance for this is one of the leading causes of cancer mortality in women. Prevention of cervical cancer by vaccine and by Pap smear screening is a major advance in public health, along with the prevention of liver cancer by hepatitis B immunization. Circumcision is now recommended by WHO for primary prevention of transmission of HPV (see Chapters 5 and 6).

Control of Sexually Transmitted Infections

In areas where a full range of diagnostic services is lacking, a "syndromic approach" is recommended for the control of STIs. The diagnosis is based on a group of symptoms and treatment on a protocol addressing all the diseases that could possibly cause those symptoms, without expensive laboratory tests and repeated visits. Early treatment without laboratory confirmation helps to cure persons who might not return for follow-up, or may place them in a noninfective stage so that even without follow-up they will not transmit the disease. STI incidence between 1950 and 2004 is shown in Table 4.9, with decline overall except around 1990, with subsequent further fall in incidence.

Screening in prenatal and family planning clinics, prison medical services, and in clinics serving prostitutes, homosexuals, or other potential risk groups will detect subclinical cases of various STIs. Treatment can be carried out cheaply and immediately. For instance, the screening test for syphilis costs $0.10 and the treatment with benzathine penicillin injection costs about $0.40. Partner

TABLE 4.9 Reportable Sexually Transmitted Infections,[a] United States, Selected Years, 1950–2004

Disease	1950	1960	1970	1980	1985	1990	2000	2004
Syphilis (all stages)	146	69	45	31	29	54	11	12
Gonorrhea	192	145	297	445	384	278	129	114

Source: *Health United States*, 1998, 2006.
[a]The increase in syphilis in 1985–1990 and subsequent decline by more than 50 percent in reported cases includes all three stages of the disease as well as congenital syphilis. Rates are cases per 100,000 population, rounded.

notification is a controversial issue, but may be needed to identify contacts who may be the source of transmission to others.

Control of STIs through a syndrome approach based on primary care providers is being promoted by WHO. Health education directed at high-risk target groups is essential. Providing easy and cost-free access to acceptable, nonthreatening treatment is vital in promoting the early treatment of cases and thereby reducing the risk of transmission.

Promoting prevention through the use of condoms and/or monogamy requires long-term educational efforts that are now fostered by the HIV/AIDS pandemic. Increased use of condoms for HIV prevention is associated with reduced risk of other STIs. Training medical care providers in STI awareness should be stressed in undergraduate and continuing educational efforts including personal protection as caregivers.

HIV/AIDS

Human immunodeficiency virus (HIV) is a retrovirus that infects various cells of the immune system, and also affects the central nervous system. Two types have been identified: HIV1, worldwide in distribution, and the less pathogenic HIV2, found mainly in West Africa. HIV is transmitted by sexual contact, exposure to blood and blood products, perinatally, and via breast milk. The period of communicability is unknown, but studies indicate that infectiousness is high, both during the initial period after infection and later in the disease. Antibodies to HIV usually appear within 1–3 months.

Within several weeks to months of the infection, many persons develop an acute self-limited flulike syndrome. They may then be free from any signs or symptoms for months to more than 10 years. Onset of illness is usually insidious with nonspecific symptoms, including sweats, diarrhea, weight loss, and fatigue. AIDS represents the later clinical stage of HIV infection. According to the revised CDC case definition (1993), AIDS involves any one or more of the following: low CD_4 count, severe systemic symptoms, opportunistic infections such as *Pneumocystis* pneumonia or TB, aggressive cancers such as Kaposi's sarcoma or lymphoma, and/or neurologic manifestations, including dementia and neuropathy. The WHO case definition is more clinically oriented, relying less on often unavailable laboratory diagnoses for indicator diseases.

This pandemic brought home lessons of public health and hygiene that had been forgotten in a smug confidence and reliance on antimicrobial therapy and vaccines that were assumed to be capable of defeating all infectious diseases. Regrettably this is not the case, and the HIV/AIDS experience showed the price of negligence in infectious disease control of sexually transmitted infections. With no vaccine yet on the horizon, the prospects for this

disease are grim and its spread certain until an effective vaccine can be developed. However, the pattern of mortality in the United States is shown in Figure 4.5 from CDC, Atlanta, indicating the potential for prolonging survival, improving quality of life and reducing transmission. Active public health measures include education on AIDS prevention and condom promotion, and effective medical care based on antiretroviral therapy (ART) and for tuberculosis and other opportunistic infections, as well as nutritional supplementation and general care, is also gaining ground as a preventive measure in sub-Saharan Africa.

AIDS was first recognized clinically in 1981 in Los Angeles and New York. By mid-1982 it was considered an epidemic in those and other U.S. cities. It was primarily seen among men who have sex with men and recipients of blood products. After initial errors, testing of blood and blood products became standard and has subsequently closed off this method of transmission. Transmission has changed markedly since the initial onslaught of the disease, with needle sharing among intravenous drug users, heterosexual activity, and maternal–fetal transmission becoming major factors. Comorbidity with other STIs apparently increases HIV infectivity and may have helped to convert the epidemiology to a greater degree of heterosexual transmission (Box 4.15 and Figure 4.5).

The disease grew exponentially in the United States but incidence of new cases has declined since 1993. AIDS is also a major public health problem in most developed and developing countries, reaching catastrophic proportions in some sub-Saharan African countries, affecting up to 30 percent or more of the population.

HIV-related deaths were the eighth leading cause of all deaths in 1993 in the United States, the leading cause among men aged 25–44 years of age, and the fourth leading cause for women in this age group. By 2005, AIDS had been diagnosed in 984,000 persons and 550,000 had died. At the end of 2003, it was estimated that up to 1.1 million persons are HIV infected in the United States. Up to 30 percent of these people may not know they are infected. In 2005, 42,000 new diagnoses were reported.

Globally, deaths from AIDS totaled 2.8 million in 2005, with an estimated 11.7 million persons having died

---

Box 4.15 HIV/AIDS, 1981–2006

AIDS was first reported as a clinical entity by Dr. Michael Gottlieb at the University of California Los Angeles (UCLA) Hospital. He reported in *Morbidity and Mortality Weekly Report* and later in the *New England Journal of Medicine* on five cases of *Pneumocystis carinii* (now *P. jiroveci*) pneumonia with cytomegaloviremia (CMV) among young male homosexuals with evidence of immune deficiency. A few weeks later, Dr. Alvin Friedman-Kien reported 26 cases of Kaposi's sarcoma in gay men from New York and California.

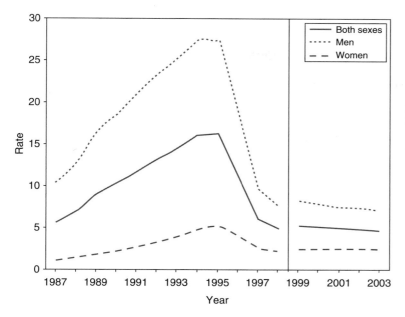

FIGURE 4.5 Age-adjusted death rates for HIV/AIDS, United States, 1987–2003. Sources: Fee, E., Brown, T. 2006. M. Michael S. Gottlieb and the identification of AIDS. *American Journal of Public Health*, 9:982–983. Stall, R., Mills, T. C. [editorial]. 2006. A quarter century of AIDS. *American Journal of Public Health*, 9:959–961. UNAIDS. 2005. AIDS Epidemic Update — December 2005. New York UNAIDS. Centers for Disease Control. 2006. Epidemiology of AIDS/HIV — United States, 1981–2005. *Morbidity and Mortality Weekly Report*, 55:589–592. Note: Definitions changed slightly in 1998.

from this pandemic up to 1997. In 2005, there were an estimated 4.1 million new cases. Due to implementation of coordinated control programs though, it is believed the pandemic expansion peaked in the late 1990s. The WHO aims to reverse the increase of HIV infection by 2015. With increased attention, training, and funding, this may be possible.

The declining incidence of new cases in industrialized nations may be the result of greater awareness of the disease and methods of prevention of transmission. Improving early diagnosis and access to care, especially the combined therapy programs that are very effective in delaying onset of symptoms, is an important part of public health management of the AIDS crisis. Until an effective vaccine is available, preventive reliance will continue to be on behavior risk reduction and other prevention strategies such as needle and condom distribution among high-risk population groups.

Throughout the world, HIV continues to spread rapidly, especially in poor countries in Africa, Asia, and South and Central America. The United Nations reports that 40 million persons are living with HIV/AIDS, 90 percent of them in developing countries, where transmission is largely by heterosexual contact. Every day, more than 8500 persons are infected including 1000 children. In Thailand, 1 person in 50 is now infected. In sub-Saharan Africa more than 1 in 40 is infected and in some cities as many as 1 in 3 people carry the virus. Estimations of new infections per year in sub-Saharan Africa range from 1 to 2 million persons, while in Asia the range is from 1.2 to 3.5 million new infected persons per year. Lessons are still being learned from the AIDS pandemic. The explosive spread of this infection, from an estimated 100,000 people in 1980 to an anticipated 40 million persons HIV

infected, shows that the world is still vulnerable to pandemics of emerging infectious diseases. Enormous movements of tourists, businesspeople, truck drivers, migrants, soldiers, and refugees promote the spread of such diseases. Widespread sexual exchange, traffic in blood products, and illicit drug use all promote the international potential for pandemics. War and massive refugee situations promote rape and prostitution, worsening the AIDS situation in some settings in Africa.

The HIV pandemic has spread throughout the world. However, there is the somewhat hopeful indication that the rate of increase has slowed in the United States. This may be an indication of a number of factors; higher levels of self-protective behavior; the most susceptible population groups have already been affected; and the spread into the general population is at a slower rate. It is also possible that this may yet prove to be only a lull in the storm, as heterosexual contact becomes a more important mode of transmission.

The UNAIDS 2006 report showed signs that combinations of several drugs from among a number of antiretroviral medications are showing promise to suppress the AIDS virus in infected people. At a current annual price of nearly $20,000 per patient, these sums are well beyond the capacity of most developing countries. Development of methods of measuring the HIV viral load have allowed for better evaluation of potential therapies and monitoring of patients receiving therapy. In developed countries, transmission by blood products has been largely controlled by screening tests, transmission among homosexuals has been reduced by safe sex practices, and transmission to newborns has been reduced by recent therapeutic advances. Safe sex practices and condom use may have helped in reducing heterosexual transmission. Further advances in therapy and prevention with a vaccine are expected over the next decade.

The HIV/AIDS pandemic is one of the great challenges to public health for the twenty-first century due to its complexity; its international spread; its sexual and other modes of transmission; its devastating and costly clinical effects; and its impact on parallel diseases such as tuberculosis, respiratory infections, and cancer. The cost of care for the AIDS patient can be very high. Needed programs include home care and community health workers to improve nutrition and self-care, and mutual help among HIV carriers and AIDS patients. The ethical issues associated with AIDS are also complex regarding screening of pregnant women, newborns, partner notification, reporting, and contact tracing, as well as financing the cost of care.

DIARRHEAL DISEASES

Diarrheal diseases are the leading cause of child mortality in the world. They are caused by a wide variety of bacteria, parasites, and viruses (Table 4.10) infecting the intestinal tract and causing secretion of fluids and dissolved salts into the gut with mild to severe or fatal complications.

In developing countries, diarrheal diseases account for half of all morbidity and a quarter of all mortality. Diarrhea itself does not cause death, but the dehydration resulting from fluid and electrolyte loss is one of the most common causes of death in children worldwide. Deaths from dehydration can be prevented by use of oral rehydration therapy (ORT), an inexpensive and simple method of intervention easily used by a nonmedical primary care worker and by the mother of the child as a home intervention. In 1983, diarrheal diseases were the cause of almost 4 million child deaths, but by 1996 this had declined to 2.4 million, largely under the impact of increased use of ORT.

Diarrheal diseases are transmitted by water, food, and directly from person to person via oral–fecal contamination. Diarrheal diseases occur in epidemics in situations

---

TABLE 4.10 Major Organisms Associated with Diarrheal Diseases

Classification	Organisms
Bacteria	*Salmonella, Shigella, Escherichia coli, Vibrio cholerae, Bacillus cereus, Campylobacter jejuni*
Virus	Enteroviruses, rotaviruses, adenoviruses, astroviruses, calciviruses, coronaviruses, small round virus group, Norwalk group
Parasites/ Protozoa	*Schistosoma, Giardia lamblia,* cryptosporidium, *Amoeba histolytica*

---

of food poisoning or contaminated water sources, but can also be present at high levels when common source contamination is not found. Contamination of drinking water by sewage and poor management of water supplies are also major causes of diarrheal disease. The use of sewage for the irrigation of vegetables is a common cause of diarrheal disease in many areas.

Salmonella

Salmonella are a group of bacterial organisms causing acute gastroenteritis, associated with generalized illness including headache, fever, abdominal pains, and dehydration. There are over 2000 serotypes of *Salmonella,* many of which are pathogenic in humans, the most common of which are *Salmonella typhimurium, S. enteritidis,* and *S. typhi.* Transmission is by ingestion of the organisms in food, derived from fecal material from animal or human contamination. Common sources include raw or uncooked eggs, raw milk, meat, poultry and its products, as well as pet turtles or chicks. Fecal–oral transmission from person to person is common. Prevention is in safe animal and food handling, refrigeration, sanitary preparation and storage, protection against rodent and insect contamination, and the use of sterile techniques during patient care. Antibiotics rarely affect disease progression and may lead to increased carrier rates and produce resistant strains; therefore only symptomatic and supportive treatment is recommended, except in systemic and life-threatening cases.

S. typhi causes typhoid fever and is estimated by WHO to kill some 500,000 persons per year and seriously affect millions of others. While treatable by ampicillin and fluid replacement, the antibiotics are becoming less effective. Two vaccines are currently available and are used in high-risk areas.

Shigella

Shigella are a group of bacteria that are pathogenic in man. The infectious dose of *Shigella* is among the lowest of all pathogens; fewer than 10 organisms are sufficient to cause disease within four groups: type A– (*Shigella dysenteriae*), type B– (*S. flexneri*), type C– (*S. boydii*), and type D– (*S. sonnei*). Types A, B, and C are each further divided into a total of 40 serotypes. *Shigella* are transmitted by direct or indirect fecal–oral methods from a patient or carrier, and illness follows ingestion of even a few organisms. Water and milk transmission occurs as a result of contamination. Flies can transmit the organism, and in nonrefrigerated foods the organism may multiply to an infectious dose. Control is in hygienic practices and in the safe handling of water and food. *Shigella* is a common cause of waterborne disease outbreaks where water supplies are contaminated and not treated adequately.

Escherichia coli

E. coli are common fecal contaminants of inadequately prepared and cooked food. Particularly virulent strains such as O157:H17 can cause explosive outbreaks of severe (enterohemorrhagic) diarrheal disease with a hemolytic-uremic syndrome and death, as occurred in Japan in 1998 with cases and deaths due to a food-borne epidemic. Sporadic, but significant epidemics occur often, mostly in developed countries where food processing and transport are common. Other milder strains cause traveler's diarrhea and nursery infections. Inadequately cooked hamburger, unpasteurized milk, and other food vectors are discussed under "Food Safety" in Chapter 8. However, food-borne disease occurs in developed countries as well as in the case of contaminated lettuce from California in 2007.

Cholera

Cholera is an acute bacterial enteric disease caused by *Vibrio cholerae,* with sudden onset, profuse painless watery stools, occasional vomiting, and if untreated, rapid dehydration, circulatory collapse, and death. Similar disease may be caused by other "cholerogenic" species of *Vibrio*. Asymptomatic infection or carrier status, and mild cases are common. In severe, untreated cases, mortality is over 50 percent, but with adequate treatment, mortality is under 1 percent. Diagnosis is based on clinical signs, epidemiologic, serologic, and bacteriologic confirmation by culture. The two types of cholera are the classic and el Tor (with Inaba and Ogawa serotypes).

In 1991, a large-scale epidemic of cholera spread through much of South America. It was imported via a Chinese freighter, whose sewage contaminated shellfish in Lima harbor in Peru (Box 4.16). Epidemics in South America, south Asia, and Iraq have caused hundreds of thousands of cases and thousands of deaths since 1991.

Prevention requires sanitation, particularly the chlorination of drinking water; prohibiting the use of raw sewage for the irrigation of vegetable crops; and high standards of community, food, and personal hygiene. Crucial treatment is prompt fluid therapy with electrolytes in large volume to replace all fluid loss with oral rehydration therapy (ORT). Tetracycline shortens the duration of the disease, and chemoprophylaxis for contacts following stool samples may help in reducing its spread. A vaccine is available but is of no value in the prevention of outbreaks.

Viral Gastroenteritis

Viral gastroenteritis can occur in sporadic or epidemic forms, in infants, children, or adults. Some viruses, such as the rotaviruses and enteric adenoviruses, affect mainly infants and young children, and may be severe enough to

---

Box 4.16 The Cholera Pandemic in South America, 1991–1998

In the 1980s, Peruvian officials stopped the chlorination of community water supplies because of concern over possible carcinogenic effects of trihalomethanes, a view encouraged by officials of the U.S. Environmental Protection Agency (EPA) and the U.S. Public Health Service. In January 1991, a Chinese freighter arrived in Lima, Peru, and dumped bilge (sewage) in the harbor, apparently contaminating local shellfish. Consumption of raw shellfish is a popular local delicacy (ceviche) and is associated with cases of cholera seen in local hospitals.

Contamination of local water supplies from sewage resulted in the geometric increase in cases, and by the end of 1992, the Pan American Health Organization (PAHO) reported an epidemic of 391,000 cases and 4002 deaths. The epidemic spread to 21 countries, and in 1992 there were a further 339,000 cases and 2321 deaths spreading over much of South America, continuing in 1999.

In the United States, 102 cases of cholera were reported in 1992; of these, 75 cases and 1 death were among passengers of an airplane flying from South America to Los Angeles in which contaminated seafood was served. In 1993, 91 cases of cholera were reported in the United States which were unrelated to international travel. These occurred mostly among persons consuming shellfish from the Gulf coast with a strain of cholera similar to the South American strain, also possibly introduced in a ship ballast. Cholera organisms are reported in harbor waters in other parts of the United States.

Sources: Anderson, C. 1991. Cholera epidemic traced to risk miscalculation. *Nature*, 354:255. Centers for Disease Control. 1993. Update cholera Western — hemisphere, 1992. MMWR, 42:89–91; Centers for Disease Control. 1993. Isolation of *Vibrio cholerae* O1 from Oysters — Mobile Bay, 1991–1992. *MMWR*, 42:91–93.

---

cause hospitalization for dehydration. Others such as Norwalk and Norwalk-like viruses affect older children and adults in self-limited acute gastroenteritis in family, institution, or community outbreaks.

Rotaviruses

Rotaviruses cause acute gastroenteritis in infants and young children, with fever and vomiting, followed by watery diarrhea and occasionally severe dehydration and death if not adequately treated. Diagnosis is by examination of stool or rectal swabs with commercial immunologic kits. In both developed and developing countries, rotavirus is the cause of about one-third of all hospitalized cases for diarrheal diseases in infants and children up to age 5. Most children in developing countries experience this disease by the age of 4 years, with the majority of cases between 6 and 24 months. In developing countries, rotaviruses are estimated to cause over 1 million deaths per year. The virus is found in temperate climates in the cooler months and in tropical countries throughout the year. Breastfeeding does not prevent the disease but may reduce its severity. Oral rehydration therapy is the key treatment. A live attenuated vaccine was approved by the FDA in 1998 and adopted in the 1999 U.S. recommended routine vaccination programs for infants.

Adenoviruses

Adenoviruses, Norwalk, and a variety of other viruses (including astrovirus, calcivirus, and other groups) cause sporadic acute gastroenteritis worldwide, mostly in outbreaks. Spread is by the oral–fecal route, often in hospital or other communal settings, with secondary spread among family contacts. Food-borne and waterborne transmission are both likely. These can be a serious problem in disaster situations. No vaccines are available. Management is with fluid replacement and hygienic measures to prevent secondary spread.

Parasitic Gastroenteritis

Giardiasis

Giardiasis (caused by *Giardia lamblia*) is a protozoan parasitic infection of the upper small intestine, usually asymptomatic, but sometimes associated with chronic diarrhea; abdominal cramps; bloating; frequent, loose, greasy stools; fatigue; and weight loss. Malabsorption of fats and vitamins may lead to malnutrition. Diagnosis is by the presence of cysts or other forms of the organism in stools, duodenal fluid, or in intestinal mucosa from a biopsy. This disease is prevalent worldwide and affects mostly children. It is spread in areas of poor sanitation and in preschool settings and swimming pools, and is of increasing importance as a secondary infection among immunocompromised patients, especially those with AIDS.

Waterborne giardia was recognized as a serious problem in the United States in the 1980s and 1990s, since the protozoa are not readily inactivated by chlorine, but require adequate filtration before chlorination. Person-to-person transmission in day-care centers is common, as is transmission by unfiltered stream or lake water where contamination by human or animal feces is to be expected. An asymptomatic carrier state is common. Prevention relies on careful hygiene in settings such as day-care centers, filtration of public water supplies, and the boiling of water in emergency situations.

Cryptosporidium

Cryptosporidium parvum is a parasitic infection of the gastrointestinal tract in man, small and large mammals, and vertebrates. Infection may be asymptomatic or cause a profuse, watery diarrhea, abdominal cramps, general malaise, fever, anorexia, nausea, and vomiting. In immunosuppressed patients, such as persons with AIDS, it can be a serious problem. The disease is most common in children under 2 years of age and those in close contact with them, as well as in homosexual men. Diagnosis is by identification of the *Cryptosporidium* organism cysts in stools. The disease is present worldwide. In Europe and the United States, the organism has been found in 1 to 4.5 percent of individuals sampled. Spread is common by person-to-person contact by fecal–oral contamination, especially in such settings as day-care centers. Raw milk and waterborne outbreaks have also been identified in recent years. A large waterborne disease outbreak due to *Cryptosporidium* occurred in Milwaukee in 1986 as described in Chapter 9. Management is by rehydration and prevention is by careful hygiene in food and water safety.

Helicobacter pylori

Helicobacter pylori, first identified in 1986, is a bacterium causally linked to gastrointestinal ulcers and gastritis, contributing to high rates of gastric cancer (Chapter 5). It is an important example of the link between infection and chronic disease. This has enormous implications for prevention of cancer of the stomach, chronic peptic ulcers, and large-scale use of hospitals and other medical resources (see Chapter 5).

A Program Approach to Diarrheal Disease Control

The control of diarrheal diseases requires a comprehensive program involving a wide range of activities, including good management of food and water supplies, education in hygiene, and, particularly where morbidity and mortality are high, education in the use of oral rehydration therapy (ORT).

Oral rehydration therapy (ORT) is considered by UNICEF and WHO to have resulted in the saving of 1 million lives each year in the 1990s. Proper management of an episode of diarrhea by ORT (Box 4.17), along with continued feeding, not only saves the child from dehydration and immediate death, but also contributes to early restoration of nutritional adequacy, sparing the child the prolonged effects of malnutrition.

The World Summit for Children (WSC) in 1990 called for a reduction in child deaths from diarrheal diseases by one-third and malnutrition by one-half, with emphasis on the widest possible availability, education for, and use of ORT. This requires a programmatic approach. Public health leadership must train primary care doctors, pediatricians, pharmacists, drug manufacturers, and primary care health workers of all kinds in ORT principles and usage. They must be backed by the widest possible publicity to raise awareness among parents.

Box 4.17 WHO Formula for Oral Rehydration Therapy (ORT)

Ingredients	Amounts in grams/liter	Ions	Concentration (millimoles/liter)
Sodium chloride, NaCl	3.5	Sodium	90
Trisodium citrate, dihydrate, or sodium bicarbonate NaHCO$_3$	2.9 or 2.5	Citrate*	20 citrate**
Potassium chloride, KCl	1.5	Potassium	10 of potassium, 80 of chloride
Glucose (anhydrous)	20.0	Glucose	111

*Note: or 2.5 grams sodium bicarbonate.
**or 30 millimole bicarbonate.
Sources: World Health Organization. 1992. *Readings on Diarrhoea: Student Manual*, Geneva: World Health Organization. Heymann, D. L. (ed.). 2004. *Control of Communicable Diseases Manual*, 18th ed. Washington, DC: American Public Health Association.

Oral rehydration therapy is an important public health modality in developed countries as well as in developing countries. Diarrheal disease may not cause death as frequently in developed countries, but it is still a significant factor in infant and child health and, even under the most optimal conditions, can cause setbacks in the nutritional state and physical development of a child. Use of ORT does not prevent the disease (i.e., it is not a primary prevention), but it is excellent in secondary prevention, by preventing complications from diarrhea, and should be available in every home for symptomatic treatment of diarrheal diseases.

An adaptation of ORT has found its place in popular culture in the United States. A form of ORT, marketed as "sports drinks," is used in sports where athletes lose large quantities of water and salts in sweat and insensible loss from the respiratory tract. The wider application of the principles of ORT for use in adults in dry hot climates and in adults under severe physical exertion with inadequate fluid/salt intake situations requires further exploration.

Management of diarrheal diseases should be part of a wider approach to child nutrition. The child who goes through an episode of diarrheal disease may falter in growth and development. Supportive measures may be needed following the episode as well as during it. This involves providing primary care services that are attuned to monitoring individual infant and child growth. Growth monitoring surveillance is important to assess the health status of the individual child and the child population. Supplementation of infant feeding with vitamins A and D, and iron to prevent anemia are important for routine infant and child care, and more so for conditions affecting total nutrition such as a diarrheal disease.

ACUTE RESPIRATORY INFECTIONS

In the developing world, respiratory infections account for over one-quarter of all deaths and illnesses in children.

As diarrheal disease deaths are reduced, the major cause of death among infants in developing countries is becoming acute respiratory infections (ARIs). In industrialized countries, ARIs are important for their potentially devastating effects on the elderly and chronically ill. They are also the major cause of morbidity in infants in developed countries, causing much anxiety to parents even in areas with good living conditions. Cigarette smoking, chronic bronchitis, poorly controlled diabetes or congestive heart failure, and chronic liver and kidney disease increase susceptibility to ARIs. ARIs place a heavy burden on health care systems and individual families. Improved methods of management of such chronic diseases are needed to reduce the associated toll of morbidity, mortality, and the considerable expenses of health care.

Acute respiratory infections are due to a broad range of viral and bacterial infections. Secondary bacterial infections progress to pneumonia with mortality rates of 10–20 percent. Acute viral respiratory diseases include those affecting the upper respiratory tract, such as acute viral rhinitis, pharyngitis, and laryngitis, as well as those affecting the lower respiratory tract, tracheobronchitis, bronchitis, bronchiolitis, and pneumonia. ARIs are frequently associated with VPDs, including measles, varicella, and influenza. They are caused by a large number of viruses, producing a wide spectrum of acute respiratory illness. Some organisms affect any part of the respiratory tract, while others affect specific parts and all predispose to bacterial secondary infection. While children and the elderly are especially susceptible to morbidity and mortality from acute respiratory disease, the vast numbers of respiratory illnesses among adults cause large-scale economic loss from work absence.

Bacterial agents causing upper respiratory tract infection include group A *Streptococcus*, *Mycoplasma pneumoniae*, pertussis, and parapertussis. Pneumonia or acute bacterial infection of the lower respiratory tract and lung tissue may be due to pneumococcal infection with

Streptococcus pneumoniae. There are 83 known types of this organism, distinguished by capsule characteristics; 23 account for 88 percent of pneumococcal infections in the United States. An excellent polyvalent vaccine based on these types is available for high-risk groups such as the elderly; immunodeficient patients; and persons with chronic heart, lung, liver, blood disorders, or diabetes.

Opportunistic infections attack the chronically ill, especially those with compromised immune systems, often with life-threatening ARIs. Mycoplasma (primary atypical pneumonia) is a lower respiratory tract infection which sometimes progresses to pneumonia. TB and *Pneumonocytis jiroveci* are especially problematic for AIDS patients. Other organisms causing pneumonias include *Chlamydia pneumoniae, H. influenzae, Klebsiella pneumonia, Escherichia coli, Staphylococcus,* rickettsia (Q fever), and *Legionella*. Parasitic infestation of lungs may occur with nematodes (e.g., ascariasis). Fungal infections of the lung may be caused by aspergillosis, histoplasmosis, and coccidiomycosis, often as a complication of antibiotic therapy.

Access to primary care and early institution of treatment are vital to control excess mortality from ARIs. In developed countries, ARIs as contributors to infant deaths are largely a problem in minority and deprived population groups. Because these groups contribute disproportionately to childhood mortality, infant mortality reduction has been slower in countries such as the United States and Russia than in other industrialized countries. The continuing gap in mortality rates between white and African-American children in the United States can, to a large extent, be attributed to ARIs and less access to organized primary care. Children are brought to emergency rooms for care when the disease process is already advanced and more dangerous than had it been attended to professionally earlier in the process. Many field trials of ARI prevention programs have proved successful, involving parent education and training of primary care workers in early assessment and, if necessary, initiation of treatment. This needs field testing in multiple settings.

Reliance on vaccines to prevent respiratory infectious diseases is not currently feasible. ARIs are caused by a very wide spectrum of viruses, and the development of vaccines in this field has been slow and limited. The vaccine for pneumococcal pneumonia has been an important breakthrough, but it is still inadequately utilized by the chronically ill because of its limitations, costs, and lack of sufficient political and public awareness, and it is too expensive for developing countries. This vaccine is recommended for infants in the United States and many industrial nations and recommended by WHO for developing countries, but as yet not widely applied in the latter. Improvements in bacterial and viral vaccine development will potentially help to reduce the burden of ARIs. A programmatic approach with clinical guidelines and education of family and caregivers is currently the only feasible way to reduce the still enormous morbidity and mortality from ARIs on the young and the elderly.

INEQUALITIES IN CONTROL OF COMMUNICABLE DISEASES

As in other fields of public health, there are wide variations or inequities between and within countries in control of communicable diseases. The differences between the industrialized countries and the developing countries are enormous. The gaps are not only in coverage, but in the content of the immunization programs. Adoption of Hib vaccine is increasing but the decade-long gap from availability to widespread global usage costs very many preventable deaths. Similarly the lag in adoption of pneumococcal pneumonia and rotavirus vaccines will prolong the time to achive the Millenium Development Goals of reducing child mortality in very many countries.

Even in the European region, there are wide differences between groups of countries as seen in Figure 4.6, comparing standardized mortality rates (3-year moving averages) for infectious and parasitic disease between long-standing members of the European Union (such as France, Germany, the United Kingdom), with the new members (since 2004, such as Hungary, Poland, and other countries of the eastern Europe) with those of the Commonwealth of Independent States (e.g., Ukraine, Russia), and finally the Central Asian Republics (e.g., Kazakhstan, Uzbekistan, Tajikistan). The trends show low and stable rates in the countries of Western and Eastern Europe, high and falling rates in Central Asia, but high and rising rates in the key countries of the former Soviet Union. While there may be artifacts of reporting, the trends are thought to be correct, and are likely to be related to many factors such as water and food safety, obsolescent immunization programs, tuberculosis and HIV control, and many other factors. Comparisons within countries will also show social and regional disparities which constitute failings of public health systems, and indicate that communicable diseases are very much part of the modern public health agenda.

The European region, which includes all of these countries, does not have a standard recommended immunization schedule and each country follows its own patterns. The western countries are generally up-to-date with the content of their programs with high coverage but this is not uniformly so in all parts of the region. The countries of the former Soviet Union are slowly updating their immunization schedules but remain largely at least a decade behind. WHO's advisory committee system on immunization has been updating their recommendations rapidly in recent years with hepatitis B, more recently Hib and pneumococcal pneumonia and adopting of the two-dose policy of MMR. As new vaccines become available, the transitional and

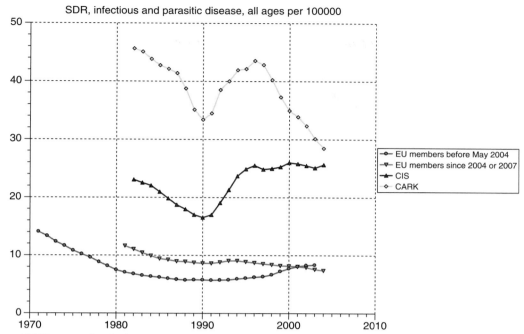

SDR, infectious and parasitic disease, all ages per 100000

FIGURE 4.6 Standardized death rates from infectious and parasitic diseases, selected regions of Europe, 1970–2005. Source: WHO European Regional Office, Health for All Database, June 2007, www.who/dk/hfadb [accessed October 3, 2007]. Note: Rates per 100,000 population and standardized to European population.

developing countries will need support to expand their programs of immunization, a key part of the drive to attain the Millennium Development Goals of reduced child mortality and control of infectious diseases (malaria, HIV, and others).

COMMUNICABLE DISEASE CONTROL IN THE NEW PUBLIC HEALTH

The success of sanitation, vaccines, and antibiotics led many to assume that all infectious diseases would sooner or later succumb to public health and medical technology. Unfortunately, this is a premature and even dangerous assumption. Despite the long-standing availability of an effective and inexpensive vaccine, the persistence of measles as a major killer of 1 million children per year represents a failure in effective use of both the vaccine and the health system. The resurgence of TB and malaria have led to new strategies, such as managed or directly observed care, with community health workers to assure compliance needed to render the patient noninfectious to others and to reduce the pool of carriers of the disease.

Successes achieved in reducing poliomyelitis, measles, dracunculiasis, onchocerciasis, and other diseases to the point of local or global eradication have raised hopes for similar success in other fields. But there are many infectious diseases

of importance in developed and developing countries where existing technologies are not fully utilized. Oral rehydration therapy (ORT) is one of the most cost-effective methods of preventing excess mortality from ordinary diarrheal diseases, and yet is not used on sufficient scale.

Biases in the financing and management of medical insurance programs can result in underutilization of available effective vaccines. Hospital-based infections cause large-scale increases in lengths of stay and expenditures, although application of epidemiologic investigation and improved quality in hospital practices could reduce this burden. Control of the spread of AIDS using combined medical therapies is not financially or logistically possible in many countries, but education for "safe sex" is effective. Community health worker programs can greatly enhance tuberculosis, malaria, and STI control, or in AIDS care, promote prevention and appropriate treatment.

In the industrialized and mid-level developing countries, epidemiologic and demographic shifts have created new challenges in infectious disease control. Prevention and early treatment of infectious disease among the chronically ill and the elderly is not only a medical issue, it is also an economic one. Patients with chronic obstructive lung disease (COPD), chronic liver or kidney disease, or congestive heart failure are at high risk of developing an infectious disease followed by prolonged hospitalization.

SUMMARY

Public health has addressed, and will continue to stress, the issues of communicable disease as one of its key issues in protecting individual and population health. Methods of intervention include classic public health through sanitation, safe water and foods, immunization, and well beyond that into nutrition, education, case finding, treatment, and changing human behavior. The knowledge, attitudes, beliefs, and practices of policymakers, health care providers, and parents are as important in the success of communicable disease control as are the technology available and methods of financing health systems. Together, these encompass the broad programmatic approach of the New Public Health to control of communicable diseases.

In a world of rapid international transport and contact between populations, systems are needed to monitor the potentially explosive spread of pathogens that may be transferred from their normal habitat. The potential for the international spread of new or reinvigorated infectious diseases constitutes threat to mankind akin to ecological and other man-made disasters.

The eradication of smallpox paved the way for the eradication of poliomyelitis, and perhaps measles, in the foreseeable future. New vaccines are showing the capacity to reduce important morbidity from rubella syndrome, mumps, meningitis, and hepatitis. Other new vaccines on the horizon will continue the immunologic revolution into the twenty-first century.

As the triumphs of control or elimination of infectious diseases of children continue, the scourge of HIV infection continues with distressingly slow progress on development of an effective vaccine or cure for the disease it engenders. Partly as a result of HIV/AIDS, TB staged a comeback in many countries where it was thought to be merely a residual problem. At the same time an old/new method of intervention using directly observed short-term therapy has shown great success in controlling the TB epidemic. The resurgence of TB is more dangerous in that MDRTB has become a widespread problem. This issue highlights the difficulty of keeping ahead of drug resistance in the search for new generations of antibiotics, posing a difficult challenge for the pharmaceutical industry, basic scientists, as well as public health workers.

The burden of infectious diseases has appeared to recede as the predominant public health problem in the developed countries but new challenges of emerging infectious diseases have come to the fore in public health, and communicable disease remains a dominating problem in the developing countries. With increases in longevity and increased importance of chronic disease in the health status of the industrial and mid-level developing nations, the effects of infectious disease on the care of the elderly and chronically ill are of great importance in the New Public Health. Long-term management of chronic disease needs to address the care of vulnerable groups, promoting the use of existing vaccines and antibiotics. Most important is the development of health systems that provide close monitoring of groups at special risk for infectious disease, especially patients with chronic diseases, the immunocompromised, and the elderly. The combination of traditional public health with direct medical care needed for effective control and eradication of communicable diseases is an essential element of the New Public Health. The challenge is to apply a comprehensive approach and management of resources to define and reach achievable targets in communicable disease control.

Control of communicable diseases is fundamental pillar of public health. The new capacities of vaccines and other methods of control develop slowly, and the advent of effective vaccines for HIV, malaria, and tuberculosis will bring untold benefit to the global community. The challenges of natural dispersion of communicable disease can be made more threatening because of the advent of bioterrorism and the emergence of new diseases or spread of those previously localized in a previously less mobile globalized world. The challenges, the potential for harm, and the benefits that can be achieved in this aspect of public health are enormous.

ELECTRONIC RESOURCES

Access to e-mail and the Internet are vital to current practice of public health and nowhere is this more important than in communicable diseases. There are many such information sites and these will undoubtedly expand in the coming years. Several sites are given as examples. The Internet has great practical implications for keeping up-to-date with rapidly occurring events and new epidemiologic and other scientific advances in this field.

Centers for Disease Control. 2006. Reported Tuberculosis in the United States, 2005. Atlanta, GA: DHHS, http://www.cdc.gov/tb/surv/surv2005/PDF/TBSurvFULLReport.pdf

Eurosurveillance Weekly. Infectious disease early warning system, http://www.eurosurveillance.org/ [accessed May 4, 2008]

Morbidity and Mortality Weekly Reports. http://www.cdc.gov [accessed May 4, 2008]

Rollback Malaria Partnership, http://www.rollbackmalaria.org [accessed June 3, 2008]

Weekly Epidemiologic Bulletin. http://www.who.ch/programmes/emc/news.htm (ProMed subscription available at http://www.isid.org/promedmail/subscribe.lasso) [accessed May 5, 2008]

World Health Organization. 2008. Avian influenza, http://www.who.int/csr/disease/avian_influenza/en/ [accessed May 4, 2008]

World Health Organization. 2008. BCG vaccine, http://www.who.int/biologicals/areas/vaccines/bcg/en/ [accessed May 4, 2008]

World Health Organization. 2008. Cholera, http://www.who.int/topics/cholera/control/en/ [accessed May 4, 2008]

World Health Organization. Dengue. http://www.who.int/csr/disease/dengue/en/ [accessed May 4, 2008]

World Health Organization. 2008. Immunization, vaccines, and biologicals, http://www.who.int/immunization/en/ [accessed May 4, 2008]

World Health Organization. 2008. Leprosy, http://www.who.int/immunization/sage_page/en/index.html [accessed May 4, 2008]

World Health Organization, MDG6: Combat HIV/AIDS, Malaria, and other diseases, http://www.who.int/topics/millennium_development_goals/diseases/en [accessed June 3, 2008]

World Health Organization, Yellow Fever, 2008, http://www.who.int/topics/yellow_fever/en

World Health Organization. 2008. Malaria, http://www.who.int/topics/malaria/en/ [accessed May 4, 2008]

World Health Organization. Neglected tropical diseases, http://www.who.int/neglected_diseases/en/ [accessed May 4, 2008]

World Health Organization. 2008. Lassa Fever, http://www.who.int/csr/disease/lassafever/en/ [accessed 3 May 2008]

World Health Organization. Zoonoses, http://www.who.int/zoonoses/en/

World Health Organization. Global malaria control and elimination, http://www.who.int/malaria/pages/elimination/malariaeliminationrec.html; http://www.who.int/malaria/docs/ReportGFImpactMalaria.pdf; http://rbm.who.int/unitedagainstmalaria/malaria/ [accessed May 3, 2008]

RECOMMENDED READINGS

Brooks, G. E., et al. 2007. *Jawetz, Melnick and Adelberg's Medical Microbiology*, 24th ed. Stamford, CT: Appleton & Lange.

Centers for Disease Control. 1999. Achievements in public health, 1900–1999: Control of infectious diseases. *Morbidity and Mortality Weekly Report*, 48:621–629.

Centers for Disease Control. 2006. *Community Guide to Preventive Services.* Multiple interventions implemented in combination are recommended to increase coverage of targeted vaccine coverage. Community Guide@CDC.gov [accessed October 30, 2007]

Centers for Disease Control. 2007. Summary of notifiable diseases — United States, 2005. *Morbidity and Mortality Weekly Report*, 54:2–92.

Centers for Disease Control. 2006. Vaccine preventable deaths and the global immunization vision and strategy, 2006–2015. *Morbidity and Mortality Weekly Report*, 55:511–515.

Centers for Disease Control and Prevention. 2007. Recommended immunization schedules for persons aged 0–18 years — United States, 2007. *Morbidity and Mortality Weekly Report Recommendations and Reports*, 5:55:Q1–4.

Cohen, J., Powderly, W. G. 2004. *Cohen & Powderly: Infectious Diseases*, 2nd ed. New York: Mosby.

Colditz, G. A., Brewer, T. F., Berkey, C. S., Wilson, M. E., Burdick, E., Fineberg, H. V., Mosteller, F. 1994. Efficacy of BCG vaccine in the prevention of tuberculosis. Meta-analysis of the published literature. *Journal of the American Medical Association*, 271:698–702.

Cook, G. C. 2002. *Manson's Tropical Diseases*, 21st ed. London: W. B. Saunders.

Heymann, D. L. 2004. *Control of Communicable Diseases Manual*, 18th ed. Washington, DC: American Public Health Association.

Mandel, G. L. 1994. *Principles and Practice of Infectious Diseases*. Edinburgh: Churchill-Livingstone.

Milstein, J., Cash, R. A., Wecker, J., Wikler, D. Development of priority vaccines for disease-endemic counties: Risks-benefits. *Health Affairs*, 24:718–728.

National Center for Health Statistics. 2006. *Health, United States, 2006. With Chartbook on Trends in the Health of Americans*. Hyattsville, MD: U.S. Department of Health and Human Services.

Plotkin, S. A., Orenstein, W. A. 2003. *Vaccines*, 4th ed. Atlanta: W. B. Saunders.

Stern, A. M., Markel, H. 2005. The history of vaccines and immunization: Familiar patterns, new challenges. *Health Affairs*, 24:611–621.

Thacker, S. B. 2006. Epidemiology and Public Health at CDC. *Morbidity and Mortality Weekly Report*, 55(SUP02):3–4.

Tierney, L. M., McPhee, S. J., Papadakis, M. A. 2006. *Current Medical Diagnosis & Treatment 2007*, 46th ed. New York: McGraw-Hill.

World Health Organization. 2007. *WHO Interim Protocol: Rapid Operations to Contain the Initial Emergence of Pandemic Influenza*, updated October 2007 http://www.who.int/csr/disease/avian_influenza/guidelines/draftprotocol/en/print.html [accessed October 30, 2007] .

World Health Organization. Hepatitis, http://www.who.int/csr/disease/hepatitis/whocdscsredc2007/en/index5.html

World Health Organization. Hepatitis B, WHO http://www.who.int/mediacentre/factsheets/fs204/en/

CDC. http://www.cdc.gov/ncidod/diseases/hepatitis/recs/index.htm, http://www.cdc.gov/ncidod/diseases/hepatitis/resource/PDFs/disease_burden.pdf

CDC. Hepatitis C, http://www.cdc.gov/mmwr/preview/mmwrhtml/00055154.htm; http://www.newscientist.com/channel/health/dn13539-hepatitis-c-is-first-target-for-new-therapy.html

WHO. http://www.who.int/mediacentre/factsheets/fs164/en/

World Health Organization. TB. WHO http://www.who.int/tb/en/

WHO. Strengthening the Global Response to XDR-TB. http://whqlibdoc.who.int/hq/2007/WHO_HTM_TB_2007.387_eng.pdf; http://www.who.int/tb/features_archive/drsreport_launch_26feb08/en/index.html, February 2008 [accessed May 3, 2008]

BIBLIOGRAPHY

Ahmed, R., Oldstone, M. B., Palese, P. 2007. Protective immunity and susceptibility to infectious diseases: Lessons from the 1918 influenza pandemic. *Nature Immunology*. 8:1188–1193.

American Academy of Pediatrics. 1999. Combination vaccines for childhood immunization: Recommendations of the Advisory Committee on Immunization Practices. *Pediatrics*, 103:1064–1077.

American Academy of Pediatrics, Committee of Infectious Diseases. 1999. Poliomyelitis prevention: Revised recommendations for use of inactivated and live oral poliovirus vaccines. *Pediatrics*, 103:171–172.

American Academy of Pediatrics Committee on Infectious Diseases. 2007. Recommended immunization schedules for children and adolescents — United States, 2007. *Pediatrics*, 2007 Jan; 119(1), 207–208.

Anderson, C. 1991. Cholera epidemic traced to risk miscalculation. *Nature*, 354:255.

Angelini, R., Finarelli, A., Angelini, P., Po, C., Petropulacos, K., Macini, P., Fiorentini, C., Fortuna, C., Venturi, G., Romi, R., Majori, G., Nicoletti, L., Rezza, G., Cassone, A. An outbreak of chikungunya fever in the province of Ravenna, Italy. *EuroSurveillance Weekly*. 2007;12(9):E070906.1. http://www.eurosurveillance.org/ew/2007/070906.asp#1 [accessed October 3, 2007]

Ashmore, J., Addiman, S., Cordery, R., Maguire, H. 2007. Measles in north east and north central London, England: A situation report. *Eurosurveillance Weekly*, 12:E070920.2. http://www.eurosurveillance.org/ew/2007/070920.asp#2 [accessed October 3, 2007]

Brooks, G. E., Butel, J. S., Morse, S. A. 2004. *Jawetz, Melnick and Adelberg's Medical Microbiology*, 23rd ed. Stamford, CT: Appleton & Lange.

Centers for Disease Control. 1992. Update: International Task Force for Disease Eradication, 1990 and 1991. *Morbidity and Mortality Weekly Report*, 41:40–42.

Centers for Disease Control. 1993. Isolation of *Vibrio cholerae* O1 from oysters—Mobile Bay, 1991–1992. *Morbidity and Mortality Weekly Report*, 42:91–93.

Centers for Disease Control. 1993. Diphtheria outbreak—Russian Federation, 1990–1993. *Morbidity and Mortality Weekly Report*, 42:840–841, 847.

Centers for Disease Control. 1993. Resurgence of pertussis—United States, 1993. *Morbidity and Mortality Weekly Report*, 42:952–953, 959–960.

Centers for Disease Control. 1993. Update cholera—Western hemisphere, 1992. *Morbidity and Mortality Weekly Report*, 42:89–91.

Centers for Disease Control. 1994. Addressing emerging infectious disease threats: A prevention strategy for the United States. Executive summary. *Morbidity and Mortality Weekly Report*, 43(RR-5):1–18.

Centers for Disease Control. 1994. Rift Valley Fever—Egypt 1993. *Morbidity and Mortality Weekly Report*, 43:693, 699–700.

Centers for Disease Control. 1994. Rubella and congenital rubella syndrome—United States, January 1, 1991–May 7, 1994. *Morbidity and Mortality Weekly Report*, 43:397–401.

Centers for Disease Control. 1994. Update: Outbreak of Legionnaire's disease associated with a cruise ship, 1994. *Morbidity and Mortality Weekly Report*, 43:574–575.

Centers for Disease Control. 1996. The role of BCG vaccine in the prevention and control of tuberculosis in the United States: A joint statement by the Advisory Council for the Elimination of Tuberculosis and the Advisory Committee on Immunization Practices. *Morbidity and Mortality Weekly Report*, 45(RR-4):1–18.

Centers for Disease Control. 1996. Compendium of animal rabies control, 1996: National Association of State Public Health Veterinarians. *Morbidity and Mortality Weekly Report*, 45(RR-3):1–9.

Centers for Disease Control. 1997. Case definition for infectious conditions under public health surveillance. *Morbidity and Mortality Weekly Report*, 46(RR-10):1–55.

Centers for Disease Control. 1997. Tetanus surveillance—United States, 1991–1994. *Morbidity and Mortality Weekly Report*, 46(SS-2): 15–25.

Centers for Disease Control. 1998. Impact of the sequential IPV/OPV schedule on vaccination coverage — United States, 1997. *Morbidity and Mortality Weekly Report*, 47:1017–1019.

Centers for Disease Control. 1998. National, state and urban area vaccination coverage levels among children aged 19–35 months — United States, July 1996–June 1997. *Morbidity and Mortality Weekly Report*, 47:108–116.

Centers for Disease Control. 1998. Primary and secondary syphilis—United States, 1997. *Morbidity and Mortality Weekly Report*, 47:493–497.

Centers for Disease Control. 1998. Progress toward elimination of *Haemophilus influenzae* type b disease among infants and children in the United States, 1987–1997. *Morbidity and Mortality Weekly Report*, 47:993–998.

Centers for Disease Control. 1998. Recommendations and reports — Vaccine use and strategies for elimination of measles, rubella, and congenital rubella syndrome and control of measles: Recommendations of the Advisory Committee on Immunization Practices. *Morbidity and Mortality Weekly Report*, 47(RR-8):1–59.

Centers for Disease Control. 1998. Varicella related deaths among children — United States, 1997. *Morbidity and Mortality Weekly Report*, 47:365–368.

Centers for Disease Control. 1999. Progress toward global poliomyelitis eradication. *Morbidity and Mortality Weekly Report*, 48:416–421.

Centers for Disease Control. 1999. Ten great public health achievements — United States, 1900–1999. *Morbidity and Mortality Weekly Report*, 48:241–243.

Centers for Disease Control. 1999. Achievements in public health, 1900–1999. Impact of vaccines universally recommended for children — United States, 1990–1998. *Morbidity and Mortality Weekly Report*, 48:243–248.

Centers for Disease Control. 2003. Progress toward global eradication of dracunculiasis, January – June 2003. *Morbidity and Mortality Weekly Report*, 52:881–883.

Centers for Disease Control. 2004. 150th anniversary of John Snow and the pump handle. *Morbidity and Mortality Weekly Report*, 53:783.

Centers for Disease Control. 2004. Increases in fluoroquinolone-resistant *Neisseria gonorrhoeae* among men who have sex with men — United States, 2003, and revised recommendations for gonorrhea treatment. *Morbidity and Mortality Weekly Report*, 53:335–338.

Centers for Disease Control. 2005. Pertussis — United States, 2001–2003. *Morbidity and Mortality Weekly Report*, 54:1283–1286.

Centers for Disease Control. 2006. Effects of measles-control activities — African Region, 1999–2005. *Morbidity and Mortality Weekly Report*, 55:1017–1021.

Centers for Disease Control. 2006. Key facts about avian influenza (bird flu) and avian influenza A (H5N1) virus. *CDC Fact Sheet*, http://cdc.gov/flu/avian [accessed October 3, 2007]

Centers for Disease Control. 2006. Measles — United States, 2005. *Morbidity and Mortality Weekly Report*, 55:1348–1351.

Centers for Disease Control. 2006. Mumps epidemic — United Kingdom, 2004–2005. *Morbidity and Mortality Weekly Report*, 55:173–175.

Centers for Disease Control. 2006. Preliminary FoodNet data on the incidence of infection with pathogens transmitted commonly through food — 10 states, United States, 2005. *Morbidity and Mortality Weekly Report*, 55:392–395.

Centers for Disease Control. 2006. Preventing tetanus, diphtheria, and pertussis among adults: Use of tetanus toxoid, reduced diphtheria toxoid and acellular pertussis vaccine: Recommendations of the Advisory Committee on Immunization Practices (ACIP) and Recommendation of ACIP, supported by the Healthcare Infection Control Practices Advisory Committee (HICPAC), for Use of Tdap Among Health-Care Personnel. 55(RR-17):1–33.

Centers for Disease Control. 2006. Recommended childhood and adolescent immunization schedule — United States, 2006. Harmonized childhood and adolescent immunization schedule. *Morbidity and Mortality Weekly Report*, 54:Q1–Q4.

Centers for Disease Control. 2006. Progress toward tuberculosis control and determinants of treatment outcomes — Kazakhstan, 2000–2002. *Morbidity and Mortality Weekly Report*, 55:11–15.

Centers for Disease Control. 2006. STD-prevention counseling practices and human papillomavirus opinions among clinicians with adolescent patients — United States, 2004. *Morbidity and Mortality Weekly Report*, 55:1117–1120.

Centers for Disease Control. 2006. Update: Influenza activity — United States and worldwide, May 21–September 9, 2006. *Morbidity and Mortality Weekly Report*, 55:1021–1023.

Centers for Disease Control. 2006. Vaccine preventable deaths and the global immunization vision and strategy, 2006–2015. *Morbidity and Mortality Weekly Report*, 55:511–515.

Centers for Disease Control. 2006. Primary and secondary syphilis — United States, 2003—2004. *Morbidity and Mortality Weekly Report*, 55:269–273.

Centers for Disease Control. 2007. National, state, and local area vaccination coverage among children aged 19–35 months — United States, 2006. *Morbidity and Mortality Weekly Report*, 31:56: 880–885.

Centers for Disease Control. 2007. Vaccination coverage among children in kindergarten — United States, 2006–07 school year. *Morbidity and Mortality Weekly Report*, 17:56:819–821.

Centers for Disease Control. 2007. West Nile virus activity — United States, 2006. *Morbidity and Mortality Weekly Report*, 56:556–559.

Centers for Disease Control. 2008. Updated recommendation from the Advisory Committee on Immunization Practices (ACIP) for use of 7-valent pneumococcal conjugate vaccine (PCV7) in children aged 24–59 months who are not completely vaccinated. *Morbidity and Mortality Weekly Report*, 57:343–344.

Centers for Disease Control. 2008. Measles — United States, January 1–April 25, 2008. *Morbidity and Mortality Weekly Report*, 57 (May 1, 2008 Early Release):1–4.

Cinatl, J. Jr, Michaelis, M., Doerr, H. W. 2007. The threat of avian influenza A (H5N1). Part I: epidemiologic concerns and virulence determinants. *Medical Microbiology, Immunology*, 196:181–190.

Diseases of Environmental and Zoonotic Origin Team, ECDC. 2007. Chikungunya in Italy: Actions in and implications for the European Union. *Eurosurveillance Weekly*, 12:E070906.2.

Editorial. 2007. *Helicobacter pylori*: Primary antimicrobial resistance and first-line treatment strategies. *Euro Surveillance Weekl*, 12(7) [Epub ahead of print].

Elderd, B. D., Dukic, V. M., Dwyer, G. 2006. Uncertainty in predictions of disease spread and public health responses to bioterrorism and emerging diseases. *Proceedings of the National Academy of Sciences*, 17;103:15693–15697.

Elliman, D. A., Bedford, H. E. 2007. MMR: Where are we now? *Archives of Diseases of Children*. 2007 Jul 11 [Epub ahead of print].

Gayer, M., Legros, D., Formenty, P., Connolly, M. A. 2007. Conflict and emerging infectious diseases. *Emerging Infectious Diseases*, http://www.cdc.gov/EID/content/13/11/1625.htm [accessed October 3, 2007]

Girou, E., et al. 2002. Efficacy of handrubbing with alcohol based solution versus standard handwashing with antiseptic soap: Randomised clinical trial. *British Medical Journal*, 325:36.

Goodman, R. A., Foster, K. L., Trowbridge, F. L., and Figuero, J. P. (eds.). 1998. Global disease elimination and eradication as public health strategies: Proceedings of a conference held in Atlanta, Georgia, USA, 23–25 February 1998. *Bulletin of the World Health Organization*, 76(Suppl 2):1–161.

Halstead, S. B. 1992. The 20th century pandemic: Need for surveillance and research. *World Health Statistics Quarterly*, 45:292–298.

Hargreaves, S. 2007. Infectious disease surveillance update. *Lancet Infectious Diseases*, (07)70154–8.

Hill, D. R., et al. 2006. Oral cholera vaccines: Use in clinical practice. *Lancet Infectious Diseases*, 6:361–673.

Humphreys, H. 2007. Control and prevention of healthcare-associated tuberculosis: The role of respiratory isolation and personal respiratory protection. *Journal of Hospital Infection*, 66:1–5.

Influenza Team (ECDC). 2007. Human influenza A/H5N1 ("pre-pandemic") vaccines: Informing policy development in Europe. *Eurosurveillance Weekly*, 12:E070920.3.

Klevens, R. M., et al. 2007. Estimating health care-associated infections and deaths in U.S. hospitals, 2002. *Public Health Reports*, 122:160–166.

Ledrans, M., Quatresous, I., Renault, P., Pierre, V. Outbreak of chikungunya in the French Territories, 2007: Lessons learned. *Eurosurveillance Weekly*, 12:E070906.3. http://www.eurosurveillance.org/ew/2007/070906.asp#3 [accessed October 3, 2007]

Lee, G. M., Santoli, J. M., Hannan, C., Messonnier, M. L., Sabin, J. E., Rusinak, D., Gay, C., Lett, S. M., Lieu, T. A. 2007. Gaps in vaccine financing for underinsured children in the United States. *Journal of the American Medical Association*, 298:680–682.

Levine, O. S., et al. 2006. Pneumococcal vaccination in developing countries. *Lancet*, 367:1880–1882.

Lombard, M., et al. 2007. A brief history of vaccines and vaccination. *Review of Science and Technology*, 26:29–48.

Lopalco, P. L. 2007. Measuring the impact of PCV7 in the European Union: Why it is a priority. *Eurosurveillance Weekly*, 12:E070614.6.

Malfertheiner, P., Megraud, F., O'Morain, C., Bazzoli, F., El-Omar, E., Graham, D., et al. 2007. Current concepts in the management of *Helicobacter pylori* infection: The Maastricht III Consensus Report. *Gut*, 56:772–781.

Matsumoto, C., et al. 2000. Pandemic spread of an O3:K6 clone of *Vibrio parahaemolyticus* and emergence of related strains evidenced by arbitrarily primed PCR and toxRS sequence analyses. *Journal of Clinical Microbiology*, 38:578–585.

Meya, D. B., McAdam, K. P. 2006. The TB pandemic: An old problem seeking new solutions. *Journal of Internal Medicine*, 261:309–329.

Mulholland, E. K. 2006. Measles in the United States. *New England Journal of Medicine*, 355:440–443.

Muto, C. A., et al. 2003. SHEA guideline for preventing nosocomial transmission of multidrug-resistant strains of *Staphylococcus aureus* and enterococcus. *Infection Control and Hospital Epidemiology*, 24:362–386.

Peiris, J. S., de Jong, M. D., Guan, Y. 2007. Avian influenza virus (H5N1): A threat to human health. *Clinical Microbiological Review*, 20:243–267.

Peterson, Z. A., Kremer, M. 2007. What works in fighting diarrheal diseases in developing countries? A critical review. *Working Papers, Center for International Development at Harvard University*, Boston, MA. http://www.cid.harvard.edu/cidwp/pdf/140.pdf [accessed October 3, 2007]

Pittet, D. 2005. Considerations for a WHO European strategy on healthcare–associated infection, surveillance, and control. *Lancet Infectious Diseases*, 5:242–250.

Raviglione, M. C., Snider, D. E., Kochi, A. 1995. Global epidemiology of tuberculosis: Morbidity and mortality of a worldwide epidemic. *Journal of the American Medical Asociation*, 273:220–226.

Sinha, A., et al. 2007. Cost-effectiveness of pneumococcal conjugate vaccination in the prevention of child mortality: An international economic analysis. *Lancet*, 369:389–396.

Slutsker, L., et al. 1997. *Escherichia coli* O157:H7 diarrhoea in the United States: Clinical and epidemiologic features. *Annals of Internal Medicine*, 126:505–513.

Smith, K. F., Sax, D. F., Gaines, S. D., Guernier, V., Guégan, J. F. 2007. Globalization of human infectious disease. *Ecology*, 2007 Aug; 88(8), 1903–1910.

Sonnenberg, P., Crowcroft, N. S., White, J. M., Ramsay, M. E. 2007. The contribution of single antigen measles, mumps and rubella vaccines to immunity to these infections in England and Wales. *Archives of Diseases of Children*, 92:786–789.

Spanaki, A., Hajiioannou, J., Varkarakis, G., Antonakis, T., Kyrmizakis, D. E. 2007. Mumps epidemic among young British citizens on the island of Crete. *Infection*, 35(2):104–106.

Stern, A. M., Markel, H. 2005. The history of vaccines and immunization: Familiar patterns, new challenges. *Health Affairs*, 24:611–621.

Stewart-Freedman, B., Kovalsky, N. 2007. An ongoing outbreak of measles linked to the United Kingdom in an ultra-orthodox Jewish community in Israel. *Eurosurveillance Weekly*, 12:E070920.1. http://www.eurosurveillance.org/ew/2007/070920.asp#1 [accessed September 28, 2007]

Suerbaum, S., Michetti, P. 2002. *Helicobacter pylori* infection. *New England Journal of Medicine*, 347:1175–1186.

Sullivan, T. 2004. *Helicobacter pylori* and the prevention of gastric cancer. *Canadian Journal of Gastroenterology*, 18:295–302.

Toungoussova, O. S., et al. 2006. Epidemic of tuberculosis in the former Soviet Union: Social and biological reasons. *Tuberculosis (Edinb.)*, 86(1):1–10.

Tulchinsky, T. H., et al. 1989. A ten-year experience in control of poliomyelitis through a combination of live and killed vaccines in two developing areas. *American Journal of Public Health*, 79:1648–1652.

Tulchinsky, T. H., et al. 1993. Measles control in developing and developed countries: The case for a two-dose policy. *Bulletin of the World Health Organization*, 71:93–103.

UNAIDS. 2006. *Report on the global AIDS epidemic: Executive summary*, Joint United Nations Proramme on HIV/AIDS, Geneva, Switzerland, 28 pp.

UNICEF. 2006. *The State of the World's Children*. New York: United Nations Children's Fund, Oxford Press.

U.S. Food and Drug Administration. 2007. FDA approves first U.S. vaccine for humans against the avian influenza virus H5N1. *FDA News*, 07(68):2 pp.

Watson, R. 2007. Chikungunya fever is transmitted locally in Europe for first time. *British Medical Journal*, 335:532–533.

Wells, C. D., et al. 2007. HIV infection and multidrug-resistant tuberculosis: The perfect storm. *Journal of Infectious Diseases*, 15(196): S86–107.

Wise, J. 2006. Demand for circumcision rises in a bid to prevent HIV. *Bulletin of the World Health Organization*, 84:509–511.

Wisner, B., Adams, J. 2002. *Environmental Health in Disasters and Emergencies: A Practical Guide*. Geneva: World Health Organization.

Wolfson, L. J., Strebel, P. M., Gacic-Dobo, M., Hoekstra, E. J., McFarland, J. W., Hersh, B. S. 2007. Has the 2005 measles mortality reduction goal been achieved? A natural history modelling study. *Lancet*, 20:369:165–166.

World Health Organization. 1990. *The Rational Use of Drugs in the Management of Acute Diarrhoea in Children*. Geneva: WHO.

World Health Orgaization. 1992. Update International Task Force for Disease Eradication 1991. *Morbidity and Mortality Weekly Report*, 41:40–42.

World Health Organization. 1996. Dracunculiasis: Global surveillance summary. *Weekly Epidemiological Record*, 71:141–148.

World Health Organization. 1996. Progress toward the elimination of leprosy as a public health problem. *Weekly Epidemiological Record*, 71:149–156.

World Health Organization. 1998. *Health for All in the Twenty-first Century*. EB101/8. Geneva: WHO.

World Health Organization. 1999. Integration of vitamin A supplementation with immunization. *Weekly Epidemiological Record*, 74:1–6.

World Health Organization. 1999. Rotavirus vaccines: WHO position paper. *Weekly Epidemiological Record*, 74:33–38.

World Health Organization. 2001. Global prevalence and incidence of selected curable sexually transmitted diseases: Overview and estimates. Geneva: World Health Organization.

World Health Organization. 2003. Treatment of tuberculosis: Guidelines for national programs. Geneva: World Health Organization; 2003 (WHO/CDS/TB/2003.313).

World Health Organization. 2005. Roll Back Malaria Programme. Global Strategic Plan Roll Back Malaria 2005-2015. Geneva, WHO. http://www.rollbackmalaria.org/forumV/docs/gsp_en.pdf [accessed May 4, 2008]

World Health Organization. 2006. Dracunculiasis eradication: Ministerial meeting 5 May 2006. *Weekly Epidemiological Record*, 24:239–224.

World Health Organization. 2007. Global tuberculosis control: Surveillance, planning, financing. WHO Report 2007. Geneva: World Health Organization. WHO/HNM/TB/2007.376.

World Health Organization. 2007. Meeting of the immunization strategic advisory group of experts, November 2006 — conclusions and recommendations. *Weekly Epidemiological Record*, 82:1–16.

World Health Organization. 2007. Meeting of the International Task Force for Disease Eradication, 11 January 2007. *Weekly Epidemiological Record*, 82:197–205.

World Health Organization. 2007. Pneumococcal vaccine for childhood — A position paper. *Weekly Epidemiological Record*, 82:93–104.

World Health Organization. 2007. Progress towards interrupting wild poliovirus transmission, January 2006–May 2007. *Weekly Epidemiological Record*, 82:245–260.

World Health Organization. 2007. Progress towards the 2005 international targets for tuberculosis control. *Weekly Epidemiological Record*, 82:169–180.

World Health Organization. 2007. Stop TB Partnership has provided treatment for 10 million people in 6 years. *Weekly Epidemiological Record*, 82:206–207.

World Health Organization. 2007. Dracunculiasis eradication: Certification of interruption of transmission. *Weekly Epidemiologic Record*, 82:161–163.

Xing, Z., Carters, T. J. 2007. Heterologous boost vaccines for bacillus Calmette-Guerin prime immunization against tuberculosis. *Expert Review of Vaccines*, 6:539–546.

Zuckerman, J. N., et al. 2007. The true burden and risk of cholera: Implications for prevention and control. *Lancet Infectious Diseases*, 7:521–530.

Chapter 5

Noncommunicable Conditions

Introduction
The Rise of Chronic Disease
The Burden of Chronic Conditions
Risk Factors and Causation of Chronic Conditions
Chronic Manifestations of Infectious Diseases
    Cardiovascular Diseases
Chronic Lung Disease
    Asthma
    Chronic Obstructive Pulmonary Disease
    Restrictive Lung Diseases
    Occupational Lung Diseases
Diabetes Mellitus
    Prevention of Diabetes and Its Complications
End-Stage Renal Disease
Cancer
    Prevention of Cancer
Chronic Liver Disease
Disabling Conditions
    Arthritis and Musculoskeletal Disorders
    Osteoporosis
    Degenerative Osteoarthritis
    Rheumatoid Arthritis and Gout
    Low Back Syndromes
Neurologic Disorders
    Alzheimer's Disease
    Parkinson's Disease
    Multiple Sclerosis
    Epilepsy or Seizures
    Brain and Spinal Cord Injury
Visual Disorders
Hearing Disorders
Trauma, Violence, and Injury
    Motor Vehicle Accidents
    Domestic Violence
    Suicide and Suicide Attempts
    Homicide
    Prevention of Violence
Chronic Conditions and the New Public Health
Summary
Electronic Resources

Recommended Readings
Bibliography

INTRODUCTION

Disease and health conditions are classified to ease efforts to monitor, control, prevent, and treat illness. The distinction between communicable and noncommunicable diseases changes over time with growing knowledge of causation and risk factors, as discussed in previous chapters. Communicable diseases are those caused by specific pathogens and may be transmitted from an infected to an uninfected host, but this process is influenced by the physical, social, and economic environment. Noncommunicable diseases are mostly due to degenerative, genetic, hereditary, and environmental conditions and life habits such as nutrition. Some are caused by infectious diseases, such as AIDS by HIV infection, peptic ulcers being due to *Helicobacter pylori*. Chronic conditions may have a multiple factor model of causation, especially with interaction of genetic tendencies and socioeconomic and behavioral factors.

During the twentieth century, the "diseases of modern life" or noninfectious conditions became the leading causes of morbidity and mortality in developed countries. This epidemiologic transition is occurring in many developing countries as well. Causation in noninfectious (chronic) disease is complex, and prevention must take into account multiple contributory or risk factors. Despite the complexity, and often for reasons not well understood, dramatic success has been achieved in reducing stroke and heart disease death rates in many countries over the past 20 years. Cancer and trauma death rates, key elements of noninfectious disease patterns, have proven more difficult to reduce.

The 15 leading causes of death in 2005 in the United States were the following:

1. Diseases of the heart
2. Malignant neoplasms
3. Cerebrovascular diseases (stroke)

4. Chronic lower respiratory diseases
5. Accidents (unintentional injuries)
6. Diabetes mellitus
7. Alzheimer's disease
8. Influenza and pneumonia
9. Nephritis, nephrotic syndrome, and nephrosis (kidney disease)
10. Septicemia
11. Intentional self-harm (suicide)
12. Chronic liver disease and cirrhosis
13. Essential (primary) hypertension and hypertensive renal disease (hypertension)
14. Parkinson's disease
15. Assault (homicide)

Source: Kung, H. C., Hoyert, D. L., Xu, J., Murphy, S. L. Deaths: Preliminary data for 2005. Health E-Stats. September 2007. http://www.cdc.gov/nchs/products/pubs/pubd/hestats/prelim-deaths05/prelimdeaths05.htm [accessed May 4, 2008]

Chronic conditions are not only the major causes of death in western countries, and increasingly so in the developing countries, but they also place very great demands on health systems. A major global health crisis exists due to rising levels of cardiovascular disease (CVD) and diabetes in developing countries, affecting the working population with serious economic consequences.

Cardiovascular diseases include coronary heart disease; stroke and renal failure are the leading causes of death and disability in the world. Heart disease and stroke kill 17 million people every year, compared to 3 million due to HIV and AIDS, with 80 percent of these deaths occurring in low- and middle-income countries.

A WHO/CDC joint Atlas of Heart Disease and Stroke (2004) attributes "many of these deaths to tobacco smoking, which increases the risk of dying from coronary heart disease and cerebrovascular disease 2–3 fold. Physical inactivity and unhealthy diet are other main risk factors which increase individual risks to cardiovascular diseases. One of the strategies to respond to the challenges to population health and well being due to the global epidemic of heart attack and stroke is to provide actionable information for development and implementation of appropriate policies."

At the same time, there is great potential for health promotion and primary, secondary, and tertiary prevention to reduce mortality and disability as well as the burden of disease and disability. The New Public Health stresses all aspects of prevention and care and is therefore increasingly required as a broad approach to this wide group of conditions and the needs associated with them.

THE RISE OF CHRONIC DISEASE

The change from the predominance of infectious diseases to noninfectious diseases as the leading causes of premature mortality has occurred primarily since the end of World War II. Antibiotics and vaccines, along with improved living standards, sanitation, nutrition, and safe water, brought about a reduction in mortality rates from infectious diseases and an increase in life expectancy. Infectious diseases, while still important, are no longer the primary concerns in public health in the developed countries, and similar trends are appearing in the developing countries.

The dramatic shift in causes of death from a predominance of infectious to the noninfectious diseases is seen in Table 5.1, which lists the leading causes of death in the United States from 1950 to 2002. This reflects similar changes which occurred in all the developed countries over the past half century.

Chronic diseases as the leading cause of morbidity and mortality are associated with a number of demographic and epidemiologic factors. First, the decline in infectious disease mortality resulted in greater longevity, increasing the numbers of persons surviving to ages when cancer and heart disease are common. Second, changes in lifestyle such as smoking, lack of exercise, diets rich in unhealthy fats and sugars, and risk-taking behavior, increased risk factors so that cardiovascular diseases and cancer became the leading causes of disease, disability, and death. Third, trauma and chronic diseases are major contributors to rising costs of health care and the economics of health care. Fourth, public health experience and new scientific knowledge are leading to new forms of prevention and medical treatment that are reducing the burden of disease and disability from chronic conditions.

Identification of many risk factors and methods of early detection have increased the potential for interventions to lower the prevalence of these diseases and their complications. As a result of these changes, the scope of public health has broadened. In this chapter we examine the major chronic diseases and their effects on the health of the population. We also look at the risk factors that contribute to them and at the interventions needed to reduce their prevalence in the population.

Causes of important chronic conditions are being identified with interventions and treatment that will further alter disease patterns in the years ahead. In addition, prevention methods, including the secondary and tertiary prevention carried out by health providers, are playing an important and measurable role in reducing the burden of disease, as is health promotion.

As expressed in a series of papers in *Lancet* in 2007, a global program goal for prevention and control of chronic diseases could avert 36 million deaths by 2015 with major economic benefits, using existing interventions. The evidence suggests this goal is possible and realistic with interventions directed towards whole populations and individuals at high risk. "The total yearly cost of the interventions in 23 low-income and middle-income countries is about US$5.8 billion" (as of 2005). This calls for "a serious and sustained worldwide effort to prevent and control

TABLE 5.1 Leading Causes of Death (Standardized Rates per 100,000 Population), United States, 1950–2004

	1950	1960	1970	1980	1990	2000	2004
All causes	1446	1339	1223	1039	939	869	801
Diseases of heart	587	559	493	412	321	258	217
Ischemic heart disease	—	—	—	345	250	187	150
Cerebrovascular diseases	181	178	148	96	65	54	50
Malignant neoplasms	194	194	199	208	216	200	186
Lung	15	24	37	50	59	56	53
Colon, rectum, anus	—	30	29	27	25	21	18
Breast	32	32	32	32	33	27	24
Prostate	29	29	29	31	38	30	25
Chronic lower respiratory diseases	—	—	—	28	37	44	41
Influenza and pneumonia	48	54	42	31	37	24	23
Diabetes mellitus	23	23	24	18	21	25	25
Chronic liver disease and cirrhosis	11	13	18	15	11	10	9
Unintentional injuries	78	62	60	46	36	35	38
Motor vehicle accident	25	23	28	22	19	15	15
Suicide	13	13	13	12	13	10	11
Homicide	5	5	9	10	9	6	6
HIV	—	—	—	—	10	5	5

Note: Data from death certificates; numbers rounded.
Source: *Health United States*. 2007. Table 029, http://www.cdc.gov/nchs/hus.htm [accessed May 4, 2008]

chronic diseases in the context of a general strengthening of health systems. Urgent action is needed by WHO, the World Bank, regional banks and development agencies, foundations, national governments, civil society, non-governmental organisations, the private sector including the pharmaceutical industry, and academics. We have established the Chronic Disease Action Group to encourage, support, and monitor action on the implementation of evidence-based efforts to promote global, regional, and national action to prevent and control chronic diseases" (Beaglehole et al., 2007).

THE BURDEN OF CHRONIC CONDITIONS

Chronic conditions place a heavy burden on the individual, the family, and society as a whole in terms of morbidity and mortality. Measurement of the burden of disease (discussed in Chapter 3) is a fundamental responsibility of public health agencies. Health expenditures as a percent of GDP (see Table 5.2) varies widely. The economic burden for health care of the population, whether funded by national responsibility as in most industrialized countries (see Chapters 10–13) or a mix of public and private

TABLE 5.2 Health Expenditures as Percent of GDP, Selected Countries, 1960–2004

Country	1960	1970	1980	1990	2000	2004
United States	5.1	7.0	8.8	11.9	13.3	15.3
Germany	NA	6.2	8.7	8.5	10.3	10.6
France	3.8	5.3	7.0	8.4	9.2	10.5
Canada	5.4	7.0	7.1	9.0	8.9	9.9
United Kingdom	3.9	4.5	5.6	6.0	7.3	8.1
Spain	1.5	3.5	5.3	6.5	7.2	8.1
Finland	3.8	5.6	6.3	7.8	6.7	7.5
Japan	3.0	4.5	6.5	5.9	7.6	8.0*
Mexico	NA	NA	NA	4.8	5.6	6.5
Poland	NA	NA	NA	4.9	5.7	6.5

Note: *2003 data.
Source: *Health United States*. 2007. Table 120, http://www.cdc.gov/nchs/data/huS/has/07pdf [accessed May 4, 2008]

expenditures, is an important factor in national economies. As chronic disease and aging of the population both increase, so will the burden of disease become increasingly important in economic as well as in health terms.

The cost of individual disease groups can be enormous. The American Heart Association reports the cost of cardiovascular diseases in the United States in 2006 at an estimated $403.1 billion, including direct costs of health services including physicians and other professionals, hospital and nursing home services, medications, home health care and other medical durables, as well as lost productivity resulting from morbidity and mortality (indirect costs). As a comparison, in 2004 the estimated cost of all cancers was $190 billion ($69 billion in direct costs, $17 billion in morbidity indirect costs, and $104 billion in indirect costs associated with premature mortality).

What is more difficult to quantify is the burden of disease on the individual, the family, and the community. Traditional measures of morbidity and mortality are supplemented by quality-adjusted life years (QALYs) and disability-adjusted life years (DALYs), as seen in Chapter 3, but the physical and emotional burden of caring for someone is almost impossible to quantify. The burden of chronic disease on the individual is reflected in his or her ability to function in the normal activities of daily life (ADL). The level of function of a person with a chronic condition is measured by ability to perform ADL, as seen in Chapter 6.

Chronic conditions often result in disabilities which impede optimal function in normal daily functions or activities. ADL measures the degree of independent capacity the patient has regarding personal care, household management, and socializing. They help determine the level and amount of home care needed or the type of facility the patient may need. While ADL measures the function of a patient, it does not address the emotional, physical, and financial stress on the caregiver in a family. Self perception of health status is important and is part of periodic surveys conducted by the U.S. National Center for Health Statistics Summary Health Statistics for the U.S. Population: National Health Interview Survey, 2006 Series.

RISK FACTORS AND CAUSATION OF CHRONIC CONDITIONS

The criteria for causation in infectious disease, discussed in Chapter 1, are known as the Koch-Henle postulates. The criteria for causation in chronic disease, called the Evans Criteria, are outlined in Box 5.1. They articulate the relationship of predisposing or risk factors to the causation of chronic disease, and are important in analyzing the relative contribution of factors to disease in a population.

Box 5.1 Criteria for Causation in Chronic Disease — Evans' Postulates

1. Prevalence of the disease should be significantly higher in those exposed to the hypothesized cause than in controls not so exposed.
2. Exposure to the hypothesized cause should be more frequent among those with the disease than in controls without the disease, when all other risk factors are held constant.
3. Incidence of the disease should be significantly higher in those exposed to the hypothesized cause than in controls not so exposed, as shown by prospective studies.
4. The disease should follow exposure to the hypothesized causative agent with a normal or log-normal distribution of incubation periods.
5. A spectrum of host responses should follow exposure to the hypothesized agent along a logical biological gradient from mild to severe.
6. A measurable host response following exposure to the hypothesized cause should have a high probability of appearing in those lacking this before exposure (e.g., antibody, cancer cell) or should increase in magnitude if present before exposure. This response pattern should occur infrequently in persons not so exposed.
7. Experimental reproduction of the disease should occur more frequently in animals or humans appropriately exposed to the hypothesized cause than in those not so exposed; this exposure may be deliberate in volunteers, experimentally induced in the laboratory, or may represent a regulation of a natural exposure.
8. Elimination or modification of the hypothesized cause should decrease the incidence of the disease (e.g., attenuation of a virus, removal of tar from cigarettes).
9. Prevention or modification of the host's response on exposure to the hypothesized cause should decrease or eliminate the disease (e.g., immunization, drugs to lower cholesterol, specific lymphocyte transfer factor in cancer).
10. All of the relationships and findings should make biological and epidemiologic sense.

Sources: Evans, A. S. 1976. Causation and disease: The Henle-Koch postulates revisited. *Yale Journal of Biology and Medicine*, 49:175–195.
Last, J. M. 2001. A *Dictionary of Epidemiology*, Fourth Edition. New York: Oxford University Press.

Disease causation may be complex with many contributing factors. Rarely is a single necessary factor sufficient by itself to produce the disease. Risk factors for chronic diseases are summarized in Table 5.3. The germ theory proved itself to be an effective basis for medical science and public health practice. At the same time, the environmental factors of the physical, social, and cultural environment, as discussed in Chapter 1 as a modern version of the miasma theory, are important for understanding modern public health with risk factors and social inequalities.

TABLE 5.3 Chronic Disease Risk Factors

Risk factor	CVD	Cancer	Chronic lung disease	Diabetes	Cirrhosis	Musculoskeletal disease	Neurologic disorders
Tobacco use	+	+	+	0	0	+	?
Alcohol use	+	+	0	0	+	+	+
High cholesterol	+	0	0	0	0	0	0
Hypertension	+	0	0	0	0	0	0
Diet	+	+	0	+	0	+	?
Physical inactivity	+	+	0	+	0	+	0
Obesity	+	+	0	+	0	+	0
Stress	?	?	0	0	0	0	0
Environmental tobacco smoke	?	+	+	0	0	0	0
Occupation	?	+	+	0	?	+	?
Air pollution	+	+	+	0	0	0	+
Low socioeconomic status	+	+	+	+	+	+	0

Source: Brownson, R. C., Remington, P. L., Davis, J. R. 1998. *Chronic Disease Epidemiology and Control*, Second Edition. Washington, DC: American Public Health Association.

Risk factors in noncommunicable diseases have been identified by important epidemiologic studies over the past four decades. The classic epidemiologic work of Jeremy Morris showed the difference in risk for coronary heart disease (CHD) between London bus drivers as compared to bus conductors. The conductors, who are both constantly required to climb bus stairs physically and also experience less stress as part of their job, had lower rates of CHD mortality than the sedentary, high-tension drivers. This was later followed by equally classic work of the Whitehall studies of British civil servants by Michael Marmot and others showing marked differences in the morbidity and mortality patterns of those with higher status, income, and control over their own lives, and the protective effect of physical exercise.

Smoking was identified as a risk factor in U.S. studies by Ernst Wynder and others in the 1940s and 1950s. In a longitudinal study of British physicians by Richard Doll and Bradford Hill in the 1950s, 35-year-old male cigarette smokers were shown to have less chance of surviving to age 65 (73 percent) as compared to nonsmokers (85 percent) and ex-smokers (81 percent). The U.S. Surgeon General's Report on Smoking and Health of 1964 summarized the hundreds of published studies up to that time and concluded that cigarette smoking was a major health hazard and a cause for lung cancer, coronary heart disease, chronic pulmonary lung disease, and stroke. Subsequent Surgeon General's Reports in 1983 and 1984 attributed 30 percent of

coronary heart disease deaths and 80–90 percent of chronic obstructive lung disease deaths to smoking. Smoking reduction has become one of the pillars of modern public health. Smoking declined in the U.S. among men from 28 percent in 1990 to 23 percent in 2005, for women from 23 percent to 18 percent, and among teenagers from 28 percent to 23 percent (Health United States, 2007).

The famous longitudinal heart study in Framingham, Massachusetts (1948), provided important epidemiologic data showing that hypertension, smoking, and elevated cholesterol are associated with increased risk of cardiovascular diseases (Box 5.2). The Framingham study pioneered the epidemiologic approach to gain insight into causes of cardiovascular diseases. This was a prospective cohort study to quantify risks for these diseases, both in terms of absolute and relative risk. Its observations have led to causal inferences (e.g., elevated blood pressure with increased risk of stroke).

Subsequent studies have elaborated on the Framingham findings. High blood pressure (hypertension) refers to elevated levels of systolic and/or diastolic blood pressure and is associated with increased risk of morbidity and mortality from myocardial infarction, stroke, and renal disease. Concepts of normality have been replaced with guidelines for "optimal" values of blood lipid and blood pressure for long-term freedom from cardiovascular disease. Atherogenic potential for serum total cholesterol was shown to be derived from the

Box 5.2 Timeline of Milestones from the Framingham Heart Study

1948 Start of the Framingham Heart Study

1956 Findings on progression of rheumatic heart disease reported

1960 Cigarette smoking found to increase the risk of heart disease

1961 Cholesterol level, blood pressure, and electro-cardiogram abnormalities found to increase the risk of heart disease

1965 First Framingham Heart Study report on stroke

1967 Physical activity found to reduce the risk of heart disease, and obesity to increase the risk of heart disease

1970 High blood pressure found to increase the risk of stroke

1974 Diabetes and its complications associated with development of cardiovascular disease

1976 Menopause found to increase the risk of heart disease

1977 Effects of triglycerides and LDL and HDL cholesterol described

1978 Psychosocial factors found to affect heart disease, atrial fibrillation (heart beats irregularly) found to increase the risk of stroke

1981 Major report issued on relationship of diet and heart disease; filter cigarettes found to give no protection against coronary heart disease

1986 First report on dementia related to vascular diseases

1987 High blood cholesterol levels found to correlate directly with risk of death in young men, fibrinogen increases the risk of heart disease; estrogen replacement therapy found to reduce risk of hip fractures in postmenopausal women

1988 High HDL cholesterol found to reduce risk of death; association of type "A" behavior with heart disease reported; isolated systolic hypertension found to increase risk of heart disease; cigarette smoking found to increase risk of stroke

1990 Homocysteine (an amino acid) found to be possible risk factor for heart disease

1993 Mild isolated systolic hypertension shown to increase risk of heart disease. Major report predicts survival after diagnosis of heart failure

1994 Enlarged left ventricle (one of two lower chambers of the heart) shown to increase the risk of stroke; lipoprotein A found as possible risk factor for heart disease; risk factors for atrial fibrillation described; apolipoprotein E found to be possible risk factor for heart disease; first Framingham report on diastolic heart failure published

1995 OMNI Study of Minorities starts (applicability of Framingham findings to many ethnic groups)

1996 Progression from hypertension to heart failure described

1997 Cumulative effects of smoking and high cholesterol on the risk for atherosclerosis reported; impact of enlarged left ventricle and risk for heart failure in asymptomatic individuals investigated

1998 New risk prediction formulas calculate a patient's risk for developing coronary disease over the next 10 years; work identifying a gene (angiotensin-converting enzyme deletion/insertion polymorphism) associated with hypertension in Framingham men published

2002 NEJM report links between body mass index (BMI)/obesity with an increased risk of heart failure; study shows BMI (body weight/mass) index to be an independent risk factor; third-generation study enrolls 3900 grandchildren of the Framingham heart study's original enrollees

2003 Offspring-based study published relating likelihood of heart attack three times greater in individuals with common genetic variation in an estrogen receptor

2004 Demonstration that having a parent with a cardiovascular disease history doubles personal risk of the disease

2005 Offspring study reports an increase of up to 45 percent for risk of heart attack, stroke, or arterial disease may occur in middle-aged people with a sibling who suffered a similar cardiovascular event

Source: The Framingham Heart Study, http://www.framingham.com/heart/timeline.htm [accessed May 2, 2008]

low-density lipoprotein cholesterol (LDL-cholesterol) fraction which is positively related to CHD incidence. High-density lipoprotein cholesterol (HDL-cholesterol) is inversely related to CHD, since it removes cholesterol from tissues. Risk of CHD is independently related to each of these lipoprotein fractions. Therefore, the ratio of total to HDL-cholesterol is an efficient lipid risk profile.

Cardiovascular disease is still the leading cause of death worldwide; 17 million people die of coronary and vascular diseases every year. An important portion of these cases can be attributed to behavioral factors, such as those that can often be modified, including cigarette smoking, physical inactivity, and unhealthy diet.

The global risk for cardiovascular disease includes many risk factors and is therefore difficult to assess. One of the results of research based on the Framingham study is a predictive model, valid for all populations (when adapting certain parameters). This permits not only the assessment of individual risk, but also allows addressing a population's risk and preventable fraction. Those evaluations are essential when planning and implementing health promotion programs, which are strongly needed in order to approach the global problem of cardiovascular disease worldwide.

The Alameda County study of Lester Breslow and colleagues followed a cohort of nearly 7000 people from the 1 million population) from 1965 to the present and identified seven health practices that were associated with reduced mortality and disability patterns (Box 5.3). The health practices studied were excessive alcohol intake, cigarette smoking, obesity, sleeping less or more than

---

Box 5.3 MONICA

Following the end of World War II, coronary heart disease (CHD) assumed epidemic proportions in western countries. CHD mortality began to decline in the U.S. in the early 1960s and was followed later in many western countries, after peaking in 1968. In 1978, the National Heart, Lung, and Blood Institute of the NIH organized the Bethesda conference on CHD mortality to assess whether prevention or improved acute coronary care were responsible for the decline in age-specific CHD mortality rates. There was no clear answer to these questions, but WHO took the leadership and organized the MONICA project (Multinational Monitoring of Trends and Determinants in Cardiovascular Disease) as an epidemiologic investigation system to assess trends and determinants of cardiovascular mortality, incidence, and case fatality from the mid-1980s to the mid-1990s. The study included 38 population centers in 21 countries worldwide.

Altogether some 13 million people were monitored over a 10-year period, 166,000 myocardial infarction patients were registered, and more than 300,000 men and women were sampled and examined for cardiovascular risk factors and many other health data. In western countries, where the CHD mortality declined on average 2–3 percent annually, two-thirds of this decline could be explained by a decline in CHD incidence and one-third by a decline in CHD case fatality. When the trends were examined in relation to changes in risk factors and CHD event rates in men over a period of 10 years in all MONICA populations, the greatest contribution to the observed decline was decreased smoking.

The MONICA project was in company with many similar studies worldwide, the Seven Countries Study including Finland (high), U.S., Netherlands, Italy, Yugoslavia, Greece, Japan (low) and the Framingham Heart Study, contributing to the development of chronic disease epidemiology and to prevention of cardiovascular diseases.

Sources: Turnstall-Pedoe, H. 2003. Background, Development and Organization of MONICA. Geneva: WHO, http://whqlibdoc.who.int/publications/2003/9241562234.pdf [accessed December 2007 and Web site confirmed May 2, 2008]. Keil, U. 2005. The Worldwide WHO MONICA Project: Results and perspectives, Gesundheitswesen. 2005 Aug; 67 Suppl 1:S38–45 (in German).

7–8 hours/day, physical inactivity, eating between meals, and not eating breakfast. Adjusting for age, sex, health status, and social networks, the occurrence of disability was half as great for those with good health practices as compared to poor practices. These patterns were prevalent in the cohort that followed in the 1960s and 1970s. The Alameda county studies continue to report socioeconomic, morbidity, and mortality gradients as well as the effectiveness of lifestyle changes primarily in reducing the complications of chronic diseases and in prolonging life.

CHRONIC MANIFESTATIONS OF INFECTIOUS DISEASES

Infections as causes of chronic diseases are of great importance for public health because such associations can lead to new treatments or preventive measures. Some of these associations are well established; others are reported but as yet are insufficiently cornered. Once confirmed, the search for vaccines could replicate the success of immunization in control of the acute infectious diseases of childhood.

The number of established and proposed relationships between certain organisms and chronic diseases is growing. The relationship of hepatitis B with chronic hepatitis, cirrhosis, and hepatic carcinoma provides the justification for wide-scale immunization to protect individuals, especially those in developing countries who are especially at risk for hepatitis B infection. Hepatitis C also causes cirrhosis and cancer of the liver, and while there is still no vaccine, there are treatments which can eliminate the chronic infection in more than half of cases, depending on the subtype of the virus. Human papillomaviruses (HPVs) are associated with cervical carcinoma. Screening for cancer of the cervix is an important public health modality, but education in hygienic practices and the control of sexually transmitted infections can help to reduce the spread of these organisms. An HPV vaccine based on genetic engineering technology is now being used for prevention of cancer of the cervix with the ultimate potential of eradicating this disease in conjunction with Pap smear programs. Varicella virus is associated with herpes zoster and post-herpetic neuralgia. The varicella vaccine now recommended for routine childhood immunization may in time eliminate this problem, although adult immunization for this virus will be an issue for prevention of this very painful and discomfiting burden for older persons.

Bacterial infections can also have long-term sequelae. Examples include peptic ulcer disease and chronic gastritis from the bacterium *Helicobacter pylori* (Box 5.4). In 1991, four case–control studies quantified and confirmed that *H. pylori* is associated with stomach cancer. This included demonstration of antibodies and growth of the organism before and after diagnosis of cancer. One study showed the relation of severity of infection and risk of stomach cancer. Currently, the association of *H. pylori* infection with cancer of the stomach and peptic ulcer and its complications is accepted, with clear evidence of dramatic reduction of cancer of the stomach, and in surgical needs for peptic ulcer treatment in recent decades.

Box 5.4 *Helicobacter Pylori*, Peptic Ulcers, and Stomach Cancer

Mortality in the United States from duodenal ulcer declined from 4/100,000 in 1950 to just over 1 in 1980. Gastric ulcer mortality remained constant at 1 per 100,000. Hospitalizations for duodenal ulcer for males declined from 28 to 8 per 10,000 population from 1965 to 1982, but still some 5000 deaths occurred per year (2 per 100,000) from bleeding, perforation, or obstruction from peptic ulcers. In 1975, costs associated with this group of diseases were estimated at $1.3–$2.6 billion. The introduction of powerful hydrogen ion antagonists in 1977 contributed to further decline in hospitalization rates from peptic ulcers. Many factors were considered important in causation for this disease, including aspirin, alcohol, smoking, coffee, stress, occupation, and genetic tendencies.

Helicobacter (Campylobacter) pylori, first reported in *Lancet* in 1983 as a curved bacterium associated with gastritis, is present in more than half the world's population. Once acquired, it persists and can cause disease decades after infection, acting as a "microbial parasite." The organism was shown in the pylorus of the gastric aspirates of 58 of 100 patients with peptic ulcer by Australian physicians Barry Marshall and J. R. Warren in 1984. Marshall experimentally ingested the organism himself and developed nausea, stomach pain, and foul breath. They suggested that the organism might be the cause of peptic ulcer and stomach cancer. Treatment of peptic ulcer, after gastroscopy, biopsy, and demonstration of the organism, followed by a short course of antibiotics is now standard treatment.

An ecological study in the United States showed high correlation between areas with high prevalence of the organism and high prevalence of cancer of the stomach. In 1998, H. *pylori* was reported to have been found in surface water, suggesting a major reservoir and method of transmission of this organism. Chlorination apparently kills the organism.

Marshall and Warren, both Australians, were awarded the Nobel Prize for physiology and medicine in 2005 for this achievement. Finding the cause and cure for peptic ulcer has radically changed the pattern of stomach cancer epidemiology and the nature of surgical practice worldwide. The relationship of the community and environmental aspects of *Helicobacter pylori* and chronic disease raises many issues related to prevention of infection by water treatment and probably in the near future vaccines to protect the population from this widespread infection.

Source: http://nobelprize.org/medicine/laureates/2005/press.html, last modified January 27, 2006 [accessed April 3, 2006]

Standard treatment of gastric and duodenal ulcers now focuses on gastroscopy and biopsy, tests for *H. pylori*, and treatment with antihistamine to reduce acid production (H₂ inhibitors) and inexpensive antibiotics, with follow-up by breath tests for the organism. This has further reduced hospitalizations and surgery for chronic peptic ulcer complications and expensive long-term drug treatment with hydrogen ion antagonists or inhibitors. *H. pylori* is accepted as the major cause of peptic ulcers, now a treatable condition, thus giving new life to the issue of infection as a cause of chronic disease.

Similarly, acute infections with STIs can lead to long-term sequelae such as sterility. Untreated syphilis can lead to long-term neurologic deterioration and valvular heart disease. These and other examples of established relationships and others that are still unproven hypotheses are shown in Table 5.4. The number of examples of infectious diseases causing chronic disease will increase with development of the biological sciences, as in the example of discovery of prions as transmitters of such diseases as scrapie in sheep, bovine spongiform encephalitis (BSE) in cattle, and Creutzfeldt-Jakob disease (CJD) in humans.

Since the 1990s, evidence of *Chlamydia pneumoniae* infection as a cause of coronary heart disease has provided fresh impetus to the search for new approaches to prevention of the leading cause of death in industrialized and many developing countries. Initial reports from Finland, the United Kingdom, and Italy were followed by supportive studies in the United States. The association has not yet been substantiated according to current criteria for causation of the Koch-Henle postulates, and more investigation is needed.

Cardiovascular Diseases

Cardiovascular disease (CVD) refers to a group of diseases of the heart and blood vessels, including coronary or ischemic heart disease, hypertension, and cerebrovascular disease (stroke). These are associated with atherosclerosis, excess fats in the diet and lipids in the body, and often with impairment of endocrine functions related to glucose metabolism and diabetes mellitus. CHD is a disease of the lining and blockage of the arteries that supply the muscles of the heart. The coronary arteries may become blocked with plaques and thromboses, thus cutting off the blood and oxygen supply (ischemia; i.e., ischemic heart disease) and leading to death (necrosis) of heart muscle, an acute myocardial infarction (AMI), also called *heart attack*. Other diseases of the heart include rheumatic diseases with damage to valves of the heart and other diseases of the heart muscle. Depending on the extent of heart muscle damage, a patient may go into congestive heart failure (CHF) due to weakened function of the heart as a pump and resulting congestion of fluids in the lungs and other tissues.

CVD is blockage or bleeding from blood vessels supplying the brain, causing death of areas of brain tissue,

TABLE 5.4 Infectious Disease Causing Chronic Diseases, Established and Hypothesized Relationships

Causative infectious disease	Outcome chronic diseases	Causal relationship
STIs	Sterility	Established
Syphilis	Neurologic and cardiac valvular diseases	Established
Streptococcal infection	Glomerulonephritis	Established
Acute rheumatic fever, streptococcal infection	Rheumatic valvular heart disease	Established
Measles	Subacute sclerosing panencephalitis (SSPE)	Established
HIV	AIDS, Kaposi's sarcoma	Established
Hepatitis B and C chronic	Cirrhosis of liver and cancer of liver	Established
Helicobacter pylori	Peptic ulcer and stomach cancer	Established
Human papilloma-virus (HPV)	Cancer of the cervix	Established
Epstein-Barr virus	Burkitt's lymphoma in Africa	Established
Schistosomiasis	Bladder cancer in developing countries	Established
Bovine spongiform encephalitis (BSE)	Creutzfeldt-Jakob disease (CJD)	Established
Chlamydia infection	Coronary heart disease and stroke	Hypothesized
Fungal infection	Polycystic kidneys	Hypothesized
Borna disease virus (BDV)	Schizophrenia	Hypothesized
Measles	Crohn's disease	Hypothesized
Hantavirus	Hypertensive renal disease	Hypothesized

otherwise known as a *cerebrovascular accident* (CVA) or *stroke*. If this is the dominant side of the brain; that is, the left side for right-handed people, then depending on the area of permanent damage after recovery, there are varying amounts of motor and mental limitations. Partial occlusion may cause transient ischemic attacks (TIAs) resulting in brief loss of motor and mental function. In both AMIs and

CVAs, the area of dead tissue will be surrounded by tissue that is inflamed and damaged. Treatment is intended to restore blood flow and minimize the inflammation or swelling (edema). Immediate care is vital to minimize damage and has a direct effect on maximizing recovery.

CVDs are the leading causes of death in the developed and in many developing countries. They have been called the diseases of "modern living" because etiologically they are due to factors such as smoking, lack of physical exercise, and a diet rich in animal fats. These are collectively known as "lifestyle" risk factors and are amenable to change by the individual at risk and through community interventions. The most important lesson since the 1950s in public health is that these risk factors and diseases can be reduced dramatically by suitable public health interventions.

Precise measurement of the prevalence of these diseases through epidemiologically sound, community-based prevalence studies is difficult because of lack of standard diagnostic criteria and terminology and limitation of resources. Measures such as mortality rates and hospitalization data may not give true prevalence rates, but they do provide important time trends and inter-area comparisons that are valid for planning public health interventions. CVDs have common pathophysiological features with diabetes, related to nutrition, exercise, and other lifestyle factors. Cerebrovascular and coronary heart diseases have important clinical differences, risk groups, as well as in screening, measurement, and interventions required.

Trends in Cardiovascular Disease Mortality

Mortality from CHD and from cerebrovascular disease increased in most western countries from the 1920s, reaching their peak levels in the 1950s. CHD mortality began to decline some years later than cerebrovascular disease and continues to decline. Cerebrovascular disease has continuously declined since the turn of the twentieth century, but the rate of decline has slowed since 1990.

There has been a dramatic decline for death rates from cardiovascular and cerebrovascular diseases in the United States from 1950 to 2002, as there has been in most western countries. Age-adjusted death rates for ischemic heart disease declined in the United States by 59 percent from 1950 to 2002, and stroke mortality by 69 percent, while mortality from all causes fell by 26 percent. The rates of decline of cardiovascular diseases, however, were not uniform in all regions or population groups. Average annual decline over this period was 3 percent for women and 3.8 percent for men. In the United States, rates of heart disease and cerebrovascular disease mortality for men and women, African-Americans and whites, have all declined, but at varying rates. The costs attributable to cardiovascular disease in the United States are shown in Figure 5.1.

Despite the decline in coronary heart disease in most western countries, this is still the largest cause of death

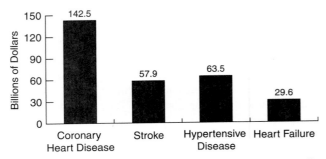

FIGURE 5.1 Estimated direct and indirect costs of major cardiovascular diseases and stroke, United States, 2006. Source: http://www.americanheart.org/downloadable/heart/1140534985281Statsupdate06 book.pdf [accessed May 2, 2008]

in the industrialized nations and increasingly in developing countries. In the United States, there are some 1.5 million heart attacks each year causing more than 500,000 deaths (Figure 5.2). Regional variation by state is wide. New York State has had higher mortality from ischemic heart disease (IHD) than the overall U.S. pattern for many decades. Within the state, with roughly half the population in upstate New York and half in New York City, there are major differences in ethnic mix, economic factors, and lifestyle. IHD is higher in New York City's population, while mortality from stroke is higher in the upstate population. IHD mortality is higher in New York City than upstate for both white and nonwhite population groups. This suggests that management of hypertension is better for the people of New York City versus the upstate population, and in the white populations compared to the nonwhite populations. Hospitalizations for AMI declined in New York State by 8 percent in 2004 following a comprehensive state ban on smoking with a direct savings of $56 million, but no reduction in stroke admissions. The potential for cost savings in CVDs from premature death,

hospitalization, and long-term care through active preventive programs is enormous.

Mortality rates from cardiovascular diseases vary widely between countries, as between regions of a country, for men and women, or for different time periods in the same country. Some countries experienced the peak of mortality from CVD in the early to mid-1950s, followed by a dramatic and sustained decline. In others, the peak was reached in the mid- to late 1960s or early 1970s followed by a more moderate rate of decline. In countries of Eastern Europe the mortality rates for CHD and strokes continued to increase in the 1990s, but may have peaked in [1994–1996] (see Figure 5.3).

Inter-country comparisons can be helpful in evaluation of public health needs and priorities. The experience of one country or region is not necessarily applicable directly to another, but the trends in cardiovascular disease mortality are now well established with declines of 40–50 percent in many countries. All countries or regions within a country that have persistent high rates should review their program priorities in public health. Figure 5.4 shows standardized death rates for a number of countries in the European Region of the World Health Organization. Finland has been a country with very high rates of cardiovascular mortality, but since the 1970s it experienced a sharp decline, as has Israel and, to a lesser degree, Sweden. Denmark and the United Kingdom did not show reductions until the late 1970s, possibly due to later adoption of new innovations, treatment of AMIs, and emphasis on dietary and smoking risk reduction.

Many studies of CVD mortality have shown racial, gender, and regional differences, but there may be many contributing factors to these differences, which are important to identify to be able to plan suitable intervention programs. Such factors include education, lifestyle and diet, access to health care, knowledge, attitudes, and practices.

Hypertension, labile or fixed, systolic or diastolic, mild or severe, for any age or sex group, is an independent contributor for CHD. Glucose intolerance or diabetes is an important risk factor for CHD. Familial history of CHD also confers excess risk, as does smoking, physical inactivity, and fatty diet. Multivariate analysis has established risk factors for intervention programs. Hypertension is more common and less well managed among the poor, African-Americans, Hispanics, and Native Americans. It is also common among native peoples in other parts of North America, Australasia, and South Sea Islanders.

Good quality medical care as early as possible during and after an AMI can reduce case fatality rates. Secondary prevention after a first AMI can reduce the risk or delay repeat AMIs and increase long-term survival. Primary prevention to reduce risk factors is also an important aspect of reducing the burden of CVD. A review of the literature and computer modeling of the experience in the United States, published in 1997, attributed less than one-third

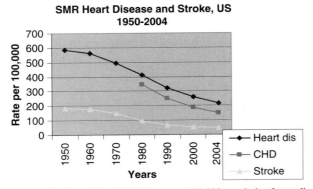

FIGURE 5.2 Age-adjusted death rates per 100,000 population for cardiovascular diseases and cerebrovascular diseases, United States, 1950–2004. Source: Health, United States, 2007, available at www.cdc.gov/nchs/data/hus/hus07.pdf [accessed November 2007 and Web site confirmed May 2, 2008]

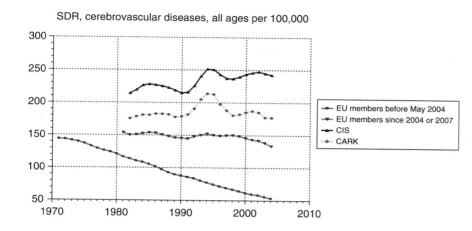

SDR, cerebrovascular diseases, all ages per 100,000

FIGURE 5.3 Standardized mortality rates (per 100,000 population) from cerebrovascular diseases, selected countries, 1970–2004. Note: CARK = Countries of Central Asia; CIS = Russian Federation, Ukraine, and other members of Community of Independent States; EU = European Union. Three-year moving averages. Source: Health for All Database, Copenhagen: WHO, November 2007, http://www.euro.who.int/ [accessed May 2, 2008]

of the reduction of mortality rates between 1980 and 1990 to primary prevention, while improved treatment accounted for half the reduction, with secondary prevention, such as routine use of aspirin and beta blockers following AMI, accounting for the rest.

Diffusion of medical interventions, discussed in Chapter 15, is sometimes seen as too rapid, but use of simple, low-cost medical technology, such as aspirin and beta blockers, both proven to be highly effective in reducing risk or delaying second AMIs, were not adopted by a majority of practitioners in a late 1990s survey in the United States. However,

using a key informant approach of respected local physicians, the percentage of local doctors using these secondary prevention medications rose sharply.

Stroke risk factors include cardiac disease, atrial fibrillation, systolic hypertension, left ventricular hypertrophy, diabetes, excessive fat and sugar in the diet, cigarette smoking, family history of early strokes, and low socioeconomic status, as well as previous stroke or transient ischemic episodes. Reduction of stroke deaths is dependent on detection and management of hypertension and its control by changes in lifestyle and by supportive medication with long-term

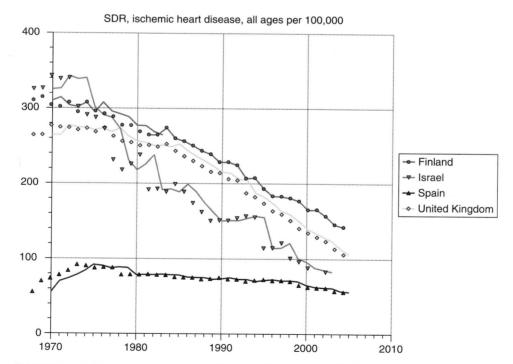

SDR, ischemic heart disease, all ages per 100,000

FIGURE 5.4 Ischemic heart disease, standardized mortality rates (per 100,000 population), selected countries, [European Region], WHO, 1970–2005. Note: Three-year moving averages. Source: Health for All Database, Copenhagen: WHO, November 2007, www.euro.who.int/hfadb [accessed November 2007 and confirmed May 2, 2008]

management and follow-up. Where stroke mortality is high, public health and medical services need to cooperate in developing education, screening, and management programs to reduce risk factors.

The long-term benefit of public health, community awareness, and improved medical care for management of elevated cholesterol can be seen in the reduction in cholesterol levels in the U.S. population as determined in NHANES surveys (see Chapter 8) since the 1960s (Figure 5.5).

With declining mortality, the long-term problem of congestive heart failure (CHF) resulting from myocardial infarction and hypertension is increasing. CHF affects 5 million people in the United States with 550,000 new cases diagnosed each year. CHF primarily affects the elderly; aging of the population and the prolonging of the lives of cardiac patients by modern and better medical treatment led to the increase in the incidence of this disease, reaching nearly 10 in 1000 people over the age of 65. It is the underlying reason for 12–15 million office visits and 6.5 million hospital days each year. Mortality rates in patients with CHF remain high; in 2006, 287,000 people died from CHF as a primary cause. Additionally, costs associated with CHF are high: CHF is the most common Medicare diagnosis-related group (DRG), and more Medicare dollars are spent for diagnosis and treatment of CHF than for any other diagnosis. It has been estimated that in 2006, the total direct and indirect cost of CHF in the United States was equal to $29.6 billion.

Prevention of Cardiovascular Diseases

The decline in cardiovascular disease mortality common in the industrialized countries over the past 25 years has been attributed to many factors, without clear evidence of the relative importance of each factor. This decline in mortality does not necessarily indicate a decline in prevalence or severity (case-fatality rates) of the disease. It may be the result of reduced prevalence of risk factors and improved access to and quality of care with secondary and tertiary

prevention including better medications, resuscitation, and emergency care delaying mortality. Higher standards of living, leisure and recreation, greater awareness of healthful nutrition and availability of appropriate foods at reasonable cost, and wider community and individual awareness of risk factors have all played a role in the reduction of mortality from CVD.

The public health implications of these alternative explanations are substantial. Current data do not allow for clear distinctions of the contribution of each to the decline in mortality. Prudent public health would continue to place stress on all of these and attempt to strengthen the trend, particularly in state or local areas where higher-than-average rates prevail. Prevention of noncommunicable diseases includes organized efforts of primary prevention (e.g., education, smoking cessation), secondary prevention (e.g., screening), and tertiary prevention (e.g., emergency medical services).

Primary prevention of CVD involves a spectrum of health promotion activities related to reduction of specific risk factors. This includes reduction of smoking and obesity, better dietary habits, and healthful foods, with increased physical activity. Prevention requires a public health approach based on creating population-based and individual changes in the factors that contribute to heart attack or stroke, along with good medical supervision (Figure 5.6).

Implementation of primary and secondary preventive activities to reduce the burden of CVD involves building institutional support, education of the public, community-based risk factor reduction activities, a healthy working environment, adequate information systems to monitor morbidity and risk factors, and a well-informed medical community. The U.S. Institute of Medicine (IOM) addressed the fragmented U.S. emergency care system and called for creation of coordinated, regionalized, and accountable emergency care systems that include "protocols for the treatment, triage, and transport of pre-hospital patients." Medical interventions include identification and

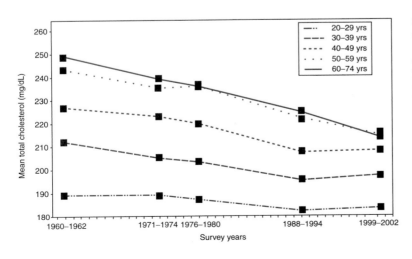

FIGURE 5.5 Trends in mean total cholesterol among adults aged 20–74 years, by age group — United States, 1960–1962 to 1999–2002. Note: Graph points represent serum total cholesterol levels at the midpoint of the survey years for the National Examination Survey conducted during 1960–1962 and the National Health and Nutrition Examination Surveys conducted during 1971–1974, 1976–1980, 1988–1994, and 1999–2002. Source: Centers for Disease Control. 2005. Trends in mean total cholesterol among adults aged 20–74 years, by age group — United States, 1960–1962 to 1999–2002. *Morbidity and Mortality Weekly Report*, 54:1288.

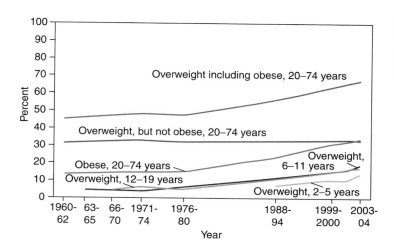

FIGURE 5.6 Overweight and obesity by age group, United States, 1960–2004. Source: *Health United States*, 2007, Chartbook on Trends in the Health of Americans, November 2007. www.cdc.gov/nchs/data/hus/hus07.pdf [accessed November 2007 and Web site confirmed May 2, 2008]

aggressive treatment of hypertension and diabetes, counseling to promote healthful dietary habits, a tobacco-free lifestyle, regular physical activity, and a supportive psychosocial environment. These are both individual and population-wide issues. Widespread use of a multidrug regimen for prevention of cardiovascular disease (a statin, aspirin, and two blood-pressure-lowering medicines), has been proposed that over a 10-year period, this regimen may avert some 18 million deaths from cardiovascular disease largely in those under 70 years of age (Lim et al., 2007). The medical care provider is involved in screening, treatment, and advising patients of the importance of reducing risk factors both before and after the onset of symptoms. The client has to take personal responsibility for many aspects of prevention

---

Box 5.5 Hypertension — Secondary Prevention

1. Hypertension @case-finding and control contribute greatly to the decline in cardiovascular mortality, especially due to stroke and heart failure. A 2 mm Hg decrease in the diastolic blood pressure in a population results in a decline of more than 5 percent in a population's risk of developing cardiovascular heart disease.

2. Screening for hypertension is an essential function of primary caregivers throughout pregnancy and for all adults.

3. Management of hypertension is a combination of diet, smoking cessation, stress management, weight loss, salt restriction, and if necessary diuretics and other antihypertensive medications.

4. Community, patient, and family education regarding hypertension, its risk factors, complications, and self-management along with medical care is fundamental to reducing the burden of individual and population morbidity and mortality from stroke, and all the associated diseases of this condition.

5. The Seventh Report of the Joint National Committee on Prevention, Detection, Evaluation, and Treatment of High Blood Pressure (2004) recommends:

   a. In the over 50 age group, systolic blood pressure (BP) greater than 140 mm Hg is a more important cardiovascular disease (CVD) risk factor than diastolic BP;

   beginning at 115/75 mm Hg, CVD risk doubles for each increment of 20/10 mm Hg;

   b. Prehypertensive individuals (systolic BP 120–139 mm Hg or diastolic BP 80–89 mm Hg) require health-promoting lifestyle modifications to prevent the progressive rise in blood pressure and CVD;

   c. For uncomplicated hypertension, a thiazide diuretic should be used in drug treatment for most, either alone or combined with drugs from other classes such as beta-blockers, angiotensin-converting enzyme (ACE) inhibitors, angiotensin-receptor blockers, or calcium channel blockers;

   d. Two or more antihypertensive medications will be required to achieve goal BP (<140/90 mm Hg, or <130/80 mm Hg) for patients with diabetes and chronic kidney disease;

   e. Patients whose BP is more than 20 mm Hg above the systolic BP goal or more than 10 mm Hg above the diastolic BP goal need therapy using two agents, one of which usually will be a thiazide diuretic;

   f. Hypertension will be controlled only if patients are motivated to stay on their treatment plan. Positive experiences, trust in the clinician, and empathy improve patient motivation and satisfaction.

Source: Adapted from Seventh Report of the Joint National Committee on the Prevention, Detection, Evaluation, and Treatment of High Blood Pressure (JNC 7): Resetting the Hypertension Sails. *Circulation*. 2003, 107:2993–2994.

as a vital factor in screening for and sustained management of high blood pressure, reduction in elevated blood lipids, reduction of obesity, screening and management of diabetes, as well as in smoking cessation and healthier exercise and dietary regimens.

Community-based programs have been designed to change life habits as a method of reducing the risks of CHD and have become accepted parts of public health. Studies of community intervention programs such as the Minnesota Five Cities and the Stanford projects showed a small direct response when the target groups were broadly defined, but other studies focusing on specific target groups, such as prevention of teenage smoking, were more effective. Social behavior studies, economic status indicators, educational levels, ethnicity, and other factors all relate to risk behavior and help to define high-risk target groups and methods of approach, and the effect of social pressures for risk behavior as in dangerous driving, smoking, and binge drinking.

Governments and manufacturers may directly or indirectly promote smoking or alcohol consumption. Where governments permit advertising of alcohol and cigarettes in the media or fail to restrict tobacco use in public places there is a tacit approval of the practice. Manufacturers promote sales of their products among the young and the poor, and in developing countries where defense mechanisms such as awareness of health risks are lower than in a middle-class population. When governments profit from taxes derived from tobacco and alcohol sales and consumption, there is a major conflict of interest between health issues and government revenues.

A 1992 conference on heart health in Victoria, British Columbia, called on all community agencies to join forces in eliminating this modern epidemic by adopting new policies, making regulatory changes, and implementing health promotion and disease prevention programs directed at entire populations. Health personnel were encouraged to work on health promotion with the community, including the media, the education system, the social sciences, professional associations, government agencies, the private sector, international and private voluntary organizations, and community health coalitions.

Numerous countries have recognized the importance of lifestyle factors and their impact over the health status of the population, and notably on the occurrence of heart diseases, representing a major burden of disease on the society. In order to facilitate the implementation of intervention programs to reduce lifestyle-related risk factors, the international CINDI (Countrywide Integrated Non-communicable Diseases Intervention) program was initiated by the World Health Organization. CINDI was started in 1990 to address the wide gap between eastern and western Europe in the prevalence of chronic disease and to address lifestyle-related behaviors (such as smoking, high blood pressure, high blood cholesterol, obesity, and excessive alcohol consumption), which are associated with the major chronic diseases such as cardiovascular disease, cancer, chronic obstructive pulmonary disease and diabetes. The member states collaborate on the implementation of an integrated approach to chronic disease prevention. The objectives are to promote measures for integrated disease prevention and health promotion in order to reduce morbidity by reducing common risk factors with inter-country collaboration in 29 countries.

Countries with very high rates of CVD should develop aggressive intersectoral intervention programs at the national and community levels. Concerted national efforts of the ministry of health, other relevant ministries, the media, NGOs, food manufacturers, and local communities are needed in order to reverse the high rates of CVD in some U.S. states and northern and eastern Europe. Legislation and litigation against tobacco manufacturers may affect the sale and pattern of smoking, but the problem is a major public health issue.

This topic should also be placed high on the national agenda in the developing countries. It is particularly relevant to those at the midlevel of development where an epidemiologic transition is under way, and CVDs are emerging as the leading cause of death. In developing countries, the problem of CVD is hidden beneath more acute issues of high morbidity and mortality from infectious diseases, and those associated with maternal and child health. But there is a pattern of increasing CVD morbidity and mortality affecting the growing middle class, with changes in diet and smoking. Age-specific mortality rates from CVD in Egypt are very much higher than in most European countries (see Chapter 3), and CVD mortality is now the principal cause of death in many developing countries. Smoking is promoted for commercial reasons in developing countries, without countervailing health hazard warning labels, limits on advertising, and other restrictive legislation. This warrants attention by health authorities in developing countries and application of lessons learned in the industrialized countries over the past 30 years (Box 5.6).

The global pandemic of cardiovascular disease is made up of a number of trends. One is a major decline in mortality in the industrialized countries, while stroke and coronary heart disease mortality are increasing in developing countries. The downward trend of the past 4 decades due to health promotion success in reduced smoking and improved treatments in the western countries may not sustain as women may be more vulnerable than in the past. The very high rates of cardiovascular mortality in Russia in recent decades has not yet reached a sustained downturn although there may be early signs of change in 2006. There is growing mortality from stroke in countries of Western Pacific Region of WHO, including China, with high rates of smoking, in association with growing wealth and dietary change. This is also related to a widespread

---

Box 5.6 Victoria Declaration on Heart Health: Conference Recommendations for Preventing Heart Disease, 1992

"Recognizing that both scientific knowledge and widely tested methods exist to prevent most cardiovascular disease, the Advisory Board of the International Conference on Heart Health calls upon

- health, media, education and social science professionals, and their associations –
- the scientific research community –
- government agencies concerned with health, education, trade, commerce and agriculture –
- the private sector –
- international organizations and agencies concerned with health and economic development –
- community health coalitions –
- voluntary health organizations –
- employers and their organizations –

to join forces in eliminating this modern epidemic by adopting new policies, making regulatory changes and implementing health promotion and disease prevention programs directed at entire populations." **The Advisory Board**

A programmatic approach on prevention of cardiovascular disease includes the following:

1. Education: educate the public, health providers, community groups, and governments in risk factor reduction.
2. Food policy: reduce fat content of milk and meat products and reduce salt content of processed foods, working with ministries of agriculture, industry, and commerce, as well as with producers and manufacturers.
3. Reduce smoking: increase cost of cigarettes through taxation; ban advertising; ban smoking in work and public places; devote some revenue from cigarette taxes to health promotion and education against smoking.

4. Promote physical exercise: promote personal and community attitudes and facilities encouraging participation in regular physical activity.
5. Reduce obesity: encourage individual and community-based health promotion.
6. Community-based initiatives: promote healthy lifestyle, including smoking cessation and fitness promotion; raise consciousness of health self-care issues; teach cardiopulmonary resuscitation (CPR).
7. Medical care: promote primary prevention techniques including screening for risk factors, management of hypertension, and patient counseling for risk factor reduction and stress management.
8. Screening: screen for risk factors such as diabetes, elevated blood lipids, and hypertension, and counsel as to findings and implications.
9. Emergency and hospital care: reduce case fatality rates, perform CPR, transport rapidly to designated medical centers with intensive care with current standards of antithrombotic agents (aspirin, streptokinase, or others) at district hospitals and ballooning, stents, or coronary artery bypass procedures at referral hospitals.
10. Rehabilitation: promote maximum recovery and function at work and in personal life; adopt preventive approaches to stop the pathological process.

The Victoria Conference on Heart Health was followed by similar conferences in Singapore in 2000 and the Biannual World Congress of Cardiology was held in Barcelona in 2006 and brings the latest scientific information on heart disease to cardiologists internationally.

Source: International Heart Health Society, http://www.internationalhearthealth.org/index.htm [accessed May 2, 2008]

---

increase of diabetes and obesity. Health promotion has a crucial role in tackling these trend shifts because they mean life or death to millions of people. The challenge is indeed a great one.

CHRONIC LUNG DISEASE

Chronic lung disease (CLD) is an important, diverse, and mostly preventable group of diseases which cause extensive morbidity and mortality. In 2002, CLD was the fourth leading cause of death in the United States. CLD can largely be prevented with good primary care and education for self-care.

When associated with acute respiratory infection, such as influenza, bronchitis, or pneumonia, CLD can result in lengthy hospitalization and premature death. Cough, shortness of breath, restricted exercise tolerance (e.g.,

climbing stairs), and difficulty sleeping are frequent symptoms, with impaired clearance of sputum and reduced lung capacity.

Asthma

Asthma is an intermittent, reversible condition of airway obstruction in response to various stimuli, resulting in wheezing and shortness of breath due to variable airflow. Usually first appearing in children up to age 5 years, it affects an estimated 14–15 million persons in the United States, including 4.8 million of those (6.9 percent) aged 0–18 years. It is the most common chronic disease among children. In 2005, asthma affected between 5 and 15 percent of persons over age 18 years. Among women, the rates were 11 percent for Hispanics, 13 percent for white non-Hispanics, and 14 percent for black non-Hispanics, while rates for men were 5 percent, 10 percent, and 10 percent, respectively.

During the 1980s, asthma prevalence and mortality increased in the United States and other countries. In the United States between 1982 and 1991, age-adjusted death rates from asthma per 1 million persons aged 5–34 years increased by 40 percent overall, but more for females. Annual mortality rates were more than fivefold higher in the African-American population than among whites. In 1994–1995, asthma was the cause of death for 11,274 persons (2.1 per 100,000 population).

Although the specific etiology of asthma is unknown, it is associated with familial, infectious, allergic, environmental, and psychosocial factors. Risk factors include animal allergens (usually from pets), household dusts and mites, primary and secondary tobacco smoke, outdoor allergens, and pollutants. Air pollution may be a factor in the increase in asthma; 63 percent of cases live in areas where pollution exceeds recommended levels. Ozone pollution, the result of hydrocarbons and nitrogen oxide emissions from motor vehicles or other sources mixed in the presence of sunlight, causes increased wheezing, coughing, and chest tightness, especially among susceptible children who play outdoors in polluted environments. The higher rates of asthma among African-Americans may be related to lower socioeconomic conditions and residence in inner city communities. Conversely, keeping children indoors for safety reasons in poorly ventilated older homes may contribute to increasing asthma.

Identification and removal of antigens, the use of bronchodilators and corticosteroids, and treatment of infections as they occur are the major methods of management. Education plays a key role in asthma control for patients, their families, school personnel, and the general public. Mortality may be increased by medication overuse, substance abuse, and tobacco use. Health care providers need continuing education regarding this condition and its management, especially during pregnancy, and regarding the possibilities and problems associated with medication usage.

Chronic Obstructive Pulmonary Disease

The term *chronic obstructive pulmonary disease* (COPD) represents advanced stages of chronic respiratory disease with airflow impairment due to chronic bronchitis affecting the smaller airways. This includes a variety of conditions resulting from damage to lung tissue, chronic narrowing of the respiratory tract, and obstruction of airflow. The term includes chronic bronchitis, emphysema, and other causes of COPD. Chronic bronchitis affects 20 percent of the adult male population of the United States. Chronic lung disease increased from 100,000 deaths in 1994–1995 to 124,816 deaths in 2002.

Chronic bronchitis and asthma can lead to emphysema, with fixed expansion and rigidity of lung tissue and reduced oxygen exchange capacity. Increasing shortness of breath, cough, loss of exercise ability, sleeplessness, repeated infections, hospitalizations, and death are common. The triad of emphysema, asthma, and bronchitis produces respiratory cripples who may function with limitations until a respiratory infection causes hospitalization and ultimately death.

Tobacco use is the greatest cause of COPD. Reduction of smoking is crucial to prevention, as is reduction of indoor pollution in the workplace and home, and outdoor pollution (see Chapter 9). Secondary and tertiary prevention should include annual influenza vaccination and pneumococcal vaccine usage. Careful monitoring of the patient at home for changing symptoms and early signs of infection would prevent long and costly hospitalization.

Mortality rates from COPD has not declined markedly (Table 5.1) in the United States because of aging of the population. Smoking reduction with increasingly aggressive legislation restricting smoking in public places including bars and restaurants, and wider use of influenza and pneumococcal pneumonia vaccines are also major factors in improving the quality and duration of life of people with COPD.

Restrictive Lung Diseases

Restrictive lung diseases are characterized by reduced lung volume, either because of an alteration in lung parenchyma or because of a disease of the pleura, chest wall, or neuromuscular apparatus. They are characterized by reduced total lung capacity, vital capacity, or resting lung volume. If caused by parenchymal lung disease, restrictive lung disorders are accompanied by reduced gas transfer, which may be marked clinically by respiratory distress on exercise. Other causes may include drugs and other treatments including cancer drugs and radiation. Inorganic dust exposure (such as silicosis, asbestosis, talc, pneumoconiosis, berylliosis, hard metal fibrosis, coal worker's pneumoconiosis) can also cause restrictive lung disease, as well as exposure to organic dust (e.g., farmer's lung, bird fancier's lung, and mushroom worker's lung, with hypersensitivity pneumonitis).

Occupational Lung Diseases

Occupational lung diseases are related to particular occupational exposures in two main categories, diseases of lung tissue and of the airway. Pulmonary fibrosis with restricted lung volume decreases lung diffusion capacity on pulmonary function testing showing increased interstitial pulmonary markings on chest x-rays. An example is silicosis and pneumoconiosis, with increased risk of tuberculosis, a major part of the burden of respiratory disease in the developing world. Obstructive airways disease is also a common pattern of occupational lung disease, which may be reversible (occupational asthma) or become irreversible (chronic bronchitis with or without obstruction or emphysema or COPD) and causes disturbed pulmonary function. The global burden of diseases related to occupational factors was estimated at 4–10 million cases per year, with approximately 3–9 million cases per year in developing countries.

Occupational lung diseases are a group of conditions associated with workplace exposures to dusts and vapors, which act as irritants, carcinogens, or immunologic agents. Primary prevention by reducing exposure levels and secondary prevention by close medical follow-up of exposed workers and ex-workers for many years after exposure when the associated diseases become apparent are an important part of occupational health. The risk of such exposures causing serious disease is accentuated by cigarette smoking and environmental pollution (see Chapter 9).

Coal Worker's Pneumoconiosis

Frequently called *black lung disease*, coal worker's pneumoconiosis (CWP) is due to prolonged exposure to coal dust for 10 years or more, with diagnosis by chest x-ray or biopsy. Anthracosis is the asymptomatic accumulation of coal pigment without cellular reaction, and is similar to accumulation found in varying degrees among most urban dwellers and tobacco smokers. Inhaled coal dust becomes a problem when the body's natural defense and processing of the dust becomes overwhelmed and, subsequently, overreactive. It has declined markedly in the United States as a result of reduced mining workforce and improved regulation and legislation (Federal Coal Mine Health Safety Act).

Silicosis

Silicosis is due to chronic inhalation of crystalline silica. The disease progresses over 20–40 years from cough and sputum production to crippling COPD due to massive fibrosis of lung tissue. Some 2000 cases are reported in the United States annually. Persons suffering from silicosis are at increased risk for tuberculosis and should undergo routine testing.

Asbestosis

Asbestosis is caused by asbestos exposure. It is a fibrous deterioration of lung tissue associated with increased rates of cancer of the lung (especially among smokers) and mesothelioma, a rapidly lethal form of cancer specific to occupational or community exposure to asbestos products.

Byssinosis

Byssinosis is both an acute and chronic disease caused by exposure to cotton dust, flax, or hemp resulting in shortness of breath, chest tightness, and chronic cough. Exposure over periods of 10 years or more causes reduced pulmonary function and COPD.

Occupational Asthma

Occupational asthma is bronchial restriction following exposure to agents to which the individual has become sensitized. Asthma with airflow obstructive symptoms may be severe and may be chronic. The variety of occupations at risk ranges from electronic workers to hairdressers, involving people exposed to a wide variety of dusts, chemical agents, and animal antigens.

DIABETES MELLITUS

WHO estimates prevalence of diabetes to be over 170 million people with 3 million deaths per year, with projected estimates of 366 million diabetics in 2030. The number of Americans with diagnosed diabetes is projected to increase 165 percent, from 11 million in 2000 (prevalence of 4.0 percent) to 29 million in 2050 (prevalence of 7.2 percent). Much of the increase in cases will be due to changes in demographic composition, to population growth, but one-third due to increasing prevalence rates.

Diabetes mellitus is a common chronic condition with disturbed carbohydrate metabolism resulting from deficiency in production of insulin in the pancreas or impaired function of the insulin receptors, resulting in increased levels of blood glucose. Diabetes is the sixth leading cause of death in the United States. It affects some 21 million Americans, or 7 percent of the population (2005), of whom 15 million were diagnosed and an estimated 6 million undiagnosed and unaware they have the disease. It is a major disease in its own right and an important risk factor for cardiovascular diseases, including coronary heart disease, stroke, and peripheral vascular disease, as well as damage to nerves, kidneys, eyes, and other organs. There are two major types of diabetes: type 1 or insulin-dependent diabetes mellitus (IDDM), and type 2 or non-insulin-dependent diabetes mellitus (NIDDM). NIDDM patients represent approximately 95 percent of all diabetes patients. NIDDM was previously known as "adult-onset diabetes," and is closely related to diet and other lifestyle factors; while IDDM was regarded as "juvenile diabetes." However, this distinction has become somewhat blurred in recent years, as NIDDM is detected more and more often in children as well (together with the increasing prevalence of obesity in children), and as the mean age of NIDDM patients declines steadily.

Diabetes prevalence has increased in the United States from 1980 through 2005, the number of Americans with diabetes increased from 5.6 million to 15.8 million. Over 700,000 new cases of diabetes are diagnosed annually, including some 12,000 among schoolchildren. Prevalence rates were approximately twice as high among African-Americans as among whites in the United States. The number of diagnosed diabetics among the black population increased from some 0.9 million to 2.5 million in 2005. Native Americans and Hispanic people also have high rates of diabetes and its complications.

Diabetes, which was ranked among the 10 leading causes of death since the 1930s, was the seventh leading cause in 2004, with over 73,000 deaths directly attributed to the disease. Additionally, diabetes is a major contributor to cardiovascular disease, the leading cause of blindness and end-stage renal disease (ESRD), and lower limb amputations.

It is the leading cause of ESRD (kidney failure requiring dialysis or transplantation) in the United States, accounting for 44 percent of new cases of treated ESRD in 2002. According to WHO, the number of diabetics in the United States was 17.7 million in 2002 and is expected to rise to over 30 million by 2030. Visual impairment is approximately twice as common among diabetics compared to non-diabetics over age 50. Lower extremity peripheral arterial disease and neuropathy are chronic conditions mainly affecting the elderly and people with diabetes. It can result in ulcer, infection, gangrene, or amputation of the lower limb and affects diabetics approximately twice as much as the elderly without diabetes.

The Canadian Community Health Survey 2002–2003 showed that the prevalence of self-reported hypertension, heart disease, and stroke among diabetics was 4 to 6 times higher than among similar age–gender groups without diabetes. The CDC estimates the direct cost of medical care for diabetes may be on the order of $50 billion per year. The addition of indirect costs may double that figure (Figure 5.7).

The discovery of insulin and its use by Frederick Banting and Charles Best in Toronto in 1921 (Nobel Prize 1923 to Banting, Best, and McLeod) gave patients with IDDM (then called *juvenile diabetes* and later *type 1 diabetes*) the opportunity to live full lives provided they were carefully managed. The more recent developments of simpler monitoring, improved insulin preparations, and insulin pump devices have made management easier and more effective for the type 1 diabetic. Insulin preparations that may be inhaled are under clinical trial. Oral hypoglycemic medications provide effective help for type 2 diabetics in combination with weight reduction, physical fitness, and careful dietary management.

WHO estimates that there are some 170 million people with diabetes globally, 90 percent with type 2 diabetes (adult onset). Diabetics are at very high risk for ischemic heart disease, stroke, kidney, eye, and peripheral vascular disease, as well as direct effects of diabetes. Controlling

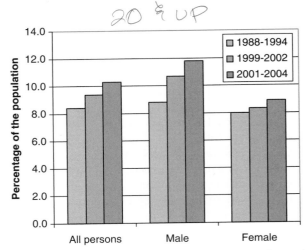

FIGURE 5.7 Diabetes among adults 20 years of age and over, by Gender, United States, 1988–1994, 1999–2002, and 2001–2004. Note: Includes self-report of physician diagnosis and undiagnosed diabetes, defined as a fasting blood glucose of at least 126 mg/dL and no reported physician diagnosis. Source: *Health United States.* 2006. Table 055, www.cdc.gov/nchs/data/hus/hus06.pdf [accessed November 2007 and Web site confirmed May 2, 2008]

high blood pressure could reduce stroke death rates in people with diabetes by 75–90 percent, and coronary heart disease by 25–50 percent. Diabetic eye disease is the most common cause of blindness, with 25,000 new cases annually, 90 percent of which could be prevented by early diagnosis and treatment. Diabetic lower extremity disease causes 57,000 leg, foot, and toe amputations per year; more than half of these could be prevented, with cost savings of $600 million annually.

Direct and indirect costs associated with diabetes in the United States amount to more than $92 billion annually, and account for 27 percent of Medicare costs or $30 billion annually. Community-based intervention programs include improved prevention, education, and services designed to reduce complications of diabetes. Such programs should be directed at populations with high prevalence rates of diabetes, and epidemiologic monitoring to identify high-prevalence populations. Using national mortality and hospitalization databases, it is possible to compare regional patterns of diabetes and its complications including coronary heart disease, cerebrovascular disease, eye disease, and reduced blood supply to lower limbs. Identifying population groups with higher incidence/prevalence of these conditions permits development of targeted community-based intervention programs to prevent obesity and diabetes through improved early diagnosis and management of diabetes and related conditions. Obesity among children and young adults is becoming a pandemic constituting a great threat to the health situation of the coming decades in developed as well as developing countries.

Prevention of Diabetes and Its Complications

Prevention is feasible for diabetes at the health promotion as well as primary, secondary, and tertiary preventive levels:

1. Health promotion involves education of the public to a high level of awareness of diabetes, its risk factors, and its complications, reduction in dietary intake of excess calories, fats and sugars, with physical activity to promote fitness and weight appropriate for height (i.e., BMI <25). This includes national and local authorities' policies to regulate and work with food manufacturers and marketing to promote healthy food consumption and avoid factors contributing to unhealthy diets for schoolchildren and in work settings, promoting substitution of health food products in marketing practices.

2. Primary prevention identifies obesity as a high-risk condition factor for preclinical diabetes, and promotes its prevention through good nutritional practices and regular exercise to reduce the risk of type 2 diabetes.

3. Secondary prevention to prevent complications of diabetes is aided by early case finding and management. Routine screening for diabetes will uncover early cases for whom treatment can reduce the severity of complications. Screening is recommended for:
   a. persons with family history of diabetes;
   b. patients with cardiovascular, renal, and eye diseases;
   c. during pregnancy;
   d. follow-up for women with glucose intolerance in pregnancy, or history of infants weighing over 4000 g;
   e. obese persons.

   For persons with abnormal blood glucose levels, follow-up management includes instruction and monitoring to prevent serious complications. Diet, exercise, regular urine and blood sugar testing, personal hygiene, and foot care should be followed closely in ambulatory and home care. In addition, medication, usually oral hypoglycemic agents, is sometimes needed to control blood sugar levels in type 2 diabetes. Insulin is always needed for control of type 1 diabetes. Social and medical support may be needed for persons unable to care adequately for themselves, especially elderly, poor, and poorly educated people.

4. Tertiary prevention aims to restore function and prevent further deterioration. For those with established peripheral vascular disease and potential gangrene, foot care can delay amputations; for amputees, rehabilitation care is essential to prevent total dysfunction. Close follow-up of diabetics by ophthalmologists promotes prevention of diabetic retinopathy by early treatment with photocoagulation (Box 5.7).

Box 5.7 Diabetes Prevention

Diabetes is a serious and costly chronic disease and is one of the leading causes of death in the United States. The effectiveness of community-based interventions has become highly appreciated for addressing lifestyle factors, such as dietary habits and physical activity.

The Diabetes Prevention Program (DPP) was a major clinical trial to determine the effectiveness of diet, exercise, and the oral diabetes drug metformin in the prevention or delay of the onset of type 2 diabetes in people with impaired glucose tolerance (IGT). The trial included participants from 27 clinical centers in the U.S. and included 3234 study participants 25 years old or older. All were overweight and had IGT, major risk factors for the development of type 2 diabetes. The study compared three intervention groups:
- Intensive lifestyle modification (healthy diet, moderate physical activity of 30 minutes a day 5 days a week)
- Standard care plus the drug metformin
- Standard care plus placebo

The study showed that diet and exercise sharply reduced the chances that a person with IGT would develop diabetes. Metformin also reduced risk, although less dramatically. The public health implications are clearly in favor of health promotion and lifestyle change for those at risk of diabetes or with IGT.

Development of community-based intervention programs to provide primary and secondary prevention of diabetes provides socioecological interventions to reduce diabetes disparities among ethnic groups found to be at higher risk. This included African-American, Native Americans, Cambodian, Chinese, Filipino, Korean, Latino/Hispanic, Samoan, and Vietnamese groups.

The Comprehensive Diabetes Intervention Research Project of the University of Arizona addresses prevention and control of diabetes among Mexican Americans living along Arizona's U.S.-Mexico border, for whom the risk of diabetes is five times higher and deaths from diabetes are 50 percent greater than for other Americans.

Similarly, the Canadian Public Health Agency has developed and implemented a series of community-based diabetes prevention programs. These projects are addressed to different groups: schoolchildren, youth, the elderly, immigrant communities, low-income families, and persons with chronic mental illness. Each of these groups is addressed via culturally adapted programs aimed at the primary and secondary prevention of diabetes through lifestyle risk factor reduction.

Sources: Knowler, W. C., Barrett-Connor, E., Fowler, S. E., et al. 2002. Reduction in the incidence of type 2 diabetes with lifestyle intervention or metformin. *New England Journal of Medicine*, 346:393–403.
Garvin, C. C., Cheadle, A., Chrisman, N., Chen, R., Brunson, E. 2004. A community-based approach to diabetes control in multiple cultural groups. *Ethnic Disease*, Summer;14 (3 Suppl 1):S83–92.
Centers for Disease Control. Comprehensive Diabetes Intervention Research Project, http://www.cdc.gov/prc/research-projects/core-projects/comprehensive-diabetes-intervention-research.htm [accessed May 2, 2008]
University of Arizona: Southwest Center for Community Health Promotion, http://www.cdc.gov/prc/center-descriptions/university-arizona.htm [accessed May 2, 2008]

The St. Vincent Declaration, developed at a World Health Organization–sponsored conference in 1990, set targets for reducing the complications of diabetes mellitus for the European region of the World Health Organization member countries. The European region of the WHO has promoted development of national diabetes programs (NDPs) in countries from western Europe to those in central Europe and the Central Asian Republics of Kyrgyzstan and Azerbaijan. Each is created by broad-based working groups in each country with representatives of ministries of health, diabetes nursing and patient's groups, as well as medical experts in the field. Similar initiatives launched in 1997 in the United States are promoting the National Diabetes Education Program (sponsored by the CDC and NIH) with participation of the American Diabetes Association and other NGOs. Such programs are intended to promote public and health care provider awareness of the seriousness of the problem and promote integrated approaches to care and improved access to diabetes care.

END-STAGE RENAL DISEASE

End-stage renal disease (ESRD) is defined as reduced renal function (to less than 10 percent of normal capacity) requiring dialysis or kidney transplantation for survival. ESRD follows severe kidney damage from infection, glomerulonephritis, hypertension, drug reactions, or diabetes.

About 40 in 100,000 persons in the United States have ESRD; almost 100,000 persons are on chronic dialysis, with more than 20,000 persons having had renal transplantation. Almost half of these are due to diabetes mellitus and most follow a long period of chronic renal failure. Prevalence varies widely between different countries and among ethnic groups in the United States. High rates occur among some ethnic groups such as Mexican and Hispanic Americans, African-Americans, North American Indians, Maoris, and Australian aborigines, probably due to high rates of diabetes and hypertension in these groups, with disparities as large as 256/1,000,000 new cases annually in whites compared with 982/1,000,000 in African-Americans. The higher rates among African-Americans as compared to whites may be mostly due to socioeconomic status and associated risk factors, such as poor infant nutrition, chronic stress, high body mass index, low levels of physical activity at work and recreationally, hypertension, lack of health insurance coverage, and lack of access to preventive health care not directly due to ethnic differences *per se*.

Prevention efforts to reduce the prevalence of ESRD should include the following:

1. Identification and effective treatment of streptococcal throat infections;
2. Careful use of medications with potential for renal damage;
3. Prompt treatment of urinary tract infection;
4. Screening for the early detection of diabetes mellitus and hypertension;
5. Proper monitoring and control of diabetes mellitus and hypertension.

ESRD is an important public health issue because it is partly preventable and it is a large consumer of health care resources. The total cost of the U.S. program for financing treatment of ESRD through Medicare was $17.9 billion in 1999 and is projected to reach $28.3 billion by 2010. Kidney transplantation is the most cost-effective method of treatment, about one-third the cost of long-term hemodialysis, but is seriously restricted by lack of donors. Continuous ambulatory peritoneal dialysis is a cost-effective option, especially where donors are limited. Usually done by the patient or the family at home, peritoneal dialysis allows the patient to engage in normal daily activities, and eliminates costly hospitalization and hemodialysis equipment needs.

Diabetes mellitus is the leading cause of ESRD (i.e., kidney failure requiring dialysis or transplantation) in the United States. CDC reports that it accounted for 44 percent of new cases of treated ESRD in 2002. Between 1984 and 2002, the crude and the age-adjusted incidence of treatment for ESRD attributable to diabetes (ESRD-DM) per 1,000,000 population increased dramatically. Incidence increased from 29.7 to 151.3 per 1,000,000 population (a 409 percent increase), and the age-adjusted incidence increased from 31.8 to 150.7 per 1,000,000 population (a 374 percent increase) (CDC, 2006).

Ethical issues in ESRD occur in both developed and developing countries. Some developing countries with specialty medical centers use donor kidneys purchased from poor persons, providing kidneys for medical tourists. In developed countries, ESRD prevalence is increasing as the population ages, as this is an age-related condition. Hemodialysis is a lifesaving procedure, and as costs of health care increase, important ethical issues arise as use of hemodialysis is increasing and in relation to the effective alternative of transplantation limited mainly by short supplies of organ donors.

CANCER

Cancer is the second leading cause of death in the industrialized countries, and is rapidly emerging as a major factor in the epidemiology of developing countries. In the United States, the lifetime probability of developing cancer is 1 in 3. From 1930 to 1990, there was an increase in age-adjusted cancer mortality rates, but they have been declining since then (Figure 5.8). The rise in lung cancer rates accounts for the increase, while incidence of new cases for some, mainly stomach, uterine, and colorectal cancers has declined. Breast cancer mortality has remained relatively constant (see Box 5.8).

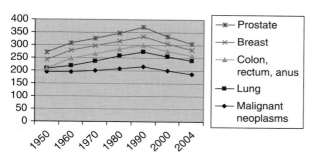

FIGURE 5.8 Age-adjusted death rates from malignant neoplasms, total, prostate, breast, colon, and lung, United States, 1950–2004. Source: *Health United States.* 2007. Table 029, www.cdc.gov/nchs/data/hus/hus07.pdf [accessed May 2, 2008]

Box 5.8 Breast Cancer Declining Mortality

Breast cancer mortality rates in the United States were steady from the 1950s all through the 1980s and even rose in 1990. However, when comparing age-adjusted death rates from breast cancer in 1990 and 2002, a reduction of 23 percent is noted (see Figure 5.9).

Investigation of the reasons for this reduction in breast cancer mortality using seven independent statistical models was developed by the Breast Cancer Working Group of a consortium of investigators sponsored by the U.S. National Cancer Institute (Cancer Intervention and Surveillance Modeling Network or CISNET). Their results showed that mammographic screening and adjuvant treatment have contributed almost equally to this fall in mortality. Their estimation was that in the absence of screening and adjuvant therapy, the death rate from breast cancer between 1975 and 2000 would have increased by about 30 percent because of increased incidence of the disease.

Sources: *Health United States.* 2005. Table 029, http://www.cdc.gov/nchs/data/hus/hus05.pdf [accessed May 2, 2008]

Spurgeon, D. 2005. Fall in mortality from breast cancer is due almost equally to screening and adjuvant therapy. *British Medical Journal,* 331:984.

According to WHO, cancer occurs because of changes of the genes responsible for cell growth and repair. These changes are the result of the interaction between genetic host factors and external agents which can be categorized as physical carcinogens such as ultraviolet (UV) and ionizing radiation; chemical carcinogens such a asbestos and tobacco smoke; biological carcinogens such as infections by virus (hepatitis B virus and liver cancer, human papillomavirus and cervical cancer) and bacteria (*Helicobacter pylori* and gastric cancer) and parasites (schistosomiasis and bladder cancer); and contamination of food by mycotoxins such as aflatoxins (products of *Aspergillus* fungi) causing liver cancer.

Cancer is responsible for much suffering and heavy economic demands on society in terms of health services, loss of work, and premature mortality (Figure 5.8). The potential for prevention of cancer is so important that any public health program must include it as part of its duties.

Many advances have been made in clinical management and in understanding of the causes of some cancers. There is strong and increasing evidence of specific risk factors which are amenable to public health intervention. The most striking example is smoking, directly related to lung cancer, and is a factor in bladder and cervical cancers. Exposure to chemical carcinogens, such as asbestos, is of major public health interest, as is prevention of liver cancer by immunization against hepatitis B. Recent research indicates that dietary factors are important contributors to colorectal, breast, and possibly lung and cervical cancers as well. The findings linking *H. pylori* infection to peptic ulcers, gastritis, and stomach cancer have been instrumental in declining incidence of these conditions internationally and the possibility of an effective vaccine for *H. pylori* would greatly enhance this benefit, joining hepatitis B and human papillomavirus as vaccines to prevent cancers.

In the United States, rates of all cancers among African-Americans are higher than whites (relative risk of 1.1), while cancer mortality is still higher (1.3:1). Cancer rates also vary widely among other ethnic groups, with Japanese, Filipino, Native American, and Mexican Americans having low age-adjusted rates. Japanese Americans have much higher cancer rates than Japanese of Japan, strongly suggesting that lifestyle differences, such as diet and smoking, are the causation or trigger events.

There are important social class differences in cancer mortality in the United States. In the United States, the relative risk (RR) for African-American men to die of lung cancer was 0.8 compared to whites in 1950, but by 1990 the RR had increased to 1.5. These differences, like those of higher rates of CHD mortality among African-American males, were thought to be race-related, but the differences are largely explained by socioeconomic or social class differences, possibly perhaps related to different smoking, occupational, and dietary patterns and less access to medical care. This is borne out by studies by mortality rate differences in different social classes seen in the United Kingdom and the Scandinavian countries.

Lung cancer is the leading cancer cause of death in the United States with an age-adjusted death rate per 100,000 population of 55 in 2002 out of 194 per 100,000 total cancer deaths. Lung cancer is also the most common cancer worldwide, 2.3 times more common in men than women, but has surpassed the breast as the leading cancer site in women. Survival rates are low (13 percent 5-year survival in the United States), so that screening for lung cancer is of dubious value. Primary prevention to reduce the number of new smokers and encourage cessation of smoking is the most crucial public health activity for lung cancer prevention.

Cancer patterns for women in the United States show a slight decrease in total cancer age-adjusted mortality rates between 1950 and 2002 (from 182 per 100,000 in 1950 to 163 in 2002), while lung cancer increased dramatically, as seen in Figure 5.9. Over the same period, breast cancer among these women remained almost constant. In 1987, lung cancer surpassed breast cancer as a cause of death among women; however, breast cancer remains a greater cause of potential years of life lost, since it affects younger women than lung cancer. In 2004, CDC reports that lung cancer was the leading cause of death from cancer both for men (31.3 percent of all cancer deaths among men) and women (25.6 percent of all cancer deaths among women). The second leading cause for women was breast cancer (15.3 percent) and for men was prostate cancer (10.1 percent). Colon cancer was the third leading cause of death from cancer both for men (9.4 percent) and women (10.1 percent) (Figure 5.10).

The single largest preventable cause of cancer in the world today is tobacco. It causes 80–90 percent of all lung cancer deaths, and about 30 percent of all cancer deaths in developing countries. It is currently responsible for the death of 1 in 10 adults worldwide (about 5 million deaths each year). Half the people who smoke today (about 650 million people) will eventually die of tobacco-related diseases, so that strong legislative and taxation policy and other efforts to reduce tobacco consumption and exposure are vital elements of public health.

In Canada, cancer mortality rates for 1970–2001 for men and women in Canada have remained stable for total cancer. There has been a fourfold increase in lung cancer mortality for women, but declining mortality from colorectal, uterus, cervix, and stomach cancer. Mortality from cancer of the breast and ovary remained essentially the same.

Cancer incidence rates vary widely between countries and within countries. A comparison of total cancer mortality between some countries in the WHO European Region show very marked differences, as seen in Figure 5.11, with declining rates since the mid-1970s in some countries (e.g., Finland), and since the late 1980s in others (e.g., United Kingdom), and rising and then stabilizing rates in some (e.g., Poland). Even neighboring Scandinavian countries show quite marked differences in total, lung, and breast cancer standardized death rates. The explanations of such differences are not at all clear, but there may be differences both in risk factors — for example, diet, environment, and smoking — as well as in treatment. Such international comparisons can generate hypotheses for further investigation.

Lung cancer mortality declined in countries such as Finland and the United Kingdom, while mortality has increased and stabilized in France and Poland, as seen in Figure 5.12. For men, lung cancer rates are more than double the rates for women. Since 1980, male mortality rates from lung cancer declined in the United Kingdom and Finland by approximately half. Rates in Poland doubled

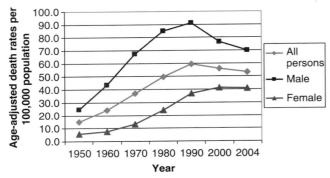

FIGURE 5.9 Age-adjusted death rates from malignant neoplasm of the lung, by sex, United States, selected years, 1950–2004. Source: *Health United States.* 2006. Table 029, www.cdc.gov/nchs/data/hus/hus06.pdf [accessed November 2007 and Web site confirmed May 2, 2008]

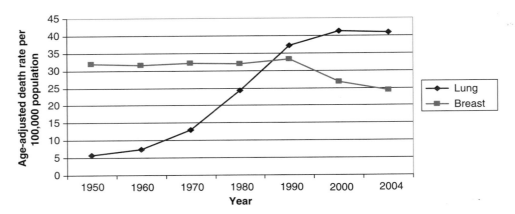

FIGURE 5.10 Age-adjusted cancer death rates for females from breast and lung cancer, United States, selected years, 1950–2004. Note: Rates per 100,000 population, standardized to the 1970 U.S. population. Lung includes trachea and bronchus. Source: *Health United States.* 2006. Table 029, www.cdc.gov/nchs/data/hus/hus06.pdf [accessed November 2007 and Web site confirmed May 2, 2008]

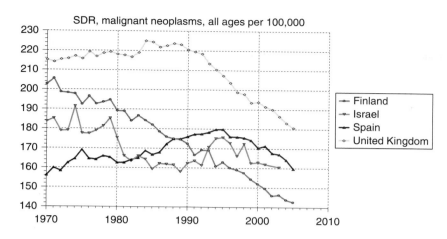

FIGURE 5.11 Standardized mortality rates (per 100,000 population) from all malignant neoplasms, selected European countries, 1970–2005. Source: Health for All Database, Copenhagen: WHO, November 2007, http://www.euro.who.int/ [accessed November 2007 and Web site confirmed May 2, 2008]

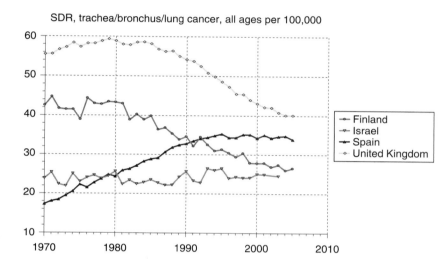

FIGURE 5.12 Standardized mortality rates (per 100,000 population) from cancer of lung, trachea, and bronchus, selected European countries, 1970–2005. Source: Health for All Database, Copenhagen: WHO, November 2007, http://www.euro.who.int/ [accessed November 2007 and Web site confirmed May 2, 2008]

to a peak in 1990 and have since declined slightly, while rates in France increased by some 40 percent to a peak in 1990, and since have also declined slowly. Female lung cancer mortality rates are still well below male rates, but increasing. These rates vary widely across Europe (from 41 per 100,000 in Denmark to around 4 per 100,000 in central Asian republics).

Prevention of Cancer

Primary prevention of cancer requires a reduction of major risk factors: smoking (lung and bladder), fatty diet (breast and colorectal), carcinogenic chemical exposures (mesothelioma, lung, leukemia, lymphoma), exposure to infectious carcinogenesic agents (hepatic), multiple sexual partners (cervical cancer), and excess sunlight exposure (melanoma). Lung and skin cancer, occupation-related cancers, and some gastrointestinal cancers are most amenable to primary prevention by reduced smoking and exposure

to sunlight, asbestos, radon, and other carcinogens. Immunization against hepatitis B is protective against most liver cancers (Table 5.5).

Secondary prevention in clinical services or in screening programs for high-risk groups focuses on early case finding. The medical care system has a major role to play, with a variety of activities, such as screening for breast, prostate, cervical, and colorectal cancer. Occupational, environmental, and social factors all play important roles in cancer. As a result, intersectoral preventive activities are needed for cancer prevention.

Cancer prevention involves reaching the total population to increase awareness of risk factors and promoting change in them. Health care providers have a vital role to educate the public, to decide when and how to screen for disease, and to cross socioeconomic barriers in reaching patients.

More research will bring important breakthroughs in this field. Identification of genes that increase risk or cause cancer, infective agents associated with carcinogenesis,

TABLE 5.5 Prevention of Cancer

Modality	Method
Self-care	Smoking cessation Sun protection Healthy body weight Physical exercise Sexual safety (e.g., reduce sexual risk behavior of multiple partners, promote the use of condoms)
Screening	Cancer of cervix (e.g., Papanicolau smears) Colorectal screening (e.g., occult blood and colonoscopy) Breast self-examination, mammography
Diet	Reduce fat and red meat intake Increase fruit, vegetable, and grain consumption Moderate alcohol consumption
Vaccination	Liver cancer — hepatitis B immunization and catch-up Cancer of cervix — human papilloma-virus (HPV)
Medical examination	Breast examination Rectal examination Skin examination
Medical management	Peptic ulcers for *Helicobacter pylori* Treatment of STIs HIV care
Toxic and carcinogenic exposures	Environmental and occupational health standards and enforcement: Control asbestos products, pesticide use, radon

medications which stop or delay progress of cancer, and other research appear very promising. The evidence for a causal relationship between nutrition and cancer is increasingly convincing.

CHRONIC LIVER DISEASE

Chronic liver disease and cirrhosis together were the seventh leading cause of death in the age group 25–64 in the United States in 2002. Cirrhosis is also a major contributing cause of morbidity, causing 1 percent of all hospital admissions. Patients with liver failure are high users of hospital care, mainly because of the serious complications, such as gastrointestinal bleeding from esophageal varices. This is a group of diseases related to chronic alcohol consumption and chronic viral hepatitis infection (mainly hepatitis B and C). The risk of cirrhosis is high among long-term heavy users of alcohol and is related to amounts consumed daily. Other nutritional factors, such as vitamin B deficiency, may be secondary contributing factors. Death rates from

chronic liver disease and cirrhosis is twice as high in men than it is in women (12.9 deaths per 100,000 and 6.3 per 100,000, age-adjusted, respectively).

Hepatitis B infection, transmitted in blood products, in body fluids, and in household contacts, is common in many countries. An estimated 2 billion persons are infected, and along with 350 million carriers they are at high risk for cirrhosis of the liver and primary liver cancer. Hepatitis C prevalence worldwide is reported by the WHO as 170 million cases, with infection rates varying from under 1 percent in Canada, Australia, and parts of western Europe, to 1–2.4 percent in the United States and much of Europe, India, and most of South America, and to rates of up to 10 percent of the population in China, much of Africa, and South America. Aflatoxin exposure in foods is a major contributor to chronic liver disease in developing countries. Liver disease varies widely by country. This is probably due to a combination of widespread hepatitis combined with high alcohol intake and nutritional and environmental factors such as widespread pesticide exposure. The high rates of mortality in Romania and France show wide divergence possibly explainable by alcohol consumption and nutritional and hepatitis prevalence changes primarily.

Liver cancer and liver cirrhosis are major public health problems, with hepatitis B being the cause of 60–80 percent of primary liver cancer, especially in developing countries of sub-Saharan Africa, east and southeast Asia, and the Pacific basin. About 2 billion people globally have been infected with the hepatitis B virus. Out of these, more than 350 million have chronic (lifelong) infections. Those chronically infected are at high risk of death from cirrhosis of the liver and liver cancer, diseases that kill about 1 million persons each year. Therefore, the WHO recommends prevention by inclusion of hepatitis B vaccine in routine infant vaccination programs and catch-up immunization of other age groups.

Hepatitis C, only discovered in 1988, is estimated to affect 170 million persons globally and 3 to 4 million persons are newly infected each year. Chronic carriers are also at risk for liver cirrhosis and primary liver cancer. Screening of blood and blood products for hepatitis C should be standard practice worldwide; however, the virus is commonly spread through unsanitary intravenous drug usage. There is still no vaccine for hepatitis C. Prevention of cirrhosis focuses on reducing daily consumption of alcohol and promoting universal immunization against hepatitis B. Needle exchange programs reduce transmission of hepatitis among intravenous drug users.

Alcohol consumption in moderation is of benefit, but long-term high levels of consumption, like binge drinking, bring health hazards in chronic liver disease. Environmental and occupational exposures to pesticides and other toxic chemicals are also factors in chronic liver disease, which is a major consumer of health care resources. Liver transplantation is effective for some in developed countries but lack of adequate numbers of donors limits the effectiveness of this modality as does its high cost.

DISABLING CONDITIONS

As the population ages, many disabling conditions place a greater burden on the individual, on the health system, and on the need for adaptation of social and health policies to meet the needs of this population group. Figure 5.13 shows the growing burden of such conditions in the U.S. population by age group.

Arthritis and Musculoskeletal Disorders

Arthritis and musculoskeletal conditions are among the most common causes of physical disability, visits to doctors, and hospitalizations. Arthritis is the leading cause of disability in the United States, affecting as much as half of the population over age 65. It was associated with total direct and indirect costs of $86.2 billion in the last Medical Expenditure Panel Survey (1997). Arthritis and other rheumatic conditions (bursitis, lupus, fibromyalgia) are among the most common chronic conditions, affecting an estimated 43 million persons in the United States; by 2020, rheumatoid arthritis (RA) is expected to affect 60 million persons. It constitutes the leading cause of disability and is two to three times more common in women than men, also occurring in children. Substantial increases in the cost of arthritis are expected in the coming years due to the aging of the population and the increasing use of expensive treatments and prescription drugs.

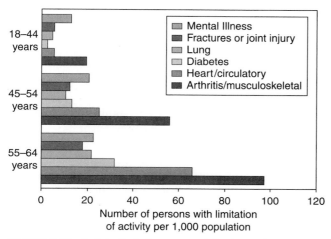

FIGURE 5.13 Activity limitation among adults due to chronic conditions, United States, 2004–2005. Source: *Health United States.* Chartbook 2007, availabe at www.cdc.gov/nchs/data/hus/hus07.pdf [accessed November 2007 and Web site confirmed May 2, 2008]

Osteoporosis

Osteoporosis is a bone disorder resulting from reduction of bone tissue density due to mineral and vitamin deficiency (calcium, vitamin D, and fluoride) which leads to weakening of the skeleton in older adults with fractures most commonly of the spinal cord, hip, and wrists. It is a major cause of disability and death, as well as costly institutional care. It appears to be more common in northern countries and in cultural groups where women are completely covered due to religious practices and have darker skin coloring, reducing sun exposure and vitamin D production.

According to the U.S. National Osteoporosis Foundation, osteoporosis is a major health threat for approximately 44 million Americans, or 55 percent of the people 50 years of age and older in 2006. Ten million individuals are estimated to already have the disease (80 percent of them are women) and almost 34 million more are estimated to have low bone mass, placing them at increased risk for osteoporosis. Osteoporosis results in more than 1.3 million fractures annually in the United States. The annual costs of medical care for fractures in 1986–1996 among this group increased from $7 billion to $14 billion. The annual cost of osteoporosis hip fractures in 2002 was $18 billion, and the cost continues to rise. Postmenopausal women are the primary group at risk, due to low bone mass. The lifetime risk of a woman to suffer an osteoporosis-related fracture is about 40 percent, and the chance of developing a hip fracture equals her combined risk of developing cancer of the breast, uterus, or ovary combined (U.S. National Osteoporosis Foundation, 1999, Web site http://www.nof.org/). Screening of those at risk by x-ray assessment of bone density is now being replaced by a relatively inexpensive, portable ultrasound instrument for use in ambulatory clinics. Postmenopausal women are at high risk for vitamin D deficiency and osteoporosis with factures of hip, spine, and other bones.

Age, gender, genetics, lifestyle (especially nutrition), and menopausal status are the major risk factors for osteoporosis, as well as lack of adequate calcium and vitamin D intake, appropriate exercise, smoking, and excessive alcohol intake. Fracture of the hip remains a serious threat to life despite improvements in surgical management. Osteoporosis management includes lifestyle issues as well as new advances in medical therapeutics. Osteoporosis is amenable to primary prevention and reducing risk factors. Primary prevention is directed toward adolescent, young adult, and perimenopausal women to assure adequate physical activity, avoidance of smoking, adequate dietary intake of calcium and vitamin D, reduction of excess alcohol consumption, and prevention of falls. For postmenopausal women, prevention should also include home safety measures, bone density screening, and hormone replacement therapy to inhibit bone resorption.

Degenerative Osteoarthritis

Osteoarthritis is a degenerative disorder, increasing in prevalence as a population ages. It is especially common in knees in women and in hips for men. It is strongly correlated with both obesity and increasing age. Reduced capacity in performing one or more activities of daily living may affect as many as 21 percent of people 18 years of age and older and 43 percent of persons over 55 years of age. This results in a heavy burden on the economy from medical costs and reduced productivity.

Reduction of obesity is the major preventive modality, requiring personal counseling and a climate of community attitudes that promotes healthy and sensible weight reduction. Treatment focuses on relief of symptoms and increasing mobility. Surgical replacement of hip or knee followed by physiotherapy and other rehabilitation measures are well-established procedures to prevent deterioration and dependency and improve quality of life.

Rheumatoid Arthritis and Gout

Rheumatoid arthritis (RA) is an autoimmune disease causing chronic inflammation of joints with stiffness, pain, deformity, and limitations of activities of daily living, affecting as many as 1 percent of adults. It is two to three times more common in women than men, and can also occur in children. RA is a generalized disease primarily affecting the joints but additionally other body systems including respiratory and gastrointestinal tracts, and is associated with excess mortality from infectious diseases and lymphomas. Supportive medical and other care should include immunization (e.g., influenza and pneumococcal pneumonia) as well as treatment of the primary condition mainly with anti-inflammatory agents.

Gout is a metabolic disorder, causing deposition of uric acid crystals in and around joints, especially those in the foot. Gout is also associated with high lead exposure in certain occupational groups, such as painting, plumbing, and ship building. National health survey data in the United States report a prevalence rate of 6 and 13 per 1000 for women and men, respectively. Reduction of lead exposure and early diagnosis and management of gout improve outcome and quality of life. Follow-up care is essential in management.

Low Back Syndromes

Low back pain from muscle injury, and abnormalities of the vertebrae, disks, or joints, with or without compression of the lumbar nerves (sciatica), cause direct and indirect costs to the U.S. economy of $15–25 billion annually. It is especially common among industrial workers, equally for men and women and mainly in the 20–40 year age group. Recovery is usual (90 percent), but chronic states often lead to surgical interventions which are expensive and often inappropriate. Prevention involves industrial engineering and education to reduce back strain, especially in heavy lifting work.

NEUROLOGIC DISORDERS

Neurologic disorders are an important burden on the affected individual and society in terms of disability, loss of productivity, premature mortality, and in health costs. According to the National Institute of Neurological Disorders and Stroke, neurologic disorders strike an estimated 50 million Americans each year, exacting an incalculable personal toll and an annual economic cost of hundreds of billions of dollars in medical expenses and lost productivity. This group of diseases includes stroke, epilepsy, multiple sclerosis, Parkinson's disease, and Alzheimer's disease, other neurodegenerative disorders, and autism. Causes of neurologic disorders include acute and chronic trauma to the brain and peripheral nerves, infection, and chemical poisoning.

Alzheimer's Disease

Alzheimer's disease is a brain disorder occurring later in life, possibly related to a genetic disorder. It is the leading cause (50–60 percent) of dementia among adults. It usually occurs after the age of 50 years, and more commonly in women than in men (1.6:1). There is no primary prevention as yet, but there is some evidence of benefit from hormone replacement and vitamin supplement therapy. Case management requires support for the family caregivers from community health resources. Other dementias, or organic brain syndromes, are due to cerebrovascular disease, Parkinson's disease, Creutzfeldt-Jakob disease, and AIDS.

Parkinson's Disease

Parkinson's disease is common after age 50, with characteristic tremor, stiff walking gait, slowness of movement, and muscular rigidity. Rates vary from 108 to 347 per 100,000 in northern Europe to 44–81 per 100,000 in Asia, Africa, and the Mediterranean. Genetic susceptibility is suspected. Treatment aimed at improving functional status is important along with support for caregivers, again to promote maximum independent function and avoid institutionalization. As the muscular rigidity affects chest muscles, respiratory infections are common, so that immunizations for influenza and pneumonia are valuable.

Multiple Sclerosis

Multiple sclerosis (MS) is a disorder of the myelin sheath of neurons, leading to impairment of vision, weakness, tremor, uncoordination, and loss of sensation and bladder and bowel control. It most commonly occurs between the ages of 20 and 50, and primarily in areas distant from the equator; for example, Canada, northern Europe, and Australasia. There is supportive evidence that this disease may be genetic, but other evidence points to an infectious agent origin. Prevention in MS focuses on good case management to reduce complications and promote rehabilitation. New medications are thought to reduce relapses and increase the length of remissions in this disease. Immunizations for influenza and pneumonia are recommended.

Epilepsy or Seizures

Epilepsy is characterized by uncontrollable convulsions starting abruptly, with or without warning symptoms, and with or without loss of consciousness. These are due to disturbances of cerebral function due to abnormal electrical activity in the brain. Isolated seizures can occur in anyone with brain hypoxia, hypoglycemia, and in children with fevers.

The World Health Organization estimates that epilepsy affects one in every 130 persons, or 40 million persons worldwide, 80 percent of whom are in developing countries, with some 2 million new cases per year. Genetic factors, infections, and brain injuries are among the major causes of epilepsy. These include infection and toxicity in the prenatal period (e.g., maternal cocaine use); asphyxia and trauma during birth; postnatal infections causing febrile convulsions; infections of the central nervous system (e.g., meningitis and encephalitis); parasitic disease (e.g., malaria, schistosomiasis); and brain damage by alcohol, trauma, and toxic substances (e.g., lead, pesticides). Prevention of epilepsy is an important reason for good quality prenatal care, safe delivery, complete infant and childhood immunization, control of fever in children, control of infectious and parasitic diseases, reduction of brain injury (e.g., home and automobile accidents), and genetic counseling.

Brain and Spinal Cord Injury

About 1.5 million Americans sustain a traumatic brain injury (TBI) each year. Of those injuries, about 75 percent are concussions or other forms of mild TBI. Although termed *mild*, those injuries may yet have long-term and even permanent effects. TBI is associated with 50,000 deaths and with the onset of more than 80,000 long-term or lifelong disability cases each year in the United States. The economic impact, including direct and indirect cost, is estimated to be over $56 billion. Head injury causes prolonged hospitalization and often irreversible brain damage. The main causes are automobile injuries (40–50 percent), falls (25–30 percent), and sports or recreation injuries (5–10 percent).

Spinal cord injuries are extremely common results of automobile accidents (50 percent of cases), recreational injuries (15 percent), and falls (15 percent). If the injury is at the cervical spine level it causes quadriplegia and if lower in the spine, paraplegia. This condition is most common among young adults aged 15–24. The estimated number of cases annually in the United States is 110,000–200,000, requiring extensive hospital and rehabilitation services.

Brain and spinal cord injuries are prevented by safer motor vehicles and work and home environments. Helmets for bicyclists and motorcyclists, football and hockey players, and other sports with records of head injury can reduce the incidence and seriousness of head trauma. Safe transportation of the injured and management in emergency departments of hospitals may help reduce the extent of brain damage from the trauma.

VISUAL DISORDERS

Blindness is defined as visual impairment sufficient to prevent the person from performing work for which sight is essential. The WHO estimates global blindness at 45 million and 180 million visually impaired, with the vast majority in developing countries where blindness prevalence rates commonly exceed 1 to 2 percent of the population. Globally, the major causes of blindness are as follows (in millions): cataract (19), glaucoma (6.4), trachoma (5.6), childhood blindness (1.5), onchocerciasis (0.3), and other causes (10).

The major causes of blindness in the United Kingdom for persons over 65 years of age include acute macular degeneration (AMD) (47 percent), cataract (20 percent), glaucoma (12 percent), diabetic retinopathy (2 percent), and others (19 percent). A study in England showed male rates increasing from 3 per 100,000 population in the 15–29 year age group to over 400 in the 75 and over group, and from 2 to 475 per 100,000 among females in these age groups. In 2003, 214,000 AMD cases of blindness were reported and the number is expected to increase to 239,000 by 2011. Currently there are between 172,000 and 245,000 with other forms of AMD.

Vitamin A deficiency is a common cause of visual impairment in children under age 5 in developing countries, causing blindness in 500,000 children per year and visual impairment in millions more. International efforts are being made to prevent this by vitamin A supplements. Onchocerciasis (river blindness), responsible for 1 million blind inhabitants of Africa and a smaller number in Latin America, is responding to improved control measures, with important economic as well as health benefits in many countries in west Africa.

Untreated gonorrhea, syphilis, measles, cataracts, and glaucoma are also important causes of blindness in developing countries. Trachoma, which when untreated leads to marked conjunctivitis and eyelid deformities causing conjunctival abrasions, can lead to blindness. This is widespread in the Middle East, Africa, and in some parts of Latin America. It is a disease of poverty, crowded living conditions, and lack of sanitation. A high percentage of cases of blindness are totally preventable by basic public health measures. Simple, inexpensive treatments are cost-effective and readily applied on a wide scale where there are planned governmental or NGO programs.

Prevention of blindness requires careful treatment of diabetes, screening of and treatment for glaucoma, cataract removal, and care of eyes using sunglasses in high sunlight areas. Public health measures also include use of safety glasses, shatterproof windows, seat belts and air bags in cars, prevention of infectious causes (STIs, measles, rubella), early treatment of eye diseases, and proper control of oxygen in incubators to prevent congenital retinal atrophy.

Blindness and visual impairment are among the 10 most common causes of disability in the United States and are associated with shorter life expectancy and lower quality of life. This is a target for *Healthy People 2010*, as summarized in Box 5.9.

Box 5.9 Strategic Approach to Assess, Evaluate, and Act in Prevention of Blindness

1. Define prevention of blindness as a national health target.
2. Set intervening subtargets such as eye care for all diabetics.
3. Mobilize key national partners, including NGOs, for collaboration with state and local health authorities.
4. Develop a national plan of action and evaluation.
5. Establish baseline and continuing monitoring of blindness.
6. Identify and alleviate health disparities in access and use of needed services.
7. Focus interventions on high-risk populations, seeking untreated diabetics and hypertensives with follow-up referral and care.
8. Population-based health initiatives for education, screening, and care (e.g., infants and children).
9. Inform and educate vulnerable groups regarding self-care and medical care needed.
10. Collaborate with professional and community groups to promote common program.
11. Develop efforts to improve awareness and competency of providers and at-risk persons to provide and utilize needed care.
12. Apply public health research methods to follow up on prevalence and new innovations in prevention and care of visual disability.

Source: Centers for Disease Control. 2006. Improving the Nation's Vision Health. A Coordinated Public Health Approach. Midcourse Review 2010. Atlanta: CDC.

HEARING DISORDERS

Hearing loss is an important disabling condition. Those who are deaf without speech can learn to communicate by hand and finger signs or writing. Those with minimal hearing may learn to lip-read and to speak. Hard-of-hearing persons have some useful hearing but require supplemental lip-reading. The psychological stress of deafness on the individual, family, and the community should be considered in developing prevention programs. Detection and correction of hearing loss can have a profound effect on an individual's well-being.

Hearing loss is probably the most common disabling medical condition, affecting large percentages of the population. The WHO estimates that 121 million people worldwide have disabling hearing impairment. In the United States, 10 million people have noise-induced hearing loss; another 20 million are exposed to hazardous noise levels in their place of employment. Prevention programs can reduce the burden of this problem.

Hearing loss may be conductive or sensory due to a neurologic defect. Conductive loss is due to obstruction in the ear canal or in the middle ear. These cases can be treated mechanically or surgically. Neurosensory hearing loss is due to damage of specialized hearing cells in the inner ear due to aging, noise trauma, infection (e.g., measles or mumps), birth defects, metabolic disorders, autoimmune disorders, side effects of medications, or unknown causes. Some causes of hearing loss can be prevented by adequate vaccination of children and limiting the use of medications that can cause hearing loss to situations when there are no valid alternatives, with monitoring of blood levels.

Noise control, especially in the workplace, is important in the prevention of hearing loss. Noise measurement in occupational settings includes decibel levels, frequency of sound waves (cycles per second or hertz), and loudness as perceived by the listener, and time or duration of exposure. Community noise levels of aircraft near airports, motor vehicle traffic, gardening equipment, and rock music for teenagers are difficult to control, but should be considered in urban planning requirements. Preventive programs include modifying machinery, erecting sound barriers, and using protective ear devices. Public health programs should be implemented in schools with the use of mobile hearing units as well as education in methods of reducing ear damage from excess noise. Infants should be screened for hearing ability before the age of 3 months, with appropriate follow-up management of treatment and education.

TRAUMA, VIOLENCE, AND INJURY

Trauma, or external injury, is a broad category that includes accidents, poisonings, suicide, homicide, and violence.

In many countries, trauma is the leading cause of death because of its greater frequency among the young and the middle-aged. It is often the leading cause of years of potential life lost (YPLL) in most developed countries and has become a major focus of intervention in modern public health program development. The possible interventions are many, requiring programmatic approaches as recommended in *Healthy People 2010* targets. There is high potential for beneficial effects of reduced injury and death primarily among young people (Box 5.10).

Trauma morbidity can be reduced by public health measures including primary, secondary, and tertiary prevention. Primary prevention reduces risk factors that are associated with trauma by such measures as enforcement of laws against alcohol abuse with driving, motorcycle helmet, car seat belt, and speed limitation laws. Secondary prevention involves early and adequate medical care at the scene of an accident and rapid transportation to a hospital trauma center. Prevention of consequences of the trauma by such intervention as cardiopulmonary resuscitation, maintaining an airway, stopping bleeding, and treatment of shock at the accident site can reduce case fatality rates. Tertiary prevention involves effective and early rehabilitation by which the degree of disability and long-term management are made more effective, as in cases of head injury (Table 5.6).

An accident is a sudden, unintended event that may be associated with human injury. The term does not imply that the event is not potentially anticipated, since there may be neglect of fundamental safety and preventive procedures and therefore increased risk of the event occurring. If a driver is drunk and/or exceeding the speed limit, the crash and deaths constitute a criminal liability. If a plant operator allows or requires employees to work in dangerous conditions without adequate safety measures, then the event is not accidental but one that could have been anticipated and probably prevented, and also constitutes criminal liability.

Injuries and deaths inflicted intentionally include homicide, rape, assault, battery, child abuse, and suicide. Public health is concerned with both intentional as well as unintentional injury. Attempts to minimize their occurrence and effects require complex interactive programs that identify and target high-risk groups; for example, for motor vehicle accidents and for suicides, through preventive-oriented programs, planning, organization, public education, training, and rescue operations.

Intervention to prevent violence is designed to break specific cycles. Violence prevention involves increased awareness by teachers, police, social workers, health professionals, and the public at-large in spotting potential and actual signs of violence, especially of abused children or women. Other forms of violence prevention include gun control, preventing weapons from entering schools, "hotlines" for victims to telephone to seek help, shelters for potential and actual victims, self-defense training, rapid response of police, enforcing drinking restrictions, and promotion of supervised teenage recreational activities. The patterns of mortality from various causes of trauma in the United States are seen in Figure 5.14.

Box 5.10 Interventions to Prevent or Mitigate Motor Vehicle Injuries

1. Strategic program involving all levels of government, judiciary, ministries of transport, local authorities, and police;
2. Mandatory seat belt (front and rear seats) legislation and enforcement;
3. Testing and enforcement of alcohol standards for drivers;
4. Administrative suspension of driving licenses;
5. Motorcycle and bicycle helmets mandatory and enforced;
6. Enforcement of speed limits of 55 miles/hour (90 km/hour) on intercity roads (50 mph or 80 kph for trucks);
7. Enforcement of minimum age drinking laws;
8. Mandatory child safety seats;
9. Driver and passenger air bags;
10. Improve vehicle design and standards;
11. Improve road design and standards;
12. Education and public policy commitment;
13. Graduated licensing for teenagers;
14. Enhance pedestrians' safety, especially for the elderly;
15. Public transport for disabled and mobility needs of the elderly;
16. Develop public health surveillance systems of dangerous roads, emergency department visits, hospitalizations, disabilities, and deaths;
17. Promote research on MVA-related injury and death and contributory factors such as emergency care on site, transportation, specialized trauma centers;
18. Pedestrian crosswalks, traffic circles (roundabouts);
19. Speed cameras;
20. Center lane barriers for highways;
21. Seek and enforce severe fines and incarceration for repeat offenders;
22. Improved public transport;
23. Define, monitor, and police dangerous roads; and
24. Promote insurance costs by risk categories and individual experience.

Source: Modified from Centers for Disease Control. 1999. Achievements in Public Health, 1900–1999. Motor-Vehicle Safety: A 20th Century Public Health Achievement. *Morbidity and Mortality Weekly Report*, 48;369–374.

TABLE 5.6 Classification of Injuries and Primary Prevention Measures

Form of injury	Regulatory control measures	Prevention/Education
Motor vehicle accidents	Mandatory seat belt, child safety seats, and speeding laws; Police enforcement of speeding and drunk driving laws; Lower permissible blood alcohol levels; Raise age for driving permits; Lower speed limits on interurban highways; Cancellation of driving privileges in high-risk groups (e.g., repeat offenders); Car safety examination, enforcement, air bags, structural standards; Pedestrian-safe crosswalks, traffic circles (roundabouts); Speed cameras; Mandatory helmets for motorcyclists and bicyclists; Regulation of insurance premiums related to risk factors	Driver education; Alcohol, drug awareness campaigns; Pedestrian safety education
Falls	Safety devices for children and the elderly; Non-skid carpets, rails in homes of elderly, bathroom and stairs railings; Hip protectors, Monitor medication use	Education and awareness campaigns
Burns	Regulators on home heating systems, regulation for manufacturers of appliances; Flame-resistant toys, children's clothing, bedding products; Building code standards for electrical wiring, smoke detectors, doors opening outwardly	Promote fire prevention awareness and occupational safety guidelines, education for care of thermal injuries
Poisoning	Manufacturers' labeling; Childproof lids on medication and household chemicals	Education regarding dangers of household medications, chemicals, safe storage, labeling and closure
Domestic violence	Mandatory reporting by medical and social workers of injuries to children and women that may originate in domestic violence and abuse; Police and medical alertness; Shelters for abused persons; imprisonment and therapy for abusers	Education in reporting of domestic violence
Occupation-related injury	Enforcement of safety standards, employer criminal liability for unsafe conditions and injuries; Monitoring small workshops and large industries, building sites, fisheries, lumbering sites	Worker and employer education
Sports injury	Mandatory use of safety equipment, helmets for motorcycle, bicycle riders, in sports, professional and amateur; Consumer Product Safety Commission monitoring recreational sports/play equipment; Licensing of coaches and sports facilities	Teaching safe sports practices in gymnastics and contact sports; Good coaching and adjudication
Suicide	Raising awareness of the general population, especially health workers, teachers, social workers, risk groups, parents; Reporting of high-risk persons to school and other authorities	Hot lines
Drowning	Water safety commission enforces guidelines for supervision of recreational swimming places, fencing of swimming pools; Lifeguards trained in rescue and CPR at public beaches and swimming pools	Swimming and boating safety awareness; Swimming education in schools, summer camps, CPR

Sources: NCHS, http://www.cdc.gov/nchs/nhcs.htm (accessed May 5, 2008)
Adams, P. F., Lucas, J. W., Barnes, P. M. 2008. Summary health statistics for the U.S. population: National Health Interview Survey, 2006. National Center for Health Statistics. *Vital Health Statistics*; 10: 1–75.

Motor Vehicle Accidents

The most common cause of loss of life from trauma is motor vehicle accidents. The CDC reports "Motor-vehicle–related injuries kill more children and young adults than any other single cause in the United States and are the leading cause of death from unintentional injury for persons of all ages. Approximately 41,000 persons in the United States die in motor-vehicle crashes each year. Moreover, crash injuries result in approximately 500,000 hospitalizations and 4 million emergency department visits annually."

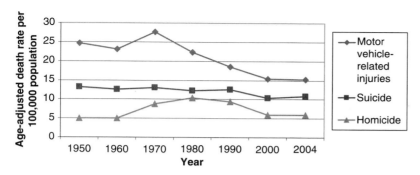

FIGURE 5.14 Age-adjusted mortality from selected causes of trauma, total population, United States, selected years, 1950–2004, rates per 100,000 population. Source: *Health United States,* 2007, Table 029, www.cdc.gov/nchs/data/hus/hus07.pdf [accessed November 2007 and Web site confirmed May 2, 2008]

Motor vehicle injuries are a problem not only in developed countries, but also in developing countries where road crash rates are extremely high. Alcohol use may be less restricted than in developed countries, road safety less advanced, cars poorly maintained, and drivers less experienced.

Minimizing the effects of trauma from car crashes involves a complex of legislation, enforcement, education, and technical development initiatives. Passenger restraints (seat belts, safety child car seats, crash air bags) and helmets for motorcycle and bicycle riders have demonstrated great potential for reducing death and injury from transportation-related crashes.

Prevention of crashes caused by speeding, alcohol abuse, fatigue, or negligence involves dealing with the human factor, based on strict enforcement and education. Laws and enforcement should ensure that vehicles are in proper working order, especially lights and brakes. Governments must ensure adequate road lighting, elimination of barriers to visibility, and well-maintained roads. The Centers for Disease Control in a review considered the reduction in motor vehicle accident mortality to be one of ten great achievements of the twentieth century (see Figure 5.14). There were many factors involved in this reduction including seat belt usage, child safety measures, improved car design, improved policing and public attitudes reducing drunk driving, improved roads and lighting, and others (Box 5.10).

Strict driver testing especially for persons under age 25 and those over 75 should be enforced. Statistically, women are safer drivers than men and should be encouraged to drive buses, trucks, and military transports. Maintaining speed limits to less than 55 miles per hour (90 kilometers per hour; 50 mph or 80 kph for trucks) reduces accident fatality rates, as do non-rigid central and side road barriers.

Law enforcement is a major factor in reducing road crash deaths and injuries. Compulsory seat belt laws are major contributors to reduced severity of injuries, and enforcement increases compliance. Enforcement of speed limits and the incarceration of drunk drivers have a strong deterrent effect and reduce death and severe injury rates. Strong police enforcement, heavy fines, criminal liability, and loss of driving permits for offenders should be permitted.

The physical relationship of speed with extent of injury is based on laws of mechanical energy or momentum. A person traveling at 55 miles per hour (90 km/hour) has four times the kinetic energy of a person traveling at 30 miles per hour (45 km/hour). Injury is more severe at higher speed. Permitting speed limits over 55 miles per hour (90 km/hour) or nonenforcement of the speed limits results in more serious injuries. This has been documented in studies of mortality patterns, which rose when speed limits were raised and declined when speed limits were lowered. Countries, especially those with limited resources, should reduce speed limits and invest in public transport and improved police enforcement including prison for serial offenders, road and vehicle improvements, and driver education. Speed cameras widely used with significant fines for speeding and use of cell phones in cars are also vital to control the tragic waste of life involved in motor vehicle crashes.

Emergency care has made important strides in the past several decades. This has undoubtedly contributed to declining mortality rates from trauma. Central trauma units serving larger populations are more likely to have a broad multiservice potential and get better results from a larger volume of cases and experience. Organized ambulance services with well-trained and supervised paramedics taking patients to central trauma centers have been shown to be effective in reducing mortality and collateral damage rates. These programs are of vital importance in achieving goals for injury control and reducing deaths and serious complications of trauma. Improved care during the early minutes following violence, whether from terrorism or motor vehicle crashes, or at the earliest stages of acute myocardial infarction can be improved by continuous training of "first responders" including police, firemen, ambulance personnel, and the general public.

Domestic Violence

Family or domestic violence is more readily identified and brought to public attention than in previous generations, so that apparent increases may be due to better reporting. Some factors that determine these intentional injuries include socioeconomic status, alcohol use, and family history of members who may have been abused themselves. Data are not readily available for incidence of child abuse, sexual abuse, or spousal abuse. Increased public awareness in recent years has led to increased reporting. Prevention requires strong public concern, police and court intervention with enforced therapy, and/or imprisonment for repeat offenders.

Suicide and Suicide Attempts

Internationally, there are wide variations in suicide rates, with the majority of cases occurring among adolescent men and the elderly. From 1981 to 2004, suicide rates in Canada among young adults (15–19 years) declined from 21.2 to 14.7 (30.7 percent) for men and rose from 3.8 to 4.7 (23.6 percent) per 100,000 population for women. Male suicide rates in Central and South American countries are below 10 per 100,000 population; in the United States and Canada the rates are less than 20 per 100,000. In Europe, overall suicide rates are lower in the Mediterranean countries (Italy, Spain, and Greece) and above 40 per 100,000 in France, Switzerland, Sweden, Denmark, Finland, Austria, and Belgium. Suicide rates in Hungary are above 100 per 100,000 population for men and 30 per 100,000 for women.

Suicide was the twelfth leading cause of death in the United States in 2004. Among persons aged 15–19 mortality from suicide remained stable, from 13.2 per 100,000 in 1950 to 10.2 in 2002 (Health United States, 2005, Table 29), but this masks an age/sex difference with a major decline in all age groups over age 45, and a nearly threefold increase in the 15–24 year age group.

For each ten suicide attempts there is one successful suicide, with men more common in the latter group. It is estimated that 30 percent of suicides are the result of mental disorders, with the remainder due to decisions regarding life circumstances, low self-esteem, binge drinking, and situational depression. Suicide among young people can become a communicable condition by emulation among people who are alienated from home and society and the fashion is as attractive as awareness of irreversibility is low.

Restriction of access to methods of suicide, such as firearms, is believed to reduce successful suicide, and restriction of publicity about suicides is thought to reduce suicide rates. Threats of suicide should be taken seriously; health care providers, teachers, counselors, and religious leaders should be instructed in suicide prevention and how to assist people through periods of depression. Mental health and supportive counseling must be part of any health care system because the suicidal individual requires immediate attention and care. Telephone counseling by volunteers on hotlines has proved to be valuable in suicide prevention.

Homicide

Homicide has become one of the major causes of death in some countries, such as in Colombia. In young adult males between the ages of 15 to 24 years in the United States, homicide is the fourth leading cause of death. The epidemiologic analysis of murders shows a relation to drug traffic, both involving rich drug competitors and street-level violence for control of the street traffic. Random violence among schoolchildren is a frequent event, as are drive-by or "road-rage" shootings, often resulting in child deaths. Murders associated with rival gangs and random violent crime with murder is now common in many former Soviet countries. Gang violence in U.S. cities is matched by concern in rural areas where murderous rampages by adolescents with access to and training in the use of deadly weapons occur with increasing frequency. Gun control legislation has made some minor gains, but weapons remain accessible to large sectors of the U.S. population.

Sources of data for epidemiologic analysis are available from national and local police, but hospital emergency department records may be more useful because more than 50 percent of violent crime goes unreported to police. Murder rates have declined in the United States since 1990 (from 9.4 per 100,000 in 1990 to 6.1 in 2002) with the decline often attributed to improved economic conditions, low unemployment, stricter punishment laws, and successful community policing. The United States remains well above other industrialized countries in homicide rates, but well below rates in Russia.

Prevention of Violence

In many countries, violence is one of the leading causes of death, especially among teenagers and young adult males. Domestic violence leading to homicide is one of the most common causes. Prevention of violence and violence-related injuries is a major public health concern because of the large scale of loss of life and personal injury as well as the long-term damage to society. Interventions involve the whole of society, not the health system alone. Nevertheless public health has an advocacy role to play.

Violence inspired by religious, nationalistic, or other political motives is a fact of life in both developed and developing countries, sometimes occurring with shocking ferocity. These events, whether bombs in buses, subways, aircraft, or buildings, or "ethnic cleansing" warfare, cause enormous physical and psychological trauma that must concern health care providers and public health personnel.

Dramatic events involving school violence and mass homicide by teenagers have occurred in the late 1990s in small communities all over the world, which are for the most part low crime areas, such as in schools or university campuses, in the United States and also in Australia, Canada, and Scotland. A 2007 CDC report of suicide rates between 1990–2003 showed a decline of 28.5 percent in rates among the 10–24 age group, but an increase among boys and girls aged 15–19 between 2003 and 2004. In 2004, 161,000 teens and young adults received emergency medical care for self-inflicted injuries.

CDC reports that 8.5 percent of U.S. students have carried weapons to school and 7.4 percent of students reported being threatened with weapons during the past 12 months. This is often attributed to a high level of violence portrayed in the media, through films, television, or the Internet. The Internet has become a forum for politically extreme and neo-Nazi ideals, as well as an avenue for illegal weapons dealing. The consensus is that this gives troubled youth the opportunity to act out fantasies of murder and mayhem with the goal of either satisfying a need for revenge, acting on an ideological hatred, or achieving notoriety. Within the context of high rates of risk-taking behavior in adolescent males, greater vigilance on the part of parents, health care providers, the educational system, and the community in seeking out warning signs, along with significant restrictions on gun sales, would help to rectify this serious problem. A 2007 CDC review of school-based programs for the prevention of violent and aggressive behavior concludes that many different strategies documented are effective at all school levels. They have beneficial effects, beyond the benefits for reduced violent or aggressive behavior, including reduced truancy and improvements in school achievement, less "problem behavior," activity levels, attention problems, social skills, and internalizing problems (e.g., anxiety and depression).

CHRONIC CONDITIONS AND THE NEW PUBLIC HEALTH

The burden of chronic conditions is an important factor in the health status of increasing aging populations, with many implications for the need of health promotion, prevention, and health service systems. Health promotion, primary, secondary, and tertiary prevention are parts of good standards of clinical and public health practice in the wide context of the New Public Health. The linkage between clinical care and public health is vital, needing strengthening of incentives and administrative changes that will foster wide implementation of existing methods of care to sustain the chronically ill in their own homes in the community and avoid or delay institutional care. Economics are an inevitable element of health policy, and the search for cost-effective ways to prevent disease and disability and to care for patients is central to progressive public health policy.

Chronic conditions often lead to an acute crises resulting in hospitalization, or long-term dependency care or death. Improved preventive care can alleviate or defer crises exacerbating a long-standing disease process. The health care system has a responsibility to prevent these acute events to give those with chronic diseases good quality of life, but also to avoid filling hospital intensive care units with patients whose balance was upset by a medical crisis not addressed early enough. The onus for prevention and management falls on all components of a health system.

The burden of chronic disease falls more heavily on the poor in industrialized countries and in developing countries as well. Obesity, diabetes, cardiovascular diseases, and cancer all have socioeconomic contributory causes related to poor nutrition and self-care, and even in those countries with universal access systems, there are gradients in utilization especially of preventive care. It may be more difficult and complex to prevent cardiovascular diseases, accidents, or diabetic complications than to treat their results, but the benefit to the individual and the community is far greater. The New Public Health includes health promotion and care of the ill in a context of limited resources and advancing medical technology, preserving individual dignity and rights, and ethical concerns.

SUMMARY

Chronic conditions are major public health problems in most industrialized countries, and are rapidly becoming so in developing countries. Cardiovascular disease and cancer are the major causes of death in most western countries, but the leading cause of years of potential life lost is trauma. Increasing longevity, improved nutrition, social support, and medical care are creating an increasingly elderly population living longer and healthier than previous generations. The public health challenge is to promote healthy middle-aged and elderly populations by reducing risk factors through health promotion and effective medical care.

The New Public Health involves working partnerships between clinical services and public health to prevent and control chronic conditions, and to prevent or delay onset of their complications. Cancer and trauma are also amenable to prevention. Dramatic lowering of mortality and morbidity from cerebrovascular and coronary heart diseases has been accomplished by this approach. The potential for prevention for increasing the well-being of those affected by these conditions should be a central element of national health policy.

ELECTRONIC RESOURCES

American College of Cardiology. http://www.acc.org/ [accessed May 3, 2008]

American Diabetes Association. http://www.diabetes.org/ [accessed May 3, 2008]

Arthritis Foundation. http://www.arthritis.org/ [accessed May 3, 2008]

Centers for Disease Control. 2006. Heart Failure Fact Sheet, http://www.cd.gov/dhdsp/library/fs_heart_failure.htm

Centers for Disease Control. 2007. Diabetes Data and Trends. End Stage Renal Disease, http://www.cdc.gov/diabetes/statistics/esrd/fig4.htm [accessed May 3, 2008]

CINDI. Updated August 1, 2005. http://www.euro.who.int/CINDI [accessed March 14, 2006]

Hypertension JCN7. 2003. Clinical Guidelines, http://www.nhlbi.nih.gov/guidelines/hypertension/express.pdf [accessed November 14, 2007]

Institute of Medicine. 2006. Hospital Based Emergency Care at the Breaking Point, http://www.iom.edu/?id=48896 [accessed May 4, 2008]

Kung, HC, Hoyert, DL, Xu, J, Murphy, SL. Health E-Stats. Sept 2007. Deaths: Preliminary data for 2005, http://www.cdc.gov/nchs/products/pubs/pubd/hestats/prelimdeaths05/prelimdeaths05.htm [accessed May 4, 2008]

National Cancer Institute. http://www.cancer.gov [accessed May 3, 2008]

National Eye Institute. http://www.nei.nih.gov/ [accessed May 4, 2008]

National Heart, Lung and Blood Institute, National Institutes of Health. 1996. Congestive heart failure in the United States: A new epidemic, http://www.nhlbi.nih.gov/health/public/heart/index.htm [accessed May 4, 2008]

National Institute of Arthritis and Skin Diseases. http://www.nih.gov/niams

National Institute of Deafness and Other Communication Disorders. http://www.nidcd.nih.gov/ [accessed May 4, 2008]

National Institute of Diabetes and Digestive and Diseases of the Kidney (NIDDK). http://www.niddk.nih.gov/ [accessed May 4, 2008]

National Institutes of Health. http://www.nih.gov/icd/

National Institutes of Health Consensus Program. http://consensus.nih.gov/ [accessed May 5, 2008]

National Institutes of Health, National Heart, Lung and Blood Institute (NHLBI). http://www.nhlbi.nih.gov/ [accessed May 5, 2008]

National Institutes of Health, National Heart Lung and Blood Institute. 2004. National High Blood pressure education program publication No. 04 5230, http://www.nhlbi.nih.gov/guidelines/hypertension/jnc7full.pdf [accessed May 4, 2008]

National Institute of Neurological Disorders and Stroke. http://www.ninds.nih.gov/about_ninds/ninds_overview.htm [accessed May 3, 2008]

National Osteoporosis Foundation. http://www.nof.org/

United Kingdom National Heart Forum. http://www.heartforum.org.uk/AboutCHD_Economicburden.aspx [accessed May 5, 2008]

World Health Organization. November 2007. European Region, Health for All database. See website for instructions to download HFA database, http://www.euro.who.int/hfadb May 4, 2008

World Health Organization International Society for Hypertension. 2007. www.who.int/cardiovascular_diseases/guidelines/hypertension/en/

International Diabetes Federation. http://www.worlddiabetesday.org/ [accessed May 5, 2008]

World Health Organization. 2006. Definition and Diagnosis of Diabetes Mellitus and Intermediate Hyperglycemia, Report of WHO/IDF Consultation, http://www.who.int/diabetes/publications/Definition%20and%20diagnosis%20of%20diabetes_new.pdf [accessed May 4, 2008]

World Health Organization. 2006. Atlas of Heart Disease and Stroke, http://www.who.int/cardiovascular_diseases/resources/atlas/en/ [accessed May 2, 2008] and http://www.who.int/occupational_health/topics/en/oehtf15.pdf [accessed May 4, 2008]

World Health Organization. 2008. A world where people can breathe freely, http://www.who.int/respiraory/gard/FlyerENblue.pdf 2008 [accessed May 4, 2008]

RECOMMENDED READINGS

Abegunde, D. O., Mathers, C. D., Adam, T., Ortegon, M., Strong, K. 2007. The burden and costs of chronic diseases in low-income and middle-income countries. *Lancet*, December 4, 2007 [Epub ahead of print].

Baker, J. L., Olsen-Lina, W., Sorensen, T. I. A. 2007. Childhood body-mass index and the risk of coronary heart disease in adulthood. *New England Journal of Medicine*, 357:2329–2337.

Bibbins-Domingo, K., et al. 2007. Adolescent overweight and future adult coronary heart disease. *New England Journal of Medicine*, 357:2471–2479.

Braunwald, E. 1997. Shattuck lecture — Cardiovascular medicine at the turn of the millennium: Triumphs, concerns, and opportunities. *New England Journal of Medicine*, 337:1360–1369.

Breslow, L., Breslow, N. 1993. Health practices and disability: Some evidence from Alameda County. *Preventive Medicine*, 22:86–95.

Carroll, M. D., Lacher, D. A., Sorlie, P. D., et al. 2005. Trends in serum lipids and lipoproteins of adults, 1960–2002. *Journal of the American Medical Association*, 294:1773–1781.

Centers for Disease Control. 1999. Achievements in public health, 1900–1999. Decline in deaths from heart disease and stroke, 1900–1999. *Morbidity and Mortality Weekly Report*, 48:649–656.

Centers for Disease Control. 2006. Improving the nation's vision health. A coordinated public health approach. Midcourse report for 2010 by CDC.

Centers for Disease Control. 2006. Visual impairment and eye care among older adults — Five states, 2005. *Morbidity and Mortality Weekly Report*, 55:1321–1325.

Centers for Disease Control. 2007. Effects on violence of laws and policies facilitating the transfer of youth from the juvenile to the adult justice system. A report on recommendations of the Task Force on Community Preventive Services. *Morbidity and Mortality Weekly Report*, 56(RR-09):1–11.

Centers for Disease Control. 2007. Increases in age-group-specific injury mortality — United States, 1999–2004. *Morbidity and Mortality Weekly Report*, 56:1281–1284.

Centers for Disease Control. 2007. National Diabetes Awareness Month — November 2007. *Morbidity and Mortality Weekly Report*, 56:1129.

Centers for Disease Control. 2007. Prevalence of actions to control high blood pressure — 20 states, 2005. *Morbidity and Mortality Weekly Report*, 56:420–423.

Centers for Disease Control. 2007. Increases in age-group-specific injury mortality — United States, 1999–2004. *Morbidity and Mortality Weekly Report*, 56:1281–1284.

Centers for Disease Control. 2007. The effectiveness of universal school-based programs for the prevention of violent and aggressive behavior. A report on recommendations of the Task Force on Community Preventive Services. *Morbidity and Mortality Weekly Report*, 56(RR-07):1–12.

Centers for Disease Control. 2007. Unintentional poisoning deaths — United States, 1999–2004. *Morbidity and Mortality Weekly Report*, 56:93–96.

Centers for Disease Control. 2007. *CDC National Vital Statistics Report*. 15 No. 4, August 21, 2007, revised October 10, 2007.

Crosson, F. J., Madvig, P. 2004. Does population management of chronic disease lead to lower costs of care? *Health Affairs*, 23:76–78.

Davey-Smith, G., Shipley, M. J., Batty, G. D., Morris, J. N., Marmot, M. 2000. Physical activity and cause-specific mortality in the Whitehall study. *Public Health*, 114:308–315.

Elders, M. J., Perry, C. L., Erikson, M. P., Giovano, G. A. 1994. The Report of the Surgeon General: Preventing tobacco use among young people. *American Journal of Public Health*, 84:543–547.

Han, T. S., Sattar, N., Lean, M. 2006. ABC of obesity: Assessment of obesity and its clinical implications. *British Medical Journal*, 333:695–698.

Hoffman, C., Rice, D., Sung, H-Y. 1996. Persons with chronic conditions: Their prevalence and costs. *Journal of the American Medical Association*, 276:1473–1479.

Julius, S. 2007. Blood pressure lowering only or more? Has the jury reached its verdict? *American Journal of Cardiology*, 6100:32J–37J.

Kannel, W. B., Wolf, P. A. [editorial]. 1992. Inferences from secular trend analysis of hypertension control. *American Journal of Public Health*, 82:1593–1595.

Khan, N. A., Hemmelgarn, B., Padwal, R., et al. 2007. The 2007 Canadian Hypertension Education Program recommendations for the management of hypertension: Part 2 — Therapy. *Canadian Journal of Cardiology*, 1523:539–550.

Kivimäki, M., Ferrie, J. E., Brunner, E., Head, J., Shipley, M. J., Vahtera, J., Marmot, M. G. 2005. Justice at work and reduced risk of coronary heart disease among employees: The Whitehall II Study. *Archives of Internal Medicine*, 165:2245–2251.

Law, M., Wald, N., Morris, J. 2003. Lowering blood pressure to prevent myocardial infarction and stroke: A new preventive strategy. *Health Technology Assessment*, 7:1–94.

Lopez, A. D., Mathers, C. D., Ezzati, M., Jamison, D. T., Murray, C. J. 2006. Global and regional burden of disease and risk factors, 2001: Systematic analysis of population health data. *Lancet*, 27(367):1747–1757.

Ludwig, D. S. 2007. Childhood obesity — The shape of things to come. *New England Journal of Medicine*, 357:2325–2327.

Marshall, S. M., Flybjerg, A. 2006. Prevention and early detection of vascular complications of diabetes. *British Medical Journal*, 333:475–480.

Nathan, D. M., Cleary, P. A., Backlund, J. Y., et al. 2005. Diabetes Control and Complications Trial/Epidemiology of Diabetes Interventions and Complications (DCCT/EDIC) Study Research Group. Intensive diabetes treatment and cardiovascular disease in patients with type 1 diabetes. *New England Journal of Medicine*, 353:2643–2653.

Newman, A. B., Simonsick, E. M., Naydeck, B. L., Boudreau, R. M., Kritchevsky, S. B., Nevitt, M. C., Pahor, M., Satterfield, S., Brach, J. S., Studenski, S. A., Harris, T. B. 2006. Association of long-distance corridor walk performance with mortality, cardiovascular disease, mobility limitation, and disability. *Journal of the American Medical Association*, 295:2018–2026.

Signorello, L. B., Schlundt, D. G., Cohen, S. S., Steinwandel, M. D., Buchowski, M. S., McLaughlin, J. K., Hargreaves, M. K., Blot, W. J. 2007. Comparing diabetes prevalence between African Americans and whites of similar socioeconomic status. *American Journal of Public Health*, 97:2260–2267.

Sytkowski, P. A., D'Agostino, R. B., Belanger, A., Kannel, W. B. 1996. Sex and time trends in cardiovascular disease incidence and mortality: The Framingham Heart Study, 1950–1989. *American Journal of Epidemiology*, 143:338–350.

Whitworth, J. A. World Health Organization, International Society of Hypertension Writing Group, 2003. World Health Organization (WHO)/International Society of Hypertension (ISH) Statement on Management of Hypertension. *Journal of Hypertension*, 21:1983–1992.

Winkleby, M. A., Taylor, C. B., Jatulis, D., Fortmenn, S. P. 1996. The long-term effects of cardiovascular disease prevention: The Stanford Five-City Project. *American Journal of Public Health*, 86:1773–1779.

BIBLIOGRAPHY

Advisory Board, International Heart Health Conference. 1992. *The Victoria Declaration on Heart Health*. Ottawa: Health and Welfare, Canada.

Beaglehole, R., Ebrahim, S., Reddy, S., Voûte, J., Leeder, S. Chronic Disease Action Group. 2007. Prevention of chronic diseases: a call to action. *Lancet*, 370:2152–2157.

Blaser, M. 1997. *Helicobacter pylori* persistence and injury of the human stomach. In Brown, F., Burton, D., Doherty, P., Breslow, L., Beck, J. L., Morgenstern, H., Fielding, J. E., Moore, A., Carmel, M., Higa, J. (eds.). Development of a health risk appraisal for the elderly (HRA-E). *American Journal of Health Promotion*, 11:337–343.

Breslow, L. 2004. *A Life in Public Health: An Insider's Retrospective*. New York: Springer Publishing Company.

Brownson, R. C., Ballew, P., Brown, K. L., Elliott, M. B., Haire-Joshu, D., Heath, G. W., Kreuter, M. W. 2007. The effect of disseminating evidence-based interventions that promote physical activity to health departments. *American Journal of Public Health*, 97:1900–1907.

Brownson, R. C., Remington, P. L., Davis, J. L. 1998. *Chronic Disease Epidemiology and Control*, Second Edition. Washington, DC: American Public Health Association.

Canadian Task Force on Preventive Health Care (formerly Canadian Task Force on Periodic Health Examination). 1994. *The Canadian Guide to Clinical Preventive Health Care*. Ottawa: Health Canada, updated 2005, http://www.ctfphc.org/ [accessed December 29, 2007]

Centers for Disease Control. 1993. Mortality trends for selected smoking related cancers and breast cancer — United States, 1950–1990. *Morbidity and Mortality Weekly Report*, 42:857–866.

Centers for Disease Control. 1993. Public health focus: Physical activity and the prevention of coronary heart disease. *Morbidity and Mortality Weekly Report*, 42:669–672.

Centers for Disease Control. 1993. Public health focus: Prevention of blindness associated with diabetic retinopathy. *Morbidity and Mortality Weekly Report*, 42:191–195.

Centers for Disease Control. 1993. Surveillance for diabetes mellitus, United States, 1980–1989. *Morbidity and Mortality Weekly Report*, 42:(SS-2):1–20.

Centers for Disease Control. 1995. Asthma — United States, 1982–1992. *Morbidity and Mortality Weekly Report*, 44:952–955.

Centers for Disease Control. 1996. Asthma mortality and hospitalization among children and young adults — United States. *Morbidity and Mortality Weekly Report*, 45:350–353.

Centers for Disease Control. 1999. Achievements in public health, 1900–1999. Motor vehicle safety: A 20th century public health achievement. *Morbidity and Mortality Weekly Report*, 48:369–374.

Centers for Disease Control. 1999. Tobacco use — United States, 1900–1999; and Cigarette smoking among adults — United States, 1997. *Morbidity and Mortality Weekly Report*, 48:986–993, 993–996.

Centers for Disease Control. 2000. Reducing falls and resulting hip fractures among older women. *Morbidity and Mortality Weekly Report*, 49(RR-02):1–12.

Centers for Disease Control. 2001. Task Force on Community Preventive Services of Motor-Vehicle Occupant Injury: Strategies for increasing use of child safety seats, increasing use of safety belts, and reducing alcohol-impaired driving: A report on recommendations of the Task Force on Community Preventive Services. *Morbidity and Mortality Weekly Report*, 50(RR-07):1–13.

Centers for Disease Control. 2004. Prevalence of visual impairment and selected eye diseases among persons aged ≥50 years with and without diabetes — United States, 2002. *Morbidity and Mortality Weekly Report*, 53:1069–1071.

Centers for Disease Control. 2005. Incidence of end-stage renal disease among persons with diabetes, United States, 1990–2002. *Morbidity and Mortality Weekly Report*, 4:1097–1100.

Centers for Disease Control. 2005. Lower extremity disease among persons aged ≥40 years with and without diabetes — United States, 1999–2002. *Morbidity and Mortality Weekly Report*, 54:1158–1160.

Centers for Disease Control. 2005. *Traumatic Brain Injury in the United States Emergency Department Visits, Hospitalizations, and Deaths 1995–2001.* National Center for Injury Prevention and Control, January 2005.

Cole, P., Amoateng-Adjepong, Y. [editorial]. 1994. Cancer prevention: Accomplishments and prospects. *American Journal of Public Health*, 84:8–10.

Dobrossy, L. (ed.). 1994. *Prevention in Primary Care: Recommendations for Promoting Good Practice.* CINDI 2000. Copenhagen: WHO Regional Office for Europe.

Doll, R., Peto, R. 1976. Mortality in relation to smoking: 20 years observation on male British doctors. *British Medical Journal*, 2:1525–1536.

Editorial. 2006. The burden of chronic kidney disease. *British Medical Journal*, 332:563–564.

Frank, J. W., Cohen, R., Yen, I., Balfour, J., Smith, M. 2003. Socioeconomic gradients in health status over 29 years of follow-up after mid-life: The Alameda county study. *Social Science and Medicine*, 57:2305–2323.

Guide to Clinical Preventive Services. 1996. *Report of the U.S. Preventive Services Task Force.* Baltimore: Williams & Wilkins.

Hunink, M. G. M., Goldman, L., Tosteson, A. N. A., Mittleman, M. A., Goldman, P. A., Williams, L. W., Tsevat, J., Weinstein, M. C. 1997. The recent decline in mortality from coronary heart disease, 1980–1990: The effect of secular trends in risk factors and treatment. *Journal of the American Medical Association*, 277:535–542.

Jeannette, E. 2001. Osteoporosis: Part I. Evaluation and assessment. *American Family Physician*, 63:897–904.

Jousilahti, P., Tuomilehto, J., Korhonen, H. J., Vartianinen, E., Puska, P., Nissinen, A. 1994. Trends in cardiovascular disease risk factor clustering in eastern Finland: Results of 15-year follow-up of the North Karelia Project. *Preventive Medicine*, 23:6–14.

Kiberd, B. A., Clase, C. M. 2002. Cumulative risk for developing end-stage renal disease in the US population. *Journal of American Society of Nephrology*, 13:1635–1644.

Kotz, K., Deleger, S., Cohen, R., Kamigaki, A., Kurata, J. 2004. Osteoporosis and health-related quality-of-life outcomes in the Alameda County Study population. *Prevention of Chronic Disease*, 1(1):A05.

Leigh, J., Macaskill, P., Kuosma, E., Mandryk, J. 1999. Global burden of disease and injury due to occupational factors. *Epidemiology*, 10:626–631.

Lim, S. S., Gaziano, T., Gakidou, E., Reddy, KS, Farzadfar, F, Lozano, R., Rodgers, A. 2007. Prevention of cardiovascular disease in high-risk individuals in low-income and middle-income countries: Health effects and costs. *Lancet*, 370:205–206.

Manson, J. E., Tosteson, H., Ridker, P. M., Satterfield, S., Hebert, P., O'Connor, G. T., Buring, J. E., Hennekens, C. H. 1992. The primary prevention of myocardial infarction. *New England Journal of Medicine*, 326:1406–1416.

Marmot, M., Elliott, P. (eds.). 1994. *Coronary Heart Disease Epidemiology: From Aetiology to Public Health.* New York: Oxford Medical Publications.

Mo, F., Pogany, L. M., Li, F. C., Morrison, H. 2006. Prevalence of diabetes and cardiovascular comorbidity in the Canadian Community Health Survey 2002–2003. *Scientific World Journal*, 6:96–105.

National Institute of Health Consensus Statement. 1994. *Helicobacter pylori in peptic ulcer disease. NIH Consensus Statement*, February 7–9, 12: 1–23.

Nichols, E. S., Peruga, A., Restrepo, H. E. 1993. Cardiovascular disease mortality in the Americas. *World Health Statistics Quarterly*, 46:134–150.

Perry, C. L., Kelder, S. H., Murray, D. M., Klepp, D. I. 1992. Community-wide smoking prevention: Long-term outcomes of the Minnesota Heart Health Program and the Class of 1989 Study. *American Journal of Public Health*, 82:1210–1216.

Speizer, F. E., Horton, S., Batt, J., Slutsky, A. S., (eds.). 2006. Respiratory diseases of adults. In *Disease Control Priorities in Developing Countries*, Second Edition. New York: Oxford University Press, pp. 681–694.

Suhrcke, M., Rocco, L., McKee, M., Mazzucco, S., Urban, D., Steinherr, A. 2007. *Economic Consequences of Non Communicable Diseases and Injuries in the Russian Federation.* Copenhagen: WHO European Region Observatory on Health Systems and Policies.

Uemura, K., Pisa, Z. 1988. The trends in cardiovascular disease mortality in industrialized countries since 1950. *World Health Quarterly*, 41:155–178.

United States National Center for health Statistics. National Vital Statistics Report, Volume 53, Number 5 (October 2004).

United States Public Health Service. 1964. The Surgeon General's Report on Smoking and Health 1964. Washington, DC: U.S. Department of Health, Education, and Welfare.

United States Renal Data System. 2001. Excerpts from the USRDS 2001 Annual Data Report: Atlas of End-Stage Renal Disease in the United States, 2001. *American Journal of Kidney Disease*, 38:S1–S248.

United States Senate Appropriations Subcommittee on Labor, Health, and Human Services. 1996. *Action Alert, The Nation's Health.* Washington, DC: American Public Health Association.

Waller, J. A. 1994. Reflections on a half century of injury control. *American Journal of Public Health*, 84:664–670.

Watson, C., Alp, N. J. 2008. Role of *Chlamydia pneumoniae* in atherosclerosis. *Clinical Science*, 114:509–531.

Waller, P. F. 2002. Challenges in motor vehicle safety. *Annual Review of Public Health*, 23:93–113.

World Health Organization. 1992. Epidemiology and public health aspects of diabetes. *World Health Statistics Quarterly*, 45:314–381.

World Health Organization, Regional Office for Europe. 1990. *Is the Law Fair to the Disabled?* European Series No. 29. Copenhagen: World Health Organization Regional Office for Europe.

Yelin, E., Cisternas, M. G., Pasta, D. J., Trupin, L., Murphy, L., Helmick, C. G. 2004. Medical care expenditures and earnings losses of persons with arthritis and other rheumatic conditions in the United States in 1997: Total and incremental estimates. *Arthritis and Rheumatism*, 50:2317–2326.

Zack, M. M., Moriarty, D. G., Stroup, D. F., Ford, E. S., Mokdad, A. H. 2004. Worsening trends in adult health-related quality of life and self-rated health — United States, 1993–2001. *Public Health Reports*, 119:493–505.

Zahran, H. S., Kobau, R., Moriarty, D. G., Zack, M. M., Holt, J., Donehoo, R. Centers for Disease Control and Prevention (CDC). 2005. Health-related quality of life surveillance — United States, 1993–2002. *Morbidity and Mortality Weekly Report*, 28:54; Surveillance Summary (4):1–35.

Zaza, S., and Thompson, R. S., (eds.). 2001. The Guide to Community Preventive Services: Reducing injuries to motor vehicle occupants; systematic reviews of evidence, recommendations from the Task Force on Community Prevention Services, and expert commentary. *American Journal of Preventive Medicine*, 2001:21(Suppl 4).

Chapter 6

Family Health

Introduction
The Family Unit
Maternal Health
   Fertility
   Public Health Concerns of Fertility
   Family Planning
   Maternal Mortality and Morbidity
Pregnancy Care
   High-Risk Pregnancy
Labor and Delivery
   Safe Motherhood Initiatives
   Care of the Newborn
   Care in the Puerperium
Genetic and Birth Disorders
   Rhesus Hemolytic Disease of the Newborn
   Neural Tube Defects
   Cerebral Palsy
   Intellectual Disability
   Down Syndrome
   Cystic Fibrosis
   Sickle-Cell Disease
   Thalassemia
   Phenylketonuria
   Congenital Hypothyroidism
   Fetal Alcohol Syndrome
   Tay-Sachs Disease
   G6PD
   Familial Mediterranean Fever
Infant and Child Health
   Fetal and Infant Mortality
Infancy Care and Feeding
Anticipatory Counseling
Documentation, Records, and Monitoring
The Preschooler (Ages 1–5 Years)
School and Adolescent Health
   Smoking
   Alcohol Abuse
   Drug Abuse
   Sexual Risk Behavior
   Dietary Risk Behavior
   Physical Activity
   Violence and Gang Behavior

Adult Health
Women's Health
Men's Health
Health of Older Adults
   Health Maintenance for Older Adults
Summary
Electronic Resources
Recommended Readings
Bibliography

INTRODUCTION

This chapter provides an introduction to health in the family context. The family structure provides an important foundation for physical and emotional health of the individual and the community. Marital and family status and interactions among family members affect each person's health and the well-being of the community and nation. The family exists and functions within the context of cultural, economic, legal, and social patterns unique to each society, with important commonalities of the role of the family in health. However, each person is a unique individual who passes through life stages with changing health needs and support systems not only in the family, but also in peer groups and society more widely.

Family health issues relate to phases involving fertility and pregnancy, infancy, childhood, adolescence, adulthood, and old age as well as the relationships among family members. Each phase has specific health risks in which prevention and other health services play an important role. This chapter addresses the health needs of family members at different stages of life. No single population group is isolated from another. Poor pregnancy outcome affects mother, child, family, and the community. The family of the person who is chronically ill, injured, or killed at work or on the road suffers economically, emotionally, and socially. The public health and medical care systems must be sensitive to the special needs of the family by providing appropriate health promotion, disease prevention, medical care, and support programs for each member of the family and the family as a whole.

The New Public Health includes traditional elements of public health, such as maternal and child health, but within a larger context of total family needs. In developed and developing countries, single teenage pregnancies entail a multitude of family and societal problems. In societies where women's rights are restricted or repressed, the health of women and children suffers measurably. The role of public health is then to advocate societal change and provide direct responses to the health needs generated by societal patterns, whether this relates to high rates of morbidity for women and children due, for example, to lack of primary care services, or whether the problem is lack of health insurance coverage for children.

THE FAMILY UNIT

The family is the basic social support unit in virtually all human societies, providing the basis for childbearing and child rearing. It has important roles in stability of basic physiological and psychological needs as well as the economic basis of the members of the unit. The family provides the key environment for the emotional needs, socialization, mutual help, and nurturing needed by adults as well as children.

The family is usually a group of two or more people related by marriage, common agreement, birth, or adoption who reside together in the same household, and may consist of one, two, or more generations. The family unit includes people living together, engaged in sexual relations, in fertility, and in rearing of children through the many stages of development before they reach independent adulthood. It also includes caring for elderly parents and relatives, as well as maintaining close contact with adult siblings and children, themselves in the process of childbearing and child rearing.

The nuclear family typically includes a male and female couple related by marriage, or living together by common consent, with or without children. Increasingly, family units include single or divorced mothers, or fathers, living alone with children, or with a parent or other person related or unrelated by marriage. The definition is being widened to include couples of the same sex, with or without children from previous heterosexual relationships, from artificial insemination, or through adoption.

The extended family is multigenerational and consists of the nuclear family and relatives of both parties, whether or not they are living in close geographic proximity. The extended family provides a broader basis of mutual support. It still exists in western societies, but frequently in an altered form. Multigenerational families consisting of single, divorced, or widowed women with children are now common, in association with high divorce rates, single parenthood, and increasing longevity, especially for women. Same-sex households can leave children of one gender without role models.

In developing countries, families often consist of large numbers of children born to poorly educated parents living in poverty. The father or less commonly the mother may be absent for long periods while working in a distant place. This can create serious health hazards for all family members. In societies where death of adults occurs from civil wars, famine, or infectious diseases such as AIDS, raising of children by single parents, neighbors, older siblings, or grandparents is common. Abandonment of children is also common in such situations. WHO estimates that more than 15 million children under 18 years have been orphaned as a result AIDS, of whom over 12 million live in sub-Saharan Africa.

Divorce and single parenthood are often associated with relative poverty, creating additional stresses in the functioning of the family unit. The absence of one parent, usually the male, also strains the family. Abusive parenting, including psychological and sexual abuse and other violent behaviors, creates a burden on children in their most vulnerable years, with long-term psychological damage.

In the United States, there has been a decline in the percentage of children in two-parent family settings. In 1980, 86 percent of all family households in the white population were two-parent families, declining to 68 percent in 2003 (U.S. Census Bureau, 2003). Among the African-American population, 56 percent of all families were in two-parent family homes in 1980, declining to 46 percent in 2000, while female-headed black households increased from 40 percent to 45 percent. For family groups with children under age 18 years, the percentage of one-parent family groups in black homes was 65 percent in 2000, as compared to 24 percent in white homes and 37 percent in Hispanic homes. This is a continuing trend. This represents a high percentage of African-American homes and children at high risk for poverty, insecurity, and inadequate health insurance. This results in lack of access to family doctors and regular care for some 25 percent of American children aged 6 or younger, with poor utilization of immunization and frequent use of hospital emergency departments. In Canada, single-parent families make up 13 percent of all families, but all are included in universal health care. In the United Kingdom, just over one-quarter of all births occur to unwed mothers, including cohabiting parents, but all are included in the National Health Service so that society ensures children access to care despite their family circumstances.

Changes in family structure affect the health status of adults as well as children. Increases in longevity of women, more than men, together with high divorce rates, produce an excess of elderly women in relatively good health who live alone. The capacity of women to adapt to this condition is positively affected by the strong social support systems in the industrialized countries where universal health insurance, national pensions, pensions from prior work of spouses, and legal protection of economic rights in divorce have produced a relatively well-protected elderly population group.

Diseases of aging from middle age onward include cardiovascular and other degenerative conditions affecting men at an age nearly 10 years younger than women. Cancers are the second leading cause of death. The epidemiologic and demographic transitions affect the family in many ways. With aging of the population, care of the elderly has become an increasing factor in family life, not only in provision of health care, but in the economics of health, for the family and the society. Where health insurance fails to fully cover long-term care, serious financial burdens are placed on middle-aged children of elderly parents. With aging of the population, it is not uncommon for middle-aged or elderly children to care for very elderly parents. Where society does bear the cost, the importance of prevention and control of chronic disease becomes essential to avert breakdown of the health system.

MATERNAL HEALTH

Women's health issues relate to their many roles: as family caregivers, individuals, workers, wives, grandmothers, mothers, and daughters. These demand lifelong responsibilities for knowledge, self-care, and family leadership in health-related issues, such as nutrition, hygiene, education, exercise, safety, fertility, child care, and care of the elderly. Changes in social roles of women create extra demands and risks in health.

Fertility

Fertility is the natural potential and ability to conceive and have children through normal sexual activity. Infertility is defined as inability to conceive after a year of regular intercourse without use of contraceptives. Childbirth and care of infants and children have been traditional concerns of public health because of the related vulnerability and historically large-scale loss of life. Improved nutrition and living conditions as well as proper care in pregnancy and at delivery have combined to reduce maternal mortality to a rare event in developed countries. In least developed countries, however, maternal mortality rates can be over 400 and as high as 1800 per 100,000 live births, while in industrialized countries it is under 10 per 100,000 live births. The difference is found in poverty and poor sanitary conditions as well as lack of adequate professional health care. It is also the result of the poor general health status of women, including education, age at fertility, previous number of births, the time or space between pregnancies, anemia and poor nutritional status, and lack of adequate facilities and standards of the health system. Poor health prior to pregnancy, little or no prenatal care, and lack of professional care during delivery in developing countries is common especially for the rural population and the urban poor.

Proper nutrition, communicable disease control, and education ensure that females reach the age of fertility physically and intellectually prepared for childbirth and child rearing. Literacy for women contributes to improved infant survival rates. The literate mother has a greater ability to address health issues, using written materials instead of depending solely on community traditions, and is better able to cope with the complexities of a health care system. Greater education is also likely to lead to a better chance of employability and greater family income.

Fertility-related health issues include preparation for and timing of pregnancy, and professional care during and following pregnancy. Using health education and prevention services, public health provides the necessary resources through which these issues may be addressed. Cessation of smoking, alcohol, drug use, and risk-taking behavior at least for the duration of the pregnancy is an important part of protection of the fetus from harm.

Fertility patterns are affected by social, cultural, religious, and other factors, including the technology available for birth control. With the availability of safe and effective birth control measures and increasing education and job opportunities, fertility patterns have changed in the industrialized countries. Ethnic differences can be substantial within a country. For African-American women, the crude birth rate fell from 22.4 per 1000 in 1990 to 16.0 in 2004, compared to 15.8 and 13.5 per 1000, respectively, for white women. The crude birth rate for Hispanic/Latino women declined from 26.7 to 22.9 during the same period.

Fertility rates in many developed countries have declined considerably since the peak "baby boom" that occurred in the years following World War II. This has generated concern that the birth rate is lower than that needed to sustain the population even at current levels. Developing countries have the opposite problem, in that high birth rates with declining morbidity are placing a heavy economic, social, and medical burden on the community while the cultural and religious ethos produces little support for family planning.

Globally, despite declining birth rates, the world's population grew from 1 billion in 1804 to 2 billion in 1927, 3 billion in 1960, and to 6.7 billion in 2007. It is anticipated to increase to 7.3 billion in 2015 and 8 billion in 2028. By 2050, the world population is projected to be 9.2 billion (United Nations Population Division, 2006 Revision).

Family planning and spacing of pregnancy are vital issues in developing countries where the burden of frequent pregnancies contributes to high maternal and infant mortality rates. The traditional means of fertility limitation in these countries was through prolonged breastfeeding, which has declined with the promotion of commercial baby formulas. Improvements in socioeconomic conditions, with female education, infant and child survival, and the strengthening of family planning services, are likely to provide the needed impetus for increased contraceptive use and fertility decline in developing countries.

Public Health Concerns of Fertility

Provision of birth control along with prenatal, delivery, and post-pregnancy care are among the central roles of any health care service. High levels of maternal mortality of the past in the industrialized countries are still present in developing countries. Traditions of high fertility rates, unattended deliveries, unsafe practices of traditional birth attendants (TBAs) and female genital mutilation greatly contribute to the poor health status of women in developing countries.

Safe prenatal care and birth practices, along with better nutrition and general health status of women, reduced maternal morbidity and mortality dramatically in the twentieth century. But the health gap remains enormous between the industrialized and least developed countries. Fertility rates have been falling in most parts of the world as individual education levels improve. Economic incentives in rural areas for more children are replaced by incentives for fewer children, as urbanization fostered by a search for better standards of health and living conditions occurs (Table 6.1).

As the technology of birth control becomes widely available, better educated women have the power to control their own fertility. There is an encouraging pattern of fertility decline in all parts of the world, including sub-Saharan Africa, which had increasing fertility rates well into the 1980s. Public health promotes spacing of pregnancy to improve health outcomes for both the mother and the newborn, with wider implications for family health.

In many societies, infertility is a problem associated with significant personal distress and social stigma. Sexually transmitted infections (STIs) are a major cause of infertility so that prevention and treatment of STIs are important aspects of managing infertility. Treatment of infertility is associated with high cost, not only in expenditures, but also in emotional trauma. Modern services to treat infertility, including stimulation of ovulation, *in vitro* fertilization, and surrogate parenting, raise many ethical and financial issues. Despite these and other problems of multiple births and high rates of very low birth weight infants with associated perinatal and developmental problems, infertility treatment is very much a part of modern health care systems.

Family Planning

Family planning enables a woman to determine the time, spacing, and frequency of pregnancy with a range of methods for preventing or expelling a conception while maintaining a normal sex life. Adoption is another aspect of family planning. The technology for birth spacing has been revolutionized, so that there are now safe and effective contraceptive methods widely available at reasonable cost in industrialized and developing countries (Table 6.2). Birth control is a public health issue and responsibility. Low levels of awareness or inaccessibility to birth control contribute to high birth rates. It also leads to use of abortion for birth control, which may complicate subsequent pregnancies. High fertility rates or pregnancies late in life pose extra hazards both to the mother and the newborn and may be a burden to the family unit.

TABLE 6.1 Total Fertility Rates by World Demographic Regions,[a] Selected Years, 1970–2005

Region/Group	1970	1980	1990	2005
Sub-Saharan Africa	na	6.7	6.3	5.4
India	5.6	4.9	3.7	2.9
China	5.3	2.8	2.2	1.7
South Asia	na	5.2	4.0	3.1
Latin America and Caribbean	5.1	4.2	3.2	2.5
Middle East and North Africa	6.2	5.9	5.1	3.1
CIS/CEE and Baltic States	2.2	2.2	2.3	1.7
Industrialized countries	2.5	1.9	1.7	1.6
World	4.7	na	3.2	2.6

[a]Total fertility rates (see Chapter 3 and Glossary). CIS/CEE refers to the former Soviet Union and Countries of Eastern Europe.
Source: UNICEF. The State of the World's Children 2007.

TABLE 6.2 Percentage of Married Women Using Some Form of Contraception, World Regions, 1960–1965 to 1997–2005

Region	1960–1965	1997–2005
Sub-Saharan Africa	5	24
Middle East and North Africa	2	53
South Asia	15	46
Latin America and Caribbean	11	71
East Asia and Pacific	17	79

Source: UNICEF. State of the World's Children, 1994, 2007.

Modern birth control methods include pharmacological and chemical prevention of conception (hormone pills, hormonal implants or injections, spermicides) and physical methods (male and female condoms and intrauterine devices [IUDs]). Traditional methods of prevention, such as breastfeeding, the rhythm method, and *coitus interruptus*, are less reliable than current methods, mainly the pill, tubal ligation, female condoms, and the IUD. Birth control measures for use by males include condoms or vasectomy. Condoms promote safer sexual contact, but both methods rely heavily on education and awareness levels.

Spacing of pregnancies by modern methods of birth control is a fundamental right of women. In many areas of the world, birth control remains a religious and political issue as well as a power struggle between the sexes. The Roman Catholic Church is strongly opposed to "artificial" birth control (although it accepts the less reliable rhythm and withdrawal methods), as are many political regimes for national demographic reasons. However, in many countries with a predominately Catholic population, the birth rate has fallen and in some is below the population replacement level. Use of birth control has increased dramatically over the recent decades in Latin America and in East Asia. Family planning has not been well understood or available in the former Soviet republics. They lack a supply and widespread awareness of modern contraception, relying on frequent abortion instead. While the absolute number of abortions has declined in Russia (1990–2005 from 2,000 to 1,000 per 1,000 live births), abortion remains 4 times higher than European Union rates, and a primary means of fertility control. Abortion continues to be an important cause of preventable morbidity and mortality.

Despite the wide availability of contraception, teenage pregnancies account for a relatively large percentage of total births in the United States. This is a serious public health problem, being more prevalent in lower socioeconomic groups with problems of single mothers with welfare dependency and associated with poor health outcomes. Teenage unmarried pregnancies are a complex of social, educational, and labor force problems not easily addressed. It is a growing problem in other industrialized countries as well. In developing countries, teenage and even child marriage produces a wide range of maternal health problems of physical and emotionally immature mothers trapped in a life role with no chance to become educated or employed.

In the United States, progress toward the health target for a reduction in teen pregnancies is being made, which declined in the age group up to age 19 from 61 per 1000 females in 1990 to 42 births per 1000 females in 2004. Abstinence among females before 15 years of age and those aged 15–17 years has increased. Both abstinence and use of condoms have increased among teen males. Recent emphasis on abstinence education programs focused on prevention of a first pregnancy and the emerging availability of emergency contraception along with fear of HIV have contributed to this phenomenon, as indicated in achievements of the Mid-Course Review of *Healthy People 2010* (http://hdr.undp.org/hdr2006/statistics, Chapter 9).

Maternal Mortality and Morbidity

A maternal death as defined by the WHO is death of a woman while pregnant or within 42 days following termination of pregnancy from any cause related to or aggravated by the pregnancy or its management, but not from accidental or incidental causes. Maternal deaths are subdivided into two groups:

1. Direct—deaths resulting from obstetric complications of the pregnant state;
2. Indirect—deaths resulting from preexisting disease or conditions not directly due to obstetric causes.

In past centuries, childbirth was life-threatening. In 1664, a report of Hotel Dieu Hospital in Paris indicated a maternal death rate of 33 percent among hospitalized parturient women. In England between 1660 and 1680, 1 woman in 44 was reported to have died in childbirth. Semmelweiss in Vienna in the 1850s reported death rates of 2 percent in women during or following delivery as compared to rates of 10 percent in medically supervised wards. This was due to lack of precautions against transmission of puerperal fever through unhygienic practices of the medical staff. Semmelweiss's work led to the control of streptococcal septicemia, a major cause of maternal death (Chapter 1).

In the United States, maternal mortality declined from 670 per 100,000 live births in 1930 (levels similar to those of least developed countries today) to just under 21.5 in

TABLE 6.3 Maternal Mortality Rates, by Selected Population Groups and Years, United States, 1950–2004

Group	1950	1960	1970	1980	1990	2000	2004
White	61.1	26.0	14.3	6.6	5.4	7.5	9.3
Black	na	103.6	60.9	22.4	22.4	22.0	34.7
Total	83.3	37.1	21.5	9.2	8.2	9.8	13.1

Source: *Health, United States*, 2007. Table 43, http://www.cdc.gov/nchs/data/hus/hus07.pdf#summary
[accessed May 5, 2008]
Notes:
1. Crude rates per 100,000 live births. Data from 1999 according to ICD 10.
2. Health target for 1990, not more than 5/100,000 overall and for all local and ethnic group rates.
3. Black or African American.

1970, 8.2 in 1990, rising to 9.8 in 2000 and 13.1 in 2004 (Table 6.3). Maternal mortality rates for African-Americans remained more than four times those of the white population. This dramatic decline in maternal mortality, representative of the trend in most industrialized countries, can be attributed to many factors including better standards of living and nutrition, improved medical care, and advances in obstetrical knowledge and training. The transfer of deliveries into maternity wards of general hospitals contributed to the improved safety of maternity.

During this period, fertility rates fell as birth control became universally available, permitting the spacing of pregnancies and the occurrence of pregnancy by choice, an important factor contributing to lower maternal morbidity and mortality rates. Low rates of maternal mortality are also due to greater access to medical care, as more Americans became insured or could pay privately, and more widespread development of quality of care programs. On the negative side of this issue, the existing and growing gaps between African-American and white populations also reflect greater poverty in the former and lack of health access because of the high percentage of uninsured and poorly insured in the United States (see Chapter 13).

The gap in maternal mortality rates between African-American and white populations remains high. The relative risk of maternal mortality in African-Americans and whites has remained over 4 to 1 since 1970, despite a decline of over 48 percent and 51 percent for African-American and white women, respectively. There remains a wide gap related to age and pregnancy preparation as well as adequacy of care, which continues to be a serious public health problem, especially for poor African-American teenage pregnant women. The rise in maternal mortality in the period since 1990 may be due in part to lack of universal health insurance, increase in the uninsured population, and lower standards of health insurance benefits for many of the insured.

Internationally, an estimated 529,000 women died in 2000 from complications of pregnancy and childbirth, 99 percent of whom were in developing countries. Maternal mortality rates range from over 800 per 100,000 (i.e., 1 percent for all births) in least developed countries in 2000 to over 400 in mid-level developing countries, 64 per 100,000 in the former Soviet countries, and fewer than 15 per 100,000 in industrialized countries, with some at 5 per 100,000 (e.g., European Union countries before 2004, Israel).

The World Bank and WHO/UNICEF estimate that the extension of prenatal, delivery, and postpartum care to 80 percent of the world's population would reduce by 40 percent the burden of disease associated with unsafe childbirth. They suggest that an appropriate intervention program should include information, education, communication, transportation, community-based obstetrics, and district hospital facilities. High maternal mortality rates are associated with high birth rates and delivery by untrained personnel (Table 6.4) (WHO/UNICEF/UNFPA/World Bank 2007 press release and http://www.un.org/millenniumgoals/docs/MDGafrica07).

The Millennium Development Goals accepted by the United Nations in 2000 call for a reduction in maternal mortality by 3/4 by the year 2015 (MDG #5). Reviews of progress up to 2007 in the African region indicate "countries with the highest initial levels of mortality have made virtually no progress over the past 15 years." A UN review in 2007 states: "Maternal health remains a regional and global scandal, with the odds that a sub-Saharan African woman will die from complications of pregnancy and childbirth during her life at 1 in 16, compared to 1 in 3,800 in the developed world" (UN, 2007). The basic issue is that lowering maternal mortality requires cultural and political changes with investment in training and infrastructure. Without appropriate political commitment at the national and international levels to this as a priority issue this MDG cannot be achieved. Despite agreement on the MDGs, failure to progress in reducing maternal mortality is in part because other issues such as HIV and TB have received much more donor attention from the international community (Chapter 16).

Maternal mortality reviews are a common method of investigating maternal deaths in order to ascertain

TABLE 6.4 Women's Health Indicators by Level of Country Development

Level of development	Crude birth rate		Births by skilled personnel		Maternal mortality rate	
	1990	2005	1990–1997	1997–2005	1990	2000
Least developed	43	37	28	35	1100	890
Developing	29	23	55	60	470	440
CIS/CEE	18	14	93	93	85	64
Industrialized	13	11	99	99	13	13

Notes: CIS/CEE refers to the former Soviet Union and countries of Eastern Europe.
Crude birth rate = number of live births per 1000 population.
Maternal mortality rate = maternal deaths per 100,000 live births.
Rates for 2000 were adjusted by UNICEF.
Source: UNICEF. State of the World's Children. 1999, 2007.

preventable causes. Such reviews can help to identify factors leading to maternal deaths and provide important lessons for medical, obstetrical, and community health staff. They are requirements for hospital accreditation in Canada and the United States (see Chapters 13 and 16).

PREGNANCY CARE

The goals of prenatal, delivery, neonatal, and infancy care are to provide the mother and child with the optimal conditions and supervision to ensure the best possible outcome of the pregnancy, preserving the health and well-being of the mother and providing the newborn with the greatest chance of survival and optimal development. Since pregnancy is fraught with potential problems for the mother and the fetus, professional prenatal care and equally important self-care by the pregnant woman are necessary. Pre-pregnancy and prenatal care should be early and complete for mother and child.

Public health programs have the responsibility to assure prenatal care for the entire population either through direct provision of care or by obstetrical services in managed care or private practice settings. Pregnancy is often a planned event, so that preparation for pregnancy is feasible. Preparation for pregnancy includes a general examination; inquiry as to possible genetic problems; STI and HIV testing; nutrition status assessment and counseling; folic acid and iron supplements; smoking, alcohol, and drug use cessation; and other counseling as required. The parents should be provided with reading material on pregnancy and parenting. Good prenatal care presupposes a basic program of visits with extra care for those at special risk.

Prenatal care in the community includes early presentation, high-risk assessment and referral, and continuous care throughout the pregnancy. Early diagnosis of pregnancy is important in permitting the woman to attend prenatal care as early as possible, hopefully in the first trimester. Early presentation for prenatal care provides the opportunity to assess the health of the mother and to advise her on appropriate nutrition and self-care. In addition, it establishes a working relationship between the mother and the caregiver. Both the mother and the father should become involved in prenatal preparation including prenatal classes and exercises. Early detection of potential complications offers better outcomes of care and genetic disorders can be diagnosed early.

Prenatal care should be built around three principles, as suggested by the American College of Obstetrics and Gynecology (ACOG) and a Delphi Group Panel, U.S. Department of Health and Human Services, in the late 1980s. These principles are:

1. Early and continuous risk assessment;
2. Health promotion;
3. Medical and psychological intervention as needed.

The ACOG also recommends pre-pregnancy consultation to discuss pregnancy and infant care issues. The ACOG suggests that prenatal care for a normal pregnancy should include a total of 13–15 doctor visits, prenatal classes on physical care, and preparation for delivery, breastfeeding, and infant care. The mother's routine checkups should include monitoring of her nutritional state, weight gain, blood pressure, emotional well-being, and potential complications. The frequency of visits and the content of normal prenatal care for healthy women vary widely among countries and are sometimes considered "excessive." However, good care should not be taken for granted. The achievement of low maternal mortality and morbidity along with low rates of risk for the newborn from low birth weight and associated developmental problems should reinforce the importance of close supervision throughout a pregnancy.

In some countries, such as Israel and France, prenatal care is separated from regular primary care and is offered in maternal and child health centers (MCH) or women's clinics. In others, such as Holland and the United Kingdom, family practitioners provide prenatal and obstetrical services. In still other countries, such as the United States and Canada, specialist obstetricians and general practitioners both provide prenatal and delivery services. In Israel, public health nurses are the major providers of prenatal care and midwives do most hospital deliveries. Whether prenatal care is part of primary care services or a separate public health service, the goals must include universal coverage starting before or very early in pregnancy, risk assessment with referral and care, and ready access to specialized care. Health education provided in prenatal classes for the pregnant woman and her partner offers additional counseling and peer group support.

In developing countries, improved access to prenatal care and increased use of mid-level health workers in "baby-friendly" birth centers, coupled with training and supervision of traditional birth attendants, spacing of pregnancies, risk assessment, and referral, are all needed to lower present high rates of maternal and perinatal mortality. As discussed in the context of maternal mortality, progress in this has been extremely disappointing over the past several decades even in countries with substantial economic growth and increased governmental revenues such as Nigeria and India.

The pregnant woman should be provided with a complete medical record of her prenatal care during scheduled visits, so that when she arrives in the maternity unit, the provider of care will have adequate information. Even if she has not been seen previously by the new caregiver, the record grants a measure of continuity beneficial to both. The patient can feel that her record of previous care is available when needed, and the provider has the use of all previous documentation on the patient. A public health system should ensure adequate records systems for maternity care in order to reduce unnecessary complications and mortality in the birth process.

Maternal mortality rates in many developing countries remain very high with little improvement over the past several decades, despite much discussion of the topic by international and local health agencies. This lack of improvement such as in India is attributed by some authors to lack of political commitment and related program planning. WHO reports that the maternal mortality ratio in 2005 was highest in developing regions, with 450 maternal deaths per 100,000 live births, in stark contrast to ratios of 9 per 100,000 live births in developed regions and 51 maternal deaths per 100,000 live births in the countries of the Commonwealth of Independent States (CIS). Moreover, the small drop in the global maternal mortality ratio reflects mainly declines in countries with relatively low levels of maternal mortality. According to the Director General of WHO in the 2007 news release, "Countries with the highest initial levels of mortality have made virtually no progress over the past 15 years" (http://

www.who.int/medWHO news release 2007 /2007/pr56/en/index.html).

High-Risk Pregnancy

Low-risk pregnancies are those in healthy women between the ages 18 and 34, who present at least once in the first trimester, who have had no more than three previous normal live births, no previous stillbirths or obstetrical complications such as gestational diabetes or preeclampsia, no history of drug or alcohol abuse, and no major medical conditions such as hypertension or kidney disease. Such women should be followed in a routine prenatal care program.

High-risk pregnancies (HRP) include very young and older women as well as those with previous or current medical and obstetrical complications. HRPs should be identified as early as possible so that the patient can be given special care for her benefit and especially for the well-being of the fetus and newborn. Identification and management of high-risk factors initially and throughout pregnancy improve pregnancy outcomes for the mother and the newborn. Some predictors of HRP are age, previous obstetric difficulties, malnourishment, poverty, many previous pregnancies, early or late maternal age, women who attend STI clinics, and use of cigarettes, alcohol, or drugs. Risk factors may include social and economic factors such as adverse family circumstances, housing, financial status, and working conditions.

The medical and obstetrical history provides evidence of previous risks such as frequent abortion, complications in pregnancy, or medical conditions that could affect the mother during the pregnancy or at the time of delivery. Pregnancy under age 16 or 17 or over age 35 should automatically define the pregnancy as being at higher than normal risk. Grand multiparity (i.e., more than five previous births) or a first pregnancy (primigravida) should also be considered as an extra risk for the mother but more so for the newborn.

A scoring system provides a set of standards or guidelines for risk assessment to assist the primary care provider in early detection and referral of patients on the basis of a reasonably objective set of criteria for high-risk factors. Detailed guidelines are needed to implement this kind of standard for high-risk pregnancy and monitoring is of value to improved pregnancy care. The form and guidelines developed should take into account local risk factors, such as high consanguinity rates in some societies or chronic malnutrition in the population. Scoring systems produce a cumulative risk assessment by adding those factors which by themselves would not mean the pregnancy is high risk, but taken together indicate potential problems. The actual cutoff points for age, parity, and education levels can be adjusted to conditions in each country, but the principle of national standards is important.

The HRP assessment and referral form outlined in Table 6.5, developed and used in a rural primary care

TABLE 6.5 High-Risk Pregnancy Scoring and Referral Form

I. Personal Data	II. Social/Personal	III. Obstetrical History
Name	Age (<17, >35 years)	Gravida (1 or >5)
Identity Number	Education (<6 years)	Para (0 or >5)
	Marital status	Abortions (>2)
Date of Birth	Economic status	Miscarriages (1+)
Address	Consanguinity (blood relation)	Fetal deaths (1)
	Smoking	Bleeding in T3
City/Town/Village	Alcohol	Stillbirths (1)
	Drug use	Previous cesarean (1)
Clinic	Home conditions	Preterm deliveries
Date last menstrual period	Summary	Birth weights
Date of first visit		Infant deaths (1)
	_____	Toxemia (1)
	_____	Birth defects (1)
	_____	Summary
	A B C	_____
		_____ A B C

IV. Medical History	V. Present Pregnancy	VI. Summary of Risk Factors
Diabetes	Number of present pregnancy	I. Personal A B C
Hypertension	Time first visit T1 T2 T3	
Renal disease	Weight before pregnancy	II. Social A B C
Heart disease	(<50 kg)	
Chronic chest disease	Height <145 cm	III. Obstetrical History A B C
Blood disorder	Bleeding	
Endocrine disease	Preeclampsia	IV. Medical History A B C
Phlebitis	Rh	
Other	Multiple pregnancy	V. Present pregnancy A B C
Summary	Abnormal presentation	
_____	Summary _____ A B C	VI. Total score A B C
_____ A B C		

Summary of Reasons for Referral: _____

Date: _____ Signature: _____

Position: _____

Report of High-Risk Clinic: _____

_____

_____

Date: _____ Signature: _____

Position: _____

---

setting, was adopted throughout a governmental health system in a developing area (West Bank and Gaza). In this case A = low risk, B = medium risk, and C = high risk. Clear definitions and staff instruction are required. In some cases one C requires referral (e.g., hypertension), or several Cs or Bs mean mandatory referral. The format and scoring system can vary, such as scoring each topic from 1 to 10 with systematic compilation of the score. Referral to HRP clinics needs not only a clear indication or reasons for referral, but also thorough assessment by specialists and feedback to the referring physician or other primary care provider. The HRP clinic should send the patient back to the referring center with a report of findings and a clear recommended plan of action. The HRP clinic may continue to follow the patient because of the risk factors, but the referring provider should have this information and help in its implementation.

A well-developed HRP assessment, referral and follow-up system contributes to improved outcomes for both mothers and newborns, preventing costly long-term consequences of maternal and infant morbidity and mortality. Its role in preventing complications during and following delivery is well justified on medical, public health, and economic grounds.

The issues may be different in developing and developed countries, but the principles are similar. In the United

States, there are substantial population groups that cannot get prenatal care for financial or bureaucratic reasons, including those who may be at highest risk. In Russia, the HRP approach is essential if the country is to bring down its high maternal mortality rates. As previously noted, countries with both universal access and well-developed obstetrical care and risk assessment have low rates of maternal mortality (Australia, Belgium, Denmark, Israel, Norway). Maternal mortality does respond to improved access to medical care, as seen in its rapid decline in Italy and Portugal during the 1970s following introduction of their national health services.

LABOR AND DELIVERY

Delivery is a highly personal service but with important public health aspects. Assurance of safe delivery affects both mother and newborn. Over the past century, the place of delivery has evolved from the private home to maternity or lying-in hospitals to maternity units in general hospitals, and to "mother and baby friendly" birth centers (in hospital settings). Delivery in maternity units in hospitals in developed countries has made an important contribution to reduced maternal and neonatal mortality and should be considered a prime goal in public health and primary care programs.

Hospital delivery is encouraged in many countries by giving maternity grants paid by national insurance or social security to the mother for covering costs of delivery in a hospital. This was done previously in France and Israel before the advent of national health insurance in order to promote hospital delivery. France continues this tradition by providing grants to women who come to prenatal care early and frequently during pregnancy. In developing countries, the place of delivery needs to be addressed, making use of existing resources with a goal of reducing the current extremely high rates of maternal morbidity and mortality.

Dangerous emergencies can occur at any stage of labor for which cesarean section, blood transfusions, general anesthesia, and fetal resuscitation or treatment of respiratory distress syndrome may be necessary. Delivery in a general hospital obstetric department is safer than at home or in a free-standing maternity home. Electronic fetal monitoring during labor does not necessarily reduce fetal intracranial damage, retardation, and cerebral palsy, as compared to clinical fetal auscultation, but provides reassurance to the anxious mother and assists the harried staff of a busy obstetrical department.

Intrapartum care should be made as homelike as possible and include the husband or other person desired by the pregnant woman. Care during labor and delivery requires information from the prenatal period, by transmission of a pregnancy care record, carried by the woman to the delivery room. This record summarizes her medical, obstetrical, and prenatal care, especially parity, date of last menstrual period, risk factors, and prenatal care findings. A mother-carried record is preferable, since access to prenatal clinic records may be limited and may not include care rendered by other providers, such as emergency departments or private physicians.

As noted above, some countries, including some with very low maternal mortality rates, use qualified midwives to perform routine deliveries while obstetricians are responsible for supervision and complex deliveries. Licensing of qualified midwives may be necessary in the United States to meet current obstetrical needs, especially where there is a shortage of obstetricians. Qualification for midwives is usually an additional year of study for registered professional or academic level nurses (Chapter 15).

Fetal monitoring during labor (intrapartum) by auscultation has been standard practice since the work of Evory Kennedy in Dublin in the 1830s. Since the late 1950s, electronic fetal monitoring has become widespread and reassuring to parents and of help to staff on a busy obstetric ward. Hospital maternity wards should be "mother, family, and baby friendly," promoting a homelike atmosphere with both midwives and medical support.

Attendance by trained anesthetists and pediatricians should be considered for routine cases and mandatory for high-risk pregnancies. The presence of a pediatrician or pediatric nurse practitioner at the time of birth gives greater assurance of appropriate early care for the newborn, especially important in cases of low birth weight, respiratory distress, or other complications at birth. In countries with low overall infant mortality, early neonatal deaths are now a major element in that mortality. Further reduction in infant mortality rates requires greater attention to this risk period when resuscitation and other immediate lifesaving procedures may be needed.

In the United States in recent years, there has been a modest trend to return to home deliveries with highly trained midwives and guaranteed backup services. Home deliveries are declining in the United Kingdom with an increase in specialist obstetric departments in general hospitals. Home births may be an adequate alternative to delivery in the hospital for low-risk pregnancies, but only when carried out by well-trained professional midwives with the backup of emergency medical teams able within minutes to deal with the complications that can occur with any delivery. Home birth is a trend that could lead to increases in maternal and infant death as an unfortunate side effect. Delivery in maternity homes or free-standing birth centers is an alternative to maternity units in general hospitals, but should be avoided in developed countries. The homes and centers lack full facilities in the event of complications unanticipated before the onset of labor. The place and methods of delivery may vary but should take into account not only the wishes of the pregnant woman but the safety of the mother and the newborn as well. High-risk mothers, including all primiparas, should always be delivered in a hospital setting.

Safe Motherhood Initiatives

Safe motherhood is an initiative of the United Nations launched in 1987 with the goal of ensuring that women go through pregnancy and childbirth safely, and give birth to healthy children. Every year, hundreds of thousands of women die or suffer serious complications from pregnancy and childbirth. Safe motherhood begins before conception with proper nutrition and a healthy lifestyle. It continues with planned pregnancy, appropriate prenatal care, prevention of complications when possible, and early and effective treatment of complications when they occur. It proceeds with labor at term with adequate care but without unnecessary interventions, followed by delivery of a healthy infant and provision of a healthy postpartum environment that supports the physical and emotional needs of the woman, infant, and family. This requires the support of many diverse sectors of the nation, the family, educators, employers, health care providers, community groups, and others. It is an investment in the physical, emotional, social, and economic well-being of women, their children, and their families and, by extension, the nation.

Home births are still common in developing countries. They are unattended or attended to by traditional birth attendants (TBAs). This may be the only choice the mother has, in which case the public health authorities should train and supervise the TBAs. Special attention should be given to hygiene, prevention of tetanus, and identification and referral of high-risk cases. The World Health Organization promotes guidelines for integrating TBAs into the health care system in developing countries. Setting standards and supervising TBAs are vital to reduce the current large-scale loss of mothers and newborns in developing countries. Primiparas should be referred to a district hospital for delivery, as should other high-risk cases including grand multiparas, women over age 35, multiple pregnancies, late presentation for care, and those with previous complicated deliveries, malnutrition, or chronic illness, as well as those with previous infant deaths.

Public health agencies should ensure hospital delivery wherever possible, and ensure that hospitals are staffed with well-trained personnel. Equally important is assurance that early and consistent prenatal care is utilized by all pregnant women. District hospital facilities should be equipped for obstetrical and newborn emergencies, including surgical, anesthesia, blood bank, and resuscitation staff and facilities for normal and cesarean deliveries. All care by community and district care centers and by TBAs should be documented in appropriate record systems developed by the public health authorities.

District and community-based obstetric service centers should be developed to provide prenatal risk assessment and care, working with TBAs in both a supportive and supervisory capacity. Community-based delivery facilities should be equipped and staffed to perform normal deliveries and safe abortions and handle emergencies such as preeclampsia, hemorrhage, and manual removal of placenta. TBAs should be trained to arrange evacuation of high-risk mothers and babies to district hospitals. Professional midwives or nurses should be appointed to supervise TBAs including visits to the villages and direct one-on-one training. TBAs performing deliveries on high-risk patients, except in absolute emergencies or non-preventable situations, should be subject to a penalty enforced by the public health authority.

Supervision of TBAs involves a process of regulation, education, and incentives to meet higher standards of practice. The health authority is empowered to license and regulate all health providers and to discipline offenders who practice without a license or do not meet current standards of practice. TBAs can be motivated to participate in training programs with incentives including provision of essential equipment and supplies. TBAs should be required to keep a log of deliveries with basic details and outcomes recorded in a state-issued log book. If TBAs are illiterate, they can be required to engage a literate person to keep such records, for regular inspection by the TBA supervisor. Training should include prenatal and delivery danger signs, hygiene in delivery, neonatal resuscitation and infancy care, and high-risk situations. TBAs can also be trained in regard to STIs, HIV, anemia, iodine deficiency, and obstetrical or medical risks.

Care of the Newborn

The newborn is totally dependent and vulnerable to many factors that may inhibit optimal growth and development. This dependency begins before conception and continues during the prenatal period, the delivery, and up to the end of the first year of life. Care should be given to assure warmth and suction to reduce the chance of aspiration and assure breathing immediately following delivery. Many of the risk situations are preventable, and the drastic decline in infant mortality seen over the past century indicates this potential.

Survival of newborns falls gradually as birth weight declines. Newborns weighing less than 2500 grams (5 pounds 8 ounces) are considered low birth weight, with normal birth weight between 2500 and 4000 grams (5 pounds 8 ounces to 8 pounds 13 ounces). In developed countries the low birth weight rate is usually below 7 percent and in developing countries often over 10 percent. Babies born weighing less than 1500 grams (3 pounds 5 ounces, very low birth weight) and under 1000 grams (2 pounds 3 ounces, ultra low birth weight) have lower rates of survival and are often left with permanent disabling conditions. A major purpose of good prenatal care is to ensure continuance of the pregnancy beyond 36 weeks and as close to full term as possible in order to promote full maturity and development of the fetus *in utero* and ensure greater chance of survival of the newborn. To reduce the factor of multiple births, the term *low birth weight* is sometimes used for single births only.

The newborn deserves the best care available. This requires immediate assessment and immediate response to resuscitation needs or other necessary interventions. The Apgar score, developed by Virginia Apgar in the early 1950s, provides an important standard method of evaluation of the newborn and its need for investigation and intensive care (Table 6.6). Scores of less than 6 at birth may indicate a need for resuscitation and specialized attention, and low scores after 2 minutes indicate an infant at risk. The Apgar score is a useful tool in assessing the newborn, but should not be considered an absolute marker for subsequent outcome and status of the infant.

Where home births take place, the potential for transport of the newborn with difficulties to a neonatal care center should be part of the preparation for low birth weight, respiratory distress, meconium inhalation, and other conditions for which the newborn requires highly trained medical and nursing personnel and equipment to improve survival and quality of outcomes. Early detection and intervention reduce unnecessary morbidity and mortality of the newborn.

In countries with high levels of neonatal mortality, such as those in the Central Asian Republics, there may be high levels of attendance for prenatal care, as was emphasized in the Soviet health system (Chapter 13), but women's health may be compromised by poor nutrition, anemia, iodine deficiency, and delivery in maternity homes detached from hospitals. Neonatal deaths may be largely due to respiratory causes with infection a likely cause, so that prophylactic antibiotics may be indicated. Vitamin K for newborns is also indicated on a routine basis.

All newborns should receive well-trained professional care including resuscitation if needed and routine eye care to prevent ophthalmic infection with gonococcus. Vitamin K should be given routinely either by injection or orally, in order to prevent hemorrhagic disease of the newborn (HDN). As many as 20 percent of newborns have low levels of prothrombin, a blood clotting factor, and are therefore subject to potentially serious and even fatal bleeding, which is preventable by intramuscular vitamin K (Box 6.1).

Screening for HIV infection if the mother's HIV status is unknown and for signs of drug or alcohol effects in newborns is important in high-risk settings. While these may be seen as part of clinical practice, it is up to modern public health agencies to establish the norms and training systems needed to foster high standards of care.

Complete examination by a pediatrician searches for birth defects and normal development and lays the basis for continued care in early infancy. The newborn should be examined for length, head circumference, birth weight, gestational assessment, and jaundice, with results recorded. Examination also includes assessment of the heart, vascular, and neurologic systems. Screening for congenital disease or abnormalities such as congenital cardiac defects, Down syndrome, congenital dislocated hips (CDH), club feet, undescended testes, phenylketonuria (PKU), hypothyroidism, hemolytic anemias (thalassemia and sickle-cell disease), galactosemia, and other inborn errors of metabolism should occur in this period with follow-up of any suspicious or positive findings. Early diagnosis of abnormal conditions allows the child to benefit from suitable management, which can save the child and family from the burdens of serious retardation, painful and frequent hospitalizations, high costs of medical care, other complications, and early death.

The 1996 recommendations of the Genetics Committee of the American Academy of Pediatrics for newborn screening were updated in 2006 to take into account

TABLE 6.6 Apgar Score

Sign/Score	0	1	2
Heart rate, beats/minute	Absent	Slow (<100)	Greater than 100
Respiratory effort	Absent	Slow, irregular	Good, crying
Muscle tone/motion	Limp	Some flexion of extremities	Active motion
Reflex irritability	No response	Grimace	Cough or sneeze
Body color	Blue, pale	Body pink, extremities blue	Completely pink

Note: A widely used system of scoring an infant's physical condition 1 and 2 minutes after birth. The maximum score is 10; those with low scores require immediate attention. The test may be repeated at 5 or more minutes in order to monitor recovery, but the long-term significance is under debate.
Sources: Kendig, J. W. 1992. Care of the normal newborn. *Pediatrics in Review*, 13:262–276.
Committee on Fetus and Newborn. 2006. The Apgar Score. American Academy of Pediatrics. *Pediatrics*, 117:1444–1447.

Box 6.1 Vitamin K and Hemorrhagic Disease of Newborn (HDN) or Vitamin K Deficiency Bleeding (VKDB)

Vitamin K for prevention of HDN in newborns was the subject for which Henrick Dam and Edward Doisy were jointly awarded the Nobel Prize for Medicine in 1943. In 1961, the American Academy of Pediatrics (AAP) recommended vitamin K by IM injection routinely for all newborns. Routine use is widely practiced, but still controversial in Europe and other parts of the world. The AAP recommendation was reaffirmed in 2003 to prevent early and late HDN (VKDB) in newborns.

Early and late HDN occurs in about 1 percent of births, with late HDN mainly in breast-fed infants (rate 4.4 to 7.2 per 100,000 live births) due to immaturity of the liver and inefficient production of prothrombin due to lack of vitamin K production in the gut.

The problem of HDN is neglected internationally due to misinterpretation of a now long-refuted study linking vitamin K and lymphoma in children. The problem of early and late HDN is being reported in developing countries, such as Turkey and India, where this preventive measure is not practiced routinely, and where low birth weight and sole breastfeeding, important risk factors for late HDN, are common.

Sources: Presentation Speech, Nobel Prize in Physiology or Medicine, 1943, http://www.nobelprize.org/nobel_prizes/medicine/laureates/1943/press.html [accessed October 2007 and Web site confirmed May 4, 2008].
Committee on Fetus and Newborn, Policy Statement. 2003. Controversies concerning Vitamin K and the newborn. *Pediatrics*, 112:191–192.
Tulchinsky, T. H., Patton, M. M., Randolph, L. A., Meyer, M. R., Linden, J. V. Mandating vitamin K prophylaxis for newborns in New York State. *American Journal of Public Health*, 1993;83:1166-1168.

Box 6.2 Recommended Standards for Routine Care of the Newborn: The American Academy of Pediatrics

1. Resuscitation by trained personnel should be available at all high-risk pregnancy deliveries, transfer to neonatologist if necessary;
2. Suction mouth and nose and establish airway;
3. Clean, warm, and cover newborn, give to mother;
4. Stimulate the newborn by rubbing back and soles of feet;
5. Oxygen via Ambu bag with face mask, if needed; examine chest for adequate ventilation;
6. Apgar score at 1 and 2 minutes following delivery;
7. Apgar score at 5 minutes following delivery;
8. If prolonged resuscitation is needed, further Apgar scores should be taken and recorded;
9. Establish bonding and breastfeeding by giving newborn to mother as soon as possible;
10. Vitamin K is routinely given to prevent hemorrhagic disease of the newborn;
11. Eye care is given with antibiotic ointment to prevent chlamydia or gonococcal eye infection;
12. Cord blood is taken for serology;
13. Umbilical cord is cut and tied, applying antiseptic or antibiotic;
14. Hepatitis B vaccine is given, the first of three doses;
15. Newborn should be loosely wrapped, not swaddled;
16. Newborn should preferably stay with mother ("rooming in");
17. Father's visits and participation in care are to be encouraged;
18. Breastfeeding is strongly encouraged;
19. Complete examination by a pediatrician or primary care physician;
20. Family and sibling visits are desirable.

Sources: Kendig, J. W. 1992; and American Academy of Pediatrics, op cit. American College of Obstetrics and Gynecology. 1992. *Guidelines for Perinatal Care*, Third Edition. Washington, DC: ACOG, and Elk Grove Village, IL: American Academy of Pediatrics.
Blackmon, L. Committee on Fetus and Newborns. American Academy of Pediatrics. 2004. Hospital stay for healthy term newborns. *Pediatrics*, 113:1434–1436.

---

technological advances and greater knowledge of inherited diseases as well as ethical issues (Box 6.2). The newborn screening system consists of newborn testing and timely follow-up of abnormal results, timely diagnostic testing and disease management. It is important to coordinate with the medical provider and genetic counseling and to conduct continuous evaluation and improvement of the newborn screening system. The disorders reviewed in the newborn screening fact sheets are available at www.pediatrics .org/cgi/content/full/118/3/e934. The March of Dimes reports that 90 percent of newborns in the United States are screened for inherited genetic disorders.

Establishment of breastfeeding is a vital event for mother and newborn and should be encouraged by all health personnel. This is important for bonding between mother and infant and for the optimal development of the infant. It should continue for at least the first half year of life especially and longer if adequately supplemented by solid foods. Breastfeeding is discussed under nutrition (Chapter 8) and in globalization of health (Chapter 16).

Care in the Puerperium

The puerperium is the 6- to 8-week period from the time of delivery of the placenta up to the resumption of normal ovulation. In the nursing mother, the puerperium may be prolonged because of the effects of lactation on hormonal balance. This is an important time for the new mother and newborn to recover from the birth process and for their physiological and psychological adjustment. Adequate follow-up of the mother and infant soon after delivery helps to prevent complications and promotes optimal health for both.

Immediate postpartum concerns focus on assurance of complete expulsion of the placenta, contraction of the uterus, absence of bleeding from any site, or phlebitis.

Basic examinations to protect the new mother's health include a breast examination, determination of the presence of any cervical or perineal tears or infections, and a hemoglobin test. Management of any residual problems should be part of the care process. The mother is provided with information on a variety of different topics including breastfeeding, nutrition, spacing for the next pregnancy, and birth control. She is also given any needed support and counseling to address the potential for postpartum depression which can become a debilitating disorder.

The mother's recovery from the pregnancy and birth process should be aimed at restoring her strength and normal family life. Postpartum depression is common and may need to be addressed by the care provider. Concern to lose weight rapidly may contradict the mother's nutritional demands, which are increased if she is breastfeeding. Iron supplements should continue in the puerperium and while breastfeeding to restore the iron lost to her during pregnancy, delivery, and breastfeeding. If necessary, Rh immune globulin to prevent future maternal–fetal blood immune reaction should be handled by obstetricians. Also if necessary, rubella immunization should be given to the postpartum mother to prevent possible future congenital rubella syndrome (Chapter 4).

Postnatal care should include a home visit by a public health nurse to assess and assist the adaptation of mother and infant and the home setting. A medical examination toward the end of the puerperium is an important element of pregnancy care because it is the first opportunity for complete assessment of the newborn after discharge from the hospital. The key concerns for the totally dependent newborn in this period are breastfeeding and bonding, hygiene and cleanliness, weight gain, vitamin supplements, beginning of immunization, and acceptance into the family. First immunizations for hepatitis B, BCG (where used), and poliomyelitis begin as close to birth as possible and the routine immunization program begins in the first 6 weeks (see Chapter 4). For home deliveries, postnatal home visits should assess and weigh the newborn, administer vitamin K, initiate the vaccination program and routine screening tests, as well as counsel on breastfeeding and infant care.

Counseling by the health care provider is very important to the new mother whether this is her first pregnancy or one of many. Family planning to promote spacing of pregnancies should be discussed as part of the health provider's attention and educational efforts. The need for support and advice is high in this period to reduce anxiety and to address many issues relating to maternal and infant health. Training of health providers working with delivery and newborn care should stress this role along with the professional skills of managing the delivery and immediate neonatal care.

GENETIC AND BIRTH DISORDERS

Worldwide, every year there are 7.8 million children born with serious genetic defects and millions of others are born with postconception defects. Approximately 3.3 million children under 5 years die each year from birth defects (Table 6.7). In the United States, an estimated 150,000 infants are born annually with birth defects with $8 billion spent on their care, and 8000 infants die. Prevention and management of birth defects and genetic disorders are therefore of public health concern, not only because of their high prevalence in many specific population groups, but also because of the financial and social cost to the family and community. Genetic disorders are increasingly both treatable and preventable. Screening, community education, and individual patient genetic counseling are likely to become even more important with the rapid development of gene therapy.

Prevention of birth defects requires screening for genetic diseases, congenital diseases, and Rh incompatibility, and education and folic acid supplementation before and during pregnancy. Prevention also requires reduced exposure to teratogenic chemicals at home, at work, or in the community, and prevention of some infections in pregnancy (Box 6.3). Rh incompatibility prevention depends on screening of pregnant women and management where incompatibility between the maternal and fetal blood can

TABLE 6.7 Causes of Birth Defect Conditions

Cause	Conditions
Genetic	Phenylketonuria, thalassemia, sickle cell disease, congenital hypothyroidism, cystic fibrosis, Down syndrome, autism, fragile X syndrome, other inborn errors of metabolism
Toxic	Thalidomide, fetal alcohol syndrome, drug abuse babies
Nutritional	Folic acid deficiency and neural tube defects, iodine and iron deficiency
Infection	Rubella syndrome, congenital syphilis, cytomegalovirus, HIV, gonococcal conjunctivitis
Physiologic	Vitamin K deficiency, hemorrhagic disease, Rh hemolytic anemia
Multiple causation or unknown cause	Low birth weight, mental retardation and developmental disability (MRDD)
Trauma	Violence, accidents, burns

Source: March of Dimes Foundation, http://www.marchofdimes.com/ [accessed May 8, 2008]

Box 6.3 Birth Defect Prevention

1. Community-wide prevention
   a. Iodine fortification of salt to prevent iodine-deficiency brain damage;
   b. Folic acid fortification of flour to prevent neural tube defects;
   c. Folic acid supplements for all women in age of fertility
   d. Rubella immunization before pregnancy to prevent congenital rubella syndrome;
   e. Treatment of STIs to prevent congenital syphilis, gonococcal, CMV, and HIV infections;
   f. Hepatitis B immunization to prevent maternal–fetal transmission;
2. Prenatal prevention
   a. Cessation of smoking, alcohol, and drug intake to prevent fetal damage;
   b. Pre-pregnancy and pregnancy nutrition counseling to promote optimal fetal development;
   c. Genetic counseling for parents at risk (e.g., consanguineous marriages);
   d. Avoidance of unnecessary medication;
   e. Iron, folic acid, and multivitamin supplements (and iodine if salt is not fortified);
   f. Risk assessment and referral to a high-risk care program;
   g. Prenatal care protocol starting in first trimester;
   h. Fasting blood sugar to screen for gestational diabetes;
   i. Management of chronic diseases (e.g., diabetes, cardiac disease) before and during pregnancy;
   j. Prevention of exposure to teratogenic agents before and during pregnancy (e.g., anesthetic agents);
   k. Prevention of exposure to lead and mercury ingestion;
   l. Hemoglobin, Rh testing, Coombs' test, urinalysis;
   m. Adequate weight gain (11–14 kg);
   n. Sonography for suspected multiple gestation, structural, or placental abnormalities;
   o. Fetal monitoring, fetal size, position, and heart rate;
   p. Safe delivery with minimal sedation and early resuscitation to prevent brain damage at birth.

Source: U.S. Preventive Services Task Force, 1996.

develop. Screening for congenital diseases such as phenylketonuria (PKU) and hypothyroidism must be followed up for confirmation of suspected cases and managed appropriately with suitable standards of care. Other birth defects, such as congenital hip dislocation, which may occur in as many as 50 per 1000 live births, and congenital cataracts are sought on clinical examination of the newborn, with hearing tested within 3 months. Findings should be referred for follow-up care in pediatric services.

The range of birth defects is wide and includes chromosomal and developmental abnormalities as well as metabolic, hematologic, musculoskeletal, cardiovascular, and neurologic disorders. Genetic disorders are transmitted by abnormal chromosomes from one or both parents. Dominant genes are those which cause the genetic disorder in a heterozygote; that is, a person who has one copy of the abnormal and one copy of the normal gene. If only one parent is affected, there is a 50 percent probability of offspring being affected. In recessive or homozygous conditions, both parents are carriers of the abnormal gene. Each child has a 25 percent chance of being affected, a 50 percent chance of being a carrier, and a 25 percent chance of being unaffected. If only one parent is a carrier of the homozygous gene, the disorder is not passed on although the child may also be a carrier.

Rhesus Hemolytic Disease of the Newborn

Rhesus hemolytic disease of the newborn is a serious condition of breakdown of red cells in the newborn due to an incompatibility between fetal and maternal blood. It occurs when the mother is Rh negative and the fetus Rh positive (inherited from the father). Fetal red blood cells can mingle with the maternal blood during delivery, causing the mother to produce antibodies to the fetal blood. This may affect subsequent pregnancies, causing hemolysis of fetal blood, producing anemia, jaundice, brain damage, or death. This was the cause of many stillbirths and neonatal deaths, until the development of exchange transfusions in newborns reduced the death rate. In the 1970s, anti-D immunoglobulin was introduced. This is given to Rh negative women following birth of an Rh positive baby to prevent the mother from developing antibodies which would affect the next pregnancy. Treatment with anti-D immunoglobulin should eliminate this disease, a major victory of preventive care.

Neural Tube Defects

Neural tube defects (NTDs) are defects of the neural tube, the precursor of the brain and spinal cord. This group includes a range of abnormalities from anencephaly, a markedly defective development of the brain which usually results in death within a few hours of birth, to spina bifida, a defective closure of the vertebral column. NTDs vary in degree of severity but many require extensive surgical and medical care. NTDs occur in 1 per 10,000 births in North America. The incidence has declined as a result of screening in pregnancy and primary prevention through folic acid supplementation during pre-pregnancy and in early pregnancy. Screening for NTDs became possible in the early 1970s with amniocentesis and testing of amniotic fluid and later in blood tests for alpha-fetoproteins (AFP). Ultrasound can also detect NTDs.

During the 1980s, British investigators showed that folic acid given before pregnancy greatly reduces the chance of this abnormality developing. Pre-pregnancy care is not common, and compliance with recommended folic acid supplementation is under 30 percent, so that the addition of folic acid to flour was adopted by the U.S. Food and Drug Administration (FDA) in 1996 and elsewhere for primary prevention of this disorder. In April 1998, the FDA and CDC recommended pre-pregnancy and pregnancy supplementation of folic acid, 400 mg/day, to augment the folic acid in bread flour in order to assure 100 percent RDA intake (see Chapter 8) (Box 6.4).

In 1998, the U.S. FDA and the Canadian Federal Department of Health mandated fortification of flour and cereal grain products not only with folic acid but also vitamin B complex and iron. Follow-up studies show a decline in NTDs and the subject of increasing the level of fortification to reduce NTDs even further is actively being debated. CDC expects fortification of bread and supplements for women in the age of fertility to substantially reduce the annual incidence of NTDs in the United States. Since mandatory fortification, there have been reports of large reductions in the NTD incidence rate in Canadian provinces (Newfoundland, Nova Scotia) and the United States as well as Chile, and the subject is under active development in many other countries (e.g., United Kingdom, Israel, the Palestinian Authority).

Cerebral Palsy

Cerebral palsy (CP) is a group of neurologic disorders occurring in about 1 in 400 births, causing motor disabilities. It may be associated with mental retardation, seizure disorders, motor spasticity, or sensory problems. CP is related to low birth weight (<2500 grams), and especially very low birth weight (<1500 grams), as well as intracranial hemorrhage, Rh incompatibility, intrauterine and birth trauma, maternal exposure to heavy metals such as mercury, and other unidentified factors. About 20 percent of CP cases are due to intrauterine fetal hypoxia. Preventive measures include reducing low birth weight births by improving maternal nutrition, smoking cessation during pregnancy, improving prenatal care and care in labor and delivery, giving vitamin K at birth, and reducing infant trauma. Use of professionally trained midwives reduces risk of CP. Prevention is limited by lack of identification of many causative factors. Increasing rates of low birth weight raise the risk of more neurologic damage including cerebral palsy.

Intellectual Disability

Intellectual disability (formerly called *retardation*) is one of the commonest birth and developmental disorders in industrialized countries and may be due to a wide variety of causes. These include Down syndrome, fetal alcohol syndrome, genetic conditions infections (such as congenital cytomegalovirus), or defects that affect the brain (such as hydrocephalus or cortical atrophy) which occur before birth. Low birth weight and asphyxia during delivery or soon after birth, and conditions occurring during childhood such as serious head injury, stroke, or infections such as meningitis.

Severe intellectual disability — intelligence quotient (IQ) of less than 50 — occurs in between 3 and 5 per 1000 newborns. Retardation may be related to prenatal factors and is often associated with CP and seizures. More than one-third of cases are attributed to chromosomal

---

Box 6.4 Neural Tube Defects and Fortification of Flour: U.S. and Canadian Experience

Neural tube defects (NTDs) such as spina bifida and anencephaly affect some 4000 births and pregnancies per year in the United States (1500 births with spina bifida and 1000 with anencephaly) and some 1500 NTD abortions per year. NTDs are associated with 1.3 percent of all infant deaths, second only to cardiac conditions as cause of neonatal deaths. NTD survivors suffer from spinal cord abnormalities with serious physical and mental disabilities. They may reach adulthood, but lifetime medical and indirect costs are estimated at $350,000 per case and a total economic burden of nearly $500 million.

Spina bifida and NTDs in Newfoundland between 1976 and 1997 occurred at a mean rate of 3.40 per 1000 births. There was no significant change in the average rates between 1991–1993 and 1994–1997. The rates of NTDs fell by 78 percent after the implementation of folic acid fortification, from an average of 4.36 per 1000 births during 1991–1997 to 0.96 per 1000 births during 1998–2001 (RR 0.22). The average dietary intake of folic acid due to fortification increased in women aged 19–44 years and in seniors. There were significant increases in serum and RBC folate levels for women and seniors after mandatory fortification. Among seniors, there was no evidence of vitamin B_{12} deficiencies, and no evidence of masking hematologic manifestations of vitamin B_{12} deficiency.

Very few countries (U.S., Canada, Chile) have adopted folic acid fortification of flour and cereal grain products as of mid-2007. WHO and the EU have thus far failed to recommend flour fortification as an essential public health program.

Sources: Committee on Genetics, American Academy of Pediatrics. Folic acid for the prevention of neural tube defects. *Pediatrics*, 104:325–327. American College of Obstetrics and Gynecology. Periconceptional folic acid and food fortification in the prevention of neural tube defects. Scientific Advisory Committee, Opinion Paper 4. April 2003.
Wald, N. 2004. Folic acid and the prevention of neural-tube defects. *New England Journal of Medicine*, 350:101–103.
DeWals, P., Tairo, F., Van Allen, M. I., et al. 2007. reduction in neural-tube defects after folic acid fortification in Canada. *New England Journal of Medicine*, 357:135–142.

abnormalities. Prenatal diagnosis, widespread use of amniocentesis in pregnant women over age 35, and termination of affected pregnancies have contributed to reducing retardation. Phenylketonuria, congenital rubella, congenital hypothyroidism, and maternal infections with toxoplasmosis and cytomegaloviruses can all cause severe retardation. Pregnancy complications such as toxemia, urinary tract infections, and anemia increase the risk of retardation. Mild disability (IQ of 50–70) may relate to both perinatal and postnatal factors, including low birth weight or asphyxia at birth. Prevention requires well-organized prenatal, perinatal, and postnatal care.

Down Syndrome

Down syndrome is a relatively common genetic disorder, occurring in 1 per 650 live births in the United States. The risk increases rapidly with increasing maternal age; for mothers over age 40 it occurs in 1 per 40 births. Although this condition occurs more commonly in older mothers, the majority of cases are in younger women. It is the most common cause of mental retardation in industrialized countries. In cases at risk, prenatal screening is done by amniocentesis, chorionic villus sampling, and fetal blood sampling, allowing for the parental choice of termination of pregnancy should the chromosomal abnormality be detected. Chorionic villus sampling can be performed in the first trimester, while amniocentesis can be performed in the second trimester. Chorionic villus sampling has itself been reported to be associated with minor birth defects and is currently under review. Pregnant women over age 35 should be screened as early as possible in pregnancy. Biochemical markers will help to increase screening with less invasive procedures. The total incidence of new births of Down syndrome cases fell by 50 percent between 1960 and 1978 in the United States, probably because of reduced fertility among older women, widespread use of birth control, and prenatal screening.

Down syndrome babies are mentally retarded, often with congenital cardiac defects and gastrointestinal obstruction. These patients now survive well into their 30s, and half to their 50s. Institutional care for a lifetime is estimated to cost $300,000, although emphasis has moved to self-care in the community. Education and training for Down syndrome children have been emphasized in recent years. With vocational and life skills training, many are able to live independently or in groups in the community rather than in institutions. Those who are integrated into the community often live longer and have a better quality of life as compared to those raised in institutions.

Cystic Fibrosis

Cystic fibrosis (CF) is the most common lethal genetic disease in the U.S. white population, occurring in 1 per 3500 for North Americans of Northern European descent, but 1 per 15,000 for black and 1 per 7000 for Hispanic newborns. The disease is a recessive genetic defect. The gene is present in 5 percent of the white population and 2 percent of African-Americans. The defect in homozygotes causes production of abnormally thick mucus in the lungs, intestines, and other glands.

The clinical disturbance is chronic obstructive lung disease, repeated infections, and destruction of lung tissue. This leads to frequent hospitalization and death, even with the best of care. In 2005 in the United States, the predicted median age of survival rose from 20 years in 1970 to 32 in 2000 and to 36.5 years in 2005. Prenatal screening can be done with chorionic villus sampling or amniocentesis. Newborn screening is now routine in most states in the United States and in other industrialized countries because of demonstrated benefits of early treatment on growth, nutrition status, cognitive development, reduced lung infection, and hospitalization improved quality of life and survival of CF patients. Early diagnosis and support and education of parents can significantly improve the duration and quality of life of the affected person. Treatment is complex and costly, involving a multidisciplinary approach. Gene therapy techniques, new vaccines, and other therapies including lung transplantation are important areas of new development in case management for CF.

Sickle-Cell Disease

Sickle-cell disease is an amino acid defect of red blood cells which affects 1 in every 346 African-American newborns and is even more common in Africa, where it may have a protective effect against malarial parasites. The disease is caused by a recessive gene affecting hemoglobin structure, usually with a benign course in the carrier state, but with episodic pain due to small blood vessel closure in the homozygous state. Clinical symptoms appear in the second half of the first year of life. The patient develops moderately severe anemia, with an increased susceptibility to severe bacterial infections (meningitis, pneumonia, septicemia) and growth retardation. Screening of newborns has been recommended since 1972 in the United States; in practice, however, screening is done selectively with 20 states screening 57 percent of African-American infants. The U.S. target is to achieve 95 percent coverage. Identification of cases and carriers is important in order to assure prompt care in crises, for preventive use of penicillin which reduces infection rates, and to ensure they receive influenza and pneumococcal vaccines. Carriers should receive genetic counseling related to marriage and pregnancy.

Thalassemia

Beta thalassemia is a recessive genetic disorder of hemoglobin structure. Beta thalassemia minor is usually without clinical significance. Beta thalassemia major, the homozygous state, is characterized by hemolytic anemia (i.e., early breakdown of red blood cells). Beta thalassemia major, also known as Cooley's anemia or Mediterranean anemia, is widespread throughout the Middle East, southern Europe, and across southern India and southeast Asia. It is ultimately fatal for those afflicted, but with current standards of treatment, including blood infusions and chelating agents (i.e., iron binding for excretion) to reduce iron overload and hemochromatosis, patients survive into their 30s (Box 6.5).

WHO estimates that 240,000 deaths, 290,000 new cases, and a total of 2.32 million cases of thalassemia and sickle-cell disorder occurred in 1995. The disorders of hemoglobin are widespread, originally localized in the Mediterranean areas, spreading to southeast Asia for the thalassemias and sub-Saharan Africa for sickle-cell disease. Because of migration, these diseases are now spread worldwide, with 10 percent of the population at risk in the United States.

Prevention of thalassemia is one of the success stories of genetic public health. Preventive approaches to this disease since the 1970s produced dramatic results in reducing the number and rate of new cases of beta thalassemia major in Sardinia, Cyprus, Greece, the United Kingdom, Canada, and other locations where this disease has been endemic among peoples of Mediterranean origin. Premarital screening, health education in schools, and access to prenatal screening are all part of a program to reduce new cases of this disease. WHO stresses the public health importance of this disease and recommends adoption of demonstrably successful preventive approaches to member states having this problem.

Screening for congenital anemias should take place at birth and at school age, because clinical cases may not appear until several years after birth. Gene carriers need to know at an age when they can understand the limitations this may place on them. In each pregnancy when both parents are carriers, chorionic villus sampling or amniocentesis should be carried out to determine if the fetus is affected. Abortion is currently recommended if the fetus is affected with thalassemia major, but bone marrow transplantation is showing promise in the treatment of new cases. The success of primary prevention in reducing new cases of thalassemia provides a model for application to disorders affecting other population groups, such as sickle-cell anemia.

Phenylketonuria

Phenylketonuria (PKU) is an inborn error of metabolism transmitted by a recessive gene. It occurs in approximately 1 per 13,500 births in North America, but this includes a wide range from 1 per 5000 in Scottish Irish parents to 1 per 300,000 in African-Americans. It involves a mutation in DNA which causes inadequate production of an

---

Box 6.5 Eradication of Thalassemia Major in Cyprus and Sardinia

Virtual eradication of thalassemia major has been achieved in formerly endemic areas. In Cyprus, 14 percent of the population was of carrier status for beta thalassemia and 1 percent of the Cyprus population (1 in 158 births) had the disease. Due to an intervention program initiated in the 1970s, only rarely are new cases of beta thalassemia major born in Cyprus and other Mediterranean locations such as Sardinia and Greece. This has been achieved by a long-term preventive program consisting of public education, screening for carriers, and counseling. Marriage between carriers is reduced by a community education program, and, when marriage does occur, careful screening of all pregnancies and termination of affected pregnancies reduces the number of thalassemia major births. The success of the Cyprus approach provides a model of control of a genetic disorder via a combination of health education, screening, and community support.

Education and planning for routine newborn screening to detect hematologic diseases, inborn errors of metabolism and other congenital conditions are part of modern care of newborns in developed countries. The benefits are in early case detection and treatment to prevent the damage done to brain, (by phenylketonuria and congenital hypothyroidism), lungs (by cystic fibrosis) and other body systems including the hematological system, as in the case of thalassemia and sickle cell anemia.

With rapidly increasing knowledge and science associated with the Human Genome Project, the potential in this field will become increasingly large. There are social and ethical issues associated with screening and ethnic focus of educational programs, but the success of thalassemia control is a great achievement of public health in the twentieth century and its application in the twenty-first century is being applied to other important genetic disorders such as Tay-Sachs and more recently cystic fibrosis. In the future it may be even more important in relation to genetic factors in cancer, cardiovascular disease, and other common conditions. This will raise many ethical and political issues which have to be addressed in the context of different settings, religious beliefs, and cultural norms.

Sources: Angastiniotis et al. 1986. How thalassemia was controlled in Cyprus. *World Health Forum*, 7:291–297.
World Health Organization. 2002. *Genomics and World Health: Summary*. Geneva: World Health Organization.
Centers for Disease Control. Birth Defects 2005, http://www.cdc.gov/ncbddd/bd/genetics_screen.htm and National Office for Public Health Genomics. 2007.
CDC Activities, http://www.cdc.gov/genomics/activities/newborn.htm [accessed May 6, 2008]

enzyme needed to metabolize phenylalanine, an amino acid. The fetus is unaffected because the maternal metabolism handles the excess phenylalanine. However, if the newborn lacks the enzyme needed, phenylalanine accumulates in the blood, leading to brain and neurologic disorders and severe retardation. If discovered early, PKU can be managed with a special low phenylalanine diet with no significant retardation.

Screening of all newborns for this condition is widespread in developed countries and reduces the number of cases requiring long-term institutional care. Where screening is not universal, pregnant women with previous PKU children should be tested during pregnancy and the newborn tested at birth. Women with PKU should maintain a special diet before and during pregnancy to help prevent mental retardation and birth defects. CDC considers phenylketonuria (PKU) screening is a major public health success, with an estimated 3000–4000 reproductive-aged women with PKU identified by newborn screening and placed on a diet that prevented severe mental retardation. Persons with PKU should stay on the special low alanine diet throughout life, but this severely restricted diet is often discontinued during adolescence. Babies born to women with PKU not following the diet are at high risk for mental retardation and birth defects, due to the mother's condition. These effects can be prevented if women follow the PKU diet before and during pregnancy (CDC, 2002, 2008).

Congenital Hypothyroidism

Congenital hypothyroidism (CH) is a relatively common congenital disorder occurring in about 1 per 3500–4000 live births in a white population with a range of 1–20 per 10,000 live births in different population groups, depending on genetic makeup and consanguinity rates. This genetic disorder causes an inefficient development of the thyroid and may be confused with iodine deficiency disorders common in areas with deficient iodine in water and soil. Therapy with thyroid replacement in CH cases prevents the severe intellectual impairment that otherwise occurs. Screening within 48 hours of birth should uncover cases for long-term follow-up and management.

Fetal Alcohol Syndrome

Fetal alcohol syndrome is caused by the toxic influence of alcohol on the fetus via the placenta. Alcohol consumption prior to and during pregnancy increases the risk of behavioral and cognitive disorders. Exposure of the fetus to maternal alcohol consumption, especially binge drinking (i.e., more than five drinks on one occasion), is a cause of fetal growth retardation and anomalies of the central nervous system. Fetal alcohol syndrome is completely preventable, as it only occurs when the mother drinks alcohol.

Tay-Sachs Disease

Tay-Sachs disease is an inborn error of metabolism associated with progressive mental deterioration and loss of vision by 4–8 months of age and death by 3–4 years. It affects 1 in 2500 births among Jews of eastern European origin. Tay-Sachs disease can be prevented by screening these risk groups before marriage, at the time of marriage, or early in pregnancy. Prenatal screening by amniocentesis or chorionic villus sampling will confirm whether the fetus is affected. In such cases, termination of pregnancy is recommended. These screening programs have reduced the incidence of new cases.

G6PD

Glucose-6-phosphate dehydrogenase (G6PD) deficiency is a genetic disorder common in Mediterranean populations, including Jews of North African or Middle Eastern origin (Sephardic Jews), Greeks, southern Italians, southeast Asians, and southern Chinese. It results in episodes of hemolytic anemia due to an infection, reactions to certain foods (e.g., fava beans), or reactions to oxidant drugs such as sulfonamides, antipyretics, and antimalarials. The degree of hemolysis varies with the agent and the degree of the enzyme deficiency. Identifying the condition helps the patient avoid exposure to hemolytic inducing agents and start prompt treatment in crises.

Familial Mediterranean Fever

Familial Mediterranean fever is a recessive hereditary condition found in Arabs, Armenians, and Sephardic Jews with periodic fevers and pains in the chest, abdomen, and joints. Control is by genetic counseling.

INFANT AND CHILD HEALTH

Public health has long played a major leadership role in improving the health of children by provision of care and regulation of conditions to prevent disease, provide early and adequate care of illness, and promote health. Pediatrics developed as a clinical specialty under the leadership of Abraham Jacobi who opened the first pediatric clinic in New York City in 1860 based on German models. The first children's hospital in the United States was opened in 1865 in Philadelphia and continues to operate to the present time.

The American Medical Association, recognizing that women and children had health needs apart from those of the general population, established a section to address those needs in 1879, leading to the founding of the American Pediatric Society in 1888. Pediatrics emerged as a separate specialty from general medicine with emphasis on the treatment of children's diseases and birth disorders, the prevention of infectious diseases, and infant nutrition. Well-child care was pioneered in the United States based on milk stations, adapted from those in France ("Gouttes de Lait"), in the poor immigrant neighborhoods in New York City, leading to the development of public health nursing and the visiting nurse function.

The first textbook of pediatrics was written in 1869 by J. L. Smith, professor of Children's Diseases at Bellevue Hospital Medical College in New York. It was followed by Holt's *Diseases of Infancy and Childhood* in 1896. Now known as *Randolph's Pediatrics*, this authoritative text was published in its 21st edition in 2003. The U.S. federal government established the Children's Bureau in 1913 to collect data on maternal and infant mortality. This later became the Bureau of Maternal and Child Health, which was subsequently empowered by federal legislation to provide grants to states for maternal and child health services. The American Academy of Pediatrics (AAP) has come to the forefront of advocacy for improved child health standards, pioneering in development of clinical guidelines and in promoting professional standards in a wide range of child health topics, from breastfeeding to cystic fibrosis screening, in parallel to the American College of Obstetrics and Gynecology (ACOG).

In the wider context of health internationally, maternal and child health are among the major priorities with special focus on primary health care. Primary health care, promoted since the Alma-Ata Conference, has placed emphasis on infant immunization, diarrheal disease control, breastfeeding and nutrition practices, and the prevention of deficiency disorders. The WHO, UNICEF, and many NGOs have provided leadership in the development of primary health care in the developing world. This return to basics in health care has been of benefit to the industrialized world as well, as health care costs soared during the 1970s and 1980s. In the United States, public health had to cope additionally with the lack of health insurance for a substantial portion of the population, and even in the early twenty-first century, federal initiatives in health are focused on providing health benefits to about 15 percent of the population who lack health insurance of the total U.S. child population of 73 million.

Fetal and Infant Mortality

A *live birth* is defined by the World Health Organization and the United States National Center for Health Statistics as a completed expulsion or extraction from its mother of a product of conception, irrespective of the duration of the pregnancy, which after separation, breathes or shows any other evidence of life such as heartbeat, umbilical cord pulsation, or definite movement of voluntary muscles, whether or not the umbilical cord has been cut or the

placenta is attached. Each product of such a birth is considered live born. The delivery can be described as spontaneous vaginal, forceps, vacuum extraction, cesarean section, or vaginal delivery after previous cesarean.

A fetal death or stillbirth is a death prior to the complete expulsion or extraction from its mother of the product of conception, irrespective of the duration of pregnancy. The death is indicated by the fact that after such separation, the fetus does not breathe or show any other evidence of life, such as beating of the heart, pulsation of the umbilical cord, or definite movement of voluntary muscles. For statistical purposes, tabulations for fetal deaths are for a stated or presumed gestation of 20 weeks or more and for 28 weeks or more gestation; the latter are known as late fetal deaths. Other indicators of infant mortality are seen in Box 6.6.

The infant mortality rate (IMR) is a generally accepted indicator of the health status of a population for internal regional and international comparisons, because it represents the cumulative effect of many socioeconomic, environmental, and health service factors. The industrialized countries have IMRs under 9 per 1000 live births and many as low as 5 or even 4 per 1000. In these countries, most infant deaths are a result of congenital anomalies and perinatal conditions associated with prematurity and the neonatal period, usually occurring during the first week of life (early neonatal deaths).

In developing countries, most infant deaths are due to acute respiratory infections, diarrheal diseases, and prematurity (LBW), with measles and tetanus remaining large-scale causes of infant death (Box 6.7). Developing countries generally have IMRs over 30 per 1000, with high neonatal and postneonatal mortality. Preventive health measures, such as immunization, breastfeeding and nutritional supplementation, good management of acute respiratory infections, and use of oral rehydration for diarrheal diseases, are effective mainly in the postneonatal period. Neonatal mortality rates can be reduced by reducing the number of low birth weight births, providing good maternal nutrition and prenatal care, minimizing birth injury, and giving good care immediately following birth. Together, these constitute the child survival package of services that are one of the main thrusts of public health. LBW is reportedly rising in industrialized countries such as Britain, possibly due to use of fertility treatments and nutritional and lifestyle issues. This raises concerns because of the much higher rates of morbidity and mortality among LBW infants especially in the neonatal period. LBW has become a major target program for the March of Dimes, an important NGO in the United States.

The United States emphasizes targeted public health programming with attempts to provide care to people without access to prepaid medical care. Federal health officials set a goal to reduce the country's infant mortality rate and then expanded access to prenatal and infant care

---

Box 6.6 Derivation of Rates in Infant Mortality

1. $Total\ Stillbirth\ Rate = \dfrac{Number\ of\ stillbirths}{Total\ stillbirths\ plus\ live\ births\ (per\ annum)} \times 1000$

2. $Perinatal\ Mortality\ Ratio = \dfrac{No.\ stillbirths\ +\ no.\ deaths\ in\ the\ first\ week\ of\ life}{Total\ live\ births\ plus\ stillbirths\ (per\ annum)} \times 1000$

3. $Early\ Neonatal\ Mortality\ Rate = \dfrac{Number\ of\ deaths\ in\ first\ week\ of\ life}{Live\ births\ (per\ annum)} \times 1000$

4. $Late\ Neonatal\ Mortality\ Rate = \dfrac{Number\ of\ deaths\ between\ the\ first\ 7\ and\ 28\ days}{Total\ live\ births\ (per\ annum)} \times 1000$

5. $Neonatal\ Mortality = \dfrac{Number\ of\ deaths\ in\ first\ 28\ days\ of\ life}{Total\ live\ births\ (per\ annum)} \times 1000$

6. $Post\text{-}Neonatal\ Mortality = \dfrac{Number\ of\ deaths\ between\ 28\ and\ 364\ days\ of\ life}{Total\ live\ births\ (per\ annum)} \times 1000$

7. $Infant\ Mortality\ Rate = \dfrac{Number\ of\ deaths\ from\ birth\ to\ 364\ days}{Total\ live\ births\ (per\ annum)} \times 1000$

TABLE 6.8 Mortality Rates for Infants,[a] United States, by Race and Percentage Change, 1970–2004

Group	1970	1980	1990	2000	2004	% decline 1970–2004
Neonatal white	13.8	7.5	4.8	3.8	3.8	72
Neonatal black	22.8	14.1	11.6	9.4	9.1	60
Postneonatal white	4.0	3.5	2.8	1.9	1.9	53
Postneonatal black	9.9	7.3	6.4	4.7	4.7	53
Total infant, white	17.8	11.0	7.3	5.7	5.7	68
Total infant, black	32.6	21.4	16.9	13.5	13.2	60
Total infant mortality	26.0	20.0	8.9	6.9	6.8	74

[a]Rates per 1000 live births.
Sources: Health United States, 2001, http://www.cdc.gov/nchs/data/hus/tables/2001/01hus023.pdf [accessed May 5, 2008].
Health United States, 2007: with Chartbook on the Health pf Americans http://www.cdc.gov/nchs/hus.htmhttp://www.cdc.gov/nchs/hus.htm [accessed May 5, 2008].

---

in order to reach this goal. From 1970 to 2004, the IMR of the United States fell from 26.0 to 6.8 per 1000 live births, a decline of 74 percent.

The U.S. health target established in 1979, when the IMR was 15 per 1000, was to reduce the IMR to no more than 9 per 1000 live births by 1990, and this was achieved. However, the rates are substantially different for white and African-American infants, reflecting socioeconomic differences and lesser access to medical care for the African-American population (Table 6.8). Although mortality rates have declined for all ethnic groups, the African-American infant mortality rates remain more than double the rates for white infants.

INFANCY CARE AND FEEDING

The new infant is dependent on a healthy, caring mother with support of the family and health providers. Physical and emotional warmth, cleanliness, and bonding with the mother are important in care and feeding (see Chapter 8). Healthful feeding in infancy is a vital issue in infant care and for primary care services, and is as important to infant survival and well-being as infectious disease control and immunization. Therefore, it has an important place in maternal education and in health provider orientation (Box 6.8).

Breastfeeding should be established early and encouraged as the sole feeding for 4–6 months, because of the

Box 6.7 Leading Causes of Infant Mortality

Developed countries — IMR Under 20/1000
1. Congenital anomalies
2. Sudden infant death syndrome
3. Prematurity/low birth weight
4. Respiratory distress syndrome
5. Newborn effects of maternal complications

Developing countries — IMR 20–30/1000
1. Acute respiratory infections
2. Diarrheal diseases
3. Prematurity/low birth weight
4. Measles
5. Newborn effects of maternal complications
6. Neonatal tetanus

Box 6.8 Recommended Infant Feeding and Supplementation

```
                              Iron syrup
|------------------------------------->
              Vitamins A+D
|-------------------------------------->
                      months
|----|----|----|----|----|----|----|----|----|----|----|
1 2 3 4 5 6 7 8 9 10 11 12
|----------------------->-------->  - - ->
      Breastfeeding
|------------------------------->
                      Add cereals, vegetables,
                      fruit, meat, eggs gradually
                      |----------------------------->
                                  Eat from family table
```

months of age when rapid brain growth and psychomotor development are at their peak. Iron supplementation should be given to all infants (from 4 to 12 months for normal birth weight babies and from 2 months for those with low birth weight) as iron-fortified formulas, supplemental iron, and/or iron-enriched foods. If the child is not receiving iron-enriched formula, iron syrup, preferably with vitamin C, is given daily from 4 to 12 months at a level of 7–15 mg/day according to weight. Secondary prevention is by screening all infants at age 9–12 months to determine if hemoglobin is normal or if further care or investigation is needed.

Following cessation of breastfeeding, infant formula enriched with iron should be given preferably up to 1 year. Whole cow's milk should not be used until the infant is close to 1 year of age. Cereals fortified with iron and vitamins are also important in infant nutrition along with fruit, vegetables, meat, and eggs for optimal growth and development and prevention of vitamin and mineral deficiencies. Later in the first year, the infant should be eating at the family table.

The physical and emotional love of the mother, father, and family provide the security and stimulus needed by the infant to develop physically, intellectually, and emotionally. Stimulus activities such as playing with and speaking to the baby help promote the normal development of the infant.

Immunization is discussed in detail in Chapter 4, and the current U.S. pediatric immunization schedule is presented. While there are many vaccination schedules and much controversy on this topic, the central importance of coverage of the infant with vaccination on schedule is of paramount importance for a successful child health program, including both infancy and school-age follow-up.

ANTICIPATORY COUNSELING

An important element of the role of the pediatrician, family physician, or other health provider is counseling parents on what is to be expected at each stage of the child's development. This includes warnings of potential difficulties that might be experienced and how to cope with them. The American Academy of Pediatrics publishes detailed guideline questions for the practitioner addressing these issues from early infancy to the teenage years (see http://www.aap.org/).

DOCUMENTATION, RECORDS, AND MONITORING

Standardized, user-friendly child health records for the primary care clinic help to promote national standards of child care practice, especially for preventive health care (Table 6.9). They help guide the busy practitioners to

ideal composition of breast milk nutritionally and for immunologic protection of the infant. Duration of breastfeeding should ideally continue to 1 year or more. Vitamins A and D should be given daily from about 2 weeks after birth to prevent clinical and subclinical micronutrient deficiency conditions for both breast- and formula-fed infants. Reports of clinical rickets and evidence of vitamin D deficiency in children and teenagers has led the American Academy of Pediatrics in 2003 to recommend routine vitamin D supplements for children up to teenage years, in addition to the vitamin D fortification of milk. Immigrant children and women covered for religious reasons are particularly subject to vitamin D deficiency, but children spending more time on computers indoors and less time playing outside are also at risk.

Prevention of iron-deficiency anemia (IDA) is important because of its effects on the infant between 6 and 24

TABLE 6.9 Components of a Child Health (Preschool and School) Health Record

I. Personal Data Mother: name, age, family status, education, occupation Father: name, age, family status, education, occupation Address, telephone no. Social security (ID) number Home situation: parent(s), sibs	II. Pregnancy/Birth Gestation: Place of birth: Date of birth: Birth weight: Delivery by: Condition at birth and at 1 and 5 minutes of age: Apgar	III. Risk Factors LBW: Previous care: Birth order: Family status: Social situation: Home conditions: Family medical history: Overall risk assessment:
IV. Growth Patterns: NCHS Growth curves in graphic chart form Weight for age Height for age Weight for height Head circumference	V. Immunizations: Dose, date, reactions BCG Hepatitis B DPT OPV IPV MMR *Haemophilus influenzae* type B Varicella	VI. Nutrition, age when added Breastfeeding: duration Formula: Vitamins A+D: Iron: Supplements: solid foods, cereals, fruit, vegetables, eggs, meat
VII. Intercurrent Illnesses Diarrheal: age, duration Respiratory: age, duration Hospitalizations: Surgery: Others:	VIII. Development Hearing Vision Responsiveness Relates with family Grasp, Smile, Grab Roll over Sit, Crawl Stand, Walk Talk: words, sentences Other milestones	IX. Medical Examinations Birth: 6 Weeks: 3 Months: 6 Months: 12 Months: 24 Months: 36 Months: 48 Months: Preschool: School ages:
X. Summary of Risk Factors Social Family Genetic Medical Other	XI. Laboratory Results PKU, Hypothyroidism Hemoglobin Others	XII. Continuing Notes Document, developments, routine examinations, illnesses, and care given

improve legal protection and professional peer group standards. Design of records systems and training of staff in their use should be given careful attention in any health care system, including private medical practice.

Well-constructed child health records set standards of care expected of the health provider. They furnish continuity of care when the contact with the health system may be episodic, or when the child is seen by different providers. They provide the caregiver with important time trends which may have clinical or epidemiologic importance. Some child records are designed to continue through to adolescence, while others focus on the child up to school age, followed by a different record. Parent-carried child health records ("road to health cards") provide the parents with all data needed for care during childhood and should continue through to adolescence, and should be designed to include all basic information on one folded hard paper record, preferably with a plasticized envelope or cover pages. Many designs are possible, but timelines and visually easy formats enhance ease of use by providers.

Maintenance of a child health record by the parents is important for its successful use.

Physical development in terms of weight and height for age and psychosocial developmental indicators should be carefully documented during infancy at the time of visits for immunization. Growth patterns are good health markers. The growth curves recommended by the World Health Organization are those developed by the U.S. National Center for Health Statistics (NCHS), based on a healthy U.S. white middle-class population. This provides an optimal or "gold standard" for use internationally, as opposed to local standards which may represent less than optimal child health and development. The NCHS is developing a revised set of standard growth curves based on data from a wider population sample. Most importantly, it provides an external standard for comparison of change over time in the local setting.

While there is controversy on use of international or local standards (discussed in Chapter 8), there is full agreement on the need for growth monitoring and early intervention where an infant or a group of infants show signs of growth faltering or failure-to-thrive (FTT). Growth faltering occurs not only due to food insufficiency, but also as a result of intercurrent illnesses, which themselves can be more severe when nutrition status is compromised, such as the relationship of high case fatality rates from measles and other infections when vitamin A deficiency is present. FTT is a medically urgent situation requiring full evaluation and intervention as rapidly and consistently as possible.

THE PRESCHOOLER (AGES 1–5 YEARS)

The preschool child is undergoing rapid growth and development. The American Academy of Pediatrics recommends health assessments at 12, 15, and 18 months, and at 2, 3, 4, and 5 years (Steering Committee on Quality Improvement and Management and Committee on Practice and Ambulatory Medicine, 2008). The examination includes assessment of development status and physical examination as well as parental concerns and skills. Hearing should be checked along with signs of strabismus and general medical examination. The examination includes behavioral assessment based on both interview and observation. The care provider should use the visits to assess parental concerns and convey information on the child's health needs. Abnormal findings on physical or developmental examination should lead to referral to a clinical or child development assessment service. Hemoglobin (or hematocrit) and urinalysis should be checked at least once from 18 months to 5 years.

Counseling on common child care problems (e.g., nighttime crying), safety issues, and stimulation of the child should be discussed with the parent at each of these visits. Dangers of accidents are serious at these ages. Prevention of accidental falls, poisoning, aspiration of food, electrocution, burns, and scalding requires diligence and safety planning in the home. Parents should be educated on how to explain safety to their children as well as how to childproof their homes (using childproof latches on cabinets and stairways, covering electrical outlets, safe storage of common household chemicals, and removal of poisonous houseplants). Signs of child abuse or neglect, especially injuries without adequate explanation, should be reported to police or welfare agencies. Assessment and counseling regarding appetite problems, toilet training, separation anxiety, demanding or obstinate behavior, relationship with parents and siblings, nutrition, self-feeding, and psychomotor development are all part of well-child care.

The physical and emotional well-being of the child up to school age is a key determinant of his or her health and capacities afterward. Completion of immunization, adequate nutrition with essential vitamin and mineral supplementation, and prompt adequate treatment of intercurrent infections are essential. Warmth and responsiveness, including speaking and reading to the child to promote interest in books, pictures, and music are part of needed stimulation and play. Even warm and caring parents need guidance, as well as advice and support from the health provider, to promote optimal development for the child.

The community and national levels have a substantial role to play in promotion of safety and health for toddlers. Mandatory safety practices for children can save many lives and reduce child morbidity. These include such factors as correct use of safe child car seats, childproof medication or poisonous material containers, lead abatement programs, regulation of child clothing standards to prohibit flammable materials, toy safety standards, supervision of nursery schools, and teacher qualifications.

Child abuse is an increasingly recognized hazard at all ages of childhood and in all socioeconomic groups. Preventive measures include public awareness through education and now mandatory reporting of signs of abuse by teachers, social workers, and doctors. Punishments of abusive parents or guardians, including imprisonment and mandatory treatment, are increasingly applied, but hidden physical, mental, or sexual abuse is far more common than previously recognized. Home, school, and playground safety involves a wide range of community activities ranging from mandatory use of helmets for bicycle riding and many sports, proper use of car seats and seat belts for children in motor vehicles, to measures to reduce access of children to firearms.

In most industrialized societies, a large proportion of women work outside the home, so that child care includes preschool centers in which health factors can be both positive and negative. The potential for spread of infectious diseases among child care center attendees is high, and outbreaks of food-borne disease can also occur. In the United States, children in licensed day care centers have three to four times as much diarrheal disease as children not in organized child care centers. A Finnish study (1984–1989) showed that between 41 percent and 85 percent of colds, otitis media, and pneumonia in 1-year-old children were due to the children attending day care centers as opposed to family day care. On the other hand, children attending well-supervised preschool day care centers benefit from interacting with teachers and other children to develop social skills and to receive nutritious meals and professional attention. Preschool care centers should provide education in hygiene, nutrition, and health services including updating of immunization and developmental assessment. Services such as these may help parents to assess and meet the emotional and health needs for their children.

The value of preschool education, especially for children from socioeconomically deprived situations, was demonstrated by follow-up studies of children who had been in the U.S. Head Start program during the 1960s War on Poverty program. Disadvantaged children participating in Head Start were found to perform significantly better in school and less likely to have a criminal record or to be on social welfare than matched children not in the program. This resulted in benefits from reduced costs for special education and public assistance.

Preschool examination and screening for developmental, cognitive, and behavioral problems can be helpful to parents and teaching personnel. A variety of screening tests are available and used, but they are controversial as to their validity and predictive value. They may be culture-specific and not necessarily valid in all societies without adaptation. Support, stimulus, and guidance can be culture-specific but are needed by the growing child.

SCHOOL AND ADOLESCENT HEALTH

The hazards that face schoolchildren include many public health issues. A central goal of public health regarding this young population is life skill training to promote healthful practices and avoidance of risk behavior. Public health efforts to reduce the toll of teenage premature loss of life (Table 6.10) must be in concert with educational, social, and police authorities. Efforts to convey a message to these risk groups should involve popular figures of the sports and popular music fields recruited to stress a positive image of teen adjustment and transition to adulthood. Research is needed to develop educational approaches for high-risk teenagers and young adults.

A school health program should ensure a safe and healthful environment for the children, including sanitation, safety from violence, temperature control, and physical facilities for study and recreation. School-aged children need physical activity for their healthful development as an integral part of a comprehensive school health education program. School meal services contribute to good child and adolescent nutrition, learning, and performance (Box 6.9).

Health monitoring includes ensuring full immunization at school entry and maintenance of its adequacy by appropriate booster doses at the elementary, secondary, and postsecondary stages. Monitoring of growth, vision, hearing, scoliosis, and skin testing for tuberculosis are frequent elements of school health programs.

The potential for learning about health in the school setting should be exploited by the public health system by encouraging the educational authorities to include health as part of the regular curriculum and preparing the teachers on how and what to teach in this area. During the early school years, the areas of study can include safety (fire and traffic), oral health (fluorides, brushing, and flossing), healthy eating habits, hygiene, how the body functions, and healthy living (sleep and exercise). Topics for continuing education of teachers include assessment of child abuse, suicidal or violent tendencies, nutritional status, mood disorders, and substance abuse. Violence prevention has become an essential topic in educational systems at the primary as well as secondary levels.

Secondary school health programs include attention to personal behavior issues, including personal and family communication, sexual relationships, the right to say no (to peer pressure in situations ranging from alcohol and drug use to sexual activity), parental responsibilities, birth control, and STI and HIV prevention. Accident and violence prevention is also vital in this population, including safety in sports and work, use of vehicles, and overall safety as a life habit.

Preadolescent and adolescent children are subject to social pressures that promote early sexual activity, smoking, substance abuse, eating disorders, accidents, suicide, and violence. The school is an important part of their life and has the potential to serve a socially educational role for attitudes and practices that may be lifelong. An activist approach to school health promotion should be integral to the school curriculum. School health committees of students, teachers, and parents can serve to promote awareness and program development, with community participation.

TABLE 6.10 Total Mortality Rates for Children and Young Adults,[a] United States, and Percentage Change, 1970–2004

Age group	1970	1980	1990	2000	2004	% change 1970–2004
1–4	85	64	47	32	30	73
5–14	41	31	24	18	17	64
15–24	128	115	99	82	80	38

[a]Rates per 100,000 population in each age group (rounded).
Source: *Health United States*, 2006.

Box 6.9 Content of a School Health Program

1. Healthful and safe school environment;
2. Health as topic in school curriculum;
3. Emergency or urgent health care in the school;
4. Physical education/fitness;
5. Education for special or disabled children within regular school program;
6. Training and supervision of teaching staff in health content of teaching program;
7. Nutrition education and eating disorders prevention and case detection;
8. School lunch and breakfast programs;
9. Family life and sex education/parenthood/birth control;
10. STI and HIV education and prevention;
11. Violence and sexual assault prevention and reporting;
12. Safety orientation — safety and prevention of fires; trauma as pedestrians; car, motorcycle, and bicycle operators or passengers; accidents at home; swimming and water safety;
13. Awareness of hazards of substance abuse, including smoking;
14. Personal hygiene and grooming;
15. Physical education and fitness habits;
16. Training in first aid/cardiopulmonary resuscitation;
17. Communicable disease control;
18. Mental and emotional health — suicide and violence prevention;
19. Understanding the health of a community;
20. Understanding health in the family — e.g., care of the elderly;
21. Understanding the environment and health;
22. Promotion of voluntarism in health.

Source: Modified from Committee on School Health. American Academy of Pediatrics, http://www.aap.org and http://www.schoolhealth.org/cshcap .htm, and American School Health Association, http://www.ashaweb.org/ [May 4, 2008]

Special needs for those with physical or mental disabilities should be identified and recognized as part of the regular duties of the school. Teaching staff should be trained and facilities modified to accommodate children with special problems and to be able to recognize physical and social ill-health, in a wide range of areas from nutrition to child abuse. The health of school-age children, adolescents, and young adults has improved in most countries as sanitation, vaccination, and nutrition programs have developed. Mortality rates overall have declined for all these age groups, but more so for preschoolers and primary school levels. Teenage and young adult health has also improved, but the problem of trauma in its various forms asserts itself and causes much loss of young life.

Adolescence is a difficult and dangerous period of transition from childhood to adulthood. The teenager feels insecure, yet demands independence and responsibility for his or her own well-being. The mix provides for a stormy personal and family transition, with important physical and mental health dangers. Violence, accidents, and suicide are major causes of death in this age group, especially for males. Sexual activity causes dangers of teenage pregnancy, hepatitis C, STIs, and HIV infection, especially when illicit drug use is involved. Developing individual responsibility in each of these categories is required of each individual, with the support of the family, the educational system, the public health system and the community.

Table 6.10 shows a decline in mortality of adolescents and young adults (ages 15–24) in the United States. Although mortality rates declined for both African-American and white persons in this age group, a sharp rise in mortality from homicide occurred, especially in African-American youths in the 1990s, which declined dramatically up to 2004, while homicide and suicide increased among white males (Table 6.11). Rates of mortality indicate the factors of motor vehicle accidents, homicide, and suicide still constitute major health problems associated with risk-taking behavior of teenage and young adult males (see Chapter 5).

Health targets for the United States are unchanged for 2010. They include reducing suicide mortality to 11 per 100,000 in the age group 15–24, primarily by health promotion and education, but also by community policing, and stricter weapons prevention programs in schools. Suicide is the seventh leading cause of death of males in Canada, nearly as high as motor vehicle deaths and more than deaths from diabetes. Male young adults are the highest risk group for suicide in North America and Europe, where rates in some countries, such as Hungary, Finland, and Sweden, are extremely high. Smoking, alcohol, and drug abuse are among the most common contributing factors for disease and early death in modern societies. These personal behaviors or "lifestyle factors" which cause personal ill health and early death constitute an enormous burden on society. They usually begin in late childhood, preadolescence, or adolescence, and their health effects continue into adulthood. Youth risk behavior is an important aspect of monitoring the health status of a population, as reported by periodic survey data (CDC, 1996). The mid-course report of *Healthy People 2010* provides an indication of the progress toward achieving the health targets. There are persistent variances between segments of the population.

Smoking

Smoking is a major contributor to cardiovascular diseases, cancer of the lung, and other diseases. The U.S. Surgeon General's Report on Smoking and Health issued in 1964 highlighted the accumulating evidence of the central role of smoking in the causation of chronic diseases. Subsequent

TABLE 6.11 Suicide, Homicide, and Motor Vehicle Accidents, White and Black Males Aged 15–24, Rate/100,000 Population, United States, 1970–2004

Group	1970	1980	1990	2000	2004	% change 1970–2004
White males						
MVA	75	74	53	41	40	−47
Homicide	8	15	15	10	10	+25
Suicide	14	21	23	18	18	+29
Black males						
MVA	58	35	36	30	26	−55
Homicide	98	83	137	85	78	−20
Suicide	11	12	15	14	12	+9

Note: Rates per 100,000 population. Homicide includes "legal intervention."
Source: *Health United States*, 2006.

reports addressed the continuing consequences of smoking and the prevalence of smoking among different population groups. Nearly all first use of tobacco occurs before age 18, and most adolescent smokers are addicted to nicotine. In 1995, one-sixth of students in grades 9–12 in the United States were regular cigarette smokers. There are identifiable psychosocial risk factors in smoking. Community-wide projects showed that public health efforts can successfully reduce adolescent use of tobacco. Many health-risk behaviors among U.S. high school students nationwide have decreased steadily from 1990 to 2007 (2007 Youth Risk Behavior Surveillance). However, many high school students continue with risk behaviors with similar rates across the country. Such behaviors include: violence; tobacco, alcohol, and other drug use; risky sexual behaviors; unhealthy dietary behaviors; and low levels of physical activity. Some 72 percent of all deaths amoung youth and young adults result from: motor vehicle accidents (30 percent), other unintentional injuries (15 percent), homicide (15 percent), and suicide (12 percent). Other consequences include some 757,000 pregnancies among 15–19 year olds, 9.1 million case of STIs, and some 5089 cases of human HIV/AIDS occurring among persons aged 15–24 years (CDC, 2008).

Alcohol Abuse

Alcohol use has been part of human culture since early civilization. Its use in excess causes serious health problems to the individual and society. The effects of alcohol abuse can be acute, chronic, or relate to the dependence itself. Binge drinking (five or more drinks per occasion) is a threat to the drinker, friends, and the community by associated violence, suicide, and motor vehicle and work

accidents. Mortality from all causes is elevated in alcoholics. In the United States, alcohol-related deaths were 5 percent of all deaths in 1988, representing 1.5 million YPLL (years of potential life lost) to age 65 and 3 million total YPLL. Alcohol affects every organ in the body, and alcohol use is a compounding risk factor for a wide variety of diseases including hypertension, stroke, coronary heart disease, and cancers of the liver, esophagus, larynx, lung, stomach, large colon, and female breast. It increases reproductive disorders including amenorrhea, anovulation, early menopause, and poor outcome of pregnancy with low birth weight or fetal alcohol syndrome in newborns. In 1997, 52.7 percent of high school seniors reported drinking alcohol, with 31.3 percent at a heavy level of consumption (5 or more drinks in a row, once or more during the previous 2 weeks). Between 1998 and 2004, increasing numbers of young people disapproved of substance use and abuse. Among high school seniors, 23 percent reported never using alcohol.

Drug Abuse

Drug abuse, while not a new problem, has become extremely widespread in most industrialized countries. Adolescence is a period of experimentation, intense social pressure, and risk-taking behavior. With widespread experience and availability of illicit drugs and other intoxicants in the United States, drug-related deaths among young people increased from 3.8 per 100,000 in 1987 to 4.1 in 1989. Cocaine-related emergency department episodes in the United States increased from 29,000 in 1985 to 80,000 in 1990 and to 142,000 in 1995, of which 2000 were for ages 6–17 and 22,000 for ages 18–25 (*Health United States*, 1998). Adolescents use such drugs as glue (or other inhalants), marijuana, and "hard drugs" including heroin and crack cocaine. In 1997, 23.7 percent of U.S. senior high school students, and 10.2 percent of eighth graders had used marijuana at least once in the previous month. Other drug use reported by high school seniors in the past month in this survey included cocaine (2.3 percent) and inhalants (2.5 percent). In 2004, 600,000 fewer American teens used drugs; 49 percent of high school seniors had never used drugs.

Drug addiction is a compulsion to use a substance and obtain it by any means with a need to increase the dosage to obtain the desired effect. There can be a physiological and/or psychological dependence on the effects of the substance. The addiction has a detrimental effect on the family and the community through increased, stress, violence and crime. Drug use behavior may be classified as experimental, recreational, circumstantial, intensified, and compulsive. The experimental version is most common and motivated by curiosity and a desire to experience an altered mood state. Recreational use is voluntary but patterned in social settings where the user is not dependent on the drug. Circumstantial use may be related to certain situations, such as amphetamine use by students during exams or truck drivers on long drives. Intensified drug use occurs daily, motivated by stress and/or the wish to maintain level of performance, by use of sedatives, barbiturates, or tranquilizers. The individual remains integrated within the social and economic context. Compulsive use is of high frequency, with a dominant psychological and/or physical dependence. These are hard-core drug users whose lives are controlled by the dependence on and financing of the habit. This leads to even more serious health problems, caused by exchange of sex for drugs, lack of condom use, sharing of needles, poor hygiene, and poor nutrition. Transmission of hepatitis, STIs, and HIV are common. Arrest, imprisonment, injury, or death by violence in drug-related crimes is also common.

Sexual Risk Behavior

Teenage sexuality risks include unplanned pregnancy and parenthood, STIs, and inadequate preparation for adulthood. The educational role in prevention of these is a vital element of a school health program. National efforts are required to control exploitation of children for prostitution, to provide care for runaway, homeless, and street children, and to regulate standards of care for orphans or disabled children in institutions. These problems are particularly acute in South American and southeast Asian countries. The HIV/AIDS epidemic is taking an enormous toll among teenagers and young adult sex workers, truck drivers, military personnel, and increasingly in the general community.

Teenagers participate in sexual intercourse, with 53 percent of U.S. high school students reporting having had sexual intercourse at least once in their lives, 18 percent having had four or more sexual partners, and 38 percent currently sexually active. Among these, just over half used condoms and 17 percent used birth control pills. By 2004, more students were using condoms; the percentage of those engaged in sex and sex with multiple partners declined. Risks of STIs and HIV are commonly associated with high-risk sexual behavior and drug abuse.

Dietary Risk Behavior

Only 27 percent of students had eaten the recommended dietary daily allowances (RDA) level of fruits and vegetables during the previous 24 hours in a 1995 U.S. survey. In addition, 28 percent thought of themselves as fat and 41 percent were currently trying to diet. Vomiting or laxative use to lose weight was nearly 8 percent among girls and over 2 percent among boys. Dietary deficiency conditions, especially iron-deficiency anemia, are common in adolescent girls due to dieting behavior and extra need for iron to compensate for menstrual losses.

The Mid-Term Review of *Healthy People 2010* shows increasing prevalence of overweight and obesity among children and adolescents aged 6–19 years from 11 percent to 16 percent between 1988 and 1994 and 1999 and 2002, whereas the target was a reduction of 5 percent. During the survey periods, the age-adjusted proportion of adults aged 20 years and older at a healthy weight decreased from 42 percent to 33 percent, while the proportion of adults who were obese increased from 23 percent to 30 percent. This phenomenon of increased obesity is a worldwide phenomenon and associated with increasing prevalence of diabetes with occurrence at younger ages than in the past.

Addressing the increase in overweight and obesity requires both public health and individual approaches. With growing food abundance and sedentary lifestyles and working conditions contribute to this pandemic. The contributing factors to overweight and obesity are complex, including psychological, behavioral, cultural, and socioeconomic, as well as genetic, metabolic, environmental issues. Slowing this trend requires attention by the health system in partnership with other sectors of society providing accessible and affordable healthy food choices, opportunities for regular physical activity and a supportive environment to facilitate individual behavior change. For obese adults, even modest weight loss (for example, 10 pounds) has health benefits. (CDC, Mid Term Review Health U.S. 2010).

Eating disorders are a common form of behavioral disorder especially among teenage girls influenced by the fashion model body type. They enter a vicious cycle of anorexia nervosa or bulimia which can be fatal, or in less severe forms extremely traumatic to the teen and family. These conditions may affect as many as 5 percent of college-age females in the United States, and they are an occupationally associated condition among both male and female athletes and professional ballet dancers or models. Overeaters who take up obesity as a lifestyle may also suffer the melancholies of this vulnerable age group. Prevention includes education and early case finding with suitable counseling and group therapy.

Physical Activity

Nearly two-thirds of U.S. students had engaged in strenuous physical activity on 3 or more days of the previous week, 60 percent were enrolled in physical education classes, and half were involved in team sports at their school. Participation in moderate physical activity has declined. Obesity in children and teenagers is increasing. Among boys aged 6–8 years, the percentage of those overweight increased from 1963–1965 to 15.4 percent. In 1988–1994, The Mid Term review indicates that moderate physical activity among students in grades 9 through 12 declined from a baseline of 27 percent in 1999, based on the percentage of students who participated in physical activity classes to 25 percent in 2003, while the target had been to increase to 35 percent. Vigorous physical activity (22-7) and participation in daily physical education in schools (22-9) among students in grades 9 through 12 also moved away from their targets. For girls, the increase in those overweight was similar to that of boys, except for the age group 15–17 where the increase was less. These patterns seem to be associated with increases in the passive activity of television watching and less activity in spontaneous or organized sports.

The prevalence of overweight and obesity among American children and adolescents rose to 16 percent by 2002, more than three times the desired target. Total fat consumption and sodium consumption remain high. Prevention-oriented policies and programs are the preferred means of intervention addressing social acceptability of physical activity and availability of healthy foods in school settings in place of junk foods and high sugar content beverages.

Violence and Gang Behavior

Teenage violence in developing and industrial countries is a growing public health hazard. Teenage alienation, availability of drugs, weapons, and violent cultural icons provide breeding grounds for gangs and violence, or individual violent outbursts which can lead to random or organized shootings or terrorist outrages.

Youth violence refers to harmful behaviors that may start early and continue into young adulthood. These range in severity and include bullying, slapping, punching, sexual harassment, ethnic slurs, victimization, weapon use, rape, and sadly frequent incidents of mass murder in school or university settings. School authorities, students, and teachers are increasingly aware of these acts and the threat they pose. Preventive factors include security services, alert mental health services, and cooperation between student and school authorities.

ADULT HEALTH

The health of adults between 25 and 64 years of age is important to the well-being of both the family and society. The higher rate of death for males in this age group creates a social imbalance with a predominance of females in the elderly age group. Premature death or disability of males often results in loss of a major part of family income, as well as productivity in the society. Much of the excess loss of males can be prevented by currently available knowledge and techniques in clinical medicine and public health. This has already been seen in reduced age-specific mortality from heart disease and stroke. Adult males should be targeted by coordinated efforts of both preventive and curative services to reduce risk and promote health.

Risk factors that contribute to disease and mortality are both intrinsic and extrinsic. Intrinsic factors include age, sex, and genetics. Extrinsic factors include personal lifestyle and the environment. Improved health for the adult population requires a wide range of primary and secondary prevention programs focused on extrinsic factors that can reduce personal risks and rates of mortality from chronic diseases and trauma (see Chapter 5). Programs to reduce morbidity and mortality from the chronic diseases include the following:

1. Smoking cessation;
2. Healthful diet (i.e., low in fat; moderate in carbohydrates; high in fruit, vegetables, and fiber);
3. Vitamin and mineral supplementation (e.g., calcium, iron, vitamins A, B, C, D, E, and others);
4. Daily physical activity (i.e., 30 minutes of moderately intense physical exercise 5 days per week);
5. Road safety (e.g., seat belts, moderate speed enforcement, non-use of alcohol with driving);
6. Workplace safety (e.g., safety helmets and shoes, preventive work site inspections);
7. Reducing violence (e.g., gun control, adolescent recreation activity programs);
8. Medical examinations for selective screening (e.g., diabetes, glaucoma, mammograms, Pap smears);
9. General screening for and management of hypertension;
10. General and high-risk group screening for and management of cancer;

11. Management of chronic diseases such as diabetes mellitus and glaucoma;
12. Support services for mental health and social stress;
13. Good management and prevention of infectious diseases;
14. Immunization with dT for all adults, influenza and pneumococcal vaccines for high-risk groups (i.e., persons over 65 and those with chronic diseases);
15. Social and recreational activity and support groups;
16. Employment, economic opportunity, and social equity;
17. Prostate-specific antigen (PSA) screening;
18. Fundoscopy to detect macular degeneration.

Detailed recommendations and clinical guidelines by disease classification are continuously updated and posted at http://www.guideline.gov/ and are referred to in various chapters of this book (Box 6.10).

The health care system has the responsibility of promoting the primary prevention and screening programs needed to reduce the burden of disease as it affects the adult population.

WOMEN'S HEALTH

Women's health is a matrix of many factors relating to fertility, sexuality, and societal conditions that affect health. Traditionally, in the years between menarche and menopause, women's health needs have largely related to fertility. With higher levels of education and participation in the workforce and other societal changes, women's health increasingly consists of issues beyond those of fertility (Table 6.12).

Since the 1980s, women's health needs have received increased attention from the public health sector. Women are the largest consumers of health care, as well as the largest group of health providers. Women live longer than men, visit physicians 25 percent more than men, and are hospitalized 15 percent more frequently. Lung cancer has replaced breast cancer as the first cause of women's cancers in many industrialized countries. Despite declining mortality rates, cardiac disease in women is more common than previously thought. Indeed, stereotypes regarding women as having symptomatic coronary heart disease less frequently than men may lead to the underassessment of the severity of this problem.

The life cycle of women between menarche and menopause involves psychosocial and physiological functions of fertility. Menarche, premenstrual syndrome (PMS), and menopause are all associated with physical and psychological stress of clinical importance as well as public health significance.

When education and awareness of women's physiological needs are accepted by the family, co-workers, and society in general, there is a greater likelihood of a more

---

Box 6.10 Adult Health Screening Recommendations

Initial and regular physical exams and preventive medical visits are recommended for all adults.

Dental Annual exam

Vision Initial screening and regular testing after age 64 or 40 for minorities

Hypertension Initial and biannual blood pressure screening

Diabetes Underserved minorities, chronic disease, personal or family history of elevated blood glucose, overweight (body mass index >25), or other risk factors

Cholesterol Initial screening and every 5 years after age 35 for men or 45 for women

Osteoporosis Bone densitometry for women once after age 60–65

Sexually Transmitted Infections Regularly for sexually active persons

Tuberculosis PPD screening for people with risks, including chronic disease, residence, or employment in institutional settings and travel to areas with elevated prevalence

Colorectal Cancer Initial colonoscopy and yearly fecal occult blood test starting at age 50

Cervical Cancer Pap test every 1–3 years for sexually active women**

Breast Cancer Regular breast self-examination and yearly mammography starting at age 40

Prostate Cancer Digital rectal exam and prostate-specific antigen (PSA) test should be offered to men yearly beginning at age 50 or earlier with risk factors

*Persons with chronic disease, personal or family history, and other risk factors should receive preventive services earlier and more often, based on medical recommendations.

**Immunization against common strains of human papillomavirus (HPV) is recommended for females ages 11–26, to greatly reduce future risk of cervical cancer.

Sources: National Guideline Clearinghouse, http://www.guideline.gov/ [accessed October 2007 and confirmed May 4, 2008]

CDC, Vaccines and Preventable Diseases, http://www.cdc.gov/vaccines/vpd-vac/default.htm

HHS. *Pocket Guide to Good Health*, for adults and children, http://www.ahrq.gov/consumer/index.html [accessed October 2007 and confirmed May 4, 2008]

American Cancer Society Guidelines for the Early Detection of Cancer, http://www.cancer.org/docroot/PED/ped_0.asp

---

appropriate response by women suffering from these effects. Women's other health issues of public health significance include iron-deficiency anemia, eating disorders, and nutrition (see Chapter 8).

Women's health is related both to biological and social factors. Postmenopausal women are subject to diseases for which they had previously been protected by hormones, such as cardiovascular diseases and osteoporosis with its accompanying fractures. The higher incidence of mental illness among women, in particular anxiety, depression, and eating disorders, relates to self-image, economics,

TABLE 6.12 Women's Health Risk and Prevention by Age Group

Age	Major health risk factors	Screening/Preventive activities
18–34	Nutrition, iron-deficiency anemia, calcium, and micronutrient deficiencies	Nutrition counseling, supplements
	Eating disorders, bodyself-image	Education, family and social support
	Marriage and parenthood	Birth control education and access
	Pregnancy and child care	Safe sex education
	Single parenthood	Prenatal care and high-risk assessment
	Physical fitness	Safe delivery
	Self-reliance, self-determination	Nutrition, exercise, weight control
	Sexuality and birth control	Self-defense
	STIs and HIV	Smoking cessation
	Safety, unwanted sexual advances, assault, rape	Premarital counseling
	Safety from domestic abuse, physical and mental	Physical exercise
	Physiologic demands of menstrual cycle, PMS	Job training, academic work
	Smoking, alcohol and drug use	Support groups
	Poverty and economic status	Shelters for abused, raped women
		Police, social service, and teacher sensitization
35–44	Late pregnancy	Counseling and support groups
	Child-rearing problems	Screening for breast cancer, cervical cancer
	Marital stress, separation, divorce	Physical exercise, weight control, fitness
	Diet	Screening for obesity, hypertension, cholesterol
	Physical fitness	Diet counseling, weight control
	Domestic abuse	Preparation for menopause (counseling, hormonal therapy)
	Obesity	Immunization (dT)
	Cancers (breast, cervix, lung)	
	Menopause	
45–64	Menopause	Periodic physical examinations
	Cancer (lung, breast)	Screening for diabetes, hypertension, bone density
	Diabetes, hypertension	Screening for breast, cervical, colonic cancer
	Physical fitness	Screening for thyroid and vitamin status
	Depression	Counseling for life crises
	Change in employment status	Immunization (dT), influenza
	Cardiovascular diseases	Physical activity
	Widowhood, divorce	Retirement planning

Source: Adapted from U.S. Preventive Services Task Force, 2005. Updates of U.S. recommendations are posted at http://www.ahrq.gov/clinic/uspstfix.htm

and society's varying pressures on women. Postpartum and menopausal depression, the pressures of child rearing and work, and long periods of loneliness as a result of family breakdowns and widowhood all contribute to an excess of mental disorders. As women have become a substantial part of the workforce, employment and career are major sources of income, self-esteem, and stress. Loss of employment can have negative effects on health for both financial and psychological reasons.

Screening for specific female cancers such as breast, ovary, uterus, and cervix are important public health programs, leading to early diagnosis and treatment. Sexual abuse, violence, single parenting, and widowhood are major factors that determine the content of health care services needed for women in the community. Females have been underrepresented in chronic disease research, which has tended to center on male subjects. Generalizations regarding treatment or prevention of a disease may not always be appropriate when they are based on male-only subjects. This has been especially problematic in heart disease, often presenting without clear symptoms, and stroke-related issues, with hypertension undermanaged.

Women's health issues relate to their role in society in terms of education, employment, and equal pay for work of equal value. Attitudes, customs, and laws and their enforcement regarding women's rights are extremely variable in different societies around the world, and the public health implications are widespread. In developed countries, a majority of people living in poverty are women and their children.

In developing countries, women's health involves a wide range of problems including high rates of pregnancy and maternal mortality, STIs and AIDS, and lack of access to family planning. In some parts of Africa, up to one-third of pregnant women are HIV-positive. WHO estimates that about one-quarter of the world's women are subject to violence and abuse in their own homes, with rates over 50 percent in Thailand, Papua New Guinea, and Korea, and as high as 80 percent in Pakistan and Chile. WHO also estimates that between 85 and 115 million women in developing countries have undergone female genital mutilation (FGM) and some 2 million young girls suffer this procedure annually. FGM can cause hemorrhage, shock, and infection in the short term. Long-term effects include pelvic inflammatory disease, infertility, psychological damage, and sexual dysfunction. It may also contribute to obstructed labor, a common contributor to maternal mortality. Prevention of FGM will require basic changes in social attitudes toward women, their education, and place in society.

The Beijing Conference on Women in 1995 included health and reproduction among those issues of most concern to women. Despite major differences in points of view between conservative religious representatives and liberal delegates on such issues as birth control and sexuality, a consensus emerged that women have rights to control their own bodies in these crucial areas. Violence and lack of education and economic development were also highlighted and recognized as impinging on the health status of women in both developed and developing societies.

Women's health is affected by socioeconomic status, education, information, societal equality, job opportunities, and women's input into health and social policies. Women are the primary caregivers in the family unit in most societies, but societal patterns may adversely affect a woman's health. Women's place in society affects many issues, ranging from fertility to public support for child care for working mothers. Some of these issues are politically contentious, with political, religious, and other social implications. Public health is closely involved in these issues, with responsibilities of advocacy, documentation, and innovative leadership.

MEN'S HEALTH

Health care for the adult male has not received as much attention as women's health issues. Men are more vulnerable than women to early mortality, have an average life expectancy of some 8 years less than women, and are more subject to risk for early disease of a wide variety of conditions and disabilities from cardiovascular disease, cancers of specific sites, violence and trauma, suicide, and occupational hazards (Table 6.13).

Some contributing factors to the lower life expectancy for males than females may include less attention to self-care and professional care, less social and psychological adaptability, more risk-taking behavior, and difficulty in transition from middle-age to elderly status. Emotional and psychosocial stresses on the middle-aged male (ages 45–64) may be very important in the morbidity and mortality patterns of a population.

Employment and career traditionally are central to the male's self-esteem. Feelings of competition with more aggressive younger colleagues, male and female, in a youth-oriented society may adversely affect the middle-aged male. Personal failure on the job or unemployment due to obsolescent industries, skills, or major economic change may result in stagnation at work, loss of self-esteem, and intrafamily strife that may lead to breakdown of physical and mental health. Social and economic changes produce loss of jobs and high rates of mortality among middle-aged men. Soaring rates of mortality from all types of chronic diseases in Russia and other states of the former Soviet Union in the early 1990s reflect the social impact on physical health in a society in rapid transition. Further research in the psychological, sociologic, and anthropologic spheres is required.

Public health and clinical services should be aware of and address self-image, social role, and occupational

TABLE 6.13 Men — Health Risks and Prevention by Age Group

Age	Major health risk factors	Screening/Preventive activities
20–34	Trauma, accidents, homicide, violence	High-risk behavior — health education, recreation
	Suicide	Enforcement of drinking-driving, seat belt, motorcycle and bicycle helmet laws
	Alcohol, drug abuse	Sexual practices education
	Sexual problems, STIs	Antismoking education
	Emotional problems	Peer group support
	Smoking	Exercise, fitness, and diet
	Fatty diet — cholesterol	Premarital screening/counseling for genetic disorders
	Lack of exercise	Preparation for fatherhood
	Marriage and parenthood	
35–44	Diet	Screening for obesity, inactivity, BP, cholesterol
	Hypertension	Immunization — dT
	Elevated cholesterol	Exercise, diet, and weight control
	Lack of exercise	Counseling for life crises (occupational, marital, family)
45–64	Cardiovascular disease risk factors	Periodic physical examination
	Hypertension, diabetes	Screening for hypertension, CVD, diabetes, glaucoma
	Cancer — colorectal, prostate	Screening for colorectal, prostate cancer
	Glaucoma	Counseling in life crises, group therapy
	Benign prostatic hypertrophy	Immunization — dT, influenza
	Change in employment status	Early retirement planning
	Depression	Physical activity

Source: Adapted from *Guide to Clinical Preventive Services*, 2007, http://www.ahrq.gov/clinic/pocketgd.htm [accessed May 6, 2008]
Recommendations of the U.S. Preventive Services Task Force, http://www.ahrq.gov/clinic/pocketgd.htm [accessed May 6, 2008]
Guide to Clinical Preventive Services, Second Edition. Report of the U.S. Preventive Services Task Force. Baltimore: Williams and Wilkins.
Canadian Task Force on the Periodic Health Examination. 1994. *The Canadian Guide to Clinical Preventive Health Care*. Ottawa: Health and Welfare Canada.
Agency for Healthcare Research and Quality, 2008. Recommendations of the US Preventive Services Task Force, ahrq. http://www.ahrq.gov/clinic/USpstfix.htm [accessed May 5, 2008]

issues related to men's health. The health care provider should utilize health education and promotion within the community to reduce risk factors in the male population.

HEALTH OF OLDER ADULTS

Improved health of the population leads to increased longevity and an increase in the elderly population in good physical, mental, social, and financial condition in industrialized countries. With the changes in disease patterns in the population, more people are living and remaining relatively disease-free longer. Many life-threatening conditions that occurred during middle age are now postponed to much later in life. As a result, it is not uncommon for the older adults to be well and free from major disease processes.

Adults 65 years and older comprise the fastest-growing group in the population of the industrialized countries. As the mortality rates in adult and middle-aged years have declined, more people survive into the over 65, the over 75, and the over 85 years old groups. These groups constitute 3.5 percent of the population in developing countries and up to 20 percent in developed countries. In the developed countries, this proportion is growing and will reach 30 percent in some countries, with the group over 75 years of age being the most rapidly growing segment of the population. The aging of the population has important implications for health services because the elderly are large consumers of health care in the context of a shrinking workforce.

Internationally, there are wide variations in life expectancy (Table 6.14) and in care programs for the elderly. The tradition of caring for the frail elderly in the family context is changing in the industrialized societies where both men and women work, housing conditions may be crowded, and tolerance for older adults reduced. This, together with a growing elderly population, has increased the reliance on long-term care in institutional settings. Offsetting this trend in western countries has been the development of adequate pensions, social benefits, universal health care, home care, and a delay in the onset of debilitating effects of disease to a later time in life.

Biological aging is measured by various functional abilities and performance of the individual, not necessarily reflected by chronologic age. Aging carries with it social, occupational, psychological, financial, and physical changes. These all directly impact health and the appropriate health and support needed by the elderly to sustain themselves. These factors can also interact with one another, causing or compounding health problems of the elderly.

Physical and mental deterioration is associated with aging, frequently resulting in a heavy burden on the family unit and on social and health support systems. Chronic diseases, such as diabetes, cardiovascular diseases, and cancer, are increasingly common with advancing age. Physical limitations can affect social interaction and mental status, with depression affecting physical abilities. Alzheimer's disease causes serious mental deterioration in people over age 50 with rates increasing with age. Parkinson's disease produces progressive physical muscular rigidity and limitations of movement (see Chapter 5).

Health Maintenance for Older Adults

Public health attempts to promote well-being by encouraging a healthful lifestyle and access to good health care. This involves nutrition, physical fitness, recreation, work or daily activities, a positive family life, social and religious participation, and an active sexual life, all in keeping with the physical and emotional capability of the individual. With rapid growth of the population of older adults, health targets for the United States and European countries have addressed prevention and health maintenance for this group with specific targets as indicated in Boxes 6.11 and 6.12.

Good nutrition, vital for a healthy old age, can be impaired by financial, social, and psychological factors. In many industrialized countries, the elderly are protected by good pensions from their employment and social security, but there are also those with inadequate pensions or other financial support who live in poverty. Isolation, loneliness, passivity, and malnutrition become a lifestyle which produces illness. Many elderly live on very limited budgets allowing little for food purchases. In addition, some have poor dental health. These are common contributing factors to the "tea and toast" syndrome of semi-starvation that becomes a way of life for this population group. Even a well-off, physically capable elderly person, living alone, may lack the appetite and incentive to prepare a properly balanced diet. The situation is especially difficult when the person is restricted in activity and unable to move about. Vitamin deficiency conditions are common, especially vitamins B and D, the latter mainly in winter months. Individual assessment by any caregiver should bear in mind the potential for low-level malnutrition among the elderly, especially those who are physically or mentally frail, or have severe dental problems. Community or voluntary programming to assist the elderly in nutrition and activity is a crucial contributor to maintaining the person in an independent life situation. The balanced diet for the elderly person is the same as for the adult, but must take into account the lower energy output in activity and less exposure to the sun, especially in winter.

Vitamin and mineral supplements are recommended for older adults. Fortification of basic foods is perhaps the most effective public health intervention for prevention of micronutrient deficiency in this and other vulnerable population groups. Food fortification not only makes essential nutrients readily available to the entire population but also supplies a significant portion of the recommended daily allowances for the elderly, including iron, iodine, and vitamins B, C, D, and E. Daily vitamin supplementation is increasingly common and is gaining support in geriatric and nutritional professional circles. In 1998, the U.S. FDA recommended routine vitamin B complex supplements for the elderly. Annual immunization for influenza and periodic immunization against pneumococcal pneumonia (every 6 years) and dT (every 10 years) should be standard care for the elderly. There is also a growing consensus that older adults should be taking routine aspirin (low dose) and possibly statins for cholesterol control.

Physical fitness is an important preventive measure in preparation for advanced age and to maintain health when reaching that state. Regular physical activity, individually or in groups, tailored to the individual's capacity, is helpful in promoting good appetite, sound sleep, and good physical appearance. It can help prevent lethargy, apathy, and atrophy. Mutual support in group physical activity is part of regular socializing and mental well-being. The individual, even disabled elderly persons, can also carry out physical exercise seated or lying down.

Social and family networks are important to the elderly and contribute to their sense of acceptance and well-being.

TABLE 6.14 Life Expectancy at Birth and at Age 65, by Sex, Selected Countries, 2002

| Country | Male life expectancy in years | | Country | Female life expectancy in years | |
	At Birth	At Age 65		At Birth	At Age 65
Hong Kong	78.6	17.8	Japan	85.2	23.0
Japan	78.3	18.0	Hong Kong	84.5	22.0
Switzerland	77.8	17.4	Spain	83.5	20.5
Sweden	77.7	16.9	France	83.0	21.3
Israel	77.5	17.3	Switzerland	83.0	21.0
Australia	77.4	17.4	Italy	82.9	20.6
Canada	77.2	17.2	Australia	82.6	20.8
Italy	76.8	76.8	Canada	82.1	20.6
Singapore	76.5	16.0	Sweden	82.1	20.0
Norway	76.4	16.2	Austria	81.7	19.7
New Zealand	76.3	16.7	Finland	81.5	19.6
Costa Rica	76.2	17.8	Norway	81.5	19.7
England and Wales	76.2	16.3	Israel	81.4	19.7
Netherlands	76.0	15.6	Germany	81.2	19.7
Austria	75.8	16.3	Belgium	81.1	19.7
France	75.8	17.0	New Zealand	81.1	20.0
Spain	75.8	16.5	Singapore	81.1	19.2
Northern Ireland	75.6	15.9	Costa Rica	81.0	20.5
Germany	75.4	16.2	England and Wales	80.7	19.2
Greece	75.4	16.7	Greece	80.7	18.8
Ireland	75.2	15.3	Netherlands	80.7	19.3
Belgium	75.1	15.8	Portugal	80.5	19.0
Finland	74.9	15.8	Puerto Rico	80.5	—
Denmark	74.8	15.4	Northern Ireland	80.4	18.9
Cuba	74.7	16.8	Ireland	80.3	18.6
United States	74.5	16.6	United States	79.9	19.5
Portugal	73.8	15.0	Denmark	79.5	18.3
Scotland	73.5	15.1	Cuba	79.2	19.3
Chile	72.9	15.4	Chile	78.9	18.8
Czech Republic	72.1	14.0	Scotland	78.9	18.1
Puerto Rico	71.6	—	Poland	78.9	17.9
Poland	70.4	14.0	Czech Republic	78.7	17.4
Slovakia	69.9	13.3	Slovakia	77.8	17.0
Bulgaria	68.9	13.1	Hungary	76.7	17.0
Hungary	68.4	13.1	Bulgaria	75.6	15.8
Romania	67.4	13.0	Romania	74.8	15.8
Russian Federation	58.9	10.9	Russian Federation	72.0	15.1

Source: *Health United States*, 2006, table 26, p. 174–175.

Box 6.11 Potential Adverse Changes Affecting the Elderly

1. Social/Occupational/Economic Status
   a. Retirement;
   b. Widowhood and bereavement;
   c. Relocation;
   d. Loss of friends and family;
   e. Loss of financial security;
   f. Loss of professional status and self-esteem;
   g. Poor nutrition;
   h. Physical inactivity.
2. Physiological
   a. Hormonal changes;
   b. Onset of non-insulin–dependent diabetes;
   c. Hypertension;
   d. Thyroid dysfunction;
   e. Osteoporosis;
   f. Decreased absorption (e.g., vitamin B).
3. Pathophysiological
   a. Chronic disease of one or more organ systems;
   b. Medical/surgical conditions, medication, or support services;
   c. Disabilities limiting mobility, activities of daily living.
4. Mental
   a. Loneliness and depression;
   b. Loss of memory;
   c. Senility and agitation;
   d. Isolation from children, family, friends.

Source: Derived from National Institute of Aging: Improve Health and Quality of Life of Older People, http://www.nia.nih.gov/ [accessed October 2007 and confirmed May 4, 2008]

Box 6.12 European and United States Health Objectives for the Elderly

WHO Europe
1. Ensure equity in health by reducing the gap in health status between countries and groups within countries;
2. Add life to years by ensuring full development and use of physical and mental capacity to derive full benefit and to cope with life in a healthy way;
3. Add health to years by reducing disease and disability;
4. Add years to life by reducing premature deaths, thus increasing life expectancy for adults over age 65.

United States
1. Vigorous exercise — 20 percent will engage in vigorous exercise three times a week for 30 minutes each occasion;
2. Muscle tone and endurance — 50% will participate in physical activities that promote and develop muscle tone and endurance;
3. Flexibility — 50 percent will regularly participate in physical activities that promote flexibility;
4. All will have more knowledge, positive attitude, and greater practice of physical exercise on a regular basis.

Source: *Healthy People* 2010, Midcourse Review.
Centers for Disease Control. 2001. Increasing physical activity: A report of the recommendations of the Task Force on Community Preventive Services. *Morbidity and Mortality Weekly Report*, 50(RR-18):1–16.
Stahl, T., Wismar, M., Ollila, E., Lahtinen, E., Leppo, K. 2006. *Health in All Policies: Prospects and Proposals*. Finland: Ministry of Health and Social Affairs, and European Observatory on Health Systems and Policies.

Familial and social relations have direct health benefits, including reduced hospitalization and long-term institutionalization. Mental health is strongly affected by the older person's self-perception of his or her role in the family and society. Recreation and social activity, including sexual relations, are part of the life of an older person. Recreational and social facilities designed to stimulate often socially isolated older persons to participate in recreational and support activity is dependent on development of a complex of adequate support systems such as rehabilitation, social security, transportation, and recreation facilities, all user-friendly and readily accessible to the elderly user (Box 6.13).

Prevention services for the elderly begin well before age 65. A health maintenance approach to preserve the well-being involves preparation for a healthy old age, and requires self-care and a preventive-oriented approach begun in earlier years. This involves early case finding of chronic disease and care of existing diseases in order to prevent debilitating complications. Continuous contact and support of the elderly by health providers can prevent or alleviate complications from medication errors, misunderstanding of health needs, inadequate nutrition, and social isolation.

Communication is vital to the life and health of the older adult. The availability of an emergency communication system may be lifesaving, as well as providing the frail person with independence and confidence in activities, personal security, and familial and other social relations. A telephone can be used for personal contact and social support as well as for emergencies and contact with medical personnel, providing a sense of security as well as contact with family and others outside the home.

Transportation for seniors or the disabled enables them to have access to medical care, social activities, shopping, and other activities of daily living. Bus companies operated by municipalities often make special arrangements. They may provide special routes including home calls and lower costs of public transportation for older persons. Ramps allow easy access for wheelchairs and walkers to public, residential, and commercial buildings and public transportation, making continued participation in community

Box 6.13 Community Health Needs of Seniors

Community programs for seniors should promote a wide range of knowledge and self-care and support services needed to prevent premature onset and progression of debilitation due to chronic disease and deterioration of general functioning of the older person, with the goal of helping them to function as independently as possible, consistent with safe and healthful conditions of a caring society.

1. Preventive self-care: healthy nutrition, regular exercise, and exposure to the sun;
2. Social contact: regular contact with family, friends, and social support systems (e.g., church, ethnic, recreation, and social clubs);
3. Health education: to promote community awareness, and knowledge among the elderly;
4. Medical care services: preventive, diagnostic, treatment, hospitalization, and rehabilitation care;
5. Nutritional support programs: to provide assistance with counseling, home-delivered "meals-on-wheels," and group meals in senior citizen centers to promote socialization;
6. Injury prevention programs: inspect homes and provide safety devices such as non-slip carpets, railings, shower aids, bath mats;
7. Medical devices loan service: a service to provide and maintain medical aid devices, such as wheelchairs and kitchen and bathroom assisting devices;
8. Home care: organized nursing, physiotherapy, shopping, cleaning, and other services to assist disabled or frail elderly persons to remain at home;
9. Hospital care: accessible but kept as short as possible to avoid infections and other complications;
10. Nursing homes: accredited facilities providing nursing and other care for elderly persons not able to live independently and requiring daily nursing care;
11. Supervised housing: group housing with supervisory nursing care, communal meals, recreational activities;
12. Recreation and occupational therapy: provided at community centers or in the home;
13. Volunteer work programs: volunteer traffic control, public garden maintenance, mutual help organizations, and community health workers;
14. Emergency call service: beeper service in case of emergency;
15. Security and safety measures installations: bathtub grips, smoke detectors, safety locks, and window bars;
16. Home help service: volunteer services to assist in maintaining independent homes by regular shopping, cleaning, and maintenance;
17. Home health aide service: cleaning, bathing, light housework, shopping, laundry, meals-on-wheels;
18. Mutual call service: telephone monitor services to maintain contact with seniors living alone;
19. Security: against abuse and violence;
20. Financial security: pensions and social support systems.

life possible for the disabled and seniors. Newer buses now have ramps for easier access and are referred to as *kneeler buses.*

Relocation and transition are part of the life of the older person. Retirement from work brings with it the potential for rest and recreation or the possibility of isolation and depression. Training people for transition in life may be as important as their physical well-being. Death and bereavement are also part of this process of transition and require organized community, as well as family, support.

Organized community systems for assistance and interventions, based on community or social networks, require public health activities and support. Identification of needs, professional support for resource development, as well as direct provision of services are all part of public health. Public health plays an advocacy and promotion role, but many services will be provided by other agencies, such as hospital-based home care programs or nongovernmental voluntary organizations that provide support services for the elderly.

Finding methods of assisting seniors to remain functional and adjusted in their own homes and in the community is essential to preserve the capacity of the health system to meet the needs of this population group. Such measures can be simple helping services, such as shopping or housecleaning. They may be devices to help in activities of daily living such as adaptations to stoves, wall grips for bathtub and toilet, or safety measures such as banisters, slip-free carpets, alarm systems, heating devices, safety locks, and police protection from vandals and burglars.

Finances in old age may be a serious burden for the younger generation. Children may need to provide financial assistance to the older members of their family who may have an inadequate pension or national social security system allocation. This can affect family relationships, living conditions, nutrition, medical care, social contact, and many other aspects of life. National social security systems have been developed in most industrialized countries to provide income security for seniors, through contributions from wages during the working years. Many social security systems are under pressure financially and politically. Increased life span and fewer births have created a situation in which the labor force may become smaller than the dependent members of society. Crises in social security systems may jeopardize the standards of living and security of the elderly in countries which today provide good income and other support systems for this population group.

Older adults in many countries comprise a significant proportion of the population and constitute a powerful political force. This has created a situation in which there is pressure on politicians to take into account the special needs of this population group. Seniors are a formidable group, strong in the political process, in professional organizations, and in municipal, provincial, and national political organizations. This "gray power" is a factor in western European and North American politics where the over-65 age group constitutes 15 percent or more of the total population, and an even larger percentage of the adult and politically active population.

The economic aspects of aging of the population are their pension support and increased health care needs. On the other hand, the older adults are consumers whose accumulated savings or social benefits employ many persons as producers and care providers. They also constitute a social asset to a country, not only as part of extended family networks, but also in their potential for volunteer work in the community, and the transmission of cultural heritage. Grandparents are important to cultural and social well-being of their grandchildren and to their society.

Despite, or perhaps because of, their use of health resources, seniors are healthier than ever before, with many continuing in the labor or volunteer workforces. Further, they are an important consumer group and tend to have the time, money, and health to play important economic and social roles. A healthy older population should not be seen as a burden but as a vibrant contributing part of the family and society generally. The New Public Health seeks to improve the health status of older adults and to ensure the provision of adequate support and health care services to assist them to function independently, as long as possible, in their own homes. Society and the New Public Health should promote volunteer and self-help networking among seniors to provide support during critical times such as illness or injury, following hospitalization, or periods of emotional stress and depression.

SUMMARY

Individuals live much of their lives in some form of family unit, but they also pass through stages of vulnerability as individuals and as population groups with common health problems. Traditionally public health has paid great attention to some groups because of their particular vulnerability, as in maternal and child health. The benefits to society as a whole are great where such programs are well developed. Middle-aged men and women are important target groups for primary and secondary preventive programming in preparation for old age. The elderly also need special attention in public health, because of increasing numbers, changes in society, and the need to find effective ways of promoting their health as a supplement or, it is hoped, in place of costly services for unattended health problems. Preventive care for the elderly to sustain health can prevent unnecessary or premature dependency on high-cost medical or nursing care facilities.

Each age group, from the newborn to the elderly, has specific problems and concerns that need to be addressed by the health care system. There are many medical, economic, and ethical issues involved in these aspects of public health. Public health needs to continuously monitor the health and social condition of the family as a key part of its overall individual care and population-oriented responsibility. Failure to do so leads to excess premature mortality and a costly burden on medical and hospital care to repair damage already done.

The New Public Health approaches the family unit both as a resource and a target group needing preventive and curative services at different stages of life. Family members may be cared for by different service providers, but there is a functional and economic relationship among those services. Inadequate prenatal care will increase the chance of poor results in infant health that can have long-term effects on the potential of the child. Inadequate support for the family struggling to cope with a chronically ill child or parent can result in unnecessary and damaging institutionalization. Lack of health promotion in nutrition, safety, and other community health issues may produce premature death of a parent, most often the male, with serious economic and social as well as emotional consequences for remaining members of the family.

The holistic approach of the New Public Health when applied to family health also addresses social and economic issues that affect or prejudice family function. Unemployment, underemployment, and poverty promote family distress and crises with long-term consequences for all members. Planning and resource allocation need to address this complex matrix of health in the family and societal context.

This approach is based on unifying factors, such as integration of various service systems and new kinds of linkages between records or a new health provider, to assist the family to cope with normal family health events and the additional burdens of chronic disease. The family physician in an ideal sense should be supported in a team approach with a family nurse to help monitor and support families. Together, they can assist the family in coping with health problems of individuals within the family.

Health promotion and prevention services are all part of this complex. Their application varies with the societal organization to address health issues and the priorities placed on resource allocation and public policy. National health targets help to define specific program content needs. The Millennium Development Goals similarly set targets for the global health community but progress in many areas has been much less than hoped for, especially with no improvement in reducing horrendous maternal mortality rates. Progress is uneven and new challenges emerge, such as growing prevalence of obesity and diabetes, with potential to reverse progress made in public health in the twentieth and the early part of the twenty-first centuries. Important challenges in family health remain as essential issues of the New Public Health.

ELECTRONIC RESOURCES

United States of America

Agency for Healthcare Research and Quality. 2008. The Guide for Clinical Preventive Services. U.S. Preventive Services Task Force, http://www.ahrq.gov/clinic/prevenix.htm[accessed May 8, 2008]

American Academy of Pediatrics, http://www.aap.org/ [accessed May 4, 2008]

American College of Obstetrics and Gynecology. http://www.acog.org/ [accessed May 4, 2008]

Beijing Conference on Women. 1995. http://www.un.org/womenwatch/daw/beijing [accessed May 4, 2008]

British Columbia Reproductive Health Program. http://www.rcp.gov.bc.ca/guidelines_alpha.htm#2 [accessed May 4, 2008]

Census 2000. http://www.census.gov/main/www/cen2000.html [accessed May 7, 2008]

Centers for Disease Control and Prevention (CDC). www.cdc.gov [accessed May 4, 2008]

Centers for Disease Control. 2002. Barriers to Dietary Control Among Pregnant Women with Phenylketonuria - United States, 1998–2000, http://www.cdc.gov/oc/media/mmwrnews/no20215.htm#mmwr [accessed May 3, 2008]

Centers for Disease Control. 2005. Intellectual Disability, http://www/cdc.gov/ncbddd/dd/mr3.htm [accessed May 5, 2008]

Centers for Disease Control. National Center on Birth Defects and Developmental Disabilities (NCBDDD). 2008. http://www.cdc.gov/ncbddd/ [accessed May 6, 2008]

Centers for Disease Control. National Center for Health Statistics, 2008, http://www.cdc.gov.nchs.nvss.htm

Centers for Medicare and Medicaid Services (CMS). http://www.cms.hhs.gov/ [accessed May 4, 2008]

Congenital hypothyroidism. 2008. http://www.ahrq.gov/clinic/uspstf08/conhypo/conhyprs.htm [accessed May 4, 2008]

Cystic Fibrosis Foundation. 2008, http://www.cff.org/AboutCF/testing/newbornscreening [accessed June 5, 2008]

Down Syndrome and Associated Medical Disorders, National Institute of Child Health and Human Development. http://www.nichd.nih.gov/publications/pubs/downsyndrome.cfm [accessed May 3, 2008]

Family Planning 9 3-26. http://www.healthypeople.gov/Data/midcourse/html/focusareas/FA09TOC.htm

Health United States. 2006. http://www.cdc.gov/nchs/data/hus/hus06.pdf [accessed May 4, 2008]

Healthy People 2010. http://www.healthypeople.gov/document [accessed May 4, 2008]

Healthy People 2010 Midcourse Review. http://www.healthypeople.gov/Data/midcourse/html [accessed May 4, 2008]

HIV 13 3-27. http://www.healthypeople.gov/Data/midcourse/html/focusareas/FA09TOC.htm

Injury and Violence Prevention 15 3-40. http://www.healthypeople.gov/Data/midcourse/html/focusareas/FA15TOC.htm

March of Dimes Birth Defects and Genetics: Cystic Fibrosis. http://www.marchofdimes.com/pnhec [accessed May 4, 2008]

March of Dimes Foundation. http://www.modimes.org/ [accessed May 4, 2008]

Maternal, Infant and Child Health 16 3-38. http://www.healthypeople.gov/Data/midcourse/html/focusareas/FA16TOC.htm

Mid-Course Review of *Healthy People 2010*. http://hdr.undp.org/hdr2006/statistics.

National Center for Chronic Disease Prevention and Health Promotion. http://www.cdc.gov/nccdphp/ [accessed May 8, 2008]

National Center for Health Statistics (NCHS). www.nchs.gov [accessed May 4, 2008]

National Guideline Clearinghouse. http://www.guidelines.gov [accessed May 4, 2008]

National Heart, Lung, Blood, Institute: Who Is at Risk for Cystic Fibrosis. http://www.nhlbi.nih.gov/health/dci/Diseases/cf/cf_risk.html [accessed May 4, 2008]

National Institute of Child Health and Human Development. http://www.nichd.nih.gov/ [accessed May 4, 2008]

National Institutes of Health: Fact Sheet — Mental Retardation, Down Syndrome and Associated Medical Disorders. http://www.nichd.nih.gov/publications/pubs/downsyndrome.cfm [accessed May 4, 2008]

National Newborn Screening and Genetics Resource Center (NNSGRC). 2006. Executive Summary. http://www.acmg.net/resources/policies/NBS/NBS_Exec_Sum.pdf [accessed May 4, 2008]

National Vital Statistics Report. http://www.cdc.gov/nchs/products/pubs/pubd/nvsr/nvsr.htm [accessed May 4, 2008]

National Vital Statistics Reports. March 15, 2007. http://www.cdc.gov/nchs/products/pubs/pubd/nvsr/nvsr.htm [accessed May 8, 2008]

Nutrition and Overweight 19 3-25. http://www.healthypeople.gov/Data/midcourse/html/focusareas/FA19TOC.htm

Pediatrics, http://pediatrics.aappublications.org/ [accessed May 4, 2008]

- The Apgar Score [2006]
- Levels of Neonatal Care [2004]
- Hospital Stay for Healthy Term Newborns [2004]
- Advanced Practice in Neonatal Nursing [2003]
- Safe Transportation of Newborns at Hospital Discharge [1999]
- Agency for Healthcare Research and Quality Hospital Discharge of the High-Risk Neonate — Proposed Guidelines [1998]
- Use and Abuse of the Apgar Score [1996]
- Safe Transportation of Premature and Low Birth Weight Infants [1996]
- Role of the Primary Care Pediatrician in the Management of High-Risk Newborn Infants [1996]

Phenylketonuria. 2008. http://www.ahrq.gov/clinic/uspstf08/pku/pkurs.htm

Physical Activity and Fitness 22 3-26. http://www.healthypeople.gov/Data/midcourse/html/focusareas/FA22TOC.htm

Sexually Transmitted Diseases 25 3-29. http://www.healthypeople.gov/Data/midcourse/html/focusareas/FA25TOC.htm

Substance Abuse 26 3-42. http://www.healthypeople.gov/Data/midcourse/html/focusareas/FA26TOC.htm

Thalassemia. http://www.cdc.gov/ncbddd/hbd/thalassemia.htm [accessed May 8, 2008]

Tobacco Use 27 3-40. http://www.healthypeople.gov/Data/midcourse/html/focusareas/FA27TOC.htm

United Nations. 2007. Africa and the Millennium Development Goals 2007 Update, http://www.un.org/millenniumgoals/docs/MDGafrica07.pdf [accessed May 5, 2008]

U.S. Census Bureau. http://www.census.gov/ [accessed May 4, 2008]

U.S. Census Bureau. America's Families and Living Arrangements 2003. http://www.census.gov [accessed May 4, 2008]

United States Department of Agriculture (USDA). www.usda.gov [accessed May 4, 2008]

International

UNICEF Monitoring and Statistics. http://www.unicef.org/statistics/index_24304.html [accessed May 4, 2008]

UNICEF Child Mortality. http://childinfo.org/areas/childmortality/ [accessed May 4, 2008]

UNICEF Maternal Mortality. http://childinfo.org/areas/maternalmortality/

UNICEF Monitoring the Situation of Children and Women. http://www
.childinfo.org/ [accessed May 4, 2008]

UNICEF 2007. State of the World's Children 2005. http://www.unicef.org/
sowc05 and http://www.unicef.org/sowc07 [accessed May 4, 2008]

United Nations, Department of Social and Economic Affairs, Population
Division. World Contraceptive Use – 2005. www.unpopulation.org
[accessed May 4, 2008]

United Nations Population Division. http://www.un.org/esa/population/
unpop.htm [accessed May 4, 2008]

United Nations Population Fund. http://www.unfpa.org/ [accessed May 4,
2008]

United Nations. 2007. Africa and the Millennium Development Goals 2007
Update, http://www.un.org/millenniumgoals/docs/MDGafrica07.pdf
[accessed May 5, 2008]

United Nations Millennium Development Goals. 2007. WHO/UNICEF/
UNFPA/World Bank 2007 press release, http://www.un.org/
millenniumgoals/docs/MDGafrica07.pdf [accessed May 4, 2008]

World Health Organization, Division of Child Health and Development.
http://www.who.int/child_adolescent_health/en/ [accessed May 8,
2008]

World Health Organization, Division of Women's Health and Development.
http://www.who.int/topics/womens_health/en/ [accessed May 6, 2008]

World Health Organization, Family Planning and Population. http://
www.who.org/rht/fpp/ [accessed May 4, 2008]

World Health Organization, Standards for Maternal and Neonatal Care
Steering Committee. Standards for Maternal and Neonatal Care.
http://www.who.int/makingpregnancy_safer/publications/en/ [accessed
May 4, 2008]

World Population Prospects: The 2006 Revision Population database.
http://esa.un.org/unpp [accessed May 6, 2008]

RECOMMENDED READINGS

American Academy of Pediatrics: Work Group on Breast-Feeding. 1997.
Breast-feeding and the use of human milk. Pediatrics, 100:1035–1039.

American Academy of Pediatrics. 2003. Controversies concerning vitamin K and the newborn. Pediatrics, 112:191–192.

American College of Preventive Medicine. 1999. Public policy statement:
Folic acid fortification of grain products in the U.S. to prevent neural
tube defects. American Journal of Preventive Medicine, 16:264–267.

Angastiniotis, M., Kyriakidou, S., Hadjiminas, M. 1986. How thalassemia was controlled in Cyprus. World Health Forum, 7:291–297.

Baker, J. P. 1994. Women and the invention of well child care. Pediatrics, 94:527–531.

Bol, K. A., Collins, J. S., Kirby, R. S., the National Birth Defects Prevention Network. 2006. Survival of infants with neural tube defects in
the presence of folic acid fortification. Pediatrics, 117:803–813.

Caravella, S., Clark, D., Dweck, H. S. 1987. Health codes for newborn
care. Pediatrics, 80:1–5.

Centers for Disease Control. 2004. Newborn screening for Cystic
Fibrosis. Evaluation of benefits and risks and recommended for state
newborn screening programs. Morbidity and Mortality Weekly
Report, 53 (RR-13):1–36.

Centers for Disease Control. 2007. Youth risk behavior surveillance—
United States, 2007. Morbidity and Mortality Weekly Report,
Surveillance Summaries, 57, SS4:1–131, http://www.cdc.gov/
HealthyYouth/yrbs/pdf/yrbss07_mmwr.pdf [accessed June 6, 2008]

Heise, L. 1993. Violence against women: The hidden health burden.
World Health Statistics Quarterly, 46:78–85.

Hessol, N. A., Fuentes-Afflick, E. 2005. Ethnic differences in neonatal
and postneonatal mortality. Pediatrics, January;115(1):e44–51.

Hilgartner, M. W. [editorial]. 1993. Vitamin K and the newbon. New
England Jouranl of Medicine, 329:957–958.

Hoyert, D. L., Mathews, T. J., Menacker, F., Strobino, D. M., Guyer, B.
2006. Annual summary of vital statistics: 2004. Pediatrics. 2006
January; 117(1):168–83 and erratum Pediatrics, 2006 June;117
(6):231–38.

Kaye, C. I., Committee on Genetics, Accurso, F., La Franchi, S., Lane, P. A.,
Hope, N., Sonya, P. G., Bradley, S., Michele, A. L. P. 2006. Newborn
screening fact sheets. Pediatrics, 2006;118:934–963

Kendig, J. W. 1992. Care of the normal newborn. Pediatrics in Review,
13:262–268.

Lozoff, B., Brittenham, G. M., Wolf, A. W. 1987. Iron deficiency anemia
and iron therapy effects on infant developmental test performance.
Pediatrics, 79:981–995.

Paneth, N. [editorial]. 1990. Technology at birth. American Journal of
Public Health, 80:791–792.

Rodriguez-Trias, H. [editorial]. 1992. Women's health, women's lives,
women's rights. American Journal of Public Health, 82:663–664.

Rowland, D. 1992. A five-nation perspective on the elderly. Health
Affairs, 11:205–215.

Singh, G. K., Kogan, M. D. 2007. Persistent socioeconomic disparities in
infant, neonatal, and postneonatal mortality rates in the United
States, 1969-2001. Pediatrics, 119:928–939.

WHO, UNICEF, and UNFPA. 2004. Maternal Mortality in 2000: Estimates Developed by WHO, UNICEF and UNFPA. Geneva: WHO,
UNICEF, and UNFPA.

World Health Organization. 2006. Integrated Management of Pregnancy
and Childbirth. Standards for Maternal and Neonatal Care. Geneva:
WHO. http://www.who.int/mediacentre/news/releases/2007/pr56/en/
index.html [accessed May 4, 2008]

BIBLIOGRAPHY

AbouZahr, C., Wardlaw, T. 2001. Maternal mortality at the end of the
decade: What signs of progress? Bulletin of the World Health Organization, 6:561–573.

American Academy of Pediatrics, Committee on Fetus and Newborn.
2006. The Apgar score. Pediatrics, 117:1444–1447.

American Academy of Pediatrics, Committee on Nutrition. 1961. Vitamin K compounds and the water-soluble analogues: Use in therapy
and prophylaxis in pediatrics. Pediatrics, 28:501–507.

American Academy of Pediatrics/American College of Obstetricians and
Gynecologists. 1992. Guidelines for Perinatal Care. Third Edition.
Washington, DC: ACOG, and Elk Grove Village, IL: American
Academy of Pediatrics.

American Academy of Pediatrics: Committee on Environmental Health.
1998. Screening for blood lead levels. Pediatrics, 101:1072–1078.

American Academy of Pediatrics: Committee on Nutrition. 1989. Iron
fortified formulas. Pediatrics, 84:1114–1115.

American Academy of Pediatrics: Committee on Nutrition. 1992. The use
of whole cow's milk in infancy. Pediatrics, 89:1105–1109.

American Academy of Pediatrics, Committee on Psychosocial Aspects of
Child and Family Health. 2001. Policy Statement — The prenatal
visit. Pediatrics, 107:1456–1458.

American Academy of Pediatrics. 2004. Policy Statement — Hospital stay for healthy term newborns. *Pediatrics*, 113:1434–1436.

American College of Obstetricians and Gynecologists, American Academy of Pediatrics: Policy Statement. 2006. The Apgar score. *Pediatrics*, 117:1444–1447.

American College of Obstetricians and Gynecologists. 1989. *Standards for Obstetric and Gynecologic Services*. Seventh Edition. Washington, DC: ACOG.

American Health Resources and Quality. 2005. *Recommendations of the U.S. Preventive Services Task Force*. Washington, DC: AHRQ.

Bailey, D. B., Skinner, D., Warren, S. F. 2005. Newborn screening for developmental disabilities: Reframing presumptive benefit. *American Journal of Public Health*, 95:1889–1893.

Belsey, M. A. 1993. Child abuse: Measuring a global problem. *World Health Statistics Quarterly*, 46:69–77.

Beresford, S. A. 1994. How do we get enough folic acid to prevent some neural tube defects? *American Journal of Public Health*, 84:348–350.

Blackmon, L. Committee on Fetus and Newborns. American Academy of Pediatrics. 2004. Hospital stay for healthy term newborns. *Pediatrics*, 113:1434–1436.

Bör, O., Akgün, N., Yakut, A., Sarhu, F., Köse, S. 2000. Late hemorrhagic disease of the newborn. *Pediatrics International*, 42:64–66.

Bren, L. 2004. Cervical cancer screening. *FDA Consumer Magazine*, January–February Issue.

Canadian Paediatric Society. 2002. Recommendations for the prevention of neonatal ophthalmia. *Paediatrics and Child Health*, 7:480–483.

Canadian Paediatric Society (CPS), and the Committee on Child and Adolescent Health, College of Family Physicians of Canada. 1997. Routine administration of vitamin K to newborns: A joint position statement, Fetus and Newborn Committee. *Paediatrics & Child Health*, 2:429–431(reaffirmed March 2004).

Canadian Task Force on the Periodic Health Examination. 1994. *The Canadian Guide to Clinical Preventive Health Care*. Ottawa: Health and Welfare Canada.

Cao, A. 2002. Carrier screening and genetic counselling in beta-thalassemia. *International Journal of Hematology*, 76:Supplement II:105–113.

Centers for Disease Control. 1982. New issues in newborn screening for phenylketonuria and congenital hypothyroidism: A commentary from the Committee on Genetics of the American Academy of Pediatrics. *Morbidity and Mortality Weekly Report*, 31:185–189.

Centers for Disease Control. 1995. Economic costs of birth defects and cerebral palsy — United States, 1992. *Morbidity and Mortality Weekly Report*, 44:694–699.

Centers for Disease Control. 1995. Surveillance for anencephaly and spina bifida and the impact of prenatal diagnosis — United States, 1985–1994. *Morbidity and Mortality Weekly Report*, 44:SS-4:1–13.

Centers for Disease Control. 1996. Youth risk behavior surveillance — United States, 1995. *Morbidity and Mortality Weekly Report*, 45: SS-4:1–84.

Centers for Disease Control. 1997. Children with elevated blood lead levels attributed to home renovation and remodeling activities — New York, 1993–1994. *Morbidity and Mortality Weekly Report*, 45:1121–1123.

Centers for Disease Control. 1998. Recommendations to prevent and control iron deficiency anemia in the United States. *Morbidity and Mortality Weekly Report*, 47:RR-3:1–30.

Centers for Disease Control. 1998. Tobacco use among high school students — United States, 1997. *Morbidity and Mortality Weekly Report*, 47:229–233.

Centers for Disease Control. 1999. Prevalence of selected maternal and infant characteristics. Pregnancy risk assessment monitoring system (PRAMS). *Morbidity and Mortality Weekly Report*, 48:SS-5:1–43.

Centers for Disease Control. 2001. Increasing physical activity: A report of the recommendations of the Task Force on Community Preventive Services. *Morbidity and Mortality Weekly Report*, 50(RR-18):1–16.

Centers for Disease Control. 2002. Annual smoking-attributable mortality, years of potential life lost, and economic costs — United States, 1995–1999. *Morbidity and Mortality Weekly Report*, 51:300–303.

Centers for Disease Control. 2004. Spina bifida and anencephaly before and after folic acid mandate — United States, 1995–1996 and 1999–2000. *Morbidity and Mortality Weekly Report*, 53:362–365.

Centers for Disease Control. 2006. Improved national prevalence estimates for 18 selected major birth defects, United States, 1999–2001. *Morbidity and Mortality Weekly Report*, 54:1301–1305.

Christianson, A., Howson, C. P., Modell, B. 2006. *March of Dimes Global Report on Birth Defects*. New York: March of Dimes Foundation.

Committee on Fetus and Newborn, Policy Statement. 2003. Controversies concerning vitamin K and the newborn. *Pediatrics*, 112:191–192.

Committee on Psychosocial Aspects of Child and Family Health, 1988, American Academy of Pediatrics. 1988. *Guidelines for Health Supervision I, II*. Elk Grove Village, IL: American Academy of Pediatrics.

Cooke, R. E. 1993. The origin of the National Institute of Child Health and Human Development. *Pediatrics*, 92:868–871.

Danielsson, N., Hoa, D. P., Thang, N. V., Vos, T., Loughnan, P. M. 2004. Intracranial haemorrhage due to late onset vitamin K deficiency bleeding in Hanoi province, Vietnam. *Archives of Diseases in Childhood — Fetal and Neonatal Edition*, 89:F546–550.

Editorial. 2006. Why should preterm births be rising? *British Medical Journal*, 332:924–925.

Fields, J. M. 2003. America's families and living arrangements. Current Population Reports, P20–553, U.S. Census Bureau, Washington, DC.

Gartner, L. M., Greer, F. R. American Academy of Pediatrics. 2003. Prevention of rickets and vitamin D deficiency: New guidelines for vitamin D intake. *Pediatrics*, 111:908–911.

Grouse, S., Waitzman, N. J., Romano, P. S., Mulinare, J. 2005. Government, politics, and law reevaluating the benefits of folic acid fortification in the United States: Economic analysis, regulation and public health. *American Journal of Public Health*, 95:1917–1922.

Guala, A., Guarino, R., Zaffaroni, M., Martano, C., Fabris, C., Pastore, G.: Neonatal Piedmont Group. 2005. The impact of national and international guidelines on newborn care in the nurseries of Piedmont and Aosta Valley, Italy. *BMC Pediatrics*, 5:45.

Haddad, J. G. [editorial]. 1992. Vitamin D—Solar rays, the milky way or both? *New England Journal of Medicine*, 326:1213–1215.

Heer, N., Choy, J., Vichinsky, E. P., et al. 1988. The social impact of migration on disease: Cooley's anemia, thalassemia, and new Asian immigrants. *Annals of the New York Academy of Sciences*, 850:509–511.

Hughes, J. G. 1993. Conception and creation of the American Academy of Pediatrics. *Pediatrics*, 92:469–470.

Ijland, M. M., Pereira, R. R., Cornelissen, E. A. Incidence of late vitamin K deficiency bleeding in newborns in the Netherlands in 2005: Evaluation of the current guideline. *European Journal of Pediatrics*, March 1; [Epub ahead of print].

Kaye, C. I., Committee on Genetics, Accurso, F., La Franchi, S., Lane, P. A., Northrup, H., Pang, S., Schaefer, G. B. 2006. Introduction to the newborn screening fact sheets. *Pediatrics*, 118:1304–1312.

Kendig, J. W. 1992. American Academy of Pediatrics. Care of the normal newborn. *Pediatrics in Review*, 1992;13:262–276.

Kozak, L. J., Lees, K. A., DeFrances, C. J. 2006. National Hospital Discharge Survey: 2003 annual summary with detailed diagnosis and procedure data. *Vital Health Statistics*, 13:1–206.

Lale-Say, S., Raine, R. 2007. A systematic review of inequalities in the use of maternal health care in developing countries: Examining the scale of the problem and the importance of context. *Bulletin of the World Health Organization*, 2007;85:733–820. http://www.who.int/bulletin/volumes/85/10/06-035659/en/index.html [accessed May 4, 2008]

Lawn, J. E., Cousens, S., Zupan, J. 2005. 4 million neonatal deaths: When? Where? Neonatal Survival 1. *Lancet*, 365:891–900.

Lawn, J. E., Cousens, S. N., Darmstadt, G. L., et al. 2006. 1 year after The Lancet Neonatal Survival Series — Was the call for action heard? *Lancet*, 367:1541–1547.

Liu, J., Wang, Q., Gao, F., He, J. W., Zhao, J. H. 2006. Maternal antenatal administration of vitamin K(1) results in increasing the activities of vitamin K-dependent coagulation factors in umbilical blood and in decreasing the incidence rate of periventricular-intraventricular hemorrhage in premature infants. *Journal of Perinatal Medicine*, 34:173–176.

Liu, S., West, R., Randell, E., Longerich, L., O'Connor, K. S., Scott, H., Crowley, M., Lam, A., Prabhakaran, V., McCourt, C. 2004. A comprehensive evaluation of food fortification with folic acid for the primary prevention of neural tube defects. *BMC Pregnancy Childbirth*, 27:4–20.

McNinch, A., Busfield, A., Tripp, J. H. 2007. Vitamin K deficiency bleeding in Great Britain and Ireland; British Paediatric Surveillance Unit Surveys, 1993–94 and 2001–02. *Archives of Diseases of Children*, 2007 May 30; [Epub ahead of print].

Mavalankar, D. V., Rosenfield, A. 2005. Maternal mortality in resource-poor settings: Policy barriers to care. *American Journal of Public Health*, 95:200–203.

Oski, F. A., Honig, A. S., Helu, B., Howanitz, P. 1983. Effect of iron therapy on behavior performance in nonanemic, iron-deficient infants. *Pediatrics*, 71:877–880.

Pooni, P. A., Singh, D., Singh, H., Jain, B. K. 2003. Intracranial hemorrhage in late hemorrhagic disease of the newborn. *Indian Pediatrics*, 2003 Mar; 40(3):243–248.

Puckett, R. M., Offringa, M. 2000. Prophylactic vitamin K for vitamin K deficiency bleeding in neonates (Cochrane Review). *The Cochrane Database of Systematic Reviews*, Issue 4. Art. No.CD002776.

Rajakumar, K. 2003. Vitamin D, cod-liver oil, sunlight, and rickets: A historical perspective. *Pediatrics*, 112:132–135.

Royston, E., Armstrong, S. (eds.). 1989. *Preventing Maternal Deaths*. Geneva: World Health Organization.

Russell, R. B., Green, N. S., Steiner, C. A., Meikle, S., Howse, J. L., Poschman, K., Dias, T., Potet, Z. L., Davidoff, M. J., Damus, K., Perrini, J. R. 2007. Cost of hospitalization for preterm and low birth weight infants in the United States. *Pediatrics*, 2007 July:120, http://www.ncbi.nlm.nih.gov/sites] [accessed at PubMed May 4, 2008]

Scott, D., Grosse, S. D., Boyle, C. A., Botkin, J. R. 2004. Newborn screening for cystic fibrosis: Evaluation of benefits and risks and recommendations for state newborn screening programs. *Morbidity and Mortality Weekly Report, Recommendations and Reports*, 53 (RR-13):1–36.

Sellwood, M. W., Huertas-Ceballos, A. 2007. NICE guidance for postnatal infant care. *Archives of Diseases of Children and Fetal Neonatal Education*, 2007 September 5; [Epub ahead of print].

Stahl, T., Wismar, M., Ollila, E., Lahtinen, E., Leppo, K. 2006. *Health in All Policies: Prospects and Proposals*. Finland: Ministry of Health and Social Affairs, and European Observatory on Health Systems and Policies.

Steering Committee on Quality Improvement and Management and Committee on Practice and Ambulatory Medicine. 2002. Principles for the development and use of quality measures: policy statement. *Pediatrics*, 121: 411–418

Stopp, G. H. 1994. *International Perspectives on Healthcare for the Elderly*. New York: Peter Lang.

Thompson, R. S., Rivara, F. P., Thompson, D. C. 1989. A case control study of the effectiveness of bicycle safety helmets. *New England Journal of Medicine*, 320:1361–1367.

Tulchinsky, T. H., Patton, M. M., Randolph, L. A., Meyer, M. R., Linden, J. V. 1993. Mandating vitamin K prophylaxis for newborns in New York State. *American Journal of Public Health*, 83:1166–1168.

UNICEF. 2005. *The State of the World's Children 2005*. New York: United Nations Children's Fund, Oxford University Press.

United States Preventive Services Task Force. 1996. *Guide to Clinical Preventive Services*, Second Edition. Report of the U.S. Preventive Services Task Force. Baltimore: Williams & Wilkins.

Waitzman, N. J., Romano, P. S., Scheffler, R. M. 1994. Estimates of the economic costs of birth defects. *Inquiry*, 31:188–205.

Weisberg, P., Scanlon, K. S., Li, R., Cogswell, M. E. 2004. Nutritional rickets among children in the United States: Review of cases reported between 1986 and 2003. Vitamin D and health in the 21st century and beyond. *American Journal of Clinical Nutrition*, 80:1697S–1705S.

Wharton, B., Bishop, N., Rickets. 2003. *Lancet*, 362(9393):1389–1400.

Williams, C. D., Baumslag, N., Jelliffe, B. 1994. *Mother and Child Health: Delivering the Services*, Third Edition. New York: Oxford University Press.

Williams, L. J., Rasmussen, S. A., Flores, A., Russell, S., Kirby, R. S., Edmonds, L. D. 2005. Decline in the prevalence of spina bifida and anencephaly by race/ethnicity: 1995–2002. *Pediatrics*, 116:580–586.

World Health Organization. 1983. Community control of hereditary anaemias. *Bulletin of the World Health Organization*, 61:63–80.

World Health Organization. 1992. *The Prevalence of Anemia in Women: A Tabulation of Available Information*, Second Edition. Geneva: World Health Organization.

World Health Organization. 2002. *Genomics and World Health: Summary*. Geneva: World Health Organization.

Yoon, P. W., Rasmussen, S. A., Lynberg, M. C., Moore, C. A., Anderka, M., Carmichael, S. L., Langlois, P. H., Edmonds, L. D. 2001. The National Birth Defects Prevention Study. *Public Health Reports*, Supplement 1, Volume 116:32–40.

Zipursky, A. 1999. Prevention of vitamin K deficiency bleeding in newborns. *British Journal of Haematology*, 1064:430–437.

Chapter 7

Special Community Health Needs

Introduction
Mental Health
  Historical Changes in Methods of Treatment
  Mental Health Epidemiology
  Mental Disorder Syndromes
  Controversies in Mental Health Policies
  Community-Oriented Mental Health
  Prevention and Health Promotion
Mental Disability
Oral Health
  Fluoridation
  Periodontal Disease
  Dental Care
  Oral Cancer
Physical Disability and Rehabilitation
Special Group Health Needs
Gay and Lesbian Health
Native Peoples' Health
Prisoners' Health
Migrant Population Health
Homeless Population Health
Refugee Health
Military Medicine
Health in Disasters
Summary
Electronic Resources
Recommended Readings
Bibliography

INTRODUCTION

The New Public Health emphasizes the importance of seeing the individual with special needs, and the special needs of groups or the total population in the context of the community and national health systems. Special community health needs may be defined according to the health risks and health losses related with special social, professional, economic conditions, or disabling chronic illnesses. Groups targeted for special community health needs may be defined according to particular population groups, conditions, and health risks. These may be related to characteristics of age, gender, socioeconomic status, location of residence, occupation, ethnicity, religion, or disabling conditions. Examples include all children and children with special needs; groups of patients with chronic diseases, such as HIV/AIDS, diabetes or end stage renal disease, where individual care may be supplemented by group activities. Other special needs include population groups such as prisoners, or population-wide problems such as mental dysfunction, and oral health, or those exposed to or potentially exposed to terrorism, natural disasters, warfare and genocide. These are issues that affect or may affect part or all of a population which must be addressed with health promotion and community approaches based on planning and strategic targets. They require a wide array of programs in addition to medical care, such as education, outreach, screening, and risk reduction programs. Research and evaluation require quantitative and qualitative research methods.

Special community health needs may change over time or be constantly present through life spans or generations. Special communities include a range of vulnerable groups which need special attention in public health. This includes native peoples, refugees, transient and homeless populations, military personnel and their families, and imprisoned persons. There are also special needs related to dental health and emergency health conditions which require organization of special services for the population as a whole. Traditionally these special needs were dealt with through separate services which segregated them from the general health care system. Even if the initial reasons for separation are no longer relevant, tradition or vested interests of the systems sometimes continue this separation, but there is a growing trend to integrate such services within primary and secondary care services (see Box 7.1).

By the fact of birth in a special group or at any point in life, every person can become a member of a community with special health needs. At times the existing health system may be too slow to address emerging health needs and there is frequently a problem of an inadequate access to care due to different barriers, which may be economic, geographic, social, prejudice, or lack of awareness. People are not always confident of the health system being able to help or they may not be ready to demand needed services. This is why study of special needs groups and services needs to be

Box 7.1 Vulnerable Groups and Factors of Mental Dysfunction

Children	Genetic factors, birth injury, fetal alcohol syndrome, low birth weight
	Nutritional deficiency: iodine, iron
	Poverty and psychosocial deprivation
	Abuse, violence: actual or exposure to it in the home or environment
	Infectious (e.g., viral encephalitis)
	Toxic exposure (e.g., lead)
Adolescents	Sexual maturation and associated stress
	Family stress, abuse, violence
	Peer pressure
	School, career, occupational expectations
	Fear of failure
	Body image: fear of obesity
	Poverty
Adults	Poverty
	Women in relation to fertility, pregnancy
	Parenthood, especially single status
	Abuse, physical, sexual, and psychological violence against women
	Occupation-related stress
	Fear of aging, menopause
	Loss of reproductive function and virility
	Job and status loss, loss of self-esteem
Elderly	Loss of spouse, friends, home
	Poverty and isolation
	Retirement and loss of occupational status
	Deterioration of mental and physical powers
	Poor nutrition: "tea and toast syndrome"
	Abuse, violence
	Loss of independence
	Fear of dying and long process of deterioration

part of the evaluation of health service delivery for access, outreach, effectiveness, and efficiency for community leaders and health managers.

This chapter addresses these special health needs, affecting particular subgroups and the population as a whole. Some of the interventions needed to protect the health of particular groups are directed specifically toward the group at risk. Other activities are directed at the general population because all are at risk, such as for development of mental health problems. The New Public Health advocates attention to the needs of these often less privileged populations, and seeks to assure adequate attention to their health care through the framework and proactive role of the community.

MENTAL HEALTH

Mental health is defined by the World Health Organization (WHO) as a state of well-being in which every individual realizes his or her own potential, can cope with the normal stresses of life, can work productively and fruitfully, and is able to make a contribution to her or his community. There is no single consensus on the definition of mental illness or mental disorder, and the phrasing used depends on the social, cultural, economic, and legal contexts in different countries.

The Special Series on Mental Health published by *Lancet* in September 2007 noted that: "An estimated 14% of the global burden of disease is due to neuropsychiatric disorders (NPDs). NPDs are the most important contributors to morbidity among the non-communicable diseases (NCDs) — more than heart disease, stroke and cancer — mainly due to the chronically disabling nature of depression, alcohol- and substance-use disorders, and psychoses. However, their true burden is likely to be underestimated because of inadequate appreciation of the connection between mental disorders and other health conditions." The series will deal with "the interface of mental health with other public health priorities; the scarcity, inequity and inefficiency of global mental health resources; the evidence for the treatment and prevention of mental disorders in low and middle income countries; the performance of individual nations on a range of mental health indicators; and the barriers to improving mental health services. The Series ends with a stirring Call for Action to scale up a basic, evidence based package of services for mental disorders, strengthen the monitoring of core mental health indicators, and invest in priority research on mental disorders."

Some 1.5 billion people of the total world population of 6.6 billion, three-quarters of whom live in developing countries, suffer from one or more NPDs, with impairment of their health and functioning in the societal context, varying from relatively mild to totally disabling. NPDs affect not only the patient but the family and the society, even more so in co-morbidity with other diseases, such as HIV, TB, and other chronic diseases. Along with the growing recognition of the widespread nature of mental illness in all societies, there is progress in understanding biological, behavioral, neurologic, and sociologic factors in this group of conditions. As a result, there is progress in development of new methods of preventing and managing mental and neurologic illness with effective medications and other methods of intervention, so that this area of health has become an important element of the New Public Health.

Short- or long-term mental and emotional problems may affect everyone to some degree during his or her life (Box 7.1). These include a wide spectrum of conditions: anxiety, depression, isolation and loneliness, psychotic states, obsessive-compulsive disorders, behavioral and eating disorders, drug abuse, delinquency, suicide and violence, alcoholism, and intra-family physical and mental abuse. These conditions affect the physical and social well-being of the patient, the family, and the community.

Services based on waiting for the presentation with full clinical manifestation often lead to long-term institutionalization as opposed to an early crisis intervention or community-based approach. In the United States, mental disorders are diagnosed based on the *Diagnostic and Statistical Manual of Mental Disorders*, fourth edition (DSM-IV, published in 1994, revised in 2000 and for review in 2010), developed in coordination with WHO and the International Classification of Disease (ICD-10) (Table 7.1).

Psychiatric conditions have traditionally been classified as psychotic or neurotic and by mode of presentation. Psychoses are major mental illnesses characterized by severe symptoms such as delusions and hallucinations. These are divided into organic and functional psychoses, the organic being caused by a demonstrable physical abnormality, while the functional has no physical disease demonstrated. The functional psychoses include schizophrenia and affective or mood disorders.

Neurotic conditions can vary in severity, but they generally reflect an exaggerated response, such as anxiety or obsessive thoughts, to normal life events. These are sometimes divided into anxiety neuroses, obsessive-compulsive neuroses, hysteria, and depression. Mental retardation and personality disorders have traditionally been considered separately from mental illness because they begin in early life or adolescence, whereas mental illness has a recognized onset after a period of normal functioning in adult life, but they are increasingly seen as part of a whole spectrum of community health problems and program needs (Box 7.2).

TABLE 7.1 International Classification of Diseases (ICD-10) Diagnostic Categories for Mental and Behavioral Disorders, F00–F99[a]

Diagnostic category	ICD-10 code	Clinical features
Organic, symptomatic, mental disorders	F00–09	Mental disturbance due to brain damage, toxicity, or trauma (e.g., Alzheimer's, vascular dementia)
Mental and behavioral disorders due to psychoactive substance use	F1–19	Alcohol, opiates, sedatives, cannabis, cocaine, hallucinogens, volatile substances, multiple drug use, and other psychoactive substances
Schizophrenia, schizotypal, and delusional disorders	F20–29	Delusional psychotic disorders and schizoaffective disorders
Mood (affective) disorders	F30–39	Manic, bipolar, and depressive disorders
Neurotic, stress-related, and somatoform disorders	F40–49	Phobias, anxiety disorders, obsessive compulsive disorders, stress reactions, dissociation, somatoform, and other disorders
Behavioral syndromes associated with psychological disturbances and physical factors	F50–59	Eating, sleeping, sexual, behavioral, and other disorders
Disorders of adult personality and behaviors	F60–69	A variety of specific personality disorder syndromes, including habit and impulse disorders, gender disorders, sexual preference disorders (pedophilia, voyeurism, etc.)
Mental retardation	F70–79	Mild, moderate, severe, and profound
Disorders of psychological development	F80–89	Speech and language, scholastic, motor, and development disorders (e.g., autism)
Behavioral and emotional disorders with onset usually occurring in childhood or adolescence	F90–98	Hyperkinetic conduct, emotional and social dysfunction disorders
Unspecified mental disorders	F99	

[a]The previous American Psychiatric Association (APA) classification of mental disorders included the major categories alcohol or drug abuse or dependence, phobias, major depression, obsessive-compulsive disorders, antisocial personality, panic disorders, cognitive impairment, schizophrenia, mania, and somatization. The present classification (DSM-IV) was developed in cooperation with the WHO and is close to the ICD-10 classification, modified in 2000. Future version scheduled for 2010 or later.
Source: American Psychiatric Association. 1994. *DSM-IV: Diagnostic and Statistical Manual of Mental Disorders*, Fourth Edition, in cooperation with the World Health Organization.

Historical Changes in Methods of Treatment

Traditionally mental illness has been stigmatized with superstition, brutal management, and treatment in isolation from the community. The methods of treating the mentally ill consisted of removal from the community into long-term institutionalization, using legal and physical restraints, and various severe forms of physical shock treatment. In the late eighteenth century, pioneering reforms by Vincenzo Chiarugi in Italy, William Tuke (and the Quakers) in England, and most influentially by Philippe Pinel in France, led to stopping the practice of chaining, starving, and beating patients in mental asylums. Psychiatric asylums, however, grew to be large isolated facilities for institutional care of the insane, usually under appalling conditions, and were the standard care of the mentally ill until well into the twentieth century.

In the United Kingdom during the nineteenth century, local authorities were encouraged to build mental asylums and to supervise the notorious private asylums. In the Mental Treatment Act of 1930, community psychiatric clinics were established providing some alternative to the grimness of mental hospitalization. In 1948, the U.K. mental hospitals came under the National Health Service.

By the 1950s, mental hospital beds in the United States and many other countries equaled the number of acute care beds. Hospitalization for a mental illness was long-term (often lifelong), with repeated readmissions and custodial treatment with little optimism for improvement or release. Therapeutic measures relied heavily on custodial care with heavy sedation, insulin, electroshock, and even lobotomy as common forms of therapy. Reduction of tertiary syphilis and other organic causes of mental illness and development of the psychotropic drugs made possible major changes in custodial policies. A British classification in the 1970s divided hospital admissions into depressive illness, schizophrenia, personality disorder, neurosis, and mania.

Mental health services based mainly on custodial care were felt to be costly and ineffectual. Large custodial mental hospitals of the past consumed significant health resources. In many countries they were mainstays of mental health services with over 5 beds per 1000 population and large total expenditures. They had few cures and produced much long-term damage in the form of institutionalized patients unable to return to normal society.

The growing strength of advocates for mental health reform, called the mental hygiene movement, and passage of the National Mental Health Act in 1946 in the United States instigated new directions in mental health care. The U.K. Mental Health Act of 1959 encouraged a rapid reduction in mental hospital beds from 152,000 in 1952 to 98,000 in 1975, and to 59,000 by the beginning of the 1990s. During this period, there was rapid increase in psychiatric units in general hospitals for short-term admissions. The concern that mentally ill persons are contributing to the increase in the homeless population in the United Kingdom and the United States is leading to a reappraisal of mental health policies.

Since the 1960s, release from institutional care and the return of large numbers of patients to the community required support by effective medication, follow-up services in the community, and other forms of therapy, as well as backup hospitalization for short- or even long-term care. Psychotherapeutic medications do not cure mental illness but can moderate or control symptoms enabling the person to function despite mental problems and coping difficulty. The objective of management is relief of symptoms and restoration of maximum possible function, as in other chronic conditions such as diabetes.

Where community services are inadequate, this policy can contribute to increasing numbers of homeless mentally ill persons who are unable to cope in modern society. Hospital admission rates for mental health in the United States during the period 1969–1993 declined in state and county mental hospitals but increased in short-term admissions to acute care general hospitals and private mental hospitals. The bed-to-population ratio in the United States has declined, as shown in Table 7.2.

TABLE 7.2 Mental Health Beds for 24-Hour Hospital and Residential Treatment, by Type of Organization: United States, Selected Years 1986–2002, Rates per 100,000 Civilian Population

Type of organization	1986	1992	1998	2002
All organizations total beds	111.7	107.5	99.5	73.3
State and county mental hospitals	49.7	36.9	25.6	19.9
Private psychiatric hospitals	12.6	17.3	12.4	8.7
Non-federal general hospital psychiatric services	19.1	20.7	20.2	14.0
Department of Veterans' Affairs medical centers	11.2	8.9	6.3	3.4
Residential treatment centers for emotionally disturbed children	10.3	11.9	11.9	13.6
All other organizations	8.8	11.7	23.1	13.9

Source: *Health, United States*, 2006, Table 11.

Reform of mental health care in the United Kingdom initiated in 1999 increased funding for specialist mental health services with emphasis on supervised community treatment. Fewer hospitalizations were needed through establishing over 700 new mental health teams in the community providing early intervention or intensive support using modern drug treatments previously rationed. National patient surveys show that 77 percent of community patients rate their care as good, very good, or excellent. The U.K. community mental health program has been regarded as a successful new approach by WHO, and as one of the most progressive mental health services in Europe.

Mental Health Epidemiology

Mental disorders at the diagnosable level are common in the United States affecting an estimated one in four adults, or 58 million people The main burden of illness is concentrated in about 6 percent of the population who suffer from a serious mental illness. Mental disorders are the leading cause of disability in the United States for ages 15–44. Many suffer from more than one mental disorder with nearly half (45 percent) of those with any mental disorder meeting criteria for two or more disorders. Co-morbidity between mental and physical illness constitutes a major issue for health services and for human suffering.

In the mid-1950s, a Midtown Manhattan study defined the population as follows: well, 19 percent; mild symptoms, 36 percent; moderate to marked symptoms, 45 percent; and severe to incapacitated, 10 percent. Studies carried out in other western countries show rates of clinical depression of between 4.5 percent and 7.2 percent in Finland and in cities such as Athens, Canberra, and Camberwell (United Kingdom). In 1985, mental illness accounted for 29 percent of all hospital bed occupancy in the United Kingdom and 4

percent of hospital admissions. Studies in the United Kingdom indicate that between 25 percent and 30 percent of patients visiting general practitioners have important or exclusively psychiatric causes for their presenting condition, even if the symptoms are primarily somatic.

The National Comorbidity Survey studied lifetime and 12-month prevalence of psychiatric disorders (DSM III) on a national probability sample of the adult population of the United States in 1990–1992. Nearly 50 percent of respondents reported at least one lifetime mental disorder, and at least 30 percent in the previous 12 months. The common disorders were major depressive episodes, alcohol dependence, and social or simple phobias. More than half of all persons reporting lifetime disorders had three or more disorders, accounting for 14 percent of the total sample. Less than 40 percent of those with lifetime disorders were ever treated professionally. Women had higher rates of affective and anxiety disorders; men had higher rates of substance abuse and antisocial personality disorders. The prevalence of psychiatric disorders was higher than expected, with a high proportion not getting professional care. These findings suggest a need for widened outreach and integration of care for psychiatric needs within general primary care. Two million people visited a hospital emergency department for mental disorders in the United States in 2002 (National Hospital Ambulatory Medical Care Survey: 2002, NCHS, CDC). CDC (2007) notes that depression takes an enormous toll on functional status, productivity, and quality of life, and is associated with elevated risk of heart disease and suicide. In addition, the rate of treatment for depression is increasing dramatically in the United States. The annual economic burden of depression alone (including direct care, mortality, and morbidity costs) totals almost $44 billion. The increasing burden and cost of depression have stimulated numerous investigations into population-based strategies to prevent the occurrence of major depression and to encourage more effective treatment to limit its course and prevent recurrence.

Prevalence studies are fraught with difficulties in survey methods with recall and other biases. There are also difficulties with response rates, diagnostic criteria, and representative sampling, as well as in survey instrument design. Mental health epidemiology draws on data from national health surveys and registries of ambulatory and hospital care. Studies require collaboration among social scientists, epidemiologists, biostatisticians, anthropologists, geneticists, and other disciplines in order to increase knowledge of contributory factors to mental illness and to promote the search for causes and treatments.

In a 2007 report of an updated comorbidity study, the prevalence of any personality disorder in the United States was reported at 9.1 percent of the population. Specific prevalence rates for borderline personality disorder and antisocial personality disorder were estimated at 1.4 percent and 0.6 percent, respectively. Thirty-nine percent of respondents with a personality disorder received treatment for problems related to mental health or substance use at some time during the previous 12 months. On average, respondents made two visits seeking mental health treatment. Even though the majority of cases was seen by a psychiatrist or other mental health professional, respondents were more likely to receive treatment from general medical providers than mental health specialists. People with personality disorders are very likely to have co-occurring major mental disorders, including anxiety disorders (e.g., panic disorder, post-traumatic stress disorder), mood disorders (e.g., depression, bipolar disorder), impulse control disorders (e.g., attention deficit hyperactivity disorder), and substance abuse or dependence. The association between personality disorders and major mental disorders affects functioning and help-seeking behaviors.

The WHO estimates that at least 5 percent of the population of the European region suffers from serious diagnosable mental disorders (neuroses and functional psychoses), although prevalence estimates vary widely from study to study. At minimum, an additional 15 percent of the population suffers from less severe but partially incapacitating forms of mental distress. These affect their well-being and create the threat of more serious mental problems, such as severe depression, chronic psychiatric conditions, or psycho-emotional problems, and life-threatening behavior, such as suicide, violence, and substance abuse. Western European countries have seen radical changes in psychiatric care over the 1980s and 1990s, with a process of "deinstitutionalization" that combines discharge of patients previously kept in long-term institutions with their reintegration into society. Development of community-based treatment and support services to promote independent living, vital social and employment skills, as well as backup support services have been part of this process.

The U.S. National Institute of Mental Health (NIMH, http://www.nimh.gov) estimates that schizophrenia affects 2 million Americans, while 10 million are affected by bipolar (manic-depressive) disorders. Other conditions (phobias, post-traumatic stress disorders, and obsessive-compulsive disorders) together affect some 30 million, and a similar number are affected by Alzheimer's disease and other brain disorders. Eating disorders probably affect millions of teenagers, while substance abuse and associated comorbidity affect millions more. With a growing trend in managed care systems in the United States, and integration of mental health into general health systems under district health services in other countries, cost-effectiveness in care is increasingly a topic of concern for health economists and health systems managers.

The mission of the NIMH is to conduct research on the brain, behavior, and genetics, to develop new diagnostic and treatment methods, testing them in real-world settings. It focuses on the neurosciences, especially at the molecular level (Julius Axelrod awarded Nobel Prize, 1970). The interaction of biological and developmental factors in mental illness is the subject of continuing study with great importance in impact for development of new forms of therapy and diagnostic–prognostic instruments. The shift in emphasis from in patient to ambulatory care has been controversial but the vast scope of mental ill health requires strengthening of community care including residential care for the severely disabled (Box 7.2).

WHO regards mental disorders as one of the major global public health problems, with tens of millions of cases accounting for approximately 10 percent of years of healthy life lost, or disability-adjusted life years (DALYs). WHO reports that in 2002, 154 million people globally suffered from depression and 25 million people from schizophrenia; 91 million people were affected by alcohol use disorders, and 15 million by drug use disorders. A WHO report indicates that in 2005, 50 million people suffered from epilepsy and 24 million from Alzheimer's disease and other dementias. Millions of others suffer from neurologic disorders, with 326 million people suffering from migraine; 61 million from cerebrovascular diseases; 18 million from neuroinfections or neurologic sequelae of infections, sequelae of nutritional disorders and neuropathies (352 million), and neurologic sequelae secondary to injuries (170 million) (Box 7.3).

Mental ill health is as relevant in developing as in industrialized societies, as a result of increasing life expectancy, and because of suffering from complex interactions between the biological, psychological, and social factors (e.g., aging, poverty, war and trauma, human rights violations, limited education, gender discrimination, and malnutrition).

The challenge of mental illness is partly one of changing priorities in health to include this set of conditions with equal priority to physical illness. The burden of mental health as measure by DALYs and social and economic cost to families and society through illness, lost productivity, and personal financial outlays is ranked among the highest of disease entities, although the mortality rates are lower. A hidden burden is the social stigma and loss of human rights so commonly associated with mental illness. Rejection by society, families, employers, and even caregivers compounds the isolation, humiliation, and pain suffered by the mentally ill, as well as their loss of earning power and independence.

Mental Disorder Syndromes

Mental disorders present with a wide range of symptoms including personality change, confused thinking, abnormal anxiety, fear or suspiciousness, withdrawal from social contact, suicidal thoughts or actions, sleeplessness, change in eating patterns, outbursts of anger and hostility, alcohol or drug abuse, or simply incapability of dealing with daily activities, such as school, job, or personal needs. Mental disorders include a heterogeneous group of disorders ranging from exaggerated response to stressful events to altered mentation from specific neurologic or genetic abnormalities (U.S. Department of Health and Human Resources, 1999). Chronicity is a problem which has, in the past, required long-term hospitalization. However, with improved medications and management in the community, hospitalization has been reduced as the primary method of treatment.

Organic Mental Syndromes

Prevention of prenatal damage to the fetus from exposure to infectious diseases (e.g., syphilis, rubella, toxoplasmosis); toxicity from alcohol, drugs, smoking, and nutritional deficiencies (iodine, iron, vitamins B and D); and low birth weight are part of pregnancy care with major potential for prevention of brain damage. Screening of newborns for phenylketonuria (PKU), congenital hypothyroidism, and many genetic defects of metabolism; infants for iron deficiency; and children for blood lead levels are all public health measures of importance for prevention of organic brain damage (Chapter 6).

Organic mental disorders in older adults can produce a range of symptoms, such as decline of memory, comprehension, learning capacity, language, and judgment, including the ability to think and calculate, or severe dementia. Alzheimer's disease is the prominent condition in this category, but other causes of organic origin include traumatic and toxic brain damage, cerebrovascular accident, Parkinson's disease, alcoholism, Creutzfeldt-Jakob disease, HIV, postencephalitic disorders, syphilis and other dementias, and mental disorders due to physical brain disease. The WHO estimates 24.3 million persons suffer from such conditions internationally, with 4.6 million new cases annually. Cognitive impairment in the elderly is less than 5 percent under age 75, but over 40 percent for those above age 80.

Prevention of brain injury, strokes, encephalitis due to vaccine-preventable diseases, management of alcoholism, and adequate nutrition reduces the prevalence of the dementias. Awareness and recognition of cognitive impairment in the elderly is an important function in primary care and specialized geriatric and psychiatric services. Because of increasing longevity, organic brain syndrome cases can place a great burden on families and the health care system. Support services for families caring for persons affected by organic brain syndromes are part of a comprehensive health program, and should include short-term respite care, home care, and long-term care services. Research into organic brain syndromes should be of high priority because of the personality destruction of an increasingly large group of the population, and the resultant effects on the individual and the family, as well as the costs of health services for this group.

Substance Abuse

Substance abuse (mental and behavioral disorders due to psychoactive drug use) is intoxication with a substance that causes physical or psychological harm, impaired judgment, or dysfunctional behavior leading to disability and harming interpersonal relationships. Dependent syndromes are characterized by the presence of three of the following: a compulsion to use the substance, physiological withdrawal symptoms, tolerance of its effects, preoccupation with the substance, and persistence in using the substance despite negative effects. Substance abuse is associated with serious problems, including death from overdosage, crime to support the habit, STIs, AIDS and hepatitis transmission, imprisonment, social ostracism, and long-term brain damage.

Drug dependence syndromes are estimated by the WHO to affect 28 million persons. Between 100,000 and 200,000 deaths occur from overdosage annually. Inhalation of volatile solvents (i.e., sniffing of glue, paint thinners, gasoline, and aerosols) among preadolescents is widespread and causes death and serious brain damage. Cannabis use is extremely widespread. New chemical formulations ("designer drugs") are appearing with death and brain damage effects. Use of opiates, cocaine, and psychotropic drugs affects all levels of society in developed and developing countries, influenced by urbanization and other social stresses and promoted by powerful economic and political interests in the international drug trade. Attempts to eliminate or control drug traffic have many similarities to international efforts to ban the slave trade in the late eighteenth and early nineteenth centuries, in that with both instances some governments covertly foster the trade, while others attempt to stop it.

Prevention should target vulnerable groups, especially young people, street children, and female drug users. Methadone substitution is widely used to reduce drug dependency. Needle exchange programs have been successful in reducing the spread of HIV and hepatitis B and C among intravenous drug users but are sometimes criticized as encouraging drug use. Detoxification and long-term treatment and follow-up programs are costly and frustrating, but they are better than the alternatives of disease, crime, social breakdown, and imprisonment now common in this group.

Alcohol abuse affects 120 million people internationally. Alcohol abuse, chronic alcoholism, and associated diseases of cirrhosis, cancers of various sites, and social breakdown are widespread in many countries, with resulting morbidity and mortality from trauma, violence, and child and spouse abuse. Alcohol abuse during pregnancy is associated with stillbirth, premature birth, low birth weight, and fetal alcohol syndrome (FAS), which is an increasing problem worldwide due to widespread use of alcohol by pregnant women. Alcohol and drug abuse continues to be a major public health concern in the United States. The Department of Health and Human Services (DHHS) 2005 national survey of Americans age 12 or older: 6.6 percent (16 million) reported heavy drinking; 22.7 percent (55 million) reported binge drinking; and 8.1 percent (19.7 million) reported illicit drugs within the month prior to the survey.

WHO defines polydrug abuse as the concurrent or sequential abuse of more than one type of drug with dependence on at least one. This type of abuse has been increasingly reported in emergency room admissions. A 2002 DHSS survey in the United States found that 56 percent of all admissions to publicly funded treatment facilities were for multiple substance abuse: 76 percent abused alcohol, 55 percent abused marijuana, 48 percent abused cocaine, 27 percent abused opiates, and 26 percent abused other drugs (Kedia 2007).

Strategies for reducing alcohol abuse include raising the price and reducing the availability of alcohol, especially to adolescents, setting a minimum age for alcohol purchase, legislation and enforcement to curb driving while under the influence of alcohol, restrictions on promotion, marketing, and advertisement of alcohol, public education and awareness programs, individual counseling, group therapy, and self-help groups, as well as inpatient, outpatient, and rehabilitation programs.

Schizophrenia

Schizophrenia is a group of chronic conditions with episodes of psychotic illness with delusional hallucinatory thought or behavior disorders. It usually manifests itself around age 20, with distorted thinking, perception, and judgment. Symptoms may include excitement, withdrawal, or a catatonic state. Hallucinations can be visual or auditory.

The disorder appears in episodes lasting a few months with interval periods of normality but is a chronic illness. It occurs in both sexes and all social classes and affects about 1 percent of the adult population. It is characterized by chronicity with periodic need for hospitalization. Management with current medications improves the outlook for many patients.

WHO estimates there are 45 million schizophrenics in the world with 33 million in developing countries. In the European Region mental ill health accounts for some 20 percent of the burden of disease and mental health problems affect one in four people at some time in life. Nine of the ten countries with the highest rates of suicide in the world are in the European Region (WHO, 2007). In the United States, there are roughly 87,000 annual acute care inpatient admissions of Medicaid patients for the treatment of schizophrenia. These admissions include a total of approximately 930,000 hospital days at a total cost of $806 million. The average length of stay in U.S. short-stay nonfederal hospitals for discharges with a first listed diagnosis of schizophrenia declined slightly between 1997 (11.9 days) and 2004 (10.8 days), while the number of schizophrenia discharges has increased from 262,000 in 1997 to 331,000 in 2004.

In the United States, the direct cost of treatment is estimated to be close to 0.5 percent of the GNP. While thought to be biological in origin, this condition is affected in course and outcome by social and cultural conditions. Acute care and long-term follow-up require well-integrated community and hospital services. Neuroleptic drugs such as chlorpromazine, introduced in the early 1950s, greatly reduce symptoms and enable patients to function in the community, especially with family support and suitable community-based services, with periodic short-term hospitalization if needed.

Mood Disorders

Elevated or depressed mood states affect some 340 million persons around the world at any given time, ranking as the fourth leading cause of the total burden of disease in developing counties. In the United States, the yearly cost of depression is estimated at $44 billion, equivalent to the cost of care of all cardiovascular diseases (CDC, 2007). WHO projects depression as second among three leading causes of burden of disease globally in the coming decades.

Mood disorders range from manic to depressed conditions, often involving both mania and depression in sequence (manic-depressive or bipolar disorders). Depression is probably the most common affective disorder, affecting approximately 5 percent of the population at any one time.

Diagnosis of depression involves four or more of the following symptoms and signs: loss of interest or pleasure in normal activities, lack of emotional response, sleep disturbances (early waking, sleeplessness, or excessive sleepiness), depression that is worse in the morning, loss of appetite, loss of 5 percent of body weight in a month, loss of libido, and a psychomotor retardation (or agitation). Severity ranges from mild to severe.

Depressive episodes may be recurrent and are a major risk factor for suicide and social breakdown. Severe depression is common in the elderly. Seasonal affective disorder (SAD) is a syndrome related to darkness during winter months and is common in northern countries with long winter nights and a closed-in lifestyle. It is associated with high rates of alcohol abuse, heart disease, and suicide.

Approximately 20.9 million American adults, or about 9.5 percent of the U.S. population age 18 and older in a given year, suffer from mood disorders. The median age of onset for mood disorders is 30 years. Depressive disorders are often associated with anxiety disorders and substance abuse.

Economic distress, unemployment, discrimination, and practices which limit women's rights are all contributors to mood disorders. Women are more commonly affected by mood disorders, with depression especially common among married women with children, related to social isolation and social devaluation of the role of the housewife. Postpartum depression can be a precursor to chronic depressive illness with a need for support and referral to prevent chronicity.

Antidepressant drugs provide an important advance in management of mood disorders, but should be accompanied by professional monitoring. In 1970, lithium was approved by the FDA for treatment of manic episodes, based on NIMH research, bringing major benefits to many persons suffering from bipolar disorders. Lithium and many newer antidepressant treatments are available but require close monitoring and supportive care by health providers. They help to reduce the disabling quality of mood disorders, reducing the economic burden of these conditions on society as well as the personal suffering of patients and families.

Primary care providers need to work jointly with mental health specialists in the same setting, to ease referral and consultation with patients unlikely to present to a separate mental health setting. Awareness of mood disorders among primary caregivers is of paramount importance in addressing this problem with understanding, patience, supportive therapy, and referral if chronic or leading to social breakdown at work or at home. Psychotherapeutic skills are important for primary care, but require support of specialized services. Community recognition is needed to address the widespread prevalence of domestic violence associated with mood disorders.

Neurotic (Anxiety and Dissociative) Disorders

This group includes a wide range of symptomatology and degrees of severity, including panic disorders, phobias, obsessive-compulsive disorders, anxiety conditions, and post-traumatic stress disorder. Specific phobias include fear of crowds, public places, traveling, social situations, objects, animals, and closed spaces. Panic conditions are discrete episodes of intense fear, starting abruptly with physical symptoms that are inconsistent with the perceived threat of a specific situation or trigger event. Obsessive-compulsive disorders are repetitive, unpleasant obsessions, causing distress or interfering with normal functioning. They include compulsive behavior such as frequent hand washing, hair brushing, cleaning, and counting. Some obsessive-compulsive patients find relief with the new generation of medications. Stress reactions may be acute or occur long after the trigger event.

Post-Traumatic Stress Disorders

Post-traumatic stress disorders (PTSDs), with flashbacks and dreams of the event reawakening painful memories, usually occur within 6 months of the stressful event or time. These may include manifestations of depression or other affective or behavioral disturbances. PTSD, originally described as related to combat experience in the Vietnam and Iraq wars, is now also recognized as occurring in reaction to catastrophic events, such as violence, genocide, torture, disasters, and sexual abuse.

Public and professional awareness and sensitivity are part of health system response, in preparation for psychological support as part of disaster planning or in response to catastrophic events such as hurricanes, earthquakes, terrorist bombings (e.g., World Trade Center in New York City 2001, Madrid 2004), or mass murder by deranged persons (e.g., Virginia Tech 2007). Veterans constitute a large population group highly subject to PTSD.

Approximately 7.7 million American adults age 18 and older, or about 3.5 percent of people in this age group in a given year, have PTSD. PTSD can develop at any age, including childhood, with a median age of onset of 23 years.

About 19 percent of Vietnam veterans experienced PTSD at some point after the war. The disorder also frequently occurs after violent personal assaults such as rape, mugging, or domestic violence; terrorism; natural or human-caused disasters; and accidents.

Dissociative (Conversion) Disorders

Dissociative (conversion) disorders involve amnesia, stupor, or trancelike conditions, which have no physical cause but which are related to a specific trigger event. Preoccupation with a variety of physical symptoms, not explainable by detectable physical disorders, and refusal to accept medical reassurance are called *somatoform disorders*. Symptoms vary but usually involve gastrointestinal, cardiovascular, dermatologic, genitourinary, or pain symptoms.

Behavioral Syndromes with Physiological Disturbance

Behavioral syndromes include eating disorders, anorexia nervosa (self-starvation) and bulimia nervosa (self-induced vomiting or purging), abuse of nonaddictive substances (e.g., vitamins, antacids), sleep disorders, sleepwalking, and night terrors. Sexual dysfunction syndromes include loss of sexual desire or enjoyment, sexual aversion, failure of sexual response (male and female), orgasmic dysfunction, premature ejaculation, and painful intercourse.

Prevention requires public discussion, especially in the media, and awareness among family and caregivers concerning teenage psychological adjustment problems and life stress situations. The potential for conversion of normal anxieties into serious, even life-threatening disorders, such as bulimia and anorexia nervosa, are especially important in adolescent care. Social norms, such as promotion of extremely thin models, who are themselves prone to eating disorders, encourage teenage emulation. Middle-aged men, who may be under stress from insecurity of employment or loss of status, may present with sleep or sexual dysfunction and are at risk for physical disease such as premature coronary heart disease events. Unemployment, loss of marriage partner, or financial distress may trigger excessive psychological and physical responses that can be life-threatening. Primary care services need to be oriented to detect and cope with potentially serious behavior syndrome situations in vulnerable groups and provide continuing support and referral services.

Personality Disorders

Personality disorders include deviations of perception and interpretation of people and events, self-images, and affect (i.e., mood or responsiveness). They are associated with difficulty in control of impulses, gratification of needs, and manner of relating to people and situations. The symptoms are persistent, inflexible, and cause distress not explainable by other mental disorders. Personality disorders range from the paranoid, or excessively suspicious, to the schizoid (i.e., emotional detachment, flattened affect, unresponsiveness, solitary life with fantasy and introspection). Emotionally unstable personality disorder involves impulsive behavior with anger and violence and difficulty maintaining a course of action not offering immediate rewards.

Borderline personality disorder includes impulsive qualities with disturbed self-image, emotional crises, threats or acts of self-harm, and chronic feelings of emptiness. The health system has a role in recognition and support of this group of disorders as it does for any significant physical illness; the consequences of such conditions can be serious for the individual, the family, and society. Early recognition may help to initiate professional care and support to help these persons through critical life periods. Mutual and family support groups with professional assistance may be of value.

Disorders of Psychological Development

Psychological developmental disorders include some level of impairment of speech, language, sound categorization, visual perception, attention, and activity control. They include childhood autism; that is, abnormal social communication, attachments, and play appearing before age 3 with lack of spontaneity and social and emotional reciprocity, and failure to develop according to mental age of peers. Early recognition and referral for specialist help require awareness and cooperation between education and health systems.

Behavioral and Emotional Disorders of Childhood and Adolescence

Hyperkinetic disorders include inattention, overactivity, and impulsivity. They include a variety of attention disorders such as attention deficit disorder (ADD) and attention deficit hyperactive disorder (ADHD). Conduct disorders are characterized by aggressive behavior, repetitive behavior with temper tantrums, lying, stealing, use of dangerous weapons, and other unacceptable behavior.

Estimates suggest that 10–12 percent of children and adolescents suffer from mental disorders, including autism, hyperactivity, depression, developmental delay, behavior disorders, and emotional disturbances. A high percentage of children and adolescents who are dysfunctional with mental disorders do not receive appropriate therapy.

Suicide

Suicide is the most serious outcome of mental disorders and in the United States is the ninth leading cause of death among men, and the third leading cause of death among 15–24-year-olds, with 4316 suicides; the total of completed suicides in 2004 was 32,439 for all ages (*Health, United States*, 2007). School systems and health care providers must be alert to the potential for teenage suicide. Provision of easily accessible care and specialist and family support services is needed at crucial times in the history of a disorder or disease that can lead to emotional breakdown or suicide. According to the latest WHO Mortality Database, for the age group 15–19 from 90 of 130 member states, suicide accounted for 9.1 percent of the 132,423 deaths reported. The mean suicide rate for the latest year was 7.4 per 100,000, with higher rates for males (10.5) than females (4.1). In the 90 countries reporting, suicide was the fourth leading cause of death for males and third for young females.

The global death toll from suicide — at almost one million people per year — accounts for half of all violent death worldwide, and estimates suggests self-inflicted fatalities could rise to 1.5 million by 2020. The highest suicide rates are found in Eastern Europe, and lowest in Latin America, Muslim countries, and some Asian nations, according to the Violence Prevention Alliance of WHO. More men than women choose to die through suicide, with the exception of rural China and some parts of India.

Learning Disabilities

Learning disabilities are defined by the U.S. National Institute for Neurological Disorders and Stroke as disorders which affect the ability to understand or use spoken or written language, do mathematical calculations, coordinate movements, or sustain direct attention. Such disabilities occur in very young children, but they may be recognized only when the child reaches school age. This requires special education with specially trained educators able to carry out diagnostic educational evaluation assessing the child's academic and intellectual potential and level of academic performance. Then the basic approach is to teach learning skills by building on the child's abilities and strengths, while correcting and compensating for disabilities and

weaknesses. Other professional services may be needed, including speech and language therapists, psychological therapy and medications to enhance attention and concentration, with wide degrees of severity from an isolated and mild learning difficulty to complex multiple problems.

The U.S. Department of Health and Human Services reports on mixed progress and regression on some Health Targets for *Healthy People 2010*, including expansion of primary care facilities providing mental health treatment, juvenile justice residential facilities that provide mental health screening, and state plans addressing mental health for elderly persons. Three objectives moved away from their targets: suicide, adolescent suicide attempts, and homeless persons with mental health problems who received services. Disordered eating behaviors among adolescents in grades 9–12 and state tracking of consumer satisfaction demonstrated no change toward or away from their targets.

Controversies in Mental Health Policies

During the 1960s, depopulation of mental hospitals and reduction in traditional psychiatric hospitals took place in the United States, Canada, and the United Kingdom. Canada reduced its psychiatric bed capacity by 32,000 beds while increasing general hospital psychiatric beds. The general hospital units tended to treat mild psychiatric conditions, such as mild depression, while resources were not put into needed community-based programs to serve the more severely disabled.

Inadequate follow-up in the community increases the risk of the ex-psychiatric patient suffering exacerbation of symptoms, despondency, homelessness, alienation, imprisonment, and suicide. Community mental health programs should include case management, rehabilitation, housing programs, and other support services. These require well-organized systems to avoid poor coordination, losing patients from follow-up, lack of accountability, ineffective case management, high readmission rates, and inadequate linkages between hospital and community care systems.

Reviews of mental health programs in the past several decades have raised a number of issues, including concern for the civil rights of the mentally ill, the destructive effects of long-term institutional care, the availability of knowledge from the neurosciences, new medications that enable treatment in the community, and the high cost of institutional care. The search for evidence-based medical practice, managed care systems, and capitation payment for mental health care are part of continuing debate in mental health policy.

Mental health policy promoting reduced hospitalization and more community care has come under much criticism. The major concern there has been a low level of investment in community based services and residential help for the mentally ill so that they live in poverty and neglect, in nursing homes, in prisons, or on the streets. The revolving door between institutions of other kinds or emergency departments of hospitals is a dramatic reminder of the need to find more sustainable support systems for the mentally ill. This debate will continue with no prospect of an early solution except a complex of well coordinated service systems meeting the multiple needs of this major public health problem.

Community-Oriented Mental Health

Advances in drug therapy, together with concern for the negative effects of institutionalization and the low cure rates of large psychiatric hospitals, prompted the development of community-oriented mental health (COMH) programs. During the 1960s, this model for delivery of community mental health services took the form of community mental health centers, but more recently there has been a trend to increase the involvement of primary care providers in mental health care.

Mental health services should be integral parts of other health and social services in the community. Patients may seek help at any one of the services, and appropriate care of the patient requires the intervention of many services. This requires development of intersectoral cooperation in health services and staff training in the elements of community mental health needs and services (Box 7.4).

Community-oriented mental health is a programmatic approach providing links between primary health care with hospitals and long-term services, and it includes social support, rehabilitation, and preventive services. These services need to be developed and function as a network in order to provide the individual and the community with the appropriate level of care needed at a particular point or particular stage of an illness.

The development of COMH may require different approaches in a metropolitan area as compared with a small city or rural area. Services should be tailored to the setting and be comprehensive in scope to help the patient in acute need as well as one in need of long-term support services, preferably within the community. Crises in the form of patient or family breakdown may require short-term hospitalization in a community hospital setting, as an integral part of the service.

Community mental health involves primary care providers willing and able to recognize and manage mental health problems. Backup services of psychiatrists, psychologists, social workers, and perhaps paraprofessional community health workers trained to serve the mentally ill, particularly in minority groups or others with special needs, are needed. This requires active educational and organizational efforts to link traditional mental health services with primary care. Primary care physicians and other health personnel must be oriented and trained in the various mental dysfunctional

---

Box 7.4 Community Mental Health Policy

1. Target populations, including a commitment to care for the significantly disabled in the community to the maximum extent possible;
2. General hospital psychiatric units linked to community support, with greater diversification to include holding beds (e.g., 48-hour observation), day beds, crisis intervention, respite care services, and detoxification units;
3. Psychiatric hospitals provide for major psychiatric and behavior disorders (e.g., dementia, brain damage, paranoid, and severely regressive schizophrenia); the supply of such beds is continuing to be reduced, but this requires well-funded and -coordinated community-based services;
4. Continuity of care is essential to ensure adequate maintenance of the chronically mentally ill with highly individualized care, needs assessment, planning, and monitoring;
5. Attention to comorbidity (e.g., drug abuse with underlying psychopathology);
6. Integration of mental health services with linkage to primary and other health care;
7. Consumerism and advocacy groups help define needs of the mentally ill in peer and family support, self-help groups, income maintenance, retraining and job placement, adequate housing, and social support;
8. Replacing dependency with independence-promoting programs;
9. Psychiatric epidemiology provides greater evidence of prevalence of mental ill health and can compare the costs and benefits of different treatment approaches;
10. Community orientation includes outreach programs, supported housing, home services, patient and family support services with attention to the cultural and organizational needs of minority and other special needs groups;
11. Case management with multidisciplinary health care teams, with specialist and hospital backup;
12. Long stay accommodation in group residences with therapeutic services in the community in homelike units, day care.

Source: WHO, Mental Health Program, http://www.int/mental_healthpolicy/services/en [accessed May 7, 2008]

---

conditions, such as mood and anxiety symptoms and disorders. Primary care providers and managed care programs need to assess the needs of risk groups, such as menopausal women and pre-retirement persons, as well as needed modalities of care that cannot be provided by mental health services alone. At the same time, traditional preventive and treatment services cannot function without backup mental health services. Each depends on the other. Building a community-oriented network approach to mental health may require finding new ways of working together, with multidisciplinary staffing either in the same setting or in close coordination.

Prevention and Health Promotion

Prevention requires working with risk groups and with groups with common problems where self-help or a support group may be the most effective form of therapy. This will encompass a wide range of resources which take into account that social, physical, and mental health issues are interactive. Identification of persons at risk for breakdown, such as suicide and violence, requires high levels of awareness by teachers, doctors, police, social workers, military personnel, employers, and the general public. Support groups take many forms, including Alcoholics Anonymous and other groups based on its format, Al-Anon for family members of an alcoholic, drug rehabilitation programs, overeater and binge eater groups, Schizophrenics Anonymous, National Association of the Mentally Ill (NAMI) representing families of the mentally ill, and bereavement groups.

Prevention programs attempt to reduce the possibility that mental dysfunction will proceed to more severe forms in which serious functional breakdown occurs. Social factors are vital in community mental health practices. Prevention in mental health involves many aspects of patient/client care on the primary, secondary, and tertiary levels. Health promotion in the mental health field involves making the public aware of a healthy lifestyle, including activity, rest, recreation, and socialization, and the dangers of substance abuse. Specific programs for activity and socialization should be targeted at the young and elderly. Secondary prevention involves early detection and supportive treatment, and tertiary prevention requires the management of long-term mental illness with adequate support systems in the community.

Primary prevention includes prevention of nutritional deficiencies (i.e., iodine, iron, and vitamins), reduction of environmental contamination (e.g., lead), reducing social and educational deprivation (e.g., education and recreation for youth in high crime/drug abuse neighborhoods), promoting toddler and preschool enrichment programs, promoting family support systems (e.g., for single parent families), providing vocational and employment assistance (e.g., for teenage mothers), providing social support for the aged and disabled (e.g., social security), targeting family and social violence (e.g., domestic, antifemale, and school violence), and raising public and professional awareness and cooperating (e.g., education and health systems).

Secondary prevention requires crisis intervention by trained professionals/paraprofessionals at the primary care level; social relief and support for abuse victims (e.g., shelters); early diagnosis of mental health disorders including screening, early referral to adequate care by police, courts, schools, hospitals, armed forces, and employers; effective treatment and follow-up including outreach and use of paraprofessionals in COMH teams; increased supply and accessibility of crisis intervention services at the community level; use of existing facilities such as hospitals for detoxification and crisis intervention; defining treatment goals and quality of care; and ensuring continuity of follow-up.

Tertiary prevention requires maintaining contact with clients to monitor medications, mood, function, family, and social relations; providing support, referral, assistance; assuring continuity in care and follow-up; assisting client contact with social support networks; protecting/rehabilitating the patient/client, family, and community regarding the effects of the disorder (e.g., drug addiction); providing support groups and structured rehabilitation; funding community-based group transitional residences, halfway houses; promoting independence and self-support; and training and providing continuing education to caregivers, clients, families, and the community.

Mental health research is opening new approaches to management of clinical psychiatric conditions, along with a wider recognition in the community that mental illnesses are real and treatable conditions. Psychotherapeutic medications and psychotherapies are effective, helping people with mental illness by correcting abnormal brain function. More research is needed to further define the relationship between the brain and behavior. New knowledge also helps establish the role of genetic and environmental factors in shaping brain and behavioral function. Great progress being made in the neurosciences and genetics helps biomedical scientists to study the normal and pathological function of the brain and devise new methods of treating mental disorders. This will bring new hope to millions of persons and their families suffering the debilitating effects of mental illness.

MENTAL DISABILITY

Mental disability, retardation, impairment, or cognitive disability is characterized by significantly lower than average scores and performance on tests of mental ability or intelligence and by limitations in daily activities. This includes a range of disabilities including:

1. Intellectual impairment: Low levels of intelligence as measured by developmental tests (IQ).
2. Learning disability or dysfunction: Specific learning disorders, such as dyslexia, which are unrelated to intelligence, require specialist assessment and educational support.
3. Mental disability: This is graded by social maladaptation, which relates to acceptance and accommodation of the disabled in society.

The term *mental disability* includes a wide range of causes and associated conditions. Among them are organic neurologic impairment, genetic endowment, and developmental or educational deprivation. Other factors that greatly influence the social adjustment of the mentally disabled include structure and orientation of services, professional attitudes and training, employment and training practices, social and cultural expectations, family and kinship structures, as well as legislation in relationship to health, welfare, education, and employment.

Many conditions that result in mental disability are preventable. Prevention can be categorized into three areas:

1. Prevention before conception includes good health habits for the future mother with nutrition, fitness, and other preparation. This includes taking folic acid tablets (along with fortification of flour), iron if anemic; screening and genetic counseling for genetic disorders as indicated by risk groups such as ethnicity, intra-familial marriage, and previous birth defects, multiple spontaneous abortions or infant deaths. Smoking cessation, nonuse of alcohol and drugs, possible occupational, toxic, radiation or chemical or other environmental hazards should be taken into account. Immunization for rubella, screening for HIV/TORCH, and sexually transmitted diseases are part of preparation for pregnancy, as is spacing between pregnancies (see Chapter 6).

2. Preventive processes during fetal life and birth include prevention targeted at the nutritional status of pregnant women, including folic acid and iron supplements, prevention of iron-deficiency and macrocytic anemia, and iodine deficiency before and during pregnancy; minimizing harm to the fetus by eliminating alcohol intake, drugs, and smoking during pregnancy; screening to detect fetal anomalies associated with mental retardation; good prenatal and obstetrical care to avoid fetal damage, anoxia, and birth trauma; good neonatal care to avoid anoxia and vitamin K deficiency–induced intracranial hemorrhages; fetal monitoring in labor.

3. Preventive processes after birth include vaccination against communicable diseases that may cause brain damage; screening for PKU, congenital and other inborn errors of metabolism, and hypothyroidism followed by case management; good infancy care and nutrition to avoid nutritional and emotional deprivation that can reduce psychomotor development; and strategies to reduce accidents and their impact.

Improved knowledge of nutrition and risk factors in pregnancy and technological developments are providing improvements in such areas as antenatal detection, treatment of fetal abnormalities, and genetic risk identification, so that mental retardation from preventable causes can be reduced in frequency. Down syndrome can be reduced in terms of new cases by family planning and reduced late maternal age pregnancies and use of currently available screening methods of amniocentesis. Down syndrome cases can also achieve higher levels of learning and social adaptation than was thought possible even a decade ago.

Adaptation of mentally disabled persons to living and functioning in the community is part of rehabilitation.

Political, community, and parental acceptance and support are essential to establish and operate successful community group living and working arrangements for this population group. Public health can address issues relating to adaptation in noninstitutional surroundings within the community. This requires advocacy and education among employers and the general public, and active promotion by the health community.

ORAL HEALTH

Oral health as defined by the World Health Organization is "a state of being free from chronic mouth and facial pain, oral and throat cancer, oral sores, birth defects such as cleft lip and palate, periodontal (gum) disease, tooth decay and tooth loss, and other diseases and disorders that affect the oral cavity. Risk factors for oral diseases include unhealthy diet, tobacco use, harmful alcohol use, and poor oral hygiene" (WHO, 2008).

The term *stomatology* is used for dental health in some countries, derived from *stoma*, the Greek word for mouth. Oral health is an important element of general health status, affecting everyone. Poor dental health can cause pain and interfere with nutrition. Oral disease can progress to loss of teeth and costly treatment, yet much is preventable. In extreme cases, dental illness can result in osteomyelitis, brain abscesses, systemic infection, and death. Public health has a key role to play in oral health because it has important preventive methods available to it. These include health education for oral hygiene, fluoridation of community water supplies, and availability and assurance of quality standards in the dental and paradental professions.

Ancient societies faced dental problems and sought explanations and treatments based on practical experience, with explanations that "nematodes" grew in teeth. Egyptian mummies show evidence of dental treatment. Hippocrates and later Ibn Sinna (Avicenna) discussed hygiene for prevention of dental disease.

In 1672 Russian *Lekars* or military doctors were trained in dental treatment for the army. Dental care was part of general medical care, but at the end of the seventeenth century it separated into its own field. Pierre Fauchard, a French surgeon (1678–1761), described 130 diseases of the mouth and teeth and is considered the father of modern dentistry.

In 1736, artificial golden dental caps were first used. Silver amalgam was first used for filling cavities in 1819. Horace Wells introduced nitrous oxide for dental anesthesia in 1844, which, along with ether, was quickly adopted by the medical community for general surgery. Special cements for filling cavities were developed in 1855. The dental drill was invented in 1870. Four years later, fluoride was discovered to prevent tooth decay. In 1942, the U.S. Dental Health Officer reported on studies in 13 cities which determined a relationship between fluoride use and reduced caries, with excessive levels of fluoride causing fluorosis. In 1946, a study was done in Grand Rapids, Michigan, Kingston, New York, and Brantford, Ontario, whereby fluoride was administered through the community water systems. The results were then compared to cities wherein fluoride was not dispensed. The study showed that fluoridated drinking water reduced caries by 48–78 percent. By 1999, 144 million Americans were using fluoridated community water, with fluoride levels adjusted to the optimal 0.7–1.2 parts per million. The CDC considers fluoridation to be one of the great public health achievements in the United States in the twentieth century.

In developing countries, as standards of living rise, increased use of sugars and candies causes deterioration in children's dental health. Use of sugar in water or tea as a baby pacifier is common in some parts of the world and should be discouraged. Education of parents to prevent this should be part of health promotion.

The major issues in dental public health are dental caries, periodontal disease, malocclusion, and oral cancer. All of these contribute to loss of dentition with effects on general health status. The important interventions to reduce caries and periodontal disease are fluoridation of community water supplies, education in oral hygiene, reduction of sugar intake in children, regular dental care, and dental sealants. Feeding babies sugared water or tea and allowing babies to go to sleep with a bottle in the mouth can cause serious tooth decay. Self-care should include regular brushing teeth after meals and dental flossing, and preventive dental checkups constitute a major aspect of dental health and should be part of school and home health education.

Fluoridation

Fluoridation of community water supplies reduces the number of caries and extractions in both children and adults by some 60 percent. Fluoride is present naturally in most water supplies. When adjusted to a level of 1 part per million, it prevents dental decay. This is one of the most effective public health interventions available. Fluoridation must be a controlled procedure because excessive fluoride in drinking water can cause fluorosis, which results in staining and fragility of the teeth. Good public health practice should therefore ensure adequate fluoride levels to reduce dental caries but avoid excessive levels of fluoridation.

Other methods of giving fluoride include tablets, fluoride rinses, fluoride in toothpaste, or fluoride enrichment of salt or milk. Fluoride tablets are a useful preventive adjunct to child health in areas without adequate fluoride in the water, but the most cost-effective method is by fluoridation of water supplies. The cost of fluoridation of water supplies depends on the size of the community served, but

in the United States this varies from $0.12–$0.21 per person per year in communities over 200,000 population, to $0.60–$5.41 for smaller communities.

Fluoridation has been the subject of political and emotional controversy over many decades and continues to be so, with organized opposition in many countries. This has interfered with the institution of this preventive health measure in many locations. But over time fluoridation is being more widely implemented. Fluoridation has been adopted by most of the major cities in the United States. Currently, 67 percent of Americans on public water systems have fluoridated water, and for U.S. communities, every dollar spent on community water fluoridation results in a savings of $38.00 in costs to repair a decayed tooth.

Progress in the reduction of the prevalence of dental caries has been occurring over the past several decades in countries where fluoridation has been adopted on a wide scale. Worldwide, approximately 210 million persons consume drinking water with optimal levels of fluoride. Health education, fluoride tablets or rinses when fluoridation is not available, and access to dentists for assessment, cleaning, and treatment are also an essential part of public health. More recently, topical application of fluoride and use of plastic dental sealants, developed in the 1970s, to reduce exposure of pits and fissures in teeth to cariogenic organisms has been shown to be highly effective. The U.S. CDC considers community water fluoridation to be safe and effective in preventing tooth decay, and includes fluoridation among its selection of the ten great achievements of public health of the twentieth century (CDC, 2007).

Periodontal Disease

Disease of the oral tissues that support the teeth is described as periodontal disease, a bacterial infection that destroys the attachment fibers and supporting bone that hold the teeth in the mouth. Left untreated, these diseases can lead to tooth loss. The accompanying inflammation of the gums or gingival tissue extends to the periodontal ligament and causes the loss of the supporting bone. If control is not imminent, teeth become loose and must be extracted. The major factor in the development of periodontal disease is poor oral hygiene, especially plaque formation. Research suggests that up to 30 percent of the population may be genetically susceptible to gum disease. Despite aggressive oral care habits, these people may be six times more likely to develop periodontal disease. Identifying these people with a genetic test before they even show signs of the disease and early intervention may help them keep their teeth for a lifetime. Prevention of this disease should include such measures as tooth brushing and flossing, along with use of antiplaque rinses. This improves oral hygiene and is protective against chronic infection, which may have systemic effects.

Dental Care

Expenditures for dental services increased from less than $2 billion in 1960 to $70.3 billion in 2002 — a more than 35-fold increase. The U.S. Centers for Medicare and Medicaid Services (formerly the Health Care Financing Administration) have projected the level of expenditures to the year 2013 to $126.3 billion, an average annual increase of 5.5 percent. Expenditures per capita for dental services increased from $10.86 in 1960 to $244.20 in 2002, an average annual increase of 7.7 percent. However, the percentage share of dental care in total health expenditures declined from 7.1 percent in 1960 to 5.2 percent in 1980 and to 4.3 percent in 2004.

With the dramatic decline in dental caries in the child population of industrialized countries since the 1960s, there has been a tendency to increase use of costly dental procedures. At the same time there has been a reassessment of the numbers of dentists required, whereas in the 1960s there was a continuous chorus of demand to increase the size and number of dental schools in western countries. Dental care is costly, and oversupply of manpower does not necessarily lower prices by competition.

The supply of dentists is affected by demand for services, and between 1985 and 1996 the number of dental schools in the United States declined from 60 to 53, and the number of dental school enrollments of new students in the United States declined from 6132 in 1960 to 4612 in 2005.

Visits to dentists for preventive care, cleaning, or restorative work is limited by economic factors and shortage of personnel in many countries. In the United States, rates of annual dental visits for persons over age 25 increased from 54 percent in 1963 to 61 percent in 1993, but nearly 25 million persons did not get dental care, mainly due to cost. In 2004–2005, dental-related illness accounted for 6.4 million days of disability, 14.3 million days of restricted activity, and a total of 20.9 million days of lost work in 1988. Many countries with universal health plans do not include dental care as a covered benefit, largely because of its high cost.

Dental care programs in schools and through community organizations should include education, fluoride rinse programs where the water supply is inadequately fluoridated, and regular dental checkups and care. Dental sealants applied to children's teeth after eruption of the first (age 6–8) and second molars (age 12–14) is a safe and highly effective means of preventing caries and should be used especially in areas without fluoridation.

In countries with dental health as a part of the national health program, such as the United Kingdom, priority on prevention by fluoridation of community water supplies and school fluoride rinse programs supplemented by dental sealants could reduce the need for restorative dentistry and allow for reducing dental manpower and costs. Dental nurses providing treatment, oral hygiene, and education in school-based programs have been successfully used in New Zealand and parts of Canada, bringing dental care to children otherwise unchecked and inaccessibility due to costs by private dentistry.

Oral Cancer

The control of oral cancer is primarily through early detection, as well as educating the public and training quality professional dentists and other health providers regarding its epidemiology and symptomatology. All suspicious lesions should be biopsied, especially in high-risk patients such as males above the age of 40 who smoke and/or drink heavily. Health promotion emphasizing the elimination of risk factors is most important. Risk factors include smoking (especially pipe and cigar smokers), heavy drinking of alcohol, use of chewing tobacco, and poor oral hygiene.

PHYSICAL DISABILITY AND REHABILITATION

The UN estimates that there are 650 million disabled persons in the world, with 80 percent of them in developing countries. They constitute some 10 percent of any given country's population, the equivalent in size of the population 65 years of age and over in developed countries, but they receive much less attention in health service and social service systems. Despite progress made, the disabled are still not on the social agenda in most countries; only 45 countries have disability discrimination law. The UN and WHO have undertaken promotion of political and professional awareness of the scope of the problem and unmet needs in the area, with a stress on prevention and rehabilitation.

Many people with disabilities need long-term care. These needs will only partially be met in a hospital environment; most care will be delivered within the community by families. Professional health care providers such as physiotherapists, occupational and speech therapists, as well as medical and nursing personnel should be part of the support services. The provision of high-quality rehabilitation services in a community should include the following:

1. Conducting a full assessment of people with disabilities and suitable support systems;
2. Establishing a clear care plan;
3. Providing measures and services to deliver the care plan.

With so many different disabling situations, this field is by its very nature both extensive and complex. In most countries, services and legislation for the disabled have developed in a piecemeal fashion, often as the result of lobbying by social reformers. In more recent years this lobbying has been by organized groups of the disabled themselves.

The rights of the disabled were addressed internationally in a series of international declarations including the 1975 in the United Nations' Declaration of the Rights of Disabled Persons; WHO's 1981 International Year of Disabled Persons; and the United Nations' Decade of Disabled Persons 1983–1992. Reviews of legislation and programs for the disabled in member countries of the WHO European Region set a benchmark in progress in the field. The UN Convention on the Rights of Persons with Disabilities was adopted in 2006. This Convention is the first human rights treaty negotiated in the twenty-first century by the United Nations. It is to be a binding treaty and it addresses issues for disabled people that have been seriously neglected internationally and in most countries.

Principles that should be common include a widespread recognition that prevention of disability is one of the most essential elements of a comprehensive approach, incorporating the classic public health definitions: primary prevention (i.e., stopping the disease or injury from occurring at all), secondary prevention (i.e., if the event occurs, stopping or reducing the severity of complications which lead to disability), and tertiary prevention (i.e., restoring the injured person to maximum feasible physical, psychological, and social functioning). Prevention of disabilities from birth injury, prematurity, failure to thrive, as well as injury in accidents are among the most important issues to address (Box 7.5).

Box 7.5 Principles of a National Public Health Program for Disabilities

1. Political commitment of national, state, local governments;
2. Multisectoral involvement of relevant government agencies, NGOs, media, professional, judiciary, church, academic, private sector;
3. Use of legislative, regulatory, and moral persuasion powers;
4. Comprehensiveness of approach;
5. Prevention-oriented approach;
6. Restoration-oriented approach for physical, psychological, and social function;
7. Promotion of full participation and equality of disabled in society;
8. Promotion of acceptance of the disabled in society;
9. Promotion of community versus institutional care;
10. Reallocation of resources and expenditures to community-based services;
11. Local service responsibility;
12. National coordination, support, and financing;
13. Participation of the disabled in decision making regarding their needs;
14. Promotion of education, employment, housing, and support services;
15. Defined legal, financial, and social rights;
16. Social benefits and compensation without litigation.

Many countries have developed a comprehensive range of laws, activities, and programs for the disabled that include prevention of disability, reduction of the severity of complications, and physical, social, and employment rehabilitation, as well as support or compensation systems. In most countries there are difficulties coordinating policies, services, and support systems.

The disabilities commonly dealt with in most countries include those due to medical, accidental, or occupational ill health or injury: veterans and civilian war victims, the hard of hearing, the blind, the mentally disabled, mentally ill, and others.

National programs in addition to medical, rehabilitative, and preventive health services include income maintenance and compensation, legal rights, normal and special education, vocational training, work in protected or normal settings, housing and the physical environment, communication and transport, leisure and sport, and self-help groups.

Prevention of accidents in motor vehicles, at work, or due to toxic exposures is an essential part of a national public health agenda. Every year, 235,000 people in the 25 European Union member states die as a result of an accident or violence. Annually, more than 50 million people receive medical treatment for an injury, from which an estimated 6.8 million are admitted to hospital. Injury presents the fourth major cause of death in Europe, after cardiovascular diseases, cancer, and respiratory diseases. In children and adolescents, it even ranks as the number-one killer.

The United Kingdom stresses not only extensive social security legislation and support systems, but also the vital role of prevention. This ranges from nutrition and vitamin/mineral enrichment of basic foods to smoking reduction, prevention of low birth weight and perinatal mortality/morbidity, infectious disease control, workers' health and safety, accident prevention, and mental health and disability legislation and compensation. The National Health Service has moved toward decentralized management through district health systems with national support. Organization of services for the disabled has evolved in this direction as well, reflecting a return to the principles of the local administration of the Elizabethan Poor Laws.

The Danish program for the disabled has evolved through social legislation over the past 100 years, especially the Public Assistance Act of 1933. A wide range of national legislation and programs have been implemented for the disabled. As in the United Kingdom, the Danish trend has also been to return responsibility for management of social and health programs from the central government to the county and local governmental levels, with support and service backup from national organizations and the national government.

The Finnish low birth weight rate of 4 percent, down from 5 percent over the past 10 years, is due in part to the 99 percent attendance at their maternal and child centers. Finnish pregnant women average 12 prenatal and postnatal visits. A maternity benefit of $125 is paid only to those who attend prenatal care before the end of the fourth month of pregnancy.

France has an impressively comprehensive program which also places great importance on prevention. As an example, there is much stress on prevention of low birth weight and associated morbidity by requiring early prenatal care for maternity benefits and 20 obligatory examinations for children from birth up to the age of 6 years, including 9 during the first year of life. Screening for metabolic disorders at birth is now augmented by screening for thalassemia, hemophilia, and sickle-cell disease. Great importance is placed on integration of disabled children in regular schools and the social integration of disabled adults. Income support systems are tailored to promote these objectives. For example, in order to qualify for birth and child financial benefits, prenatal and infant visits to preventive care services are obligatory in France.

In Russia, the disabled have traditionally been isolated from the mainstream of society. With over 13 million disabled persons (2006), only 15 percent of disabled people in Russia are involved in any kind of productive work or business, compared to 35 percent in the United States and 80 percent in China. The responsibility for the disabled in Russia is shared jointly by the Ministry of Labor and Social Affairs and the Ministry of Health. The major issues in disability prevention relate to the high rates of binge drinking, violence, suicide, homicide, poisonings, and industrial and road accidents, with associated injury, disability, and mortality.

Progress is being made in developing national programs in prevention, care, and services for the disabled. Sharing experiences from one country to another can help in defining how health and social services can be more effectively arranged both to prevent disability and to integrate the disabled into society. Review of legislation and service organizations as well as the incorporation of new medical, therapeutic, and technological innovations are part of that process.

SPECIAL GROUP HEALTH NEEDS

In every society there are groups of people with special health needs. Indeed, a society is often judged on how it cares for its native peoples, its prisoners, its homeless, and others. They may be set aside from the mainstream of the society as a whole by historic, ethnic, legal, or economic circumstance. They may require special attention to deal with their problems of health because they are either more dependent, vulnerable, or less able to access services based on traditional medical practice, or simply because their needs are greater or more specialized than those of the general population. The New Public Health has an advocacy and pioneering role here, as it does for special needs of the whole population.

GAY AND LESBIAN HEALTH

Gay, lesbian, or bisexual orientation includes some 10 percent of the population of the United States Accurate rates are difficult to determine; societal stigmatization of homosexuality may lead to underreporting, whereas nonrandom convenience studies (e.g., patients at STI clinics) may overestimate true prevalence. Furthermore, sexual orientation may not always correlate with sexual behavior (e.g., 70 percent of gay men report having sex with married men and 45 percent of lesbians report having sex with men). Regardless of the exact figures, gay men, lesbians, and bisexuals represent a significant proportion of the population. In heterogeneous societies, such as the United States, the gay/lesbian populations are part of the diversity of the population-at-large, in terms of ethnicity, religion, socioeconomic status, and geography. Thus, health care needs deriving from homosexual behavior must be addressed in their appropriate social context.

Specific health problems related to same-sex behavior include high incidence of STIs, gastrointestinal infections, and hepatocellular and anal cancer among gay men, related to oral–anal, oral–genital, and receptive sexual intercourse. Health problems of particular concern to lesbians include high incidence of breast, ovarian, and endometrial cancer related to low rates of parity and low rates of breastfeeding and oral contraceptive use. Lesbians tend to visit gynecologists less frequently, and may therefore not receive important tests such as Pap smears or mammograms. Antigay violence is another serious health issue for both gays and lesbians, as are high rates of smoking, alcoholism, substance abuse, depression, suicide, and cardiovascular disease, all possibly related to the stress of living in a homophobic society.

Among the most significant and relatively easy to improve health issues for gay men and lesbians is the lack of preventive services and early treatment due to negative experiences with the health care establishment and fear of stigmatization by the medical community. Studies have shown that only 10–40 percent of primary care physicians routinely take a sexual history from a new adult patient and only 30 percent feel comfortable with gay patients. Over 60 percent of Gay and Lesbian Medical Association members surveyed felt that homosexuals who revealed their sexual orientation risked substandard care as a result. The overall outcome is a failure to screen, diagnose, and treat important medical problems of gay and lesbian patients.

The AIDS epidemic has brought sexual behavior issues to the fore, especially for the homosexual population. However, with only an estimated 10 percent of U.S. students receiving comprehensive sex education and fewer than 5 percent of primary care physicians routinely obtaining information regarding high-risk activities, gay and lesbian organizations themselves have often had to fill the vacuum. Significant successes have been recorded (e.g., widespread adoption of condom use among gay men in the 1980s), but educational programs face enormous ongoing challenges.

One target group of particular concern is the gay and lesbian adolescent population. According to the American Academy of Pediatrics (1993), these teens are severely hindered by societal stigmatization and prejudice, limited knowledge of human sexuality, a need for secrecy, a lack of opportunity for open socialization, and limited communication with healthy role models. Subjected to overt rejection and harassment at the hands of family members, peers, school officials, and others in the community, they may seek, but not find, understanding and acceptance by parents and others. Such rejection may lead to isolation, runaway behavior, homelessness, domestic violence, depression, suicide, substance abuse, and school or job failure. Heterosexual or homosexual promiscuity may occur, including involvement in prostitution (often in runaway youths) as a means to survive.

Thus, with the new generation as with the previous one, ostracism by society leads to those very behaviors which put both homosexuals and heterosexuals at increased risk for social isolation and the spread of diseases, including hepatitis B and AIDS. Breaking this vicious cycle will require a concentrated multidisciplinary effort, in which public health workers, primary care providers, and gay and lesbian organizations all have pivotal advocacy roles to play. There is a process of change in social attitudes to gay and lesbian people which may reduce barriers to preventive and curative services, but still exposure to sexually transmitted diseases is a significant health risk due to unsafe sexual lifestyle, particularly related to multiple partners and failure to use condoms and other safe sex choices.

NATIVE PEOPLES' HEALTH

The United States, Canada, Australia, and other countries provide health care for native populations as special categories under direct provision of care by federal government agencies. These populations are often segregated from the general population, with different schools, legal systems, and health care systems.

The Canadian Indian population in the sixteenth century was approximately 222,000 persons, but disease and famine accompanying European immigration reduced this to 102,000 by 1867 (the year of Canada's confederation). By 1941, the number of aboriginal people was between 100,000 and 122,000, but by 1988 had reached 443,884. There has been a renewal of tribal growth and increase of the Indian population of Canada to 748,371 persons in 2005, following disillusionment with the experience of absorption into the general society, which was marked by tragedies of epidemic alcoholism, violence, drug abuse,

prostitution, and social breakdown. New wealth has benefited many reserves. They have shopping centers, gambling casinos, hotels, and other successful enterprises servicing nearby white communities. However, this has not reduced the flow to the cities where native people are subject to severe stresses with alcohol and drug abuse, child neglect, and many acute and chronic diseases.

Health status indicators in Canada's native peoples indicated widespread alcoholism in the form of binge drinking among all age groups from children and adolescents upward. Glue sniffing, diabetes, hypertension, and hospitalization rates are about twice the Canadian averages. Over the past several decades, there has been a decline in mortality rates among Canadian Indian and Inuit people, but the gap with the overall population remains high. A nutrition survey carried out in 1973 and subsequent follow-up studies showed marked levels of micronutrient deficiency conditions among Canadian Indians and Inuit, with inadequate intakes of vitamins A and D, iron, calcium, and high levels of iron-deficiency anemia and rickets. In 1990, the relative mortality rates as compared to the total population varied from 5:1 in infancy to 1.1:1 for those over age 65. The major causes of death were, in order of decreasing frequency, injury and poisoning, diseases of the circulatory system, neoplasms, and diseases of the respiratory system. Suicide rates for those aged 15–24 were five to six times the Canadian rates for this age group.

In 1997, infant mortality in the Canadian aboriginal population was twice the national average, and life expectancy at birth was 8 years less than the national level. Chronic illness, including diabetes, cardiovascular disease, cancer, and end-stage renal disease, as well as infectious diseases including tuberculosis, STIs, and hepatitis are all much more common among native peoples than the general population. Poverty, physical violence, and alcohol and drug abuse are widely prevalent.

Indian and Inuit peoples come under direct federal responsibility for health and are part of hospital and medical services operated by each province under Canada's universal health insurance program. Services not insured under provincial health plans are provided by the federal department, Health Canada. Recent trends to decentralize management of services to the tribal council have been matched by proposals for constitutional change making aboriginal communities recognized levels of government. This has, however, not been adopted. The social problems of unemployment, poor education, alcohol abuse, family breakdown, violence, and welfare syndromes plague Canadian Indians who move to the cities perhaps even worse than those who remain on the reserves.

The violent record of the United States in its wars against the Indians was followed by forcible concentration on poor land reservations and lethal administrative practices far harsher than the Canadian record. However, provision of care and prevention of social breakdown may be even less effective in Canada than in the United States in the latter part of the twentieth century.

The history of U.S. government provision of health care to American Indian and Alaskan natives is equally troubling. Since the early nineteenth century, care for Indians was provided by the federal government; namely, by military doctors up to 1849, then by the Bureau of Indian Affairs as part of the Department of the Interior until moved to the U.S. Public Health Service in 1954. The Indian Health Service (IHS), formed as a national health service for Indians, is administered through regional offices, providing primary and specialty care with increasing emphasis on community participation and control.

The U.S. population of Native Americans and Eskimos in 1996 totaled 2.3 million persons and is expected to reach 4.3 million and just over 1 percent of the population by 2050 (U.S. Census Bureau, 2008). The IHS provides care to some 60 percent of this population. Gains in health status in reduced infant and general mortality still leave the native peoples in poorer health than the general U.S. population. Poor nutrition, unsafe water supplies, isolation and poor transportation, inadequate waste disposal, and high rates of obesity, alcohol abuse, and violence shorten life expectancy. The IHS operates a network of 43 hospitals and over 110 health centers and health stations. Expenditures for American Indian health care, however, are 60–65 percent of per capita national average expenditures.

Life expectancy for Native Americans improved from about 60 years in 1950 to 73.2 years in 1989–1991. Improved infant mortality and a reduction in deaths due to tuberculosis, gastroenteritis, and acute respiratory disease have been accompanied by a reduction in alcohol- and violence-related mortality. Rates of diabetes (type 2) are high and increasing, as compared to the non-Indian population. Fetal alcohol syndrome (FAS) is a large problem for Native Americans, with an estimated incidence of 2.7 per 1000 live births. The rate of FAS for American Indians is higher than that for any other racial or ethnic group in the United States. Increasing tribal management of health services has been accompanied by increasing tribal activity to reduce alcohol abuse and violence. Health goals for 2000 included reducing rates of cancer, infant mortality, and alcoholism.

The aboriginal populations, initially decimated by famine and acute infectious disease and currently by alcohol abuse, violence, and diabetes-related conditions, have a health status that remains well below the health of both Canadian and American populations. Federal administration and separateness from general health services have been subject to much criticism in both countries. Integration with other health services and decentralization of management with community involvement are current trends. Federal financing of this service has stabilized, while increasing local management leads to increased expenditures by this population group in the private medical market.

The prospects of the native population in the United States, Canada, and elsewhere will depend on social and economic development, with a large measure of self-government. In 1999, the Canadian government established a self-governing Inuit (Eskimo) territory which has authority for health, social services, education, taxation and economic rights (e.g., minerals), and many other areas accorded to Canadian provinces. This will provide an important test for this concept.

Similar problems appear among aboriginal populations and minority groups in other countries, including Australia, New Zealand, and Peru and other South American countries. Aboriginal populations in Papua, New Guinea, Taiwan, central Africa, Russia, and Australia suffer from conditions of high rates of cardiovascular disease, diabetes, end-stage renal disease, alcoholism, glue sniffing and drug abuse, rheumatic fever, infant and child mortality, and general social breakdown. The poor health of native peoples is a blight on the records of many countries without clearly defined health responsibility, authority, and programming. The issue of health of aboriginal peoples is a serious challenge to the New Public Health in finding ways to provide adequate preventive and curative care, and more importantly to reduce social gaps engendering apathy and social decay.

PRISONERS' HEALTH

Since the mid-1980s, there have been extraordinary changes in the condition of jails and the health of prison inmates in the United States. The increase of urban decay, illicit drug use, and poverty and associated epidemics has had a great impact on incarcerated Americans. Prison medical services have been transformed into outposts faced with the challenge of meeting near impossible demands.

Imprisonment grew in the United States from 1.6 million prisoners in 1997 to 2.3 million in December 2006 (Bureau of Justice Statistics) held in federal or state prisons or in local jails. This was an increase of 2.9 percent from 2005, but this was less than the average annual growth of 3.4 percent since 1995. Incarceration rates rose from 411 at the end of 1995 to 501 prison inmates per 100,000 U.S. residents at the end of 2006.

Incarceration rates by gender and race show increases in the number of women in state or federal prisons by 4.5 percent from 2005, reaching 112,498, and the number of men rose 2.7 percent, totaling 1,458,363 in 2006. At year-end 2006, there were 3042 African-American male prison inmates per 100,000 African-American males, compared to 1261 Hispanic male inmates per 100,000 Hispanic males and 487 white male inmates per 100,000 white males. Approximately 6 percent of white males are incarcerated and 23 percent of African-American males are under the supervision of the correction system. About 47 percent of

U.S. prisoners are African-American, and a large number are young and poor (http://www.ojp.usdoj.gov/bjs/prisons.htm [accessed January 24, 2008]).

In 1970, concern for the poor state of prisoner health led the American Medical Association and the American Public Health Association to conduct reviews of prison health services, recommending guidelines which provided state prison services with standards to meet and resulted in improved services in many states. Prison health services took on a more professional role, with less reliance on part-time practitioners. Voluntary accreditation of services by the American Council on Health Services Accreditation serves as a valuable external review procedure promoting transparency in the quality of care.

Persons in correctional institutions are at increased risk for tuberculosis because of the high prevalence of HIV and hepatitis B infections and latent TB, overcrowding, poor ventilation, and the frequent transfer of inmates within and between institutions. The recent emergence of multi-drug-resistant TB as an important opportunistic infection of HIV-infected persons underscores the need for improving infection control practices. The increase in TB in Russia since 1990 is in part related to large-scale release of prisoners following perestroika in the 1980s. Specific control measures in correctional facilities should include the following:

1. Regular and systematic screening of inmates and staff for HIV and TB. Those who test positive should be eligible for preventive therapy.
2. Rapid identification, isolation, and treatment of suspected cases of TB.
3. Directly observed therapy as well as rigorous follow-up and record keeping to ensure completion of treatment.
4. Follow-up to assure continuity of care both inside and outside the correctional facility.

Prisoners are also at risk for sexually transmitted infections. Both male and female prisoners may be subject to physical or sexual assault or intimidation by fellow prisoners and staff. Prison medical services should have screening and treatment capacity as well as powers to assure protection of vulnerable prisoners, such as segregation of young prisoners from potentially violent or long-term inmates.

In the 1990s, U.S. federal legislation mandating life sentences for repeat offenders ("three strikes" law) increased the number of elderly prisoners ineligible for parole. As larger numbers of inmates spend longer periods in prison, there is a need to regard prisoners' health as part of that of society as a whole.

Short-term and long-term health risks for prison staff require the attention of prison health services. Alcoholism, smoking, obesity, and job stress are linked to family breakdown and early mortality from cardiovascular diseases especially. Helping staff deal with the stress and latent violence on the job should work to reduce personal risk and propensity to staff-initiated violence.

The need for mental health services in prison is great and growing. The deficiency of community mental health services has accelerated the movement of individuals from the street to the penitentiary. A study by the National Association for Mental Health found that 25 percent of prisoners admitted were psychotic, with another 14 percent manifesting some psychotic symptoms. Mental illness may be responsible for a large part of violent behavior and violent crime. There is a need to combine jail treatment facilities with rehabilitation programs. A study in New Zealand compares prevalence of mental disorders among prisoners with people living in the community as shown in Table 7.3.

Ethical issues in prisoner health include torture, rape, murder, starvation, unethical medical experimentation, and inadequate medical and psychiatric care. The Nuremberg and Helsinki codes of medical conduct in medical experimentation are internationally accepted standards. The Geneva Convention standards related to war and military occupation conduct are also accepted, often ignored, but still a basis for judgment. Abuse of prisoners in civil war situations, such as in Bosnia in the mid-1990s, reached genocidal levels with massacres, starvation, and rape applied widely. The health community in any country or conflict situation has a professional obligation to monitor care of prisoners and play an advocacy role in times of peace and of civil war.

TABLE 7.3 Comparative Prevalence of Psychiatric Disorders in Prisoners and in People Living in the Community

Disorder	Prevalence	
	In prisoners (%)	In community (%)
Any psychiatric disorder	80	31
Psychosis	7	0.7
Affective disorder	23	9
Anxiety disorder	38	11
Substance abuse disorder	66	18
Personality disorder	43	9

Source: Butler, T., Allnutt, S., Cain, D., et al. 2005. Mental disorder in the New South Wales prisoner population. *Australia New Zealand Journal of Psychiatry*, 39:407–413.

MIGRANT POPULATION HEALTH

In many countries, large numbers of persons and families move with seasons for farm or other temporary work. They lack stable social environments and support systems and are often dependent on exploitative employers willing to provide only subsistence wages, and a possibly hostile surrounding community. This group includes migrant and seasonal farm workers, miners, foresters, and building workers who often live away from their families for extensive periods.

In the United States, this phenomenon is widespread. The Migrant Health Program of the U.S. Department of Health and Human Services estimates the size of the population of migratory workers at 1.5 million, and another 2.5 million persons work as seasonal farm workers. The majority are married with children, so the total size of the population including dependents is much larger. Most of these families are living below the poverty line. Many are illegal immigrants, subject to exploitation and abuse by employers or contractors. They are exposed to poor sanitation and housing conditions, pesticides, and infectious diseases.

Migratory workers in the United States are of mixed ethnic origin, including white Americans, Hispanics, Haitians, and Jamaicans. Immigrant groups, such as Haitians, Mexicans, and Filipinos, have high rates of tuberculosis, and their work in agriculture may lead to disease transmission to other farm workers. Migrant farm workers are approximately six times more likely to develop TB than the general population of employed adults.

Migrant farm workers may bring their families with them to poor housing encampments operated by their employers with unsanitary conditions and limited access to food and other necessary supplies. Arrangements for health care, schooling, and recreation are likely to be inadequate. Migrant workers in urban settings are frequently single men, who may have left families at home and who migrate from declining rural regions in the same country or who are brought into a foreign country for specific work in agriculture, construction, or mining. Women workers are often brought in for housekeeping or personal caregiver jobs, or as sex workers. Professional migrant workers in nursing medicine, computers, and other skills are also common in many parts of the world. People living in a foreign culture as contract workers are subject to abuse and exploitation as well as social isolation with associated mental stress, violence, and STIs.

Migrant workers are more likely to live in poverty, and to suffer unsanitary and crowded living conditions, with lack of medical care. In addition, they generally lack the necessary documentation for establishment of legal residence status. Access to health insurance and regular medical care, especially for children and women is often one of the most pressing issues, especially for illegal immigrants. Nutritional problems of iron deficiency and stunted growth are common as well as micronutrient deficiencies. Dental disease is common in migrant families, and access to dental care is usually lacking. Injury and deaths from work accidents are high among farm workers. Agricultural workers are the occupational group having one of the highest mortality rates (37 per 100,000 agricultural workers) from work-related injuries (see Chapter 9).

High prevalence of syphilis, HIV infection, and TB among migrant workers underscores the need for public health professionals who are trained to respond to health care needs within the migrant worker population. These studies led to development of cross-training for public health workers on STIs (including HIV infection), TB, and other communicable diseases among migrant farm workers.

Migration for work or to seek a better life is an international phenomenon. In Europe, large numbers of "guest workers" provide unskilled labor in agriculture and the building trades primarily to the wealthy economies which over several decades imported workers from poor countries in North and sub-Saharan Africa or Eastern Europe. This has created long-term social and political problems for workers and their families, who are often subject to ethnic resentment, violence, and cultural adaptation difficulties. Public health agencies should supervise living conditions and arrange preventive and treatment services with screening for common diseases in this population group.

Migrants are a vulnerable group with many health needs, ranging from infectious diseases and chronic diseases to mental health. They may face discrimination and language and cultural barriers, and work at menial jobs in agriculture, building trades, and caregiving. They leave their homes moving from rural to urban centers in their own countries or to new countries to improve their economic conditions, sending money home to help poor families and hoping to stay or return with savings to reestablish themselves in a better condition. This can be a rewarding or tragic process. Some conditions are akin to slavery as in trafficking of women for sex work. Exposure to disease and abuse is common.

Addressing the health needs of migrant populations is in the self-interest of host countries to prevent the spread of infectious diseases, such as HIV/AIDS and multidrug-resistant tuberculosis, usually acquired in the host country. This is an important population group requiring special attention in the New Public Health.

HOMELESS POPULATION HEALTH

Homelessness due to poverty and lack of low-rental permanent housing has become a common problem in urban centers in industrialized countries and has become a longstanding issue in developing countries. It is increasingly capturing attention in the media because of the contrasting lifestyles of the homeless and the wealthy population.

Depending on how "homeless" is defined, the WHO estimates that there are between 100 million and 1 billion homeless people worldwide (World Health Organization, 1997). An estimate suggests that as much as 7.4 percent of the U.S. population has experienced homelessness at some time in their lives. Long-term street dwellers are at great risk for health problems.

Health care, especially preventive, is not a priority for homeless people. Instead, many of their resources, such as time and money, go toward securing a safe, warm, and dependable place to sleep, toward purchasing food, and often toward drugs and alcohol.

The global pandemic of homelessness is the result of the lack of affordable housing, unemployment, and the deterioration of social services and support. The homeless do not "choose" to live on the streets; rather, they are often victims of circumstance. It is often argued that homelessness is the result of deinstitutionalization from mental institutions, and some estimates of the percentage of homeless individuals suffering from mental illness range from 25 to 50 percent. It is difficult to gauge whether homelessness causes mental illness or vice versa, but it is clear that the two combine to produce a vicious cycle in which individuals face social isolation, inadequate nutrition, and hygiene barriers to housing and employment, which together serve to perpetuate illness and insecurity.

Surveys of health problems faced by the homeless show that rates of both chronic and acute diseases are extremely high. With the exception of stroke, cancer, and obesity, homeless people are far more likely to suffer from every category of chronic health complication. Because of the transient nature of the population, conditions which require vigilant care and attention, such as TB, are rampant. This, of course, has grave dangers for the public at-large. Along with prisons, homeless shelters account for a large percentage of TB cases in the industrialized countries. Other health problems include death by freezing or violence, frostbite, leg ulcers, burns, respiratory infections, STIs including HIV, and trauma from muggings, beatings, and rapes. Homelessness precludes good nutrition, shelter, warmth, security, personal hygiene, and basic first aid. The homeless have poor access to care until the medical condition is dire.

As the members of this group are often uninsured, access to health care services is compromised by cost and distance. One of the most effective means of serving this population is through mobile health units. In Washington, DC, among other U.S. cities, "health vans," coupled with mobile "soup kitchens," operated by charitable organizations, provide nutritional and medical services to the homeless. If conditions are severe enough to warrant more substantial care, the individual is encouraged to enter a city or private homeless facility, or may be referred for medical care. These provide medical care, a hot shower, a place to stay the night, and hot meals. The role of NGOs and charitable organizations is important in helping the homeless to survive. Clearly, though, these are only temporary solutions to the millions of homeless people worldwide living in dangerous conditions.

Homelessness of female-headed families with young children may occur from family breakdown due to an abusive spouse, mother's use of drugs and alcohol, pregnancy, migration, or the collapse of an economic base of the family with eviction from a stable home. Homeless runaway children or teenagers are vulnerable to serious social and health consequences, including violence, sexual abuse, substance abuse, suicide, and high-risk behavior, resulting in unwanted pregnancies, STIs, or HIV infection. In many developing countries, homelessness and poverty provide a source for sale of babies, organs, and child prostitution.

The United Nations has estimated the population of street children worldwide at 150 million, with the number rising daily. These young people are more appropriately known as *community children*, as they are the offspring of our communal world. Ranging in age from 3 to 18, about 40 percent of those are homeless — as a percentage of world population, unprecedented in the history of civilization. The other 60 percent work on the streets to support their families. They are unable to attend school and are considered to live in difficult and dangerous circumstances. Increasingly, these children are the defenseless victims of brutal violence, sexual exploitation, abject neglect, chemical addiction, and human rights violations. Released prisoners and the mentally ill are overrepresented in the homeless population, but this is also an issue among U.S. veterans.

Homelessness is a public health problem, mainly to the homeless themselves, but also affects the community around them. Solutions, like those for many other social problems, are complex and costly. The root issues lie at conquering the common characteristics of homelessness: poverty and the inadequate supply of permanent housing. Reintegrating the homeless into mainstream society requires community-based treatment and social services. This means client engagement, case management, housing options, long-term follow-up, and support. Vocational training and employment opportunities may be the key. Until such help is available, homelessness will continue to cause much suffering as well as strain the charitable and public health systems which try to care for this population.

REFUGEE HEALTH

The UN definition, which the United States accepts, considers a refugee "any person who is outside any country of such person's nationality . . . and who is unwilling or unable to return . . . because of persecution or a well-founded fear of persecution on account of race, religion, nationality, membership in a particular social group, or political opinion." An estimated 35 million persons have fled their countries as refugees or have been displaced internally at the end of 2006 — 32.9 million people. Of some 20 million cross-border refugees, about 14 million are in Africa, southwest Asia, or the Middle East. The mass exodus of people from their homelands can be the result of ethnic or religious persecution, political conflict, war, civil strife or political conflict, environmental degradation, or conflict over economic resources. Annually, millions of persons abandon their homes, farms, regions, or countries in search of food and water, employment, or security.

The ability of a country to cope with a sudden mass refugee situation depends on the function of the state infrastructure and basic services, the security situation, and prior planning for disasters. Appropriate and timely response depends on local strengths in these areas, with international assistance as a backup for needed supplies and services. Effective relief services must be primarily based on the establishment of security, shelter, clothing, sanitation, safe water and food supplies, family reintegration, personal preventive services, and treatment services. Education, family planning, and longer-term health services are part of any relief program when the refugee situation continues.

The United Nations High Commissioner for Refugees (UNHCR) monitors and coordinates refugee assistance needs and relief services. The 1951 Geneva Refugee Convention established UNHCR to protect the human rights of vulnerable persons displaced by war and civil strife from being forced to return to a country where they may be persecuted. UNHCR helps in civilian repatriation, integration in countries of asylum, or resettlement in third countries. Its world network provides shelter, food, water, and medical care in the immediate aftermath of refugee situations (UNHCR, 2008). At the end of 2006, UNHCR estimated global refugee population at 9.9 million people.

The acute phase of a refugee situation, discussed later in the section on man-made or natural disaster relief, may become a long-term situation which imposes a different set of problems. The drama of the acute phase may become an international media event with many countries and NGOs offering assistance, relief supplies, and financial contributions. Longer-term refugee situations require help which places more responsibility on the UNHCR and NGOs such as the International Red Cross, Médicins sans Frontières, Catholic Relief Services, and others (see Chapter 2). Long-term refugee centers require all the basic services of any population group, with the additional factor of the risk of collapse of fragile sanitation, nutrition, or health services following political change or further disaster. The Kosovo crisis of 1999 brought together international efforts of the UNHCR, the North Atlantic Treaty Organization, and many international NGOs, with bilateral and public assistance with finances and essential survival material. But the immediate solution to the massive displacement was resolved through military and political action to force the home country to allow repatriation of the refugees, although long-term political solutions are hard to foresee.

The UN Environmental Program Report in 1985 defined *environmental refugees* as "people forced to leave their traditional habitat, temporarily or permanently, because of marked environmental disruption (natural and/or triggered by people) that jeopardized their existence and seriously affected the quality of their life." The Framework Convention on Climate Change (Bali, 2007) estimated the possibility for 50 million people being environmentally displaced by 2010, with a long-term projection of about 200–250 million people by 2050 (Biermann and Boas, 2007, http://unfccc.int/2860.php [accessed May 8, 2008]).

MILITARY MEDICINE

Armed forces are part of national responsibility for defense and security. This includes provision of preventive and medical care for the military and their families. Armed forces seek to recruit people in apparent good health, but conditions of service may cause disease both in peace and in wartime. Military medicine emphasizes prevention to maintain the health of personnel who will be placed in hazardous conditions for disease and injury.

Military medicine has played an important role in the development of surgery and related skills and public health. Roman military success was aided by skill in preventive health measures through hygiene and camp discipline, no less than in care of wounds, basic military organization, and discipline. Armies and navies depend on medical and nutritional support to maintain health and ability to perform missions assigned. The mandatory use of limes for British sailors following the epidemiologic breakthrough of James Lind established the necessity of nutritional support and discipline to maintain function and competence of the individual and the unit. The American army adopted mandatory vaccination soon after Jenner's method became known. Innovations in public health by military personnel during the nineteenth century were numerous, from Ronald Ross's work on the malaria parasite to the conquest of yellow fever by the U.S. Army in Cuba in 1901.

Death from disease outweighed deaths in battle in most armies until the early part of the twentieth century, as seen in data from the U.S. Army from the Civil War to the Vietnam War (Table 7.4). Awareness of the predominance of disease in army casualties was a major contribution of Florence Nightingale from the Crimean War. Military medicine has contributed to the development of emergency medical care, bringing enormous benefit for the civilian sector, including innovations in medical adaptations from military technology such as ultrasound.

Protection of troops from disease requires assurance of immunization to prevent diseases that could easily be transmitted in barracks or shipboard living conditions.

TABLE 7.4 Number and Percent of Deaths from Disease and Battle Injury, United States Army, 1860–1975

War	Disease (D)	Battle injury (BI)	D/BI%
Civil War (North) 1861–1865	199,720	138,154	145
Spanish American War, 1898	1939	369	525
Philippines, 1899–1902	4356	1061	410
World War I, 1917–1918	51,447	50,510	102
World War II, 1941–1945	15,779	234,874	7
Korean War, 1950–1953	509	27,704	2
Vietnam War, 1961–1975	1433	30,900	5

Source: Adapted from Legters and Llewellyn, 1992. Military medicine, In Last, J. M., Wallace, J. B. (eds.). *Public Health and Preventive Medicine*. London: Prentice Hall.

This may include updating childhood vaccinations with boosters of diphtheria and tetanus and vaccinating against hepatitis B, influenza, polio, measles, mumps, rubella, meningococcal meningitis, hepatitis A, anthrax, and others. Antimalarials and other preventive measures are used depending on location of service.

Nutrition; protection from food-borne, waterborne, and vector-borne diseases; and prevention of training and motor vehicle accidents, suicides, exposure to toxic materials, contact with STIs, and HIV exposure are among the many issues of peacetime armed forces. Violence and brutality are frequent issues in the training period, as are suicides. Officer and noncommissioned officer vigilance, accountability, with medical and psychological surveillance are required.

Prevention of war is the surest method of preventing related military and civilian death and injury. In war, battle casualty prevention depends on armaments, skill in their application, discipline, and leadership. Medical support in the field, evacuation and triage, and rapid transfer with life support systems to medical centers are crucial to keep fatality to injury ratios low and save many young lives. Organization of medical services from the medic on the battlefield to the evacuation post and the base hospital requires skilled organization and management with assurance of adequate supplies of everything from water and food to diagnostic and treatment resources.

Casualty management to conserve fighting strength is based on prevention and collection, treatment, and evacuation of casualties in a manner that supports morale of the troops and prevents complications or death following injury. Triage or sorting is based on providing the best care one can for the most patients under the circumstances. Triage categories include the following: urgent cases require airways, chest tubes, hemorrhage control, and replacement therapy to prevent immediate death; immediate cases are life-threatening wounds temporarily stable but requiring surgery within a short time and a good chance of recovery; delayed cases are injuries that can be cared for successfully 8–16 hours after the injury; minimal, or superficial, wounds require minor surgery, fracture setting, or observation; expectant cases are mortal wounds with little chance of survival.

Special U.S. veteran hospitals and health services are operated by the Department of Veterans Affairs. The VA deals with long-term disabilities from consequences of the injuries and trauma, Agent Orange exposure during the Vietnam War, Gulf War syndrome, and post-traumatic stress disorder (PTSD). Although each successive war brings renewed attention to this syndrome, it wasn't until the Vietnam War that PTSD was first identified and named.

Conditions of potential atomic, biological, and chemical warfare require special preparation and, above all, prevention. The potential exists, and preparation for its effects on the health of combat units and civilian populations are part of this special field of military medicine and, indirectly, public health. New phases of historical and political developments brought up war with terrorism and bio-preparedness. Military medical systems work with civilian public health to reduce threats of bioterrorism through strengthening public health preparedness and bio-defense research.

HEALTH IN DISASTERS

Natural and man-made disasters are frequent occurrences (Table 7.5) and have important implications for public health. Natural disasters are naturally occurring extreme events that can cause excess morbidity, mortality among the population, and damage to the physical environment.

These include earthquakes, floods, hurricanes, droughts, blizzards, and volcanic eruptions. The term *disaster* also includes a wide range of events including war, industrial accidents, terrorist incidents, and others. Disasters are often classified into "man-made" and "natural" and "sudden" or "slow" onset. The distinctions are often blurred, since a natural disaster may be the result of inappropriate policy decisions such as a drought occurring in an area made vulnerable by political or policy actions. In semiarid areas with a delicate food balance and chronic undernutrition, a modest natural disaster may tip the balance and cause wide-scale

TABLE 7.5 Selected Disasters — Man-Made and Natural, 1976–2008

Location	Type	Effects
Serveso, Italy 1976	Chemical factory explosion	17,000 evacuated
Cambodia 1979	Genocide, political	1–2 million deaths
Bhopal, India 1984	Chemical leak	2000 deaths, 70,000 evacuated
Mexico 1985	Earthquakes	10,000 deaths, 60,000 homeless
Colombia 1985	Volcano	23,000 deaths, 200,000 homeless
Chernobyl, Ukraine 1986	Nuclear reactor meltdown	30 dead, 100,000 evacuated; long-term effects (cancer, birth defects) unknown
Bosnia, 1993–1995	Civil war, genocide	Tens of thousands of casualties, mass rape and murder, breakdown of civilian services
Rwanda 1994	Genocide, tribal	Up to 500,000 deaths
Kobe, Japan 1995	Earthquake	6000 deaths
Caribbean 1995	Hurricane Gordon	11,000 deaths in Haiti, Cuba, Jamaica, Dominican Republic
China 1998	Flood	4150 dead, 18.4 million displaced, and 180 million affected
Nicaragua 1998	Hurricane Mitch	10,000 dead, 120,000 homeless
Kosovo 1999	Ethnic cleansing	1 million persons forcibly displaced, with mass murder, community destruction
Tsunami 2005	Tidal wave of 11 countries bordering Indian Ocean	Over 225,000 people killed and missing
New Orleans, USA, 2005	Hurricane Katrina	1836 killed, most of New Orleans destroyed
Darfur 2002–2008	Genocide	200,000–400,000 killed, 2.5 million displaced
Republic of Congo 2006–2008	Civil war	Hundreds killed monthly
Kenya 2008	Political-ethnic conflict	250,000 Kenyans displaced, 350 killed
Brazil 2008	Floods	49,506 people affected
Myanmar 2008	Cyclone Nargi, floods	Over 100,000 killed; 1.6 million homeless

Sources: Relief Web Open Forum, http://www.reliefweb.int/rw/dbc.nsf/doc100?Open Forum [accessed May 10, 2008]
UN News Center, http://www.un.org/apps/news/ [accessed January 23, 2008 and May 10, 2008]

suffering and malnutrition continuing over an extended period of time.

The accumulated experience of disaster relief management has been enhanced in the past several decades by improved technology in communications and transportability of air-mobile supplies of tents, blankets, food, chlorinators, generators, heavy equipment for lifting debris, as well as field hospitals, and medical supplies such as sterilizing and trauma equipment, antibiotics, vaccines, and oral rehydration therapy.

Disasters require a highly professional and aggressive response based on intersectoral cooperation. This can be a matter of life and death to large groups of people.

Disasters can occur in the midst of an urban metropolis or in a remote jungle. The details will differ, but meeting basic human needs is common to both settings. Limiting further death and injury requires attention to protection and security and safe water, food, and shelter as first priorities. Organized prevention in the form of sanitation and management of the most common health problems such as diarrheal diseases, acute respiratory diseases, measles immunization, and other diseases associated with poor sanitation (e.g., hepatitis, typhoid, cholera, and gastroenteritis) is needed. Dysentery, malnutrition, and respiratory infection outbreaks can kill large numbers of debilitated children and the elderly. PTSD can be dealt with as part of a community-oriented support program with involvement of the population affected as health aides.

Epidemiologic monitoring of death, injury, and disease is usually difficult because of the chaotic events associated with disasters. Monitoring of the process and epidemiologic patterns provides information for intervention and lessons which can be important in planning and management of future crises. Monitoring admissions to hospitals indicates morbidity patterns for specific diseases such as typhoid fever and viral hepatitis. Such was the case during the 1980 earthquake in southern Italy causing water contamination. Cholera is also a serious danger in disaster situations as occurred in the Rwanda refugee disaster. Contamination of limited water supplies makes this one of the most urgent aid issues, with pumps and portable chlorinators as basic equipment needs. Tents and food distribution facilities may need to be protected by armed guards.

Intervention by aid agencies should include the potential for vaccination against measles and poliomyelitis and sustaining regular DPT vaccination, along with vitamin and iron supplements and insecticide impregnated bed nets if the return to normal conditions is delayed over months. Typhoid and salmonellosis are combated by sanitary and case management approaches. The absolute necessity of potable water and oral rehydration therapy on a large scale was demonstrated in the Rwanda disasters of 1993–1994.

Over longer periods, nutrition monitoring of child weight and height for age may be done on sample populations to assess the effects and changing situation. Disaster relief should include the possibility for vitamin and iron supplements for children to offset the damage of food deprivation. Provision of food staples is a high priority along with shelter and water, but the bulk shipment required may overwhelm even the ability of developed countries to cope with local disasters.

Secondary damage from unsanitary conditions can spread infectious disease. This is aggravated by food deprivation and lack of water and sanitary facilities. The large numbers of refugees from civil war and ethnic massacres seen in the 1994 Rwanda situation resulted in massive loss of life. International or local relief efforts may be overwhelmed and hampered by lack of coordination. Planning refugee camps in disaster situations must take into account natural drainage, water sources, access roads, separation of sanitary facilities including latrines and garbage disposal with safe areas for food and water distribution.

Better coordination of storm warnings could have reduced deaths at sea; warnings to remain off the roads and curb use of vehicles in danger areas may have reduced road crash deaths. Emergency information to promote safe evacuation, safe havens, risk avoidance, and assurance of support services are part of public health action in emergency situations, in conjunction with other local, state, and federal agencies.

In late 1996, a massive refugee crisis emerged in Rwanda and Zaire, with complex ethnic and political origins, which resulted in death on a massive scale due to genocidal military actions as well as dehydration, disease, and starvation. International intervention involving military field hospitals, aid, and health organizations were sporadic and insufficient. NGOs such as Médicins sans Frontières and UN aid organizations are often the first on the ground, but without strong international political and military backing, all efforts will be limited. Following the tsunami of 2005, U.S. armed forces provided immediate aid with troops, helicopters, and medicine. Bilateral governmental aid is important and often brings essential services and supplies not available locally. Medical interventions with field hospitals capture media attention, but more fundamental efforts to promote basic organization for shelter, safe water, and sanitation receive fewer resources and less support.

Disasters during 2005 caused 99,425 deaths, of which 84 percent were due to October's south Asian earthquake. Hurricane Katrina, which hit America's Gulf Coast in 2005, killed around 1300 people and was followed by Hurricane Stan, which killed over 1600 people in Guatemala. In 2005, natural disasters globally affected 161 million people and cost around $160 billion (U.S.). Between 1996 and 2005, disasters killed over 934,000 people, with 2.5 billion people affected across the globe. Earthquakes, floods, heatwaves, and other natural and technological hazards cause thousands of deaths and billions of dollars in economic losses each year. In the WHO European Region between 1990 and 2006, 1483 events were recorded, killing 98,119 people and affecting more than 42 million others, with an estimated economic loss of over €130 billion. Increasing evidence indicates that climate change, environmental changes such as the depletion of stratospheric ozone, and increasing interconnectedness resulting from changes in trade, travel, and technology may threaten human health through complex and interdependent mechanisms. Health systems will need to include these evolving threats in comprehensive efforts to protect health.

Prevention of human injury and death in disasters must take into account experience in a given area. Earthquakes in California are a serious threat and have led to strict building codes that reduce the damage and secondary loss of life. Monitoring can give warning of potential disasters such as hurricanes, and preparation of evacuation plans and facilities makes a major difference in disaster management. Zoning laws and building code enforcement can prevent many deaths when earthquakes strike in urban areas. Flood control measures in areas traditionally threatened are worthwhile investments to prevent the massive damage to property and homes. Storage of appropriate equipment and supplies and training of human resources for such emergencies is part of prudent public administration, and is important for public health as well.

Wars, genocide, and terrorism are part of past and recent experience in many parts of the world. These too are public health disasters requiring interventions to provide security, basic needs of victims and refugees, as well as limiting the spread or continuation of the human tragedies. Incitement to violence and terror are early warning signs of potential ethnic conflict or genocide and by themselves constitute a public health call to duty.

Documentation of disaster experience is essential to improve efforts in future situations, and although lessons may be absorbed slowly, they are part of the development of public health. The needs of addressing disasters include:

1. Treat and evacuate the injured;
2. Limit further death, injury, and disease;
3. Assure safety/protection/security/public order to coordinate interventions of police, army, official health agencies, local and international NGOs;
4. Provide shelter, safe water, food, and warmth;
5. Provide sanitary facilities and prevent environmental hazards;
6. Promote epidemic control/prevention and treatment of communicable diseases (e.g., diarrheal disease, acute respiratory infections, measles, hepatitis, malaria);
7. Provide ongoing medical care for the injured and sick;
8. Mobilize and coordinate all official and voluntary local, state, national, and international aid;
9. Prevent malnutrition, provide micronutrient supplements;
10. Provide maternal and child care for pregnancy, delivery, infancy, and childhood;
11. Monitor disease and epidemiologic surveys;
12. Mobilize health aides among the affected population;
13. Promote early restoration of normal functions (e.g., family, health care, work);
14. Prevent post-traumatic stress disorder;
15. Assess, evaluate, monitor, and report the process and lessons learned;
16. Survey, document, publish, and follow up;
17. Educate (e.g., provide temporary preschool and school activities);
18. Employ able-bodied persons (e.g., promote participation in refugee care activities);
19. Promote rapid return to homes and rehabilitation;
20. Review experience and plan for potential future disasters.

Disaster planning is an essential component of public health agencies at the national state/provincial/regional levels and at local health authority levels. Coordination with police, army, civil defense, fire service, local and state disaster planning, hospitals, and many other local agencies as well as international relief agencies is fundamental to coping with disasters. Preparation of protocols for operating procedures with drills may make a large difference in outcomes. Although disaster situations are unique, they have common features relevant to planning, adapted from documentation of trial, error, and experience.

SUMMARY

A society is often judged on how it treats its minorities, its poor, its prisoners, and its refugees as much as how it cares for the main population groups. All such groups need special attention because they are people in need, but also because they can affect the health of others, as in the case of tuberculosis spread to prison guards in the United States, and the spread of tuberculosis in the general population by ex-prisoners in the former Soviet Union.

Public health agencies are often the advocates and pioneers in implementing programs for such groups. These may be special needs of the whole population or needs of special groups in the population. The public health approach emphasizes prevention in all its phases, and preparation for foreseen and unexpected emergency health situations in coordination not only with health service systems but also many other agencies in society.

Preparation for handling emergencies is an important challenge to all elements of a health system, from sanitation and pest control to tertiary care neurosurgery. In such a situation, there will be little question of the need for all elements to work together to treat the injured and prevent further damage. This requires preparation as well as improvisation at the time of the event. Mental and dental health are two areas where community care and prevention are established as effective measures but generally not well linked to the mainstream of organization and funding of health activities.

At-risk groups such as refugees, prisoners, minorities, migrants, or those suffering from natural and man-made disasters provide living proof of John Donne's famous phrase, "No man is an island unto himself alone." Today's prisoner or refugee can become tomorrow's free citizen infected with HIV or tuberculosis. The migrant worker may bring a disease from the home country or acquire it in the host nation and bring it back to the original place of residence. Disasters or terrorism can affect anyone. The role of health care providers and public health agencies is to prepare for and respond to such challenges.

Special community health needs are sometimes perceived with a stigma or social or political neglect. Failure to address such needs can endanger the public health and only an adequate response to a special community need can be a safeguard to population health. The New Public Health seeks to use all potential interventions and resources of prevention and care effectively for what have been treated as marginal populations or issues.

ELECTRONIC RESOURCES

American College of Mental Health Administration (ACMHA). http://www.acmha.org/ [accessed January 24, 2008]

American Dental Association. http://www.ada.org/

American Psychiatric Association. *Diagnostic and Statistical Manual of Mental Disorders*, http://www.psych.org [accessed May 7, 2008]

http://allpsych.com/disorders/dsm.html [accessed May 10, 2008]

Centers for Disease Control. 2007. Community Water Fluoridation Overview, http://www.cdc.gov/fluoridation/ [accessed May 9, 2008]

Centers for Disease Control. Community Mental Health Services, http://www.thecommunityguide.org/mental/ [accessed June 12, 2008]

Centers for Disease Control. 2008. Immigrant, Refugee and Migrant Health, http://www.cdc.gov/ncidod/dq/refugee/ [accessed May 8, 2008]

Centers for Disease Control. 2008. Refugee Health Guidelines Overseas and Domestic Malaria Guidelines for Refugees from sub-Saharan Africa, http://www.cdc.gov/ncidod/dq/refugee/rh_guide/index.htm [accessed June 12, 2008]

Centers for Disease Control. Community water Fluoridation: Guidelines and Recommendations, http://www.cdc.gov/fluoridation/guidelines/index.htm [accessed June 12, 2008]

Indian Health Service. U.S. Department of Health and Human Services, http://www.ihs.gov/ [accessed May 8, 2008]

Lancet Series on Global Mental Health. http://www.thelancet.com/online/focus/mental_health/collection [accessed May 8, 2008]

Migrant Health. http://aappolicy.aappublications.org/ and http://www.bphc.hrsa.gov [accessed May 8, 2008]

National Hospital Ambulatory Medical Care Survey. 2002, http://www/cdc/gov/nchs/about/major/ahcd/adata.htm [accessed June 12, 2008]

National Institute of Dental and Craniofacial Research, http://www.nidcr.nih.gov/ [accessed May 7, 2008]

U.S. National Institute of Health. 2006. NIH Almanac 2006–2007. NIH Publication No. 06-5. http://www.nih.gov/about/almanac/ [accessed January 5, 2008]

U.S. National Institute of Mental Health. http://www.nimh.nih.gov/ [accessed May 8, 2008]

U.S. National Institute of Mental Health. The numbers count — mental disorders in America, http://www.nimh.gov/health/ publications [accessed May 7, 2008]

National Institute of Neurological Disorders and Stroke. http://www.ninds.nih.gov/ and http://www.ninds.nih.gov/disorders/learningdisabilities/learningdisabilities.htm [accessed May 8, 2008]

Native Peoples' Health. http://www.cdc.gov/ and http://www.aaip.com/ [accessed May 8, 2008]

Prisoner's Health (U.S.). http://www.ojp.usdoj.gov:80/bjs/pub/pdf/cpius95.pdf [accessed May 8, 2008]

Tsunami, www.searo.who.int/en/Section23/Section1108/Section1835.htm [accessed May 8, 2008]

United Kingdom Mental Health Services. http://www.dh.gov.uk/en/Publicationsandstatistics/Publications/PublicationsPolicyAndGuidance/DH_072730 [accessed May 8, 2008]

United Nations High Commissioner for Refugees. http://www.unhcr.ch [accessed May 8, 2008]

United Nations Environment Programme (UNEP), 2008, http://www.unep.org/ [accessed June 12, 2008]

United Nations Refugee Agency, 2008, http://www.unhcr.org/protect.html [accessed June 12, 2008]

United States Department of Veteran's Affairs. http://www.va.gov/ [accessed May 8, 2008]

United States National Center for Health Statistics. http://www.cdc.gov/nchs/fastats/indfacts.htm [accessed May 8, 2008]

U.S. Department of Health and Human Services. *Healthy People 2010*, http://www.healthypeople.gov/data/midcourse/html/focusareas/FA18ProgressHP.htm [accessed May 8, 2008]

U.S. Census Bureau. 2008. The American Indian, Eskimo, and Aleut Population, http://www.census.gov/population/www/pop-profile/amerind.html [accessed May 9, 2008]

U.S. Department of Justice. http://www.ojp.usdoj.gov/bjs/prisons.htm [accessed January 23, 2008]

U.S. Federal Emergency Management Agency. http://www.fema.gov/ [accessed January 23, 2008]

U.S. Department of Health and Human Services. Agency for Healthcare Research and Quality. http://www.ahrq.gov/ [accessed May 9, 2008]

U.S. National Institute of Mental Health. 2008-05-10, http://www.surgeongeneral.gov/library/mentalhealth/home.html [accessed May 6, 2008]

WHO. 1992. *ICD-10 Classification of Mental and Behavioural Disorders — (The) Clinical Descriptions and Diagnostic Guidelines.* Geneva: World Health Organization, (easier to find) http://www.who.int/classifications/icd/en/bluebook.pdf [accessed May 8, 2008]

WHO Mental Health. http://www.who.int/mental_health/en/ [accessed May 8, 2008]

WHO Migrant Health. http://www.who.int/dg/speeches/2007/20070928_lisbon/en/ [accessed January 23, 2008]

WHO. *World Health Report. Mental Health: New Understanding, New Hope, 2001*, http://www.who.int/whr/2001/en [accessed January 20, 2008]

WHO European Region. 2007. The challenges of mental health in Europe, http://www.euro.who.int/mentalhealth [accessed May 8, 2008]

WHO. *World Health Report 2007 — A safer future: Global public health security in the 21st century*, http://www.who.int/whr/2007/en/index.html [accessed January 24, 2008]

WHO. 2008. Mental health program, http://www.who.int/mental_health/en/ [accessed May 7, 2008]

WHO. Oral health, http://www.who.int/topics/oral_health/en/ [accessed May 9, 2008]

Women's Health, http://www.ahcpr.gov/research/womenres.htm [accessed May 8, 2008]

RECOMMENDED READINGS

Breakey, W. R. (editorial). 1997. It's time for the public health community to declare war on homelessness. *American Journal of Public Health*, 87:153–154

Butcher, J., Samarasekera, U., Wilkinson, E., Shetty, P. Special report. 2007. *Lancet*, 370:117–124.

Centers for disease Control. 1999. Achievements in public health, 1900–1999: Fluoridation of drinking water to prevent dental caries. *Morbidity and Mortality Weekly Report*, 48:933–940.

Centers for Disease Control. 1999. Achievements in public health, 1900–1999: Improvements in workplace safety — United States, 1900–1999. *Morbidity and Mortality Weekly Report*, 48:461–469.

Centers for Disease Control. 1999. Achievements in public health, 1900–1999. Motor-vehicle safety: A 20th century public health achievement. *Morbidity and Mortality Weekly Report*, 42:905–909.

Centers for Disease Control. 2005. World Mental Health Day — October 10, Notice to readers: 2005. *Morbidity and Mortality Weekly Report*, 54:1000.

Centers for Disease Control. 2006. Alcohol and other drug use among victims of motor-vehicle crashes — West Virginia, 2004–2005. *Morbidity and Mortality Weekly Report*, 55:1293–1296.

Centers for Disease Control. 2006. Assessment of health-related needs after Hurricanes Katrina and Rita — Orleans and Jefferson Parishes, New Orleans Area, Louisiana, October 17–22, 2005. *Morbidity and Mortality Weekly Report*, 55:38–41.

Center for Mental Health Services. *Mental Health, United States, 2004*. Manderscheid, R. W., Berry, J. T. (eds.). DHHS Pub. No. (SMA) 06-4195. Rockville, MD: Substance Abuse and Mental Health Services Administration, 2006.

Chilvers, R., Macdonald, G. M., Hayes, A. A. 2006. Supported housing for people with severe mental disorders. *Cochrane Database Systematic Review*; 18(4):CD000453.

Cookson, S., Waldman, R., Gushulak, B., MacPherson, D., Burkle, F., Paquet, C., Kliewer, E., Walker, P. 1998. Immigrant and refugee health. *Emerging Infectious Diseases*, 4:Special Edition; July–September 1998.

Durkin, P. W. (ed.). 1996. Beyond mortality — Residential placement and quality of life among children with mental retardation. *American Journal of Public Health*, 86:1359–1360.

Editorial. 2007. Mental health: Neglected in the UK. *Lancet*, 370:104.

Freudenberg, N., Daniels, J., Crum, M., Perkins, T., Richie, B. E. 2005. Coming home from jail: The social and health consequences of community reentry for women, male adolescents, and their families and communities. *American Journal of Public Health*, 95:1725–1736.

Gatherer, A., Moller, L., Hayton, P. 2005. The World Health Organization European Health in Prisons project after 10 years: Persistent barriers and achievements. *American Journal of Public Health*, 95:1696–1700.

Horton, R. 2007. Launching a new movement for mental health. *Lancet*, 370:806, http://www.thelancet.com/online/focus/mental_health/collection [accessed May 7, 2008].

Institute of Medicine. 2002. *Crossing the Quality Chasm: Report for Behavioral Health*. Washington, DC: U.S. Academies of Science, National Academies Press.

Jorgensen, J. G. 1996. Recent twists and turns in American Indian health care. *American Journal of Public Health*, 86:1362–1364.

Kedia, S., Sell, M. A., Relyea, G. 2007. Mono- versus polydrug abuse patterns among publicly funded clients. *Substance Abuse Treatment, Prevention and Policy*, 2:33.

Kessler, R. C., McGonagle, A., Zhao, S., Nelson, B., Hughes, M., Eshelman, S., Wittche, H. U., Kemder, K. S. 1994. Lifetime and 12 month prevalence of DSM-III-R psychiatric disorders in the United States: Results from the National Comorbidity Study. *Archives of General Psychiatry*, 51:8–19.

Kunitz, S. J. 1996. The history and politics of U.S. health care policy for American Indians and Alaska natives. *American Journal of Public Health*, 86:1464–1473.

Lenzenweger, M. F., Lane, M. C., Loranger, A. W., Kessler, R. C. 2007. DSM-IV personality disorders in the National Comorbidity Survey Replication. *Biological Psychiatry*, 62:553–564.

Neugebauer, R. 1999. Mind matters: The importance of mental disorders in public health's 21st century mission. *American Journal of Public Health*, 89:1309–1311.

Noji, E. K. 2005. Public health issues in disasters. *Critical Care Medicine*; 33(1 Suppl):S29–33.

Phelan, J. C., Link, B. G. 1999. Who are the "homeless": Reconsidering stability and composition of the homeless population. *American Journal of Public Health*, 89:1334–1338.

Pols, H., Oak, S. 2007. War and military mental health: The US psychiatric response in the 20th century. *American Journal of Public Health*, 97:2132–2142.

Restum, Z. G. 2005. Public health implications of substandard correctional health care. *American Journal of Public Health*, 95:1689–1691.

Spiegel, P., Sheik, M., Gotway-Crawford, C., Salama, P. 2002. Health programmes and policies associated with decreased mortality in displaced people in postemergency phase camps: A retrospective study. *Lancet*, 360:1927–1934.

Ustün, T. B. 1999. The global burden of mental disorders. *American Journal of Public Health*, 89:1315–1318.

VanRooyen, M., Leaning, J. 2005. After the tsunami — Facing the public health challenges. *New England Journal of Medicine*, 352:435–438.

Waring, S. C., Brown, B. J. 2005. The threat of communicable diseases following natural disasters: A public health response. *Disaster Management Response*, April–June;3(2):41–47.

Yano, E. M., Simon, B. F., Lanto, A. B., Rubenstein, L. V. 2007. The evolution of changes in primary care delivery underlying the Veterans' Health Administration's quality transformation. *American Journal of Public Health*, 97:2151–2159.

BIBLIOGRAPHY

Auerbach, J. 2008. Lesbian, gay, bisexual, and transgender public health: progress and challenges. *American Journal of Public Health*, 98:970.

American Academy of Pediatrics. 1993. Homosexuality and adolescence. *Pediatrics*, 92:631–634.

American Medical Association Council on Scientific Affairs. 1996. Health care needs of gay men and lesbians. *Journal of the American Medical Association*, 275:1354–1359.

American Psychiatric Association. 1994. *DSM-IV: Diagnostic and Statistical Manual of Mental Disorders*, Fourth Edition. Washington, DC: The American Psychiatric Association.

Biermann, F., Boas, I. 2007. Preparing for a warmer world. Towards a global governance system to protect climate refugees. Global Governance Working Paper No 33 – November 2007, http://www. glogov.org/images/doc/WP33.pdf [accessed May 7, 2008]

Braddock, D. 1992. Community mental health and mental retardation services in the United States: A comparative study of resource allocation. *American Journal of Psychiatry*, 149:175–183.

Buckner, J. C., Trickett, E. J., Corse, S. J. 1985. *Primary Prevention in Mental Health: An Annotated Bibliography*. Bethesda, MD: National Institutes of Mental Health.

Eaton, W. W. 1985. Epidemiology of schizophrenia. *Epidemiologic Reviews*, 7:105–126.

Fisher, W. H., Geller, J. L., Altaffer, F., Bennett, M. B. 1992. The relationship between community resources and state hospital recidivism. *American Journal of Psychiatry*, 149:385–390.

Fitzgerald, G. J. 2008. Chemical warfare and medical response during World War I. *American Journal of Public Health*, 98:611–625.

Garrison, C. Z., McKeown, R. E., Valois, R. F., Vincent, M. L. 1993. Aggression, substance use and suicidal behaviors in high school students. *American Journal of Public Health*, 83:179–184.

Hveman, M. J. 1996. Epidemiologic issues in mental retardation. *Current Opinion in Psychiatry*, 9:305–311.

International Helsinki Federation for Human Rights. 1999. *Annual Report on United States, 1997*. Vienna, Austria: IHF.

Kemp, D. A. (ed.). 1993. *International Handbook on Mental Health Policy*. Westport, CT: Greenwood Press.

Leaning, J., Briggs, S. M., Chen, L. C. 1999. *Humanitarian Crises: The Medical and Public Health Responses*. Cambridge, MA: Harvard University Press.

Leaning, J. 2004. Diagnosing genocide — the case of Darfur. *New England Journal of Medicine*, 19;351:735–738.

Leff, H. S., Lieberman, M., Mulkern, V., Raab, B. 1996. Outcome trends for severely mentally ill persons in capitated and case managed mental health programs. *Administration and Policy in Mental Health*, 24:3–23.

Legters, L. J., Llewellyn, C. H. 1992. Military medicine (Chapter 71). *In* Last, J. M., Wallace, R. B. *Macxy-Rosenau-Last, Public Health and Preventive Medicine*, Thirteenth Edition. Norwalk, CT: Appleton & Lange.

Logue, J. N. 1996. Disasters, the environment, and public health: Improving our response. *American Journal of Public Health*, 86:1207–1210.

Mathers, C. D., Loncar, D. 2006. Projections of Global mortality and burden of disease from 2002 to 2030. Public Library of Science Medicine. November;3(11):e442.

Marcus, S. C., Olfson, M. 2008. Outpatient antipsychotic treatment and inpatient costs of schizophrenia. *Schizophrenia Bulletin*, January; 34(1):173–180 [epub June 19, 2007].

Mayer, K. H., Bradford, J. B., Makadon, H. J., Stall, R., Goldhammer, H., Landers, S. 2008. Sexual and gender minority health: what we know and what needs to be done. *American Journal of Public Health*, 98:989–995.

Mechanic D. 2007. Mental health services then and now. *Health Affairs*, 26:1548–1550.

Menu, J-P. (ed.). 1996. Emergency and humanitarian action. *World Health Statistics Quarterly*, 49:165–242

National Center for Health Statistics. *Health, United States, 2006, with Chartbook on Trends in the Health of Americans*. Hyattsville, MD, http://www.cdc.gov/nchs/data/hus/hus06.pdf

National Center for Health Statistic. *Health, United States, 2007, with Chartbook on Trends in the Health of Americans*. Hyattsville, MD, http://www.cdc.gov/nchs/data/hus/hus07.pdf

Nease, D. E., Volk, R. J., Cass, A. R. 1999. Investigation of the severity-based classification of mood and anxiety symptoms in primary care patients. *Journal of the American Board of Family Practice*, 12:21–31.

O'Neill, D. M. 1996. Measuring changes in resource use in state hospitals, 1969–1990: The effect of alternative deflators. *Administration and Policy in Mental Health*, 24:3–23.

Post, B. 1997. The underserved and desperate: Native health, its time for action. *Canadian Medical Association Journal*, 157:1655–1656.

Ramstedt, M. 2001. Alcohol and suicide in 14 European countries. *Addiction*, 96 (Suppl 1):S59–S75.

Sandhaus, S. 1998. Migrant health: A harvest of poverty. *American Journal of Nursing*, 98:52–54.

Stastny, P., Penney, D. 2008. Lost luggage, recovered lives. *American Journal of Public Health*, 98:986–998.

Steele, C. B., Meléndez-Morales, L., Campoluci, L. R., DeLuca, N., Dean, H. D. *Health Disparities in HIV/AIDS, Viral Hepatitis, Sexually Transmitted Diseases, and Tuberculosis: Issues, Burden, and Response, A Retrospective Review, 2000–2004*. Atlanta, GA: Department of Health and Human Services, Centers for Disease Control and Prevention, November 2007. http://www.cdc.gov/nchhstp/healthdisparities/ [accessed May 8, 2008]

U.S. Census Bureau. 2006. Statistical Abstract of the United States, http://www.census.gov/compendia/statab/ [accessed May 9, 2008]

U.S. Department of Justice. 2008. Bureau of Justice Statistics. Corrections statistics, http://www.ojp.usdoj.gov/bjs/correct.htm# [accessed May 10, 2008]

U.S. Department of Health and Human Services. 1999. *Mental Health: A Report of the Surgeon General*. Rockville, MD: U.S. Department of Health and Human Services, Substance Abuse and Mental Health Services Administration, Center for Mental Health Services, National Institutes of Health, National Institute of Mental Health, http://www.surgeongeneral.gov/library/mentalhealth/home.html [accessed May 10, 2008]

Wasserman, D., Cheng, Q., Jiang, G. X. 2005. Global suicide rates among young people age 15–19. *World Psychiatry* 4:114–120.

World Health Organization. 1985. *Targets for Health for All: Targets in Support of the European Regional Strategy for Health for All*. Copenhagen: WHO Regional Office for Europe.

World Health Organization. 1991. *Health for All Targets: The Health Policy for Europe. European Health for All Series No. 4*. Copenhagen: World Health Organization Regional Office for Europe.

World Health Organization. 1997. *The World Health Report 1997. Conquering Suffering, Enriching Humanity*. Geneva: WHO, http://www.who.int/whr/1997/en/ [accessed May 10, 2008]

World Health Organization. 1998. *The World Health Report, 1998: Life in the 21st Century: A Vision for All*. Geneva: WHO.

World Health Organization. 2007. There is no health without mental health. Fact sheet no. 220, September, Geneva, www.who.int/entity/mediacentre/factsheets/fs220/en/ [accessed May 7, 2008]

Zaza, S., Briss, P. A., Harris, K. W. 2005. *The Guide to Community Preventive Services: What Works to Promote Health*. Oxford University Press, New York.

Chapter 8

Nutrition and Food Safety

Introduction
Development of Nutrition in Public Health
Nutrition in a Global Context
Nutrition and Infection
Functions of Food
    Composition of the Human Body
Human Nutritional Requirements
    Carbohydrates
    Proteins
    Fats and Oils
    Vitamins
    Minerals and Trace Elements
Growth
Measuring Body Mass
Recommended Dietary Intakes
Disorders of Undernutrition
    Underweight: Protein-Energy Malnutrition (PEM)
    Failure to Thrive
    Marasmus
    Kwashiorkor
    Vitamin A Deficiency
    Vitamin D Deficiency (Rickets and Osteomalacia)
    Vitamin C Deficiency
    Vitamin K Deficiency (Hemorrhagic Disease of the
        Newborn)
    Vitamin B Deficiencies
    Iron-Deficiency Anemia
    Iodine-Deficiency Diseases
    Osteoporosis
    Eating Disorders
Diseases of Overnutrition
    Overweight/Obesity
    Cardiovascular Diseases
    Cancer
Nutrition in Pregnancy and Lactation
Promoting Healthy Diets and Lifestyles
Dietary Guidelines
Vitamin and Mineral Enrichment of Basic Foods
    Controversy in Food Enrichment
Food and Nutrition Policy
    The Evolution of a Federal Role
    Nutrition Issues in Development Policies

The Role of the Private Sector and NGOs
The Role of Health Providers
Nutrition Monitoring and Evaluation
    Standard Reference Populations
    Measuring Deviation from the Reference Population
Food Quality and Safety
Nutrition and the New Public Health
Summary
Electronic Resources
Recommended Readings
Bibliography
Nutrition and Food Technology Journals

INTRODUCTION

Nutrition has a direct effect on growth, development, reproduction, and both physical and mental well-being. It is one of the most important factors for the health of an individual or a community, and is, consequently, a fundamental issue in modern public health. People require nutrients such as carbohydrates, fats, and protein to provide heat and energy to regulate body processes (water, minerals, fiber, vitamins, proteins, and essential amino acids), and build and renew body tissue (water, proteins, and mineral salts). The nutritional status of an individual and society is influenced by the supply, quality, distribution, access to, and cost of foods. It is also affected by knowledge, attitudes, beliefs, and practices regarding essential and balanced nutrition. Food policy at the individual and governmental levels, eating habits, as well as economic and technical factors all contribute to the nutritional state of the public and the individual person.

Improved nutrition has made a major contribution to better health in recent ages. In the twentieth century, nutrition emerged as a basic and applied science. Knowledge of the elements of proper nutrition and its role in prevention of deficiency diseases or chronic diseases has played a vital part in the development of modern public health. And, despite rapid population growth, food production and average food consumption have improved steadily worldwide.

Nevertheless, malnutrition is widely prevalent throughout the world. Developed countries struggle with problems of inappropriate and excessive nutrition which can lead to chronic diseases along with diseases of nutritional deficiencies. Rising standards of living in many developing countries have brought diseases of modern living to prominence as communicable diseases are increasingly brought under control. Subpopulations within rich and poor nations alike suffer from a broad spectrum of nutritional diseases.

Public health attempts to ensure all groups in the population have adequate, but not excessive, intake of the basic food groups, essential vitamins and minerals for growth, health maintenance, and physical activity. This includes recommendations of daily human needs for nutrition and energy, which vary according to age, sex, body size, level of activity, individual health status, and environmental conditions. It also requires monitoring of the nutritional status of the population and its subgroups.

DEVELOPMENT OF NUTRITION IN PUBLIC HEALTH

The steady improvement in life expectancy seen in Europe and North America in the seventeenth to nineteenth centuries probably had as much to do with improved nutrition as with improved sanitation. Pioneering epidemiologic studies of James Lind in the mid-eighteenth century, the first recorded clinical epidemiologic experiment, and that of Joseph Goldberger in the early twentieth century, opened up the field of nutritional epidemiology. They each established proof of deficiency conditions that met the Koch–Henle criteria of causation in epidemiology.

Just as Snow's work on cholera preceded Koch's discovery of the *Vibrio cholerae* organism by 30 years, so Lind's work on scurvy preceded the isolation of ascorbic acid by more than a century. Antoine Lavoisier (1743–1794) in Paris developed basic concepts of metabolism, measuring oxygen consumption and carbon dioxide production, and is called the "father of the science of nutrition." Justus von Liebig (1803–1873) demonstrated that fat, protein, and carbohydrates are burned in the body and developed methods of analysis of composition of foods, body tissues, urine, and feces. In the mid-nineteenth century, scientists in France, Germany, and later in Britain and the United States made rapid advances in the chemistry of oxygen, carbon dioxide, calcium, iodine, and iron. In 1897, Christian Eijkman demonstrated the origin of beriberi by showing the disease in populations eating polished rice and its absence when rice was eaten with its husk (Nobel Prize, 1929).

In the early years of the twentieth century, animal studies showed that diets consisting of only pure protein, carbohydrates, and fats led to failure to thrive, sickness,

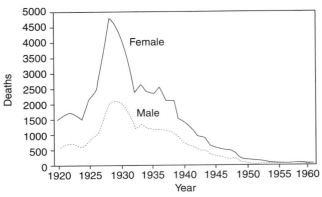

FIGURE 8.1 Reported pellagra deaths, by gender and years, United States, 1920–1960. Source: Centers for Disease Control. 1999. Achievements in Public Health, 1900–1999: Safer and healthier foods. *Morbidity and Mortality Weekly Report*, 48:905–913.

and death. This led Kasimir Funk to the idea of "vital amines" missing in the diet, later termed *vitamins*, and the isolation of chemical substances with antineuritic and antiberiberi properties. The delineation of the vitamins necessary to prevent disease and promote health produced immeasurable improvement in human health where the knowledge was applied.

The development of applied nutritional public health interventions in the United States was led by the federal Departments of Agriculture and Education; the Department of Health and Human Services also contributed greatly to this evolution (Figure 8.1). The investigation of pellagra, the establishment of national recommended dietary allowances, national food supplementation, and school lunch programs established nutrition as a central issue in public health in the United States. In many parts of the world, application of the knowledge of the fundamental importance of essential vitamins and trace minerals has still to be implemented as applied nutritional public health measures.

NUTRITION IN A GLOBAL CONTEXT

Worldwide malnutrition affects one in three people. It affects all age groups, but is most devastating for children. It is associated with poverty, inadequate access to health education, clean water, and good sanitation. Malnutrition contributes to half of all childhood deaths. This is due to poor feeding practices, contaminated water, diarrheal and respiratory illnesses, as well as food deficiencies. Developing countries also suffer from a growing burden of chronic diseases associated with changing dietary patterns, increased consumption of fats, particularly saturated fat, and refined carbohydrates. These patterns are combined with an increasingly sedentary lifestyle (WHO and FAO, 2005).

The people most affected by poverty are women, children, and the elderly, and poor nutrition is one of the key mediators of poverty and poor health. The effects of maternal malnutrition on newborns can be lifelong. Micronutrient deficiencies, despite their preventability by low-cost public health measures, are still widespread. Yet, while millions are starving or in a chronic state of undernutrition, there is enough food in the world to feed everyone, so that national strategies for combating the effects of poverty are needed based on good governance, suitable legislation, participation by all in society, and provision of effective basic services. Poverty, unequal distribution of income and food, ignorance, low education levels, large family size, acute and chronic illness, and government inaction are major contributors to world hunger (Table 8.1).

The cumulative effects of poverty, population, environmental degradation, overfarming of land, erosion of topsoil, harmful agricultural practices, and inadequate storage and transportation are vital in global health.

Malnourished people, especially children, are susceptible to diseases including diarrhea and acute respiratory infections, many of which are vaccine-preventable.

NUTRITION AND INFECTION

The infectious disease–malnutrition cycle causes millions of deaths of children yearly from preventable causes. Control of measles and wide use of oral rehydration therapy alone would save hundreds of thousands of lives and improve the nutritional status of millions of children. Malnutrition and infection interact, each exacerbating the other. Infections such as measles may have a case fatality rate of 2 per 1000 in an industrialized country but 50 or more per 1000 in a developing country where there are widespread deficiencies of essential nutrient elements

TABLE 8.1 Nutritional Conditions in Developing and Developed Countries

Issues	Factors in developing countries	Factors in developed countries
Poverty	Lack of resources or access to healthy foods; traditional crops and foods inadequate for healthy diet	General prosperity but poverty in significant minorities; supplementation programs; e.g., WIC, food stamps, school lunches offset deprivation
Educational deprivation	Lack of knowledge of good nutrition	High awareness of good nutrition, but obesity common, especially among the poor
Micronutrient deficiencies	Widespread deficiencies of iron, iodine, and vitamins A, B, C, and D; inadequate food processing technology	Food fortification with iodine, iron, vitamins A, B, C, D
Malnutrition-infection cycle	Poor sanitation; lack of control of vaccine-preventable diseases, parasitic, diarrheal and respiratory diseases; HIV, tuberculosis, malaria all harm nutrition status	High levels of sanitation, vaccination coverage, and hygiene with benefit to nutritional state; high-risk subgroups — HIV, tuberculosis, drug abusers, the homeless
Food security	Harmful agriculture practices; overgrazing and overfarming; soil erosion, small plots; inadequate and wasteful storage and distribution facilities; food costly relative to incomes	Science-based technological agriculture; high productivity, abundant produce, variety, food cheap relative to family income
Unhealthy diets and noncommunicable diseases	Undernutrition with poverty, inadequate calories per person; middle-class overnutrition with high rates of cardiovascular diseases, diabetes	Overnutrition with excess animal fat intake; high rates of cardiovascular disease; pockets of poverty and nutritional inadequacy; cardiovascular disease mortality falling
Breastfeeding (capacity to care)	Common and prolonged but with poor supplementation practices	Increasing and well supplemented
Food quantity, quality, variety, and cost	Contamination, waste and destruction common; lack of supply and variety of vegetables, fruit, protein; costly, wasteful	Good supply and quality; good distribution, and marketing systems; labeling, regulation, and supervision; comparatively inexpensive
Monitoring	Monitoring supply, distribution, intake of food; growth, anemia, and intake studies needed	Nutrition surveys
Need for national policies and objectives	Prevent deforestation and land loss; rural poverty, lack of education; lack of credit and agricultural support; poor and harmful farming practices; pesticides used dangerously	Government promotes agriculture, research, support systems, transport, and marketing; limit use of pesticides; industrialized farming

Source: Adapted from *International Conference on Nutrition: Nutrition the Global Challenge*. World Health Organization and Food and Agricultural Organization, 1992.

such as vitamin A. Conversely, even a relatively common infection can adversely affect the nutritional state, growth pattern, and resistance to further infection of a child. The time required to make up nutrient losses from infection following recovery may be two to three times as long as the duration of the infection itself.

Normal growth in weight and height indicate that a child is more likely to resist infection or prolongation of infections that occur. Child health survival strategies rest on the twin pillars of infectious disease control and nutritional adequacy. Nutritional deficiency may also be a factor in reduced resistance to infection in the elderly and the immunocompromised, as both social groups may be socially and economically marginalized in industrialized societies.

FUNCTIONS OF FOOD

Consumed food substances provide varying levels of energy and essential requirements for growth and maintenance of body functions. Exercise and moderate eating habits maintain body weight and reduce the risk for chronic diseases associated with excess body fat, such as diabetes, hypertension, cardiovascular disease, and some cancers.

Nutrients have specific roles within the body, but their functions are interdependent. The diet of the individual determines their availability. Consequently, it is important to identify the sources of proteins, carbohydrates, fats, vitamins, and minerals available in common foods in a target community. The six important nutrient groups are macronutrients (carbohydrates, fats, and proteins), the micronutrients (trace minerals and vitamins), and water. The macronutrients provide energy, essential and other amino acids, and essential fatty acids. Micronutrients are required for utilization of that energy.

The body processes foods into simpler forms in order to absorb them by digestion through a continuous mechanical and chemical process in the digestive tract. Foods are first ground up by chewing of food, requiring good dentition. Mixing of the food with saliva and swallowing brings the partially digested food to the stomach and small intestine, where it is acted on by enzymes. These enzymes break the food down into smaller and smaller fragments which can then be absorbed through the walls of the small intestine to enter the bloodstream. Disease of the gastrointestinal tract can interfere with this process.

Pancreatic enzymes are released into the small intestine as proteases (which split proteins), amylases (which split polysaccharides), and lipases (which break down fats). The pancreas also produces insulin, vital for control of blood sugars. Carbohydrates are absorbed as sugar and stored to provide energy in the liver and muscles as glycogen, which is released into the bloodstream to sustain sugar levels. The liver also stores fat-soluble vitamins and manufactures enzymes, cholesterol, proteins, vitamin A, blood coagulation factors, and bile salts which are released into the intestine to help in absorption.

Composition of the Human Body

The human body is composed by weight of approximately 62 percent water, 17 percent protein, 13 percent fat, 6 percent minerals, and 2 percent carbohydrates. Body composition can vary with stress and nutritional status. Food deprivation uses body stores. During starvation, depletion of carbohydrates is made up by synthesis from reserves of fat and protein. Depletion of up to 10 percent of total body water can occur without serious risk, but in small children the margin of safety is smaller.

HUMAN NUTRITIONAL REQUIREMENTS

Calculation of the appropriate amounts of nutrients for age and sex groups is a complex task. Energy from food is converted into mechanical work (up to 25 percent) and growth, dissipated as heat, and used in maintaining the basic functions and temperature of the body. The international unit of energy is a joule, but the more commonly used measure in nutrition is the kilocalorie, or calorie (1 kcal = 1 cal = 4.1868 kilojoules, kJ). Technically, 1 calorie is the amount of energy required to raise the temperature of 1 gram of water 1° centigrade. The calorie measure used commonly to discuss the energy content of food is actually a kilocalorie or 1000 real calories. This is the amount of energy required to raise 1 kilogram of water (about 2.2 pounds) 1° centigrade.

Need for energy intake varies with body size and is increased by activities of work and recreation. Persons with a sedentary lifestyle and average body size will need less food intake than those with moderate or high levels of activity or greater body size to maintain a status quo.

Both deficiency and excessive dietary intake of any nutrient can cause disease or death. The range of intake needed for optimal physiological function depends on age, body size, gender, and activity level, as well as on pregnancy, disease, or injury. The range of intake needed for optimal health emphasizes that there are subclinical phases of basic undernutrition and overnutrition. Table 8.2 lists the essential nutrients and their functions in the human body.

For nutrients and energy sources, there is a range of intake that confers optimal physiological function. Below this range, deficiencies can cause disease or death. Excessive intake, in some cases, also can lead to toxicity. The optimal range varies for each nutrient and is affected by many individual and environmental factors.

TABLE 8.2 Essential Nutrients and Their Functions

Dietary components	Types	Functions
Carbohydrates	Energy (4 Kcal/gram) Sugar and starches	Provides efficient source of energy; water-soluble, easily transported and available in tissue fluids; should comprise 40–85% of energy intake
Proteins	Energy (4 Kcal/gram) Essential amino acids	Provides amino acids and building material for all body cells, especially muscle and bone; should comprise 10–15% of energy intake
Fats and oils	Energy (9 Kcal/gram) Essential fatty acids (linoleic and linolenic acids)	Concentrated energy source; transports fat-soluble vitamins; enhances flavor; should comprise 25% of diet, mostly non-animal sources
Minerals	Arsenic, boron, calcium, chromium, copper, fluoride, iodine, iron, magnesium, manganese, molybdenum, nickel, phosphorus, selenium, silicon, vanadium, zinc	Essential for building healthy body tissue and fluids, blood, hormones, electrolyte balance; bones and teeth
Vitamins	Fat-soluble (A, D, E, K); water-soluble	Healthy body tissues of bone, muscle, blood, central nervous, and immunologic systems
Fiber	Vegetable matter	Food bulk and prevention of cancer
Water		Fluid and body tissue balance; conveyer of food and water-soluble vitamins

Source: Adapted from Surgeon General's Report on Nutrition and Health, 1988, and Institute of Medicine, Dietary Recommended Intakes, 2004, http://www.iom.edu/Object.File/Master/21/372/0.pdf and http://www.iom.edu/Object.File/Master/7/300/Webtablemacro.pdf (accessed May 10, 2008)

Carbohydrates

Carbohydrates are a major source of energy (4 kcal per gram) used for metabolic processes and for producing cellular substances including enzymes and cell membranes. Carbohydrates are classified as monosaccharides or di-saccharides (simple carbohydrates) or as polysaccharides (complex carbohydrates).

Monosaccharides are found as glucose and fructose in fruits and honey. They are simple sugars that can be absorbed in the gut without any digestive process. Disaccharides are made up of two monosaccharides and are commonly found in fruits and vegetables, including sugar beets, sugar cane, and as lactose in dairy products.

Polysaccharides are larger molecular structures of monosaccharides linked together. Disaccharides and polysaccharides must be broken down into monosaccharides before they can be absorbed. Excess glucose is stored in the liver as glycogen and muscles (not fat tissue). Diets rich in complex carbohydrates (polysaccharides) and fiber are associated with reduced risk of cancer and cardiovascular diseases.

Proteins

Proteins are large molecules made up of chains of amino acids that are broken down by the digestive process into their component units (1 g protein yields 4 kcal). There are 20 common amino acids in biological materials needed by the body. Humans lack the ability to synthesize the nine amino acids, and it is therefore essential to obtain them in protein from animal sources or combinations of foods, such as legumes and cereals, in the diet. Young children and adolescents need protein for their growth spurts. Proteins function in the body as structural components of cells and tissues, enzymes which act as catalysts for chemical reactions, and hormones which act as chemical messengers. Lack of protein and calories in the diet is called protein energy malnutrition (PEM).

Fats and Oils

Foods of animal and plant origin include a variety of substances known as fats and oils (lipids) that are soluble in organic solvents but not in water. Dietary fats are broken down in the gut for absorption into the body to provide energy and fatty acids needed for many physiological functions. They provide a concentrated form of energy (9 kcal per gram as compared to 4 for proteins and carbohydrates, and 7 for alcohol). They also provide essential fatty acids needed for production of hormones, cell membranes, and other substances. Linoleic and linolenic acids are essential fatty acids which cannot be synthesized in the body but can be retrieved from animal fats and plant sources like walnuts and flaxseed oil.

Fats or lipids stored in body fat tissue insulate and protect vital organs, insulate the body against heat loss, and provide energy during periods of reduced consumption or greater body need of energy in periods of growth, illness, or injury. Fats provide for production of bile acids needed for absorption of fat-soluble vitamins (A, D, E, and K).

Dietary fats consist of mixtures of saturated, monounsaturated, and polyunsaturated fats, depending on their chemical structure. The degree of saturation of fatty acids is based on the number of double bonds in the side chains of molecules made up of carbon, hydrogen, and oxygen atoms. Fats from animal sources (meat, poultry, fish, and dairy products) are mostly saturated fats (i.e., contain no double bonds). Fats from plant sources such as sunflower, olive, or peanut oils are monounsaturated (i.e., contain side chains with one double bond) and are preferable to saturated or polyunsaturated fats. Whereas coconut and palm oils are high in saturated fats, fish are an excellent source of unsaturated fats, including omega-3 fatty acids, known to reduce harmful cholesterol levels.

Foods of animal origin contain high amounts of saturated fats and cholesterol. Cholesterol, which can also be synthesized in the body, is needed for synthesis of sex hormones, vitamin D, and cell membranes. Excess dietary intake of saturated fats and cholesterol increases the risk for atherosclerosis, cardiovascular diseases, diabetes, and some forms of cancer.

Vitamins

Vitamins are organic compounds that are essential in small amounts for specific functions of the body for health, growth, reproduction, and resistance to infection (Table 8.3). They differ in physical and chemical properties and in biological functions. Vitamins function in highly specialized metabolic processes. They cannot be synthesized in sufficient quantity by the body alone and must be obtained from the diet or from supplements.

Vitamins are classified based on their solubility, either in fat or water. Fat-soluble vitamins (A, D, E, and K) are found in high concentrations in the fatty portions of food. Excretion of excess intake of this type of fat-soluble vitamins is minimal. Vitamin C and those of the B-complex group are water soluble and should be supplied in adequate amounts in the daily diet, as they are easily excreted.

Storage of vitamins in the body for the water-soluble vitamins is limited, so that regular sources are even more essential than for the fat-soluble vitamins, which are stored in body fat and the liver. Deficiencies of either in the diet leads to depletion of body stores, followed by non-specific symptoms (fatigue, confusion, weakness, neuritis, and reduced resistance to infection) before classic deficiency can be recognized clinically. A deficiency condition of even one vitamin can jeopardize health.

Vitamins are present in natural foods, and an appropriate diet should supply most vitamin needs; however, since eating such a balanced diet is often problematic, food enrichment or supplementation is necessary, especially for vulnerable groups (e.g., children, elderly, adolescents, and institutionalized patients). Enrichment means replacing nutrients in foods to levels found in the natural product before processing. For example, enrichment of white flour should replace the 22 natural elements that are normally present in whole grains but are removed in processing, including B vitamins, vitamin D, calcium, and iron salts.

Minerals and Trace Elements

Minerals are distributed in a variety of foods, but they usually are present in limited amounts. Diets must contain a sufficiency and variety of foods to meet daily requirements. Eighteen known minerals are required for body maintenance and regulatory functions. Of these, dietary allowances (RDA) have been established for seven: calcium, iodine, iron, magnesium, phosphorus, selenium, and zinc. Other active minerals in the body include sodium, chloride, potassium, chromium, cobalt, copper, fluoride, manganese, molybdenum, sulfur, and vanadium. Sodium, potassium, chloride, and calcium are particularly crucial for electrolyte balance in the blood and body tissues. Essential trace minerals also include boron, silicon, nickel, and arsenic for optimal growth and membrane function. The body needs a small but continuing intake of these elements for its structure and function. If metabolic needs are not met, a deficiency ensues. Deficiency disorders vary with the mineral element involved, the duration and extent of dietary intake deficiency, and depletion of body stores (Table 8.4).

GROWTH

Growth is not a steady progression, but a process during which nutrition requirements are determined by a genetic timetable, affected by nutritional intake and health status. Optimal growth occurs only if the organs and tissues receive the nutrients needed for synthesis of proteins and other molecules. Insufficient energy and protein (protein energy malnutrition, or PEM) are common in developing countries or deprived populations as causes of failure to thrive, stunting, and wasting. Deficiencies of essential minerals and vitamins also adversely affect growth and development. Iodine deficiency slows thyroid hormone production and causes adverse developmental effects. Lack of micronutrients such as vitamins A or D or minerals such as iron, iodine, calcium, and phosphorus adversely affects growth and development of epithelial cells, bone, and red blood cells. Measurement of growth and development is one of the most important health status indicators of individual and population health.

TABLE 8.3 Essential Vitamins

Vitamins	Source and activity in body	Deficiency condition
Water-Soluble	Absorbed in intestines, excreted in urine, so very large amounts are required to produce overdosage. Body needs adequate daily intake or tissue depletion occurs within weeks or months; they are essential for enzymes to catalyze biochemical reactions in energy production, biosynthesis, and nervous system development and maintenance	
Vitamin B complex	Thiamin, riboflavin, niacin, pyridoxine, cobalamin, and folic acid; sources are whole-grain cereals, legumes, leafy vegetables, meat, and dairy foods	Loss of memory, mental confusion; occurs with chronic illness, alcoholism, dietary restriction; may lead to serious clinical conditions, such as amnesia, dementia, heart failure, neurologic disorders, death
Vitamin B_1 (thiamin)	Part of enzyme systems for release of energy from carbohydrates	Beriberi, anorexia, emotional lability, depression, fatigue, constipation; cardiomyopathy, cardiac failure; polyneuritis; Wernicke's encephalopathy; Korsakoff's psychosis, amnesia, dementia, death
Vitamin B_2 (riboflavin)	Enzymes for metabolism of protein and carbohydrates	Dry skin and mucous membranes disorders, stomatitis, photophobia, blurred vision, polyneuritis
Vitamin B_3 (niacin, nicotinamide, nicotinic acid)	Maintains normal gastrointestinal and nervous system	Pellagra; gastrointestinal, skin, and neurologic changes; depression, psychosis, neuropathy, dermatitis, diarrhea, dementia, death
Vitamin B_5 (pantothenic acid)		
Vitamin B_6 (pyridoxine)	Part of enzyme process in protein metabolism	Irritability, depression, muscle weakness, cardiomyopathy, liver damage; prevents neuropathy in isoniazid therapy for TB
Vitamin B_7 (biotin)	Skin condition	Acne
Vitamin B_9 (folic acid, folate)	Red blood cell formation; homocysteine metabolism	Megaloblastic anemia of pregnancy; neural tube defects (spina bifida and anencephaly); elevated homocysteine, possible link to coronary heart disease and mental deterioration with aging
Vitamin B_{12} (cobalamin)	Found only in foods of animal origin; essential for red blood cell formation	Macrocytic anemia; peripheral neuropathy, pernicious anemia, mental retardation
Vitamin C (ascorbic acid)	Source fruits and vegetables; needed to form, maintain intercellular substances	Scurvy, poor bone and cartilage formation, anemia, stunting, infections, bleeding
Fat-Soluble	Found mainly in fat component of food; absorbed, transported along with fat; requires bile and dietary fats; stored in body fats and takes longer to deplete than water-soluble vitamins	
Vitamin A (retinol)	Found in yellow vegetables; essential for epithelial cells of mucous membranes; regulation of vision in dim light	Night blindness, loss of color vision, dryness, corneal ulceration and scarring, blindness; poor bone and tooth formation; susceptibility to infection and poor survival rates from infectious diseases
Vitamin D (calciferol)	Produced in skin by exposure to sun; found in enriched foods (milk products); enhances calcium and phosphorus utilization, bone growth, thickness, and density	Rickets, stunting, soft bones, bowed legs, carious teeth impaired development of bone length; osteomalacia, soft bones, fractures, muscle pain in adults; contributes to osteoporosis
Vitamin E (tocopherol)	Found in leafy green vegetables, legumes, nuts; protects fat from oxidation (antioxidant) and red cell breakdown	Low birth weight; hemolysis of the newborn; degenerative disorders
Vitamin K, phylloquinone (vitamin K_1), menanquinone (vitamin K_2)	Spinach, cabbage, cauliflower; formation of prothrombin	Hemorrhagic disease of newborn, prolonged clotting time

Note: Coenzymes including folate (vitamin B_{90}), pantothenate (vitamin B_5), cobalamin (vitamin B_{12}), biotin (vitamin B_8), and molybdenum cofactor (Moco) play essential roles in energy transfer and many vital metabolic processes.

Source: Adapted from Canada's Food Guide Handbook; Passmore and Eastwood, 1986, and *The Surgeon General's Report on Nutrition and Health*, 1988.

TABLE 8.4 Essential Minerals

Minerals	Activity in body	Effect of deficiency	Food source
Calcium	Builds and maintains bone structure and teeth; muscle and cardiac function; blood coagulation; neuromuscular irritability	Poor bone and teeth formation; rickets in children, osteoporosis in elderly	Fortified milk, hard cheese, egg yolk, cabbage, clams, cauliflower, soybeans, spinach
Iron	Constituent of hemoglobin, muscle and bone; carries oxygen in red blood cells	Iron-deficiency anemia, fatigue, poor linear growth in infants, psychomotor deficiency affecting school and work performance	Liver, red meat, turkey, legumes, egg yolk, peaches, apples, raisins, prunes, molasses
Iodine	Constituent of thyroxine, normal thyroid function needed for growth and mental development	Stunting, retardation, cretinism	Iodized salt, seafood
Phosphorus	Builds and maintains bones, teeth, cells, body fluids	Poor teeth, stunting, rickets	Milk, cheese, egg yolk, meat, legumes, cereals, nuts, vegetables
Chloride	Needed to maintain acid-base and osmotic balance in body fluids	Lost in diarrhea, with dehydration; shock and death in children or elderly from diarrheal diseases	Milk, salt, fish, cheese
Copper	Needed for the central nervous system, hemoglobin formation	Anemia, liver function, metabolism of ascorbic acid	Seafood, nuts, whole-grain cereals, liver, meat, legumes, vegetables
Fluoride	Strengthens tooth enamel, bone formation	Dental caries and osteoporosis in the elderly	Fluoridated drinking water, fluoride rinses
Magnesium	Constituent of bones and teeth, enzymes; needed for cardiac and neurologic functions	Cardiac arrhythmia, nervous irritability	Same as diet for calcium and phosphorus
Sodium	Intracellular and extracellular fluid balance, muscle and nerve irritability	Fluid loss, circulatory collapse	Table salt, meat, milk
Potassium	Electrolyte balance, cardiac arrhythmia	Fluid loss, circulatory collapse, and muscle irritability	Vegetables, cereals, fruits, bananas, melons
Selenium	Antioxidant	Changes in biochemical systems, cardiomyopathy, carcinogenesis, liver damage	Meat, seafood, cereals, grains

Source: Adapted from the *Surgeon General's Report on Nutrition*, 1988.

MEASURING BODY MASS

The body mass index (BMI) is a standard method of measuring body size. It encompasses in one number height and weight, summarizing nutrition status. BMI is useful clinically for the individual patient and as a description of nutritional status of a community population based on survey data. It is calculated as follows:

$$BMI = \text{body weight in kilograms}/(\text{height in meters})^2$$

or

$$BMI = \text{weight in pounds} \left[\times 703/(\text{height in inches})^2\right]$$

Obesity is defined for males and females as a BMI above 30; overweight, BMI = 25–30; undernutrition is defined as a BMI below 18.5.

Body weight is taken as usual weight and expressed as a percentage of desirable weight. The BMI is a convenient measure. Table 8.5 shows the WHO interpretation for the BMI ranges. These provide a useful categorization which may need to be augmented by clinical and other anthropometric measures and have become a guideline for proper nutrition for the health-conscious individual. According to the WHO criteria, about two-thirds of U.S. adults (66 percent), 61.6 percent of women and 70.5 percent of men, are overweight or obese (BMI ≥25) (National Center for Health Statistics, 2006).

TABLE 8.5 WHO Interpretation for Body Mass Index (BMI)

BMI <16.00	grade 3 thinness
BMI 16.0–16.99	grade 2 thinness
BMI 17.0–18.49	grade 1 thinness
BMI 18.5–24.99	normal range for an individual
BMI 25.0–29.99	grade 1 overweight
BMI 30.0–39.99	grade 2 overweight
BMI ≥40.00	grade 3 overweight

Source: WHO. 1995. Physical Status: The use and interpretation of anthropometry. WHO *Technical Report Series No. 854.* WHO: Geneva.
BMI Categories:
Underweight = <18.5
Normal weight = 18.5–24.9
Overweight = 25–29.9
Obesity = BMI of 30 or greater

Moreover, according to the same criteria mentioned above, 31.4 percent of U.S. adults age 20 and older are obese (BMI ≥30). The prevalence has steadily increased over the years among both genders, all ages, all racial and ethnic groups, and all educational levels. From 1960 to 2004, the prevalence of overweight increased from 45 to 66 percent in U.S. adults ages 20 to 74 (National Center for Health Statistics, 2006). The prevalence of obesity during this same time period more than doubled among adults ages 20 to 74 from 13 to 32 percent, with most of this rise occurring since 1980. Figure 8.2 shows the increase in overweight and obesity rates in the United States from 1960–2004.

Many studies show measurable increases in mortality associated with obesity. Individuals who are obese have a 10–50 percent increased risk of death from all causes, compared with healthy weight individuals (BMI 18.5 to 24.9). Most of the increased risk is due to cardiovascular causes. In the United States, obesity is associated with about 112,000 excess deaths per year in the population relative to healthy weight individuals.

RECOMMENDED DIETARY INTAKES

In 1941, the Committee of Food Nutrition of the National Research Council in the United States developed recommended daily allowances (RDAs, later renamed *daily dietary allowances*) in response to a request from the U.S. Council of National Defense, which was concerned over possible food shortages and nutritional ill effects on the health of the population during World War II.

The RDAs were developed by the Food and Nutrition Board of the Institute of Medicine of the U.S. Academies of Sciences established in 1940 to study adequacy and safety of food supply, and to recommend standards of adequate nutrition. They represent the levels of essential nutrients needed to adequately meet the nutritional needs of virtually all healthy persons. They are used for planning for national emergencies and for the needs of institutionalized or socially deprived persons. The RDAs are modified in light of expanding knowledge of nutrition.

These serve to remind us of what should be included in our regular diet, and assist in preventing excessive intake of essential nutritional elements that could be potentially harmful. The U.S. National Academy of Sciences (NAS) revised the RDAs in 1989 and again in 1997. In 1997, the Food and Nutrition Board of the National Academy of Sciences created the Dietary Reference Intakes (DRIs). With the creation of DRIs, the NAS changed the way nutritionists and nutrition scientists evaluate the diets of healthy people.

From 1941 until 1989, the RDAs' primary goal was to prevent diseases caused by nutrient deficiencies. They were established and used to evaluate and plan menus that

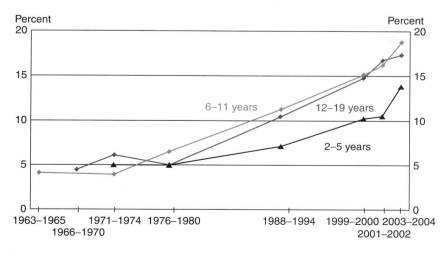

FIGURE 8.2 Trends in child and adolescent overweight, United States, 1963–2003. Source: National Center for Health Statistics. Prevalence of Overweight and Obesity Among Children and Adolescents, United States, 2003–2004. Available at http://www.cdc.gov/nchs/products/pubs/pubd/hestats/overweight/HealthEstat1206.gif [accessed May 9, 2008]

would meet the nutrient requirements of groups as well as other applications such as interpreting food consumption records of populations, establishing standards for food assistance programs, and establishing guidelines for nutrition labeling. Technically speaking, the RDAs were not intended to evaluate the diets of individuals, but they were often used this way.

In the early 1990s, the Food and Nutrition Board of the National Academy of Science undertook the task of revising the RDAs, and a new set of nutrient reference values was born—the Dietary Reference Intakes (DRIs). There are four types of DRI reference values: the estimated average requirement (EAR), the recommended dietary allowance (RDA), the adequate intake (AI), and the tolerable upper intake level (UL). (See Box 8.1 for definitions of these values.) The primary goal of having new dietary reference values was to not only prevent nutrient deficiencies, but also to reduce the risk of chronic diseases such as osteoporosis, cancer, and cardiovascular disease.

RDA values were reviewed and published as DRI values and released in stages. The first report, *Dietary Reference Intakes for Calcium, Phosphorus, Magnesium, Vitamin D and Fluoride*, was published in 1997. Since then, additional reports have been released which address folate and other B vitamins, dietary antioxidants (vitamins C, E, selenium, and the carotenoids), and the micronutrients (vitamins A, K, and trace elements such as iron, iodine, copper, selenium, zinc, and others), macronutrients such as dietary fat and fatty acids, protein and amino acids, carbohydrates, sugars, and dietary fiber, as well as energy intake and expenditure.

The DRIs serve as nutritional guidelines and are useful in standardizing food manufacturing package labeling and education practices. They have become an essential part of both clinical nutrition and public health nutrition standards internationally. The RDIs are average figures, and, although age- and sex-specific, they do not take into account periods of illness, injury, or physical stress, nor the elderly. There are, as in many other areas of public health, different points of view on the importance of DRIs. Intake of enriched foods along with additional vitamin or mineral supplements may have the potential for excessive intake, if multiple sources of intake are used, as in the case of vitamin A. Tables 8.6 and 8.7 include recommendations for regular diet and for the prevention of hypertension (see Chapter 5).

The NAS sets DRIs, which are used by government, industry, researchers, and clinicians for individual and public health purposes. As an example, the NAS recently recommended folic acid and vitamin B supplements for routine use to prevent birth defects and micronutrient deficiency conditions among vulnerable groups in the U.S. population.

DISORDERS OF UNDERNUTRITION

Undernutrition includes at least 25 different deficiency diseases resulting from lack of one of the essential nutrients, proteins, vitamins, fats, or minerals (Table 8.8). A person is more likely to suffer from multiple deficiencies as opposed to deficiency in only one nutrient. Deficiency disorders may be subclinical and not show characteristic clinical features.

The WHO and UNICEF estimate that severe acute malnutrition affects an estimated 20 million children under the age of 5 worldwide (WHO, 2007). Malnourished children can suffer developmental disability of varying degree, including reduced physical and/or mental ability, often associated with reduced strength (measured as hand-grip strength), impaired cognitive function, and reduced occupational activity.

Malnourished children are subject to illness and complications or death from diseases that would otherwise be less dramatic. Much of this is due to early cessation of breastfeeding, lack of adequate food supplies, ignorance, and failure of public health systems to fortify foods or provide supplements to vulnerable groups. But the main cause is political chaos or repression, with indifference to poverty and its consequences (UNICEF, 1998).

Failure of the political systems to sustain or promote conditions to ensure adequacy of food supply or distribution systems may cause widespread undernutrition or famine.

Box 8.1 Dietary Reference Intakes: Definitions

Dietary Reference Intakes (DRIs) revision of RDAs qualitatively defining nutrient requirements as the lowest continuing intake level of a nutrient that will maintain a specific indicator of adequacy in an individual, as defined by a respected national governmental agency such as the FDA in the United States, United Kingdom, Canada, or other government.

Recommended Dietary Allowance (RDA) the average daily dietary intake level that is sufficient to meet the nutrient requirement of nearly all (97 to 98 percent) healthy individuals in a particular life stage and gender group.

Adequate Intake (AI) a recommended intake value based on observed or experimentally determined approximations or estimates of nutrient intake by a group (or groups) of healthy people, that are assumed to be adequate; used when an RDA cannot be determined.

Tolerable Upper Intake Level (UL) the highest level of daily nutrient intake that is likely to pose no risk of adverse health effects for almost all individuals in the general population. As intake increases above the UL, the potential risk of adverse effects increases.

Estimated Average Requirement (EAR) a daily nutrient intake value that is estimated to meet the requirement of half of the healthy individuals in a life stage and gender group; used to assess dietary adequacy and as the basis for the RDA.

TABLE 8.6 Recommended Dietary Allowances (RDAs) by Age and Sex, United States, 2005

| Gender | Age (years) | Activity level[b,c,d] | | |
		Sedentary[b]	Moderately active[c]	Active[d]
Child	2–3	1000	1000–1400[e]	1000–1400[e]
Female	4–8	1200	1400–1600	1400–1800
	9–13	1600	1600–2000	1800–2200
	14–18	1800	2000	2400
	19–30	2000	2000–2200	2400
	31–50	1800	2000	2200
	51+	1600	1800	2000–2200
Male	4–8	1400	1400–1600	1600–2000
	9–13	1800	1800–2200	2000–2600
	14–18	2200	2400–2800	2800–3200
	19–30	2400	2600–2800	3000
	31–50	2200	2400–2600	2800–3000
	51+	2000	2200–2400	2400–2800

[a]These levels are based on Estimated Energy Requirements (EER) from the Institute of Medicine Dietary Reference Intakes macronutrients report, 2002, calculated by gender, age, and activity level for reference-sized individuals. "Reference size," as determined by IOM, is based on median height and weight for ages up to 18 years and median height and weight for that height to give a BMI of 21.5 for adult females and 22.5 for adult males.
[b]*Sedentary* means a lifestyle that includes only the light physical activity associated with typical day-to-day life.
[c]*Moderately active* means a lifestyle that includes physical activity equivalent to walking about 1.5 to 3 miles per day at 3 to 4 miles per hour, in addition to the light physical activity associated with typical day-to-day life.
[d]*Active* means a lifestyle that includes physical activity equivalent to walking more than 3 miles per day at 3 to 4 miles per hour, in addition to the light physical activity associated with typical day-to-day life.
[e]The calorie ranges shown are to accommodate needs of different ages within the group. For children and adolescents, more calories are needed at older ages. For adults, fewer calories are needed at older ages.
Source: U.S. Department of Health and Human Services Dietary Guidelines for Americans, 2005, http://www.health.gov/DietaryGuidelines/dga2005/document.

TABLE 8.7 Comparison of Selected Nutrients in the Dietary Approaches to Stop Hypertension (DASH) Eating Plan,[a] the USDA Food Guide,[b] and Nutrient Intakes Recommended per Day by the Institute of Medicine (IOM)[c]

Nutrient	DASH eating plan (2000 kcals)	USDA food guide (2000 kcals)	IOM recommendations for females 19–30
Protein, g	108	91	RDA: 46
Protein, % kcal	21	18	AMDR: 10–35
Carbohydrate, g	288	271	RDA: 130
Carbohydrate, % kcal	57	55	AMDR: 45–65
Total fat, g	48	65	–
Total fat, % kcal	22	29	AMDR: 20–35
Saturated fat, g	10	17	–
Saturated fat, % kcal	5	7.8	ALAP[d]

Continued

TABLE 8.7 Comparison of Selected Nutrients in the Dietary Approaches to Stop Hypertension (DASH) Eating Plan,[a] the USDA Food Guide,[b] and Nutrient Intakes Recommended per Day by the Institute of Medicine (IOM)[c] —Continued

Nutrient	DASH eating plan (2000 kcals)	USDA food guide (2000 kcals)	IOM recommendations for females 19–30
Monounsaturated fat, g	21	24	–
Monounsaturated fat, % kcal	10	11	–
Polyunsaturated fat, g	12	20	–
Polyunsaturated fat, % kcal	5.5	9.0	–
Linoleic acid, g	11	18	AI: 12
Alpha-linolenic acid, g	1	1.7	AI: 1.1
Cholesterol, mg	136	230	ALAP[d]
Total dietary fiber, g	30	31	AI: 28[e]
Potassium, mg	4706	4044	AI: 4700
Sodium, mg	2329[f]	1779	AI: 1500, UL: <2300
Calcium, mg	1619	1316	AI: 1000
Magnesium, mg	500	380	RDA: 310
Copper, mg	2	1.5	RDA: 0.9
Iron, mg	21	18	RDA: 18
Phosphorus, mg	2066	1740	RDA: 700
Zinc, mg	14	14	RDA: 8
Thiamin, mg	2.0	2.0	RDA: 1.1
Riboflavin, mg	2.8	2.8	RDA: 1.1
Niacin equivalents, mg	31	22	RDA: 14
Vitamin B_6, mg	3.4	2.4	RDA: 13
Vitamin B_{12}, µg	7.1	8.3	RDA: 2.4
Vitamin C, mg	181	155	RDA: 75
Vitamin E (AT)[g]	16.5	9.5	RDA: 150
Vitamin A, µg (RAE)[h]	851	1052	RDA: 700

Estimated nutrient levels in the DASH Eating Plan and the USDA Food Guide at the 2000-calorie level, as well as the nutrient intake levels recommended by the Institute of Medicine for females 19–30 years of age.

[a]DASH nutrient values are based on a 1-week menu of the DASH Eating Plan. NIH publication No. 03-4082. www.cnhlbi.nih.gov.

[b]USDA nutrient values are based on population-weighted averages of typical food choices within each food group or subgroup.

[c]Recommended intakes for adult female 19–30, RDA = Recommended Dietary Allowance; AI = Adequate Intake; AMDR = Acceptable Macronutrient Distribution Range; UL = Upper Limit.

[d]ALAP = As Low As Possible while consuming a nutritionally adequate diet.

[e]Amount listed is based on 14 g dietary fiber/1000 kcal.

[f]The DASH Eating Plan also can be used to follow at 1500 mg sodium per day.

[g]AT = mg d-α-tocopherol

[h]RAE = Retinol Activity Equivalents

Source: U.S. Department of Health and Human Services Dietary Guidelines for Americans, 2005, http://www.health.gov/DietaryGuidelines/dga2005/document.

TABLE 8.8 Terms and Syndromes of Undernutrition

Term/Syndrome	Condition
Protein-energy malnutrition (PEM)	Person does not get enough food, calories, or protein needed for normal growth and energy or normal human activities; low weight and height for age; fatigue, poor work or school capacity
Kwashiorkor ("deposed child")	Calorie, vitamin, and protein deficiency commonly affecting children aged 2–4 when the next child is born and insufficient food, or triggered by an infection (e.g., measles); depigmented, flaky skin and hair; poor appetite; low serum albumin; edema or swelling of abdomen; enlarged liver in a child who is thin in trunk, arms, and legs
Marasmus	A wasting disease due to lack of energy and protein; usually occurs between 6 and 18 months with combination of maternal neglect and non-breastfeeding; occurs in urban slums; infant looks like a wizened old person, skin and bones, with recurrent infections; causes permanent impairment of brain development; needs warmth, attention, adequate nutrition, maternal education, and prevention of infectious diseases
Specific deficiency conditions	Vitamin A deficiency — xerophthalmia Vitamin B deficiency — pellagra Folic acid deficiency — anemia and congenital birth defects Vitamin C deficiency — scurvy Vitamin D deficiency — rickets Vitamin K deficiency — hemorrhagic disease of newborn Iron-deficiency anemia (IDA) Iodine-deficiency disorders (IDD)
Low birth weight (LBW)	Birth weight of 2500 grams or less; %LBW is a good indicator of nutrition status of a population
Low weight for age (wasting)	Indicator of current calorie and protein undernutrition (chronic or acute or both)
Low height for age (stunting)	Low height for age compared to reference population indicates chronic undernutrition at some time during the growth years, but may not be currently undernourished
Low weight for height	Thin person whose weight is low for his or her height, indicating current undernutrition
Infection-malnutrition cycle	Interaction of nutrition status, susceptibility to infection with effects of infectious disease on nutrition status

Source: Williams, C. D., Baumslag, N., and Jelliff, D. 1994. *Mother and Child Health: Delivering the Services*, Third Edition. New York: Oxford University Press.

Political and military conflict, often associated with drought and famine, may produce the conditions in which food production, storage, and marketing break down and prove inadequate for growing populations. Hunger is particularly common in populations living in sub-Saharan Africa, the Indian subcontinent, and Southeast Asia, with chronic poverty of undernourished populations in all communities throughout the world.

Malnutrition is a pathological state caused by a relative or absolute deficiency or excess of one or more essential nutrients, the clinical results being detectable by physical examination or biochemical, anthropometric, or physiological tests. The types of malnutrition identified by Derrick and Patrice Jelliffe included underweight/starvation, overweight/obesity, specific deficiency, and imbalance.

Underweight: Protein-Energy Malnutrition (PEM)

The state of being underweight, or PEM, is a nutritional deficiency resulting from either inadequate energy (caloric) or protein intake and manifesting in either marasmus or kwashiorkor. The main characteristics are weight loss and wasting of body fat and muscle mass, low weight for height, and low height for age. Severe forms include a spectrum of failure to thrive, marasmus, or kwashiorkor in children and starvation in adults. Weight loss in adults may be the result of loss of appetite, fasting, anorexia nervosa, persistent vomiting, inability to swallow, incomplete absorption, and increased base metabolic rate such as in prolonged fever, hyperthyroidism, cancer, diabetes mellitus, or other medical conditions. Chronic underweight in developed countries can occur in high-risk population groups because of poverty, illness, or ignorance of appropriate diets.

In young children, PEM may be due to an infection and/or a lack of food. In developing countries, undernutrition due to poverty or failure of the food supply is the most prevalent public health problem, particularly in infants and young children. A malnourished child is more vulnerable to infection, with lowered resistance and reduced immunity. A child with an infection then becomes even more undernourished and may suffer long-term growth retardation as a result.

In starvation there is a compensatory reduction in metabolic rate, slowed and weak pulse, lowered blood pressure, loss of body fat, muscle wasting, decreased muscle tone, loss of skin elasticity, mental dullness, and easy fatigue. The symptoms of specific deficiencies in vitamins and minerals are likely to be minimal. Recovery of weight loss after the start of feeding is slower in adults than in children. Starvation is more likely to affect children, women, and the elderly, with infants and toddlers especially vulnerable.

Failure to Thrive

Failure to thrive, or growth retardation, describes the failure of growth in keeping with age compared to a standard growth pattern. This is particularly common in inadequately fed babies in developing countries. Failure to recover growth following an illness such as diarrhea, acute respiratory infection, or measles is also common.

Marasmus

Marasmus is a severe failure to thrive condition due to marked deficiency of caloric and protein intake and general deprivation. It is characterized by wasting of body tissues, particularly muscles and subcutaneous fat, and is usually a result of severe restrictions in energy intake. It occurs in the 3–9-month age group, commonly due to early weaning and inadequate feeding with resultant starvation of the infant. The child appears wasted and irritable with depletion of subcutaneous fat and muscle tissue which are burned up to maintain blood glucose.

Kwashiorkor

Kwashiorkor, a severe form of undernutrition, is usually the result of severe restrictions in protein intake and is characterized by edema (particularly ascites or abdominal swelling). It occurs in infants and children up to about age 6. This widespread protein deficiency syndrome occurs in young children who have been weaned, often after the birth of a new child, to a diet high in carbohydrates and low in protein, and the condition is often aggravated by an infectious disease. Kwashiorkor is characterized by retarded growth and development, apathy, gastrointestinal irritability, depigmentation of hair, edema resulting in a swollen abdomen, fatty infiltration of the liver, and dry skin. Treatment consists of establishing adequate dietary intake and balance. Untreated, the mortality of this condition is high. Both types of malnutrition can be present simultaneously (marasmic kwashiorkor) and mask malnutrition due to the presence of edema.

Malnutrition is often a synergistic factor underlying deaths in children in many developing countries but can also directly result in death. The following classification of malnutrition of infants and children is that recommended by WHO in managing the malnourished child.

Deficiency conditions for one or more of the essential nutrients at a clinically detectable or subclinical level are important in public health. In developing countries, incidence varies according to urban/rural residence and social class, but among the majority of rural and urban poor they are major factors in excess morbidity and mortality. In developed countries, the predominant nutritional problem is obesity, but pockets of deficiencies exist among special population groups (Table 8.9).

Even in developed countries, vitamin and mineral deficiency can be widespread. The Canadian National Nutrition Survey in the early 1970s showed vitamin and mineral deficiencies in different parts of the country, especially in native people, teenagers, women, and the elderly. Many developed countries have taken steps to reduce these problems through improved standards of living and the implementation of policies of vitamin and mineral enrichment of basic foods. The history of pellagra in the southern United States in the early part of the twentieth century illustrates the serious damage vitamin deficiencies can cause and the remedy by fortification of flour with vitamin B_1 shows the effectiveness of public health interventions.

Vulnerable population groups include pregnant and lactating women, infants and toddlers, and the elderly, especially those in poverty groups. Other groups at risk include alcoholics, persons with chronic or frequent infection such as AIDS patients, persons with chronic diseases, and those who restrict themselves to certain foods such as vegetarians. Vegetarianism is compatible with good health, provided it is done with good nutritional advice, but when coupled with pregnancy and breastfeeding carries the risk of anemia and other deficiency conditions especially for the breastfed infant. The largest vulnerable populations are the poor populations in sub-Saharan Africa, southeast Asia, and Latin America.

According to WHO Guidelines on Food Fortification (WHO, Allen L., et al., 2006), 0.8 million deaths (1.5 percent of the total) can be attributed to iron deficiency each year, and a similar number to vitamin A deficiency. In terms of the loss of healthy life, expressed in disability-adjusted life years (DALYs), iron-deficiency anemia results in 25 million DALYs lost (2.4 percent of the global total), vitamin A deficiency in 18 million DALYs lost (1.8 percent of the global total), and iodine deficiency in 2.5 million DALYs lost (0.2 percent of the global total). These are important public health issues especially because they are preventable by currently accepted and inexpensive interventions.

Vitamin A Deficiency

Vitamin A is essential for normal vision and ocular function, because of its primary role in forming visual pigment. The dietary sources of vitamin A are animal products, such

TABLE 8.9 Prevalence of the Three Major Micronutrient Deficiencies by WHO Region[a]

WHO region	Anemia total population		Iodine intake deficiency		Insufficient vitamin A (Preschool children)	
	No. (millions)	% total	No. (millions)	% Total	No. (millions)	% Total
Africa	244	46	260	43	53	49
Americas	141	19	75	10	16	20
Southeast Asia	779	57	624	40	127	69
Europe	84	10	436	57	No data available	No data available
Eastern Mediterranean	184	45	229	54	16	22
Western Pacific	598	38	365	24	42	27
Total	2030	37	1989	35	254	42

[a]Based on the proportion of the population with hemoglobin concentrations below established cut-off levels.
[b]Based on the proportion of the population with urinary iodine <100 μg/L (micrograms per liter).
[c]Based on the proportion of the population with clinical eye signs and/or serum retinol.
Sources: Allen, L., de Benoist, B., Dary, O., Hurrell, R. 2006. *Guidelines on Food Fortification with Micronutrients.* Food and Agricultural Organization of the United Nations, World Health Organization.
World Health Organization. 2001. *Iron Deficiency Anaemia: Assessment, Prevention, and Control. A Guide for Programme Managers.* Geneva: World Health Organization, (WHO/NHD/01.3).
de Benoist, B., et al., (eds.). 2004. Iodine status worldwide. WHO *Global Database on Iodine Deficiency.* Geneva: World Health Organization.
World Health Organization. 1995. *Global Prevalence of Vitamin A Deficiency. Micronutrient Deficiency Information System Working Paper No. 2.* Geneva: World Health Organization, WHO/NUT/95.3.

as egg yolk, liver, dairy products, and breast milk, and plants containing carotenoids, such as dark green leafy vegetables, yellow and reddish fruits, and red palm oil. Symptoms of vitamin A deficiency include growth retardation, alterations in the differentiation and morphology of epithelial and mesenchymal tissues, and impaired vision. WHO estimates 3 million children have some form of xerophthalmia and, on the basis of blood levels, another 250 million preschool children are vitamin A deficient and it is likely that in vitamin A–deficient areas a substantial proportion of pregnant women are vitamin A deficient, and that 250,000 to 500,000 vitamin A–deficient children become blind every year, half of them dying within 12 months of losing their sight. WHO and UNICEF estimate that vitamin A deficiency affects very large numbers of children:

Deficient intake (subclinical), 562 million
Deficient intake (clinical susceptibility to infection),
    231 million
Night blindness, 13.5 million
Xerophthalmia, 3.1 million
Severe eye damage/blindness, 0.5 million

Vitamin A deficiency decreases resistance to infections and increases the severity, complications, and risk of death from various diseases. It also results in night blindness and xerophthalmia (drying out of the eyes, leading to scarring). This nutritional deficiency is widespread among children in developing countries, and the problem is exacerbated by a tendency by some to withhold vegetables from children for cultural or other reasons. Vitamin A deficiency is associated with increased mortality from measles, while vitamin A has both protective and therapeutic roles in measles treatment. High doses of vitamin A should be given to susceptible populations and children during measles outbreaks.

Vitamin A deficiency is a public health problem in more than half of all countries, especially in Africa and Southeast Asia, hitting hardest young children and pregnant women in low-income countries. For children, lack of vitamin A causes severe visual impairment and blindness, and significantly increases the risk of severe illness, and even death, from such common childhood infections as diarrheal disease and measles. For pregnant women in high-risk areas, vitamin A deficiency occurs especially during the last trimester when demand by both the unborn child and the mother is highest. The mother's deficiency is demonstrated by the high prevalence of night blindness during this period. The impact of vitamin A deficiency on mother-to-child HIV transmission warrants further investigation.

Preventive measures for populations at risk for early or marginal vitamin A deficiency can markedly reduce the risk of mortality from intestinal and respiratory infections and measles. Methods for improving vitamin A status include periodic distribution of large-dose capsules appropriate for age, fortification of readily consumed dietary staples, and increased intake of vitamin A–rich foods. Adding vitamin A, along with vitamin D, to margarine, milk, and dairy products was practiced in the United States, Canada, and the United Kingdom during World War II. The practice is mandatory in Canada and nearly universal in the United States.

The basis for lifelong health begins in childhood. Vitamin A is a crucial component. Since breast milk is a natural source of vitamin A, promoting breastfeeding is the best way to protect babies from vitamin A deficiency. WHO has a goal of worldwide elimination of vitamin A deficiency and its tragic consequences, including blindness, disease, and premature death. To successfully combat vitamin A deficiency, short-term interventions and proper infant feeding must be backed up by long-term sustainable solutions. The combination of breastfeeding and vitamin A supplementation, coupled with promotion of vitamin A–rich diets and food fortification, holds the promise of achieving this goal.

For vitamin A–deficient children, the periodic supply of high-dose vitamin A in swift, simple, low-cost, high-benefit interventions has also produced remarkable results, reducing mortality by 23 percent overall and by up to 50 percent for acute measles sufferers. Planting these "seeds" between 6 months and 6 years of age can reduce overall child mortality by 25 percent in areas with significant vitamin A deficiency. However, because breastfeeding is time-limited and the effect of vitamin A supplementation capsules lasts only 4–6 months, they are only initial steps toward ensuring better overall nutrition and not long-term solutions.

Cultivating leafy vegetables in a home garden is a community-based phase necessary to achieve long-term results. Food fortification takes over where supplementation leaves off. Food fortification (e.g., sugar in Guatemala), maintains vitamin A status, especially for high-risk groups and needy families.

For vulnerable rural families, for instance, in Africa and Southeast Asia, growing fruits and vegetables in home gardens complements dietary diversification and fortification and contributes to better lifelong health.

In 1998 WHO and its partners — UNICEF, the Canadian International Development Agency, the United States Agency for International Development and the Micronutrient Initiative — launched the Vitamin A Global Initiative. In addition, over the past few years, WHO, UNICEF, and others have provided support to countries in delivering vitamin A supplements. Linked to sick-child visits and national poliomyelitis immunization days, these supplements have averted an estimated 1.25 million deaths since 1998 in 40 countries.

Cumulative evidence since the mid-1980s has reinforced the importance of vitamin A in combating infections, so that food fortification is recommended especially in developing countries or for populations subject to high infection rates, such as HIV-positive persons. Routine vitamin A supplementation for children is now recommended by WHO in conjunction with immunization programs (EPI Plus, see Chapter 4). Massive doses of vitamin A are used to treat children with measles with great benefit in reducing mortality and complication rates.

Fortification of sugar with vitamin A has been implemented in a number of countries in South and Central America. The Philippines has required fortification of margarine with vitamin A during 1998. Indonesia is using a variety of techniques to increase supplementation and fortification to reduce deficiency conditions. Supplementation policies providing vitamin A supplements to children and postpartum mothers carried out in 78 countries in 1996 is continuing and spreading with WHO and UNICEF promotion.

Vitamin A toxicity may occur when more than three times the RDA is consumed, usually from medication overdose, especially among pregnant women, alcoholics, or persons with chronic liver conditions. This can result in hyperkeratinosis (i.e., orange skin), nausea, vomiting, birth defects, and neurologic, gastrointestinal, and dermatologic symptoms.

Vitamin D Deficiency (Rickets and Osteomalacia)

Rickets from vitamin D deficiency was a common disorder in many parts of the developed world well into the twentieth century (Box 8.2). It constitutes one of the important diseases of infancy because of its serious complications, including disorders of long bone growth, bowing of legs, pelvic deformity, and, in extreme forms in infants, tetany and convulsions. The vitamin D content of human milk is extremely low, and the exclusively breastfed infant not adequately exposed to sunlight may develop clinical rickets. In adults, malabsorption or poor dietary intake of vitamin D can result in osteomalacia and osteoporosis with fragile bones and frequent fractures. Among the elderly, who are closed in especially during winter months, and among homebound or institutionalized patients, vitamin D deficiency is common.

There are few natural sources of vitamin D, but deficiency is preventable by exposing the skin to the ultraviolet rays of the sun. Prevention became common through the practice of giving children cod liver oil, a successful antirachitic measure, until replaced by use of vitamin D supplements for infants. In colder climates or in foggy locations where exposure to the sun may be limited, rickets was common. However, seasonal deficiency of vitamin D can occur even in locations with ample sunlight.

Rickets remained widespread until fortification of milk was introduced in the 1940s during World War II in

Box 8.2 Rickets and Vitamin D Deficiency

Rickets is described in ancient medical writings from the first and second centuries CE as bone deformities in infants by Soranus and Galen, Roman physicians of that era. In 1650, Francis Glisson, a Cambridge physician, published a treatise on rickets, describing its clinical features and suggesting treatments.

In 1870, as many as one-third of the poor children of cities such as London and Manchester suffered from obvious rickets. At the turn of the twentieth century, rickets was rampant among infants living in industrialized, polluted cities of North America and Europe. In 1919, Edward Mellanby, an English physician, clearly established the role of diet in the cause of rickets via animal experiments. Elmer McCollum, an American nutritional biochemist, developed a method of biological analysis of nutritive value of foods and eventually discovered vitamin D. In 1919, while investigators in Germany showed that exposure to sunlight cured rickets and that it acted by altering fats to produce vitamin D.

As late as 1921, McCollum claimed that probably half of the children in the United States had rickets. Rickets was at that time the most common nutritional disease of children, affecting approximately 75 percent of infants in New York City. Cod liver oil and irradiated foods were introduced to prevent rickets and widely used in Europe and North America.

Rickets remained widespread until fortification of milk was introduced in the 1940s in Britain and North America. Rickets prevalence in the United Kingdom, especially in the industrial cities in northern England and Scotland, declined dramatically. In Canada, fortification of milk with vitamin D was routine during the 1940s, but stopped in the 1950s and 1960s. Significant deficiencies in certain population groups led to mandatory fortification in Canada starting from 1979, with subsequent disappearance of rickets.

Rickets in infants continues to be reported in the United States and other countries due to inadequate vitamin D intake and decreased exposure to sunlight. In 2003, because of reappearance of rickets and low levels of vitamin D, the American Academy of Pediatrics (AAP) recommended all infants, especially breastfed, have a minimum supplement of 200 IU of vitamin D per day from first 2 months of life. They also recommend 200 IU vitamin D per day be continued in childhood and adolescence because sunlight exposure is not easily determined for a given person. New vitamin D intake guidelines are based on U.S. National Academy of Sciences recommendations.

Osteoporosis has its highest occurrence among postmenopausal women and results in the loss of bone mass, often leading to fractures, including those of the hip and spine. Hypovitaminosis D and related abnormalities in bone metabolism and strength are common in the elderly in Europe, but are reported in all elderly populations. In 1995, there were 382,000 hip fractures in 15 countries of the European Union with an estimated total care cost of about $9 billion. In the United Kingdom, hip fractures were found to place the greatest demand on resources and have the greatest impact on elderly patients because of increased mortality, long-term disability, and loss of independence.

The problem of vitamin D deficiency should be addressed with a multiple strategy approach, including both supplementation (with calcium) targeted to higher-risk groups, and food fortification in order to reach the general population.

Sources: Centers for Disease Control. 1999. Achievements in Public Health, 1900–1999: Safer and healthier foods. *Morbidity and Mortality Weekly Report*, 48:905–913.
American Academy of Pediatrics. 2003. Section on Breastfeeding and Committee on Nutrition. Prevention of rickets and vitamin D deficiency: New guidelines for vitamin D intake. *Pediatrics*, 111:908–910.
Rajakumar, K., Greenspan, S. L., Thomas, S. B., Holick, M. F. 2007. Solar ultraviolet radiation and vitamin D: A historical perspective. *American Journal of Public Health*, Aug 29; [Epub ahead of print].

Britain and North America. Rickets prevalence in the United Kingdom, especially in the industrial cities in northern England and Scotland, declined dramatically. The 1971 Canadian National Nutrition Survey found significant vitamin D deficiencies in certain age, sex, ethnic, and geographic population groups. Although antirachitic procedures (cod liver oil and adding vitamin D to milk) were routine during the 1940s, the practice waned in the 1950s and 1960s. An increase in hospitalizations for rickets in Montreal followed the abandonment of vitamin D milk enrichment.

Addition of vitamin D to milk, nearly universal in the United States, was made mandatory in Canada in 1979 and rickets disappeared. No incidents of vitamin D toxicity were reported. Vitamin D toxicity, due to human error in formula preparation or from multiple source supplementation, can cause failure –to thrive, nausea, vomiting,

and weakness. Canada's mandatory milk and other food fortification policy continues and was renewed in 2006.

Even in a sunny climate, there are seasonal variations in vitamin D availability, and social customs of keeping infants overly wrapped and out of the sun may cause rickets in some population groups. Studies of vitamin D levels among institutionalized elderly found low levels, and among elderly in the community vitamin D levels were reported to be low in the winter months.

Prevention of vitamin D deficiency conditions includes routine supplements of vitamin A and D for infants from 1–12 months (and up to 24 months). Fortification of baby formulas and cereals is common. Standard textbooks of pediatrics and of public health take the same position with respect to rickets (vitamin D deficiency): vitamin D deficiency is a problem in childhood and in the elderly that cannot be addressed by providing vitamins to high-risk groups, so that milk fortification is needed. Fortification of milk products is one of the most important public health nutrition measures used.

In 2003, the American Academy of Pediatrics recommended vitamin D supplementation with 200 International Units daily for infants and children up to adolescence because of increasing reports of vitamin D deficiency at the level of clinical rickets and in low vitamin D levels in older children. Risk factors include prolonged sole breastfeeding, mothers with dark skin in northern climates, in winter, and when women are totally covered for religious reasons and not adequately skin exposed to solar radiation. Vitamin D deficiency conditions are being reported in the literature in countries including the United Kingdom which do not fortify milk nor recommend routine vitamin D supplements.

Deficiency of vitamin D remains a major public health problem that requires attention both in food fortification and supplementation for groups at risk for deficiency, infants, children, teenagers, adults, and the elderly. Failure to promote its application widely internationally has been an important lapse of the international health community.

Vitamin C Deficiency

Scurvy is a deficiency disease due to lack of vitamin C (ascorbic acid) in the diet. It was common among seamen (see Chapter 1) and others deprived of fresh fruits and vegetables. Vitamin C deficiency causes skin lesions, weakness, fatigue, weight loss, muscle pains, susceptibility to infection, hemorrhages, debility, and even death. It can occur at any age due to inadequate diet. Infantile scurvy formerly appeared in bottle-fed infants, but with the advent of vitamin-fortified infant formulas this has become rare in industrialized countries. Infants should have a source of vitamin C from the first month of life, such as in orange juice.

Fortification of fruit juices and flavored drinks is a common practice to assure adequate intake of vitamin C. As a water-soluble vitamin, there is no risk of excess intake from multiple sources in the course of a normal diet.

The use of vitamin C in prevention/treatment of the common cold and respiratory infections remains controversial, with ongoing research. For cold prevention, more than 30 clinical trials including over 10,000 participants have examined the effects of taking daily vitamin C (200 mg or more). Overall, no significant reduction in the risk of developing colds has been observed. Also other uses for vitamin C have been proposed, but few have been conclusively demonstrated as being beneficial in scientific studies. In particular, research in asthma, cancer, and diabetes remains inconclusive, while no benefits have been found in the prevention of cataracts or heart disease.

Vitamin K Deficiency (Hemorrhagic Disease of the Newborn)

A deficiency of vitamin K occurs in the normal newborn, and may cause impaired production of prothrombin, a key blood clotting factor. Lack of prothrombin shortly after birth can cause hemorrhagic disease of the newborn (HDN or vitamin k deficiency bleeding disorder; see Chapter 6). Secondary HDN can occur weeks later. To prevent the deficiency, a single injection of vitamin K is administered by injection, or orally at birth and vitamin K is added to fortified baby formulas and cereals. Administration of vitamin K to newborns is now instituted in most U.S. hospitals and should be a priority for prevention of neonatal mortality.

Vitamin B Deficiencies

Niacin Deficiency (Pellagra)

Niacin (nicotinic acid) is essential for specific oxidation–reduction reactions in the body. A deficiency of niacin causes diarrhea, dermatitis, and dementia. Pellagra was established as a nutritional deficiency (and not infectious) condition in investigation of the condition in orphanages and hospitals in the southern United States in 1917–1922 (see Chapter 1). Pellagra is prevented with adequate dietary intake of niacin or niacin substitutes contained in vitamin-enriched bread, liver, meat, fish, poultry, potatoes, green vegetables, peanuts, and cereals. Niacin also exerts beneficial effects by reducing blood lipids and raising high-density lipoproteins (HDL), slowing atherosclerosis and coronary artery lesions.

In 1998, the National Science Foundation issued a recommendation for multiple vitamin supplements for all adults.

Thiamine (B₁) Deficiency (Beriberi)

Thiamine deficiency causes derangement of carbohydrate metabolism. "Dry beriberi" results in neurologic symptoms and death. "Wet beriberi" is characterized by cardiac symptoms and disorders, including heart failure and death. This disease was common among prisoners of war in Japanese camps during World War II due to diets mainly of polished rice. Prevention requires a diet containing liver, glandular organs, yeast, wheat germ, whole wheat or cereals, unpolished rice, milk, legumes, soybeans, and peanuts.

Alcoholic (Korsakoff's syndrome) or nutritional dementia (Wernicke's encephalopathy) may be a combination of B complex deficiency and constitute important public health problems. Wernicke's encephalopathy together with Korsakoff's syndrome are termed *cerebral beriberi*. These diseases are of particular concern for people who consume excess alcohol (>4 oz/day). Habitual alcoholics may develop severe vitamin B deficiency leading to dementia and hemorrhagic stroke. Enrichment of bread, breakfast cereals, and other flour-based products is practiced in Canada and in many countries in Latin America on a mandatory basis, and in the United States, where it is mandatory when the term "enriched" is used.

Folate Deficiency

Folic acid is required for normal blood formation and neurologic health. Its deficiency is common among low socioeconomic groups, especially during pregnancy, infancy, and childhood (Box 8.3). Deficiency is common in alcoholics, due to malnutrition or impaired absorption. Alcoholics with good diets are less likely to develop folate deficiency than those with poor eating habits. Folate deficiency constitutes a major health risk for alcoholics to develop severe neurologic damage in the spinal cord or in the optic or peripheral nerves.

Folic acid deficiency has been demonstrated to be a cause of neural tube defects (NTDs), an important congenital disorder (see Chapter 6). This condition ranges in severity from anencephaly to disabling defects of the spinal cord and is largely preventable by prenatal supplements of folic acid or food fortification, which became mandatory in the United States, Canada, and Chile in 1998 with subsequent decline in NTD incidence. In many countries, including Europe, folic acid supplementation does not receive adequate attention. As a major prevention measure of birth defects, all women of reproductive age should receive folic acid supplements. In addition, adequate intake is important for children, men, and women of all ages in maintaining proper homeostasis and health.

Folic acid supplementation may also be important in prevention of coronary heart disease by lowering homocysteine levels, but this requires further substantiation (see Chapter 5).

Vitamin B_{12} Deficiency

Vitamin B_{12} deficiency causes enlargement of red blood cells with poor hemoglobin content (macrocytic anemia). Vitamin B_{12} deficiency may be due to malabsorption in the stomach (pernicious anemia) or from long-term deficiency of vitamin B_{12} intake from vegetarian diets which exclude eggs and dairy products. Vitamin B_{12} deficiency and folic acid deficiency are often comorbid, with serious deleterious effects for women of childbearing age, the elderly, and alcoholics. Pregnant women and lactating mothers, especially if vegetarian, need vitamin B_{12} supplements. Vitamin B_{12} deficiency can result in degeneration of the spinal cord, optic nerves, cerebral tissue, and peripheral nerves. Prevention is by promotion of healthy nutrition with foods rich in B complex vitamins and supplementation for infants in fortified cereals.

Box 8.3 Folic Acid and Neural Tube Defects

In the 1930s, Lucy Wills showed folate was a nutrient needed to prevent anemia of pregnancy, using brewer's yeast from spinach leaves. Folic acid was first synthesized in 1946. It is a water-soluble B vitamin, important in new cell formation via DNA synthesis. It is available in leafy green vegetables such as lettuce, spinach, and beans and in peas and liver. Folic acid levels in population studies in the U.S. showed widespread deficiency in the 1990s NHANES studies.

In the 1990s, international studies showed that folic acid supplements before pregnancy resulted in reduced rates of neural tube defects (NTDs) among newborns. When this was confirmed by the British Medical Council, recommendations for folic acid supplements for women in the age of fertility were adopted by many governmental and professional organizations such as the CDC, the American College of Obstetrics and Gynecology, American Academy of Pediatrics, and many others. In 1996, the U.S. Food and Dug Administration and the Canadian Federal Department of Health adopted mandatory fortification of flour, implemented in 1998, as did Chile.

Studies have shown reduced rates of NTDs in Canada and the United States and flour fortification has become widely recommended for other countries. No country in the European Union has yet adopted this measure. Some controversy along with apathy obstructs fortification by claims that potential harmful effects may occur by masking vitamin B_{12} deficiency and neurologic damage.

The benefit of folic acid fortification along with supplementation for high-risk groups (including women and infants) to achieve intake of 400 micrograms per day has been demonstrated, but cumulative evidence is mounting regarding other benefits such as prevention of cardiovascular disease, dementia, and cancer.

This is a case in which evidence of benefit far outweighs theoretical possibilities of harm, and delayed implementation causes significant damage to newborns and others. The case for action is strong but reaction to implementation is also strong. The knowledge-action gap is still wide.

Sources: Dietary Supplement Fact Sheet: Folate. Office of Dietary Supplements. NIH Clinical Center, National Institutes of Health, http://ods.od.nih.gov/factsheets/folate.asp, updated August 22, 2005 [accessed January 14, 2007].
De Wals, P., Fassiatou, T., Van Allen, M., et al. 2007. Reduction in neural tube defects after folic acid fortification in Canada. *New England Journal of Medicine,* 357:135–142.

Iron-Deficiency Anemia

Iron-deficiency anemia is the most common nutritional deficiency condition in the world, widespread in both developing and developed countries. Malnutrition is a pathological state caused by a relative or absolute deficiency or excess of one or more essential nutrients, the clinical results being detectable by physical examination or biochemical, anthropometric, or physiological tests.

In 2007, WHO stated:

"Iron deficiency is the most common and widespread nutritional disorder in the world. As well as affecting a large number of children and women in developing countries, it is the only nutrient deficiency which is also significantly prevalent in industrialized countries. The numbers are staggering: 2 billion people — over 30 percent of the world's population — are anemic, many due to iron deficiency, and in resource-poor areas, this is frequently exacerbated by infectious diseases. Malaria, HIV/ AIDS, tapeworm and hookworm infestation, schistosomiasis, and other infections such as tuberculosis are particularly important factors contributing to the high prevalence of anemia in some areas. Iron deficiency affects more people than any other condition, constituting a public health condition of epidemic proportions. More subtle in its manifestations than, for example, protein-energy malnutrition, iron deficiency exacts its heaviest overall toll in terms of ill health, premature death and lost earnings.

Iron deficiency and anemia reduce the work capacity of individuals and entire populations, bringing serious economic consequences and obstacles to national development. Overall, it is the poor and least educated who are most affected by iron deficiency, and it is they who stand to gain the most by its reduction. This concern is reflected in the Millennium Development Goals of reduction of poverty and of hunger."

Iron plays a critical role in the transport of oxygen in human blood. Iron exists in two major forms in food. The first is heme iron, which is found only in animal sources. It is readily available, as absorption is not hindered by other constituents of the diet. The second type is inorganic iron, whose absorption is strongly influenced by factors present in foods ingested at the same time. The composition of a meal can affect the amount of iron absorbed. Consumption of foods containing heme iron will improve the absorption of non-heme iron. Vitamin C enhances the absorption of non-heme iron, but substances like tannin from tea combine with the iron in the intestine to inhibit its absorption.

Iron-deficiency anemia as a public health problem has received considerable attention since the 1980s by international organizations such as UNICEF and the World Health Organization. The issue is especially vital for developing countries, but even in developed countries iron enrichment of staple foods is an important public health measure. Iron-deficiency anemia particularly affects infants, pregnant women, the adolescent and adult populations, as well as the elderly.

In developing countries, two-thirds of children and women of childbearing age are estimated to suffer from iron deficiency; one-third or more of them have the more severe form of the disorder, anemia. Symptoms include listlessness and fatigue. Low levels of iron are associated with often irreversible damage to brain development. Studies of the effects of iron-deficiency anemia have shown diminished psychomotor performance suggestive of low levels of brain dysfunction or damage resulting from iron deficiency, even in the absence of anemia. The intellectual development of children and the physical activity of both adults and children are impaired in those suffering from anemia. This may be complicated by lead toxicity, which is common in iron-deficient children, who are more susceptible to lead absorption. There is a direct correlation between low hemoglobin levels and the prevalence of diarrheal and respiratory diseases which result from an impaired immune system. Lack of recovery in many children even after iron supplementation underscores the importance of preventing iron deficiency through dietary manipulations and health education.

The International Society for Prevention of Iron Deficiency Anemia promotes preventive approaches, such as iron fortification of basic foods (bread, sugar, salt) and routine supplementation for infants and pregnant women, to prevent what it defines as the most widespread nutritional deficiency in both developed and developing countries. Globally, nearly 2 billion people are anemic and 3.6 billion are iron deficient.

Iodine-Deficiency Diseases

Iodine is an essential element in nutrition. Insufficient iodine in natural sources causes clinical or subclinical thyroid disorders. Deficient supply of iodine damages fetal development and produces fetal hypothyroidism. Low levels of circulating thyroid hormones and urinary iodine, cause varying degrees of brain damage in infants, including cretinism. Excess prevalence of goiter, or enlargement of the thyroid gland, indicating suboptimal thyroid function, is present in large areas of the world where there are low levels of ground and surface water iodine.

Iodine deficiency is the world's most prevalent cause of brain damage. In 2003, some 2 billion people suffered from iodine deficiency: 436 million in Europe, 624 million in Southeast Asia, 365 million in the Western Pacific, 260 million in Africa, 229 million in the Eastern Mediterranean, and 75 million in the Americas (Anderson et al., 2007). David Marine and David Cowie, following a series of studies from 1910 onward, pioneered the idea of prevention of iodine deficiency by iodination of commercial table salt in 1924 (Morton's Iodized Salt, see Chapter 1). By 1930, most of the salt consumed in the United States was iodized, and goiter had largely disappeared even in previously endemic areas.

Iodination of salt as a preventive measure has become standard public health practice in many countries since World War I. It has been compulsory in Canada since 1979, and widespread in Western Europe, although not universal and not always at effective levels.

In the 1980s, the World Health Organization expressed growing concern with the widespread nature of iodine-deficiency disorders in large areas of the world, affecting an estimated 2.3 billion persons especially in China, the former Soviet Union, Southeast Asia, and many developing countries. In 1986, the World Health Assembly called on all nations to introduce iodination of salt or other appropriate technology to reduce this silent pandemic.

Australian scientist and public health advocate Basil Hetzel demonstrated that insufficient iodine during fetal development adversely affected brain development and prompted WHO and other international organizations to respond to the issue of iodine-deficiency disorders.

Iodine deficiency was described as follows in a 1993 editorial in the *New England Journal of Medicine:* "The most important effects of iodine deficiency are on the developing central nervous system, and they form a continuum from mild intellectual impairment to full blown cretinism." Prevention of iodine deficiency is best achieved through the iodination of salt on a national scale at a level of 1 part iodine to 10,000–20,000 parts of salt. A global effort for prevention of iodine-deficiency disorders is under way.

UNICEF, the International Council for Control of Iodine Deficiency Disorders, the European Thyroid Association, Kiwanis International, some countries (e.g., Canada), and the World Bank have called for national and international action to control this widespread public health problem. The World Summit for Children called for universal iodizing of salt with a target of 95 percent iodination in each country by 1995. By 1994, 94 countries had national plans for iodizing salt, with 58 countries, including almost 60 percent of the world's children, on schedule. Progress is being made, but inertia and complacency are still barriers to achieving this goal.

The World Health Organization Geneva, in a 2004 report on iodine status, stated that the number of countries where iodine deficiency is a public health problem has halved over the past decade. Therefore, the main strategy — universal salt iodination — has been successful. The number of countries where iodine deficiency is a public health problem was reduced to 54 in 2003 from 110 in 1993. Of the 54 iodine-deficient countries, 40 are mildly iodine deficient and 14 moderately or even severely iodine deficient. WHO says that of the 126 countries for which data were available in 2003, iodine intake is now adequate in 43, and sustained efforts are still required to strengthen salt iodination programs worldwide. Also, UNICEF estimates that globally 66 percent of households now have access to iodized salt. Still many countries such as in the Central Asian Republics and in South America have difficulties with smuggling of non-iodized salt, and poor compliance with national programs of iodination. A study carried out in northern France in 2003 reports 24 percent iodine deficiency in schoolchildren that is mostly mild to moderate but still unsatisfactory. Iodine deficiency continues to be a major public health problem in many parts of the world in the twenty-first century.

Osteoporosis

Osteoporosis is an important chronic condition, as discussed in Chapter 5, with serious health consequences, particularly among older women in terms of fractures of the hip, the spine, and forearm. Hip fracture mortality is between 12 and 20 percent, and many survivors are institutionalized as a result of complications. Osteoporosis is preventable to a large degree by adequate calcium and vitamin D supplementation and exposure to exercise and sunlight from the early years of life. Fluoridation of water may have beneficial effects as fluoride in appropriate amounts increases bone strength. Physical exercise, particularly weight bearing, helps to increase bone mass.

Because of the increasing population of elderly persons, especially women, osteoporosis and its complications are important challenges for preventive, curative, and rehabilitative services as well as an issue for health promotion in terms of dietary and other self-care and preventive measures in individual medical practice as well as in public health. Food enrichment with calcium and vitamin D play an important role in reducing the severity of this condition currently affecting some 44 million Americans, 10 million with the disease and 34 million with low bone density and at risk for osteoporosis (National Osteoporosis Foundation, United States 2008).

Exposure to sun produces vitamin D but this is countered by concerns that solar exposure increases skin cancer, and use of sunscreens reduces ultraviolet light exposure of the skin, reducing vitamin D production. Sun exposure is also affected by seasonal variation and is dependent on the latitude, skin color, total body coverage by clothing worn for cultural and religious reasons, as are dark-skinned people living in high latitude countries, particularly in winter, when people tend to remain indoors and cover up when outside.

Eating Disorders

Eating disorders are an important health risk, primarily for teenage girls and young women, and an occupation-related health risk associated with sports, ballet, and modeling. These disorders can be communicated between people by group pressure of fashion, precedent, and close contact among teenage girls and young women. Fashions in a society have great influence on vulnerable adolescents who can enter a cycle of self-denial or purgation that may be extremely destructive and even fatal. In all eating disorders, a vital component of treatment and public health guidance should be behavioral and family therapy, as eating practices are highly influenced by family development and conceptions about image.

Anorexia Nervosa

Anorexia nervosa is a self-imposed severe dietary restriction that may be chronic or acute. This condition is most common among teenage girls and occupational groups including models and ballet dancers. Eating disorders reportedly affect some 5 percent of female college students in the United States. Anorexia is often associated with laxative and diuretic abuse or excessive exercising. Specific features common to anorexia nervosa are abnormal sensitivity to being fat and fear of losing control over the amount of food eaten, severe restriction of food intake and refusal to accept food, marked weight loss, cessation of menstruation, damage to dentition and chronic tooth pain, and more serious consequences including liver and heart muscle damage. The psychological features of rigidity, perfectionism, and fear of obesity preceding the condition are usually resistant to treatment. Hospitalization is commonly required with strict regimens of antidepressant and behavior therapy. Anorexia nervosa may lead to death by self-starvation or suicide.

Bulimia or Binge Eating

A binge-eater has a compulsion to eat large quantities of food within a short period of time, usually 2 hours or less, with deliberate vomiting to expel the food to lose or maintain his or her weight. Many features are common to both bulimia and anorexia nervosa. The physical effects of this form of eating disorder can be severe (such as esophageal damage from repeated vomiting) but usually less so than anorexia nervosa. The presence of depression may require antidepressant therapy supported by psychotherapy techniques or behavior modification therapy and hospitalization. Like anorexia, this disorder is most common in societies that promote images of beauty as thinness through advertisement and related social pressures affecting the psychologically vulnerable with poor self-images.

DISEASES OF OVERNUTRITION

Though media attention tends to focus on the extreme forms of malnutrition, the more common nutritional health concerns in industrialized countries are those relating to specific deficiencies and dietary imbalances. In the United States, the *Surgeon General's Report on Nutrition and Health, 1988,* addressed the problems of nutrition in a developed country in the 1980s, with an emphasis on a type of malnutrition which could be described as overnutrition. The report also addressed the issues of specific vitamin and mineral deficiency states and food fortification. Diseases of overnutrition, discussed in Chapter 5, are of increasing importance in developing countries as well, as the diseases of improving standards of living affect growing middle classes and as social patterns, food, and eating habits switch from traditional styles toward western diets including excess calories, especially from fatty foods and red meat.

Overweight/Obesity

Economic development leads to changes in a population's diet. Dietary habits identified with an affluent or western lifestyle are characterized by an excess of energy-dense foods rich in fat and simple sugars with a relative deficiency of complex carbohydrate food. Modest improvements in the economic state of a country are associated with an epidemiologic shift characterized by a rise in the incidence of chronic diseases of middle and later adult life. In developing countries, these typically coexist with the traditional and persistent problems associated with nutritional deficiencies. Obesity in the industrialized countries is often associated with poverty, and its significance increasingly recognized as a major public health issue for the industrialized countries but also in developing countries.

Obesity is an excess of fat in the body caused when energy consumed has been greater than the energy expended. Nutrient excess may occur in the short term or over a long period (chronic nutrient toxicity). Excessive weight for body size (BMI, see section entitled "Measuring Body Mass") is a public health problem because it is associated with premature death and is a risk factor for coronary heart disease, diabetes mellitus, hypertension, asthma, and gastrointestinal disorders. While there is evidence of a genetic predisposition for obesity, diet and other environmental factors, such as a sedentary lifestyle, play a major role in excess body fat. Increasing obesity at all ages is a widespread phenomenon in the industrialized countries and in developing countries.

Restriction of caloric intake and increased exercise are advisable for those individuals seeking to lose excess weight. Obese individuals tend to be inactive; attempts to increase physical exercise have well-documented health benefits. Drugs and fad diets are treatment techniques that have the potential for damaging one's health, and any weight loss tends to be temporary. Surgical treatment may produce permanent weight loss but often with serious side effects, such as chronic diarrhea. Nutritional counseling and assistance in development of nutritious and low-fat diets will be more useful, especially when accompanied by exercise.

Obesity is extremely resistant to treatment, and thus primary prevention of obesity is a major public health target. This requires health education of mothers and very young children in proper nutritional practices. Eating and physical activity habits learned in childhood are very difficult, although not impossible, to change. Information about food choices should highlight the need to reduce fat and increase dietary fiber and complex carbohydrates. Food labeling can serve to inform the individual of the caloric and fat content of items intended for consumption. The individual person as well as governments, schools, parents, the media, and communities all have responsibilities in health promotion for prevention of obesity.

Diabetes mellitus, discussed in Chapter 5, is a chronic metabolic disorder found throughout the world. It develops in individuals who lack sufficient insulin production or whose insulin is impaired in function. As a result, they have reduced capacity to utilize glucose derived from carbohydrate foods or from body stores of glycogen. Type I, also called insulin-dependent diabetes (IDDM) or juvenile diabetes, appears in childhood or adolescence due to failure of insulin production and is not diet-generated, but its management requires dietary controls as well as insulin. Type II, or mature onset, nutritional, or non-insulin-dependent, diabetes (NIDDM) occurs in middle adulthood or in the elderly, mostly due to excess dietary caloric intake, with saturated fats and low intake of dietary fiber, resulting in decreased insulin sensitivity and abnormal glucose tolerance. Genetic predisposition and possibly fetal influences result in a high prevalence of type II diabetes in some ethnic groups such as native Americans, Inuit, and Australian aborigines, but prevalence is magnified by societal factors including poor diet, lack of physical exercise, and alcohol abuse. This interacts with other risk factors, such as body weight, to instigate the onset of disease.

Type I and type II diabetes both require careful nutrition. Since the discovery of insulin by Frederick Banting and Charles Best in Toronto in 1921, type I diabetes has been managed by regulation of blood sugar with daily insulin injections, with monitoring of blood sugar, diet, and exercise. Type II diabetes is controllable by diet and exercise but may also require medication to lower blood sugar. Approximately 80 percent of patients with type II diabetes are obese. Modern public health emphasizes weight reduction for obese diabetic patients to control their blood glucose and reduce their high risk of coronary heart disease and stroke. Nutrition plays a key role in the causation and management of type II diabetes.

Cardiovascular Diseases

As discussed in Chapter 5, a strong relationship exists among diet, lifestyle, and the risk of cardiovascular disease (CVD), and especially between saturated fat intake and incidence of this disease. Cholesterol levels can be reduced by lowering dietary intake of foods high in cholesterol and saturated fats and by increasing foods with fiber and monounsaturated fats, such as olive oil and avocado.

Mortality from coronary heart disease and strokes has declined dramatically in western countries, but strokes still occur in some 500,000 persons in the United States each year, with some 150,000 deaths. The American Heart Association estimates the direct and indirect costs for cardiovascular disease in 2007 at $431.8 billion. Estimates of the incidence of strokes based on the Framingham study underestimate the true incidence because the study is primarily of a white middle-class population; more representative population-based studies report higher rates because strokes are more common in African-American and Hispanic populations.

Hypertension and diabetes are major risk factors for heart disease and stroke. Hypertension is associated with imbalanced nutrition, with excess fats and low intake of fruits and vegetables. Primary prevention is largely dietary, with weight loss, salt restriction, smoking cessation, and increased physical activity. Nutrition and diet play a major role in clinical care of the patient with cardiovascular diseases as well as in public health policy related to food, nutrition, and population awareness. Alcohol, especially red wine, in moderation, has a protective effect against coronary heart disease, but in excess contributes to increased rates. Clinical and health promotion intervention must also promote regular screening for elevated resting blood pressure, associated risk factors, and careful clinical management of hypertension to reduce risk of CVD. Antihypertensive medications play an important role in prevention of known complications of this condition.

Cancer

The relationship between specific nutrients and cancer is less well established than between diet and cardiovascular disease (Chapter 5). The relationship of food consumption data and cancer rate is supported by studies including the following: changing cancer rates in ethnic groups after migration with changes in dietary patterns; case–control studies of cancer patients and controls; prospective studies of populations with known dietary habits; and animal experimental data.

Cancers in which diet is implicated as an etiological factor include cancers of the oral cavity and pharynx, larynx, gastrointestinal tract, breast, liver, pancreas, lung, endometrium, cervix, and prostate (Table 8.10). It is currently accepted that there is a relationship between diet and specific cancer sites. The strongest dietary associations are high fat intake with cancers of the prostate and

TABLE 8.10 Association between Selected Dietary Factors and Specific Cancer Sites

Cancer site	Body weight	Fat	Fiber	Fruits/Vegetables	Alcohol	Smoked, salted, pickled
Lung	0	0	0	P	0	0
Breast	C	C	0	0	C	0
Colon	0	C*	P	P	0	0
Prostate	0	C*	0	0	0	0
Bladder	0	0	0	P	0	0
Rectum	0	C	0	P	0	0
Endometrium	C*	0	0	0	0	0
Oral	0	0	0	P	C	0
Stomach	0	0	0	P	0	C*
Cervix	0	0	0	P	0	0
Esophagus	0	0	0	0	C*	C

Note: C — contributes to development of cancer; C* — greater contribution to the development of cancer; P — protective effect against cancer; 0 — no known effect.
Sources: World Health Organization, 1990; and *The Surgeon General's Report on Nutrition and Health*, 1988.

colon; high body weight with cancer of the endometrium; alcohol with cancer of the esophagus; and smoked, pickled, or salted foods with cancer of the stomach. Widespread consumption of betel nuts in India and other South Asian regions is a major risk factor for cancers of the alimentary tract. The protective effect of a diet high in fruit, whole grains, and vegetables on colon and other cancers are possibly due to vitamin A and C and their antioxidant effects.

Some epidemiologists attribute 30–40 percent of cancer in men and 60 percent in women as diet-related. Since the 1960s, there has been a growing consensus that diet plays an important, although still undefined, role in carcinogenesis. The "Mediterranean diet," which is low in total and saturated fat, high in green and yellow vegetables and citrus fruits, and low in alcohol and salt-pickled, smoked, and salt-preserved foods, is consistent with a lower risk of many of the currently major cancers.

Dietary guidelines for prevention of cancer currently focus on the following:

1. High intake of fruits, vegetables, and whole grains;
2. Low intake of salt-cured, pickled, and smoked foods;
3. Low animal fat consumption (with <30 percent of total caloric intake of fat from all sources);
4. Moderate alcohol consumption;
5. Moderate caloric intake and physical activity to reduce obesity.

Public health nutrition includes a wide-ranging program of health education, and health promotion programs are effective primary prevention. Governments, especially departments of agriculture and finance, should take steps to ensure fruit and vegetable supply at low cost to the consumer. Decisions affecting the supply and pricing of foods interact with the interests of farmers, producers, agribusiness, transport and storage, as well as marketers of food to the consumer and the food production industry. However, they exist within a social context in which public opinion and purchasing power affect the type of produce offered and available.

NUTRITION IN PREGNANCY AND LACTATION

Pregnant women and lactating women are in effect feeding two people. They must eat enough to meet their own requirements at a time of increased need, as well as provide for the growth of the fetus and infant. During pregnancy, a woman needs 300 extra calories and an additional 20 g of protein per day. During breastfeeding, the mother needs an extra 500 calories and 20 g of protein above her dietary needs based on height, weight, and activity levels (see Table 8.6). The needs for (many) vitamins and minerals are similarly higher during pregnancy and lactation.

Adequate nutrition and weight gain during pregnancy are important for the development of the fetus and for the woman's health. Weight gain of a mother is a good predictor of birth weight of her infant. Since infant birth weight is a determinant of potential for survival and future development, the recommended weight gain for the mother is important to achieve. Women who are of normal weight for their height or slightly overweight women have better pregnancy outcomes than those who are underweight.

Both caloric intake and nutritional quality need to be considered. A pregnant woman needs to increase her intake of folic acid, iron, and certain trace elements. This can be done through supplementation as well as food fortification. Other factors influencing weight gain include smoking, strenuous physical work, and chronic illness. The social pressure on women to be thin may make it difficult for some to allow themselves adequate weight gain. For further discussion on nutritional needs during pregnancy, see Chapter 6.

Mineral and vitamin supplementation is also important in pregnancy and lactation; iron, iodine, selenium, folic acid, and vitamins A, B, and C are especially important when these additional needs cannot be met from diet alone. As noted in Chapter 6, the need for folic acid supplements precedes pregnancy in order to prevent neural tube defects in the fetus, since the neural tube develops in the first weeks after conception when a woman may not realize she is pregnant. Since not all pregnancies are planned and pre-pregnancy preparation is not common, fortification of flour with folic acid has been made mandatory in Canada and the United States in order to prevent neural tube defects. Food enrichment is discussed later in this chapter.

Lactating women should continue iron, folate, and multiple vitamin supplementation. Return to pre-pregnancy weight is often a major preoccupation but should be secondary to meeting the calorie, micronutrient, and fluid needs of lactation and energy requirements of caring for a newborn.

PROMOTING HEALTHY DIETS AND LIFESTYLES

Nutrition plays a central role in health of an individual and a population, making this a major function of public health. Nutrition is also a very personal matter and requires understanding on the part of each individual as well as the society which deals with food and agricultural policies, costs and infrastructure, cultural standards, food enrichment, and many other aspects of nutrition. Education of the public in nutrition is a part of creating a social awareness of nutrition and its role in health (U.S. Department of Agriculture).

Nationwide and community-based education to promote healthy eating patterns is an essential part of health promotion. Education for health can be promoted by departments or ministries of health, education, and agriculture, as well as by nongovernmental organizations.

Availability, quality, variety, and cost of foods depend on national policy, economics, and personal preferences, knowledge, and community patterns. A nutrition program should provide consumers with information regarding food selection and should consider why people make their choices.

Healthy eating behavior is part of all stages of life. Beginning in kindergarten and grade school, children can be taught about the nutritional values of foods. School curricula and teacher training should provide sufficient background in nutrition to provide children with guidance in food selection and promote desirable food habits. Child and adult education programs offer opportunities to emphasize the value of nutritious, balanced, and adequate diets for child and maternal health.

Dietary diversification programs need to take into account cultural beliefs about appropriate foods for different age groups, economic barriers to a healthy diet, and the availability of food types. The use of mass media to inform populations of food choices should not be overlooked.

Food support programs should be used to promote healthful nutrition. Community promotion of healthful nutrition can be implemented through schools and through health and social services especially for vulnerable groups, including women at the age of menopause, single-parent families, the elderly, the homeless, and the immunocompromised.

DIETARY GUIDELINES

Dietary guidelines provide a basis for individual and community education regarding healthful nutrition. The U.S. Department of Agriculture's (USDA) food pyramid is found at http://www.mypyramid.gov/index.html. The pyramid of recommended daily nutrition (Figure 8.3) provides a personal eating plan with the foods and amounts suitable for each person. The pyramid food patterns are designed for the general public ages 2 and over. They are not therapeutic diets for specific health conditions or for pregnancy or lactation. Those with a chronic health condition should consult with a health care provider to find a dietary plan appropriate for them. Special dietary requirements should be taken into account for adolescents, pregnant and lactating women, and the elderly. The recommended dietary guidelines will be revised periodically.

VITAMIN AND MINERAL ENRICHMENT OF BASIC FOODS

Health education regarding good nutrition practices alone is not sufficient to prevent significant manifestations (borderline or clinical) of vitamin and mineral deficiency

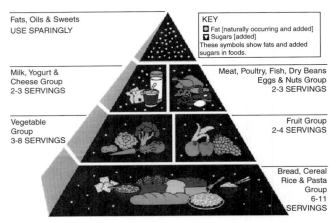

FIGURE 8.3   Food pyramid. U.S. Recommended daily servings of food. Source: Food pyramid provided for the public domain by the USDA Center for Nutrition Policy and Promotion (CNPP). 2005. [accessed May 9 2008]. Note: For a vegetarian diet pyramid see http://www.vrg.org/nutrition/adapyramid.htm [accessed May 9, 2008]

states in vulnerable sections of the population even in wealthy countries. It is a public health responsibility to ensure that all people get an adequate basic vitamin/mineral daily intake even if their food budget, or access to knowledge, is limited. This can best be assured through the appropriate vitamin and mineral enrichment of basic foods, such as bread, milk, and salt. Food enrichment to provide basic essential trace elements is standard practice widespread in North America and the United Kingdom. It must become a basic component of modern public health in nations where it has not yet been implemented.

In the United Kingdom, enrichment of white flour with two vitamins, thiamine and niacin, as well as two minerals, iron and calcium, has been compulsory since World War II. In Canada, as a result of a 1971 national nutrition survey, the federal government's Food and Drug Directorate passed regulations which made it illegal to sell milk products without vitamin A and D, bread without iron, vitamins B, and niacin, or salt without iodine, each to specified levels. In the United States, enriched bread is permitted and defined in FDA regulations, and enrichment is very widely practiced, with most breads enriched in keeping with these regulations.

In the early 1990s, studies in the United Kingdom and elsewhere showed the protective effects of folic acid, if taken prior to and during early pregnancy (periconceptional), against neural tube defects (NTDs), mainly anencephaly and spina bifida. This important finding raised new possibilities for prevention of birth defects by nutritional means. For planned pregnancies, this can be done through education and physician advice and taking of folic acid supplements prior to and during pregnancy. However, this is not practical for the majority of women as approximately half of all pregnancies are unplanned.

The alternative or supplementary approach is addition of folic acid to a common food such as bread and other flour-based foods to reach the population at risk to prevent these fatal or difficult to manage birth defects (Figure 8.4). As a result, folic acid fortification of enriched flour is mandatory in the United States, Canada, and Chile. Folic acid fortification of flour has been proposed in the United Kingdom, but not in most countries of Europe.

Controversy in Food Enrichment

Food enrichment is a well-established practice in the United States, Canada, and Latin America and is increasingly being adopted in the developing countries, but is controversial in Europe. Opposition to food enrichment is largely based on interpretation of food fortification as an invasion of personal rights and a lack of professional support.

Food fortification is a cost-effective method of ensuring nutrients to large segments of the population without requiring changes in food consumption. Fortification has been used since the 1920s in industrialized countries to restore micronutrients lost in food processing, especially the B vitamins, or adding elements absent in the environment such as iodine. Fortification, inexpensive and harmless to others, has played a key role in eradication or control of diseases associated with deficiencies in these vitamins.

Is food enrichment an infringement of personal rights, or a manifestation of "paternalism versus liberalism"? It is the most liberal countries, including Canada, the United States, that have led in food enrichment for the common good, whereas the tradition-bound, more centralistic, less liberal countries have persisted in ignoring this issue. Chlorination and fluoridation of community drinking water are relevant precedents with positive experience in the western world. The philosophy of health promotion is based not only on prevention of disease, but also on ensuring an optimal health environment particularly for the vulnerable groups in society (see Chapters 7 and 15).

Mandatory enrichment should be supported by mandatory labeling requirements, monitoring of levels of vitamin and mineral enrichment, and continuing health education on appropriate eating habits. National monitoring of the nutrition status of the population should be carried out in sentinel population groups through appropriate measurements (i.e., hematologic, biochemical, and anthropometric).

Although malnutrition is widespread throughout the world, it is most common in the form of micronutrient deficiencies. These are not only health problems as defined by the World Health Organization and UNICEF affecting well-being, but also they affect the economic growth potential of a population. The World Bank has increasingly accepted the importance of public health nutrition as a basis for infrastructure and economic growth, with nutrition interventions being among the most cost-effective public health measures. These include

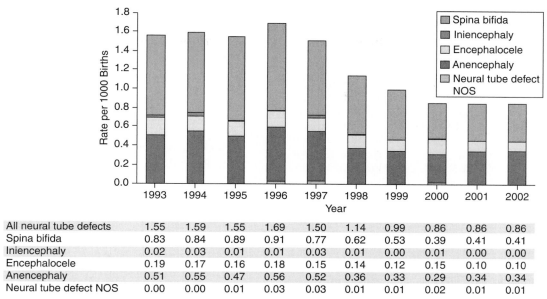

	1993	1994	1995	1996	1997	1998	1999	2000	2001	2002
All neural tube defects	1.55	1.59	1.55	1.69	1.50	1.14	0.99	0.86	0.86	0.86
Spina bifida	0.83	0.84	0.89	0.91	0.77	0.62	0.53	0.39	0.41	0.41
Iniencephaly	0.02	0.03	0.01	0.01	0.03	0.01	0.00	0.01	0.00	0.00
Encephalocele	0.19	0.17	0.16	0.18	0.15	0.14	0.12	0.15	0.10	0.10
Anencephaly	0.51	0.55	0.47	0.56	0.52	0.36	0.33	0.29	0.34	0.34
Neural tube defect NOS	0.00	0.00	0.01	0.03	0.03	0.01	0.01	0.02	0.01	0.01

FIGURE 8.4 Decline in neural tube defects in Canada before and following fortification of flour with folic acid. Source: De Wals, P., Fassiatou-Tairou, F., Van Allen, M. I., Uh, S. H., Lowry, B., Sibbald, B., Evans, J. A., Van den Hof, M. C., Zimmer, P., Crowley, M., Fernandez, B., Lee, N. S., Niyonsenga, T. 2007. Reduction in neural-tube defects after folic acid fortification in Canada. *New England Journal of Medicine*, 357:135–142. With permission.

nutrition education (including promoting breastfeeding), micronutrient fortification, micronutrient supplementation, food supplementation, food price subsidies, and control of parasitic diseases (World Development Report, 1993).

FOOD AND NUTRITION POLICY

A national food policy aimed at prevention of conditions related to nutritional deficiency or excess can only succeed with continuous intersectoral cooperation (Box 8.4). Regulation of food products by public health authorities involves national and provincial governments, as well as the local health authority. The food processing industry must be educated to compliance and self-regulation as part of a corporate culture working within adequate government regulation and educated consumer demand, with the competition of the marketplace serving a positive role. Active support is needed from government departments or ministries, food manufacturers and marketers, community leaders, health professionals, educators, women's groups, and the media in informing the public of the key issues.

The Evolution of a Federal Role

The U.S. federal government has been a force for improved nutrition for the past century. This included extension or outreach education programs as well as in

food and agricultural research and development and in food supplementation programs. The federal role in nutrition research began in 1887 with the establishment of a nutrition laboratory which was the forerunner of the National Institutes of Health. The USDA instituted food monitoring programs in 1893 and extension services to promote nutrition education in 1914.

In 1917, national concern for nutrition increased as a result of research findings that large numbers of draftees were being rejected from military service because of poor nutritional status (e.g., goiter). The U.S. Food Administration was established to supervise food supply during World War I. National initiatives in nutrition increased during the Great Depression (1929–1936) with use of surplus agricultural commodities for food relief. This initiative later developed into routine school lunch programs on a national level continuing to the present time.

In 1927, the U.S. Food and Drug Administration (FDA) was established. During the 1930s, the USDA conducted the first nationwide food consumption surveys, and in 1939 a federal food stamp program was established. In the 1950s and 1960s, food stamp and lunch programs were expanded. In 1972, the USDA established the Women, Infants, and Children Program (WIC) providing food supplements to needy pregnant women and children at risk. Programs to provide meals for the elderly were also developed and the school lunch program expanded to include breakfasts in needy communities.

Box 8.4 Food and Nutrition Responsibilities of National and International Authorities

International agencies Establish and promote the science and practice of food production, processing, distribution, and marketing for adequacy of nutrition internationally. Agencies concerned include the United Nations Food and Agriculture Organization (FAO), the World Health Organization, United Nations Development Programme, the World Bank and many bilateral governmental aid agencies; for example, USAID (United States), DFID (United Kingdom), CIDA (Canada), and many others.

National authority Establish national standards and regulatory requirements in such areas as food safety, labeling, fortification, and food content; responsible for regulating imported foods; establishing standards and enforcement of safety, packaging, and labeling; promotion of nutrition education and policies by government; coordination with the other departments of government (e.g., agriculture, trade, industry, and international agencies).

State/Provincial authority Legislate ordinances which reinforce national standards; licensing of food establishments; supervisory and regulatory functions. Coordination with agricultural and industry departments of government; nutrition education; liaison with private food production, processing, and marketing industry.

Local authority Supervision of local food producers; inspection of meat and dairy products and certain processed foods; monitor school lunch programs; provision of nutrition services and education; periodic inspection of public eating places. Promotion of nutrition education, breastfeeding, and healthy weaning practices. Provision of well-child care including growth monitoring and nutrition education. Good nutrition in schools and control of unhealthy and fattening foods in schools and other child activity settings.

A 1940 report on malnutrition in the United States led to a national nutritional conference and to recommended dietary allowances (RDAs). In 1941, the FDA issued enrichment standards for flour and bread with vitamin B complex and iron, and in 1942 addition of vitamins A and D to milk and margarine was mandated. Following World War II, in 1946 a national School Lunch Program was established. In 1965, as a part of the War on Poverty, the Head Start program was initiated, the Food Stamp Act was passed by Congress, followed by the Child Nutrition Act and School Breakfast Program in 1966. The National Health and Nutrition Examination Surveys carried out periodically since 1971–1974 (NHANES I), NHANES II (1976–1980), and NHANES III (1988–1991) and 2005–2006 are vital to documentation of the nutritional status of the U.S. population (Table 8.11).

The WIC supplementary food programs for women, infants, and children and the elderly were initiated in 1972. WIC provides federal grants to states for supplemental foods, health care referrals, and nutrition education for low-income pregnant, breastfeeding, and non-breast-feeding postpartum women, and to infants and children up to age 5 who are found to be at nutritional risk. WIC distribution sites are also active in promoting immunizations and education regarding pregnancy and child nutrition.

Labeling of foods with added nutrients or nutrient claims was mandated. Enrichment of infant cereals and formulas with vitamins B_1, B_2, B_6, and iron, with vitamin C in drinks and dessert powders, was required. In 1979, a Nutrition Policy Board and Surgeon General's publication *Healthy People* established nutritional objectives for the nation. In 1988, Dr. Everett Koop, the Surgeon General of the United States, issued the *Surgeon General's Report on Nutrition and Health*, a landmark document on nutrition in national health policy. Federal regulations were issued in 1993 under the Nutrition Labeling and Education Act, and in 1994 regulations were issued for labeling of meat and poultry. This was followed in 1996 by federal requirements for addition of folic acid to enriched flour, and in 1998 federal agencies recommended vitamin B supplements for older persons.

Nutrition Issues in Development Policies

Countries or regions that are poor and suffer from high rates of borderline or overt malnutrition must address the nutritional needs of the population as an essential part of planning for development. Economic development and nutritional status are directly related, as a malnourished population is not productive and cannot learn well. Efficient planning and policymaking require all governments to establish a cycle of assessment, intervention, and monitoring of the existing nutritional situation.

Piped or clean water and public sanitation and improved food safety help to reduce diarrheal disease which, especially in children, aggravates malnutrition. Microfinancing (small loans) and technical support for development of small agricultural, food production, and marketing businesses in rural and poor urban areas can improve the food supply and food quality.

Successful nutrition policy ultimately depends on consumer knowledge and access to quality foods at costs that are affordable. Fortification is an important supplement, but education of the producer, processor, marketer, and consumer is no less vital. Consideration must be given not only to the supply of food and nutrients but also to the short- and long-term effects of agricultural policies. Food security policies must aim to guarantee all families access to their minimum food requirements.

The cost to the health system of sickness associated with over- or undernutrition alone justifies expenditures on educational effort to reduce to burden of disease or disability associated with inappropriate nutritional status.

TABLE 8.11 NHANES Publications 2007

Lab 06 Heavy Metals (L06HM_c 2003-2004)	Immunization (2003-2004)
Lab 06 Urinary Mercury (L06UHG_c 2003-2004)	Lab 06 Ferritin and Transferrin Receptor (2003-2004)
Lab 06 Methylmalonic acid and Homocysteine (L06MH_c 2003-2004)	Lab 04 Volatile Organic Compounds in Blood and Water (1999-2000)
Lab 24 Polyfluorinated compounds (L06PFC_c 2003-2004)	Lab 39 Erythrocyte Protoporphyrin and Selenium (2003-2004)
Lab 26 Priority Pesticides and Organophosphate Pesticides (2001-2002)	Urine Collection (Pregnancy) (1999-2000)
Lab 28 Persistent Organochlorines, Furans, and Co-Planar PCBs (2001-2002)	Dermatology (DEX C 2003-2004)
Lab 26 Priority Pesticides and Organophosphate Pesticides (1999-2000)	NHANES III 27A — Surplus Sera Laboratory Component: Cystatin C
Lab 02 Hepatitis (2003-2004)	Lab 06 Nutritional Biochemistries (2001-2002)
Lab 02 Hepatitis (2001-2002)	Lab 06 Vitamin C (2003-2004)
Lab 02 Hepatitis (1999-2000)	Lab 24 Polyfluorinated Compounds (2003-2004)
Lab 24 Environmental Phenols (L24 C 2003-2004)	Lab 02 Hepatitis files for 1999-2000, 2001-2002, and 2003-2004
Lab 28 Polybrominated Diphenyl Ethers (2003-2004)	Lab 06 Nutritional Biochemistries (1999-2000)
Lab 05 Chlamydia and Gonorrhea (L05 B 2001-2002)	Lab 06 Nutritional Biochemistries (2001-2002)
Lab 05 *Chlamydia trachomatis* and *Neisseria gonorrhoeae* (2003-2004)	Depression Screener (2005-2006)
Lab 11 C-Reactive Protein (2001-2002)	Hepatitis C Follow Up Questionnaire (2005-2006)
Food Security (2003-2004)	Housing Characteristics (2005-2006)
Dietary Supplement Use (2003-2004)	Prostate Specific Antigen Follow Up Questionnaire (2005-2006)
The 1999-2000, 2001-2002, and 2003-2004	Prostate Specific Antigen Follow Up Questionnaire (2003-2004)
NHANES III 22A — Surplus Sera Antibody to Human Herpes Virus 8	Vision (2005-2006)
Lab 43 Vitamin B$_6$ (2003-2004)	Blood Lead and Cadmium (PBCD_D 2005-2006)
Varlab 1999-2000	Ferritin (2005-2006)
Dietary Supplement Use (2003-2004)	Blood Mercury (2005-2006)
Social Support (2003-2004)	Transferrin Receptor (2005-2006)
Alcohol Use (2003-2004)	Dioxins, Furans, and PCBS (2003-2004)
Drug Use (2003-2004)	Non-Dioxin Polychlorinated Biphenyls (2003-2004)
	Organochlorine Pesticides (2003-2004)
	Volatile Organic Compounds in Blood and Water (2001-2002)

Source: U.S. NCHS, NHANES, http://www.cdc.gov/nchs/about/major/nhanes/whatsnew.htm [accessed May 10, 2008]

The Role of the Private Sector and NGOs

The private sector grows, processes, and markets food and thus plays a vital role in nutrition and food safety. This includes the farmer producer, but equally the food processor/marketing industry. The private sector responds to the consumer but can also create demand for new products with a health component by labeling, advertising, and marketing practices.

Manufacturers of salt initiated iodination of salt in the United States, with widespread manufacture and marketing of iodized salt to prevent goiter in 1924. Baby food manufacturers enriched their products with vitamins and minerals without being required to by law. Where governments are reluctant to mandate food enrichment, private industry may already have made the decision to do so. This can be a valuable marketing tool, and where government regulations stipulate permissible enrichment, private industry fills an important public health function by pursuing enrichment policies.

Nongovernmental organizations have played important roles in nutrition especially in the form of women's organizations. This was seen in the early part of the twentieth century in rural women's organizations in North America which promoted better nutrition through education and consciousness-raising. Lobbying of government and private industry for constructive nutrition policies is also important for promoting better nutrition for vulnerable groups. Concern for nutrition of the elderly is an area where NGO support for nutritional assistance programs can be very effective. Women's magazines have also played an important role in disseminating information on healthful diets and raising interest in the relationship of nutrition and health.

The Role of Health Providers

Providers of care play a pivotal role in education of patients regarding appropriate nutrition during normal healthy periods and certainly during illness. The provider of health care has the trust of the patient and the professional knowledge to convey information to the patient in a time and manner that may be more effective than any other form of education. This includes the person at risk for development of ischemic heart disease, hypertension, and other diseases associated with nutritional factors, before and even more so after the disease process has become clinically apparent.

Malnutrition is associated with an increase in morbidity from infectious and chronic diseases. Conversely, disease, whether infectious or chronic, can cause a worsening of underlying malnutrition. Health providers need to be sensitive to these interactions and aware of their consequences. Prevention and management of infectious diseases can have an important effect on the nutritional status of children at risk for protein energy malnutrition, growth faltering, or failure to thrive. Identification of children at risk for malnutrition can result in additional attention and instruction of the mother regarding appropriate nutrition. Counseling and education for healthful nutrition according to age and needs is a key element of all patient care, as it is a public health issue.

NUTRITION MONITORING AND EVALUATION

Ongoing monitoring of the nutritional status of representative samples of the target population is an essential part of evaluating the health of a population This may be done through population samples or at selected sentinel centers. Nutrition monitoring systems, such as the U.S. National Health and Nutrition Examination Survey (NHANES), enable governments to monitor trends in food intake of individuals sampled and trends in nutritional status and general health of the population. Study results can direct policymakers in dietary standards and recommendations for manufacturers, state and local governments, and the general public (Table 8.12).

TABLE 8.12 Standards and Samples for Monitoring Nutrition Status of Population

Standards/ samples	Factors
Reference standards	International standards preferable to local standards to compare and monitor changes over time with optimal "ideal growth" population (gold standard) as a standard measure for comparison
Representative samples	Epidemiologically representative sample of total population at risk
Sentinel populations	Monitor population in particular program such as food supplement program (e.g., WIC) or attendees of a mother and child preventive care program
Sentinel centers	Selected places of contact with population in which monitoring can be added to basic program such as MCH centers, emergency departments, doctors' offices

Surveys of food purchases and consumption through direct observation, and interviews of family members can provide valuable nutrition monitoring data, as can interviews with food marketing networks (Table 8.13). Food supply and consumption tables can be calculated from total national data to estimate average consumption. Anthropometric measurements of height and weight, skin-fold, and body mass indexes provide valuable data on nutrition status, as do hematologic (iron, ferritin) and biochemical indicators (cholesterol, lipids, fasting blood sugar, vitamin levels). Large-scale surveys are costly, so sentinel center studies in selected representative locations may be more practical and can provide time trends in successive cross-sectional studies.

Standard Reference Populations

Anthropometric measures were first developed in the nineteenth century by Richer using skin-fold thickness as an index of fatness. During World War I, Matiega developed ways of measuring the composition of the human body to assess the physical efficiency of soldiers. Anthropometry includes composition of the body at the atomic, molecular, cellular, tissue, and whole body levels. Whole body studies involve measurements of body weight and stature, including skin folds, circumferences, and bone measures. These studies have clinical as well as public health importance. The association between weight loss and disease outcome has become a useful predictor of prognosis in individual patient care.

Reference populations are used for comparison purposes and have been found to reflect the "optimal" growth of children who are not excessively ill and who receive what is currently thought to be "good nutrition." The data may be presented in chart form and as percentiles. Recorded weights or heights at specific ages for the individual or group of children are seen graphically at different ages.

The U.S. National Center for Health Statistics growth patterns data derived from a white middle-class population sample is increasingly used internationally as the standard reference population since 1977, but these were updated and reissued in 2000 based on revised curves and BMIs for children, including about half breastfed children, with data from the NHANES survey and new statistical analysis.

The original NCHS standard was adopted and used by the World Health Organization since the late 1970s as an international standard reference population. The WHO concluded that child growth potential is similar across national and ethnic lines if nutrition is appropriate and intercurrent illness is not excessive. The standards were used internationally for several decades.

The NCHS/WHO reference standard was criticized, however, as the data used to construct the reference covering birth to 3 years of age came from a longitudinal study

TABLE 8.13 Measures for Monitoring Nutrition Status of Population

Assessment	Indicators
Indirect measures	
Food supply and consumption	Total available food supply by categories of foods divided by the total population
Education levels and cultural practices	Literacy, years of schooling, and especially female education levels are directly and indirectly related to adequacy of infant and child care mediated through improved family incomes and competency in dealing with the challenges of child care; traditional dietary patterns and individual choices
Household income and food purchase surveys	Per capita GNP, average household incomes, and other indicators of purchasing power are derived from census data and special surveys, including purchasing practices
Food balance sheets	An indirect measure of nutrition state by aggregating national data on amounts of food products adding imports, subtracting exports; the balance is converted to calories, deriving per capita consumption which are reasonable estimates of actual average intake
Infant mortality, LBW	Where these rates are high, it is assumed that malnutrition is widespread
Parasitosis	High rates of parasite infestation contribute to malnutrition
Direct measures	
Clinical assessment	Hospitalizations with primary or secondary diagnoses serving as index cases of extent of nutritional problems such as rickets, diabetes
Anthropometric measures and surveys	Weight and height for age and weight for height using WHO/NCHS standards now recommended for use as the international "gold standard"; skin-fold thickness of upper arm is also widely used to indicate protein/calorie status of individual child or the population of children; body mass index is another widely used measure; WHO issues new child health growth curves 2007
Biochemical measures and surveys	Examination of body fluids including blood, urine are measures of nutrients or indicators of complex metabolic processes; anemia is measured by hemoglobin, hematocrit, serum iron, or ferritin; thyroid function tests and iodine levels in urine measure iodine status; vitamin C in cells and serum, serum levels of vitamins B and D also are used; these are expensive tests not readily available for surveys; surrogate cheap field tests such as zinc protoporphyrin can serve as indicators of blood lead and/or anemia; routine fasting blood sugars on pregnant women can give an idea of prevalence of glucose intolerance
Dietary assessment	Dietary surveys in institutions or households are usually based on recall of intake over past 24 hours or 7 days, or on dietary records weighing food served and uneaten

of children of European ancestry from a single community in the United States. Further, measurements were taken every 3 months, which is inadequate to describe the rapid and changing rate of growth in early infancy. The statistical methods available at the time the NCHS/WHO growth curves were constructed were too limited to correctly model the pattern and variability of growth. WHO experts agree with the principle and importance of international growth standards but they undertook to develop new growth charts consistent with "best" health practices and that would show optimal child growth in all countries rather than describe how they grew in one country.

The WHO Multicentre Growth Reference Study (MGRS) produced the new international standards based on healthy children of different ethnic origins living under conditions likely to favor achievement of their full genetic growth potential. The mothers of the children selected for the construction of the standards engaged in fundamental health-promoting practices — namely, breastfeeding and not smoking. The new WHO standards were derived from an international study of 8440 children from a diverse set of countries, including Brazil, Ghana, India, Norway, Oman, and the United States. This provides a sample with ethnic or genetic variability in addition to cultural variation in how children are nurtured to strengthen their universal applicability. The new WHO standards explicitly identify breastfeeding as the biological norm and the breastfed child as the normative model for growth and development. They also include markers for six gross motor developmental milestones. WHO released the new growth charts in 2006, (under review by a joint committee of the U.S. CDC and the American Academy of Pediatrics) and they include:

1. Height and weight to calculate body mass index (BMI);
2. Knee height to predict adult stature;
3. Circumferential measures, including head, mid–upper arm, mid-thigh, and mid-calf;

4. Skin-fold measures, including biceps, triceps, subscapular area (1 cm below lower angle of scapula), suprailiac (mid-axillary 2 cm above iliac crest), thigh, and calf.

Such combinations as these are used to calculate the fat and fat-free content of body mass, which is useful for clinical assessment and special studies of groups. Fat-free body mass (FFM) indicates the muscle mass for nutrition studies or in patients in long-term nutrition follow-up and for teaching purposes.

Measuring Deviation from the Reference Population

Comparing the growth pattern of a study population to a standard provides a comparison similar to measuring blood cholesterol and comparing it to a normal range. A simple method of surveying a population for comparison with a standard, such as the NCHS/WHO reference population, is to record heights and weights of children presenting for immunization on a standard growth chart and compare the clustering of those observed to the international standard (Figure 8.5). This provides personnel in primary care a method of monitoring the population they serve. Table 8.14 lists classification systems used for monitoring child growth and defining malnutrition.

Z-scores are a widely used method of showing anthropometric data compared to a standard (Figure 8.6). A Z-score shows deviation from the mean in terms of standard deviations. Thus, the birth weight or growth pattern of children

may be compared to a norm in standard deviations and not absolute numbers. Observations representing 68 percent of a normal distribution curve are included within one standard deviation, while 95 percent are included within 1.96 standard deviations (see Chapter 3). Z-scores simplify graphic summation of the data in anthropometric studies.

FOOD QUALITY AND SAFETY

Food-borne disease is a major public health issue in all societies. In developing countries, it is one of the major causes of morbidity and mortality, and even in the most advanced countries great care must be taken by governments, the food industry, and the consumer to prevent food-borne disease. Both government and the food industry have a vested interest in providing nutritious and safe food to the public at reasonable cost. Other concerns of private industry are attractiveness of food and its packaging, long shelf life, low rates of loss of food or spoilage, contamination, and deterioration. Because the food producer and the processing/marketing industry wish to maximize their profit, self-policing is not sufficient, and a strong governmental regulatory agency is necessary to protect the public interest.

Fruits and vegetables contaminated with enteric pathogens are common causes of disease. Ingestion of contaminated products results in frequent outbreaks of gastroenteric disease, and contributes to the background level of diarrheal diseases not necessarily identified as outbreaks.

TABLE 8.14 Classification Systems Used Internationally for Monitoring Child Growth

Classifications	
Gomez (Weight-for-Age)	Median weight-for-height of those 90% or more of median considered normal; those 76–90% called mildly malnourished (1st degree); those 61–75% of median, moderately malnourished (2nd degree); those 60% or less as severely malnourished (3rd degree)
Waterlow (Weight-for-Height *and* Height-for-Age)	Height-for-age identifies past malnutrition, and weight-for-height indicates current malnutrition; uses NCHS/WHO reference population
NCHS/WHO	Weight and height for age, weight for height and head circumference standards based on NCHS survey finding of primarily white middle-class children from Iowa; this is accepted by WHO as an international standard
Shakir Arm Circumference	A well-nourished child has same arm circumference between 1–6 years; Shakir used tape measurement of arm circumference as a simple screening device for severely undernourished preschoolers; effective for case finding of moderate to severe malnutrition
WHO Malnutrition Classification, 1999	Moderate malnutrition — no edema, 2–3 SD below weight for age and height for age Severe malnutrition — edema, -3 SD below weight for age (severe wasting), and -3 SD below height for age (severe stunting)
WHO, 2007	New international growth curves for children up to 18 years

Source: *Management of Severe Malnutrition: A Manual for Physicians and Other Senior Health Workers*. Geneva: World Health Organization, 1999, http://whqlibdoc.who.int/hq/1999/57361.pdf and World Health Organization. *The WHO Child Growth Standards, 2008*, http://www.who.int/growthref/en [accessed June 13, 2008]

Note: SD = observed value minus median reference value divided by standard deviation for reference population.

Animals used for human food can be a source of transmission of communicable disease as well as chronic disease. This may be in the form of infectious diseases when food is contaminated with pathogenic organisms during growth, transportation, preparation, storage, or handling before consumption. Food poisoning or diarrhea can result from *Salmonella* or *Campylobacter* on chicken meat or eggs. Animals raised for food can harbor diseases such as brucellosis that can be transmitted to humans through consumption of unpasteurized milk products from infected animals. People can become seriously affected by toxins in food from animal origin, including meat, milk, and seafood. Diseases such as brucellosis, anthrax, and Rift Valley fever can be spread from infected domestic animals by direct contact of persons handling meat products. Table 8.15 lists various food-borne agents capable of affecting human health.

Animal husbandry practices are now part of public health concerns along with food processing, transport, and cooking. A large number of cases of dangerous *E. coli* infections resulting from contaminated meat in the United States, Japan, and many other countries show the potential

danger even in technologically advanced industrial countries. Use of animal materials rendered for animal feed is common practice instead of more expensive alfalfa. This raises serious health risks because of the *Salmonella* and *Campylobacter* they contain. In August 1997, tainted beef found in products of one company resulted in tens of thousands of tons of hamburger meat being condemned in the United States. A ban was imposed on British beef by the European Union in the late 1990s as a result of the bovine spongiform encephalopathy (BSE or mad cow disease) linked to Creutzfeldt-Jakob disease cases in humans (see Chapters 4 and 5) due to animal feed using contaminated animal parts.

Hazards associated with foods require constant vigilance on the part of government, the food industry, and consumers. Cooperation between government and food producers, processors, and manufacturers is one of mutual interest as well as a regulatory relationship. Private industry has a major role to play in improving food supply, and nutritional quality both for commercial and legal liability reasons. The consumer is also responsible for safe food handling practices, in terms of hygiene, storage, and

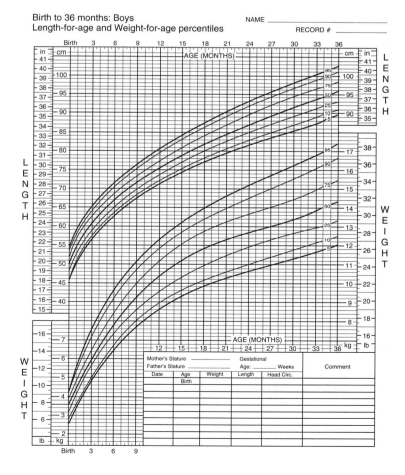

FIGURE 8.5 Standard Growth Curve — Boys Age 0–36 Months, Length for Age and Weight for Age, Centers for Disease Control, 2000. Source: Centers for Disease Control, NCHS, 2000. http://www.cdc.gov/nchs/about/major/nhanes/growthcharts/Powerpt.htm, January 11, 2007 [accessed May 9, 2008]

Weight-for-age GIRLS

Birth to 5 years (Z-scores)

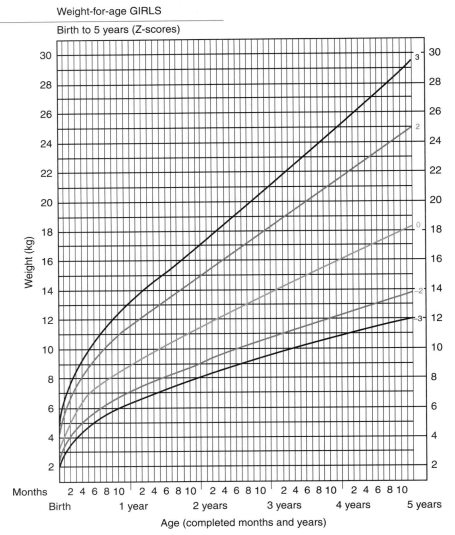

FIGURE 8.6 Recommended Growth Curve — Girls Age 0–5 Years, Weight for Age, WHO, 2007. Source: WHO, 2007. http://www.who.int/nutrition/media_page/Launching.pdf [accessed May 10, 2008]

usage. Sources of health-related problems exist at the points of production, processing, transporting, storing, sale, preparation, and serving of food products. Outbreaks of food-borne diseases serve as a reminder of the need for careful inspection of fresh produce, meat and dairy products, and processed foods on both a national and local basis.

Hazards that occur in many agricultural regions include contamination by pathogens. It is not uncommon in developing countries for foods to be exposed to pathogens by the use of untreated or inadequately treated sewage water for irrigation, malfunctioning sewage systems, or deliberate use of "nightsoil" for fertilization. Such fruits and vegetables eaten raw can expose the consumer to a host of infectious diseases transmitted via the fecal–oral cycle. Destruction and contamination of food by ruminants, rodents, or other pests during growth, harvesting, storage,

transportation, or processing of food can be major factors in safety and adequacy of food supply. Even in highly developed agriculture, contamination of food can occur at many stages in its processing. Imported and domestic food contamination caused 6647 outbreaks of foodborne diseases with 128,370 persons reported in 1998–2002. In 2006, CDC reported 1247 foodborne disease outbreaks (viral, bacterial, and chemical) in the United States with 25,659 cases (CDC, 2006).

A second major concern is the hazard posed by ingestion of chemicals used in food production. Modern agriculture has come to rely on extensive use of chemical fertilizers and pesticides. These accumulate in the soil over time and form a residue on the produce grown. Pesticide residues on orchard and field crops and in processed foods include chlorinated hydrocarbons, dieldrin, and

TABLE 8.15 Food-Borne Disease Agents, Organisms, Chemicals, and Vectors

Agent	Organism/Chemical	Food vector
Prions	Bovine spongiform encephalitis (BSE)	Food of animal origin for cattle transmitting Creutzfeldt-Jakob disease (CJD) to humans
Viruses	Poliomyelitis, hepatitis A and E	Vegetables, fruit irrigated with sewage; shellfish
Bacteria	*Campylobacter, Clostridia, E. coli, Salmonella, Vibrio cholerae, Listeria*	Raw and processed foods contaminated with human, bird, or animal excreta
Molds	Mycotoxins, aflatoxins, ochratoxins	Stored nuts and cereals
Protozoa	Amoebae, *Giardia, Cryptosporidia*	Vegetables, fruit, water, unpasteurized milk
Helminths, parasites	*Ascaris, Fasciola, Taenia, Trichinella, Trichura, Opisthorchiasis*	Vegetables, undercooked meat, raw fish from areas with contaminated water; untreated sewage or irrigation water
Chemicals, agricultural, and veterinary	Pesticides, fungicides, fertilizers, hormones, antibiotics	Contaminated food from abuse of agents; widespread antibiotic use may engender allergies and organism resistance
Heavy metals	Mercury, cadmium, tin	Fish, seafood from water contaminated by industrial waste
Radionuclides	Radioactive materials	Foods contaminated from atmospheric fallout

Source: Adapted from International Conference on Nutrition, Food and Agriculture Organization, World Health Organization, 1992.

DDT. Pesticides can cause serious problems of toxic exposure to farm workers, contamination of rain runoff to surface waters or groundwater, and contamination of inadequately washed farm produce. Safety requires consumer self-protection by selection of foods and careful washing before use, but education in seeking effective alternatives and reducing use of pesticides is important primary prevention (see Chapter 9).

Food additives are substances used in production, processing, treatment, packaging, transportation, or storage of food. These include simple additives such as salt, baking soda, vanilla, or yeast, but also complex ones. Additives are used in food to maintain product consistency, to improve nutritional value (vitamins and mineral enrichment), to maintain palatability and retard spoilage, to provide leavening and prevent acidity/alkalinity, and to enhance flavor or desired color. All additives are subject to FDA regulation, and some have been banned for harmful effects, including cyclamates (sweetener), cobalt salts (beer), tar derivative used for food coloring, some pesticides, fungicides, and herbicides, and polyvinyl chloride in plastic containers.

Growth enhancers, including antibiotics and hormones to promote rapid and vigorous animal growth and milk production, can adversely affect milk and meat products. Prophylactic use of antibiotics may contribute to increasing the numbers of resistant organisms which can potentially endanger public health.

Food additives are substances intentionally added to foods to modify their taste, color, texture, nutritive value, appearance, and resistance to deterioration. Some have been in use for a long time, such as sugar, salt, and spices. Natural additives are important in processed foods; others have positive public health benefit, such as in added vitamins and minerals.

Some additives are of concern due to their potential toxic or carcinogenic effects, and more so when these additives are for cosmetic value only. In 1958, the Food Additives Amendment was added to the U.S. Food, Drug, and Cosmetics Act. This amendment requires food manufacturers to comply with FDA rulings on food additives. In the same year, the famous Delaney Clause banned any additive that has been shown to be carcinogenic in any amount from use as a food additive. As a result, there was large-scale testing of additives. Those additives already in use were categorized as "generally regarded as safe" or GRAS. Some of these were subsequently tested and found to be carcinogenic and removed from general use by FDA ruling (Box 8.5).

Methods of preserving foods, some as old as civilization itself, were developed to protect the quality of foods and ensure supply beyond the immediate period of harvesting of crops or foods of animal origin. They are meant to prevent microbial or chemical deterioration of food, without causing harmful effects to the safety, nutritional, or aesthetic value or shelf life of the foods. Techniques developed to protect and preserve the quality of food beyond the immediate period following production are vital to any population group. They are meant to prevent microbiological and chemical deterioration, without harmful

Box 8.5 Food Additives Removed from Use by FDA Ruling

1. **Cyclamates** artificial sweeteners banned in 1970;
2. **Cobalt salts** used to improve stability of beer foam, but associated with heart attacks;
3. **Polyvinyl chloride** use in plastic liquor bottles, banned when this carcinogen was found in liquor;
4. **Food coloring** some banned because they contained coal tar derivatives shown to be carcinogenic;
5. **Pesticides, fungicides, and herbicides** EPA found many of these to be carcinogenic; some were banned in the United States, but continue to be produced and exported.

TABLE 8.16 Major Food Processing and Preservation Technologies

Drying	Radiant (solar), spray, air, freeze- or vacuum-drying
Fermentation	Yogurt, cheese, soy sauce, wine and beer, vinegar, sauerkraut
Chemical treatment	Salting, pickling, sugaring, smoking, fumigating, or by food additives
Thermal treatment	Cooking (many forms), blanching, pasteurizing, aseptic filling with high-temperature short-time, retorting
Cold treatment	Refrigerating, freezing
Radiation	Microwave, ionizing

chemical, toxic, or carcinogenic effects. Some of the major categories of food preservation technology are shown in Table 8.16.

While government and industry have major roles in ensuring safe food, final responsibility rests with the consumer. Public education, knowledge, attitudes, and practices determine the safety of food served to the family. The consumer in a market economy can influence and even determine which goods to purchase, creating demands felt in the marketplace that guide producers and processors to change the content, packaging, labeling, advertising, and prices. Consumers can affect the market by refusing to purchase goods known to be problematic or lacking nutritional value, an effective part of quality control. Consumer advocacy can also play a role in food quality, as manufacturers and governments are sensitive to public expression of concern through the media on food quality. The Food and Drug Act setting up regulatory mechanisms in the United States was enacted in response

to newspaper articles and novels (such as *The Jungle*, by Upton Sinclair) exposing the food quality in the United States in 1906. Consumer boycotts have also made producers very sensitive to public concerns.

Concern for public health aspects of food have increased with serious incidents of food poisoning from new variants of *E. coli* and *Salmonella* being resistant to available antibiotics, and as large sectors of the population are vulnerable to infection, especially immunocompromised persons such as HIV carriers and splenectomized patients. In the summer of 1997, 25 million pounds of ground beef were recalled in the United States because of suspected contamination with *E. coli* O157:H7 (FDA, 1998).

In the 1990s, despite safe water supplies, good hygiene, and food technology, food poisoning increased in many industrialized countries. Current estimates suggest that between 5 and 10 percent of the population of industrial countries are affected annually. In the United States, there are between 3.3 and 12.3 million cases and up to 3900 deaths per year from food poisoning. Food-borne and waterborne diseases are perhaps the most important public health problems of developing countries, with 1.5 billion cases and 3 million deaths annually from diarrheal disease. Consumer resistance to use of irradiation of food remains high, despite its being widely regarded as probably the safest and most reliable method of protecting food supplies from contamination. This resistance may decline as more instances of food contamination are reported in the public media.

Irradiation to improve food safety, genetic engineering to produce food crops of higher quality, and freedom from need for pesticides and chemical fertilizers are producing a new revolution in agriculture and food safety. They are also controversial and under critical review for safeguards and labeling requirements to protect consumer freedom to choose. Nevertheless, these are important new technologies which will be vital to meeting the needs of developing countries, as well as those of the industrial world in promoting good nutrition and food safety.

NUTRITION AND THE NEW PUBLIC HEALTH

Population and individual health are closely related to nutrition status. The individual chronically ill patient may experience a serious decline in his or her condition as a result of inadequate daily nutrition. The cost of hospitalization alone for such persons may outweigh the support services that should assist that person in maintaining adequate nutrition. Overt or subclinical deficiency conditions can cause widespread health damage if national health policies fail to take into account current knowledge and practices for their prevention. Failure to implement such policies engenders undernutrition with important health consequences. The success of developed countries in controlling micronutrient deficiency conditions has been achieved partly by social and economic development and improved pediatric care, but also by enlightened food policy with food fortification. Adoption of population-oriented food policies is crucial for developing countries to cope with basic undernutrition for large vulnerable parts of the population.

The complexity of food and nutrition policy is demonstrated in the following figure (Figure 8.7). The interaction of food supplies and prices with governmental policy and other related factors impinge on individual family food purchases and individual consumption. Education and culture play important roles as do the economics of food and global factors such as use of grains for fuel production that causes a severe shortage of corn and grain for animal production and for even more basic foods.

Similarly, overnutrition and its associated diseases, such as diabetes, cardiovascular diseases, and cancer, are the cause of widespread pathology and use of costly health services, whereas education and other programs of intervention can have important preventive effects. These cannot be addressed solely as individual patient problems, but require a parallel population approach which involves a broad set of food and nutrition policies. The New Public Health takes a holistic approach to prevention

and management of disease, as well as promotion of optimal health. Nutrition is central to that task, requiring well-planned interventions at the national, community, and individual level.

SUMMARY

While much of the glamour and stress in public health achievements over the past century have involved sanitation and control of communicable diseases, the contribution of improved nutrition and food safety to better health has been enormous. Mortality from all causes, including infectious diseases, began to decline in the seventeenth century, well before organized infectious disease control emerged, and that was, in part, a result of the agricultural revolution and improved food production.

Today, nutrition is one of the major public health issues of developed countries. The problems are largely of overnutrition or inappropriate balance and excessive caloric intake. In developing countries, mass nutritional deprivation is perhaps the most important public health problem, resulting in wide-scale deficiencies of calories, protein, and essential vitamins and minerals. Mid-level developing countries are experiencing "nutrition transition." Cardiovascular mortality rapidly increased during the 1940s to the 1960s in the industrialized countries

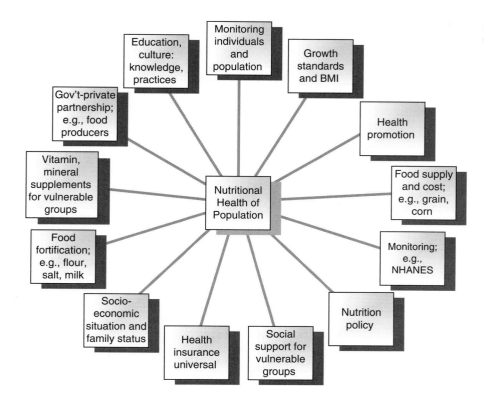

FIGURE 8.7 Factors in nutritional health of a population

due to excessive animal fat intake as living standards increased for part of the population and has declined rapidly since with more awareness of healthy diet and reduced cigarette smoking as well as with improved medical care (see Chapter 5). In developing countries, however, rapid nutritional change is affecting large sections of the population, while undernutrition remains widespread for the majority.

Since the 1960s, the people in most industrialized countries have paid a great deal of attention to their personal nutrition, prompted by concern in the media and the health professions. The dietary approach to preventive health care has contributed to the dramatic drop in death rates from coronary heart disease, stroke, and cancer of the stomach in most western countries.

The success of folic acid fortification of flour in reducing neural tube birth defects has renewed interest in the issue of food fortification. WHO has come forward with important new initiatives in this field and in setting new growth standards for application internationally. The rapid growth of the problem of obesity and diabetes is one of the challenges to public health in the coming decades.

Monitoring nutrition status and developing national policies to assure adequate and high-quality food are major governmental functions. The economic, agricultural, and marketing systems all have roles to play in ensuring population nutritional health. Food fortification and promotion of healthful dietary habits are vital to prevent disorders due to deficient and excess nutrition. The supply and quality of food for growing populations with rising expectations will depend on new science and technology such as genetic engineering and food irradiation, despite current controversy and manipulation of legitimate public concerns. The clinician deals with the individual patient and public health with the individual and the community. This duality is the New Public Health in which nutrition plays a central role.

ELECTRONIC RESOURCES

Centers for Disease Control Growth Standards. http://www.cdc.gov/growthcharts/who_standards.htm [accessed May 11, 2008]

Centers for Disease Control. NCHS, 2000. *Standard Growth Curve — Boys Age 0–36 Months, Length for Age and Weight for Age*, http://www.cdc.gov/nchs/about/major/nhanes/growthcharts/Powerpt/htm [accessed May 9, 2008]

Centers for Disease Control. 2006. *Summary Statistics for Foodborne Outbreaks, Number of Foodborne Outbreaks by Etiology*, http://www.cdc.gov/foodborneoutbreaks/documents/2006_Line_List/2006_Line_List.pdf [accessed June 13, 2008]

Food and Agriculture Organization (FAO) home page. http://www.fao.org/ [accessed May 10, 2008]

Health Canada, Review of Canada's Policies on the Addition of Vitamins and Minerals. April 1999. Canadian Food Inspection Agency, http://www.hc-sc.ca/food-aliment/ [accessed May 11, 2008]

Institute of Medicine. National Academies of Sciences. Food and Nutrition. 2008. http://www.iom.edu/CMS/3708.aspx [accessed May 10, 2008].

Institute of Medicine. Nutrition Board. *Dietary Reference Intakes: 2004*, http:///www.iom.edu/Object_File/master/21/372/0.pdf and http://www.iom.edu/?id=18495&redirect=0 [accessed June 13, 2008]

National Center for Health Statistics. NHANES. 2008, http://www.cdc.gov/nchs/otheract.htm [accessed May 10, 2008]

National Center for Health Statistics. National Health and Nutrition Survey III. Body Measurements (Anthropometry), http://www.cdc.gov/nchs/data/nhanes/nhanes3/cdrom/nchs/manuals/anthro.pdf [accessed May 10, 2008]

OECD-FAO. 2007. Agricultural Outlook 2007–2016. Organization for Economic Cooperation and Development and Food and Agriculture Organization of the United Nations. Paris.

U.S. Department of Agriculture. Nutrient data for specific foods, http://www.ars.usda.gov/ba/bhnrc/ndl and http://fnic.nal.usda.gov/nal_display/index.php?info_center=4&tax_level=2&tax_subject=256&level3_id=0&level4_id=0&level5_id=0&topic_id=1342&placement_default=0 [accessed May 10, 2008]

U.S. Department of Agriculture. 2008. National Agricultural Library home page, http://www.nal.usda [accessed May 9, 2008]

USDA's "Mypyramid.gov," http://www.mypyramid.gov/index.html [accessed May 9, 2008]

U.S. Department of Health and Human Services and Department of Agriculture. 2005. *Dietary Guidelines for Americans*, http://www.health.gov/DietaryGuidelines/dga2005/document/ [accessed June 13, 2008]

U.S. Environmental Protection Agency (EPA) home page, http://www.epa.gov/ [accessed May 9, 2008]

U.S. NHANES. http://www.cdc.gov/nchs/nhanes.htm [accessed May 9, 2008]

U.S. Women Infants and Children, http://www.fns.usda.gov/wic/ [accessed May 9, 2008]

World Health Organization, Child growth standards, http://www.who.int/childgrowth/en/ and http://www.who.int/growthref/en/ [accessed May 9, 2008]

World Health Organization. Vitamin and mineral nutrition information systems, http://www.who.int/nutrition/databases/micronutrients/en/ [accessed May 9, 2008] and http://www.who.int/nutrition/topics/vad/en/ [accessed May 9, 2008]

World Health Organization. 2001. *Iron Deficiency Anemia: Assessment, Prevention, and Control. A Guide for Program Managers*. WHO, Geneva, http://www.who.int/nutrition/databases/micronutrients/en/index.html

World Health Organization. 2007. WHO child growth standards: Length/Height-for-age, weight-for-age, weight-for-length, weight-for-height and body mass index-for age: Methods and development, http://www.who.int/bookorders/anglais/detart1.jsp?sesslan=1&codlan=1&codcol=15&codcch=660# [accessed May 9, 2008]

World Health Organization. 2008 Iron deficiency anemia, http://www.who.int/nutrition/topics/ida/en/index.html [accessed May 9, 2008]

RECOMMENDED READINGS

American Academy of Pediatrics. 2003. Section on Breast Feeding and Committee on Nutrition. Prevention of Rickets and Vitamin D Deficiency: New Guidelines for Vitamin D Intake, *Pediatrics*, 111:908–910.

American Heart Association. *Heart Disease and Stroke Statistics: 2007 Update*. Dallas, Texas.

Anderson, S. H., Vickery, C. A., Nicol, A. D. 1986. Adult thiamine requirements and the continued need to fortify processed cereals. *Lancet*, 2:85–89.

Anderson, M., de Benoist, B., Darnton-Hill, I. and Delange, F. 2007. Iodine deficiency in Europe: a continuing public health problem. WHO/UNICEF, Geneva and New York.

Allen, L., Bruno de Benoist, B., Dary, O., Hurrell, R. 2006. *Guidelines on Food Fortification with Micronutrients*. Food and Agricultural Organization of the United Nations. Geneva: World Health Organization.

Backstrand, J. R. 2002. The history and future of food fortification in the United States: A public health perspective. *Nutrition Reviews*, 60:15–26.

Centers for Disease Control. 1999. Achievements in public health, 1900-1999: Safer and healthier Foods. 1999. *Morbidity and Mortality Weekly Report*, 48:905–913.

Centers for Disease Control. 2002. Iron deficiency — United States, 1999–2000. *Morbidity and Mortality Weekly Report*, 51:897–899.

Centers for Disease Control. 2005. Use of dietary supplements containing folic acid among women in childbearing years — United States. *Morbidity and Mortality Weekly Report*, 54:955–958.

DeMaeyer, E. M., Dallman, P., Gurney, J. M., Hallberg, L., Sood, S. K., Srikantia, S. G. 1989. *Preventing and Controlling Iron Deficiency Anemia Through Primary Health Care*. Geneva: World Health Organization.

De Wals, P., Fassiatou, T., Van Allen, M., et al. 2007. Reduction in neural tube defects after folic acid fortification in Canada. *New England Journal of Medicine*, 357:135–142.

Dunn, J. T. (editorial). 1992. Iodine deficiency — The next target for elimination? *New England Journal of Medicine*, 326:267–268.

Dwyer, J., Picciano, M. F., Raiten, D. J., Members of the Steering Committee. 2003. Estimation of usual intakes — what we eat in America — NHANES. *Journal of Nutrition*, 133:609S–636S.

Flegal, K. M., Graubard, B. I., Williamson, D. F., Gail, M. H. 2005. Excess deaths associated with underweight, overweight, and obesity. *Journal of the American Medical Association*, 293:1861–1867.

Foster, P. 1992. *The World Food Problem: Tackling the Causes of Undernutrition in the Third World*. London: Boulder Adamantine Press.

Gartner, L. M., Greer, F. R. 2003. Prevention of rickets and vitamin D deficiency: New guidelines for vitamin D intake. *Pediatrics*, 111:908–910.

Gorstein, J., Akre, J. 1988. The use of anthropometry to assess nutritional status. *World Health Statistics Quarterly*, 41:48–58.

Hanley, D. A., Davison, K. S. 2005. Vitamin D insufficiency in North America. *Journal of Nutrition*, 135:332–337.

Hetzel, B. S. 2005. Towards the global elimination of brain damage due to iodine deficiency — the role of the International Council for Control of Iodine Deficiency Disorders. *International Journal of Epidemiology*, 34:762–764.

Institute of Medicine. 1998. *Dietary Reference Intakes for Thiamin, Riboflavin, Niacin, Vitamin B₆, Folate, Vitamin B₁₂, Pantothenic Acid, Biotin, and Choline*. Washington, DC: National Academy Press.

Institute of Medicine (IOM). 1998. *Prevention of Micronutrient Deficiencies: Tools for Policymakers and Public Health Workers* (Christopher P. Howson, Eileen T. Kennedy, and Abraham Horwitz, eds.). Washington, DC: National Academy Press.

Institute of Medicine. 2001. *Dietary Reference Intakes for Vitamin A, Vitamin K, Arsenic, Boron, Chromium, Copper, Iodine, Iron, Manganese, Molybdenum, Nickel, Silicon, Vanadium, and Zinc*. Washington, DC: National Academy Press.

Institute of Medicine. 2004. *Dietary Recommended Intakes 2004*, http://www.iom.edu/object.file/master/21/372/o.pdf and http://www.iom.edu/object-file/master/7/300/webtablemacro.pdf [accessed May 10, 2008]

King, D. E., Mainous, A. G., III, Geesey, M. E. 2007. Turning back the clock: Adopting a healthy lifestyle in middle age. *American Journal of Medicine*, 120:598–603.

Mokdad, A. H., Ford, E. S., Bowman, B. A., Dietz, W. H., Vinicor, F., Bales, V. S., Marks, J. S. 2003. Prevalence of obesity, diabetes, and obesity-related health risk factors, 2001. *Journal of the American Medical Association*, 289:76–79.

Nathoo, T., Holmes, C. P., Ostry, A. 2005. An analysis of the development of Canadian food fortification policies: The case of vitamin B. *Health Promotion International*, 20:375–382.

National Academy of Sciences — National Research Council. 1989. *Recommended Dietary Allowances*, Tenth Edition. Food and Nutrition Board. Washington, DC: National Academy Press.

National Institutes of Health. National Heart, Lung, and Blood Institute. 1989. *Clinical Guidelines on the Identification, Evaluation, and Treatment of Overweight and Obesity in Adults — The Evidence Report*, www.nhlbi.nih.gov/guidelines/obesity/ob_gdlns.htm [accessed May 9, 2008]

Ogden, C. L., Carroll, M. D., Curtin, L. R., McDowell, M. A., Tabak, C. J., Flegal, K. M. 2006. Prevalence of overweight and obesity in the United States, 1999–2004. *Journal of the American Medical Association*, 295:1549–1555.

Otten, J. J., Hellwig, P. J., Meyers, L. D. (eds.). 2006. *Dietary Reference Intakes: The Essential Guide to Nutrient Requirements*. Institute of Medicine. Washington, DC: National Academies Press.

Park, Y. K., Sempos, C. T., Barton, C. N., Vanderveen, J. E., Yetley, E. A. 2000. Effectiveness of food fortification in the United States: The case of pellagra. *American Journal of Public Health*, 90:727–738.

Pouessel, G., Bouarfa, K., Soudan, B., Sauvage, J., Gottrand, F., Turck, D. 2003. [Iodine nutritional status and risk factors for iodine deficiency in infants and children of the French North Department] (in French). *Archives Pediatrie*, 10:96–101.

Rajakumar, K., Greenspan, S. L., Thomas, S. B., Holick, M. F. 2007. Solar ultraviolet radiation and vitamin D: a historical perspective. *American Journal of Public Health*, 97:1476–1754.

Rose, D., Oliveira, V. 1997. Nutrient intakes of individuals from food-insufficient households in the United States. *American Journal of Public Health*, 87:1956–1961.

Scrimshaw, N. 1995. The new paradigm of public health nutrition. *American Journal of Public Health*, 85:622–624.

Sidel, V. 1997. The public health impact of hunger. *American Journal of Public Health*, 87:1921–1922.

World Health Organization/Food and Agricultural Organization. 2005. *Vitamin and Mineral Requirements in Human Nutrition*, Second Edition. Geneva: World Health Organization.

BIBLIOGRAPHY

Barkan, I. D. 1985. Industry invites regulation: The passage of the Pure Food and Drug Act of 1906. *American Journal of Public Health*, 75:18–26.

Beghin, I., Cap, M., Dujardin, B. 1988. *A Guide to Nutritional Assessment*. Geneva: World Health Organization.

Bentley, T. G. K., Willett, W. C., Wenstein, M. C., Kuntz, K. M. 2006. Population-level change in folate intake by age, gender, and race/ethnicity after folic acid fortification. *American Journal of Public Health*, 96:2040–2047.

Calvo, M. S., Whiting, S. J., Barton, C. N. 2005. Vitamin D intake: A global perspective of current status. *American Society for Nutritional Sciences*, 135:310–316.

Cantorna, M. T., Zhu, Y., Froicu, M., Wittke, A. 2004. Vitamin D status, 1,25-dihydroxyvitamin D3, and the immune system. *American Journal of Clinical Nutrition*, 80:1717S–1720S.

Centers for Disease Control. 1989. CDC criteria for anemia in children and child bearing-aged women. *Morbidity and Mortality Weekly Report*, 38:400–404.

Centers for Disease Control. 1994. Daily dietary fat and total food energy intakes — Third National Health and Nutrition Examination Survey, Phase 1, 1988–1991. *Morbidity and Mortality Weekly Report*, 43:116–117, 123–125.

Centers for Disease Control. 1994. Epidemic neuropathy — Cuba 1991–1994. *Morbidity and Mortality Weekly Report*, 43:183–189.

Compston, J. E. 1998. Vitamin D deficiency: Time for action. *British Medical Journal*, 317:1466–1467.

de Benoist, B., et al. (eds.). 2004. *Iodine Status Worldwide. WHO Global Database on Iodine Deficiency*. Geneva: World Health Organization.

Delange, F., Burgi, H. 1989. Iodine deficiency disorders in Europe. *Bulletin of the World Health Organization*, 67:317–325.

de Onis, M., Blössner, M. 2003. The World Health Organization Global Database on Child Growth and Malnutrition: Methodology and applications. *International Journal of Epidemiology*, 32:518–526.

de Onis, M., Garza, C., Habicht, J. P. 1997. Time for a new growth reference. *Pediatrics*, 100:E8.

de Onis, M., Garza, C., Onyango, A. W., Borghi, E. 2007. Comparison of the WHO child growth standards and the CDC 2000 growth charts. *Journal of Nutrition*, 137(1):144–148.

Doet, E. S. et al. 2008. Current micronutrient recommendations in Europe: towards understanding their differences and similarities. *European Journal of Nutrition*, 47 (Suppl 1):17–40.

Dunn, J. T., Van der Haar, F. 1990. *A Practical Guide to the Correction of Iodine Deficiency*. Amsterdam: International Council for the Control of Iodine Deficiency Disorders.

Editorial. 1986. Prevention and control of iodine deficiency disorders. *Lancet*, 2:433–434.

Edejer, T. T., Aikins, M., Black, R., Wolfson, L., Hutubessy, R., Evans, D. B. 2005. Achieving the millennium development goals for health: Cost effectiveness analysis of strategies for child health in developing countries. *British Medical Journal*, 19:331(7526):1177.

Food and Agriculture Organization. 1987. The Fifth World Food Survey. Rome: Food and Agriculture Organization of the United Nations.

Food and Agricultural Organization, 2006. *The State of Food Insecurity in the World*. Rome.

Gartner, L. M., Greer, F. R. 2003. Prevention of rickets and vitamin D deficiency: New guidelines for vitamin D intake. *Pediatrics*, 111:908–910.

Gartner, L. M., Greer, F. R., and AAP Section on Breastfeeding and Committee on Nutrition. 2003. Prevention of rickets and vitamin D deficiency: new guidelines for vitamin D intake. *Pediatrics*, 111:908–910, http://pediatrics.aappublications.org/cgi/reprint/113/1/180.pdf

Gennari, C. 2003. Calcium and vitamin D nutrition and bone disease of the elderly. *Gerontology*, 49:273–278.

Gerstner, H. 1997. Vitamin A — Functions, dietary requirements and safety in humans. *International Journal of Vitamin and Nutrition Research*, 67:71–90.

Gibson, R. S. 1990. *Principles of Nutritional Assessment*. New York: Oxford University Press.

Grote, J. 2006. Vitamin D deficiency. *Journal of the American Medical Association*, 295:1002.

Health Protection Branch, Health and Welfare Canada. 1974. *Addition of Vitamin and Mineral Nutrients to Foods*, 426:1–4. Ottawa: Health and Welfare Canada.

Henkel, J. 1998. Irradiation: A safe measure for safer food. *FDA Consumer*, 32:12–17.

Hintzpeter, B., Mensink, G. B., Thierfelder, W., Müller, M. J., Scheidt-Nave, C. 2007. Vitamin D status and health correlates among German adults. *European Journal of Clinical Nutrition*, May 30 [Epub ahead of print].

Holick, M. F. 2004. Sunlight and vitamin D for bone health and prevention of autoimmune diseases, cancers, and cardiovascular disease. *American Journal of Clinical Nutrition*, 80:1678S–1688S.

Holick, M. F. 2006. Resurrection of vitamin D deficiency and rickets. *Journal of Clinical Investigation*, 116:2062–2072.

Krebs-Smith, S. M., Cook, D. A., Subar, A. F., Cleveland, L., Friday, J. 1995. US adults' fruit and vegetable intakes, 1980 to 1991: A revised baseline for *Healthy People 2000* objectives. *American Journal of Public Health*, 85:1623–1629.

Lynch, M., Painter, J., Woodruff, R., Braden, C. 2006. Surveillance for foodborne-disease outbreaks, United States, 1998–2002. *Morbidity and Mortality Weekly Report*, 55(SS10):1–34.

Marshall, J. R. (editorial). 1995. Improving American diet — Setting public policy with limited knowledge. *American Journal of Public Health*, 85:1609–1611.

McKenna, M. J. 1992. Differences in vitamin D status between countries in young adults and the elderly. *American Journal of Medicine*, 93:69–77.

Mertz, W. 1981. The essential trace elements. *Science*, 213:1332–1338.

Morley, J. E., Mooradian, A. D., Silver, A. J., Heber, D., Alfin-Slater, R. B. 1988. Nutrition in the elderly. *Annals of Internal Medicine*, 109:890–904.

Müller, O., Krawinkel, M. 2005. Malnutrition and health in developing countries. *Canadian Medical Association Journal*, 173:279–286.

National Academy of Sciences — National Research Council. 1968. *Recommended Dietary Allowances*, Seventh Edition. Washington, DC: Food and Nutrition Board.

National Center for Health Statistics. 2006. *Chartbook on Trends in the Health of Americans. Health, United States, 2006*. Hyattsville, MD: Public Health Service.

National Center for Health Statistics, *Health, United States 2007, with Chartbook on Trends in the health of Americans*. Hyattsville, MD, http://www.cdc.gov/nchs/data/hus/hus07.pdf

National Osteoporosis Foundation, 2003, *Clinician's Guide to Prevention and Treatment of Osteoporosis*. Washington, DC.

National Research Council. 1989. *Diet and Health: Implications for Reducing Chronic Disease Risk*. Washington, DC: National Academy Press.

Nishida, C., Uauy, R., Kumanyika, S., Shetty, P. 2004. The Joint WHO/FAO expert consultation on diet, nutrition and the prevention of chronic diseases: Process, product and policy implications. *Public Health Nutrition*, 7:245–250.

North American Menopause Society. 2002. Management of postmenopausal osteoporosis: Position statement of the North American Menopause Society. *Menopause*, 9:84–101.

Otten, J. J., Hellwig, J. P., and Meyers, L. D. (eds.). National Academy of Sciences — National Research Council. 2006. *The Dietary Reference Intakes. The Essential Guide to Nutrition Requirements*. Washington, DC: National Academy Press.

Passmore, R., Eastwood, M. A. 1986. *Davidson and Passmore: Human Nutrition and Dietetics*, Eighth Edition. Edinburgh: Churchill Livingstone.

Pinstrup-Andersen, P., Pelletier, D., and Alderman, H. (eds.). 1995. *Child Growth and Nutrition in Developing Countries: Priorities for Action.* Ithaca, NY: Cornell University Press.

Raiten, D. J., Picciano, M. F. 2004. Vitamin D and health in the 21st century: Bone and beyond. *American Journal of Clinical Nutrition*, 80:1673S–1677S.

Rajakumar, K. 2003. Vitamin D, cod-liver oil, sunlight and rickets: A historical perspective. *Pediatrics*, 112:132–135.

Sachan, A., Gupta, R., Das, V., Agarwal, A., Awasthi, P. K., Bhatia, V. 2005. High prevalence of vitamin D deficiency among pregnant women and their newborns in northern India. *American Journal of Clinical Nutrition*, 81:1060–1064.

Shils, M. E., Olson, J. A., Shike, M. (eds.). 1994. *Modern Nutrition in Health and Disease,* Eighth Edition. Philadelphia: Lea & Febiger.

Taha, W., Chin, D., Silverberg, A. I., Lashiker, L., Khateeb, N., Anhalt, H. 2001. Reduced spinal bone mineral density in adolescents of an ultra-orthodox Jewish community in Brooklyn. *Pediatrics*, May; 107(5):E79.

UNICEF. 2006. *The State of the World's Children, 2006.* New York: Oxford University Press.

United States Department of Agriculture. 1995. *Dietary Guidelines for Americans*, Fourth Edition. Washington, DC: USDA.

United States Department of Health and Human Services. 1988. *The Surgeon General's Report on Nutrition and Health 1988.* DHHS (PHS) Publ. No. 88-50210. Washington, DC: United States Department of Health and Human Services.

Utiger, R. D. 1998. The need for more vitamin D. *New England Journal of Medicine*, 338:828–829.

van der Meer, I. M., Karamali, N. S., Boeke, A. J., Lips, P., Middelkoop, B. J., Verhoeven, I., Wuister, J. D. 2006. High prevalence of vitamin D deficiency in pregnant non-Western women in The Hague, Netherlands. *American Journal of Clinical Nutrition*, Aug;84(2):350–353.

Wang, Y., Moreno, L. A., Caballero, B., Cole, T. J. 2006. Limitations of the current World Health Organization growth references for children and adolescents. *Food and Nutrition Bulletin*, 27(4 Suppl Growth Standard):S175–188.

Weber, P., Bendich, W. A., Schach, W. 1995. Vitamin C and human health — A review of recent data relevant to human requirements. *International Journal of Vitamin and Nutrition Research*, 66:19–30.

Weisberg, P., Scanlon, K. S., Li, R., Cogswell, M. E. 2004. Nutritional rickets among children in the United States: Review of cases reported between 1986 and 2003. *American Journal of Clinical Nutrition*, 80(6 Suppl):1697S–1705S.

West, K. P. (ed.). 1993. *Bellagio Meeting on Vitamin A Deficiency and Childhood Mortality.* New York: Helen Keller Foundation.

Whiting, S. J., Calvo, M. S. 2005. Dietary recommendation for vitamin D: A critical need for functional end points to establish an estimated average requirement. *American Society for Nutritional Sciences*, 135:304–309.

Willett, W. 1990. *Nutritional Epidemiology.* New York: Oxford University Press.

Willett, W. C. 1994. Diet and health: What should we eat? *Science*, 264:532–537.

Williams, C. D., Baumslag, N., Jelliffe, D. B. 1994. *Mother and Child Health: Delivering the Services*, Third Edition. New York: Oxford University Press.

Woolf, A. D., Åkesson, K. 2003. Preventing fractures in elderly people. *British Medical Journal*, 327:89–95.

World Bank. 1993. *World Development Report, 1993: Investing in Health.* New York: Oxford University Press.

World Health Organization. 1990. *Diet, Nutrition, and the Prevention of Chronic Diseases.* Technical Series Report 797. Geneva: World Health Organization.

World Health Organization. 1995. *Global Prevalence of Vitamin A Deficiency.* Geneva: World Health Organization (WHO/NUT/95.3).

World Health Organization. 1995. *Global Prevalence of Vitamin A Deficiency. Micronutrient Deficiency Information System Working Paper No. 2.* Geneva: World Health Organization (WHO/NUT/95.3).

World Health Organization. 1995. *Physical Status: The Use and Interpretation of Anthropometry Report of a WHO Expert Committee.* Technical Report Series, No. 854. Geneva: World Health Organization.

World Health Organization, 1999. *Management of Severe Malnutrition: A Manual for Physicians and Other Senior Health Workers.* Geneva.

World Health Organization. 2001. *Iron Deficiency Anaemia: Assessment, Prevention, and Control. A Guide for Programme Managers.* Geneva: World Health Organization (WHO/NHD/01.3).

World Health Organization. 2003. *Diet, Nutrition and the Prevention of Chronic Diseases: Report of a Joint WHO/FAO. Expert Consultation.* WHO Technical Report Series 916. Geneva: World Health Organization.

World Health Organization. 2006. *Child Growth Standards: Length/ Height for Age, Weight for Age, Weight for Length, Weight for Height and Body Mass Index for Age. Methods and Standards.* Geneva: World Health Organization.

World Health Organization and Food and Agricultural Organization. 1992. *Major Issues for Nutrition Strategies, 1992. International Conference on Nutrition.* Italy: FAO and WHO.

World Health Organization/Food and Agricultural Organization. 2005. *Vitamin and Mineral Requirements in Human Nutrition*, Second Edition. Geneva: FAO and WHO.

WHO Regional Office for Europe. 2000. *Comparative Analysis of Progress on the Elimination of Iodine Deficiency Disorders.* Copenhagen: WHO. http://www.euro.who.int/document/e68017.pdf[accessed May 9, 2008]

NUTRITION AND FOOD TECHNOLOGY JOURNALS

American Journal of Clinical Nutrition (American Society of Clinical Nutrition)
European Journal of Clinical Nutrition
Food and Nutrition Digest (Kansas State University)
Food Technology
International Journal of Vitamin and Nutrition Research
Journal of Food Sciences
Journal of Nutrition (American Society for Nutritional Sciences)
Nutrition

Chapter 9

Environmental and Occupational Health

ENVIRONMENTAL HEALTH
Introduction
Environmental Issues
Geographic and Environmental Epidemiology
Environmental Targets
Global Environmental Change
   Climate Change
   Environmental Impact on Health Burden of Disease
Community Water Supplies
   Waterborne Diseases
Sewage Collection and Treatment
Toxins
   Toxic Effects on Fertility
   Toxic Effects of Lead in the Environment
Agricultural and Environmental Hazards
Air Pollution
   The External Environment
   Methyl Tertiary Butyl Ether
Indoor Pollution
   Radon Gas
   Outdoor–Indoor Pollutants
   Biological Pollutants
   Sick Building Syndrome
Hazardous or Toxic Wastes
   Minimata Disease
   Toxic Waste Management
Radiation
   Ionizing Radiation
   Non-Ionizing Radiation
Environmental Impact
   Emergency Events Involving Hazardous Substances
   Man-Made Disasters, War, Terrorism
   Preventing and Managing Environmental Emergencies
Environmental Health Organization
OCCUPATIONAL HEALTH
Introduction
Development of Occupational Health
The Health of Workers
The Burden of Occupational Morbidity and Mortality
   Occupational Health Priorities in the United States
International Issues in Occupational Health

National and Management Responsibilities
   Standards and Monitoring
Occupational Health Targets
Toxicity at the Workplace and in the Environment
   Lead
   Asbestos
   Silica
   Cotton Dust (Byssinosis)
   Vinyl Chloride
   Agent Orange
Workplace Violence
Occupational Health in Clinical Practice
Inspecting the Place of Work
Risk Assessment
Preventing Disasters in the Workplace
Occupation and the New Public Health
Summary
Electronic Resources
Recommended Readings
Bibliography — Water Quality and Waterborne Disease
Bibliography — Occupational and Environmental Health

ENVIRONMENTAL HEALTH

INTRODUCTION

One of the eight Millennium Development Goals (MDGs) adopted by the United Nations in 2001 is "to ensure environmental sustainability" with the following specific targets:

- Target 9: Integrate the principles of sustainable development into country policies and programs; reverse loss of environmental resources
- Target 10: Reduce by half the proportion of people without sustainable access to safe drinking water
- Target 11: Achieve significant improvement in lives of at least 100 million slum dwellers by 2020 (UNDP, 2008)

The MDGs call for international cooperation to prevent environmental degradation resulting in global warming.

A safe environment is fundamental to health; clean water is as important as shelter and food in a hierarchy of health and survival needs. Access to safe water has increased, but globally 19 percent of the burden of disease among children 0–1 years is from diarrheal disease largely due to contaminated water, while 10 percent is due to malaria and another 10 percent to malnutrition, intestinal infestation, and childhood disease clusters related to poor environmental conditions.

Overshadowing other environmental issues are climate change and global warming as a result of both natural and man-made phenomena. The result could be massive threats to public health through spread of diseases related to climate such as malaria; desertification of highly vulnerable zones of the world; disruption of water and food supplies; and wide-scale natural disaster phenomena of rising sea levels with permanent flooding of coastal areas, hurricanes, and ecologic changes of unpredictable severity. A wide consensus of scientific opinion raises the level of concern over such disastrous effects such that governments and the public seem to be ready to act to reduce fossil fuel consumption and other root causes of greenhouse gases.

Safe water supplies and waste management are fundamental and still problematic aspects of public health and community hygiene. Contamination by biological, chemical, physical, or other disease-causing agents in the external environment and the workplace are major public health and political concerns of the twenty-first century. Since the 1960s, a high degree of consciousness has developed regarding these problems. Air, water, ground, and workplace pollution are issues of concern to the public, the business sector, the media, and governmental and nongovernmental organizations, and are part of the general culture of our times. The growth of the concepts of right-to-know, consumerism, and advocacy in public health has led to greater sensitivity to these issues in many countries.

Occupational health developed as a separate area of concern from environmental health, but in recent years these have come to address issues in common. It is included as the second part of this chapter because of common advocacy, professionalism, technology, and regulatory approaches. The level of public response to environmental threats is illustrated by the groundswell of public opinion against environmental decay.

Issues have become more complex and go beyond the prevention of disease and traditional public health. While the resources needed to reduce the environmental neglect from inadequate sanitation and high levels of pollutants in the air, water, and soil are costly, the burden to society of environmental decay can be even greater in the long term.

Twentieth-century advocacy groups and reformers have made major contributions to public policy, which are akin to the achievements of their predecessor reformers of the eighteenth and nineteenth centuries in the areas of abolition of slavery, humane treatment of prisoners and the mentally ill, improvements in working conditions in factories and mines,

and public health sanitary improvements (see Chapter 1). Globalization, industrialization, and fossil fuel dependency have become an accelerated threat to the global environment. It is increasingly accepted not only by the scientific community, but by governments as well as the business community and the general public, that human society must order its affairs so that its use of natural resources does not deplete or overwhelm the self-sustaining capacity or natural regenerative powers of the environment. Environmental health is a central issue in the New Public Health in that it is the root cause of much disease and death that is preventable and degrades the environment with irreversible loss to society.

ENVIRONMENTAL ISSUES

The World Summit on Sustainable Development held in Johannesburg in 2002 called for world leaders: "aiming to achieve, by 2020, that chemicals are used and produced in ways that lead to the minimization of significant adverse effects on human health and the environment." It was noted in the specific recommendations that both technical and financial assistance will be needed for developing countries and economies in transition to build their capacity (Johannesburg, 2002). The U.S. Institute of Medicine in 2007 called for cooperative participation of industry in "green chemistry" and voluntary compliance both at home and internationally eliminating double standards in industrialized and developing countries, and complying with a robust regulatory environment to achieve less industrial, air, and global environmental pollution (IOM, 2007). In 2007, the executive board of the World Health Organization (WHO) received a report by the secretariat of global warming which calls for action to raise awareness, wide-scale preventive measures, and preparedness for the consequences of global climate change (WHO, 2007) (Box 9.1).

The environment and society interact and are mutually dependent. The ecological issues that face the world include those that can be addressed locally and nationally and others that require concerted international cooperation. Local action is part of global responsibility. Local issues require close cooperation among different agencies of government at all levels, with local authorities, supported by state and national levels. Nongovernmental organizations, the media, and voluntary groups all have important roles in promotion of a healthful environment. Unrestrained population growth and rising standards of living in many developing countries with attendant demand for consumption standards of developed countries undermine local and international efforts to maintain a balance between nature and human society. At the same time, the industrialized countries are beginning efforts to reduce polluting standards, but the time available to prevent global warming is very short.

At the same time, global society must face those environmental, social, and health issues that relate to poverty and high population growth in the poorest countries.

Source: Campbell-Lendrum, D., Woodruff, R. 2007. Climate change: Quantifying the health impact at national and local levels. A. Prüss-Üstün, C. Corvalán (eds.). WHO *Environmental Burden of Disease Series* No. 14. Geneva: World Health Organization, http://libdoc.who.int/publications/2007/9789241595674_eng.pdf [accessed June 14, 2008]

Box 9.1 Global Environmental Challenges for the Twenty-First Century

1. Global warming;
2. Inequalities between industrial and nonindustrial countries;
3. Population growth;
4. Social, economic, and political inequalities nationally;
5. Deforestation;
6. Water supplies/shortages;
7. Food production and distribution;
8. Energy and resource depletion;
9. Soil erosion/desertification;
10. Air pollution;
11. Chemical/toxic wastes;
12. War/nuclear threats/terrorism/armament costs;
13. Ozone depletion;
14. Economic growth.

Among the long-range issues confronting many countries are water supplies and their quality, which are endangered by overuse and the pollution of groundwater sources. Air and soil pollution, deforestation, and desertification require local, national, and international multisectoral cooperative planning and intervention.

Public consciousness regarding these issues has increased during the past several decades. Environmental concern has become an essential part of accepted public philosophy in many developed countries. Its place in developing countries is often of low priority, coming after the struggle to expand economically as well as the severe problems of population growth and basic services. Economic growth and health status are closely related to agriculture, food supplies, and distribution systems, as well as preservation of agricultural land and rational use of energy. As was the case in numerous countries during their industrial development and urbanization, many eastern European countries prioritized industrialization over all other issues and subordinated environmental concerns, so that accumulated environmental degradation is part of the long-range burden of post-Soviet societies.

GEOGRAPHIC AND ENVIRONMENTAL EPIDEMIOLOGY

Geographic epidemiology is defined as the description of spatial patterns of disease incidence and mortality. It is part of descriptive epidemiology that generally describes the occurrence of disease according to demographic characteristics of the population at risk and in terms of place and time. Snow's description of cholera in London in 1854 and many other observational studies supported hypotheses that turned out in practice to be the case, even though the direct causal relationships were not demonstrable at the time.

Geographic epidemiology helps to generate hypotheses that can then be tested by more rigorous methods. Environmental and occupational epidemiology uses a wide range of research methods to the study of disease in relation to environmental or work-related conditions. In practical everyday public health, the findings of a common point source of disease, injury, or death may lead directly to contaminated water, toxic exposure at a work site, a risk condition, or polluted air of a city. While these may need case–control or other more formal studies for confirmation, the findings of known risk factors on routine surveillance should be sufficient to lead to adequate public health intervention by appropriate regulatory authorities.

Epidemiologic studies may describe in quantitative terms the relationship between the frequency of disease and the degree of exposure to a particular agent. Such studies are subject to errors in the measurement of exposure. Measurement of exposure by place of residence or work is only an approximation. Moreover, within the same community there will be wide variation in actual exposure levels to the toxic agent. The agent may affect different populations or subgroups differently. There may be genetic and social factors at play as well. In cases where there is a long time lapse between exposure and resultant disease, and many independent variables, it may be very difficult to attribute the disease to a specific exposure, as in the case of asbestos exposure and mesothelioma, but cumulative evidence is such that the relationship is established and action to eliminate asbestos usage is warranted.

ENVIRONMENTAL TARGETS

In 1985, the European Region of the WHO issued consensus statements on health targets for 2000; some of these are listed in Table 9.1. These targets represent a broad societal commitment to stop environmental degradation. These have been reissued in various forms and increasingly represent a wide commitment to slow global warming, as this has become the center stage issue of environmental health in recent years.

GLOBAL ENVIRONMENTAL CHANGE

Public health has traditionally placed high priority on sanitation, housing, and urban planning in the battle to reduce the burden of infectious disease. The sanitary movement of the nineteenth century had an enormous impact on the

TABLE 9.1 Selected European Region WHO Environmental Health Targets and Issues

Target	Issue
Multisectoral policies to protect the environment	Coordination between agencies at international, national, regional, and local levels
Raising public awareness of global climate and the health effects of environmental health	Promotion of public/private consortia for environmentally friendly policies to reduce greenhouse effects, and promote alternative fuels, chemicals, construction, agriculture policies
Monitoring and control mechanisms for environmental hazards	Chemicals, ionizing radiation, noise, biological agents, consumer goods, risk assessment
Adequate supplies of safe drinking water	Quantity, quality of water; international, national programs, ground and surface water surveillance, quality control; water management standards
Protection against air pollution	Legislative administrative and technical measures to control indoor and outdoor pollution
Reduced risk of food contamination including harmful additives	Legislative administrative and technical measures to control food contamination and additives, and production, storage, transport, sale, and use
Eliminate risks of hazardous wastes	Effective legislative, administrative, and technical measures for surveillance and control of dumped wastes
Healthy and safe urban environment	Housing and urban planning standards, waste disposal, potable water supply, recreation, open spaces, traffic control, waste disposal, and sanitation
Protection against work-related risks	Protection against biological, chemical, physical hazards; worker education, industry self-monitoring and government regulation

Source: Commission of the European Communities. 2004. *The European Environment Action Plan 2004–2010*, http://ec.europa.eu/environment/health/pdf/com2004416.pdf [accessed May 13, 2008]
Campbell-Lendrum, D., Woodruff, R. 2007. Climate change: Quantifying the health impact at national and local levels. A. Prüss-Üstün, C. Corvalán (eds.). WHO Environmental Burden of Disease Series No. 14. Geneva: World Health Organization, http://libdoc.who.int/publications/2007/9789241595674_eng.pdf [accessed June14, 2008]

control of communicable diseases. The lessons learned in disinfection of potable water supplies and treatment of solid and liquid wastes are still not applied universally. This includes not only the less-developed countries, but middle-level developing and industrialized countries as well. The threat of local, national, and international disasters, including the re-emergence of cholera in South America and Russia and *Giardia* as a major outbreak in the United States, returned the classic issues of water quality to center stage in modern public health. Plague in India in 1994 and Rift Valley fever in Egypt in 1977 and then in various parts of the Middle East (Saudi Arabia, Yemen) where it is now probably endemic and resurgence of malaria and dengue fever highlight the problems of vector control in modern public health. In the 1990s, West Nile fever appeared for the fist time in the northeast United States and is now endemic in many parts of the United States. In 2003, SARS was spread from China to Toronto, causing the city to be under virtual quarantine. In 2007, Chikungunya fever appeared in France, Italy, and in Hong Kong in 2008, brought by travelers from Asia and possibly becoming endemic with spread via *Aedes albopictus* mosquitoes which are common in many parts of Western and Central Europe.

The concept of environmental health has been widened in recent decades by the spectrum of global changes to the environment as a result of man-made environmental pollution and natural events such as volcanic eruptions. The greenhouse effect is the warming of the global environment through retention of solar energy heating of the earth by increasing the greenhouse gases in the earth's surrounding atmosphere. Disposal of toxic and radiologic waste constitutes a very difficult public health challenge in many countries. Land degradation, loss of topsoil, deforestation, groundwater depletion, and acidification of water and soil are all challenges in environmental health in the twenty-first century. The effects of global environmental changes cannot be predicted with certainty, but there is scientific consensus on the serious and imminent dangers to the environment and human society which requires both global and local preventive action and public health crisis response capacities.

Poverty, low levels of education, and rapid population growth in the poorest countries with limited food production potential stand in contrast to high levels of consumption and energy use and low rates of population growth in the industrialized countries. Many environmental issues involve more than one country, partly because of transportation of waste products or hazardous materials from one country to another, by wind, water, or deliberately by man. Economic concerns include the destruction of fishing stocks, damage to forests, and more global concerns of ozone depletion, global warming, and ocean pollution. Intersectoral cooperation within a country, and international cooperation and regulation to reduce pollution of common waters in seas, lakes, and rivers shared by more than one country are part of a broad New Public Health agenda.

Climate Change

There is widespread consensus that the warming of the earth is a result of man's activities and it is an emerging risk factor for health for the spread of infectious diseases and disruption of food and fresh water supplies. The effects of global warming may include serious weather disruptions and changes in ecology that could threaten human, plant, and animal life on earth. The policymaking uses of estimates of health impacts are the identification of groups at risk for specific diseases and the use of scarce resources. This helps to target resources in controlling the emissions of greenhouse gases.

The Human Development Report 2007/2008 sees climate change as the defining human development challenge of the 21st century. Failure to respond to this challenge will stall and reverse international efforts to reduce poverty. The poorest countries and most vulnerable people will suffer the most damaging setbacks, but no country will be immune to the impact of global warming. Increased exposure to droughts, floods, and storms is already destroying opportunity and reinforcing inequality. Overwhelming scientific evidence indicates that the world is moving towards the point of irreversible ecological catastrophe. Avoiding the impact of the most damaging climate changes requires global action the decades ahead. The financial resources and technological capabilities exist, but implementation requires a sense of urgency, public interest, and political will to make deep cuts in greenhouse gas emissions. Achievement of Millennium Goal 7 (to ensure environmental sustainability) rests very much on addressing the issues of global warming.

Environmental Impact on Health Burden of Disease

Environmental hazards contribute to a wide range of diseases. WHO reports that as much as 25 percent the burden of disease worldwide is from preventable environmental exposures with more than 13 million deaths annually and nearly one-third of mortality and morbidity in developing countries. It is responsible for more than 33 percent of illness in children under 5 worldwide and as many as 4 million of their lives could be saved mostly in developing countries by preventive environmental measures. Safe household water storage and hygienic measures, cleaner and safer fuels, better buildup of environment, less air pollution, better home and workplace management and use of

TABLE 9.2 Diseases with the Largest Total Annual Health Burden Globally from Environmental Factors

Disease	DALYS (annual, in millions)	Disease-specific burden
Diarrhea	58	94% of all diarrhea
Lower respiratory infections	37	41% of all cases
Unintentional injuries (excluding road)	21	44% of all cases
Malaria	19	42% of all cases
Chronic obstructive lung disease	12	42% of all cases
Perinatal conditions	11	11% of all cases

Source: World Health Organization. 2006. *Preventing Disease Through Healthy Environments — Towards an Estimate of the Environmental Burden of Disease.* Geneva: WHO.

toxic substances, and better water resource management would reduce diseases such as diarrhea, respiratory infections, malaria, West Nile virus, and others (Table 9.2).

Environmental factors affect the developing countries most suffering from poor water supplies, low levels of sanitation, housing standards, education (especially of girls), and highest rates of poverty. It is these topics which are addressed in the Millennium Development Goals. Most of those diseases with large numbers of deaths are amenable to change with available policies, technologies, and preventive and public health measures. This could result in 2.6 million fewer deaths from cardiovascular diseases, 1.7 million fewer deaths annually from diarrhea, 1.5 million fewer from respiratory diseases, 1.4 million fewer from cancers, and close to 1 million fewer from external injuries (motor vehicle accidents, poisonings, and others).

Sulfur and nitrogenous oxides from fossil fuel electric power plants can travel long distances after being released from tall chimneys. Such pollutants falling as acid precipitation have led to the destruction of forests in countries of Central and Eastern Europe. Acid rain generated in one European country may fall in another, affecting waters and animal life in addition to forests. Reduction of acid rain has been achieved during the 1980s in North America by greater selectivity in fossil fuels, and the result is reduced damage to forests and water sources.

The release of various organic solvents, called *chlorofluorocarbons* (also known as freons or CFCs), used in cooling systems, refrigerators, and consumer aerosol products, cause damage to the earth's ozone layer. This permits entry of ultraviolet (UV) light that was formerly excluded into the earth's atmosphere. UV light causes a rise in skin cancer and cataracts in humans. Substitution for freons is vital to reduce damage to the ozone layer. This can be achieved on an individual level by use of water-based paints and chemical products in daily life. The search for substitutes for refrigerants and toxic chemicals to replace those that damage the environment and exposed workers, has, along with regulation, become the hallmark of environmental and occupational health.

Greenhouse gases are built up in the atmosphere by carbon dioxide emissions largely due to increasing carbon dioxide and other gases produced from excessive and inefficient use of fossil fuels along with wide-scale destruction of forests which are protective through natural conversion of CO_2 to water. These gases block infrared radiation from the earth's surface, leading to trapping of heat. This effect resembles the use of glass or plastic covers to retain heat in a greenhouse. This global warming effect may have long-term serious consequences for the earth's thermal balance. The effects on the polar ice caps can lead to global changes in the level of oceans. Reduction of the greenhouse effect requires international, national, and individual effort, and especially an environmental consciousness and action by governments, the media, the scientific and business communities as well as the general public.

Hazardous wastes are being exported from developed to developing countries. This is potentially solvable by heightened national awareness and stronger international conventions with publicity and fines imposed by international courts against offending firms or nations. In a global economy, all of these factors link up with effects on the physical environment as well as on working conditions and many social/political factors, such as the widening gap between rich and poor.

COMMUNITY WATER SUPPLIES

Fresh water is vital for all living organisms and is becoming an increasingly scarce resource. Waterborne diseases are still among the major causes of death in developing countries, which often lack adequate supplies of water. In both developed and developing countries, pollution control, reuse of wastewater, and water planning are vital to the national economy and public health.

The International Decade for Drinking Water and Sanitation in the 1970s and early 1980s promoted national, bi-national, and international efforts to improve community water supplies, sanitation, drainage, education, and hygiene. Implementation of appropriate technology for maintaining water and sanitation infrastructure was emphasized. Safety of community drinking water, as defined by WHO, requires a combination of standards and protection of raw water sources from contamination. Treatment of community water supplies requires sedimentation, coagulation, filtration,

chlorination, and continuous monitoring. High standards of construction and maintenance of water distribution systems are needed whether at the village well or the municipal water supply system. Filtration removes solid and suspended particles, improving the quality of surface source water, and disinfection by chlorination effectively kills most microorganisms.

Coverage and protection of reservoirs and canals is also beneficial in improving the security of water sources and in preventing contamination from natural sources, including birds, animals, and vegetation. Community water regulation and enforcement require both physical treatment and disinfection to protect the public against microbiological, chemical, and other health hazards. Agricultural runoff of pesticides and animal wastes are also important contaminants of water sources.

The Clean Water Act (CWA) of 1977 amended the 1972 Federal Water Pollution Control Act. At that time, some of the Great Lakes were seriously polluted, as were many of the major rivers of the country. The CWA set new U.S. national standards and regulatory mechanisms at federal, state, and local levels of government. It increased regulatory powers to "restore and maintain the chemical, physical, and biological integrity of the Nation's waters." Regulations under the act permitted effective action against industrial and other polluters and allowed for controls to be established where multiple municipalities were involved in a river or regional water system. This led to steady improvement in water quality of lakes, rivers, and groundwater sources throughout the country, but in 1999, 40 percent of U.S. waterways were below EPA standards. A Clean Water Restoration Act proposed in the U.S. Congress in 2007 is intended to clarify federal jurisdiction and standards of water supervision. However, it is controversial because of alleged federal infringement of state responsibilities and has not been passed as of 2008.

Concern about potential carcinogenic effects of trihalomethanes may cause withdrawal of mandatory chlorination of surface waters. The absence of adequate disinfection with chlorine increases the risk of serious waterborne disease outbreaks such as the wave of cholera epidemics in South America during the 1990s. New standards may require time to be implemented because of prevailing conservative professional and public attitudes and the cost of treatment plants. In Israel, for example, opinion gradually shifted to mandatory chlorination policy. This was due to a number of factors: increased public and news media awareness of drinking water quality, a greater recognition at the leadership level of the Ministry of Health of the importance of preventive and environmental factors in enteric disease, an increasing presence of younger, better trained sanitary engineers willing to challenge previously accepted dogmas, and persuasive documentation of the impact of contaminated community water supplies on the infectious disease burden of the country.

Waterborne Diseases

Waterborne disease may be so common as to escape detection in point outbreak form. This seems to be the case in many countries, where hepatitis (especially hepatitis A and E) is endemic and where incidence of gastroenteritis from *Shigella* and *E. coli* remains high. In industrialized countries, waterborne disease outbreaks have become uncommon events because of high levels of water management (Box 9.2). Water contamination and enteric disease can occur from organisms for which routine testing is not currently practiced. For example, testing for rotaviruses (which cause enteric disease) and organisms such as *Campylobacter* and *Giardia* is not done routinely,

but water is tested if there is a suspicion of contamination. Safe water requires physical treatment as well as disinfection of all surface water sources. Effective water management can reduce the burden of gastroenteric disease even in a relatively developed country.

Israel, developing rapidly in the 1960s, built a national water carrier to distribute unfiltered, chlorinated water to communities and for agriculture. Local groundwater was not necessarily chlorinated, and sewage system development was inadequate. In the 1970s and 1980s, Israel experienced large numbers of waterborne disease outbreaks. A 1985 outbreak resulted from the contamination of groundwater sources by a sewage pipe which accidentally broke during roadwork, resulting in 9000 cases of shigellosis, 49 cases of typhoid fever, and 1 death. Introduction of mandatory chlorination in 1988 in Israel resulted in a substantial improvement in the quality of community water supplies and greatly reduced the incidence of waterborne disease outbreaks and the total burden of diarrheal disease. At the same time, there was a marked reduction in the overall burden of enteric diseases in the country, including hepatitis A; however, primarily food-borne salmonellosis continues to increase.

In the United States during 1995–1996, there were 22 waterborne disease outbreaks due to contaminated drinking water with 2567 cases, largely due to *Giardia*. In 1993, *Cryptosporidium* contamination of water sources caused waterborne disease outbreaks in Milwaukee and elsewhere. Some waterborne organisms such as *Giardia* and *Cryptosporidium* constitute a special risk for persons with compromised immune systems, including cancer patients being treated with chemotherapy, HIV-positive persons, and patients following organ transplantation being treated with immunosuppressants. During the period of 1991 to 2002 there were 2007 waterborne disease outbreaks in the United States with 433,947 cases of illness reported. Problems in the distribution system were the most commonly identified deficiencies under the jurisdiction of a water utility, underscoring the importance of preventing contamination after water treatment. Contaminated ground water provides support for the Ground Water Rule of 2006. Outbreaks of *Giardia* and *Cryptosporidium* in the United States have raised concerns because these organisms are not efficiently eliminated by standard water treatment, and are not routinely tested in regular water-sampling monitoring.

In 1993, the largest reported waterborne disease outbreak in U.S. history occurred in Milwaukee with approximately 403,000 ill persons, of whom 4400 required hospitalization. Attack rates were as high as 50 percent in some parts of the city. *Cryptosporidium*, in addition to being transmitted from person to person and from animals to humans, can also be transmitted in swimming pools. *Cryptosporidium* is reportedly present in 65–87 percent of surface water samples tested in the United States.

Early detection by laboratory diagnosis requires preparation of laboratories for identification of these organisms.

Box 9.2 International Water Management Standards

1. International: United Nations and World Health Organization promoted the International Decade for Drinking Water and Sanitation and promulgated clear standards of water quality for community water supplies (1958, 1963, 1971, 1984, and 1997).
2. National, state, and local authorities: policy commitment, funding, and professional departments for supervision of community water systems.
3. Municipal water systems: water management and testing varies according to the quality of the source water and methods of treatment including:
   a. High standards of acceptability of source surface water;
   b. Physical treatment — coagulation and filtration;
   c. Disinfection by chlorination — routine and mandatory;
   d. Maintaining and monitoring of residual chlorine;
   e. Construction and maintenance of water storage and distribution systems;
   f. Monitoring of enteric disease;
   g. Investigation of suspect waterborne disease outbreaks;
   h. Continuous monitoring by bacteriologic and chemical testing;
   i. Assurance of safe distance between sewage and water pipes;
   j. Integrity of water distribution system against inflow.
4. Village wells:
   a. Protection of wells from human and animal wastes;
   b. Regular or periodic chlorination;
   c. Supervision by trained and supervised village health workers.
5. Sanitary education: at all levels of society including governments, NGOs, intersectoral cooperation, public, medical and other professional communities, and in schools.

Source: World Health Organization. 2006. *Guidelines for Drinking Water Quality. First Addendum to Third Edition. Recommendations.* Geneva: World Health Organization. http://www.who.int/water_sanitation_health/dwq/gdwq3rev/en/index.html [accessed May 14, 2008]

Regular testing of the community water supply at its origin and within the supply system is essential to monitor water safety. The presence of coliform bacteria indicates fecal contamination and potential hazards, warning sanitation officials that other more dangerous organisms, such as dysentery bacilli or enteric viruses such as hepatitis, may be present. Testing for *Cryptosporidium*, *Giardia*, and viruses is difficult, costly, and insensitive; therefore, routine testing is not done. Chlorination and filtration may not be sufficient to prevent waterborne disease transmission of these organisms. This is a problem for sanitary control. New methods of testing and disinfection of water supplies need to be devised. At present, filtration and chlorination remain the basic methods of assuring safe community water supplies, supplemented by boiling of suspect water during outbreaks of disease.

Standard water treatment processes remove solid and suspended material, bacteria, and odors from water and have been outstandingly successful in reducing waterborne disease. New concerns over chemical contamination of community water supplies have become prominent in recent decades. Heavily polluted waters have been linked to neurologic damage and cancers of the bladder, intestinal tract, liver, and kidney.

The U.S. Safe Drinking Water Act of 1974, as amended in 1996, establishes criteria for monitoring of public water systems for microbiologic, chemical, and other contaminants (Box 9.3). The act defines maximum contaminant levels (MCLs) for specified chemical pollutants. The U.S. Environmental Protection Agency (EPA) sets MCLs for pollutants, out of hundreds of organic, inorganic, biological, and radiologic contaminants detected in water supplies around the country. This is an area of public health concern that still requires much epidemiologic and sanitary engineering research.

Right-to-know laws, a critical investigative press, and an environmentally conscious public are fundamental to prevent serious ecological degradation. Environmental activism has made important contributions to public health, but such activism can be a two-edged sword. One example is the excessive zeal focused on the environmental impact of chlorination and its by-products. Trihalomethanes, produced by the combination of chlorine and nitrogenous material (chloroform, bromoform, bromochloromethane, and chlorodibromomethane) in unfiltered contaminated surface water, can reach levels that are carcinogenic and suspicions of teratogenicity have been disproven (Niewenhuijsen et al., 2008). Levels within accepted norms are considered safe; total trihalomethane levels under 0.10 mg/L (EPA, 2008).

Some opposition to disinfection by chlorination has been based on this concern and led to the spread of cholera in South America during the 1990s (see Chapter 4). The offset of benefits against risks has resulted in current professional consensus that this is not a reason to cease

---

Box 9.3 Water Contaminants Under U.S. Environmental Protection Agency Regulation

Microbiological contaminants turbidity, total coliforms, viruses, *Giardia lamblia*, *Cryptosporidium*, *Legionella*;

Disinfection by-products bromate, chlorite, haloacetic acids, total trihalomethanes;

Disinfectants chloramines and chlorine (as Cl_2), chlorine dioxide (as ClO_2);

Volatile organic chemicals trichloroethylene, tetrachloroethylene, carbon tetrachloride, vinyl chloride, benzenes, ethylene, and ethane compounds;

Synthetic organic compounds pesticides (lindane, endrin, 2,4-D, chlordane), carbon tetrachloride, carbofuran, chlorobenzene, diadipate, dichloromethane, dichloropropane, diphthalate, dinseb, dioxin, diquat, endrin, epichlorohydrin, ethylbenzene, ethylene and dibromide, glyphosate, heptachlor and epoxide, hexachlorobenzene, methoxychlor, oxamyl, polychlorinated biphenyls (PCBs), pentachlophenol, simazine, styrene, terachloroethylene, toluene, toxaphene, silvex, trichloroethylene, vinyl chloride, xylene;

Inorganic chemicals antimony, arsenic, asbestos, barium, beryllium, cadmium, chromium, copper, cyanide, fluoride (>4 mg/L), lead, mercury, nitrates, nitrites, selenium, thallium;

Radiologic contaminants alpha and beta particle activity, natural uranium, radium, radon.

Note: Maximum contamination levels (MCLs) are set by the EPA for the listed contaminants.
Source: www.epa.gov/safewater/contaminants/index.html#micro [accessed May 13, 2008]

---

chlorination. Rather, it provides additional justification for physical treatment of raw water before chlorination to reduce the nitrogenous material content and thereby reduce the combination with chlorine which produces trihalomethanes, improving water potability and clarity.

Developmental programs including local and large-scale dam projects can have negative health effects by providing a hospitable environment for vectors for diseases such as malaria, schistosomiasis, and onchocerciasis, resulting in resurgence of diseases once controlled. Planning of development projects must take into account the potential ecological effects and the needed control measures to prevent greater health damage than benefit.

SEWAGE COLLECTION AND TREATMENT

Sewage collection and treatment, along with filtration and disinfection of drinking water, have made enormous contributions to improved public health, perhaps even more than the use of modern medicines and vaccines. Collection of sewage prevents surface environmental contamination as well as seepage into groundwater and contamination of local water sources. Sewage contains bacteria, viruses, protozoa, and other pathogens that can cause serious disease; treatment entails killing the pathogenic organisms present in the sewage. The purpose of sewage treatment is to improve the quality of wastewater to a level where it can be discharged into a waterway or prepared for reuse for agriculture without damaging the aquatic environment or causing human health problems in the form of waterborne disease.

Primary treatment of community wastewater begins with the removal of solids from the wastewater. This is done by several mechanical processes of screening and sedimentation. The wastewater is passed through screens to remove large solid objects and then through grinders to further break up the solid wastes. The wastewater then flows at reduced velocity through a grit chamber where sand, gravel, and other inorganic materials settle out. Air is injected into the tank to remove trapped gases and to maintain an aerobic environment. The wastewater then flows into secondary settling tanks where further sedimentation of solid particles takes place. Primary treatment removes just over half of the suspended material and particles in preparation for secondary treatment.

Secondary treatment of wastewater is based on biological treatment assisted by mechanical methods, accelerating the natural decomposition of organic wastes. Aerobic microorganisms are used in the presence of an abundant oxygen supply to decompose the organic material into carbon dioxide, water, and minerals. The wastewater is sprayed over trickling filters or beds of crushed stone covered with a slime containing various types of microbes. These microbes absorb the organic material and act to break it down into its various components. The sewage is then processed by the activated sludge method, carried out by introducing bacteria-containing sludge into a tank of wastewater along with compressed air. The waste is then agitated and mixed for 4–10 hours. The microbes are adsorbed to suspended particles and oxidize the organic material. After this process, the sludge, consisting of masses of bacteria, settles out into the tank. The sludge is then removed and recycled into the next tank of wastewater.

Following primary and secondary treatment, the suspended material and the biochemical oxygen demand (BOD) are reduced by approximately 90 percent. This process depends on temperature, which affects the metabolic rate and activity of the organisms needed to break down the suspended organic material. Secondary treatment is most effective in removing protozoa, worms, and bacteria, but less effective against viruses, heavy metals, and other chemicals. Since 1988, all sewage plants in the United States are required by federal regulations to provide at least secondary treatment.

Tertiary treatment is required if the wastewater is to be recycled for the purposes of agricultural irrigation, recreation, or community use. Tertiary treatment includes a combination of physical, chemical, and biological processes to reduce the particles and BOD to less than 1 percent of those of the original wastewater. The process includes chemical coagulation, filtration, sedimentation, activated carbon adsorption, oxygenation ponds and aerated lagoons, osmosis, ion exchange, foam separation, and land application. All of these processes remove different pollutants present in the wastewater, especially tiny particles of suspended organic matter. They also remove synthetic chemicals, ammonia, nitrates, phosphates, and dissolved organic materials. Recycled wastewater is an important source of water in a world running short of water. Desalination is also becoming an attractive option as costs of treatment are reduced and competitive with other forms of water management. Another important potential is in new technology to use evaporated water in the air as a source of household water supply.

Disposal of the sludge remaining after sewage treatment by incineration or ocean dumping is environmentally problematic. Use of the sludge for compost in agriculture or gardening is increasing, but contamination may enter the food chain and create another hazard. Sludge disposal should be carefully regulated.

Disinfection is the final stage, accomplished by introducing chlorine into the water so that there is a residual level of chlorine to protect the water from contamination in the water storage and distribution system. In many countries or regions, lack of sufficient local water supplies for community agriculture and industrial uses necessitates recycling of wastewater as part of the process of water conservation. Supplementation of water sources by desalination and recycling will be increasingly important as population growth, increasing standards of living, and pressures of agricultural and industrial contamination of water sources increase. New technology in membrane filtration offers hope to improve the efficiency and economics of this sector of the ecological sciences.

The disposal of solid waste has been a challenge from prehistoric to modern times and will only increase in the future. With the growth of cities, disposal of refuse took on more significant health importance. In biblical times, Jerusalem burned its garbage in a valley outside the city walls (*Gehennam*, a term later adopted for "Hell"). The Greek city-states had ordinances against dumping refuse in or near cities, providing waste disposal sites for this purpose. In medieval European cities, garbage as well as human and animal wastes were discarded into the streets and areas surrounding the home. In the thirteenth century, Parisians were forbidden to throw waste on the streets and had to dump it outside the walls of the city. In 1388, the English parliament prohibited waste disposal in public waterways. During the industrial revolution, medieval cities evolved into working-class slums. Crowding, poor housing, and poor sanitation forced municipal governments to organize measures to reduce the nuisance and health hazards of solid waste.

Waste management continues to be a problem as greater amounts are generated by the affluent lifestyles of the population of industrialized countries. In developing countries, where rural to urban population shift is under way on a massive scale, rapid population growth, crowding, and slums increase the burden of solid waste disposal. Since the 1980s, return, recycling, and reuse of waste products entered the popular culture in some countries and are beginning to have an impact on reducing landfill needs. Recycling of paper, plastic, and glass bottles and metals contributes to reduced solid waste for disposal and has become an economically attractive activity. Biogas methods are improving so that animal wastes can be used for production of methane gas to be used for energy produced for home or general use.

Waste management is becoming more controversial. There is confusion about the differences between issues such as hospital waste, industrial waste, and toxic household waste and a lack of trust by the community in governments and scientific communities. Hence there is great need for continuing community education and communication by government agencies and community leaders.

Sources of solid waste include agriculture, mining, industry, and urban waste. In the United States, 95 percent of solid wastes are from agriculture, mining, and industry. The remainder is from household waste, which generates 150–180 million tons of solid waste annually. This is the equivalent of 4 pounds of refuse per person per day. Municipal waste collection and disposal are serious problems involving high costs and a serious public health burden if not done well.

Waste management involves a variety of techniques, including reusing and recycling, composting, incineration, and land refill. Each has its advantages and disadvantages. These techniques are part of the engineering of community infrastructure. Seawater dumping is still practiced in some countries, but increasing global concern about the effects of such practices on the ecology of the lakes and oceans makes this solution unacceptable. Landfill is the most prevalent method of solid waste disposal. It involves spreading garbage in layers 8–10 feet deep and covering them with a thin layer of soil. This method is adequate if well planned and supervised and has the benefit that methane gas produced by anaerobic decomposition can be recovered for use. The problems of seepage of toxic materials and potentially explosive gas accumulation require careful assessment of landfill sites and limit the potential of landfills to serve as a sustainable, feasible option. Limited possibilities for suitable landfill locations in large urban concentrations make this method of disposal a serious urban planning problem. Sanitary landfill is expensive because of the cost of collection and transportation, the land value, and the human resources required. Landfill under sanitary conditions requires compaction of waste and covering by well-spread yet compacted earth far from ground and surface water. Sites must be fenced to prevent scavenging by people, animals, and off-hours dumpers. It should have well-paved and well-drained access roads. The landfill should be located away from residential areas and should be well maintained and tidy. It should be seeded, in completed areas, with grass and trees to control erosion and must be maintained by well-trained sanitarians.

Composting or conversion of waste products into topsoil can be applied at the household and municipal levels. By-products of wood and food processing can be composted and used to reduce soil pollution from petroleum-based products. This involves separation of non-biodegradable materials from biodegradable materials and treatment of the materials to break down organic waste. Decomposition at high temperatures (140° F) kills flies, weed seeds, and potentially pathogenic organisms. In closed systems with forced draft aeration, this process can be accomplished in a few days; however, with passive methods it takes many months. After further treatment of "curing" and screening or grinding, an excellent soil conditioner can be produced that can be used to enhance agricultural or horticultural work such as in nurseries, public gardens, and parks. Incineration is attracting wide interest, but its use is limited by high capital cost and the possible release of potentially toxic materials such as dioxin and heavy metals into the atmosphere. Meticulous maintenance requirements are needed to properly mix the materials for clean burning at high temperatures. In addition, there is the residual problem of disposal of ash itself that is toxic. Waste-to-energy incineration reduces the volume of waste products by 80–90 percent and produces energy that can generate electricity and replace fossil fuels. In Japan and Western Europe, 30–40 percent of solid waste is incinerated in waste-to-energy plants.

Using garbage as feed for pigs is no longer acceptable because of the problem of meat contamination with trichinosis (pork tapeworm). However, the practice is returning on an experimental basis with scraps ground and steamed prior to being used as animal feed. Control of the use of animal parts for animal feed is now being re-evaluated and more intensively regulated following the bovine spongiform encephalopathy (BSE) experience in the United Kingdom and Europe in the 1990s (Chapter 4).

Recycling and waste reduction are methods gaining wide support. Reducing use of disposables (e.g., packaging materials, disposable diapers) requires an ecologically conscious public and municipal, nongovernmental, or volunteer collection systems. Scrap metal, paper, glass, and plastic recycling can be commercially successful. Industry and commercial enterprises can be convinced to reduce use of bulky packaging materials and to adopt "ecologically friendly" practices. Plastics and rubber tires are also recyclable in economically valuable ways. Ecological consciousness is fundamental to the success of such practices.

The location and management of landfills requires professional management with a fair and transparent process and involvement of the community, with a focus on replacing poor-quality landfill practices. Epidemiologic surveillance programs should only be undertaken after a

TABLE 9.3 Hazardous Waste Disposal Sites: Summary of Epidemiologic Evidence

Health effect	Level of evidence	Health effect	Level of evidence
Early fetal deaths (spontaneous abortion)	Inadequate	Oro-facial birth defects	Inadequate
Late fetal deaths (stillbirths)	Inadequate	Musculoskeletal birth defects	Inadequate
Intrauterine growth retardation	Inadequate	Genitourinary birth defects	Limited
Small for gestational age	Inadequate	Gastrointestinal birth defects	Inadequate
Birth weight adjusted for gestation length	Inadequate	Chromosomal abnormalities (structural)	Inadequate
Term birth weight	Inadequate	All childhood cancers	Inadequate
Low birth weight (not adjusted for gestation length)	Limited	Leukemia	Inadequate
Preterm birth, gestation length	Inadequate	Lymphoma	Inadequate
Total birth defects	Limited	Reproductive system development	Inadequate
Central nervous system birth defects	Limited	Thyroid function	Inadequate
Cardiovascular birth defects	Limited	Kidney function	Inadequate

Sources: Adapted from Wigle, 2004.
World Health Organization, European Office. 2007. *Population Health and Waste Management: Scientific Data and Policy Options.* Copenhagen: WHO.

feasibility analysis and with suitable protocols. The chemical exposure pathways and the effects on at-risk segments of the population should be considered. The adverse effects on health from factors such as noise, odor, negative impacts on property values, and the community views should all be considered. A 2007 WHO review of the epidemiology evidence of hazardous waste effects on health shows the epidemiology is weak or limited (Table 9.3).

Production of both steel and aluminum from virgin ore are both very polluting and energy intensive. As such, recycling of iron, steel, and aluminum in the United States in 1991 was a substantial part of total new production of these metals. Over 11 million vehicles are recycled, supplying 37 percent of all ferrous scrap. Use of recycled iron and steel reduces air pollution by 86 percent, water pollution by 76 percent, and solid wastes by 105 percent, as compared to production from new ores. Similar benefits accrue from recycling of aluminum scrap. Community waste collection and recycling has become widely practiced in many countries. In the United States, the rate of recycling municipal waste products has doubled to over 32 percent of total waste, saving some 64 million tons of household waste from landfill or incinerator disposal.

TOXINS

A toxin is a substance in the environment with the potential for causing human disease or injury. Toxicology is the study of such substances and their effects on humans. All chemicals are toxic under some conditions, depending on the dose, concentration, and threshold or sensitivity of a given species for that substance. The range of chemical toxins and methods of classifying them are shown in Table 9.4.

The factors that affect the toxicity of an agent, in addition to the extent and duration of exposure, include host factors (e.g., age, gender, fitness level, previous exposure), environmental factors (e.g., temperature, air flow), and the nature of the toxic agent (e.g., physical and chemical properties) (Box 9.4). Toxicology is an important part of environmental and occupational health; further reference will require a specialized text and appropriate Internet websites (see bibliography).

Toxic Effects on Fertility

Toxins can adversely affect fertility, pregnancy, and early or later child development. Reproductive potential can be adversely affected by reduced male reproductivity, such as by exposure to the pesticide dibromochloropropane (DBCP). Other chemicals have been implicated in increased abortion rates among exposed pregnant women. Birth defects or teratogenesis occurred with exposure to thalidomide. Other chemicals relate to low birth weight and toxicity in newborns. Exposures to chemicals such as lead produce brain damage in children.

TABLE 9.4 Classification of Toxic Agents in Environmental and Occupational Health

Classification	Subgroups	
By structure: organic; inorganic	Organic: aromatics (e.g., benzenes), polyaromatics, amines, ethers, ketones, alcohols	Inorganic: anions, cations, heavy metals, metalloids (e.g., selenium)
By chemical type	Organochlorines, organophosphates, halogenated aliphatic hydrocarbons, halogenated ethers, polychlorinated biphenyls, monocyclic aromatic hydrocarbons, phthalate esters, polycyclic aromatic hydrocarbons, nitrosamines, metals and inorganics, others	
By source	Natural: plants, bacteria, fungi	Synthetic: industrial reagents, products, or by-products; pharmaceuticals
By use	Pesticides, solvents, paints, dyes, coatings, detergents, cleansers, pharmaceuticals	
By action	Enzyme damage, metabolic poisoning; macromolecular bindings (e.g., DNA); cell membrane damage; sensitization, irritation	
By target organ	Affecting nervous system, blood, kidneys, liver, lungs, skin, metabolic processes Reproductive system and genetic effects, teratogens; carcinogens	

Sources: Adapted from Last, J. M. *Public Health and Preventive Medicine*, Thirteenth Edition. op. ci, 1992.
Samiullah, Y. *Prediction of the Environmental Fate of Chemicals*, 1990, op. cit.

Box 9.4 Basic Concepts of Toxicology

Bioavailability the ability of a substance that enters the body to be liberated from its environmental matrix (water, tissue, soil) and to enter the circulation of the host.

Dose–response relationship the relationship between the quantity of a toxicant received by the host and the probability of an effective concentration at the vulnerable site.

Intermediary metabolism the metabolic changes that a chemical undergoes once it reaches the cell of the body, usually in the liver. The substance may be detoxified to benign compounds, or may be converted to biologically harmful metabolites. The toxic substance acts on a cellular or subcellular level to disrupt the living organism. Some toxic agents are metabolic poisons; others act on cell membranes, interfere with chemical reactions, or bind to nucleic acids.

Susceptibility the ability of a living thing to be harmed by an agent, which may be influenced by age, sex, genetic disposition, nutrition, prior exposure, immune state or general health, stress, location at work, airflow, temperature, and humidity.

Threshold the lowest dose of a chemical that has a detectable effect.

Toxic effect damage to an organism as measured in terms of loss, reduction, or change of function, clinical symptoms, or signs. Effects may be adverse in one person and not in others.

Teratogens are substances that cause birth defects, diseases, or abnormalities in the embryo or fetus either by disturbing maternal homeostasis or by acting directly on the fetus. Birth defects historically were attributed to retribution for sin, witchcraft, or moral or physical defects in the mother. Scientific knowledge of genetic disorders has grown since the 1940s, and many agents have been shown to cause birth defects. Such agents act on fetal development and not on genetic DNA, so that a threshold effect is assumed; that is, the effect occurs only if the causative exposure is above a certain threshold. Some currently known teratogenic agents and their effects are shown in Table 9.5.

Toxic Effects of Lead in the Environment

In the United States in the 1920s, tetraethyl lead use in fuel was promoted to improve automobile performance. This caused a long struggle between public health regulatory agencies and the automobile industry. Industry won, and leaded gasoline was used well into the 1960s and is still available in many parts of the world. Alice Hamilton investigated the widespread use of lead in industry during the 1920s and successfully lobbied for legislative changes to increase surveillance and improve safety by reduced exposure (Box 9.5). Community exposure to lead was identified as a public health problem in the 1960s when trace quantities were found in food, beverages, soil, and air. It was found that the main sources of community exposure were from leaded fuels for cars and lead-based paints manufactured from the 1920s to the 1960s.

Children are particularly vulnerable to these environmental contaminants. Clinical effects appeared particularly in children and at lower blood concentration levels than previously thought to be significant. "Acceptable" levels were lowered and lead abatement programs introduced.

TABLE 9.5 Some Teratogens and Their Effects on the Fetus and Newborn

Teratogen	Effects on fetus and newborn
Maternal Infections	
Rubella	Congenital rubella syndrome, deafness, cataracts, heart defects
Syphilis, herpes simplex	Mental retardation, microcephaly
Cytomegalovirus	Infected kidney, liver, lungs
Toxoplasmosis	Central nervous system lesions
HIV	HIV neonatal transmission
Others — varicella, mumps, parvovirus	
Nutritional deficiency	
Protein deficiency	Abortion, prematurity, low birth weight
Folic acid deficiency	Anencephaly, spina bifida
Ionizing radiation	
X-rays or nuclear radiation or fallout	Central nervous system disorders, microcephaly, mental retardation
Drugs	
Alcohol	Mental retardation, microcephaly, facial defects
Cocaine	Prematurity, retardation, addiction
Thalidomide	Phocomelia (i.e., small deformed limbs)
Dilantin, valproic acid	Heart malformations, cleft palate, retardation, microcephaly
DES (diethylstilbestrol)	Vaginal cancer in girls, genital deformities in boys
Anesthesia	Miscarriages, structural deformities
Barbiturates	Heart defects, microcephaly, retardation
Chemicals and heavy metals	
Methyl mercury, lead, cadmium	Miscarriages, mental retardation, neurologic disorders
Dioxin	Physical deformities, miscarriage
Cigarette smoke — direct and "secondhand smoke"	Miscarriage, prematurity, low birth weight

Source: Adapted from Nadakavukaren, A. 1990. *Man and Environment*, Third Edition. Prospect Heights, IL: Waveland Press, p. 160. Boston Children's Hospital. http://www.childrenshospital.org/az/Site474/mainpageS474P0.html [accessed May 14, 2008]

Box 9.5 Alice Hamilton and Tetraethyl Lead

Alice Hamilton, a pioneering researcher and public health advocate in the 1910s and 1920s, demonstrated workplace hazards to toxic substances such as white phosphorus used in match production, lead additives to gasoline, and radium in watch dials. Tetraethyl lead (TEL) was produced and promoted by DuPont, despite being identified as hazardous.

Hamilton and others strenuously opposed its use, but TEL use expanded, and with it her research on behalf of state and federal government commissions. Environmental lead toxicity increased until the 1970s when further research revealed the extent of the problem and its public health effects, especially on children. Hamilton's work set standards for toxicology research in occupational and environmental health that led to the regulatory successes of the 1970s in the United States.

Source: Rossner, P., Markowitz, G. 1985. A "Gift of God"? The public health controversy over leaded gasoline in the 1920s. *American Journal of Public Health*, 331:161–167; and CDC. 1999. Improvements in workplace safety—United States, 1900–1999. *MMWR*, 48:461–469.

These programs were especially needed in urban slum areas where children were exposed to lead-based paints in older homes and heavy urban traffic and were found to have high blood lead levels (BLLs) with risk of brain damage as a result. In 1992, the American Academy of Pediatrics adopted lower BLLs levels as danger signs of lead toxicity sufficient to cause brain damage in children. Between 1991 and 1994 in the United States, 4.4 percent of children 1–5 years of age had elevated blood lead levels (exceeding 10 µg/dL). Current professional opinion is that there is no safe level of blood lead and that levels under 10 µg/dL are also harmful to the brains of young children (Box 9.6).

There may be no harm-free blood level of lead, and current recommendations include routine testing of infants and young children as well as exposed workers, along with environmental measures to reduce emission levels and industrial or home use of lead or lead-containing products. This topic has received much attention in the United States and Canada but less so in other countries. CDC lead control programs have targets of eliminating elevated blood lead levels in the United States by 2010. This

---

Box 9.6 Lead Abatement in the United States, 1977–2000

Reduction of elevated blood lead levels (BLLs >10 µg/dL) is one of the targets of Healthy People 2010. Studies in the United States based on the national representative surveys NHANES II (National Health and Nutrition Examination Survey) showed that elevated BLLs in 88.2 percent of children aged 1–5 years in 1976–1980 (13.5 million). The rate of elevated BLL fell to 8.6 percent in 1998–1991 (1.7 million children), to 4.4 percent in 1991–1994 (850,000 children), and 2.2 percent in 1999–2000 (434,000 children).

This reduction was due to a number of factors, including the following:

1. Reduction in use of lead in gasoline since 1976;
2. Reduced use of food and soft drink cans containing lead solder;
3. Reduced use of lead-based house paint;
4. Promulgation of national standards for lead exposure in industry;
5. A ban on the use of lead soldering on household plumbing;
6. Screening of children as part of routine child care and intervention where elevated BLLs found;
7. Lead abatement by county health departments through removal of lead-based paint in older housing;
8. Increased provider and parental awareness of lead-induced permanent brain damage hazard;
9. Strong positions of the Centers for Disease Control, American Academy of Pediatrics, state and county health departments, and child advocacy organizations;
10. Increased public awareness;
11. Regulation and elimination of use of lead paint on child toys;
12. Reduction of house lead paint exposure by removal, painting over, and aluminum covers.

Source: Centers for Disease Control. 2008. Children's blood lead levels in the United States, http://www.cdc.gov/nceh/lead/research/kidsBLL.htm [accessed May 14, 2008]

---

program focuses on assisting states and municipalities in lead poison prevention programs. WHO recommends preventive measures including:

- Environmental standards that remove lead from gasoline, paint, and plumbing;
- If lead pipes cannot be removed, cold water should be flushed through in the morning before drinking;
- Enforcement of occupational health standards;
- Surveillance of potentially exposed population groups, especially the vulnerable ones (small children, pregnant women, workers);
- Water treatment;
- Removing lead solder from food cans;
- Use of lead-free paint in homes;
- Screening of children for blood levels over acceptable limits and referral for medical care as necessary.

The health-based guideline for lead in drinking water is 0.1 milligrams per liter (WHO, 1993). If high levels are detected in a supply, alternative supplies or bottled water may be necessary to protect young children (WHO, http://www.who.int/water_sanitation_health/diseases/lead/en/print.html [accessed May 13, 2008]).

AGRICULTURAL AND ENVIRONMENTAL HAZARDS

Pesticide and herbicide use to increase agricultural production is a worldwide phenomenon. Resistance to widely used chemicals has developed so that there is a continuing search for new chemicals. Excess use affects the ecosystem by the buildup of pesticides in the food chain and in groundwater, and the long-term effects may be serious.

Short-term exposure to agricultural chemicals may result in acute poisoning, especially in developing countries where it is estimated to affect some 3 million persons with 220,000 deaths annually. Suspected concentration of pesticides in breast fat tissues may be linked with excess breast cancer risk. Pesticide use in North America and the former Soviet countries is high but has declined since the 1980s, while their use in western Europe is increasing. Widespread use of pesticides in developing countries is often poorly supervised, and pesticide poisoning episodes are common.

The use of herbicides and pesticides within the recommended limits of the *Codex Alimentarius* (joint foods standards manual of the Food and Agriculture Organization, or FAO, and the WHO) and methods recommended by the International Code of Conduct on the Distribution and Use of Pesticides are considered safe. Current recommended practice is to reduce the amounts of pesticide and herbicide use, accompanied by care and safe use and storage practices to reduce the chance of acute poisonings. Alternative agricultural methods, using little or no chemicals, are the subject of wide research and experimentation.

AIR POLLUTION

The External Environment

Air pollution is contamination of the air by smoke, solid material, or chemicals that cause health and ecological damage to the community and the environment. It includes the oxides of sulfur and nitrogen spread locally and over long distances, domestically and internationally. The effects are increasingly important as the demand and use of fossil fuels have grown for the internal combustion engine, heating, and power generation. Coal fuel used in homes created the terrible air pollution of nineteenth- and early twentieth-century London (Box 9.7). This has subsided since the 1950s with reduction of brown soft coal usage in individual homes. Use of coal fuel energy plants in central and eastern Europe has created a gray zone of air pollution carried long distances, destroying forests and creating serious damage to the human environment and health hazards to large population groups. Similarly, extensive damage has occurred in Canadian forests from acid rain originating in the United States.

Large modern fossil fuel plants built near population centers may use high chimneys to disperse the effluent. This reduces exposure of the adjacent population but contributes to long-distance pollutant effects, carrying sulfur and nitrogen oxides to forests and bodies of water creating sulfuric and nitric acid or acid rain. Acid precipitation affects rivers, streams, and lakes, many already burdened with sewage effluent and pesticide runoff, damaging the ecosystem and animal and plant life. The effects on human health are not easily measurable in a directly attributable way, but environmental damage affects the quality of life. International transmission of environmental damage is seen in nearly a quarter of Europe's forests from acid rain originating in eastern European countries with poor emission control standards.

The range of damage measured by the percentage of dead and dying trees varies from over 24 percent in central and western European countries (Denmark, Norway, the Netherlands, and Germany) to over 50 percent in some eastern European countries (the Czech Republic and Poland). In the 1979 Convention on Long-Range Trans-Boundary Pollution, European states agreed to reduce emissions that could cross international boundaries by 30 percent by 1993.

Air pollutants can enter the food chain by contaminating fish, fowl, and livestock. Changes in the acidity of water can create further harmful effects by corrosion of water pipes, affecting the lead, mercury, aluminum, cadmium, or copper content of drinking water. Acidified metals may cause chronic conditions such as chronic obstructive lung disease and asthma as well as specific chemical toxicity. Toxicity levels are difficult to measure

epidemiologically; regulation of source emissions are set as proxy measures for preventable exposure to unhealthful contaminants.

Particulate matter in air pollution has both physical and chemical effects on the nasopharynx and respiratory tract. Excess cancer of the respiratory tract and chronic obstructive lung disease can be demonstrated in exposed populations. A variety of syndromes are associated with specific respiratory irritants, such as coal dust (miners' lung), cotton dust (byssinosis), and others among exposed occupational groups. A study of regional cancer rates in Israel in the 1980s showed an excess of cancers of the nasopharynx and respiratory tract in persons living in an area exposed to high levels of silicate materials in emissions from a local cement plant. Geographic cancer epidemiology in the United Kingdom shows higher levels of many diseases in terms of standard mortality rates (SMRs) in urban or other polluted areas in Britain, correlating excess morbidity with excess air pollution (Chapter 3).

The London "killer fog" incident in 1952, implicated in up to 4000 deaths, raised international concern over the deadly effects of critical levels of pollution as well as the long-term effects. In Britain, this led to controls on use of soft coals for home fires and gradual reduction of the Victorian levels of smog that had fouled ambient air quality in British industrial and commercial centers. A similar inversion in 1948 in Donora, Pennsylvania, affected over 40 percent of the population of 14,000 with 20 deaths. A smog crisis in New York City in 1966 occurred a month before the third National Conference on Air Pollution, followed by a series of smog crises in California.

Localized air pollution is largely generated by the automobile as well as by general industry. Pollution of urban areas with lead, sulfur dioxide (SO_2), and nitric oxide (NO) has been reduced where catalytic converters and unleaded gasoline are compulsory. However, the beneficial effect is reduced simply by the increase in the number of automobiles, as shown in the southern California experience. During the 1960s and 1970s, there was a growing sense of crisis in environmental pollution in the United States. Until the 1970s, solid and liquid industrial wastes were dumped or discharged indiscriminately with volatile chemicals contaminating water sources and the air. Pollution of lakes and rivers and poor air quality in the cities led to a series of federal legislative acts, as well as the establishment of the Environmental Protection Agency (EPA) in 1970, including the Motor Vehicle Air Control Act of 1967, the Air Quality Act of 1967, the more effective Clean Air Act of 1970, the Clean Water Act of 1977, the Safe Drinking Water Act of 1974 (amended 1996), and the Water Quality Act of 1987.

Traffic congestion in modern cities exposes car occupants and pedestrians to exhaust fumes containing particulate matter and air pollutants. Pollutants may act to compound the ill effects of other risk factors such as smoking. Los Angeles is subject to heavy pollution and temperature inversions, producing harsh conditions for those prone to chronic bronchitis, asthma, and chronic obstructive lung disease.

A study in Los Angeles showed that an increase of 10 parts per million of carbon monoxide (CO) levels in the air was associated with a 37 percent increase in hospital admissions. Other large cities that have not implemented such pollution controls, such as Mexico City, continue to have serious air pollution levels.

Where the numbers of cars and trucks increase but pollution control measures are not required, the pollution can have alarming effects on child health in the form of increased blood lead levels and respiratory tract damage from other chemical and particulate pollutants. CO blocks red blood cell uptake of oxygen and can reduce the oxygen carrying capacity of blood. In vulnerable groups, such as children, the elderly, pregnant women, and the immunosuppressed, this can have serious deleterious effects on psychomotor function. Polycyclic hydrocarbons released from car emissions and other sources are carcinogens. Nitrogen oxides (NO_x) affect the terminal respiratory tract alveoli, increasing susceptibility to lower respiratory tract infection in children. Ozone (O_3) and secondary pollutants affect ultraviolet (UV) light absorption, increasing skin cancer incidence. Ozone can travel hundreds of kilometers, causing clinical effects close to and well away from the site of the traffic. Carbon dioxide (CO_2) affects global warming with potentially important effects on world climate and water supplies. The health and environmental effects of air pollutants pose a severe challenge to the global community.

Emission control through regulation and new technology should be seen in the context of overall transportation policy. Policy in transportation has long-term effects in determining degrees of air pollution, land use, and trauma from motor vehicle crashes. A full accounting of the costs of morbidity and mortality associated with air pollution and traffic accidents should be included in cost-effectiveness studies of rail versus road transport, especially in crowded urban communities and in countries with limited land space.

The U.S. Federal Clean Air Act of 1970 established air quality standards for major pollutants such as nitrogen oxides (NO_x), CO, SO_2, ozone, asbestos, dioxin, and other toxic air contaminants (TACs). Improving enforcement, especially of automobile emissions, has led to improved air quality in many parts of the country. Though federally legislated, implementation is at the state level. Standards are set for ambient air quality, automobile emissions, and emission by stationary facilities, such as power plants and factories. Such standards are also being implemented in many other countries.

The Clean Air Act Amendments of 1990 listed 189 hazardous air pollutants (HAPs) for which Congress mandated the EPA to issue standards. These include asbestos, dioxin, diesel, and many other potentially toxic agents, including latex, which has been identified as a factor in causing asthma. The FDA is continuing to develop standards for other HAPs. The Clean Air Act as amended provides for state agencies to regulate local air districts.

The California Air Quality Management Board regulates regional air quality management boards (e.g., southern California) that carry out a certification process of local industry. This board has powers to sanction changes in industrial practices in any given industry by attributing its component of ambient air pollution, with the potential for closing an offending industry. As a result, California was able to reduce air pollution dramatically since the mid-1980s, with only one major smog alert occurring in 1997 compared to 66 in 1987 in Los Angeles. Beginning in the late 1990s, technological innovations have become standard in some new automobiles, thus further reducing emissions and increasing air mileage per gallon of gasoline. Other innovations incorporate hydrogen fuel cells and hybrid and electric vehicles that will release nearly zero pollutants.

Consistent with the Clean Air Act, the U.S. goal is to reduce the proportion of persons exposed to air that does not meet the standards with a target of zero for 2010. Ozone affects 43 percent, particulate matter 12 percent, carbon monoxide 20 percent, nitrogen dioxide and sulfur dioxide 2 percent, and lead less than 1 percent of the population.

Diesel air pollutants became the subject of scrutiny by the state air resources board which carried out a meta-analysis and defined diesel pollution as a health hazard. This decision requires use of best available control technology (BACT) to reduce emissions from the defined "acceptable risk" of 10 excess cases of cancer per million population. In the case of diesel emissions the excess rate was determined to be 100 times in excess of the acceptable rate. Industry opponents raise the specter of tremendous economic effects of such decisions, but the BACT approach minimizes this potential harm to the economy. At the same time, the process of identifying the issue spurs industry to seek out technological solutions that are compatible with greater efficiency in the long run.

Methyl Tertiary Butyl Ether

Methyl tertiary butyl ether (MTBE) is a synthesis of methanol and isobutylene, developed as an additive to gasoline to improve octane performance, and was widely adopted in the United States, especially in California, in place of ethanol used in other states. Ethanol is a farm product that was encouraged by the U.S. Department of Agriculture as a new economic opportunity for farmers and relatively nontoxic agent environmentally. Some gasoline producers opted to use MTBE instead, as it is produced and promoted by the chemical industry.

MTBE is a volatile, ether-based chemical agent that when found in drinking water gives a bad taste. MTBE was adopted widely without adequate testing for potential toxic effects and has come under scrutiny. Evidence of carcinogenesis in rats raised concerns that MTBE may have the same effect in humans, especially to drivers, gasoline station attendants, and refinery workers exposed to high levels. The effectiveness of MTBE in promoting clean burning of gasoline and reducing exhaust pollutants has also been questioned. MTBE has been found in 3.4 percent of water districts in California. Some 50 percent of drinking water wells in Santa Monica, California, were closed due to MTBE contamination in 1995.

The MTBE case is an example of EPA-sanctioned use of a harmful chemical substance widely used in industrial settings as a gasoline additive, thus present in automobile emissions. This replaced a more environmentally safe substance from the farm industry even before environmental concerns brought the issue under public scrutiny. Since 1996, the American Public Health Association has called on the FDA to ban MTBE as a hazardous chemical, to place restrictions on the use of gas-powered boats on lakes and rivers, to return to ethanol-based fuel additives, and to inspect underground storage facilities for gasoline to reduce leakage and contamination of ground water. In 2000, the EPA included MTBE under the Toxic Substances Control Act and efforts to reduce MTBE contamination of ground and surface water supplies (EPA, 2007).

INDOOR POLLUTION

Contaminants within private dwellings may be a greater health hazard than external pollution. Indoor pollution particularly affects women, the very young, the ill, and the elderly because they usually spend more time within the home. Increased insulation, window layers, sealed doors, and smoking all contribute to increased concentration of indoor pollutants, including benzene, formaldehyde, carbon monoxide, and radon gas, as well as bacteria, fungi, and viruses. Smoking is a widespread habit, and passive smoking, or inhalation of smoke generated by other persons, is a long-term health hazard.

Wood and its waste products, vegetable matter, and animal dung are sometimes referred to as *bamboo fuels*. These are not efficient fuels in terms of heat produced per unit mass as compared to fossil fuels. Approximately half of the world's population depends on such fuels for their daily needs. These fuels are used extensively in rural areas of developing countries because they are cheap and widely available, but they require much time to gather and lead to deforestation with other damage to the environment. Often primitive stoves are used, which result in creating fire hazards and high levels of continuous daily indoor pollution due to poor ventilation.

The dangers associated with use of bamboo fuels include fires, smoke inhalation, and chronic indoor pollution. These fuels release many chemical compounds including suspended particulate matter, carbon monoxide, nitrogen and sulfur oxides, aldehydes, hydrocarbons, benzene, phenols, and complex hydrocarbons. Women in India show high rates of right heart failure (cor pulmonale) from cooking stove fumes. Technological development of more efficient wood stoves would reduce the problem; however, other forms of energy are more efficient and less damaging to health in the home and to the environment.

Indoor pollution from materials used in construction is a serious health problem. Asbestos in the home may contribute to mesothelioma and lung cancer. Lead paint in the home increases the hazard of lead toxicity among young children, which is associated with brain damage. Unsafely packaged household chemical solvents and mold in the home contribute to poisonings as well as asthma morbidity and mortality.

Radon Gas

Radon is a very heavy gas that produces harmful alpha particles as a by-product. Radon originates in the natural radioactive decay of uranium from soil and rocks such as granite, shale, and phosphate and is present as a gas in ground crevices, dissolved water, or dispersed open air. It seeps into homes via basement cracks and into well water and point sources. Radon was first detected in homes in the United States in 1984 near Philadelphia. Early investigations showed in-home radiation exposure as high as the equivalent of 455,000 chest x-rays. Further investigation revealed that sections of eastern Pennsylvania, New Jersey, and New York lie over uranium-rich geological formations that result in high levels of radon contamination.

The U.S. Environmental Protection Agency in 1988 advised that all homes be checked for radon levels. Inexpensive home radon detectors are available that meet EPA standards. In 1988, the EPA estimated that radon contributes to between 7000 and 30,000 cases of lung cancer per year, or up to 10 percent of all lung cancer deaths in the United States. Hundreds of thousands of U.S. citizens receive as much radiation as did persons living near the Chernobyl plant at the time of the nuclear accident in 1986. Cigarette smoking has a synergistic effect, enhancing the radon-related risk of lung cancer by a factor of 10. Radon reduction can be carried out in high-risk homes by carefully planned sealing of identified sources, ventilation, and fans for high radon basements.

Outdoor–Indoor Pollutants

Carbon monoxide, nitrogen oxides, chemicals, and particulate matter are common outdoor pollutants that can accumulate in homes with kerosene and wood stoves, attached automobile garages, or cigarette use. Passive smoking can expose the nonsmoker to benzene and other carcinogens. Formaldehyde is produced from insulation material, plywood, and floor coverings, especially in mobile homes. Chemical fumes from household products, such as disinfectants, solvents, hair sprays, furniture polish, and dry cleaning solvent, also pollute the home atmosphere as well as provide a potential for childhood poisonings.

Carbon monoxide poisoning from home heaters where there is inadequate ventilation causes 100 deaths per year in the United Kingdom. Some deaths from carbon monoxide poisoning may be attributed to heart disease and can only be diagnosed affirmatively by measurement of carbon monoxide in the air or blood carboxyhemoglobin levels.

Biological Pollutants

Bacteria and fungal spores can enter a building and infect inhabitants, usually through the ventilation system, as is the case with Legionnaires' disease (see Chapter 4). Occupants of a building may suffer from allergies due to fungal spores, mites, animal dander, and feces of roaches or mites. These allergies are more likely to occur in buildings using humidifiers or vaporizers with stagnant water, which favor bacterial and fungal growth.

Sick Building Syndrome

The term *sick building syndrome* is used to describe a common symptomatology (headache, eye and nose irritation, dizziness, fatigue, wheezing, or recurrent respiratory infections) among persons working in a specific building. This may occur as a result of a poor ventilation system that fails to provide sufficient fresh air relief from specific microbiological pollution of a ventilation or humidifier system, vehicle exhaust entering a ventilation intake vent, ozone emissions from photocopying machines, formaldehyde in wood paneling or furniture, or cigarette smoke.

Nonresidential buildings are often sealed with ventilation provided by mechanical means, so that such conditions can occur when the ventilation system is inadequate. Building codes should specify minimum levels of outside air admission, acceptable levels of oxygen, carbon monoxide, and carbon dioxide, odor dilution, and adequacy of ventilation equipment.

HAZARDOUS OR TOXIC WASTES

Toxic materials used in industrial processes can cause ill effects in workers exposed to the material at the site of production and in storage, transport, and use of the materials. They can also cause harmful effects to persons living near the material, as well to the environment. Case studies of serious environmental pollutants and their effects on health demonstrate the problems involved.

Hazardous wastes are defined as any discarded material that may pose a substantial threat to human health or the environment when improperly handled. They include toxic wastes such as arsenic, heavy metals, and pesticides which can cause acute or long-term health problems; ignitable wastes include organic solvents, oils, plasticizers, and paint waste; and corrosive wastes (with a pH of under 2 or over 12.5) which can eat away metal containers or living tissue.

Reactive wastes include obsolete munitions and acids that react with water or air to produce explosions or toxic fumes. Radioactive and infectious wastes from hospitals are also hazards to public health. Hospital wastes took on new importance with the dangers of transmission of hepatitis B, HIV, and drug-resistant microorganisms in contaminated materials. The problem caught worldwide attention in the late 1980s when waste material from hospitals washed onto beaches in the United States.

Prevention and waste site remedies have gained wide attention by industry; federal, state, and local government; and media and public concern generated by episodes such as the Love Canal of the late 1970s which served to mobilize public awareness of environmental health in the United States. In the 1890s, Mr. William T. Love built a canal bypassing Niagara Falls with the intent of building an industrial city using inexpensive hydroelectric power. The project failed and the canal was abandoned and the land sold at public auction. In 1942, the Hooker Chemical Company (subsidiary of the Occidental Petroleum Co.) received permission to use the canal to dump chemicals from its several plants in the area. Up to 1953, 21,000 tons of chemical wastes (acids, alkalis, solvents, chlorinated hydrocarbons, etc.) were disposed of at the site, when it was covered by landfill. Despite warnings, the land was sold and over 1000 homes, apartments, and schools were constructed along the covered canal.

Beginning in the 1950s, local residents complained of foul odors and chemicals oozing from the covered canal. In 1978, pressure from local congressmen and news media prompted investigation by the U.S. Environmental Protection Agency and the New York State Department of Health. Over 200 different chemicals were identified, including dioxin and 12 known or suspected carcinogens, mutagens, and teratogens. The New York State Commissioner of Health proclaimed an imminent health peril and called for evacuation of pregnant women and children under age 2. Over 1000 families were evacuated and 300 homes demolished at public expense. Work at the site to contain the chemicals and prevent seepage and groundwater contamination cost over $180 million. An initial investment of $2 million dollars by the Hooker Chemical

Company at the time of the disposal could have prevented the health damages.

Epidemiologic studies of the exposed residents showed that they experienced statistically significant elevated rates of miscarriage, birth defects, and chromosomal abnormalities, but the studies' methods and conclusions remain controversial. This episode focused national concern on the approximately 16,000 hazardous waste sites throughout the United States. In 1980, Congress established a superfund program, funded by federal tax on the chemical and petroleum industries, to locate, investigate, and clean up the worst sites in the country.

Minimata Disease

Minimata disease is a chronic neurologic disorder caused by methyl mercury, a heavy metal with many industrial uses. The disease was first reported near Minimata Bay in Japan in 1968 when mercury oxide was being discharged from a chemical plant into the waters of the bay. It was converted to an organic form, methyl mercury, by organisms in the mud and slime of the bay floor. Mercury poisoning of fish is a recurrent phenomenon where industrial wastes discharged into rivers, lakes, and the sea enter the food chain, and humans are affected through fish consumption. As of March 2001, 2265 victims had been officially recognized (1784 of whom had died) and over 10,000 had received financial compensation. Minimata disease is one of four major pollution diseases of Japan caused by environmental pollution due to improper handling of industrial wastes by Japanese corporations. Compensation, cleanup, and damages cost hundreds of millions of dollars. This episode also served to mobilize international public opinion to the dangers of toxic waste disposal as a health hazard. In 1999, people living in remote areas of Brazil were found to have methyl-mercury poisoning, probably from fish contaminated by methyl mercury used to purify gold.

Toxic Waste Management

Pollution prevention at the workplace has become part of management processes as industry responded to increasing federal and state regulation and as the public demand for greater corporate responsibility led to increasing punitive litigation. In 1986, the Federal Office of Technology Assessment published a comprehensive work on the topic entitled *Serious Reduction of Hazardous Waste*. An Organization of Economic Cooperation and Development (OECD) 1992 publication called on workers to play a greater role in pollution prevention. The chemical industry responded with the idea of total quality environmental management (TQEM), adopting pollution prevention as integral to industrial management. Companies like 3M,

Monsanto, and Rhone-Poulenc and industrial associations (the Chemical Manufacturers Association) undertook environmental prevention policies. Community activism helped industry to respond positively and openly to environmental hazards in their communities. The search for safe alternatives to toxic chemical waste management can be costly, but inevitably saves a company large expenditures in fines, litigation, and damage to corporate image. The issue is now very broadly shared among government, private industry, workers, and community, involving planners, scientists, engineers, regulators, residents, as well as environmental organizations and consumers. The EPA is currently promoting waste minimization of persistent bioaccumulative and toxic (PBT) chemicals from industrial sources. This includes source reduction and recycling aimed at reducing hazardous waste products (EPA, 2008). In the OECD Key Indicators for 2008, the quantity of municipal waste generated in the OECD area exceeded 650 million tons in 2006 or 570 kg per inhabitant.

RADIATION

In 1895, Wilhelm Roentgen's discovery of x-rays revolutionized medical science. Ionizing radiation includes particulate radiation of alpha and beta particles, as well as electromagnetic x-rays and gamma rays. Alpha particles are easily stopped by a thin sheet of paper, while beta and gamma radiation can penetrate barriers both inside and outside of the body. Ionizing radiation can dislodge atoms or parts of atoms and destroy chemical bonds. This can adversely affect living organisms, especially vulnerable fetal cells, resulting in mutations or carcinogenesis.

Ionizing Radiation

Ionizing radiation includes high-energy electromagnetic radiation, such as x-rays and gamma rays, which are of shorter wavelength and higher energy than ultraviolet or visible radiation. It also includes high-energy particles such as electrons, neutrons, protons, and alpha particles. Excessive exposure to these forms of radiation has early and late effects depending on dose and the tissue exposed. Early effects of exposure to high doses of radiation may be fatal due to acute damage to the gastrointestinal, erythropoietic (blood-forming), and central nervous systems. Late effects include malignant disease such as leukemia and birth defects. The principal sources of radiation exposure for the general public include natural background radiation from radon (55 percent), outer space (8 percent), the Earth (8 percent), inhaled or ingested materials (11 percent), medical exposures (15 percent), discharges from nuclear sources (1 percent), and consumer products (3 percent). Variation in background exposure from natural sources, home construction materials, and geographic location can be quite high.

Shorter exposures with high dosage are far more serious than long-term low-dose exposure. Radiation sickness of people exposed to radiation from the atomic bomb explosions at Hiroshima and Nagasaki and nuclear accidents ranged in severity with a variety of short- and long-term responses. The long-term responses have been less severe than originally feared.

Ionizing radiation of humans can act as a mutagen, a carcinogen, and a teratogen. It can cause cataracts, impaired fertility, premature aging, and skin damage. Radiation-induced cancer can occur as little as 2 to 5 years after exposure, or following a latency period of up to 25 years after exposure. Greater risk occurs for those exposed *in utero*. X-ray–induced disease from excess exposure, faulty equipment, or human error is a hazard of medical care. There is perhaps no safe exposure to ionizing radiation beyond atmospheric background, and any extra exposure should be limited, with prudent exposure to x-rays and limited exposure to atomic radiation from domestic or military uses.

Non-Ionizing Radiation

There are two types of non-ionizing radiation: optic and some electromagnetic fields. Optic radiation includes ultraviolet and infrared. Electromagnetic fields, such as those induced by microwave or radio frequencies, are described in terms of wavelengths or frequency. The harmful effects of non-ionizing radiation are of three main types: photochemical (sunburn or snow blindness), thermal, and electrical.

The health effects of ultraviolet (UV) radiation include increasing incidence of squamous and basal cell carcinoma and melanoma of the skin, a highly malignant cancer. This kind of radiation is associated with excess exposure to the sun, which in addition to these skin cancers, causes skin and eye burns, cataracts, reduced immunity, and damage to blood vessels. Infrared radiation exposure over long periods is associated with increased risk for cataracts, impaired fertility, and tissue damage.

Long-term exposure to cellular phone use, high voltage power lines, and radio and radar transmitters is suspected to be associated with increased risk of cancer, but this has not yet been proven. Microwave exposures at high levels can damage vulnerable tissues, but the level of dangerous exposure has not yet been conclusively determined. Lasers are pulsed electromagnetic waves used increasingly in medicine and industry. Lasers not intended for medical use (and misused medical lasers) can cause irreparable retinal damage and severe burns.

Use of low-dose irradiation in production, processing, and handling of foods to prevent food safety hazards is being widely supported by professional organizations. It provides an important adjunct to sanitation and good manufacturing practices to reduce morbidity and mortality associated with food-borne diseases even in industrialized countries. More than 40 years of research and use in the United States and many other countries have demonstrated the effectiveness and safety of low-dose irradiation. This is rapidly becoming an essential part of public health protection from food-borne disease in the United States and internationally, although public acceptance is still problematic.

ENVIRONMENTAL IMPACT

The U.S. National Environmental Policy Act (NEPA), passed in 1970, made protection and restoration of the environment matters of national policy. NEPA required all federal agencies to take environmental considerations into account in decision-making processes and program implementation. *Environmental Impact Statements* are required for major construction and public works programs, delineating positive impact, possible adverse effects, alternatives, and any irreversible effects. This legislation resulted in changes in many national projects and promoted a governmental regulatory approach to supervision, control, and prevention of pollution with materials and processes that could harm human health and the environment.

Emergency Events Involving Hazardous Substances

Since World War II, there has been a rapid increase in the number of chemicals developed and used worldwide. More than 60,000 chemicals are available, with some 600 new substances produced every year, an unknown number of which are hazardous. The health effects resulting from the release of a hazardous substance are often unknown. A hazardous substance release is defined as the uncontrolled or illegal release or threatened release of chemicals or their hazardous by-products.

Reportable events are defined as those events in which the substances need to be removed or cleaned up. Plant management is liable for damages due to negligence in both civil and criminal law. Where community exposure occurs from negligence, accident, or natural disaster, a public health emergency response is required, based on prior preparation.

In a 4-year period in the United States, 34,575 toxic chemical accidents were reported, of which 2186 involved deaths, injuries, or evacuations. These accidents released 680 million pounds of toxic chemicals into the environment. Almost two-thirds of these incidents involved one of 15 chemicals with polychlorinated biphenyls (PCBs) heading the list, followed by anhydrous ammonia, sulfuric acid, chlorine, hydrochloric acid, ethylene glycol, sulfur dioxide, radioactive materials, and hydrogen sulfide. The potential for chemical disasters requires a fundamental preventive approach by industry with supervision by federal and state regulatory agencies. In the United States, the major federal agencies involved are the Environmental Protection Agency (EPA) and the Occupational Safety and Health Administration (OSHA).

Even in industrialized countries, monitoring for heavy metals is problematic. The U.S. goal is to increase the number of states and territories that monitor for diseases from heavy metal environmental hazards. All states monitor for lead poisoning, while 20 monitor for pesticides, 14 for mercury, 10 for arsenic, 10 for cadmium poisoning, and 35 for birth defects. The goal is to monitor for exposure to pesticides in humans by measuring urine concentrations of metabolites. Linked health effect, exposure, and hazards data for environmental public health surveillance were used by 15 states in 2004.

Internationally, a number of major disasters have occurred in recent decades. In Seveso, Italy, in 1976, an explosion in a chemical factory resulted in 17,000 persons being evacuated and many terminations of pregnancy among exposed women. In 1984, a sudden release of highly toxic methyl isocyanide from a chemical plant in Bhopal, India, caused thousands of deaths, and blinding and permanent injury of several thousands more, requiring evacuation of an estimated 300,000 persons living in adjacent neighborhoods. While the transfer of hazardous occupations and industries to less-developed areas is a growing issue, the tragedy at Bhopal led to greater recognition by policymakers and the public that the potential for toxic accidents is everywhere, at any time, and any place, and not just in developed countries (Box 9.8).

Nuclear and chemical disasters have become a major element in disaster planning for corporate, investor, and occupational and environmental health agencies, as well as for communities adjacent to chemical production, storage, or transportation. Emergency responses to chemical, radiation, or biological catastrophes involve specialized expertise, based on common principles of prevention, monitoring, and crisis management. These include prior emergency planning, speed, coordination of civil and military resources, skilled professional teams providing information to the public, logistical, medical, and laboratory support, on-site case management and evacuation, investigation of causes, and continuous teamwork among all involved agencies.

The *Exxon Valdez*, a large oil tanker that ran aground in Alaska in 1989, spilling large amounts of crude oil in Prince William Sound, served as a precedent case in acknowledging that personal and fiscal responsibility for cleanup and other costs to reduce the environmental

---

Box 9.8 Nuclear Accidents: Three Mile Island (1979) and Chernobyl (1986)

Three Mile Island In 1979, the nuclear plant at Three Mile Island in Pennsylvania suffered a near disaster that devastated the plant but did not release nuclear material. It led to review of safety procedures and heightened public concern as to the overall safety of nuclear energy facilities. There was no loss of life or radiation release.

Chernobyl In 1986, a nuclear energy plant located at Chernobyl in the Ukraine suffered a meltdown that breached the integrity of the containment vessel and caused a massive explosion of the reactor. Design problems and a series of staff errors led to loss of control of the reactor with power levels soaring to 120 times the normal, rupturing of the fuel rods, and vaporizing the cooling system. A steam explosion then blasted open the 100-ton concrete slab covering the reactor, starting uncontrollable fires. Despite valiant attempts by emergency personnel and staff, the fires could not be controlled immediately. Air dropping of sand, lead, clay, and limestone led to fire control, but the heat of the reactor and radiation could not be reduced for many days. Immediate deaths numbered 33 individuals, mostly among the firefighters, with 237 suffering acute radiation poisoning. 135,000 people were evacuated from a 19 square mile area.

The nuclear fallout material carried in a 1900-foot plume, including iodine-131, cesium-137, and xenon isotopes, spread across much of Europe. Fallout reached some 20 countries, and an international public health threat of major proportions occurred. Ten years after the incident, there was a highly significant increase in thyroid cancer cases in children in the three affected countries: Ukraine, Belarus, and Russia. The long-term effects in terms of increased cancer and birth defects are hard to assess, but current estimates are of 500 (1–2%) additional cancer cases among 100,000 persons exposed to 10–20 rads. The actual increase in incidence of thyroid cancer, other cancers, and birth defects and the general impact on health will only be determined by careful epidemiologic follow-up of the exposed population over many years. The economic impact of the disaster is estimated at over $19 billion, and replacement of the plant reaches a similar sum. Close to the tenth anniversary of the Chernobyl disaster, a second nuclear leak nearly occurred due to human error. The Ukrainian government reopened the second reactor in 1999. International assistance in technical and financial aspects of nuclear energy in the Ukraine is in process.

Sources: International Conference One Decade after Chernobyl. 1996. Sponsored by WHO, International Atomic Energy Agency, and other international agencies. Austria. International Atomic Energy Association. http://www.iaea [accessed May 13, 2008]; World Nuclear Association. http://www.world-nuclear .org/info/chernobyl/inf07.htm [accessed May 13, 2008]; United States Nuclear Regulatory Commission. 2007. http://www.nrc.gov/reading-rm/doc-collections/ fact-sheets/3mile-isle.html and http://www.nrc.gov/reading-rm/doc-collections/fact-sheets/fschernobyl.html [accessed January 28, 2008]

damage lies with the company that owns the ship. Cleanup efforts required enormous amounts of money, and the spill became a *cause célèbre* for the environmental movement. The incident highlighted the importance of monitoring of seagoing chemical and fuel vessels. New cleanup techniques have been researched and applied in recent years. In response, the Coalition for Environmentally Responsible Economics (CERES) of investment fund advisors and social advocates established the "CERES principles" demanding environmental monitoring of corporations regarding energy use, public disclosure, damage compensation, sustainable use of natural resources, and environmental representatives on boards and in management of corporations.

Man-Made Disasters, War, Terrorism

The man-made disaster of war has used chemical, biological, and nuclear methods of destruction as well as traditional methods of warfare including economic blockade. War's offspring, terrorism, has used chemical armamentaria and may use biological or even nuclear destruction sooner or later. When disasters occur, lessons can be learned to improve services for the next disasters, be it natural or man-made (see Chapter 7).

Since World War I, poison gas has been used as a weapon against both frontline troops and civilian populations. This practice continued almost to the end of the twentieth century. Poison gas was used with deadly efficiency by the Nazis in the Holocaust of the Jews in World War II. Egypt used poison gas in its war in Yemen in the 1960s, and in the 1980s Iraq targeted Kurdish civilian villages killing thousands of men, women, and children. Defoliants (Agent Orange) used widely during the Vietnam War are believed to have long-term effects on Vietnamese civilians and on exposed military personnel.

During the Gulf War of 1991, the potential use of poison gas in long-range rockets on civilian population targets was narrowly averted. Several years later, thousands of U.S. service personnel reported a variety of neurologic symptoms and general fatigue. By 1996, these cases were acknowledged by the Department of Defense as possible long-term sequelae of accidental exposure by troops to toxic agents following destruction of Iraqi chemical weapons, or due to antidotes taken for potential gas warfare exposure (Soman). In 1995, a chemical attack with an extremely dangerous chemical warfare agent (sarin) was carried out by an extremist cult in Japan on subway passengers in Tokyo, resulting in 12 deaths and 3000 injuries with hundreds of hospitalized persons.

Terrorist bombing incidents occurred in many parts of the world during the 1990s. The 1995 bomb detonated by domestic terrorists in a federal building in Oklahoma City in the United States killed over 160 persons. Terrorist bombings of a U.S. military housing complex in Saudi Arabia, U.S. embassies in Africa in 1998, Moscow apartment buildings in 1999, and Israeli public bus lines caused large numbers of deaths and injuries. Each incident resulted in national concern over the threat of terrorist action causing mass casualties. Destruction of pipelines and oil fields caused extensive environmental damage in the aftermath of the Gulf War in 2001.

The largest terrorist attack was the hijacking of commercial civilian aircraft and crashing two into the World Trade Center in New York City on September 11, 2001, causing some 3000 deaths and massive fear and outrage initiating a worldwide war on terror. The Islamic jihad movement led by Osama Bin Laden created a world system of semi-independent cells operating with terrorist motives and methods creating large loss of life in a train bombing in Madrid in 2004 and in London in 2005. The fear of such events increased when people realized that many bombs could be made with easily obtainable chemicals or explosives, with the potential for nuclear or "dirty bombs" causing destruction of human life and property on a truly massive scale.

Unmarked land mines cause continuous loss of life and limbs, often among farmers and children. Millions of land mines are present in many areas of conflict, and cleanup is dangerous and costly. Between 2003 and 2005, there were an estimated 7000 land mine deaths and casualties a year worldwide. The most were in Iraq, Afghanistan, Cambodia, and Colombia. Land mines limit land and water use and have serious economic consequences for farmers. An international movement to ban the use of land mines gained international prominence with support by Princess Diana and by awarding the 1997 Nobel Peace Prize to Jody Williams, founder of this movement. Prevention is achieved by raising awareness and political action to prevent land mine use and support efforts for land mine clearance.

The potential for intentional, negligent, or accidental disasters, whether man-made or natural, is a real and present danger requiring health officials to coordinate with civil defense and military authorities to prepare disaster plans and continuously prepare for such events. Planning can greatly reduce the number and severity of casualties of toxic chemical disasters.

Preventing and Managing Environmental Emergencies

Public health has an important role to play in prevention, management, and mitigation of the effects of man-made and natural disasters. The U.S. Congress passed the Emergency Planning and Community-Right-to-Know Act (EPCRA) of 1986 following the 1984 Bhopal disaster. This legislation established state and local agencies for managing chemical emergencies. It requires facilities that handle hazardous chemicals to make information available to the public and preparedness for possible chemical accidents. This involves a holistic approach integrating technology, procedures, and management practices. The first responsibility lies with management, which must have a high level of awareness and commitment to accident prevention and safe practices. The range of industries at risk is very broad in modern societies. It includes local dry cleaners and furniture manufacturers as well as the chemical industry. The right to know extends from governments, professional societies, trade associations, labor unions, the research community, the news media, and environmentalists as well as the general public. The right to know has become the need to know.

Environmental emergencies occur from release of chemicals or radiation into the air. Inhalation and fallout effects downwind of the site depend on weather conditions and dispersion of the smoke plume. Clinical management of exposed civilians and emergency personnel is an activity of health management that involves organizing triage and transportation services at the site of the disaster. The decision to evacuate civilians is often made with limited information but must take into account the potential for exposure during evacuation, weighed against the protective effect of sealing homes and staying indoors.

The approach to management of environmental health problems requires a continuum of interrelated activities and phases ranging from prevention, preparedness, detection, response, and recovery. It includes measures to rebuild infrastructure and also lives and livelihoods affected by the emergency. Prevention is ultimately the most cost-effective and cost-beneficial means of dealing with potential environmental health problems. The principles of disaster and environmental preparedness include:

- Planning
- Coordination with sister agencies
- Preparation
- Research
- Adaptation with science and technology
- Training
- Monitoring
- Supply
- Detection
- Prevention
- Event
- Response
- Revision based on lessons learned

The appropriate team to handle such a situation involves public health, occupational health, and epidemiology investigators as well as police, fire services, civil defense, armed forces, chemical warfare units, and psychological staff. Post-disaster recovery planning is part of the planning process (disaster planning is discussed in Chapter 7). Long-term effects include post-traumatic stress disorder (PTSD) that can result in serious psychological dysfunction

in affected individuals. PTSD can be alleviated by early psychological support for victims of mass disasters at the site and at evacuation or follow-up centers and should be part of emergency care planning.

Rapid risk assessment involves weighing the hazard, exposure potential, dose–response, as well as both short- and long-term risks. Command centers and designated leaders are needed to maintain control of the multitude of needs for information, coordination between agencies, and the distribution of resources to areas of greatest need. Long-term epidemiologic assessment may be necessary for legal and compensation purposes, as well as for training and preparation for future events.

Advocacy is a public health function, and environmental and safety issues are areas where advocacy can bring important public benefit. Leadership in defining public health problems and in defining necessary action to reduce risk factors, or short- or long-term ill effects, requires skill in interpretation of epidemiologic events and studies, providing perspective for policymakers for addressing those issues.[1]

ENVIRONMENTAL HEALTH ORGANIZATION

The *World Health Organization Commission on Health and Environment Report* (1992) represents a consensus documentation of international environmental health issues. This commission, chaired by Simone Weil of the European Parliament, included many distinguished scientists, professional leaders, and international organizations. The report represented a strong international consensus on joint action to prevent and clean up environmental degradation that had occurred in Europe over several decades.

National organization for environmental health can take various forms. In the past, it was common for ministries of health to have environmental health departments, but in recent years this has increasingly moved to ministries of the environment.

Since 9/11, there is increased governmental and public concern regarding possible emergencies along with environmental decay and incidents of food or waterborne disease. The possibilities of natural and man-made disasters in the environment call for renewed efforts to prepare emergency plans and conduct suitable training, and exercises with public health, hospital, and primary care centers, as well as ambulance, fire, police, and military services is of great importance (see Chapter 10) (Box 9.9).

---

[1]Information and technical assistance bulletins are available by contacting the following: Emergency Planning and Community-Right-to-Know Information Service, U.S. Environmental Protection Agency, OS 120, 401 M Street SW, Washington, DC. See website http://www.epa.gov/ and Department of Homeland Security at http://www.dhs.gov/index.shtm [accessed May 14, 2008]

Box 9.9 Emergency Procedures for Hazardous Substances, Chemical or Radiation Disasters

1. Plan and prepare local public health, hospital, and other first responders for possible chemical or radiation disasters or attacks;
2. Establish early warning reporting and communication procedures;
3. Contain and reduce spread of toxin;
4. Inform the community to remain inside homes or other buildings;
5. Notify municipal, state, and federal emergency organizations;
6. Minimize exposure by ensuring the potentially exposed population remains inside, with the affected areas sealed off and quarantined or by limited evacuation;
7. Identify, decontaminate, and triage exposed persons;
8. Measure exposure and reaction;
9. Determine causative agents and antidotes;
10. Initiate antichemical procedures for exposed persons including removal of clothing, showers, antidote;
11. Coordinate on-site triage and evacuation for medical care;
12. Ensure medical or hospital care for exposed persons;
13. Provide accurate information to the public;
14. Promote health and supportive care at evacuation sites;
15. Investigate — professional and criminal;
16. Compensate the injured or displaced;
17. Pursue civil and criminal charges against negligent management persons and corporations;
18. Provide documentation and recommendations from lessons learned;
19. Review procedures and revise disaster plan operation;
20. Promote public and professional discussion.

Sources: Khan, A. S., Levitt, A. M., Sage, M. J. 2000. Biological and chemical terrorism: Strategic plan for preparedness and response: Recommendations of the CDC Khan Strategic Planning Workgroup. *Morbidity and Mortality Weekly Report*, 49(RR-4)1–26.
http://www.epa.gov/superfund/programs/er/hazsubs/ [accessed May 13, 2008]
http://www.epa.gov/emergencies/index.htm [accessed May 13, 2008]

Because of concern for environmental decay and fragmentation of government regulation efforts, the United States established the EPA in 1970 as the head federal agency reporting to the president to coordinate the administration of a wide range of environmental health problems. The EPA sets standards and regulations for a variety of legislation pertaining to the environment, such as air and water pollution, solid and hazardous waste management, noise, public water supplies, pesticides, and radiation. Despite the growth of the EPA and its control of a superfund to reduce toxic and other waste sites, interagency coordination is complex. In the U.S. federal government, a wide variety of agencies located in different government departments have responsibilities related to

the environment (Table 9.6). The substantial environmental progress made in the United States in the past 25 years is outlined in Box 9.10.

OCCUPATIONAL HEALTH

INTRODUCTION

The CDC considers that improvement in workers' health and safety was one of the ten great achievements of public health in the United States in the twentieth century. The U.S. National Safety Council reports from 1933–1997 indicate that deaths from unintentional work-related

TABLE 9.6 U.S. Federal Government Agencies with Environmental Responsibilities

Environmental Protection Agency (Independent)	Consumer Product Safety Commission (Independent)
Council on Environmental Quality (Executive Office)	Public Health Service (HHS)
Nuclear Regulatory Commission (Independent)	Centers for Disease Control (HHS)
Office of Environmental Safety and Health (Dept. Energy)	Food and Drug Administration (HHS)
Office of Environmental Management (Dept. Energy)	Agency for Toxic Substances and Disease Registry (HHS)
Office of Environmental Policy and Assistance (Dept. Energy)	Occupational Safety and Health Admin., OSHA (Dept. Labor)
Office of Surface Mining Reclamation and Enforcement (Dept. Interior)	Mine Safety and Health (Dept. Mines)
Bureau of Land Management (Dept. Interior)	Fish and Wildlife Service (Dept. Interior)
Center for Environmental Health, CDC (HHS)	Soil Conservation Service (Dept. Agriculture)
National Institute of Occupational Safety and Health, NIOSH (CDC)	Department of Homeland Security

Note: The department under which each agency falls is listed in parentheses. HHS = Department of Health and Human Services.

Box 9.10 Environmental Milestones in the United States, 1970–2006

1970 President Richard Nixon creates EPA; mission to protect the environment and public health; Congress amends the Clean Air Act to set national air quality, auto emission, and antipollution standards.

1971 Congress restricts use of lead-based paint in residences and on cribs and toys.

1972 EPA bans DDT, a cancer-causing pesticide, and requires extensive review of all pesticides.

U.S. and Canada agree to clean up the Great Lakes, which contain 95 percent of America's fresh water and supplies drinking water for 25 million people.

Congress passes the Clean Water Act, limiting raw sewage and other pollutants flowing into rivers, lakes, and streams.

Only 36 percent of the nation's assessed stream miles were safe for fishing and swimming; in 2006 about 60 percent are safe for such uses.

1973 EPA begins phasing out leaded gasoline; OPEC oil embargo triggers energy crisis, stimulating conservation and research on alternative energy sources. EPA issues its first permit limiting a factory's polluted discharges into waterways.

1975 Congress establishes fuel economy standards and sets tailpipe emission standards for cars, resulting in the introduction of catalytic converters.

1976 Congress passes the Resource Conservation and Recovery Act, regulating hazardous waste from its production to its disposal.

President Gerald Ford signs the Toxic Substances Control Act to reduce environmental and human health risks; EPA begins phase-out of cancer-causing PCB production and use.

1977 Clean Air Act Amendments to strengthen air quality standards and protect human health.

1978 Residents discover Love Canal, New York, is contaminated by buried leaking chemical containers; federal government bans chlorofluorocarbons (CFCs) as propellants in aerosol cans; CFCs destroy the ozone layer, which protects the earth from harmful ultraviolet radiation.

1979 EPA demonstrates scrubber technology for removing air pollution from coal-fired power plants. This technology is widely adopted in the 1980s.

Three Mile Island nuclear power plant accident near Harrisburg, Pennsylvania, increases awareness and discussion about nuclear power safety. EPA and other agencies monitor radioactive fallout.

1980 Congress creates a superfund to clean up hazardous waste sites. Polluters are made responsible for cleaning up the most hazardous sites.

1981 National Research Council report finds acid rain intensifying in the northeastern United States and Canada.

1982 Congress enacts laws for safe disposal of nuclear waste. Dioxin contamination forces the government to purchase homes in Times Beach, Missouri. The federal government and the responsible polluters share the cleanup costs.

A PCB landfill protest in North Carolina begins the environmental justice movement.

1983 Cleanup actions begin to rid the Chesapeake Bay of pollution stemming from sewage treatment plants, urban runoff, and farm waste.

EPA encourages homeowners to test for radon gas, which causes lung cancer; more than 18 million homes tested for radon. Approximately 575 lives are saved annually due to radon mitigation and radon-resistant new construction.

1985 Scientists report that a giant hole in the earth's ozone layer opens each spring over Antarctica.

1986 Congress declares the public has a right to know when toxic chemicals are released into air, land, and water.

1987 The United States signs the Montreal Protocol, pledging to phase out production of CFCs.

Medical and other waste washes up on shores; beaches closed in New York and New Jersey.

1988 Congress bans ocean dumping of sewage sludge and industrial waste.

1989 *Exxon Valdez* spills 11 million gallons of crude oil in Alaska's Prince William Sound.

1990 Clean Air Act Amendments require states to show progress in improving air quality.

EPA's Toxic Release Inventory tells the public which pollutants are being released from specific facilities in their communities.

Number of chemicals listed in EPA's Toxic Release Inventory nearly doubled, from 328 in 1990 to 644 in 1999.

Pollution Prevention Act signed emphasizing the importance of preventing, not just correcting, environmental damage.

National Environmental Education Act signed, for educating the public to ensure scientifically sound, balanced, and responsible decisions about the environment.

1991 Federal agencies begin using recycled content products. EPA launches voluntary industry partnership programs for energy-efficient lighting and for reducing toxic chemical emissions.

1992 EPA launches the Energy Star® Program to help consumers identify energy-efficient products.

1993 EPA reports secondhand smoke contaminates indoor air, posing serious health risks to nonsmokers. Today, more than 80 percent of Americans protect their children from secondhand smoke exposure at home.

A *Cryptosporidium* outbreak in Milwaukee, Wisconsin's drinking water sickens 400,000 people and kills more than 100.

Federal government uses its $200 billion annual purchasing power to buy recycled and environmentally preferable products.

Continued

Box 9.10 Environmental Milestones in the United States, 1970–2006—Cont'd

1994   EPA Brownfields Program to clean up abandoned, contaminated sites to return them to productive community use.

        EPA issues new standards for chemical plants to reduce toxic air pollution by more than half a million tons each year, equivalent of removing 38 million vehicles annually.

1995   EPA launches an incentive-based acid rain program to reduce sulfur dioxide emissions.

        EPA requires municipal incinerators to reduce toxic emissions by 90 percent from 1990 levels.

1996   Public drinking water suppliers required to inform customers about chemicals and microbes in their water; funding made available to upgrade water treatment plants.

        Vast majority of American households now have safe drinking water.

        EPA requires that home buyers and renters be informed about lead-based paint hazards.

        Food Quality Protection Act signed to tighten standards for pesticides used to grow food, with special protections to ensure that foods are safe for children to eat.

1997   Executive Order issued to protect children from environmental health risks, including childhood asthma and lead poisoning.

        EPA issues tough new air quality standards for smog and soot, an action that would improve air quality for 125 million Americans.

1998   Clean Water Action Plan announced to continue making America's waterways safe for fishing and swimming.

1999   New emissions standards for cars, sport utility vehicles, minivans and trucks, requiring them to be 77–95 percent cleaner in the future.

        EPA announces new requirements to improve air quality in national parks and wilderness areas.

2000   EPA establishes regulations requiring more than 90 percent cleaner heavy-duty highway diesel engines and fuel.

2002   Small Business Liability Relief and Brownfields Revitalization Act were signed to reclaim and restore thousands of abandoned properties.

2003   Healthy Forests Restoration Act signed to prevent forest fires and preserve nation's forests.

        Over 4000 school buses to be retrofitted through the Clean School Bus USA program, removing 200,000 pounds of particulate matter from the air over next decade.

        Clear Skies legislation and alternative regulations proposed to create a cap and trade system to reduce SO_2 emissions by 70 percent and NO_X emissions by 65 percent below current levels.

2004   New, more protective, 8-hour ozone and fine particulate standards go into effect across the country.

        Clean Air Rules of 2004 are proposed that will make people healthier.

        EPA requires cleaner fuels and engines for off-road diesel machinery such as farm or construction equipment.

2005   EPA issues the Clean Air Interstate Rule and the Clean Air Mercury Rule.

2006   WaterSense is a program to raise awareness about the importance of water efficiency, ensure the performance of water-efficient products and provide good consumer information. Clean Water Restoration Act of 2007 to clarify federal jurisdiction in surface waters in process in Congress.

Sources: http://www.epa.gov/earthday/history.htm [accessed May 10, 2008]; http://infotrek.er.usgs.gov/traverse/f?p=NAWQA:HOME:5755038819145367 [accessed May 10, 2008], http://epa.gov/watersense/ [accessed June 14, 2008]

---

injuries declined 90 percent, from 37 per 100,000 workers to 4 per 100,000, a "reduction of the number of deaths from 14,500 to 5,100; during this same period, the workforce more than tripled, from 39 million to approximately 130 million" (CDC, 1999).

Occupational health is the promotion and maintenance of the highest levels of physical, mental, and social well-being of workers in all occupations by preventing departures from health, controlling risks, and adapting of work to people and people to their jobs (International Labour Organization and WHO, 1950). Diseases related to occupations, always an essential part of public health, increasingly relate to environmental health, but to other fields as well. The worker is also a member of a family and a breadwinner, so that the health of the worker is related to family health. The worker is concerned not only with what happens at the place of employment but also with hazardous agents he or she might accidentally bring home. The retired or laid-off worker is worried about well-pensioned and honorable retirement. Occupational health in this wider context has an important place in the New Public Health.

DEVELOPMENT OF OCCUPATIONAL HEALTH

Occupational health is one of the oldest sectors of public health, dating back to Roman times. Documentation of occupational diseases was begun in 1700 by Ramazzini. Historic examples of work-related health hazards and diseases include scurvy among sailors, cancer of the scrotum specific to chimney sweeps in nineteenth-century England, black lung in coal miners, mercury poisoning in hat makers, byssinosis in cotton mill workers, and mesothelioma in asbestos workers. The list is long and extends to musculoskeletal injuries and hepatitis B in hospital workers, spinal disorders in typists, and medial neuritis in computer workers (carpal tunnel syndrome). Interventions vary widely from the banning of asbestos usage to modifying the office work environment through better chairs, exercise breaks, and ergonomic training of workers.

During the early part of the nineteenth century, the harsh working conditions of children, women, and other workers led to parliamentary action to regulate mines and factories, improving conditions generally. The first factory inspectors in the United Kingdom were appointed in 1833 to administer the provisions of the Factory and Workshops Acts. In 1898, Thomas Legge became the first medical doctor appointed to the post of Chief Factory Inspector in the United Kingdom. He articulated the basic public health approach to workers' health and established the principle that management is responsible for the health of the employees. These issues are termed Legge's Axioms and are still relevant to the field of occupational health today (Table 9.7).

Government responsibility for setting standards, monitoring, intervening, and regulating compensation grew slowly over the past century. Case reports, epidemiologic studies, and advocacy regarding the effects of lead, asbestos, vinyl chloride, silica, and dust fibers led to steps to reduce the hazards to workers and provided the professional support for legislative initiatives. International standards developed by the League of Nations, the International Labour Organization (ILO), and other international organizations promoted development of this field.

THE HEALTH OF WORKERS

The health of workers is subject to normal health threats for the adult population, but there are specific threats to health associated with the work situation. Workers have lower death rates from the general population because they are demographically different from the general population and even epidemiologically different from a population matched for age and sex. This is due to the fact that there is a process of selection of workers that excludes the severely ill and disabled from employment. The selection process continues with attrition of unhealthy persons from the workplace. This is termed the *healthy worker effect* and is a factor to be considered in occupational health studies and practice. Death rates or other population-based norms from the general population may be inappropriate for comparison if this effect is not taken into account. Case matching or control studies may be needed to accommodate this phenomenon. Other population groups such as immigrants or refugees go through similar selection, where only the healthy may be included or survive.

THE BURDEN OF OCCUPATIONAL MORBIDITY AND MORTALITY

In the United States, the workforce is made up of 135 million persons with 71.4 percent between the ages of 25 and 54. Premature disease, injury, and death related to occupational

---

TABLE 9.7 Thomas Legge's Axioms on Workers' Health and Modern Equivalent

Legge's axioms	Modern version
1. Unless and until the employer has done everything — everything means a good deal — the workman can do next to nothing to protect himself.	Don't blame the victim; the health of workers is the responsibility of management.
2. If you can bring an influence to bear, external to the worker, that is one over which he can exercise no control, you will be successful; if you cannot or do not, you will never be wholly successful.	Structural change is best.
3. Practically all industrial lead poisoning is due to inhalation of dust and fumes.	If you stop the exposure, you stop the poisoning.
4. All workmen should be told something of the danger of the material with which they come into contact and not be left to find out for themselves — sometimes at the cost of their lives.	Workers have the right to know of potential hazards to their health at their place of employment.

Sources: Hunter, D. 1969. *The Diseases of Occupations*, Fourth Edition. London: The English Universities Press Ltd.
Harrington, J. M. 1999. 1998 and beyond — Legge's legacy to modern occupational health. *Annals of Occupational Health*, 43:1–6.

exposures are a major burden on the economy and the health system. Over the period from 1980–1994, a total of 88,622 workers died in the United States from work-related injuries and an additional 60,000 died from occupational diseases. The estimated annual cost of occupational injuries is $128–$155 billion.[2]

Deaths due to work injuries declined from 8.9 per 100,000 workers in 1980 (7400 deaths) to 5.6 per 100,000 in 1989 (5714), 4.3 per 100,000 (5290 deaths) in 2000 and 4.3 per 100,000 (5734 deaths) in 2005. In 2006, there were 5703 work-related injury deaths in the United States (about 16 deaths per day), and more than four million new nonfatal injuries. Deaths from work-related disease are estimated at 100,000 annually. The largest numbers of deaths occur in the following industries: construction (22.6 percent), transportation/communications/public utilities (18.3 percent), and manufacturing (14.0 percent). The decrease in occupation-related deaths from 1980–2005 is related to the cumulative effect of increased awareness and regulation of worksite dangers and toxins, as well as new technology and mechanization, changes in the economy, and workforce distributions.

Even though occupational deaths have been declining, permanent impairment suffered on the job grew during the 1980s. During 1987, permanent impairments were suffered by 70,000 workers, and the number of totally disabled increased to 1.8 million. Further, there is an increasing rate of lost workdays. The decline in mortality and increase in work injury may be due to improving care of the injured or to a real increase in the number of injuries.

In 2001 in the United States, there were 5.7 nonfatal injuries with lost workdays per 100 employees in the private sector, a reduction of 34 percent from 1992. There is a trend toward substantial improvements in the more dangerous occupations such as agriculture, fishing and forestry, mining, construction, and manufacturing during the 1990s and early twenty-first century, as seen in Table 9.8.

The ten most frequent work-related diseases and injuries in the United States are:

1. Lung disease;
2. Musculoskeletal injuries;
3. Cancers;
4. Severe trauma;
5. Cardiovascular disorders;
6. Disorders or reproduction;
7. Neurotoxic disorders;
8. Noise-related hearing loss;
9. Dermatologic conditions;
10. Psychological strain and boredom.

Occupational Health Priorities in the United States

Priorities for research in occupational health are focusing on work-related anxiety and neurotic disorders, hearing loss, musculoskeletal disorders, injuries (fatal and nonfatal), poisoning, respiratory conditions, and skin disorders. These are the most common occupational health issues and costliest to the economy (CDC Worker Chartbook, 2004). Injury surveillance in the United States is maintained by the CDC's National Institute of Occupational Safety and Health.[3]

INTERNATIONAL ISSUES IN OCCUPATIONAL HEALTH

Occupational health has become an international issue as the global economy transfers manufacturing from one country to another with great speed and ease (Box 9.11). This is often motivated by lower wages and also by lower occupational and environmental regulatory controls and less stringent or nonexistent legal protection against toxic exposures and child labor in developing countries. Transfer of occupational hazards from industrialized to non-industrialized countries has become an issue in international cooperation and trade agreements. Developed countries have stricter environmental regulations and worker organization than developing countries that are anxious for job-producing industry at any price.

In 2000, risk factors were responsible worldwide for 37 percent of back pain, 16 percent of hearing loss, 13 percent of chronic obstructive pulmonary disease (COPD), 11 percent of asthma, 8 percent of injuries, 9 percent of lung cancer, and 2 percent of leukemia. These risks at work caused 850,000 deaths worldwide and resulted in the loss of about 24 million years of healthy life. Needlestick injuries accounted for about 40 percent of hepatitis B and hepatitis C infections and 4.4 percent of HIV infections in health care workers. Exposure to occupational hazards accounts for a significant proportion of the global burden of disease and injury, which could be substantially reduced through application of proven risk prevention strategies.

---

[2]The National Electronic Injury Surveillance System (NEISS) is maintained by the federal Consumer Product Safety Commission (CPSC). It has collected data since 1981 for surveillance of work-related injuries treated at 65 of 91 hospital emergency departments selected from a stratified sample of all hospitals in the United States. The CDC uses these data to summarize injury and illness data from the workforce (*Morbidity and Mortality Weekly Report*, 47:302–306, 1988, and *Morbidity and Mortality Weekly Report*, 56:393–397, 2007).

[3]*Health, United States, 2007, with Chartbook on Trends in the Health of Americans*. Centers for Disease Control. Workplace Safety and Health, http://www/cdc/gov/workplace and Centers for Disease Control. Workers' Health Chartbook. N10SH Publication No. 2004–146. Work Health Chartbook 2004, http://www2a.cdc.gov/niosh-Chartbook/content.asp [accessed June 15, 2006]

TABLE 9.8 Occupational Injury Death Rates (per 100,000 Workers) by Industry, United States, Selected Years 1985–2005

Industry	1985	1990	1993	2003	2005	% Change 1985–2005
Total civilian workforce	5.8	4.6	4.2	4.0	4.0	−31.3
Mining	30.0	30.0	25.4	26.9	25.6	− 7.5
Agriculture, fishing, forestry	23.7	18.0	18.5	31.9	32.5	+37.1
Construction	16.6	14.0	11.8	11.7	11.1	−33.1
Transportation, communication, and public utilities	15.7	10.4	10.1	17.8	17.7	+1.2
Public administration	6.4	3.8	4.2	2.7	2.4	−62.5
Manufacturing	4.0	4.0	3.6	2.5	2.4	−0.40
Wholesale trade	2.8	3.6	3.6	4.2	4.6	+64.2
Retail trade	2.7	2.8	2.9	2.1	2.4	−0.11
Services	1.8	1.5	1.4	na	na	—

Source: Health, United States, 1998 and 2007.

Box 9.11 Occupational Health Issues in the Global Economy

1. Technology transfer from industrial to developing countries or areas within a country;
2. Child labor in developing and developed countries;
3. Pesticide overuse, toxicity, and food contamination;
4. Ecological damage from toxic waste spills and waste disposal;
5. Toxic waste transfer from industrial to developing countries;
6. High-technology industrial toxic wastes;
7. Nuclear energy, accidents, and wastes;
8. Technological and professional common interest between occupational and environmental health;
9. Poor safety and control standards in former Soviet and developing countries;
10. Poor wages, psychological stress, boredom, and shift work;
11. Management negligence and lack of accountability for workplace safety;
12. Governmental negligence and corruption in developing regulatory role;
13. Inadequate health and safety measures in developing countries;
14. Widening income gap between upper- and lower-income groups.

NATIONAL AND MANAGEMENT RESPONSIBILITIES

In the United States, workers' health benefits cost more than the steel to make a car. As a result, there is a growing interest on the part of management and of workers in promoting workers' health through improved nutritional monitoring of canteens and cafeterias, antismoking activities, and physical fitness programs. The management interest in a healthier workforce to contain rising health care costs is part of the modern corporate culture. The primary responsibility, however, legally and morally, lies with management, in addition to protecting the worker by monitoring risks, providing a safe environment, and providing care at the time of injury (Box 9.12).

Occupational injuries and illnesses are social as well as engineering and management concerns. Compensation, litigation, class-action suits, and union action are all associated with increasing awareness of toxic and trauma effects on workers, and court decisions regarding management liability. The field is made more complex because some occupational illness may occur long after the exposure: silicosis, asbestos-related mesothelioma, and asbestosis may develop after a long latency of up to 20–30 years following exposure. Follow-up of exposed workers may be difficult, and issues such as compensation may

Box 9.12 Principal Tasks of Occupational Health

1. Anticipation: dealing with potential disease and injury to include preparation for prevention as facilities are planned or renovated;
2. Surveillance and monitoring assuring timely and accurate identification, reporting, and recording of occupational disease and injury; medical surveillance: passive or active and industrial hygiene and safety;
3. Right-to-know: for workers, health professionals, community at large;
4. Epidemiologic analysis: analyzing collected data — linking exposure to outcome data helps to locate trends, clusters, associations, and causes of disease and injury for more in-depth investigation and prevention;
5. Exposure reduction: minimizing toxic exposure, to prevent approaching or exceeding established limits;
6. Substitution: substituting less toxic substances;
7. Awareness: promoting awareness at government, management, community, worker, and consumer levels;
8. Government regulation: on-site supervision by regulatory agencies; publication of standards of exposure and "good practice";
9. Compensation: compensating for illness and loss of life related to work accidents, toxicity, and stress;
10. Management–worker cooperation: recognizing that worker participation in health and safety is of mutual benefit.

Source: Weeks, J. L., Levy, B. S., Wagner, G. R. (eds.). 1991. *Preventing Occupational Disease and Injury*. Washington, DC: American Public Health Association.

Standards and Monitoring

Monitoring of occupational health involves a set of activities designed to increase the safety and protection of the workers. It involves a number of parallel services to promote the health of the individual worker and the safety of the work environment and should be coordinated in an overall strategy.

In the United States prior to 1970, prevention of occupational injuries, death, and disease was governed by state and local government or market forces. Federal initiatives to raise standards of occupational health and safety were mandated in the Occupational Safety and Health Act of 1970, which established two government agencies to implement the act, the Occupational Safety and Health Administration (OSHA) and the CDC's National Institute of Occupational Safety and Health (NIOSH). OSHA is responsible for promulgation and enforcement activities, within the U.S. Department of Labor. OSHA sets standards based on consensus derived from professional organizations in consultation with labor, industry, and health authorities, meant to promote safety and reduce risk for employees and set performance standards for employers. NIOSH was established to conduct research related to the objects of the act for occupational disease, particularly those derived from exposure to toxic physical and chemical agents.

The act provides an environment for regulation and study of occupational health issues including public petitions, court decisions, and new research findings used to formulate priorities for standards development. Monitoring is done by a combination of federal, state, and local health authorities with participation of professional and industrial organizations. The legal responsibility for worker safety and health is placed with the employer (Table 9.9), but worker awareness and participation in safety programs are vital to a successful approach.

also be complicated. Occupational health involves a governmental regulatory function and legislated responsibility to protect workers from toxic or physical risks at the work site.

TABLE 9.9 Management and Governmental Responsibilities in Workers' Health

Management responsibility	Governmental responsibility
Substitute less dangerous materials	Legislation — substitute, ban, define legal responsibility (civil and criminal compensation)
Enclose/separate	Regulation to set and enforce standards for toxic emissions and controls
Process exhaust	Litigation — civil suits vs. compensation
General ventilation	Test environment and workers with notification of test results
Good housekeeping	Label hazardous materials, labeling, disposal
Monitor health of workers	Monitor health of workers
Personal protection	Educate managers and workers
Investigate	Research — scientific and operational
GMP (good manufacturing practices)	Regulate and compensate for income loss and health damage

---

TABLE 9.10 U.S. Health Targets in Occupational Health for 2000

Subject	Previous	Target
Reducing death from *work-related* injuries	6/100,000 full-time workers (1983–1987)	<4/100,000
Reduce *work-related injuries* resulting in medical treatment, lost time from work, or restricted work activity	<7.7/100 full-time workers in 1987	<6/100
Reduce *hepatitis* B *infection* by increasing immunization levels to >90% among occupationally exposed workers	6200 cases in 1987	<1250 cases
Increase worksites with 50 or more employees that mandate use of *occupant protection systems* (such as seat belts), during work-related motor vehicle travel		>75%
Reduce proportion of workers exposed to average daily *noise levels* that exceed 85 decibels		<15%
Eliminate exposures which result in workers having *blood lead concentrations* greater than 25 μg/dL of whole blood		
Implement statewide *occupational safety and health plans* for the identification, management, and prevention of work-related diseases and injuries	10 states in 1989	50 states
Establish exposure standards adequate to prevent the *major occupational lung diseases* to which their worker populations are exposed, including byssinosis, asbestosis, coal workers' pneumoconiosis, and silicosis		50 states
Increase the proportion of worksites with 50 or more employees that have implemented *programs on worker health and safety*		>70%
Increase the proportion of worksites with 50 or more employees that offer *back injury* prevention and rehabilitation programs	28.6% in 1985	>50%
Establish either public health or labor department programs that provide *consultation and assistance to small businesses* to implement safety and health programs for their employees		50 states
Increase the proportion of *primary care providers* who routinely elicit occupational health exposures as a part of patient history and provide relevant counseling		>75%

Source: U.S. Surgeon General, Healthy People 2000: National Health Promotion and Disease, http://www.surgeongeneral.gov/

OCCUPATIONAL HEALTH TARGETS

The U.S. Surgeon General's report *Healthy People 2000* formulated a number of targets for occupational health and safety issues (Table 9.10). These are national targets that are also being adopted by state departments of health and have organizational as well as legal implications. The midcourse review identifies the progress made since establishment of the goals and targets. The findings are difficulties in systematic evaluation of actions, and in establishing that some medical conditions are related to the job. The review also found lack of awareness of prevention methods and risks are acceptable for some conditions otherwise preventable. There is renewed effort on activities such as research, surveillance, renewed preventive measures, and information dissemination and training.

TOXICITY AT THE WORKPLACE AND IN THE ENVIRONMENT

Toxic substances are widely used in industry, not only in manufacturing but also in services such as laboratories, and they constitute a major concern of both occupational and environmental health. Extensive information on toxic substances is published by the World Health Organization and the Centers for Disease Control.[4]

Much of the concern of occupational health has been on detection, prevention, and reduction of exposure to toxic materials at a workplace, but more recently concern has increased with regard to contamination of the

---

[4]Toxic Substances and Disease Registry, Division of Toxicology, Centers for Disease Control, 1600 Clifton Road N.E., Atlanta, GA 30333.

surrounding environment. The scientific knowledge of toxins used in occupational settings and their sources, uses, effects, actions, and target organs is extensive.

Factors that affect the toxicity of an agent include the extent and duration of exposure, as well as host factors such as age, gender, fitness, previous exposure, and compounding risk factors such as smoking and nutritional status. Environmental factors include temperature and air flow as well as the physical and chemical properties of the toxic agent. A number of examples of toxic substances and the history of control measures for them illustrate the complexity of this problem.

Lead

Lead is a mineral with thousands of applications because of its plasticity and its softness. Lead poisoning has been a worker hazard since ancient times. Lead enters the body through inhalation and ingestion, affecting the gastrointestinal, nervous, hematologic, and circulatory systems. It is associated with intestinal colic, encephalopathy, delirium, and even coma in its acute forms. Chronic forms of plumbism or lead poisoning cause mental dullness, headache, memory loss, neurologic defects (wrist drop), anemia, and a blue line on the gums.

Lead toxicity has been a traditional health problem of glaziers and potters because of lead use in the manufacturing process. Wines or rum produced and stored in lead containers or in pewter (lead–tin alloy) utensils were known to be associated with the "dry gripes" in the seventeenth and eighteenth centuries. The Devonshire colic, described in 1776 by George Baker, was widespread for more than 100 years in parts of England where cider was made and stored in lead containers.

Lead toxicity and excess exposure in the workplace remains a problem in the United States. Lead-induced hypertension, neuropathy, carcinogenesis, reproductive damage for men, and abortion for women are the major toxic effects. The 1995 blood lead surveillance by the CDC's NIOSH Adult Blood Lead Epidemiology and Surveillance Program, which monitors elevated blood levels among adults, reported a continuing hazard of work-related exposures as an occupational hazard in the United States. Studies of lead exposure in industrial settings in the United States have shown widespread exposure above permissible exposure limits. This includes the traditional high-exposure industries such as primary and secondary lead smelting, battery and pigment manufacturers, brass/bronze foundries, and 47 other industries. Highest-exposure jobs throughout industry were painters.

Occupational exposure continues to be an important source of lead toxicity. OSHA standards promulgated in 1978 came at a time when lead prices dropped, reducing the number of producers and the degree of compliance overall.

Concern for lead toxicity evolved from a strictly occupation-related toxicity to an environmental one in which both the exposed worker and the general population are adversely affected by this widely used metal. In 1997, the CDC adopted a blood lead level (BLL) standard of <10 micrograms per deciliter (<10 µg/dL), a level at which a negative effect on cognitive development is recorded. Between 1976 and 1980 and 1980 and 1991, geometric mean BLLs of persons aged 1–74 in the United States declined from 12.8 µg/dL to 2.9 µ/dL, and even further in 1991–1994 to 2.3 µg/dL (NHANES surveys).

Despite major improvements (see Box 9.6), some 1.7 million children ages 1–5 in the United States still have BLLs above 10 µg/dL. Further progress in BLLs will require reduction in lead hazards in housing and reduced contact with lead-contaminated dust, house paint lead, and work-site exposure. Work-related and environmental lead exposures continue to be public health problems in the United States, requiring continued diligence on the part of pediatricians and internists as well as occupational and public health workers.

The Upper Silesia region of Poland with its capital Katowice and a population of 4 million is the site of many nonferrous metal plants, especially using lead and zinc. In the Katowice district there are four such plants, two of which are more than a century old and have a high output of atmospheric lead, and two built in the 1960s with inadequate pollution control equipment. Although emissions of lead and cadmium from one plant reportedly fell during the late 1980s, high levels of blood lead and cadmium are found in children, and soil contamination is extensive, including high levels of contamination of vegetables. This problem is widespread in eastern Europe.

Reduction in lead exposures has been achieved in the United States by a combination of legislation and professional and social pressures, resulting in the adoption of lead-free gasoline, removal of lead from paints, and its substitution in many industrial practices. Awareness and active lobbying by public health–minded groups has had a beneficial effect in reducing lead toxicity in the community and workplace. The American Public Health Association is concerned that 4.4 percent of U.S. children ages 1–5 have BLLs above 10 g/dL, and is promoting a wide-ranging program of further abatement of lead paint hazards including litigation against manufacturers as well as community-based prevention and health education programs.

Asbestos

Asbestos-related disease is an occupational and public health problem that grew from the rapid increase in the use of asbestos during World War II. It left a legacy of death and disease that only became apparent many years later. Fibrotic lung disease resulting from asbestos exposure was called *asbestosis* by W. E. Cooke in 1927. A subsequent British government investigation of the subject reported to Parliament that inhalation of asbestos dust over a period of years results in the development of a serious type of fibrosis of the lung and recommended dust suppression measures. This was followed by many case reports and the wide recognition of the health hazards associated with asbestos exposure. During World War II, the U.S. Navy issued minimum requirements for safety in shipyards contracting for naval work, involving some 1 million workers.

The first reports of an association between asbestos and lung cancer began to appear in the 1930s. Studies by Irving Selikoff in 1965 in New York reported high rates of lung cancer in several large population groups of ex-shipyard workers. Selikoff and colleagues also showed a synergistic relationship between asbestos exposure and cigarette smoking (Table 9.11); namely, a greater risk of lung cancer with heavier smoking and a reduction in risk following cessation of smoking.

The U.S. Toxic Substances Control Act of 1976 placed the responsibility for harmful chemicals, including asbestos, on those who would profit from their sale. The long time lag between the first reports of asbestos-related disease followed by definitive studies and implementation of control measures raised questions as to the way in which occupational health functions. As a result of these studies and the regulatory responses by federal legislators, there was a fourfold reduction in asbestos usage in the United States from 1972–1982. In 1986, the U.S. Asbestos Hazard Emergency Response Act reinforced federal regulation of asbestos usage.

Asbestos exposure is accepted as the cause of mesothelioma, a highly malignant cancer of the chest or abdominal lining. The latency period may be 20–30 years or more, and the risk of the disease is related to the extent of exposure. The exposure may occur in asbestos-cement production, shipyard workers, garage workers exposed to brake linings, plumbers, and construction workers using asbestos-based products. During the 1980s, concern was expressed that asbestos was being exported to developing countries lacking the regulatory mechanisms of the developed world. In the 1990s, there was still some concern that asbestos products manufactured in developing countries are being imported to developed countries. In 1999, the European Union effectively banned the use of asbestos products.

Silica

Silicosis is one of the oldest known occupational diseases, affecting miners in particular. It was described in ancient Greece and Rome as the "fatal dust." Silica occurs in minerals and rocks throughout the world either as free silica or combined in quartz, flint, or sandstone. Mining; tunneling; stone cutting; quarrying; iron and steel works; sandblasting; brick making; polishing of stone, glass, and metals; and many other industries cause the exposure of workers to inhalation of silica dust (Box 9.13).

Silicosis is a condition of massive fibrosis of the lungs resulting from prolonged inhalation of silica dust. It is classified as a pneumoconiosis, a general inflammatory fibrotic lung condition caused by inhalation of dust particles. This condition can progress through mild symptoms to shortness of breath, with radiologic evidence of pulmonary consolidation and concomitant tuberculosis. Silica has not been proved to be a carcinogen.

Studies of hard coal miners in the nineteenth century documented the effects of silicosis. By 1918, English workers could receive disability compensation for silicosis

TABLE 9.11 Lung Cancer Death Rates (Age Standardized) for Workers Exposed to Asbestos Dust and Cigarette Smoking with Controls

Group	Asbestos exposure	Cigarette smoking	Death rate	Mortality difference	Mortality ratio
Control	No	No	11.3	0	1.0
Asbestos workers	Yes	No	58.4	+47	5.2
Control	No	Yes	122.6	+111	10.9
Asbestos workers	Yes	Yes	601.6	+590	53.2

Note: Rates per 100,000 man-years, age standardized, on 12,051 asbestos-exposed workers followed prospectively between 1967 and 1976. Controls included 73,763 similar men in a prospective study of the American Cancer Society for the same decade. The number of lung cancer deaths is based on death certificate information.
Sources: From Hammond, E. C., Selikoff, I. J., Seidman, H. 1979. Asbestos exposure, cigarette smoking, and death rates. *Annals of the New York Academy of Science*, 330:473–490.
Selikoff, I. J. 1986. Asbestos-associated disease. In J. M. Last (ed.). *Maxcy-Rosenau: Public Health and Preventive Medicine*, Twelfth Edition. Norwalk, CT: Appleton–Century–Crofts.

Box 9.13 Ramazzini on Silicosis, 1700

"We must not underestimate the maladies that attack stonecutters, sculptors, quarrymen and other such workers. When they hew and cut marble underground or chisel it to make statues and other objects, they often breathe in the rough, sharp, jagged splinters that glance off, hence they are usually troubled with cough, and some contract asthmatic affections and become consumptive."

Source: Ramazzini, B., *De Morbis Artificum Diatriba*, 1700, as quoted in Hunter, D. 1969. *The Diseases of Occupations*, Fourth Edition. London: English University Press.

---

and tuberculosis. In the 1920s and 1930s in the United States, studies showed silicosis in cement production workers, anthracite miners, tunnel workers, lead–zinc miners, and other hard rock miners. In the mid-1930s, an estimated 700 U.S. workers died as a result of construction of the Hawk's Nest Tunnel in Gauley Bridge, Fayette County, Virginia, leading to compensation laws covering workers with silicosis. At present there is still a controversy as to legally enforceable standards, and the problem remains difficult to prevent.

Cotton Dust (Byssinosis)

Cotton dust has been a common cause of chronic obstructive lung disease among long-term workers in textile industries, widespread in the United States until the 1960s. OSHA promulgated new standards in 1978 based on assessment of the potential of improved ventilation and filtration, and improved machinery use. The industry at that time was in the process of replacing old equipment with modern and more automated machines, which gave improved production speed, more effective use of floor space, reduced labor input, and a higher quality product, along with lower dust levels. The technical and economic feasibility of the higher standard was correct, and compliance by industry exceeded early expectations at about one-third of anticipated costs.

Vinyl Chloride

Vinyl chloride is a colorless, flammable gas with a faintly sweet odor. It is an important component of the chemical industry because of its flame-retardant properties, low cost, and many end product uses. It is also a carcinogen causing liver, brain, and lung cancer, as well as spontaneous abortion. Vinyl chlorides are dangerous primarily when inhaled or ingested. Vinyl chloride usage increased since the 1930s and more dramatically after the end of World War II until the 1970s. In the 1960s, polyvinyl chloride (PVC) was shown to be associated with Raynaud's phenomenon and later with malignancies, including hemangiosarcoma of the liver.

The carcinogenicity of PVC was established as a result of the review of all evidence in 1974 by the U.S. Office of Technology Assessment and OSHA. Scientists concluded that there was no safe level of exposure to vinyl chlorides. OSHA adopted 1 part per million as the maximum possible dose. While the risk assessment issues are still controversial, reduction of exposure to workplace carcinogens such as vinyl chloride is the accepted standard of modern occupational health.

Despite vigorous opposition by the industry to this reduction in permissible emission level (PEL), full compliance was achieved within 18 months by improving ventilation, reducing leaks, modifying reactor designs and chemical pathways, and using greater automation of the process. Even more effective was a major improvement in the production of PVC using less vinyl chloride. The costs to industry of reducing exposure levels were less than 25 percent of anticipated costs because of unanticipated innovations in the production process.

Agent Orange

Agent Orange is an herbicide used widely by the U.S. armed forces in the Vietnam War to defoliate large areas of that country. This agent includes dioxin and is carcinogenic. High levels of dioxin have been found in breast milk, adipose tissue, and blood of the Vietnamese population. Even though sampling has not been systematic, studies carried out between 1984 and 1992 show that high levels of dioxin-like contaminants (TCDD) or 2,3,4,8-tetrachloro-dibenzo-*p*-dioxin are seen in blood samples of the Vietnamese population exposed to Agent Orange during the war.

Studies of effects among American veterans of the Vietnam War have not produced convincing evidence of long-term effects. Additional studies will be needed to verify effects such as increased cases of cancer or birth defects. However, court and compensation decisions have been made in favor of veterans exposed to Agent Orange despite inconclusive epidemiologic evidence of its ill effects on health.

WORKPLACE VIOLENCE

Violence is endemic in many societies and affects many organizations and institutions. Violence has become a leading cause of fatal injuries in the workplace. Violence in the health setting has an extensive history, with the first documented case in 1849 when a patient fatally assaulted a psychiatrist in a mental health care facility. Since then there have been many other studies reporting assaults, hostage taking, rapes, robbery, and other violent acts in the health care and community settings. During the 1990s, homicide became the leading occupational cause of death for females and the second leading cause, after motor vehicle accidents, for men in the United States.

Murder of convenience store employees in the United States and other countries has become a major occupational problem. NIOSH issued new guidelines for addressing this problem in April 1998. Response from the association of operators has opposed standards such as installation of bulletproof glass and television monitors and nighttime double-staffing that have been demonstrated to reduce violence and death in armed robberies.

Shocking incidents of violence and homicide have occurred in which bombs and handguns were used in assassinations of health workers in clinics carrying out abortions, while assault and murder of health workers occurred in hospitals and other settings. The U.S. target of work-related homicides for the year 2010 is 0.4 per 100,000 workers over 16 years of age.

Homicide at work has only recently been addressed as an occupational hazard, and research in this area is in its infancy. No universal standards exist to protect workers from work-related violence, and no policy has been created to protect workers. Preventing violence in the workplace is essential and must be addressed at the national level. Currently the California Occupational Safety and Health Authority has promulgated guidelines with emphasis on preventing violence before it occurs, by developing an effective policy to ensure workplace safety. Management and workers' organizations as well as the health system share responsibility. Prevention of drug, alcohol, and sexual abuse or exploitation at work are vital to prevention of workplace violence.

OCCUPATIONAL HEALTH IN CLINICAL PRACTICE

The clinical physician should be aware of the patient's occupation and previous work history. The inclusion of questions related to workplace factors of current or past employment (see Table 9.12) may be crucial in investigation of patients and without which it might be impossible to find the cause of the disease. The health care provider should be aware of industries in the community and their

TABLE 9.12 Factors for Walk-Through Inspection of Worksites

Marker	Observations of conditions, safety arrangements, and effects on workers
Sensory effects	Eye irritation, poor lighting, noise levels, metallic taste in air, visible fumes, exhaust, temperature (heat/cold)
Safety devices	Use of hard helmets, welding masks, safety shoes and clothing, ear protectors, eye and face protectors, first aid facilities, respirators, monitoring procedures
Storage	Hazardous chemical substances closets; unlabeled bottles, containers
Toilets	Cleanliness, fixtures, soap, toilet paper, waste disposal bins
Worker hygiene	Changing place, showers, lockers, clothing change
Eating place	Separate tables, cleanliness, wash-up facilities
Workers' ages	Children, teenagers, elderly, pregnancy
Workers' complaints	Headache, fatigue, dizziness, nausea, breathlessness, skin problems
Worker morale	Worker morale is reflected in turnover and absenteeism
Worksite layout	Safety in movement of supplies, products, ventilation
Medical service	On-site staff, first aid, evacuation procedures
Emergency procedures	Spills, contamination, terrorist attack, communications, reporting, evacuation, staff training
Hazard control	Labeling, process recording, worker records, periodic screening
Cleanliness	Removal of waste products, oil or chemicals on floors, machines, tables
Vents, fans	Exhaust of fumes, odors, dust
Worker–management cooperation	Mechanisms for worker and management to consult and share responsibility to reduce hazards and improve performance

Source: Adapted from Weeks, J. L., Levy, S. S., Wagner, G. R. (eds.). 1991. *Preventing Occupational Disease and Injury.* Washington, DC: American Public Health Association.

potential hazards. The clinician is particularly important because he or she may be the first to see index cases of toxicity. This requires simple questions, such as the following: What is your job or hobby? What do you do at work? Are you exposed to any chemicals at work or at home? Are there others at work with similar exposure and similar symptoms? How long have you been exposed to these chemicals? The clinical suspicion is the key to finding a potential toxic cause to a set of symptoms, and may lead to a wider public health problem.

INSPECTING THE PLACE OF WORK

The public health authority responsible for health at the place of work may be under the authority of a Ministry of Labor or under a public health authority. Site inspection provides a guide to management and workers for safety and health issues. Noncompliance with federal, state, or local standards should lead to regulatory action to correct deficiencies and should include, if necessary, punitive damages to management. Examination of the work site involves on-site observations as listed in Table 9.12. The inspection should be documented and made available to management, workers, and follow-up inspections.

RISK ASSESSMENT

Identifying and quantifying occupational and environmental risks is difficult, but clinical or public health observations, supplemented by epidemiologic analysis, can identify toxic or carcinogenic factors that can be reduced or eliminated by public health intervention. High levels of awareness by clinicians of potential health effects from environmental or occupational exposures can help identify index cases just as in infectious disease, leading to an investigation and removal of the cause. Similarly, epidemiologic small area analysis can identify populations at high risk for cancers or other toxic effects, giving a localization for further investigation.

Establishment of dose–response relationships requires well-conducted observational studies. Some studies may be so insensitive as to dismiss risks which are at low levels of statistical significance, but still represent preventable risks that can be sufficient to warrant compensation. This occurred in the case of veterans in the United States who were exposed to Agent Orange in Vietnam in the 1960s and those suffering effects attributed to toxic exposures in the Gulf War in 1991 and Iraq War (Second Gulf War) of 2003–2008.

Regulatory and compensation decisions must often be made in the face of inconclusive or contradictory evidence from epidemiologic studies. In the 1960s, the FDA used the Delaney Clause applied to food additives or coloring in which any degree of ill effect noted in animal studies

was enough to disqualify a drug from acceptability, but this has not become an accepted legal standard. The topic remains one of controversy and contradiction, with cases providing precedents that affect future court and regulatory decisions. The contribution of epidemiology to resolving such issues also remains controversial.

PREVENTING DISASTERS IN THE WORKPLACE

A disaster in a workplace can affect the workers and the surrounding community. The major responsibility for prevention is with management, but the worker and society also have roles in the process. Prevention involves education of workers and management, and constant vigilance. Government has the overall responsibility to legislate and enforce standards, safe conditions of work, and control of toxic materials and ensure fair compensation for injury or disease. The simple qualitative observations listed in Table 9.13 can provide a useful picture of the disaster management capacity of a work site. These observations can be made by management, health professionals, and workers' representatives to monitor and promote improved worker health and safety.

TABLE 9.13 Markers and Indicators of Disaster Management Capability in an Industrial Setting

Markers	Indicators
Engage key stakeholders	Ongoing consultative mechanism to develop and implement plan
Administrative	Occupational health disaster plan; access to first aid; frequent disaster drills; close supervision of subcontractors
Investigation	Thorough investigation of complaints, leaks, and spills
Monitoring workers	Monitor worker injuries, illnesses, toxic levels; use of safety measures
Technological	Fail-safe monitoring devices; real-time monitoring; minimal on-site storage; automatic alarm/shut-down devices; local incineration/neutralization
Transportation	Vehicle and container standards; driver training, fatigue, alcohol and drug abuse, traffic offenses
Information/ feedback	Workers' information; right-to-know of workers and community; community disaster plan

Source: Richter, E. D., Deutsch P., Adler, J. 1992. Recognition and use of sentinel markers in preventing industrial disaster. *Pre-hospital and Disaster Medicine*, 7:389–395.

The principle of "good worksite practice" is parallel to good manufacturing practices required by food and drug authorities. It is based on the concept that current standards of acceptable safety involve standards of facilities, staffing, and operational criteria. The healthful and safe work site should be maintained and accredited on that basis.

OCCUPATION AND THE NEW PUBLIC HEALTH

Social class, often defined by occupation and education, is a key determinant of health status. A population of unskilled workers has much higher rates of coronary heart disease, strokes, and cancer and their children have much higher rates of mortality and morbidity than higher skilled workers or business and professional people. The evidence points to a feeling of having less control over one's own life as a major consideration. The worker who has little say in determining his or her own activities may be subject to higher stress on the job, such as on the production line, or in job security, advancement, and wages. Loss of work is a key factor in increasing the vulnerability of men especially to a variety of life-threatening conditions, including suicide, alcoholism, violence, cardiovascular disease, and others. The phenomenon of downsizing, or reducing the workforce, affects production workers disproportionately, but also reaches middle- and upper-management levels, so that the danger of losing a position at an age when finding new employment is unlikely may become a real health hazard. Awareness and responsiveness to a variety of risks associated with employment and occupation is part of health responsibility. Prevention may predominate in some situations, screening for case finding in others, and clinical management in others.

SUMMARY

Environmental and occupational health are increasingly prominent elements of the New Public Health along with concern for the ecology of the world, especially since the 1960s. The problems of this field have become more complex in the past several decades with numerous global ecological concerns emerging. These include global warming, hazards associated with nuclear accidents on the scale of Chernobyl, and chemical disasters occurring frequently in all parts of the world. Other massive environmental issues such as desertification, destruction of forests, and massive air pollution are health concerns, but also societal issues in general. Concern for the environment and the worker often clash with desire for economic growth, especially in the poorer countries as they try to cope with rapidly increasing populations and increasing expectations for a better life.

Important progress has been made in management of water, waste products, toxic wastes, and air quality standards especially since the 1970s. Workers' health and safety have improved dramatically over the past century in the industrial countries. Some of these gains are at the cost of moving hazardous materials and working conditions to newly industrializing or developing countries in the global economy. Even a vigilant health sector is, by itself, incapable of dealing with the problems of the environment and of occupational health. It requires many levels and agencies of government as well as the support of public opinion. The role of the public health community is to act in the professional and advocacy roles with intersectoral cooperation to address these complex and vital issues. Epidemiology provides tools to measure mortality, morbidity, or physiologic change that may occur as a result of environmental damage, but these may not be sufficiently rapid nor sensitive. Both epidemiology and testing technology are improving steadily, providing hope for standards that one would expect to contribute to a cleaner, safer, and more aesthetically pleasing environment.

The New Public Health includes a broad range of health targets of personal and community care for all segments of the population as outlined in *Healthy People 2010* in the United States. The environment affects all, but the poor for many reasons more so. Work or lack of satisfactory work occupies a large portion of the time and energy of a person. It is also the location of many activities of daily life, including diet and physical activity. Policy employers and workers all need to take this into account in developing worksite conditions, management, access to health services, life habits, nutrition, and planned activities. It is not only safety and risk reduction for the benefit of the employer and the worker to preserve and protect the health of workers. This is increasingly so since more people are employed in knowledge-based industries and with the aging of society, fewer workers are available for unskilled jobs as well. For these reasons, health targets cut across all aspects of society including environment and workplace health. The issue include:

- Physical activity
- Overweight and obesity
- Tobacco use
- Substance abuse
- Responsible sexual activity
- Mental health
- Injury and violence
- Environmental quality
- Immunization
- Access to Health Care

The New Public Health includes long-standing public health issues of the environment and occupational health, but widens the field to include clinical services, the community, and the individual. All need to be involved in healthy public policy, in case finding, and in documenting the results of workplace and environmental risks. For a society there are choices to be made in creating a less toxic and hazardous environment. Choices, for example, are between private and public transportation, between jobs in industries with toxic emissions, or between producing energy from fossil fuels or nuclear sources. The search for substitutes for toxic materials and raising the level of social consciousness are needed to reduce the gross pollution that was the price of industrialization over the twentieth century. Equally challenging is the need to prepare and deal with natural and man-made disasters that may involve conventional explosives or biological, chemical, and even nuclear methods of destruction. Avoiding the impact of the most damaging climate changes requires global action in the decades ahead. The financial resources and technological capabilities exist but implementation requires a sense of urgency, public interest, and political will to make deep cuts in greenhouse gas emissions. Achievement of the Millennium Development Goals to reduce poverty and ensure environmental sustainability rests on how global society addresses the issues of global warming. The price of unrestrained pollution and man-made destruction is too great to bear. Investment in a healthy environment is a health, economic and quality-of-life issue for each community and for the entire planet.

ELECTRONIC RESOURCES

Agency for Toxic Substances and Disease Registry. http://www.atsdr.cdc. gov/atsdrhome.html; PCBs, www.atsdr.cdc.gov/HAC/PCB/b_pcb_cvr. html; full list of hazardous substances, http://www.atsdr.cdc.gov/cercla/ 97list.html [accessed May 14, 2008]

American Public Health Association. http://www.apha.org/science/

American Public Health Association, Policy Statements, http://www. apha.org/science.policy.html [accessed May 14, 2008]

Campbell-Lendrum, B., Woodruff, R. Climate change: Quantifying the impact at national and local levels in: A. Prüss-Üstün, C. Corvalán (eds.). *Environmental Burden of Disease Series 14.* WHO. http:// libdoc.who.int/publications/2007/9789241595674_eng.pdf/ [accessed May 14, 2008]

Case Studies in Environmental Medicine. http://www.atsdr.cdc.gov/ csem/csem.html [accessed May 14, 2008]

Centers for Disease Control. National Institute for Occupational Safety and Health. 2004. Worker Health Chartbook, http://www.cdc.gov/ niosh/docs/chartbook/ [accessed May 14, 2008]

Centers for Disease Control. 2004. Worker's Health Chartbook. N1OSH Publication No. 2004-146, www.cdc.gov/niosh/nas/mining/pdfs/2004-146.pdf

Centers for Disease Control. 2008. *Children's Blood Lead Levels in the United States,* www.cdc.gov/nceh/lead/research/kidsBLL.htm [accessed May 14, 2008]

Centers for Disease Control. 2004. *Healthy People 2010.* Midcourse Review. 2004. Occupational Health and Safety; and Environmental Health, http://www.healthypeople.gov/data/midcourse/html [accessed May 14, 2008]

Chemical Agent Briefing Sheets (CABS). http://www.atsdr.cdc.gov/cabs/

Collegium Ramazzini, http://www.collegiumramazzini.org/ [accessed May 14, 2008]

Commission of the European Communities. 2004. The European Action Plan 2004–2010, http://ec.europa.eu/environment/health/pdf/com2004416. pdf [accessed May 13, 2008]

Craun, M. F., Craun, G. F., Calderon, R. L., Beach, M. J. 2006, *Journal of Water and Health,* 04, Suppl 2:19–30.

Environmental Protection Agency (EPA). http://www.epa.gov/ [accessed May 14, 2008]

Environmental Protection Agency. Unified Air Toxic Website, Office of Air Quality, Planning and Standards (EPA). http://www.epa.gov/ttn/ atw/index.html [accessed May 14, 2008]

Environmental Protection Agency. Office of Air and Radiation. http://www. epa.gov/oar/ and http://www.epa.gov/air/data/index.html [accessed May 14, 2008]

Environmental Protection Agency. NO$_x$, What is it and where does it come from http://www.epa.gov/air/urbanair/nox/what.html [accessed May 12, 2008]

Environmental Protection Agency, Wastes, http://www.epa.gov/osw [accessed May 13, 2008]

EPA. 2008. http://www.epa.gov/osw/ [accessed May 13, 2008]

History of the Clean Air Act. January 2008. http://www.epa.gov/air/caa/ caa_history.html [accessed May 14, 2008]

International Conference One Decade after Chernobyl. 1996. Sponsored by WHO. International Atomic Energy Agency. Austria. International Atomic Energy Association, www.iaea.org/Publications/Documents/Infcircs/1996/inf510.shtml [accessed May 13, 2008]

Introduction to the Clean Water Act. March 2008. http://www.epa.gov/ watertrain/cwa/ [accessed May 14, 2008]

Liang, J. L., Dziuban, E. J., Craun, G. F., Hill, V., Moore, M. R., Gelting, R. J., Calderon, R. L., Beach, M. J., Roy, S. L. Centers for Disease Control and Prevention (CDC). http://www.ncbi.nlm.nih.gov/ pubmed [accessed May 14, 2008]

MTBE. http://www.epa.gov/mtbe/ [accessed May 14, 2008]

National Center for Environmental Health, Agency for Toxic Substances and Disease Registry. http://www.atsdr.cdc.gov/ and http://www.cdc. gov/nceh/information/org_chart.pdf [accessed May 14, 2008]

National Institute of Environmental Health Sciences (NIEHS). http:// www.niehs.nih.gov/ [accessed May 14, 2008]

Occupational Safety and Health Agency. http://www.osha.gov/ [accessed May 14, 2008]

OECD. http://www.oecd.org/dataoecd/60/47/40501197.pdf [accessed May 13, 2008]

U.S. National water Quality Assessment Data Warehouse. http://infotrek. er.usgs.gov/traverse/f?p=NAWQA:HOME:5755038819145367 [accessed May 15, 2008]

United Nations Millennium Development Goals. 2001. http://www.un. org/millenniumgoals/# [accessed May 14, 2008]

United Nations. 2007. Fighting climate change: Human solidarity in a divided world, http://hdr.undp.org/en/media/hdr_20072008_en_ complete.pdf

U.S. Environmental Protection Agency. Office of Air Quality and Standards Air Quality Assessment Division. Latest findings on national air quality: Status and trends through 2006, http://www.epa.gov/air/ airtrends/2007/ [accessed May 14, 2008]

United Nations Development Programme, Millennium Development Goals, 2007. Progress Report, http://www.undp.org/mdg and http:// mdgs.un.org/unsd/mdg/news.aspx?article1D=34 [accessed June 14, 2008]

United Nations Environment Programme. World Summit on Sustainable Development, 2002, http://www.unep.org/wssd [accessed June 14, 2008]

United Nations, Johannesburg Summit, document post August 24, 2006, http:www.un.org/jsummit/html/documents/documents.html [accessed June 14, 2008]

United States Environmental Protection Agency. *Disinfection By-Products Health Effects*, http://www.cpa.gov/enviro/htm/icr/dbp_health.html [accessed June 14, 2008]

United States Environmental Protection Agency, *Water Sense*, http://www.epa.gov/watersense/ [accessed June 15, 2008]

United States Nuclear Regulatory Commission. 2007, http://www.nrc.gov/reading-rm/doc-collections/fact-sheets/3mile-isle.html and http://www.nrc.gov/reading-rm/doc-collections/fact-sheets/fschernobyl.html [accessed January 28, 2008]

United States Surgeon General, http://www.surgeongeneral.gov/publichealthpriorities.html [accessed June 15, 2008]

Universities of Michigan and Sheffield Landmine Monitor. 2006. http://www.icbl.org/lm/2005/intro/survivor.html#Heading4 [accessed May 14, 2008]

WHO Europe. 2007. Report of a WHO workshop, population health and waste management scientific data and policy options, www.euro.who.int/document/E91021.pdf [accessed May 14, 2008]

WHO. 2006. Preventing disease through healthy environments: Towards an estimate of the environmental burden of disease, http://www.who.int/quantifying_ehimpacts/global/en [accessed May 14, 2008]

World Nuclear Association, http://www.world-nuclear.org/info/chernobyl/inf07.html [accessed May 13, 2008]

RECOMMENDED READINGS

Advisory Committee on Childhood Lead Poisoning Prevention. 2007. Interpreting and managing blood lead levels <10 µg/dL in children and reducing childhood exposures to lead. Recommendations of CDC's Advisory Committee on Childhood Lead Poisoning Prevention. *Morbidity and Mortality Weekly Report*, 56(RR-08):1–14.

Bhatia, R. 2007. Protecting health using an environmental impact assessment: A case study of San Francisco land use decision making. *American Journal of Public Health*, 97:406–413.

Bunn, F., Collier, T., Frost, C., Roberts, I., Wentz, R. 2003. Traffic calming for the prevention of road traffic injuries: Systematic review and meta analysis. *Injury Prevention*, 9:200–204.

Centers for Disease Control. 1995. *Vibrio cholerae* O1—Western hemisphere, 1991–1994, and *V. cholerae* O139—Asia, 1994. *Morbidity and Mortality Weekly Report*, 44:215–219.

Centers for Disease Control. 1999. Achievements in public health, 1900–1999: Control of infectious diseases. *Morbidity and Mortality Weekly Report*, 48:621–629.

Centers for Disease Control. 1999. Achievements in public health, 1900–1999: Fluoridation of drinking water to prevent dental caries. *Morbidity and Mortality Weekly Report*, 48:933–940.

Centers for Disease Control. 1999. Achievements in public health, 1900–1999. Improvements in workplace safety—United States. *Morbidity and Mortality Weekly Report*, 48:461–469.

Centers for Disease Control. 1999. Achievements in public health, 1900–1999: Motor-vehicle safety: A 20th century public health achievement. *Morbidity and Mortality Weekly Report*, 48:369–374.

Centers for Disease Control. 2000. Strategic planning workgroup. *Morbidity and Mortality Weekly Report*, 49(RR-04):1–14.

Centers for Disease Control. 2004. 150th anniversary of John Snow and the pump handle. *Morbidity and Mortality Weekly Report*, 53:783.

Centers for Disease Control. 2007. Interpreting and managing blood lead levels <10 µg/dL in children and reducing childhood exposures to lead: Recommendations of CDC's Advisory Committee on Childhood Lead Poisoning. *Morbidity and Mortality Weekly Report*, Recommendations and Reports, RR-8, 56:1–16.

Centers for Disease Control. 2007. *Health United States*. Washington, DC: Department of Health and Human Services.

Centers for Disease Control. 2007. Nonfatal occupational injuries and illnesses — United States, 2004. *Morbidity and Mortality Weekly Report*, 56:393–397.

Concha-Barrientos, M., Nelson, D. I., Fingerhut, M., Driscoll, T., Leigh, J. 2005. The global burden due to occupational injury. *American Journal of Industrial Medicine*, 48:470–481.

Desai, M. A., Mehta, S., Smith, K. R. 2004. *Indoor smoke from solid fuels: Assessing the environmental burden of disease at national and local levels*. Geneva: World Health Organization (WHO Environmental Burden of Disease Series, No. 4).

Dolbokova, D., Krzyzanowski, M., and Lloyd, S. (eds.). 2007. *Children's health and the environment in Europe: A baseline assessment*. Copenhagen: WHO European Region.

Grandjean, P., Landrigan, P. J. 2006. Developmental neurotoxicity of industrial chemicals. *Lancet*, 2006 Dec 16;368(9553):2167–2178.

Greenberg, M. R. 2007. Contemporary environmental and occupational health issues: More breadth and depth. *American Journal of Public Health*, 97:395–397.

Khan, A. S., Levitt, A. M., Sage, M. J. 2000. Biological and chemical terrorism: Strategic plan for preparedness and response: Recommendations of the CDC Khan Strategic Planning Workgroup. *Morbidity and Mortality Weekly Report*, 49(RR-4):1–26.

Kleinman, M. T., Sioutas, C., Froines, J. R., Fanning, E., Hamade, A., Mendez, L., Meacher, D., Oldham, M. 2007. Inhalation of concentrated ambient particulate matter near a heavily trafficked road stimulates antigen-induced airway responses in mice. *Inhalation Toxicology*, 19(Suppl 1):117–126.

Kouznetsova, M., Huang, X., Ma, J., Lessner, L., Carpenter, D. O. 2007. Increased rate of hospitalization for diabetes and residential proximity of hazardous waste sites. *Environmental Health Perspectives*, 115:75–79.

Lahiri, S., Levenstein, C., Nelson, D. I., Rosenberg, B. J. 2005. The cost effectiveness of occupational health interventions: Prevention of silicosis. *American Journal of Industrial Medicine*, 48:503–514.

Ma, J., Kouznetsova, M., Lessner, L., Carpenter, D. O. 2007. Asthma and infectious respiratory disease in children — Correlation to residence near hazardous waste sites. *Paediatric Respiratory Reviews*, 2007 Dec; 8:292–298.

MacKenzie, W. R., Hoxie, N. J., Proctor, M. E., Gradus, M. S., Blair, K. A., Peterson, D. E., Kazmierczak, J. J., Addiss, D. G., Fox, K. R., Rose, J. B., Davis, J. P. 1994. A massive outbreak in Milwaukee of *Cryptosporidium* infection transmitted through the public water supply. *New England Journal of Medicine*, 331:161–167.

Nadakavukaren, A. 1990. *Man and Environment*, Third ed. Prospect Heights, IL: Waveland Press.

Nelson, D. I., Concha-Barrientos, M., Driscoll, T., Steenland, K., Fingerhut, M., Punnett, L., Prüss-Üstün, A., Leigh, J., Corvalan, C. 2005. The global burden of selected occupational disease and injury risks: Methodology and summary. *American Journal of Industrial Medicine*, 48:400–418.

Paneth, N., Vinten-Johansen, P., Brody, H., Rip, M. 1998. A rivalry of foulness: Official and unofficial investigations of the London cholera epidemic of 1854. *American Journal of Public Health*, 88:1545–1553.

Peterka, M., Peterkova, R., Likovsky, Z. 2007. Chernobyl: Relationship between the number of missing newborn boys and the level of radiation in the Czech regions, Department of Teratology, Institute of Experimental Medicine. *Environmental Health Perspectives*, 115:1801–1806.

Prüss-Üstün, A. 2006. *Preventing Disease Through Healthy Environments: Towards an Estimate of the Environmental Burden of Disease.* Geneva: WHO.

Resnik, D. B., Wing, S. 2007. Lessons learned from the Children's Environmental Exposure Research Study. *American Journal of Public Health*, 97:414–418.

Rossner, D., Markowitz, G. 1985. A "Gift of God"?: The public health controversy over leaded gasoline in the 1920s. *American Journal of Public Health*, 75:344–352.

Schulte, P. A. 2005. Characterizing the burden of occupational injury and disease. *Journal of Occupational and Environmental Medicine*, 47:604–622.

Solomon, S., Qin, D., Manning, M., Chen, Z., Marquis, M., Averyt, K. B., Tignor, M., and Miller, H. L. (eds.). Intergovernmental Panel on Climate Change IPCC. 2007. Summary for Policymakers. In: *Climate Change 2007: The Physical Science Basis.* Contribution of Working Group I to the Fourth Assessment Report of the Intergovernmental Panel on Climate Change. Cambridge and New York: Cambridge University Press.

United States Department of Health and Human Services. *Health, United States, 2007, with Chartbook on Trends in the Health of Americans.* Washington, DC.

World Health Organization. 2006. *Guidelines for Drinking Water Quality. First Addendum to Third Edition. Recommendations.* Geneva: World Health Organization. http://www.who.int/water_sanitation_health/dwq/gdwq3rev/en/index.html [accessed May 14, 2008]

BIBLIOGRAPHY — WATER QUALITY AND WATERBORNE DISEASE

Calderon, R. L., Craun, G. F. 2006. Estimates of endemic waterborne risks from community-intervention studies. *Journal of Water and Health*, 4 Suppl 2:89–99.

Centers for Disease Control. 1993. Update: Cholera — Western hemisphere, 1992. *Morbidity and Mortality Weekly Report*, 42:89–91.

Centers for Disease Control. 1994. Assessment of inadequately filtered public drinking water — Washington, DC, December 1993. *Morbidity and Mortality Weekly Report*, 43:661–663.

Centers for Disease Control. 2007. Ground Water Awareness Week, March 11–17, 2007. *Morbidity and Mortality Weekly Report*, 56:199.

Centers for Disease Control. 2007. National Drinking Water Week — May 6–12, 2007. *Morbidity and Mortality Weekly Report*, 56(17), 426–427.

Centers for Disease Control. 2007. National Drinking Water Week — May 6–12, 2007, World Water Day — March 22, 2007. *Morbidity and Mortality Weekly Report*, 56:228–229.

Committee on Earth–Atmosphere Interactions. *Understanding and Responding to Multiple Environment Stresses.* Nation Research Council. 2007. National Academies Press, Washington DC, http://www.nap.edu/catalog/11748.html

Craun, G. F., Calderon, R. L. 2006. Workshop summary: Estimating waterborne disease risks in the United States. *Journal of Water and Health*, 4 Suppl 2:241–253.

Cutler, D., Miller, G. 2005. The role of public health improvements in health advances: The twentieth-century United States. *Demography*, 42:1–22.

Esrey, S. A., Potash, J. B., Roberts, L., Shiff, C. 1991. Effects of improved water supply on ascariasis, diarrhoea, dracunculiasis, hookworm infection, schistosomiasis, and trachoma. *Bulletin of the World Health Organization*, 69:609–621.

Hurst, C. J. 1991. Presence of enteric viruses in freshwater and their removal by the conventional drinking water treatment process. *Bulletin of the World Health Organization*, 69:113–119.

Institute of Medicine. 2007. *Understanding Multiple Environmental Stresses: Report of a Workshop Committee on Earth-Atmosphere Interactions: Understanding and Responding to Multiple Environmental Stresses*, National Research Council, Washington DC.

Last, J. M., Wallace, R. B. 1992. *Public Health and Preventive Medicine*, Thirteenth Edition. Norwalk, CT: Appleton and Lange.

Liang, J. L., Dziuban, E. J., Craun, G. F., Hill, V., Moore, M. R., Gelting, R. J., Calderon, R. L., Beach, M. J., Roy, S. L. Centers for Disease Control and Prevention. 2006. Surveillance for waterborne disease and outbreaks associated with drinking water and water not intended for drinking — United States, 2003–2004. *Morbidity and Mortality Weekly Report*, 55(12):31–65.

Nieuwenhuijsen, M. J., Toledano, M. B., Bennett, J. et al. 2008. Chlorination disinfection by-products and risk of congenital anomalies in England and Wales. *Environmental Health Perspectives*, 116:216–222.

Ritter, L., Solomon, K., Sibley, P., Hall, K., Keen, P., Mattu, G., Linton, B. 2002. Sources, pathways, and relative risks of contaminants in surface water and groundwater: A perspective prepared for the Walkerton inquiry. *Journal of Toxicology and Environmental Health*, 65:1–142.

Schoenen, D. 2002. Role of disinfection in suppressing the spread of pathogens with drinking water: Possibilities and limitations. *Water Research*, September;36(15):3874–3888.

Tulchinsky, T. H., Burla, E., Brown, A., Goldberger, S. 2001. Safety of community drinking-water and outbreaks of waterborne enteric disease outbreak: Israel, 1976–97. *Bulletin of the World Health Organization*, 78:1466–1473.

United Nations. 2008. *International Year of Sanitation 2008.* New York: United Nations.

United Nations Development Programme Report 2007/2008. *Fighting Climate Change: Human Solidarity in a Divided World.* UNDP, New York.

Wigle, D. T., Arbuckle, T. E., Walker, M., Wade, M. G., Liu, S., Krewski, D. 2007. Environmental hazards: evidence for effects on child health. *Journal of Toxicology and Environmental Health*, Part B Critical Reviews, 10:3–39.

World Health Organization. 1993. *Guidelines for Drinking Water Quality,* Volume 1. *Recommendations*, Second Edition. Geneva: WHO.

World Health Organization. 1994. *Operation and Management of Urban Water Supply and Sanitation Systems: A Guide for Managers.* Geneva: WHO.

World Health Organization. 1996. *Guidelines for Drinking Water Quality,* Volume 2. *Health Criteria and Other Supporting Information*, Second Edition. Geneva: WHO.

World Health Organization. 1997. *Guidelines for Drinking Water Quality,* Volume 3. *Surveillance and Control of Community Water Supplies*, Second Edition. Geneva: WHO.

World Health Organization. 1998. *Guidelines for Drinking Water Quality, Addendum to Volume 2. Health Criteria and Other Supporting Information.* Geneva: WHO.

World Health Organization, *Water and Sanitation: Facts and Figures, 2002,* http://www.who.int/water_sanitation_health?General/factsandfigures.htm [accessed May 14, 2008]

World Health Organization. 2006. *Guidelines for Drinking-Water Quality,* Third Edition, *Incorporating First Addendum:* Volume 1 — *Recommendations.* Geneva: World Health Organization.

BIBLIOGRAPHY — OCCUPATIONAL AND ENVIRONMENTAL HEALTH

American Public Health Association. 1997. Policy Statement 9704. Responsibilities of the lead pigment industry and others to support efforts to address the national child lead poisoning problem, http://www.apha.org/science.policy.html

American Public Health Association. 1998. Policy Statement 9806. Preventing adverse occupational and environmental consequences of methyl tertiary butyl ether (MTBE) in fuels, http://www.apha.og/science.policy.html

Attfield, M. D., Castellan, R. M. 1992. Epidemiological data on US coal miners' pneumoconiosis, 1960 to 1988. *American Journal of Public Health,* 82:964–970.

Centers for Disease Control. 1994. Occupational injury deaths — United States, 1980–1989. *Morbidity and Mortality Weekly Report,* 43:262–264.

Centers for Disease Control. 1994. Surveillance for emergency events involving hazardous substances — United States, 1990–1992. *Morbidity and Mortality Weekly Report,* 43(SS–2):1–6.

Centers for Disease Control. 1999. Outbreak of West Nile like encephalitis, New York, 1999. *Morbidity and Mortality Weekly Report,* 48:845–849.

Centers for Disease Control. 1999. Achievements in public health, 1900–1999: Improvements in Workplace Safety — United States, 1900–1999. *Morbidity and Mortality weekly Report,* 48:461–469.

Centers for Disease Control. 2000. Outbreak of Rift Valley Fever, Saudi Arabia, August–October, 2000. *Morbidity and Mortality Weekly Report,* 45:905–908.

Centers for Disease Control. 2001. Biological and chemical terrorism: Strategic plan for preparedness and response. *Morbidity and Mortality Weekly Report,* 49(RR-4).

Centers for Disease Control. 2001. New York City Department of Health response to terrorist attack, September 11, 2001. *Morbidity and Mortality Weekly Report,* 50:821–822.

Centers for Disease Control. 2001. Ongoing investigation of anthrax — Florida, October 2001. *Morbidity and Mortality Weekly Report,* 50:877.

Centers for Disease Control. 2003. Recognition of illness associated with exposure to chemical agents — United States, 2003. *Morbidity and Mortality Weekly Report,* 52:938–940.

Centers for Disease Control. 2006. Adult blood lead epidemiology and surveillance — United States, 2003–2004. *Morbidity and Mortality Weekly Report,* 55:876–879.

Centers for Disease Control. 2006. Rapid community needs assessment after Hurricane Katrina — Hancock County, Mississippi, September 14–15, 2005. *Morbidity and Mortality Weekly Report,* 55:234–236.

Centers for Disease Control. 2006. Worker illness related to ground application of pesticide — Kern County, California, 2005. *Morbidity and Mortality Weekly Report,* 55:486–488.

Centers for Disease Control. 2007. Advanced pneumoconiosis among working underground coal miners — Eastern Kentucky and Southwestern Virginia, 2006. *Morbidity and Mortality Weekly Report,* 56:652–655.

Centers for Disease Control. 2007. Chikungunya fever diagnosed among international travellers to the United States, 2006. *Morbidity and Mortality Weekly Report,* 56:276–277.

Centers for Disease Control. 2007. Fatal occupational injuries and illnesses — United States, 2005. *Morbidity and Mortality Weekly Report,* 56:297–301.

Centers for Disease Control. 2007. Non-fatal occupational injuries and illnesses — United States, 2004. *Morbidity and Mortality Weekly Report,* 56:393–397.

Deutsch, P. V., Adler, J., Richter, E. D. 1992. Sentinel markers for industrial disasters. *Israel Journal of Medical Sciences,* 28:526–533.

Doll, R. 1992. Health and the environment in the 1990s. *American Journal of Public Health,* 82:933–941.

Dwyer, J. H., Flesch-Janys, D. 1995. Agent Orange in Vietnam. *American Journal of Public Health,* 85:476–478.

Editorial. 1992. Environmental pollution: It kills trees but does it kill people? *Lancet,* 340:821–822.

Edling, C. (editorial). 1985. Radon exposure and lung cancer. *British Journal of Industrial Medicine,* 42:721–722.

Elliott, P., Cuzick, J., English, D., Stern, R. (eds.). 1992. *Geographic and Environmental Epidemiology: Methods for Small Area Studies.* World Health Organization, Regional Office for Europe, Oxford University Press.

Environmental Protection Agency. 1989. Why accidents occur: Insights from the Accidental Release Information Program. *Chemical Accident Prevention Bulletin,* Series 8 number 1, July, Washington, DC.

Froines, J. R., Baron, S., Wegman, D. H., O'Rourke, S. 1990. Characterization of airborne concentrations of lead in U.S. industry. *American Journal of Industrial Medicine,* 18:1–17.

Ginsberg, G. M., Tulchinsky, T. H. 1992. Regional differences in cancer incidence and mortality in Israel: Possible leads to occupational causes. *Israel Journal of Medical Sciences,* 28:534–543.

Gottlieb, R. (ed.). 1995. *Reducing Toxins: A New Approach to Policy and Industrial Decisionmaking.* Washington, DC: Island Press.

Hammond, E. C., Setikoff, I. J., Seidman, H. 1979. Asbestos exposure, cigarette smoking, and death rates. *Annals of the New York Academy of Science,* 330:473–490.

Harrington, J. M. 1999. 1998 and beyond — Legge's legacy to modern occupational health. *Annals of Occupational Health,* 43:1–6.

Harris, P., Harris-Roxas, B., Harris, E., Kemp, L. 2007. *Health Impact Assessment: A Practical Guide.* Sydney: Centre for Health Equity Training, Research and Evaluation (CHETRE). UNSW Research Centre for Primary Health Care and Equity, University of New South Wales.

Hunter, D. 1969. *The Diseases of Occupations,* Fourth Edition. London: The English Universities Press Ltd.

Institute of Medicine, National Academies of Sciences. *2007. Global Environmental Health in the 21st Century: From Governmental Regulation to Corporate Social Responsibility.* Workshop Summary. Roundtable on Environmental Health Sciences, Research, and Medicine. Board on Population Health and Public Health Practice. Washington, DC: National Academies Press.

Landrigan, P. J. 1992. Environmental disease — A preventable epidemic. *American Journal of Public Health,* 82:941–943.

Maisonet, M., Correa, A., Misra, D., Jaakkola, J. J. 2004. A review of the literature on the effects of ambient air pollution on fetal growth. *Environmental Research,* 95:106–115.

McMichael, A. J. 1993. Global environmental change and human population health: A conceptual and scientific challenge for epidemiology. *International Journal of Epidemiology*, 22:1–8.

National Institute of Occupational Safety and Health. 2004. *Worker Health Chartbook 2004* (NIOSH Publication No. 2004-146). Washington, DC: National Institute of Occupational Safety and Health.

Nicholls, G. 1999. The ebb and flow of radon. *American Journal of Public Health*, 89:993–995.

Pavia, M., Bianco, A., Pileggi, C., Angelillo, I. F. 2003. Meta-analysis of residential exposure to radon gas and lung cancer. *Bulletin of the World Health Organization*, 81:732–738.

Peden, M., Scurfield, R., Sleet, D., Mohan, D., Hyder, A. A., Jarawan, E., Mathers, C. 2004. *World Report on Road Traffic Injury Prevention*. Geneva: World Health Organization.

Prüss-Üstün, A., Corvalán, C. 2006. *Preventing Disease Through Healthy Environments: Towards an Estimate of the Environmental Burden of Disease*. Geneva: WHO.

Racioppi, F., Eriksson, L., Tingvall, C., Villaveces, A. 2004. *Preventing Road Traffic Injury: A Public Health Perspective for Europe*. Copenhagen: WHO Regional Office for Europe.

Richter, E. D., Berman, T., Friedman, L., Ben-David, G. 2006. Speed, road injury and public health. *Annual Reviews of Public Health*, 27:125–152.

Rosen, G. 1993. *A History of Public Health*, Expanded Edition. Baltimore, MD: Johns Hopkins University Press.

Selikoff, I. J. 1986. Asbestos-associated diseases. In Last, J. M. (ed.). *Maxcy-Rosenau: Public Health and Preventive Medicine,* Twelfth Edition, pp. 523–525. Norwalk, CT: Appleton–Century–Crofts.

Sinclair, U. 1906. *The Jungle*, Classic Series, 1965. New York: Airmont Co.

Snow, J. On the mode of transmission of cholera. in 1849. In *Snow on Cholera*. New York: The Commonwealth Fund, 1936. http://www.ph.ucla.edu/epi/snow.html [accessed May 14, 2008]

Stayner, L. 1999. Protecting public health in the face of uncertain risks: The example of diesel exhaust. *American Journal of Public Health*, 89:991–993.

Stern, C., Young, O. R., and Druckman, D. (eds.). 1992. *Global Environmental Change: Understanding the Human Dimensions*. Washington, DC: National Academy Press.

Tulchinsky, T. H., Ginsberg, G. M., Shihab, S., Goldberg, E., Laster, R. 1992. Mesothelioma mortality among former asbestos-cement workers in Israel, 1953–90. *Israel Journal of Medical Sciences*, 28:543–547.

Valent, F., Little, D., Bertollini, R., Nemer, L., Barbone, F., Temburlini, G. 2004. Burden of disease attributable to selected environmental factors and injury among children and adolescents in Europe. *Lancet*, 363:2032–2039.

United States Surgeon General, *Healthy People 2000:National Health Promotion and Disease.*

Weeks, J. L., Levy, B. S., and Wagner, G. R. (eds.). 1991. *Preventing Occupational Disease and Injury*. Washington, DC: American Public Health Association.

World Health Organization, European Region. 2000. *Air Quality Guidelines for Europe,* Second Edition. Copenhagen: World Health Organization Regional Office for Europe, (WHO Regional Publications, European Series, No. 91).

World Health Organization. 2002. *World Health Report 2002 — Reducing Risks, Promoting Healthy Life*. Geneva: World Health Organization.

World Health Organization. 2003. *Guidelines for Safe Recreational Water Environments*. Volume 1: *Coastal and Freshwaters*. Geneva: World Health Organization.

World Health Organization. 2004. *Evaluation of the Costs and Benefits of Water and Sanitation Improvements at the Global Level*. Geneva: World Health Organization.

World Health Organization. 2004. *World Health Report 2004 — Changing History*. Geneva: World Health Organization.

World Health Organization. 2005. *Ecosystems and Human Health — Health Synthesis*. Geneva: World Health Organization.

World Health Organization. 2005. *Water-Related Diseases: Malnutrition*. Geneva: World Health Organization.

World Health Organization. 2006. *Fuel for Life, Household Energy and Health*. Geneva: World Health Organization.

World Health Organization, European Regional Office. 2007. *Population Health and Waste Management: Scientific Data and Policy Options*. Copenhagen: World Health Organization, European Region.

World Health Organization. 2008. Climate change and health: Report by the Secretariat. Geneva: WHO. http://www.who.int/gb/ebwha/pdf_files/EB122/B122_4-en.pdf [accessed May 13, 2008]

World Health Organization. UNICEF. 2004. *Meeting the MDG Drinking Water and Sanitation Target: A Mid-Term Assessment of Progress*. Geneva: World Health Organization.

Yoder, J. S., Beach, M. J. 2007. Cryptosporidiosis surveillance — United States, 2003–2005. *Morbidity and Mortality Weekly Report*, 56(SS-07):1–10.

Zirm, K. L., Mayer, J. 1989. *The Management of Hazardous Substances in the Environment*. London: Elsevier Applied Sciences.

Chapter 10

Organization of Public Health Systems

Introduction
Government and Health of the Nation
Federal and Unitary States
Checks and Balances in Health Authority
Government and the Individual
Functions of Public Health
Regulatory Functions of Public Health Agencies
Methods of Providing or Assuring Services — Direct or Indirect?
Nongovernmental Roles in Health
Disasters and Public Health Preparedness
Medical Practice and Public Health
Incentives and Regulation
Promotion of Research and Teaching
Accreditation and Quality Regulation
National Government Public Health Services
State Government Public Health Services
Local Health Authorities
Monitoring Health Status
National Health Targets
Universal Health Coverage and the New Public Health
Hospitals in the New Public Health
Hospital Classification
Supply of Hospital Beds
The Changing Role of the Hospital
Regulation of Hospitals
The Uninsured as a Public Health Challenge
Summary
Electronic Resources
Recommended Readings
Bibliography

INTRODUCTION

Formal structures to ensure public health evolved over the centuries as local authorities addressed fundamental societal needs for sanitation, safe water and food supply, business licensing and other issues. These structures developed in response to the challenges of industrialization and urbanization along with growing scientific and applied methodologies for disease prevention and health promotion. Nongovernmental charitable, religious, and advocacy organizations pioneered many services which were part of addressing the broad spectrum of public health needs. With the widening range of public responsibilities, state and national governments took on increasing roles of leadership. This included financial support and professional development of public health and, in parallel, medical care systems to meet the growing public expectation for good health. These challenges remain important for current and future needs of both individual and population health. In the United States the high and rising cost of health care, and lack of universal insurance coverage are continuing political and public health issues, while many other industrialized countries have better health outcomes such as longer life expectancy (Chapters 11 and 13).

This chapter examines the organization of public health and health care delivery services, illustrating how separate systems of service coexist and interact. Each system evolved in its own organizational and financing format, yet they come together, as medical care and prevention become more mutually interdependent. Traditional public health systems must increasingly develop intersectoral cooperation with other components of the health care industry, as well as with government and related fields, such as agriculture, business, social welfare, education, police, and community organizations.

Governments have legislative, regulatory, and taxation powers set out in constitution and law for common action for the public good, including powers to promote health and to restrict individual actions that may jeopardize the health of others. City-states in ancient Greece provided sanitation for the entire community and medical care for the poor. The Elizabethan Poor Laws in Britain in the early seventeenth century established the responsibility of the local authority for health and welfare. Subsequent developments brought local, state, and national government into sanitation, disease control, and other aspects of public health and health planning. Later this extended to assuring provision of comprehensive health care on a social-equity basis for all or to meet the specific needs of vulnerable groups within a society.

Societies have learned to prevent disease by social action and have learned that individual health depends on such action. Governments are involved in that process, whether the governmental structure is based on democratic and free market principles, or is centrally managed with a command economy. Society has accepted some limitations on individual rights for the public good. This limits the individual from attacking and harming another person, or damaging goods whether private or public. A person is restricted from throwing garbage in the street, and industry is prohibited from polluting the environment or endangering its workers.

Public health policy, legislation, and action involve common measures to protect the individual and the community. Such measures may take the form of mandatory reporting of an infectious disease, chlorinating and fluoridating community water systems, regulating food and drug industries, requiring children to be immunized prior to entry to school, or to fine or imprison industry managers whose negligence causes death and injury, or whose industry pollutes the environment.

Achievement of public health goals requires organization (Box 10.1). Public health organization requires a formal structure for a defined population in which finance, management, scope, and content are defined in law and regulations. It includes services contributing to people's health and health care to be delivered in many settings such as homes, communities, educational institutions, workplaces, hospitals, and clinics. Public health also addresses the policy, legislative, and regulatory functions of societal health including the physical and psychosocial environment. A health system is organized at various levels, starting at the most peripheral, the community or primary level. It includes district, regional, state, and national levels as well as international aspects. International strategies for health and national health systems should be seen as investments that produce health gain rather than merely management of existing medical care institutions and services.

Function and structure are interdependent. Structure should evolve from the desired function; that is, to achieve national goals and objectives for health. This is fulfilled through legislative, regulatory, financing, and service functions, which provide the underpinnings to meet health needs in any country. Some countries provide universal health care through a governmental system. Others legislate financing of health care, while another approach focuses on financing for certain population subgroups, such as the elderly and the poor, placing greater emphasis on provision of facilities and research in health care.

This chapter describes public health organization primarily using examples from the United States, including federal, state, and local public health authorities. In contrast to most industrialized countries, the United States lacks universal health care. As a result, health care is provided through a mix of independent, private, and public agencies. While this is sometimes described as a "non-system," it is in fact a complex network of interactive services. Yet, it lacks universality, leaving many individuals without access to even basic private health care. As a result, public health organizations in the United States play a very important role in providing essential services for people or needs not otherwise met. Yet there are socioeconomic, ethnic, and regional variation and inequalities in insurance coverage and resource allocation, leaving substandard access and outcomes for many in the U.S. health system. Partly to compensate for this fragmentation of health care in the United States, public health has played a leadership role particularly in advocacy, development, and achievement in promoting health.

GOVERNMENT AND HEALTH OF THE NATION

Public health involves a wide variety of issues that should be directly under governmental responsibility as they require legislation, enforcement, and taxing powers. These include, for example, environment, nutrition, food and drug control, sanitation, immunization, traffic laws, firearms control, and health education. Many of these functions are promoted by nongovernmental organizations (NGOs), with delegated governmental regulatory powers.

Financing and allocation of public funds for health care are an important means of influencing health activities. This may mean direction of public funds to support research, teaching facilities, and provision of services. National governments may directly provide services, but increasingly this is being decentralized to lower levels of government (regional, district, municipal) or to nongovernmental health care providers. Academic, professional, and public advocacy organizations play important roles in the New Public Health, such as in personnel training, education, research, and professional standards setting. These functions can be diffused to a variety of professional, consumer, and academic institutions. This enables governments to act through direct regulatory functions. Governments can also act indirectly, setting standards and norms through financial and other incentives or sanctions. This includes an organized system of accountability, accreditation, licensing activities, and quality guidelines.

Box 10.1 What Is a Public Health System?

"A network of public, private, and voluntary entities that contribute to the health and well-being of a community."

Source: WHO World Health Report 2004.

Federal and Unitary States

Public health requires a basis in law, public administration, and financing. The form of government may differ from country to country, some being federal, others unitary.

In a federal system, three levels of government, federal, state, and local, have separate but overlapping responsibilities for public health. Federal states have constitutions conceived and written in a historical period when state rights were emphasized and health care was perceived mainly as a private activity between patient and doctor. Consequently, primary responsibility for health rested at the state or provincial level. However, because of greater resources at the national level, federal government roles have increased in the health field over the years. National governments have a responsibility to ensure equity of social policy. A growing federal role has been a historical process common to many countries. At a minimum, the federal level is responsible for national health policy, planning and setting national health targets. The United States, Canada, Russia, Argentina, and Nigeria are examples of countries with federal forms of government.

A unitary state is a form of government that has a central national level, and local governments, but no intermediary legislating level. This includes countries such as the United Kingdom and governments based on the French Napoleonic Code, including most Spanish-speaking countries. In these countries, the central government has great responsibility for health, but here, too, local government is still a major factor in sanitation and local public health. The powers of regional and local authorities are derived from the national structure. Public health grew initially at the local level with regulations for sanitation, business premises and product licensing, food safety, and the like. In the United Kingdom, the national government promoted local public health organization, later organizing personal health services programs for the entire population in the centrally controlled National Health Service.

Diffusion of authority is common to all health systems to differing degrees, mainly based on historical precedents. In recent years, national health authorities have largely been responsible for overall policy, law, financing, standards, monitoring, research, and assurance of services to meet national health goals. Management of services is generally decentralized, with responsibility at the state, regional, and local health authority or institutional level. Diffusion or sharing of responsibility from each level of authority is common in current planning to cope with the wide range of activities and interests that make up the health sector of a society. Nongovernmental agencies often precede governmental authority in the field, and their presence and participation make up important elements of the health complex, whether as providers of services, as advocates, or as fundraisers for programs that a government cannot manage to include in its "basket of services."

Local authorities often delegate administration of services to independent institutions or other agencies. Diffusion of responsibilities occurs in different degrees in administration of services, in education, in training, as well as registry of health professionals including the related professional and accreditation organizations. Diffusion also occurs in research, in intersectoral cooperation between governmental agencies, along with NGOs, or advocacy groups, and in academic as well as research facilities. Legislation may initiate and direct changes in health programs using regulatory and financing measures, but implementation also requires a broad spectrum of participation of individuals and organizations of consumers, providers, and other health interest groups. Health is not an isolated service, but a reflection of the social values and standards and economic development of a society, with a large degree of interdependence and interaction between health agencies and other governmental and non-governmental elements of that society.

Checks and Balances in Health Authority

The balance between government intervention and private organization, between regulation and self-governance, is not easy to define or to achieve in health. Historically, elements of health care developed at different times and with different degrees of political, economic, and public support. The accumulated experience of modern public health indicates that all elements of health need to be considered as part of a spectrum of services (Figure 10.1). Weakness in one area threatens the well-being of the totality. Poor levels of nutrition and sanitation breed disease, for which treatment is more expensive and less effective than prevention. At the same time, low medical care standards due to inadequate training, motivation, resources, and supervision can lead to low standards of health among large segments of the population.

Public health services are in some cases partially integrated with curative service systems and the public–private mix may be beneficial in delivery of specific preventive services, such as screening and immunization. However, when such services are provided by medical care providers, there will always be a need for special provision to uninsured or noncompliant population groups. Organization for public health services, whether integrated into a total care system or separate from curative service systems, requires a combination of centralized and decentralized responsibilities. The overriding national responsibility may be met best by setting policy goals and standards, while assuring regional and social equity. Decentralization allows local authorities direct operational responsibility, with resources and accountability. Diffusion of responsibility means that many agencies operate at different levels of the national entity. Each has its own sphere of interest, and

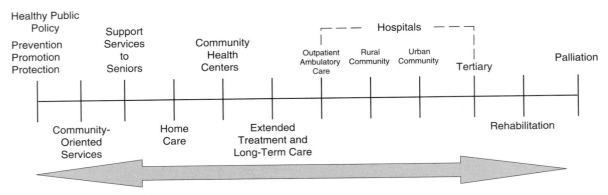

FIGURE 10.1   A continuum of health services.

these link together to form a working whole, with checks, balances, and cooperation among them.

A centralized health organization that controls policy, administration, financing, services, personnel training, research, and regulation may lack checks and balances needed to prevent authoritarian control. Formerly highly centralized health systems are now seeking decentralization as a means of infusing additional funding, a sense of localism, pride, privacy, and quality in their health systems. They are combining this with universal access and regional, ethnic, and social equity. Comprehensiveness and cost constraint are the challenges of organization of public health systems.

There are advantages to a federal structure in the division of responsibilities for health. The senior level of government serves as the overall policy level with financing and regulatory roles. By its very nature, the state level is closer to the community and represents regional interests, while the local government is closest to the community and, with state and federal backing, can serve to promote community health interests subject to state and federal guidelines and accountability.

The New Public Health seeks a balance and cooperation between government-operated health services and the diffused network of private, often competing, organizations, working together to use resources effectively to achieve common health targets that meet the needs of the individual and the population as a whole.

Government and the Individual

Conflicting ideas as to the overall role government should play affect public health in many ways. In 1869, John Stuart Mill, the founder of modern economics, wrote in the introduction to *On Liberty,* "the only purpose for which power can rightfully be exercised over any member of a civilized society, against his will, is to prevent harm to others. His own good, either physical or moral, is not a sufficient warrant." This philosophy has been adapted to a recognized and essential role for government in public health, as in education and other essential services.

The institutions of basic sanitation and community hygiene have had to contend with such individualistic ideas. The issue of governmental interference in "private matters," such as in health, is not new and is actively debated in industrialized western societies, in the post-Soviet countries, and in developing nations alike. Laissez-faire economists promote the idea of minimal governmental involvement in all economic affairs including social services such as health.

During the nineteenth and increasingly in the twentieth centuries, it became apparent and imperative for protection and promotion of health that the state intervene to set and enforce public health measures in all societies. At the other extreme, disillusionment occurred when governments assumed total responsibility for health and total central management of health services. Most countries have their own balance between the two extremes. Paradoxically, the most decentralized and privatized of all national health systems, that of the United States, has been proactive in, and has emphasized development of, national and professional standards, monitoring, setting national targets and regulation in health, and is in the process of profound change from individual care toward managed care systems.

FUNCTIONS OF PUBLIC HEALTH

The American Public Health Association (APHA), founded in 1872, periodically issues policy statements on the mission and essential services of public health organizations. These guidelines help government to provide or assure provision of services through other agencies. The 1994 APHA statement of the overall vision and the mission of public health in America was endorsed by the Association of State and Territorial Health Officials, the National Association of County and City Health Officials, the Institute of Medicine, the Association of Schools of Public Health, the U.S. Public Health Service, and others. Periodic review and revision, with consensus among the many professional organizations concerned with public health, help maintain relevance for local and central public

---

Box 10.2 Mission and Essential Services of Public Health, American Public Health Association (APHA)

Public health responsibilities or mission
1. Prevent epidemics and spread of disease.
2. Protect against environmental hazards.
3. Prevent injuries.
4. Promote and encourage healthy behaviors.
5. Respond to disasters and assist communities in recovery.
6. Assure quality and accessibility of health services.

Essential public health services
1. Monitor health status to identify community health problems.
2. Diagnose and investigate health problems and health hazards in the community.
3. Inform, educate, and empower people about health issues.
4. Mobilize community partnerships and action to solve health problems.
5. Develop policies and plans that support individual and community health efforts.
6. Enforce laws and regulations that protect health and ensure safety.
7. Link people to needed personal health services and assure provision of health care when otherwise unavailable.
8. Assure an expert public health workforce.
9. Evaluate effectiveness, accessibility, and quality of health services.
10. Research for new insights and innovative solutions to health problems.

Source: Essential Public Health Services Work Group. 1997. *Morbidity and Mortality Weekly Report*, 46:150–152; http://www.apha.org/ [accessed May 16, 2008]

---

health organizations. The mission and essential services of public health in the United States are shown in Box 10.2.

For many of the responsibilities legislated for public health agencies at the national, state, provincial, or local health authority levels, a combination of methods and approaches is needed. Regulatory functions are those based on the legal authority of a public health agency to set and enforce standards. Setting health targets, policies and financing, and national or state standards is important in promoting new program initiatives. Health promotion includes not only direct and formal teaching, but promotion of awareness of public health problems to the general public, health care providers, and other agencies. Services may be provided directly or may be funded and supervised by the public health agency. Direct service is the provision of services to the public, especially useful in areas where universal coverage is essential (e.g.,

immunizations), or for high-risk groups not able to access other services (e.g., prenatal care for the poor).

Intersectoral cooperation is the coordination with other agencies of government, NGOs, or service providers to work toward common objectives that will improve public health. This is an area where public health advocacy is important in that the public health authority tries to engage other agencies, as in development of water and sewage systems or policing of highways to reduce road accidents deaths and related morbidity. NGOs, voluntary organizations, and advocacy groups have in the past and will in the future play a vital role in developing health programs.

Regulatory Functions of Public Health Agencies

Regulatory function in public health is based on a legal mandate to protect the public from health hazards and to assure certain standards for provision of care. Whatever degree of decentralization occurs, there are key central standards in public health that must be maintained at the federal level in essential areas such as nutrition, sanitation, food and drug control, and others over which the individual citizen or health provider has no direct control. The regulatory function covers a wide range of public health activities (Box 10.3).

Methods of Providing or Assuring Services — Direct or Indirect?

Whether a governmental agency provides or assures the provision of services varies from country to country. Canada's health insurance program is operated by the provinces with federal cost sharing. In Scandinavian countries, the counties, which have many of the characteristics of provinces, operate most local health services. In centrally managed economies, such as former Soviet countries, health services have been operated with a high degree of central control. The trend internationally toward decentralization of management of services is under critical review and a mix of centrally managed and decentralized services will likely be the trend in coming decades.

Only government can perform many public health functions because certain services require legislative, taxing, and regulatory powers, or because they are directed at the total population. Central coordination is required for key public health functions such as epidemiology and disease control, monitoring population health, and others discussed in the previous section.

In keeping with specific health targets formulated by national or international public or professional bodies, local, state, or national health authorities directly provide certain basic public health services, such as those of specialized

Box 10.3 Examples of Regulated Aspects of Public Health in the United States

1. Regulation and processing data from birth and death certificates and other data sources from local, state, and national authorities — National Center for Health Statistics (NCHS);
2. Business premises and product licensing approval: local health authorities;
3. Building code compliance: local health authorities under state and federal codes;
4. Sanitation and environmental health: municipal, state, and national agencies, such as the Environmental Protection Agency (EPA);
5. Regulation of health professionals: state boards;
6. Licensing and certification of health facilities: local, state, and federal authorities;
7. Communicable disease control: local, state, and federal authorities with the Centers for Disease Control (CDC);
8. Food safety: local, state, and federal standards and inspections by the Food and Drug Administration (FDA);
9. Pharmaceutical standards: safety, efficacy, labeling, and manufacturing standards by the FDA;
10. Occupational health and safety: local, state, and federal standards and inspections within the Occupational Safety and Health Administration (OSHA), for standards, regulation, and enforcement, and the National Institute for Occupational Safety and Health (NIOSH), for research.

laboratories. In the United States, public health agencies provide services not otherwise available to high-risk or otherwise underserved population groups. Many of these developed under special funding by higher levels of government to promote specific programs such as immunization, lead abatement, prenatal care, and HIV testing. They are generally services that are often not adequately covered by health insurance systems or by private practitioners and health care systems.

Immunization may be provided as a governmental service, which is the case in Israel, or by private or managed care providers, as in the United Kingdom and the United States. Even in countries with well-developed primary care systems, there may be need for additional special services, such as screening for cancer of the cervix, hypertension, or congenital disease. Health education, a function of all levels of government and nongovernment health services, involves those activities centered on raising consciousness and knowledge in the health professions, the public, or vulnerable target groups, cutting across virtually all public health activities.

Financial incentives in the form of grants or other categorical funding may be directed to programs to promote specific public health services, research, or education. Financial incentives are used widely in seeking solutions to particular problems, such as incentive payments to physicians for achieving performance indicators or national health targets such as full immunization, or Pap smears and mammography for target population groups in the United Kingdom. National goals may be set in a consultative process, taking into account their importance to the health of the nation. They must also address economic and human resource capacity to organize and deliver relevant programs to meet goals stated with the potential impact evaluated. Incentive or categorical funding is often a useful method to introduce a new set of activities, to strengthen a weak area of public health, or to promote a shift in emphasis in the health system.

NONGOVERNMENTAL ROLES IN HEALTH

Both the government and the private sector, including both not-for-profit and for-profit service systems, have vital roles to play in public health and health care. The private sector includes service providers; professional organizations; universities; and consumer, volunteer, and advocacy groups. Because of the private sector's contribution to service delivery, professional standards, and education of health personnel, it can make a major contribution to any health system.

NGOs may be able to innovate through voluntary action and programming to meet areas of need with which formal health systems may have difficulty. In the United States, the March of Dimes (Box 10.4) is an outstanding example of a volunteer organization and its contribution in the development of the Salk polio vaccine in the 1940s, subsequently in the care of people affected by polio and, more currently, in prevention of birth defects. There are many organizations raising funds for promotion of research and services for specific health concerns, ranging from diabetes to multiple sclerosis.

Voluntary organizations can often initiate services that the public sector cannot. Examples are numerous, but this may suffice: In Jerusalem, a father and son established a voluntary organization in memory of the wife and mother (Yad Sarah) in 1976 to provide a wide range of free, loaned medical devices and services, from wheelchairs to home meals, to day care centers and emergency call systems. The mission of Yad Sarah is to help the elderly and handicapped to function in their own homes. Subsequently, branches were established in 70 cities all over Israel. Other organizations established similar projects in over 25 cities of the former Soviet Union, and plans are in progress for a similar organization in New York City.

Box 10.4 The March of Dimes (MOD)

Founded in 1938 with the participation of President Franklin Delano Roosevelt, a 1921 victim of polio himself, the March of Dimes played a major role in providing care for polio-stricken children and the search for a vaccine to prevent the disease. Thousands of volunteers helped to raise funds and to organize wide-scale clinical trials of the Salk vaccine in the 1950s. Following the eradication of polio in the United States, the March of Dimes shifted its focus to major health problems of children: birth defects, low birth weight, infant mortality, and lack of prenatal care.

MOD's 2005 Global Report on Birth Defects states: "Every year an estimated 8 million children — 6 percent of total births worldwide — are born with a serious birth defect of genetic or partially genetic origin. Additionally, hundreds of thousands more are born with serious birth defects of post-conception origin due to maternal exposure to environmental agents. At least 3.3 million children less than 5 years of age die annually because of serious birth defects and the majority of those who survive may be mentally and physically disabled for life.

The organization promotes and funds activities to reduce birth defects and infant mortality by measures to prevent low birth weight (to 5 percent or less), and to increase the number of women receiving prenatal care in the first trimester (to 90 percent). It funds work to promote genetic research including gene therapy, testing, counseling, and gene mapping. MOD promotes work on the Human Genome Project with genes related to immune disorders, mental retardation, leukemia, improved blood tests for newborn screening, and improved perinatal care for cerebral palsy and respiratory distress of the newborn. MOD works actively to promote use of folic acid among women in the age of fertility to reduce risks of neural tube defects, and supports comprehensive newborn screening for all babies (for at least 29 conditions, as "for each of these conditions, screening has a documented benefit to the child, and a reliable test that enables early detection is available").

Source: March of Dimes Birth Defects Foundation. 2006. http://www.marchofdimes.com/ [accessed May 12, 2008]

Box 10.5 Planning Assumptions for Emergency Mass Critical Care

1. Mass casualties from bioterrorist attacks or accidental, chemical, or biological releases may occur without warning and could result in hundreds, thousands, or more critically ill victims.
2. National, state, and local health authorities should prepare, direct, and coordinate activities in planning and managing such critical situations as illness due to pandemic, natural disaster, and other man-made or natural disaster situations, utilizing all public and private resources for such events.
3. Mass illness (or injury) from a pandemic may produce large numbers of critically ill patients requiring acute respiratory care.
4. Mass critical illness will place great stress on local community hospitals which will have a key role in decreasing morbidity and mortality rates after a bioterrorist attack or pandemic disaster situation.
5. Substantial pre-event planning is required for mass critical care with new approaches to triage and care.
6. Any hospital will have limited ability to divert or transfer patients to other hospitals in such an event.
7. Currently deployable medical and epidemiologic teams of the U.S. federal government will have a limited potential for increasing a hospital's immediate ability to provide critical care to large number of victims of a bioterrorist attack.
8. Hospitals will need to depend on non-federal sources or reserves of medications and equipment necessary to provide critical care to the seriously ill for the first 48 hours following discovery of the bioterrorist attack, or during a pandemic.

Sources: Khan, A. S., Levitt, A. M., Sage, M. J., et al. 2000. Biological and chemical terrorism: Strategic plan for preparedness and response. Recommendations of the CDC Strategic Planning Workgroup. *Morbidity and Mortality Weekly Report*, 49(RR-04):1–14.
Belson, M. G., Schier, J. G., Patel, M. M. 2005. Case definitions for chemical poisoning. *Morbidity and Mortality Weekly Report*, 54(RR-01):1–24.
Rubinson, L., Nuzzo, J. B., Talmor, D. S., et al. 2005. Augmentation of hospital critical care capacity after bioterrorist attacks or epidemics: Recommendations of the working group on emergency mass critical care. *Critical Care Medicine*, 33:E1–E13.

DISASTERS AND PUBLIC HEALTH PREPAREDNESS

After September 11, 2001, preparedness for terrorism became a high priority for federal, state, and local governments (Box 10.5). With federal funding and other support, communities have strengthened their ability to respond to public health emergencies. Collaborative relationships developed for bioterrorism preparedness have proven useful in addressing other threats, such as health impacts of natural disasters and infectious disease outbreaks. The primary role in disaster response is increasingly recognized as a local responsibility. Funding constraints, inadequate surge capacity, public health workforce shortages, competing priorities, and jurisdictional issues all continue to hamper adequate preparation and response as witnessed by the Katrina hurricane aftermath in New Orleans. The U.S. federal government has responded with an investment of some $5 billion since 2001 to upgrade the public health system's ability to prevent and respond to large-scale public health emergencies, whether caused by terrorism or by natural agents.

The public health system will continue to face demands for emergency preparedness and health protection in the face of natural disasters and terrorism. The challenges are to use focused, risk-based resource allocation, regional planning, technological upgrades, workforce restructuring, improved integration of private-sector assets, and better performance monitoring.

Disaster preparedness requires activities and readiness at all levels of government, and by first responders (police, firefighting, and ambulance services) as well as by health care institutions. Activities include preparation of essential supplies, organizational guidelines, staff training and orientation, as well as adequate funding to meet these needs. Since disasters with mass casualties may appear in many forms, the response teams need flexibility and capacity for improvisation. Coordination between different levels of government can be difficult with lines of command and lateral communication unclear and potentially disastrous. Preparation for treatment of mass casualties of bioterrorism requires similar resources to a situation of pandemic and mass illness due to a new variant of SARS or avian influenza (Box 10.6).

MEDICAL PRACTICE AND PUBLIC HEALTH

Public health and clinical services are both vital and interplay to improve individual and population health. Ready access to high-quality health care services is a basic right and a requirement of good public health. This calls for

---

Box 10.6 Lessons from Recent Disasters and Threatened Pandemics — 9/11, SARS, Tsunami, Katrina, and Avian Flu, Burma and China, 2008

The twenty-first century began with the 9/11 massive Islamic terrorist attack on New York City's World Trade Center in Manhattan by hijacked civilian aircraft causing over 2500 deaths and many injuries. This event stirred world-wide repercussions and was followed by very deadly terrorist strikes in Madrid, London, Bali, Mumbai, and many other parts of the world. These attacks caused national and international reactions including calls for disaster preparedness with stress on local capacity for response to man-made and natural disasters, with emphasis on basic "first responder" service capacity.

During 2004, a threatened pandemic of severe acute respiratory syndrome (SARS) started in China and was transmitted via an infected person to Toronto, Canada, in a short time. The Canadian provincial and municipal authorities were taken by surprise and lacked adequate federal mechanisms for addressing the problem. Provincial and municipal authorities managed the epidemic by hospitalization and isolation of all suspected cases with quarantining of hospitals involved. As a result of review of this experience, Canadian governmental authorities developed new federal institutions in part modeled on the U.S. Centers for Disease Control and established a federal Ministry of Public Health with direct authority to increase the federal presence in epidemic control.

In 2004–2005, three huge natural disasters occurred in different parts of the world, showing the crucial importance of disaster preparedness and response organization, preparation, and intergovernmental coordination. The tsunami in Thailand and region, Hurricane Katrina in Louisiana and especially New Orleans, and the earthquake in northern Pakistan showed the crucial need for coordination and speed as well as preparation for natural disasters by all levels of government working with voluntary organizations for rescue and relocation needs.

In 2006, the H5N1 influenza virus, also called "bird flu," threatened to become a new world pandemic of a scope similar to the influenza pandemic of 1917–1918. National and world public health organizations mobilized under the leadership of the WHO implementing monitoring and control measures. These largely rest on identification of cases among wild and domestic birds and rapid identification, isolation, and treatment of human cases. Culling of domestic agricultural birds took place to restrict transmission of the virus H5N1, which could produce a human pandemic of epic proportions if transmitted from birds to humans and then by human-to-human transmission.

In May 2008, a cyclone disaster in Burma (Myanmar) killed many tens of thousands of people with some 1.5 million left homeless, and destitute, and vulnerable to secondary disasters from new floods, exposure, famine, and infectious diseases. The response from its military government has been criminally negligent, preventing foreign aid to reach the people in need. China was struck by a massive earthquake and series of aftershocks which killed an estimated more than 100,000 people and devastated many cities, towns, and villages. The governmental response was immediate and effective, accepting limited foreign assistance, unable to cope with the enormity of the calamity, but probably limiting the secondary disasters from famine and infectious diseases.

These experiences and threatened pandemics have brought public health organizations and key public health functions into the spotlight of national thinking in many countries after many years of financial cutbacks and administrative neglect or "outsourcing" to private providers. This public awareness may be fleeting, and should be utilized to help strengthen public health infrastructure capacity and workforce development.

Sources: National Academies of Science. http://www.nationalacademies.org and http://newton.nap.edu/catalog/11619.html [accessed May 14, 2008]
Centers for Disease Control. Public Health Emergency Response Guide for State, Local, and Tribal Public Health Directors, http://www.bt.cdc.gov/planning/responseguide.asp [accessed May 14, 2008]
Myanmar disaster, http://www.searo.who.int/LinkFiles/Myanmar-Cyclone_sitrep_170508.pdf [accessed May 18, 2008]

high-quality organization and the availability of professionals to provide both clinical and preventive care. The phenomenon of private payment to physicians working in public sector health systems is widespread, as is that of physicians in public service who practice privately after official working hours. This is almost impossible to stop, but can be regulated.

In the United Kingdom, private practice by specialists employed by the hospitals is permitted and encouraged, allowing faster access to hospital care for private patients. This is often seen as a built-in injustice in the National Health Service (NHS). In Israeli teaching hospitals, a private medical service is organized using senior physicians on the hospital premises, with a percentage of the generated funds going to the hospital.

Fee-for-service payment practice of medicine is still a common mode of practice in the United States and Canada, even though each of the two countries has different methods of financing services. Canada's national health insurance program is based on private fee-for-service practice of medicine, but it bans extra billing by physicians, which could threaten equity of access for all population groups, as part of federal criteria for support of provincial health plans.

The United States has a mixed situation of private health coverage, mainly through employer-subsidized insurance, Medicare for those over age 65, and Medicaid for the poor and people with disabilities. This combined system has proven inadequate on a societal level; some 40 million people lack health insurance and another 15 million have poor levels of coverage, with further difficulties for those who change jobs and lose their health insurance coverage. Growth of managed care plans is occurring as private medical practice is declining in the United States. Operated as for-profit or as not-for-profit programs, managed care plans provide lower cost and more comprehensive coverage than traditional insurance plans.

Medical care outside hospitals as reviewed for eight countries (Australia, Denmark, England, Finland, New Zealand, and Sweden) by Ettelt et al. in 2006 pointed out wide variation in patterns of organization. In Finland, this emerges as multispecialist and general practitioner networks with integration into single centers providing medical service. In Germany, polyclinics, once rejected, are making a comeback.

Reform of health care is going on in many countries. This requires incentives to promote ambulatory and community outreach services, through incentives and integration of hospital and long-term care. Managed care is important in the United States, and the model is relevant in other countries because of the linkage of restraint of unnecessary use of hospital and unreferred specialist services, placing emphasis on primary care and preventive care (see Chapters 11–13).

INCENTIVES AND REGULATION

Incentives and disincentives are important tools in health policy and management. Governments are responsible for assuring adequate supplies and quality of health facilities and personnel to meet the needs of the population. They also are responsible for assuring that financing of the system is adequate and efficient. These responsibilities include the use of public authority to ensure a balanced and high-quality system of care equitably available to the people of all regions and social classes. Whether services are owned and administered by government, non-profit agencies, or private auspices, the public authority is responsible and accountable for ensuring that the health needs of the population are met.

The appropriate balance among different elements of health systems serving the same regional or district population is an important public health planning issue. Health facilities such as hospitals and long-term and community care facilities are licensed and regulated by the appropriate public health authority. This regulatory power is necessary, but not sufficient without financing arrangements to combine incentives and disincentives (i.e., a "carrot and stick" approach, Box 10.7).

The ratios of hospital beds and medical personnel per thousand population are crucial determinants of health economics, so that national and state health authorities must use their regulatory powers to contain supply and distribution. Excess labor supply of medical specialists is a problem in many mid-level developing countries, such as in Latin America. Regulatory or financial powers, as well as financial controls, can be used to reduce the oversupply of specialists and to redirect doctors to underserved areas of a country and primary care.

Box 10.7 Carrots and Sticks

"Carrot and stick" is a phrase used to refer to the act of simultaneously rewarding "good" behavior while punishing "bad" behavior. An older interpretation is the use of a carrot dangling on a stick in front of an uncooperative mule, so that the encouragement is constant, but the satisfaction is permanently elusive.

The combination provides financial mechanisms and regulation to encourage health facilities to develop, in keeping with national, state, or local needs. In developed countries, this may mean closure of excess hospital beds and reallocation of resources to community-based health services, as is under way in the United Kingdom, many European countries, Canada, the United States, and others. In Russia and many former Soviet countries, the incentives and requirements produced a heavily hospital-oriented health system with lower priorities to community-based services.

A federal government authority can act to promote health programs by setting financial incentives and disincentives. The categorical grant approach provides funds for a specific purpose or cost sharing for a program that meets defined guidelines. The Canadian health insurance plan is based on provincial plans meeting federal guidelines to qualify for a share of the costs.

National health insurance was included in the proposed Social Security legislation during the Roosevelt administration but excluded from the Social Security Act of 1935, because it created opposition to the major elements of the Act. Following the end of World War II, in 1946, the proposed Wagner-Murray-Dingell Bill for national health insurance failed to reach the floor of Congress, dying in committee, under severe pressure from the American Medical Association and the health insurance industry. A portion of that proposal emerged, however, as the Hill-Burton Act to provide federal assistance to local agencies to build or upgrade hospitals. The Hill-Burton model is a relevant approach to problem solving in a federal state using a categorical grant mechanism to promote what is seen as a health priority. Such an approach may be useful to strengthen a weak health program such as immunization and maternity care in a developing country. It may be used to change the balance in supply of services and resources. A system of incentives or cost-sharing arrangements can provide capital funding; for example, to reduce total bed capacity and to promote integration of maternity, mental health, geriatric, and tuberculosis facilities into general hospitals. A "downsize and upgrade" conditional grant would provide for renovation and transition to an approved program of facilities to modernize hospital services. The federal grant system, pioneered by the Hill-Burton Act (Box 10.8) would encourage the local authority to apply and match part of the funding, and meet federal criteria and guidelines for this process.

In countries where health systems were highly centralized, such as in Britain's National Health Service and in former Soviet health systems, decentralization and diffusion of power were promoted by financing mechanisms. The Canadian national health insurance plan was implemented with fiscal cost-sharing incentives by the federal government to induce the provinces to agree to federal conditions of universal coverage, comprehensiveness, portability, and public administration as criteria for the provincial plans. When the federal government moved from a fixed percentage of expenditures to block grants, it lost some control over detailed management of provincial plans, but it retains a strong voice in requirements for equity, portability (i.e., transferability of insured benefits from one province to another), public administration, and prohibiting extra billing by providers for insured services. As federal shared cost program funding declined as a share of total provincial health costs, the provinces were under pressure to reform mainly by reducing the

---

Box 10.8 The Hill-Burton Act

The Hill-Burton Act, adopted by the U.S. Congress in 1946 as the Hospital Survey and Construction Act, provided a federal–state–local partnership that channeled large federal grants to assist development of hospitals and standards for construction (Hospital Survey and Construction Act, 1946, Title VI of the Public Health Service Act). This affected 4000 communities in 6800 projects to modernize hospitals suffering from lack of investment from the depression and World War II period. Initially it covered hospitals, but later was expanded to extended care, rehabilitation facilities, and public health centers. In 1975, this was further expanded to grants, loan guarantees, and interest subsidies for health facilities. Facilities assisted under Title XVI were required to provide uncompensated services in perpetuity.

It brought national standards and financing to local hospitals. The program helped raise standards of medical care throughout the United States in the 1950s and 1960s. It led to an increase in numbers of hospitals in underserved areas and the renovation of obsolete facilities. It promoted desegregation in the southern United States and provided a mechanism for treatment of the uninsured in the nation's hospitals.

The program also succeeded in limiting the buildup of an excess of hospital beds, setting standards at 4–4.5 acute care hospital beds per 1000 population (more for rural areas), without an increase in the total supply of beds. While it favored middle-class communities because it required local financial contributions, it also channeled federal monies to poor communities, thus raising standards of hospitals and equity in access to quality care. It required hospitals assisted by federal funding to provide a reasonable volume of free or reduced-cost care to the poor (see Chapter 11) and emergency treatment of the uninsured. In setting upper limits on hospital beds, it limited hospital expansion and contributed to a continuing process of improving and shortening hospital stays.

The program had a number of basic failings, including the promotion of the hospital as the main center of health care, leaving community care out of the main flow of added funds. It led to an increase in the proportion of health expenditures going to hospital care. Expenditures for hospital care in the United States as a percentage of total health expenses increased from 34.5 percent in 1960 to a high of 41.5 percent in 1980, but declined to 35.4 percent in 1995. In the 1980s, the Hill-Burton Act was expanded to promote clinic and primary care facilities.

Source: Department of Health and Human Services. http://www.hhs.gov/ocr/hburton.html (updated June 2000); http://www.hrsa.gov/hillburton/compliance-recovery.htm [both accessed May 14, 2008]

---

hospital bed supply and promoting community-wide health service organization.

Unregulated chronic care facilities operated by private interests resulted in proliferation of poor-quality facilities and sometimes extremely low levels of care in many communities in the United States. Public health authorities were powerless to interfere except in cases of gross neglect or poor sanitary facilities. The introduction of

Medicare for the elderly and Medicaid for the poor provided federal and state agencies with the power to set minimum standards for care facilities, by requiring all facilities serving Medicare patients to be accredited by a nongovernmental agency accepted by the federal health authorities. This has become a standard requirement throughout the United States, and the Canadian provincial health insurance plans also apply economic sanctions on unaccredited hospitals or other inpatient facilities.

Another measure to increase regulation of health care facilities was the requirement for any hospital proposing expansion or renovation to seek state approval through a Certificate of Need (CON). The CON, as used in the United States under state health legislation, makes approval by the state contingent on demonstrating need and sources of funding which comply with state regulations. This can be linked with incentive grants but can also be used as a simple regulatory mechanism. The CON approach by state departments of health was only partially successful in limiting unbridled ambitious expansion of hospital facilities. In the 1980s and especially the 1990s, competition and changes in payment systems have resulted in hospital closures and downsizing in the United States.

Promotion of Research and Teaching

Medical research and education are the basis for future developments in health care. They foster new health science developments, such as vaccines or, through initiatives such as the Human Genome Project, treatment for genetic and chronic diseases. This contributes to the development of medical schools, but also safeguards, guarantees, and increases their quality, raising standards of care. Research in public health depends on the basic and clinical sciences, but equally on epidemiology and documented experience of field programs.

In the United States, the National Institutes of Health (NIH), starting with the National Cancer Institute in the 1930s, have done much to encourage high-quality medical education and research. The NIH granting system has been a major factor in promoting standards of medical education by financing research and teaching faculties in medical schools throughout the United States. NIH funding has played a major role in moving the United States into the forefront of the biomedical sciences since World War II. There are currently 27 separate National Institutes of Health including centers and divisions (Box 10.9).

A combination of collegial competition, the free publication and exchange of research studies, views in peer-reviewed journals, and professional meetings in government agencies promotes scientific and applied progress in the medical sciences. The private sector manufacture of drugs and medical devices contributes to the continued development of medical and public health sciences. National centers of

---

Box 10.9 U.S. National Institutes of Health, Centers and Divisions, and Internet Addresses, May 13, 2008

Institutes general home page http://www.nih.gov/
National Cancer Institute (NCI) www.cancer.gov/
National Eye Institute (NEI) www.nwei.gov
National Heart, Lung, and Blood Institute (NHLBI) www.nhlbi.nih.gov/
National Human Genome Research Institute (NHGRI) www.genome.gov/
National Institute on Ageing (NIA) www.nia.nih.gov/
National Institute of Alcohol Abuse and Alcoholism (NIAA) www.niaa.nih.gov/
National Institute of Allergy and Infectious Diseases (NIAID) http://www3.niaid.nih.gov/
National Institute of Arthritis and Musculoskeletal and Skin Diseases (NIAMS) www.niams.nih.gov
National Institute of Biomedical Imaging and Bioengineering http:///www.nibib.nih.gov/
National Institute of Child and Human Development (NICHD) www.nichd.nih.gov/
National Institute of Deafness and Other Communication Disorders (NIDCD) www.nidcd.nih.gov/
National Institute of Dental and Craniofacial Research (NIDCR) www.nider.nih.gov/
National Institute of Diabetes and Digestive and Kidney Diseases (NDDK) http://www2.niddk.nih.gov/
National Institute of Drug Abuse (NIDA) www.nida.nih.gov/
National Institute of Environmental Health Services (NIEHS) www.niehs.nih.gov/
National Institute of General Medical Sciences (NIGMS) www.nigms.nih.gov/
National Institute of Mental Health (NIMH) www.nimh.gov
National Institute of Neurological Disorders and Stroke (NINDS) www.ninds.nih.gov/
National Institute of Nursing Research (NINR) www.ninr.nih.gov/
Centers and divisions
National Institute of Health Clinical Center (CC) http://clinicalcenter.nih.gov
Center for Information Technology (CIT) www.cit.nih.gov/
National Library of Medicine (NLM) and MEDLARS* www.nlm.nih.gov
National Center on Minority Health and Health Disparities http://www.nchmd.nih.gov/
National Center for Research Resources (NCRR) www.ncrr.nih.gov
National Center for Complementary and Alternative Medicine (NCCAM) http://nccam.nih.gov
John Fogarty International Center (FIC) www.fic.nih.gov
Center for Scientific Review (CSR) http://cms.csr.nih.gov/

Source: National Institutes of Health, Bethesda, Maryland. May 2008, http://www.nih.gov/
*MEDLARS = Medical Literature Analysis and Retrieval System.

excellence in public health in other countries include the Pasteur Institute in France and Cambridge Laboratories in the United Kingdom. They receive national funding and have a critical mass of high-quality researchers. Federal funding of medical teaching centers supports development and maintenance of academic standards for undergraduate medical education.

Federal or external granting mechanisms can be used to promote schools of public health and health administration that are needed to prepare the next generation of health leaders, academics, and researchers. Research may be initiated in response to requests for proposals by scientists in university or research institutes, or in the governmental or private sector. A competitive grant system can be useful to upgrade medical education and university academic standards by promoting research and graduate education.

Accreditation and Quality Regulation

Public health authorities have sufficient powers to regulate health facilities. However, in practice, accreditation based on professional guidelines and systems outside the governmental structure (see Chapter 15) plays an important role in quality of health care provider organizations, as an important adjunct to the official regulatory approach of health departments.

The Joint Commission on Hospital Accreditation (JCHA) started in the United States in 1913, and included Canada from 1951–1959 when the latter established its own accreditation system. The JCHA was established by a consortium of the American College of Surgeons, the American Hospital Association, and other voluntary professional bodies. It carries out voluntary peer review of hospitals throughout the United States. The commission established minimum standards in 1918, and has gone on to develop extensive guidelines based on physical, organizational, and professional criteria, to protect the safety and rights of the patient, standards of care, and efficient organization of services. Accreditation involves a process of external review of the facilities, organization, staffing, and related functions including staff qualifications, continuing education, medical records, quality assurance, and others (Chapter 15).

The JCHA review was initially conducted on the basis of a voluntary request by the institution, but accreditation has become virtually mandatory for the economic survival of a hospital in the United States and Canada. Since 1965, Medicare and Medicaid accept accreditation as compliance with federal standards for the purpose of payment, and refuse to pay for services in an unaccredited hospital. The renamed Joint Commission for Accreditation of Healthcare Organizations (JCAHO) has gone on to develop standards for accreditation of facilities for the mentally retarded

(1969), psychiatric facilities (1970), long-term care facilities (1971), ambulatory facilities (1975), hospices (1983), managed care programs (1989), and home care and ambulatory care in 1990. There is a growing emphasis on action plans for quality improvement for rural hospitals, health care networks, laboratories, and public health programs. The JCAHO has become active in promoting accreditation organizations in other countries such as the United Kingdom and Australia.

The New York State Department of Health has its own mandatory regulatory system for hospitals and long-term care facilities. While perhaps unpopular with providers and business leaders, regulation is viewed by health officials as essential to the maintenance of quality standards and prevention of professional and human rights abuses. Israel, during the 1990s, established a national system of inspection of private long-term care facilities, which has improved standards of facilities and care. While opponents see this as excessive state interference, in principle they accept the need for state regulation. Resultant improvements in quality of care measures have justified prudent regulation and oversight of health care facilities. These models could be useful for raising standards in other health care systems.

NATIONAL GOVERNMENT PUBLIC HEALTH SERVICES

National governments can use their financial power to promote programs directly to the state, provincial, or local governmental level or indirectly through nongovernmental agencies. The latter include universities, voluntary teaching hospitals, and private NGOs. Direct or indirect funding may be used to diffuse and promote national standards, such as in medical education and research. Governments can also ensure regional equality of services by cost sharing or grants that favor poorer regions of the country. National governmental health agencies are responsible for external relations, including those with international bodies such as the United Nations, the World Health Organization, the Food and Agriculture Organization, the International Labour Organization (Chapter 16), as well as with parallel ministries of health in other countries, and other national agencies in the same country.

Prior to and after World War II, most western industrialized countries developed some form of national health program. In North America, health care was provided through private insurance, largely union-negotiated, employment-based health plans. The attempts by President Harry Truman to bring in a national health insurance plan in 1946 were unsuccessful in the United States. As a result, federal support for health was channeled into many categorical programs by funding state and county public health services and research and teaching facilities.

It included establishment of the Centers for Disease Control and the National Institutes of Health (NIH). This promoted high levels of competitive, peer-reviewed programs throughout the country, but failed to assure universal access to health care (Chapter 13).

In all forms of government, the national responsibility for health has led to specialized public health services as well as supervisory and regulatory functions (Box 10.10). These include provision of vital support services, such as public health reference laboratories, epidemiology and communicable disease control activities (e.g., national epidemiologic publications, airport, and port surveillance), national health statistics, approval and supervision of drugs and biologicals, research and teaching facilities, and cooperation among federal, state, and local authorities. Standards bureaus and agencies provide the guidelines, monitoring, and/or supervision of health care at the lower levels of government and in the nongovernmental and private sectors.

The federal government entered the public health arena in areas where only a national jurisdiction could function. The Marine Hospital Service was established in 1798 to provide care for U.S. and foreign seamen, becoming the U.S. Public Health Service in 1889. Under the organizational structure of the U.S. Department of Health and Human Services, the USPHS provides direct care in many areas of American society, including Native American reservations, physician shortage areas, the U.S. Coast Guard, and penal institutions. The federal Food and Drug Act of 1906 updated frequently, protects the consumer from adulterated foods and ineffective or dangerous medicines. The Social Security Act has provided pensions for the elderly and the disabled since 1935. In 1965, the Social Security Act was extended to include Medicare as a federal program providing health insurance for the elderly. In the same year, Medicaid was also established, providing health care for the poor, set up as a cost-sharing program with state and local authorities. The history of development of public health in the United States reflects advancing scientific knowledge, societal demands for better health, and the evolution of interactive organization at federal, state, and local levels. In some respects, public health in the United States has provided professional leadership in the field internationally; in other respects, the United States has lagged behind other industrialized countries.

The Department of Health, Education, and Welfare was established in 1953 under a cabinet-level officer of the executive branch of the Eisenhower administration. This brought together a variety of federal agencies and programs, and subsequent reorganization led to emergence of the Department of Health and Human Services (HHS). The present organizational structure of HHS is presented in Figure 10.2. The federal role in direct regulation and funding of projects deemed to be in the national interest helps to promote state and local health authority response to public health problems. The categorical grant system has been instrumental in advancing specific areas of activity, such as maternal and child health, which remain a major activity of both state and local public health departments. The initiatives of the Health Care Financing Administration (HCFA) in promoting changes in methods of paying for hospital care through diagnosis-related groups (discussed in Chapters 12 and 13) helped reduce hospital lengths of stay, days of care, and the hospital bed to population ratio.

The Surgeon General of the Public Health Service is also the Assistant Secretary for Health and provides important professional leadership to the public health movement in the United States. Dr. C. Everett Koop, an outstanding surgeon general, who served for many years during the Reagan administration, exemplified this kind of leadership role. The American Public Health Association initially opposed the appointment of Dr. Koop, a pediatric cardiac surgeon, but within a short period he showed a degree of professional and moral leadership that served as a beacon recognized by the public health profession. Dr. Koop is now honored as one of the great leaders in U.S. public health history, responsible for many accomplishments, most notably increased awareness of the deadly effects of tobacco use and his advocacy for HIV/AIDS research and treatment funding.

The Centers for Disease Control and Prevention (CDC) plays a continuous role in dispersing epidemiologic data and evaluation throughout the country and the world (see Chapter 4). The training program of Epidemic

---

Box 10.10 Key Functions of a Federal or National Ministry or Department of Health

1. National health planning
2. National health financing
3. National health insurance
4. Assurance of regional equity
5. Defining goals, objectives, targets
6. Setting standards and quality of care
7. Promotion of research in quantity and quality
8. Operating or delegating professional standards/licensing
9. Environmental protection
10. Food and drug standards, licensing
11. Epidemiology of acute and chronic disease
12. Health status monitoring
13. Medical/pharmaceutical industrial development
14. Health promotion
15. Nutrition and food policy
16. National reference laboratories
17. Social assistance
18. Social security

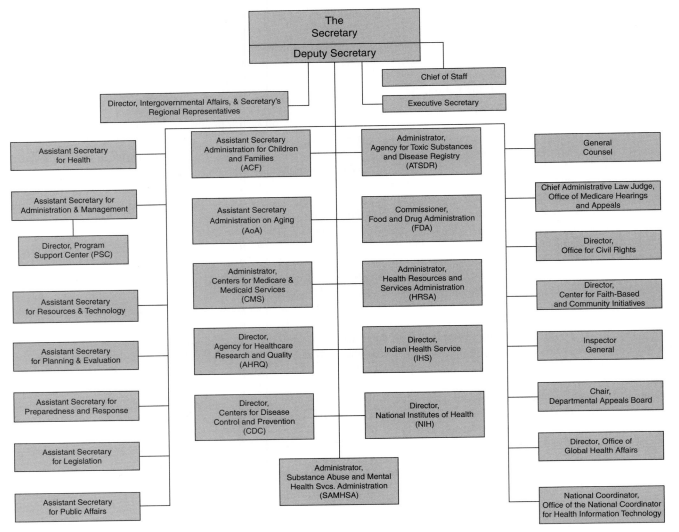

FIGURE 10.2　U.S. Department of Health and Human Services (HHS). Note: The Assistant Secretary for Health is also the Surgeon General of the United States. Source: U.S. Department of Health and Human Services, http:// www.hhs.gov/about/orgchart.html [accessed May 16, 2008]

Intelligence (EIS) officers for federal, state, and local health departments continues to provide high-quality medical epidemiologists capable of developing leadership in this field.

Other agencies of the federal government control health-related programs, including the Departments of Agriculture, Defense, Environment, Interior, Labor, and Transportation. The Department of Agriculture operates a National School Lunch program and a food stamp program to supplement food purchasing power for the working poor. The Department of Labor operates the Occupational Safety and Health Administration. The Environmental Protection Agency is an independent federal agency responsible for air and water quality, pollution control, pesticide regulation, solid waste control, radiation and toxic substance hazard control, and noise abatement.

STATE GOVERNMENT PUBLIC HEALTH SERVICES

The state or provincial level of government has a leading role in health in most federal countries. State or provincial governments have responsibility to ensure adequacy in organization, set standards and targets, assist financially, and provide professional and technical support services to local health departments. State functions, such as financing and in some cases direct services and monitoring health status, are listed in Box 10.11.

State or provincial departments of health are complex organizations with many responsibilities for financing, regulating, inspecting, and assuring health-related issues. In the United States, responsibilities include administration of health insurance for the poor under Medicaid; in

---

Box 10.11 Functions of a State/Provincial Ministry or Health Department

1. Coordinate with other government departments: governmental planning and priorities; education, social welfare, labor, agriculture, mental health, and financing of universities;
2. Establish standards, finance, develop, advise, and supervise local health departments;
3. Legislate and regulate health-related matters: preparation, assistance, and enforcement;
4. Plan and set health priorities and targets;
5. Provide epidemiologic and laboratory services to local health departments and conduct biological surveys;
6. Maintain and publish vital statistics, epidemiology, and health information systems;
7. Develop standards and monitor quantity, quality, and distribution of diagnostic and treatment services;
8. Ensure occupational health supervision;
9. Ensure environmental health monitoring and supervision;
10. License and discipline health professionals and health care institutions;
11. Provide occupational and personal health services to state employees;
12. Coordinate with related state services: social services, mental retardation, drug and rehabilitation, and prison services;
13. Ensure mental health services are part of mainstream health;
14. Coordinate with national and other state/provincial health authorities;
15. Monitor health status indicators of state/province and local authorities;
16. Health education;
17. Promotion of quality of care in long-term care and hospitals, and in primary care;
18. Communicable and infectious disease control;
19. Prepare and train for natural and man-made disasters as well as health emergencies, including potential mass epidemics and bioterrorism;
20. Legislate and promote positive health behaviors such as smoking restriction and environments in schools, workplaces, and public spaces.

Canada, the provinces administer universal health insurance plans. States may initiate programs that are shared with local health authorities and with federal cost sharing, or respond to federal initiatives and seek funding for a wide variety of programs through federal requests-for-proposals for maternal and child health or other categorical grants.

The New York State Department of Health (DOH) has a strong tradition of regulation in chronic care facilities, laboratories, and hospitals, and in environmental health. The various regulatory functions of the department make it a powerful determinant of the operation of health care in the state. Among its functions are granting certificates of need, regulation of reimbursement methods for hospital care (see Chapter 13), establishing health standards and surveillance systems, rural health systems, and many other activities. This state DOH is active in screening programs for congenital and infectious diseases of the newborn, laboratory certification, and quality assurance. An AIDS Institute is responsible for prevention, screening, and AIDS care programs. The Center for Community Health operates a wide range of public health programs, from epidemiologic surveillance of infectious diseases, to prenatal and newborn care among the underserved, to community health worker programs, to nutrition monitoring and many other intervention programs focused on high-risk groups or topics. Environmental epidemiology and monitoring are also strong in the state, which experienced the Love Canal incident (see Chapter 9). Box 10.12 is the 1996 configuration of the New York State DOH. This is not necessarily typical but does show the wide range of activities, including state, federal, and local initiatives.

In New York State, selected public health functions are the responsibility of other government departments or agencies. Box 10.13 displays the range of public health responsibilities in other agencies.

The New York State Department of Health is unique in that it is a cosponsor with the State University of New York (SUNY) of a School of Public Health at Albany, which involves departmental personnel as faculty and students in internships in branches of the DOH. While not necessarily representative of other states, this health department represents the broad scope of public health at the state level of government (Box 10.14).

LOCAL HEALTH AUTHORITIES

Historically, the local health authority (LHA) is responsible for sanitation and the provision of direct care to the poor and high-risk population groups. Boards of Health were established in Philadelphia in 1794 and in New York City in 1796 for these purposes.

The city or county local public health department is the official public health agency closest to the population served. The LHA provides a range of direct supervisory sanitation functions to ensure compliance with local, state,

Box 10.12 New York State Department of Health: Organization Chart

Commissioner

Health Facility Management
Office of Managed Care
Government Affairs and Strategic Planning
School of Public Health
Office of Minority Health

Office of public health (OPH)	Office of health systems management (OHSM)
AIDS Institute HIV care and community services Prevention	**Health Care Financing** Elderly pharmaceutical coverage Financial management and institutional support Health economics Long-term care reimbursement Primary and acute care reimbursement
Center for Community Health Family and local health Nutrition Epidemiology Chronic disease prevention and control	**Health Care Standards and Surveillance** Home care Hospital services Professional medical misconduct Emergency medical services Funeral directing Controlled substances Standards health facility planning
Center for Environmental Health Environmental protection Environmental health assessment Occupational and environmental health Epidemiology District offices	**Architecture and Engineering** Health facility planning Long-term care initiatives Facility and service review Financial analysis and review Project management
Laboratory Center Molecular medicine Infectious diseases Environmental disease prevention Laboratory quality certification Laboratory operations Genetic disorders	

Source: New York State Department of Health, http://www.health.state.ny.us/ [accessed May 15, 2008]

Box 10.13 Agencies with Public Health Responsibilities (New York State)

Agencies	Responsibilities
Department of Education	School sanitation, health education, licensure of physicians and other health professionals
Department of Labor	Health and safety of workers, in-plant pollution and radiation control
Department of Environmental Conservation	Control of pesticides, rabies control, air pollution, sewage and solid waste control
Department of Social Services	Medicaid (program for the poor)
State University of New York	School of Public Health, student health services
Department of Mental Hygiene	Mental institutions and community services
Narcotics Addiction Control Commission	Treatment facilities, research, education
Department of Agriculture	Licensure of meat dealers and slaughterhouses, inspection of restaurants, school-meal regulation of food additives
Department of Corrections	Operation of prison hospitals and clinics, TB case finding
Department of Motor Vehicles	Highway safety promotion

Box 10.14 Albany County Health Department

Albany County Department of Health Programs and Services

Albany County is located generally in the vicinity of Albany, New York, the capital of New York. Albany is also the county seat of Albany County with a population of 297,414 in 2005. The county health department is a team of public health and administrative professionals that has built community comprehensive programs focusing on prevention of disease, protection from environmental hazards, and promotion of good health. Individuals, community groups, schools, and businesses can all benefit from these services, always being assured that any personal or environmental health issue will be handled in a professional manner.

Programs and services available from the Department of Health

- Administration and Engineering Programs
- Birth to Five Program
- Bright Beginnings Program
- Communicable Disease Program
- Community Health and Food Protection
- Community Health Worker Program
- Dental Service
- Heart Disease and Cancer Screening Program
- HIV Testing and Counseling
- HIV/AIDS Community Education Program
- Home Care Assessment Program
- Home Health Care Program

- Immunization Program
- Infant and Child Health Assessment Program (ICHAP)
- Lead Poisoning Prevention Program
- Neighbor Notification Law
- Smokers' Quitline
- Perinatal Hepatitis B Program
- Public Health Nursing Program
- Rabies
- Residential Public Health Programs
- Sexually Transmitted Diseases (VD) Clinic
- Travel Immunization Program
- Tuberculosis Program

Source: Albany County Health Department. http://www.albanycounty.com/departments/health/programs_services.asp?id=171 [accessed May 15, 2008]

---

and federal sanitary codes. The local public health department may also provide direct services, usually personal preventive services, such as those for uninsured pregnant women, funded by the local government authority or by higher levels of government. In the United States, the local public health department is the agency attempting to ensure services to persons inadequately served by voluntary or federal and state insurance plans. Programs may be funded by cost sharing or may be based on categorical or block grants from state or federal governments.

Even though there has been massive growth in the involvement of higher levels of government in public health, the LHA remains the major force for public health at the community level (Box 10.15). In the United States and Canada, the LHA is organized in the form of city or county health departments.

In new health initiatives, such as Health for All, district health systems, and Healthy Cities, the LHA is involved in a wider set of programs for health of its population. In recognition of the objectives of these programs, formerly highly centralized systems, such as those of the United Kingdom, the Scandinavian countries, developing nations, and republics of the former Soviet Union, are being decentralized to LHAs, with varying degrees of central funding, planning, and direction.

In 1940, the American Public Health Association adopted a recommended standard of six basic responsibilities of the LHA:

Box 10.15 Health Responsibilities of a Local Health Authority

Registration and vital statistics
Epidemiology of infectious diseases
Health education, health promotion
Environmental protection, sanitation
Control of communicable diseases, STIs, HIV, TB
Preventive prenatal, infant, toddler care
Allocation of resources
Planning and management of services
Licensing and supervision of health facilities
Hospitals and home care
Care of disabled
Rehabilitation and long-term care
Coordination of health services
Intersectoral cooperation
Mental health
Social assistance
Nutrition
Community participation

1. Vital statistics;
2. Communicable disease control: childhood diseases, TB, STIs, and tropical diseases;
3. Environmental sanitation: water, food processing and marketing, sewage, garbage, sanitary condition of places of business, public eating places, and workplaces;

4. Laboratory services;
5. Maternal, child, and school health;
6. Health education.

In 1950, the APHA adopted an expanded list of program of responsibilities for the LHA, which included the above plus the following:

1. Chronic disease control;
2. Housing and urban planning;
3. Accident prevention;
4. Coordination with other agencies;
5. Surveillance of total health status; births, deaths, chronic disease, morbidity data, surveys, reporting of morbidity, and evaluation of community needs;
6. Education of the public and professional community regarding health status and needs;
7. Supervisory and regulatory activities including health services providers;
8. Personal health services: direct provision and supportive services, varying from comprehensive service programs to services for those in need;
9. Planning of health facilities, urban planning and renewal;
10. Special diagnostic services: including STIs, TB, cancer, child development, and dental care.

Cooperation between the different levels of government is vital to define and achieve national health objectives. Each level of government has a unique role to play. Decentralized administration of public health without national financing and policies will not achieve the full potential of public health and will produce inequities between different regions of a country. National governments are responsible for setting policies, priorities, and goals with definable health targets. State and provincial governments are direct providers and supervisors of public health standards, while local authorities are those directly responsible for sanitation, local planning, and direct services to reduce public health risks.

MONITORING HEALTH STATUS

As discussed in Chapter 3, public health depends on information, just as an army depends on intelligence in order to modify approaches in accordance with changing circumstances and need. Collection, collation, and analysis of this information is vital for informed health policy and must be available to all concerned with health for analysis and policy debate. All levels of government are engaged in health status monitoring, with geographic information systems (GIS), a multisource database related to health indicators for the population of a geographic region, helping to identify localized problems for intervention.

The responsibility of gathering vital statistics lies largely at the local government level, as does the reporting of infectious diseases and other events. Initial collation of the data occurs at this level, and information is then passed on to state health authorities and subsequently to the national level. The gathering of information is a strongly developed tradition in the industrialized countries, and the United States has in many ways done this effectively. In the United States, the CDC serves as a national leadership and reference center, not only for infectious diseases, but also for chronic diseases such as cardiovascular disease, nutrition, diabetes, perinatal epidemiology, and many other conditions.

Health statistics provide the ongoing data needed for monitoring the health status of populations. They provide routine diagnostic and population-based monitoring data that supply valuable epidemiologic information on congenital conditions, STIs, tuberculosis, and HIV infection. Centers of excellence of all kinds, funded or administered directly by federal or state government or by the NIH mechanism, provide tertiary level medical care and conduct biomedical and epidemiologic research, making important contributions to the information pool needed to promote quality analysis and health care.

The national health authority is responsible for the central collation and analysis of health information on the epidemiology of infectious and chronic diseases, vital statistics, utilization of services, and monitoring of national and regional variations in health. This information is only of value if gathered, processed, and published so that it is readily available to health administrators, planners, epidemiologists, care providers, and the public. Census data provide the population denominators for calculation of rates of death and disease incidence or prevalence.

Inexpensive technology of personal computers with modems, as well as telephones and facsimiles, enables local public health agencies to receive real-time information through Internet connections for continuous health profiles of their communities. Sources of data include the following:

1. Vital statistics and national centers for health statistics;
2. Epidemiologic reports of infectious and reportable diseases, including STIs;
3. State, national, and international reporting centers for disease control;
4. Census data;
5. Special disease registries (e.g., cancer);
6. Hospital discharge information systems;
7. Public health laboratories;
8. Poison control centers;
9. Central medical libraries with Medline;
10. Registries of medical, nursing, and dental professionals.

Geographic epidemiology has been important in the history of public health. Fragmentation of information systems has delayed applying modern information technology to multiphasic evaluation and integrating data from multiple sources. This provides information to identify a basic framework of standards and policies for both public- and private-sector participation. A common framework of policies and standards would be strengthened by information sharing among regional and other health networks including academic, service sectors, and community organizations, as well as the media. An example of such an information system is the Health for All database developed by WHO European Region, which is also available for use within countries to show interprovincial or intra-county variations in health status indicators (available at http://www.who.dk/hfadb).

NATIONAL HEALTH TARGETS

The U.S. Public Health Service has set national health targets since 1979. These are increasingly accepted at all levels of the national public health complex. Targets highlight areas of concern that require effort by all levels of government and the health care system. They also serve an educational role for health providers and the community.

Some of the progress made in reducing morbidity and mortality from epidemiologically important diseases is the result of that wider awareness and a growing concept of "self-care." *Healthy People 2010* is a set of health objectives for the United States. It is important as a guideline for states, communities, professional organizations, and others to help them develop programs to improve health. This initiative began in 1979 with the Surgeon General's Report, *Healthy People, and Healthy People 2000: National Health Promotion and Disease Prevention Objectives*. These were developed through a broad consultation process, with available scientific knowledge and are monitored by measurable indicators over time.

Healthy People 2010 Midcourse Review, issued by the National Center for Health Statistics and the CDC, showed progress being made toward over 450 separate objectives in 28 focus areas designed to prevent disease and injury and to promote health in the United States. Of the 281 objectives with tracking data, some 10 percent of the goals have been met and progress made in another 49 percent. For 20 percent targets there were regressions, and for 20 percent mixed results or no change. The annual publication *Health, United States* provides annual updating of a wide range of health statistics as in Box 10.16.

Another approach to national health promotion developing in Europe relates to decision-making in public health (Box 10.17). The European Union lacks many of the institutions available to a federal state such as the United States. It is attempting to find ways to compensate such as in establishing a European Centers for Disease Control to promote pan-European cooperation in communicable and other

---

Box 10.16 *Healthy People* 2010

Healthy People 2010 provides a framework for prevention for the United States. It is a statement of national health objectives designed to identify the most significant preventable threats to health and to establish national goals to reduce these threats.

The Leading Health Indicators are:

- Physical Activity
- Overweight and Obesity
- Tobacco Use
- Substance Abuse
- Responsible Sexual Behavior
- Mental Health
- Injury and Violence
- Environmental Quality
- Immunization
- Access to Health Care

Source: Department of Health and Human Services. Healthy People, http://www.healthypeople.gov/ [accessed May 15, 2008]

---

disease control with guidelines and common policies of health promotion. This effort is in its early stages, but has been advanced by concern over the threats of pandemics such as SARS and avian influenza.

The concepts of prevention and health promotion are integral to setting and attaining health targets. The methods of public health are increasingly moving toward wider responsibilities in terms of health monitoring and organization to reach the stated goals and objectives. The New Public Health provides a conceptual basis for this process.

UNIVERSAL HEALTH COVERAGE AND THE NEW PUBLIC HEALTH

Because the United States lacks a universal coverage national health insurance program, it is commonly cited in the literature that the United States has a "non-system." This is misleading; the United States has a very complex and unfinished health system, with a major deficiency in lack of universal access health insurance. Yet, the United States is a world leader in public health, not only in the development of new vaccines, but in implementation of important advances in preventive and health promotion, such as fluoridation of community water supplies. The United States has the costliest health system with total expenditures reaching 16 percent of GDP in 2006, but it lags behind many other countries in important indicators of health status (see Chapter 13). Still, the United States has other indirect public health programs that support poverty groups, including a universal school lunch program and WIC (Women, Infants and Children) food supplementation

Box 10.17 Effective Decision Making for Public Health Policy

Public health does not take place in a vacuum. It requires a societal commitment that places health in a high social priority for funds and public policy. Allin, Mossalios, McKee, and colleagues examined public health policy in eight countries (Denmark, Finland, France, Germany, the Netherlands, Sweden, Australia, and Canada). The authors discussed the following key issues for strong public health policy.

1. Political commitment and support at all governmental levels (national, state, local).
2. Intersectoral cooperation between government agencies and with NGOs.
3. Preparation of the population (e.g., societal acceptance of smoking restriction legislation).
4. Health law developed and codified with appropriate enforcement capacity.
5. Promotion of individual and population behavior changes consistent with "healthy lifestyle" and supportive socio-economic context such as in alleviating poverty and inequities in health.

6. Adequate infrastructure and resources for organized public health structures at all levels of government with sufficient, well-trained personnel and programs.
7. Independence from political control so that the voice of public health can operate to identify and meet challenges in population health and not be submerged under a clinically oriented health system.
8. Organization, funding, and support for research to provide the skills and material to evaluate health of the population and identify new risk factors and associations.
9. Health policies that are realistic and targeted to measurable goals with identification of priorities and feasible programs to meet these objectives.
10. Development of training and research environments and capacities consistent with the standards and culture of public health at the highest international standards.

Source: Modified from Allin, S., Mossalios, E., McKee, M., Holland, W. 2004. Making decisions in public health: A review of eight countries. *European Observatory on Health Systems and Policies.* Copenhagen: European Observatory, http://www.euro.who.int/observatory; document available, at http://www.euro.who.int/document/E84884.pdf [accessed May 15, 2008]

for pregnant women and toddlers in need. Further, the U.S. health system is a complex interactive set of organizations, subject to system changes, that has pioneered many innovations in health sciences, health care administration, and public health.

Publicly administered universal access elements exist, even if underfunded. The middle class is protected by employment-based health insurance, the elderly by Medicare, and the poor by the federal–state–locally administered Medicaid program. The failure to adopt national health insurance providing equitable access to health care continues to be a major obstacle to improving health of the vulnerable poor and marginalized sectors of society. Public health services at all levels of government spend much energy and resources trying to cover deficiencies resulting from inequities in access to services.

Managed care plans in which financial incentives are in play to promote ambulatory and preventive care and decreasing use of hospital care increasingly cover the U.S. population. Collaboration between organized public health and medicine, long-standing antagonists in the United States, took a new direction in the mid-1990s with development of a "new paradigm" of cooperation. The American Medical Association and the American Public Health Association agreed to work together to promote networking in the form of collaborative local programs to resolve unmet health needs of the community. This

mutual awareness represents recognition of the importance of both clinical medicine and public health. Intersectoral dialogue helps identify the potential for cooperation in the context of the dramatic changes taking place in the United States in health care organization. The health insurance coverage of people under age 65 in the United States is shown in Figure 10.3.

In other countries such as the United Kingdom and the Scandinavian countries, organization of health services

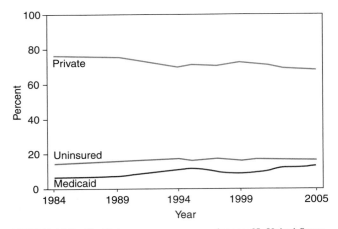

FIGURE 10.3 Health insurance coverage under age 65, United States, 1984–2005. Source: Centers for Disease Control and Prevention, National Center for Health Statistics, *Health, United States, 2007.*

moved to district health systems in which public health is a full partner with clinical services, and where prevention is integral to the economics and function of a population-based program. The managed care evolution in the United States since the 1990s may well promote a new level of cooperation between clinical medicine and public health. Integration of services financed by Medicare and Medicaid, with federal waivers of eligibility conditions for age and poverty, may allow a new approach based on residence in areas of need. Expanding Medicaid will occur largely through enabling enrollment into managed care programs of large numbers of eligible persons not currently enrolled.

Downsizing the hospital sector, constraint in health costs, increasing enrollment in managed care, focusing on health targets, and increasing coverage through managed care will constitute a national health program evolving toward some form of the New Public Health. The United States has been very innovative in financing systems to promote efficiency in use of services, and other countries have begun to apply those lessons in their national health insurance plans. The United States will benefit from examining the reforms going on in many countries, including Canada and European countries, as their health systems also evolve. The American public health community, including the schools of public health, has capacity and experience with professional leadership and advocacy, and it can make a great contribution toward adaptation of the New Public Health.

Expenditures on population-based public health services in 50 states are 3.0 percent of total health expenditures. Most expenditures by state health departments (66 percent) were for personal care services, mostly for people ineligible for health insurance or with benefits excluding preventive care. This represents the predominant priority for hospital and ambulatory care services based on insured or personal outlay for services. While much of ambulatory care involves preventive services, the relatively low expenditure for community-oriented public health activities reflects traditional values and underevaluation of the potential impact of community-oriented approaches to health promotion. The health reforms going on in most countries, especially those in transition from the Soviet system, require a shift of priorities for expenditures from a hospital orientation to a community orientation. This is a difficult process with many political implications, especially loss of jobs in many communities.

HOSPITALS IN THE NEW PUBLIC HEALTH

The hospital is an important element of the New Public Health. The individual institutional health facility, whether a general rehabilitation center, a nursing home, or a mental or other special hospital, is characterized as a hospital. Each has a defined role, administrative structure, budget, and *modus operandi* as a unique organization. Hospitals evolved under municipal, religious, voluntary, governmental, university, private, or other sponsorship. Hospitals have traditionally been separate administrative units from other health services, although often with a strong connection to medical and paramedical training programs. The organizational structure is often based on the history of the organization, and may need adaptation to address the facility's mission, resources, and role as part of a larger community health system.

Hospitals are among the largest employers in any country, with some three-quarters of all health personnel and, depending on the country and its traditions and reform processes, between 38 percent and 75 percent of total health expenditures. The magnitude of the hospital sector and the key role it plays in the health service system make it vital to rationalize services, preventing duplication, bed surpluses, overemphasis on specialized services versus primary care, and depersonalization of patients and workers.

The modern hospital is the most costly and visible element of a health system to the public; it employs the most personnel and it provides care for the seriously ill. Hospital management is therefore an important factor in managing the total health system. While health care is an organizational system, the component facilities such as hospitals are also living organizational entities that require structure, management, and planning.

The supply and utilization of community general hospitals have changed over the past half century, from 3.6 beds per 1000 population in 1960, rising to 4.5 in 1980, and declining steadily since to 2.7 beds per 1000 in 2005. At the same time, occupancy rates have also declined from 75 percent in 1960 to 67 percent in 2005. A similar trend has occurred in most industrialized countries and more recently even in the former Soviet countries (see Chapters 11 and 13).

Under managed care systems, the hospital will try to satisfy two parties: the patient and the managed care system with its economic constraints. These two parties may have different objectives and methods of assessment of the functioning of the institution and the community it serves. The insured patient, in his or her role as a hospital patient and with the option to change health plans, will be able to exert some influence on the quality of care he or she receives. Similarly, the managed care system can judge the quality of care rendered by a hospital and express dissatisfaction by choosing an alternate provider.

The mission of a hospital is to provide high-quality care and service to the patient within the limits of current standards of knowledge and resources. In addition, there are many other objectives of the hospital as an organization.

These objectives include professional and economic survival as an institution, teaching functions, research, and publication. The hospital provides an important contribution to the community, providing employment, financial stability and solvency, prestige, education research, and a system of access to health care.

To meet these diverse goals and objectives, hospitals have become complex organizations with an extensive division of labor (see Chapter 12). This involves many different professional areas, as well as "hotel services and facilities" such as the provision of food, laundry, housekeeping, supplies, and the financial and personnel administrative functions. As a large organization of great complexity, a hospital must have a formal, quasi-bureaucratic structure with clear lines of authority and responsibility. Given that, however, the modern hospital cannot function under a traditionally authoritarian, paternalistic pattern of administration. Coordination of the many complex skills brought together in a hospital requires lateral coordination between departments and staff at all levels or the machine simply will not function. As a result, the hospital is highly dependent on the motivation and integrity of its staff, and their ability to network with others in different departments or professional levels freely and without excessive bureaucratic constraints.

Nevertheless, basic acceptance of authority, discipline, and rigid requirements to maintain standards of care are still essential to hospital function and predictability of performance. A great demand on hospitals is efficiency, so that waste, duplication of service, poor maintenance and function of facilities and equipment, corruption, negligence, or thievery cannot be tolerated by the organization. The modern hospital has formal bureaucratic lines of authority, and hundreds or perhaps thousands of examples of informal networks and sometimes formal organizations to carry out the daily work of patient care, while meeting the other needs of the hospital and "good standards of care" with efficiency in use of resources. There are many checks and balances in the structure with multiple lines of authority and responsibility, and sometimes even tension between administrative and professional elements.

Hospital Classification

Hospitals are institutions whose primary function is to provide diagnostic and therapeutic medical, nursing, and other professional services for patients in need of care for medical conditions. Hospitals have at least six beds, an organized staff of physicians, and continuing nursing services under the direction of registered nurses. The WHO considers an establishment a hospital if it is staffed continuously by at least one physician, can offer inpatient accommodation, and can provide active medical and nursing care.

Any hospital bed that is set up and staffed for care of inpatients is counted as a bed in a facility. A bed census is usually taken at the end of a reporting period. The WHO defines a *hospital bed* as one regularly maintained and staffed for the accommodation and full-time care of inpatients and situated in a part of the hospital that provides continuous medical care. A bed is measured functionally by the number and quality of staff and support services that provide diagnostic and treatment care for the patient in that bed.

Hospitals include those operated on a not-for-profit and those on a for-profit basis. Most are operated as not-for-profit facilities as public services provided by government, municipalities, religious organizations, or voluntary organizations. In the United Kingdom, hospitals formerly operated by the NHS have been transformed into public trusts to operate as not-for-profit public facilities. In the Scandinavian countries, hospitals and other local health services are operated by the county health department. Private, for-profit hospitals, though increasing, are still a minority of general hospitals but include a large proportion of chronic care facilities.

In the United States, Canada, and Israel, long-term care for the elderly and infirm is largely provided by private for-profit facilities. In these countries, private facilities arose because of inadequate public resources for direct provision of services. As payment systems evolved, private operators were encouraged to enter the field. Government supervision and regulation have diminished the abuses and exploitation that occurred in the 1960s, but the standards of care can be compromised by the profit motive. There are, however, good examples of large-scale operations of long-term care facilities run by private organizations that are efficient and provide good standards of care. As illustrated in Box 10.18, hospitals are also defined by the types of services provided, the population served, and average length of stay.

Supply of Hospital Beds

The supply of hospital beds is measured in terms of hospital beds per 1000 population. This varies widely between and within countries. Historically, hospital development was initiated by church or religious groups, municipalities or voluntary charitable societies, or by local, state, or national governments without national planning criteria. In all health systems, regardless of administration and financing methods, the supply of hospital beds and their utilization are fundamental to health economics and planning.

The hospital bed is often a political issue. In some countries, the hospital has been traditionally regarded as a center of refuge from harsh conditions of life, climate, and social conditions. This is especially the case in rural areas with lesser access to health care. Pressures for more

Box 10.18 Types of Hospitals

Short-stay hospitals are those in which more than half of the patients are admitted to units in the facility with an average length of stay of fewer than 30 days. These include teaching, general, community, and district hospitals providing a broad range of services, as well as specialized hospitals that focus on special categories of patients by age, sex, or medical condition.

Long-stay hospitals are those in which more than half of the patients are admitted to units in the facility with an average length of stay of more than 30 days. These may include special hospitals and may be jointly managed with short-stay hospitals.

Nursing homes are establishments with three or more beds that provide nursing or personal care to the aged, infirm, or chronically ill. They employ one or more registered or practical nurses and provide nursing care to at least half of the residents.

Skilled nursing homes provide more intensive nursing care, as defined by nursing care hours per patient day.

Hostels are residential facilities attached to a medical center for overnight stay of patients undergoing outpatient investigation or care.

Hospices are facilities related to a medical center especially organized to provide a humane, personalized, and family-oriented setting for care of dying patients.

Non-profit hospitals are operated by a government, voluntary, religious, university, or other organization whose objectives do not include financial profit.

Proprietary hospitals are operated for profit by individuals, partnerships, or corporations.

General hospitals provide diagnoses and treatment for patients with a variety of medical conditions or for more than one category of medical discipline (e.g., general medicine, specialized medicine, general surgery, specialized surgery, and obstetrics). This excludes hospitals which provide a more limited range of care.

Community hospitals serve a town or city and are usually short-stay (fewer than 30 days average length of stay) general hospitals.

District hospitals are general hospitals that serve a population of a defined geographic district and have, as a minimum, four basic services: general medicine, surgery, obstetrics and gynecology, and pediatrics.

Teaching hospitals are those operated by or affiliated with a medical faculty in a university or institute.

Special hospitals are single-category inpatient care facilities such as a children's, maternity, psychiatric, tuberculosis, chronic disease, geriatric, rehabilitation, or alcohol and drug treatment center which provide a particular type of service to the majority of their patients.

Tertiary care hospitals are referral and teaching hospitals; a secondary level hospital is a community or district hospital providing a wide range of services; and a primary level hospital is a limited service community hospital in a rural area.

Source: *American Hospital Association*. 2006. *and Health, United States*, http://www.aha.org/aha/about/ [accessed May 15, 2008]

beds may come from physicians or from the public. Political figures tend to favor more hospitals because they provide jobs in a community, signify access to medical care, and create a public sense of well-being. The addition or closing of hospital beds is one of the difficult and controversial issues in health planning and health politics. However, if politicians are responsible for paying the hospital operational costs, they must take into account that operational costs will equal capital costs in about 2 years. It is also difficult to close redundant or uneconomic hospital beds, because this means a loss of jobs in the community unless combined with transfer of personnel to other services, a painful procedure itself.

The hospital bed is a functional economic unit with accompanying staff and fixed costs, so it has important economic implications for the health system. The cost per bed is measured by the total expenditure of the hospital divided by the number of beds. Building and operating costs, on average, are such that the cost of construction of a hospital unit is usually equal to the cost of operating the bed over 2–3 years. The decision to build a bed obliges the health system to indefinitely fixed costs even if that bed is unused as a result of regulation or reduced utilization from professional or economic incentives. Hospital planning is no longer left to the initiative of the facility itself even in the most competitive, market economy–oriented health system.

The tendency to build excess hospital beds and the resultant costs of maintaining them were common to both developed and developing countries in the 1950–1980s. Excess supply is associated with high utilization rates and long lengths of stay. Ironically, most non-emergent diseases may be better treated on an outpatient basis, as hospital-associated infections and disease, such as deep vein thrombosis (DVT), increase length of stay and morbidity and mortality and raise health costs dramatically. Where there is no incentive for the hospital or physician to increase efficiency, patients tend to linger in the hospital. This results in higher overall costs of health care and is associated with medical mishaps, including falls in the hospital, errors in care, drug errors, anesthetic mishaps, and nosocomial infections. Excess bed capacity can be managed in a number of ways. Essentially, it requires conversion of bed stock and staff to other purposes or closure of obsolete facilities.

Especially since the 1980s, many countries are reducing excess hospital bed utilization by shortening the length of stay, increasing the efficiency in diagnostic procedures, decreasing unwarranted surgical procedures, and adopting less traumatic procedures (e.g., breast conserving surgery for breast cancer, and endoscopic surgery). Ambulatory services replace inpatient care for many types of surgeries, including most eye, ear, nose, and throat surgery, and for medical care for oncology, hematology, mental health, and many internal medical problems (Chapter 11).

Development of alternatives to hospital care, such as organized home care, assist in earlier discharge of patients from acute care hospitals by providing services to the patient at home, such as nursing, physiotherapy, intravenous care, change of dressings, or removal of stitches following surgery. Rehabilitation facilities provide appropriate low-cost alternatives to lengthy recovery periods after surgery such as hip or knee replacements. Long-term care facilities provide services for geriatric patients requiring extensive nursing care. These patients may not benefit from lengthy stays in acute care hospitals, and need access to alternatives to hospital care. Closure or reduction of beds is important to assure that savings in one area of service are transferred to a common financing system to provide funding for those alternative services. This may require investment in these extended community services before savings are realized from reduced hospital utilization. While hospitals are vital for acute care in life-threatening disease, preventive capacity is optimized by decentralizing and taking medical care to the community. Hospital size, number, and beds must be balanced using an economic- and public health–focused approach.

The capitation system of payment provides incentives for district health or managed care systems to limit admissions and lengths of stay. Sweden succeeded in reducing the percentage of GNP spent on health care during the 1980s by reducing hospital bed supplies, while maintaining the improvement of health status indicators. Managed care systems and diagnosis-related groups (DRGs) have the same effect in the United States. District health system capitation is leading to reduced hospital bed supplies in the United Kingdom. This is a complex and controversial issue, but managing the numbers of hospital beds is essential especially in view of aging populations with chronic diseases, and the highly intensive and expensive kinds of care needed by many patients (Chapter 11).

The Changing Role of the Hospital

Hospitals are more technologically oriented and more costly to operate. Under the influence of rising costs, incentives for alternative forms of care have led to the development of home care, ambulatory services, and linkages with long-term care. Forces acting on the hospital as an organization and economic unit place the hospital in a context where community-based care is an essential alternative that requires organizational and funding linkage to promote integration.

As a key element of any health system, the hospital will undergo changes as technology and health management sciences advance. Managing health systems with fewer hospital days requires reorganization within the hospital to provide the support services for ambulatory, diagnostic, and treatment services, as well as home care.

The interaction between the hospital-based and community-based services requires changes in the management culture and community-oriented approaches. Involvement of all the staff in quality of the service has become part of this management (Box 10.19).

In countries which operate hospitals as part of the Ministry of Health or National Health Service, there is a growing tendency to transfer hospital ownership and operation to not-for-profit agencies, or trusts as free-standing economic units, or integrated within service programs of district health authorities. Competition for patients and payment for services such as by a DRG system will increase competition and the need for excellence in

---

Box 10.19 Hospital Mergers in Los Angeles County — The University of California, Los Angeles (UCLA), Health Sciences Center and Community Outreach

Los Angeles is a large, multi-ethnic, and rapidly growing metropolitan city of over 9 million people in southern California. The hospital bed-to-population ratio has been 3.5 per 1000 population during the 1980s and 1990s. Payment by DRGs in the 1980s, and growing membership in managed care, led to reduced hospital bed occupancy with 45 percent of beds occupied in 1996. In 1998, the vast majority of insured Angelinos belonged to managed care programs. As a result, many hospitals are selling to for-profit hospital chains, or are under threat of closure, some being converted to long-term or ambulatory care facilities.

As an example, the UCLA network includes the Santa Monica Hospital, a 337-bed acute-care facility serving the health care needs of Los Angeles and Santa Monica since 1926. The UCLA network includes community clinics (Brentwood, Malibu, Santa Monica, Westwood, and others). This provides a wide population for the tertiary care center in competition with other tertiary care centers in Los Angeles. The UCLA Health Sciences Center is a teaching hospital owned and operated by the university.

In the mid-1990s this center developed contracts to provide hospital care to many managed care programs. In order to broaden its community service base, the center purchased several community hospitals and established affiliation agreements with medical group practices in adjacent areas of the city. This enabled the center to ensure its catchment population in a highly competitive market. The emphasis is increasingly on developing contractual arrangements with primary care medical services. The Health Sciences Center is replacing the hospital due to damage in the 1994 earthquake and will do so with a substantially lower number of beds. This is the survival strategy adopted to ensure its continuing role as a major teaching and community service hospital in the changing medical market in the twenty-first century.

Source: UCLA Medical Center. Health Sciences. 1998. http://healthcare.ucla.edu/santa-monica/about [accessed May 15, 2008]

hospital care and its management for the financial survival of the facility. There is a trend in the United Kingdom, Israel, and many countries in transition from the Soviet and post-colonial health systems toward less centralized management and greater competition in health care. The trend to include hospitals in district health authorities, as in the Nordic countries, as part of geographic managed care programs is another important policy direction of health reform. Some Nordic countries, however, are reversing this trend and re-establishing centralized management of district hospitals.

In the United States, hospital networks are developing in the for-profit and not-for-profit sectors with integration of management and other cost savings in scale of purchasing and operation. Integration of health services can be lateral, integrating related services and the medical providers of these services, or vertical (integrating different types of services and different levels of health prevention, such as acute with long-term care, and community care services) (Figures 10.4 and 10.5). Contracts with managed care organizations for hospital care have replaced the previous system under which the insured patients' hospitalization depended on whether the attending doctor had privileges or worked on staff. The for-profit hospital corporations,

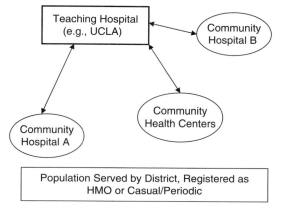

FIGURE 10.4   Integration of health services, UCLA Medical Center.

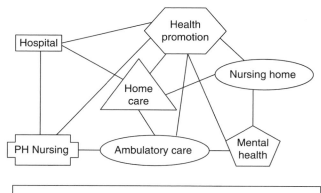

FIGURE 10.5   Vertical integration of health services.

along with similar managed care organizations, have brought health care to the stock market with profits larger than many other sectors of the private economy. Hospital mergers may then be seen in the context of any other business merger or corporate development.

Regulation of Hospitals

Governments have the responsibility and authority to assure standards of health for the population. Licensing of health facilities is one of the traditional methods used to ensure the public interest and prevent harmful practices in patient care facilities. The licensing procedure is the basis for regulation of standards and the content of the service, as well as for controlling health care expenditures (Chapter 11).

Governments have a number of methods to regulate hospitals. One method is through control of the funding mechanism. This allows room for negotiation and influence on standards and level of satisfaction with care; the second is regulation of the number of hospital beds as the licensing and standards authority; the third is control of capital expenditures; a fourth method is to link payment for insured patients to accreditation of the hospital. The level of government responsibility for regulation varies from country to country, usually depending on the constitutional division of responsibility between the different levels of government and the size of the country. Generally, the state and local authorities have the greatest influence because of their proximity. Where government agencies operate hospitals directly, there is a conflict of interest in the form of self-regulation.

The combination of governmental roles of financing, operating, and regulating hospitals in a highly centralized health system may appear to have some advantages, but separation of these conflicting functions is important in promoting a high-quality service. The separation of financing and regulation from operation of services is a widening trend in national health systems.

Governmental regulation may be used to make nongovernmental accreditation virtually mandatory. The accreditation agency's standards are accepted by government as a requirement for hospitals and long-term care facilities in the United States. This is use of an NGO inspection system as a proxy for governmental standards and frees the government from the necessity of establishing large-scale regulatory and inspection systems.

THE UNINSURED AS A PUBLIC HEALTH CHALLENGE

The 40 million Americans without health insurance and the large number of citizens with poor coverage constitute a serious challenge for public health in the United States.

County and municipal health departments focus a great deal of their activities and attention on this population, who are largely poor and in need of health care. The coverage of the elderly and the very poor under Medicare and Medicaid has given a base of protection to these groups, but the near-poor and the near-elderly are still highly vulnerable, especially when job layoffs are a major part of the economic condition. While this problem is becoming more acute with a growth in the number of uninsured following the failure to enact national health insurance in 1994, there are increasing federal and state initiatives to widen coverage for Medicaid and especially target the approximately 20 percent of children who are uninsured. The United States, despite still being the only industrialized country lacking universal health insurance, has established and led in the development of public health programs that have had positive health effects, such as expanding the content of routine immunization of children and adults, school lunch programs, a wide range of categorical health programs to promote prenatal care, lead screening and exposure reduction, mammography, Pap smears, and other preventive services.

No prediction can be made safely as to when or if the United States will adopt national health insurance, or alternatively state programs to mandate health insurance coverage for all. This will remain a potent political issue in coming elections. Delay in establishment of a universal health insurance program remains a continuing burden on the full realization of America's national health potential, for its individual citizens and for the nation as a whole. In order to improve health in the United States in the twenty-first century, the political echelons at federal and state levels will need to find a suitable formula for universal health coverage. Public health professionals have to engage the public, the business community and public policymakers to promote this process toward achievement of individual and community health as well as a healthy workforce.

There are alternative approaches to universalization of U.S. health insurance. One would be a federal-state subsidy program for private health insurance. This would be administratively costly and difficult to monitor for quality of coverage and service. A second approach would be enlargement of the federally administered Medicare program. This is an attractive option as the administration already exists, is generally well appreciated, and could be expanded to include other groups such as low-income groups not eligible for Medicaid. Another possible approach could be federally initiated costs haring for state-level universal coverage meeting federal criteria, based on the Hill-Burton model and to some extent the Canadian health insurance system. This may begin with universal coverage for children and expand to the general population. In the 1990s, the state of Oregon developed an excellent universal coverage system, which consisted of a combination of public funding and regulated private administration of health insurance. The Oregon Health Plan (OHP) was revolutionary, as it did not compete directly with the political influence of established private institutions. There are caveats though. Because neighboring states did not have comparable coverage, migration occurred, based on hope for better care. The public primary care network was unable to expand to meet demands. Necessary increases in local business taxes contributed to regional economic difficulty during a broader national recession, causing many employment providers to leave the state. While the Oregon Health Plan is far less extensive and in danger, it presents a model for development of universal health care without threatening the political and economic power of established private-sector health systems. If instituted on a larger regional scale, economic viability is much more likely.

SUMMARY

Public health is organized at local, state, and national levels to define and work toward achievement of health targets. A balanced health care system requires resources to be rationally allocated to the different preventive, curative, or environmental elements of health. Resources must be directed to all vulnerable groups in the population, recognizing that some groups have greater needs than others. At the same time, issues that affect all, such as nutrition, sanitation, housing, and socioeconomic conditions, affect the poor and the elderly disproportionately. Sound public policy must also take into account the need to ensure adequate quality of care by health care providers and institutions, through developing and regulating standards, licensing procedures, and quality assurance mechanisms.

Impressive progress has been made in public health in the United States over more than two centuries since the establishment of the U.S. Marine Hospital Service, and the evolution of public health systems in the twentieth century. Despite lack of a national health system, the United States has been a leader in formulating administrative mechanisms to improve the efficiency of health care.

The evolution of health care in the United States in the 1990s toward managed care is causing a large-scale reorganization of hospitals with both vertical and lateral integration; that is, formation of networks of hospitals and linkage of hospitals with primary care. Adjustment to meet the health care organization environment of the twenty-first century requires changes for hospitals. These include downsizing, development of ambulatory and home care services, and linkages with primary care services to ensure a "catchment population." The competitive factors in which primary care providers and the community have a role in determining a hospital's utilization, occupancy, and ultimately its survival will help build a more community-oriented health system.

In the United States, public health has been separated from and is unequal to medical services. The advent of managed care for a large portion of the population creates a professional and economic challenge for both sides. Organized public health in the United States needs to seek a closer liaison with managed care to promote a more comprehensive New Public Health approach. Managed care organizations need to develop health promotion and at the same time ensure the interests of the patient to successfully promote their long-term economic interests, and vice versa. If public health remains outside the issues of organization and financing of personal care services, the isolation of public health in the United States will deepen.

The New Public Health is a comprehensive approach to health care, stressing the interdependence of medical and hospital services with prevention and health promotion. Clinical medicine, management of health services, and community health approaches are interactive in many forms, in the United States and elsewhere. In northern Europe and the United Kingdom, district health systems are responsible for and are budgeted on a per capita basis to ensure community health and all levels of personal care services. In the United States, lack of universal health access and central payment systems for all has, paradoxically, promoted development of managed care systems linking all levels of health care. However, public health remains detached from this process, being organized and financed separately.

Health impact assessment applied systematically at the community level is important for determining priorities and use of evidence-based public health. Health impact assessment is an approach to assessing both the health burden from conditions in sectors other than health and the potential of health improvements by modifying those conditions. It combines procedures, methods, and tools by which a policy, program, or project may be judged as to its potential effects on the health of a population and the distribution of those effects within the population. This can facilitate intersectoral action to improve health by evaluating health effects of actions in other sectors such as agriculture, education, economic policy, transportation, and housing. Systematic review is a formal process that identifies all of the relevant scientific studies on a topic, assesses their quality, individually and collectively, and sums up their results. Systematic reviews make it easier for practitioners and policymakers to understand all of the relevant information that is available, how it was collected and assembled, and how the conclusions and recommendations relate to the information that was reviewed. This range of techniques and tools can serve to assure that an intervention or policy will be appropriate and feasible in particular settings.

The New Public Health approach seeks to link those activities of local, state, and national government that promote the health of a population, including the provision of personal care in hospitals, community, and long-term care settings. The New Public Health approach also seeks to link those activities of local, state, and national governments with nongovernmental services that have similar goals.

ELECTRONIC RESOURCES

Agency for Health Care Policy Research. http://www.ahcpr.gov/ [accessed May 15, 2008]

American Hospital Association, http://www.aha.org/aha/about/index.html [accessed May 15, 2008]

American Public Health Association. http://www.apha.org/ [accessed May 15, 2008]

Bureau of Primary Care (DHHS). http://www.bphc.hrsa.gov/ [accessed May 15, 2008]

Centers for Disease Control and Prevention (DHHS). http://www.cdc.gov/[accessed May 15, 2008]

Department of Health and Human Services (DHHS). http://www.hhs.gov/ [accessed May 15, 2008]

Health Care Financing Administration (DHHS). http://www.os.dhhs.gov/about/opdivs/hcfa.html [accessed May 15, 2008]

Health Care Indicators, United States, http://www.cms.hhs.gov/Medicare-ProgramRatesStats/03_HlthCrInds.asp [accessed May 15, 2008]

Healthy People 2010, http://www.healthypeople.gov/Publications/Cornerstone.pdf [accessed May 15, 2008]

Healthy People 2010, Midcourse Review, http://www.healthypeople.gov/Data/midcourse/pdf/ ExecutiveSummary.pdf [accessed May 15, 2008]

Hill-Burton Obligated Facilities, Health Resources and Services Administration. http://www.hrsa.gov/hillburton/hillburtonfacilities.htm [accessed May 15, 2008]

Indian Health Services (DHHS). http://www.ihs.gov/ [accessed May 15, 2008]

Institute of Medicine, National Academy of Science. http://www.iom.edu/ [accessed May 15, 2008]

Joint Commission on Accreditation of Healthcare Facilities, http://www.jcaho.org/index.htm [accessed May 15, 2008]

March of Dimes Birth Defects Foundation, 2006, http://www.marchofdimes.com [accessed May 12, 2008]

National Center for Health Statistics. http://www.cdc.gov/nchswww/

National Institutes of Health. http://www.nih.gov/ [accessed May 15, 2008]

UCLA Medical Center, Health Sciences. 1998, http://www.healthcare.ucla.edu/santa-monica/about [accessed May 15, 2008]

U.S. Department of Health and Human Services. *Healthy People 2010.* http://www.healthypeople.gov/ [accessed May 15, 2008]

RECOMMENDED READINGS

Afifi, A., Breslow, L. 1994. The maturing paradigm of public health. *Annual Review of Public Health*, 15:223–235.

Allin, S., Mossalios, E., McKee, M., Holland, W. 2004. Making decisions in public health: A review of eight countries. *European Observatory on Health Systems and Policies*. Copenhagen: World Health Organization, http://www.euro.who.int/Document/E84884.pdf

Alpert, J. J. 1998. (Editorial). Serving the medically underserved. *American Journal of Public Health*, 88:347–348.

Beitsch, L. M., Brooks, R. G., Menachemi, N., Libbey, P. M. 2006. Public health at center stage: New roles, old props. *Health Affairs*, 25:911–922.

Belson, M. G. Schier, J. G., Patel, M. M. 2005. Case definitions for chemical poisoning. *Morbidity and Mortality Weekly Report*, 54 (RR–01):1–24.

Centers for Disease Control. 1996. Historical perspectives of CDC. *Morbidity and Mortality Weekly Report*, 45:526–530.

Centers for Disease Control. 1997. Estimated expenditures for essential public health services — selected states, fiscal year 1995. *Morbidity and Mortality Weekly Report*, 46:150–152.

Centers for Disease Control. 1999. Ten great public health achievements — United States, 1900–1999. *Morbidity and Mortality Weekly Report*, 48:241–243.

Centers for Disease Control. 1999. Achievements in public health, 1900–1999: Changes in the public health system. *Morbidity and Mortality Weekly Report*, 48:1141–1147.

Centers for Disease Control. 1997. Estimated expenditures for essential public health services — selected states, fiscal year 1995. *Morbidity and Mortality Weekly Report*, 46:150–152.

Cranmer, H. H. Hurricane Katrina. 2005. Volunteer work — logistics first. *New England Journal of Medicine*, 353(15):1541–1544.

Dausey, D. J., Lurie, N., Diamond, A. 2005. Public health response to urgent case reports. *Health Affairs*, [Epub ahead of print].

Department of Health and Human Services, http://www.hhs.gov/ocr/hburton.html and http://www.hrsa.gov/hillburton/compliance-recovery.htm [accessed May 14, 2008]

Effelt, S., Nolte, E., Mays, N., Thomson, S., McKee, M., et al. 2006. *Policy brief: Health care outside hospital: accessing generalist and specialist care in eight countries. European Observatory on Health Systems and Policies.* Copenhagen: WHO European Region.

Greenough, P. G., Kirsch, T. D. 2005. Hurricane Katrina. Public health response — assessing needs. *New England Journal of Medicine*, 353:1544–1546.

Kennedy, E. M. 2005. The role of the federal government in eliminating health disparities. *Health Affairs*, 24:452–458.

Khan, A. S., Levitt, A. M., Sage, M. J., et al. 2000. Biological and chemical terrorism: strategic plan for preparedness and response. Recommendations of the CDC Strategic Planning Work Group. *Morbidity and Mortality Weekly Report*, 49 (RR–04):1–14.

Oberle, M. W., Baker, E. L., Magenheim, M. J. 1994. *Healthy People 2000* and community health planning. *Annual Review of Public Health*, 15:223–235.

Robinson, L., Nuzzo, J. R., Talmor, D. S., et al. 2005. Augmentation of hospital critical care capacity after bioterrorist attacks or epidemics: Recommendations of the Working Group of Emergency Mass Critical Care. *Critical Care Medicine*, 33:E1–E13.

Saltman, R., Bankkauskaite, V., and Vrangbaek, K. (eds.). 2007. Decentralization in health care, *European Observatory on Health Systems and Policies Series*. Berkshire, UK: World Health Organization, Open University Press.

BIBLIOGRAPHY

American Public Health Association. 1991. *Healthy Communities 2000: Model Standards*. Third Edition. Washington, DC: APHA.

Atkins, D., Siegel, J., Slutsky, J. 2005. Making policy when the evidence is in dispute. *Health Affairs*, 24:102–113.

Bates, D. W. 2005. Physicians and ambulatory electronic health records. U.S. Physicians are ready to make the transition to EHRs — which is clearly overdue, given the rest of the world's experience *Health Affairs*, 24:1180–1189.

Benjamin, G. C. 2006. Putting the public in public health: New approaches. *Health Affairs*, 25:1040–1043.

Carlson, M. J., DeVoe, J., Wright, B. J. 2006. Short-term impacts of coverage loss in a Medicaid population: Early results from a prospective cohort study of the Oregon Health Plan. *Annals of Family Medicine*, 4:391–398.

Centers for Disease Control. 1991. *Profile of State and Territorial Public Health Systems: United States, 1990*. Atlanta: U.S. Department of Health and Human Services.

Centers for Disease Control. 1997. Estimated expenditures for essential public health services— Selected states, fiscal year 1995. *Morbidity and Mortality Weekly Report*, 46:150–152.

Centers for Disease Control. 2001. Recommendations and Reports. Updated Guidelines for Evaluating Public Health Surveillance Systems. Recommendations from the Guidelines Working Group. *Morbidity and Mortality Weekly Report*, 50(RR-13):1–35.

Fielding, J. E., Briss, P. A. 2006. Promoting evidence-based public health policy: Can we have better evidence and more action? *Health Affairs*, 25:969–978.

Gebbie, K. M., Turnock, B. J. 2006. The public health workforce, 2006: New challenges. *Health Affairs*, 25:923–933.

Gollust, S. E., Jacobson, P. D. 2006. Privatization of public services: Organizational reform efforts in public education and public health. *American Journal of Public Health*, 96:1733–1739.

Green, L. W., Ottoson, J. M. 1994. *Community Health*, Seventh Edition. St. Louis: Mosby.

Halamka, J., Overhage, J. M., Ricciardi, L., Rishel, W., Shirky, C., Diamond, C. 2005. Exchanging health information: Local distribution, national coordination. As more communities develop information-sharing networks, a coordinated approach is essential for linking these networks. *Health Affairs*, 24(5):1170–1179.

Hurley, R. E., Pham, H. H., Claxton, G. A widening rift in access and quality: Growing evidence of economic disparities. *Health Affairs*, December 6 [Epub ahead of print].

Institute of Medicine. 1988. *The Future of Public Health*. Washington, DC: National Academy Press.

Katz, A., Staiti, A. B., McKenzie, K. L. 2006. Preparing for the unknown, responding to the known: Communities and public health preparedness. *Health Affairs*, 25:946–957.

Lasker, R. D. (ed.), and Committee on Medicine and Public Health. 1997. *Medicine and Public Health: The Power of Collaboration*. American Public Health Association and American Medical Association. New York: New York Academy of Medicine.

Lurie, N., Wasserman, J., Nelson, C. D. 2006. Public health preparedness: Evolution or revolution? *Health Affairs*, 25:935–945.

Mill, J. S. 1869. *On Liberty*. London: Bartleby Co., www.bartleby.com/130 [accessed May 15, 2008]

Miller, C. A., Moos, M.-K. 1981. *Local Health Departments: Fifteen Case Studies*. Washington DC: American Public Health Association.

National Center for Health Statistics. 2006. *Health, United States, with Chartbook on Health of Americans 2006*. Hyattsville, MD: NCHS, http://www.cdc.gov/nchs/data/hus/hus06.pdf [accessed May 15, 2008]

National Center for Health Statistics. 2007. *Health, United States, with Chartbook on Health of Americans 2007*. Hyattsville MD: NCHS, http://www.cdc.gov/nchs/data/hus/hus07.pdf [accessed May 15, 2008]

Patrick, D. L., Erickson, P. 1993. *Health Status and Health Policy: Quality of Life in Health Care Evaluation and Resource Allocation*. New York: Oxford University Press.

Pickett, G., Hanlon, J. J. 1990. *Public Health Administration and Practice*, Ninth Edition. St. Louis: Times Mirror/Mosby College Publishing.

Reschovsky, J. D., Staiti, A. B. 2005. Access and quality: Does rural America lag behind? *Health Affairs*, 24:1128–1139.

Robertson, J. E. Jr., Boyd, J., Hedges, J. R., Keenan, E. J. 2007. Strategies for increasing the physician workforce: The Oregon model for expansion. *Academic Medicine*, 82:1158–1162.

Rockenschaub, G., Pukkila, J., Profili, M. C. 2007. *Toward Health Security: A Discussion Paper on Recent Health Crises in the European Region WHO.* 2007. Copenhagen: World Health Organization, European Region, http://www.euro.who.int/Document/E90175.pdf [accessed July 19, 2007]

Rohrer, J. E. 1996. *Planning for Community-Oriented Health Systems.* Washington, DC: American Public Health Association.

Rubinson, L., Nuzzo, J. B., Talmor, D. S., et al. 2005. Augmentation of hospital critical care capacity after bioterrorist attacks or epidemics: Recommendations of the working group on emergency mass critical care. *Critical Care Medicine*, 33E1:E13.

Saha, S., Solotaroff, R., Oster, A., Bindman, A. B. 2007. Are preventable hospitalizations sensitive to changes in access to primary care? The case of the Oregon Health Plan. *Medical Care*, 45:712–719.

Salinsky, E., Gursky, E. A. 2006. The case for transforming governmental public health. *Health Affairs*, 25:1017–1028.

Simon, P. A., Fielding, J. E. 2006. Public health and business: A partnership that makes cents. *Health Affairs*, 25:1029–1039.

Turnock, B. J. 2004. *Public Health: What It Is and How It Works.* Third Edition. Boston: Jones and Bartlett.

United States Department of Health and Human Services. 1992. *Healthy People 2000: National Health Promotion and Disease Prevention Objectives.* DHHS Publication PHS 91-50212. Washington, DC: U.S. Department of Health and Human Services.

United States Department of Health and Human Services. 1994. *Healthy People 2000 Review, 1994.* DHHS Publication No. (PHS) 94-1256-1. Washington, DC: U.S. Department of Health and Human Services.

Whitener, B. L., Van Horne, V. V., Gauthier, A. K. 2005. Health services research tools for public health professionals. *American Journal of Public Health*, 95:204–207.

Williams, S. J., Torens, P. R. (eds.). 2007. *Introduction to Health Services,* Seventh Edition. Clifton Park, NY: Cengage Delmar Learning.

Chapter 11

Measuring Costs: The Economics of Health

Introduction
Economic Issues of Health Systems
  Investing in Health
  National Health Care Spending
Basic Concepts in Health Economics
Supply, Need, Demand, and Utilization of Health Services
  Normative Needs
  Felt Need
  Expressed Need
  Comparative Need
  Demand
  Supply
  Grossman's Demand Model
Competition in Health Care
Elasticities of Demand
Measuring Costs
Economic Measures of Health Status
Cost-Effectiveness Analysis
Cost-Benefit Analysis
**Basic Assessment Scheme for Intervention Costs and
  Consequences**
The Value of Human Life
Health Financing — The Macroeconomic Level
Costs of Illness
  Costs and Variations in Medical Practice
  Cost Containment
Medical and Hospital Care — Microeconomics
  Payment for Doctor's Services
  Payment for Comprehensive Care
Health Maintenance and Managed Care Organizations
District Health Systems
Paying for Hospital Care
Capital Costs
Hospital Supply, Utilization, and Costs
Modified Market Forces
Economics and the New Public Health
Summary
Electronic Resources
Recommended Readings
Bibliography

INTRODUCTION

Health economics is an important element of health policy, both at the strategic (macroeconomics) and tactical levels (microeconomics).[1] Macroeconomics in health deals with overall financing and allocation of health resources, while microeconomics compares alternative approaches to dealing with specific health issues. Monetary resources for health are limited in all countries, and difficult choices have to be made in their allocation. Management of health care requires an understanding of use of resources, priorities, and trade-offs in health.

All professional health care providers and planners need a working knowledge of the fundamentals of health economics and how regulation and economic incentives and disincentives affect the supply, demand, and ultimately the cost of health services. This knowledge helps one to understand and appreciate how health care, while beneficial in terms of reduced morbidity and mortality, also has a cost in terms of resources used, and how health can be improved while facing constraints of limited resources.

Health economic analysis provides a set of tools for management and decision making in the selection of priorities. It can add a measurable empirical element to policy formation as a necessary, though not sufficient, instrument for health policy decisions. Sometimes, there is a conflict between health economics and professional, ethical, and moral issues in solving the everyday problems of preventive and curative services.

The high and rising cost of health services is increasingly coming under public scrutiny and economic analysis. The total expenditures and how they are allocated are central issues in all health systems. The achievement of a balance between these issues is part of everyday health management, and therefore of the New Public Health.

---

[1]In standard economics, *macroeconomics* is defined as the aggregate of economic activity; *microeconomics* is the theory of how individual organizations (e.g., suppliers) and consumers behave.

ECONOMIC ISSUES OF HEALTH SYSTEMS

Health care expenditures vary widely among different countries, ranging from under 4 percent to over 16 percent of GDP, and are rising faster than general inflation in the industrialized and many mid level developing and transitional countries. In many developing nations total health expenditures are far below levels needed to produce a sustainable infrastructure and meet the ongoing crises of HIV, malaria, tuberculosis, maternal and child health, as well as rising mortality from cardiovascular diseases and diabetes.

In the United States, expenditures for health care have risen from 7.2 percent of GDP in 1965 to over 16 percent of GDP in 2007, and it is projected to be 20 percent of GDP just 10 years from now (Table 11.1). Despite this, and a strong leadership position in medical sciences and research, a substantial portion of the population lacks any or adequate health insurance and the United States is well behind many other less wealthy countries in terms of key health indicators (Chapter 13).

Basic economic problems in health affect different countries through underinvestment, overinvestment, and misallocation of health resources. The World Bank's 1993 *World Development Report: Investing in Health* discussed the economic aspects of addressing health needs, mainly, but not exclusively, in developing countries (Box 11.1). This justified economic investment in health care focusing on less costly and more effective interventions. This approach has influenced international donors and the World Bank as a lending agency. The World Bank has been widely criticized both externally and internally for health project lending policies that have been oriented to medical services without adequate emphasis on population health issues, promoting privatization in health care and inadequate external evaluation of projects.

In 2006, the World Bank Group, a consortium supported by WHO, the Gates Foundation, and other donor agencies, published its second edition of *Disease Control in Developing Countries*. This reviews progress made and guidelines for addressing current public health issues in developing countries, again with an economic justification that a healthier population is essential for economic growth.

The United Nations-approved Millennium Development Goals (for 2015) and their ongoing review provide a globally accepted framework linking poverty reduction and health indicator progress, such as reducing child and maternal mortality. The complexity of achieving these goals is discussed elsewhere, with little chance of achieving the targets by 2015. Nevertheless, they provide a universally accepted set of defined goals and specific targets to monitoring of progress with global indicators.

The Commission on Macroeconomics and Health (CMH) launched by WHO Director-General Dr. Gro Harlem Brundtland in 2000, authored by Jeffrey Sachs at Harvard University, analyzed the impact of health on development. The commission recommended specific interventions for health sector investments that could positively impact economic growth and equity in developing countries for poverty reduction and economic development.

The United Nations Millennium Development Goals along with other international initiatives and many national programs strongly promote the concept that health is a prerequisite for development and economic growth. They call for international and national policy responses and financing mechanisms to address global problems such as poverty, AIDS, malaria, tuberculosis, environmental sanitation, and the threat of avian influenza, as well as chronic diseases such diabetes and obesity.

TABLE 11.1 National Health Expenditures and Average Annual Percent Change, United States, Selected Years, 1960–2005

	1960	1970	1980	1990	2000	2005
	Amount (billions)					
GDP (billion $)	526	1039	2790	5803	9817	12456
National health expenditures per capita	148	356	1102	2813	4790	6697
	Percent distribution					
National health expenditures (NHE) (percent of GDP)	5.2	7.2	9.1	12.3	13.8	16.0
Average percent increase in NHE from previous year shown	na	10.5	13.0	10.9	5.9	6.9
Private expenditures (as % NHE)	75.3	62.4	58.1	59.8	55.9	54.6
Public expenditures (as % NHE)	24.7	37.6	41.9	40.2	44.1	45.4

Source: Health, United States, 2007, table 121.

Box 11.1 Essential Cost-Effective Health Services for Developing Countries: World Bank

Public health
1. Immunizations EPI Plus = DPT, polio, measles, hepatitis B, yellow fever, vitamin A, iodine supplements (later expanded further);
2. School-based health services;
3. Information for family planning and nutrition;
4. Programs to reduce tobacco and alcohol consumption;
5. AIDS prevention.

Clinical services
1. Pregnancy-related care;
2. Family planning;
3. Tuberculosis control;
4. STI control;
5. Management of diseases of children — ARI, diarrheal diseases, measles, malaria, malnutrition.

Notes: EPI = Expanded Programme of Immunization (Chapter 4).
ARI = Acute respiratory infection.
STIs = Sexually transmitted infections (Chapter 4).
Source: World Bank. 1993. *World Development Report, 1993: Investing in Health: World Development Indicators.* New York: Oxford University Press, p. 117.
Disease Control Priority Project, World Bank Group, http://www.dcp2.org/pubs/DCP [accessed June 21, 2008]

Capacity for health economics analysis is important to examine health intervention alternatives and priorities (Figure 11.1). The importance of health as an international issue also requires attention of non-health sector bodies, including addressing global warming, water shortages, and agricultural and trade policies in the interests of the world as a whole.

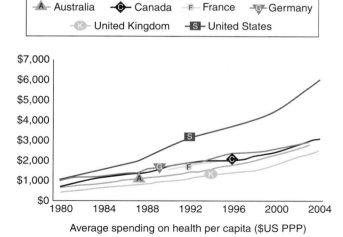

FIGURE 11.1 Trends in international comparisons in spending per capita on health, 1980–2004. Source: The Commonwealth Fund.

Investing in Health

The wide acceptance of the vital role of health in economic development is based on the concept that a healthy population is not only a well-meaning social goal based on human rights, but, like an educated population, is essential to the development of a strong economy. Healthier populations are better workers and contributors to economic growth. Healthier children learn better in schools and thus, also have prospects for contributing to economic development of their country.

These reports address investment in health as follows:

1. Good health is a crucial part of well-being;
2. Spending on health can be justified on purely economic grounds;
3. Improved health contributes to economic growth:
   a. It reduces production loss by worker illness;
   b. It permits use of natural resources that have been inaccessible because of disease;
   c. It increases the enrollment of children in school and makes them better able to learn;
   d. It frees for alternative uses resources that would otherwise have to be spent on treating illness;
4. Sound policy in financing and resource allocation is essential to achieve good health.

The World Bank report calls on governments to increase spending on health and also to foster an atmosphere in which families and communities do the same. It calls for competition and diversity in health care, based on an economic rationale, with health development based on "a basket of essential public health and clinical services" (see Box 11.1).

This report is a landmark document in international health, as important as the Declaration of Alma-Ata, Health for All 2000, and the more recent Millennium Development Goals (see Chapter 2). These reports continue to press the policy that investment in health is an efficient contributor for economic development, and thereby places health among the priorities for national and international financial investment. Health is essential for productivity and economic growth, and allocation of limited resources to health should promote quality and efficiency in care using less relatively unproductive services, such as excessive hospital services, placing greater emphasis on quality, health promotion, and primary care.

In 2007, the prestigious journal *Science* published a lead article describing the technology-implementation gap in which "many evidence-based innovations fail to produce results when transferred to communities in the global south, largely because their implementation is untested, unsuitable, or incomplete. For example, rigorous studies have shown that appropriate use of insecticide-treated bed nets can prevent malaria, yet on average fewer than 10 percent of children in 28 sub-Saharan African countries regularly sleep with this protection" (Madon, 2007). This article calls for implementation sciences to address this in a scientific manner with promotion of research to identify barriers to adoption of currently available technology. The same concern is expressed regarding the availability and provision of antiretroviral (ARV) treatment to stop the epidemic of HIV in Africa and to save countless lives with affordable quality drugs now available to those countries, but using public health strategies. Of the estimated 6.5 million people in need of antiretroviral treatment, as opposed to clinical methods not yet available, 1.65 million were reported to have access to ARV treatment in low- and middle-income countries (WHO, June 2006). At the same time in the highest prevalence region, basic public health infrastructure is weak with very high rates of mortality from preventable maternal and child deaths, tuberculosis and malaria.

National Health Care Spending

The Commonwealth Fund reported in 2007 that health spending in the United States has continued to climb but at a lower rate of increase and has remained at a nearly constant percentage of GDP per capita from 2003–2005. This report called for steps to reduce the rate of cost increase to the level of growth of the economy through: "increasing transparency and public reporting of cost and quality information, rewarding quality and efficiency, and expanding the use of information technology and systems of health information exchange."

Growth per annum in national health spending is projected to slow to 7.4 percent, from a peak of 9.1 percent in 2002. In 2005, the United States spent 16 percent of its gross domestic product (GDP) on health care. Medicare Part D drug coverage in 2006 produced a shift in spending but had little net effect on aggregate spending growth. Health spending is now more than four times U.S. spending on defense, and is expected to grow faster than the economy. Projections indicate that health spending will reach 20 percent of GDP by 2015. At the same time the severe problem of nearly 47 million uninsured Americans remains a political and social dilemma for the United States.

BASIC CONCEPTS IN HEALTH ECONOMICS

All societies have limited resources and must, according to politically determined priorities, provide funds for health care in competition with funds for education, defense, agriculture, and others. The availability of limited funds requires making choices. These choices reflect the overall political commitment to health and should, as far as possible, be based on an objective assessment of costs and benefits of available options.

The components of economic evaluation in health care are seen in Figure 11.2. Expenditure of resources (in terms

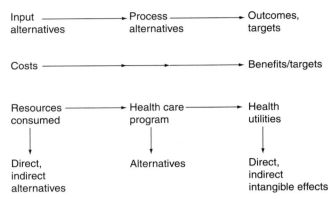

FIGURE 11.2 Economic models of health evaluation: resources-programs-benefits. Source: Commonwealth Fund, www.commonwealthfund.org/ [accessed May 22, 2008]

of financial and human resources), both direct and indirect, is targeted to a health program. This is expected to produce health benefits or utilities, which can also be both direct and indirect. Health benefits may be expressed in terms of a direct reduction in morbidity and mortality, or as improved productivity and quality of life.

Measurement of both input and output is an essential part of health management. Health inputs include resources such as expenditures on buildings, hospital or nursing home beds, equipment, personnel, home care, ambulatory care, and preventive programs. Other elements of health costs, not directly related to the provision of health services but resulting from it, include patient's travel time, loss of work time for both patients and caregivers, loss of full functioning years of life, and loss of quality of life. The "input-output" theory of health economics may sound simplistic, but it provides a useful marker when examining the benefits and costs of a specific health intervention. Alternatives can be examined and analysis of their cost-effectiveness made, in order for decision makers to select the most suitable ones.

SUPPLY, NEED, DEMAND, AND UTILIZATION OF HEALTH SERVICES

Supply and demand are fundamental concepts of economics, particularly of a market economy. *Demand* refers to the quantity of a product or service wanted by buyers, and this relates to the price of the service and the availability of supply. *Supply* is the quantity of the service or product available. Demand is affected also by price, which if too high will reduce demand for that specific product. In a traditional market situation, promotion, supply, price, and demand are interactive.

In health care, price and demand are not the same as products on a free market. Supply may be limited by government regulation and price is offset by third-party payment through insurance mechanisms. In most industrialized countries, these are established by government and cover the entire population. In the United States, health insurance coverage is through a combination of governmental and private insurance arrangements, with a substantial population without adequate or any such coverage. So the supply of services and the method of payment are important economic factors in demand and utilization of health care.

Need and demand for medical services are not necessarily the same. Need for medical care exists when an individual has a disorder or risk of such, with symptoms, illness, or disability, for which he or she believes there to be an effective, acceptable, and beneficial treatment or cure. Need also refers to preventive care which may not be a pressing issue for the individual (i.e., immunization, smoking reduction). Demand for medical care exists when the individual considers that he or she has a need and is willing to spend resources including money, time, energy, loss of work, travel, and inconvenience to receive care. Utilization of services occurs when the individual acts on this demand or need and receives health services.

Normative Needs

Normative needs are those services determined by experts to be essential for a specific need or condition for a specific population group. These include many standard guidelines for both preventive and clinical health care, such as prenatal care, immunization, child care for infants and toddlers, management of diabetes and hypertension, and screening for breast and prostate cancer. "Evidence-based public health" consists of summation of the published literature and reports from countries with successful application of public health "best practices." There are very often legitimate differences of opinion about public health issues, based on alternative interpretations of information, incomplete evidence, or lack of access to international sources and literature.

As scientific knowledge advances, new information is not absorbed into decision-making processes as rapidly as needed, especially in developing and transitional countries. Professional value judgments may be traditions or biases in medical opinion, and not adequately responsive to advances in clinical, technological, and epidemiologic evidence, and evidence from best practices in leading countries. Normative needs, such as cancer screening by Pap smears and mammograms, should be under continuous review by professional panels, with representatives from epidemiological clinical, and public health services, as well as managers and consumers of health care. These reviewers must consider the available literature and experience with such programs worldwide. Disciplines

such as health economics, sociology, health education, and urban planning add to the understanding of contributing factors to a disease and its presence and effects on a population and how to address it. A current problem requiring much multidisciplinary consideration is the obesity epidemic with related diabetes and its complications. Each professional field can contribute to interpretation and decision-making about standards for addressing such problems in the health system.

The individual characteristics of people seeking care, including factors such as age and sex, help determine the type and amount of health services needed. For example, a woman of 40 may not need a mammography as frequently as a woman over 50 years of age. An infant may need to be seen for preventive care assessment more often than a 3-year-old. A male aged 45 needs his blood pressure checked more often than a 25-year-old, and a teenager needs more attention paid to prevention of risk-taking behavior than a 35-year-old.

Clinical guidelines are in common use in many countries providing norms for care based on consensus by professional groups or health insurance providers and have been adopted in countries such as the United States, the United Kingdom, and Israel. Norms are also used for payment purposes, but this method, as used in the Soviet Semashko health system, proved to be rigid and difficult to alter when health conditions change. A norm system for hospital beds proved to be a major barrier to adaptation to changing scientific information and population needs, as well as expectations of the clients, beneficiaries or patients. Where economic incentives (and disincentives) can be used to help promote selected health priorities and are limited resources, as occurs in all countries, then choices must be made. This often comes into conflict with public expectations in health care systems.

Felt Need

Felt need is the subjective view of the patient or the community, which may or may not be based on actual physiological needs. Though subjective, felt need is a prerequisite to whether a person actually undertakes to seek care. There is a growing recognition of the importance of sharing health information with the population to increase the possibility that rational choices will be made (the health-belief model). Greater public knowledge is vital to acceptance of preventive programs such as immunization and compliance with treatment regimens for chronic diseases. Felt needs also affect health planning. A community or donor may, for example, feel that a community needs a new hospital, whereas the same resources might better be spent on developing primary care or health educational services that have a greater impact on the health of the population. Even in an author-itarian society, public opinion may direct decision makers to make irrational choices, such as placing an excessive portion of health expenditures on high levels of hospital bed supply or to adequately address noncommunicable diseases or nutrition issues.

Expressed Need

Expressed need is a felt need that is acted on, such as in visiting a clinic or general practitioner. Felt needs may not be acted on because economic, geographic, social, or psychological barriers may inhibit a person from seeking or receiving care. Accessibility may be limited because the individual cannot afford to pay the fee. A service may be free, but not readily accessible due to such obstacles as distance, language, religious or cultural barriers, difficulty in arranging an appointment, or a long waiting period. As a result, the person seeking care may not be able to receive it and may delay interfacing with the health system until a more urgent, and often more costly, problem arises. Dis-tance, time, and cost of travel, inconvenience, and loss of wages may affect the seeking of service, more so for pre-ventive care than urgent surgical conditions, for example, even if the service is free of charge. Elderly persons may sometimes avoid turning their felt needs into actions as they may not feel comfortable with the fact that they are ill, or may not wish to become a burden. Altering the supply, location, type of service, and its availability can change these factors, thus improving equity of access.

Unexpressed need may be the result of lack of knowl-edge, awareness, access, or taboos from religious, cultural, or even political factors which, for example, may prevent a woman from utilizing birth control even when further pregnancies may jeopardize her life. Lack of knowledge may also interfere with appropriate use of available clinical or preventive health care. This may necessitate outreach ser-vices to access such people as migrant laborers, immigrants, refugees, IV drug users, commercial sex workers, and other groups whose social circumstances place them at risk for disease, but whose access to appropriate preventive and curative services may be very limited.

Comparative Need

Comparative need is a term that relates needs of similar population groups, as in two adjacent regions with the same age/sex/ethnic mix and socioeconomic status. One region may have a certain service, such as fluoridation of the com-munity water supply, while the comparison community does not. The population of the second community is objec-tively in need of that service according to the best current professional and scientific evidence. There are no definable absolutes in the extent of demand for health care, but there are accepted basic standards that are part of world standards at a particular point in time for health promotion, preven-tion, or health care. These are derived from trial and error as much as from science and must be continuously reex-amined in light of new information, as well as the measur-able benefits and costs derived from them.

Demand

Demand is based on individual and community expecta-tions (Figure 11.3). Economists consider this to be a part of the economic demand theory of laissez-faire[2] in which the individual is seen as the best judge of his or her need. The individual may feel that he needs a service, but expert opinion may say that this is not a reasonable demand. A patient may ask a physician for an antibiotic to treat a viral

---

[2]The term *laissez-faire* is used by economists to refer to minimal or absent interference by government in economic affairs; for instance, the free market of goods and services. Laissez-faire is short for "laissez faire, laissez aller, laissez passer," a French expression meaning "let do, let go, let pass."

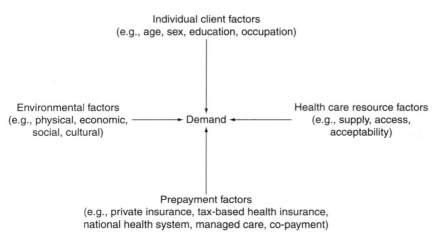

FIGURE 11.3 Factors in demand for health services.

infection which would not help and may even cause harm. A community hospital may wish to increase its bed supply or purchase a highly technical piece of medical equipment due to consumer demand. A patient may feel that denial of a referral by a doctor in a managed care plan is infringement of his rights, but there may be a legitimate and ethical reason for the refusal. Doctors may wish to have the prestige and convenience of certain equipment locally, but economic and planning assessments may say that this is not justified on economic or medical grounds. Such analyses, however, are not immutable, as costs of a procedure may change as technology, clinical experience, and costs of a technology change. The wide-scale adoption of hepatitis B vaccine was greatly influenced by the fall in price of the vaccine, originally costing $100 per immunization and currently less than $5 per dose, so what was once difficult to include in free immunization budgets has now become highly cost-effective even in poor countries. The new human papillomavirus vaccine costs nearly $100 per dose so that its inclusion will be problematic until the price comes down. Antiretroviral treatment drugs were not accessible in developing countries until prices were lowered and international aid programs encouraged their use to prevent HIV transmission from mother to child and in other patterns. Priorities are therefore very much affected by prices and such considerations are unavoidably part of health care planning.

Supply

Demand may also be induced by the supply or provision of care. Making available more hospital beds may increase their use beyond justifiable need, or it may lead to an expectation or demand by patients or their families for an unnecessarily long stay in the hospital. Providing some services at no cost to patients may induce people to utilize those services more than they really objectively require for health reasons according to current best standards. An inappropriate or excessively frequent use of a service may be promoted, and used by the upper middle class, while there may be a lack or scarcity of other important services for the poor due to selection of priorities and inequitable allocation of resources. Sometimes the interest of the health care providers is such that they may act to promote the use of services because payments are received for each service rendered (fee for service). This occurs, for example, in situations where a greater supply of surgeons results in unnecessary cases and rates of surgery being performed.

Grossman's Demand Model

A frequently used economic demand model is that described by Michael Grossman in 1972. This method looks at health within the framework of a production function; that is, health status (output) is a result of health care activities (input) by the environment, the individual, and the health services system. Individual demand for health care is affected by many factors, such as socioeconomic, educational, and cultural barriers or incentives to health care, as well as age, gender, and health status.

In this model, everyone inherits a stock of health when born. Health depreciates over time, however, and investment is required to sustain it. As people age, there is an increase in the rates of illness and death and in utilization of health services. The rate at which a person's stock of health depreciates over time is represented in a health depreciation–time curve. The stock of health can be sustained by investment to maintain health, such as investment in health-promoting activities (e.g., recreation and fitness facilities) and health services.

Change in health is thus a function not only of medical care received, but also of exercise, good housing, nutrition, smoking restraint, and societal factors that are difficult to quantify. Over their life cycle, people will attempt to offset to an increasing extent the rate of depreciation in their health by increasing their expenditures or use of medical care services. The production function depends on environmental and behavioral variables, such as education, that alter the efficiency of the production process. Personal choices affect health, depending on the way the individual allocates resources to its production, for example, the amount of leisure time dedicated to jogging or the decisions whether to eat fatty foods or to smoke cigarettes.

Health is also an investment good. Being unhealthy brings discomfort, a reduced sense of well-being, and a measurable loss of income from reduced work hours or performance. Health as a consumption good means that a health-related activity will improve the quality and enjoyment of life, prevent discomfort or illness, or improve appearance, as in cosmetic surgery. Older people use more hospital and ambulatory medical care and other services than younger people as their health declines, and they are subject to more disease. There are also factors within the health services system or in an insurance system related to the way individuals act. Personal lifestyle may influence the process of services delivery/provision, including access to services and quality of care. For example, an insurance plan may refuse to accept an enrollee on grounds of age, personal habits (e.g., smoking), or preexisting medical conditions, or may not cover preventive services.

In Grossman's model of health demand, rising income may adversely affect health because of an increase of unhealthy or risk-taking behavior. Together with the rise in per capita income, excessive consumption of fatty foods, smoking, alcohol abuse, and motor vehicle accidents increased as well, and led to increasing death rates from cardiovascular diseases and trauma in the industrial countries in the 1940s and 1950s. This also occurred in the 1980s and 1990s in the growing middle class of developing countries. Very poor populations, whose basic health problems are inadequate food intake and infectious diseases, benefit from rising incomes in developing countries if family incomes rise to allow more expenditures for food and better nutrition, with positive effects on health status.

A consumption good has been interpreted by economists as use of resources in a manner which most benefits the individual and society, based on the best currently available evidence. Free market economists, and institutions such as the World Bank, once considered health care as ineffective consumption and recommended that it would be better to invest these resources in economically "productive" development *per se*. At the other end of the political spectrum, Marxist economics looked at health spending as nonproductive consumption, in comparison to investments in heavy industry or infrastructure. In contrast, democratic socialists, such as those in the Nordic countries, have long adopted the view that investment to improve health is justified on both social and economic grounds, with health, like education, being a social right and foundation of society. This concept gained international acceptance with the Health for All program for fundamental social justice and human rights, as expressed in the Alma-Ata Declaration of 1978. The World Bank's 1993 report, *Investing in Health,* and *WHO's Macroeconomics of Health in 2000* all take a utilitarian view that spending money on health and education is a sound investment in economic growth, and not a drain on the economy.

COMPETITION IN HEALTH CARE

Health services were historically established as private, charitable, religious, or governmental public services. In recent years, even in countries in which government finances the services, a market-oriented approach has evolved, giving the consumer choice to utilize competing health services (Box 11.2).

Reforms in the United Kingdom's National Health Service in the 1990s included the primary care provider who holds the funds for the people who select him or her for their primary care. The fund holding by general practitioners was stopped and transferred to primary care trusts (PCTs). These are established in all parts of England and receive budgets directly from the Department of Health, and since 2002, PCTs have taken control of provision of local health care, under supervision of standards and performance monitoring of Regional Health Authorities.

In the United States, managed care health plans receive insurance payments per capita and select hospitals, physicians, or other services on a competitive basis. In some other countries (e.g., Israel, Colombia, Philippines, Jordan)

Box 11.2 Trends in Health Reform

- Downsize the hospital sector
- Develop alternatives to hospitalization
  - Home care
  - Ambulatory medical and surgical care
  - Palliative care
- Develop primary health care including outreach services
- Linkage between insurance and service
- Linkage between services — vertical and lateral integration
- Define basket of services
- Generic drugs
- Clinical guidelines
- Technology assessment

---

clients select health plans or sick funds that are afterwards responsible for all care and costs, seeking competitive benefits from providers such as hospitals or medical practitioners.

Health care organizations, whether hospitals or primary care services, must provide quality service to the community in order to meet their mission and to survive as institutions (Box 11.3). The patient or consumer is, in principle, at the center of the process. However, market analysis is essential regarding health care program costs, contents, level of satisfaction of individuals and the community, as well as health care providers' (individual or institutional) satisfaction and performance indicators. Since 2000, many countries have tried to introduce free market ideas of competition into their health systems to move away from state-run health systems. Introduction of free market mechanisms in a national health program compromises universal entitlement to health care, creating different levels of services for those who can afford to pay and those who cannot. A two-level system introduces greater degrees of inequity between urban middle class and the urban and rural poor.

To fulfill the potential of the U.S. health system, it is important to continue to seek savings and better value for the high investment in health. This requires creating more efficient and effective health promotion, health care, and insurance systems with universality and equity as major goals. Strategies with the potential to achieve savings, slow growth in spending, and improve health system performance include:

1. Increased public information and greater competition between health care organizations;
2. State mandated universal health coverage, access, affordability, and equity;
3. Reduction of insurance administrative overhead;
4. Financial incentives to promote efficient and effective care;

Box 11.3 Obligations of Health Service Providers

1. **Availability** provides continuous service coverage, 24 hours every day and on holidays;
2. **Accessibility** the client can readily reach the service within reasonable travel and waiting time;
3. **Accountability** the system and provider explain programs, decisions, and actions to the client;
4. **Affordability** services are provided at reasonable cost to consumer and insurer;
5. **Acceptability** user-friendly in style and manner of staff to the patient and family;
6. **Accredited** undergoes external evaluation and implements recommendations of accrediting agencies;
7. **Equality** provides fair and equivalent access to services needed regardless of age, sex, ethnic origin, religions, social, or political identification, place of residence, prior medical condition, ability to pay;
8. **Efficiency** produces effective results with a minimum of waste, expense, or unnecessary effort in use of human, financial, and other resources;
9. **Ethical standards** meets current professional, societal, ethical, and legal standards;
10. **Consumer's rights** ensures the patient and consumer are informed of their rights and alternatives of care, before medical or administrative decisions are made;
11. **Financial soundness** is able to meet financial obligations;
12. **Goals and objectives** defined, written, reviewed, and used as a basis for planning and monitoring;
13. **Innovativeness** is open to new methods in clinical and preventive approaches to health;
14. **Quality promotion:** to set high standards of facilities and services in accordance with current professional criteria and standards of leading providers;
15. **Community acceptability** meets community expectation with participation in promoting health standards;
16. **Comprehensiveness** linked to a broad range of services provided or arranged to meet client needs.

---

5. Expanded efforts in health promotion and training in public health;
6. Patient- and risk group-centered primary care;
7. Health information technology; and
8. Categorical grants to improve equity in health.

Managed care systems depend on reducing unnecessary hospital care and increasing use of ambulatory and home care services. But this has become a part of the entire system of health care in the United States because of reduced hospital bed supply and payment by DRGs. Services previously provided on an inpatient basis are being provided equally or even more successfully on an outpatient basis. The manager or chief executive officer of a health facility needs to guide its transition from a passive receiver of the sick to an institution with emphasized ability to meet health needs in the community, while at the same time controlling costs in order to remain competitive and financially viable.

ELASTICITIES OF DEMAND

Classically, *elasticity of demand* means that demand is related to price of a product. Demand may increase more rapidly than the fall in price (elastic), or may not change even if the price falls (inelastic), or may fall in direct relation to price change (unitary). But demand in health care is influenced by many other factors determined by the consumer, the provider, the supply, the location of services, and the knowledge and motivation of the consumer. Cost to the consumer is a factor in choosing to purchase goods or seek services. If the price goes up, then demand will decline, or vice versa. In other words, demand is not an absolute but can be affected by supply, price, and type of payment required for the service.

Classical market mechanisms fail in health care because there are so many intervening factors, including the high degree of information asymmetry between provider and the consumer, the role of third-party payers such as national or private health insurance, and regulatory roles of government. Universal access through national systems of coverage are also not enough guarantee of equity, as access is often unequal and because the consumer may not be sufficiently informed of the potential benefit of care or prevention before making his choice. Cost containment can be addressed using many "market" factors in health care, such as the supply of facilities that affect demand; reducing costs of a specific vaccine, drug, or surgical procedure; and method of payment for doctors and hospital services. All of these affect the economics of health. However, where consumers have no choice of service provider, there is risk of exploitation by providers and consumers alike.

In classic market economic theory, the individual is seen as the best judge of his or her own needs and decides what to buy (consumer sovereignty). It assumes that consumers purchase services or a health plan based on factors such as cost and quality, as one would do when purchasing a refrigerator. Individual decisions are made on the basis of personal perception, information, priorities, and resources. Proponents of the market approach in health care suggest that it provides the consumer with more control and choice, and indirectly raises the competition for consumer demand and the quality of services while lowering costs.

Opponents to this view claim that this market approach fails in medical care because the issue is more complex than purchaser and provider. The market mechanism in health care is as much determined by supply, access, and method of payment as by consumer choice. Supply of services creates demand, as does prepayment. Consumers rely on their doctors to advise them; this is known as "the agency relationship." Physicians make or recommend decisions for their patients on the basis of both patient needs and the supply of services. Payment for services by a third party, such as an insurance plan or the government, where the consumer may provide little or no direct payment, can lead to the provision of unneeded services, particularly if the doctor has incentives of the fee-for-service payment system.

Someone may want to purchase additional or different services according to their wishes for expected benefits (marginal utility), such as reduced waiting time for an elective surgical procedure. The marginal cost is the cost of an extra unit of the commodity used. In classic economic theory, the consumer decides to purchase a service when the marginal utility is equal to or greater than the marginal cost; that is, the added benefit is worth the additional cost. This free market approach in health care may result in under provision of vital preventive health services, especially to the population in greatest need. If medical care services operated in a free market, then consumers would not necessarily take into account benefits to persons other than themselves (externalities). For example, if vaccines are available purely on a free market basis, many persons would not purchase the vaccinations, due to lack of resources or lack of awareness of the importance of the preventive measure. This would increase the risk to the population at large by reducing herd immunity. Externalities should be taken into account in public policy decisions, valuing the benefits to society as a whole.

Where there are many providers, they will compete with each other in principle by offering services at lower prices in order to attract clients. However, this rarely occurs in medical services where fees or salaries are set by collective bargaining. In terms of hospital and insurance services, the existence of monopolies (i.e., only one provider) or oligopolies (i.e., too few providers) prevents price competition and often results in collusion to fix prices. A monopsony is a situation when there is only one purchaser, so that the provider may be subjected to pressures to lower prices or meet additional demands of the buyer, such as a managed care organization purchasing hospital services.

In a highly privatized system of health care, such as that in the United States up to 2008, demand for care is rationed through fees, co-payment by the consumer, or by limitations set by the indemnity insurance plan and lack of insurance benefits for many. Indemnity insurance plans require co-payments and deductibles, so that the insured person has to take additional insurance to cover expenditures or pay part of the charges. Medicare beneficiaries in the United States have two alternatives: to pay part of their health care expenditures and have a greater variety of service providers to choose from, or to join a managed care option which covers their service costs, but with a restricted list of physicians. Therefore, the beneficiary must choose between extra payments versus limited choice of service providers.

In a public system of health care, demand is rationed through limitations on supply of services by a requirement that the patient wishing to see a specialist be referred by a general practitioner who plays the role of a "gatekeeper," so that the consumer's choice is limited to the options decided by the general practitioner. This has been adopted by many health systems, including managed care plans in the United States. This is a controversial limitation on the consumer, who may wish to consult other specialists, but it may be essential to avoid frivolous "shopping" for care that drives up the cost of health care.

The market approach is based on choices being made by consumers on the basis of anticipated benefits in terms of improved health or health care, or reassurance. The employer must offer the employee several options for health insurance. In a fee-for-service indemnity plan, allowing the consumer greater choice of physicians, the employee pays additional monthly premiums, as compared to a managed care option. This carries with it a measure of inequality because of differences within the population, both in health needs and in the ability to purchase or utilize needed services.

Insurance for health care means sharing the risk and paying from your place of employment to protect yourself against the costs of illness which are unpredictable. The insuring body, whether public or private, can predict needs based on population data. When governments provide an insurance system, they seek to share the risks among the total population, and not only the highest risk groups such as the elderly and the poor, as is done in the United States, where the government has taken responsibility for these groups, leaving the general public to arrange insurance privately or through collective bargaining at their workplace, who are generally a healthier age group and lower risk population group. Where the government insures the highest risk groups it is a form of adverse selection of the highest risk population groups. Yet other risk groups are left without coverage or basic information on health insurance. Sharing risk is the basis of insurance, but those left outside of insurance protection or with inadequate coverage are at risk for major economic loss due to illness. Those with insurance may also seek extra coverage or co-insurance to protect themselves against costs of uninsured health needs, such as drugs or dental care.

Often those with the greatest need are those least able to get the desired or required services. Persons who have no source of income cannot be consumers or make decisions to purchase health services on the free market. Rather, they are dependent on free charitable services. This problem may be addressed by a number of economic alternatives, such as providing low-income individuals with free health insurance under Medicaid or with vouchers to purchase services.

Decisions or actions of patients are influenced not only by cost and access, but also by knowledge and attitudes toward care. A person unaware of modern birth control, or living in a society that discourages or limits its use on religious or political grounds, is not able to make an informed decision about its use. Market mechanisms in health work in different ways in different health systems. Even where care is a free service and it is seen as a right for everyone, there are financial and human resource limitations in the supply of services. Market mechanisms have a major role in the reform of many health systems by giving consumers the choice of provider, even if this means choice of health care plan such as with managed care systems. Financial incentives are used to promote quality care, such as immunization or preventive procedures such as Pap smears, or disincentives to reduce unnecessary or wasteful services, as in the limitation of hospital bed supply.

MEASURING COSTS

Costs in health can be analyzed in various ways. Direct costs of services are those paid by the patient or by the insurer or sick fund on behalf of the patient, including costs of the hospital or other provider. Indirect costs of an illness to the patient, his family, and society, include loss of income due to time off work or lowered productivity as expressed in work or school absence, and poor quality of product or learning capacity.

Opportunity cost refers to the resources used that could have been applied to other uses. Hospital land and building costs, for example, could be allocated for other purposes, such as primary health care facilities or facilities outside the health sector such as after-school programs for children. Increasing the proportion of the GNP spent on health care may limit society's ability to spend money on education and other important social programs.

Social costs include indirect expenditures for health and illness such as the total value of lost production and costs of social support for a person whose health and work capacity have been impaired by illness. Private costs include out-of-pocket expenditures that an individual makes to purchase health care plus related expenses such as payments for health insurance, loss of wages, purchase of pharmaceuticals, and co-payments for health services.

ECONOMIC MEASURES OF HEALTH STATUS

Economic analysis assesses not only input, as in costs and resources, but also output, as in extension of years of life and reduction of disability, morbidity, and mortality.

Greater functional levels that improve the quality and quantity of life are output measures of health care. This should be part of an economic evaluation of the use of national or personal resources. Disability-adjusted life years (DALYs) and quality-adjusted life years (QALYs) are measures of the total burden of disease as a guide to population health status (both of death and disability).

DALYs are calculated as the present value in years of disability-free life that might be lost as a result of premature death and disability occurring due to a disease in a particular year. QALYs measure life expectancy adjusted by changes in quality of life measured by assessing two or more aspects of health, such as pain, disability, mood, or capacity to perform self-care or socially useful activities such as paid employment or housework. DALYs and QALYs are constructed by using expert evaluation to estimate the degree of impairment (normal, impaired, or incapacitated) from specific diseases. These include impairments such as loss of ability to communicate; sleep disturbance; pain; depression; and sexual, eating, and mobility dysfunctions.

The value of the health status of an individual can be expressed in numerical values for comparisons. The values of the total scores are then added together and the overall score calculated out of a maximum value for comparison. This allows a measure of health status and may be used for the purpose of comparing the effectiveness of alternative interventions. This is subjective depending on the perception of the assessor, and interobserver variability may be high. While such measures do not include all factors which may contribute to development of a disease, they contribute to using the economic impact of disease as part of health planning by pooling mortality and economic indicators.

DALYs and QALYs are useful economic measures to provide a common basis for comparison of different cases, settings, and time changes by using mortality, disability, and quality of life as measures. They are used as proxy health status indicators to analyze different approaches to health policy, to justify specific interventions, and determine priorities. The World Bank, WHO, and other organizations are examining alternative indicators to link health and its underlying determinants of the total burden of disease and disability, and to refine the process of establishing priorities for research and decision making for interventions.

Gains in life expectancy from preventive or curative medical interventions may be measured from published data sources. The gain in life expectancy calculated for a patient who survived cardiac arrest by placement of an implantable pacemaker is calculated as 36–46 months or who had bone marrow transplantation for recurrence of non-Hodgkin's lymphoma is calculated at 72 months. For preventive measures, the average gains for the total population appear smaller. Cervical cancer screening, for example, increases the life expectancy of all women by 3 months, but for a woman whose cancer of the cervix was detected early, the gain is an average of 25 years. This methodology relies on published studies, and can be of great value in comparison and analysis of alternative strategies and health care priorities.

COST-EFFECTIVENESS ANALYSIS

Cost-effectiveness analysis (CEA) in health care is the net gain in health or in reducing the burden of disease from a specific intervention in relation to its cost. It is used to determine the least expensive way of achieving the goal, by comparing alternative methods of intervention in order to make a choice. The most cost-effective method is the one that achieves the same goal using the fewest resources. A low cost per DALY gained indicates a high degree of cost-effectiveness, and therefore an intervention that should be of high priority, given limited resources.

Alternative methods of treatment may also be compared, such as use of medication compared to surgery, day surgery as compared to inpatient care, or treatment in the community as compared to hospital inpatient care. As seen in Table 11.2, expanded programs of immunization are the most cost-effective services for developing countries where vaccine-preventable diseases are the major causes of loss of DALYs for the population. Other cost-effective programs include preventing iodine and vitamin A deficiency and treating intestinal worms, even though these are relatively lower causes of lost DALYs.

The Disease Control Priorities Project (2006) examined 25 priority health conditions in low- and middle-income countries, assessing their public health significance and the cost-effectiveness of various clinical and public health interventions. Such analyses help to construct a basket of essential services on the basis of comparative cost-effectiveness. Highly cost-effective interventions include vitamin A supplementation, measles control, and directly observed chemotherapy for TB. A high-cost but highly effective intervention is chemotherapy for leukemia in children under age 15. This intervention is justified as benefits are high. The same chemotherapy in a 75-year-old would present a low DALY value. CEA studies examine issues such as day surgery versus inpatient surgery, operations versus medications (e.g., for peptic ulcers and coronary

TABLE 11.2 Main Causes of Burden of Disease in Children in Developing Countries and Cost-Effective Interventions, 1990

Disease	No. DALYs lost (millions)	% Total DALYs lost	Intervention	Cost effectiveness ($ per DALY)
1. Vaccine-preventable childhood diseases	65	10	EPI	12–30
2. Vitamin A deficiency	12	2	EPI Plus	12–30
3. Iodine deficiency	9	1	Iodine supplementation	19–37
4. Intestinal helminths	17	3	School health	20–34
5. Diarrheal diseases	92	14	Integrated management of sick child (MISC)	30–100
6. Protein-energy malnutrition	12	2	MISC	20–150
7. Perinatal morbidity and mortality	96	15	Family planning Prenatal and delivery care	30–100
8. Respiratory infections	98	15	MISC	30–100
9. Malaria	31	5	MISC	30–100
10. Congenital malformations	35	5	Surgery	High
11. Others	193	28	—	—
Total	660	100	—	—

Notes: MISC = medically integrated services for children.
Ranking is by intervention cost-effectiveness.
EPI = expanded program of immunization.
EPI Plus = EPI with the addition of vitamin and iodine supplements (see Chapter 4).
Source: Bobadilla, J.-L., Cowley, P., Musgrove, P., Saxenian, H. 1994. Design, content and financing of an essential national package of health services. *Bulletin World Health Organization*, 72:653–662.

heart disease), public versus individual dental prevention (e.g., fluoride versus dental hygienist care), and community versus institutional care.

A comparison of life years gained for patients with end-stage renal disease in the United States found that renal transplantation was less expensive ($3600 per year of life gained) compared to home dialysis ($4200 per life year gained) and hospital dialysis ($116000 per life year gained). Moreover, transplantation provides a higher quality of life. This was perhaps the first example of "cost–utility" analysis, where life years gained were weighted according to quality of life. This can be expressed as the cost-effectiveness per QALY. Policy decisions based on such findings are constrained by difficulty in obtaining sufficient donor kidneys and lack of personnel and facilities suitable to carry out transplantation effectively.

Surgical removal of the gallbladder and even cancers of the colon are now done with endoscopy instead of the traditional abdominal cholecystectomy. Endoscopy is less traumatic with faster recovery, so the patient is discharged from the hospital on the next day and returns to work within a day or two, while the patient with abdominal cholecystectomy requires a much longer hospital stay and recuperation at home before returning to work. The newer procedure is easier on the patient and safer. Cost–benefit analysis must take into account not only the medical and hospital costs, but also the social costs of lost work time for the patient and caregivers. Computed tomography (CT) scanning, once considered costly and for special use, has become a valuable part of investigation of many conditions and is used frequently, often in place of costly, dangerous, and less effective procedures previously used to investigate many conditions.

Care of the infirm elderly, up to a certain level of disability, in a private home with outside help, including meal preparation and delivery, nursing, physiotherapy, and social worker visits, is less costly than care of the same patient in an institution. Home care promotes earlier hospital discharge and recuperation at home. These assessments must take into account social costs and transfer of costs of services provided in an institution, such as food, laundry, heating, and electricity to the patient's family. Home settings promote improved recovery, avoidance of hospital infections, and a general feeling of well-being of the patient. For persons with severe illness or many disabilities, requiring a higher degree of nursing and/or medical care, institutions are more cost-effective than home care. Respite care sometimes provides support for a family in caring for a patient with multiple disabilities, delaying more costly institutional care.

Research comparing treatment of psychiatric patients in a large mental hospital, in the psychiatric ward of a general hospital, and in a day treatment center shows day treatment center care to be least costly, but some measure of the severity of illness and need of care has to be added

---

Box 11.4 Management for Cost-Effectiveness

- Cost-containment AND moderation (of increase) in health expenditures
- Priorities shift toward health promotion and disease prevention
- Cost-effective health initiatives
- Decentralized management
- National policy, monitoring, and standards
- Information systems/monitoring
- District health profiles
- Increase in primary care
- Outreach services
- Increase home care
- Increase in long-term care facilities
- Increase non-admission surgery, long-term care
- Health information systems
- Managed care
- DRGs (diagnosis-related groups) for payment for hospital care

---

to this assessment. Planning mental health services and facilities with reduced hospitalization requires adequate resources for mental health care in the community in order to prevent chronic mental patients from becoming part of the homeless population as has happened in many large cities.

Sometimes the least costly method is the least effective. For example, a study showed that prevention of pregnancy by the withdrawal method is least costly but is far less effective than use of the birth control pill. Abortion as a method of birth control may be less costly than use of the pill, but, in addition to the ethical issues, it produces complications and contributes to excess morbidity and mortality in subsequent pregnancies, both for the mother and the newborn.

Cost-effectiveness analysis takes into account both the cost and effectiveness of interventions, as a measure of value for cost, but does not answer the question of whether or when the intervention should be done. Management of health systems is part of the New Public Health. It reflects the priorities and capacity development of a broad spectrum from health promotion through prevention to institutional care. Box 11.4 summarizes management issues related to operational aspects of cost containment and substitution of high and inappropriate costs by lower-cost services which address actual patient and community needs.

COST-BENEFIT ANALYSIS

Cost-benefit analysis (CBA) compares the expense of a specific program to its expected monetary yield or savings. Costs include direct expenditures as well as the indirect costs of loss of productivity and loss of contribution to society. Direct benefits include prevention of premature loss of life through reductions in morbidity and mortality and the associated savings in medical care costs, such as hospitalization, doctors' services and drugs used, and the attachment of economic value to this. Indirect benefits include savings to the patient's family in terms of expenditures to visit the patient (transportation costs) or time away from work to look after a sick child or relative. Other indirect benefits also accrue to society, in terms of savings in reduction of lost work time by the patient or his or her family during an illness.

The assessment of costs and benefits involves three stages: enumeration, measurement, and explicit valuation. Assessing a particular treatment, or enumeration, requires measurement of change in health status, in the cost of use of resources, as well as in the patients' productive output. Economic appraisal depends on determination of the many factors needed in managing a public health problem and its expected outcomes. Explicit valuation, or estimation of the cost of a variable, is based on determination of the economic value of these factors. Many factors need to be taken into account and simplified; consequently, the estimations of the costs are approximations rather than exact figures.

A CBA study of a phenylketonuria (a congenital metabolic disorder) program in the United States showed the total cost, including screening 660000 newborns, confirmation tests, the special diet for those affected, and administration of the program, to be $1.39 million. The benefits gained were $1.26 million for medical and other services, and $1.05 million for prevented loss of productivity, for a total of $2.31 million. The benefit/cost ratio was 2.31/1.39 = 1.66. For each dollar invested, the gain to society was $1.66. According to CBA studies on the benefit-to-cost ratios of vaccines, the addition of a second dose of measles vaccine was found to have a higher benefit-to cost-ratio (CBA = 4.5/1) in both developed and developing countries as compared to one dose of hepatitis B or *Haemophilus influenzae* type b (Hib) vaccine on an immunization schedule. Hepatitis B vaccinations were found to have high benefit-to-cost ratios, even in countries with intermediate levels of endemicity (CBA = 4.5/1). For Hib vaccine, the social benefits were found to exceed the costs to society; these include indirect social benefits such as reduced need for special education for brain-damaged children. The benefit-to-cost ratio, if viewed solely from the point of view of the health sector, was lower. Such a result might prevent the health system from adopting a beneficial program if the benefit-to-cost study is too narrowly applied.

The decision to adopt a specific program may include a CBA but is often made on other grounds, including public and professional opinion as well as political factors. A CBA can give a prioritized ranking to alternative interventions, and thereby help in the decision-making process. Ranking according to the relative costs and benefits can help a health ministry to choose among putting resources into a high-technology hospital, home care, expansion of an immunization program, or investment in primary care services.

Both CBA and CEA include initial as well as ongoing costs, but they must take into account that the future value of money will be less than the present value, referred to as *discounting*. The costs as well as benefits to be derived from the project must be calculated as they accrue, so that a portion of the effect is observed the next year, and a portion the following year. The cumulative discounted value is called the *net present value* (NPV). Cost-utility analysis (Box 11.5) is commonly used.

In recent years, the gold standard for assessing the feasibility and the ranking of potential interventions has been the cost per QALY. The costs and QALYs lost under a null scenario, where no intervention is provided, are used as a baseline.

Box 11.5 Cost–Utility Analysis

This is a method of analysis in which the outcome of a program or intervention is measured by outcomes such as the cost per Quality Adjusted Life Years (QALYs). Cost–utility analysis (CUA) integrates economic information and epidemiologic information as to the proposed intervention.

The primary outcome measure is the incremental cost-effectiveness ratio (ICER) defined as net cost per QALY, where:

$$ICER = \text{net costs (intervention cost} - \text{treatment costs)}$$

and

$$QALY = \text{Decrease in morbidity and mortality as measured by quality} - \text{adjusted life years gained) due to the project.}$$

CUA, CEA, and CBA are used for different purposes. CUA has been criticized by some economists because this method is based on the incremental cost calculation, and it does not take into account the overall health system and population. It is, however, widely used under the pressing and frequent need for CEA of health projects. Each health project could be evaluated on the basis of its incremental cost per extra QALY provided to patients, as in the case of new drugs for treatment of advanced cancers such as of breast or colon.

Second, a calculation is made as to the costs and QALY losses associated with the current scenario of provision (e.g., Pap smears once every 5 years). This will enable the average cost-effectiveness ratio (ACER) in terms of costs per QALY to be calculated (from the differences in costs and QALYs between the current and null scenarios). Finally, a calculation will be made as to the costs and QALY losses associated with the various proposed scenarios (e.g., Pap smears annually and human papillomavirus vaccination). This allows for calculation of the incremental cost-effectiveness ratio (ICER) in terms of costs per QALY comparing the costs and QALYs between the proposed and the current scenarios.

Projects whose cost per QALY exceed three times the GNP per capita are deemed not to be cost-effective. Those whose cost is between 1 and 3 times the GNP per capita and less than the GNP per capita are cost-effective and very cost-effective, respectively. Projects whose net cost is negative, as a result of savings in treatment costs exceeding the project costs, are deemed to be cost saving.

The choice of a cost-effectiveness threshold is a value judgment that depends on several factors:

1. Who the decision maker is and what the purpose is of the project;
2. How a decision maker values health outcomes and money, how he or she is willing to substitute one for the other, and what his or her attitude is about risk;
3. The resources available.

On a societal basis, the use of antiretroviral therapy for HIV infection illustrates vividly the dependence of the cost-effectiveness threshold on resources: although antiretroviral therapy may be considered cost-effective in the United States, a cost-effectiveness threshold of $50000 per life year gained is completely implausible in the developing world, where per capita health spending may be less than $10 annually. Resources and the cost-effectiveness threshold tend to rise and fall together, all other factors being equal.

BASIC ASSESSMENT SCHEME FOR INTERVENTION COSTS AND CONSEQUENCES

Assessment of effectiveness and costs of an intervention has become a basic part of policymaking in health. An approach called Basic Assessment Scheme for Intervention Costs and Consequences (BASICC) has been widely promoted by the U.S. Centers for Disease Control. BASICC is a more complex approach that looks at the efficacy of the intervention, the cost, including direct outlays, productivity costs (i.e., loss of time for work or recreation), and intangible costs (i.e., pain and suffering).

Costs include fixed costs, or those that do not vary according to the quantity of the service provided but also include a portion of rent, utilities, and equipment allocated to the program. The average costs are the total cost of a program divided by the total units of output produced. Variable costs are those that vary according to the level of service provided, such as the number of nursing visits required for a home care patient. Marginal costs are those additional costs to basic program costs such as expansion of staff or facilities to accommodate extra activities.

BASICC focuses on intervention costs and direct cost savings in terms of medical care. Net costs can be summarized as the cost of the intervention and its side effects for n persons minus the direct costs of the expected number of cases averted for the same n persons, calculated as follows:

Net cost = (cost of program + cost of side effects)
— cost of adverse health outcomes averted

The steps of BASICC include the following:

1. Describe the program, its objectives, target population, effectiveness of intervention, external constraints, resources required, management of the program, implementation strategy, and scientific evidence of effectiveness;
2. Define the burden of the disease, its incidence, and prevalence without the program;
3. Define outcomes anticipated in terms of improved quality of life, reduced incidence or severity of the disease, and prevention of premature death;
4. Measure efficacy of the intervention, taking into account that interventions are rarely 100 percent successful in practice because of compliance and effectiveness of the intervention;
5. Measure intervention costs per unit;
6. Measure direct medical costs of outcome averted by the intervention;
7. Assess resources required for the intervention, which include fixed, variable, and total, as well as unit costs.

THE VALUE OF HUMAN LIFE

The main anticipated benefit of a health intervention is the saving of human life. Placing an economic value on life is useful in calculating the benefits of specific interventions or perhaps for compensation to the family of a person who loses his or her life as a result of, for example, negligence by a doctor or plant manager.

The value of a human life in economic terms was first calculated by William Petty in 1699 while developing his idea of political arithmetic. In 1876, William Farr used life tables to calculate economic equivalents. More recently, economists have introduced other methods of quantifying the value of human life, such as calculating the value of human capital, willingness to pay for services, years of life saved, and more precise evaluations involving QALYs and DALYs saved.

Ethical and political conflicts surround the issue of calculating the economic value of human life. A materialistic approach would evaluate human life based solely on the value of contributions that the individual might make to society. A humanistic approach would place virtually unlimited value on a human life according to the ethical principle that saving one life is as important as saving all human beings (Sanctity of Human Life, see Chapter 1).

By placing infinite value on human life, society provides doctors with precious resources to save one human, without considering that this may be at the expense of other lives. For example, the cost of a heart transplant which may add quality and years to one person's life may be equivalent to the cost of a program on prevention of heart disease that might save many more lives. International agencies spend hundreds of millions of dollars to eradicate polio, a much feared, crippling, but usually non-lethal disease, while measles, often thought to be a common benign disease, kills over hundreds of thousands of children annually. The valuing of human life is not meant to fuel ethical argument, but rather to provide a measurement tool for the planning of priorities and litigation needs of health planners.

In health economics some arbitrary measures are used in order to demonstrate alternative ways of using limited resources. The implicit social value (ISV) of life (Box 11.6) rates a program by the lives it saves, and assumes that, in a democratic society, all lives have the same intrinsic value. Governments made some decisions showing inconsistency in the appraisement of ISV. A United Kingdom government decision not to introduce childproof drug containers implied an estimated value of a life saved at less than $5000, while the same government decided to change a building code that implied a valuation of $50 million per life saved. Estimated costs per person year of life saved may vary among specific public health interventions: annual mammography for women aged 40–49 is estimated to cost $62,000 per life year saved, as compared to $2700 for a program of mammography

every 3 years for women aged 60–65. A smoking cessation advice program for men aged 50–54 yields a cost-effectiveness of $990 for a year of life saved (Brownson, Remington, and Davis, 1998). The question of routine mammograms in reducing cancer mortality has been questioned and the issue is again being reviewed, but until better screening methods are developed it continues to be recommended by leading professional organizations. Smoking cessation has a demonstrated strongly positive benefit-to-cost ratio. Economic analyses such as the ISV are very much part of decision making in health systems management.

Early economists valued life in terms of loss of net output to society, or the future loss of earnings minus the future loss in consumption resulting from the premature death of an individual. This human capital method is still widely used because of the simplicity of its calculations. However, it does not take into account the grief of the family. It places a negative value on the life of a pensioner who is no longer a worker and "producer" in society, and gives no value to work done in the household, such as cooking, home maintenance, and rearing children. Nor does it give value to the intangible social and psychological benefits of the multigenerational family for all its members.

Another approach to valuation of life is based on court awards for compensation. It is a highly subjective method, often based on the court's interpretation and judgment of degree of contributory negligence, such as whether the injured person in a car accident was wearing a seat belt at the time, or may be based on assumed earning power of the individual in the years of work lost.

A major method is the willingness-to-pay approach, where valuations of life are based on what individuals are willing to pay for reductions in their probability of dying. For example, how much would persons pay for new car tires, or how much extra would they pay in order to travel on an airline with a better safety record? How much will a patient be ready to pay above his or her insurance coverage to have a world-famous surgeon operate on him or her as opposed to accepting the surgeon available within the health service? Such measurement is difficult to perform and is often based on asking questions about hypothetical situations. Answers are also influenced by the income level of the respondent, by their attitude toward risk, and by the probability of death.

The issue is not only theoretical. If the cost to prevent HIV transmission to newborns is $3000, and the number of cases of HIV-positive pregnant women who may transmit the virus in a developing country is such that a very large part of the national budget for health may go to this purpose, while there are insufficient funds for basic immunization, then choices need to be made, and they may be painful ones. All societies must make choices in priorities and in allocation of resources. Choosing to build large superhighways and neglecting public mass transit is a

---

Box 11.6 Implicit Social Value of Life — An Example

The implicit social value (ISV) of life is summarized in the following equation:

$$ISV = \frac{\text{sum of costs} - \text{sum of benefits}}{\text{sum of life years saved}}$$

As an example, in the United Kingdom, home dialysis was provided in the 1960s despite the fact that the social costs were $5000 more than the calculated social benefits. In other words, society was willing to pay $5000 to keep one member of society alive for 1 year. From this decision, we can infer an ISV of at least $5000.

decision which assumes certain social values, but will cost lives and health because of downstream effects such as increased pollution, motor vehicle injuries, and deaths.

HEALTH FINANCING — THE MACROECONOMIC LEVEL

Financing health care has evolved from personal payment at the time of service delivery to financing through health insurance (prepayment) by employer/employee at the workplace. This has evolved in most industrialized countries toward governmental financing through social security or general taxation, supplemented by private and nongovernmental organizations (NGOs) (Table 11.3), and personal out-of-pocket expenditures. Ultimately, every country faces the need for governmental funding of health care either for the total population or at least for vulnerable groups such as the elderly and the poor, as in the United States, where governmental funding comes to nearly 50 percent of total health expenditures. Government funding is necessary also for services that insurance plans avoid or are inefficient in reaching, such as community-oriented services and groups at special risk such as infants and women (see Chapter 13).

Health financing involves not only methods of raising money for health care, but also allocation of those funds. National health expenditures are derived from government and nongovernment sources and are used to finance a wide array of programs and services. There is competition for funds in any system, and the way in which money is allocated affects not only the way the services are provided but also setting of priorities, as indicated in the "laws" of health economics in Box 11.7.

The economic consequences of decisions made in resource allocation are major determinants of health care economics. Each country has to cope with similar issues in reforms to correct for changing health needs and the economic results of former decisions (see Chapter 13). A comparison of total national health expenditures is seen in Table 11.4. The United States has consistently been

---

TABLE 11.3 Sources of Financing Health Services

Public sources	Private sources	International cooperation
Federal, state and local government general revenues, mainly from taxes: income, excise, resources, business, inheritance, value added, capital gains, property, special taxes	Private health insurance	United Nations affiliates
	Personal expenditures	Foundations
	Private donations, wills	Religious organizations
Social Security payroll tax	Private foundations	Other NGOs
Compulsory health insurance	Voluntary community service	World Bank
Lotteries		Government bilateral aid
Dedicated taxes: cigarettes, alcohol, gambling	User fees	

Box 11.7 "Laws" of Health Resource Allocation

1. **Sutton's Law** Willy Sutton was a bank robber and when asked by a reporter why he robbed banks, he replied: "Well, that's where the money is." This expression is used to indicate that health services emphasize those aspects which are better financed. If more funds are available for treatment services, and preventive care is relatively underfunded, then treatment will have greater emphasis than prevention.

2. **Capone's Law** Al Capone, a well-known gangster, planning the division of Chicago among his colleagues, said: "You take the north side and I'll take the south side," i.e., let's divide things up according to our mutual interest. This expression in the health context is taken to mean that planning may reflect interests of providers, as opposed to that of the general public. An alternative use of the concept is that macroeconomics planning may serve a general interest at the expense of the individual patient.

3. **Roemer's Law** "Hospital beds, once built and insured, will be filled." The supply of hospital beds is a key determinant of utilization, especially where the public has health insurance benefits covering hospitalization. This "law" was modified by the experience of changing payment systems with incentives to reduce utilization. Following the introduction of the diagnosis-related group (DRG) method of payment in the United States in the 1980s, there has been a reduction in hospital bed supply and occupancy. Incentives to control both hospital bed supply and utilization are crucial elements of health planning in most industrialized countries.

4. **Bunker's Law** "More surgeons; more surgery." A greater supply of surgeons generates more surgery. This has also been modified as managed care and gatekeeper functions limit referrals and self-referral to specialists, and as professional organizations and governments limit training positions and licensing for such specialists.

TABLE I1.4 Per Capita Health Expenditures as Percentage of Gross Domestic Product for Selected Industrial Countries and Years, 1960–2004/2005

Country	1960	1970	1980	1990	1995	2000	2004/2005
United States	5.0	6.9	8.7	11.9	13.3	13.1	15.3
France	3.8	5.4	7.1	8.6	9.5	9.3	11.1
Germany	—	6.2	8.7	8.5	10.6	10.6	10.7
Canada	5.4	7.0	7.1	9.0	9.2	8.9	9.8
Sweden	—	6.9	9.1	8.4	8.1	8.4	9.1
Denmark	—	—	9.1	8.5	8.2	8.4	9.1
United Kingdom	3.9	4.5	5.6	6.0	7.0	7.3	8.3
Japan	3.0	4.5	6.5	5.9	6.8	7.6	8.0

Source: *Health, United States*, 2006, and Organization of Economic Cooperation and Economic Development, Health Data 2007, OECD Health Data 2007, http://www.oecd.org/ (health at statistics portal) [accessed May 17, 2008]

the highest spender on health care, but has succeeded in reducing the rate of cost increase in the 1990s. Canada also experienced high rates of cost increase in health during the 1970s and 1980s, but managed to reduce the rate of increase and moved from the second leading country in per capita health expenditures to fourth place after the United States, Germany, and France.

Health care expenditure involves money spent from all sources for the entire health sector, regardless of who operates or provides the services. The methods of financing health care include tax-supported, social security-supported, employer-employee financed, charitable organizations, or consumer payment at the time of service. The total of expenditures for health care and how those funds are spent are the most fundamental issues in health economics and planning. Allocation of resources requires a skillful planning process to balance spending on different sub-sectors of the system and to assure equity between regions and various socioeconomic groups in society.

What is the "right" amount of health care financing? This is a political decision which reflects the social and economic value placed on health by a nation. These attitudes affect such issues as how well medical and other health care staff are paid in comparison to other professions, and the supply of physical and human resources for health care in a given society. Virtually all developed countries have recognized the importance of national health and the role of financing systems to make health care universally available. Some basic principles and recommendations for successful health care financing policies are outlined in Box 11.8.

These rules are not absolute and solutions vary from country to country, as discussed in Chapter 13. But it is important to stress that the system of financing greatly affects the services provided. The United Kingdom

Box I1.8 Health Care Financing

- Universal coverage through social security or tax-based system
- Financing within national means for social benefits
- Adequate overall financing (>6 percent GNP)
- Shift from supply-side planning to costs per capita
- Performance or output measures
- Categorical grants to promote national objectives and specific health target programs
- Increase financing at national, state, and local government levels (7–10 percent GNP)
- Health insurance as a supplement
- Define "basket of services" and consumer rights
- Reduce acute care beds to <3.0/1000 population
- District health authorities with capitation funding
- Incentives for improved performance measures including:
  - Preventive health measures
  - Health promotion
- Disincentives for excess hospitalization, surgery
- Incentives for integration of services

continues to operate its National Health Service at a relatively low percentage of GDP as does Japan (Table 11.4).

There are large differences in levels of expenditures on health between countries. In the established market economies, on average 9.3 percent of GDP goes to health, while the former socialist economies expend 3.6 percent, and developing countries generally under 4.5 percent.

Per capita health expenditures also vary widely. The total per capita expenditure on health, whether as percent of GDP or as dollars per capita, does not reflect the efficiency with which the resources are used. Many countries not only have low overall levels of health expenditures but also allocate those meager resources inefficiently.

Regardless of how efficiently money is allocated, countries spending less than 4 percent of GNP on health will have poorly developed health care. Those spending between 4 and 5 percent of GNP may try to have universal coverage, but often achieve this through low staff salaries, inadequate equipment, and spreading limited resources too thinly. This is accentuated when a disproportionately large hospital system and excessive supply of physicians create a siphoning effect on health care spending, or when resources are concentrated in cities while most of the population is rural.

Developed countries that spend between 8 and 16 percent of the GNP on health care have made a value judgment. They have placed health care among the vital priorities in their societies. In those countries with high health care expenditures, such as the United States, physicians' incomes are very high, even when compared with other highly paid professionals. Where financing is centralized in a single paying agency, administrative costs are less than in countries with multiple funding sources. Canada's provincial health insurance plans operate with administrative overheads of less than 5 percent, compared to some 30 percent in U.S. private health insurance.

The World Health Organization issued a *Global Strategy for Health Development* which stressed the importance of efficiency in use of resources as a vital element of health development. The WHO recommends preferential allocation to primary and intermediate care services, especially for currently underserved rural populations. In most countries, reallocation of resources is necessary to strengthen primary care and to adopt new technology and health programs, shown to be cost-effective in terms of costs as well as anticipated benefits.

Where there are multiple sources of health financing, it is more difficult to develop effective national planning. Regulation and supplemental funding by government are needed/required to prevent inequity between socioeconomic groups and between urban and rural populations. When multiple agencies are involved in health insurance or direct government granting systems for specific services, there are gaps (inadequate coverage or access) in services, usually for politically, geographically, and socially disadvantaged sectors of the population, who may have the greatest needs. Under such circumstances, public health services very often become oriented to provision of basic services for persons excluded from health benefits because of lack of health insurance. This places a great financial burden on public health services, which are generally underfunded in comparison to clinical services. Such countries often bring in national health insurance for the disadvantaged groups (e.g., the elderly and the poor). These insurance plans may pay less well than private insurance for the middle class and organized workers. This applies to the United States and to many mid-level developing countries (see Chapter

13). The United States addresses this issue by promotion of national (and state) health targets, guidelines, accreditation systems, and strong professional organizations and medical centers with high levels of research capacity, but still lacks universal coverage for health care. However, the United States provides some 100 billion dollars in "uncompensated care" for the poor and large families, and these costs are built into the budgets. Hence, this care is subsidized by the insured person.

Where financing of health care is centralized, a potential exists for rational allocation of resources. But this depends on adequacy of total financing and rational allocation policies to promote equitable access to services and a balance between one service sector and another. Allocation of monies within the total health expenditures means selection from many alternatives. Misallocation of resources between sectors within the health sector can lead to a wasteful and even counterproductive health system, such as excessive funding of tertiary care while primary care is lacking (Table 11.5).

Where funds are allocated to regional or local health authorities, the potential for shifting resources to meet local needs should be greater. But this may be limited by lack of data or lack of analysis on a local or district basis to highlight priority areas of need. Where there is a highly decentralized management system, some centralized functions are essential to promote national health needs and equity between regions of the country. These include setting policy and standards, monitoring health status indicators, and determining health targets with funding to promote national priorities.

The range of services or programs requiring funding for a population group is indicated in Table 11.5. In 2004 in the United States, there were 45.8 million uninsured, or 15.7 percent of the civilian population. Medicaid includes those enrolled in the State Children's Health Insurance Program (SCHIP). Medicaid and SCHIP covered 37.5 million low-income individuals (12.9 percent of the U.S. population), primarily children, pregnant women, elderly, and disabled people. Medicare, which insures the elderly, the disabled, and patients with end-stage renal disease, had 39.7 million enrollees or 14 percent of the population in 2004. In 2002, Medicare expenditures were $256.8 billion. Both are parts of the Social Security Act (SSA) Amendments of 1965, Medicare (SSA Title XVIII) is under federal administration, while Medicaid (SSA Title XIX) is shared between federal, state, and local administration. Figure 11.4 shows the breakdown of health insurance coverage in the United States in 2004, with 16 percent of the population without any health coverage and a similar number with low levels of health insurance protection. Nearly 47 million Americans lack health insurance, and another 16 million have low insurance coverage. The Institute of Medicine reports that lack of adequate insurance is the cause of an estimated 20,000 excess deaths per year in the United States (Commonwealth Fund, 2008).

TABLE 11.5 Major Categories of Health Expenditures

1. **Institutional care**
   Teaching hospitals, general hospitals, mental and other special hospitals, long-term nursing care, residential care, hospice

2. **Pharmaceuticals and vaccines**

3. **Ambulatory care**
   Primary care, family practice, pediatric, prenatal, and medical specialist; medical, diagnostic, and treatment; ambulatory and day hospital clinics; surgical, medical, geriatric, dialysis, mental, oncologic, drug and alcohol treatment

4. **Home care**

5. **Elderly support activity/service centers**

6. **Categorical programs**
   Immunization, maternal and child health, family planning, mental health, STIs, HIV, tuberculosis, screening for birth defects, cancer, diabetes, hypertension

7. **Dental health**

8. **Community health activities**
   Healthy communities, health promotion in the community for risk groups; smoking restriction, promotion of physical fitness and healthy diet; environmental and occupational health; nutrition and food safety, safe water supplies, special groups

9. **Research**

10. **Professional education and training**

Sources: Testimony of A. Bruce Steinwald before the Subcommittee on Health, Committee on Energy and Commerce, House of Representatives, July 25, 2006.
Medicare physician payments: Trends in service utilization, spending, and fees prompt consideration of alternative payment approaches. Statement of Director, Health Care, at http://www.gao.gov/new.items/d061008t.pdf [accessed May 17, 2008]

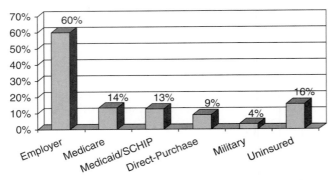

FIGURE 11.4 Sources of insurance coverage in the United States, 2004. Source: Overview of the uninsured in the United States: An analysis of the 2005 Current Population Survey. U.S. Department of Health and Human Services, Office of the Assistant Secretary for Planning and Evaluation (ASPE), September 22, 2005; ASPE tabulations of the 2005 Current Population Survey, http://aspe.hhs.gov/health/reports/05/unin-sured-cps/index.htm [accessed May 17, 2008]

COSTS OF ILLNESS

Direct expenditures for health care in the United States by type of illness are measured in periodic National Medical Expenditure Surveys of the civilian, non-institutionalized population, covering thousands of persons and homes for self-reported expenditures. The largest items of health care expenditure were for cardiovascular disease, followed by injury, then neoplasms.

Following especially high rates of increase in health expenditures during the 1980s in the United States, measures were taken to restrain growth in health care costs, leading to a slower rate of increase. In part this was due to growth of managed care and incentives for lower hospital utilization, and by shifts in payment schedules for hospital and ambulatory care. In the United States, out-of-pocket expenditures as percentage of total health expenditures declined from 55 percent in 1960 to 15 percent, while private health insurance increased from 21 percent in 1960 to 36 percent in 2005; government expenditures for health grew from 24 percent in 1960 to 45 percent (34 percent federal and 11 percent state and local) in 2005 (Health, United States, 2007, Table 125).

From 1990–2004, acute care hospital beds in the United States, declined from 3.7 to 2.7 per 1000 population, while occupancy rates were stable at 67 percent. The rate of growth of health expenditures slowed from annual increases of 9.1 percent in 2002 to 6.6 percent in 2005; however, health expenditures are expected to reach 20 percent of GDP by 2015. The economic effect of adoption of any possible universal health plan in the coming decade will be difficult to forecast, but it will not be entirely additional to current expenditures, as much episodic health care is provided by hospitals to the uninsured.

Costs and Variations in Medical Practice

Increasing costs of health care, waste, variations, and fraud in medical practice inevitably come under scrutiny whether prepayment is in the private or public sector. Variations due to the different needs of various population groups may be justified. However, if an epidemiologic analysis reveals no apparent reasons for the variations, then they become administrative problems which require other approaches. Comparing the quantity and quality of services between population groups is part of epidemiologic and administrative health practice. This approach, when supported by review of relevant current literature on methods of treatment, provides a basis for what is termed *evidence-based medical practice*.

Analysis of medical practice by examination of medical and hospitalization data may show quite startling differences between different cities, regions, and countries. What has come to be called "small area analysis" looks at patterns of practice and tries to determine what may be the cause of such differences. For example, no evidence exists of benefit from higher rates of some types of surgeries, such as hysterectomy, cholecystectomy, and tonsillectomy. Further, there is a cost attached to a surgical procedure that includes a certain mortality rate from anesthetic mishaps and other iatrogenic complications; that is, caused by medical care itself. For example, cholecystectomy rates in the early 1990s in Canada were 600 per 100,000 population, 370 in the United States, and 122 in the United Kingdom. These studies concluded that excess supply of surgeons and the fee-for-service type of payment may lead to an excess of unnecessary and potentially harmful surgical procedures. The cost implications for a health care system are high and can be calculated.

In the United States, health maintenance organizations (HMOs) have shown the capacity to provide comprehensive care over long periods to large population groups with relatively low hospital utilization. HMOs and for-profit managed care coverage increased dramatically in the 1990s. For the period between 1993 and 1994, hospital admissions, average length of hospital stay, and days of care for HMO members (non-Medicare) as well as for Medicare beneficiaries, were well below comparable rates for fee-for-service-based insurance plans.

Technological innovations using simpler, less costly, less invasive, and less risky procedures have led to important changes in health care standards. Continuous evaluation of criteria for "good practice" leads to change based on new knowledge, experience, consensus or leading opinion and meta-analysis, and is essential to quality promotion in health care (see Chapter 15).

Cost Containment

High public and professional expectations from health care can, along with increasing demands of an aging population, costly medical technology, and oversupply of highly technological medical services, lead to a rapid rise in health care costs and a fiscal crisis in many countries. Cost containment became important as the costs in all health systems increased at rates well above economic growth during the 1970s and 1980s. Governments everywhere sought ways to restrain cost increases. Cost effectiveness and cost-benefit analyses have become a part of the planning and management review of ongoing or new interventions in health for both operational and capital expenditures as critical tools of health service planning for rational decision making to restrain health cost increases. Because hospitals are the major consumers of health care expenditures (between 40 and 60 percent in different countries), the emphasis on cost containment has been placed on reducing hospital utilization and developing alternative services or programs of ambulatory and community care.

Cost containment and high-quality health care can coexist. Indeed, cost-containment measures (Table 11.6)

TABLE 11.6 Example Health Service Programs Promoting Cost Containment

Program	Mode of operation
Home care	Reduces length of stay following medical or surgical hospital treatment; reduces the incidence of nosocomial (hospital-acquired) infections; helps elderly or chronically ill to remain at home rather than enter a long-term care facility
Long-term care facilities	For persons unable to be cared for in the family setting reduces length of hospital stay
Regulatory limitations	Limit supply of beds; limit medical services
Ambulatory or day care surgical, medical, mental	Reduces stay in hospital, with fewer secondary infections and iatrogenic complications
Prevention	Prevention (primary, secondary, and tertiary) reduces hospitalization for vaccine-preventable disease, cardiovascular disease, diabetes, and their complications
Environmental health	Chlorination of community water supplies prevents diarrheal diseases and hospitalizations; fluoridation reduces dental disease
Health promotion	Interventions to reduce trauma from motor vehicle accidents; restriction of smoking leads to less lung cancer and coronary heart disease
Diagnosis-related groups (DRGs)	Payment promotes reduced length of hospital stay
Health maintenance organizations and managed care organizations (HMOs and MCOs)	Promote alternatives to hospitalization and long stays, lowers hospital utilization; incentives for employers, employees, and governments to enroll beneficiaries in less costly managed care organizations; capitation provides incentive to prevent illness and institutional care, strengthen ambulatory and preventive care
District health system or regionalization	Promotes rationalization of services, elimination of excess facilities and duplication; promotes greater community orientation in service complex

are associated with greater precision in care and more appropriate use of resources than previous patterns of care. Some measures relate to substitution of lower-cost care for more costly services such as home care for acute hospital care. Others relate to changes in professional services; for example, outpatient surgery in place of inpatient care, and shorter hospital length of stay following myocardial infarction.

Countries with public funding of health care systems are especially concerned with establishing cost containment in order to reduce the rate of increase in health costs. In Canada, governments have shifted their concern from assurance of access to care to cost containment.

Health expenditure grew by 12.5 percent annually in the 1980s in Canada, well above growth of the economy. Canada's cost-containment approaches include controls on fees, regionalization to reduce duplication, excess hospital supply and utilization, as well as increased oversight in utilization of medical care. During the 1990s, these measures succeeded in slowing the rate of cost increase (Chapter 13).

MEDICAL AND HOSPITAL CARE — MICROECONOMICS

Resource allocation policy, made at the national, regional, health insurance, or sick fund level, must address many specific factors affecting the way services are provided and paid for. Incentives and disincentives for efficient care include how doctors and hospitals are paid, and how services are organized. Payment for doctor's services includes fee-for-service, case payment, capitation, salary, or a combination of these methods. Each has its historical roots, its advantages and disadvantages, as well as proponents and opponents.

Payment for Doctor's Services

Fee-for-service is payment for each unit of service, such as a visit or surgical procedure. Payment for a complete service covering the whole period of an illness or another type of care such as obstetrical care including prenatal care and delivery, or other, is called *case-payment*. Fee-for-service is historically the common method of paying for doctor's services and is still the norm in Canada, Germany, and other countries. In some places, payment may be according to a fixed-fee schedule negotiated between the insurance mechanisms, whether public or private, and the doctors' representatives. Fee schedules are often weighted toward medical specialists who have greater prestige than primary care physicians.

Fee-for-service tends to promote an overabundance of the more expensive kinds of care, including surgery, often without real need. This is especially so when the patient is fully covered by health insurance and is therefore better able to pay for the service than the person without insurance. Some insurance systems require participation of the user in the co-payment or user fees or charges. This is often promoted by the idea that it restrains the consumer from seeking unnecessary care, as well as helping cover costs, while opponents justly reply that user fees affect the poorer sector of any population disproportionately and discourage preventive care.

Capitation is payment of doctors by a fixed sum of money for the individual registered for care for a specified period of time. This can apply to a comprehensive health service, as in managed care organizations, as well as to general practitioner services, as in the United Kingdom. Compared with salaried service, this method allows a greater degree of personal identification of the patient with the doctor. It has been in use in the United Kingdom since the introduction of national health insurance in 1911. The recent introduction of incentive fees for full immunization or screening programs has improved performance in these areas.

The United Kingdom's budget holder system initiated in the late 1980s which pays a group of general practitioners for their registered patients has been replaced by per capita funding of primary care trusts. Hospitals are being increasingly paid on a diagnosis-related group (DRG) basis.

Salary payments for doctors and other health workers are common in hospitals even where fee-for-service or capitation is the prominent method of payment. This has advantages for the physician in predictability of income, with less incentive to promote unnecessary servicing. Salary payment may be combined with incentive payments for additional services.

The method of payment for doctors has an important impact on the way in which medical services are used. Empirical evidence indicates that fee-for-service promotes excessive use of the system, including unnecessary surgical procedures, while salaried services are often criticized for diminished identification with patients and, perhaps, underservicing. Increasingly mixed systems of payment are emerging, with capitation as a predominant method.

Payment for Comprehensive Care

Per capita budgeting is a system of payment based on a defined population registered for care with a specific health service system providing a comprehensive range of services, such as a district health system or a managed care organization. Capitation payment covers responsibility for total care, so that economies in hospital care can be applied to cost-effective alternatives such as strong ambulatory care, home care, and long-term care. The population may be enrolled either on a voluntary basis, as in health maintenance organizations (HMOs), prepaid group practice systems and managed care systems, or on a geographic basis as in regional or district health systems.

In some financing systems, the per capita payment takes into account the age and sex distribution of the region, locality, or the registered population. It applies national hospital utilization rates for different categories of age and sex. The capitation method provides an incentive against unnecessary admissions and decreases length of hospital stay, but it is not in a hospital's best interest to discharge a patient prematurely because of the potential for litigation and because the patient may later return in need of more care, adversely affecting hospital costs.

Capitation values may be adjusted by applying regional standard mortality rates (SMRs), as in the United Kingdom, to account for age, sex, and morbidity differences. The British NHS is paying many of its general practitioners by a combination of capitation and DRG systems discussed later.

HEALTH MAINTENANCE AND MANAGED CARE ORGANIZATIONS

Health maintenance organizations (HMOs) are integrated health insurance and provider systems, responsible for hospital, ambulatory, and preventive care for an enrolled population. It is a system of prepaid health care in which the insured person joins or becomes an enrolled member of a health plan that has received a fixed per capita payment from the insurer to provide comprehensive health care for a defined period of time. This approach, which

was developed in the United States, creates non-profit organizations sponsored by industry, unions, and cooperative groups. Formerly called *Prepaid Group Practice*, these plans were developed by Kaiser Permanente in California during World War II and later in many other parts of the country.

Since the 1973 HMO Act, the HMO has become part of the accepted mainstream of health care in the United States. Some large HMOs operate their own hospitals, utilizing 1.5 beds per 1000 population, well below U.S. averages, even when adjusting for age and selection factors. They operate with 1.2 doctors per 1000 enrollees, as compared to 4.5 per 1000 for fee-for-service health care systems. Doctors working in HMOs may be paid on salary or capitation in a staff and group HMO or on a fee-for-service basis in an independent practice association (IPA) or a preferred provider organization (PPO) (Box 11.9).

Health care in the United States has been influenced by the HMO experience and that of other health insurers using HMO-like cost-control measures which limit unrestricted fee-for-service practice. The HMO or managed care approach to health care organization is less costly, largely because of better management of patients in the community and lower hospital utilization patterns.

The major increase in enrollment in managed care took place in the 1990s, much of it in for-profit managed care. Managed care was successful in taking a large part of the market share of health insurance because of its advantages of lesser cost and more comprehensive coverage than traditional fee-for-service health insurance.

Managed care health plans undertake responsibility for the comprehensive care of enrolled members. Managed

---

Box 11.9 Managed Care Organization Models

Managed care plans are health insurance plans that contract with health care providers and medical facilities to provide care at reduced costs. They provide a network of services and are responsible for the quality of care and comprehensiveness of services according to the contract with the insured persons.

1. **Health maintenance organizations** The HMO is a health system providing insurance and service to enrolled members. The traditional **group model HMO** is based on the prepaid group practice in which the HMO employs or contracts with physician groups to provide comprehensive care. Payment is on a capitation payment basis to enrolled members, usually in health centers operated by the HMO and in hospitals owned or contracted with the HMO. Group HMOs may be partnerships that share in the incentive payments. **Staff model HMOs** are plans which employ physicians and other providers in HMO-owned facilities. Network model HMOs contract with multiple physician groups including single or multi-specialty medical groups.

2. **Preferred provider organization (PPO)** A formally organized entity, usually of physicians, hospitals, pharmacies, laboratories, or other providers, which contracts to provide care to HMO members on an agreed (discounted) fee schedule or capitation basis. Each provider works independently but agrees to contracted conditions, including utilization review. The beneficiary has choice of providers within the panel.

3. **Individual practice association (IPA)** The providers may include individual practice physicians who contract to provide services to HMOs, and may also provide services to members of other health insurance plans.

4. **Point of service (POS) Plan** This type of plan allows choice of HMO or PPO services at any time.

Source: U.S. National Library of Medicine and the National Institutes of Health, Managed Care, 24 March 2008 accessed 14 May 2008.
National Center for Health Statistics. 2007. http://www.cdc.gov/nchs/datawh/nchsdefs/hmo.htm [accessed May 14, 2008]
Health, United States, 2007.

care systems are being promoted by private employers, by insurance companies, by states for Medicaid beneficiaries, and by the federal government PPOs.

Since the 1970s, managed care has become the predominant form of health care in most parts of the United States. More than 70 million Americans are enrolled in health maintenance organizations (HMOs) and almost 90 million part of preferred provider organizations (PPOs). Enrollment in HMOs peaked in 2001, declined substantially since, but managed care remains a dominant type of health care and coverage. Medicaid managed care grew rapidly in the 1990s. In 1991, 2.7 million beneficiaries were enrolled in some form of managed care and by 2004, that number had grown to 27 million. Of the total Medicaid enrollment in the United States in 2005, some 63 percent receive Medicaid benefits through managed care. All states (except Alaska, New Hampshire, and Wyoming) have all, or a portion of their Medicaid population enrolled in an MCO. States can make managed care enrollment voluntary, or require certain populations to enroll in an MCO. For 2006, the breakdown of enrollment by plan type was as follows: 20 percent HMO, 60 percent PPO, 13 percent point of service providers (POSs), 4 percent high-deductible health plan (HDHP), and 3 percent conventional indemnity plans. HCFA regulates HMOs and has instituted guidelines for reporting and quality assessment in an accreditation approach to quality assurance (see Chapter 15). There has been some backlash against managed care with negative publicity regarding restrictions in referrals and other client concerns (National Conference of Sate Legislators).

Managed care, especially in the for-profit sector, is under criticism in medical and public health organizations and journal editorials, as well as in the media and state and federal legislatures. It is alleged that the system promotes denial of access to specialists and other needed care because of the economic incentives built into the capitation system, especially when administered by for-profit companies. The economic benefits are generally accepted. The controversy focuses on the incentives to underservice and on loss of choice by the consumer in for-profit managed care systems. The quality and ethical issues of managed care are discussed further in Chapter 15. Legislative efforts at state and federal levels to define patient's rights, grievance procedures, and minimum baskets of service have been under way in Congress, with a narrow (50–47) defeat in late 1998, when President Bill Clinton actively promoted a Patient's Bill of Rights which is likely to reappear in future congressional sessions and broad political debates.

Opponents to the managed care approach argue that lower HMO hospital utilization may in part be due to lower costs due to a bias by selection of healthy members, and that HMOs may underservice patients in order to reduce costs, or increase physician incomes or profits.

Available evidence supports HMO experience as providing high-quality medical care at lower cost than competing open-ended, fee-for-service insurance systems. The leveling off of expenditure for health in the United States during the 1990s is largely attributable to the move from fee-for-service care plans to managed care of a large percentage of the population. Managed care is also emerging in other countries, in the Sick Funds in Israel and in some European countries, in Latin America (Argentina, Brazil, Mexico, Chile, Peru, and others), as well as in the Philippines, all seeking to restrain cost increases, while extending health care to a greater part of their populations.

DISTRICT HEALTH SYSTEMS

In the United Kingdom and in the Scandinavian countries, a comprehensive service model has existed in the form of district health systems for many years. The residents of a district have their health benefits provided by or contracted out by the district. In principle, the geographic unit of service allows for efficiency in transfer of resources and patients from one service to another, based on need, and not on the financial interests of the insurance system or the provider.

The Scandinavian countries have a long tradition of management of health facilities at the county level with budgets derived from a combination of local taxation and national grants. In reforms since the 1980s, integration of various services into district health systems with reduced hospital bed supplies has resulted in a leveling off of cost increases for health.

In the United Kingdom, methods of budget allocation to health regions were the subject of a long and detailed study by the Regional Allocation Working Party (RAWP) in the 1980s. The decision was made that the optimal method of allocating funds to district health authorities would be by per capita grants with adjustment by the SMRs of the district. This adjustment takes into account age and morbidity differences between different districts and promotes an equitable approach to resource allocation. Reforms since the 1990s focused on building workforce and physical capacity, and issues such as waiting times which have fallen, clinical outcomes for cancer and heart disease have improved, and NHS facilities have been modernized. Since 2003, a new stage of reform promised more choice for patients: more freedom to innovate and improve services; competition on quality with financial incentives to improve care and promote sound financial management; national standards and regulation to guarantee quality, safety, and equity are being emphasized, along with improved information management and technology to support reforms and deliver better, safer care.

Health reform in some provinces in Canada includes reducing the per capita hospital bed supply along with regionalization and integration of services in regional or district health systems. Many provinces have adopted district health boards which amalgamated hospital, nursing home, and public health boards. Per capita funding allows transfer of funds from hospital care to other sector services such as home and community care. The provinces have managed to level off health expenditure increases to rates less than the growth of GDPs.

Regionalization of hospital and other services is another approach to rationalization of health care and cost control. In communities with excessive hospital beds and competing services, regionalization provides a method of rationalization, with voluntary or mandatory elimination of wasteful, competing departments or investigative units such as *in vitro* fertilization, cardiac surgery units, advanced imaging devices (MRIs), or excess bed capacity. In the United States, efforts were made to regionalize certain services, such as the introduction of perinatal care systems in the 1980s, but did not lead to wider application of this approach. In the 1990s, hospital networks in both the for-profit and not-for-profit sectors have expanded aggressively to increase market share and vertical integration for service and management cost-efficiency as part of the managed care dominance of the U.S. health insurance market.

PAYING FOR HOSPITAL CARE

Hospitals are the most costly component of a health service. Traditionally, hospitals were paid on a *per diem* or flat rate per patient-day. The *per diem* may be determined by using actual costs or by national, state, or regional averages. The daily operating costs are divided by the number of beds, with perhaps adjustment for teaching or research functions. The *per diem* based on actual costs per patient in specific units in a hospital, such as intensive care, may be higher or lower than the budget provides for that specific service.

The *per diem* method of payment encourages long lengths of stay, rewards hospitals with low technology, and if based on national or regional averages may penalize hospitals with high levels of staffing and technology, such as teaching hospitals. When the service is insured, there is no financial incentive for shortening the patient's hospital stay. The *per diem* method is associated with inefficient use of facilities, such as admission to the hospital for diagnostic tests or prolonging a stay for additional testing or care that could be provided in alternative and less costly ways. The provider has an incentive to hospitalize and provide prolonged care to a relatively healthy patient, while the sicker patient is a financial liability, as are teaching and research functions, unless funded separately. This system lacks incentives to improve efficiency by developing alternative ambulatory or day care services, and it

punishes more efficient hospitals which reduce length of stay or occupancy rates.

Fee-for-service payment for each service supplied in a hospital favors unnecessary marginal care, long lengths of stay, high admission rates, and the provision of duplicative or unnecessary services. This method was common in the United States with its multiple insurance systems but is increasingly being replaced by DRG payment (discussed later in this section). Fee-for-service payment provides and incentive to overservice with no incentive to reduce costs or admissions of length of stay.

Historical budgeting is remuneration based on the previous year's budget, adjusted for inflation and the cost of new services. The budget may be reviewed line by line by the paying authority or be on a global or block budget basis, which frees the hospital to make internal reallocations within the overall allotment. Payment can include a capital fund for renovation. This method is often used when a hospital is directly operated by the Ministry of Health. As opposed to the *per diem* payment system, this method should theoretically provide some incentive to reduce length of stay and to search for efficiency in the use of hospital resources.

Payment by norms means financing according to nationally fixed standards of numbers of beds, staffing, and other measures. This method as practiced in the Soviet health system provided national incentives to maintain high hospital bed to population ratios, long length of hospital stay, little investment in improving ambulatory care, low salaries, and generally low quality of care. Reform in post-Soviet countries requires cancellation of these historic norms, reducing excess hospital bed capacity and adoption of incentives for efficiency in health care (Chapter 13).

As a result of concern over high costs and utilization rates, alternative methods of payment have developed in the United States since the 1960s. The diagnosis-related group (DRG) system was adopted in 1983 by the U.S. Health Care Financing Administration (HCFA) as the basis for payment for hospitalization of Medicare patients. The DRG system has been the basis for paying for hospital care in the United States since 1999, and it is increasingly being used in other industrialized countries, such as the United Kingdom and Israel, and some developing countries, such as the Philippines.

The DRG system is a prospective payment system for hospital care reimbursement, and the system pays the hospital according to 495 treatment classifications of diagnoses or procedures, each with a fixed hospital payment rate. This provides an incentive to reduce length of stay, more efficient use of diagnostic and treatment services, and reduced overall bed capacity. As a result, hospital outpatient services increased rapidly in the United States while bed occupancy rates and the hospital beds to population ratio declined steadily over the 1990s. The DRG system does not lead to fewer admissions and may encourage falsification of diagnostic criteria or increasing the diagnostic severity of case definition to increase revenues ("DRG creep").

Different hospital budgeting methods have advantages and disadvantages. Payment by DRGs is most likely to promote rational use of hospital care. Regional budgets allocated on a per capita basis with hospital payment by DRGs may be the most effective way of achieving a balance between ambulatory and hospital care, combining regional equity and incentives for efficient use of diagnostic and treatment services. Prospective payment systems must be associated with quality assurance mechanisms, a vital issue in health management (Chapter 15).

CAPITAL COSTS

The capital cost to build or renovate a health facility is based on long-term considerations but has important effects on current operating costs. The cost of operating a new health care facility may equal the capital cost in 2–3 years. Capital costs may be financed by public or private donations, risk-capital investment, or government-guaranteed loans. Government regulatory agencies may approve a capital project of construction or equipment of a hospital under a certificate of need (CON) procedure and then agree to a grant mechanism to provide funds to match local contributions or to budget or adjust rates to include repayment of long-term loans for capital costs. This occurred both in the United States under the Hill-Burton Act (see Chapter 10), and in Canada under the National Health Insurance System. Where hospitals are operated independently of government, they may borrow or raise money privately through long-term bonds or low-interest loans. Repayment can be built into the operating costs and amortization of the loan over many years.

When government finances capital costs, it has greater control over the direction, distribution, and supply of hospital facilities. Government norms may encourage an increased bed supply by encouraging hospital construction, or maintenance of high numbers of beds that may not be used or may be of poor quality. Norms may also be used to set upper limits or provide incentives to reduce bed supply. One of the common elements of cost-containment strategies in many industrialized countries is reduction in hospital bed supply, which is occurring without apparent harm to the quality of care. Hospital bed reduction is partly offset by transfer of long-stay patients to home care programs or to nursing homes with a transfer of capital and operating costs. Overall, maintaining quality of care is not compatible with supporting a large and costly number of beds to population ratio (the converse of the aforementioned Roemer's law, Box 11.7) because of the excessive resources required to maintain these beds at the expense of other needed services in the community.

Health expenditures by category of services in the United States are shown over the period 1960–2005 in Table 11.7. In 2005, hospital care accounted for about 30 percent of total expenditures, a reduction from nearly 40 percent in 1980. Physician and other clinical services took 21 percent of total expenditures, a small decline from 1990, prescription drugs 13 percent, dental care 4.4 percent, nursing homes just over 6 percent, governmental and private health insurance administration 7.2 percent, governmental public health activities 2.8 percent, research 2.0 percent, and structures and equipment 4.4 percent.

HOSPITAL SUPPLY, UTILIZATION, AND COSTS

Acute care hospital bed to population ratios in the United States increased from the 1940s to the 1980s and declined thereafter. The supply and utilization of hospital beds are changing as economic incentives increase pressure to find less costly forms of care, and as ambulatory and community-oriented care is perceived to be more effective in many instances. In the United States, hospital utilization, average length of stay and percentage occupancy show a decline mainly during the period 1980–2004. Hospital staff per 1000 patient days increased from 226 in 1960 to 583 in 1991, reflecting increased support and technical services and greater severity of illness of those hospitalized. Increased staffing, technological innovations, and expensive medications increased the cost of patient care in hospitals. Table 11.8 shows the trend in hospital bed supply and percentage occupancy in acute care nonfederal general hospitals in the United States from 1960–2004.

There has been a trend to decrease the hospital bed supply and utilization in the United States during the 1980s, 1990s, and 2000s. Despite aging of the population, the trend to lower overall hospital utilization has resulted from the following: changing morbidity patterns, ambulatory services in place of inpatient care, adoption of the DRG system of payment reducing length of stay, greater stress on health economics and cost containment in medical considerations, more efficient methods of care, greater health consciousness in the general population, and improved self-care and prevention.

Mortality from coronary heart disease has decreased markedly during this period; however, admission rates for heart disease overall have not declined, while total days of care fell by 38 percent. This is in part due to changing patterns of care, with shorter length of stay and a more aggressive rehabilitation approach to myocardial infarction, and emphasis on ambulatory care. Medical treatment during the acute myocardial infarction stage is more effective than previous treatments, with technology

TABLE 11.7 National Health Expenditures Percent Distribution, United States, 1960–2005

	1960	1970	1980	1990	2000	2005
National health expenditure (billions $)	28	75	254	714	1,353	1,988
			Percent distribution			
Total national health expenditures	100.0	100.0	99.9	100.1	100.0	100.0
Investment (structure and capital equipment)	6.9	7.8	5.7	4.9	4.7	4.4
Research	2.5	2.6	2.1	1.8	1.9	2.0
Personal health care percent						
Hospital care	33.3	36.8	39.8	35.2	30.8	30.8
Physician and clinical services	19.4	18.7	18.5	22.0	21.3	21.2
Dental care	7.1	6.2	5.2	4.4	4.6	4.4
Other personal care	3.6	2.7	2.7	3.8	5.6	5.8
Nursing homes	2.9	5.4	7.5	7.3	7.0	6.1
Home health care	0.2	0.3	0.9	1.8	2.3	2.4
Prescription drugs and medical products	18.0	14.0	10.1	10.4	12.6	13.0
Government administration and net cost of private health insurance percent	4.4	3.7	4.8	5.5	6.0	7.2
Public health percent (government)	1.5	1.9	2.5	2.8	3.2	2.8

Source: Health, United States, 2007, from table 124.

TABLE 11.8 Acute Care Hospital Bed Supply and Utilization, United States, 1980–2004

Facilities	1960	1970	1980	1990	2000	2004
Beds per 1000 population	3.6	4.3	4.5	3.7	2.9	2.8
Discharges per 1000 population	na	na	173	125	113	118
Average length of stay (ALS)	na	na	7.7	7.2	5.8	5.6
Total days of care per 1000 population	na	na	1197	819	558	569
Percent occupancy	75	77	75	67	64	67

Notes: Includes community hospitals; does not include federal hospitals.
na = data not available
Source: Health, United States. 2006. http://www.cdc.gov/nchs/dat/hus/hus06.pdf [accessed May 17, 2008]

such as streptokinase, angioplasty, stents, and other interventions. All of this has been accompanied by a steady fall in mortality rates (Chapter 5).

Many western European countries began to reduce their hospital beds in the 1980s, as seen in Figure 11.5. Sweden and Finland reduced their hospital capacity by 53 percent and 36 percent, respectively, and Western Europe as a whole by 26 percent. Countries of Eastern Europe and the former Soviet Union have high but declining hospital bed to population ratios (i.e., beds per 1000 population) and

hospital inventories. Some have increased health spending, but some (Russia) are still relatively low in overall expenditures on health care per capita.

Reducing hospital utilization and bed supply creates a problem of staff and resource reallocation. Hospital facilities themselves can sometimes be converted to other purposes, as outlined in Box 11.10. Often, the most constructive use of obsolete hospital facilities is to transfer them out of the health sector, since the land may be of greater value than its continuing use for health care purposes.

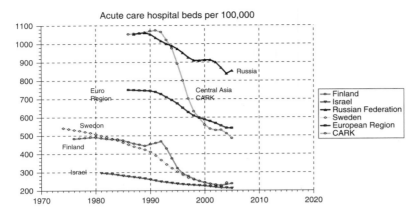

FIGURE 11.5 Acute care hospital beds, selected countries, 1970–2005. Source: World Health Organization, Regional Office for Europe, Health for All database, November 2007. Copenhagen: WHO.

Box 11.10 Managing Excess Hospital Bed Capacity

1. **Convert to long-term care (LTC) facility** extended care, rehabilitation, chronic care, or elderly persons housing;

2. **Close maternity homes** replace with maternity units in district general hospitals;

3. **Develop acute psychiatric units in general hospitals** conversion of excess acute care beds, closure of long-term psychiatric beds, development of community services and group residential facilities;

4. **Develop acute geriatric units in general hospitals** short-term care, with home and LTC facilities;

5. **Develop tuberculosis units in general hospitals** short-term investigation and therapy, with closure of long-term beds in TB hospitals and strengthening community care systems;

6. **Develop detoxification units for alcohol and drug abuse** general hospitals with community facilities support;

7. **Convert to special needs shelters** homeless persons, abuse or rape crisis shelters;

8. **Convert to ambulatory care facilities** use inpatient facilities and staff for outpatient, day hospital services;

9. **Develop hospices** care of terminal patients;

10. **Convert to other socially useful functions** community centers, schools, or vocational training;

11. **Demolish and dispose of obsolete facilities** land value may pay for part of new health programs.

Box 11.11 Classical Market Forces

- Supply
- Demand
- Competition in cost, quality
- System macro-efficiency
- Vertical integration
- Lateral integration
- System micro-efficiency
- Incentives
- Disincentives
- Reputation

---

of a manager, provider, and policy planner for a strategic role in health systems. Some of these modifiers are governmental regulatory factors, such as in supply of hospital beds. Others are related to access to services and the amount and method of payment and other factors that affect needs for care, as well as the quality and efficiency of a service.

Market mechanisms are modified by regulations, incentives, and other factors used to promote a balance of preventive, curative, and rehabilitative services, including health promotion to improve health and help the individual seek and find the most appropriate care at any point in time. These factors are summarized in Box 11.12.

MODIFIED MARKET FORCES

Classically, market forces are seen as a means of empowering the purchaser to seek the least expensive and/or best goods or services from competing providers. The classical market forces are presented in Box 11.11.

In health care, there are modifying factors that affect market forces. Understanding the modifiers of market forces summarized in Table 11.9 is part of the preparation

ECONOMICS AND THE NEW PUBLIC HEALTH

The New Public Health has a vital interest in methods of financing of health services, the economics of health care, and allocation of resources, whether in terms of money, human resources, or fixed assets. No other approach has taken the broad view of promotive, preventive, curative, and long-term care services. The balance and interdependence among these elements of the health system are

TABLE 11.9 Market Forces and Modifying Factors in the Economics of Health

Determinants in health demand	Modifiers of demand	Examples
Classical market factors	Supply	Providers combine to restrict supply
	Demand	Prepayment increases effective demand
	Competition in cost, quality	Managed care versus fee-for-service plans
	System macro-efficiency	District health systems and HMOs
	Vertical integration	Multiservice systems for defined populations
	Lateral integration	Multihospital networks for efficiency
	System micro-efficiency	Quality improvement, computerization, staff attitudes, scheduling, and service hours
	Incentives: disincentives	Budget, method of payment: block, *per diem*, DRG
	Reputation	Consumer, community, provider satisfaction
Regulatory factors	Regulate supply	Reduce hospital beds/1000 and personnel
	Regulate demand	Gatekeeper functions
	Regulate price	User fees
		Fee control, income capping, salary or capitation payment for doctors
	Regulate method of payment	DRGs, block budgets for hospitals
	Health promotion issues	Food enrichment, safety factors, seat belts
Health and societal factors	Differing population needs	Demographic and epidemiologic transitions
	Social inequities	Reduce social gaps, universal access, risk groups
	Improve infrastructure to reduce needs	Sewage, water treatment, road safety
	Socioeconomic improvements	National and family incomes
	Public social policies	Social security, pensions, compensation
	Health as a national and local priority	Health system reform
	Health promotion	Involve community and providers in prevention
	Improve KABP (knowledge, attitudes, beliefs, and practices)	Providers, beneficiaries, consumers' rights, needs, responsibilities
System determinants	Shift in resource allocation	Balance of institutional and community care
	Technological innovations	New vaccines, drugs, diagnostic equipment, ORS, community health workers
	Regulation and accreditation	Home care, generic drugs, nurse practitioners
	Substitution	External accreditation, internal review systems, patient choice, continuous quality improvement
	Total quality management	

Box 11.12 Regulatory Factors

- Regulate supply
- Regulate demand — gatekeeper, user fees
- Regulate price
- Regulate benefits
- Regulate method of payment
- Health promotion issues
- Accreditation of providers

influenced by financing systems and choices among alternative ways of expending resources.

Economic analysis, like epidemiologic assessment, is a vital tool in health planning and management, particularly in evaluating allocation of resources within a health system. Failure to carry out such assessments results in inefficiency in health planning. Methods of economic analysis, described briefly in this chapter, are part of the armamentarium of the New Public Health. Both the new and "veteran" health professionals need a basic understanding of the economic issues involved in setting priorities; service organization; utilization of services; and the complex of related ethical, political, social, and management issues.

SUMMARY

The basic view of Health for All promoted by the WHO at the Alma-Ata Conference in 1978 was this concept adopted anew in the *Macroeconomic Health Report* of 2001, and more recently on the thirtieth anniversary of Alma-Ata. It was also adopted in the United Nations Millennium Development Goals of 2001. The incorporation of an economic rationale for investment in health expounded in the World Bank Human Development Report of 1993 and later updated in the Disease Control Priorities Project of 2006 has strengthened the concept that investment in health is a part of and necessary for economic development. This is the principle that health is a human right for everyone, and that management of resources is crucial to achieve improved health. This combination of overall public health commitment with an economic justification and management is the essence of the New Public Health.

Advancement in its application requires political commitment, and funding adequate for a balanced program including health promotion, primary health care, hospital care, and long-term care, all essential elements of an effective health system. Health promotion and primary care are the most cost-effective interventions in improving the health status of the population. Where there has been an excessive emphasis on institutional care, there is real potential for transfer of resources and emphasis within the health system as part of the process of raising primary care and health promotion standards. This is the essential direction of health reform in many countries which had overemphasized hospitalization for health care.

Innovations in health care financing and administration, such as HMOs, managed care, capitation payment, fund-holding general practitioners, district health systems, and DRGs are all part of the search for more efficient ways of using resources and limiting cost increases. Innovations in achieving better individual and population health include a wide array of technologies, organizational, and other improvements in medical care and prevention care, health promotion including legislation and regulatory public health functions, smoking reduction and greater awareness of healthy lifestyle, and outreach and home care services for people at high risk. Many new technologies such as endoscopic and outpatient surgery, early and more effective care for cardiovascular disease, simple and inexpensive diagnosis and cure for peptic ulcers, new vaccines for prevention of cancer, and other innovations are making an impact on health systems.

The effectiveness of new health innovations and the economics of health in society are critically important elements of the New Public Health. As populations age and as technology advances, costs will inevitably increase, but this can be moderated by resource allocation to prevent or delay the onset of complications among chronically ill persons in the community.

Health must compete with other government programs for resource allocation. In the New Public Health, health care is an investment in human capital for the development of a country, as well as an ethical obligation of a society to its individual members. The New Public Health is involved in management of health care in all its aspects so that an understanding of basic issues in health economics is as vital to its practice as is an understanding of communicable disease or any other element of the broad panorama of health.

ELECTRONIC RESOURCES

American Public Health Association, Policy Statement 9615(PP), Supporting national standards of accountability for access and quality in managed health care, advocacy/policy statements, http://www.apha.org/ [accessed May 17, 2008]

American Public Health Association, Policy Statement 9716(PP), The issue of profit in health care, advocacy/policy statements, http://www.apha.org/ [accessed May 17, 2008]

American Public Health Association, Policy Statement 9802, Managed care and people with physical/mental /disabilities, advocacy/policy statements, http://www.apha.org/ [accessed May 17, 2008]

Centers for Disease Control, Vaccine costs, http://www.cdc.gov/vaccines/programs/vfc/cdc-vac-price-list.htm [accessed May 17, 2008]

Commonwealth Fund. www.commonwealthfund.org [accessed May 17, 2008]

Disease Control Priority Project (DCPP). 2006, http://www.dcp2/page/main/About.html [accessed May 18, 2008]

Health Care Financing Administration. http://www.os.dhhs.gov/about/opdivs/hcfa.html [accessed May 15, 2008]

Managed Care. http://www.ahrq.gov/research/managix.htm [accessed May 17, 2008]

Managed Care On-Line. http://www.managedcaredigest.com [accessed May 17, 2008]

National Conference of State Legislators. 2008, http://www.ncsl.org/programs/health/managed.htm [accessed June 23, 2008]

Organization of Economic Cooperation and Development, OECD Health Data 2007, http://www.oecd.org/health/ [accessed May 18, 2008]

Organization of Economic Cooperation and Development, OECD Health Data. 2007. Statistics and Indicators for 30 Countries. OECD, User's Guide, statistics/health portal, http://www.oecd.org/ and http://www.oecd.org/els/health/workingpapers [accessed May 15, 2008]

The Commonwealth Fund: A Private Organization Working Toward a High Performance Health System, 2004–2008, http://www.commonwealthfun.org/ [accessed June 22, 2008]

United Kingdom Department of Health. http://www.dh.gov.uk/en/Publicationsandstatistics/Publications/PublicationsPolicyAndGuidance/DH_4070242 [accessed May 15, 2008]

United States Department of Health and Human Services, Office of the Assistant Secretary for Planning and Evaluation (ASPE). 2005. ASPE tabulations of the 2005 Current Population Survey, http://aspe.hhs.gov/health/reports/05/uninsured-cps/index.htm [accessed May 17, 2008]

UK Annual Report 2007, http://www.dh.gov.uk/en/Publicationsandstatistics/Publications/AnnualReports/DH_074767.

United States. Steinwald Before the Subcommittee on Health, Committee on Energy and Commerce, House of Representatives, July 25, 2006. Medicare Physician Payments: Trends in Service Utilization, Spending, and Fees Prompt Consideration of Alternative Payment Approaches. Statement of Director, Health Care, http://www.gao.gov/new.items/d061008t.pdf [accessed May 15, 2008]

World Bank Health Reform Online, http://www.worldbank.org/healthreform/

World Bank Group. 2006. Global burden of disease and risk factors, http://www.dcp2.org/pubs/GBD

World Health Organization, European Region, Health for All Database, http://www.euro.who.int/ and http://www.euro.who.int/information sources [accessed May 15, 2008]

Health Care Spending in the United States and OECD Countries, http://www.kff.org/insurance/snapshot/chcm050206oth2.cfm and http://www.kff.org/insurance/snapshot/chcm010307oth.cfm [accessed May 15, 2008]

RECOMMENDED READINGS

Anderson, G. F., Ponllier, J.-P. 1999. Health spending, access and outcomes: Trends in industrialized countries. *Health Affairs*, May/June:178–192.

Bobadilla, J.-L., Cowley, P., Musgrove, P., Saxenian, H. 1994. Design, content and financing of an essential national package of health services. *Bulletin of the World Health Organization*, 72:653–662.

Borger, C., Smith, S., Truffer, C., Keehan, S., Sisko, A., Poisal, J., Clemens, M. K. 2006. Health spending projections through 2015: Changes on the horizon. *Health Affairs*, 25:61–73.

Centers for Disease Control. 1995. Assessing the effectiveness of disease and injury prevention programs: Costs and consequences. *Morbidity and Mortality Weekly Report*, 44(RR-10):1–10.

Centers for Disease Control. 1995. Economic costs of birth defects and cerebral palsy — United States, 1992. *Morbidity and Mortality Weekly Report*, 44:694–699.

Desvarieux, M., Landman, R., Liautaud, B., Girard, P.-M. 2005. The INTREPIDE Initiative in Global Health. Antiretroviral therapy in resource-poor countries: Illusions and realities. *American Journal of Public Health*, 95:1117–1122.

Evans, R. G., Lomas, J., Barer, M. L., et al. 1989. Controlling health expenditures: The Canadian reality. *New England Journal of Medicine*, 320:571–577.

Haberer, J. E., Garrett, B., Baker, L. C. 2005. Does Medicaid managed care affect access to care for the uninsured? *Health Affairs*, 24:1095–1105.

Kahn, J. G., Kronick, R., Kreger, M., Gans, D. N. 2005. The cost of health insurance administration in California: Estimates for insurers, physicians, and hospitals. *Health Affairs*, 24:1629–1639.

Kuttner, R. 2008. U.S. Market based failure – a second opinion on U.S. health costs. *New England Journal of Medicine*, 358:549–551.

Leroy, J. L., Habicht, J.-P., Pelto, G., Bertozzi, S. M. 2007. Current priorities in health research funding lack an impact on the number of child deaths per year. *American Journal of Public Health*, 9:219–223.

National Center for Health Statistics, Health, United States 2007 with Chartbook on the Health of Americans, Hyattsville, MD, http://www.cdc.gov/nchs/data/hus/hus07.pdf [accessed June 24, 2008]

Sachs, J. D. 2001. Macroeconomics and health: Investing in health for economic development, http://www.cid.harvard.edu/archive/cmh/cmhreport.pdf

Schoen, C., Osborn, R., Doty, M. M., Bishop, M., Peugh, J., Murukutla, N. 2007. Toward higher performance health systems: adults' health care experience in seven countries, 2007. *Health Affairs*, 26:717–734.

Schoenbaum, S. C., Davis, K., Holmgren, A. L. 2007. *Health Care Spending: An Encouraging Sign?* New York: Commonwealth Fund.

Wennberg, J. E. 1990. Outcomes research, cost containment, and the fear of health care rationing. *New England Journal of Medicine*, 323:1202–1204.

WHO Commission on Macroeconomics and Health. 2001. *Macroeconomics and Health: Investing in Health for Economic Development. Report of the Commission on Macroeconomics and Health.* Geneva: World Health Organization. Follow-up reports from 2004 available http://www.who.int/macrohealth/action/update/en/

World Bank. 1993. *World Development Report, 1993: Investing in Health. World Development Indicators.* New York: Oxford University Press.

World Health Organization. Regional Office for Europe, Health for All Database, November 2007. Copenhagen, WHO.

Tan-Torres, T., Edejer, T., Baltussen, R., Adam, T., Hutubessy, R., Acharya, A., Evans, D. B., Murray, C. J. L. (eds.). 2003. *WHO Guide to Cost-Effectiveness Analysis.* Geneva: World Health Organization.

Thorpe, K. E. 2005. The rise in health care spending and what to do about it. *Health Affairs*, 24:1436–1445.

BIBLIOGRAPHY

Abel-Smith, B. 1991. Financing health for all. *World Health Forum*, 12:191–200.

Anderson, R. M., Rice, T. H., Kominski, G. F. 1996. *Changing the U.S. Health Care System: Key Issues in Health Services, Policy and Management.* San Francisco: Jossey-Bass Publishers.

Anderson, T. F., Mooney, G. (eds.). 1990. *The Challenge of Medical Practice Variation: Economic Issues in Health Care.* Hong Kong: Macmillan.

Bodenheimer, T. 1999. The American health care system: Physicians and the changing medical marketplace. *New England Journal of Medicine*, 340:584–588.

Book, E. L. 2005. Health insurance trends are contributing to growing health care inequality. *Health Affairs*. 2005 Dec 6 [Epub ahead of print].

Brownson, R. C., Remington, P. L., Davis, J. R. 1998. *Chronic Disease Epidemiology and Control*, Second Edition. Washington, DC: American Public Health Association.

Bunker, J. P. 1970. Surgical manpower: A comparison of operations and surgeons in the United States and in England and Wales. *New England Journal of Medicine*, 282:135–144.

Centers for Disease Control. 1992. A framework for assessing the effectiveness of disease and injury prevention. *Morbidity and Mortality Weekly Report*, 41(RR-3):1–12.

Centers for Disease Control. 1995. Economic costs of birth defects and cerebral palsy — United States, 1992. *Morbidity and Mortality Weekly Report*, 44:694–699.

Chassin, M. R., Kosecoff, J., Park, R. E., Winslow, C. M., Kahn, K. L., Merrick, N. J., Keesey, J., Fink, A., Solomon, D. H., Brook, R. H. 1987. Does inappropriate use explain geographic variations in the use of health services? *Journal of the American Medical Association*, 258:2533–2537.

Chernew, M. E., Goldman, D. P., Pan, F., Shang, B. 2005. Disability and health care spending among Medicare beneficiaries. *Health Affairs*, 24 Suppl 2:W5R42–52.

Creese, A. L., Henderson, R. H. 1980. Cost–benefit analysis and immunization programmes in developing countries. *Bulletin of the World Health Organization*, 58:491–497.

Department of Health. 2006. The NHS in England: The operating framework for 2007/08. London: Department of Health.

Drumond, M. F. 1985. Survey of cost-effectiveness and cost–benefit analyses in industrialized countries. *World Health Statistics Quarterly*, 38:383–401.

Drumond, M. F., Stoddart, G. L. 1985. Principles of economic evaluation of health programmes. *World Health Statistics Quarterly*, 38:355–367.

Drumond, M. F., Stoddart, G., Labelle, R., Cushman, R. 1987. Health economics: An introduction for clinicians. *Annals of Internal Medicine*, 107:88–92.

Folland, S., Goodman, A. C., Stano, M. 1997. *The Economics of Health and Health Care*, Second Edition. Upper Saddle River, NJ: Prentice-Hall.

Foote, S. B., Halaas, G. W. 2006. Defining a future for fee-for-service Medicare. *Health Affairs*, 25:864–868.

Friede, A., Taylor, W. R., Nadelman, L. 1993. On-line access to a cost–benefit/cost–effectiveness analysis bibliography via CDC WONDER. *Medical Care*, 31(Supplement):JS12–17.

Ginsberg, G. M., Berger, S., Shouval, D. 1992. Cost–benefit analysis of a nationwide inoculation programme against viral hepatitis B in an area of intermediate endemicity. *Bulletin of the World Health Organization*, 70:757–767.

Ginsberg, G. M., Tulchinsky, T. H., Abed, Y., Angeles, H. I., Akukwe, C., Bonn, J. 1990. Costs and benefits of a second measles inoculation of children in Israel, the West Bank, and Gaza. *Journal of Epidemiology and Community Health*, 44:274–280.

Ginsberg, G., Tulchinsky, T., Filon, D., Godfarb, A., Abramov, L., Rachmilevitz, E. A. 1998. Measuring costs: The economics of health benefit analysis of a national thalassemia programme in Israel. *Journal of Medical Screening*, 5:120–126.

Goldie, S. J., Gaffikin, L., Goldhaber-Fiebert, J. D., Gordillo-Tobar, A., Levin, C., Mahe, C., Wright, T. C. 2005. Alliance for Cervical Cancer Prevention Cost Working Group. Cost-effectiveness of cervical-cancer screening in five developing countries. *New England Journal of Medicine*, 353:2158–2168.

Goldman, L. 2005. Cost-effectiveness in a flat world — can ICDs help the United States get rhythm? *New England Journal of Medicine*, 353:1513–1515.

Grossman, M. 1972. *Demand for Health*. New York: Columbia University Press.

Inglehart, J. K. 1999. The American health care systems—Medicare. *New England Journal of Medicine*, 340:327–332.

Jacobs, P. 1991. *The Economics of Health and Medical Care*, Third Edition. Gaithersburg, MD: Aspen.

Jamieson, D. T., Breman, J. G., Measham, A. R., Alleyne, R., Claeson, M., Evans, D. B., Jha, P., Mills, A., Musgrove, P. 2006. *Disease Control Priorities in Developing Countries. Second Edition*. Washington, DC: The World Bank Group.

Jamieson, D., Breman, J. G., Measham, A. R., et al. 2006. Disease Control Priorities in Developing Countries. Second Edition. The World Bank Group, Washington, DC.

Kelley, J. E., Burrus, R. G., Burns, R. P., Graham, L. D., Chandler, K. E. 1993. Safety, efficacy, cost and morbidity of laparoscopic versus open cholecystectomy; a prospective analysis of 228 consecutive patients. *American Surgeon*, 59:23–27.

Kinnon, C. M., Velasquez, G., Flori, Y. A. World Health Organization Task Force on Health Economics. 1994. *Health Economics: A Guide to Selected WHO Literature*. Geneva: WHO.

Lieu, T. A., Cochi, S. L., Black, S. B. 1994. Cost-effectiveness of a routine varicella vaccination program for U.S. children. *Journal of the American Medical Association*, 271:375–381.

Lord, J., Thomason, M. J., Littlejohns, P., Chalmers, R. A., Bain, M. D., Addison, G. M., Wilcox, A. H., Seymour, C. A. 1999. Secondary analysis of economic data: A review of cost–benefit studies of neonatal screening for phenylketonuria. *Journal of Epidemiology and Public Health*, 53:179–186.

Madon, T., Hoffman, K. J., Kupfa, L., Glass, R. I. 2007. Implementation science. *Science*, 318:1728–1729.

Marquis, M. S., Rogowski, J. A., Escarce, J. J. 2005. The managed care backlash: did consumers vote with their feet? *Inquiry*, 41: 376–390.

McKee, M., Edwards, N., Atun, R. 2006. Public–private partnerships for hospitals. *Bulletin of the World Health Organization*, 84:890–896.

Mills, A. 1985. Economic evaluation of health programmes: Application of the principles in developing countries. *World Health Statistics Quarterly*, 38:368–382.

Mills, A. 1985. Survey and examples of economic evaluation of health programmes in developing countries. *World Health Statistics Quarterly*, 38:402–431.

Minicozzi, A. 2006. Medical savings accounts: What story do the data tell? *Health Affairs*, 25:256–267.

Mitchell, J. B., Haber, S. G., Hoover, S. 2005. Premium subsidy programs: Who enrolls, and how do they fare? Oregon's premium subsidy program and SCHIP serve low-income children equally well, but additional efforts are needed to enroll some populations. *Health Affairs*, 24:1344–1355.

Mooney, G. H., Drummond, M. F. 1982. Essentials of health economics: Part 1 — What is economics? *British Medical Journal*, 285:949–950.

National Center for Health Statistics. 2007. *Health, United States, 2007, with Chartbook on Trends in the Health of Americans*. Hyattsville, MD; http://www.cdc.gov/nchs/data/hus/hus07.pdf

Neumann, P. J., Rosen, A. B., Weinstein, M. C. 2005. Medicare and cost-effectiveness analysis. *New England Journal of Medicine*, 353:1516—1522.

O'Donnell, O., van Doorslaer, E., Wagstaff, A., Lindelow, M. 2007. *Analyzing Health Equity Using Household Survey Data*. Washington, DC: World Bank.

Patrick, D. L., Erickson, P. 1993. *Health Status and Health Policy: Quality of Life in Health Care Evaluation and Resource Allocation*. New York: Oxford University Press.

Pauly, M. V., Zweifel, P., Scheffler, R. M., Preker, A. S., Bassett, M. 2006. Private health insurance in developing countries. *Health Affairs*, 25:369–379.

Posa, P. J., Harrison, D., Vollman, K. M. 2006. Elimination of central line-associated bloodstream infections: Application of the evidence. *American Association of Critical Care Nurses: Advanced Critical Care*, 17:446–454.

Regopoulos, L., Christianson, J. B., Claxton, G., Trude, S. 2006. Consumer-directed health insurance products: Local-market perspectives. *Health Affairs*, 25:766–773.

Robinson, R. 1993. Cost–benefit analysis. *British Medical Journal*, 307:924–926.

Roemer, M. I. 1961. Bed supply and hospital utilization: A natural experiment. *Hospitals*, 1:35–42.

Sackett, D., et al. 1996. Evidence based medicine: What it is and what it isn't. *British Medical Journal*, 312, 7023:71–72, http://www.shef.ac.uk/scharr/ir/def.html.

Sanmartin, C., Berthelot, J. M., Ng, E., Murphy, K., Blackwell, D. L., Gentleman, J. F., Martinez, M. E., Simile, C. M. 2006. Comparing health and health care use in Canada and the United States. *Health Affairs*, 25:1133–1142.

Scanlon, W. J. 2006. The future of Medicare hospital payment. *Health Affairs*, 25:70–80.

Smith, C., Cowan, C., Heffler, S., Catlin, A. 2006. National health spending in 2004: Recent slowdown led by prescription drug spending. *Health Affairs*, 25:186–196.

Smith, S., Freeland, M., Heffler, S., McKusick, D. and the Health Expenditures Projection Team. 1998. The next ten years of health spending: What does the future hold? *Health Affairs*, Jan/Feb:128–140.

Stocker, K., Waitzkin, H., Iriart, C. 1999. The exportation of managed care to Latin America. *New England Journal of Medicine*, 340:1131–1135.

Super, N. 2006. From capitation to fee-for-service in Cincinnati: A physician group responds to a changing marketplace. *Health Affairs*, 25:219–225.

Thorpe, K. E., Florence, C. S., Howard, D. H., Joski, P. 2005. The rising prevalence of treated disease: Effects on private health insurance spending. *Health Affairs*, Jan-Jun;Suppl:W5-317–W5-325.

Vayda, E. 1973. A comparison of surgical rates in Canada and in England and Wales. *New England Journal of Medicine*, 289:1224–1229.

Weinstein, M. C., Stason, W. B. 1977. Foundations of cost-effectiveness analysis for health and medical practices. *New England Journal of Medicine*, 296:716–721.

Wennberg, J. F., Gittelsohn, A. 1973. Small area variations in health care delivery. *Science*, 182:1102–1108.

White, C. C., Koplan, J. P., Orenstein, W. A. 1985. Benefits, risks and costs of immunization for measles, mumps and rubella. *American Journal of Public Health*, 75:739–744.

White, C. 2008. Why did Medicare spending growth slow down? *Inquiry*, 27:793–802.

Wilensky, S., Rosenbaum, S., Hawkins, D., Mizeur, H. 2005. State-funded comprehensive primary medical care service programs for medically underserved populations: 1995 vs. 2000. *American Journal of Public Health*, 95:254–259.

Wolff, J. L., Agree, E. M., Kasper, J. D. 2005. Wheelchairs, walkers, and canes: What does Medicare pay for, and who benefits? *Health Affairs*, 24:1140–1149.

World Health Organization, Regional Office for Europe. 1996. *European Health Care Reforms: Analysis of Current Strategies, Summary*. Copenhagen: WHO.

Wright, B. J., Carlson, M. J., Edlund, T., DeVoe, J., Gallia, C., Smith, J. 2005. The impact of increased cost sharing on Medicaid enrollees. *Health Affairs*, 24:1106–1116.

Chapter 12

Planning and Managing Health Systems

Introduction
Health Policy and Planning as Context
The Elements of Organizations
Scientific Management
Bureaucratic Pyramidal Organizations
Organizations as Energy Systems
  Cybernetics and Management
Target-Oriented Management
  Operations Research
  Management by Objectives
Human Relations Management
  The Hawthorne Effect
  Maslow's Hierarchy of Needs
  Theory *X*–Theory *Y*
Network Organization
Total Quality Management
Changing Human Behavior
Empowerment
Strategic Management of Health Systems
Health System Organization Models
  Functional Model
  Corporate Model
  Matrix Model
Skills for Management
The Chief Executive Officer of Health Organizations
Community Participation
Integration — Lateral and Vertical
Norms and Performance Indicators
Health Promotion and Advocacy
Philanthropy and Volunteerism
New Organizational Models
New Projects and Their Evaluation
Systems Approach and the New Public Health
Summary
Electronic Resources
Recommended Readings
Bibliography

INTRODUCTION

Planning and management are changing in the era of the New Public Health with new advances in prevention and treatment of disease, in population health needs, in new technologies such as genetic engineering, new immunizations that prevent cancers and infectious diseases, prevention of noncommunicable diseases, environmental and nutritional health, and health promotion to reduce risk factors and improve healthful living for the individual and the community. Modern and successful public health also must address social, economic, and community determinants of health and the promotion of public policies and individual behaviors for health and well-being. The social capital or norms that promote cooperation among people are the basis of what is called a *civil society* (i.e., the totality of voluntary, civic, and social organizations and institutions of a functioning society alongside the structures of governmental and commercial institutions) using technologies available in medicine and the environment to promote the health and well-being of a population. This includes security against the effects of threatened terrorism, growing social isolation, and inequities in health. Management in health can learn much from concepts of business management that have evolved to address the economic and human resource aspects of a health system, at the macro level or individual unit of service at the micro level.

The New Public Health is not contained within one organization, but rather reflects the collected efforts of national, state, regional, and local governments, many organizations in the public and nongovernmental sectors, and finally efforts of individual or group advocates and providers and the public itself. The political level is crucial for adequate funding, legislation, and promotion of health-oriented policy positions and in public health management. The responsibility for public health management is shared across all parts of society. This includes individuals, communities, business, and all levels of government.

The New Public Health identifies and addresses community health risks and needs. Planning is critical to the process of keeping a health system sustainable and adaptable and in creating adequate responses to new health threats. Health needs measurement and documentation are vital to design and adapt an effective program and to measure impact. Data on the targeted issues must be available.

Health is a hugely expensive and expansive complex of services, facilities, and programs provided by a wide range of professional and support service personnel making up one of the largest employers of any sector in a developed country. Services are increasingly delivered by organized groups of providers. But all health systems operate in an environment of economic constraints, imposing a need to seek efficiency in the use of resources. How organizations function is of great importance not only for their economic survival, but also, and equally important, for the well-being of the clients and providers of care.

An organization is two or more persons working together to achieve a common goal. Management is the process of defining the goals and making effective use of an organization to attain those goals. Even very small units of a human organization require management. Management of human resources is vital to the success of an organization, whether in a production or service industry. Health systems may vary from a single structure to a network of many organizations. No matter how organizations are financed or operated, they require management.

Management in health care has much to learn from approaches to management in other industries. Elements of theories and practices of profit-oriented sector management can be applied to health services even if they are operated as non-profit enterprises. Physicians, nurses, and other health professionals will very likely be involved in the management of some part of the health care system, whether a hospital department, a managed care system, a clinic, or even a small health care team. At every level, management always means working with people, using resources, providing services, and working toward common objectives.

Health providers require preparation in the theory and practice of management. A management orientation helps to understand the wider implications of clinical decisions and the provider's role in helping the health care system achieve goals and targets. Students and practitioners of public health need preparation in order to recognize that the health care system is more complex than the direct provision of individual services. Similarly, policy and management personnel need to be familiar both with individual and population health needs and related care issues.

HEALTH POLICY AND PLANNING AS CONTEXT

Health has evolved from an individual one-on-one service to complex systems organized within financing arrangements mostly under government auspices. As a governmental priority, health may be influenced by political ideology, sometimes reflecting societal attitudes of the party in power and sometimes apparently at odds with its general social policy. Following Bismarck's 1881 introduction of national health insurance for workers and their families, funded by both workers and employers, most countries in the industrialized world introduced national health plans. Usually, this has been at the initiative of socialist or liberal political leadership, but conservative political parties have preserved national health programs once implemented. Despite the new conservatism since the 1990s with its preeminent ideology of market forces, the growing roles for national, state, and local authorities in health have led to a predominantly government role in financing and overall responsibility for health care, even where there is no universal national health system, as in the United States. The United Kingdom's National Health Service initiated by a Labor government in 1948 has survived through many changes of government including the conservative Margaret Thatcher period in which many national industries and services were privatized.

Health policy is a function of national (government) responsibility overall for health, but implementation is formulated and met at state, local, or institutional levels. The division of responsibilities is not always clear-cut but needs to be addressed and revised both professionally and politically within constitutional, legal, and financial constraints. Selection of the direction to be taken in organizing health services is usually based on a mix of factors, including the political view of the government, public opinion, and rational assessment of needs as indicated through epidemiologic data, cost–benefit analysis, the experience of "good public health practice" from leading countries, and recommendations by expert groups. Lobbying on the part of professional or lay groups for particular interests they wish to promote is part of the process of policy formulation and has an important role in planning and management of health care systems. There are always competing interests for limited resources, some in the health field itself and others outside of the health sector.

The political level is vitally involved in health management in establishing and maintaining national health systems, and in determining the place of health care as a percentage of total governmental budgetary expenditures, in allocating funds among the competing priorities. These competing priorities for government expenditures include defense, roads, education, and many others, as well as those within the health sector itself. A political commitment to health must be accompanied by allocation of resources adequate to the scope of the task. Thus, health policy is partially determined by society and is not a prerogative of government, health care providers, or any institution alone.

As a result of long struggles by trade unions, advocacy groups, and political action, well-developed market economies have come to accept health as a national obligation and essential to an economically successful and well-ordered society. This has led to the implementation of universal access systems in most of the industrialized countries. Once initiated, national health systems require high levels of resources, because the health system is labor-intensive with relatively high salaries for health care professionals. In these countries, health expenditures consume between 7 and 15 percent of GNP.

Some industrialized countries, notably those in the former Soviet bloc, lacking mechanisms for advocacy, including consumer and professional opinion, tended to view health with a political objective of social benefits, and also as a "nonproductive" consumer of resources rather than a producer of new wealth. As a result, budget allocations and total expenditures for health as a percentage of GNP were well below those of other industrialized countries. Salaries for health personnel were low compared to industrial workers in the "productive" sectors. Furthermore, industrial policy did not promote modern health-related industries, as compared to the military or heavy industrial sectors. Typically, such countries allocate between 3 and 5 percent of the GNP to health, in part because of the lingering idea of health as a nonproductive investment. The former socialist countries have been increasing expenditures on health since 2002. The developing countries generally spend under 4 percent of GNP because health is addressed as a relatively low political priority.

Financing of health care and resource allocation require a balance among primary, secondary, and tertiary care. Economic assessment, monitoring, and evaluation are part of determining the health needs of the population. Regulatory agencies are responsible for defining goals, priorities, and objectives for resulting services. Targets and methods of achieving them help provide the basis for implementation and evaluation strategies. Planning requires written plans that include a statement of vision, mission objectives, target strategies, methods, and coordination during the implementation. Designation and evaluation of responsibilities, resources to be committed, participants and partners in the process are part of the continuous process of management.

The dangers of taking a "wrong" direction may be severe, not only in terms of financial costs, but also in terms of high levels of preventable morbidity and mortality. Health policy is often as imprecise a science as medicine itself. The difference is that inappropriate policy can affect the lives and well-being of very large numbers of people, as opposed to an individual being harmed by the mistake of one doctor. There may be no "correct" answer, and there are numerous controversies along the path. Health policy remains more an "art" than the more quantitative and seemingly precise field of health economics.

Societal, economic, and cultural factors as well as personal habits have long been accepted as having an important impact on vulnerability to coronary heart disease. But other factors such as the degree of control over one's life, as suggested in studies of British civil servants, religiosity, and the effects of migration on families left behind are part of the social gradients and inequalities seen in many disease entities with consequent excess morbidity and mortality in some contexts, such as in Russia and the Ukraine.

Health policy, planning, and management are interrelated and interdependent. Any set goal should be accompanied by planning how to attain it. A policy should state the values on which it is based, as well as specify sources of funding, planning, and management arrangements for its implementation. Examination of the costs and benefits of alternative forms of health care helps in making decisions as to the structure and the content of health care services. This may involve both internal structures (within one organization) and external linkages (intersectoral cooperation with other organizations). The methods chosen to attain the goals become the applied health policy.

The World Health Organization's Health for All (HFA) strategy (1977) was directed to the political level and intended to increase governmental awareness of health as a key component of overall development. To some degree it succeeded despite its expansive aspirations, and even after 30 years, its objectives remain worthwhile even in well-developed health systems. Within health, primary care was stressed as the most effective investment to improve the health status of the population. In 1993, the World Bank's *World Development Report* adopted the Health for All strategy and promoted the view that health is an important investment sector for general economic and social development. However, economic policies promoting privatization and deregulation in the health sector threaten to undermine this larger goal in countries with national health systems, and delay achievement of universal access in the United States, where some 47 million Americans in 2006 were uninsured for health care (16 percent of the population), an increase from 45 million in 2004 (15 percent) (see Chapter 13).

In the New Public Health, health promotion, preventive, and clinical care are all part of public health because the well-being of the individual and the community requires a coordinated effort from all elements of the health spectrum. Establishing and achieving national health goals requires planning, management, and coordination at all levels. The achievement of health advances depends on organizations and structured efforts to reach health goals such as those defined above and more recently by the United Nations in the Millennium Development Goals (see Chapter 2). This requires some understanding of organizations and how they work.

THE ELEMENTS OF ORGANIZATIONS

The study of organizations developed within sociology, but has gradually become a multidisciplinary activity involving many other professional fields, such as economics, anthropology, individual and group psychology, political science, human resources management, and engineering. Organizations, whether in the public or private sector, exist within an external environment, and utilize their own structure, participants, and technology to achieve goals. In order for an organization to survive and thrive, it must adapt to the physical, social, cultural, and economic environment.

Organizations participating in health care establish the connection between service providers and consumers, with the goal of better health for the individual and the community. The technology for this includes the legislation, regulation, professionalism, instrumentation, medications, vaccines, education, and other modalities of intervention for prevention and treatment. The social structure of an organization may be "formal" (structured stability), "natural" (groupings reflecting common interests), or "open" (loosely coupled, interacting, and self-adjusting systems to achieve goals).

Formal systems are deliberately structured for the purposes of the organization. Natural systems are less formal structures where participants work together collaboratively to achieve common goals defined by the organization. Open systems relate elements of the organization to coalitions of partners in the external environment to achieve mutually desirable goals. In the health system, structures should focus on prevention and treatment of disease and improvement in health and well-being of society. The social structure of an organization includes values, norms, and roles governing the behavior of its participants.

Government, business, or service organizations, including health systems, require organizational structures, with a defined mission and set of values, in order to function. An organizational structure needs to be tailored to the size and complexity of the entity and the goals it wishes to achieve. The structure of an organization is the way in which it divides its labor into distinct tasks and coordinates them. The major organizational models, which are not mutually exclusive, and may indeed be complementary, are the pyramidal (bureaucratic) and network structures. The bureaucratic model is based on a hierarchical chain of command with clearly defined roles. In contrast, the matrix or network organization brings together professional or technical persons to work on specific programs, projects, or tasks. Both are vital to most organizations to meet ongoing responsibilities and to address special challenges.

SCIENTIFIC MANAGEMENT

Some classic organization theory concepts help to set the base for modern management ideas as applied to the health sector. Scientific management was pioneered by Frederick Winslow Taylor (1856–1915). His work was pragmatic and based on empirical engineering, developed in observational studies carried out for the purpose of increasing worker, and therefore system, efficiency. Taylor's industrial engineering studies of *scientific management* were based on the concept that the best way to improve worker productivity was by designing improved techniques or methods used by workers. This theory viewed workers as instruments to be manipulated by management and assumed that efficient, rationally planned methods would produce better industrial results and industrial peace as the tasks of managers and workers would be better defined.

Time and motion studies analyzed work tasks to seek more efficient methods of work in factories. Motivation of workers was seen to be related to payment by piecework and economic self-interest to maximize productivity. Taylor sought to improve the productivity of each worker and to make management more efficient in order to increase earnings of employers and workers. He found that the worker was more efficient and productive if the worker was goal-oriented rather than task-oriented. This approach dominated organization theory during the early decades of the twentieth century.

Resistance to Taylor's ideas came from both management and labor; the former because it seemed to interfere with managerial prerogatives and the latter because it expected the worker to function at top efficiency at all times. However, Taylor's work had a lasting influence on the theory of work and organizations.

BUREAUCRATIC PYRAMIDAL ORGANIZATIONS

The traditional pyramidal bureaucratic organization, existing since the dawn of human society, was analyzed by sociologist Max Weber between 1904 and 1924. This form of organization is classically seen in the military and civil services, but also in large-scale industry, where discipline, obedience, and loyalty to the organization are demanded, and individuality minimized. Leadership is assigned by higher authority, and is presumed to have higher knowledge than members lower down in the organization. This form of organization is effective when the external and internal environments, the technology, and functions are relatively well defined, routine, and stable.

The pyramidal system (Figure 12.1) has an apex of policy and executive functions, a middle level of management personnel and support staff, and a base of the people who produce the output of the organization. The flow of information is generally one way, from the bottom to the top level, where decisions are made for the detailed performance of duties at all levels. Lateralizing the information systems so that essential data can be shared to help

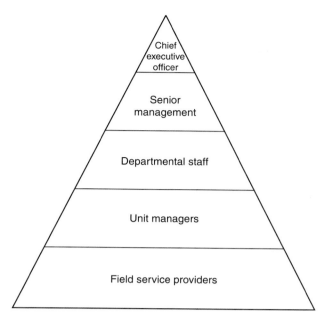

FIGURE 12.1 Pyramid structure of organizations.

staff at the middle and field or factory floor levels of management is generally discouraged because this may promote decentralized rather than centralized management. Even these types of organizations have increasingly come to emphasize small-group loyalty, leadership initiative, and self-reliance.

The bureaucratic organization has the following characteristics:

1. There is a fixed division of labor with clear jurisdiction and based on assignments, which are subject to change by the leader;
2. There is a hierarchy of offices, with each lower functionary controlled and supervised by a higher one;
3. A documented, stable set of rules governs decisions and actions;
4. Property and rights belong to the office, not the person in the office;
5. Officials are selected on the basis of qualifications; salaries and benefits are based on technical competencies;

6. Employment is viewed as a tenured career for officials, after an initial trial period.

The bureaucratic system, based on formal rationality, structure, and discipline, is widely used in production, service, and governmental agencies, including military and civilian departments and agencies.

ORGANIZATIONS AS ENERGY SYSTEMS

Health systems, like other organizations, are dynamic and require continuous management, adjustment, and systems control. Continuous monitoring and feedback, evaluation, and revision help to meet individual and community needs. The input–process–output model (Figure 12.2) depends on feedback systems to make the administrative or educational changes needed to keep moving toward the selected objectives and targets.

Organizations use resources or inputs that are processed to achieve desired results or outputs. The resource inputs are money, personnel, information, and supplies. Process is the accumulation of all activities taken to achieve the results intended. Output, or outcome, is the product, its marketing, its reputation and quality, and profit. In a service sector such as health, output or impact can be measured in terms of reduced morbidity and/or mortality, improved health, or number of successfully treated and satisfied patients at affordable costs. The management system provides the resources and organizes the process by which it hopes to achieve the established goals.

Program implementation requires systematic feedback for the process to work effectively. When targets are set and strategy defined, resources, whether new or existing, are placed at the service of the new program. Management is then responsible to use the resources to best achieve the intended targets. The results are the outcome or output measures which are evaluated and fed back to the input and process levels.

Health systems consist of many subsystems, each with an organization, leaders, goals, targets, and internal information systems. Subsystems need to communicate within themselves, with peer organizations, and with the macro (health) system. Leadership style is central to this process. The surgeon as the leader of the team in the operating

FIGURE 12.2 Organizations as energy systems.

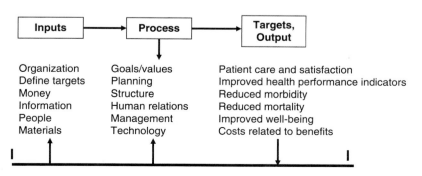

room is dependent on the support and judgment of other crucial persons on the team, such as the anesthesiologist, the operating room nurse, pathologist, radiologist, and laboratory services, all of whom lead their own teams. The hospital director cannot function without a high degree of decentralized responsibility and a creative team approach to quality development of the facility.

Health systems management includes analysis of service policy, budget, decision making in policy, as well as operation, regulation, supervision, provision, maintenance, ethical standards, and legislation. Policy formulation involves a set of decisions made in pursuit of a course of action for achieving selected health targets, such as those in the Millennium Development Goals (MDGs) or continuing to update *Healthy People 2010* health targets in the United States (see Chapter 2).

Cybernetics and Management

Cybernetics, a term coined by Norbert Wiener, refers to systems or organizations which are dependent on each other to function, and whose interdependence requires flexibility of response. Cybernetics gained wide credence in engineering in the early 1950s, and feedback systems became part of standard practice of all modern management systems. Its later transformations appeared in operating service systems, as information for management. Application of this concept is entering the health sector. Rapid advances in computer technology, by which personal computers have access to Internet systems and large amounts of data, have already enhanced this process. In mechanistic systems, the behavior of each unit or part is constrained and limited; in organic systems, there is more interaction between parts of the system. The example used in Figure 12.3 is the use of a thermostat to control the temperature and function of a heater according to conditions in the room. This is also described as a feedback system.

Cybernetics opens up new vistas on the use of health information for managing the operation of health systems.

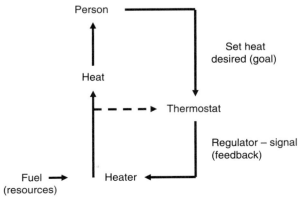

FIGURE 12.3 Cybernetic feedback control organization.

A database for each health district would allow assessment of current epidemiologic patterns, with appropriate comparisons to neighboring districts or regional, state, and national patterns.

Data would need to be processed at state or national levels in comparable forms for a broad range of health status indicators. Furthermore, the data should be prepared for on-line availability to local districts in the form of current health profiles. Thus, data can be aggregated and disaggregated to meet the management needs of the service, and may be used to generate real targets and measure progress toward meeting them. A geographic information system may demonstrate high rates of a disease in a region due to local population risk factors, and thus become the basis for an intervention program.

In the health field, the development of reporting systems based on specific diseases or categories has been handicapped by lack of integrative systems and a geographic reporting approach. The technology of computers and the Internet should be used to process data systems in real time and in a more user-friendly manner. This will enable local health authorities and providers to respond to actual health problems of the communities.

Health is a knowledge-based service industry, so that knowledge management and information technology are extremely important parts of the New Public Health, not only in patient care systems, but in mobilization of evidence and experience in best practices for policies and management decision making. The gap between information and action is wide and presents an ethical as well as a political challenge. Regions with the most severe health problems lack trained personnel in assessment and exploitation of current state of the art practices and technology in many practical public health fields including immunization policy and in management of risk factors for stroke.

Knowledge and evidence exist, but the capacity to access and interpret that information is commonly lacking in many countries so that very large numbers of people die of preventable diseases even when there are, in total, sufficient resources to address the challenges. International guidelines are vital to help countries adopt current standards and make use of the available knowledge for public policy. This often comes down to lack of political support or openness to international norms. As examples, there is no common recommended immunization program yet for the European Union, so badly needed for the new access and potential new members, and stroke mortality is more than two to three times higher in CIS countries than in countries of Eastern Europe. Systems management requires access and use of knowledge to bridge this gap.

Adaptation of knowledge to address local problems is essential in a globalized world, if only to prevent international spread of threatened pandemics or, more simply, spread of unhealthful lifestyles to countries in mid-development of a middle class along with a massive poor population. Application of knowledge and experience successful in leading countries can help foster innovation and create experience that may help to generate a local renewal process. Knowledge management relies on collaboration to apply current potential to the severe health problems many countries face.

TARGET-ORIENTED MANAGEMENT

The management of resources to achieve productivity and measurable success has been characterized and accompanied by development of systems of organizing people to create solutions to problems or to innovate toward defined objectives.

Operations Research

Operations research is a concept developed by British scientists and military personnel in search of solutions for specific problems of warfare during World Wars I and II. The approach was based on the development of multidisciplinary teams of scientists and personnel. The development of the Anti-Submarine Detection Investigation Committee (ASDIC) for underwater detection of submarines during World War I characterized and pioneered this form of research. The common story of the camel being a horse designed by a committee is amusing, but such organization works. The famous Manhattan Project whereby the United States assembled a powerful research and development team which produced the atomic bomb during World War II is a prime example.

Team- and goal-oriented work was very effective in problem solving under the enormous pressure of wartime needs. It also influenced post-war approaches to developmental needs in terms of applied science in such areas as the aerospace and computer industries. The computer hardware and software industries are characterized by innovation conceived by informal working groups with a high level of individual competence, peer group dynamism, and commitment to problem solving. This is how the "nerds" of Macintosh and Microsoft beat the "suits" of IBM in innovation and introduction of the personal computer. Similar startup groups, such as Google, successfully exploited the Internet and much of the product of California's Silicon Valley as well as its imitators in many parts of the country and worldwide.

In the health field, innovation in organization developed prepaid group practice which became the health maintenance organization (HMO), and later the managed care organization, now a major, if controversial, factor in health care provision in the United States. Other examples may be found in multidisciplinary research teams working on vaccines or pharmaceutical research, and in the increasingly multidisciplinary function of hospital departments and especially highly interdependent intensive care or home care teams.

Management by Objectives

The business concept of management by objective (MBO), pioneered in the 1960s, has become a common theme in health management. MBO is a process whereby managers of an enterprise jointly identify its goals, define each individual's areas of responsibility in terms of the results expected, and use these measures as guides for operating the unit and assessing the contributions of its members.

The common goals and then the individual unit goals must be established, as well as the organizational structure molded to help achieve these goals. The goals may be established in terms of outcome variables, such as defined targets for reduction of infant or maternal mortality rates. Goals may also be set in terms of intervening or process variables, such as achieving 95 percent immunization coverage, prenatal care attendance, or screening for breast cancer and mammography. Achievements are measured in terms of relevancy, efficiency, impact, and effectiveness.

The MBO approach has been subject to criticism in the field of business management because of its stress on mechanical application of quantitative outcome measures and because it ignores the issue of quality. This approach had great influence on the adoption of the objective of Health for All by the World Health Organization, and on the U.S. Department of Health and Human Services' 1979 health targets for the year 2000 (*Healthy People 2000*), now continuing for 2010. Targeting diseases for eradication may contribute to institution building by developing experience and technical competence to broaden the organizational capacity. On the other hand, categorical programs or target-oriented programs can detract from the development of more comprehensive systems approaches. An example of this is a perception that the issues of polio control with reliance on national immunization days may have distracted planning and resource allocation for the buildup of the essential public health infrastructure, such as the basic immunization system so fundamental to child health. A suitable balance between comprehensive and categorical approaches requires very skilled management. The Millennium Development Goals (MDGs) agreed to by the United Nations as targets for the year 2015 provide a set of measurable objectives and a formula for international aid and for national development planning to help the poorest nations, with the wealthy nations providing aid, education, debt relief, and economic development through fairer trade practices.

HUMAN RELATIONS MANAGEMENT

Management is the activity of coordinating and integrating organizational resources, including people, money, materials, time, and space. The purpose is to achieve defined/stated objectives as effectively and efficiently as possible. Whether in terms of producing goods and profits or in delivering services effectively, management deals with human motivation and behavior because workers are the key to achieving goals. Knowledge and motivation of the individual client and the community are also essential for achieving good health. Thus management must take into account the knowledge, attitudes, beliefs, and practices of the consumer as much as or more than those of the people working within the system, as well as the general cultural and knowledge level in the society, as reflected in the media, political opinions, and organizations addressing the issues.

Management, like medicine, is both a science and an art. The application of scientific knowledge and technology in medicine involves both theory and practice. Similarly, management practice involves elements of organizational theory, which, in turn, draws on the behavioral and social sciences and quantitative methodologies. Sociology, psychology, anthropology, political science, history, and ethics contribute to the understanding of psychosocial systems, motivation, status, group dynamics, influence, power, authority, and leadership. Quantitative methods including statistics, epidemiology, survey methods, and economic theory are also basic to development of systems concepts. Comparative institutional analysis helps to develop principles of organization and management, while philosophy, ethics, and law are part of understanding individual and group value systems.

Organizational theory, a relatively new discipline in health, as an academic study of organizations, addresses health-related issues using the methods of economics, sociology, political science, anthropology, and psychology. The application of organizational theory in health care has evolved and become an integral part of training for, and the practice of, health administration. Related practical disciplines include human resources (HR) and industrial and organizational psychology. Translation of organizational theory into management practice requires knowledge, planning, organization, assembly of resources, motivation, and control. Health organizations have become more complex and costly over time, especially in their mix of specializations in science, technology, and professional services. Organization and management are particularly crucial for successful application of the principles of the New Public Health, as it involves integration of traditionally separate health services. Delegation of responsibilities in health systems, such as in intensive care units, is fundamental to success in patient care with nurses taking increasing responsibility for management of the severely ill patient with multiple system failure.

The Hawthorne Effect

Elton Mayo of the Harvard School of Business carried out a series of observational studies at the Hawthorne, Illinois, plant of the Western Electric Company between 1927 and 1932. Mayo and his industrial engineer, along with psychologist colleagues, made a major contribution to the development of management theory. Mayo began with industrial engineering studies of the effect of increased lighting on production at an assembly line. This was followed by other improvements in working conditions, including reduced length of the working day, longer rest periods, better illumination, color schemes, background music, and other factors in the physical environment. These studies showed that production increased with each of these changes and improvements. However, the researchers discovered, to their surprise, that production continued to increase when the improvements were withdrawn. Further, in a control group where conditions remained the same, productivity also grew during the study period. This led Mayo to the conclusion that worker performance improved because of a sense that management was interested in them.

Traditionally, industrial management viewed employees as mechanistic components of a production system. Previous theory was that productivity was a function of working conditions and monetary incentives. What came to be known as the Hawthorne effect showed the importance of social and psychological factors on productivity. Formal and informal social organizations among management and employees were recognized as key elements in productivity, now called *industrial humanism*. Research methods adapted from the behavioral sciences contributed to scientific studies in industrial management. Traditional theories of the bureaucratic model of organization and management were modified by the behavioral sciences. This led to the emergence of the systems approach, or scientific analysis to analyze complex structures or organizations, taking into account the mutually interdependent elements of activities, interactions, and interpersonal relationships between management and workers.

The Hawthorne effect in management is in some ways comparable to the placebo effect in clinical research and health care practice. It is also applied to clinical practice whereby medical care provided by doctors is measured for specific "tracer conditions" to assess completeness of care according to current clinical guidelines. Review of clinical records has been shown to be a factor in improving performance by doctors in practice, such as in treatment of acute myocardial infarction, management of hypertension, or in completeness of carrying out preventive procedures such as screening for cancer of cervix, breast, or colon (see Chapters 3 and 15). The fact of being studied is a factor in improved performance.

Maslow's Hierarchy of Needs

Abraham Maslow's hierarchy of human needs was an important contribution to management theory. Maslow

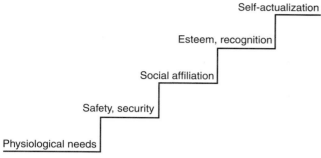

FIGURE 12.4 Maslow's hierarchy of needs.

(1908–1970) was an American psychologist, considered "the father of humanism" in psychology. Maslow defined a prioritization of human needs (Figure 12.4), starting with those of basic physical survival, and at higher levels, human needs include social affiliation, self-esteem, and self-fulfillment. Others in the hierarchy include socialization and self-realization; later revisions include cognitive needs.

This concept is important in terms of management because it identifies human needs beyond those of physical and economic well-being. It relates them to the social context of the work environment with needs of recognition, satisfaction, self-esteem, and self-fulfillment. It opened many positive areas of management research, not only in the motivation of workers in production and service industries, but also in the motivation of consumers.

Maslow's hierarchy of human needs which contributed to the idea that workers' sense of well-being is important to management. Maslow's theory played an important role in application of sociologic theory to client behavior, just as the topic of personal lifestyle in health became a central part of public health and clinical management of many conditions, such as in risk factor reduction for cardiovascular diseases. This concept fits well with the epidemiologic studies referred to in the introduction such as those showing strong relationships with sociopolitical factors as well as socioeconomic conditions.

Theory *X*–Theory *Y*

Theory *X*–Theory *Y* (Table 12.1), developed by clinical psychologist and professor of management Douglas McGregor in the 1960s, examined two extremes in management assumptions about human nature that ultimately affect operations of organizations. Organizations with centralized decision making, a hierarchical pyramid, and external control are based on certain concepts of human nature and motivation. McGregor's theory, drawing on Maslow's hierarchy of needs, describes an alternative set of assumptions that credit most people with the capacity for self-direction.

Management by direction and control (Theory *X*) assumes workers are lazy, unambitious, uncreative, and motivated only by basic physiological needs or fear. The Theory *Y* model provides a more optimistic view of human potential, assuming that, if properly motivated, people can be self-directed and creative at work, and that the role of management is to unleash this potential in workers. Many other theories of motivation and management have been developed to explain human behavior and how to utilize inherent skills to produce a more creative work environment, reduce resistance to change, reduce unnecessary disputes, and ultimately create a more effective organization.

Variants of the human motivation approach in management carried the concept further by examining industrial organization to determine the effects of management practices on individual behavior and personal growth within the work environment. Theory *X* assumes that management produces immature responses on the part of the worker: passivity, dependence, erratically shallow interests, short-term perspective, subordination, and lack of self-awareness. In contrast, the other end of the immaturity–maturity spectrum was the mature worker, with an active approach, independent mind, capable of a broad range of responses, deeper and stronger interests, a long-term perspective, and a high level of awareness and self-control. This model has been tested in a variety of industrial settings, showing that giving workers the opportunity to grow and mature on the job helps satisfy more than

TABLE 12.1 Theory *X*–Theory *Y*

Theory *X*	Theory *Y*
1. Work is inherently distasteful to people.	1. Work is as natural as play in favorable conditions.
2. Most people are not ambitious, have little desire for responsibility, and prefer to be directed.	2. Self-control is indispensable in achieving goals.
3. Most people have little capacity for creatively solving organizational problems.	3. The capacity for creativity in solving organizational problems is widely distributed in the population.
4. Motivation occurs only at physiological and safety levels.	4. Motivation exists at the social, esteem, and self-actualization levels, as well as the physiological and security levels.
5. Most people must be closely controlled and often coerced to achieve organizational objectives.	5. People can be self-directed and creative at work if properly encouraged.

basic physiological needs and allows them to use more of their potential in accomplishing organizational goals.

In *The Motivation to Work* (1959), U.S. clinical psychologist Frederick Herzberg wrote of his motivation–hygiene theory. He developed this theory after extensive studies of engineers and accountants, examining what he called *hygiene factors* (i.e., administrative, supervisory, monetary, security, and status issues in work settings). His motivating factors included achievement, recognition of accomplishment, challenging work, and increased responsibility with personal and collective growth and development. He proved that the motivating factors had a great positive effect on job satisfaction.

These human resource theories of management helped to change industrial approaches to motivation from "job enrichment" to a more fundamental and deliberate upgrading of responsibility, scope, and challenge of work, by letting workers develop their own ways of achieving objectives. Even when the theories were applied to apparently unskilled work, such as plant janitors, the workers were changed from an apathetic, poorly performing group into a cohesive, productive team, taking pride in their work and appearance. This approach gave members of the team the opportunity to meet their human self-actualization needs by taking greater responsibility for problem solving, and it resulted in less absenteeism, higher morale, and greater productivity with improved quality.

Rensis Likert, who along with McDougal and Herzberg helped to pioneer the "Human Relations School" in the 1960s, applied human resource theory to management systems and styles. He classified his theory into four different systems as follows:

System 1. Management has no confidence or trust in subordinates, and avoids involving them in decisions and goal setting, which are made from the top down. Management is task-oriented, highly structured, and authoritarian. Fear, punishment, threats, and occasional rewards are the principal methods of motivation. Worker–management interaction is based on fear and mistrust. Informal organizations within the system often develop that lead to passive resistance of management and are destructive to the goals of the formal organization.

System 2. Management has a condescending relationship with subordinates, with some degree of trust and confidence. Most decisions are centralized, but some decentralization is permitted. Rewards and punishments are used for motivation. Informal organizations become more important in the overall structure.

System 3. Management places a greater degree of trust and confidence in subordinates, who are given a greater degree of decision-making powers. Broad policy remains a centralized function.

System 4. Management is seen as having complete confidence in subordinates. Decision making is dispersed, and communication flows upward, downward, and laterally. Economic rewards are associated with achieving goals and improving methods. Relationships between management and subordinates are frequent and friendly, with a sense of teamwork and a high degree of mutual respect.

Case studies showed that a shift from management system 1 toward system 4 radically changed the performance of production, cut manufacturing costs, reduced staff turnover, and increased staff morale. Furthermore, workers and managers both shared a concern for the quality of the product or service and the competitiveness and success of their business. The health industry includes highly trained professionals and paraprofessional workers who function as a team with a high degree of cohesion, mutual dependence, and autonomy, such as a surgical or emergency department team.

NETWORK ORGANIZATION

The network, or task-oriented working group, is basically an organization of relationships rather than authority, sometimes called an *adhocracy*. This is a more organic form of organization, best suited to be effective for adaptation when the environment is complex and dynamic, when the workforce is largely professional, and when the technology and system functions change rapidly. Complexities and technological change require information, expertise, flexibility, and innovation, strengths best promoted in free exchange of ideas in a mutually stimulating environment.

In a network organization, leadership may be formal or informal, assigned to a particular function, which may be temporary, medium-term, or permanent, to achieve a single defined task or develop an intersectoral program. The task force is usually for short-term specific assignment; a working group, often for a medium-term project, such as integrating services of a region; and a committee for permanent tasks such as monitoring an immunization program.

Significant advantages of this form of organization are the challenge and the sharing of information and responsibility, which give professionals challenges and job satisfaction by providing the opportunity to demonstrate their creativity. Members of the task force may each report within their own pyramidal structure, but as a group they work to achieve the assigned objective. They may also be interdisciplinary or interagency working groups to review the "state of the art" in this particular issue as documented in reports, professional literature, and to coordinate activities, review previous work, or plan common future activities.

An ongoing network organization may be a government cabinet committee to coordinate government policy and the work of the various government departments, or a joint chiefs of staff to coordinate the various armed services. This approach is commonly used for task groups where interdisciplinary teams of professionals meet to coordinate functions of a department in a hospital, or where a multidisciplinary group of experts is established with the specified task of a technical nature.

Network organizational activity is part of the regular functions of a health professional. Informal networking is a day-to-day activity of a physician in consultations with colleagues and also a part of more formalized network groups. The hospital department must to a large extent function as a network organization with different professionals working as a team more effectively than in a strictly authoritarian pyramidal model. A ministry of health may need to develop a joint working group with the ministry of transport, the police, and those responsible for standards of motor vehicles in order to seek ways to reduce road accident deaths and injuries. If a measles eradication project is envisioned, a multidisciplinary and multi-organizational team, or a network, should be established to plan and carry out the complex of tasks needed to achieve the target (Figure 12.5).

In a health service context, a task group to determine how to eradicate measles locally might be chaired by the deputy chief medical officer. Members might include the chief district nurse, an administrative and budget officer, pharmacist, the chief of the pediatric department of the district hospital, a primary school administrator, a health educator, a medical association representative, the director of laboratories, the director of the supply department, a representative of the department of education, a representative of a voluntary organization interested in the topic, and others as appropriate.

Most organizational structures are mixed, combining elements of both the pyramidal and the network structures. It is often difficult for a rigid pyramidal structure to deal with parallel bodies in a structured way, so that the network approach is necessary to establish working relations with outside bodies to achieve common goals. A network is a democratic functional grouping of those professionals and organizations needed to achieve a

Chairperson/Facilitator/Coordinator

A	B	C	D
E	F	G	H
I	J	K	L

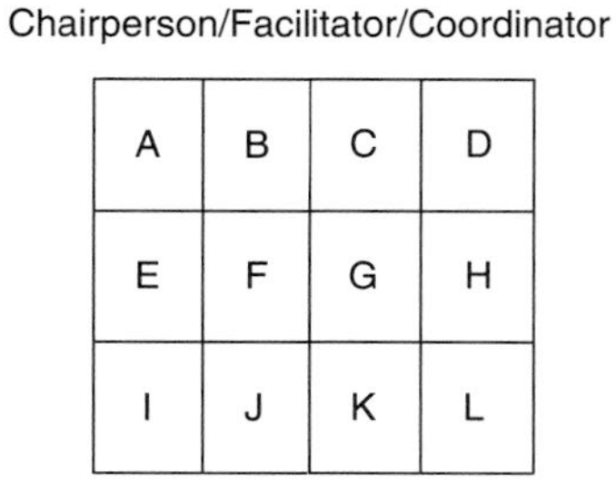

FIGURE 12.5 Network organization structure. Note: The letters represent the participation by individuals or organizations as a task group to achieve a defined goal, as set out in terms of reference and a time frame.

defined target, sometimes involving persons from many different organizations. The application of this concept is increasingly central in health care organization as multilevel health systems evolve in the form of managed care or district health systems. These are vertically integrated management systems involving highly professional teams and units whose interdependence for patient care and financial responsibility are central elements of the New Public Health.

TOTAL QUALITY MANAGEMENT

In the United States during World War II, W. Edwards Deming, a physicist and statistician, developed a system of economic and statistical methods of quality control in production industries. Following the war, Deming was invited to teach in Japan and moved from the university to the level of industrial management. Japanese industrialists adopted his principles of management and introduced quality management into all industries, with astonishingly successful results within a decade. The concept, later called *total quality management* (TQM), has since been adopted widely in production and service industries.

In the Deming approach to management of companies, quality is the top priority and is the key responsibility of management, not of the workers. If management sets the tone and involves the workers, quality goes up, costs come down, and both customer satisfaction and loyalty increase. This means enhancing the pride of the worker, listening to his or her ideas, and avoiding a punitive inspection approach. Removal of fear and building mutual participation and common interest is the responsibility of leadership. Training is one of the most important investments of the company. The differences between traditional management and the TQM approach are shown in Boxes 12.1 and 12.2. In societies with growing economies, the role of an educated workforce becomes greater as information technology and services, such as health, become larger parts of the economy and require professionalism and self-motivating workers.

The TQM approach integrates the scientific management and human relations approaches by giving workers credit for intellectual capacity and expects them to use it to analyze and improve the tasks they perform. Even more, this approach expects workers at all levels to contribute to better quality in the process of design, manufacture, and even marketing of the product or the service.

The TQM ideas were revolutionary and successful when applied in business management in production industries. The TQM concept is much in discussion in the service industries. The World Health Organization has adapted TQM to a model called Continuous Quality

---

Box 12.1 Standard Management Theory

1. Quality is expensive;
2. Inspection is the key to quality, and control experts and inspectors can assure this;
3. Systems are designed by outside experts — no input is needed from workers;
4. Work standards, quotas, and targets can help productivity;
5. People may be hired when needed and laid off when not needed;
6. Rewards and punishments will lead to greater productivity and creativity;
7. Buy at the lowest cost;
8. Change suppliers frequently, based on price alone;
9. Profits are based on keeping costs down and revenue high;
10. Profit is the most important indicator of a company.

Box 12.2 Total Quality Management

1. Quality leads to lower costs;
2. Inspection is too late — worker involvement in quality services eliminates defects;
3. Quality is made by management;
4. Most defects are caused by the system, not the worker;
5. Eliminate all work standards and quotas in industry;
6. Fear leads to disaster;
7. Make workers feel secure in their jobs;
8. Judgment, punishment, reward for above- or below-average performance destroys teamwork essential for quality production;
9. Work with suppliers to improve quality and costs;
10. Profits are generated by loyal customers — running a company for profit alone is like driving a car by looking in the rearview mirror.

Improvement (CQI), with the stress on mutual responsibilities throughout a health system for quality of care. The application of TQM and CQI approaches is discussed in Chapter 15, including the external regulatory and self-development TQM approaches.

CHANGING HUMAN BEHAVIOR

Human behavior is individual but takes place in a social context. Changing individual behavior is needed to reduce risk factors for many diseases. Change can be threatening; it requires alteration, substitution, transformation, or modification of purposes, procedures, methods, or style.

Implementation of plans usually requires some change, which often meets resistance. The resistance to change may be professional, technical, psychological, political, emotional, or a mix of all of these. The manager of a health facility or service has to cope with change and gather the support of those involved to participate in creating or implementing the change effectively.

The behavior of the worker in a production or service industry is vital to the success of the organization. Equally important is the behavior of the purchaser or consumer of the product or service. Diagnosing organizational problems is an important skill to bring to leadership in health systems. Even more important is the ability to identify and alter the variables that require change and adaptation to improve performance of the organization. High expectations are essential to produce high performance and improved standards of service or productivity. Conversely, low expectations not only lead to low performance, but produce a downward spiraling effect. This applies not only within the organization, but to the individuals and community served, whether in terms of purchase of goods produced or in terms of health-related behavior.

People often resist change because of fear of the unknown. Participation in the process of defining problems, formulating objectives, and identifying alternatives is needed to bring about changes. Change in organizational performance is complex, and this is the test of leadership. Similarly, change at the individual level is essential to achieve the goals of the group, whether this is in terms of the functioning of a health care service unit, such as a hospital, or whether it is an individual's decision to change from smoking to nonsmoking status. The health of an individual and a population both depend on the individual health team member's motivation and experience.

The behavior of the individual is important to his own and community health. Even small steps in the direction of a desirable change in behavior should be rewarded as soon as possible (i.e., reinforcing positive performance in increments). Behavior modification is based on the concept that change of behavior starts with the feelings and attitudes within the individual, but can be influenced by knowledge, peer pressure, and legislative standards. Change involves a number of elements to define "where you are at":

1. Knowledge: what is the level of adequate health information?
2. Attitudes: what is the person's perception of that information?
3. Behavior of the individual: what does the individual actually do?
4. Behavior of the group: what are the social norms and acts?
5. Behavior of the organization: what does the health system do to change these factors?

Change in behavior is vital in the health field. The health belief model (Chapter 2) is widely influential in psychology and health promotion. The belief-intervention approach involves programs meant to reduce risk factors or a public health problem. It may require change in organizational behavior, with involvement and feedback to the people who determine policy, those who manage services, and the community being served.

EMPOWERMENT

In the 1980s, major industries in the United States were unable to compete successfully with the Japanese in the consumer electronics and automobile industries. Management theory began to place greater emphasis on empowerment as a management tool. The TQM approach stresses teamwork and involvement of the worker in order to achieve better quality of production. Comparatively, empowerment went further to involve the worker in operation, quality assessment, and even planning of the design and production process. Results in production industries were remarkable, with increased efficiency, less absenteeism, and greater searching for ideas to improve quality and quantity of production, with the worker as a participant in the management and production process.

The concept of empowerment entered the service industries with the same rationale. The rationale is that improvement in quality and effectiveness of service require the active physical and emotional participation of the worker. Participation in decision making is the key to empowerment. This requires management to adopt new methods that allow the worker, whether professional or manual, to be an active participant. Successful application of the empowerment principles in health care extends to the patient, the family, and the community, emphasizing patients' rights to informed participation in decisions affecting their medical care, and the protection of privacy and dignity.

Diffusion of powers occurs when management of services is decentralized. Delegation of powers to professional, NGO, and advocacy organizations is part of empowerment in health care organizations. Governmental powers to govern or promote areas such as licensure, accreditation, training, research, and service can be devolved to local authorities or NGOs by delegation of authority or transfer of funds. Organizational change may involve decentralization. Institutional changes such as amalgamation of hospitals, long-term care facilities, home care programs, day surgery, ambulatory care, and public health services are needed to produce a more effective use of resources. Integration of services under community leadership and management should encourage transfer of funds within a district health network from institutional care to community-based care. Such changes

are a test of leadership skills to achieve cultural change within an organization, which requires behavior change and involvement of health workers in policy and management of the change process.

STRATEGIC MANAGEMENT OF HEALTH SYSTEMS

Strategic management emphasizes the importance of positioning the organization in its environment in relation to its mission, resources, consumers, and competitors. It requires development of a plan of action or implementation of a strategy to achieve the mission or goal of the organization within acceptable ethical and legal guidelines. Articulation of these is a key role of the management level of an organization. Defining the mission and goals of the organization must take into account the external and internal environment, resources, and operational needs to implement and evaluate the adequacy of the outcomes. The strategy of the organization matches its internal approach with external factors, such as consumer attitudes and competing organizations. Strategy is a set of methods and skills of the health care manager to attain the objectives of a health organization, including:

1. providing high-quality care at current professional standards
2. innovating to avoid obsolescence;
3. developing good internal and external professional relationships;
4. utilizing human resources effectively;
5. ensuring accountability and accreditation within the local and national environment;
6. promoting the service to improve market share;
7. managing financial, human, and other resources efficiently; and
8. promoting the public and professional reputation of the institution.

Policy is the formulation of objectives and priorities. *Strategy* refers to long-range plans to achieve stated objectives, indicating the problems to be expected and how to deal with them. Strategy does not identify all actions to be taken, but it includes evaluation of progress made toward a stated goal. While the term has traditionally been used in a military context, it has become an essential concept in management, whether of industry, business, or health care. Tactics are the methods used to fulfill the strategy. Thus strategic management by objectives is applicable to the health system, incorporating definition of goals and targets, and the methods to achieve them (Box 12.3).

Change in health organizations may involve a substantial alteration in the size or relationships between existing, well-established facilities and programs (Table 12.2).

Box 12.3 The Strategic Management Process

1. Policy and planning
   a. Define mission, goals, objectives
   b. Surveillance
   c. Analysis of external environment
   d. Analysis of internal environment
   e. Assessment of capabilities
   f. Evaluating strategic choices, short-range
   g. Developing strategic planning, long-range
   h. Guiding the implementation process
   i. Communicating policy direction
2. Implementation
   a. Motivation: clearly communicate the goals and plans of the organization
   b. Differentiate between short- and long-term goals
   c. Ensure that staff know their responsibilities
   d. Ensure provision of adequate resources
   e. Promote sense of staff involvement
   f. Modify structure to meet needs
   g. Delegate authority, assign responsibility
   h. Promote interdepartmental coordination and interpersonal relations
   i. Promote capacity to deal with change
   j. Review policies in keeping with progress toward goals
   k. Promote understanding of change and resistance to change
3. Monitoring
   a. Evaluation of effectiveness
   b. Evaluation of outcome, lessons learned
   c. Revise strategic plan
   d. Redeploy resources in keeping with lessons learned

A strategic plan for health reform in response to the need for cost containment, redefined health targets, or dissatisfaction with the status quo requires a model or a vision for the future. This requires a strategic plan and a well-managed program. Opposition to change may occur for psychological, social, and economic reasons, or because of fear of loss of jobs or changes in assignments, salary, authority, benefits, or status. Downsizing in the hospital sector, with buildup of community health services, is one of the major issues in health reforms in many countries. This can be accomplished over time by naturally occurring vacancies, or attrition due to retirement, or by retraining and reassignment. All of this requires skilled leadership.

HEALTH SYSTEM ORGANIZATION MODELS

The New Public Health is an integration or coordination of many participating health care facilities and health-promoting programs. It is evolving in various forms in different places as networks with administrative and financial interaction between participating elements. Each organization provides its own specific services or groups of services. How they function internally and how they interact functionally and financially are important aspects of the management and outcomes of health systems. The health system functions as a network with formal and informal relationships; it may be very broad and loosely connected as in a highly decentralized system, with many lines of communication, payment, regulation, standards setting, and levels of authority.

TABLE 12.2 Transformation of Health Care Paradigms

Old paradigm	New paradigm
Emphasis on in-patient care	Emphasis on continuum of care
Emphasis on treating illness	Emphasis on maintaining, promoting wellness
Responsibility for the individual patient	Accountable for defined population
Specialists rewarded more than generalists	Greater economic parity between providers
Surgery rewarded more than medical services	Prevention rewarded versus surgery
Goal to fill beds	Provision of care at appropriate level of care
Separate organization, funding of hospital and other services	Integrate health delivery system
Managers run an organization, department	Managers promote market share
Managers coordinate services	Managers promote intersectoral cooperation

Source: Modified from Shortell, S. M., Kaluzny, A. D. 2005. *Health Care Management, Organization Design and Behavior.* Fifth Edition. Albany, NY: Delmar Publishing.

The relationship and interchange between different health care providers has functional and economic elements. This may be best shown by an example. A pregnant woman who is healthy and prepared for pregnancy physically and emotionally, and who receives comprehensive prenatal care is less likely than a woman whose health is neglected to develop complications and require prolonged hospital care as a result of childbirth. The cost of good prenatal care is a fraction of the economic cost of treating the potential complications and damage to her health or that of the newborn. A health system is responsible to ensure that women in the reproductive age take folic acid tablets orally before becoming pregnant, that she has had family planning services so that the pregnancy is a desired one, that the space between pregnancies is adequate for her health and that of her baby, and that she has adequate prenatal care. An obstetrics department should be involved in assuring or providing the prenatal care, especially for high-risk cases, and delivery should be in hygienic and professionally supervised settings.

Similarly for care of children and the elderly, there are a wide range of public health and personal care services that make up an adequate and cost-effective set of services and programs. The economic burden of caring for the sick child falls on the hospital. When there is a per capita grant to a district, then the hospital and the primary care service have a mutual interest in reducing morbidity, and hence mortality. This is the principle of the HMOs and district health systems discussed elsewhere. It is also a fundamental principle of the New Public Health.

Health care organizations differ according to size, complexity, ownership, affiliations, types of services, and location. Traditionally a health care organization provides a single type of service, such as an acute care hospital providing episodic inpatient care, or a home health care agency. In present-day health reforms, health care organizations, such as an HMO or district health system, provide a population-based, comprehensive service program. Each organization must have or develop a structure suited to meet its goals, both in the internal and external environments. The common elements that each organization must deal with include governance of policy, production or service, maintenance, financing, relating to the external environment, and adapting to changing conditions.

Functional Model

A functional model of an organization perhaps best suited to the smaller hospital is the division of labor into specific functional departments; for example, medical, nursing, administration, pharmacy, maintenance, and dietary, each reporting through a single chain of command to the CEO (Figure 12.6). The governing agency, which may be a local non-profit board or a national health system, has overall legal responsibility for the operation and financial

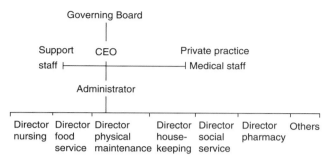

FIGURE 12.6 Functional model of organization.

status of the hospital, as well as raising capital for improvements.

The medical staff may be in private practice and work in the hospital with their own patients by application for this right as "attending physician," according to their professional qualifications, or the medical staff may be employed by the hospital similar to the rest of the staff. Salaried medical staff may include physicians in administration, pathology, anesthesia, and radiology, so that even in a private practice market system many of the medical staff are hospital employees. Increasingly, hospitals employ "hospitalers," who are full- or part-time physicians, to provide continuity of inpatient and emergency department services, augmenting the services of staff or private practice physicians. This is in part related to the increasing numbers of female physicians who run their homes and families as well as practice medicine and who find this mode of work more attractive than full-time private practice.

This is the common arrangement in North American hospitals. The governing board of a "voluntary," nongovernmental, not-for-profit organization with municipal and community representatives may be appointed by a sponsoring religious, municipal, or fraternal organization.

Corporate Model

The corporate model in health care organization (Figure 12.7) is often used in larger hospitals or where mergers with other hospitals or health facilities are taking place. This requires the CEO to delegate responsibility to other members of the senior management team who have operational responsibility for major sectors of the hospital's functioning.

A variation of the corporate model is the divisional model of a health care organization based on the individual service divisions allowing middle management a high degree of autonomy (Figure 12.8). There is often departmental budgeting for each service, which operates as an economic unit; that is, balancing income and expenditures. Each division is responsible for its own performance, with powers of strategic and operational decision-making

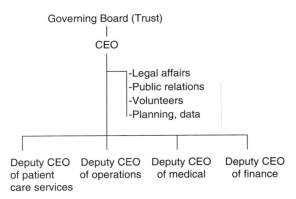

FIGURE 12.7   Corporate model of health care organization.

authority. This model is used widely in private corporations, and in many hospitals in the United States. With increasing complexity of services, this model is also employed in corporate health systems in the United States, with regional divisions.

Matrix Model

The matrix model of a health care organization is based on a combination of pyramidal and network organization (Figure 12.9). This model is suited to a public health department in a state, county, or city. Individual staff persons report in the pyramidal chain of command, but also function in multidisciplinary teams to work on specific programs or projects. A nutritionist in the geriatric department is responsible to the chief of nutrition services but is functionally a member of the team on the geriatric unit. In a laterally integrated health maintenance organization or district health system, specialized staff may serve in both institutional (i.e., hospital) and community health roles (Figure 12.9).

The organizational structure appropriate to one set of circumstances may not be suitable for all. Whether the

payment system is by norms, per diems, historical budget, or per capita in a regional or district health system structure (see Chapters 10 and 11), the internal operation of a hospital will require a model of organization appropriate to it. Hospitals need to modify their organizational structure as they evolve, and as the economics of health care change.

SKILLS FOR MANAGEMENT

Leadership in an organization requires the ability to define the goals or mission of the organization and to develop a strategy and define steps needed to achieve these goals. It requires an ability to motivate and engender enthusiasm for this vision by working with others to gain their ideas, their support, and their participation in the effort. In health care as in other organizations, it is easier to formulate plans than to implement them. Change requires not only the ability to formulate the concept of change, but also to modify the organizational structure, the budgeted resources, the operational policies, and, perhaps most importantly, the corporate culture of the organization.

Management involves skills that are not automatically part of a health professional's training. Skilled clinicians often move into positions requiring management skills in order to build and develop the health care infrastructure. In some countries, hospital managers must be physicians, often senior surgeons. Clinical capability does not transfer automatically into management skills to deal with personnel, budgets, and resources. Therefore, training in management is vital for the health professional.

The manager needs training for investigations and fact-finding as well as the ability to evaluate people, programs, and issues, and set priorities for dealing with the short- and long-term issues. Negotiating with staff and outside agencies is a constant activity of the manager, ranging from the trivial to major decisions with wide implications. Perhaps the most crucial skill of the manager is

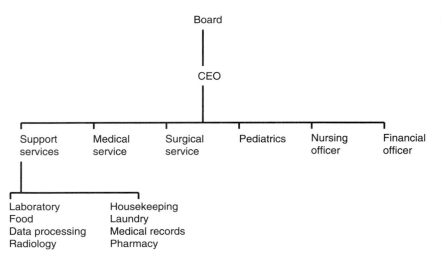

FIGURE 12.8   Divisional model of health care organization.

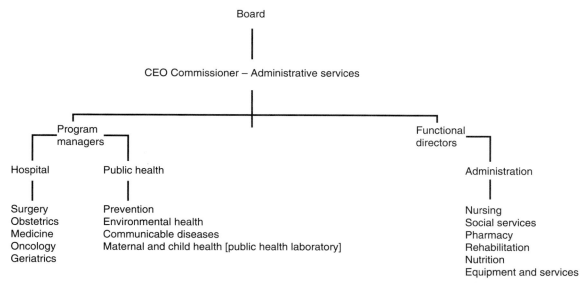

FIGURE 12.9 Matrix model of health care organization.

communication: the ability to convey verbal, written, or unwritten messages that are received and understood and to assess the responses as an equal part of the exchange.

Interpersonal skills are a part of management practice. The capable manager can relate to personnel at all levels in an open and equal manner. This is essential to help foster a sense of pride and involvement of all personnel in working toward the same goals and objectives, and to show that each member of the team is important to meeting the objectives of the organization. At the same time, the manager needs to communicate information, especially as to how the organization is doing in achieving its objectives.

The manager is responsible for organizing, planning, controlling, directing, and motivating. Managers assume multiple roles. A role is an organized set of behaviors. Henry Mintzberg described the roles and skills needed by all managers: informational, interpersonal, and decisional roles. Robert Katz identified three managerial skills that are essential to successful management: technical, human, and conceptual. The distribution of these skills between the levels of management is shown in Figure 12.10.

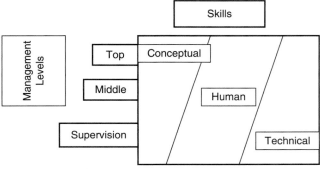

FIGURE 12.10 Distribution of skills within an organization.

THE CHIEF EXECUTIVE OFFICER OF HEALTH ORGANIZATIONS

Hospital directors in the past were often senior physicians, often called the *superintendent*, without training in health management. The business manager CEO has become common in hospital management in the United States. During the 1950s when the CEO was called *administrator*, he or she worked under the direction of a board of trustees who raised funds, set policies, and were often involved in internal administration.

Where the CEO was a non-physician, the usual case in Northern American hospitals, often a conflict existed with the clinical staff of the hospital. In some settings, this led to appointment of a parallel structure with a full-time chief of medical staff with a focus on clinical and qualitative matters. In European hospitals, the CEO is usually a physician, often by law, and the integration of the management function with the role of clinical chief is the prevalent model.

Over time, the CEO role has changed to one of a "coordinator" as the cost and complexity of the health system have increased. The CEO is now more involved in external relations and less in the day-to-day operation of the facility. The CEO is a leader/partner but *primum inter pares,* or first among equals, in a management team that shares information and works to define objectives and solve problems. This de-emphasizes the authoritarian role and stresses the integrative function.

The CEO is responsible for the financial management of operational and capital budgets of the facility, which is integral to the planning and future development of the facility. Budgets include four main factors: income, fixed or regular overhead, variable or unpredictable overhead, and capital or development costs, all essential to the survival and development of the organization.

The key role of top management is to develop a vision, goals, and targets for the institution, to maintain an atmosphere and systems to promote the quality of care, financial solidity, and to represent the institution to the public. The overall responsibility for the function and well-being of the program is with the CEO and the governing board of directors.

COMMUNITY PARTICIPATION

Community participation in management of health facilities has a long-standing and constructive tradition. The traditional hospital board has served as a mechanism for community participation and leadership in promoting health facility development and management at the community level. The role of hospital boards evolved from primarily a philanthropic and fund-raising one to a greater overall responsibility for policy and planning function working closely with management and senior professional staff. This occurred as operational costs increased rapidly, as government insurance schemes were implemented, and as court decisions defined the liability of hospitals and reinforced the broadened role of governing boards in malpractice cases and quality assurance. Centrally developed health systems such as the United Kingdom's National Health Service have promoted district and county health systems with high degrees of community participation and management, both at the district level and for services or facilities.

INTEGRATION — LATERAL AND VERTICAL

Rationalization of health facilities increasingly means organizational linkages between previously independent facilities. Mergers of health facilities are common events in many health systems. In the United States, mergers between hospitals, or between facilities linked to HMOs or managed care systems, are frequent events. Health reform in many countries is based on similar linkages. Governmental approval and alteration of financing systems are needed to promote linkages between services to achieve greater efficiency and improve patient care (see Chapters 10 and 11).

Lateral integration is the term used for amalgamation among similar facilities. This is like a chain of hotels, and in health care involves two or more hospitals, usually meant to achieve cost savings, improve financing and efficiency, and reduce duplication of services. Urban hospitals, both not-for-profit as well as for-profit, often respond to competition by purchasing or amalgamating with other hospitals to increase market share in competitive environments. This is often easier for hospital-oriented CEOs and staff to comprehend and manage, but it avoids the issues of downsizing and integration with community-based services.

Vertical integration describes organizational linkages between different kinds of health care facilities to form integrated, comprehensive health service networks. This permits a shift of emphasis and resources from inpatient care to long-term, home, and ambulatory care. This is the managed care or district health system model. Community interest is a factor in promoting change to integrate services, which can be a major change for the management culture, especially of the hospital.

The survival of a health care facility may depend on integration with appropriate changes in concepts of management. In the 1990s, a large majority of California residents moved to managed care programs because of the high cost of fee-for-service indemnity health insurance and because of federal waivers to promote managed care for Medicare and Medicaid beneficiaries. Independent community hospitals without a strong connection with managed care programs were in danger of losing their financial base.

Hospital bed supplies were reduced in the United States from 4.5 beds per 1000 population in 1980 to 2.9 in 2000 and 2.7 in 2005. Occupancy rates also fell from 75 percent in 1980 to 6.4 percent in 2000 and rose slightly to 67 percent in 2005. Hospital discharges also fell during these years from 173 per 1000 population in 1980 to 113 and 116, respectively, in 2000 and 2005 while days of care fell from 1297 to 537 and 554, respectively, in 2000 and 2005 (Health U.S., 2007). These data are monitored by the National Hospital Discharge Survey and CDC's National Center for Health Statistics. The lower hospital bed supply and utilization since the 1980s and 1990s reflects the adoption of payment by DRGs in which the insurance system pays by diagnostic group, rather than on a per diem basis. There was a shift to stronger ambulatory care as occurred throughout the industrialized countries despite an aging of the population. These trends were largely due to greater emphasis on ambulatory surgery and other care, and major medical centers responded with strategic plans to purchase community hospitals and develop affiliated medical groups and contract relationships with managed care organizations to strengthen their "market share" service population base for the future. The new payment environment and managed care also promoted hospital mergers (lateral integration) and linkages between different levels of service, such as teaching hospitals with community hospitals and primary community care services (vertical integration).

Vertical integration is important not only in urban areas, but can serve as a basis for developing rural health care in both developed and developing countries. The district hospital and primary care center operating as an integrated program can provide a high-quality program. Hospital-centered health care, common in industrialized countries, channels most of the available funds for health to the hospital, thus preventing adequate funding to primary health care. To prevent this, developing countries can adopt the primary care approach, which is now the basis of managed care systems, and try to limit hospital care by improving ambulatory and preventive care services.

NORMS AND PERFORMANCE INDICATORS

Norms are useful to promote efficient use of resources and promote high standards of care, if based on empirical standards proved by experience, trial and error, and scientific observation. Norms may be needed even without adequate evidence, but should be tested in the reality of observation, experience, and experiment. This requires data and trained observers free to examine, report, and publish their findings for open discussion among colleagues and peers in proceedings open to the media and the general public.

Normative standards of planning are the determination of a number per unit of population that is deemed to be suitable for population needs; for example, the number of beds or doctors per 1000 population, or length of stay in hospital. Many organizations based on the bureaucratic model used norms of resources as the basis for planning and allocation of resources including funding (see Chapter 11). This led to payment systems which encouraged greater use of that resource. If a factory is paid by the number of workers and not the number and quality of the cars produced, then management will have no incentive to introduce efficiency or quality improvement measures. If a district or a hospital is paid by the number of beds, or by days of care in the hospital, there is no incentive to introduce alternative services such as same-day or outpatient surgery and home care.

Performance indicators are measures of completion of specific functions of preventive care such as immunization, mammography, Pap smears, and diabetes and hypertension screening. They are indirectly measures of economy, efficiency, and effectiveness of a service and are being adopted as better methods of monitoring a service and paying for it, such as by paying a premium. General practitioners in the United Kingdom receive additional payments for full immunization coverage of the children registered in their practices. A block grant or per capita sum may be tied to indicators which reflect good standards of care or prevention, such as low infant, child, and maternal mortality. Incentive payments to hospitals can promote ambulatory services as alternatives to admissions and reduce lengths of stay. Limitations of financial resources in the industrialized countries and even more so in the developing countries make the use of appropriate performance indicators of great importance in the management of resources.

HEALTH PROMOTION AND ADVOCACY

Social marketing is the systematic application of marketing alongside other concepts and techniques to achieve specific behavioral goals for a social good. Initially focused on commercial goals in the 1970s, the concept became part of health promotion activities to address health issues where there was no current biomedical approach such as in smoking reduction and in safe sex practices to prevent the spread of HIV.

Social marketing was based initially on commercial marketing techniques but now integrates a full range of social sciences and social policy approaches using the strong customer understanding and insight approach to inform and guide effective policy and strategy development. This has become a part of public health practice and policy setting to achieve both strategic and operational targets. A classic example of the success has occurred with tobacco reduction strategies in many countries using education, taxation, and legislative restrictions. Other challenges in this field include risk behavior such as alcohol abuse through binge drinking, unsafe sex practices, and dietary practices harmful to health.

PHILANTHROPY AND VOLUNTEERISM

Philanthropy and volunteerism have long been important elements of health systems through building hospitals, mission houses, and food provision, and other prototype initiatives on a demonstration basis. This approach has been instrumental in such areas as improved care and prevention of HIV, immunization in underdeveloped countries, global health strategies, and maternal and child health services.

During the late twentieth and early twenty-first centuries, a new "social entrepreneurship" was initiated and developed by former President Bill Clinton, Microsoft's Bill Gates, and George Soros. This promoted integration and consortia for promotion of AIDS and malaria control in many developing countries. GAVI (Global Alliance for Vaccine and Immunization) is a U.S.-based organization which links international public and private organizations and resources to extend access to immunization globally. It includes UNICEF, WHO, bilateral donor countries, the vaccine industry, the Gates Foundation, and other major donors. GAVI has made an important contribution to advancing vaccine coverage and adding important new vaccines in many developing countries and regions. These organizations focus funds and activities on promoting improved care and prevention of HIV, TB, and malaria along with improved vaccination for children, reproductive health, global health strategies, technologies, and advocacy. It brings publicity and raises consciousness at political levels where resource allocations are made. To some extent, this is a modern version of missionary work widely done to spread concepts of religion, and in this case the promotion of "civil societies" in a globalized world of free trade, democracy, and peace.

These efforts included promoting improved large-scale marketing of antiretroviral drugs for treatment of HIV infection, including price reduction so that developing countries could include antiretroviral treatment, especially to reduce mother-to-infant transmission. This effort branched out into malaria-preventing bed net production, distribution of low-cost pharmaceuticals, marketing drugs for the poor, desalination plants, and solar roof units, in Africa mainly.

A private sector movement has become attractive to big business in an attempt to appear to be proactive in environmental consciousness. The automobile industry is facing both public concern for improved gas mileage and public demand for larger cars. Hybrid cars using less fuel have been successfully introduced into the market for low-emission, fuel-efficient cars. Public opinion is showing signs of moving toward promoting environmentally friendly design, marketing, and purchasing practices in energy consumption, conservation practices, and public policy. Public opinion and the price of fuel will play a major part in driving governments to legislate energy and conservation policies to address global warming and damage to the environment with many negative health consequences. However, such changes must work with public opinion because of the sensitivity of consumers to the price of fuel. In addition, when food crops, such as corn, are used to produce ethanol for energy to replace oil, then food prices rise and consumers suffer and respond vigorously.

NEW ORGANIZATIONAL MODELS

New models of health care organization are emerging and developing rapidly in many countries. Partly this is a search for more economical methods of delivering health care and is also the result of the target-oriented approach to health planning that seeks the best way to define and achieve health objectives. The developed countries seek ways to restrain cost increases, and the developing countries seek effective ways to quickly and inexpensively raise health standards for their populations. The new organizational models that try to meet these objectives include district health systems and managed care systems, described in greater detail in Chapter 11. Critical and basic elements of a health system organization are shown in Figure 12.11.

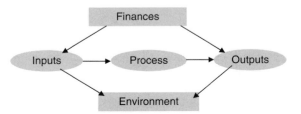

FIGURE 12.11 Basic elements of an organizational system. Source: Modified from Gibson et al., 2003. *Organizations: Behavior, Structure, Process.* New York: McGraw-Hill/Irwin.

NEW PROJECTS AND THEIR EVALUATION

In any organization or system of health service, there will be identification of issues and decisions to launch new endeavors or projects to advance the state of the art or to address unmet needs, or to meet competition. In developing and developed countries, many NGOs provide services which are directed by their funding source and its priorities, often running parallel as a result of uncoordinated activities. Public health agencies can play a leadership and regulatory role in coordination of activities and directing new programs to areas of greatest national needs. The public health agency may also seek funding to launch new pilot or specific needs programs. The agency may introduce a new vaccine into a routine immunization program in phases pending government approval and funding to incorporate it as a routine immunization program based on evaluation of the initial phase. An example is the introduction of *Haemophilus influenzae* type b (Hib) vaccine in Albania in 2006 funded by GAVI for 5 years based on a study and proposal including a cost-effectiveness study (Bino, S., Ginsberg, G. — personal communication, 2007).

Proposals for projects need to be prepared in keeping with the vision, mission, and objectives of the organization. A project proposal should include why the project is important, what its specific goals and objectives are, available or new resources, and the time frame needed (Box 12.4). It should describe the means proposed to accomplish the goals, and how the proposed program will impact the community.

During the planning process, data will be collected on past projects, amount funded, and special project proposal guidelines of the intended donor agencies. Relevant data can be obtained from vital and health statistics, past and current programs, and outcomes. Partnering with other agencies (such as NGOs) or institutions may augment the capacity and appropriateness of the strengths, weaknesses, opportunities, and threats (SWOT) of the potential partner agencies or personnel.

A proposal for a project should include title page, introduction, general aim and objectives, expected results, activities, work plan, monitoring and evaluation, and a budget. The first part of the proposal that the sponsoring/funding agency will see is the title page. A title page should be easy to read and convey the project name and the implementing agency's address. The executive summary of the proposal includes the time frame, country, region, district or target population of the program, the total budget, the amount requested, the equivalent shared by the implementing organization, the submission date of the proposal, and the evaluation according to objectives of the program.

The introduction outlines the current state of the problem and the case for action. It should describe existing

Source: Centers for Disease Control. 1999. Framework for program evaluation in public health. *Morbidity and Mortality Weekly Report*, 48(RR-11):1–40.

Box 12.4 Program Evaluation Information Needs

The following utility standards ensure that an evaluation will serve the information needs of intended users:

1. **Stakeholder identification** Persons involved in or affected by the evaluation should be identified so that their needs can be addressed.
2. **Evaluator credibility** The persons conducting the evaluation should be trustworthy and competent in performing the evaluation for findings to achieve maximum credibility and acceptance.
3. **Information scope and selection** Information collected should address pertinent questions regarding the program and be responsive to the needs and interests of clients and other specified stakeholders.
4. **Values identification** The perspectives, procedures, and rationale used to interpret the findings should be carefully described so that the bases for value judgments are clear.
5. **Report clarity** Evaluation reports should clearly describe the program being evaluated, including its context and the purposes, procedures, and findings of the evaluation so that essential information is provided and easily understood.
6. **Report timeliness and dissemination** Substantial interim findings and evaluation reports should be disseminated to intended users so that they can be used in a timely fashion.
7. **Evaluation impact** Evaluations should be planned, conducted, and reported in ways that encourage follow-through by stakeholders to increase the likelihood of the evaluation being used.

programs which address that issue, with proposed collaboration, and expansion or improvement of programs but avoiding duplication of services. Background information needs to relate the project to the priorities of the prospective funding organization. The objectives should follow the acronym "SMART": Specific, Measurable, Achievable, Relevant, and Time-Based. This term originally used for computer disc self-management has been adapted as a current form of management by objectives from the 1950s and 1960s.

The project objectives should be feasible and the expected results of the project should be based on the stated objectives. The funding organization wants to know what will be the expected product of the program in measurable process (e.g., immunization coverage) or outcome indicators (e.g., reduced child mortality). Projections will be based on the intended activities and known outcomes of other past programs with similar goals in the same or other countries (environmental scan). This should be supported by a review of local and international literature on the topic. The activities section of a proposal should include a time line of the intended actions and a description of activities based on best practices. The expected outcomes, monitoring and evaluation, and justification are all part of the presentation (see Box 12.5).

The proposed funding agency expects convincing evidence of how this program will be effective, efficient, practical, and realistic. This information is presented in the activities section, which also needs to address the resources that will be needed to implement the program such as vaccine purchase, management costs, cold chain equipment, syringes, alcohol swipes, transportation, management, training and supervision costs, such as the budget, supplies, and human capital. After completing the activities section, a realistic and achievable work plan and time frame is required.

Well-planned projects have monitoring and evaluation criteria. Monitoring follows the performance of the program, documenting successes, failures, and lessons learned, as well as expenditures. Evaluation guidelines of the program define the methods used to assess the impact of the project and if the project was carried out in an effective and efficient manner and may be required periodically through the life of the project.

The most difficult issue is sustainability. A project funded by an NGO is usually time limited to 3–4 years and the survival of the program usually depends on its acceptability and the capacity of government to continue it. Thus evaluation becomes even more crucial for the follow-up of even successful short-term projects. Harm reduction programs for HIV among drug users in the Russian Federation funded in its initial period of years by the Open Society Institute (Soros Foundation) have been continued and expanded with funding by the Russian Federation Ministry of Health (Bobrick, A. — personal communication, 2007).

Public health work within departments or ministries of health or local health authorities operate at a disadvantage in comparison with other health activities, especially hospitals, pharmaceuticals, diagnostics, and medical care. The competition for resources in a centrally funded system is intense and the political and bureaucratic battles for funds may pit new immunization agents or health promotion programs against new cancer treatment drugs or MRI scanners, and this is very often a difficult struggle. Presentation of program proposals for new public health interventions requires skill and professionalism, good timing, and requires help of informed public and professional opinion. The political level decides on allocation of resources in a tax-based universal system, while even in a social security (Bismarckian) system where funding is through an employee-employer payroll deduction,

---

Box 12.5 Project Proposal Summary

Title page Name of project; principal persons and implementing organizations; contact person(s); time frame; country (state, region); target group of project; estimated project cost; date of submission.

Introduction Provides project background including the health issue(s) to be addressed, a situational analysis of the health problem, the at-risk and target populations, and existing programs in the community. Includes an international and national literature review of the topic with references.

Aim of the project Intended accomplishment of the project.

Objectives Specific, Measurable, Achievable, Relevant, and Time-based (SMART).

Expected results Based on the objectives — what will be produced and delivered.

Activities Actions and time frame in keeping with the objectives and expected results.

Work plan Time line of all activities, including preparation, training, pilot, and implementation stages.

Budget Estimated cost of expenditures including human resources, activities, running cost, and overhead for project and evaluation.

Monitoring and evaluation A plan for periodic follow-up of project activities (including time line and measures) in order to determine if the objectives are being met, positive or negative outcomes, and use of resources. How efficient and effective is the project?

Reporting Report of the project, with publication in peer-reviewed journal if possible.

Justification Why is this project important and timely and how will it contribute to the health of the community?

Sources: Adapted from Centers for Disease Control. 1999. Framework for program evaluation in public health. *Morbidity and Mortality Weekly Report*; 48(RR-11):1–40.
Interagency Resource Management unit (IRM), World Health Organization Regional Office for Africa. 2000. *Standard Format for Afro Project Documents*, p. 1–33. http://www.afro.who.int/irm/reports/project_eng.pdf [accessed December 20, 2007, and May 21, 2008]

---

additional funding from government is essential to keep up with the continuing flow of new modalities of treatment or prevention.

Public health is handicapped in portraying the costs and benefits of important interventions so that it often ends up with insufficient resources for new programs, not including the staffing, administrative cost (e.g., office space, phone service, transportation costs) which are part of any public health program. Portraying the cost of the new proposed program should be based on the total population served, not just the specific target population for a new program (i.e., it should be represented as a per capita cost). Similarly, projected benefits should extrapolate the results from other areas, such as pandemic or avian flu or SARS, and the likely impact on the target geographic area and its population.

SYSTEMS APPROACH AND THE NEW PUBLIC HEALTH

Management in health is for the purpose of improving health and not merely the maintenance of an institution. Separate management of a variety of health facilities serving a community has derived from different historical development and funding systems. In competition for public attention and political support, public health suffers in comparison to hospitals, new technology and drugs and other competitors for limited resources. The experience of successes in reducing mortality from noninfectious as well as infectious conditions comes largely from public health interventions. Medical care is also an essential part of public health, so that management and resource allocation within the total health sector are interactive and mutually dependent. The New Public Health looks at all services as part of a network of interdependent services, each contributing to health needs, whether in hospital care or in enforcing public health law regarding motor vehicle safety and smoking restriction in public places.

Separate management and budgeting of a complex of services results in a disproportion of funds, staff, and attention going to traditional services such as hospitals and fails to redirect resources to more cost-effective and patient-sensitive kinds of services, such as home and preventive care. However, reducing the supply of hospital beds and implementing payment systems with resources for early diagnosis and incentives for short stays have changed this situation quite dramatically in recent decades. The effects of incentives and disincentives built into funding systems are central issues in determining how management approaches problem solving and program planning, and are therefore important considerations in promoting health.

The management approach to resolving this dilemma is professional vision and leadership to promote the broader New Public Health. Thus managers of hospitals and other health facilities need broad-based training in a New Public Health in order to understand the interrelationships of services, funding, and population health. Managers who persist in an obsolescent paradigm with the traditional emphasis, regardless of the larger picture, may find the hospital noncompetitive in a new climate where economic incentives promote downsizing institutions and upgrading health promotion. Defensive, internalized management will become obsolete, while forward-looking management will be the pioneers of the New Public Health. This may be seen as a systems approach to improve population and individual health, based on strategic planning for immediate needs and adaptation of health systems in the longer-term issues in health.

SUMMARY

Health care is one of the largest and most important industries in any country, consuming anywhere from 3 to 16 percent of GNP, and still growing. It is a service, not a production industry, and is vital to the health and well-being of the individual, the population, and the economy. Because it employs large numbers of skilled professionals and many unskilled persons, it is often vital to the economic survival of small communities, as well as for a sense of community well-being.

Management includes planning, leading, controlling, organizing, motivating, and decision making. It is the application of resources and personnel toward achieving targets. Therefore, it involves the study of the use of resources, and the motivation and function of the people involved, including the producer or provider of service, and the customer, client, or patient. This cannot take place in a vacuum, but is based on the continuous monitoring of information and its communication to all parties involved. These functions are applicable at all levels of management, from policy to operational management of a production or a service system. Creative management of health systems is vital to the functioning of the system at the macro level, as well as in the individual department or service. This implies effective use of resources to achieve objectives, and community, provider, and consumer satisfaction. These are formidable challenges, not only when money is available in abundance, but even more so when resources are limited and difficult choices need to be made.

Modern management includes knowledge and skills in identification and measuring of community health needs and health risks. Critical needs are addressed in strategic planning with measurable impacts and targets. Public health managers should have skills gained in marketing, networking, data management, managing human resources and finance, engaging community partners, and communicating public health messages.

Many of the methods of management and organization theory developed as part of the business world have become part of public health. This includes defining mission, values and objectives of the organization, strategic planning and management, management by objectives, human resource management (recognizing individual and professional values), incentives–disincentives, regulation, education, and economic resources. Ultimately the mission of public health is the saving of human life and improving its quality, and to do this as efficiently as possible with the highest standards of professionalism and community involvement as possible.

The scope of the New Public Health is broad. It includes the traditional public health programs, but equally must concern itself with managing and planning comprehensive service systems and measuring their function. Selection of targets and priorities is often determined by the feasible rather than the ideal. The health manager, either at the macro health level or managing a local clinic, needs to be able to conceptualize the possibilities of improving health of clients and the population in his or her service responsibility with current and appropriate methods. Good management means designing objectives based on a balance between the feasible and the desirable. Public health has benefited greatly from its work with the social sciences and assistance from management and systems sciences to adapt and absorb the new challenges and technologies in applied public health. The New Public Health is not only a concept; it is a management approach to improve the health of individuals and the population.

ELECTRONIC RESOURCES

Agency for Healthcare Research and Quality. http://www.ahcpr.gov/ [accessed May 21, 2008]

American College of Healthcare Executives. http://www.ache.org/ [accessed May 21, 2008]

American College of Medical Quality. http://www.acmq.org/ [accessed May 21, 2008]

American Hospital Association. http://www.aha.org/aha/about/index.html [accessed May 21, 2008]

Centers for Disease Control, Management and Analysis Services Office. http://www.cdc.gov/maso/mapb.htm [accessed May 21, 2008]

Glossary of Managed Care Terms. http://www.pohly.com/terms.html [accessed May 21, 2008]

The Joint Commission. http://www.jointcommission.org/ [accessed May 21, 2008]

National Association of Public Hospitals and Health Systems. http://www.naph.org/ [accessed May 21, 2008]

National Center for Health Statistics. http://www.cdc.gov/nchs/pressroom/01news/99hospit.htm [accessed May 21, 2008]

World Health Organization. http://www.who.int/management/en/

http://www.who.int/management/district/overall/en/index.html [accessed May 21, 2008]

RECOMMENDED READINGS

Arias, D., C. 2008. C. Everett Koop: The nation's health conscience. *American Journal of Public Health*, 98:396–399.

Bobak, M., Murphy, M., Rose, R., Marmot, M. 2007. Societal characteristics and health in the former communist countries of Central and Eastern Europe and the former Soviet Union: A multilevel analysis. *Journal of Epidemiology and Community Health*, 61:990–996.

Centers for Disease Control. 1999. Framework for program evaluation in public health. *Morbidity and Mortality Weekly Report*, 48(RR-11): 1–37.

Centers for Disease Control. 2000. Biological and chemical terrorism: Strategic plan for preparedness and response. Recommendations of the CDC Strategic Planning Workgroup. Recommendations and Reports. *Morbidity and Mortality Weekly Report*, 49(RR-04):1–14.

Katz, R. 1974. Skills of an effective administrator. *Harvard Business Review*, September–October, 90–110.

McGlynn, E. A., Asch, S. M., Adams, J., Keesey, J., Hicks, J., DeCristofaro, A., Kerr, E. A. 2003. The quality of health care delivered to adults in the United States. *New England Journal of Medicine*, 348:2635–2645.

McMahon, R., Barton, E., Piot, M., Gelina, N., Ross, F. 1992. *On Being in Charge: A Guide to Management in Primary Health Care*, Second Edition. Geneva: World Health Organization.

Marmot, M. 2007. Commission on Social Determinants of Health. Achieving health equity: From root causes to fair outcomes. *Lancet*, 29;370:1153–1163.

Mintzberg, H., 1973. *The Nature of Managerial Work*. New York: Harper and Row.

Satcher, D., Higginbotham, E. J. 2008. The public health approach to eliminating disparities in health. *American Journal of Public Health*, 98:400–403.

Schuster, R. J., Terwoord, N. A., Tasosa, J. 2006. Changing physician behavior to measure and improve clinical outcomes. *American Journal of Clinical Quality*, 21:394–400.

Scott, R. W. 1992. *Organizations: Rational, Natural and Open Systems*, Third Edition. Englewood Cliffs, NJ: Prentice-Hall.

Suba, E. J., Murphy, S. K., Donnelly, A. D., Furia, L. M., Huynh, M. L. D., Raab, S. S. 2006. Systems analysis of real-world obstacles to successful cervical cancer prevention in developing countries. *American Journal of Public Health*, 96:480–487.

Trochim, W. M., Cabrera, D. A., Milstein, B., Gallagher, R. S., Leischow, S. L. 2006. Practical challenges of systems thinking and modeling in public health. *American Journal of Public Health*, 96:538–546.

Woodall, J. P. 1988. Epidemiological approaches to health planning, management and evaluation. *World Health Statistics Quarterly*, 41:2–10.

BIBLIOGRAPHY

Aday, L. A., Begley, C. E., Lairson, D. R., Slater, C. H. 1993. *Evaluating the Medical Care System: Effectiveness, Efficiency, and Equity*. Ann Arbor, MI: Health Administration Press.

Adler, P. S. 1993. Time-and-motion regained. *Harvard Business Review*, January–February:97–108.

Aguayo, R. 1990. *Dr. Deming: The American Who Taught the Japanese About Quality*. New York: Simon & Schuster.

Bar Yam, Y. 1996. Improving the effectiveness of health care and public health: A multi-scale complex systems analysis. *American Journal of Public Health*, 96:459–466.

Barron, W. M., Grsek, C., Weber, D., Cerese, J. 2005. Critical success factors for performance improvement programs. *Joint Commission Journal of Quality and Patient Safety*, 31:220–226

Cohen, J. 1990. Health policy, management and economics. *In* T. Lambo, S. B. Day (eds.). *Issues in Contemporary International Health*. New York: Plenum.

Darr, K., Rakich, J. S. 1989. *Hospital Organization and Management*, Fourth Edition. Owings Mills, MD: National Health Publishing.

Donabedian, A. 1976. *Aspects of Medical Care Administration: Specifying Requirements for Health Care*. Boston: Harvard University Press.

Duncan, W. J. P., Ginter, P. M., Swayne, L. E. 1995. *Strategic Management of Health Care Organizations*. Cambridge, MA: Blackwell Business.

Dutton, D. B. 1979. Patterns of ambulatory health care in five different delivery systems. *Medical Care*, 17:221–243.

Ellencweig, A. Y. 1992. *Analysing Health Systems: A Modular Approach*. Oxford: Oxford University Press.

Etzioni, A. 1964. *Modern Organizations*. Englewood Cliffs, NJ: Prentice Hall.

Fajans, P., Simmons, R., Ghiron, L. 2006. Helping public health systems innovate: The strategic approach to strengthening reproductive health policies and programs. *American Journal of Public Health*, 96:435–440.

Fleming, S. T., Scutchfield, F. D., Tucker, T. C. 2000. *Management Epidemiology*. Washington, DC: Health Administration (AUPHA) Press.

Gibson, J. L., Donnelly, J. H., Konopaske, R., Ivancevich, J. 2005. *Organizations: Behavior, Structure, Processes*. Irwin/McGraw Hill, Boston, MA.

Glynn, J. J., Perkins, D. A. (eds.). 1995. *Managing Health Care: Challenges for the 90s*. London: W. B. Saunders.

Griffith, J. R. 1993. *The Moral Challenges of Health Care Management*. Ann Arbor, MI: Health Administration Press.

Jex, S. M. 2002. *Organizational Psychology: A Scientist–Practitioner Approach*. New York: John Wiley & Sons.

Joffe, M., Mindell, J. 2006. Complex causal process diagrams for analyzing the health impacts of policy interventions. *American Journal of Public Health*, 96:480–487.

Katz, R. 1974. Skills of an effective administrator. *Harvard Business Review*, September–October, 90–101.

Katzenbach, J. R., Smith, D. K. 1993. *The Wisdom of Teams: Creating the High-Performance Organization*. Cambridge, MA: Harvard Business School Press.

Kotler, P., Roberto, N., Lee, N. 2002. *Social Marketing: Improving the Quality of Life*. Thousand Oaks, CA: Sage Press.

Lied, T. R., Kazandjian, V. A. 1998. A Hawthorne strategy: implications for performance measurement and improvement. *Clinical Performance and Quality Health Care*, 6:201–204.

Loeffel, G., Blumental, D. 1989. The case for using industrial quality management science in health care organizations. *Journal of the American Medical Association*, 262:2869–2873.

McGregor, D. 1960. *The Human Side of Enterprise*. New York: McGraw-Hill.

Maslow, A. 1954. *Motivation and Personality*. New York: Harper.

Maslow, A. H., Sakoda, J. M. 1952. Volunteer-error in the Kinsey study. *Journal of Abnormal & Social Psychology*, 47(2):259–262.

May, E. 1945. *The Social Problems of an Industrial Civilization*. Boston: Graduate School of Business Administration, Harvard University.

Midgley, G. 2006. Systemic intervention for public health. *American Journal of Public Health*, 96:466–472.

Mintzberg, H. 1973. *The Nature of Managerial Work*. New York: Harper and Row.

Mintzberg, H. 1994. The fall and rise of strategic planning. *Harvard Business Review*, January–February, 107–114.

Oleske, D. M. (ed.). 1995. *Epidemiology and the Delivery of Health Care Services: Methods and Applications*. New York: Plenum Press.

Parsons, H. M. 1974. What happened at Hawthorne? *Science*, 183:93

Reinke, W. A. (ed.). 1988. *Health Planning for Effective Management*. New York: Oxford University Press.

Robbins, S. P. 1990. *Organization Theory: Structure, Design and Applications*, Third Edition. Englewood Cliffs, NJ: Prentice-Hall.

Shortell, S. M., Kazluny, A. D. (eds.). 2005. *Health Care Management: Organization Design and Behavior*, Fifth Edition. Albany, NY: Delmar Publishing.

Sloan, M. D., Chmel, M. 1991. *The Quality Revolution and Health Care: A Primer for Purchasers and Providers.* Milwaukee, WI: ASQC Quality Press.

Smith, D. 1972. Organizational theory and the hospital. *Journal of Nursing Administration,* 2:19–24.

Tarimo, E. 1991. *Towards a Health District: Organizing and Managing District Health Systems Based on Primary Health Care.* Geneva: World Health Organization.

Taylor, C. E. 1992. Surveillance for equity in primary care: Policy implications from international experience. *International Journal of Epidemiology,* 21:1043–1049.

Taylor, F. W. 1947. *Scientific Management.* New York: Harper & Brothers, reprint.

Turnock, B. J. 2004. *Public Health: What It Is and How It Works,* Third Edition. Sudbury, MA: Jones and Bartlett Publishers.

United States Department of Health and Human Services. *Health, 2007, with Chartbook on Trends in the Health of Americans.* Washington, DC.

Vaughn, J. P., Morrow, R. H. 1989. *Manual of Epidemiology for District Health Management.* Geneva: World Health Organization.

World Health Organization. 1994. *Information Support for New Public Health Action at the District Level:* Report of a WHO Expert Committee. Technical Support Series 845. Geneva: World Health Organization.

Chapter 13

National Health Systems

Introduction
Health Systems in Developed Countries
  Evolution of Health Systems
The United States
  Federal Health Initiatives
  Medicare and Medicaid
  The Changing Health Care Environment
  Health Information
  Health Targets
  Social Inequities
  The Dilemma of the Uninsured
  Summary
Canada
  Reform Pressures and Initiatives
  Provincial Health Reforms
  Health Status
  Summary
The United Kingdom
  The National Health Service
  Structural Reforms of the National Health Service
  Reforms Since 1990
  Social Inequities
  Health Promotion
  Health Reforms
  Primary Care Trusts
  Summary
The Nordic Countries
  Sweden
  Denmark
  Norway
  Finland
Western Europe
  Germany
  The Netherlands
Russia
  The Soviet Model
  Epidemiologic Transition
  Post-Soviet Reform
  Future Prospects
  Summary
Israel
  Origins of the Israeli Health System

  Health Resources and Expenditures
  Health Reforms
  Mental Health
  Healthy Israel 2020
  Summary
Health Systems in Developing Countries
  Federal Republic of Nigeria
  Summary
Latin America and the Caribbean
  Colombia
Asia
  China
  Japan
Comparing National Health Systems
  Economic Issues in National Health Systems
Reforming National Health Systems
Summary
Electronic Resources
  Canada
  China
  Europe
  International
  Israel
  United States
Recommended Readings
Bibliography
  Canada
  China
  Colombia
  Developing Countries
  Europe
  Israel
  Japan
  Latin America and the Caribbean
  Nigeria
  Nordic Countries, Germany, and the Netherlands
  Russia
  United Kingdom
  United States

INTRODUCTION

Assuring access to quality health care for all is a basic principle of the New Public Health. There are many personal or community risk factors which affect health status, and medical care is a vital part of the broad spectrum of health needs. But the individual and the population very much need the many effective methods of preventing and treating many diseases and stopping the progress of others. Despite its value, medical care, by itself, is not sufficient to produce a high standard of health for the population. Availability and access to care must be seen in the context of the individual and of society in conditions which increase risk of disease and measures to reduce those risks in order to prevent disease and promote health. This interrelates with human resources for health (Chapter 14), financing and economics (Chapter 11), organization (Chapter 10), technology, law, ethics (Chapter 15), and globalized health (Chapter 16).

WHO defines a health system as "The people, institutions and resources, arranged together in accordance with established policies, to improve the health of the population they serve, while responding to people's legitimate expectations and protecting them against the cost of ill-health through a variety of activities whose primary intent is to improve health. It is a set of elements and their relationship in a complex whole, designed to serve the health needs of the population. Health systems fulfill three main functions: health care delivery, fair treatment to all, and meeting health expectations of the population" (WHO, 2000).

Most industrialized countries have implemented national health programs as health insurance systems or national health services. Each was developed in the political context of the country and continues to evolve with its own and international experience. Developing countries are also struggling to achieve universal access to care and health for all by expanding primary health care and social security plans providing benefits to workers. As they move up the scale of economic development, developing countries also address the problem of how to decrease morbidity and mortality, to achieve equity in access to health care as well as how to expand the funding basis for health care through national health insurance. Some countries experiencing rapid economic development lag behind in applying increased national wealth to improve health status due to lack of focused political commitment, trained policy analysts and a trained public health workforce (Chapters 14 and 16).

Each national health system has its own characteristics and problems. System management requires continuous evaluation based on well-developed information systems, trained health management personnel, and involvement of society through professional organizations and advocacy groups. There is no single defined "gold standard" plan to provide universal access to health care that is suitable for all countries. Each country develops and modifies a program of national health appropriate to its own cultural needs and resources available. But there are evolving sets of patterns in health care, so that countries can and do learn from one another (Box 13.1).

Barriers to care can be geographic, cultural, social, and psychological as well as financial. Removing financial barriers to care is necessary but not sufficient for optimal health and to address the health problems of an individual and of a society. Equity in access to quality health care, with minimal interregional, socioeconomic, and demographic differences, is vital to good public health standards. Special attention to high-risk groups in the population is not less essential, whether the group is based on age, gender, occupation, risky lifestyle, location of residence, or ethnicity, religion, sexual orientation, economic status, or other factors that increase susceptibility to

---

Box 13.1 Key Elements of National Health Systems

1. Health targets;
2. Demographic, epidemiologic, economic monitoring;
3. Health promotion;
4. Universal coverage by public insurance or service system;
5. Access to a broad range of health services;
6. Strategic planning for health and social policies;
7. Improvement in health status indicators;
8. Recognition of special needs of high-risk groups and related issues;
9. Portability of benefits when changing employer or residence;
10. Equity in regional and sociodemographic accessibility and quality of care;
11. Adequacy of financing;
12. Cost containment;
13. Efficient use of resources for a well-balanced health system;
14. Consumer satisfaction and choice of primary care provider;
15. Provider satisfaction and choice of referral services;
16. Public administration and regulation;
17. Promotion of high-quality service;
18. Promote patient and staff safety;
19. Comprehensive primary, secondary, and tertiary levels of care;
20. Well-developed information and monitoring systems;
21. Continuing policy and management review;
22. Promotion of standards and accreditation of services, professional education, training, research;
23. Governmental and private provision of services;
24. Decentralized management and community participation;
25. Assurance of ethical standards of care for all;
26. Conduct health systems research;
27. Preparation for mass casualties from disasters and bioterrorism.

disease, premature death, or disability. The services needed are not only those that are available on patient demand, but also those services that reach out to the entire population, especially to people at high risk who are often least able to seek appropriate care.

A program that provides equal access for all may not achieve its objective of better health for the population unless it is accompanied by other important governmental activities. These include enforcement of environmental and occupational health laws, food, nutrition, and water standards, improved rural care, higher educational levels, and provision of health information to the public. Additional national programs are needed to promote health generally and to reduce specific risk factors for morbidity and mortality. Responsibility for health lies not only with medical and other health professionals, but also with society and its executive instruments, governmental and voluntary organizations, as well as the individual, the family, and peer groups.

Individual access to an essential basket of services as a prepaid insured benefit is integral to a successful national health program. Each country addresses this issue according to its means and traditions, but the benefits should be examined as to which are most effective at least cost to meet epidemiologic and demographic needs. Payments for heart transplantation may be beyond the means of a health system, but early and aggressive management of acute myocardial infarction is effective in saving lives at modest cost. Improved diets, smoking reduction, and physical fitness are even more effective and less costly. Prevention is cost-effective and should be integral to the development of service priorities within the basket of services.

Globalization affects health systems around the world not only in the ease of spread of infectious diseases, but in increased access to modern preventive, diagnostic and treatment modalities. Access to antiretroviral drugs is changing the face of AIDS in many developing countries with support of international and bilateral donors. The same is true for vaccines so that the spread of Hib vaccine will save many children's lives in the coming years. Information technology, migration of medical professionals, and internalization of educational standards are all global health issues affecting national health systems (Chapters 14, 15, and 16). Health systems are facing similar problems in population health, with aging of populations, rising obesity and diabetes, and rising costs of care. Health systems research capacity is important in each country as it attempts to cope with rapid changes in population health and individual health needs within limited resources.

In this chapter, national health systems are presented representing major models of organization, which may be influential in health care system formulation in both developing and developed countries, and in countries restructuring their health services. Health care systems and financing are under pressure everywhere, not only to assure access to health for all citizens, but also to keep up with advancing medical technology, and contain the increase in costs to sustainable levels. Because a health system is judged by more than its costs and measures of medical services, this chapter includes indicators of health status of the population, including morbidity and mortality. This topic has developed a complex terminology of its own, and some of the key words are defined in this and other chapters in this text and in the glossary.

Finally, health systems are meant to improve health and quality of life, as measured by quantitative and qualitative methods (see Chapter 3). Since 2000, the Human Development Index provides a standard method of comparison which combines many health indices into a summary figure. These include life expectancy at birth, GDP per capita, child mortality, and others. Table 13.1 provides some of these key indications discussed in this chapter for some industrialized as well as mid-level and other developing countries.

HEALTH SYSTEMS IN DEVELOPED COUNTRIES

Evolution of Health Systems

The tradition of prepayment of health care goes back to ancient times, when municipal doctors were employed by local authorities to provide care for the poor and the slaves. In the Middle Ages, the Church provided charitable care for the poor. In the medieval and Renaissance periods, guilds provided prepaid health care to members and their families. These later evolved into the friendly (benevolent) societies, as mutual benefit programs that provided for burials, pensions, and payment for health services for members. In the twentieth century, these developed through collective bargaining into health insurance plans through private or professionally sponsored insurers, and labor union–sponsored health plans. Governmental responsibility for health systems evolved in public health and health protection systems in the nineteenth and twentieth centuries, and continues to evolve to face new challenges and preventive and treatment capacities.

Social Insurance

Otto von Bismarck, Chancellor of Germany, introduced the first national health insurance plan for workers. It followed previous legislation in Germany establishing workmen's compensation on railroads (1838) and compulsory miners' benevolent societies (1854). Workmen's compensation and other benefits were extended in 1871 to many workers in other industries, such as those in domestic service, workers in mines, factories, and quarries, and seamen. Bismarck's 1883 compulsory health insurance legislation was intended to improve the health of workers and their families, and especially of potential army recruits, as well as to stave off the political advancement

TABLE 13.1 Human Devlopment Index Ranking of Selected Countries, 2005

	HDI Rank	LE in Years at Birth	GDP Per Capita ($PPP)	GDP Spent on Health (percent)	Under 5 Mortality Rate	Maternal Mortality Ratio (adjusted 2000)
Norway	2	79.8	41,420	9.1	4	7
Canada	4	80.3	33,375	9.8	6	7
Sweden	6	80.5	32,525	9.1	4	3
Japan	8	82.3	31,267	8.0	4	6
Netherlands	9	79.2	32,864	9.2	5	6
Finland	11	78.9	32,153	7.5	4	7
United States	12	77.9	41,890	15.2	7	11
Denmark	14	77.9	33,973	9.1	5	3
United Kingdom	16	79.0	33,238	8.3	6	8
Germany	22	79.1	29,461	10.7	5	4
Israel	23	80.3	25,864	8.0	6	4
Russia	67	67.0	10,845	5.2	18	28
Colombia	75	72.3	7,304	7.3	21	120
China	81	72.5	6,757	4.7	27	45
Nigeria	158	69.1	1,128	3.9	194	1,100

Note: LE–Life Expectancy, HDI–Human Development Index, GDP-Gross Domestic Product.

Sources: Human Development Reports 2007/08, http://hdrstats.undp.org/indicators/and World Health Statistics 2008, http://www.who.int/whosis/whostat/2008/en/index.html [accessed June 25, 2008]

of the socialist parties. The program was based on the principle of social insurance, involving payroll deductions at the place of work, with contributions from the employer and employee, to cover medical care, unemployment benefits, and pensions for workers.

The Bismarckian model established state social insurance with prepayment by workers and their employers. It utilized Sick Funds (*Krankenkassen*) as insurers to provide payment to the physician, hospital, or other provider. In the years before World War I, many countries in Central and Eastern Europe implemented similar health plans. In the period between the World Wars, national health insurance programs were developed in many countries in the industrialized world. In Europe, most countries developed models based on the Bismarckian approach, with compulsory contributions by workers and their employers to a national social security system. This then finances approved Sick Funds that pay for services usually through private medical practice with fee-for-service payment. The Bismarckian model is used widely in Europe and Israel. This model has also influenced post-Soviet health reforms in countries of Eastern Europe.

In 1911, the Liberal government of Great Britain, initiated by David Lloyd-George, Chancellor of the Exchequer, influenced by the German compulsory health insurance scheme, introduced the National Health Insurance Act. It was compulsory for all wage earners between ages 16 and 70 who made payments along with their employers and a state contribution. This two-part plan provided a contributory system for unemployment insurance and for medical care against illness for workers and their families. General practitioner services were paid on a capitation basis rather than on a salary, preserving their status as self-employed professionals. Initially this plan covered one-third of the population, but coverage increased to one-half by 1940. Administration was through approved mutual benefit societies (the Friendly Societies), some based on insurance companies and others by professional associations and trade unions. European countries and Japan gradually developed compulsory health insurance following World War I, and completed universal coverage following World War II.

The Social Security model of health insurance for urban workers also became prominent in many countries in Latin America. Social security plans are financed by mandatory contributions of workers and employers, and administered by the state. The Social Security Act of 1935 in the United States was instituted to alleviate the social distress of the Great Depression. This social experiment of the "New Deal" of President Franklin Roosevelt provided cash benefits for widows, orphans, and the disabled, as well as pensions for the elderly, and provided a base for future reform including health insurance. Since 1965, this legislation has provided the basis for U.S. medical and hospital coverage of the elderly under Medicare and the poor under Medicaid. Later proposals for national health insurance in the United States have also largely been based on the federal Social Security funding system.

National Health Service

In some countries, the state directly assumed the responsibilities for both social security and health care. The welfare state took on measures such as unemployment and disability insurance, special disability benefits for the blind, widows, orphans, and the elderly through pensions, and child benefits to raise levels of child care and nutrition through general governmental revenues from taxation and other sources.

In 1918, following the Russian revolution, the new Soviet Union (U.S.S.R.) introduced its national health plan for universal coverage within a state-run system of health protection. The Soviet model, designed and implemented by Nikolai Semashko, provided free health care for all as a government-financed and -organized service. It brought free health services to the U.S.S.R., with a system of primary and secondary care, with universal and equitable access to care through district organization of services. It achieved control of epidemic and endemic infectious diseases and expanded services into the most remote areas of the country.

In the early days of World War II, the British government established a national Emergency Medical Service to operate hospitals in preparation for the large-scale civilian casualties expected. The plan established national health planning and rescued many hospitals from near bankruptcy resulting from the effects of the Great Depression in the United Kingdom. During World War II, a postwar social reconstruction program was developed by William Beveridge, at the behest of the wartime government of Winston Churchill. The Beveridge Report of 1942, *Social Insurance and Health Services*, outlined the nature of the future welfare state including a national health service, placing medical care in the context of general social policy for the total population.

The wartime government coalition approved the principle of a national health service, which had wide public support, despite opposition from the medical association. In 1948, the Labour government of Clement Attlee implemented the National Health Service (NHS), a nationally financed, universal coverage system providing free care by general practitioners, specialists, hospitals, and public health services.

National Health Insurance

The Canadian system of tax-based national health insurance is based on provincial health plans meeting federal government requirements for cost sharing. The program evolved from provincial initiatives led by Tommy Douglas, premier of the Province of Saskatchewan. Initiated in 1946, provincial plans provided universal insured hospital services under provincial public administration, later followed by medical and other services.

Developed over the period 1946–1971, the plans were promoted by federal governmental cost sharing, political support, and national standards. The plans were initially financed by taxation and premiums, but later by general tax revenues alone. The Canadian "Medicare" plans are publicly administered by the provinces with federal standards, cost sharing, and comprehensive coverage. Care is provided by private medical practitioners on a fee-for-service basis under negotiated medical fee schedules. Hospitals are operated by non-profit voluntary or municipal authorities, with payment by block budgets. This Medicare-type plan was later adopted in a number of other countries including Australia. Figure 13.1 indicates health expenditures by the member countries of the Organization for Economic Cooperation and Development.

THE UNITED STATES

In 2006, the population of the United States reached 300 million, with a GDP per capita of $43,800 and a child mortality rate of 7 per 1000 live births (34th in world ranking). In 2004, the United States ranked 46th in average life expectancy from birth and 42nd in infant mortality; and U.S. life expectancy (total) at birth was 77.9 years in 2005, ranking 31st among developed countries. The United States is near the middle of the 32 OECD nations in death rates from all causes, mainly heart disease, cancer, and stroke. With declining birth rates and increasing longevity, the population is aging with over 12 percent aged 65 and over. In 2006–2007, the United States stood 12th among the leading nations in the Human Development Index, although steadily improving in HDI since 1975 and ranking above the average for OECD countries.

The United States has a federal system of government, with 50 states each having its own elected government with legislative, judicial, enforcement, and taxing powers. The U.S. Constitution gives primary responsibility for health and welfare to the states, but direct federal services are provided to the armed forces, veterans, and Native Americans. However, the federal government has established a major leadership role in health by development of national standards, regulatory powers, and information systems. It also serves as a major agency for financing research, health services, and training programs.

Federal Health Initiatives

In 1798, the federal government under President John Adams established the U.S. Marine Hospital Fund to provide for prepaid care sick merchant seamen. It later became the Marine Hospital Service and then the U.S. Public Health Service as a commissioned corps and a uniformed service headed by the U.S. Surgeon General (1873).

% GDP

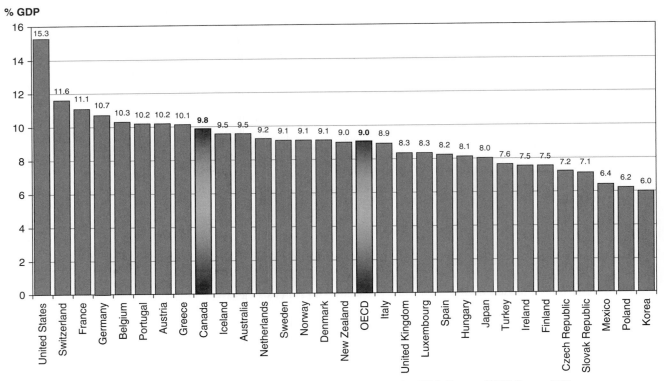

FIGURE 13.1   Health expenditure as a share of GDP, OECD countries 2005. Source: OECD Report 2007, http://www.oecd.org/dataoecd/46/33/38979719.pdf [accessed May 24, 2008]

In the late nineteenth and early twentieth centuries, the federal Department of Agriculture Extension Service promoted nutrition and hygiene education throughout the rural areas of the country. Later legislation provided federal grants to establish state, municipal, and county health departments. Health hazards in poor standards of food and drugs, nursing homes, lack of care for the elderly and the poor, dangerous automobiles, environmental pollution, and deficiencies in health services led to government intervention to protect the public interest. The Food and Drug Control Act of 1906 was promulgated to regulate and control commerce. In 1921, the Sheppard-Towner Act established the federal Children's Bureau that administered grants to assist states to operate maternal and child health programs, which were later incorporated into the Social Security Act.

In 1927, the Committee on the Costs of Medical Care, a commission funded by several private foundations, recommended that the United States implement a universal national health program based on medical group practices with voluntary prepayment. From the 1920s, labor unions won health insurance benefits through collective bargaining, which became the main basis for prepayment for health care in the United States until today. These initiatives were slowed due to the Great Depression from 1929–1939, but resumed after World War II. The Social Security Act of 1935 increased social support for millions with disabilities or occupational

injuries, widows and orphans, and the elderly, alleviating some of the worst effects of the Depression.

During World War II (1941–1945) millions of Americans in the armed forces and their dependents, previously with limited access to prepaid health care, were enrolled in a national plan for free health care (Emergency Maternity and Infant Care for the Wives and Children of Servicemen). At the same time, health benefits through voluntary insurance for workers were vastly expanded in place of wage increases, which were forbidden by federal wartime regulation. At the end of the war, millions of veterans were eligible for health care through the Veterans Administration (VA), which established a national network of federal hospitals and primary care services for this purpose.

In 1946, President Truman attempted to bring in national health insurance but the legislation (the Wagner-Murray-Dingell Bill) failed in the U.S. Congress. One section of the bill was approved, enabling the federal government to initiate a program of categorical grants to upgrade hospital facilities around the country under the Hill-Burton Act. Another provided massive federal funding for health to strengthen the National Institutes of Health (NIH), established after World War II, to promote research and strengthen public and private medical schools, teaching hospitals, and research facilities. In the 1950s, the federal government also established the Centers for Disease Control (CDC) as well as increased public health grants providing assistance for state and local public health activities.

From the 1940s through the 1960s, voluntary health insurance became the major method of prepayment for health care needs, mostly through employment contracts. The private insurance industry developed rapidly, with minimal governmental regulation to ensure fair pricing and payment. During the 1970s and 1980s, employers became concerned about the costs of health insurance for their workers and pressed the government to act to restrain health care costs. Federal initiatives included public insurance for the elderly and poor, promoting efficiency in payment for hospital care, and later promotion of health maintenance organizations (HMOs) and managed care.

Medicare and Medicaid

In the mid-1960s, despite the growth of voluntary and employment-based health insurance, a large percentage of the elderly and poor American population lacked health insurance. In 1965, President Lyndon B. Johnson introduced Medicare for the aged (over age 65), disabled persons, and persons on renal dialysis as Title XVII of the 1935 Social Security Act. This brought some 10 percent of the population a limited form of national health insurance.

Medicaid, Title XIX of the Social Security Act also enacted in 1965, provided federal cost sharing for acceptable state health plans for the poor, with local authority participation. These two plans together brought some 25 percent of Americans into public systems of health insurance. Limitations included variable definitions of poverty in each state, and co-payments for Medicare beneficiaries. Medicare covers hospitalization, including skilled nursing home care, medical and appliances and other benefits with co-payments. In 2006, a drug benefit program was added.

In 1997 Title XXI of the Social Security Act the State Children's Health Insurance Program (SCHIP) was initiated to provide federal funds to assist approved state plans to extend health insurance for children. This

program provides health coverage for families that are not eligible for Medicaid due to their income status but who cannot afford to purchase independent insurance. While funding for SCHIP is provided by both federal and state governments, each state runs its own SCHIP program under the broad guidelines of the federal government and the specific guidelines created by the state. This program was originally set to expire in 2007 but is being proposed for reauthorization for 5 more years in the president's fiscal year 2008 budget. It was vetoed by President George W. Bush in October, 2007. A congressional effort to override the veto failed by 13 votes (273–156, with two-thirds approval required) 15 days later. The subject of extending health insurance coverage for the uninsured population, and particularly for children, is a subject of ongoing political activity in 2008.

In 2006, 67.9 percent of the United States population was covered under private health insurance, mostly employment-based, 13.6 percent under Medicare, 12.9 percent under Medicaid, while 15.8 percent were uninsured. Medicare and Medicaid brought many previously uninsured persons under health insurance coverage. Public funding for health care in the United States includes Medicare, Medicaid, and SCHIP, research and medical education, and promotion of community health centers and services in impoverished or underserved areas (see Table 13.2). The percentage of public funding in the United States rose from under 25 percent of total health expenditures in 1960 to approximately 45 percent of total health expenditures in the years 1995–2004. The population enrolled in Medicare increased from 19 million in 1966 to 40.3 million in 2006, including 6.8 million disabled persons under the age of 65. The Medicaid enrolled population increased from 28.2 million in 1991 to 49.3 million in 2006. This contributed to increasing health expenditures in the public sector, a concern for both critics and supporters of public health care programs.

In 2006 about 4.5 million individuals were enrolled in the SCHIP. Congress created SCHIP with the goal of significantly reducing the number of low-income uninsured children. Under certain circumstances, states may also

TABLE 13.2 Health Expenditures, United States, 1990–2005

Health expenditures	1990	1995	2000	2005
Total (billion $)	714	1017	1353	1988
Annual percent change	11.8	5.6	7.0	6.9
Percent of gross domestic product spent on health	12.3	13.7	13.8	16.0
Health expenditures (percent distribution)	100	100	100	100
Private	59.8	54.3	55.9	54.6
Public	40.2	45.7	44.1	45.4

Source: Health, United States, 2007

cover adults, and in June 2006 about 349,000 adults were enrolled. Each state receives an annual allotment of federal funds, available as a federal match based on the state's expenditures. Generally, states have 3 years to use each fiscal year's allotment, after which unspent federal funds may be redistributed. Congress initially authorized SCHIP for 10 years, from 1998 through 2007.

In 1997 Congress created the Medicare + Choice plan (now known as Medicare Advantage) which gives Medicare enrollees the choice of various health plans. It was also created in the hopes of controlling Medicare costs. The Medicare Prescription Drug Improvement and Modernization Act of 2003 was signed into law by President George W. Bush in an effort to make prescription drugs affordable to Medicare enrollees.

Medicare costs are increasing at a faster rate than the economy. Financing of Medicare comes from two trust funds: the Hospital Insurance and the Supplementary Medical Insurance. Taxes paid by employees and employers support the Hospital Insurance (HI) trust fund which finances inpatient care. This trust fund is expected to be depleted by the year 2019. The Supplementary Medical Insurance (SMI) is supported by general income tax revenues and enrollee premiums. The SMI covers physician services, outpatient and hospital services, and prescription drugs. The federal government faces the challenge of making appropriate reforms in Medicare in order to avoid consuming more federal revenues and taking from other federal programs, especially as the baby boom generation becomes eligible for Medicare benefits.

Medicaid is financed through shared responsibilities of the federal government and state governments. Medicaid spending increased to $295.9 billion in 2004 as compared to $205.7 billion in 2000. In 2003, the Jobs and Growth Tax Relief Reconciliation Act was passed by Congress to help states meet some of their Medicaid budget shortfalls. Although Medicaid has had a substantial increase in adult and children enrollees, much of its costs go toward the disabled and elderly populations. A reduction in employer-based insurance also contributes to the growth in the demand for Medicaid. Medicaid faces growing expenses and new challenges in cost containment, as does the U.S. health system generally.

The Changing Health Care Environment

From the 1960s through the 1990s, rapid cost increases were attributed to many factors. These included an increasing elderly population; high levels of morbidity in the poor population; the spread of AIDS; rapid innovation and costly medical technology; specialization; high laboratory costs; and large-scale public investment in medical education and research and health facility construction.

Other equally important factors were high levels of preventable hospitalizations; an institutional orientation of the health system; high administrative costs due to multiple private billing agencies in the private insurance industry; high incomes for physicians, especially for specialists; and high medical malpractice insurance costs. The pressure for cost constraint came from government, industry, and the private insurance industry.

Most hospitals are owned and operated by nonprofit agencies, including federal, state, and local governments, voluntary organizations, and religious organizations. Privately owned hospitals operating for profit increased from 7.8 percent of community, short-term hospital beds in 1975 to 12.7 percent in 1996. Private medical practice, with payment by fee-for-service, was the major form of medical care until the 1990s. HMOs and other forms of managed care have grown rapidly to become the predominant method of organizing health care in the United States.

Prepaid group practice (PGP) originated from company-provided contract medical care, especially in remote mining camps. The Community Hospital of Elk City, Oklahoma, established in 1929, is considered the first real medical cooperative or prepaid group practice. Later, many rural cooperatives were formed to provide prepaid medical care. Union-sponsored health services were developed to provide medical care in poor mining areas in the Appalachian Mountains, as well as in an urban cooperative in Washington, D.C., in 1937. In the 1940s, New York City sponsored the Health Insurance Plan of Greater New York to provide prepaid medical care for residents of urban renewal and low-income housing areas. This was later supported by organized union groups such as municipal employees and garment industry workers.

Prepaid group practice became best known in the Kaiser Permanente network developed for workers of Henry J. Kaiser Industries, at the Boulder Dam and Grand Coulee Dam construction sites in the 1930s. This experience was applied in Kaiser's rapidly growing industries in the San Francisco Bay area. Kaiser Permanente health plans now provide care for millions of Americans in many other states. Initially opposed by the organized medical profession and the private insurance industry, PGP gained acceptance by providing high-quality, less costly health care. This became attractive to employers and unions alike, and later to governments seeking ways to constrain increases in health costs.

Since the 1970s, the generic term *health maintenance organization* (HMO) has been used, especially by the federal government seeking to promote this concept. The HMO concept, linking health insurance and medical care in the same organization, was promoted through the HMO Act by President Richard Nixon in 1973. The HMO including both HMOs and other forms of prepaid insurance plans, now called *managed care*, has become an accepted, if criticized, part of medical care in the

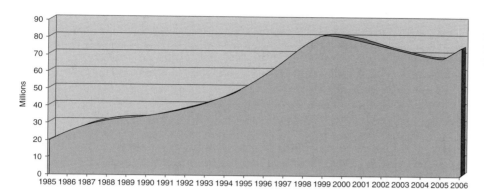

FIGURE 13.2 National HMO Enrollment. Source: http://www.mcareol.com/factshts/factnati.htm [accessed March 6, 2008]

United States and an important alternative to fee-for-service, private practice medicine (Figure 13.2).

In order to encourage more efficient use of hospital care, the method of payment was changed during the 1980s. In 1983, a prospective payment system, called *diagnosis-related groups* (DRGs), was adopted for Medicare, with payment by categories of diagnosis (HCFA, 1998). This replaced the previous system of paying by the number of hospital days, or *per diem*. DRGs encourage hospitals to diagnose and treat patients effectively and expeditiously and to discharge them as quickly as their condition allows. Payment for Medicare and Medicaid patients was shifted to this method. In many states this has also become standard for patients with private health insurance. Between 1980 and 1990, because of the DRG payment system and HMOs or managed care systems, which promote alternatives such as home and ambulatory care, hospital utilization was reduced in the United States. While total costs of health care increased in this period, without reduction of hospital utilization the increase would have been considerably higher.

During the late 1980s, the term *managed care* was introduced, expanding from HMOs of the Kaiser Permanente type to include both non-profit and for-profit systems. Managed care plans of the HMO type operate their own clinics and staff (i.e., the staff model). Other managed plans operate on a not-for-profit or a for-profit basis. These are independent practice associations (IPAs), which operate with physicians in private practice, or preferred provider organizations (PPOs), which cover care with doctors and other providers associated with the plan providing services to the enrolled members or beneficiaries at negotiated prices (Chapter 12). Managed care enrollment of Medicare beneficiaries at the end of 2006 with 65 percent overall varied widely by state, with the majority of most states over 45 percent.

Following the failure of the Clinton national health insurance proposal in 1994, managed care experienced tremendous growth as employers sought to provide their employees with comprehensive coverage at reasonable costs. Managed care systems have been able to cut costs in health care in ways that the U.S. government could not.

In 1996, 74 percent of insured American workers were enrolled in managed care plans, as compared to 55 percent in 1992. In California, with a long tradition of HMOs such as Kaiser Permanente, enrollment at the end of 2006 was 65 percent of the total state population. In the United States as a whole, in addition to the nearly 58 million persons enrolled in HMOs, another 91 million persons are enrolled in PPOs, with 25 percent of Medicaid and 10 percent of Medicare beneficiaries in managed care plans.

The search for cost containment led to the development of a series of important innovations in health care delivery, payment, and information systems. HMOs have demonstrated that good care provision can be operated efficiently with lower hospital admission rates than care provided on a fee-for-service basis. The managed care systems brought about profound changes in health care organization in the United States. The number of plans declined from 572 in 1990 to 412 in 2004 as a result of mergers. Of total enrollment in managed care, a total of 149 million persons (Table 13.3) or 51 percent of the insured population and 49 percent of the total U.S. population.

Managed care received much negative publicity due to private for-profit insurance operators apparently making tight restrictions on access to care to reduce costs, and this has made the capitation payment method widely criticized by supporters of open-ended fee-for-service. Proponents of managed care point to successful high-quality programs such as the Harvard Pilgrim Managed Care plan which has been very successful in growth and sustaining high-quality care within reasonable cost parameters. They have pioneered in utilization review, preventive practices as part of regular medical care, computerization of medical records, and quality promotion. The topic of managed care will be a central issue for the 2008–2012 federal government in the search for universal coverage health insurance at affordable costs.

Hospitals and other specialty services are competing for contracts with managed care organizations and establishing community service systems of their own in order to compete for "market share" of insured clients. In many locales, excess hospital beds have become an economic

TABLE 13.3 Managed Care Coverage by Type of Health Insurance Plan, United States, 2007

Numbers of Enrollees by Insured System (millions)	Total U.S.	U.S. (percent)	Managed Care	Managed Care (percent)
Medicare	44.2	14.6	8.7	20.0
Medicaid	58.9	19.4	38.3	65
Commercial	152.9	50.5	101.9	66.6
Uninsured	47	15.5	0.0	0.0
Total	303	100.0	148.9	49.1

Source: Managed Care Fact Sheet, Managed Care National Statistics, 2008, http://www.medicarehmo.com/mcmnu.htm [accessed June 23, 2008]

burden, forcing many hospitals to downsize or sell out to larger hospital chains. Hospitals have responded by establishing contracts with managed care organizations and by reducing bed capacity; others have closed as they were unable to compete for sufficient patient flow.

Federal and state legislative initiatives are attempting to define patients' rights in managed care because of public complaints with limitations of managed care. In response to widespread complaints regarding managed care restricting access to specialty services and shortened hospital stays, in 1998 the U.S. Congress passed a bipartisan-sponsored law that requires minimum 48-hour maternity stays. Many other pieces of legislation to protect consumers' rights and choice of doctor have been proposed in Congress and in state legislatures.

Health Information

The United States has developed extensive information systems of domestic and international importance. The CDC publishes the *Morbidity and Mortality Weekly Report* (*MMWR*), which sets high standards in disease reporting and policy analysis. The U.S. National Center for Health Statistics (NCHS), the Health Care Financing Administration (HCFA), the U.S. Public Health Service (PHS), the Food and Drug Administration (FDA), the National Institutes of Health (NIH), and many nongovernmental organizations (NGOs) carry out data collection, publication, and health services research activities important for health status monitoring. National nutrition surveillance (NHANES, see Chapter 8) and other systems of health status monitoring are reported in the professional literature and in publications of the CDC. National monitoring of hospital discharge information facilitates the understanding of patterns of utilization and morbidity. These information systems are vital for epidemiologic surveillance and managing the health care system.

The 1965 U.S. Surgeon General's *Report on Smoking*, linking smoking and lung cancer, had a major impact on public knowledge and behavior. The 1988 *Surgeon General's Report on Nutrition* was another landmark document, setting international standards in public health. More recent reports of the Surgeon General include *Women and Smoking: A Report of the Surgeon General* (2001), *Youth Violence: A Report of the Surgeon General* (2001), *Mental Health: Culture, Race, and Ethnicity, A Supplement to Mental Health: A Report of the Surgeon General* (2001), *Bone Health and Osteoporosis: A Surgeon General's Report* (2004), *The Health Consequences of Smoking: A Report of the Surgeon General* (2004), *The Health Consequences of Involuntary Exposure to Tobacco Smoke: A Report of the Surgeon General* (2006), *Children and Secondhand Smoke Exposure — Excerpts from* The Health Consequences of Involuntary Exposure to Tobacco Smoke: *A Report of the Surgeon General* (2007).

The Surgeon General continues to promote awareness of knowledge on important public health issues including physical activity and health, mental health, oral health, youth violence, bone health and osteoporosis, underage drinking, sexual health, and other topics. Reduction in smoking and obesity continue to be priority issues. Media coverage of health events and topics is very extensive, and public levels of health knowledge grow steadily but vary widely by social class and educational levels.

The CDC created the National Center for Public Health Informatics (NCPHI) in 2005 to provide leadership and coordination of shared systems and services, to build and support a national network of integrated, standards-based, and interoperable public health information systems. This is meant to strengthen capabilities to monitor, detect, register, confirm, report, analyze, provide feedback and alerts on important health events. This will enable partners to communicate evidence for decisions with health impacts. Electronic medical and personal health records are now widely used that both protect patient privacy and confidentiality and serve legitimate clinical and public health needs. U.S. health costs are rising, now representing 16 percent of GDP. At the same time acute and chronic health threats challenge United States, and other countries' capacities to address them with efficient and effective disease prevention.

Health Targets

Despite rapid increases in health care expenditures during the 1970s and 1980s, improved health promotion activities, and rapidly developing medical technology, the health status of the American population has improved less rapidly than that in other western countries. Infant mortality in the United States in 1996 remained higher than that in 21 other countries. Even the rate of infant mortality of the white population of the United States was higher than that of 16 countries which spent much less per person and a lesser percentage of GNP per capita on health care.

The 1979, the U.S. Surgeon General's Report *Healthy People* set forth a series of national health targets for a wide variety of public health issues. The program defined 226 objectives in 15 program areas within the three categories of prevention, protection, and promotion. These goals and objectives were formulated based on research and consultation by 167 experts in different fields who participated in a conference by the U.S. Public Health Service. Consensus was based on position papers, studies, and conferences involving the national governmental health authority, the National Academy of Science Institute of Medicine, and professional organizations, such as the American Academy of Pediatrics and the American College of Obstetrics and Gynecology. Many private individuals and organizations contributed to this effort, including state and local health agencies, representatives of consumer and provider groups, academic centers, and voluntary health associations.

These targets (Table 13.4) are periodically assessed as performance indicators of the U.S. health system and then updated. Progress made during the 1980s included major reductions in death rates for three of the leading causes of death: heart disease, stroke, and unintentional injuries. Infant mortality decreased, as did the incidence of vaccine-preventable infectious diseases.

Healthy People 2000, published in 1992 by the Surgeon General, detailed 332 specific health targets, in six groups, for the year 2000, in the areas of health promotion, health protection, preventive services, surveillance and data systems, and age-related and special population groups (see Chapter 11). The final reviews of *Healthy People 2000* showed significant decreases in mortality from coronary heart disease and cancer.

TABLE 13.4 Categories and Program Target Areas for Health Objectives, United States, 1979–1995

Categories	Specific groups or activities with measurable targets
Health promotion	Improve physical activity and fitness Nutrition; reduced deficiency and excess dietary conditions Reduced tobacco use and exposure Reduce alcohol and drug abuse Promote family planning Promote improved care for mental illness, mental disorders Reduce violent and abusive behavior Promote educational and community-based programs
Health protection	Reduce unintentional injuries Improve occupational safety and health Reduce environmental toxic exposure Improve food and drug safety Improve oral health
Preventive services	Reduce maternal and infant illness and risk factors Reduce risk factors for heart disease and stroke Reduce risk factors and morbidity from cancer Improve case finding and management of diabetes and other chronic disabling conditions Reduce incidence of HIV infection Reduce incidence of sexually transmitted infections Improve immunization coverage to control infectious diseases Increase use of clinical preventive services
Surveillance and monitoring	Improve surveillance and data systems at federal, state, and local levels
Age-related objectives	Improve services and utilization of services for children, adolescents and young adults, adults, and the elderly
Special population objectives	Improve access to and care of people with low income: African-Americans, Hispanics, Asians and Pacific Islanders, American Indians and other Native Americans

Source: Public Health Service. 1992. *Healthy People 2000: National Health Promotion and Disease Prevention Objectives.* United States Department of Health and Human Services. Boston: Jones and Bartlett; Developing Healthy People 2020, http://www.healthypeople.gov/ [accessed June 24, 2008]

In 2000, the DHHS released *Healthy People 2010* with two main goals: "increase the quality and years of healthy life" and to "eliminate health disparities." These goals focus on 28 specific areas developed by over 350 national membership organizations and 250 state health, mental health, substance abuse, and environmental agencies. A midcourse review of *Healthy People 2010* shows that 60 percent of the objectives are either being met or moving forward. The United States is moving toward the goal to "increase the quality and years of healthy life" although there are still clear gender, race, and ethic discrepancies. Reducing health disparities continues to be a challenge in the United States.

Many states have adopted use of these targets as their own measures of health status and performance. Annual publications by the U.S. Public Health Service, in cooperation with the National Center for Health Statistics, make available a wide set of data for updating health status and process measures relating to these national health goals. The value of working toward health targets is widely accepted, and Healthy People 2020 is in preparation (http://www.healthypeople.gov).

Health promotion has received wide public, governmental, and professional support in the United States over the 1980s and 1990s, and continues in the twenty-first century. In part, this reflects a long tradition of education on health matters in the rural agricultural sector and school health education. Nutrition and antismoking consciousness has grown in part because of wide media attention to many important epidemiologic studies.

Consumer advocacy has been a potent factor for change in the United States in the twenty-first century, and especially since the 1960s. It has contributed to strengthened governmental regulation in a wide area of public health-related fields (Chapter 2). These include automobile safety features and emission control, environmental standards, Mothers Against Drunk Driving (MADD), nutritional labeling and vitamin and mineral fortification of basic foods, and legal action against cigarette manufacturers. Food fortification, pioneered in the United States, is not mandatory as in Canada, but is nevertheless nearly universal, and mandatory for those foods labeled "enriched" (see Chapter 8). This is accepted in the general population based on advocacy, informed public opinion, and an innovative, highly competitive food industry. Despite much public controversy, fluoridation of community water supplies covers 67 percent of the population, a higher coverage than most industrialized countries.

Advocacy groups can also promote regression in public health measures, as with groups currently fighting against immunization on the grounds of exaggerated concerns over reactions to vaccines. Some opposed to abortion have greatly affected public policy and promote sometimes violent activities against proponents and providers of abortions. Research and wide media coverage of health issues encourage a high level of individual and community consciousness of health-related issues and a climate receptive to health promotion.

Social Inequities

Lack of universal access and the empowerment it brings encourages an alienation or non-engagement with early health care, promoting inappropriate reliance on emergency department care and hospitalization in response to undertreated health needs. With large numbers of uninsured persons and many lacking adequate health insurance, access and utilization of preventive care are below the levels needed to achieve social equity in health, especially for maternal and child health and for chronic diseases such as diabetes, cancer, and heart disease.

As measured by the infant mortality rate in 2004, the rate among black mothers was twice as high as the national average. A significantly higher rate of infant mortality exists among Puerto Rican and American Indian populations as compared to the national average. Efforts to improve immunization coverage of U.S. infants to meet national health targets have been partially successful with efforts directed toward poor population groups. During the 2006–2007 school year, approximately 95 percent of all children entering kindergarten received all mandatory vaccinations. Efforts to fully immunize the preschool population on time will require special attention of the public health system for years to come. In 2002, a program called Racial and Ethnic Adult Disparities in Immunization Initiative was introduced in order to reduce the low levels of influenza and pneumococcal vaccinations among minorities aged 65 and over.

School lunch programs and nutrition support for pregnant women and children in need have reduced some of the ill effects of poverty in the United States, but lack of health insurance affects these groups severely. Chronic disease and trauma are also diseases of poverty with higher rates of morbidity and mortality in virtually all categories.

Health disparities are a complex problem that goes beyond the issue of uninsured Americans. Low-income and illegal immigrants face challenges accessing medical insurance. New immigrants to the United States who obtained citizenship after August 1996 must wait 5 years before they are eligible for Medicaid. The structure of the medical system plays an important role in an individual's ability to obtain medical attention. This includes convenience of appointment making and office hours, waiting times, and transportation. A lack of health literacy also plays a role in an individual's ability to seek medical attention. Individuals not fluent in English experience communication gaps. In 2003, it was estimated that an excess of $58 billion a year is spent on health care in the United States, as a result of low health literacy. In certain areas of the country, medical facilities are scarce. Minorities are underrepresented in medical professions. Black, Latino, and Native American populations make up approximately 6 percent of the physician workforce although these populations represent over 26 percent of the population in the United States.

Health disparities remain an important social and political issue in the United States The Office of Minority Health (OMH) of the Department of Health and Human Services was established in 1986 to address issues of health disparities among racial and ethnic minorities. One of the main goals of *Healthy People 2010* is to eliminate health disparities. In 2007, President George W. Bush announced the Affordable Choice Initiative, meant to enable states to redirect existing funds in health care in order to pay for insurance for uninsured populations.

The Dilemma of the Uninsured

Universal access is widely accepted as essential to reduce the social inequalities in health even when income gaps are high. Conversely, increasing family disposable income for the poor is an effective way of reducing the health inequities. The two are complementary and equally important in social policy in the United States.

High percentages of the population are without any, or have inadequate, health insurance. Loss of health coverage with change of place of employment and the rapidly increasing cost of private health insurance generated widespread pressure for a national health program. The business community, too, had lost confidence in voluntary health insurance as costs of health insurance mounted rapidly as a cost of employment in an increasingly competitive international business climate.

The Clinton health plan (1994) was based on federally administered compulsory universal health insurance through the place of employment, with alternative plans available to choose from at different costs. A state could opt to form its own health insurance program and even designate its own department of health to fulfill this function. Physicians could contract with health insurance plans to provide care on a fixed-fee schedule, or in HMOs, whether based on group or individual practice.

The Clinton health plan failed in Congress. Apathy or frank opposition was widespread among the majority of the population who already had good insurance benefits under their employment-based health insurance plans or Medicare. Their interest was in the status quo, and the insurance industry and organized medical community used this to defeat the bill. Federal legislation protecting workers' health rights under collective bargaining, prevented states from mandating health insurance benefits. Federal assistance and waivers for state health insurance allow states to opt for managed care for Medicaid beneficiaries. Medicare and Medicaid waivers also allow states to include these beneficiaries in state health plans, but universal access to care will require enabling legislation in Congress. At the same time, conservative attacks on public programs such as Medicare keep the issue of national

health insurance on the public agenda. State initiatives for universal health coverage are alternative approaches to achieving national universal access to health care.

Many employers have switched to promoting managed care coverage, while offering indemnity plans as options to the employees but with additional premiums. The movement to managed care became an avalanche in the 1990s, with a high percentage of the population insured at their workplace becoming members of HMOs or other forms of managed care. The swing to managed care produced major effects in the health care system, not only for doctors increasingly pressed to join HMOs or PPOs, but also for hospitals and for the consumer who had to adjust to the rules of managed care. Restrictions on access to specialists and new procedures generated public and political criticisms leading to a decline in enrollment from 1999–2005, with an increase in 2006, but this economically-driven changeover has profound effects on the United States health system.

In 1996, many states introduced legislation to regulate HMOs, of which 56 laws were enacted in 35 states. Criticisms of for-profit HMOs are appearing frequently in the popular media, and there is a growing backlash of opinion against imposed limitations on specialist referrals, emergency department visits, hospitalization, and some therapeutic interventions (e.g., bone marrow transplants for terminal cancer cases). Some of these have also generated legal suits for malpractice with large settlements. A 1998 Commission on Health Quality appointed by President Clinton produced a bill of rights for patients that called for additional information on health plans and for the right of appeal to an independent panel on health plan decisions regarding denials of coverage for emergency care or access to specialists.

The non-profit prepaid group practice type of HMO uses over 90 percent of premiums for patient care, whereas the for-profit plans spend higher proportions of premiums for administration, including very high salaries for executive staff. The growth trend of managed care will certainly continue, but perhaps with greater regulation of for-profit HMOs in order to ensure access to services based on medical criteria in the patient's interest and quality assurance.

In 2006, U.S. health care spending increased 6.7 percent to $2.1 trillion, or $7026 per person which was 16 percent of GDP. Prescription drug spending growth increased to 8.5 percent of total expenditures in part due to Medicare Part D covering prescription drugs for older adults. Most other major health care services experienced slower growth in 2006 than in prior years.

With the large percentage of uninsured in the population (16 percent), and another 15 percent underinsured, the political pressure for some form of national health insurance continues in the twenty-first century, while all other industrialized nations have some form of universal health programs. The 2008 U.S. presidential election has health insurance as a major issue with a variety of plans being presented primarily by the Democratic party.

Summary

The United States has managed to achieve many of the targets set by the 1979 Surgeon General's *Healthy People* report. At the same time, the annual average increases in health care expenditures in the United States slowed markedly from the 1986–1990 period with average annual increases of 10.7 percent, falling to under 7 percent annually in the 1995–2005 period. This is partly due to lower general inflation rates (<3 percent), but also to cost-containment measures being adopted by government insurance (Medicare, Medicaid), the health insurance industry, the growth of managed care, and rationalizing the hospital sector by downsizing and promoting lower-cost alternative forms of care.

National health insurance was delayed by congressional rejection of the Clinton health plan, but is a major issue in the 2008 presidential election. A number of possibilities exist to extend health insurance coverage: state health insurance initiatives with federal waivers and cost sharing; or a federal plan based on Medicare or the Veterans Administration program or a revised form of the Clinton plan of 1994 could dominate health care reform political initiatives to provide universal access to the more than 47 million Americans without, and an equal number with poor levels of health coverage. In the mid- to late 1990s, employers promoted managed care options for their workers, so that managed care grew rapidly through market mechanisms. State governments are acting to regulate this by legislation, such as those requiring minimum hospital stays for obstetrics cases, by limiting managed care programs from certain kinds of contracts for services, and by establishing appeals mechanisms for managed care members. Increased access to Medicaid may be fostered by states by raising the income levels defining poverty to increase health insurance coverage.

The term *non-system* is often applied to health care in the United States, with many stakeholders and providers, with high costs and poorer results than health systems in other industrialized countries. Much of this implied criticism is justified. The U.S. health system is a diffuse and incomplete system with good to outstanding quality of care for those with insurance but very inadequate for those without. Social and regional inequities in health status are still present, but not necessarily greater than in some countries with universal access to health care. Further, there are many parallel programs in the United States that have important positive public health content, such as universal school lunch programs; nutrition support for poor women, infants, and children (the WIC program); food stamps for the working poor; fortification of basic foods; free care in emergency departments and for urgent hospital care for the poor, Medicare for the elderly, and Medicaid for the poor.

Nevertheless, equitable universal access is lacking, and the system is the costliest in the world, without the best levels of health as measured by process indicators, such as immunization and prenatal care coverage, nor in outcome measures such as infant and other mortality rates. Life expectancy at birth in the United States increased by 7.9 years between 1960 and 2004, substantially less than the increase of over 14 years in Japan, or 8.9 years in Canada. In 2004–2005, life expectancy in the United States was 77.8 years, almost one year below the OECD average of 78.6 years (OECD 2007). And social inequities in these health status indicators are further evidence of failures of the U.S. health system to reach its full potential, despite its being the costliest system in the world and its high quality for those with access (see Davis, 2008).

CANADA

Canada is a federal country with ten provinces and three northern territories, a population of 32.6 million, and a GDP per capita of $35,700 (U.S.) in 2006. Life expectancy in 2005 was 80.3 years. The Human Development Index rating for Canada in 2005–2006 was 4th among the developed countries, having increased steadily since 1975 and placed above the OECD average. Canada's total health expenditures were 9.8 percent of GDP in 2005, almost one percentage point higher than the average of 9.0 percent in OECD countries, with some 70 percent of total expenditures from the public sector.

The Canadian constitution sets responsibility for health at the provincial level of government, except for the aboriginal Indian and Inuit populations, armed forces, RCMP (Royal Canadian Mounted Police), and veterans. Despite many geographic, historic, cultural, and political similarities to the neighboring United States, Canada developed its own unique national health insurance program.

Starting in the 1930s, federal grants-in-aid were given to the provinces for categorical health programs, such as cancer and public health programs. Based on this precedent, Canada's national health program is a system of provincial health insurance with federal government financial support and standards. It developed in stages between 1946 and 1971, first with hospital and diagnostic services and subsequently with medical care insurance, now collectively known as Medicare. It brought all Canadians into a system of publicly financed health care, while retaining the private practice model of medical care. Hospital care is provided mostly through non-profit, non-governmental hospitals.

The Canadian health program differs markedly from those of the United Kingdom and the United States. Each health system is an important part of the political and cultural traditions of the country. Each within its own tradition is attempting to constrain the rate of cost increases and preserve, or develop, universal coverage. Comparisons are attempted using various health indicators and can be controversial but the Canadian universal health service or insurance coverage seems to have improved the health status of the population more rapidly than similar indicators for the total U.S. population, but not necessarily for the insured population. After decades of stress on developing national health insurance, Canada became a leading innovator in health promotion.

Initiatives for national health insurance in Canada go back to the 1920s, but definitive action occurred only after World War II. The development of national health insurance was partly the result of the experience of the Great Depression of the 1930s, a strong agrarian cooperative movement, and the collective wish for a better society following the war. In 1946, the recently elected social democratic government of Saskatchewan, a large wheat-growing province of 1 million persons on the western prairies, under the leadership of Tommy Douglas, the founder of Canada's Medicare program, established a hospital insurance plan. This plan provided free hospital care for all residents of the province on a prepaid basis under public administration. Within several years, other provinces developed similar plans, and in 1956, the federal government passed legislation (the Hospital Insurance and Diagnostic Services Act) to provide a cost-sharing plan for provinces, adopting universal, publicly administered hospital insurance plans. By 1961, all ten provinces and the (then) two territories had implemented hospital insurance plans meeting federal criteria in a two-tiered national health insurance plan; that is, universal provincial health plans with federal standards and cost-sharing.

In 1961–1962, Douglas and the province of Saskatchewan again led the way by implementing a universal plan for medical services (Medicare). This was opposed by a bitter, 23-day doctors' strike which resulted in some compromises, but the universal plan came into effect, paying doctors' bills on a fee-for-service basis. Again, this was based on the principles of universal coverage, comprehensive benefits, and public administration.

Following the controversies over this plan, a federal Royal Commission on Health Services (the Hall Commission) recommended establishment of similar plans with federal cost sharing. In 1966, the federal government introduced its Medicare Act, providing federal cost sharing of approved provincial plans. Federal reimbursement to the provinces included 25 percent of national average medical care expenditures per capita and 25 percent of the actual expenditures by each individual province. This provided higher than national average rates of support to poorer provinces as well as portability between provinces. By 1971, all provinces had implemented such plans.

Reform Pressures and Initiatives

The Canadian health program established universal coverage for a comprehensive set of health benefits without changing the basic practice of medicine from individual medical practice on a fee-for-service basis. Poorer provinces were able to use the cost-sharing mechanism to raise standards of health services, and a high degree of health services equity was achieved across the country.

Rapid increases in health care costs led to a review of health policies in 1969 (the Federal–Provincial Committee on the Costs of Health Services). The resulting report stressed the need to reduce hospital beds and develop lower-cost alternatives to hospital care, such as home care and long-term care. Federally led initiatives during this period extended coverage to home care and long-term nursing home care, while restricting federal participation in cost sharing to the rate of increases in the GNP. Since then, many provincial and federal reports have examined the issues in health care and recommended changes in financing, cost sharing, hospital services, and development of primary care.

In 1974, a new approach to health was outlined by the Federal Minister of Health, Marc Lalonde, in a landmark public policy document, *New Perspectives on the Health of Canadians*. This report described the Health Field Theory in which health was seen as a result of genetic, lifestyle, and environmental issues, as well as medical care itself (see Chapter 2). As a result, health promotion became a feature of Canadian public policy, with the objective of changing personal lifestyle habits to decrease risk factors such as smoking, obesity, and physical inactivity. The pioneering work in nutrition from the National Nutrition Survey published in 1971 led to the adoption of federal mandatory enrichment regulations for basic foods with essential vitamins and minerals (see Chapter 8). This and other initiatives in the 1980s led to the Ottawa Charter on Health Promotion (see Chapter 2).

In the mid-1980s, physicians' organizations pressed for the right to bill patients above the rates paid for by Medicare, but this was forbidden by national legislation (the Medical Care Act), passed unanimously by the Federal Parliament. This act penalizes provincial governments which allow extra billing by physicians by withholding federal funding. The Canada Health Act was passed by the federal parliament in 1984. This act outlines specific principles and requirements for all Canadian provinces and territories on health care, in order to qualify for public funding. Annual reports are published outlining the status of health care for provinces and territories. Those who do not adhere to the requirements are subject to withholdings of transfers or penalties.

The Canadian health expenditures showed high rates of cost increase, approximately 12 percent annually, during the 1980s, while GDP grew at 3 percent per year. National expenditures on health rose from 5.4 percent of GDP in 1960 to 7.0 percent in 1980, 8.9 percent in 1990, stabilizing at 8.8 percent in 2000, but again rising to 9.8 percent in 2005.

During the late 1990s, the rate of increase in health care costs was reduced by politically painful measures of retrenchment, especially in hospitals. Acute care hospital bed supplies were reduced from 5.0 per 1000 population in 1975 to 4.0 in 1990, and 2.9 in 2004. In the period 1975–1991, when the rate of growth of health expenditures was averaging 11 percent, Canada was second only to the United States in percentage of GNP expended on health. In 1998 it was the fourth highest in the world (with 9.2 percent of GNP in 1996), after the United States, Germany, and France. In 2005, Canada's percentage of GDP spent on health was above the OECD average, but well below the United States (16 percent) and 7 other countries of OECD.

Of total health expenditures in 2005, hospitals accounted for 30 percent, drugs 18 percent, and physicians 13 percent. Expenditures for drugs and public health grew as percentages of total expenditures, while hospital expenditures declined and costs of physician services remained stable.

Financing of total health expenditures in 2005 was 70 percent from public sector sources, including federal, provincial, and municipal governments and workers' compensation, and holds at a stable rate since 1996. Federal government cost sharing in health expenditures has gradually declined since the 1970s, so that provincial governments are facing difficulty with continued financing at current levels, and are pressured to control rates of increase. This led many provinces to reduce the hospital bed supply, from 6.9 beds per 1000 in 1979 to 4.7 in 1995 and to 3.0 in 2003.

The Canada Health and Social Transfers (CHST) established in 1997 serves as the mechanism for federal transfer of money to provinces and territories through cash contributions and tax transfers. As long as the provinces and territories adhered to the ideology of the Canada Health Act, money could be allocated to various social programs. In 2004, the CHST was split into the Canada Health Transfer (CHT) and the Canada Social Transfer (CST). By creating the CHT and the CST, the federal government can allocate cash contributions and tax transfers in a way that is accountable and transferable to provinces and territories in order to maintain the goals and obligations set forth by the Canada Health Act. The Public Health Agency of Canada and the position of Chief Medical Officer were established in 2004, following severe criticism of public health organization in Canada for the handling of the 2003 outbreak of severe acute respiratory syndrome (SARS).

Provincial Health Reforms

In the 1970s, a growing emphasis on health promotion and development of alternatives to acute hospital care led the province of Manitoba to institute reforms in the delivery of services. It established a district health system — integrating hospitals, nursing homes, home care, preventive services, and medical practice, reaching many of the rural areas of the province over the next decades. Following the 1970s Castonguay-Nepveau Commission Report, Quebec implemented Community Health Centers (CLSCs) throughout the province. In the 1990s, Saskatchewan began development of similar integrated district health service and regionalized hospital systems. Other provinces have since followed suit, each using a unique formula to regionalize and consolidate services, and increase the provision of services in the community setting. Ontario, the economically largest province, but the last to regionalize, recently introduced 14 local health integrated networks to manage services to its 12 million inhabitants.

In most provinces, regional health authorities (RHAs) are autonomous health care organizations responsible for health administration within a defined geographic region within a province or territory. The regions have appointed or elected boards of governance responsible for funding and delivering community and institutional health services within the regions. RHAs fund and administer core services including public health; home care and community-based services; mental health services and long-term care institutions; alcohol and substance abuse programs; and hospitals. Provincial and territorial ministries collect taxes to finance the health care system and develop regional funding envelopes; regional health boards allocate funds to service organizations based on their own needs assessments and policy priorities.

Health Status

Criticism of the Canadian health system focuses on waiting time (in comparison to the United States) for diagnostic and surgical procedures, lesser access to high-tech equipment and procedures, and reduction in hospital staff positions. Such comparisons, however, are not substantiated by objective analyses, or in measurable health indicators. Waiting times are reportedly reduced in recent years and the supply of high-tech equipment such as CT scanners and MRIs has increased in comparison to other OECD countries.

Since the implementation of Medicare, Canada's position in major health status indicators improved in comparison to other countries. Infant mortality rates were higher than those of the United States up to the 1960s (28 versus 22 per 1000), but were lower in the 1990s (6 versus 7 per 1000 in 1997). Canada's infant mortality rate was 5.3 deaths per 1000 live births in 2004, lower than in the United States (6.8) and the OECD average (5.4).

In 2005, Canada's maternal mortality rate was 7 per 100,000 live births, compared to 7 for the United Kingdom, and 11 for the United States. In 2007, Canada's life expectancy at birth was 80 compared to 78 years in the United States; the Canada–U.S. gap decreased from 2.7 years in 1993 for men to 1.8 years in 2007, and the difference among women stayed at 2.9 years.

TABLE 13.5 Hospital Discharge Rates for Cerebrovascular Diseases per 100,000 Population, Selected Countries, OECD

	1990	1995	2000	2004–2005
Canada	184	192	173	147
Denmark	na	na	426	349
Finland	676	814	658	600
Germany	na	487	462	422
Japan	na	na	na	452
Netherlands	175	194	185	194
Sweden	613	616	518	463
United Kingdom	245	254	222	235
United States	259	282	268	247

na = not available.
Source: OECD Health Data 2007 — Version: October 2007, data available from 1980–2005.

Immunization coverage for infants was reported in 2005 as over 90 percent. Canada ranks among the 15 OECD countries with the lowest total mortality rate. Canada has achieved major progress in reducing tobacco consumption, with the rate of daily smokers among adults cut by half since 1980, from 34.4 percent in 1980 to 17.3 percent in 2005. Canada is 5th lowest among the OECD countries in circulatory system diseases. Hospitalization for cerebrovascular disease is lower in Canada than in many OECD countries (see Table 13.5), reflecting both prevention programs, as well as outpatient and home management of stroke patients. The main causes of death in Canada are cancer, circulatory, respiratory, digestive, and infectious diseases. Obesity is a growing health concern with approximately 25 percent of the population overweight and 11 percent of the population obese in 2003. Public health efforts have led to a decrease in tobacco use from 34 percent in 1980 to 17 percent in 2004, making Canada the lowest consumer of tobacco among OECD countries. The aboriginal Canadian populations suffer higher rates of poor health status from immunization-preventable diseases and alcohol- and tobacco-related illness (see Chapter 7).

Summary

Canada's health system successfully established universal tax-supported national health insurance in North America. Prior to universal coverage, Canada was on a similar track as the United States but since has bypassed the United States in increased longevity and lower cause-specific mortality rates for cardiovascular disease, cancer, and stroke as well as in child mortality, all at considerably lower per capita expenditures.

The Canadian health program has important lessons for health care reform internationally. Universal health insurance was implemented without changing the basic mix of services or the way they are funded, but with inadequate attention to the public health portion of the system until the SARS episode of 2003. These are all issues that need to be considered in the Canadian experience. At the same time, Canada pioneered the idea of health promotion from the 1970s in such areas as Healthy Cities, fitness, and food enrichment. This has helped to achieve positive results in reduced smoking, falling rates of cardiovascular morbidity and mortality, food fortification, and increased health consciousness of the population and the political leadership over a number of decades. Pioneering reform in integrated regional health management is now widespread across the country, based on the Manitoba model developed in the 1970s.

The Canadian health insurance model can be regarded as a success although with drawbacks. It has tended to freeze the medical private practice model, and was slow in implementing reform measures. Despite transfer of tax points from federal to provincial governments, the provinces have difficulty coping with health costs which are the largest item of provincial budgets. The principle of universal health insurance delivered by provincial plans under federal regulation and cost sharing has been preserved. The quality of care is high, and Medicare is one of the most popular public institutions in Canada of which most Canadians, including physicians, are proud. Reforms carried out during recent decades appear to be succeeding in controlling the rate of increase in costs.

The health status of Canadians is rated among the highest in the world. Despite the financial burden and the need for economic analysis with priority selection, the Medicare program remains highly popular with the Canadian public, and the federal and most provincial political leadership remains committed to universal, publicly administered health care and this is likely to continue. The Canadian model is a success from many points of view and will be of importance in reform for other countries, particularly the neighboring United States.

THE UNITED KINGDOM

The United Kingdom's population in 2005 was 58.2 million and the GDP per capita was $33,238 (PPP USD), ranking it economically as 16th, well below Scandinavia, North America, and many European countries. The United Kingdom also ranked 16th among OECD countries in the Human Development Index of 2005–2006, with total life expectancy of 79 years; the child mortality rate in 2005 was 6 per 1000 live births.

The United Kingdom is a unitary state with public health developing at the local authority level, with national health initiatives evolving slowly since the mid-nineteenth century. The National Health Service (NHS) developed and maintained high professional and technical standards, despite modest levels of funding of the service. Immunization coverage in 2005 was 97 percent for DPT, 91 percent for poliomyelitis, and 82 percent for measles (MMR), with a second dose of measles vaccine at school entry.

The British developed a unique and important model of health care as a tax-financed public service that is widely influential in other national health systems. It is popular and was successful in achieving its initial goals, and has undergone periodic reforms since its inception in 1948, surviving many changes of government and political philosophy.

The National Health Service

As described earlier, the United Kingdom developed its present NHS over many decades. This program evolved from previous milestones including reform of the Poor Laws of the eighteenth to nineteenth centuries, the Friendly Societies, the National Health Insurance Act of 1911 for workers and their families, the National Emergency Medical Service of World War II, and the (William) Beveridge Report of 1942. In 1946, under the Labour government of Clement Atlee, Parliament approved the National Health Service Act with implementation in 1948, under the leadership of Aneurin Bevan.

The NHS is financed through general tax revenues to provide comprehensive service to the entire population. The NHS was originally organized as three parallel services: the hospital service with salaried doctors, the general practitioner (and dental) services provided by independent practitioners with capitation payment, and the public health service with salaried staff. The hospital and general practice services were operated by separate public boards or councils; the public health service was administered by the local authorities.

Structural Reforms of the National Health Service

During several stages of reform in the 1970s and 1980s, the NHS was reorganized, reducing the number of administrative levels and trying to achieve integration and coordination between highly specialized and fragmented services. The 1974 reform established Regional Health Authorities, and integrated Area Health Authorities (AHAs) beneath them to replace the previous multiplicity of hospital management committees, boards of governors, and local health authority committees. The AHAs were non-elected lay bodies that absorbed public health and hospital management functions, consolidating many previously overlapping jurisdictions. Multidisciplinary management teams were introduced at the AHA and district levels, with decision making by consensus, stressing professional managerial competency.

A further reorganization in 1982 abolished the AHAs, placing the managerial responsibility at the District Health Authority level, with a stress on further decentralization of management authority to hospital and community service structures. Reviews of the NHS were conducted during the Conservative government led by Margaret Thatcher, focusing on managerial efficiency, government and business viewpoints, the growth of the private sector, consumer group advocacy issues, and protection of consumer rights.

Despite the aging of the population, the number of acute care hospital beds in the United Kingdom fell from 3.5 beds per 1000 in 1980 to 2.7 in 1990 and 2.4 in 1998; the average length of stay in acute care hospitals fell from 8.5 days in 1980 to 5.0 days in 1996. Psychiatric beds also were reduced in these years, while geriatric and nursing care beds increased from 0.5 per 1000 in 1980 to 4.3 per 1000 population in 2000. There has been an increase in medical practitioners (from 1.9 per 1000 in 1998 to 2.3 in 2004) and nurses (from 8.0 in 1998 to 9.2 per 1000 population in 2004) in the United Kingdom; the supply of physicians is below that of other European OECD countries, but the supply of nurses is above OECD levels. Access to high technology such as CT scanners and MRIs has increased in the past decade, but remains well below OECD averages. Community care services of all kinds increased during this period. Health expenditures increased from 3.9 percent GDP in 1960 to 5.6 percent in 1980, 6.0 percent in 1990, 7.3 percent in 2000, and 8.3 percent in 2005.

Reforms Since 1990

In 1990, the National Health Service and Community Care Act attempted to further rationalize management of the NHS. Three types of statutory health authorities were redefined: Regional Health Authorities (RHAs), District Health Authorities (DHAs), and the Family Health Service

Authorities (FHSAs). The RHAs and DHAs became the primary administrative levels. The FHSAs manage contracts with the general practitioners (GPs). The NHS operations in Wales, Scotland, and Northern Ireland operate under similar arrangements.

The 14 RHAs assess health needs, set strategic direction for service development, monitor quality of management and care, and allocate resources to promote cost-effective services. They also promote medical audits, specific program development (e.g., transplantation services) and provide assistance to health providers such as hospitals with management problems. RHAs do not provide services. The DHAs operate under the authority of boards similar to those of the RHA, and are the major purchasers of services from hospitals and other providers. They contract with hospitals for services based on assessed need and on satisfaction with hospital performance. They may also operate NHS hospitals or other services, such as ambulance services.

Reforms since 1990 included the introduction of competition between providers, the development of community health services, and further reduction of the supply of hospital beds. These were intended to introduce greater choice for the patient and the primary care provider (the GP), with incentives for efficiency and quality of care.

FHSAs under the 1990 Act are governed by boards similar to the RHAs and DHAs. The FHSA is responsible for contracting with GPs, general dental practitioners, optometrists, and community pharmacists. The role of FHSAs expanded to include formulation of policies, supervision of facilities and services, as well as remuneration of contracting providers. Patients register with GPs and are referred to hospitals and specialists in keeping with medical needs. GPs have traditionally been paid on a capitation basis for the patients registered with them, and the patient has the right to change GPs. Capitation is the allocation of funds per person registered as a service beneficiary for a specified period of time to cover care for a range of services. Weighted capitation is allocation per person, with adjustments made for factors such as age, sex, and regional standardized mortality rates, which reflect both need and demand for health services. Standardized mortality rates (SMRs) are used as a proxy for morbidity in capitation allocation (see Chapter 3). GPs are paid extra premiums for performance indicators; for example, specific preventive services such as immunization, Pap smears, and mammogram screening.

A major new innovation has been to allow the FHSAs to administer budgets for fund holder GPs, in the form of per capita payments including both GP and hospital services. The GPs are increasingly working in health centers, along with district public health nurses. By 1995, some one-third of GPs worked as fund holders with per capita payment by the NHS for ambulatory and hospital care. This empowered GPs to negotiate with the hospitals, reduce waiting times, and improve other health care conditions for their patients, placing the hospital in the position of having to compete for the referral work of the GP. Experiments with financing of hospital care through the GPs were designed to raise the quality of care and promote cost containment. The GP fund holder movement seems to be a successful program, although not well evaluated.

Hospitals are encouraged to become National Health Service Trusts, which are non-profit public corporations governed by boards of trustees appointed by the national government, usually representing the local authorities. Hospital Trusts must demonstrate management capacity and viability to operate as economic units. They must compete for referrals, striving for patient and GP satisfaction. Hospitals are no longer funded directly by the NHS, but derive their income from providing services to the Health Authorities, fund holding GPs, private insurance, and self-paying patients, paid for services by a DRG system. This permits them to operate as independent economic units, enabling them to charge for services, determine staff conditions, raise capital by borrowing money, and, within limits, buy or sell land or facilities.

Financing of the NHS continues to be by governmental allocations from general tax revenues. Some revenues come from other sources, including user fees, such as for prescription drugs or dental services. Operating budgets are allocated to RHAs to cover costs of hospital, community health, and primary care services. The allocation is determined on the basis of population size and adjusted by SMRs, with some local weighting factors based on service utilization. The DHAs are in turn funded by the RHAs, using similar criteria. RHAs administer GP fund holding units, whereas FHSAs are funded to pay for contracting primary care services. Capital allocations to replace and modernize facilities and equipment are based on long-term planning at the RHA level.

Market reforms in the United Kingdom are still being developed. Despite being the subject of continuous critical scrutiny in the press and at political levels, the NHS continues to have support of the general public and all political parties and has provided universal access and maintained high quality at reasonable costs. Health expenditures were increased by the Tony Blair government because of criticisms that the NHS was being operated at lower levels of expenditures than most industrialized countries, generating criticism of underfunding of important areas but without change in methods of operation of the system. In 2005, the cabinet officer responsible for the NHS put forward a proposed series of changes, including greater funding for disadvantaged areas of the country, changes in methods of financing hospitals from block budget to an incentive-based budget (yet to be defined), and greater flexibility in funding for Primary Care Trusts and GPs to innovate in developing programs, and greater choice of hospitals by patients. The public support for the NHS is an important element in its durability.

The values of the NHS are described in a 2007 review by the Nuffield Trust as including:

- Universalism (compulsory coverage)
- Equity (social justice, fairness)
- Democracy (accountability, answerability)
- Choice (autonomy, freedom)
- Respect for human dignity (honesty, consideration, fair dealing)
- Public service (public service ethos, altruism, noncommercial motives)
- Efficiency (cost-effectiveness, waste avoidance)
- Promoting desirable outcomes and processes
- Accountability

Social Inequities

Social inequities in the health status of the population of Britain, which were part of the justification for the establishment of the NHS in 1946, have persisted. The Black Report (Douglas Black, 1980) documented this problem, and subsequent reports indicate the persistence and even worsening of social inequalities into the 1990s. Changes in definitions and distribution of the population in the different social classes may explain some of the differences; however, there is a widening of the gap between the social classes, with continuous increase in the SMR of class V and continuous decline in SMRs for classes I and II (see Chapter 4). Many studies show higher mortality from all causes by social class and now health profiles for every local authority and region across England are published by the Department of Health and Public Observatories. Higher cause-specific mortality in lower socioeconomic classes is seen especially in cardiovascular disease, trauma, and cancer.

This social gap is not easily explained on the grounds of the classic health risk factors alone. The health gap correlating to economic disparities may be due to poor diet, more smoking, less physical activity, and social and working conditions with less reward, personal satisfaction, and less control of life events than for the higher social classes. There are also regional differences in SMRs in the United Kingdom; the reasons for these are not always well understood but probably relate to a variety of social, economic, personal lifestyle, and environmental risk factors.

In 1998, Donald Acheson, a senior professor of public health in the United Kingdom, reported on the Blair government-initiated inquiry into social disparities in health in Britain. His report confirmed the findings of the Black report and evaluated findings of the many studies of social gradients in health status since that report was issued. The Acheson report has been a factor in government policies in tax and welfare reform, preschool child care programs, and tobacco legislation, and in some aspects of NHS reforms.

Health Promotion

During the 1950s and 1960s, mortality from cardiovascular diseases increased in the United Kingdom, as in most industrialized countries. These rates began to decline in the 1970s in the United States, Canada, and other European countries, but remained high in the United Kingdom for another decade, with coronary heart disease mortality only declining substantially since 1985. This delay in reduction of cardiovascular disease mortality may be explained by then prevailing conservative attitudes toward treatment of acute myocardial infarction in the United Kingdom such as aggressive, intervening methods of treatment and intensive care units. The NHS was also slow in response to changing approaches to health promotion and risk factor reduction. Mortality from ischemic heart disease continued to increase up to 1978. The United Kingdom continued to have higher mortality rates from cardiovascular diseases compared to many countries in Europe. These and other public health issues, including relatively low immunization coverage levels, led to formulation of health promotion strategies in the Department of Health.

In the late 1980s and early 1990s, a number of major initiatives sought to improve prevention and health promotion activities in the United Kingdom, including greater public awareness of healthy nutrition and the risks of smoking. From 1978–1997, mortality rates from ischemic heart disease declined by 41 percent. Incentive payments to GPs resulted in a sharp improvement in immunization rates. Local authorities are required to have specialized staff to promote motor vehicle safety, which contributed to a reduction in road crash mortality of more than one-third from 1980–1995. A new Water Act and the Environmental Pollution Act of 1990 increased the supervisory and regulatory role of the national government in these areas of public health.

The Health of the Nation report (Secretary of State for Health, 1991) placed health promotion and national health targets as a major focus of a national health program. Declining mortality from the major causes of death (cardiovascular diseases, cancer, and trauma) may reflect an increasing effectiveness of health promotion activities in the United Kingdom.

Health Reforms

Between 1991 and 1998, the government led by Margaret Thatcher introduced the option of holding budgets for general practices for prescribing and elective secondary care. The Labour government of Tony Blair, elected in 1997, undertook further reform in the NHS especially in methods of financing primary care and the market forces of GP fund holding. The government increased funding for the NHS by 4 percent above inflation over the period 1999–2003 to strengthen clinical service sectors which suffered from

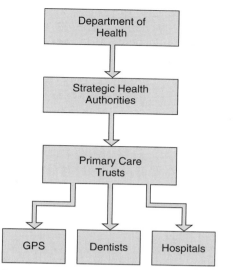

FIGURE 13.3 National health service structure, England. Source: British Broadcasting Corporation.

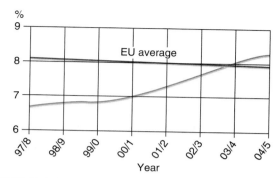

FIGURE 13.4 Trend of U.K. health spending as a percent of GDP 1997–1998 to 2004–2005. Source: DoH/HM Treasury.

excessive cutbacks in the previous decade mainly going to improvement in salaries and medical equipment.

In 1999, the Blair government initiated a new reform of the NHS, establishing primary care groups (PCGs) throughout the country with GP groups serving population groups of between 30,000 and 250,000 persons. The PCGs replaced the purchasing of services previously performed by the fund holding GPs and the health authorities.

NHS policy in England is directed from the center by the Department of Health (DH) (Figure 13.3). The DH has provided new tools to improve monitoring with a Commission for Health Improvement (CHIMP) and a National Institute for Clinical Excellence (NICE) established in 1999. GPs gained online appointment systems and patients have access to free 24-hour telephone nurse consulting service to improve access and contact with patients.

Funding for the NHS was increased since 2000 and this mainly went to increase staff salaries and capitation payments. Family doctors have benefited the most, more than doubling their incomes. In 2004, an additional payment was added to the basic capitation payment to family practitioners based on their performance of 146 quality indicators relating to clinical care for 10 chronic diseases, organization of care, and patient experience. Fund holding was reintroduced in 2005.

In 2006, the Department of Health published its Seventh White Paper entitled "Our health, our care, our say: A new direction for community services" promising a "fundamental shift" toward integrated services provided in local communities. It was intended to improve access and more local coordination among services, cost-effectiveness in reducing hospitalization, to improve quality and to save money in the longer term by emphasizing prevention now to avoid costly illnesses later. It abolished the original fund holding scheme (where GPs held budgets and bought services on behalf of

their patients) but set up PCGs, which later became primary care trusts (PCTs) with the aim of improving the quality of local services. The PCTs set up GP contracts with new quality incentives built in and from 2000 promoted a policy very similar to GP fund holding.

Hospital reform became the prominent issue in public assessment of the NHS particularly in reducing waiting times. This was to be achieved through a combination of targets and, since 2003, financial incentives to promote an incentives-based system. Payment by results (PBR) started in 2006 and provides for competition for hospital trusts, increasing local control and the range of non-hospital treatment options.

The NHS was relatively underfunded but increased funding during the Blair government has brought overall health expenditures of the NHS to above the European Union average since 2002 (Figure 13.4).

Primary Care Trusts

Local health bodies called *primary care trusts* are charged with providing and commissioning services to a geographic population. They are supervised by 10 regional strategic health authorities (SHAs). The number of PCTs was reduced to 152 in 2006 to match geographic division of other social services bodies. They are based on the registered populations of enrolled GPs in the geographic area and are responsible for primary care, hospitalizations, community health promotion, dental care, and health promotion.

PCTs are budgeted through a capitation formula with many factors taken into consideration, with adjustments for age, mortality, in- and out-migration between geographic areas, ethnic mix, prison populations, army personnel and their dependents, and so forth. Incentive payments for immunization and preventive procedures such as Pap smears and mammography have become part of the capitation base budget. Only HIV/AIDS is identified separately.

PCTs pay for hospital care from the allocations for geographic areas with resident populations and for community hospital and office visits, and so on. DRGs are used but are called *case mix groups*. Community health services are part of the HCHS. Chiropody, family planning, and screening are part of this component. Community health services include district nurses, community psychiatric nurses, health promotion programs, community dental health, and health visitors (i.e., public health nurses).

Electronic medical records are now used by almost all GPs in the United Kingdom. This has contributed to implementation of the performance measurement systems to evaluate practice on a national level system called Quality Management and Analysis System (QMAS). This is used both to calculate payments and as a public source of information on quality of care, providing a base for comparison against individual GP previous performance and in comparison to other practices locally and nationally, with data accessible at http://www.qof.ic.nhs.uk/index.asp. These also affect GP payments for performance of specified services (99 percent of practices and a high level of care with 83 percent of incentive payments claimed) in the first years of the program.

The indicators, particularly those in the clinical areas, represent a mixture of process measures and intermediate outcome measures. Generally, intermediate outcome indicators are more difficult to achieve and so represent a greater workload. Most clinical measures are process in nature (registers, improving systems), but many include intermediate measures such as lowering blood pressure, lipid and glucose levels in heart disease, stroke, hypertension, diabetes, and kidney disease patients.

Successive governments of the different political parties have supported the NHS, and despite criticism, it remains a popular institution with the British public, surviving many changes of political leadership over the past 50 years. Reform, on the average of every decade, has enabled the NHS to evolve with experience and to meet the changing economic and health needs of the country.

Changing epidemiologic patterns have also led the U.K. Department of Health and the NHS to develop health promotion strategies. This has helped to reduce high rates of mortality from cardiovascular disease and trauma. This may help to reduce social and regional inequalities in health still present after nearly a half century of universal access. NHS reforms in the early 1990s and in the 2000s promote local community participation and clients' rights with the primary care trusts, and GP satisfaction with much higher incomes and control ("gatekeeper") over use of secondary and tertiary care.

Numerous innovations in organization, incentive funding, information technology, and quality promotion with clinical guidelines appear to have had beneficial effects on access to care, shortening waiting times for primary and specialty care, and probably the quality of care as

well. It is still too early to evaluate innovations on cost-efficiency and social equity of the NHS. Health promotion such as in smoking regulation, physical activity, and dietary change to combat obesity are now active elements of the DH and NHS.

Summary

The NHS has succeeded in its mission of providing universal access in a tax-financed and relatively economical service. It has guaranteed access to health care for all, but has failed to alleviate social class inequalities in health status. This has fostered new efforts and resource allocation to needier geographical areas and to health promotion efforts as in other industrialized countries to reduce the burden of cardiovascular disease, cancer, and other diseases that affect the poor more than the well-to-do.

The Beveridge model NHS has been influential in the Nordic countries since the 1950s and in countries of southern Europe (Greece, Italy, Portugal, Spain, and Turkey) in their various reform programs since the 1970s. The NHS continues to evolve, and is an important and successful international model of health care systems, and one of the most cost-effective. It has adopted many measures of health promotion and increased efficiency of services. Despite many criticisms, the NHS remains one of the most important and respected social institutions of the United Kingdom.

THE NORDIC COUNTRIES

The Nordic countries share common principles of a "Nordic welfare model" with features of universality (right to social protection), a strong public sector, and tax funding based on legislative rights of citizens, equal treatment, and high social benefits. Church-based philanthropy and charity have not played much of a role in welfare provision. The municipal "welfare model" roots go back to the early eighteenth century, long before the emergence of organized philanthropy and charity.

The Nordic countries have working committees to focus on joint cooperative projects in the health sector and many institutions. The work concerns the common interests between these countries and related matters with the European Union.

Each of the health care systems in the Nordic countries has its own characteristics with reforms in progress in each country. Denmark, Finland, Norway, and Sweden, with social democratic governments both before and after World War II, in many ways pioneered the welfare state. They were later influenced by the United Kingdom's NHS, but with strong regional or local governmental organization and taxation have had more emphasis on a decentralized program of health services. Their achievements in social welfare and health care over many decades have been

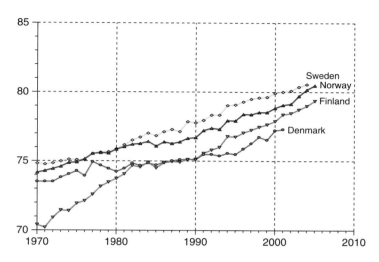

FIGURE 13.5 Life expectancy at birth in years, Nordic countries, selected years, 1970–2005. Source: Health for All Database, WHO European Region, November 2007.

widely acclaimed successful models for social protection in prosperous industrial economies. In 2005, the Human Development Index ranks for the Nordic countries — Denmark – 14, Finland – 11, Norway – 2, and Sweden – 6 — are among the highest in the world with a steady increase in life expectancy (Figure 13.5). Expenditures on health have remained less than 10 percent compared to other countries and are similar among the Nordic countries (Table 13.6). Total health expenditures per capita have increased moderately on an annual basis.

Most (77–85 percent) of the health system revenues are from public sources (Table 13.7). Commonly, between 50 and 70 percent of health system revenues are generated from personal income taxes levied at the regional (Sweden, Norway, Denmark) or municipal (Finland) levels of government. Most of the remainder comes from general revenues raised by the national government through value-added taxes or excise and personal or corporate income taxes. The national funds are distributed as block grants to minimize interregional inequities, with additional grants for medical education. National sickness funds pay for ambulatory visits. Municipal governments pay for long-term care for the elderly. Patient co-payments

provide 2–3 percent of Sweden's county health expenditures and were introduced in Finland in 1993. The user fees are not a significant hardship because of the widespread prosperity and well-established social security systems of the Scandinavian countries.

The Nordic countries have traditionally emphasized maternal and child health and have achieved very low rates of infant mortality. They have high rates of mortality from cardiovascular diseases as compared to countries in southern Europe. This is thought to be related to traditional dietary patterns with high-fat diets, along with smoking and heavy alcohol usage. These risk factors have been the subject of much successful effort at health promotion and are slowly declining. All Nordic countries have greatly reduced the supply of acute care hospital beds as shown in Table 13.8.

Sweden

In 2005, Sweden's population was 9 million, and the 2005 GDP per capita was $32,525 (PPP USD). The 2005 infant mortality rate was 3 per 1000 live births, and life expectancy at birth was 80.5 years. In that year, crude birth rate was

TABLE 13.6 Total Expenditures on Health as percent of GDP, Nordic Countries, Selected Years, 1980–2005

	1980	1985	1990	1995	2000	2005
Denmark	8.9	8.5	8.3	8.1	8.3	9.1
Finland	6.3	7.1	7.7	7.5	6.6	7.5
Norway	7.0	6.6	7.6	7.9	8.4	9.1
Sweden	9.0	8.6	8.3	8.1	8.4	9.1

Source: OECD Health Data 2007 — Version: October 2007, data available from 1980–2005.

TABLE 13.7 Public Expenditure on Health as percent of Health Expenditures, Nordic Countries, Selected Years, 1980–2005

	1980	1985	1990	1995	2000	2005
Denmark	87.8	85.6	82.7	82.5	82.4	84.1
Finland	79.0	78.6	80.9	75.6	75.1	77.8
Norway	85.1	85.8	82.8	84.2	82.5	83.6
Sweden	92.5	90.4	89.9	86.6	84.9	84.6

Source: OECD Health Data 2007 — Version: October 2007, data available from 1980–2005.

TABLE 13.8 Acute Care Beds per 1000 Population, Nordic Countries, Selected Years, 1980–2005

	1980	1985	1990	1995	2000	2004–2005
Denmark	5.3	4.7	4.1	3.9	3.5	3.1
Finland	4.9	4.8	4.3	4.0	3.2	2.9
Norway	5.2	4.7	3.8	3.3	3.1	3.0
Sweden	5.1	4.6	4.1	3.0	2.4	2.2

Source: OECD Health Data 2007 — Version: October 2007, data available from 1980–2005.

11 per 1000 population, with 100 percent of births taking place in medical facilities, and the maternal mortality rate averaging 3 per 100,000 between 2000 and 2005. Immunization coverage in infancy in 2006 was over 95 percent.

Sweden's health insurance system evolved over many decades, and became compulsory in 1955, covering compensation for medical clinic and hospital services and private ambulatory care. Swedish health care is tax-financed, with funding mainly from employers and government, but patients are charged a co-payment for services. Health expenditures as a percentage of GDP ranged between 8.1 and 8.6 percent from 1985 to 2000, increasing to 9.1 percent in 2004 and 2005. Per capita cost comparisons between primary, psychiatric, and specialized care show that specialized care is $3\frac{1}{3}$ times more than primary care and specialized psychiatric care is slightly more than 50 percent of primary care costs in 2006.

The county or municipality is the principal level of government responsible for management of health care. There are 23 county councils and three large municipalities with populations ranging from 60,000 to 1.5 million. In 1993, the counties or municipalities, which have an income tax base of financing, provided 75 percent of funding for health care, with 7 percent coming from the national government, 5 percent from national insurance, 2–3 percent from patient fees, and the remainder from miscellaneous sources. Current reforms include improved primary care coupled with

reduction in the hospital bed supply. Primary care is provided in health centers staffed by salaried GPs, nurses, and other staff serving about 15,000 clients. In 1993, private practice accounted for about 20 percent of total doctor visits. Sweden has a system of economic equalization to compensate for uncontrollable factors such as age differences and rate differences for certain costly disease conditions.

Sweden has traditionally had a very high ratio of hospital beds to population. In 1985, this included 4.6 acute care, 6.2 long-term, and 2.4 mental hospital beds per 1000 population. The acute care beds were reduced to 2.2 per 1000 in 2005. Hospitalization was a common form of care, especially in areas with sparse population and long distances to hospitals and doctors. Reduction in hospital bed supplies has been a long-term strategy in Sweden since the 1940s, and emphasized since the 1960s, with a steady reduction in medical, surgical, and community care beds, as well as psychiatric beds. Long-term social care for the elderly has been transferred to social service agencies. This was accomplished while maintaining high-quality service and improving national health indicators, such as infant mortality rates and maternal mortality rates, which are among the lowest in the world.

Recent reforms in Sweden allowed contracting out for public sector services. This strengthened the role of primary care providers, now able to select more efficient and user-friendly services. Hospitals operate as economic units, balancing revenues and expenditures, and must compete for patients in the new public market for health care. Public institutions must also compete with the private sector and, in some instances, purchase services from private providers. This has helped to reduce waiting times for operations and led to bankruptcy of inefficient or unacceptable hospitals.

Sweden, like other Nordic countries, has refocused health planning on the principle that all people should have equal access to the same conditions for good health with a renewed emphasis on vulnerable groups such as immigrants and single parents and their children. It includes a focus on avoidable hospital days for chronic long-term conditions (e.g., asthma, diabetes, heart failure, hypertension) and acute conditions (bleeding ulcers, diarrhea, and inflammatory conditions).

Denmark

In 2004, Denmark's population was 5.4 million; the GDP per capita was $33,973 (PPP USD). The 2005 infant mortality rate was 4 per 1000 live births, and life expectancy at birth was 77.9 years and the Human Development Index was 0.949, ranking 14th highest of 177 countries. The crude birth rate was 11 in 2005 with 100 percent of births occurring in medical facilities. The maternal mortality rate was 10 per 100,000 (1990–2004); the maternal mortality rate in 2000 was 3.0 adjusted. Immunization coverage in infancy in 2004 was 93 percent for polio and DPT, and 95 percent for measles.

In 1803, the predecessor of the National Board of Health was established; from 1858 local boards of health began to be set up. There is a long history of decentralized health services which are the responsibility of local towns and municipalities since the early years of the eighteenth century.

Reforms focus on ensuring continuity of care across administrative sectors, with easy access to unified prevention, primary care, and rehabilitation services. The focus is improved service for multi-problem situations, the disadvantaged, the chronically ill, and at-risk children. Denmark has not built any institutional accommodation since 1987 but has developed subsidized housing and extensive home care services for older people. The percentage of GDP spent on health increased from 8.1 percent in 1980 to 9.1 percent in 2005. Between 1980 and 2004, acute care beds declined from 5.3 to 3.1 per 1000 population.

Norway

In 2005, Norway's population was 4.6 million; the GDP per capita was $41,420 (PPP USD). The 2005 infant mortality rate was 3 per 1000 live births, and life expectancy at birth was 79.8 years. The Human Development Index for Norway in 2005 was 0.968, the second highest among 177 countries. The crude birth rate was 12 in 2005 and

100 percent of all births take place in medical facilities. The maternal mortality rate was 7 per 100,000 and immunization coverage in infancy in 2005 was over 91 percent.

Norway, with a GDP 43 percent above the average in the European Union, is one of the richest countries in the world. Health expenditures in 2005 were $4364 (PPP USD) per capita. The percentage of GDP spent on health between 1985 and 2005 rose steadily from 6.6 percent to 9.1 percent. The proportion of government expenditures spent on health is similar to that of the other Nordic countries. Norway is the only Nordic country where central government is directly involved in the decision-making process for tertiary care services. Sweden, Denmark, and Finland delegate this to regional authorities or municipalities.

A summary of the trends in health reform over the past few decades:

1970s — reducing inequities and build up health services;
1980s — cost containment and decentralization;
1990s — efficiency and leadership;
2000s — structural changes in delivery and organization with a focus on reducing inequities.

Primary care is the responsibility of the local municipalities; five regional health authorities are responsible for specialist care; with ownership of hospitals transferred to central government. Hospital services are organized as enterprises with day-to-day operations run by a general manager and executive board. National reforms have focused on responsibility for providing service, priorities, patient rights, and cost-containment.

Finland

Finland is a republic with a population of 5.3 million persons in 2004 and a 2005 GDP per capita of $32,153 (PPP USD) and a Human Development Index of 0.952 ranking 11th out of 177 countries, just above the United States. Finland has achieved one of the lowest infant mortality rates in the world, declining from 22 per 1000 live births in 1960 to 3 per 1000 in 2005. Maternal mortality averaged 6 per 100,000 live births (2000–2005). Child care is provided free by the municipalities. Immunization coverage rates include 97 percent for polio, 97 percent for DPT, and 97 percent for MMR for 1-year-old children (2004). Despite high immunization coverage for polio, Finland experienced an outbreak of polio due to use of an inadequately immunizing IPV vaccine in the 1980s. Longevity increased by 5.5 years for men and 5.1 years for women from 1971 to 1991, with a life expectancy of 78.9 years overall in 2005, an increase from 68 years in 1960.

Finland has three tiers of government. Strong municipal governments provide primary, secondary, and tertiary care services, as well as public health, education, and other social services. The states subsidize municipalities to provide these services, with management by locally elected officials. Taxes on income are shared between the municipal and national governments. Universal access to care is guaranteed to all.

Health policy is determined at the level of the national government, which regulates capital investment in health facilities and subsidizes municipalities, which are responsible for providing health and social services. State and municipal governments together collect approximately half of total taxation, high compared to other countries, reaching 46 percent of GDP. The economy was in recession during the early 1990s with a decline in GDP; as a result, the percentage of GDP spent on health care rose sharply, from 6.3 percent in 1980 to 7.7 percent in 1990, falling to 6.6 percent in 2000 and rising to 7.5 percent in 2005.

The constitution provides social protection for the people made up of preventive social and health policy, social welfare and health services, sickness, unemployment, old age, and other benefits.

Public health care services consist of primary care provided by municipal health centers and specialized hospital care. A health center can be run by more than one municipality on a cooperative basis. Primary care includes well child care, school health care, medical rehabilitation and dental care. Services may be purchased from private providers. Finland has 20 districts which provide specialized hospital care and includes a central and a regional hospital. There are five university hospitals. There is a fee paid at the time of visit to the health center for municipal services but with a cap on the annual amount the person is charged, fees for long-term care are based on the person's income.

High rates of mortality from cardiovascular diseases, injury, and suicide affect middle-aged men disproportionately. The widely known North Karelia project (see Chapter 5) to promote reduction in risk factors for heart disease stimulated national efforts and contributed to substantial reductions in mortality rates from these diseases (Figure 13.6).

Cardiovascular mortality rates declined 52 percent from 1970 to 1996, in part because of changes in diet with less meat and greater vegetable consumption. Hospital discharge rates for cardiovascular diseases declined significantly (Table 13.9). Smoking rates for men in the early 1970s reached 50 percent but have declined to 26 percent in 2005, with 18 percent of women smoking (Table 13.10). Overall alcohol consumption is low, but binge drinking is common and relates to the high suicide and trauma rates.

Finland had high hospital bed ratios up until the 1980s when it changed health policy, recognizing the limitations of hospital care and placing greater emphasis on primary care, preventive and social services, and health promotion. Hospital bed supplies are still being reduced, with shorter lengths of stay and increasing ambulatory care and outpatient care. Mental hospital beds were decreased by 50 percent during the 1980s. The total hospital bed to population ratio declined from 15.6 in 1980 to 9.3 per 1000 population in 1995. Acute care hospital beds per 1000 decreased from 4.9 to 4.0 from 1980 to 1995 and 2.9 in 2005 (OECD data report 2007).

Reform in primary care services during the 1980s has reduced inefficiency, bureaucracy, and waiting times, and raised consumer satisfaction. A combination of capitation and fee-for-service payment is used. During 1993, reforms in health care financing converted national support for municipal health services to block grants based on capitation formulas to the municipalities, which now fund both the hospitals and primary care services. This allows the municipalities greater freedom in seeking a new balance of services and redirecting resources from the hospital to the primary care sectors. Local health centers provide most medical and health-related services, including rehabilitation and addiction services. Recent health reform activities have emphasized guaranteed access to care within maximum time frames with uniform criteria for non-emergency care. Oral health

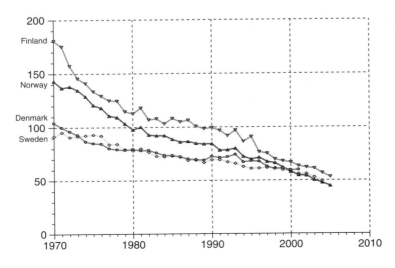

FIGURE 13.6 Nordic countries, SDR, cerebrovascular diseases, all ages per 100,000 population, 1970–2005. Source: Health for All Database, WHO European Region, November 2007.

TABLE 13.9 Hospital Discharge Rates, Cerebrovascular Diseases, Nordic Countries, Selected Years, 1990–2005

	1990	1995	2000	2003	2005
Denmark	na	na	426	384	349
Finland	676	814	658	646	600
Norway	292	382	319	351	342
Sweden	613	616	518	467	463

Source: OECD Health Data 2007 — Version: October 2007, data available from 1980–2005.

TABLE 13.10 Tobacco Consumption, Daily Smokers 15 Years and Older, percent of Population, Selected Years, Nordic Countries, 1980–2005

	1980	1990	1995	2000	2004–2005
Denmark	50.5	44.5	35.5	30.5	26.0
Finland	26.1	25.9	24.0	23.4	21.8
Norway	36.0	35.0	33.0	32.0	25.0
Sweden	32.4	25.8	22.8	18.9	15.9

Source: OECD Health Data 2007 — Version: October 2007, data available from 1980–2005.

care is supported by public funding that covers the total population.

Hospital-based physicians are permitted to practice privately. Over 90 percent of GPs work in publicly operated health facilities, but nearly one-third also conduct private practices in their off-duty time. GP satisfaction with the changes in the health system, with the combination of capitation and fee-for-service, is reportedly high.

The search for greater efficiency now includes a mix of planned and market economies in health. The strong tradition of publicly operated health services will continue despite introduction of market elements, but regional inequities may be an undesired result. Health reform in Finland continues with decentralized service management and central planning and financial support. Finland emphasizes *Health in All Policies* whereby health and social issues are included in all local and national planning (Ministry of Social Affairs and Health, Finland and European Observatory on Health Systems and Policies, 2006).

WESTERN EUROPE

The countries of continental western and central Europe pioneered national health insurance through place of employment, with the national government regulating conditions of insurance, establishing fee schedules, and setting national health policies. The generic type is termed the *Bismarckian national health insurance program*, and is characteristic of Germany, France, Holland, Belgium, Luxembourg, Austria, and Switzerland, each with distinct characteristics and mixed features of social insurance with national service elements. These have been termed *sickness insurance*, based on the solidarity principle of workers' benefits, including old-age pensions, disability benefits, and compensation for loss of working capacity. The funds have maintained a treatment-oriented approach, and only under exceptional circumstances have they undertaken disease prevention, much less health promotion.

Germany

Germany is a federal state with a century-old tradition of social protection legislation. Most aspects of management are delegated to self-governing insurers and associations of providers. The population of Germany in 2007 was 85.4 million, and life expectancy at birth in 2004 was 79.4 years; infant mortality (2004) was 4.1 per 1000 live births, and maternal mortality was 5.2 per 100,000. In 2005, the GDP per capita was $29,461 (PPP USD) and Germany ranked 22nd of 177 countries with the Human Development Index (0.935), just above the OECD average. Immunization coverage for infants was over 90 percent in 2005.

Bismarckian Health Insurance

Germany's system of national health insurance is based on Chancellor Otto von Bismarck's plan that introduced care for low-income workers financed through a social security system by employer and employee contributions. The Sickness Insurance Act of 1883 provided that all workers earning below a designated level be insured by a sick fund, with employer–employee contributions. This is also known as "statutory health insurance" (SHI) or as the "Bismarckian system" based on making health insurance mandatory for certain employees.

The Sick Funds (*Krankenkassen*) might be owned by unions or employer associations, which can operate their own health services to provide comprehensive medical and hospital services for enrolled members and their families. The Sick Funds or mutual benefit societies may also provide cash benefits for accidental injuries, burial benefits, and widow's pensions. This plan was later extended to cover virtually the entire population and remains the foundation of Germany's health and social insurance up to the present time.

In 1911, a framework for social insurance was introduced by adopting an Imperial Insurance Regulation. In 1923 the Imperial Committee of Physicians and Sickness Funds (later known as the Federal Joint Committee) was created as the authority responsible for decisions regarding benefits and the delivery of outpatient care. Later, the Sick Funds became obliged by law to provide hospital care not only to their members but also to family dependents, and coverage extended to include health care benefits for pensioners. Health care benefits were gradually extended further and in 2004 the unemployed, students, disabled, and recipients of social welfare were incorporated into the statutory health plan.

The statutory health insurance system is characterized by three main principles: solidarity — the willingness of the healthy people to pay for the sick and offering a universal and comprehensive benefit package; decentralization and organization of the health care system from the bottom up; and the principle of corporatist organization, namely representation of employees and employers on the management boards of sick funds.

In Germany, health care is governed at the national level by the Federal Assembly, the Federal Council and the Federal Ministry for Health and Social Security as the key authorities liable for passing health reforms concerning statutory health insurance. The federal government is responsible for setting the health policy for delivery of medical services. The corporatist level consists of 292 non-profit, quasi-public Sick Funds and associations of SHI-contracted physicians and dentists on the provider side. The 16 *Länder* are accountable for planning and management of the hospital sector, policy development, and implementation for social and nursing care services, including prevention and monitoring of transmissible diseases, pharmaceuticals and drugs, and environmental hygiene.

The entire German population is entitled to health care services; in 2003, 88 percent were covered by statutory health insurance (SHI), 10 percent were covered by private health insurance companies, and the remaining 2 percent of the population were covered by specific governmental schemes (military, police, social welfare, and assistance for immigrants seeking asylum). Thirty-seven percent of SHI insured were members of general regional funds (AOK), 33 percent were insured by substitute funds, 21 percent were members of company-based sickness funds (BKK), and 6 percent were covered by guild funds (IKK).

Statutory health insurance is the core of the German health care system. Outpatient care is provided by private for-profit care providers characterized with monopoly and no gatekeeping functions. Physicians and other health professionals working in hospitals or institutions for nursing care or rehabilitation are paid salaries. Private physicians and dentists are paid on a fee-for-service (FFS) basis with the fee schedule determined by the Federal Ministry of Health and Social Security. Inpatient care is delivered by a mixture of public and private providers. The Sick Funds represent the collectors, purchasers, and payers of SHI and long-term care insurance. Sick Funds are self-governed and based on mandatory membership.

As a result of amalgamations, the number of Sick Funds decreased from 1200 in 1993 to 292 in 2004. By law, they have the right to raise contributions, and to negotiate prices and quality assurance with providers of care they contract with. Sick Fund membership is mandatory for employees whose gross income does not exceed a specified upper level of the gross salary per month (in 2005) in order to prevent high-earning voluntary members from leaving the SHI. Contributions for SHI are dependent on income, and not on the risk. From 1949 until 2004, contributions were shared equally between the employees and their employers. Since 2005, the contribution rate for employees was increased to 54 percent with employers obliged to pay the remaining 46 percent. For people earning below a threshold minimum salary, the employers pay a standard rate of 11 percent contributions for all Sick Funds. Since 2004, pensioners have to pay the full contribution rate. In 1995, mandatory insurance for long-term care was introduced. The long-term care insurance scheme is run by the Sick Funds and private health insurers. There is a uniform co-payment for outpatient services and products and co-payment of €10 per inpatient day for a maximum of 28 days.

Since reunification of the eastern and western parts of Germany in 1990, several health care reforms were launched with a focus mainly on expenditure control and improving technical efficiency by enhancing managed competition and taking measures to avoid adverse effects on equity and quality. In 2004, the total government expenditure as a percentage of GDP was 47 percent, whereas the total health spending as a percentage of GDP was 10.9 percent, placing Germany among the OECD countries with highest expenditure on health.

Health Insurance Reform

In Germany, governments contribute 21 percent of total health expenditures, while employer/employee contributions make up 60 percent, out-of-pocket 11 percent, and private insurance 7 percent. Hospitals were paid on a *per diem* basis, including salaried physician services.

Traditionally, the German citizen had no right to choose the Sick Fund and was assigned to the appropriate fund based on geographic and/or job characteristics. However, since 2002, every SHI member has a choice of Sick Fund membership at any time of the year, but a minimum membership period of 18 months is required before being able to switch to another Sick Fund. The company-based funds (BKK) and the guild funds (IKK) have the right to remain closed, but if they decide to open, they are obliged to contract with all applicants. Only the farmers' and sailors' funds as well as the miners' fund remained closed with assigned membership.

In order to assure competition and to balance income-level differences in contribution rates among the funds, a risk structure compensation scheme (RSC) was launched in two stages during 1994–1995. In 2001, Disease Management Programmes (DMPs) were introduced as a new instrument to avoid "cream-skimming" among the Sick Funds, as well as providing incentives for care of the insured chronically ill. Since 2004, the Sick Funds are obliged to receive a fixed amount from the federal budget for several benefits relevant to family policies, such as maternity benefits, sick pay for parents caring for sick children, and *in vitro* fertilization, and in 2007 the scheme became "morbidity oriented."

State governments have the authority to plan hospitals. By 1985 legislation, hospital capital costs were funded by state and local governments through a certificate of need. In 2002, 54 percent of hospitals were public; most are operated by municipalities, with 38 percent by non-profit NGOs, and 8 percent by for-profit corporations. Until 2003, hospitals, except for university hospitals, traditionally provided inpatient care only. Since then, hospitals may treat patients with diseases requiring highly specialized treatment on an outpatient basis. In 2005, Germany's acute care hospital bed supply had declined from 8.4 per 1000 in 1991 to 6.3, as compared to 3.7 in France and 3.9 among the original members of the European Union in 2004. The average length of stay for acute care hospitals in 2004 was 8.6 days and bed occupancy rate was 75.6 percent in comparison with most other EU countries with an average of 6.7 days and 75.5 percent bed occupancy.

One of the major reforms in the German health care system concerns the hospital payment system. Operating costs were paid on a *per diem* basis by the Sick Funds at standard rates for all patients but differing among hospitals. There were no incentives for hospitals to reduce costs of utilization. In 1986, global budgeting was introduced for hospitals, intended to promote cost-effective services, outpatient treatment, and hospital financing for greater ambulatory care and coordination of medical care. Germany has a high hospital bed supply and low occupancy rates. In 1988, and again in 1993, health reform laws were passed trying to restrain health cost increases. This included a law limiting fee increases, the supply of physicians, and use of expensive technologies in ambulatory care. Since 2004, hospitals are reimbursed on the basis of diagnosis-related groups (DRGs); in 2005, the acute hospital cases were classified in 878 DRGs. Mandatory quality assurance carried out by external authorities was initiated since 2004 in order to provide transparency and improve quality of care.

The professional associations and hospitals have had a strong role in determining the costs of health care by negotiating high salary levels and promoting an emphasis on high technology, high levels of surgery, and overlapping services. Patients have choice of physician but may be obliged to join one of the 294 Sick Funds according to the choice of the employer or a person's professional grouping.

In 2002, 75 percent of total health expenditures were from public sources and 25 percent from private payments. In 2005, total health spending accounted for 10.7 percent of GDP in Germany, 1.7 percentage points higher than the average of 9.0 percent in OECD countries. Only the United States (15.3 percent), Switzerland (11.6 percent), and France (11.1 percent) allocated more of their GDP to health than Germany in 2005 (OECD).

Germany pioneered in social security–based health insurance. Its health system coped well with the challenge of integrating the former East German health system and population. Germany's health care standards are among the highest in the world; life expectancy rates are improving steadily but continue to be below those of France and the EU, but well above neighboring countries of eastern Europe (Figure 13.7). Mortality rates from cerebrovascular disease and heart disease are well above those of France and the original EU members, while cancer mortality is slightly lower. Health promotion approaches are not part of Sick Fund responsibilities, but are being developed in recent health reforms.

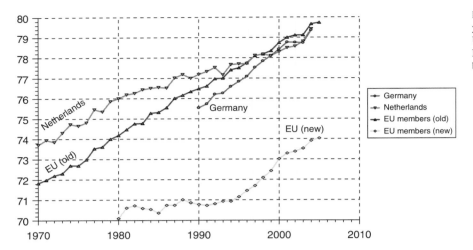

FIGURE 13.7 Life expectancy at birth in years, Germany, Netherlands, European Union, 1970–2005. Source: Health for All Database, WHO European Region, November 2007.

The Netherlands

In 2006, the Netherlands had a population of 16.3 million, with a GDP per capita of $38,725 (2005) and life expectancy of 79.4 years (2004), among the world's highest. Health expenditures in 2004 were 9.2 percent of GDP, just above the OECD average of 9.0 percent. The crude birth rate was 11.5 live births per 1000 population in 2005. The infant mortality rate declined from 18 in 1980 to 4 per 1000 live births in 2005, compared to U.S. rates of 26 and 6 per 1000, respectively. Maternal mortality was 8 per 100,000 live births in 1990–2004. Immunization coverage in infancy in 2004 was 96–98 percent for DPT, polio, and measles. The Netherlands experienced two outbreaks of poliomyelitis among nonimmunized religious groups from imported polio virus in 1987 and 1992 and a large mumps outbreak in 2008.

The health care system of the Netherlands is a combination of public and private financing, with private delivery of care. The system evolved from medieval guilds and mutual benefit associations to health insurance through employer–employee payments to non-profit sick funds or private insurance plans. By 1933, health insurance offered by such groups covered 41 percent of the population. National health insurance was introduced in 1941 (by Germany). Sick funds were established on a geographic basis covering a majority of the population. Physicians are paid on a fee-for-service basis for insurance patients and by capitation for sick fund patients.

A new health insurance system was created in 2006, replacing the former fragmented insurance system and includes occupational disease and workplace injuries. It is a private insurance system with statutory safeguards covering the total population; covering long-term nursing care, acute care, and supplementary insurance. It is described as a hybrid model between public and private insurance. The medical insurers (30 companies) are required to accept all applicants and offer the same insurance coverage under the same terms and conditions. The insured person pays a nominal premium and an income-related contribution. At year's end, those who made little use of the system get a rebate of part of the premium. The tax system levies the income-related contribution through the employers.

Municipal health services are responsible for public health services on behalf of local and regional authorities (governments). Lifestyle factors are seen as important aspects of public health policy on smoking, alcohol abuse, physical activity, nutrition, diabetes, and mental depression.

Preventive and health promotion targets for improving health include smoking, problem drinking, overweight, diabetes, and depression as key areas of reducing health inequities. More than 70 percent of care expenditures are for treatment of those with chronic diseases. The private insurance system for personal health services limits opportunities for prevention-oriented activities for lifestyle-related conditions. In the absence of objectives and targets, providers and insurers determine the types and levels of preventive health services.

Patients must have a referral from their general practitioner before seeing a specialist (i.e., the GP as gatekeeper). This helps to prevent unnecessary referrals, strengthening the role of the GP and helping to control health care costs. Most specialists are hospital-based and are paid on a fee-for-service basis. Most hospitals are not-for-profit and paid on a block budget negotiated with the private insurers. The supply of hospital beds is closely regulated by the government as is technology investment, restraining cost increases for the hospital sector.

Reform of the health system in the Netherlands emphasizes competition and market-based approaches to private insurance. Health expenditures as a percentage of GDP increased from 6.7 percent in 1972 to 8.4 percent in 1982, remained relatively stable until 1990 at 8.0 percent, and subsequently increased to 9.2 percent in 2004. The acute care hospital bed-to-population ratio was reduced from 5.5 in 1970 to 3.1 beds per 1000 population in 2005.

Mortality patterns show the Netherlands population to be at relatively high risk for cancer, but at lower risk than most northern European countries for cardiovascular disease. The Dutch health system has been successful in restraining cost increases as compared with the United States, while providing universal coverage, preserving primary care medical services, and achieving health status measures among the best in the world.

RUSSIA

The Russian Federation is the largest country in the world, stretching from Europe to the Pacific Ocean. Russia has a highly urbanized (73 percent) and educated (99.6 percent), multiethnic population of 142.2 million persons (2007) and abundant natural resources. The Russian Federation GDP per capita between 1995–2005 averaged 6.7 percent, with inflation and widespread poverty especially in rural areas. It is estimated at $14,600 PPP USD for 2007. High incomes from oil and raw materials exports have helped reduce Russia's unemployment rate to 6 percent in 2007. Immigration from neighboring countries such as the Central Asian Republics helps to moderate the depopulation trend to some extent. The Human Development Index of 0.802 places Russia in 67th position, but following a catastrophic decline in the first half of the 1990s, has recovered to the level before 1990. In 2006, life expectancy at birth was 66.6 years (female 73 and males 60 years).

Despite strong economic growth since 2000, Russia's population decline is largely due to low birth rates and premature deaths from stroke, coronary heart disease, violence, traffic accidents, and alcoholism. Cardiovascular diseases are the most frequent cause of death in Russia. Standardized mortality rates are high compared to western European countries, which have been experiencing declining mortality rates especially since the 1960s. Some 1.3 million people die of cardiovascular diseases annually in Russia particularly among less-educated men not surviving to pension age with factors such as unemployment, difficult living conditions and lifestyle (smoking and binge drinking), poor education levels, and unstable marital status among middle-aged men. Heart diseases account for 56.7 percent of total deaths, with about 30 percent involving people of working age. Mortality among Russian men rose by 60 percent since 1991, four to five times higher than in Europe. Smoking affects up to 70 percent of men and 30 percent of women in Russia. In the United States, the mortality rate from heart diseases and stroke is 116 deaths per 100,000 men 35–59 years of age, while in Russia the rate is 576 deaths from these causes. Figure 13.8 compares standardized mortality rates from cardiovascular diseases in the Russian Federation with the older and new members of the European Union and countries of Central Asia.

Excessive alcohol consumption and binge drinking also result in high motor vehicle death rates and other trauma, particularly in homicide and suicide rates which are among the highest in the world. The number of officially registered HIV-positive cases (October 2007) was reported as 396,524, for an HIV prevalence rate of 262 cases per 100,000 population. The most affected age group is between 18 and 24 years old. Currently the HIV epidemic in the Russian Federation is concentrated in high-risk population groups with the principal driving forces being injection drug use and unsafe sex. Tuberculosis case notification rates more than doubled from 34.2 to 90.4 per 100,000 population between 1990 and 2000, but decreased by 8.5 percent from 2001 to 2003, and appears to have since stabilized at 83.8 (2005). Multidrug-resistant strains of TB are present in as much as 15 percent of cases in some regions of Russia.

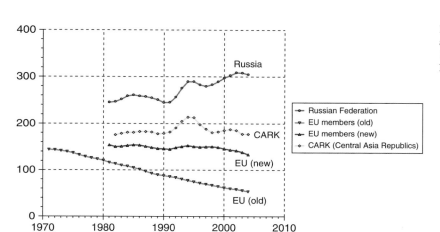

FIGURE 13.8 SDR, cerebrovascular diseases, all ages, per 100,000 population, Russia, EU members, 1970–2004. Source: Health for All Database, November 2007. Note: 3-year moving averages.

This crisis in health is related not only to the period of economic transition in the 1990s, but goes deep into the former Soviet health system. The "old" state-operated service provided free universal health care with ample, indeed excessive, resources in medical personnel, hospital beds, polyclinics, and other services, but with quantity compromising quality since the epidemiologic transition and changes in health profile of the population. The system operated as a state monopoly, with the central government controlling budgets, setting mandatory norms, and totally controlling personnel training and research. The system lacked mechanisms for epidemiologic or economic analysis and accountability to the public. Medical standards, research, and education were very isolated from the outside world with poor access to literature and professional contacts. The epidemiologic transition from predominance of infectious to noninfectious diseases was addressed by further increases in the quantity of services. Policy and funding favored hospitals over ambulatory care and individual routine checkups over community-oriented preventive approaches.

Health expenditures have increased from 2.5 percent of GDP in 1992 to 5.6 percent of GDP in 2004, while other industrialized European countries expend an average of 9 percent of GDP on health. The Russian per capita GDP declined from $3220 in 1991 to $2410 in 1996, but subsequently increased to $10,845 (PPP USD) in 2005. Russia has traditionally maintained a very high hospital bed-to-population ratio, which has been declining since 1990.

After the 1991 breakup of the Soviet Union, the Russian Federation entered a period of political, economic, and social reform with important effects on the national health system and health of the population. In 1993, a compulsory (mandatory) national health insurance (MHI) plan was adopted to augment funding and promote decentralized management of health care and movement toward a market economy in health. The health issues are, however, complex, and changing methods of financing medical care services alone may worsen the health situation by reducing access to care.

The Soviet Model

Prior to the 1917 revolution, Russia was a largely rural country with higher mortality rates than European countries. Public medical care and other social services for the rural poor majority were established in Czarist Russia in 1864 under the local district assemblies or *Zemstvos,* providing tax-financed services for medical and hospital care. Health insurance was established in 1912 based on the Bismarckian social security model, covering about 20 percent of industrial workers.

Following World War I, the 1917 October Revolution, and the Civil War, Russia was racked by mass epidemics and starvation. In 1918, reconstruction planning included the Soviet concept of health care formulated by Nikolai Semashko, based on the principles of government responsibility for health; universal access to free services; a preventive approach to the "social diseases"; quality professional care; a close relation between science and medical practice; continuity of care between health promotion, treatment of the sick, and rehabilitation; and community participation.

The state undertook to provide free medical services for all, through a governmental unified health system. The "social diseases" referred to all diseases related to the poor living and working conditions of the workers, mainly infectious and occupational diseases as well as maternal and child health problems, and were the focus of special attention and measures of prevention and control. Epidemic control was implemented on an urgent basis, especially for tuberculosis, typhoid fever, typhus, malaria, and cholera. Community prevention approaches were enforced, often with use of punishment measures. Prophylactic measures such as quarantine were implemented, urban sanitation and hygiene improved, and malarial swamps drained in the huge territories of the U.S.S.R., resulting in elimination of malaria by 1960.

Medical prevention of social diseases focused on routine checkups for the working population. From the 1920s, emphasis was placed on prevention and control of infectious diseases. In order to meet the needs of the system of providing health care throughout the country, increases in the supply of hospitals, polyclinics, doctors, and nurses were a national priority. In 1937, all insurance and hospital-based sick funds were closed, and hospitals and other health facilities nationalized and organized under district health management. Virtually all health personnel became public employees. Parallel services were provided within industries and for special categories, especially party leadership, some ministries, defense and security personnel, miners, workers in heavy industries, and transport workers.

General government revenues provided financing of health services as part of national plans for social and economic development. The central administration directly employed staff, paid salaries, and provided supplies for all health care facilities and research and training institutes. Directors of health facilities therefore administered their allotted resources, supplies, and human resources with no opportunity for program management or internal accounting of service costs. The health system was developed, financed, and managed under strong central government control, with payments based on norms such as for hospital beds and staffing. Mandatory norms for facilities and personnel were enacted by the Commissariat (later Ministry) of Health, under strict regulation of the central authorities of the Communist Party, and later by the Ministry of Finance. These norms were revised periodically at Party Congresses, with expansion of services being the major policy orientation. The policy of continuing to increase the supply of hospital beds and medical personnel was reiterated in the mid-1980s, and continued into the 1990s, but has been reduced since 2000.

During World War II, the Soviet health system was mobilized for the war effort, effectively providing care for huge numbers of military and civilian casualties. The Soviet Union lost some 25 million people in World War II. Despite the harsh conditions for both military and civilian populations, no mass epidemics occurred. External observers including Garrison in the 1920s, Sigerist in the 1940s, Field in the 1960s, and more recently Roemer (1991, 1993), as well as Russian medical historians Yeravinski, Smirnov, and others, noted the remarkable achievements of reducing epidemic diseases, meeting wartime demands, and bringing health care to the whole country. Postwar stabilization allowed restoration of health services and makeup of trained personnel lost in the conflict.

In order to assure equal access, a district health system evolved with each district required to have sanitary epidemiologic stations (SESs), hospitals, polyclinics, and specialized treatment facilities according to national norms based on population size. The SESs supervised water, sewage, air, and ground quality; conducted epidemiologic investigations of infectious disease outbreaks; and monitored child health and nutrition status. *Medsanchest* clinics located in industrial plants provided on-site medical and occupational health services, and prophylactic health centers provided a variety of medical rehabilitation services, sanatoria, and vacation benefits. Originally, polyclinics in each district were linked as outreach facilities to the district hospital with staff rotation between them to promote continuity of care and improve professional education. However, this became impractical because of rapid expansion of the number of polyclinics. Prevention of disease continued to be based on routine screening checkups for workers and other specified groups.

With universal access of the population to preventive and curative care, control of infectious disease was achieved and the health status of the population dramatically improved. A strong system of epidemiologic surveillance and control evolved and successfully defended the huge population through the challenges of Russian history of the twentieth century with the social disruption, starvation, migration, mass imprisonment, and executions in the gulags in the 1920s and 1930s. The tragic losses of some 23 million soldiers and civilians during World War II (13.7 percent of the total population) were followed by the dramatic return to society of millions of prisoners after Stalin's death in 1953. During the 1950s, the Soviet model of a state-operated health system was widely promoted and emulated in countries of Eastern Europe, Central Asia, newly independent countries in Africa, Asia, and the Middle East as well as in Latin America. It also influenced the development of the Alma-Ata approach of Health for All based on universal access to primary health care.

As a response to the increasing prevalence of chronic diseases in the mid-1960s, the Communist Party Plenum in 1983 decided to implement annual *dispanserizatzia* or checkups as a uniform program for the general population, provided in polyclinics, hospitals, and specialized clinics. The checkups and treatment involved clinical care, follow-up ambulatory or hospital care, sanatoria (i.e., rest homes), and a change of work if necessary. The screening program increased demands for hospitalization because of limited ambulatory diagnostic resources, placing the major focus of care on hospitalization and institutional care. In the mid-1980s, the Ministry of Health enunciated the continued direction of health policy as concentrating on "development of preventive medicine and improvement of health care facilities through a program for building general and specialized hospital establishments." With central control of financing, the state set mandatory norms for personnel and hospital beds, and controlled medical education to produce the human resources to operate the system. The state monopoly on health, however, led to stagnation with a bias of the system toward hospital care, without financial or epidemiologic accountability for efficiency and effectiveness. The focus on hospitalization and institutional care has begun to change and the per capita acute hospital bed supply has declined from the mid-eighties to under 8 per 1000 by 2005 (Figure 13.9).

In 2005, President Vladimir Putin established priority projects in education, health, housing, and agriculture. Priority health projects were intended to improve health status of the population, to increase accessibility, and improve the quality of medical care. It stressed strengthening primary care as well as health promotion and disease prevention activities, and projects to improve accessibility to tertiary care. These included upgrading ambulatory care, additional immunization programs, new check-up programs for infants and pregnant women, and AIDS prevention and treatment. Primary care centers are being re-equipped with cardiographs and ultrasound equipment. Salaries for general practitioners and nurses have been improved to attract young staff.

Epidemiologic Transition

Despite major improvements during the Soviet period (1917–1991), mainly due to control of infectious diseases, health status of the population dramatically deteriorated in the last quarter of the twentieth century. By 1994, life expectancy at birth for males had fallen to less than 59 years, almost 6 years fewer than in 1985 (Table 13.11).

Life expectancy improved up to the 1960s, but since has lagged well behind other countries (Figure 13.10). Very high mortality rates from cardiovascular diseases and trauma are

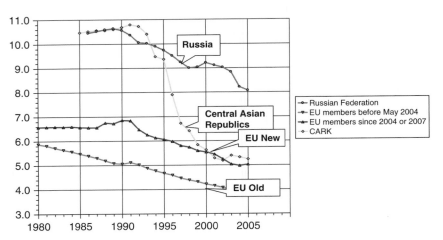

FIGURE 13.9 Acute care hospital beds per 1000 population, Russia and European Union, 1980–2005. Source: Health for All Database. WHO, European Region 2007.

primarily responsible for low and declining life expectancy. Cardiovascular disease mortality is twice as high as in OECD countries. Trauma mortality in males is 2.5–3 times that of western industrial countries. However, in the WHO European Region, even in 2003, in all age groups from 15–60 years old, the Russian Federation trauma mortality rates are the highest. However, there has been a stabilization of mortality and life expectancy between 2003 and 2006.

Even before the impact of the collapse of the Soviet system was felt in 1991, mortality rates in Russia were much higher than those in other industrialized countries. Standardized mortality rates in Russia were 1.5 times higher for total mortality. Table 13.12 shows even higher rates in categories such as cerebrovascular disease, trauma, and infectious diseases, with alcohol binge drinking and violence as major factors.

The crude birth rate declined in Russia from 17.2 per 1000 in 1987 to 8.4 per 1000 in 1999 and rising modestly to 10.2 in 2005. The total fertility rate declined from 2.0 in 1989 to 1.1 in the period 2000–2005, well below replacement levels. Infant mortality rates fell from 22 per 1000 in 1980 to 16 in 1998, and 11 in 2005, still twice the rates in western European countries. Abortion is the main method of birth control, and modern methods are not widely available or trusted. Maternal mortality in Russia declined from 68 per 100,000 live births in 1980 to 44 in 1998 but has declined sharply since 2000, reaching 25 per 100,000 live births in 2005 as compared to rates under 10 in the industrialized countries (HFA, 2007) as birth rates declined.

The decline in health status since 1990 cannot be blamed solely on the current economic crisis, or entirely on the health care system. The worsening mortality pattern

TABLE 13.11 Life Expectancy at Birth, by Gender, Russian Federation, Selected Years, 1961–2006

Year	Male	Female	Total
1965	64.3	73.4	69.5
1970	63.2	73.6	68.9
1975	62.3	73.0	68.1
1980	61.5	73.1	67.6
1985	63.8	74.0	69.3
1990	63.8	74.3	69.2
1995	58.3	71.7	64.6
2000*	59.0	72.3	65.3
2006*	60.4	73.2	66.6

Note: Population of Russia (1965–1997) Moscow, Goscomstat of Russia, 1998 (in Russian). Rounded to nearest decimal place.
*Preliminary data.
Source: Data from the State Committee of the Russian Federation on Statistics (GOSKOMSTAT), http://www.gks.ru and http://www.gks.ru/bgd/regl/b07_13/IssWWW.exe/Stg/d01/04-23.htm.

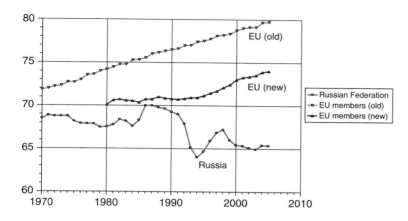

FIGURE 13.10 Life expectancy at birth in years, Russia, EU members, 1970–2007. Source: Health for All Database. WHO, European Region, November 2007.

TABLE 13.12 Age-Standardized Mortality Rate per 100,000 Population for Selected Causes of Death in Russia and Other Countries, 2002

Country	Noncommunicable diseases (Total)	Cardiovascular diseases (Total)	Cancer (Total)	Trauma (Total)
Russia	960	688	152	217
Poland	593	324	180	53
Germany	444	211	141	29
Denmark	503	182	167	40
United States	460	188	134	47
United Kingdom	434	182	143	26
Israel	399	136	133	30
Sweden	379	176	116	30
Canada	388	141	138	34
France	368	118	142	48
Japan	287	106	119	39

Note: Data are standardized to the world population.

Sources: *World Health Statistics* 2007, WHO *Statistical Information System* (WHOSIS), http://www.who.int/whosis/whostat2007/en/index.html; http://www.who.int/whosis/indicators/2007compendium/en/index.html

is due to a combination of factors: stress, alcohol abuse, smoking, violence, lack of a balanced diet, lack of modern health care technology, environmental pollution, and a general mood of anxiety and depression related to the dramatic decline in economic and political stability since 1990. Males are not surviving to pension age at 60 years.

A combination of factors encouraged a medical bias toward care of individual patients and failure to apply the successful experience of the 1930s to the control of epidemics of noncommunicable disease. The concept of prevention took on a primarily medical orientation, stressing routine checkups. Health policy continued to promote increased supplies of doctors, polyclinics, and emphasis

on hospital beds. The number of medical graduates per 100,000 population was declining but has been stable since 2000. Table 13.13 shows a comparison of human resources in Russia to other countries.

A hospital sector with a passive strategy of treatment and long hospital stays was unable to keep up with technological advances and consumed a large share of the very limited amount of funds allocated to health care.

Post-Soviet Reform

The Russian Federation continues to provide basic social security and health care for all citizens. Until 1993, when

TABLE 13.13 Human Resources (Selected) per 1000 Population, Selected Countries and Years, 2001–2004

	Physicians	Nurses	Midwives	Dentists
Canada	2.1	10.0		0.6
China	1.1	1.1	0.03	0.1
Israel	3.8	6.3	0.2	0.2
Japan	2.0	7.8	0.2	0.2
Russian Federation	4.3	8.1	0.5	0.3
United Kingdom	2.3	12.1	0.6	1.0
United States	2.6	9.4	—	1.6

Source: World Health Report, 2006 Annex 4; data rounded, http://www .who.int/whr/2006/annex/06_annex4_en.pdf | |accessed May 24, 2008|

compulsory health insurance was established, all social benefits were funded from the general budget of the government. The health insurance scheme was based on mandatory payment by employers to regional health insurance funds.

From the early 1990s, the World Bank has played a significant role in influencing policymakers in Russia promoting health insurance to partially replace state funding of health services, as well as decentralization and privatization of health and social services. Decentralization of management of most state health services and financing of health care increased regional and local health autonomy. The sudden and nearly complete decentralization of authority and funding hampered central management by the Ministry of Health in its capacity to develop new policies for public health issues. Central management of the sanitary epidemiologic service was hampered by limited funding for expanding the immunization program or to promote nutrition and other health promotion initiatives.

Decentralization of compulsory health insurance through regional systems allowed local change in health management issues and shifting from obsolescent inflated national norms for hospital beds and personnel. Epidemiologic, economic, and cost-effectiveness analysis is vital to reform in health care, especially in harsh economic conditions. Calculation of the cost of a service is fundamental. Regional and municipal authorities now have more financial responsibility and power to reallocate funds and shift priorities from institutional treatment to prevention and ambulatory care, but lack trained health management personnel to challenge old assumptions, such as the norms for hospital beds and human resources, still used as guidelines.

The deteriorating health situation is a part of a health-care system in decline. The structure continued to focus on inpatient services with less attention to ambulatory and primary health care, disease prevention, and health promotion. The per capita hospital bed supply and average length of stay are much greater than in the European Union. In addition, public funding of health care has declined considerably, and the collection of informal user fees by public health providers has reduced the access of the poor to health care. Public sector funding for health care is through the federal, regional, and municipal levels and the 3.1 percent payroll tax. International reports (e.g., WHO, OECD, UN) show that the federal level allocates 3.6 percent of expenditures to health.

Development of information systems, training of leadership personnel in modern management theory and practice, and reduction of the hospital sector with transfer of resources to primary care are needed to improve health care efficiency and quality. Access to international literature and the Internet will help to improve quality of continuing education in the health sector. Reforms based, in part, on reallocation of existing resources, will require additional funding to meet the cost of the transition and raise the quality of care.

Health reforms are essential to preserve universal access and to raise population health status, medical care, and public health to international standards. Changes in financing of health care, adoption of international health targets, and changes in manpower development programs are needed. But these depend on a new set of priorities and new standards at the national, oblast (province/state in Russia), and local health authority levels of government. Decentralization and diffusion of the overly centralized system require epidemiologic information and dialogue on health issues to raise health awareness and management practices to meet the health needs of the people.

The SESs are a force with the potential to expand their traditional roles to lead health promotion activities at the community and district levels, raising public awareness and knowledge of health issues. This will mean a change in attitude from being defenders of the old system to responders to community needs. That means redefining objectives, instituting training programs for new personnel, modernizing technology for laboratories, and environmental quality and enforcement issues.

Training for modern behavioral epidemiologic data collection, analysis, and its distribution are essential to promote knowledge of risk factors and their control. Policy, provider, and community levels need to define cost-effective programs to meet local conditions. Wide distribution of relevant data to government and the general public is needed to help change knowledge, attitudes, beliefs, and practices related to risk factors.

Future Prospects

The Russian health system has important assets with potential for change. Increased funding during the first decade of the twenty-first century is linked to economic growth and concern for the demographic decline. Allocation of resources will need to represent a reordering of priorities and allocating funds to promote primary care, upgrading salaries and equipment, and revision of the role of the existing polyclinic system. A system of grants to local authorities specifically directed toward strengthening primary care and health promotion would help to downsize and upgrade the hospital sector in terms of equipment and upgrade primary care, with the government of Russia developing new approaches toward reallocating rising expenditures and more effective management of financing the health system. There is a multichanneled system of financing in place, based on the state budgets, the compulsory insurance system, household expenditures, services by other governmental and industrial complexes, and voluntary insurance.

Bureaucratic segmentation of services presents a formidable barrier to reform. Privatization is not a solution and will not help to reduce the current burden of excess mortality in the Russian Federation. Instead, a reform process should build on the main existing structure of SESs and polyclinics but with major revision of content and quality standards.

OECD recommendations (2006) for reform in the Russian health system include the following: "closing the gap between formal commitments to the population and available resources; shifting the structure of provision towards greater reliance on integrated primary care; adopting payment schemes in the healthcare sector that encourage more cost-effective therapeutic choices; and modernizing the system of mandatory medical insurance."

Health reforms needed include refocusing on a number of key issues:

- Health promotion regarding smoking reduction, alcohol abuse, nutrition, physical activity, and trauma prevention;
- Sustained increased level of funding for health;
- Pooling of regional health budgets and health insurance;
- National standards and guidelines for a regional "basket of services" for all;
- Replacement of obsolescent financing norms of the Ministry of Finance;
- Reduced hospital bed supply with upgrading of hospital quality and greater emphasis on ambulatory and home care;
- Development of polyclinics with preventive and curative services for defined populations with capitation funding and incentives for improved quality and efficiency of services;

- Financing patient care on a capitation basis;
- Control of corruption and under-the-table payments;
- Raising standards of care and quality of services;
- Increased contact with the international community through health literature and professional meetings.

Summary

The Soviet health system brought health care to a vast, underdeveloped country. This system provides universal access, within a totally state-operated system of service. It was a source of pride to the Soviet state, and was recognized internationally as an important model because of its successes from the 1930s to 1960s. During the 1950s, the Soviet model, a state-operated health system, was widely promoted and emulated in eastern Europe and in newly independent countries in Africa, Asia, and the Middle East, as well as in Latin America. This model also influenced development of the Alma-Ata approach of Health for All based on universal access to primary health care.

From the 1960s to the 1990s, an epidemiologic transition occurred in varying degrees in the different republics and ethnic populations of the Soviet Union. This transition was characterized by sharp declining mortality from infectious diseases and rising death rates from noninfectious diseases. Life expectancy remained static during the 1970s and 1980s. In the 1990s, life expectancy declined dramatically, especially for men, during the economic and social crisis following the breakup of the Soviet Union.

This crisis in health related not only to the period of economic transition in the 1990s, but goes deep into the former Soviet health system with quantity compromising quality since the epidemiologic transition and changes in health profile of the population. The system operated as a state monopoly, with the central government in control. It lacked mechanisms for epidemiologic or economic analysis and accountability to the public. The epidemiologic transition from predominance of infectious to noninfectious diseases was addressed by further increases in the quantity of services. Policy and funding favored hospitals. Individual health was deeply affected by stress associated with great uncertainty, economic collapse, and breakdown of social safety nets. Alcohol consumption, homicide, and suicide in Russia are among the highest in the world. The challenge of health promotion and adequate prevention and treatment are not met by the existing health system. A decrease in alcohol consumption in Russia is critical for the long-term improvement in the demographic crisis.

Reform since 1991 has centered on compulsory national health insurance and decentralized management of services. In order to free resources to address health needs more effectively, reforms aimed at rationalizing the health care delivery system are needed. However, the reform movement was lacking a broad national health strategy to address the fundamental public health problems and especially the present enormous excess of preventable mortality. A new national health project reports 138,000 fewer deaths in 2006 compared to 2005 and this decline is attributed to new national initiatives in social and health policy in the Russian Federation. Rising national income and standards of living in recent years will foster this improvement, but structural and content reform of Russia's health system is important to reduce the continuing dreadful toll of preventable deaths in the country.

A 2007 joint report of the World Bank and WHO states that: "Non Communicable Diseases and Injuries (NCDIs) share common risk factors, underlying social and cultural determinants and opportunities for intervention. In some CIS countries, seven leading risk factors — high blood pressure, tobacco consumption, alcohol abuse, high blood cholesterol, overweight, low fruit and vegetable intake, and physical inactivity — account for more than 60% of the disease burden, measured in disability-adjusted life-years (DALYs). NCDIs and their causes are unevenly distributed in the population, with higher rates among the poor and vulnerable. Gaining better health for the people of the CIS countries is achievable, as it is possible to significantly reduce the burden of premature death, disease and disability through comprehensive action on underlying determinants and risk factors of noncommunicable diseases and injuries" (World Bank and WHO, 2007).

ISRAEL

Israel in 2006 had a population of 7.1 million. The GDP per capita was $25,864 in 2005, life expectancy at birth of 80.3 years, and the ranking in the Human Development Index was 23rd (HDI 2007/08). Infant mortality in 2005 was 6 per 1000 live births, and maternal mortality 5 per 100,000 (2000–2003) (HFA, 2007). Total health expenditures as a percentage of GDP were 8.7 percent in 2005, with 70 percent from public sources.

Immunization rates are high, with 2005 rates of 95 percent for DPT, 93 percent for polio, and 93 percent for measles. The content of the publicly funded program has gradually expanded to include hepatitis A and B, *Haemophilus influenzae* type b, but not yet pneumococcal pneumonia, varicella, rotavirus, or influenza vaccines for children. Despite high immunization rates, an epidemic of polio with 15 cases occurred in 1988, and measles epidemics in 1991 and 1994, leading to the adoption of improved immunization policies.

Health indicators for Israel show an advanced state of health but with important ethnic, regional, and gender inequalities. In infant mortality, Israel ranked 10th in 2005 among the advanced countries, a rise in ranking from 25th place in 1990. For longevity, Israel is second among the leading countries for males and among the top 10 for female longevity.

Origins of the Israeli Health System

Israel's health system evolved gradually over the past century. Palestine under the Ottoman Turkish Empire was a poor, disease-ridden, remote province rife with malaria, dysentery, and other infectious diseases. Immigration of Jews from eastern Europe and Arabs from surrounding countries since the 1880s led to initiation of charitable hospitals to provide care for the urban poor.

Jewish immigrants from eastern Europe formed labor brigades and mutual aid associations. Sick funds were initiated in 1912 based on mutual benefit principles derived from European models, associated with the union movement, and later with other political organizations. The sick funds grew to provide medical care insurance and services to over 95 percent of the population. They provide services through neighborhood and specialized clinics, or affiliated doctors in their own clinics, purchasing hospital care from government or NGO-operated hospitals in areas where they lack their own.

Preventive care originated in 1911 by nurses from the United States sponsored by Hadassah, an international women's organization. Following the conquest of the area by British forces from the Turks in 1917, Hadassah sent the American Zionist Medical Unit from the United States to help establish a network of health facilities in Palestine. This consisted of 44 doctors, nurses, dentists, and other personnel with equipment and financial support from Hadassah and the Joint Distribution Committee. The unit opened hospitals in many urban centers, and established nursing training and preventive care programs for immigrants and schoolchildren, as well as mother and child health stations (*Tipot Halav* or "drop of milk" stations). These were gradually located in towns, villages, and neighborhoods throughout the country, providing prenatal care and child care for infants and toddlers. They provided immunization, child development monitoring, and nutrition counseling to almost all the infants in the country, and prenatal care for most women in the country, with others going to private doctors.

The British mandate from 1917–1948 brought successful colonial administrative experience and development of basic public health law and systems, licensing of medical professions, sanitation, food and drug laws, as well as public health laboratories, malaria control, and many other features of public health, standard of the time.

From 1912–1948, the health system grew based on primary health care through the *Tipot Halav* and the labor movement's sick fund clinics in towns and villages throughout the country, providing ready access to primary care treatment and referral services.

Following the establishment of the State of Israel in 1948, massive immigration from post-Holocaust Europe and the Middle East brought an enormous burden of health problems to the country. The new Ministry of Health established regional hospitals throughout the country in abandoned British army camps, providing acute care, rehabilitative services, mental health, and chronic care services. Other hospitals are owned by the major sick funds and by NGOs. Reliance on ambulatory and primary care with regional medical and hospital centers is the basis of the Israeli health system.

Health Resources and Expenditures

Health care expenditures as a percentage of GNP increased from 7.9 percent in 1990, to 9.3 percent in 2002, then declined to 8.7 percent in 2004 (Table 13.14). Acute care hospital beds per 1000 population were reduced in Israel from 3.0 in 1980 to 2.6 in 1990, 2.3 in 2000, and 2.1 in 2006. During this period, psychiatric beds were reduced from 2.2 per 1000 in 1980 to 1.5 in 1990, 0.9 in 2000, and 0.6 in 2006. Nursing and elderly care beds increased from 1.4 in 1980 to 2.0 in 1990, 2.9 in 2000, and 3.2 in 2006.

Expenditures on hospital care increased from 34 to 41 percent of total health expenditures from 1975–1990, followed by a decrease to 38 percent in 2004 (see Table 13.15).

Along with the decline in acute care beds, average length of hospital stay (acute care), fell from 6.8 days in 1980 to 4.1 days in 2006 and bed occupancy increased from 90 percent to 95 percent. Ambulatory care and community health consume about one-third of total expenditures, an increase since 1975. Salaries in the health sector have been low compared to other sectors in the society, as has been capital investment but have increased since. Cost restraint, improving the physical infrastructure, and keeping up with technological advances in medicine are a major challenge for the future.

TABLE 13.15 Health Expenditures (percent) by Type of Service, Israel, 1985–2004

Category/Year	1985	1990	1995	2000	2004
Hospitals	42.7	40.7	40.6	37.9	38.4
Public clinics/ prevention	32.5	32.6	34.5	38.5	41.2
Other	24.8	26.7	24.9	23.6	20.4
Total	100	100	100	100	100
Government as percent of total health expenditures	na	na	72.3	69.5	67.9
Private expenditures	na	na	27.8	30.6	33.2
Social Security as percent of total government expenditure	na	na	47.1	48.5	50.0
National health expenditures as percent of GDP	6.8	7.8	8.6	8.0	8.1

Source: Statistical Abstract of Israel, 1996; and Health in Israel, 1998, http://www.who.int/nha/country/isr.pdf [accessed May 24, 2008], and Ministry of Health. Health in Israel, 2005, Selected Data. Ministry of Health, Jerusalem, and Ministry of Health, Health, Israel, and WHO European Region Health for All Database, November 2007.

Health Reforms

After many years of debate, several national commissions on health, and gradual reform of health services, Israel's national health insurance (NHI) plan was implemented on January 1, 1995. It covers the total population through the universal National Insurance social security system. The individual pays for this through a 3 percent deduction from his or her salary along with an equivalent employer's

TABLE 13.14 Health Expenditures, Hospital Resources and Utilization, Israel, 1960–2004

Resources/Utilization	1960	1970	1980	1990	2000	2004
Health expenditures as percent of GNP	5.5	5.4	6.9	7.9	8.6	8.7
Acute care beds/1000 population	3.1	3.2	3.0	2.6	2.3	2.1
Hospitalization (acute) days/1000 population/year	954	1148	997	834	785	761
Discharges (acute)/1000 population/year	115	129	139	156	175	173
Average length of stay (acute care) [days]	8.3	8.9	6.8	5.3.	4.3	4.2
Mental health beds/1000 population	2.1	2.4	2.2	1.5	0.9	0.8
Mental diseases days/1000 population	615	631	721	496	379	220
Nursing and elderly beds/1000 population	Na	Na	1.4	2.0	2.9	3.1

Source: Statistical Abstract of Israel, 1975, and Ministry of Health. Health, Israel, and WHO Euro Region Health for All Database, November 2007, and Ministry of Health. Health in Israel, 2005, Selected Data. Ministry of Health, Jerusalem.

contribution to a mandatory NHI program, which also covers old age and disability pensions, workers' compensation, and other social benefits. Each family must select membership in a Sick Fund which functions as a health maintenance organization. The National Insurance Institute transfers funds to the Capitals for Sick Fund and Health Maintenance Organizations according to a per capita formula, with a larger per capita payment for the elderly and for populations in developing towns.

The Ministry of Health supervises the sick funds, which are required by the new law to provide a basic basket of services that is very comprehensive. The Sick Funds are obliged to provide all specified services or arrange for those services it cannot itself provide. They provide comprehensive care either through their own neighborhood clinics or through affiliated private physicians who are paid on a capitation basis. Additions to the obligatory basket of services, such as new medications and diagnostic tests, are funded by the Ministry of Health. Co-payments for specialists, diagnostic testing, and pharmaceuticals are a source of cost sharing. People receiving welfare do not pay co-payments and the elderly population pays 50 percent of the quarterly ceilings. However, there are no subsidies to low-income groups, resulting in inequalities in the access to health care.

The sick funds are accountable for the services rendered, and the individual is entitled to change sick funds semiannually. Most hospital beds are operated directly by the Ministry of Health. Government hospitals are meant to be transferred to independent trusts, to operate as economic units able to allocate funds internally and compete for clients, with payment on a DRG basis. Regionalization of services will be difficult to achieve in the present configuration of the NHI law because each sick fund has its own regional organization.

Health promotion is gaining strength in Israel, and health awareness has generally increased (Table 13.16). The compulsory seat belt law has met with compliance by a large majority of car drivers, and similar legislation requiring use of helmets for motorcycle drivers is also generally implemented. Similar requirements were passed in 2007 for bicycle users. Increases in permitted speed limits on major highways have been followed by a rise in motor vehicle deaths and case fatality rates.

Non-smoking legislation banning smoking from public buildings and workplaces has helped to reduce smoking in the adult population, especially among males. Smoking is still practiced by some 23 percent of adults in 2006, a reduction from 34 percent in 1991. In 2007, smoking was banned by law in bars, restaurants, and cafés, with the owners held responsible and fined for violations. Compliance is reportedly well accepted.

Low-fat foods are now commonly available in supermarkets. Private food manufacturers fortify baby formulas and cereals with vitamins and minerals. Breakfast cereals are also enriched, but food fortification of bread, milk, and salt with essential minerals and vitamins is not practiced. Obesity is common, especially in women over age 45, and while consciousness of the importance of physical fitness is increasing, it is still not at an acceptable level.

Mortality from stroke and coronary heart disease has declined dramatically over the past 3 decades largely as a result of improved treatment of hypertension and myocardial events, but also because of a decline in smoking and greater interest in self-help to maintain health. Cardiovascular diseases have now fallen below cancer as leading causes of mortality, although prevalence rates remain high. Obesity and diabetes are growing as national health problems.

The health agenda has paid increasing attention to health promotion as well as to the structure of health services in health reform. The Ministry of Health concentrates on reform in health services and national health insurance, but needs a cohesive strategy with health targets and strengthening of health promotion strategies. Decentralized administration of services is through the long-standing sick funds which are increasingly emphasizing preventive services and a health maintenance function. Regional disparities in health resources and health status are an issue and a powerful argument for regional budgeting by the centrally managed sick funds.

Regional, social, and ethnic disparities are still important in Israel's health status; the Arab population has higher rates of infant mortality than the Jewish population, 8 versus 4 per 1000 in 2005. Large-scale immigration of Russians and Ethiopians since the 1990s brings together people with different risk factors. More than 21,000 immigrants came to Israel in 2005. Of these immigrants, 9400 came from the former Soviet Union and 3600 from Ethiopia. Traditional distribution of health resources favors the more concentrated population centers, while the more rural areas receive fewer resources per capita.

Mental Health

Mental health reform has been a controversial subject since the establishment of National Health Insurance. In 2006, the Ministry of Finance and the Ministry of Health agreed to transfer Mental Health Care to the sick funds. The sick funds will be responsible for including mental health care in the basic basket of services. A debate between various stakeholders is delaying the process. Concerns over the transfer of mental health care to the sick funds include increased demand for services, increased cost, loss of jobs, and poor working conditions. This reform is projected to be implemented during 2008.

Healthy Israel 2020

The *Healthy Israel 2020* initiative was created by the Ministry of Health to define Israeli policy in the areas of

TABLE 13.16 Health Promotion Initiatives in Israel

Topic	Action	Effects
Smoking	Nonsmoking legislation; restricted advertising; nonsmoking promotion by NGOs	Increased awareness of health effects of smoking; ban on smoking in public places, extended in 2007 to include bars, restaurants with responsibility of owners
Cancer prevention	Promoting mammography, reduced sun exposure; restricting smoking	Improving public awareness; cancer increasing, but far lower than most European rates, except for breast cancer
Nutritional health education	Mediterranean diet; high vegetable, fruit intake; low consumption of animal fats	Increasing public awareness of healthy nutrition contributes to rapidly declining cardiovascular mortality and low rates of cancer; decreasing obesity
Motor vehicle accidents	Mandatory seat belt use implemented; highway speed limits raised; inadequate highway patrols; improved emergency care and transport	Overall fatality rates declining Police activity in driver licensing, road monitoring; electronic monitoring demonstrated to be more effective; Urban round abouts, speed bumps, increased interurban speed cameras
Water quality	Mandatory chlorination since 1988 Filtration plan operational for main surface water source from 2006	Less waterborne diarrheal disease
Sewage	Increasing treatment	Reuse of wastewater increasing
Enrichment of basic foods	Law permits but does not require private manufacturers' initiatives	Breakfast cereals enriched; infant formulas and cereals enriched; basic foods (bread, salt, milk) not enriched
Food quality	Food supervision strengthened Standards at international levels	Public awareness increased; low-fat foods now widely available
AIDS/STI prevention	School health education	Widespread information on use of condoms and avoidance of transmission in drug use
Preventive health care	Universal health insurance; sick funds become health maintenance organizations	Prevention increasing in primary care for all ages; computerized medical records and clinical guidelines
Healthy Cities	Active association and networking of healthy cities	45 cities have "Healthy Cities" programs; health profiles and sustainable strategy
Health promotion	Raising professional and public consciousness of health and lifestyle	Growing public consciousness of diet, fitness, smoking as health factors
Healthy aging	Community centers for elderly	Municipal and NGO sponsorship of activity centers and programs
Healthy Israel 2020	Define health targets for 2020 with measurable indicators	Use of evidence-based approaches to health promotion addressing risk factors

Source: Adapted from Israel Center for Disease Control. Health Status in Israel, 1999; and Donchin, M., Shemesh, A. A., Horowitz, P., Daoud, N. 2006. Implementation of the Healthy Cities' principles and strategies: An evaluation of the Israel Healthy Cities network. *Health Promotion International*, 21:266–273.

disease prevention and health promotion. It will establish and prioritize objectives, quantitative targets, and interventional strategies necessary to improve health and reduce health inequities. The initiative is similar to *Health 21*, the European Region version of the WHO's *Health for All*, and to *Healthy People 2010* of the U.S. Department of Health and Human Services. It will adapt current evidence and standards of health behaviors and environmental factors affecting morbidity. The initiative involves collaboration of a broad spectrum of individuals and organizations; including representatives of government ministries, local community centers, businesses, and the Knesset. They are dealing with 21 focus areas, consulting international experts, addressing health behaviors and risk factors, using evidence-based targets and interventions. *Healthy Israel 2020* will promote policies, implementation of preventive interventions, and health-related behaviors that have the potential to increase life expectancy and quality of life, as well as to moderate the increase in health care costs.

Summary

Israel has achieved high standards of health care and health status indicators. The Israeli health system has been a quasi-national health service with over 95 percent coverage through Sick Funds for decades; and NHI in 1995 brought universal coverage to all residents of the country. The health system has helped the Israeli population achieve low rates of mortality from infectious and noninfectious diseases and life expectancies among the longest in the world.

Medical and paramedical professional education, research, and the medical and drug industries have reached high levels of excellence. The 1995 implementation of NHI through Social Insurance provides greater equity in financing and reduces political manipulation of the health system. Primary care services still separate community-based preventive and treatment facilities, but the sick funds are increasingly prevention oriented with improved standards of primary care. Regionalization of financing and service management is needed to alleviate disparities, but the Israeli health system has contributed to high standards of health for the whole population.

HEALTH SYSTEMS IN DEVELOPING COUNTRIES

In most developing countries, health services were inherited from colonial regimes and subsequently influenced by the Soviet model of health care in the 1950s and 1960s. The development of primary health care was neglected, and underfunded with excessive allocation of resources to teaching hospitals in the main population centers, leaving little for the rural majority. As a result, most developing countries are facing the need to reform their health systems.

During the 1980s, emphasis slowly moved toward primary care under the influence of the *Health for All* initiatives sponsored by the World Health Organization. Achievements during the 1980s and 1990s included greatly improved immunization coverage, wide-scale use of oral rehydration therapy (ORT), and better sanitation. There has been a decline in birth rates in most regions of the world, including sub-Saharan Africa, which had until the 1990s seemed totally resistant to birth control. National health programs emphasize primary care with immunization, ORT, promotion of breastfeeding, supplemental feeding for infants, and birth spacing.

Productivity and per capita GDP are rising in many countries in the developing world so that a combination of universal primary education and improved economic status is furthering the potential for continuing improvement in health standards. In sub-Saharan Africa, low and declining levels of economic activity reduce the likelihood of increasing funds for health care. AIDS, malaria, tuberculosis, measles, other infectious diseases, poverty, malnutrition, and high birth rates with high child mortality aggravate a poverty–population–environment (PPE) cycle, which impairs national growth potential. As western diets, lifestyles, and technology are absorbed in developing countries, they face a dramatic increase in chronic diseases such as hypertension, diabetes, stroke, coronary heart disease, and motor vehicle accidents. This, along with increasingly costly technology, places new burdens on health services.

Most developing countries spend less than 4 percent of their low national incomes on health and much of that on costly hospitals in the capital cities. Lack of adequate governmental budgetary funding raises political interest in national health insurance, especially in the mid-level developing countries. The purpose is to bring more of the population into the health care system and raise additional funds for health care beyond the little that is provided through government allocations.

In this section, following brief regional overviews, we give several examples of developing countries actively working to reform their health care systems. Health insurance is needed to increase funding for health care and provide for the growing urban employed and middle-class health needs, but at the same time, ministries of health must provide direct services to the rural poor majorities. As in developed countries, there will be no uniform approach, but sharing of lessons learned will be helpful.

Sub-Saharan Africa includes 40 countries with a total population of 713.5 million persons in 2005 and a GNP per capita of $764. Annual births of 28.7 million in 2005 represent a decline in total fertility rates from 6.6 in 1960 to 5.9 in 1997 and 5.4 in 2005. Life expectancy increased from 44 years in 1970 to 50 years in 1990 and declined to 46 in 2005. Mortality of children under age 5 declined from 244 per 1000 in 1970 to 188 per 1000 in 1990 and then to 169 per 1000 in 2005. Maternal mortality remains very high, with a rate of 940 per 100,000 in 2000, and with only 43 percent of births attended by a trained attendant (1997–2005). Immunization rates for the major childhood diseases have improved markedly and range between 65 and 75 percent with tetanus immunization of pregnant women at 39 percent. Malnutrition (acute and chronic) further damages child health status. Tuberculosis, malaria, AIDS, measles, and other infectious diseases are major contributors to high rates of morbidity and mortality. In addition, chronic diseases and trauma are increasingly important contributors to the total burden of disease.

In the face of economic decline, political chaos in many countries, and the aforementioned health problems, resource allocations for health budgets have been jeopardized in many countries. Some countries in the region devote less than $2 per capita to health budgets. Despite these challenges, progress has been made in efforts to improve sanitation and expand primary care services to the underserved rural areas and urban slums, giving hope for effective public health in the twenty-first century.

The dramatic effects of HIV/AIDS, with accompanying epidemic of tuberculosis and multidrug-resistant forms have created a public health crisis of great severity in sub-Saharan Africa, largely overwhelming their nascent health infrastructures. Yet progress is being made in immunization coverage for children and successful assistance programs to widen the impact of EPI, DOTS, oral rehydration, antiretroviral therapy, and other modalities of public health hold out hope for a better future. The loss of skilled health manpower to wealthy countries such as the United Kingdom, Europe, and North America is a serious deficit, yet signs of hope in political, economic, and social development are being seen.

Dramatic progress has been made in the eradication of poliomyelitis, dracunculiasis, and onchocerciasis. Less progress is seen in tuberculosis, schistosomiasis, and malaria control. The WHO recommends wide-ranging new efforts in prevention of cancer by immunization against hepatitis B, screening transfusion blood for hepatitis C, schistosomiasis control, smoking and alcohol control, reducing risk factors for cardiovascular diseases and diabetes, mental health, and oral health. The effects of civil war, collapse of governments in some areas, and refugee situations have had dreadful effects on public health. Since the mid-1990s, signs of stabilization in the governments of some countries (such as Nigeria, discussed below) of sub-Saharan Africa and economic progress offer new hope for the future of this potentially wealthy continent.

Federal Republic of Nigeria

Nigeria is the most populous country in Africa, and the eighth in the world with a population of 140 million in 2006, an increase from 61.5 million, with 48.3 million in urban settings and 51.7 rural, the latter living in more than 90,000 villages. Nigeria has more than 250 ethnic groups, with varying languages and customs. Religious groups are Muslim 50 percent, Christian 40 percent, and indigenous beliefs 10 percent. The number of children born in one year is around 5 million. Nigeria has 55 percent of its population living in poverty, and less than 60 percent of primary-age children attend school. In 2005, 4.4 percent of the population were infected with HIV, based on serosurveys.

The 2005 GDP per capita was $1128 (PPP U.S.$). The Human Development Index ranked Nigeria as 159 (as shown in Table 13.18) out of 177 countries in 2005. Life expectancy at birth was 43.3 years with a difference between males and females of only one year. Nigeria's child mortality rate fell from 230 in 1990 to 191 in 2006, and infant mortality was reduced from 120 to 99 in the same time period. (UNICEF, 2008). Table 13.17 provides some vital statistics indicators for Nigeria.

In 1960, Nigeria gained independence from the United Kingdom, becoming a republic in 1963. It has had a turbulent political history since then, full of violence and instability resulting in a slow rate of development. Despite the vast oil wealth discovered during the 1970s, Nigeria (a member of OPEC and the 6th largest producer of oil in the world), long ruled by repressive military regimes, saw corruption erode all levels of government functioning.

TABLE 13.17 Vital Statistics Indicators of Health Status, Nigeria

Life expectancy at birth (years) males	47.0 (2005)
Life expectancy at birth (years) females	48.0 (2005)
Children under five years of age mortality rate	194 (2005)
Infant mortality rate (per 1000 live births)	101.0 (2005)
Neonatal mortality rate (per 1000 live births)	47 (2004)
Maternal mortality ratio (per 100,000 live births)	800 (2000)

Note: World Health Organization Statistical Annual 2008, http://who.int/countries/nga/nga/en/ and http://www.who.int/whosis/database/core/core_select_process.cfm [accessed June 25, 2008]

The military in Nigeria has played a major role in the country's history since independence.

Africa's attainment of the Millennium Development Goals depends on Nigeria's success, as one in every four Africans is a Nigerian. In 2004, the government of Nigeria launched its strategy for growth and poverty reduction — the National Economic Empowerment and Development Strategy (NEEDS) and the state-level State Economic Empowerment and Development Strategies (SEEDS), which complement NEEDS. The NEEDS is based on three pillars:

- Empowering people and improving social service delivery;
- Growing the private sector and focusing on non-oil growth;
- Changing the way government works and improving governance.

However, there is a big discrepancy between stated health strategy, developed with modern knowledge and good understanding of New Public Health, and existing progress with the health status of the population. Table 13.18 provides some Human Development Index comparisons with other countries in sub-Saharan Africa.

In Nigeria, adult literacy rates for men increased from 47 percent in 1960 to 67 percent in 1995 and 76 percent in 2005; rates for women rose from 23 to 47 percent and 61 percent, respectively. According to the United Nations, Nigeria has experienced very rapid population growth with one of the highest fertility rates in the world with annual growth rates of 3.2 percent. By their projections, Nigeria will be one of the countries in the world that will account for most of the world's total population increase by 2050.

The birth rate has declined among the educated urban population, but remains high in the Muslim northern half of the country and the southern primarily Christian rural areas, with an overall total fertility rate of 6.0 children

TABLE 13.18 Human Development Indices, Selected African Countries, 2005

Country	Human development index rank	Life expectancy at birth (years)	Under 5 mortality rate/1000, 2004	Maternal mortality ratio per 100,000 births
Ghana	136	56.7	112	540
Cameroon	144	45.8	149	730
Togo	147	54.2	140	570
Kenya	152	47.0	120	1000
Nigeria	159	43.3	197	800
Benin	163	53.8	152	850
Ivory Coast	164	46.0	194	600
Chad	171	43.6	200	1100
Sierra Leone	176	40.6	283	2000
Niger	177	44.3	259	1600

Source: Human Development Report 2006 (UNDP).

per woman in 1997. In 2007, estimated fertility rates were 5.5 children born per woman and 40.2 births per 1000 population. Only 37 percent of infants are delivered by trained health personnel; 57 percent are born at home, with considerable variation by region. Infant mortality declined from 122 per 1000 in 1960 to 112 in 1997, to estimated rates of 96 (male 102; female 88) per 1000 live births (2007 estimated). Child mortality in Nigeria in 2006 was 12th highest in the world at 191 per 1000 live births, down from 207 in 1960.

The maternal mortality ratios were reported between 800–1100 in the period 2000–2005. Maternal mortality rates (Nigerian Demographic and Health Survey, 2003) showed geographic disparity from 166 in the southwest to 1549 in the northwest per 100,000 live births. In 2005, WHO estimates that 536,000 women died of maternal causes world-wide; of these, 86 percent occurred in sub-Saharan Africa and south Asia and less than 1 percent in more developed countries. Some 10 percent (59,000 women, estimated in 2005) of the worldwide losses by maternal mortality occurred in Nigeria, despite the fact that the country is only 2 percent of the world's population. Large regional differences in maternal deaths demonstrate that most of these deaths are preventable. WHO estimates that for each woman who dies from childbirth in Nigeria, another 30 suffer long-term damage to urogenital organs, often with vesico-vaginal fistula (with continuous leakage of urine through the vagina), tubal damage (resulting in infertility and ectopic pregnancy), and chronic pelvic pain. The UNFPA estimates that some 2 million women are affected by fistulas in the developing world, out of which 800,000 (40 percent) are in Nigeria,

especially in the northern part of the country where early marriage and frequent pregnancies are promoted.

The commonest causes of maternal mortality and morbidity in Nigeria are bleeding immediately after delivery (postpartum hemorrhage, 23 percent), prolonged obstructed labor, eclampsia (hypertensive disease of pregnancy, 11 percent), postpartum infection (17 percent), and unsafe abortion (11 percent), along with anemia (11 percent), malaria (11 percent), and other causes (5 percent). As the result of the restrictive abortion law in the country, women often use dangerous methods to produce abortion, with high rates of complications, often resulting in death; every day, 160 pregnant Nigerian women die from the complications of pregnancy. It has been estimated that nearly 610,000 women resort to induced abortion each year, and of this number 10,000 die. Low levels of use of contraception in the 15–25-year age group result in 60 percent of pregnancies being unwanted, with 80 percent of women with such pregnancies resorting to unsafe and illegal abortion.

About one-fifth of children who are born in Nigeria die before reaching 5 years, almost twice as high as in Ghana. Life expectancy at birth increased from 43 years in 1970 to 52 in 1997 and dropped to 43.3 in 2005 (47 years in 2007). The quality of health and vital statistics is low.

Female Genital Mutilation

Female genital mutilation (female circumcision) is among traditional practices that are deeply entrenched in Nigeria. This practice has over the years received global attention and condemnation because of its many serious physical, mental, social, economic, and political implications. The

Nigerian government observes the International Day for Zero Tolerance to Female Genital Mutilation (FGM). The fight against the harmful practice is to be marked on February 6 each year. Nigeria is one of the 27 countries in Africa where FGM is still practiced. It is estimated that between 40 and 60 percent of Nigeria's women are victims to the practice of FGM, with some regions being over 90 percent. This practice has no health benefits and harms girls and women in many ways. It involves removing and damaging healthy and normal female genital tissue, and hence interferes with the natural function of girls' and women's bodies. It poses a great burden on the women of the country and on the health system, including its adverse effects in transmission of HIV and sexually transmitted infections.

Although the Nigerian federal government has publicly condemned FGM as a harmful practice, it has not taken any legal action against it. FGM is done by largely untrained women with crude implements, with no anesthesia or antibiotics; there is usually bleeding which sometimes leads to death or anemia. Besides the direct consequences of bleeding, there is the ever-present risk of infection, especially tetanus or HIV/AIDS. Many NGOs have been established by Nigerian women to fight this issue.

A WHO multi-country study in which more than 28,000 women participated confirmed that women who had undergone genital mutilation had significantly increased risks for adverse events during childbirth, with high rates of cesarian section and postpartum hemorrhage. Genital mutilation of mothers has negative effects on their newborn babies. Consequences of genital mutilation are even more severe for the majority of Nigerian women who deliver outside a hospital setting.

Communicable Diseases

The World Bank Country Status Report on Nigeria (2005) notes that communicable diseases, often in association with malnutrition, are the major causes of mortality among children, predominantly malaria, measles, meningitis, pneumonia, yellow fever, dysentery, tuberculosis, and AIDS. Immunization coverage in Nigeria is among the lowest in Africa. Malaria causes the largest number of child deaths in Nigeria, estimated at 300,000 annually. Other major causes of childhood deaths are food- and waterborne diseases; bacterial and protozoal diarrhea, hepatitis A, and typhoid fever; respiratory disease; meningococcal meningitis; and aerosolized dust or soil contact disease. Some parts of Nigeria are highly endemic for Lhassa fever.

Highly pathogenic H5N1 avian influenza was identified among birds in this country or the surrounding region in January 2006. The potential devastation from emergence of a pandemic strain in Africa has led to a sudden shift in disease control to a public health focus with international aid funding available for pandemic preparedness, but this has led to concern over the possible distortion of priorities and damage to critical basic public health programs.

Immunization rates increased during the early 1980s but declined in the latter part of the decade. Immunization of pregnant women against tetanus was 23 percent in 1995–1997; infant immunization for bacille Calmette-Guérin (BCG) was 29 percent, 21 percent for DPT, 25 percent for polio, and 38 percent for measles. Coverage for DPT (3 doses) was 54 percent, polio 61 percent, and measles 62 percent in 2006 (UNICEF, 2008). Measles accounts for 12 percent of child deaths and AIDS is a major public health issue in Nigeria as in other sub-Saharan countries. Nigeria in 2007 was the only polio-endemic country in the African region — and one of only four in the world — but there was an 80 percent reduction in wild poliovirus in 2007. Polio control and eradication measures are ongoing with advocacy and support of community and religious leaders. This support, especially in the vaccine-averse north, along with efforts to control measles and other childhood killer diseases, creates awareness of acute flaccid paralysis (AFP) and disease surveillance, and intersectoral cooperation of governmental, private sector, and community financial and logistic support for immunization activities. Immunization Plus Days (IPDs) have been helpful but have led to deterioration of the routine immunization program in Nigeria.

WHO reports that Nigeria has the fourth highest number of TB cases in the world, and the first in Africa among the high-prevalence TB countries with more than 700,000 people infected. In 2004, there were an estimated 293 new TB cases per 100,000 total population (incidence) and an estimated 546 per 100,000 population (prevalence) (WHO, 2006).

Chronic Diseases

Although communicable diseases are major causes of mortality and morbidity in the country, chronic diseases represent a fairly large share of the burden of disease among Nigerians. Chronic disease mortality represents 25 percent total mortality. Half of the deaths are due to cardiovascular diseases, a quarter due to cancers and about a tenth due to respiratory diseases. Sickle cell anemia is the most common genetic disorder affecting Nigerians. Hypertension affects an estimated 11.2 percent (4.3 million) of Nigerians over 15 years of age. Smoking reported at that time was about 4.14 million Nigerians over the age of 15 years. Diabetes in this national survey was estimated in 1.05 million Nigerians with at least 1.5 years of disease.

Nigeria's Health System

The federal, state, and local governments support works in a three-tier system of health care. The essential features of the system are its comprehensive nature, multisectoral inputs, community involvement, and collaboration with nongovernmental providers of health care. The system is based on the 1979 constitution of the country which put health care on the concurrent legislative list of responsibilities of all three levels of government. International health quarantine and control of drugs as well as poisons are exclusively the responsibility of the federal government. A national health policy based on the philosophy of social justice and equity was developed in 1984 and adopted in October 1988. The policy was revised in 2004 as the "Nigerian Health System on Primary Health Care."

The health system inherited from the British colonial period included limited hospital care in the urban centers, and some medical training facilities. Following independence in 1960, the state-operated health system began to develop a widened network of primary care services, in parallel with state primary education. Health care expenditures in 1992 were $1.50 per capita, and health constituted 5 percent of the national budget.

The present health system is seriously underfunded and covers less than two-thirds of the population, with large parts of the rural population outside of the system. UNICEF estimated access to health services at 85 percent for the urban population, 62 percent for the rural population, and 66 percent for the total population. Curative services in hospitals and primary care clinics receive the major share of the fiscal resources. Proposed changes in allocation will allocate health resources in Nigeria as follows: 15 percent to federal government-operated specialty hospitals; 25 percent to state government-operated district hospitals; and 60 percent to local government-operated primary health care clinics, including maternal and child health, school health, and other aspects of primary health care.

Spending on health is low, with total expenditures reported as 3.5 percent of GDP with 1.5 percent from government sources (UNDP, 2008). According to Central Bank of Nigeria, the federal government health spending increased from the equivalent of $14 million (U.S.) in 1998 to the equivalent to $288 million (U.S.) in 2003. Most of the federal health spending goes to teaching and specialized hospitals and federal medical centers. Tertiary health care institutions receive more than two-thirds of the total budget allocated to health, of which about two-thirds is spent on personnel and administrative overhead. Out-of-pocket expenditure accounts for 70 percent of Nigeria's total health expenditure and represents more than 9 percent of total household expenditure.

Health care insurance is through Social Security and state National Assistance, together with special group coverage for members of the armed forces and organized urban groups such as those working in the transport sector. The public health services suffer from low salaries, lack of supplies, and inefficient administration. Private practice is common in the urban centers, serving mainly the middle class. Drugs are expensive and imported in an unrestricted fashion.

Despite medical training considered to be at a high standard, overall quality of care and efficiency in health management are low. The federal government is undertaking initiatives to broaden health insurance in order to raise revenues for health care and to increase equity of access to services. A National Health Care Fund is being established that will receive funds from federal government general revenues, rural cooperative health insurance premiums, and employed persons' health insurance. This is expected to expand coverage for basic health care to a large part of the rural and urban poor population.

Although the data are incomplete, available information from the Federal Ministry of Health Record for 2005 reported the following federal government-operated hospitals: 19 teaching and specialist hospitals, 8 psychiatric hospitals; 3 orthopedic hospitals, and 24 federal medical centers. In addition there are 59 tertiary health facilities operated by the states. The tertiary and specialized hospitals suggest there is a relatively good average availability of high-level services. In 2000, there were 3275 secondary care level facilities in the public sector, a population-to-facility ratio of around 135,000 persons per facility supplemented by 3000 facilities in the private sector. Primary care is based on over 21,585 public sector primary health care (PHC) facilities and almost 7000 private PHC facilities in 2003 with more in process for development.

There is steady increase in the numbers of health professionals produced in Nigeria to meet the health care needs. There were 31,000 physicians in 2001, with the number of newly trained nurses and midwives reported to be about 95,000 registered nurses and 70,000 registered midwives. Medical human resources include over 35,000 physicians or 2.8 per 10,000 population, but most physicians are located in the urban areas (Table 13.19).

Medical education has been given high priority with 18 fully and 5 partially accredited medical schools in the country with many more awaiting accreditation. They graduated about 2000 doctors, 5000 nurses, and 800 pharmacists in 2002/2003. Some of the universities/teaching hospitals have fully developed departments of

TABLE 13.19 Health Professionals in Nigeria, 2003

Personnel	Number	Density per 10,000 population
Physicians	34,923	2.8
Nurses	127,580	1.03
Midwives	82,726	0.67

Source: WHO 2008, estimated as of 2003 http://www.who.int/whosis/database/core/core_select_process.cfm (accessed March 6, 2008)

public health where public health physicians are trained in conjunction with the National Postgraduate Medical College of Nigeria and West African College of Physicians.

A master's degree in public health programs is also offered, but there is no school of public health. The priority given to curative services largely fails to address the basic health problems of the country which require the application of well-known and cost-effective public health programs. Increasing death rates from noninfectious diseases and trauma require attention in planning preventive and curative services for the future.

Nigeria is one of the several major health-staff–exporting countries in Africa, with nurses and physicians emigrating, both legally and illegally, mainly to Britain, which is a threat to sustainable health care delivery in Africa's most populous country. About 20,000 health professionals are estimated to emigrate from Africa annually. Data on Nigerian doctors legally migrating overseas are scarce and unreliable, but estimates are that hundreds of Nigerian-trained doctors continue to migrate annually. Internal migration from state and rural posts is a major threat to the achievement of the Millennium Development Goals (MDGs). Doctors are attracted to university teaching hospitals rather than employment by states because the salaries are far higher in the federal establishment than in state employment. Almost half of Nigerian doctors choose to work in Lagos in federal health institutions. A unified salary scale for doctors at both the state and federal government employment should be implemented to motivate doctors to stay in the state of origin or local governments to render services because that is where they are most needed. The Nigeria Medical Association for many years had been advocating for a unified salary scale called the Medical Salary Scale (MSS) to counter the maldistribution of medical doctors in the country.

Cancer

Data from the National Cancer Registry show that 100,000 new cases of cancer are currently diagnosed each year. The most common cancers in Nigeria are of the cervix, liver, breast, and lymph glands.

HIV/AIDS

Nigeria has third largest number of people infected with HIV/AIDS in the world. Violations of women's rights escalate the rate of HIV infections throughout the continent. Sexual oppression combined with a high biological receptiveness of viral transmission due to genital mutilation put women at risk. As a consequence, the violence against women threatens to destroy whole communities.

Violence

While the level of violence against Nigerian women in the home remains poorly mapped, pilot studies conclude it is "shockingly high." Up to two-thirds of women in certain communities in Nigeria's Lagos State are believed to have experienced physical, sexual, or psychological violence in the family; in other areas, around 50 percent of women say they are victims of domestic violence.

In the absence of official studies, research into the prevalence of violence in the family has been conducted by individuals and organizations. In a recent small-scale study of gender inequality in Lagos and Oyo states, 40 percent of the women interviewed said they had been victims of violence in the family, in some cases for several years.

Health Reform

The Health Reform Foundation of Nigeria (HERFON) is a non-profit, nongovernmental organization developed by a group of reform-minded Nigerians who have come together in response to the deplorable health system and health status of Nigeria. The foundation aims to support and help to sustain the health reform agenda in the country. The origin of the foundation was the Change Agent Programme (CAP) which was funded by the U.K. Department for International Development (DFID). CAP/HERFON has since trained about 200 change agents (CAs) drawn from various stakeholders across the country based on the premise that sustainable reform in Nigeria will be Nigeria-owned and led, but that change may be catalyzed through support from those leading the change. These CAs have assisted in accomplishing the various reform agendas of the government. The Health Minister of Nigeria, Professor Eyitayo Lambo was the Director of CAP from 2003–2007.

HERFON remains committed to supporting the Federal Ministry of Health (FMOH) to achieve its stated health sector reform objectives of improving the stewardship role of government; strengthening the National Health System and its management; reducing the burden of disease; improving health resources and their management; improving access to quality health services; improving consumers' awareness and community involvement; and promoting effective partnership, collaboration, and coordination. Toward this end, HERFON is supporting the Federal Ministry of Health and the states to empower the states to deliver/implement projects towards achieving health-related MDGs.

Summary

The evolution from a very limited colonial health service to a centrally managed service with serious underfunding, and then to a more universal system, reflects post-independence trends in many countries. Ethnic violence over the oil-producing Niger Delta region, interreligious relations, corruption, and inadequate infrastructure are basic issues in the country. Facing a population explosion and contracting economies, African countries went through a very difficult transition in the 1980s.

The primary care system needs strengthening to meet the challenge of preventable diseases, which have been exacerbated by a decline in immunization coverage in recent years. Decentralization of organization to increase the role of the state and local government authorities may improve community participation and efficiency of services. It may also increase revenues by providing a mechanism for local financial input.

There has been a lack of prioritization of maternal and child health in terms of resource allocation and systematic programming. The low rate of political attention given to maternal and child health in the country in part reflects the continuing adverse affects of some harmful traditional, religious, and cultural practices. The Minister of Health, Professor Adenike Grange, in November 2007 reported to the National Health Council on a seven-point agenda that places a high premium on the development of human capital, and recognizes that health and education are the twin engines that drive national development by developing human capital.

Since 1999, the Nigerian health sector has proposed reform targeted toward improving health service delivery and quality of care, but "these programmes have fallen short of making a significant impact towards improving health service delivery, due to a relatively poor emphasis on implementation, monitoring and evaluation." Professor Grange promised that the present administration would move health sector reform forward, building on the policies and frameworks that have been developed, focusing on implementation, integration, monitoring, and evaluation.

LATIN AMERICA AND THE CARIBBEAN

The Latin American region includes 20 countries, with a population of more than 487 million persons, experiencing rapid economic growth since 2000, but burdened with widespread poverty and inequality in incomes, health, and well-being. GDP per capita ranges from a low of $528 (USD) in Haiti to Chile with $8903 (USD) in 2006. Gross national income per capita in 2006 averaged $4847 (USD), as compared to $32,217 for the industrialized countries.

Life expectancy at birth increased from 60 years in 1970 to 68 in 1990 and 73 years in 2006. There are over 11 million births annually in the region (2006), but the crude birth rate has fallen from 37 to 27 to 20 per 1000 population from 1970 to 1990 and 2006, respectively. Crude mortality declined in the same years from 10 to 7 and to 6 per 1000 population. From 1970 to 1990 and

2006, the child (<5 years) mortality rate fell from 123 to 55 to 27 per 1000 live births, with average annual reductions of 4 percent and 4.4 percent in the periods 1970–1990 and 1990–2006. Infant mortality fell from 106 to 43 to 22 per 1000 live births from 1970 to 1990 and 2006, respectively. Maternal mortality in the period 2000–2006 was still high, 130 per 100,000 live births, with 86 percent of women delivered in a health facility (UNICEF, 2008).

These all indicate impressive economic and health care progress for the region. Continent-wide eradication of wild poliovirus and control of measles and other vaccine-preventable diseases have been achieved. Violence and trauma, cardiovascular diseases, tuberculosis, malaria, dengue fever, Chagas' disease, and cholera are still major public health problems.

Despite the impressive but uneven progress in health, inequalities of income and health status between and within countries are also dramatic, with widespread poverty in rural and urban slums. In order to achieve the Millennium Development Goals of reducing poverty and hunger, sustained economic growth and higher quality work will be needed. The health sector can help very much, however, by furthering the successes to date in control of infectious diseases, with attention to malaria, tuberculosis, and other endemic diseases with sanitation, extended immunization, and improved maternal and child health care. Colombia is presented as an example of the progress and challenges facing the health sector in Latin America.

Colombia

Colombia is located in the northwestern region of South America and has a land area of 1.141 million square kilometers divided into 32 departments (states), and further subdivided into 1076 municipalities. It is a mid-level developing nation with 45.6 million inhabitants, with a per capita GDP of $7304 (PPP USD) in 2005. The population is 77 percent urban (2006), with 92.8 percent literate and 52.7 percent living below the national poverty line (17.8 percent living on less than $2 per day).

Life expectancy at birth increased from 57 years in 1960 to 71 in 1997 and 72.3 years in 2005. From 1970 to 1990 and 2000, the crude mortality rate fell from 9 to 6 to 5 per 1000. The total fertility rate in 2000–2005 was 2.5 per woman. The infant mortality rate decreased from 82 per 1000 live births in 1960 to 25 in 1997 and 17 in 2005. Child mortality rates declined from 130 per 1000 live births in 1960 to 30 in 1997 and 21 in 2005. Maternal mortality remains high at 120 per 100,000 in 2005. Primary school enrollment is universal, and adult literacy rates are high (92.8 percent in 2005).

Total expenditures on health were 7.8 percent of GDP in 2004. According to the Human Development Index, Colombia's HDI in 2005 was 0.79, placing the country in the 75th position out of 177 countries surveyed (HDI 2007/08). Furthermore, Colombia has a GINI index (a measure of inequality of wealth distribution) of 58.6 (on a scale from 0 to 100, 0 being total equality and 100 total inequality).

National estimates indicate that from 1970 to 2004, the crude mortality rate fell from 9 to 4.2 per 1000; the infant mortality rate decreased from 123.2 per 1000 in 1950 to 19 per 1000 live births in 2005; maternal mortality remains high at 78.7 per 100,000 in 2004; and life expectancy at birth increased from 50.6 in 1950 to 72.6 years in 2005. The internal civil conflict has caused since 1985 the displacement of nearly 4 million Colombians, which has a devastating impact on the health profile of the Colombian population.

In Colombia in 2002, the leading causes of death included cardiovascular diseases and diabetes 27 percent, violence and trauma 19 percent, and chronic and lower respiratory infections 8 percent. Cardiovascular diseases are increasing as the associated risk factors of smoking, fatty diet, inactivity, hypertension, and diabetes are more prevalent than in the past throughout the area.

During the 1990s, Colombia's health system experienced a major reform that replaced the previous national health system and the Bismarckian social security system with a new social security system that covers standards governing the general system of pensions, professional risks, and complementary social services. The reform of the 1970s National Health System attempted to respond to the global initiative promoted by the World Bank in 1987 that aimed to consolidate health systems in different nations. As a result, Colombia implemented Law 10 in 1990, by which territorial entities became financially and administratively autonomous to operate the public hospitals circumscribed to its areas and to execute free public health activities within the frame of their local plans.

This process of decentralization was favored by the new Colombian constitution of 1991, which conferred more power to the territorial entities and defined social security as a mandatory public service that should be coordinated and controlled by the state. This mandate was enacted in 1993 under Law 60, which governs matters relating to the authority and resources of the various territorial entities (today Law 715/2001), and Law 100, which created a new scheme for the General Social Security System for Health (GSSSH). Based on the concept of universal access through a demand-oriented model, the reform seeks to implement equity of access, free choice of HMOs (*Entidades Promotoras de Salud*), institutional autonomy, decentralized administration, and national regulatory mechanisms by assigning each person a per capita unit adjusted by risk.

To assure universality and financial solidarity, the reform intends to cover all individuals under both contributory and subsidized systems based on a partnership scheme of income redistribution. The law stipulates that the employed contribute 12 percent of their salary (two-thirds of which is paid by their employer) while self-employed pay 12 percent of their declared income. On the other hand, the subsidized system is financed with the resources of the municipalities, one-twelfth of the resources collected through the compulsory system, fiscal allocations to the departments, national income assigned to the departments, resources from gambling taxes, voluntary contributions from the municipalities and departments, royalties from new oil wells, contributions from the compensation funds, value-added tax destined for social programs, tax on firearms and ammunition, and co-payments and prorated fees from the members and their families.

Funding resources are collected in the National Solidarity and Guaranty Fund. The system is directed, standardized, regulated, and controlled by the National Council of Social Security for Health, a body of the Ministry of Social Protection (previously called the Ministry of Health), composed of a professional group of the main participants in the system and the sectional health services in each state. The legal framework is supervised and evaluated by Committee VII of the Senate and the House of Representatives. The HMOs administer the provision of the services and the Health Provider Institutions (*Instituciones Prestadoras de Salud*) provide the services. The Superintendancy of Health controls and monitors the system.

The contributory system offers a comprehensive compulsory health plan (*Plan Obligatorio de Salud* — POS) under the Social Security system that includes initiatives to benefit the individual, the family, and the community in general. The compulsory health plan of the subsidized program (*Plan Obligatorio de Salud del Regimen Subsidiado* — POS-S) is territorially based, composed mainly by actions in the area of health promotion and disease prevention, and provides only 70 percent of the services offered through the contributory system. To select the subsidized population, the municipal authorities apply annually a survey that combines criteria of the Poverty Line Index and the Index of Unsatisfied Basic Needs. The HMOs mobilize the financial resources and organize health promotion activities, complementary health plans, provide the POS and the POS-S for individuals affiliated, and other medical services for the disabled or those who have an occupational disease or a work-related accident. These medical services are provided either by the HMOs through their own Health Provider Institutions or through other Health Provider Institutions (public hospitals, independent health service centers, individual or groups of health professionals) that are contracted by the HMOs.

Compared to the United States, Colombia has a lower rate of physicians per population (13.5/10,000 vs. 25.6/10,000) and a lower rate of nurses/per population

(6/10,000 vs. 9.4/10,000). In order to respond to the human resources needed by the health sector, the government implemented Law 30 and Law 115 of 1994, which authorized educational institutions to create new programs. Consequently, Colombia is experiencing an uncontrolled and hazardous growth of study programs and private vocational schools at the technical and auxiliary levels. The National Council on Human Resources Development regulates the basic formation of the health technicians, such as health promoters (*promatores de salud*), family, community health workers, and nursing assistants. Training of health technicians in rural areas recruited from the population served constitutes a great asset, because it guarantees intensive outreach and culturally sensitive health educational functions.

Use and quality control of pharmaceutical products is supervised since 1995 by the National Institute for the Surveillance of Drugs and Food (INVIMA), which follows the Good Manufacturing Practices (GMP) guidelines of the World Health Organization. At the same time, the Bureau of Pharmaceutical and Laboratory Services of the Ministry of Social Protection develops strategies to promote the development of services for pharmaceutical care and the rational use of drugs and also designs policies related to this area.

After more than a decade of implementation and the recognition of the World Health Organization as one of the most responsive models in Latin America, diverse evaluations have criticized the performance of the Colombian health system. In terms of access, different reports indicate poor improvements in coverage rates, with clear disadvantages for the unemployed and the poorest segment of the population. Other reports evidence deterioration in public health services due to the lack of commitment of the HMOs to fulfill their obligations in regard to public health. In addition, local governments and local health authorities have been unable to ensure adequate levels of public health services. For instance, immunization coverage has decreased dramatically and morbidity and mortality from malaria, tuberculosis, and other communicable diseases have increased. At the same time, the reports indicate an increase in the total expenditure for health care from 7 percent of GDP in 1990 to 10.5 percent in 1999 and declining to 7.8 percent in 2004. There is a large participation of private expenditures by families without an increase in coverage rates or improvements in the quantity and quality of services, indicating that large amounts of money are being deviated from social objectives by the HMOs. Moreover, due to the imbalanced competition between private and public providers and the enormous debt that the government has with the public hospitals, five of the largest national public hospitals have closed and ten more are in liquidation processes. In practice, the reform promoted privatization and minimal state involvement in care delivery.

In an attempt to overcome most of these problems, in 2006 the government enacted Law 52, which constitutes the first reform of Law 100. This new law is intended to increase the level of coverage, from the current 47 percent to 85 percent and to equalize the mandatory health plan for both contributory and subsidized systems. It is also meant to diminish access barriers such as co-payments and pro-rated fees for the subsidized system, and the number of weeks' waiting period required to treat chronic conditions for the contributory system. The government also pays the debts of the subsidized system and fortifies the provision of public health services through the implementation of the National Plan of Public Health. Additionally, this law proposes the creation of the Health Regulatory Commission and the Colombian Territorial Fund, to define new regulations within the system, control the use of resources, and monitor the quality of the services provided by the HMOs. The Colombian health system is in a process of continuing change meant to improve quality of health care, universal coverage, and equity.

Health reform is meant to introduce universal coverage and consumer choice of HMO-like organizations based on the concept of universal access to a market-oriented set of service alternatives. They will provide care paid on a capitation basis and be subject to accreditation and quality control with GPs as gatekeepers. Direct service development of primary health care continues as a responsibility of the Ministry of Health with some assistance by NGOs. *Promatoras* (Community Health Workers) are an important part of that strategy. This will provide an important experience in health care reorganization in a mid-level developing country, as well as a major step forward for Colombia's social security, but will not resolve the problem of providing care for the underserved rural population.

Colombia faces continuing struggles with armed conflict with rebel groups, drug trafficking, poverty, unemployment, and poor sanitation and nutrition in many sectors of the country. Rural populations are at serious social and health disadvantage and promoting health to achieve the Millennium Development Goals for the country will be a serious challenge in the coming years. The health system is an important factor in this process but is concentrated in the cities and requires a strengthening of health promotion activities with prioritization of improved sanitation, maternal and child health, and communicable disease control, as well as facing the growing burden of chronic diseases.

ASIA

UNICEF divides Asia into two groups: 1) South Asia and 2) East Asia and the Pacific (Table 13.20). The former includes India and has a total population of 1.5 billion persons, while the latter includes China and has 2.0 billion persons. Japan is excluded, being linked to the industrialized countries.

TABLE 13.20 Countries of South and East Asia, Demographic and Health Indicators

Indicators	South Asia	East Asia and the Pacific
Countries	India, Pakistan, Bangladesh, Nepal, Afghanistan, Sri Lanka, Bhutan	China, Indonesia, Malaysia, Korea (North and South), Philippines, Thailand, Vietnam, Cambodia, Hong Kong, Singapore and Others
Population (000s) 2006	1.54 billion	1.97 billion
Annual births (000s) 2006	37.9 million	29.7 million
Total fertility rate		
1970	6.8	5.8
1990	5.0	4.3
2005	3.1	3.1
Under 5 mortality rate		
1970	206	122
1990	128	58
2005	84	33
Maternal mortality rate		
2005	500	150
Life expectancy at birth		
1970	48	58
1997	61	68
2005	64	91

Source: UNICEF. State of the World's Children, 1999 and 2007.

The countries of south Asia have progressed less rapidly than those of east Asia in terms of economic, demographic, and health status indicators. Each is a diverse group of nations, but many have common problems, including infectious diseases (e.g., AIDS, tuberculosis, and malaria), poor nutrition for the majority, problems related to rapid urbanization, and the growing problem of noninfectious diseases.

China

China, with a population of more than 1.323 billion persons (2007), is in the process of rapid change and economic growth. Life expectancy increased from 61 years in 1970 to 70 years in 1997 and to 71 and 74 years for males and females, respectively, in 2005. Healthy life expectancy at birth was 63 and 65 years for males and females, respectively, in 2002. The GDP per capita increased from $300 in 1988 to $6757 in 2005 (PPP USD). Total expenditure on health per capita was $277 or 4.7 percent of GDP in 2004. The HDI for China is 0.777, ranking 81st out of 177 countries (HDI Index China, 2007).

Life expectancy at birth in 2005 was 72.5 years. Infant mortality fell from 140 per 1000 in 1960 to 85 in 1970 and 23 in 2005. The mortality rate for children younger than 5 years fell from 185 in 1970 to 27 in 2005. In 2006, China ranked 101st among 177 countries in terms of the child mortality rate with 24/1000 live births.

Fertility declined in China with the rate of births per woman falling from 4.9 (1970–1975) to 1.7 (2000–2005) while the ratio of newborn boys to girls increased from 109:100 in the 1980s to 117:100 in 2000. The crude birth rate fell from 33 per 1000 population in 1970 to 13 per 1000 in 2006. Maternal mortality in 1980–1997 declined from 60 per 100,000, to 45 per 100,000 in 2005, with 83 percent delivering in health care facilities (UNICEF). Immunization coverage in 2006 was 95–96 percent for tuberculosis and measles, DPT, polio, and measles. Fertility rates have decreased rapidly, from 2.4 births per woman in 1980 to 1.7 in the early 2000s, while the ratio of newborn boys to girls increased from 109:100 in the 1980s to 117:100 in 2000.

China has traditionally placed strong social value on health and education, with major achievements in development of a health care infrastructure during the twentieth century. Primary school education is universal. Adult literacy increased from 79–90 percent for males and from 53–73 percent for females between 1980 and 1995 and to 91 percent between 2000 and 2004. As a result of falling birth rates and mortality patterns, the population pyramid is becoming similar to that of developed countries, with a rapidly aging population. In 2004, public expenditures on health were 1.8 percent of GDP, while private expenditures were 2.9 percent of GDP. The demographic transition is contributing to China's health challenges with increasing longevity and declining mortality rates.

Ancient China had a rich tradition of medical care and vital statistics. The Confucian and Taoist streams of Chinese culture supported a "high-order" medical system, emphasizing both preventive and curative services. Classical medical texts documented an empirical base of pharmacopoeias and therapeutic traditions. The yin–yang principle of resonant harmonies between alternative structures was in contrast to the single causation emphasis of western culture. Ancient Chinese medicine based itself on treatment with herbal medicines, and at the same time included a holistic, psychosomatic perspective. Preventive medicine included attention to diet, rudimentary sanitation, personal hygiene, destruction of rabid animals, inoculation against smallpox, and an orientation to the well-being of the individual as essential to health. However, this high-order medicine was available only to the elite of a rigid feudal–bureaucratic society. The vast bulk of the rural population had to rely on folk medicine based on herbal and other traditional practices.

Western medicine was introduced to China with the advent of missionary activities in the nineteenth century. This was accepted as another eclectic element of medicine, and medical schools were opened in the early twentieth century to train medical personnel in western medicine. In the period 1911–1949, medicine and public health advanced with establishment of the national Ministry of Public Health (1927), 30 medical colleges, municipal public health departments, rural district hospitals, military medical services, a factory inspection service, and an array of public health professional departments including maternal and child health, and a large number of provincial medical centers. This brought vaccination, ophthalmic and other forms of surgery, western hospitals, clinics, and medical schools to the provinces and rural areas. The Japanese invasion and civil war that ravaged China from 1936–1948 halted this progress.

Since the establishment of the People's Republic of China under Mao Tse Tung in 1949, China placed improvement of living and health conditions among the rural population as a high national priority. Between 1949 and 1965, China's national government, acting with advisors from the Soviet Union, emphasized rapid expansion of training of mid-level health personnel—nurses, midwives, dispensers, and feldshers (Chapter 14) as well as doctors, whose numbers increased from 13,000 in 1945 to 150,000 in 1966. Hospital bed supply was also expanded rapidly so that by 1965 every county had at least one modern hospital.

In 1966, as part of the Cultural Revolution, policy review placed emphasis on traditional medicine and self-sufficiency in health care at the community level. Western medical training was reduced in scope and duration. Auxiliary or "barefoot" doctors were trained briefly in a mixture of western and traditional Chinese medicine. The barefoot doctors brought health care to the rural population living in 27,000 communes, as well as to urban neighborhoods, focusing on sanitation, family planning information, immunization, and treatment of common illnesses.

Since the 1960s, emphasis on family planning resulted in a slowing of population growth. The policy of one child per family is enforced with many sanctions. This has led to widespread social problems including female infanticide and abandonment, and a high male to female population ratio in younger age groups.

The rural population of China constitutes some 80 percent of the total population. Rural health care was based on cooperative medical services (CMS) funded by the rural communes using barefoot doctor and referral services. This provided effective preventive and curative services to the vast bulk of the rural population of China during the 1960s and 1970s. However, the economic reform in agriculture as part of the transition to a market economy, in effect, abolished the rural communes. As a result, the CMS system had no organizational or financial basis, and was replaced with fee-for-service practice by the former barefoot doctors, who became private medical practitioners during the 1980s. By 1986, only 9.5 percent of the rural population was still covered by the CMS system, in comparison to 90 percent in 1978. This has resulted in greater use of emergency service and hospitalization, with less diligence in performance of preventive health services. In some areas, the CMS model is being restored as cooperative measures under local initiatives. The national Ministry of Health and provincial/regional or municipal departments of public health are responsible for health services in their jurisdictions, with a high degree of local autonomy.

From 1979 with the adoption of market-oriented reforms and new economic policies, a new focus on modernization replaced the ideological zeal and violence of the Cultural Revolution, and has been associated with a period of rapid economic growth. Barefoot doctors were retrained and examined for licensing as village doctors. China's earlier high health standards have played a key role in the country's economic success but the benefits of economic growth have not been shared equally, with a wide gap in socioeconomic indicators between different regions and communities, between urban and rural, and migrant and resident communities within cities. Surveys show that 30–50 percent of poor people in China indicate health is the single biggest factor in their poverty, with reduced earning capacity and medical care costs.

Hospital bed ratios in China increased from 4.6 beds per 1000 population in 1985 to 6.1 in 1989 in urban areas and from 1.5 to 1.4 beds per 1000 in the rural areas during the same period. Similarly, in 1989, health professionals increased in urban areas to 12.6 per 1000 urban residents compared to 2.3 per 1000 rural residents. In 2006 nation-wide, China had 22 hospital beds, 15.5 physicians and 11.1 nurses per 10,000 population. Polyclinics, and SESs were established throughout the country; patients are charged fees for services to support the health system.

WHO reports total heath expenditures in China in 2005 at 4.7 percent of GDP; national or provincial governments covered 39 percent of total health expenditures, with private expenditures covering 61 percent (WHO, 2007). A national health survey in 2003 indicated a decline in access to health care especially in rural areas, with a declining level of coverage in private or public health insurance systems.

In the 1990s, a national campaign to eradicate poliomyelitis was conducted, showing good results with reduction of cases from 5065 in 1990 to 1191 in 1992, by supplemental oral polio vaccine national immunization days (NIDs) for children up to age 4 years. China has since joined the polio-free nations of the world. Crude mortality rates fell from 25 per 1000 in 1949 to 7.6 in 1970 and 6.8 per 1000 in 2006. Maternal mortality fell from 1500 per 100,000 births in 1949 to 95 per 100,000 in 1990 and 48 per 100,000 in 2006. Life expectancy increased from 44 years in 1960 to 73 years in 2006. Other health indicators are seen in Tables 13.21 and 13.22.

Serious health problems in China include high rates of cardiovascular diseases, and lung cancer in polluted industrial cities, with very high rates of smoking. The leading causes of death are similar to those in developed countries, but regional disparities are apparent, with rural populations having higher death rates in all categories. Urban health care has always been at an advantage in China.

At the end of 2005, there were 300,000 medical and sanitary facilities, including 60,000 hospitals and ambulatories, 2964 facilities for maternal and child care, and 1470 specialized clinics with a total 3.07 million beds. The health workforce included 4456 million people, including 1.938 million doctors and doctor's assistants and 1.34 million nurses (15.5 doctors and 11.1 nurses per 10,000 population in 2006). The same year, there were 3592 SESs with 161,000 medical staff, and 1925 sanitary control function facilities with 38,000 medical workers. Regional health care includes 40,000 ambulatories/health centers with 653,000 beds and 848,000 medical staff.

TABLE 13.21 Health Indicators, China, 1960–2006

Indicator	1960	1990	2006
IMR	140	36	20
Child mortality rate	209	45	24
Life expectancy	44	70	73

Source: UNICEF. State of the World's Children, 1999 and 2008.

TABLE 13.22 Vital Statistics, People's Republic of China, 1970–2006

Indicator/Year	1970	1980	1990	1992	2006
Crude birth rate/1000	33.4	18.2	21.1	18.2	12.1
Crude mortality rate/1000	7.6	6.3	6.7	6.6	6.8
Natural increase/1000	25.8	11.9	14.4	11.6	5.3

Source: Ministry of Public Health, Beijing, 1994; and WHO 2007.

During the last decades, one of the country's priorities was achieved with respect to human resources for health — an increase in the quantity of health personnel with 2–6 years of professional training. Consequently the availability of health services has expanded rapidly, particularly in cities and better-off rural areas. Privatization of health services has, however, created a difficult situation with half the population unable to afford health services; only 25 percent of the urban and 10 percent of the rural population have any form of health insurance.

Health workers are not evenly distributed, and the poor rural areas suffer from shortages. There are also concerns about the quality of public health professional and clinical standards in education, training, and practice. The rising costs of health services create a paradox of increase in numbers of health personnel and decline in use of health services. For profit health facilities increased during 2000–2003, despite a decline in number of patients; it is estimated that only 25 percent of the urban population and 10 percent of the rural population use any form of health care. Preventive and health promotion services are more cost-effectively delivered by nurses and other health disciplines. A national strategy for human resources planning will be needed to redefine the roles of health care practitioners, and to meet the needs of rural areas.

Emerging Infectious Diseases

At the end of 2005, there were 650,000 HIV-infected persons reported, including 75,000 clinical cases, with 25,000 previous deaths from AIDS. China had an estimated 120 million people infected with hepatitis B in 2004. The big public health challenges of SARS in 2003, and HIV along with syphilis, have increased dramatically since 2000 and are predicted to become a major epidemic fueled by millions of migrant workers with poor levels of sex education working in China's booming mega-cities. They are far from restrictive rural home environments and tend to access commercial sex workers, in part due to a shortage of young women in the population.

Emerging infectious diseases (such as SARS and avian influenza) are increasingly important because of their potential to become epidemics and pandemics; and in addition to the illness and death they bring, they can cause social instability. The SARS outbreak in 2003 affected 5327 persons in mainland China (348 deaths) and spread to other countries via air transport. Since 2003, there have been 25 reported human cases of H5N1 in China, with 16 deaths.

China's achievements in control of vaccine-preventable and other infectious diseases have been matched by success in birth control and in arranging access to medical care for a population of over 1.3 billion people. However, 69 percent of the urban population and 28 percent of the rural population live with good sanitary conditions; but 7 percent of urban and 37 percent rural populations did not have access to clean water in 2004. China has achieved better outcomes in terms of infant and child mortality and life expectancy with low health expenditures (3.5–4.7 percent of GDP) than many other developing countries. The transition to a market economy left many, especially in the rural areas, with no medical care. About 90 percent of children in the rural areas have serious health problems and low vaccination rates. Collapse of the state medicine led to a decline in the health of children in the rural areas of China at the beginning of the twenty-first century. The "one child" policy adopted since the 1960s with illegal ultrasound prenatal testing has promoted abortion of female fetuses and female infanticide with the result of a large-scale deficit of marriage-age women, and an important societal problem particularly in the rural population.

Aging of the population and the one child per family policy create a situation where the tradition of family care of the elderly will be by a couple who will have sole responsibility for four parents. This will be compounded by the rapid movement of young people to the cities for economic opportunity, so that care of the elderly will be a major problem in the coming decades. Demographic projections suggest there will be close to 350 million people older than 65 years (24.5 percent) in China in 2050. With economic growth, dietary, and lifestyle changes, vascular-related diseases are increasing rapidly. Chronic diseases cause about 80 percent of deaths and are projected to result in $550 billion (USD) of lost productivity between 2005 and 2015 due to associated deaths and disabilities.

As the country rapidly expands its economic potential, national health insurance is in an advanced stage of preparation. The Chinese experience in health status improvement for its huge population in a chaotic period is an enormous achievement considering the economic level of development in China. The country has successfully reduced fertility rates in an attempt to limit population growth and reduce infant, child, and general mortality rates, but it faces challenges not only in transforming the health system to a market economy but also from the effects of the profound demographic shift.

Obesity and smoking are becoming leading health problems in China. Currently, 23 percent of the population is overweight and 150 million people are suffering from hypertension. Diabetes prevalence is projected to double by 2030 to more than 42 million cases. Some 350 million Chinese continue to smoke, increasing their risk of developing related chronic diseases.

China has made good progress toward achieving the Millennium Development Goals since 2000, reducing childhood stunting and malaria control, but progress in reducing maternal and childhood mortality rates has shown less than projected progress, with marked variation among the provinces, with the rural population and poorer provinces at significant disadvantage. Progress in control of TB has been successful where directly observed therapy (DOT) was implemented, but in general China lags behind in this MDG. This is in part due to fee-for-service payments required by the current health system.

The Chinese government announced a new wide-ranging health initiative for the period 2006–2010 of increasing government investment in health and improving the public health and clinical service delivery system, and establishing a medical safety net for the poor. Measures taken included improving capacity in disease prevention and control; improved control of HIV/AIDS, schistosomiasis, and hepatitis B; actively preventing occupational and endemic diseases; strengthening maternal and child health care; and promoting the development of community health services. Other measures include deepening health system reform and allocating health resources rationally; better regulating pharmaceutical production/products and the market; and fostering a modern traditional Chinese medicine industry (WHO, 2005).

For the tens of millions in the countryside, health provision is patchy, with poor access and rampant corruption causing social discontent. China's Health Ministry has announced a plan to reform the health system and provide a national service for all citizens, including the rural population. The *Healthy China 2020* program would provide a universal national health service and promote equal access to public services with some comparisons with the National Health Service in Britain.

China's 11th Five-Year Plan (2006–2010) forms the current basis for the government's economic and social development efforts. The plan aims to sustain the rapid and steady development of China's "socialist market economy" while in addition achieving the "five balances" (between rural and urban development, interior and coastal development, economic and social development, people and nature, and domestic and international development) and making economic and social development more people-oriented, comprehensive, balanced, and sustainable. It includes two key quantitative targets: to achieve annual gross domestic product (GDP) growth of 7.5 percent with the goal of doubling the 2000 GDP per capita; to reduce energy consumption per unit of GDP by 20 percent, and reduce the total discharge of major pollutants by 10 percent by 2010, with promotion of a more balanced development among the different regions of the country and a better quality of life for all.

Despite over two decades of great economic growth and a large and rapidly growing wealthy urban population, China continues to have a large poor rural population. The severe earthquake of 2008 with its effects of tens of thousands of deaths and millions displaced from shattered homes, towns, and villages, revealed the weak infrastructure of the country. China is on the road to becoming an economic and political superpower. Its health system will need a great effort to keep up its role in this challenge.

Japan

Japan is a centralized industrialized democratic country with a 2004 population of 127.9 million and a per capita GDP of $31,267 (PPP USD). Longevity is among the world's highest, with a combined male and female life expectancy increase from 72 years in 1970 to 80 years in 1997 and 82.3 years in 2005. In 2005, the birth rate was 1.3 births per woman, and the infant mortality rate was 2.8 per 1000 live births, about half of the OECD average of 5.4 and one of the lowest in the world. Maternal mortality (adjusted) was 8 per 100,000 in 1990–2004. Immunization coverage in 2006 was 97–99 percent for DPT, polio, and measles vaccines.

Japan ranks high (8th) in the Human Development Index, well above its GDP 17th ranking in 2005 ($31,327). Japan had the highest life expectancy among OECD countries at 82.3 years (2005). Improved longevity has been largely due to declining death rates from heart diseases (the lowest of all OECD countries for both males and females). The birth rate and infant mortality rates in Japan have both fallen dramatically in recent decades. The IMR is just above half of the OECD average of 5.4 per 1000 live births.

Following World War II, the Japanese placed emphasis on maternal and child health, providing free maternal and child care services. Pregnant women receive maternity bonuses to encourage early prenatal care; child care services include an extensive immunization program, screening for diseases of the newborn, developmental testing, and special care for low birth weight or disabled newborns.

Japan has very low rates of heart disease, diabetes, and malignant disease mortality, but relatively high rates of stroke and trauma (motor vehicle accidents and suicides). Coronary heart disease death rates in Japan are low, 25 per 100,000 for men as compared to 118–164 in Canada, the United States, Sweden, and the United Kingdom. However, cerebrovascular death rates are higher than in these and other countries. Stomach cancer rates are higher, but lung and breast cancer mortality are lower. The Japanese diet is low in animal fat and cholesterol, which may relate to the low cardiovascular disease mortality rates, but high in smoked and salty foods, perhaps explaining the higher cerebrovascular disease and stomach cancer mortality rates.

National Health Insurance

The basic health insurance program was enacted in Japan in 1922 as an extension of the employment-related social insurance law of 1874. In 1935, health insurance was extended to all manual workers, further expanded in 1938 to self-employed persons. By the mid-1960s, virtually the entire population was covered by a health insurance plan, either through employers, local government, or trade associations. Government-managed health insurance covers employees of small businesses of fewer than 300 employees, which encompasses some 29 percent of the population. Large companies, or groups of companies, with more than 700 employees, as an alternative to the government health insurance plan, can set up independent insurance plans for their employees. This currently covers some 25 percent of the population. Mutual aid associations provide coverage for civil servants, educators, and others (approximately 10 percent of the population).

Two laws promulgated in 1972 and in 1992 provide coverage for the elderly and low-income earners (32 percent of the population). Insurance for these groups is administered by local authorities or trade associations. There are also many health laws governing a wide range of issues including nutrition, tuberculosis prevention, communicable disease control, mental health, environmental sanitation, and health planning.

Financing and Services

In Japan, 81 percent of total health expenditures are from the public sector. Japan's health service is financed by a payroll tax with rates fixed by law at 3.6–4.55 percent for employees and 4.1–4.7 percent for employers. Government subsidies for health insurance cover 65 percent of health costs, with control of costs by national obligatory fee schedules for a basket of covered services. Co-payments by patients include 10 percent for employees and 30 percent for their dependents for hospital care and outpatient care.

Health plan benefits include medications, long-term care, dental care, and some preventive services, as well as medical and hospital services. Preventive care is provided free of charge through a nationwide network of health centers with costs shared by the central and local governments. Total expenditures for health increased from 4.8 percent of GDP in 1972 to 6.8 percent in 1982, and 8 percent of GDP in 2005, well below the OECD average of 9.0 percent. Expenditures per capita in Japan were $1713 in 2005, compared to $6401 for the United States, $3326 for Canada, and $2532 for the United Kingdom.

The acute care hospital bed ratio of 8.2 beds per 1000 population (2005) is more than twice the OECD average of 3.9.

Japan has a very high hospital bed-to-population ratio, with 8.2 acute care hospital beds per 1000 population, more than double the OECD average of 3.9, with few beds designated for long-term or nursing care. Hospital utilization rates are therefore high with average lengths of stay much longer than western countries. Hospitals, generally small with an average size of 166 beds, include both acute and chronic nursing care patients. Japan has had the highest number of CT and MRI scanners per capita with 92.6 CT scanners and 40.1 MRI units per million population as compared to 20.6 and 9.8 per million population, respectively, in OECD countries. The availability of high-tech medical devices such as CT and MRI is very high compared to OECD countries (see Chapter 15). On the other hand, Japan has fewer physicians with a ratio of 2.0 per 1000 in 2005 as compared to the OECD average of 3.0, but the supply of nurses is 9.0 per 1000 as compared to the OECD average of 8.6 per 1000 population.

Patients have free choice of doctors, two-thirds of whom work as private practitioners in both public and private hospitals. About one-third of physicians are solo general practitioners, paid on a fee-for-service basis, which favors primary care. National fee schedules promote primary care by financial incentives. Physicians also dispense medicines in their private clinics, so that the Japanese consume more medications than most industrialized populations. Physician contact rates are at least double those in western countries; 12.9 contacts per capita per year, compared to 2.8 in Sweden and 5–7 in Canada, the United States, and the United Kingdom.

Japan has had very low birth and fertility rates since the 1950s. This, coupled with low mortality rates and increasing longevity, contributes to an aging of the population, posing problems for the health services in the years ahead. These include a need for geriatric facilities, nursing homes, home care, and support services for family care of the elderly. Proliferation of medical technology is a problem in the health system, and cost containment is now a major issue, with government regulation in health care likely to increase. Obesity rates have increased over recent years, but at 3 percent are well below rates in the United States (32 percent). Smoking prevalence remains very high in Japan (especially among males at 46 percent) with 30 percent of all adults reporting smoking, one of the highest of OECD countries compared to the OECD average of 24 percent (OECD, 2007; HDI, 2007/08).

Summary

The Japanese health system is highly decentralized, but regulated by the national authorities. It has achieved success in lowering mortality rates for most ages and conditions to among the lowest in the world, while restraining health care expenditures. Incentives for primary care seem to have been successful, despite the promotion of excess use of medication. Japan has a high total hospital bed ratio because it has a high percentage of elderly in its population and lacks alternative facilities for long-term care. The problem of caring for the elderly will be a challenge in the years ahead.

COMPARING NATIONAL HEALTH SYSTEMS

The major participants in national health insurance networks include governments, employers, insurers, consumers, providers, and the public. Governments have increasingly come to recognize the economic and social value of improving the health of the population (Box 13.2). They carry this out through public health measures to ensure the basic health of the nation, as well as through legislation regarding the nature of health insurance, whether it is provided through private or public insurance mechanisms. In both the original U.K. Beveridge and the Soviet Semashko models, the government directly finances and provides health care. Services in the United Kingdom are provided by independent contractors, general practitioners, and hospitals operated by free-standing hospital boards (now trusts). The Semashko model was a totally state-financed and -operated service, with decentralized management.

In the Bismarck model, health insurance is financed through social insurance, paid at the place of employment, with Sick Funds paying for services of private medical practice and nongovernment hospitals. The Canadian plan finances health services by provincial governments funded by general tax revenues with federal government financial

---

Box 13.2 Stakeholders in National Health Systems

1. Government — national, state, and local health authorities;
2. Employers — through negotiated health benefits for employees;
3. Insurers — public, not-for-profit, and private for-profit;
4. Patients, clients, or consumers — as individuals or groups;
5. Risk groups — persons with special risk factors for disease (e.g., age, poverty);
6. Providers — hospitals, managed care plans, medical, dental, nursing, laboratories, others;
7. Not-for-profit provider institutions;
8. For-profit institutions, individual providers, and groups;
9. Teaching and research institutions — universities, hospitals, institutes;
10. Professional associations, societies, academies, colleges;
11. Social security systems — with employer and employee contributions;
12. The public, the community, public opinion;
13. Political parties, philosophies, and social agendas;
14. Advocacy groups — age, disease, poverty, or public interest groups;
15. The media — advocacy and watchdog roles;
16. Economies — national, regional, and local;
17. International health organizations and movements;
18. Pharmaceutical and medical technology industries.

support, but care is provided by private practitioners and not-for-profit community-based hospitals. In all variations of health insurance systems, the place of the government as provider and insurer is important to the care received by the consumer and the general state of public health.

There are many variations in methods of assuring national access to health care. Different approaches taken in the development and current structure of health systems in the United States, Canada, United Kingdom, European, and Nordic countries, Japan, Russia, Israel, and the developing countries are given as examples in this chapter. Improved health, as measured by outcome indicators such as reduced morbidity, mortality, or sociophysiologic dysfunction, is the major underlying objective of a national health system. This is sometimes forgotten in debates that may reflect interests of groups such as insurers, providers, institutions, governments, professional groups, or even political philosophies.

A typology of national health systems based on methods of financing and administration of health services provides a framework for their classification and for comparisons (Table 13.23). Mixed models have also developed as the dynamics of health system reform evolves in many countries.

Economic Issues in National Health Systems

As discussed earlier, the costs of health care are a major issue in national health systems. This is in part due to the rising costs of technology in medicine and the increasing age of the population with associated increasing importance of chronic disease, but it is also due to the traditional emphasis on institutional care. Health expenditures for preventive care, health promotion, and environmental health are usually not well financed or analyzed in routine economic data reporting. This makes economic analysis and comparison of interventions difficult, thereby handicapping the search for cost-effective interventions.

National expenditures on health care are usually expressed in terms of U.S. dollars as a percentage of gross national product (GNP) or gross domestic product (GDP). The two economic figures are expressions of the total goods and services in a country, but GDP excludes international transfer of funds. Health care costs are also expressed directly as expenditures per capita (per person, per year), and indirectly as resources such as the number of hospital beds or medical personnel per 1000 (or 10,000) population (see Table 13.24). The percentage of

TABLE 13.23 Typology of Financing and Administration of National Health Systems

Type	Financing Source	Administration
Bismarckian health insurance through social security (e.g., Germany, Japan, France, Austria, Belgium, Switzerland, Israel)	Compulsory employer-employee tax payment to Sick Funds or through Social Security	Germany — governments regulate Sick Funds which pay private services; strong Sick Fund and doctor's syndicates; Israel's Sick Funds compete as HMOs with per capita payments for a mandatory "basket of services"
Beveridge National Health Service (e.g., United Kingdom, Norway, Sweden, Denmark, Italy, Spain, Portugal, Greece)	Government — taxes and revenues; U.K. national financing; Nordic countries combine national, regional, and local taxation	Central planning, decentralized management of hospitals, GP service, and public health; integrated district health systems with capitation financing in U.K.
Semashko national health systems (e.g., former USSR)	Government — taxes and revenues; post-Soviet national health insurance	Strong central government planning and control; financing by fixed norms per population; allocation of facilities and manpower promote increase in hospital beds and medical staff; post-1990 reforms emphasize decentralization with capitation and compulsory health insurance (i.e., payroll taxation)
Douglas national health insurance through government (e.g., Canada, Australia)	Taxation — cost sharing between provincial and federal governments	Provincial government administration; federal government regulation; medical services paid by fee-for-service; hospitals on block budgets; reforms to regionalize and integrate services
Mixed private/public system [e.g., United States; Latin America (e.g., Colombia); Asia (e.g., Philippines); and African countries (e.g., Nigeria)]	Private insurance through employment and public insurance through Social Security for specific population groups	Strong government regulation (U.S.); mixed private medical services, public and private hospitals, state/county preventive services; DRG payment to hospitals, rapid increase in managed care; extension of Medicaid coverage

TABLE 13.24 Population, GDP per Capita, Health Facilities and Health Indicators, Selected Countries and Years, 2004–2006

Countries	Population (millions) 2006	GDP per capita $ 2005#	Percent GDP for health 2005	Acute care beds/1000 2003–2005*	Average length of acute care hospital stay (days)* 2004*	Discharges per 1000 2004*	Infant mortality/ 1000 2006	Life expectancy 2006
United States	296.5	41,980	15.3	3.3	5.6	121	6.0	78
Canada	32.3	33,375	9.8	3.6	7.3	88	5.0	80
Sweden	9.0	32,525	9.1	2.8	6.0	151	3.0	81
Germany	82.5	29,461	10.7	6.3.	8.7	204	4.0	79
Finland	5.3	32,153	7.5	2.5	4.2	199	3.0	79
Denmark	5.4	33,973	9.1	3.1	3.4	169	1.0	78
Israel	6.8	25,864	8.7	2.2	4.2	173	4.0	80
United Kingdom	60.2	33,238	8.3	2.4+	5.0+	235	5.0	79
Russian Federation	143.2	10,845	6.0	8.1	12.2	220	14.0	65

*Data from Health for All Database, WHO Euro Region, June 2007.

+U.K. reports hospital acute care hospital beds and average length of stay up to 1996.

#UNDP Human Development Index GDP per capita in $PPP (purchasing power parity), http://hdrstats.undp.org/countries/country_fact_sheets/cty_fs_GBR.html [accessed February 11, 2008]

Sources: *Health, United States,* 2007. UNICEF. State of the World's Children, 2008; OECD Health Data. 2007. Hospital average length of stay and hospital discharges from: CRS Report for Congress, Sept. 2007, U.S. *Health Care Spending, Comparison With Other OCED Countries.*

GNP spent on health care often is not necessarily directly related to health indicators, such as infant mortality or longevity, as funds may be allocated to or spent on less effective and more costly care. This said, countries with low GNP per capita that spend less than 4 percent on health have poorer health indicators because there are insufficient resources to provide a basic health level for all. Underfinancing and inappropriate allocation of funds have been severe problems in most Soviet health systems and in most developing countries.

The supply of health care services remains one of the difficult and controversial topics in health planning. Economic analysis usually focuses on methods of financing in health care, and on methods of reimbursement or payment for services, placing less emphasis on the supply and quality of services. The World Bank's 1993 *World Development Report,* discussed in previous chapters, places major emphasis on the economic benefits of prevention and cost-effective measures to reduce the burden of disease. Excessive hospital utilization is not cost-effective.

Roemer's law (see Chapter 11) states that hospital utilization under insurance varies with bed supplies. Despite its essential validity, subsequent evidence shows that payment systems for hospital care can be modified so that there are incentives to prevent unnecessary admissions and to shorten hospital stays. As health costs increased rapidly, the concept of providing health care with fewer hospitalizations and more emphasis on ambulatory service became one of the essentials of health policy in many countries since the 1970s.

Health resources indicators are quite variable among the developed market economy countries. Acute care bed ratios represent the number of general short-term beds per 1000 population. A hospital bed is not only a piece of furniture; it represents a service unit and staffing, services, maintenance, food, laundry, and other services. It is therefore an economic unit with fixed and variable costs when in use or even empty. Total hospital beds per 1000 includes all institutional beds utilized for inpatient medical care, but not geriatric custodial care. Acute care hospital beds per 1000 is a more precise and comparable indicator (Table 13.24). Many countries have or are actively reducing hospital bed supplies (United Kingdom, the Nordic countries, most western European countries, the United States, and Israel), developing alternatives to hospital care, using incentive payments to promote ambulatory or day-hospital treatments.

The hospital bed supply (i.e., the acute care bed-to-population ratio) of a country reflects historical patterns, medical practice traditions, concepts, medical technology, and organization. It is also a function of financial incentives or disincentives. Reduced hospital bed supply has become part of standard health reforms in industrialized countries as more efficient care is achieved through better diagnostic facilities and ambulatory care. There is also a wide recognition that hospitals are vital for short-term acute care, but themselves are health risks from infections and there are incidents that relate to errors, disorientation of patients, and the discomfort of being away from the family environment. Because the elderly are greater consumers of health services than the young, another major factor that influences this ratio is the age distribution of the population. Investment in alternatives to hospital care and health promotion to reduce morbidity is essential to help control the rate of increase in costs of health care. This requires investment in education, legal action, screening, nutrition education, group counseling, and selective home support services, and many other elements of the broad concept of health promotion.

Important factors in determining costs of national health systems include salary or income of providers, levels of technology in the service, health planning criteria (norms), and hospital bed supply and utilization. Other factors are availability of home care and comprehensive community care services, use of integrated or regionalized models of health care delivery, methods of paying for hospital services, use of incentive payment systems to promote more efficient use of resources, and emphasis on prevention and health promotion. All of these are issues in the reform of national health systems. Table 13.24 shows a comparison of expenditure, resource utilization, and outcome indicators for selected industrialized countries. Globalization of economics and weakening of public services with trends to privatization in health care are accompanied by technologic advances, aging, and migration, all creating new challenges for a New Public Health.

No analysis of a health system can be complete without addressing the importance of poverty as a major contributing factor to morbidity and mortality. Poverty is associated with high rates of mortality from stroke, coronary heart disease, trauma, asthma, and cancer. Poverty is also related to many specific risk factors for illness: low educational levels, poor housing conditions, poor nutrition, psychological depression, cigarette smoking, alcohol and drug abuse, teenage pregnancies, single parenthood, early bereavement or abandonment, lack of prenatal care, low birth weights, family and neighborhood violence, and others. Universal access to traditional medical care may alleviate some of these effects, but it fails to address the core issues. Social policy and health programs are interdependent, each contributing to improving the quality and length of life. Health planning, including economic indicators, must take this factor into account.

REFORMING NATIONAL HEALTH SYSTEMS

Health care systems are developed in the historical and political context of each country and continue to evolve slowly to meet challenges of demographic, economic, and epidemiologic change, public awareness and expectation, and changing technology in health. Impetus for reform of a health system may derive from a need for cost restraint, universal coverage, or efficiency in use of resources, or an effort to improve satisfaction of consumers or providers (see Table 13.25). The objective of improving the health of the population is also a motive, but this is often expressed as improved access, equity, efficiency, quality of care, and outcomes.

Political and philosophical considerations for health reform often stress issues such as universal access, social solidarity, and equity in resource distribution, human resources, and hospital beds, but it is equally important to focus on targets for improving the health of the general population and special groups at risk. Philosophical and historical issues and arguments for national health insurance have included the need for social protection as a matter of national honor, but a system that fails to improve national health in terms of international outcome indicators does not meet this objective.

Debates and reforms in organizing health systems continue and increase in intensity as the political objective of Health for All meets the reality of rising costs. Efficiency in use of resources and satisfaction of the public and providers are major issues in all health systems. There is no single best means, despite claims by proponents of state-operated systems and equally ideological claims by market-force proponents. Direct importation of a total health system model is not feasible, because there are many factors contributing to the development of a health system relating to the political, social, and professional cultures of each country.

The assumption that market forces produce a better quality of health care is commonly expressed. This point of view has merit if taken in the sense that personal management of finances and choices in health care empower the individual to choose. This may be an advantage for a better educated urban population living near specialized services unavailable to others. Choice for consumers and freedom of choice (autonomy) for providers are different aspects of the market force issue. Taken together, they provide a measure of protection of the rights of the consumer and provider to choose health systems. However, they diminish the responsibility and ability of the system to reach out and provide care and preventive services, or to manage resources effectively, so that important

TABLE 13.25 Goals, Issues, Strategies, and Tactics for National Health Policies

Goals	Issues	Strategies/Tactics
National political commitment to improved health for all	Health as a government responsibility Universal access Adopt international standards Regional and social equity in access Rights to choose within health system Healthy lifestyle as national policy	Health promotion as policy Law/regulations Regulate consumers' rights in health Public information on health Advocacy groups — public, professional
Financing within national means for social benefits	Adequate overall financing (>6% GNP) Shift from supply planning to cost per capita per output Categorical grants to promote national objectives	Increase financing at national, state, and local government levels Health insurance as supplement Define "basket of services" and consumer rights Reduce acute care beds to <3.0/1000 District health authorities with capitation funding
Management for cost-effectiveness	Cost containment Cost-effective health initiatives Decentralized management National policy, monitoring, and standards Information systems/monitoring District health profiles	Incentives for primary care and outreach services Incentives for home care, long-term care facilities Increase home care, non-admission surgery, and long-term care facilities Health information systems Managed care and DRGs
Defining national health targets	Define leading causes of morbidity, mortality, and YPLL, hospitalization Regional, socioeconomic, ethnic analysis Health promotion vs. treatment philosophy Prioritization for use of available resources Use relevant international standards	Social factor analysis in health Improve health KABP Community attitudes to health promotion Promote public health, nutrition, environment, immunization policies
Monitoring health status	Reporting, data systems, information technology	Computerization of medical records, IT and public access to population-based statistics

programs particularly in health promotion and public health (such as care for high-risk groups, immunization, prenatal care, and care of the elderly) may suffer as a result. This set of rights is also sometimes in conflict with the imperative of cost control and the rapid increase in availability of new innovations in diagnosis and management whose benefits may be limited and costly, preventing other proven measures from being implemented. They may also have the undesired effect of promoting excess services such as unnecessary surgery, which has costly and potentially harmful consequences. Market mechanisms that promote individual as well as health system responsibility can make important contributions in health.

Government responsibility is to implement health promotion initiatives that may limit individual rights. This includes adding chlorine and fluoride to water; iodine to salt; and vitamin B, iron, and folic acid to flour. This is part of the substance of public health and requires people who may not directly benefit to accept this social solidarity in the interests of the need of others in the community and the community at large. A local, state, or national health authority may close a business (drug, food, manufactured product) that is hazardous to health, such as an

unhygienic restaurant or a manufacturer of lead-contaminated toys. Management of health care systems must address macroeconomic and microeconomic issues for efficiency. Communities and regions will often address health planning in terms of its impact on business, jobs, and prestige in the community, as opposed to national or regional plans and priorities.

Since the 1970s, there has been a growing stress on health promotion as a way of reducing the burden of chronic disease and the cost of health care for those diseases. This was stimulated and promoted by the Health Field Concept (Marc Lalonde, 1974), the Alma-Ata Conference on primary health care (1978 and 2000), and the WHO's Health for All concept (1978). Specific health targets in the United States (*Healthy People 2000* and *Healthy People 2010*) and in the European region of WHO (1985 and 2005) place emphasis on measurable objectives as the basis for health planning, affecting the planning process (see Chapter 2).

The 1990s was a decade of major reforms in national health systems. Industrialized countries attempted to restrain cost increases while retaining universal access. Sweden has brought down its health expenditures by reducing hospital bed supplies. The United States, building on its social security-based health insurance plan for the elderly and the poor, failed in its 1995 attempt for national health insurance, but is undergoing dramatic changes in the managed care revolution, propelled by the need to control the rate of cost increases. Canada is facing a crisis in sustaining its national health insurance system as federal withdrawal of funding leaves the provinces to finance a generous range of benefits and high levels of hospital bed supply. Israel has moved from voluntary sick funds to National Health Insurance, with the Sick Funds as managed care systems. The Eastern European countries are in a state of transition away from the pre-1990 Soviet model, adopting national health insurance and decentralized administration of services. In developing countries there is concern that directly financing services through the government will hinder development of health services, so that there is a tendency to look toward national health insurance as a way to improve funding of services and bring more people into care. China has moved toward fee-for-service in its rural health care for some 70 percent of its population. All countries are struggling to develop adequate prevention models to reduce the burden of disease that can bankrupt a national health system.

Universal access to health care does not necessarily address social inequities in health. Removal of financial barriers by itself does not guarantee good health. Many social, cultural, and environmental health risk factors are not correctable or preventable by medical or hospital care. They may be of greater importance than the medical care provided (see Chapter 3). The models presented may serve as examples for other countries, and will continue to do so. It is therefore useful to understand how they evolved, their successes and failures, and how they are continuing to develop.

There are two basic directions for reform, which are sometimes in conflict. One is the primary health care approach. This is based on tackling the basic health problems of developing countries by promoting primary health care as a public service through decentralized delivery and administration. The alternative approach, based on the market economy theory, is to promote access to health care by national health insurance, funded through employer–employee contributions or through general taxation.

The fundamental differences in these two approaches present a dilemma for the developing countries and in many ways for the developed countries as well, as they struggle to control health care costs. A health insurance approach may increase funds available for health care, but it invites increases in expenditures for care, inequities in access to care, and an emphasis on curative as opposed to preventive service. This is decidedly a medical approach, promoting hospital and physician services, with public health inadequately addressed and left to the care of private medical practitioners.

The market approach assumes that promoting competition will increase the quality of care and attention to consumer needs, but it is often associated with overutilization of costly services and drives health costs to very high levels. It is a luxury available only to the very wealthiest countries and still not providing all citizens with equal access to services. Developing countries may not have adequate funds to provide health care for all. At the same time, developed economies may not be able to fund health services on demand at levels that consumers and providers might consider ideal. This has led many countries to restrict access to specialist services and place other limitations on services, and is the basis for the managed care approach in the United States.

The public service model often leaves a national program underfunded, leading to problems of quality and morale for the provider as well as the consumer. However, a national health policy is still essential for vulnerable population groups or areas, whether in a developing or developed country. Even the most developed countries have substantial population groups living in poverty, with poor health conditions. The United States has over 42.1 million persons, 16.4 percent of the population (in 2005), without health insurance and probably an equal number with very inadequate coverage. It also has rural areas ill-served by collapsed rural health services. Study of the international experience of health care systems helps to promote international standards and criteria.

The health sector is under great pressure to constrain costs. Employer–employee contribution systems are implementing changes to control costs because health costs are partly responsible for making their industry noncompetitive in the global market. At the same time, there are inflationary pressures of the aging of the population, medical technologic innovation, and high professional and public expectations. Health system reform includes downsizing the hospital sector and building up community health care.

SUMMARY

National health systems throughout the world are in a process of change, seeking restraint in increasing costs, universal coverage, equity in access and quality, as well as efficiency and effectiveness in use of resources to achieve health targets. Many countries are looking for ways to provide universal and equitable care, while controlling costs and improving efficiency. There is no one answer to the search for a health system that works.

Social security and social welfare systems took up the task of assuring access to health services over the twentieth century. National health systems evolved to provide access to medical, hospital, preventive, and community health services. Financing of services through general taxation based on progressive income tax, resource taxes, and excise taxes may be the most equitable way of raising funds. Many countries use social security systems based on employer–employee contributions to pay for health services. Universal access is a means of assuring that the economic barrier is removed for the total population and may lead to increased access to medical and hospital services for those previously excluded. It does not, of itself, guarantee achievement of important health targets. Allocation of resources is an even more fundamental problem.

Beyond financing and resource allocation, there are many "non-tariff" barriers to health. Even in highly developed national health systems, such as that of the United Kingdom, social class, place of residence, education level, and ethnicity play important roles in morbidity and mortality rates. Factors other than medical or hospital care are vital, as classic risk factors for disease, such as diet, smoking, and physical fitness. Partly, however, social class differences in morbidity and mortality are the result of less well-defined aspects of poverty, such as depression, fear, insecurity, and lack of control over one's life. These are issues that are important to the achievement of national health goals and equity.

Health systems must be continuously evaluated. Traditional outcome indicators, such as infant and child, maternal, and disease-specific mortality rates, are important but not sufficient. Incidence of vaccine-preventable diseases, immunization rates for infants, anemia rates in infancy and pregnancy, and disabling conditions are also necessary. Newer measures such as DALYs and QALYs (see Chapters 3 and 11) may help to change the emphasis from mortality to quality of life measures as part of the evaluation. National health systems require data systems that generate information needed for this continuous process of monitoring. High-quality academic centers for epidemiologic, sociologic, and economic analysis are needed to train health leaders and managers and to carry out the studies and research vital for health progress.

Despite the structural diversity and underlying philosophical differences in national health systems, there are important common elements. They are large employers and among the large economic sectors in their respective countries. All face problems of financing, cost constraint, overcoming structural inefficiencies, and at the same time, funding incentives for high quality and efficiency.

A national health system is a complex with many parts that includes but goes well beyond medical care. The quality of the community infrastructure (sewage, water, roads, and communication), the quantity and quality of food, levels of education, and professional organization are all parts of this continuum. Narrow planning for health systems ignores this message at the risk of missing its targets of improved health indicators, such as those adopted by the United Nations as the Millennium Development Goals,

and control of the burden of noncommunicable diseases and injuries. National health systems are not only a matter of adequacy and methods of financing and assuring access to services; they address health promotion, national health targets, and adaptation to changing needs of the population, the environment, and a broad intersectoral approach to health of the population and the individual. The structure, content, and quality of a health system play a vital role in the social and economic development of a society and its quality of life.

Since the end of the Cold War in 1991, a new movement of globalization with economic and political dimensions has taken place with greater stress on human rights with direct application to health. The former socialist countries have gone through painful periods of transition. Many countries have developed free market systems with dynamic growth in national economies. Health systems have struggled to adapt but great gains in longevity and reduced mortality from preventable diseases have been made in many countries. Public and private donor partnerships have emerged to help the poorest countries cope with overwhelming health problems of HIV, tuberculosis, malaria, diarrhoeal and respiratory diseases and the vaccine-preventable diseases. The Millennium Development Goals represent an international consensus on reducing poverty and preventable mortality especially of women and children. The potential for achieving these goals depends on developing infrastructures of health systems which provide access for all and distribution to meet geographic and social inequities in health. Each country needs to develop its own system, but can learn from the experience of others. The purpose of this chapter is to highlight the unique and common features of national health systems.

ELECTRONIC RESOURCES

CANADA

Health Canada. http://www.hc-sc.gc.ca/index_e.html; http://www.cihi.ca/cihiweb/dispPage.jsp?cw_page=media_07dec2005_e; http://www.fin.gc.ca/budget03/booklets/bkheae.htm; http://www.hc-sc.gc.ca/ahc-asc/media/nr-cp/2002/2002_care-soinsbk4_e.html; http://www.hc-sc.gc.ca/ahc-asc/media/nr-cp/2002/2002_care-soinsbk5_e.html [accessed May 24, 2008]

Canadian Institute for Health Information. http://www.cihi.ca/; http://secure.cihi.ca/cihiweb/dispPage.jsp?cw_page=cihi_portal_e [accessed May 24, 2008]

Regionalization. http://secure.cihi.ca/cihiweb/en/downloads/spend_nhex-enhance_e_RegionStudy.pdf; http://www.unicef.org/infobycountry/canada.html [accessed May 24, 2008]

OECD Health Data. 2007. *How Does Canada Compare?* www.oecd.org/health/healthdata; www.oecd.org/canada; UNDP; http://hdrstats.undp.org/countries/country_fact_sheets/cty_fs_CAN.html [accessed May 24, 2008]

UNICEF. http://www.unicef.org/infobycountry/canada.html [accessed May 24, 2008]

CHINA

United Nations Development Program. Human Development Report 2007/2008 — China, http://hdrstats.undp.org/countries/data_ sheets/cty_ds_CHN.html [accessed May 24, 2008]

World Bank. 2005. Rural health in China: China's health sector — why reform is needed, http://siteresources.worldbank.org/INTEAPREG-TOPHEANUT/Resources/502734-1129734318233/BN3whyreform final.pdf [accessed May 24, 2008]

World Bank. 2005. Rural health in China: China's progress in implementing MDGs, http://siteresources.worldbank.org/INTEAPREGTO-PHEANUT/Resources/502734-1129734318233/BN2-MDG-final.pdf [accessed May 24, 2008].

World Health Organization. Regional Office for the Western Pacific. 2005. China, http://www.who.int/countries/chn/en/ [accessed May 24, 2008]

World Health Organization. Regional Office for the Western Pacific Region. China, http://www.wpro.who.int/countries/chn/national_ health_priorities.htm [accessed May 24, 2008]

WHO Regional Office for the Western Pacific. 2005. http://www.wpro .who.int/countries/chn/major_information_sources.htm [accessed May 24, 2008]

EUROPE

Campbell, S., Benita, S., Coates, E., Davies, P., Penn, G. 2007. Analysis for policy: Evidence-based policy in practice. Government Social Research Unit — HM Treasury — London, U.K., http://www.gsr.gov .uk/downloads/resources/pu256_160407.pdf [accessed May 24, 2008]

Denmark. http://www.ess-europe.de/en/denmark.htm [accessed May 24, 2008]

Finland. http://www.stm.fi/Resource.phx/eng/index [accessed May 24, 2008]

Germany, Federal Ministry of Health. http://www.bmg.bund.de [accessed May 24, 2008]

Health Policy Monitor Country Fact Sheet Finland. http://www.hpm.org/en/Country_Facts/Country_Selection/Europe/Finland.html [accessed May 24, 2008]

Human Development Index:

Denmark. http://hdrstats.undp.org/countries/country_fact_sheets/cty_fs_ DNK.html [accessed May 24, 2008]

Germany. http://hdrstats.undp.org/countries/country_fact_sheets/cty_fs_ DEU.html [accessed May 24, 2008]

Netherlands. http://hdrstats.undp.org/countries/country_fact_sheets/cty_ fs_NLD.html [accessed May 24, 2008]

Sweden. http://hdrstats.undp.org/countries/country_fact_sheets/cty_fs_ SWE.html King's Fund. 2006. Our health our care, our say. A new direction for community services: Briefing, www.kingsfund.org.uk/publications/briefings/our_health_our.html [accessed May 24, 2008]

Norway. http://www.shdir.no/ [accessed May 24, 2008]

European Observatory on Health Care Systems in Transition – Finland. 2002. Copenhagen: WHO, European Region, http://www.euro.who .int/document/e74071.pdf [accessed May 24, 2008]

Finland, Ministry of Social Affairs and Health. *Social and Health Report 2006 (for 2002–2005)*, http://www.etk.fi/Default.aspx?Lang=2 [accessed May 24, 2008]

Netherlands Ministry of Health. Welfare and Sport. http://www.minvws .nl/en/ [accessed May 24, 2008]

Organization for Economic Cooperation and Development (OECD) Health Division. 2007, http://www.oecd.org/health/healthdata [accessed May 24, 2008]

OECD Health Data 2007. How does Sweden compare? http://www.oecd .org/sweden; How does Denmark compare? http://www.oecd.org/denmark [accessed May 24, 2008]

OECD Health Data 2007. How does Germany compare? http://www .oecd.org/germany [accessed May 24, 2008]

OECD Health Data 2007, http://www.oecd.org/dataoecd/45/55/38979836.pdf [accessed May 24, 2008]

Swedish Association of Local Authorities and Regions, Quality and Efficiency in Swedish Health Care — Regional Comparisons 2007. The 2001 English report is available as a supplement to the Scandinavian *Journal of Public Health* (Health in Sweden — The National Report 2001, supplement 58, 2001).

Swedish National Board of Health and Welfare. http://www.socialstyrelsen.se/en/ [accessed May 24, 2008]

Swedish National Board of Health and Welfare. 2007. Quality and efficiency in Swedish health care, www.socialstyrelsen.se/publicerat [accessed May 24, 2008]

Swedish National Institute of Public Health. http://www.fhi.se/default 1417.aspx [accessed May 24, 2008]

Sweden: Quality and efficiency in Swedish health care, http://www .socialstyrelsen.se/en/showpub.htm?GUID={9F559A98-E4F8-432B-85B8-8A3B96908488 [accessed May 24, 2008]

Sweden: Social report 2006 — The national report on social conditions in Sweden. http://www.socialstyrelsen.se/en/showpub.htm [accessed May 24, 2008]

Swedish National Board of Health and Welfare. www.socialstyrelsen.se/publicerat [accessed May 24, 2008]

Sweden. http://www.fhi.se/shop/material_pdf/public%20health%20strat (1).pdf [accessed May 24, 2008]

United Kingdom. http://www.bbc.co.uk/history/historic_figures/beveridge_william.shtml; http://www.connectingforhealth.nhs.uk/; http://www.qof.ic.nhs.uk/index.asp; http://www.dh.gov.uk/en/Rss/Rss?Feed=DH_076655; http://www.medicalnewstoday.com/articles/35121.php; http://www.qof.ic.nhs.uk/index; http://www.qof.ic.nhs .uk/index [accessed May 24, 2008]

INTERNATIONAL

Judge, K., Platt, S., Costongs, C., Jurczak, K. 2005. Health inequalities: A challenge for Europe: A background paper on health inequalities commissioned by the UK Presidency to coincide with Presidency Conference on Inequalities in Health, http://www.dh.gov.uk/assetRoot/04/12/15/83/04121583.pdf [accessed May 24, 2008]

Organization of Economic Cooperation and Development (OECD), http://www.oecd.org/health/healthdata and http://www.oecd.org/health/healthataglance [accessed May 24, 2008]

UNICEF. 2007. The State of the World's Children 2008: Women and Children — Child Survival, www.unicef.org/sowc [accessed May 24, 2008]

UNICEF. http://www.unicef.org/infobycountry/canada_statistics.html [accessed May 24, 2008]

Human Development Index. http://hdr.undp.org/en/statistics/ [accessed May 24, 2008]

UNICEF. http://www.unicef.org/infobycountry/; http://www.unicef.org/infobycountry/canada.html; http://www.unicef.org/infobycountry/israel_statistics.html; http://www.unicef.org/infobycountry/usa.html [accessed May 24, 2008]

UNICEF. 2008. State of the world's children, http://www.unicef.org/ [accessed May 24, 2008]

World Bank. Health reform, http://www.worldbank.org/healthreform/index.htm [accessed May 24, 2008]

World Health Organization. European Region Health for All Data Set, http://data.euro.who.int/hfadb/ [accessed May 24, 2008]

World Health Organization. 2000. A quick reference compendium of selected key terms used in the World Health Report 2000, http://www.who.int/health-systems-performance/docs/glossary.htm#health_system [accessed May 24, 2008]. Note: Glossary consists of terminology as it stood with the release of the WHO 2000; changes in definitions as a result of consultation and peer review are not accounted for. From WHOTERM, the WHO Terminology Information System.

ISRAEL

Gertner Institute. http://www.health.gov.il/english/Pages_E/default.asp?maincat=2Israel [accessed May 24, 2008]

Central Bureau of Statistics. 2006, http://www.cbs.gov.il/hodaot2006n/01_06_252e.pdf; http://www.hpm.org/en/Surveys/Brookdale_Institute/09/Responsibility_for_mental_health_care_(2).html [accessed May 24, 2008]

Israel Ministry of Health. http://www.health.gov.il/english/; http://www.unicef.org/infobycountry/israel_statistics.html; http://www.health.gov.il/download/docs/units/comp/hfa/2003/chap7.pdf [accessed May 24, 2008]

UNICEF. http://www.unicef.org/infobycountry/; http://www.unicef.org/infobycountry/israel_statistics.html [accessed May 24, 2008]

UNITED STATES

America's Health Insurance Plans. http://www.ahip.org/ [accessed May 24, 2008]

American Medical Student Association. http://www.amsa.org/div/ [accessed May 24, 2008]

Health Care Financing Administration. http://www.os.dhhs.gov/about/opdivs/hcfa.html [accessed May 24, 2008]

Healthy People. http://www.healthypeople.gov/ [accessed May 24, 2008]

Healthy People 2010. http://www.health.gov/healthypeople [accessed May 24, 2008]

History of health insurance, http://eh.net/encyclopedia/article/thomasson.insurance.health.us [accessed May 24, 2008]

Human Development Index – United States, http://hdrstats.undp.org/countries/country_fact_sheets/cty_fs_USA.html [accessed May 24, 2008]

Holahan, J., Cook, A., Dubay, L. 2007. Characteristics of the uninsured: Who is eligible for public coverage and who needs help affording coverage? Kaiser Family Foundation.

Kaiser Family Foundation. http://www.kff.org/ [accessed May 24, 2008]

Kaiser Family Foundation. 2007. President's FY 2008 Budget and the State Children's Health Insurance Program (SCHIP). Washington, DC: 2007, http://www.kff.org/medicare/1066.cfm; http://www.kff.org/medicare/healthplantracker/topicresults.jsp?i=6&rt=1&ctot=&sr=&ss=&yr=5&s=a&x=47&y=7; http://www.medicarehmo.com/mchmwhat.htm; http://www.medicarehmo.com/mdmnu.htm [accessed May 24, 2008]

Medicare. http://www.medicare.gov/ [accessed May 24, 2008]

National Center for Health Statistics. http://www.cdc.gov/nchs/ [accessed May 24, 2008]

National Center for Health Statistics. 2006. *Health, United States, 2006: With Chartbook on Trends in the Health of Americans.* DHHS Publication number 76-641496. Hyattsville, MD, 2006.

United States Census Bureau. http://www.census.gov/ [accessed May 24, 2008]

United States Census Bureau. 2006. Nation's Population to Reach 300 Million on Oct. 17. United States Census Bureau Press Release. Health Insurance coverage 2007, http://pubdb3.census.gov/macro/032007/health/toc.htm; http://pubdb3.census.gov/macro/032007/health/h01_001.htm [accessed May 24, 2008]

United States Centers for Medicare and Medicaid Services. 2007. Table 121. National Health Expenditures by Type 1980 to 2004. Managed Care Enrollment in Medicaid, http://www.statehealthfacts.org/comparetable.jsp?ind=217&cat=4&sub=56&yr=1&typ=2&sort=a&o=a [accessed May 24, 2008]

United States Department of Health and Human Services. 2006 and 2007. Health, United States, http://www.cdc.gov/nchs/data/hus/hus06.pdf; http://www.cdc.gov/nchs/data/hus/hus07.pdf; http://www.healthy-people.gov/data/midcourse/html/execsummary/progress.htm; http://www.health.gov/healthypeople; http://www.cdc.gov/nchs/products/pubs/pubd/hp2k/review/highlightshp2000.htm; http://www.cdc.gov/nchs/data/hus/hus06.pdf#135; http://www.cdc.gov/nchs/data/hus/hus06.pdf#120; http://www.healthypeople.gov/About/developed.htm; http://www.healthypeople.gov/data/midcourse/pdf/ExecutiveSummary.pdf; http://www.cdc.gov/nchs/data/nvsr/nvsr55/nvsr55_14.pdf; http://www2.cdc.gov/nip/schoolsurv/nationalAvg.asp; http://www.hhs.gov/news/press/2002pres/02minorityhealth.html; http://www.uscis.gov/files/nativedocuments/M-618.pdf; http://www.medicarehmo.com/mcmnu.htm; http://www.census.gov/compendia/statab/health_nutrition/; http://www.interstudypublications.com; [accessed May 24, 2008]

United States Department of Health and Human Services. 2006. CMS Statistics.

Health Maintenance Organizations (HMO): 1990 to 2004 (Rep. No. Table 142) (2005) Nashville, TN: HealthLeaders-InterStudy.

United States Department of Health and Human Services. 2006. Healthy People 2010 Midcourse Review Executive Summary. Washington, DC: DHHS.

United States Department of Health and Human Services Surgeon General, http://www.dhhs.gov/surgeongeneral/National Center for Health Statistics; http://www.cdc.gov/nchs/ [accessed May 24, 2008]

United States Department of Homeland Security. Welcome to the United States: A Guide for New Immigrants. http://www.uscis.gov/files/nativedocuments/M-618.pdf [accessed May 24, 2008]

United States Department of Human Services, http://www.omhrc.gov/templates/browse.aspx?lvl=1&lvlID=7 [accessed May 24, 2008]

United States White House, The White House. 2003. Fact Sheet: Medicare Prescription Drug, Improvement, and Modernization Act of 2003. The White House Press Release, http://www.whitehouse.gov/news/releases/2003/12/20031208-3.html; http://www.whitehouse.gov/stateoftheunion/2007/initiatives/index.html [accessed May 24, 2008]

RECOMMENDED READINGS

American College of Physicians. 2008. Achieving a high-performance health care system with universal access: What the United States can learn from other countries. *Annals of Internal Medicine*, 148:1–21.

Connell, J., Zurn, P., Stilwell, B., Awases, M., Braichet, J. M. 2007. Sub-Saharan Africa: Beyond the health worker migration crisis? *Social Science and Medicine*, 64:1876–1891.

DeWitt, L. Social Security Administration, Historian's Office. 2003. *Historical Background and Development of Social Security*, http://www.ssa.gov [accessed May 24, 2008]

Hofmarcher, M. M., Oxley, H., Rusticelli, E. 2007. *Improved Health System Performance through Better Care Coordination.* Health Working Papers, No. 30, December 2007.

Hussey, P. S., Anderson, G. F., Osborn, R., et al. 2004. How does the quality of care compare in five countries? *Health Affairs,* 23:89–99.

OECD Health Data. 2007. Statistics and Indicators for 30 Countries. OECD 18 July 2007, http://www.oecd/health/stats [accessed May 24, 2008]

Organization for Economic Co-operation and Development (OECD) http://www.olis.oecd.org/olis/2007doc.nsf/ENGDATCORPLOOK/ NT00005926/$FILE/JT03237930.PDF [accessed May 24, 2008]

Peterson, C. L., Burton, R. 2007. U.S. health care spending: Comparison with other OECD countries: Report to Congress. September 17, 2007. Domestic Social Policy Division. Congressional Research Service, http://assets.opencrs.com/rpts/RL34175_20070917.pdf [accessed May 24, 2008]

van der Ven, W., Schut, F. 2008. Universal mandatory health insurance in the Netherlands: a model for the United States? *Health Affairs,* 27:771–781, http://content.healthaffairs.org/cgi/content/full/27/3/771 [accessed June 25, 2008]

BIBLIOGRAPHY

Anderson, G. F., Hussey, P. S., Frogner, B. K., Waters, H. R. 2005. Health spending in the United States and the rest of the industrialized world. *Health Affairs,* 24:903–914.

Banks, J., Marmot, M., Oldfield, Z., Smith, J. P. 2006. Disease and disadvantage in the United States and in England. *Journal of the American Medical Association,* 295:2037–2045.

Blackwell, D. L., Martinez, M. E., Gentleman, J. F. 2008. Women's compliance with public health guidelines for mammograms and Pap tests in Canada and the United States: An analysis of data from the Joint Canada/United States Survey of Health. *Women's Health Issues,* January 5, 2008 [Epub ahead of print]

Blendon, R. J., Brodie, M., Benson, J. M., Altman, D. E., Buhr, T. 2006. Americans' views of health care costs, access, and quality. *Milbank Quarterly,* 84:623–657.

Brekke, K. R., Nuscheler, R., Straume, O. R. 2007. Gatekeeping in health care. *Journal of Health Economics,* 26:149–170.

Breslow, L. 2005. The organization of personal health services. *Milbank Quarterly,* 83:759–777.

Brown, L. D. 2003. Comparing health systems in four countries: Lessons for the United States. *American Journal of Public Health,* 93:52–56.

Cylus, J., Anderson, G. F. 2007. *Multinational Comparisons of Health Systems Data, 2006.* The Commonwealth Fund, May 2007, http:// www.commonwealthfund.org/publications/publications_show.htm? doc_id=482648#areaCitation [accessed May 24, 2008]

Feacham, R. 2000. [editorial]. 2000. Health Systems Special Edition. Health systems: More evidence, more debate. *Bulletin of the World Health Organization,* 78:715.

Hofmarcher, M. M., Oxley, H., Rusticelli, E. 2007. *Improved Health System Performance Through Better Care Coordination.* OECD Health Working Paper No. 30. Paris: Organization for Economic Cooperation and Development.

Lasser, K. E., Himmelstein, D. U., Woolhandler, S. 2006. Access to care, health status, and health disparities in the United States and Canada: Results of a cross-national population-based survey. *American Journal of Public Health,* 96:1300–1307.

Marmot, M. 2001. Inequalities in health. *New England Journal of Medicine,* 345:134–136.

Nolte, E., McKee, C. M. 2008. Measuring the health of nations: Updating an earlier analysis. *Health Affairs,* 27:58–71.

Organization for Economic Cooperation and Development (OECD). 2005. *Health at a Glance.* Paris: OECD.

Organization for Economic Cooperation and Development (OECD). 2006. *OECD Reviews of Health Care Systems: Switzerland.* Paris: OECD.

Roemer, M. I. 1991 and 1993. *National Health Systems of the World, Volumes 1 and 2.* New York: Oxford University Press.

Sanmartin, C., Berthelot, J. M., Ng, E., Murphy, K., Blackwell, D. L., Gentleman, J. F., Martinez, M. E., Simile, C. M. 2006. Comparing health and health care use in Canada and the United States. *Health Affairs,* 25:1133–1142.

Sarti, C., Rastenyte, D., Cepaitis, Z., Tuomilehto, J. 2000. International trends in mortality from stroke, 1968 to 1994. *Stroke,* 31:1588–1601.

Schoen, C., Doty, M. M. 2004. Inequities in access to medical care in five countries: Findings from the 2001. Commonwealth Fund International Health Policy Survey. *Health Policy,* 67:309–322.

Schoen, C., Osborn, R., Huynh, P. T., Doty, M., Peugh, J., Zapert, K. 2006. On the front lines of care: Primary care doctors' office systems, experiences, and views in seven countries. *Health Affairs,* 25:555–571.

Schoen, C., Osborn, R., Doty, M. M., Bishop, M., Peugh, J., Murukutla, N. 2007. Toward higher-performance health systems: Adults' health care experiences in seven countries, 2007. *Health Affairs,* 26:717–734.

Sochalski, J., Aiken, L. H. 1999. Accounting for variation in hospital outcomes: A cross-national study. *Health Affairs,* 18:256–259.

Suhrcke, M., Rocco, L., McKee, M. 2007. Health: *A Vital Investment for Economic Development in Eastern Europe and Central Asia.* European Observatory on Health Systems. Copenhagen: WHO European Region.

Tuohy, C. H. 1999. Dynamics of a changing health sphere: The United States, Britain, Canada. *Health Affairs,* 18:114–134.

UNICEF. 2006. *The State of the World's Children, 2006: Child Survival.* New York: UNICEF.

UNICEF. 2008. *The State of the World's Children, 2008: Excluded and Invisible.* New York: UNICEF.

World Bank. *World Development Report 1993: Investing in Health.* Washington, DC: World Bank, http://files.dcp2.org/pdf/WorldDevelopmentReport1993.pdf [accessed May 24, 2008]

World Health Organization. 2005. *European Health Report 2005. Public Health Action for Healthier Children and Populations.* Copenhagen: WHO European Region.

CANADA

Canadian Institute for Health Information (CIHI). National health expenditure trends, 1975–2005, http://www.cihi.ca/cihiweb/dispPage.jsp? cw_page=media_07dec2005_e [accessed May 24, 2008]

Canadian Public Health Association. 1996. Focus on health: Public health in health services restructuring: An issue paper. Ottawa: CPHA.

Curtis, L. J., McMinn, W. J. 2007. Health-care utilization in Canada: 25 years of evidence. SEDAP Research Paper No. 190 — A Program for Research on Social and Economic Dimensions of an Aging Population, http://socserv2.socsci.mcmaster.ca/~sedap/p/sedap190 .pdf [accessed May 24, 2008]

Deber, R. B. 2003. Health care reform: Lessons from Canada. *American Journal of Public Health,* 93:20–24.

Evans, R. G. 1989. Controlling health expenditures — The Canadian reality. *New England Journal of Medicine,* 320:571–577.

Guyatt, G. H., Devereaux, P. J., Lexchin, J., Stone, S. B., Yalnizyan, A., Himmelstein, D., Woolhandler, S., Zhou, Q., Goldsmith, L. J., Cook, D. J., Haines, T., Lacchetti, C., Lavis, J. N., Sullivan, T., Mills, E., Kraus, S., Bhatnagar, N. 2007. A systematic review of studies comparing health outcomes in Canada and the United States. *Open Medicine*, 2007, 1:e27–36.

LaLonde, M. 1974. *New Perspectives on the Health of Canadians*. Ottawa: Department of National Health and Welfare.

Marchildon, G. P. 2005. *Health Systems in Transition: Canada*. Copenhagen: WHO Regional Office for Europe on behalf of the European Observatory on Health Systems and Policies, http://www.euro.who.int/Document/E87954.pdf [accessed May 24, 2008]

Nason, E. 2007. Health and Medical Research in Canada. RAND Europe's Health Research System Observatory, 2008, http://www.rand.org/pubs/documented_briefings/2008/RAND_DB532.pdf [accessed May 24, 2008]

Pineault, R., Lamarche, P. A., Champagne, F., Contandriopoulos, A. P., Denis, J. L. 1993. The reform of the Quebec health care system: Potential for innovation? *Journal of Public Health Policy*, 14:198–219.

Roos, N. P., Brownell, M., Shapiro, E., Roos, L. R. 1998. Good news about difficult decisions: The Canadian approach to hospital cost control. *Health Affairs*, 17:239–246.

Statistics Canada. 1999. *Towards a Healthy Future: Second Report on the Health of Canadians*. Ottawa.

Stewart, D. E. 2006. A broader context for maternal mortality. *Canadian Medical Association Journal*, 174:302–303.

CHINA

Akin, J. S., Dow, W. H., Lance, P. M., Loh, C. P. 2005. Changes in access to health care in China, 1989–1997. *Health Policy and Planning*, 20:80–89.

BBC NEWS. January 7, 2008. http://news.bbc.co.uk/go/pr/fr/-/1/hi/world/asia-pacific/7175501.stm [accessed May 24, 2008]

Chen, C. M. 2008. Overview of obesity in Mainland China. *Obesity Review*. 2008 Mar, 9(s1Obesity in China), 14–21.

Grogan, C. M. 1995. Urban economic reform and access to health care coverage in the People's Republic of China. *Social Science and Medicine*, 41:1073–1084.

Hong, Y., Li, X., Stanton, B., Lin, D., Fang, X., Rong, M., Wang, J. 2006. Too costly to be ill: Healthcare access and health-seeking behaviours among rural-to-urban migrants in China. *World Health and Population 2006*, 8(2):22–34.

Hougaard, J. L., Osterdahl, L. P., Yu, Y. 2008. The Chinese Health System: Structure, Problems, and Challenges. Discussion Papers, Department of Economics, University of Copenhagen.

Human Development Index — China, http://hdrstats.undp.org/countries/country_fact_sheets/cty_fs_CHN.html [accessed May 24, 2008]

Lawson, J. S., Lin, V. 1994. Health status differentials in the People's Republic of China. *American Journal of Public Health*, 84:737–741.

Zheng, X., Hillier, S. 1995. The reforms of the Chinese health care system: The Jiangxi study. *Social Science and Medicine*, 41:1057–1064.

COLOMBIA

Ayala-Cerna, C., Kroeger, A. 2002. Health sector reform in Colombia and its effects on tuberculosis control and immunization programs. *Cad Saude Publica*, 18:1771–1781.

Consultoría para los Derechos Humanos y el Desplazamiento. 2006. *Más o menos desplazados, Boletín de la Consultoría para los Derechos Humanos y el Desplazamiento*, No. 69, Bogotá: CODHES.

De Vos, P., De Ceukelaire, W., Van der Stuyft, P. 2006. Colombia and Cuba, contrasting models in Latin America's health sector reform. *Tropical Medicine and International Health*, 11:1604–1612.

Franco-Aguledo, S. 2003. Algunas lecciones de la reforma de salud y seguridad social en Colombia. *Gerencia y Políticas de Salud*, 4:58–69.

Hernandez, M. 2002. Health reform, equity and the right to health in Colombia. *Cad Saude Publica*, 18:991–1001.

Human Development Index — Colombia, http://hdrstats.undp.org/countries/country_fact_sheets/cty_fs_COL.html [accessed May 24, 2008]

Kroeger, A., Ordoñez-Gonzalez, J., Aviña, A. I. 2002. Malaria control reinvented: Health sector reform and strategy development in Colombia. *Tropical Medicine and International Health*, 7:450–458.

Ministerio de Proteccion Social/Organización Panamericana de la Salud/Organización Mundial de la Salud. 2006. *Situacion de Salud en Colombia. Indicadores Basicos 2006*. Bogotá: MPS/OPS/OMS.

Organización Panamericana de la Salud/Organización Mundial de la Salud. 2002. *Perfil del Sistema de Servicios de Salud de Colombia. 2002*. Washington, DC: PAHO.

Pan American Health Organization/World Health Organization. 2001. *Situation de la Salud en las Americas. Indicadores Basicos 2000*. Washington, DC: PAHO and 2005, http://www.paho.org/English/DD/AIS/cp_170.htm [accessed May 24, 2008]

Proyecto de Ley 052.http://www.esepolicarpa.gov.co/ArchivosWeb/REFORMASG.pdf [accessed May 24, 2008]

UNDP. 2006. Human Development Report, http://hdr.undp.org/hdr 2006/statistics/andhttp://unstats.un.org/unsd/mdg/Data.aspx?cr=170 [accessed May 24, 2008]

Vélez, A. L. 2000. Marco político y legal de la promoción de la salud. *Colombia Medica*, 31:86–95.

UNICEF, http://www.unicef.org/infobycountry/colombia.html [accessed May 24, 2008]

World Health Organization. 2006. Core health indicators, http://www.who.int/whosis/database/core/core_select_process.cfm?strISO3_select=ALL&strIndicator_select=healthpersonnel&intYear_select=latest&language=english [accessed May 24, 2008]

World Health Organization. 2007. Colombia, http://www.who.int/countries/col/en/ [accessed May 24, 2008]

World Health Organization, Mortality: Country fact sheet Columbia, http://www.who.int/whosis/mort/profiles/mort_amro_col_colombia.pdf [accessed May 24, 2008]

DEVELOPING COUNTRIES

Barnum, H., Kutzin, J. 1993. *Public Hospitals in Developing Countries: Resource Use, Cost, Financing*. Baltimore, MD: The Johns Hopkins University Press.

Pauly, M. V., Zweifel, P., Schleffer, R. M., Preker, A. S., Basset, M. 2006. Private health insurance in developing countries: Voluntary private insurance could fill in the gaps that limited public resources cannot cover. *Health Affairs*, 25:369–379.

Ron, A., Abel-Smith, B., Tamburri, G. 1990. *Health Insurance in Developing Countries: The Social Security Approach*. Geneva: International Labor Office.

Schieber, G., Maeda, A. 1999. Health care financing and delivery in developing countries. *Health Affairs*, 18:135–143.

Tarimo, E., Creese, A. (eds.). 1990. *Achieving Health for All by the Year 2000: Midway Reports of Country Experiences*. Geneva: World Health Organization.

EUROPE

LSE Health, London School of Economics and Political Science London, United Kingdom. 2008. Health system snapshots: perspectives from six countries. *Eurohealth*, 14, No. 1, http://www.lse.ac.uk/collections/LSEhealth/pdf/eurohealth/vol14no1.pdf

Mackenback, J. P., Sirbut, I., Reskam, A. J., Schaap, M. M., Menvielle, G., Leinsalu, M., Kunst, S. E. 2008. European Union Working Group on Socioeconomic inequalities in health in 22 European countries. *New England Journal of Medicine*, 358:2468–2481.

McDaid, D., Drummond, M., Suhrcke, M. 2008. How can European health systems support investment in and the implementation of population health strategies? Policy brief—Health Systems and Policy Analysis. World Health Organization, on behalf of the European Observatory on Health Systems and Policies, http://www.euro.who.int/document/hsm/1_hsc08_epb_2.pdf

WHO Statistical Information System — WHOSIS 2007. http://www.who.int/whosis/en/ [accessed May 24, 2008]

United Nations Human Development Report 2007–08. http://hdr.undp.org/en/ [accessed May 25, 2008]

ISRAEL

Brookdale Institute. 2006. Audit of Health Plan Performance. Myers-JDC-Brookdale Institute, Jerusalem, Israel.

Brookdale Institute. 2007. Redesigning primary care services in Maccabi. Myers-JDC-Brookdale Institute, Jerusalem, Israel.

Brookdale Institute. 2007. Restricting supplemental insurance services. Myers-JDC-Brookdale Institute, Jerusalem, Israel.

European Observatory on Health Systems on Health Systems and Policies — Israel 2003. Copenhagen: WHO, European Region.

European Observatory Health Systems in Transition, Israel. 2004. http://www.euro.who.int/Document/E81826sum.pdf [accessed May 24, 2008]

Ginsberg, G., Tulchinsky, T., Filon, D., Goldfarb, A., Abramov, L., Rachmilevitz, E. A. 1998. Cost-benefit analysis of a national thalassaemia prevention programme in Israel. *Journal of Medical Screening*, 5:120–126.

Gross, R., Brammli-Greenberg, S., Bennun, G. 2007. Responsibility for mental health care Rep. No. 9. Jerusalem: The Myers-JDC-Brookdale Institute.

Health for All, http://www.health.gov.il/download/docs/units/comp/hfa/2003/chap4.pdf [accessed May 24, 2008]

Healthy Israel. 2002, http://www.health.gov.il/english/Pages_E/default.asp?maincat=14 [accessed May 24, 2008]

Health Policy Monitor. http://www.hpm.org/index.jsp [accessed May 24, 2008]

Human Development Index — Israel, http://hdrstats.undp.org/countries/country_fact_sheets/cty_fs_ISR.html [accessed May 24, 2008]

International comparisons, Israel Central Bureau of Statistics. 2006. *Demographic Situation in Israel 2005*. Jerusalem: Central Bureau of Statistics Press Release.

Israel Ministry of Health. *Health in Israel, 2005, Selected Data*. Jerusalem: Ministry of Health.

Israel Ministry of Health. http://www.health.gov.il/english/ [accessed May 24, 2008]

Israel Center for Disease Control. 1999. *Health status in Israel 1999*. Jerusalem: Ministry of Health.

Israel Ministry of Health. Department of Information. http://www.health.gov.il/pages/default.asp?maincat=2&catId=558&PageId=3241 [accessed May 24, 2008]

Kahan, N. R., Waitman, D. A., Blackman, S., Chinitz, D. P. 2006. Suboptimal pneumococcal pneumonia vaccination rates among patients at risk in a managed care organization in Israel. *Journal of Managed Care Pharmacy*, 12:152–157.

Levav, I., Levinson, D., Radomislensky, I., Shemesh, A. A., Kohn, R. 2007. Psychopathology and other health dimensions among the offspring of Holocaust survivors: Results from the Israel National Health Survey. *Israel Journal of Psychiatry and Related Sciences*, 44:144–151.

Penchas, S., Shani, M. 1995. Redesigning a national health-care system: The Israeli experience. *International Journal of Health Care Quality Assurance*, 8:9–18.

Rosen, B., Goldwag, R., Thomson, S., Mossialos, E. 2003. *Health Care Systems in Transitions — HiT Summary Israel*. European Observatory on Health Systems and Policies, http://www.euro.who.int/document/E81826.pdf [accessed May 24, 2008]

Country Fact Sheet, Israel. http://www.hpm.org/en/Country_Facts/Country_Selection/Middle_East/Israel.html;jsessionid=DBA0686151FD7A62200533BE7AB47E59 [accessed May 24, 2008]

Shtarkshall, R., Soskolne, V., Bubis-Feder, P., Daoud, N. 2002. The teaching of social sciences, health behavior, and health behavior change in public health. *Public Health Reviews*, 30:201–208.

Tulchinsky, T. H. 1985. Israel's health system: Structure and content issues. *Journal of Public Health Policy*, 6:244–254.

UNDP. Human Development Index, http://hdrstats.undp.org/countries/data_sheets/cty_ds_ISR.html [accessed May 24, 2008]

UNICEF. State of the World's Children, http://www.unicef.org/infoby country/israel_statistics.html [accessed May 24, 2008]

World Health Organization Health Policy Monitor, http://www.hpm.org/en/Country_Facts/Country_Selection/Middle_East/Israel.html;jsessionid=DBA0686151FD7A62200533BE7AB47E59 [accessed May 24, 2008]

World Health Organization. Regional Office for Europe. 1996. *Highlights on Health in Israel*. Copenhagen: World Health Organization, Regional Office for Europe.

Yosefy, C., Dicker, D., Viskoper, J. R., Tulchinsky, T. H., Ginsberg, G. M., Leibovitz, E., Gavish, D. 2003. The Ashkelon Hypertension Detection and Control Program (AHDC Program): A community approach to reducing cardiovascular mortality. *Preventive Medicine*, 37:571–576.

JAPAN

Health Policy Monitor. http://www.hpm.org/en/Country_Facts/Country_Selection/Asia/Japan.html [accessed May 24, 2008]

Human Development Index — Japan. http://hdrstats.undp.org/countries/country_fact_sheets/cty_fs_JPN.html [accessed May 24, 2008]

Ikegami, N., Campbell, J. C. 1999. Health care reform in Japan: The virtues of muddling through. *Health Affairs*, 18:56–75.

Ingelhart, J. K. 1988. Health policy report: Japan's medical care system, parts 1 and 2. *New England Journal of Medicine*, 319:807–812 and 1166–1172.

Nishimura, S. 2007. Promoting health during the American occupation of Japan: the public health sections, Kyoto Military Government Team, 1945–1949. *American Journal of Public Health*, 98:424–434.

Nishimura, S. 2008. Promoting health during the American occupation of Japan: the public health sections, Kyoto Military Government Team, 1945–1949. *American Journal of Public Health*, 98:424–434

OECD Health Data. 2007. How Does Japan Compare? http://www.oecd.org/dataoecd/45/51/38979974.pdf [accessed May 24, 2008]

LATIN AMERICA AND THE CARIBBEAN

UNDP. 2005. The Millennium Development Goals: A Latin America and Caribbean Perspective, http://www.undp.org/latinamerica/docs/Regionalenglish.pdf [accessed May 24, 2008]

UNICEF. 2008. *The State of the World's Children 2008*, Child Surriral, UNICEF, New York.

NIGERIA

Aid Harmonization and Alignment. Initiatives for Nigeria: Country-Level Harmonization, http://www.aidharmonization.org/ [accessed May 24, 2008]

Anyangwe, S. C., Mtonga, C. 2007. Inequities in the global health workforce: The greatest impediment to health in sub-Saharan Africa. *International Journal of Environmental Research and Public Health*, 4:93–100.

Beaglehole, R., Sanders, D., Dal Poz, M. 2003. The public health workforce in sub-Saharan Africa: Challenges and opportunities. *Ethnicity and Disease*, 13:S24–30.

Connell, J., Zurn, P., Stilwell, B., Awases, M., Braichet, J. M. 2007. Sub-Saharan Africa: Beyond the health worker migration crisis? *Social Science and Medicine*, 64(9):1876–1891.

Federal Republic of Nigeria National Bureau of Statistics, http://www.nigerianstat.gov.ng/descr.php?recordID=9 [accessed May 24, 2008].

Human Development Report. 2007/08. Nigeria, http://hdrstats.undp.org/countries/data_sheets/cty_ds_NGA.html [accessed May 24, 2008]

National Bureau of Statistics. 2005. Federal Republic of Nigeria. *Social statistics in Nigeria*, http://www.nigerianstat.gov.ng/social_statistics/SSD%20final.pdf [accessed May 24, 2008]

Nigerian National Health Conference Proceedings (NHC2006), http://www.herfon.org/docs/NHC2006_Details_of_Proceedings.pdf [accessed May 24, 2008]

Nwokolo, E. The role of health in Nigeria's national development: A model for community-based control of major chronic and communicable diseases. Nigerian National Health Conference Proceedings (NHC2006), http://www.herfon.org/docs/NHC2006_Details_of_Proceedings.pdf [accessed May 24, 2008]

Stilwell, B., Diallo, K., Zurn, P., Vujicic, M., Adams, O., Dal Poz, M. 2004. Migration of health-care workers from developing countries: Strategic approaches to its management. *Bulletin of the World Health Organization*, 82:595–600.

United States Library of Congress. Federal Research Division. 2006. Country Profile Nigeria, http://lcweb2.loc.gov/frd/cs/profiles/Nigeria.pdf [accessed May 24, 2008]

World Health Organization. 2007. Maternal Mortality in 2005, estimates developed by WHO, UNICEF, UNFPA, and the World Bank, http://www.who.int/reproductive-health/publications/maternal_mortality_2005/mme_2005.pdf [accessed May 24, 2008]

World Health Organization, Monitoring and evaluation, http://www.who.int/reproductive-health/global_monitoring/index.html [accessed May 24, 2008]

World Health Organization. 2007. Religious leaders unite against polio, http://www.who.int/countries/nga/features/2007/sokoto/en/print.html [accessed May 24, 2008]

World Health Organization. 2007. Nigerian 51st National Council of Health, http://www.who.int/countries/nga/mediacentre/releases/2007/healthcouncil/en/index.html [accessed May 24, 2008]

NORDIC COUNTRIES, GERMANY, AND THE NETHERLANDS

Brown, L. D., Amelung, V. E. 1999. "Manacled competition": Market reforms in German health care. *Health Affairs*, 18:76–91.

Busse, R. 2005. Risk adjustment compensation in Germany's Statutory Health Insurance. *European Journal of Public Health*, 11:174–177.

Busse, R., Riesberg, A. 2004. *Health Care Systems in Transition: Germany*. Copenhagen, WHO Regional Office for Europe on behalf of the European Observatory on Health Systems and Policies. http://www.euro.who.int/Document/E85472.pdf [accessed May 24, 2008]

Elola, J., Daponte, A., Navarro, V. 1995. Health indicators and the organization of health care systems in Western Europe. *American Journal of Public Health*, 85:1397–1401.

European Observatory on Health Systems and Policies — Sweden, 2005. Copenhagen: WHO, European Region, http://www.euro.who.int/Document/E88669sum.pdf [accessed May 24, 2008]

Greiner, W. 2005. Health economic evaluation of disease management programmes: The German example. *European Journal of Health Economics*, 50:191–196.

Harrison, M., Calltorp, J. 2001. The reorientation of market-oriented reforms in Swedish health care. *Health Policy*, 50:219–240.

Health Insurance in the Netherlands. The new health insurance system from 2006, The Hague 2005.

Hecke, T. L., Erzberger, M. 2005. Continuous case management of a German statutory health insurance. *Health Care Financing Review*, 27:59–68.

Hermanson, T., Aro, S., Bennett, C. L. 1994. Finland's health care system: Universal access to health care in a capitalist democracy. *Journal of the American Medical Association*, 271:1957–1962.

Hofmarcher, M. M., Oxley, H., Rusticelli, E. 2007. *Improved Health System Performance Through Better Care Coordination*. OECD Health Working Paper No. 30. Paris: OECD.

Hurst, J. W. 1991. Reforming health care in seven European nations. *Health Affairs*, 10:7–21.

Iglehart, J. K. 1991. Germany's health care system. *New England Journal of Medicine*, 324:1750–1756.

Järvelin, J., Rico, A., Cetani, T. 2002. *Health Care Systems in Transition, Finland 2002*. Copenhagen: WHO European Region.

Petrov, I. C. 2007. The elderly in a period of transition: Health, personality, and social aspects of adaptation. *Annals of the New York Academy of Science*, 1114:300–309.

Roberts, J. L. 1996. Terminology for the WHO Conference on European Health Care Reforms: *A Glossary of Technical Terms on the Economics and Finance of Health Services*. Copenhagen: World Health Organization Office for Europe.

Saltman, B., Bankauskaite, V., and Vrangbaek, K. (eds.). 2007. *Decentralization in Health Care*. Open University. European Observatory on Health Systems and Policies. Copenhagen: WHO European Region.

Saltman, R. B., Figueras, J. 1997. *European Health Care Reform: Analysis of Current Strategies*. Copenhagen: World Health Organization, Regional Office for Europe.

Stahl, T., Wismar, M., Ollila, E., Lahtinen, E., Leppo, K. 2006. Health in All Policies: Prospects and potentials. Euopean Observatory on Health Systems and Policies/Ministry of Social Affairs and Health: Finland.

van Herten, L. M., Gunning-Schepers, L. J. 2000. Targets as a tool in health policy. Part I: Lessons learned. *Health Policy*, 53:1–11.

Van Herten, L. M., Gunning-Shepers, L. J. 2000. Targets as a tool in health policy. Part II: Guidelines for application. *Health Policy*, 53:13–23.

van de Ven, W. P., Beck, K., Buchner, F., Chernichovsky, D., Gardiol, L., Holly, A., Lamers, L. M., Schokkaert, E., Shmueli, A., Spycher, S., Van de Voorde, C., van Vliet, R. C., Wasem, J., Zmora, I. 2003. Risk adjustment and risk selection on the sickness fund insurance market in five European countries. *Health Policy*, 65:75–98.

World Health Organization. Regional Office for Europe. 1999. *Health 21: The Health for All Policy Framework for the WHO European Region*. Copenhagen: World Health Organization, Regional Office for Europe.

World Health Organization. Regional Office for Europe. 2005. *Health for All: The Policy Framework for the WHO European Region 2005 Update*. Copenhagen: WHO.

RUSSIA

Bobak, M., Murphy, M., Rose, R., Marmot, M. 2007. Societal characteristics and health in the former communist countries of Central and Eastern Europe and the former Soviet Union: A multilevel analysis. *Journal of Epidemiology and Community Health*, 61:990–996.

Centers for Disease Control. 1992. Public health assessment—Russian Federation, 1992. *Morbidity and Mortality Weekly Report*, 41:89–91.

Centers for Disease Control. 1995. Diphtheria epidemics — New independent states of the former Soviet Union, 1990–1994. *Morbidity and Mortality Weekly Report*, 44:177–181.

Danishevski, K., Balabanova, D., McKee, M., Nolte, E., Schwalbe, N., Vasilieva, N. 2005. Inequalities in birth outcomes in Russia: Evidence from Tula oblast. *Paediatric and Perinatal Epidemiology*, 19:352–359.

Danichevski, K., McKee, M., Balabanova, D. 2008. Prescribing in maternity care in Russia: The legacy of Soviet medicine. *Health Policy*, 85:242–251.

Demoscope Weekly. Institute of Demographics of the State University — Higher School of Economics.

European Observatory on Health Care Systems, Health in Transition. 2003. Russian Federation, http://www.euro.who.int/document/Obs/russum.pdf [accessed May 24, 2008]

Human Development Index Russian Federation, http://hdrstats.undp.org/countries/country_fact_sheets/cty_fs_RUS.html [accessed May 24, 2008]

Leon, D. A., Saburova, L., Tomkins, S., Andreev, E., Kiryanov, N., McKee, M., Shkolnikov, V. M. 2007. Hazardous alcohol drinking and premature mortality in Russia: A population based case-control study. *Lancet*, 369:2001–2009.

Lyons, R. F., Rudd, A. G., Alvero, C. 2008. Advances in health policy. *Stroke*, 39:264–267.

Murphy, M., Bobak, M., Nicholson, A., Rose, R., Marmot, M. 2006. The widening gap in mortality by educational level in the Russian Federation, 1980–2001. *American Journal of Public Health*, 96:1293–1299.

Nemtsov, A. 2005. Russia: Alcohol yesterday and today. *Addiction*, 100:146–149.

Pomerleau, J., McKee, M., Rose, R., Haerpfer, C. W., Rotman, D., Tumanov, S. 2008. Hazardous alcohol drinking in the former Soviet Union: A cross-sectional study of eight countries. *Alcohol and Alcoholism*, 2008 Feb 3 [epub ahead of print].

Shchepin, O. O., Tishuk, E. A., Vishnevsky, A. 2003. [Medicodemographic problems in Russian Federation]. The depopulated superpower. *Russia in Global Affairs*, No. 3, July/September 2003. http://eng.globalaffairs.ru/numbers/4/488.html [accessed May 24, 2008]

Tulchinsky, T. H., Varavikova, E. A. 1996. Addressing the epidemiologic transition in the former Soviet Union: Strategies for health system and public health reform in Russia. *American Journal of Public Health*, 86:313–320.

Vestnik Rossiyskoy Akademy Nauk, 2005, (9):3–6 [Article in Russian].

World Health Organization. 2005. *Highlights on Health in the Russian Federation*. Copenhagen: WHO European Region, http://www.euro.who.int/document/E88405.pdf [accessed May 24, 2008]

UNITED KINGDOM

Acheson, D. 2000. Health inequalities: Impact assessment. *Bulletin of the World Health Organization*, 78:75–77.

Black, D. 1980. *Inequalities in Health: Report of a Research Working Group*. London: Department of Health and Social Security.

Campbell, S., Reeves, D., Kontopantelis, E., et al. 2007. Quality of primary care in England with the introduction of pay for performance. *New England Journal of Medicine*, 357:181–190.

Coulter, A. 1995. Evaluating general practice fundholding in the United Kingdom. *European Journal of Epidemiology*, 5:233–239.

Department of Health. 2005. *Resource Allocation: Weighted Capitation Formula*, Fifth Edition. Leeds: Department of Health.

Department of Health. 2006. *The NHS in England: The Operating Framework for 2007/08*. London: Department of Health.

Department of Health. 2006. *Our Health, Our Care, Our Say: Making It Happen*. London: HMSO.

Department of Health. 2006. *Our Health, Our Care, Our Say: New Direction for Community Services*. London: HMSO.

Department of Health. Shaping Health Care for the Next Decade. 2007. *Diverging Values? The Values of the United Kingdom's National Health Services*. London: Nuffield Trust.

Department of Health. 2007. *Primary Care Trust Recurrent Revenue Allocations, 2006–07 & 2007–08. Health Service Circular: Finance Directorate, Resource Allocation Monitoring and Analysis*. Leeds: Department of Health.

Dixon, T., Shaw, M. E., Dieppe, P. A. 2006. Analysis of regional variation in hip and knee joint replacement rates in England using Hospital Episodes Statistics. *Public Health*, 120:83–90.

Doran, T., Fullwood, C., Gravelle, H., Reeves, D., Kontopantelis, E., Hiroeh, U., Roland, M. 2006. Pay-for-performance programs in family practices in the United Kingdom. *New England Journal of Medicine*, 355:375–384.

Dusheiko, M., Gravelle, H., Yu, N., Campbell, S. 2007. The impact of budgets for gatekeeping physicians on patient satisfaction: Evidence from fundholding. *Journal of Health Economics*, 26:742–762.

European Observatory on Health Care Systems. *Health Care Systems in Transition: United Kingdom 1999*. Copenhagen: World Health Organization, European Region.

Klein, R. 2006. The troubled transformation of Britain's National Health Service. *New England Journal of Medicine*, 355:409–415.

Legido-Quigley, H., McKee, M., Nolte, E., Glinos, I. A. 2008. Assuring the Quality of Health Care in the European Union: A Case for Action. World Health Organization. Regional Office for Europe. Copenhagen.

Light, D., Dixon, M. 2004. Making the NHS more like Kaiser Permanente. *British Medical Journal*, 328:763–765.

Lilford, R. J., Brown, C. A., Nicholl, J. 2004. Use of process measures to monitor the quality of clinical practice. *British Medical Journal*, 335:648–650.

McFadden, E., Luben, R., Wareham, N., Bingham, S., Khaw, K. T. 2008. Occupational social class, educational level, smoking and body mass index, and cause-specific mortality in men and women: a prospective study in the European Prospective Investigation of Cancer and Nutrition in Norfolk (EPIC-Norfolf) cohort. *European Journal of Epidemiology*, 23:449–458.

McKee, M. 2004. What can we learn from the British fundholding experience? *European Journal of Public Health*, 5:231–232.

Marmot, M. 2004. Commission on Social Determinants of Health. Achieving health equity: From root causes to fair outcomes. *Lancet*, 370:1153–1163.

Pocock, S. J., Shaper, A. G., Cook, D. G., Phillips, A. N., Walker, M. 2004. Social class differences in ischemic heart disease in British men. *Lancet*, 11:197–201.

Ramsay, S. E., Morris, R. W., Lennon, L. T., Wannamethee, S. O., Whincup, P. H. 2008. Are social inequalities in mortality in Britain narrowing? Time trends from 1978 to 2005 in a population-based study of older men. *Journal of Epidemiology and Community Health*, 62:75–80.

Rose, D., 1995. Official Classifications in the U.K. *Social Research Update. University of Surrey.* Issue 9, 1995 (available via m.brownett@soc.surrey.ac.uk).

Secretary of State for Health. 2004, reprinted 1995. *The Health of the Nation.* London: Her Majesty's Stationery Office.

Smith, G. D., Bartley, M., Blane, D. 2004. The Black Report on socioeconomic inequalities in health 10 years on. *British Medical Journal*, 301:373–377.

Whitfield, M. D., Gillett, M., Holmes, M., Ogden, E. 2004. Predicting the impact of population level risk reduction in cardio-vascular disease and stroke on acute hospital admission rates over a 5 year period — A pilot study. *Public Health*, 120:1140–1148.

UNITED STATES

Catlin, A., Cowan, C., Hartman, M., Heffler, S., and the National Health Expenditure Accounts Team. 2008. National health spending in 2006: A year of change for prescription drugs. *Health Affairs*, 27:14–29.

Davis, K. 2008. Public Programs: Critical Building Blocks in Health Reform. Invited Testimony, Senate Finance Committee, June 16, 2008, The Commonwealth Fund, http://www.commonwealthfund.org and www.commonwealthfund.org/publications/publications_show.htm?doc_id=689819

Department of Health and Human Services. *Healthy People 2000: National Health Promotion and Disease Prevention Objectives.* U.S. Department of Health and Human Services. Boston: Jones and Bartlett.

Ibrahim, S. A. 2007. The Veterans Health Administration: A domestic model for a national health care. *American Journal of Public Health*, 97:2124–2126.

Iglehart, J. K. 2007. The battle over SCHIP. *New England Journal of Medicine*, 357:957–960.

Institute of Medicine. 2007. *Crossing the Quality Chasm: A New Health System for the 21st Century.* Washington, DC: National Academies Press.

Kerr, E., Fleming, B. 2007. Making performance indicators work: Experience of the U.S. Veteran's Health Administration. *British Medical Journal*, 335:971–973.

Kuttner, R. 2007. Market-based failure — A second opinion on U.S. health care costs. *New England Journal of Medicine*, 358:549–551.

Lindenauer, P. K., Rothberg, M. B., Pekow, P. S., Kenwood, C., Benjamin, E. M., Auerbach, A. D. 2007. Outcomes of care by hospitalists, general internists, and family physicians. *New England Journal of Medicine*, 357:2589–2600.

McGlynn, E., Asch, M., Adams, A., et al. 2007. The quality of health care delivered to adults in the United States. *New England Journal of Medicine*, 348:2635–2645.

Mathews, T. J., Macdorman, M. F. 2007. Infant mortality statistics from the 2004 period linked birth/infant death data set. *National Vital Statistics Reports*, 55:14.

National Center for Health Statistics. 2007. *Health, United States: With Chartbook on Trends in the Health of Americans.* Hyattsville, MD: National Center for Health Statistics.

National Center for Health Statistics. 2007. *Health, United States, with Chartbook on Trends in the Health of Americans.* Hyattsville, MD: National Center for Health Statistics.

Oberlander, J. 2007. Learning from failure in health care reform. *New England Journal of Medicine*, 357:1677–1679.

OECD Health Data. 2007. How does the United States compare. www.oecd.org/health/healthdata [accessed May 25, 2008]

Rittenhouse, D. R., Robinson, J. C. 2007. Improving quality in Medicaid: The use of care management processes for chronic illness and preventive care. *Medical Care*, 44:47–54.

Schroeder, S. A. 2007. We can do better — Improving the health of the American people. *New England Journal of Medicine*, 357:1221–1228.

Smith, C., Cowan, C., Heffler, S., Catlin, A., and the National Health Accounts Team. 2007. *Health Affairs*, 25:186–196.

U.S. Census Bureau. International Data Base. 2007. Country summary [online], http://www.census.gov/ipc/www/idb/country/gmportal.html. [accessed August 25, 2007]

Welch, W. P., Miller, M. E., Welch, H. G., Fisher, E. S., Wennberg, J. E. 2007. Geographic variation in expenditures for physicians' services in the United States. *New England Journal of Medicine*, 328:621–627.

Chapter 14

Human Resources for Health Care

Introduction
Overview of Human Resources
Human Resources Planning
  Supply and Demand
Basic Medical Education
Postgraduate Medical Training
Specialization and Family Practice
Training in Preventive Medicine
Nursing Education
In-Service and Continuing Education
Accreditation of Medical Educational or Training Facilities
The Range of Health Disciplines
Licensure and Supervision
Constraints on the Health Care Provider
New Health Professions
  Nurse Practitioners
  Physician Assistants
  Feldshers
  Community Health Workers
Alternative Medicine
Changing the Balance
Education for Public Health and Health Management
Health Policy and Management of Human Resources
Summary
Electronic Resources
Recommended Readings
Bibliography

INTRODUCTION

Development and sustainability of the New Public Health and its ability to respond to old and emerging threats depend on the quantity and quality of human resources of the total and especially the public health workforce. The great achievements of public health of the nineteenth and twentieth centuries have led to doubling of life expectancy in the industrialized countries and the emergence of important health advances in all but the least developed countries. New challenges of globalization of diseases such as HIV/AIDS and SARS and threatened bioterrorism and pandemics of avian influenza present great challenges in the twenty-first century. At the same time we are witnessing an increasing flow of trained health professionals from poor to rich countries.

The New Public Health is concerned with the total health system and related issues. It requires an understanding of issues related to the training, supply, distribution, and management of many kinds of human resources, including the balance between personnel working in institutions and in the community. It was for these reasons that WHO World Report (2006) noted "an estimated shortage of almost 4.3 million doctors, midwives, nurses and support workers worldwide" and recommended a 10-year program to address this fundamental issue particularly for the developing world. The developed nations are facing many shortages in critical areas such as nursing personnel, but also in other skills needed to care for an increasingly elderly population, rising tides of diabetes, obesity, and their long-term sequelae.

Health systems require adequate numbers of well-trained, well-remunerated, and up-to-date providers working with adequate facilities and support systems. Health professions are made up of many disciplines working in a complex network of facilities and programs. Their range of activities includes provision of patient care in the community and in hospital or other institutional settings. They work to promote health, prevent disease, treat illness, and rehabilitate in a compassionate, ethical, professional, and cost-effective manner. This means that health care providers must be educated not only for competence and humaneness in clinical functions, but also they must be continuous learners and knowledgeable in the medical and social sciences of health including the economic aspects of health care. They must be aware of and be able to synthesize knowledge from related fields such as epidemiology, economics, and management as well as the social and behavioral sciences. The quality of the practitioner depends on the recruitment of socially motivated and talented people, on education, training, and professionalization as providers, as well as on the structure, content, and quality orientation of the health system in which they work.

Determining need and allocation of human resources is an important health planning issue. A relative over- or undersupply of one or more heath professions creates a bias or imbalance in the health system and its economics. Mid-level practitioners and community health workers are being recognized as essential to ensure access to appropriate levels of service and to provide for unmet service needs in both developed and developing countries. The ongoing pandemics of AIDS, tuberculosis, malaria, and comorbidities of infectious diseases with micronutrient deficiencies stretch beyond the limited capacities of existing resources. At the same time, political attention is unable to focus on health needs sufficiently to provide for developing and sustaining the human resources necessary to meet such challenges as the Millennium Development Goals. Many countries will be unable to meet such goals, particularly with relation to child and maternal mortality, with the financial and human resources available even with assistance from donor countries and organizations.

Globalization is shaping our understanding of the importance of the global health workforce in view of looming threats of pandemic influenza and bioterrorism, ongoing pandemics, and unprecedented migration of the skilled health workforce from developing countries to developed countries due to perceptions of more attractive incomes, professional settings, and way of life. At the same time, with aging populations and new technologies revolutionizing medicine, new generations of health professionals and health workers are needed to meet rising demands and expectations.

In seeking efficient ways of improving health, health systems have opened many new professional roles in new organizational frameworks. As definitions of health service were widened to include health maintenance, new health professions were added to the total health service spectrum. Continuing education is vital to maintain and upgrade quality in a health care system. Registration systems and databases are important to provide basic information on all relevant aspects of health personnel.

The World Health Report of 2006 states the issue as:

"The world community has sufficient financial resources and technologies to tackle most of these health challenges; yet today many national health systems are weak, unresponsive, inequitable—even unsafe. What is needed now is political will to implement national plans, together with international cooperation to align resources, harness knowledge and build robust health systems for treating and preventing disease and promoting population health. Developing capable, motivated and supported health workers is essential for overcoming bottlenecks to achieve national and global health goals." [*World Health Report 2006:* Working Together for Health]

This chapter will examine the importance of human resources for the New Public Health and the elements essential for training in relation to the quantity, quality, and changing interaction among the health professions.

OVERVIEW OF HUMAN RESOURCES

The World Health Report 2006 was centered on global problem of human resources for health. This is not only a severe problem for developing countries but an ongoing issue in the industrialized countries as well. The numbers, types, and distribution of personnel supply are major determinants of access, availability, appropriateness, and costs of health care. The training, quality, and performance of health personnel and the technology they use are all important health planning issues. Every health professional needs knowledge of the principles and current standards of public health in order to perform his/her functions, as all of health care now routinely involves prevention, teamwork, management, quality assurance, cost containment, and related ethical issues.

In many countries, the major focus of education of health personnel has been to prepare clinicians, without equal emphasis on preparation of public health policy analysts, health managers, and public health professionals. Yet the latter are especially important when health reforms are under way and when health promotion and prevention are needed to cope with changes in the health needs of a society.

The principal problems in human resources development vary from country to country but consistently include the following:

1. Inadequate funding, training positions, salaries, incentives, safety, and support systems for health workers;
2. Imbalance in training of health professionals; severe shortages of nurses and other health professionals, compared to the physician workforce;
3. Insufficient training for medical and nursing personnel in developing countries, possibly excess capacity for medical training in post-Soviet countries;
4. Excess of medical subspecialists, and insufficient incentives and training of primary care physicians, inflating health costs and compromising access to care;
5. Geographic maldistribution of vital professional categories with concentration in urban centers and poor supply in rural areas;
6. Underfinancing for public systems of health care in comparison to private, fostering poor work conditions, low remuneration, and indifferent career opportunities, with low staff morale, performance, and client satisfaction;
7. Insufficient standards and length of training of specialist physicians to produce well-qualified professional leaders;
8. Lack of orientation of all health providers to public health, an overmedicalization in the health field, and excessive influence of the pharmaceutical industry on medical education, practice, and health priorities;

9. Lack of postgraduate accredited academic centers for research and training of public health specialists in epidemiology, health-related social sciences, health system policy analysis, or health system management, compromising the ability of a health system to monitor its outcomes and resource allocation, or to evaluate program effectiveness;

10. Licensing of health providers by the government, which may allow for compromises in quality to ensure adequate numbers of graduates; conversely, delegation of licensing to professional syndicates may result in a protectionist approach, placing the interests of the profession above those of the public;

11. Compromising the quality of human resources by inadequate recruitment and educational standards, inadequate continuing examination and recertification;

12. Conflicts of academic, professional, government, or insurer interests with public and individual patient interests in training policies;

13. Poor coordination and communication between government and managerial sectors involved in health policy;

14. Globalization and migration of the skilled workforce from developing countries, where they are most needed;

15. Inadequate development of community health workforces as front line access and outreach personnel in rural and underserved urban areas as integral participants in the health system, including for underserved high need populations in developed countries.

From the 1950s to the 1970s, medical schools were opened and existing schools expanded to meet problems of access to care, perceived to be due to a shortage of doctors. It was thought that increased numbers of doctors would increase competition and lower doctors' incomes; however, medical incomes continued to rise and problems of access to care were unresolved. With growing emphasis on health economics and health promotion and disease prevention, there was a realization that excess medical personnel would not contribute to the national health, and that in some countries the excess of medical personnel had become a liability. In both fee-for-service medicine and salaried health service, increases in physician supply generate increases in health expenditures. Supply and demand market forces do not adapt well to health care, because the consumer demand is to a large extent generated by provider decisions (e.g., for return visits, investigation, or hospitalization). Fees may be fixed arbitrarily or by negotiation with a public insurance mechanism; the service is paid by a third party, and the consumer is less knowledgeable than the provider.

Each country addresses the issue of how many and what kinds of human resources to train for its own needs, related to the design and operation of its health system. During the 1970s, the province of Alberta, Canada, had relatively stable expenditures for medical services and physician-to-population ratios. During the 1980s, the province experienced a marked economic downturn and zero population growth, but the supply of physicians and services per physician increased by some 20 percent. The reduced numbers of clients per physician led to an increase both in fees and in volume of services per capita so that physician incomes were sustained. As a result, total and per capita expenditures for health care increased sharply. In many countries during the 1980s, policies were reformulated to reduce the size of medical school training entry classes.

Oversupply of medical specialists can also be a serious problem for a health system, promoting a bias toward a specialized medical orientation in health care at the expense of other more basic needs of public health, primary care, and fundamental support systems for vulnerable groups in society. An excessively specialized medical orientation fosters misallocation of limited resources by creation of tertiary care and high physician density in central cities, leaving rural and primary care underdeveloped. This is widely prevalent in developing countries such as India, Mexico, Colombia, and Latin American countries. In some countries the problem is often compounded by an inability of the health budget to employ needed numbers of physicians. Unemployment among young physicians is a substantial problem in many countries.

In developing countries, health workforce shortages are already at crisis levels. As both the populations and workforces of industrialized nations increase in age, these societies also face an increasing demand for health workers across the professional spectrum that outstrips supply. In 1996, the Association of American Medical Colleges (AAMC) in cooperation with the Council on Graduate Medical Education issued a call for a 30 percent increase in medical school capacity over the following decade. As the current workforce nears retirement age it becomes clear this may even be insufficient to maintain basic health care services. The shortages encompass nearly every field, but are most pronounced for rural, primary care, and public health; fewer than 32 percent of physicians in the United States practice as generalists. The American Association of Family Physicians predicts a need for an increase of at least 40 percent of physicians in primary care alone by 2020.

The ratio of physicians in active medical practice per 10,000 population in the United States increased from 13.5 in 1975 to 18.0 in 1985, 21.3 in 1995, and 23.8 in 2005 (Health, United States, 2007), and is expected to rise to 29.2 by the year 2020. Concerns of shortages of both primary care and specialist doctors are based on current geographic and specialty training distribution, but also the changing demography of the aging population. This is, however, based on current medical practice organization and should take into account more efficient methods of practice, such as prepaid group practice, full-time salaried physician staffing in hospitals, and increased potential for preventive care by nurse practitioners and community health workers. Debate on increasing of graduate training positions by removing the current ceiling on their funding by Medicare is in full swing. This topic is the center of vigorous debate as many argue that there is a steady increase in medical personnel, but the geographic distribution favors the major population centers over rural and underserved remote areas. Further, some medical specialties, such as anesthesia and primary care, are in severe shortages. They further argue that increasing the output of medical schools increases social disparities now present and only major reforms including national health insurance would reduce such inequities.

In the United States, mid-level health care providers fill more than half the supply of primary care clinicians. Compounded by a generalized nursing workforce shortage, nurse practitioner graduation rates are now decreasing at a rate of 4.5 percent each year. Policy studies in family medicine and primary care in the United States. predict similar declines in physician assistant graduations, reaching a 25 percent loss by 2020. Mid-level practitioners have grown in influence and have proven to be outstanding public health and primary care professionals, bringing needed health care to rural and other marginalized populations. Enhancement and support of the nurse practitioner and physician assistant professions is an essential component of any successful effort to meet health care workforce demands. As graduation rates for mid-level practitioners decrease, a crisis of unmet need arises. Current educational programs must be expanded, new programs developed, and incentives created to attract workers to public health and primary care.

While nearly all health professions face projected shortages, the situation for nurses is most severe. Throughout the world, the nursing workforce is very insufficient to meet the public's needs. In sub-Saharan Africa, simply to accomplish immediate health intervention goals, an additional 600,000 nurses would be required. In the United States, the number of registered nurses increased by 8 percent between 2000 and 2004 to a new high of 2.9 million, but this is increasingly a group approaching retirement age. The picture varies from country to country; however, every region of the world faces dramatic nursing shortages. A 2006 report of the International Council of Nurses presents policy guidelines for recruiting new workers, curtailing migration trends which have stripped impoverished regions of nurses, and improving working conditions and labor strength, a priority in retention of the current workforce.

Table 14.1 highlights the disproportion between nursing and medical professions for several developed nations. While total physician supply varies greatly, there is a global pattern of maldistribution of human resources, with greater

TABLE 14.1 Physician and Nurse Density per 1000 Population for Select Countries

Country	Physicians	Nurses
United Kingdom	2.3	12.1
Sweden	3.3	10.2
Germany	3.4	9.7
United States	2.7	9.4
Russian Federation	4.3	8.1
Israel	3.8	6.4
Greece	4.4	3.9
Mexico	2.0	0.9

Note: Rounded to one decimal place.
Source: World Health Organization. Working Together for Health: The World Health Report 2006, Annex table 4, various years 1997–2003.

than half of providers practicing in subspecialties and metropolitan areas. In terms of public health need, the shortage of both doctors and nurses is at crisis levels for rural and marginalized populations.

The achievement of the goal of Health for All through primary health care requires the effective and coordinated services of many types of health personnel within a national health system designed to reach this goal. Political policy is crucial for the preparation, composition, and work patterns of the health workforce. National expenditures on health are dependent on the political priority given to health compared to other issues that may be equally or more pressing to the governing power. A strong national health policy can nevertheless be constructed, even in a poor country, by well-defined health programs. Community and rural health policy in China during the 1950s was based on a number of elements: development of a 3-year family doctor training program for rural service, upgrading training of village doctors to physician assistant level, and incentives to encourage work in the countryside and at a grassroots level. This program was successful in raising health standards in China beyond that which might be expected from its economic level. With recent economic reforms, this system is going through profound changes (Chapter 13).

HUMAN RESOURCES PLANNING

The health infrastructure of a country includes the resources available and their organization. Human resources are essential to any health system. The supply of personnel and facilities, economic support of the system, management and policy, methods of payment of providers, and organization of the services are therefore vital in health planning (Box 14.1).

Resources available to health systems include facilities, personnel, and financial resources for health care. The organizational and financial structure of a health system determines how these resources are allocated or expended, in public as well as the private health care sectors. Both structure and methods of payment affect how services are provided. Health systems require economic support sufficient for basic and continuing education of high-quality human resources, as well as managing their appropriate and optimum use.

Regulation of health personnel includes licensure and discipline and is an important governmental function. Measures to control or limit the supply of medical practitioners, along with incentives to promote more efficient health care, are important issues in rationalizing health care systems.

Continuing education is a vital part of health manpower planning. The rapid and continuous development of medical sciences (e.g., genetics and nanotechnology) and technology requires (e.g., new vaccines and health promotion techniques) health workers to have access to continuing education to keep up with new developments. The methods of doing this should include short courses, longer formal training periods, such as the master of public health degree for health managers, and development of distance learning with wide access to Internet resources. Increased access to well-developed consensus guidelines for clinical care and for public health policy are vital for human resource planning.

Fundamental to the process of determining labor needs is knowledge of the current personnel situation (Box 14.2). Essential for this are data systems based on periodic registration or census-taking of persons practicing a health profession. Practitioners may retire, die, migrate, or leave the profession and should be taken out of registries of those actively practicing. An accurate, up-to-date picture of actual human resources provides information on specialty, geographic distribution, age, sex, and current work activities. International comparisons of professional personnel help to place a national pattern in the context of other countries with similar socioeconomic and health standards. Human resources supply should be matched to the targets and resources of a country. Alternative approaches may be needed if the supply of workers is insufficient or inappropriate to meet health needs and targets.

Assessing current personnel supply and determining future needs are specific tasks of a government agency concerned with comprehensive national socioeconomic planning. They may be assigned to a planning agency, board, commission, or committee empowered by authorities working with education systems, consistent with general health planning. Academic training centers play an important role not only in training, but in implementation of national human resource policies, so they are integral to determining policy.

Supply and Demand

A common form of quantitative human resources planning, or non-planning, is a market-oriented approach, based on the needs of the training institution and demands of trainees. The demand for training as physicians may be high, and the medical schools have an interest in training more students for financial or prestige reasons. If unregulated, the creation of new private medical schools will be based on a profit motive and not take into account the needs or capacity of the country to absorb new graduates, leading to creation of excessive training capacity, and a surplus of poorly trained doctors with little prospect of professional employment as has happened in some mid-level developing countries.

During the 1950s and 1960s, planners in many countries thought universal access to medical care would solve most health problems and more doctors would be needed to fulfill that dream. It was assumed, even in centrally planned health systems, that supply and demand would direct new graduates to underserved regions or professional specialties. However, increasing the supply of medical graduates is costly to society and results in oversupply, especially in major urban centers, with increased utilization and subspecialization of medical care.

Increasing the supply of physicians was expected to increase access to health care and to increase the numbers of doctors entering less popular fields of practice, such as primary care, and moving to underserved geographic areas. This approach has been less accepted since the 1980s; even in free market societies, it inflates the costs of health care and fails to meet needs in underserved populations or specialties. Immigration and emigration of medical personnel, or departures from active practice, are also factors in supply and distribution of health personnel. There were also concerns during the 1980s of a possible oversupply of doctors, and medical school enrollments were limited as a policy. This approach was promoted as part of the concern for rapid increases in health care costs and partially as a result of concerns in the medical profession of excess supply bringing more competition and possibly lowering incomes of physicians. In the 1980s, health care costs rose rapidly, associated with increasing specialization and a search for new organizational patterns of health care such as the health maintenance organizations (HMOs). There was also a growing realization that health needs depend more on prevention than increased supply of physicians. This led to a trend to reduce numbers of new students entering medical schools, a decrease in subspecialty training positions, and a greater reliance on immigration of doctors to the United States, Canada, and the United Kingdom. This created a growing migration of medical and nursing personnel and reliance of developed countries on importation of doctors and nurses from poorer countries, often with serious and destructive effects on poorer countries in Europe and especially in severely manpower-deficient countries in sub-Saharan Africa.

Comparing Canada to the United States and other OECD countries (Organization of Economic Cooperation and Development) shows that Canada in 2005 had 2.2 practicing physicians as compared to the OECD average of 3.0 and 10.0 qualified nurses as compared to the OECD average of 8.6 per 1000 population. In 2005, the United States had a higher practicing physician-to-population ratio (2.4) and a lower nurse-to-population ratio (7.9 per 1000 population) than Canada (OECD, 2007). As seen in Chapter 13, Canada has better life expectancy, lower mortality rates, and better access to physicians than U.S. counterparts. The difference seems to be in universal coverage still lacking in the United States, despite higher health expenditures there.

Medical and health profession schools are costly to establish and operate; they can generate high cost to a health system if they produce excess graduates. Founding new schools and maintaining existing schools at present levels of enrollment require careful consideration of the effects of the numbers of medical graduates on the health system. In either a regulated environment or a free market situation, the supply of human resources can be powerful in driving up health care costs. A period of restraint in health expenditures calls into question the wisdom of continuous increases in personnel and unlimited service as free or insured benefits. Even in free market settings such as the United States, government funding and regulatory powers are used to reduce the number of training positions in the specialties in favor of increased incentives and openings in primary care.

Table 14.2 shows trends in medical education in the United States between 1960 and 2005. The number of medical graduates increased twofold between 1960 and 1980 but has grown at a slower rate since 1990. Medical personnel per population increased by 40 percent from 1960 to 1980, and another 35 percent from 1980 to 2005. Some 25 percent of the total physician supply in the United States are graduates of international medical schools. A leveling off in production of new doctors is now occurring. Due to population expansion and aging, the American Medical Association and American Association of Medical Colleges now predict a growing shortage of physicians, particularly in primary care and underserved areas. This shortage is partially addressed by the expansion of osteopathic schools; however, the AMA and AAMC have called for a rapid increase in U.S. medical school

---

TABLE 14.2 Medical and Osteopathy Schools, Graduates, and Physician Supply, United States, 1960–2005

	1960	1970	1980	1990	2000	2005
Medical and osteopathy schools	92	110	140	142	144	145
Graduates (thousands)[a]	7.5	8.8	16.2	16.9	18.0	18.5
Doctors/10,000 population[b]	14.0	15.5	19.6	23.2	26.8	26.9

[a]Includes all graduating allopathic and osteopathic physicians.
[b]Includes all practicing and non-practicing physicians.
Source: U.S. Department of Health and Human Services. *Health, United States,* 2006 and 2007 (Table 107, 110).

admissions by 30 percent. Immigration of physicians from developing countries is providing an important source of medical manpower, but is contributing to the deficiency of physicians in the source countries.

A normative approach uses standards or norms derived in some systematic, arbitrary way. The standards may be based on empirical criteria of the number of physicians, nurses, or other health personnel required. This approach may be excessively rigid and unresponsive to changes in disease prevalence and technological changes in health service needs. Standards may also be adopted from ratios found in other countries or in other successful or "gold standard" areas of the same country.

Human resources planning may set certain goals intended to produce personnel in numbers maintaining or increasing the current supply-to-population ratio by a selected percentage, for example, by 5 or 10 percent within a chosen time period. A country wishing to increase this ratio will need to take into account new training needs and loss due to emigration. This approach is less likely to lead to an oversupply but may maintain an arbitrarily high level of human resources despite changes in epidemiologic patterns or increased efficiency of the services. For example, as tuberculosis declined, fewer tuberculosis specialists were needed, but as the disease recurs, there is a demand to improve the training and numbers of specialists in the field. Hospitals are becoming less the center of health care, and reduction in hospital beds has become part of restructuring of services. This should lead to a shift of personnel from institutional to community-based services, with provision for retraining and skilled system management.

Many countries require medical graduates to serve one or several years in rural or underserved areas. This exposes young graduates to the realities of primary care as part of their professional development and, it is hoped, infuses them with concern for the harsh realities of the living conditions of rural poverty. However, it places inexperienced young professionals in isolated locations without adequate collegial support or supervision, where they are unlikely to remain beyond a compulsory period of service. Efforts to require young graduates to work in rural areas are temporary solutions, generally frustrated by the desire of doctors and nurses to live in urban areas and practice in clinical subspecialties.

Partly in search of methods to constrain cost increases and in part searching for ways to improve access to care for high-risk groups, mid-level health provider training is increasingly accepted in human resources planning. Human resources planning should take into account the many different disciplines needed for both clinical care and public health, taking into account changing patterns of need, technology, and spread of health care responsibility among many professions.

Organization of care affects the numbers of different health workers required. Independent private practice and free choice of physician or specialist promote higher utilization patterns and create waiting lists, rapid cost increases, and an apparent shortage of personnel. Centrally controlled health systems such as the Soviet health system created inflated norms of staff-to-population ratios and low-efficiency health services (see Chapter 13).

Qualitative methods are as important as quantitative planning. Quality of training programs at the undergraduate and graduate levels, accreditation, licensure procedures, and ongoing quality assurance measures are important elements in the quality of national health systems.

BASIC MEDICAL EDUCATION

The education of medical doctors and training of specialists are, in principle, a national commitment. Governments have a responsibility to ensure an adequate number of well-trained health professionals to provide services. This is a combined function of health and education authorities, carried out by providing financial support and standards for the universities or medical training institutes where education occurs. Funding support and accreditation of educational institutions provides mechanisms for applying national or state policy for both quantity and quality of educational programs. National or provincial departments of education set guidelines and standards for funding through a university grants mechanism or commission, often based on enrollment. Standards may be set for curriculum, faculty, basic sciences, and clinical training, as part of approval for funding or through nongovernmental accreditation structures organized by the medical schools themselves (see Chapter 15).

The long tradition of multi-faculty, university-based medical education is widespread in the industrialized countries and their former colonies, now independent states. Medical training gains from an environment that promotes research and service in an academic atmosphere with its associated standards. This tradition of linking research with education and service is important in promoting quality education. Having a research climate of peer-reviewed work raises the aspirations of the institution and its faculty and sets a standard for students for their life's work. A university degree confers prestige to a profession, encourages the pursuit of peer recognition of excellence, and academic criteria for student selection, curriculum, and faculty standards. This is widely the case for medical schools, and increasingly for schools of nursing and other health professions. However, a university degree is not required for all health professions. Community colleges may more appropriately provide a multi-faculty educational environment and a broad education base for some health care jobs.

In the nineteenth century, medical training in the United States was primarily carried out by private, commercial schools of medicine with poor facilities, staffing, and standards. The Carnegie Foundation sponsored a study of medical education in the United States and Canada, carried out by Abraham Flexner, a non-physician educator, who reported in 1910 on the poor quality of these commercial schools (Box 14.3). This report promoted university-based medical schools modeled on The Johns Hopkins University, which itself was based on successful, scientifically oriented German medical schools, combined with the strong clinical orientation of British teaching hospital medical schools. Most of the 450 commercial schools in the United States closed soon after this report and were replaced by the present 126 university-based medical schools with high standards of medical education and academic research. Since the 1950s, U.S. medical schools were stimulated by large amounts of federal funds channeled into research and training through the National Institutes of Health (NIH), as well as from nongovernmental sources, including private and foundation donations.

Medical schools are service resources for the community as well as being centers of academic excellence. Their goal should be to provide a balanced education in an academic environment where teaching, research, and service interact to produce medical graduates competent and oriented to meet the needs of the population. This requires a balance among the biomedical, psychological, population-based, and sociological perspectives on health care.

Teaching methods should be designed to promote the objectives of the program. Many medical schools teach primarily by lecture to very large classes, with limited supervised clinical experience. This reduces the chance for the student to develop patient-oriented and problem-solving skills. It promotes a didactic approach to medicine, and minimizes the opportunity for the student to work with multidisciplinary teams, or see medical care as part of a diverse team. Working with students of other sciences and professions in a collegial fashion helps the medical student understand the team role of workers in the health care system.

The purpose of training medical practitioners is to have skilled professionals providing patient care and the professional leadership needed to develop and maintain high-quality health care and public health systems. In order to meet these goals, high standards are required in selection of candidates. Medical schools in the United States are graduate schools requiring a prior university degree for candidates. In other countries such as Canada and the United Kingdom, medical education includes 2 years of premedical studies followed by 4 years of medical school. Quality medical education requires continuous curriculum development and review, as well as highly qualified teachers, library access, clinical training, and examination during and at completion of training. The nature of undergraduate training will be a key factor in determining the lifelong practice habits of the providers, but equally important are the specialization period and ongoing education throughout their professional lives.

In most industrialized countries, enrollment of women and minority groups has increased dramatically in recent decades as part of social policy. While there are social and political reasons to promote access to professional schools for all segments of a population, but this should be without compromising academic standards, and should not adversely affect the quality of services provided to the patient or the population as a whole. Private medical schools are a highly lucrative business in some developing countries, which, if unregulated, may contribute to overproduction of inadequately trained doctors, compromising national efforts to promote quality of training.

Where the language of instruction is not one used internationally for scientific literature, the local medical community may be limited in access to current textbooks and peer-reviewed professional literature, domestic and especially international journals, as well as Internet access. The language of instruction in most schools of medicine is in the national language, but increasingly with English as a second language required in many European schools. Lack of English-language training prevents or hinders access to the world literature and participation in international exchanges and effectively holds back scientific progress in many countries.

During the transition period of the post-Soviet era, medical schools sought enrollment of foreign students to

---

Box 14.3 The Flexner Report, 1910

"For twenty-five years there has been an enormous over-production of uneducated and ill-trained medical practitioners in absolute disregard and without serious thought to the interests of the public. Taking the United States, physicians are four or five times as numerous in proportion as in older countries like Germany. Over-production is due to the very large numbers of commercial schools. Colleges and universities have failed to appreciate the great advance in medical education and the increased cost of teaching it along modern lines. A hospital under complete educational control is as necessary to a medical school as is a laboratory of chemistry or pathology. Trustees of hospitals, public and private, should, therefore, go to the limit of their authority in opening hospital wards to teaching. Progress for the future would seem to require a very much smaller number of medical schools, better equipped and better conducted and the needs of the public would equally require fewer physicians graduated each year better educated and better trained."

Source: Flexner, A. 1910. *Medical Education in the United States and Canada: A Report to the Carnegie Foundation for the Advancement of Teaching.* Reprinted New York: Arno Press and The New York Times, 1972. From: Introduction by H. S. Pritchett.

increase revenues, reducing the numbers of local students. However, they still produce graduates at levels well above the capacity of the system to absorb if salaries of physicians are to rise above tradesman levels. In some developing countries, private medical schools have sprung up with inadequate facilities and faculty producing large numbers of poorly trained doctors with little chance of employment in the profession in their own countries.

Curriculum reform, as in the days of Flexner's recommendations, must be an ongoing process to meet health needs of the population, in keeping with current international standards. This includes adequate attention to basic medical sciences, clinical experience and patient care, hospital and community-based training, and research. Access to libraries with adequate supply of current international literature, textbooks, and computers with Internet services is essential to maintain acceptable standards.

Reform in medical education is focusing on producing practitioners for the twenty-first century, meeting the needs of both primary care and specialized medical services. In recent years, there has been a growing concern that there has been too much emphasis on science and specialization to the detriment of primary care in training U.S. physicians. All medical students should be exposed to patient contact earlier in their training than in the past, in different health care settings, including teaching hospitals, outpatient clinics, and community clinics, as well as public health programs. They should also be familiar with community-based resources for the infirm, disabled, and poor. Training should include multidisciplinary components so the student is familiar with the professional elements of other disciplines including those in public health, health-related economics, and social sciences.

International conferences on medical education sponsored by the World Federation for Medical Education (WFME) in 1988 and 1993 attempted to define a new direction for education of physicians to promote their role in promotion of health as well as treatment and prevention of illness (Table 14.3). In 2003, WFME met again in Copenhagen, Denmark, reaffirming these targets. Quality and uniformity in standards for basic medical education were stressed as a priority at the global level. Sponsored by the WHO, UNICEF, UNESCO, UNDP, and the World Bank, these conferences established an international forum for reevaluation of medical education in the twenty-first century in the context of changes in medical and public health technology, organization of health care, and needs of the population. Change in medical education is often difficult due to competing concepts of what medical students should know, and a lack of focus on what the practicing physician should be.

Medical Education Issues

The costs of medical education are high and require public subsidization. University grant commissions are semi-autonomous bodies with financial grants from education

---

TABLE 14.3 Medical Education Issues — The Edinburgh Declaration (1988) and the World Summit on Medical Education, 1993 (Reaffirmed in 2005)

1. Conducted in relevant educational settings — hospital, community, workplace, homes
2. Curriculum based on national health needs
3. Emphasis on disease prevention and health promotion
4. Lifelong active learning
5. Competency-based learning
6. Teachers trained as educators
7. Integration of science with clinical practice
8. Selection of entrants for social commitment, intellectual attributes
9. Coordination of medical education with health care services
10. Balanced production of types of doctors
11. Multiprofessional training
12. Continuing medical education requirements
13. Students involved in planning and evaluation of medical education
14. A multiscience-base medical graduate
15. Ethical and moral basis of medical practice
16. Curriculum options for dealing with information overload
17. Postgraduate education in relation to community needs
18. Health teams and multiprofessional education
19. Community participation in medical education
20. Population-based education — care for individual patients in context of needs for a defined population

Source: Adapted from World Federation for Medical Education. World Summit on Medical Education: The Changing Medical Profession. Edinburgh, August 1993, reaffirmed at Copenhagen in 2005; Promotion of Accreditation of Basic Medical Education: A Programme within the Framework of the WHO/WFME Strategic Partnership to Improve Medical Education. WFME Office. The Panum Institute, Faculty of Health Sciences, University of Copenhangen, November 2005, http://www2.sund.ku.dk/wfme/Activities/WFME%20PROMOTION%20OF%20ACCREDITATION_291105.pdf

---

departments of governments. Thus both financial and regulatory powers are used to set criteria for standards and accreditation of faculties of medicine. This represents an important diffusion of power and responsibility from direct control by government. Regulation by accreditation of schools is also strengthened by national organizations which promote national standards of medical education.

POSTGRADUATE MEDICAL TRAINING

Undergraduate medical training provides an educational base, but is not adequate preparation for a medical practitioner. Postgraduate training of high quality and adequate duration is essential to assuring quality in health care services (Box 14.4). Specialty training requirements should be regulated by a national or state authority, or a professional body (college or association) delegated the legal right to license practitioners. This includes designation of facilities accredited for training, academic and research

Box 14.4 Standards for Postgraduate Medical Training

1. Regulated by national board with professional, governmental, and public representation for quality, admissions/enrollment and clinical and community-based training;
2. Duration of training of 4–6 years, depending on specialty;
3. Supervised independent clinical experience;
4. Accreditation of training centers based on academic and service criteria of licensing body;
5. Supervised research period in basic science laboratory or epidemiologic study;
6. Required familiarity with relevant international literature;
7. Rotation with part of training in a different medical center;
8. Demonstrated high levels of clinical ability, responsibility, knowledge, and ethical standards;
9. Examinations in mid-training with written examinations based on international standards;
10. Examinations at end of training; clinical and written examinations based on international standards;
11. State board or professional college setting examinations and certification;
12. Recertification requirements.

areas within the curriculum, clinical experience, duration of training, and requirements for examination at several stages during the training period. National standards are needed to ensure equivalent quality and permit freedom of movement for professionals. However, this may put some areas at a disadvantage by promoting a "brain drain," or loss of professionals, usually from rural to urban areas, or from poor countries to wealthy ones. The rights of an individual practitioner to select place and type of practice are limited by open positions in training centers or in practice settings.

Licensing of medical specialists is a state responsibility, but in some countries this is delegated to a professional association. In the United States, postgraduate training is under the control of state and national boards, made up of state-appointed officials and public and professional representatives. In Canada, licensing of physicians is delegated to the provincial medical associations, while postgraduate examinations and certification are under the authority of a professional body, the Canadian Royal College of Physicians and Surgeons. In the United Kingdom, licensing of physicians is under a state-appointed body, the General Medical Council, while specialty recognition is by a series of Royal Colleges.

Standards for specialty training must reflect the views of the specialty practitioners as well as the public interest. The public interest is best protected by a combination of state and professional supervisory systems with the force

of law, including the regulatory and disciplinary measures needed to maintain professional and ethical standards demanded by the public interest. The specialist-trainee requires supervised time and experience to mature as a professional. Supervised clinical experience, research, publication in peer-reviewed journals, and continuing peer review are all essential in the training process to produce specialists motivated and capable of keeping up with rapidly evolving standards of modern medicine. Clinical specialization time requirements vary widely from country to country. Eligibility for specialty boards in the United States is generally 3–4 years of recognized training after graduation, with examination by member boards of the American Board of Medical Specialties.

SPECIALIZATION AND FAMILY PRACTICE

Good medical care depends on access to primary care and appropriate referral for specialty care. Most systems utilize the primary care physician as a gatekeeper for referral to specialty care. *Laissez-faire* insured service systems allowing unreferred access to specialty care face the difficulty of maintaining primary care medicine and continuing pressures on physicians to select specialty training as their career choice (Table 14.4). Due to uncertain career outlook, few graduating physicians choose a career in family practice. The aging generalist workforce also influences trends which have led to inadequate primary care capacity, crowded clinics, unacceptably long wait time for patients, and questions of sustainability and quality assurance.

Maldistribution of medical practitioners is widespread, with rural and urban poverty areas often suffering from lack of access to primary care. Regional variation in practicing physicians ranges from high ratios in Massachusetts (38.4 per 10,000 population) to a low rate of 15.4 in Oklahoma. Specialist physicians are less likely than generalists to live in rural areas. Maldistribution by specialty is another problem in medical resource planning. Regulations to redistribute the number of training positions are

TABLE 14.4 Physician Workforce and Percentage of Primary Care Doctors, United States, Selected Years, 1950–2004

Year	1950	1980	1995	2005
Physicians per 10,000 population	14.1	19.0	21.3	23.8
% in primary care	59.0%	43.0%	43.0%	45.6%

Note: From 1970, primary care includes general primary care specialists and primary care subspecialists.
Source: U.S. Department of Health and Human Services. *Health, United States*, 2007.

now operational in the United States and common in many countries. Medical teaching centers in the United States are now under regulations which require them to include primary care in their postgraduate training programs.

National health systems address these problems with regulations to mandate and financial incentives to attract physicians to underserved areas and understaffed specialties. In the United Kingdom, as in many other European countries, the National Health Service uses the general practitioner (GP) as the key family practitioner, primary care provider for all beneficiaries, with specialty access through the GP. Managed care programs in the United States also stress and require patients to see primary care physicians. The changing economic environment of health care will be associated with changes in medical specialization more easily than the urban–rural inequities. These issues are leading to greater role delegation to nursing and new kinds of health workers.

TRAINING IN PREVENTIVE MEDICINE

Preventive medicine is recognized as a clinical specialty in the United States. Promoted since the 1970s, this specialty attempts to bring public health and clinical medicine closer together. Preventive medicinal training is one of 24 accredited clinical specialties in the United States, with doctors becoming board certified in one or more subspecialties: general preventive medicine and public health, occupational medicine, and aerospace medicine. These programs are part of the postgraduate training program system of the American Medical Association, in conjunction with the American Board of Preventive Medicine. Master's or doctoral degrees are earned in more than 23 graduate programs situated in departments of community or preventive medicine within a medical faculty.

Preventive medicine is a specialized field of medical practice composed of distinct disciplines that utilize skills focusing on the health of defined populations to maintain and promote health and well-being and prevent disease, disability, and premature death. The American Board of Preventive Medicine requires trainees to have core competencies in biostatistics, epidemiology, administration, planning, organization, management, financing and evaluation of health programs. Training also includes environmental and occupational health, and social and behavioral factors in health and disease and the practice of prevention in clinical medicine. It applies primary, secondary, and tertiary prevention measures within clinical medicine. Graduates in this field provide a supply of potential health planners, administrators, teachers of preventive medicine, researchers, and clinicians applying preventive medicine in health care settings of practice. They may also serve in governmental (local, state, national, and international) public health departments, educational institutions, organized medical care groups, in industry, other employment settings, and the community, voluntary health agencies, and professional and related health organizations. Requirements include a graduate year of training and experience in a clinical area of medicine; a year of academic training in a fundamental aspect of preventive medicine; and a practicum or year of supervised practical experience (e.g., occupational). Training of clinicians in health services research and clinical epidemiology also provides a potential career path for physicians entering one of the many fields of public health.

NURSING EDUCATION

Nursing is the backbone profession in hospital and community health care. The place of nursing in a health system reflects cultural values of the society and has an important effect on the health system. Whereas medicine is generally a high-prestige profession, in many countries nursing is of low social status, with strong cultural biases against women entering nursing. Finland and Sweden have more than four nurses per physician, while developing countries such as India and Bangladesh have between one and two nurses per physician. This represents a widespread overemphasis on medical training and an underemphasis on training of nurses in developing countries. The health system thus suffers from a lack of personnel to develop and operate primary care services, with biases to high-cost secondary and tertiary care services. Furthermore, lack of high-level professional nursing personnel prevents full development of quality secondary and tertiary care services. Lack of nursing at the professional level may be one of the biggest factors in retarding development of health services in many countries.

The number of nursing graduates more than tripled from 1950 to 2000 in the United States (Table 14.5). From 1996 to 2000, there was a decline in nursing graduates by 31 percent overall. The promotion of the academic aspects of nursing is seen in the growth of baccalaureate nursing education from 13 percent of all nursing graduates in 1960 to 26.5 percent in 2000. The decline in the number of nursing schools in the 1950s was due to closure or consolidation of individual hospital schools of nursing. North American schools for nursing education are now largely associated with university or associate degree programs in community colleges. University-based schools in the United States provide academic degree programs at the bachelor, master, and doctorate levels. Nursing education at the master and doctorate levels provides the teaching, research, and management cadres needed for a progressive health care system.

The scope of activities that professional nurses are authorized to carry out by law and custom has gradually broadened over the past several decades to include procedures previously performed only by physicians in the United States. This was partially associated with increasing academization of the nursing profession and increasing

TABLE 14.5 Nursing Schools and Graduates in the United States, 1950–2000

	1950	1960	1970	1980	1990	1996	2000
Nursing schools	1770	1137	1340	1385	1470	1508	na
Total nursing graduates (000)	25.8	30.1	43.1	75.5	66.1	94.8	79.7
Graduates with BA/Bsc (thousands)	na	4.1	9.1	25.0	18.6	32.4	26.5
Registered nurses/10,000 population	na	na	35.6	56.0	69.0	79.8[a,b]	na

na = not available.
[a]Includes bachelor of sciences degrees.
[b]Data for 1995.
Source: U.S. Department of Health and Human Services. *Health, United States*, 1998, 2006.

emphasis on bachelor's and master's degrees for nurses and PhDs for nursing teachers.

Nurse practitioners are trained to diagnose and treat illness, usually under authorization from a supervising physician. In some developing countries, especially in rural areas, auxiliary nurses, as well as professional nurses, are expected to diagnose and treat common ailments, in addition to conducting health education and primary and secondary prevention. This is vital, especially in areas without medical practitioners, and should be under supervision and guidelines of the ministry of health or other public health agency.

Nursing specialization may be at a certificate or master's level. Certificate courses are in fields where the nursing role involves highly skilled practice crucial to patient outcomes such as in intensive care or emergency department nursing. Master's programs in areas such as pediatrics, geriatrics, or adult health produce a more broadly based and independent practitioner, researcher, or educator.

The first baccalaureate program in nursing was established at the University of Minnesota in 1909; by 1980 there were 377 and in 1995 521 bachelor's programs for registered nurses. Upgrading educational standards for existing professions, such as nursing or midwifery involves considerations of the costs and effects on personnel supply as well as the desirability of raising professional standards. The advent of degree programs in nursing raised the level of prestige, leadership, research, teaching, and service of the profession. The transition from hospital apprenticeship training to university-based education (i.e., "academization") was opposed by traditional interests such as hospital management and the medical profession, but this resistance subsided with the demonstration of greater capacity in the nursing profession to take responsibility and incorporate rapid scientific and technological advances. Table 14.5 shows the reduction in schools of nursing in the 1960s followed by growth in baccalaureate nursing education in the United States. The trend of planning for 25 percent of nursing graduates to be at the baccalaureate level continues to the present time with other registered nurses being trained in 2-year community college programs.

Based on current trends, there is an anticipated growth in demand for RNs. Health Resources and Service Administration (HRSA) of the U.S. Department of Health and Human Services estimates there was a 6 percent shortage of RNs in the United States in 2000. The shortage is projected to reach 800,000 nurses in 2020 (a shortfall of 29 percent). Nursing is in competition for new students with many other career opportunities for young people. The U.S. National League of Nursing estimated that 530,000 of a total of 850,000 nurses in 1970 were active in the field. By 1980, there were 1.3 million licensed nurses, most of whom were working at least part time, increasing to over 2.0 million in 1994. The nurse-to-population ratio increased from 56/10,000 in 1980 to 69 in 1990 and 78.5 in 1994. The number of employed RNs increased during the period 1999–2005 by an average annual increase of 1.2 percent from 2.205 million to 2.368 million (Health United States, 2007). In 2002, the United States had 7.9 nurses per 1000 population compared to 8.6 per 1000 for OECD averages. The hospital bed supply in 2005 in the United States was 2.9 beds per 1000 (a decline from 4.4 beds in 1980), as compared to the 2005 OECD average of 3.9 (OECD Health Data 2007, How Does the United Sates Compare?). Nursing colleges and universities in the United States are struggling to expand enrollment levels to meet the current nursing shortage and the rising demand for nursing personnel.

Acute care hospital bed supplies have fallen over the while long-term beds for nursing care have increased (Chapter 11). However, demand for RNs is expected to grow by 2 percent to 3 percent each year. Some estimates indicate a need for 30,000 additional nurses to graduate annually in order to meet the nation's health care needs and that more than one million new and replacement nurses will be needed in the United States by 2016.

As health care copes with both an aging and healthier total population, chronic diseases require care at the primary level with increased roles (and needed retraining) for physicians and for nurses in home care and other

TABLE 14.6 Predicted Shortfalls in Nurse Supply in Some OECD Countries

Country	Number of RNs in workforce	Predicted shortfall (year)
United States	2,202,000	800,000 (2020)
United Kingdom	500,000	53,000 (2010)
Ireland	49,400	10,000 (2008)
Canada	230,300	78,000 (2011)
Australia	179,200	40,000 (2010)

Sources: Polsky, D., Sochalski, J., Aiken, L. H., Cooper, R. A. 2007. Medical migration to the U.S.: Trends and impact. LDI *Issue Brief*, April-May; 12(6):1–4.
Aiken, L. H., Buchan, J., Sochalski, J., Nichols, B., Powell, M. 2004. Trends in international nurse migration. *Health Affairs*, 23:69–77.

outreach programs of care for persons at high risk (e.g., secondary prevention for hypertension or diabetes).

National health authorities need to take professional views into account, but balance the vested interests of each profession with other factors such as the hospital bed supply and utilization, alternative forms of care (which can be labor-intensive), aging of the population, changing disease patterns, and technological changes in prevention and health care.

The medical and nursing professions were in conflict over numbers and roles of nurses. The nursing profession struggled to establish greater autonomy and academic quality while the medical association fought to maintain large numbers of nurses and a subordinate role for nursing. The conflict between "need and demand" is a matter of definition, viewpoint, and priority. Reorientation of the health system with greater emphasis on community care will provide more employment opportunities for nurses with expanded professional responsibilities (Table 14.6).

IN-SERVICE AND CONTINUING EDUCATION

Rapid changes in all fields of medical science and practice make in-service and continuing education a necessity of any health program to maintain professional standards. In-service education increases the sense of self-esteem of workers and motivates staff to better performance. It serves to reinforce knowledge, introduce new information, and is essential to facilitate change in an institution. It also provides opportunity for the supervisory staff to reinforce and raise quality standards. Introduction of new programs and technologies should be accompanied by staff orientation as part of an ongoing in-service education program.

Continuing education refers to ongoing professional education in the form of courses, conferences, workshops, and literature. Medical graduates who complete requirements for specialization must continue to upgrade their training with periodic courses in specialty areas, where rapid advances are continuous. In public health, staff may take summer courses in epidemiology at schools of public health or departments of clinical medicine. Many medical, nursing, and other professional organizations require proof of continuing education for continued licensure and for professional advancement.

Recent advances in genetics, biotechnologies, patient safety, and many other important findings and discoveries require continuing education for all health workers. Science is changing our health paradigms and professional education needs to provide continuous updating in new knowledge and professional skills.

Governments, educational authorities, professional associations, provider organizations, nongovernmental health agencies, and the general public all have strong interests in continuing education for the health professions. In-service and continuing education should be part of the working schedule of a health institution and included in budgetary planning for all levels of health personnel, from laundry room staff to hospital managers and from community health workers to medical officers of health.

ACCREDITATION OF MEDICAL EDUCATIONAL OR TRAINING FACILITIES

All facilities training health professionals should be accredited to do so by the national or provincial authority or by an agency recognized by them for this purpose. In Canada, accreditation is carried out by the Medical Council of Canada, which is also the examining body for graduates of all medical schools. Provincial licensing bodies accept the License of the Medical Council of Canada (LMCC) as the basic requirement for licensure. In the United States, the Association of American Medical Colleges provides guidelines and accreditation of existing schools and reviews applications for new schools wishing to be recognized. Medical schools are subject to the state educational boards governing higher-education facilities. State boards are responsible for examination of graduates and their licensure.

Universities or colleges establishing schools for other health disciplines are subject to the requirements of the authorities governing postsecondary education. A university wishing to establish a medical, dental, nursing, pharmacy, or other professional school would need prior approval showing the need for the facility, financial resources, and a complete proposal including curriculum, staffing, facilities, organizational affiliations, and objectives. Recruitment standards and policies, clinical affiliations, quality of library and basic sciences facilities, and budget would be scrutinized.

Staff qualifications, tenure procedures and requirements, publications and research, access to international professional literature, availability of textbooks, and students' ability to read them (i.e., in a foreign language) should be part of the accreditation process.

Many new medical schools have followed patterns set at such schools as McMaster University in Hamilton, Ontario, Canada, and Ben Gurion University in Beersheva, Israel, with a focus on preparing primary care physicians, but it is not clear to what extent they have succeeded in this objective. Curriculum review has become widespread in schools of medicine with concern that there may be an excessive emphasis on basic sciences and specialty clinical services so that the graduate has little orientation to family and community practice or public health.

THE RANGE OF HEALTH DISCIPLINES

New professions such as the nurse practitioner are developed from graduates of degree programs and require a master's level of training in an accredited program. Establishing or recognizing new health professional roles, such as nurse practitioners, optometrists, or community health workers, is dependent on and related to the needs of the health system. Development of curricula, criteria for enrollment, and site of the training program should be governed by the objectives of the program, but also should ensure wide acceptance of the new profession and potential for career advancement. Acceptance by the community is important, especially in programs intended to improve services for persons in high-risk groups with the health and social services systems. Traditional birth attendants and community health workers are categories of personnel providing health care where cultural adaptation is especially important.

Clinical medicine has evolved from primarily a medical and nursing service to involve a highly complex team of professionals. Similarly, in public health the range of professions is broad. Interdisciplinary training is important for adequate functioning of a department or service increasingly dependent on teamwork.

The complexity of modern public health and clinical services is shown in the number of different professions listed in Table 14.7. This broad range of professions in public health requires graduate studies with an interdisciplinary approach to preparation of leaders, teachers, and researchers for the field. Public health professionals work within a variety of settings. They need a wide base of training in order to understand the broad professional aspects of public health that relate to complex and rapidly changing professions and practices.

LICENSURE AND SUPERVISION

All countries have legal or regulatory systems by which newly trained health personnel are permitted to practice their profession. Requirements differ from country to country and for various types of personnel within a country. In some countries, health personnel must pass licensing examinations in addition to completing the prescribed training. In others, registration by the government is more or less automatic after the prescribed training, including the examinations, has been successfully completed. For some disciplines, such as medicine, dentistry, nursing, or pharmacy, the legal requirements for the license may be delegated to professional colleges or to state or national boards. Certification and relicensing of medical and other health care practitioners have become standard practice in the United States and some other jurisdictions to assure the health care provider meets accepted professional standards of the day and public expectations.

Examination of undergraduate students is generally by the teaching institution itself, but examination at completion of training for licensing to practice medicine should be by external examination, preferably at a national or even international level. National examinations, formulated and supervised by professional and governmental authorities, establish and maintain the standards of medical graduates. In the United States, state boards govern medical licensure and specialty certification.

Licensing of health professions in some countries such as Canada allows the health professions self-government to set standards and govern the discipline within the profession as a form of peer review. National examinations and limitations of foreign graduates are spelled out in regulation or by decisions of the governing body of the profession. Foreign schools may be accepted for equivalent status or examination requirements may be established. As many as 30 percent of doctors working for the National Health System (NHS) of the United Kingdom of Great Britain and Northern Ireland (U.K.) have obtained their primary qualifications from a country outside the European Union.

In Canada, the Medical Council of Canada, a consortium of provincial professional bodies, establishes and supervises medical graduation examinations, while licensure for medical practice is by a provincially authorized medical body. Other countries regulate medical licensure directly but delegate specialty training supervision to professional organizations. Many countries have developed national examinations for medical, dental, nursing, pharmacy, and other professional licensure to promote high-level requirements and avoid the conflict of interests of a school examining its own graduates.

Graduates of Canadian schools of nursing are accepted for licensure by some states in the United States. Medical graduates are licensed by state boards in the United States. Graduates of U.S. medical schools in one state are accepted in other states for postgraduate training but not necessarily for medical practice, although some states have agreements of reciprocity. Canadian provinces accepted graduates of British medical schools, but this

TABLE 14.7 Major Types of Health Professions in Public Health and Clinical Services

Public health professions	Clinical health professions
Public health management	Health facility administrators — accounts, materials management, human resources, admitting, dietary, purchasing, planning, legal, public relations, secretary, volunteers
Health policy analyst	
Medical officer of health	
Epidemiologist	Physician — general and specialty
Dental public health officer	Dentist
Veterinary public health officer	Nurse — administration, general, and specialists
Industrial health physician	Midwife
School health officer	Pharmacist
Health economist	Physical therapist
Medical sociologist	Physician assistant
Medical anthropologist	Psychologist
Legal officer	Communication — speech therapist, audiologist
Information scientist	Occupational therapist
Demographer	Speech therapist
Statistician, biometrician	Respiratory therapist
Health service researcher	Social worker — medical, psychiatric
Supervisor of midwives	Laboratory personnel — biochemistry, microbiology, genetic, and others
Health educator — community, public, school and industrial	Genetic counselor
	Radiologic technician
Nutritionist	Imaging technician
Computer personnel	Dental auxiliary
Public health nurse	Nutritionist
Sanitary engineer	Dietitian
Product safety engineer	Pharmacy assistant
Food technologist	Licensed nursing assistant
Biochemist, microbiologist, food scientist, toxicologist	Health record clerk
Veterinary scientist	Community health worker
Sanitarian	Vital records clerk
Community health worker	Assistant midwife
Others	Others — kitchen, laundry, cleaning staff, supply, maintenance

was restricted in the 1970s to reduce the flow of immigrant doctors. In the United Kingdom, the General Medical Council is the legislated body empowered to license local graduates and immigrant physicians.

Licensing of physicians, nurses, midwives, psychologists, optometrists, nurse practitioners, or other professionals must be based on legislation or regulation under public health law to designate the scope of permissible functions in each profession, licensing and examination procedures, as well as a code of ethics. Diffusion of power in governance of medical practice has contributed to setting high standards of practice.

Control of education and licensing by the same authority that operates the national service may tend to compromise standards. Development of multiple systems of accountability in a previously totally state-controlled system, as in Russia, will require many changes in existing practices of medical education, examinations, licensure, specialty training, and examination and discipline, as well as development of independent professional organizations, accreditation bodies (see Chapter 16), and standards of care.

CONSTRAINTS ON THE HEALTH CARE PROVIDER

Maintaining standards requires organized supervision of performance by public bodies, in written guidelines, based on accepted current standards of care. This is often based on a consensus of professional views and practices, as well as recommended guidelines of professional bodies. Caution should be taken to avoid penalizing legitimate innovations or differences of professional opinion, such as whether simple lumpectomy is sufficient care for cancer of the breast as opposed to radical mastectomy. This is part of quality assurance, discussed in Chapter 16.

Ideally, the constraints that impinge on the health care providers are the sum of training, licensing, practice, collegial relationships, and self-governance of ethical, humanitarian, and professional standards. These are under scrutiny and potential disciplinary procedure from a variety of sources, including legal responsibility and standards of care expected by the employer or institution in which the provider functions.

The provider is also under scrutiny in the eyes of the public or consumers where consumer choice is part of the system. The total effect of peer review on a continuous basis, encouraging good standards of practice in the community, helps to assure basic standards, but at the same time may promote conformity, limiting medical innovation, especially in the areas of organization of health care. Recognition of new professions can lead to conflicts of interest with self-governing professions, as has happened in the case of optometrists and nurse practitioners. Similarly, professional groups may oppose changes in health care financing and organization, both for the public good and sometimes for professional self-interests.

Application of standard medical curricula, national standard examinations, national licensing and disciplinary boards, and standards of medical practice are essential for maintaining and improving health care. External peer review is consistent with current emphasis on total quality management or continuous quality improvement. When entering medical practice, doctors seek access to hospitals where they apply for "hospital privileges" and are assessed by professional peers for experience and qualifications. Individuals are expected to be members of professional organizations in their specialty and participate in departmental staff meetings and programs of continuing education organized by professional associations, hospitals, and medical schools.

Health insurance plans monitor billing or practice patterns of physicians and investigate aberrant practice or potential fraud. This may be followed by administrative action against the offending physician or, rarely, by criminal procedure for fraud.

Monitoring of surgical procedures may point out poor practice that may lead to disciplinary procedures. Criminal conviction means suspension of license to practice. Malpractice insurance is vital to protect any physician or other health care provider against litigation and may be very costly depending on the specialty. The National Vaccine Injury Compensation Program passed by the Federal government in 1986 is designed to compensate individuals quickly, easily, and generously so as to prevent litigation, which damaged immunization in the United States. The no-fault insurance for injury protects the provider and the injured party and may be a model for a more rational system than the litigation approach.

Professional accountability is specific in each country. In the United Kingdom, the General Medical Council is empowered by the state to issue medical licenses and discipline practitioners. A Patient's Charter sets out the rights, entitlements, and standards of service the citizen may expect in health care. This, coupled with the right to change general practitioner, empowers the patient to seek redress of grievances. Complaints regarding hospital care are investigated and can be pursued through stages of investigation. Consumer satisfaction is a factor in the recent innovation of GP fund holding in the United Kingdom (see Chapter 13), where the patient may, with the GP, select among hospitals or other support services.

Health care is complex and requires a skilled and integrated team functioning with mutual trust, based on a common set of professional and ethical goals and standards. This is clear in the hospital drama as seen in popular television programs, but applies equally in the larger, real-world scale of health system organization and interaction among institutions, insurers, and public health networks. In addition to oversight by financial authorities and accreditation bodies, the scrutiny of the media, the political sector, the consumer, and the public at large is important. In short, the health provider and the health system are, and should be, under scrutiny, internally and externally.

NEW HEALTH PROFESSIONS

New professional roles emerge as the health needs of the community evolve. Public health and health management professionals as well as health care providers are all essential, tailored to the community they serve. The health system needs to assure that traditional health workers are available in numbers sufficient to provide for community and national care needs. In addition, there has been a growing realization of the need for new levels of health workers.

Mid-level health worker experience in Russia with the feldsher was important to provision of primary care in rural, underserved areas. In many developing settings, experience with community health workers has been growing, and this has also been applied in some developed countries. Nurse practitioners and physician assistants have emerged as new health professional roles in the United States to augment the medical workforce, provide health care in underserved areas, and in some cases to provide health care for targeted underserved population groups such as the elderly or diabetics.

Nurse Practitioners

In the 1960s health care providers and planners in the United States became aware of the growth of specialization in medicine and the decline of the general practitioner. The nursing profession promoted expansion of nursing roles to fill this gap. These factors and increasing costs of medical care fostered the wider role of nurses to provide medical services.

In 1965, the first program to train nurse practitioners was developed at the University of Colorado. Nurses were taught specific medical functions, not as a doctor, but as a logical extension of the nurse's traditional function of assisting patients to regain health and independence. In 1971, the federal government adopted the Nurse Training Act, which defined the role of nurse practitioners.

A nurse practitioner (NP) is a registered nurse (RN) who has advanced training and education in the medical field. NPs perform detailed physical examinations, order laboratory tests for diagnostic purposes, assess the results of such tests, write prescriptions for medications and other medical devices, make referrals, and perform other delegated medical functions within authorized limitations and medical supervision. An NP can provide care in hospitals or the community as part of a health team.

A 2004 U.S. Department of Health and Human Services National Survey of RNs showed 115,000 NPs or 3.8 percent of the estimated 3 million currently licensed. The NPs were 44 percent of all advanced practice nurses (APNs), who represent some 6.3 percent of the total RN population. Applicants and enrollees more than doubled in the United States between 1993 and 1995, and the number of graduates increased by a factor of 1.8. More recent surveys show this expansion has slowed though and NP graduation rates are currently in decline. The number of NPs may reach 10 percent of all RNs in the first decade of the twenty-first century, as primary care roles for NPs increase. These include family practice; women, adult, and school health; pediatrics; and gerontology. Specialty care tracks include neonatal and acute care, occupational therapy, psychiatric care, and others in institutional care settings.

In each state, the practice of nursing is established and regulated by nurse practice acts and common law. These acts establish educational and examination requirements, provide for licensing or regulation of individuals who have met these requirements, and define the functions of the professional nurse in general and specific terms. The criteria establish parameters within which each NP may practice. The Commission on Collegiate Nursing Education, an accreditation organization, has made great strides in standardizing NP training programs to ensure quality of the graduates.

Nurses constitute the largest single group of professionals among health personnel. The expansion of nursing roles is essential to improved quality of health care, especially in medically underserved areas. Equally important, nurse practitioners in underserved professional areas such as geriatrics and primary care have innovated programs and provided a research and theoretical basis for new directions in health care. With a growing use of clinical guidelines, the roles of the nurse practitioner may be expected to increase in the twenty-first century. To meet this goal, NP programs must work to expand and attract nurses to the profession.

Physician Assistants

In 1923, 89 percent of U.S. physicians were general practitioners. By the mid-1960s, the figure had declined to about 25 percent. Concern with the shortage of primary care physicians in the country led Eugene Stead, at Duke University in Durham, North Carolina, to develop the first physician assistant (PA) training program in 1965. PA training programs were developed for three reasons: to help alleviate a perceived shortage of primary care physicians; to compensate for geographic and specialty maldistribution of physicians; to help control escalating health care costs and to accommodate medics from the military returning to civilian life and seeking a continued role in health care.

PA programs are not part of nursing, and this has engendered conflict with nurse practitioner programs and nursing authorities. There are currently 52 accredited PA training programs in the United States. The majority of PAs in the United States work outside of public health and primary care; however, the profession is very adaptable. In addition to medical science, PAs are taught preventive health care, patient education, utilization of community health and social service agencies, and health maintenance. The main focus in basic PA training is patient care in the primary care practice setting. The first 6 to 12 months of training is devoted to preclinical studies and clinical laboratory procedures, followed by 9 to 15 months of clinical training. The curricula are reviewed regularly and modifications made in keeping with changes in the health care setting.

On completion of a PA training program, an entry-level competence examination is given by the National Commission of Certification of Physicians' Assistants. When passed, it allows PAs to append the title PA-C (physician assistant-certified) to their names. PAs must register every 2 years, documenting 100 hours of approved continuing medical education. To assure clinical competency, PAs must take a recertifying examination every 6 years.

PAs perform tasks such as history-taking, physical examination, simple diagnostic procedures, data gathering, synthesis of data for a physician, formulation of diagnoses, initiation of basic treatment, management of common acute and emergent conditions, management of stable chronic conditions, patient and family counseling, supportive functions, and prescribing privileges throughout the country. Task delegation of PAs is determined by the State Board of Medical Examiners within each state. The PA is not a substitute for the physician, nor an independent provider like a nurse practitioner. The PA is not licensed for independent practice, and the physician must assume all responsibilities and bear all the professional and legal consequences of the PA's actions.

The American Academy of Physician Assistants lists nearly 75,000 persons eligible to practice. Some two-thirds of PAs had baccalaureate degrees prior to PA training. Some one-third of respondents were employed in hospitals, another one-third in solo or group practice offices, and 9 percent in community health centers. PAs work in over 60 specialty fields, but over 50 percent are involved in general/family practice. Employment opportunities are increasing for PAs in various specialties and practice settings. A PA may be able to handle one-half to three-quarters of the clinical services provided by the supervising physician, indicating that they are productive and cost-effective in their employment settings. PA salaries are one-fourth to one-third those of physicians, so the costs and benefits of PAs are an attractive option for organized health systems, but this discipline should not be seen as an independently practicing profession. The literature on this topic reports that the quality of care offered by PAs is comparable to that given by a physician.

The PA, although initially controversial, is now a widely accepted role. The nursing profession has regarded this as a method of increasing physician incomes, an infringement of nursing roles, and a reaction on the part of physicians to the growing professionalization of nursing. The medical profession defends the PA as a way of extending the possibility of medical care to larger population groups and improving the economics of medical practice.

Feldshers

The feldsher is a unique Russian mid-level health worker. The role originated as military company-level surgeons introduced by Peter the Great in the seventeenth century. Retired army feldshers returned to rural areas not served by physicians, becoming the sole providers of rural medical care. The feldsher was adapted to the Soviet health system to provide care in rural areas still underserved by doctors, despite the increase in medical personnel.

The feldsher is trained in a course of 2–3 years following intermediate school graduation. Feldshers complement physicians in urban and rural practice, especially in small medical posts, in mobile emergency medical services, and in industrial health stations. Since the health reforms of the 1990s, the feldsher is a declining profession, which may result in serious difficulties in maintaining rural health care.

Community Health Workers

The concept of the community health worker (CHW) is not new but has found new expression in health programs in many parts of the world as part of the primary health care initiatives springing from Alma-Ata. It is an adaptation of traditional village practice of midwives and healers to modern, organized public health services. CHWs were first developed to provide care in rural areas in developing countries without access to health care. More recently, there has been an interest in the CHW model for urban community health needs where access to health is limited by geographic or socioeconomic reasons (Box 14.5).

Box 14.5 Community Health Worker (CHW) Program Models

Independent CHWs
1. Feldsher in Russia for rural health care;
2. Barefoot doctors in China for rural heath care;
3. "Where There Is No Doctor" CHWs providing all health care in remote villages in Latin America.

Categorical CHWs
1. Program-specific CHWs; e.g., malaria control in Colombia, Guinea worm disease and onchocerciasis in Africa; AIDS or TB care in New York City; immunization in Los Angeles.
2. Public health nurse extender CHWs; e.g., Albany County Health Department, New York State.

Preventive-Oriented CHWs
1. Preventive village health workers providing on-site services as part of public health system, visiting medical-nursing services and close supervision; e.g., Hebron, West Bank, Palestinian Authority.
2. "Urban villages"; i.e., urban poverty area CHWs in United States, with CHWs as part of county health department or community-based organization services such as Housing Authority of City of Los Angeles.

Mid-level CHWs include generalists in independent practice, categorical or targeted health workers, and preventive community health workers. The exact functions of these personnel, the duration of training, and the framework within which they work have varied much more than their titles. The generalist village health worker programs in some areas lack close supervision, and may seek fee-for-service practice. Community health workers, advisors, or *promatores* are commonly used in Latin American countries, often as volunteers. CHWs are recommended for rural locations in developing countries without supervisory or organized contact with professional health services, providing a wide range of diagnostic and treatment services.

Community health workers may provide services on categorical target diseases. These include malaria control, tuberculosis, directly observed therapy—short course (DOTS), providing medication under supervision to assure compliance, support services and counseling for multiproblem families in an inner-city poverty area, STI follow-up, and promotion of immunization. Prototypes of the task-oriented CHW include malaria control CHWs in Colombia and tuberculosis DOTS and AIDS case workers, TB case workers, and public health nurse assistants in New York State. In Africa, CHWs are crucial to programs for eradication of guinea worm disease and river blindness.

A CHW program with a focus on preventive services was developed in and since 1985 in Hebron, in the West Bank, first under Israeli and later under Palestinian jurisdiction. CHWs act as preventive care workers as persons of contact representing the government health service, providing on-site prenatal and child care, immunization, nutritional counseling, pregnancy care with a medical-nurse support team, first aid, and emergency or non-urgent referrals to the district hospital or to nearby medical clinics. The CHW is trained and supervised to provide primary care and outreach services in a community as part of an organized health system. Visits by supervisory professional staff are vital to the function of this system of care. The emphasis may be on preventive and community services in small villages without on-site services, or as part of an outreach and support service in a large urban setting.

Community health workers should be recruited from the community to be served and trained in settings with both didactic and field experience. Training of village CHWs may be very different from schools for training conventional health personnel. Training usually takes place in rural health centers or hospitals, to which classrooms and student living quarters have been added. The training is nearly always sponsored by the ministry of health, sometimes in collaboration with foreign agencies or international organizations. Candidates for such training are typically young people from rural families, who have been selected by their communities. During training, the student should be salaried, and on completion of training return to work in the community of origin.

In the United States, public health nurses have been traditional providers of home visiting programs or outreach programs to provide health care and education for families in need. The emergence of community-oriented primary health care broadens this strategy, using selected and trained community residents as CHWs. The CHW concept was used to carry out many Great Society programs in the United States in the 1960s and 1970s. Volunteer and paid workers worked as lay home visitors or health guides in programs in selected areas or target populations, such as pregnant women or parenting families. Other CHW programs in the United States included Navajo communities, urban health centers, rural Texas, and Alaska.

Community health workers have been trained to provide outreach and case management in the complex environment of New York State, going into the community to assist high-risk families in underserved areas of large cities and in AIDS, STI, and tuberculosis patient care, especially with DOT for TB. CHW experiences working with the homeless and mental health patients have shown positive results. More recently, CHW projects have been developed in urban poverty "villages" in Los Angeles, sites of low income, public housing projects, with backup services of on-site clinics and sponsoring medical centers.

Evaluation and cost–benefit justification of CHW programs are difficult in terms of establishing population denominators and control groups, and in determining changes in outcome measures in mortality, morbidity, and physiological indicators, such as growth patterns. This limitation is shared with many health programs, not only in primary care but perhaps even more so in highly technological medicine. The village health worker concept has had criticism as well as advocacy in recent years. The CHW as an all-purpose health provider of both preventive and treatment services may be impossible to sustain and is undesirable conceptually. The more selective approach may be more feasible and manageable, with the CHW showing the promising potential to provide a new parameter in health care.

ALTERNATIVE MEDICINE

Use of alternative medical treatment has become a widespread phenomenon in industrial countries and remains a mainstay of health care in preindustrial societies. Alternative medical care is based on the belief that illness includes physical, mental, social, and spiritual factors. Alternative medicine views health as a positive state, rather than as the absence of disease, and believes in the natural healing capacity of the human body. Its interventions are generally noninvasive and less technological than conventional medical care.

The growth in popularity of alternative medicine in part relates to a widespread disillusionment with the depersonalization and technological orientation of medical care. Other factors include medicine's failure to treat the patient as a whole person, bias toward single causes, and "magic bullet" treatments for disease. Disappointment with medical outcomes is common, with failure of cures or complications of treatment itself as iatrogenic disease. In the United States, the number of visits to providers of unconventional therapies is greater than the number of visits to primary care doctors, with high levels of out-of-pocket expenditures. Users of alternative medicine are largely in the 25–49-year age group and are among better-educated persons in upper-income categories. Alternative medicine categories include acupuncture, chiropractic, massage, commercial weight loss programs, lifestyle diets, herbal medicine, megavitamin therapy, self-help groups, energy healing, biofeedback, hypnosis, homeopathy, and folk remedies.

Changing attitudes within conventional medical care are seen in a growing tendency to accept some previously excluded professions, including optometry, chiropractic, and acupuncture, in insured benefits or even multidisciplinary health care systems as complementary or supplementary to conventional medical care. Conventional medicine is itself in a process of change with addition of many paramedical professions and awareness of limitations of the biomedical model as the sole basis for health care.

Consumer demand plays an important role in this change, as does willingness of medical practitioners to refer chronic health problems such as back and neck pain, stress and related problems, phobias and addictions, allergy and skin disorders, and hormonal and menstrual disorders. Conversely, alternative practitioners seem to be increasingly aware of limitations in treatment of conditions such as cancer, hypertension, and chronic and hereditary disorders, and the need for referral for treatment by conventional medical methods. The nursing profession in the United States has promoted use of touch therapy as an independent healing modality over the past several decades.

Absorption of "holistic methods" by medical practitioners has become widespread. Many major medical centers now include departments of alternative or holistic medicine, with acupuncture, hypnosis, and other nontraditional approaches in the roster of services. This increases legitimization, status, and incomes of practitioners, but is only achieved over long periods of conflict and opposition by orthodox medical practice. In 1992, the U.S. Congress mandated establishment of an office of Alternative Medicine within the National Institutes of Health, which in 1998 was expanded to become the National Center for Complementary and Alternative Medicine. This shows the growing acceptance of a now major element of community health. European medicine and health agencies, with long traditions of healing baths and rest cures, have been more open to complementary and alternative medicine than their North American counterparts.

CHANGING THE BALANCE

As cost containment becomes more important and many countries cut back on health expenditures in order to control rates of increase, the stress of readjustment falls on health workers. If such measures are implemented on an emergency basis rather than over time as part of transition and realignment, the process will generate hostility, defensiveness, and political opposition. If, however, long-term planning takes into account the issue of human resources in a changing balance of services, then the burden of the individual or group of health workers who suffer from the downsizing can be minimized.

During the 1960s, Canada was embarking on its national health insurance program, and leading thinkers of the day called for rapid expansion of medical and nursing schools to meet future needs of the population. They assumed that national health insurance would bring a significant portion of the population who lacked access to care to the health services, overloading the medical and hospital services.

In the mid-1990s, all provinces cut back on the hospital bed supply, creating unemployment for nurses and maintenance staff. Provinces are attempting to restrict the numbers of practicing doctors and to modify the payment systems away from fee-for-service. Inflation in health care utilization and costs has resulted in a crisis management approach, rather than structural reform to promote a more integrated population-oriented health care organization. This can lead to a great deal of public and professional dissatisfaction within a health care system. Public education is vital to raise consciousness and participation rates in preventive care such as influenza vaccination, screening for colon and cervical cancer, and many other aspects of health promotion and disease prevention.

EDUCATION FOR PUBLIC HEALTH AND HEALTH MANAGEMENT

Successful implementation of the New Public Health requires that many health disciplines work together. The training milieu should have a capacity for interdisciplinary training in a comprehensive program, including fundamental and applied research, as well as a relationship with service programs and community health assessments. To facilitate teamwork, all public health practitioners need a background of the medical sciences, epidemiology, economics, social sciences, environmental and occupational health, health systems analysis, and management theory. It is important that they be familiar with the terms and concepts of fields other than their own specialized discipline. This is more likely to be found or created in a university atmosphere and is difficult to foster in separate, categorical institutes.

University resources are essential for a school of public health to provide teachers and courses from other faculties such as schools of business administration and the social, physical, and biological sciences. Schools that are unaffiliated with a degree-granting university lack the broad academic atmosphere and requirements, as well as the connection with parent disciplines, such as economics, sociology, microbiology, and business management faculties.

Preparation of personnel for public health and health management should be at the graduate school level followed by continuing education. The U.S. Institute of Medicine's 1988 report *The Future of Public Health* defined a need for schools of public health to teach not only professional and technical skills, but also an understanding of how a particular discipline relates to public health as a whole, and the value system that is part of public health's coherence.

Training of public health physicians began in the United Kingdom in 1871 in Dublin's Trinity College, which granted the Diploma in Public Health (DPH). The program was designed to provide for the training of medical officers of health to lead the work of the boards of health that were established under public health acts and were required to have qualified medical practitioners as medical officers of health. Other universities later offered DPH or equivalent training, supervised by the College of Physicians and Surgeons, as public health became a recognized medical specialty. In 1924, the London School of Hygiene and Tropical Medicine brought together several institutes and produced a major center of training and research in public health. Since the 1991 Acheson Report on schools of public health in the United Kingdom, there has been a rapid growth of schools of public health under various names in a number of universities. This has coincided with growth of interest in the NHS in working toward health targets as opposed to simply managing health services.

The tradition of schools of public health is especially strong in the United States, where the Johns Hopkins and Harvard Schools of Public Health were founded in 1913. In 1915, the Rockefeller Foundation sponsored a national program to promote public health education at the University of Michigan, Yale University, and the University of Pennsylvania, which established graduate training for public health professionals to meet the needs of crowded urban industrial cities of the United States. These schools saw their mission as the training of public health practitioners, and secondly academics, educators, and researchers. They attempted to develop and assimilate new knowledge into public health practice. The development of public health as a multidisciplinary field made it vital to be independent from but affiliated with a medical faculty. Schools of public health in the United States produced many generations of well-trained epidemiologists, social scientists, health educators, practitioners, and leaders who were crucial for development of the field.

In the 1960s, schools of public health were criticized because of their separateness from and lack of influence on clinical medical training. In some jurisdictions, this led to closure and replacement by departments of community medicine within faculties of medicine. This occurred in the United Kingdom and in British Commonwealth countries. Canada's two schools of public health were closed, replaced by departments of social and preventive medicine or community health, developed within medical schools. Departments of community health within a medical faculty serve as only one department among many clinical or basic science departments. Such departments generally may lack prestige in the hierarchy of medical schools, in an environment promoting a narrow, medically oriented approach to public health, with an insufficiently multidisciplinary program and faculty. This model provides training at the undergraduate, master of public health (MPH), and doctoral levels. The full academic potential of a graduate school of public health is most suited to be in an independent, multidisciplinary, university-based academic center for public health research and training.

Periodically, threats of absorption of schools of public health into other sectors of the university arise. In 1994, a plan to close the University of California, Los Angeles (UCLA), School of Public Health by transfer to the School of Public Policy was halted by university and nationwide protests. Health administration is sometimes considered as better integrated within schools of business, as happened at the University of Minnesota School of Public Health.

Since the 1980s, there has been a renaissance of schools of public health in the United States with an expanding market for graduates. The Association of Schools of Public Health (ASPH) represents the 37 accredited U.S. accredited schools and many with associate member programs or schools not yet accredited through a formal review process of the Council on Education of Public Health (CEPH). Many other programs also provide postgraduate education in public health. A well-rounded graduate education program is based upon five core public health and cross-disciplinary studies (see Box 14.6).

Today, students in schools of public health come from many different backgrounds, including medicine, dentistry, nursing, engineering, economics, social sciences, statistics, mental health, veterinary sciences, and others. In 2008, ASPH member schools of public health in the United States, Puerto Rico, and Mexico have a combined 4000 faculty, 19,000 students, and 7000 graduates per year. The most popular specialties were international health and epidemiology. The career outlook for graduates of schools of public health now includes other traditional public sector positions, but also higher-paying positions in the private sector, including consulting firms and managed care organizations. In the 1996–1997 academic year, 14,007 students were enrolled in schools of public health in the United States.

In the United States, graduate schools of public health accredited by the Council on Education for Public Health (CEPH) have grown from 28 in 1999 to 40 in 2008, with an additional 70 recognized master's of public health (MPH) programs. They are all based on a multidisciplinary approach to training for public health, including technical and administrative leadership in epidemiology of communicable and noninfectious diseases, biostatistics, management of personal health services, environmental health, maternal and child health, health economics, health education, and other related fields (Box 14.7). The interdisciplinary aspect of public health is emphasized by the wide range of backgrounds and experiences of the students and their many areas of specialization. Analysis of problems with the skills of epidemiology, sociology, and other related disciplines permits the graduate to enter practice with a problem-solving approach. Another 26 university graduate programs in community health education, community health, and preventive medicine are accredited to extend public health training to most parts of the country.

In European countries, there is a growing trend to public health education at the master's, PhD, and increasingly

Box 14.6 Elements of Public Health Training

Objectives

1. Prepare target-oriented practitioners, researchers, policy analysts, and managers;
2. Provide continuing education for public health practitioners;
3. Promote public health research and policy analysis;
4. Advocate and promote health-related issues in public policy.

Requirements

1. Postgraduate training in a multi-faculty academic setting;
2. Multidisciplinary approach;
3. Problem-oriented skills training to identify targets and problem-solving management approaches;
4. Link education, research, and service in public health.

Core curriculum

1. Basic tools of social analysis: history of public health, demography, medical sociology and anthropology, biostatistics, population sampling and survey methods, political science of health systems, principles of program evaluation and health economics;
2. Health and disease in populations: vital statistics, major human diseases and zoonoses, epidemiology of diseases

and risk factors, methods of clinical diagnosis and prevention, infectious and chronic diseases, nutrition, environment, special disease and risk groups, global ecology of disease and risk factors;

3. Promotion of health and prevention of disease: communicable disease control, chronic diseases prevention, environmental and occupational health, maternal, child, adolescent, adult, and elderly health, mental health, STI/AIDS control, nutritional and dental health, health education and promotion; rehabilitation, refugee, migrant, and prisoner health, military medicine, and disaster planning;

4. Health care systems and their management: organization and operation of national health care systems, health insurance and social security, health services and workforce development, health facilities and their management, drugs and their logistics, health planning, principles of management and application to health programs, budgeting, cost control and financial management, record and information systems, health systems research, health legislation and ethics, technology assessment, accreditation and quality promotion in health care, information systems, monitoring, and research methods for management, global health.

Sources: American Association of Schools of Public Health, http://www.asph.org [accessed October 16, 2007]
Roemer, M. 1999. Genuine professional doctor of public health the world needs. *Journal of Nursing Scholarship*, 31:43–44.

Box 14.7 Core and Cross Disciplines in Public Health Education

Core disciplines

- Biostatistics
- Environmental health sciences
- Epidemiology
- Health service administration
- Behavioral science and health education

Cross disciplines

- Communication and informatics
- Diversity and culture
- Professionalism
- Program planning
- Public health biology
- Systems thinking
- Ethics
- Law

Sources: Association of Schools of Public Health, 2004, www.asph.org [accessed October 14, 2007]
ASPH Committee, 2006.

also at the bachelor's levels. The European Association of Schools of Public Health (ASPHER) represents this movement and is promoting peer review, mission and values statement, competency standards, and an accreditation

program. This is a new movement and is beginning to work in alliance with the European Public Health Association (EUPHA) to promote public health policies, research and educational standards. Similar organizations are working in other regions such as in Southeast Asia where many new schools of public health are developing, e.g., in India.

In former Soviet countries, public health training is a stream within medical training institutes or academies at the undergraduate level. These are the *Sanepid* doctors who provide communicable disease control, food and environmental hygiene, and are the main cadre of public health practitioners in the Semashko system (see Chapter 13). Postgraduate training is provided as a medical specialty through *Ordinatura, Aspirantura,* and *Candidate* levels in the medical academies. Research institutes in various fields of public health provide graduate-level training up to the doctoral (PhD and Doctor of Science) levels.

Most former Soviet countries provided specialization during basic medical training (i.e., internal medicine, pediatrics, or sanitary epidemiology). Specialized graduate training takes place in various departments of medical academies, with criteria including length of training and examinations established by the Ministry of Health. Newly established graduate schools or programs in public health are providing training at the master's of public health level in many countries of Eastern Europe (CEE), the former Soviet Union and Central Asia, including

Hungary, Poland, Romania, Macedonia, Bulgaria, Moldova, Albania, Czech Republic, Ukraine, and in the Central Asian Republics of Kazakhstan, Uzbekistan, and Tajikistan, with five developing in Russia. The master's of public health degree is not yet recognized in all these countries, but is in keeping with the Bologna Agreement (see Box 14.9) on postgraduate education in Europe as agreed to by most of these countries.

Europe recognized a need to modernize its university education standards (the Bologna Declaration and Process). It came at a time when the health workforce was becoming an increasingly urgent issue recognized by the World Health Organization and many individual countries as crucial to health policy. Global threats to population health, such as HIV, SARS, threatened bioterrorism, and the avian flu pandemic, have increased public and political recognition of the vital importance of a trained cadre of public health experts (Box 14.8).

In many countries in transition from socialist systems, the model of public health continues to be one largely focused on infectious disease and is hospital-oriented, although there is advent of compulsory health insurance and elements of privatization of services in some countries. Noninfectious disease issues are seen as clinical problems and left to medical practitioners to resolve; health promotion remains a vague concept. As a result, the populations of countries in transition suffer the tragic consequences of high rates of preventable morbidity and mortality from chronic diseases.

The importance of newer training models for the public health workforce is one of the most urgent keys to successful, well-functioning, and well-managed health systems. This is reinforced and evidenced by the events and outcomes of newly developing schools of public health in countries of Eastern Europe and Central Asia as well as the new schools now in development in India and Southeast Asia.

As mentioned, WHO addressed the "global crisis in health systems and health workforce needs" by choosing Human Resources for Health as the theme of the World Health Report — Working Together for Health (WHR) 2006. The major stress was placed on the shortage of medical and nursing and other health providers, and the issues of their training, conditions of work, and migration. There is also recognition that preparing a competent public health workforce is a key element for effective and sustainable health systems. There is no clear agreement on what the public health workforce is. Definitions for identifying the specific members of the public health workforce, their corresponding roles within the health systems as classifications and roles vary across countries.

WHR 2006 states in regard to education and training of the workforce that education of the health workforce requires attention to curricular content, pedagogical learning methods, training of teaching staff, research and service, and moreover, that "more schools of public health are needed."

Public health workforce development (PHWD) is a crucial element in increasing capacity of national health systems, allowing them to address present and future population health challenges. The development of advanced-level programs of post-diploma public health education in the central and eastern European region is an important innovation to help countries cope with public health crises of low performance of their health systems.

In Europe there has been a burgeoning of schools of public health since the 1990s and increasingly this is also occurring in countries of Eastern Europe and Central Asia. Associations of schools of public health have promoted these new schools with help from the Open Society Institute (Soros Foundation) of New York. The economic and political consolidation of Europe has highlighted the need for uniformity, standardization, and reciprocity in the European Union's systems of higher education. Following a series of meetings, a June 1999 meeting of European ministers of education issued what has come to be known as the Bologna Declaration (Box 14.9). Signatory states agreed in principle to work toward a European Higher Education Area (EHEA), to improve quality of collegiate education, and to provide for enhanced movement of students and academics within the EU.

Box 14.8 Competencies for Providing Essential Public Health Services

- Monitor health status to identify community health problems;
- Diagnose and investigate health problems and health hazards in the community;
- Inform, educate, and empower people about health issues;
- Mobilize community partnerships to identify and solve health problems;
- Develop policies and plans that support individual and community health efforts;
- Enforce laws and regulations that protect health and ensure safety;
- Link people to needed personal health services and assure the provision of health care when otherwise unavailable;
- Assure a competent public health and personal health care workforce;
- Evaluate effectiveness, accessibility, and quality of personal and population-based health services;
- Research for new insights and innovative solutions to health problems.

Source: Modified from http://www.trainingfinder.org/competencies/list_nolevels.htm (accessed October 23, 2007)

Box 14.9 Bologna Declaration

- Target date for 2010;
- Work to increase international competitiveness of European system of higher education;
- Agree to work together toward uniformity within EU universities;
- Adopt a system of comparable degrees to promote European citizens in academia;
- Adopt a system of two main university cycles, undergraduate and graduate degrees at master's or doctoral level;
- To establish a system of easily transferable academic credit between institutions and countries;
- Promote widespread student mobility, improve access for students and training opportunities;
- Promote European cooperation in quality assurance working toward compatibility;
- Promote European dimensions of higher education;
- Move Europe toward comparable degrees and cooperation in quality assurance.

Source: Declaration 19 June 1999 by European Ministers of Education in Bologna, Italy, http://ec.europa.eu/education/policies/educ/bologna/bologna.pdf [accessed October 14, 2007]

The implications for public health education are great. Creation of new schools of public health has been initiated in several countries, particularly in eastern and southeastern Europe, where the need for trained health workers is perhaps greatest. The Bologna Declaration allows for widespread recognition of the master's of public health degree and clarification of the public health worker's role as a professional. There are similar consequences for nursing, medicine, and allied health professions, which will move toward common nomenclature, educational standards, and academic exchange.

Accreditation involves external review of facilities, faculty, curriculum, student selection criteria, internships or field experience, and academic standards. An accredited school is better able to generate research and scholarship funds and is more attractive to students for future career advancement in an expanding job market. These schools, plus many other nonaccredited schools or university departments, provide operating public health and health care agencies with well-trained personnel with a wide range of undergraduate and professional experience. This enriches the field not only with practitioners, but also with researchers, administrators, and policy analysts of high quality.

Schools of public health should have close working relations with state and local health agencies, recruiting part-time faculty from service agencies and conducting research in real public health problems that confront health agencies. Schools of public health can provide important services to departments of health in research, consultation and assessment on public health issues.

Graduate schools of public health are ranked annually by surveys of deans, top administrators, and senior faculty conducted as part of continuous surveys of graduate schools in the United States by the weekly news magazine *U.S. News and World Report*. These surveys are based on annual assessments, including reputation among deans and senior faculty (40 percent), research activity (30 percent), student selectivity (20 percent), and faculty resources (10 percent). In 2007, the top ranked schools of public health included Johns Hopkins, Harvard, and the University of North Carolina (Chapel Hill). While these surveys and rankings have no official status, they are widely used as guides for student selection of graduate schools, possibly affecting research grants, fundraising, and faculty recruitment. The Council on Education for Public Health (CEPH) provides a recognized accreditation system with requirements for standardized public health curricula and core content, and the process of accreditation ensures quality graduate education across the country.

Field epidemiology is an essential component of effective public health practice, and developing such capacity is a critical step in a country's efforts to improve the health of its citizens. Since 1975, a total of 28 field epidemiology training programs (FETPs) have been established worldwide. Most FETPs have resulted from partnerships among CDC, host country health agencies, the World Health Organization, the U.S. Agency for International Development, and others. Modeled on CDC's Epidemic Intelligence Service (EIS), FETPs follow the EIS approach of combining service with training. FETPs also participate in training programs in the epidemiology and public health interventions network, which provides a venue for information sharing, program development, and quality improvement. In 2003, EIS and FETPs graduated approximately 250 field epidemiologists.

In continental Europe, training in public health was traditionally through a job-oriented, vocational approach carried out in government or independent institutes as courses for medical officers of health or hospital managers. Since the 1980s or early 1990s, European schools of public health of the broader model have been established, in Germany, Holland, France, Spain, Poland, and Romania (Table 14.8). The Association of Schools of Public Health of the European Region has developed and promotes the idea of standardization and reciprocity for master's of public health degrees in the European Union.

ASPHER has over 70 institutional members located throughout the Member States of the European Union (EU), Council of Europe (CE), and the European Region of the World Health Organization (WHIO). ASPHER's values, vision, mission, and aims address their role: ASPHER represents the academic component of development of the education, training, research, and service of

---

TABLE 14.8 Principles of Public Health Education Standardization for Europe

1. Public health as an organized system is vital for societies for long-term improvement in health of the population and to meet emergency situations of pandemics, disasters, and bioterrorism;
2. Public health depends on laws, organizational structure, resources, planning, and training;
3. Many countries in Europe lack clear differentiation of powers between different levels of government;
4. Trained workforce is crucial to an effective public health network;
5. Training, licensure, accreditation are all needed in Europe in the western and eastern countries, especially in their transition stages;
6. Development of schools of public health as multidisciplinary settings with autonomy as separate faculties of semi-autonomous status within medical faculties is vital for them to succeed in their mission of training, research, and service;
7. WHO must take a leadership role in promoting public health reform and education, networking with organizations already active in this field such as ASPHER, EUPHA, and others;
8. Ministries of Health should be proactive in developing robust structures and training programs for public health in conjunction with sister ministries and academic organizations;
9. Europe-wide standards and health targets will help individual countries cope with these challenges;
10. Europe-wide funding to promote new organization and training capacities in member countries is essential to enable this process to happen and to implement the intent of the Bologna Declaration in higher education.

Source: Dubois, C. A., McKee, M., Nolte, E. 2005. *Resources for Health in Europe*. Buckingham: Open University Press.

---

the public health workforce of Europe for the future. This includes the implementation of the Bologna Declaration, promoting equality, not only in the context of health standards, but also in education standards, in recognition of qualifications, accreditation systems, and in improving the health of the peoples of Europe. Many European countries are still in transition or development with currently poor standards of public health. Training in professional aspects of public health is vital to the educational goals of participating members, but also to the public health movement as a whole, within basic principles of ethics and values for the modern public health movement, and of the highest standards of public health practice.

Schools of public health are of special importance for developing countries because of the prime importance of public health approaches in meeting their health needs. Nigeria is a very populous country where the main health issues are those in the public health sector, yet there are over 20 medical schools and no school of public health. Departments of social or community medicine, primarily teaching medical students, are common in most medical schools, but this fails to provide students at the graduate school level with an academic environment and multidisciplinary training and specialization that the field requires to provide the professional leadership needed to meet the health challenges of their societies.

Leadership positions in health systems at the local, state, and federal levels are now held by doctors trained in clinical medicine, often with added training in public health. Few, however, have training in management and many with training focused in subspecialties of public health. On the other hand, schools of public health at the doctoral level focus on preparation of scholars in research and teaching rather than health leadership and management. Some seek preparation in master's of business administration (MBA) programs. Preparation at the PhD level of doctors in public health requires post-baccalaureate training with broad areas of knowledge: tools of social analysis, health and disease in populations, promotion of health and prevention of disease, and health care systems and their management (see Box 14.8). Some schools of public health are moving in this direction by providing special part-time programs for working health executives. This will be especially important in the preparation of leadership capable of coping with the complexities of managed care or district health programs developing in many countries facing the organizational, economic, and ethical aspects of individual and population health.

The impact of public health and preventive medicine on national health has gained prominence since the 1980s, with the idea that medical need can be reduced by decreasing the burden of illness. This led to an increase in demand for preventive medicine and public health training, as medical care costs were thought by economists to be a function of need and demand. Health education programs designed to reduce health risks and reduce costs were shown to have documented effectiveness, with claims reductions of 20 percent in some health insurance systems in the United States. Specific program features including chronic disease self-management, risk reduction, and increased self-efficacy appear important. This concept was further supported by the review by CDC of achievements of public health in the twentieth century, in which reductions in the burden of disease were directly related in large measure to public health programs. The HIV, SARS, bioterrorism, and natural pandemics heightened the sense of concern, need, and urgency to strengthen public health training in the United States and internationally.

Countries of Eastern Europe and the Commonwealth of Independent States are facing a combination of high rates of mortality from preventable diseases and pressures for reform of the health care systems. Development of schools of public health as independent schools within single or multi-facility universities should be an important priority for international aid and for national authorities.

This is a challenge of integrating experience from many countries in the industrialized world with local academic centers in the field of public health. Traditional departments of social hygiene within medical academies need to evolve to educate new generations of doctors and other health professionals to cope with challenges facing the health systems in these countries. Development of postgraduate centers of training is essential to provide the leadership and professional staffing to address the New Public Health.

For developing countries — such as India and Nigeria, with vast populations and poor health, and Asia — the need for schools of public health is even more urgent. Achieving the Millennium Development Goals in many countries will not be possible without developing and sustaining a strong cadre of well-trained public health analysts, leaders, and field workers. This will require academic centers capable of training, research and service to prepare such public health workers and to advocate policies and priorities to achieve these targets.

The essential competencies for public health are those outlined by the American Public Health Association and discussed in Chapter 10. These are skills acquired through training and experience and not part of the skills of physicians per se. They include what the APHA calls the essential public health services.

Essential Service #1: Monitor health status to identify community health problems

Essential Service #2: Diagnose and investigate health problems and health hazards in the community

Essential Service #3: Inform, educate, and empower people about health issues

Essential Service #4: Mobilize community partnerships to identify and solve health problems

Essential Service #5: Develop policies and plans that support individual and community health efforts

Essential Service #6: Enforce laws and regulations that protect health and ensure safety

Essential Service #7: Link people to needed personal health services and assure the provision of health care when otherwise needed

Essential Service #8: Assure a competent public health and personal health care workforce

Essential Service #9: Evaluate effectiveness, accessibility, and quality of personal and population-based health services

Essential Service #10: Research for new insights and innovative solutions to health problems

These are the elements essential for public health practice and a training curriculum to address those needs.

A professional, qualified, and multidisciplinary workforce, in sufficient numbers, is vital to the organization and management of effective public health systems in Europe and around the world. Such a workforce is essential to evaluate and respond to growing threats to population health, to address health inequalities between and within countries, and to develop and implement scientifically-based interventions in a timely and appropriate manner within the limits of available resources.

HEALTH POLICY AND MANAGEMENT OF HUMAN RESOURCES

Preparation for policy and management roles in public health and health systems has become widespread in schools of public health and in business schools in the United States. This trend will undoubtedly increase as managed care increases, and as intersectoral mergers or other functional arrangements become more common. Departments of management and policy or health services in schools of public health have the mission to study and seek methods of improving efficiency and effectiveness of personal and population-based health organizations. As academic fields, they share a population perspective that includes interdisciplinary faculty from economics, law, management, medicine, history, sociology, and policy analysis. They focus on societal, population, economic, and organizational perspectives.

Personnel and managerial costs are the largest single component of total health expenditures, so that management and utilization of this resource is of prime importance to a health system. Health personnel should be recruited, trained, and utilized in a manner appropriate to meet the health needs of the population. This means employing their skills under conditions that promote effective work. Human resource management includes determining which category of worker can best provide specific services, how many are required, and what organizational frameworks are required to provide needed care most effectively. This requires not only delegation of responsibility, but also the resources with accountability to carry out the tasks.

Human resources management includes determination of numbers and types of health personnel needed for the health system of the future. Delegation of responsibility from professional levels can be made to appropriately qualified paraprofessionals. This includes planning, management, financing constraints, licensing, discipline procedures, and quality control measures. Accountability for performance is essential for any system dedicated to provision of quality care and to meeting its goals of improved health outcomes.

Restricting numbers of professionals is sometimes done in the self-interest of a professional group to restrain competition. Oversupply can be costly and destructive to the public interest by misdirecting health resources in nonproductive or even harmful ways. An excess supply of surgeons generates higher rates of elective surgical procedures than necessary or safe, while shortages of primary care physicians prevent adequacy in basic health services. Poor supply or quality of nursing personnel compromises the quality of hospital care and primary care in the community. This requires long-term retraining and redeployment policies developed and implemented over time rather than spasmodic mass layoffs of nurses and other hospital workers.

Managed care in the United States and similar comprehensive service programs in other countries provide the opportunity to seek a new balance of services and introduce new roles in health care. Physicians, nurse practitioners, community health workers, and many other kinds of health care professionals and technicians will be part of the complex of health care provision when the economics of care necessitate cost-effectiveness, where prevention and treatment are part of the same complex, and where a health promotion approach is fundamental to the objectives of the organization. In turn, public health workers should take active roles informing and participating in health systems management.

The evolution of public health is discussed in Chapter 2 and its organization in Chapter 10. The preparation of public health leaders and professional staff is well described by the Association of Schools of Public Health. The ASPH, representing 37 graduate schools of public health, which provides approximately 85 percent of the public health graduates in the United States has identified these core master's in public health competencies for students upon graduation. Public Health, as a profession and a discipline, focuses on population and society's role in monitoring and achieving good health and quality of life. Public health professionals work in many settings to guarantee:

- Optimal human growth, development, and dignity across the life-span;
- Respect for community participation and preferences in health;
- Air, food, and water safety;
- Workplace, school, and recreation safety;
- Timely detection of disease outbreaks and public health threats;
- Science-based responses to public health problems;
- Health care access, efficiency, and effectiveness;
- Encouragement of healthy choices that prolong a high quality of life; and
- Design and maintenance of policies and services to meet community and individual needs for physical and mental health

Public health professionals also recognize the contributions of other disciplines, including but not limited to the health professions, Business, Economics, Education, Engineering, Law, Political Science, Psychology, Public Administration, and Sociology.

SUMMARY

Education and training of medical and allied health personnel are important issues in health care systems development, and include issues of both quantity and quality. Regular reassessment is needed lest the numbers of practitioners produced become larger or fewer than the needs of the services, and as a result health care standards may decline or the system may become excessively costly while needed health promotion is inadequate. Preparation of managers and planners skilled in data and program analysis and leadership is as important as training health care providers.

Training of health professionals should be accompanied by orientation to the broad sweep of the New Public Health, including its management and evaluation skills. New health professional roles will evolve based on individual patient and community health needs.

Management in the New Public Health is confronted with many difficult challenges in human resource policy. These include not only the quantity and quality of training but flexibility in utilization, including redeployment of personnel from institutional care settings to community health and health promotion activities. Personnel and management expenses constitute some 75 percent of costs of patient care; any program for reallocation of resources toward community and preventive care must necessarily involve health workers, not only as an economic issue, but as a qualitative one. These professional and personnel issues must be treated with care and sensitivity.

A health care system depends on the quality, ethics, pride, and professional skills of its team members. The training and retraining of such personnel are therefore fundamental considerations of the New Public Health.

ELECTRONIC RESOURCES

American Academy of Nurse Practitioners (AANP). http://www.aanp.org/ [accessed May 24, 2008]
American Academy of Physician Assistants. http://www.aapa.org/gandp/pro-issues.html#ama [accessed May 24, 2008]
American Association of Medical Colleges Center for Workforce Studies. http://www.aamc.org/workforce/ [accessed May 24, 2008]
American Board of Preventive Medicine. http://www.abprevmed.org/aboutus.cfm [accessed May 24, 2008]
American College of Preventive Medicine. http://www.acpm.org/ [accessed May 24, 2008]
American Medical Association (AMA). http://www.ama-assn.org/ [accessed May 24, 2008]
American Public Health Association (APHA). http://www.apha.org/ [accessed May 24, 2008]

Association of Schools of Public Health (ASPH). http://www.asph.org/ [accessed May 24, 2008]

Association of Schools of Public Health in the European Region (ASPHER). http://www.ensp.fr/aspher; http://www.aspher.org/ [accessed May 24, 2008]

Association of Schools of Public Health. Master's Degree in Public Health Core Competency Development Project, Version 2.3: October 2004–August 2006, http://www.asph.org/userfiles/version2.3.pdf

Association of Schools of Public Health of the European Region. http: www.asph.org/document.cfm?page=851[accessed July 3, 2008]

Association of Teachers of Preventive Medicine (ATPM). http://www .atpm.org/ [accessed May 24, 2008]

Core competencies, http://www.trainingfinder.org/competencies/list_no-levels.htm [accessed May 24, 2008]

Canada – Health Personnel Trends, http://www.cihi.ca/cihiweb/dispPage .jsp?cw_page=AR_21_E [accessed May 24, 2008]

Council on Education for Public Health (CEPH). http://www.ceph.org/ [accessed May 24, 2008]

International Council of Nurses (ICN). http://www.icn.ch [accessed May 24, 2008]

OECD Health Data 2007. How does the United Sates compare, http:// www.oecd.org/dataoecd/46/2/38980580.pdf [accessed May 24, 2008]

Public Health Agency of Canada, Core Competencies for Public Health in Canada. September 2007, http://www.phac-aspc.gc.ca/ccph-cesp/ pdfs/cc-manual-eng090407.pdf

Tulchinsky, T. H., Birt, C. A, Kalediene, R., Meijer, A. *ASPHER's Values, Vision, Mission, and Aims: A Working Paper*, December 2007, ASPHER Brussels Office, Belgium, office@aspher.org, Association of Schools of Public Health, http://www.aspher.org/media .pdf/aspher_mission_and_values_working_paper_15_dec_07.pdf

UNDP is the United Nations Development Programme, http://www.undp .org [accessed May 24, 2008]

United Nations Educational, Scientific and Cultural Organization (UNESCO). http://portal.unesco.org/en/ev.php-IRL_ID=3328&URL_ DO=DO_TOPIC&URL_SECTION=201.html [accessed May 24, 2008]

UNICEF is the United Nations Children's Fund, http://www.unicef.org/ [accessed May 24, 2008]

U.S. News and World Report ranking of U.S. schools of public health, http://grad-schools.usnews.rankingsandreviews.com/usnews/edu/ grad/rankings/hea/brief/pub_brief.php [accessed May 24, 2008]

U.S. Department of Health and Human Services. Health Resources and Services Administration. http://www.hrsa.gov/ [accessed May 24, 2008]

World Bank. http://www.worldbank.org/; http://www.worldbank.org/ibrd/ [accessed May 24, 2008]

RECOMMENDED READINGS

Feil, E. G., Welch, H. G., Fisher, E. S. 1993. Why estimates of physician supply and requirements disagree. *Journal of the American Medical Association*, 269:2659–2663.

Handler, A., Schieve, L. A., Ippoliti, P., Gordon, A. K., Turnock, B. J. 1994. Building bridges between schools of public health and public health practice. *American Journal of Public Health*, 84:1077–1080.

Kahn, K., Tollman, S. M. 1992. Planning professional education at schools of public health. *American Journal of Public Health*, 82:1653–1657.

Legnini, M. W. 1994. Developing leaders vs. training administrators in the health services. *American Journal of Public Health*, 84: 1569–1572.

Roemer, M. 1999. Genuine professional doctor of public health the world needs. *Journal of Nursing Scholarship*, 31:43–44.

Weiller, P., Hiatt, H., Newhouse, J., Brennan, T. A., Leape, L., Johnson, W. P. 1993. *A Measure of Malpractice*. Cambridge, MA: Harvard University Press.

World Federation for Medical Education. 2003. *Global Standards for Quality Improvement*. WFME Office, University of Copenhagen, Denmark.

BIBLIOGRAPHY

American Academy of Family Physicians. 2006. *AAFP Adopts New Physician Workforce Policy*. Press Release.

Barnard, K., Köhler, L. 1994. Creating a good learning environment — A review of issues facing schools of public health. In *Training in Public Health, Strategies to Achieve Competences*. Copenhagen: World Health Organization, Regional Office for Europe.

Beaglehole, R., Bonita, R. 1997. *Public Health at the Crossroads*. Cambridge: Cambridge University Press.

Beaglehole, R., Dal Poz, M. 2003. Public health workforce: Challenges and policy issues: Commentary. *Human Resources for Health*. 1:4.

Bender, D. E., Pitkin, K. 1987. Bridging the gap: The village health worker as the cornerstone of the primary health care model. *Social Science and Medicine*, 24:515–528.

Berman, P. A., Watkin, D. R., Burger, S. E. 1987. Community-based health workers: Head start or false step toward Health for All? *Social Science and Medicine*, 25:443–459.

Brotherton, S. E., Rockey, P. H., Etzel, S. E. 2005. U.S. Graduate Medical Education, 2004–2005: trends in primary care specialties. *Journal of the American Medical Association*, 294:1075–1082.

Buerhaus, P. I., Potter, V., Staiger, D. O., French, J., Auerbach, D. I. 2009. *The Future of the Nursing Workforce in the United States: Data, Trends and Implications*, Sudbury, MA: Jones and Bartlett. http://www.jbpub.com/covers/newlarge/0763756849.jpg [accessed May 24, 2008]

Bury, J., Gliber, M. 2001. *Quality Improvement and Accreditation of Training Programmes in Public Health*. European Association of Schools of Public Health, Paris Fondation Mérieux.

Cavallo, F. *Public Health Education and Training in Europe*. 1997. In L. Köhler, B. Barnard (eds.). 1997. *EU and Public Health: Future Effects on Policy, Teaching and Research*. Göteborg: Nordic School of Public Health.

Centers for Disease Control. 2003. Assessment of the Epidemiologic Capacity in State and Territorial Health Departments — United States. *Morbidity and Mortality Weekly Report*, 2003/52(43): 1049–1051.

Centers for Disease Control. 2003. Building epidemiology capacity. *Morbidity and Mortality Weekly Report*, 52:1037.

Chinn, P. L. 2006. The global nursing shortages and healthcare. *Advances in Nursing Science*, 29(1):1.

Committee on Educating Public Health Professionals. Gebbie, K., Rosenstock, L., Hernandez, L. M. (eds.). 2003. *Who Will Keep the Public Healthy? Educating Public Health Professionals for the 21st Century*. Washington, DC: Institute of Medicine, National Academic Press.

Council on Graduate Medical Education. 1994. *Fourth Report: Recommendations to Improve Access to Health Care through Physician Workforce Reform.* Rockville, MD: U.S. Department of Health and Human Services.

Council on Graduate Medical Education. 1999. *Tenth Report: Physician Distribution and Health Care Challenges in Rural and Inner-City Areas.* Rockville, MD: U.S. Department of Health and Human Services.

Dubois, C. A., McKee, M., Nolte, E. (eds.). 2006. *Human Resources for Health in Europe.* European Observatory on Health Systems and Policies. New York: Open University Press.

Evans, G. 2006. Update on vaccine liability in the United States: presentation at the National Vaccine Program Office Workshop on strengthening the supply of routinely recommended vaccines in the United States. *Clinical Infectious Disease,* 42 (Suppl. 3):S130–S137.

Fee, I., Acheson, R. M. (eds.). 1991. *A History of Education in Public Health.* Oxford: Oxford Medical Publications.

Foldsprang, A. 2008. Public health education in Europe and the Nordic countries: status and perspectives. *Scandinavia Journal of Public Health,* 36:113–116.

Fries, J. F., Koop, C. E., Sokolov, J., Beadle, C. E., Wright, D. 1998. Beyond health promotion: Reducing need and demand for medical care. *Health Affairs,* 17:70–84.

Fulop, T., Roemer, M. I. 1987. *Reviewing Human Resources Development: A Method of Improving National Health Systems.* Public Health Papers, 83. Geneva: World Health Organization.

General Medical Council. 1993. *Tomorrow's Doctors: Recommendations on Undergraduate Medical Education.* London: GMC.

George, J. T., Rozario, K. S., Jeffrin, A. E., Gerard, B. J., McKay, A. 2007. Non-European Union doctors in the National Health Service: Why, when and how do they come to the United Kingdom of Great Britain and Northern Ireland? *Human Resources for Health,* 5:6.

Gevena-Kindig, D. A., Cultice, J. M., Mullan, F. 1993. The elusive generalist physician: Can we reach a 50% goal? *Journal of the American Medical Association,* 270:1069–1073.

Goodman, D. C., Fisher, E. S. 2008. Physician workforce crisis? Wrong diagnosis, wrong prescription. *New England Journal of Medicine,* 358:1658–1661.

Goodman, J., Simmons, N. 2002. ASPHER PEER review: A discussion of its role in the joint Open Society Institute (OSI) — Association of Schools of Public Health in the European Region (ASPHER) program. In T. H. Tulchinsky, L. Epstein, C. Normand (eds.). Proceedings of the International Conference on Developing New Schools of Public Health. *Public Health Reviews,* Vol. 30, Nos. 1–4.

Goodman, D. C. 2004. Twenty-year trends in regional variations in the U.S. physician workforce, *Health Affairs,* Suppl. Web Exclusives VAR90-7.

Goodman, D. C., Fisher, E. S. 2008. Physician workforce crisis? Wrong diagnosis, wrong prescription. *New England Journal of Medicine,* 35:1658–1661.

Goodman, J., Overall, J., Tulchinsky T. 2008. *Public Health Workforce Capacity Building: Lessons Learned from "Quality Development of Public Health Teaching Programmes in Central and Eastern Europe."* Brussels: European Association of Schools of Public Health.

Handler, A., Schieve, L. A., Ippoliti, P., Gordon, A. K., Turnock, B. J. 1994. Building bridges between schools of public health and public health practice. *American Journal of Public Health,* 84:1077–1080.

Harper, D., Johnson, J. 1998. The new generation of nurse practitioners: Is more enough? *Health Affairs,* 17:158–164.

Hayes, E. 2007. Nurse practitioners and managed care: Patient satisfaction and intention to adhere to nurse practitioner plan of care. *Journal of the American Academy of Nurse Practitioners,* 19:418–426.

Health Resources and Services Administration, Bureau of Health Professions. 2002. *Projected Supply, Demand and Shortages of Registered Nurses 2000–2020.* Washington, DC: U.S. Department of Health and Human Services.

Inglehart, J. K. 2008. Grassroots activism and the pursuit of an expanded physician supply. *New England Journal of Medicine,* 358: 1741–1749.

Institute of Medicine. 1988. *The Future of Public Health.* Washington, DC: National Academy Press.

Institute of Medicine. 2003. *The Future of the Public's Health in the 21st Century.* Washington, DC: Institute of Medicine.

Institute of Medicine. 2003. *Who Will Keep the Public Healthy? Educating Public Health Professionals for the 21st Century.* Washington, DC: Board of Health Promotion and Disease Prevention.

Kahn, K., Tollman, S. M. 1992. Planning professional education at schools of public health. *American Journal of Public Health,* 82:1653–1657.

Köhler, L. 1991. Public health renaissance and the role of schools of public health. *European Journal of Public Health,* 1:2–9.

Köhler, L. 1992. The mission of public health during the next 25 years. A European perspective. *Public Health and Socio-Economic Changes at the Dawn of the 21st Century, Implication on Public Health Academic Education.* Jakarta: University of Indonesia.

Köhler, L., Bury, J., De Leeuw, E., Vaughan, P. 1996. Proposals for Collaboration in European Public Health Training. *European Journal of Public Health,* Vol. 6, No. 1.

Lasser, K. E., Himmelstein, D. U., Woolhandler, S. 2006. Access to care, health status, and health disparities in the United States and Canada: Results of a cross-national population-based survey. *American Journal of Public Health.* 2006 July; 96(7): 1300–1307.

Legnini, M. W. 1994. Developing leaders vs. training administrators in the health services. *American Journal of Public Health,* 84:1569–1572.

McKee, M., et al. 1992. Public health medicine training in the European community: Is there scope for harmonization? *European Journal of Public Health,* 2:45–53.

National Center for Health Statistics. 1998. *Health, United States,* 1998, with Socioeconomic Status and Health Chartbook. Hyattsville, MD: DHHS (PHS), 98–1232.

Oliver, R., Sanz, M. 2007. The Bologna Process and health science education: Times are changing. *Medical Education,* 41 (3):309–317.

Oulton, J. A. 2006. The global nursing shortage: an overview of issues and actions. *Policy, Politics and Nursing Practice,* 7:34S–39S.

Rimpelä, A. 1996. Postgraduate public health programmes in Nordic countries. In A. Rimpelä, L. Köhler (eds.). *Postgraduate Public Health Training in the Nordic Countries.* Göteborg: Nordic School of Public Health.

Roemer, M. 1999. Genuine professional doctor of public health the world needs. *Journal of Nursing Scholarship,* 31:43–44.

Salsberg, E., Grover, A. 2006. Physician workforce shortages: Implications and issues for academic health centers and policymakers. *Academic Medicine,* 81:782–787.

Schorr, T. M., Kennedy, M. S. 1999. *100 Years of American Nursing: Celebrating a Century of Caring.* Baltimore: Lippincott Williams and Wilkins.

Shortell, S. M., Weist, E. M., Sow, M. S., Foster, A, Tahir, R. 2004. Implementing the Institute of Medicine's recommended curriculum content in schools of public health: a baseline assessment. *American Journal of Public Health*, 94:1671–1674.

Tilson, H., Gebbie, K. M. 2004. The public health workforce. *Annual Review of Public Health*, 25:341–356.

Tulchinsky, T. H. 2002. Developing schools of public health in countries of Eastern Europe and the Commonwealth of Independent States. *Public Health Reviews*, 30:179–200.

Tulchinsky, T. H., Epstein, L., Normand, C. (eds.). 2002. Proceedings of the Jerusalem Conference on Developing New Schools of Public Health. *Public Health Review,* 30:1–4:1–392.

Tulchinsky, T. H., Varavikova, E. A. 1996. Addressing the epidemiologic transition in the former Soviet Union: Strategies for health system and public health reform in Russia. *American Journal of Public Health*, 86:313–320.

U.S. Department of Health and Human Services Public Health Service. 1997. *The Public Health Workforce: An Agenda for the 21st Century.* A Report of the Public Health Functions Project. Washington, DC: Department of Health and Human Services, Public Health Service.

Walt, G., Gilson, L. (eds.). 1990. *Community Health Workers in National Programs: Just Another Pair of Hands?* Milton Keynes, PA: Open University Press.

Workshop on Alternative Medicine. 1992. *Expanding Horizons: A Report to the National Institutes of Health in Alternative Medical Systems and Practices in the United States.* Washington, DC: U.S. Government Printing Office.

World Federation for Medical Education. 1993. World Summit on Medical Education. Edinburgh.

World Federation for Medical Education. 2003. *Global Standards for Quality Improvement.* WFME Office: University of Copenhagen, Denmark.

World Health Organization. 1987. *The Community Health Worker: Working Guide: Guidelines for Training: Guidelines for Adaptation,* Third Edition. Geneva: World Health Organization.

World Health Organization. 1993. *Reviewing and Reorienting the Basic Nursing Curriculum.* Copenhagen: World Health Organization, Regional Office for Europe.

World Health Organization. 1994. *Nursing Beyond the Year 2000: Report of a Study Group.* Geneva: World Health Organization.

World Health Organization, *World Health Report 2006*, Working Together for Health, http://www.who.int [accessed May 2008]

World Health Organization 2006. World Health Report 2006 — Working Together for Health, Geneva: WHO, http://www.who.int/whr/2006/en/ [accessed May 24, 2008]

Chapter 15

Health Technology, Quality, and Ethics

Introduction
Innovation, Regulation, and Quality Control
Appropriate Health Technology
Health Technology Assessment
Technology Assessment in Hospitals
Technology Assessment in Prevention and Health
   Promotion
Technology Assessment in National Health Systems
Dissemination of Technology
Diffusion of Technology
Quality Assurance
Adverse Events and Negligence
Licensure and Certification
Health Facility Accreditation
Peer Review
Algorithms and Clinical Guidelines
Organization of Care
Diagnosis-Related Groups
Managed Care
Performance Indicators
Consumerism and Quality
The Public Interest
Total Quality Management
Public Health Law
Environmental Health
Public Health Law Reform
Ethical Issues in Public Health
Individual and Community Rights
Ethics in Public Health Research
Ethics in Patient Care
Ethics in Public Health
Human Experimentation
Sanctity of Life Versus Euthanasia
The Imperative to Act or Not Act in Public Health
Summary
Electronic Resources
Recommended Readings
Bibliography

INTRODUCTION

Ethics and law in public health reflect the values of a society. They inevitably evolve as they face dramatic social, economic, demographic, and political changes; new health challenges; and new technological and scientific possibilities for improving health. Ethics are the foundation of the value systems of a society and thus of its health concepts. Biblical religious values such as the *Sanctity of Human Life* and *Act to Improve the World* for the sick led to the humanistic precepts of the *Universal Human Rights* and *Health for All* in the recent era (see Chapters 1 and 2).

The law sets the basic responsibilities, powers, and limitations of public health practice, with legislation and court decisions. Innovations in the technology of medical care and public health are powerful forces contributing to increased longevity, quality of life, and economic growth, but they also bring challenges to implementation impeded by additional costs of the health system and slow adaptation in countries with the greatest need. These are challenges to national and international political, organizational, and economic systems to address health with the full potential for saving lives. Determining standards of "good practice" is a continuing process with the rapid development of new knowledge, technology, and experience.

The law is a dynamic process involving old and new legislation, court decisions, and new issues not previously faced, often following rather than anticipating public health issues. Public health has had both positive and negative ethical experiences and continues to face new issues with changing population needs, technology, science, and economics.

Management of a production or a service system requires attention to the quality of personnel, as much as to the system in which they work. Their motivation and sense of participation, the scientific and technological level of the program, as well as the legal and ethical standards of individual providers and of the system as a whole, are all important to the quality of care provided.

Quality is the result of input and process, and is measured by outcome or performance indicators as well as perception of the service by the patients, the staff, and the community as a whole. *Input* refers to the institutional and financial resources for education, human resources, supplies, medications, vaccines, diagnostic capacity, and services available. *Process* refers to the use of those resources, including peer group expectations of professionalism. *Outcomes* generally include measures of morbidity, mortality, and functional status of the patient and the population. Defining and measuring achievements of national health objectives and targets, the methods of financing services, and the efficiency of organization help determine quality. Training, supply, and distribution of health personnel are all determinants of access to and quality of care. Continuous and adequate availability of essential preventive, diagnostic, and treatment services, as well as accountability and internal methods of promoting standards, are all elements of the quality of a health service for the individual, the population as a whole, and groups within the population with special needs.

The content and standards of service can be assessed through organized review by professional peers within an institution, and from outside. Peer review within an institution and external evaluation by accreditation or governmental inspection, based on cumulative evidence and the recognized current "state of the art" contribute to accountability and improved quality of care. Continuous quality improvement (CQI) among health care teams and organizations includes regular practice assessments, evidence gathering, remediation, and re-evaluation which will be discussed later in this chapter. The perception of the services by the community, along with the knowledge, attitudes, beliefs, and practices of health are all vital to improvement of health status.

Health-related technology is in a continuing state of change. Systematic review and absorption of new scientific knowledge, technology, and innovations are essential to promote and renew health care methods. Public health serves in a regulatory role to assure high-quality care to the individual and the community. New technology, whether in the form of diagnostic procedures, new drugs, devices, or vaccines, or new types of health personnel, requires evaluation for effectiveness and appropriateness to the system.

Technology assessment also involves epidemiologic and economic aspects of effectiveness. Failure to continuously monitor developments and to assimilate those that are demonstrably successful is an ethical and management failure which tragically costs many millions of lives from preventable diseases yearly, such as in delayed adoption of well-proven vaccines or tobacco restriction legislation. This is due to political failure even more than professional weakness, and constitutes one of the saddest ethical dilemmas of public health: failure to convince policymakers of the prime importance of health promotion and disease prevention in the health sector.

INNOVATION, REGULATION, AND QUALITY CONTROL

Health care technology has advanced with an increasing stream of innovation since the seventeenth-century definition of *separateness of mind and body* by Descartes and the discovery of smallpox vaccine by Jenner, to the dramatic innovations of the end of the twentieth century (Table 15.1). The pace of innovation is rapid, creating the need for regulation, quality control, and technology assessment.

National governments are responsible for assuring that pharmaceuticals, biological products, food, and the environment are regulated to protect the public. In some countries, these responsibilities are divided among ministries of trade, industry, commerce, health, and environment. In a federal system of government, there may be a division of responsibility among federal, state, and local government, but with the national government often providing national standards and leadership in this area.

Government regulation and control are meant to protect the public health. The U.S. Food and Drug Administration (FDA) is responsible to enforce the Food, Drug and Cosmetic Act, the Fair Packaging and Labeling Act, sections of the Public Health Services Act relating to biological products for control of communicable diseases, and the Radiation Control for Health and Safety Act. The FDA is part of the structure of the Department of Health and Human Services (DHHS).

Drugs and devices include all drugs, diagnostic products, blood and its derivatives, biologicals, veterinary medicines, and medicated premixed animal products. All manufacturers and distributors are required by law to register these products with the national authority in order to be allowed to market or import them. All countries need to govern the food, drugs, vaccines, and cosmetics regulated for production, importation, marketing, and use within their jurisdiction. Organizations within each government must be responsible for assuring the consumer that foods are pure and wholesome, safe to eat, and produced under sanitary conditions; that drugs and medical devices are safe and effective for their intended uses; that cosmetics are safe and made from appropriate ingredients; and that labeling is truthful, informative, and not deceptive.

The national authorities such as the FDA operate under legislation and regulations govern both domestic and imported products. They set and enforce standards, or use external agency standards as a "gold standard," meaning that products meet high standard of safety and efficacy. The FDA also investigates contents manufacturing standards under "good manufacturing practices" (GMPs), which includes regular accreditation of a producer's facilities, staffing, planning and monitoring capacity. Testing of products is carried out according to safety, potency, and toxicity using accepted reference laboratory procedures

TABLE 15.1 Health Care Innovations from the Seventeenth to Twenty-First Centuries

Period	Examples of scientific, technological, and organizational innovation
17th century	Biological basis of disease (Descartes), circulation of blood (Harvey), microscope (Leeuwenhoek)
18th century	Thermometer, lime juice supplements (Lind), vaccination (Jenner), surgical anatomy (Hunter), clinical sciences (Sydenham)
19th century	Stethoscope, anesthesia, laryngoscope, ophthalmoscope, blood pressure cuff, sanitation (chlorination and filtration of community water supplies), antisepsis, Braille printing, hygiene in obstetrics, nursing, microscopic pathology, pathological chemistry, microbiology, vaccines, x-ray, national health insurance, syringes, well-child care
1900–1930	Biomedical education, salvarsan, insulin, blood groups, vitamins, conquest of yellow fever, vitamin B, cost-benefit analysis, food fortification
1931–1945	Penicillin, streptomycin, randomized clinical trials, antimalarials, and vector controls
1946–1960	Vaccines, antihypertensives, psychotropic drugs, cancer chemotherapy, prepaid group practice
1961–1980	DNA, oral rehydration therapy, measles/mumps/rubella vaccines, cost-effectiveness analysis, open heart surgery, pacemakers, organ transplantation, computed tomography (CT), eradication of smallpox, HMOs, DRGs, district health systems, Pap smears
1981–2000	Magnetic resonance imaging (MRI), positron emission tomography (PET), endoscopic surgery, *Helicobacter pylori* control, managed care, *Haemophilus influenzae* b (Hib) vaccine, poliomyelitis eradication campaign, local eradication of beta thalassemia, rise and control of HIV infection
2000–2008	Natural disasters, bioterrorism, and pandemic avian flu preparedness; human papillomavirus (HPV) vaccine; measles control and imported epidemics

as published in the compendium *Official Methods of Analysis of the Association of Official Analytical Chemists.* When federal, state, or local investigators, in some cases known as *consumer safety officers*, observe conditions that may result in a violation of an act, they leave a written report with the manufacturers with recommendations for correcting the conditions. In more blatant cases, the authorities may issue urgent recall or seizure orders for products in violation of standards constituting a danger to public health, such as contaminated products or lead-painted children's toys. This may include contaminated foods causing food-borne disease outbreaks, which occur not infrequently in imported and domestically produced foods in the United States Supervision of food standards may also fail, as occurred in Israel in 2004 when total absence of vitamin B_1 in a soy-based baby formula imported from Germany, resulted in three deaths and permanent brain damage to others due to severe beriberi. This episode led to criminal charges of negligence in performance of duties resulting in death in 2008 against owners of the company which imported the foods and staff members of the Ministry of Health.

The FDA and its counterparts in each country are responsible for regulation of:

- **Food** food-borne illness, nutrition, dietary supplements;
- **Drugs** prescription, over-the-counter, generic;
- **Medical devices** pacemakers, contact lenses, hearing aids;
- **Biologics** vaccines, blood products;
- **Animal feed and drugs** livestock, pets;
- **Cosmetics** safety, labeling;
- **Radiation-emitting products** cell phones, lasers, microwaves;
- **Combination products.**

New drugs and biological products for human use are required to pass rigorous review before approval for marketing is granted. Applications are submitted by the manufacturer or sponsor with acceptable scientific data including test results to evaluate product safety and effectiveness for the conditions for which it is being offered. All manufacturers of drugs are required to be registered with the FDA and to meet its requirements for each drug produced and marketed, including the reporting of adverse reactions and labeling criteria. Manufacturers are required to operate in conformity with current good manufacturing practices, which include stringent control over manufacturing processes, personnel training, computerized operations, and testing of finished products. The FDA publishes guidelines to help manufacturers familiarize themselves with current standards. The *United States Pharmacopoeia*, the *National Formulary*, and the *WHO Model Formulary 2004* are the official listings of approved products.

Medical devices are also supervised by the FDA. Thousands of products for health care purposes require premarket approval, ranging from basic articles such as thermometers, tongue depressors, and intrauterine devices (IUDs), to more complex devices such as cardiac monitors, pacemakers, breast implants, and kidney dialysis machines. These products are subject to controls of good manufacturing practices, labeling, registration of the manufacturer, and performance standards.

Monitoring for efficacy and potential hazards has been strengthened since the 1970s as a result of findings of long-term carcinogenic and mutagenic effects of estrogens, and toxic effects of chloramphenicol on bone marrow. The drug thalidomide, widely used in Europe, Canada, and Australia in the 1960s, was not approved by the FDA. Thalidomide caused large numbers of serious birth defects leading to its ban in most countries. Controls of blood and blood products have been strengthened since the transmission of HIV, hepatitis B, and hepatitis C by contaminated blood products in the 1980s. The responsibility of this regulatory function is well illustrated by the 1995 criminal conviction of several senior health officials in France for failing to stop the use of blood products contaminated with HIV in the mid-1980s. Concern regarding possible carcinogenic effects of silicone breast implants led to legal action and greater controls of all implantable products. A balance between safety and well-regulated approval of new products requires a highly professional and motivated regulatory agency, well-developed procedures, and staff.

The concepts of standardization of good manufacturing practices for pharmaceutical products and written protocols for good medical practice or good public health practice are accepted norms based on best available evidence of current scientific knowledge and experience. Recommended immunization schedules, water quality, ambient air standards, food fortification, and screening programs for early stages of diabetes are examples of accepted practice that have become recommended standards of public health practice, paralleling qualitative measures developed in clinical care.

APPROPRIATE HEALTH TECHNOLOGY

The topics discussed in the growing literature and meetings of the International Society of Technology Assessment in Health Care represent the dynamic field of technology assessment. The issues range from economic evaluation of pharmaceuticals, modeling approaches, measures of quality of life, technology dissemination and impact, and outcomes measurement. The range of issues also includes finance and health insurance, health care in developing countries, informatics, teleradiology, technologies for the disabled, screening, and cost-effectiveness. Evaluations in the scrutiny of both high- and low-technology services based on a combination of clinical, epidemiologic, and economic factors are necessary. As health costs rise and populations age, medical innovation proceeds at a rapid rate, and both client and community expectations in health care continually rise.

Appropriate technology is defined by the World Health Organization as the level of medical technology needed to improve health conditions in keeping with the epidemiologic, demographic, and financial situation of each country. All countries have limited resources and so must select strategies of health care and appropriate technology to use those resources effectively to achieve health benefits.

In developing countries, training and supervision of traditional birth attendants for prenatal preparation and normal deliveries may be important ways to reduce maternal mortality in rural areas. Similarly, community health workers can provide care to underserved rural poor populations with a defined package of services that can be tailored to meet specific local needs, such as immunization, growth monitoring, nutrition counseling, and malaria and tuberculosis control.

A major example of appropriate technology has been the WHO initiatives to promote national drug formularies (NDFs) as a consensus list of essential drugs that are sufficient for the major health needs of a country, eliminating unnecessary duplication and combined products on the commercial market. The WHO called on all member states to ensure availability and rational use of drugs and vaccines and supported states wishing to select an essential list of drugs for economic procurement. Assistance with drug regulatory agencies, legislation, quality control, information, supply, and training was offered to help the member countries. Standard reference laboratories, the *International Pharmacopoeia*, and the *WHO Drug Bulletin* promote international standards and provide guidance to member states. The *WHO Model List of Essential Drugs* is a valuable tool to improve quality and cost management in national health systems.

The World Bank's *World Development Report* of 1993 defined cost-effective clusters of clinical and public health programs essential to improving health outcomes for low- and middle-income developing countries. The programs focus on those diseases which contribute heavily to the burden of disease and are amenable to relatively inexpensive interventions. This report defined interventions most able to reduce the burden of disease in low- and middle-income countries using clinical and public health interventions, as summarized in Table 15.2. The EPI Plus program, for example, prevents 6 percent of the total burden of disease in a low-income country and costs $14.60 per immunized child, or $0.50 per capita if purchased through international organizations.

In 2003, the Bellagio Study Group on Child Survival estimated that the lives of six million children could be saved each year if 23 proven interventions were universally available in the 42 countries in which 90 percent of child deaths occurred in 2000. The Millennium Development Goals set out by the United Nations can only be achieved by the year 2015 if public health infrastructures especially

TABLE 15.2 World Bank Recommended Model for Analysis of Priority Cost-Effective Health Interventions in Low- and Middle-Income Developing Countries, 1990

Minimum package service type	Burden of disease averted in countries			
	Low income		Middle income	
Public Health	% averted	$ cost per capita	% averted	$ cost per capita
EPI Plus immunization (DPT, polio, measles, BCG, hepatitis B, yellow fever + vitamin A)	6.0	0.50	1.0	0.80
Other public health programs (family planning, health, and nutrition education)	n/a	1.40	n/a	3.10
Tobacco and alcohol control programs	0.1	0.30	0.3	0.30
AIDS prevention program	2.0	1.70	2.3	2.0
School health program (including de-worming)	0.1	0.30	0.4	0.60
Sub Total (public health)	8.2	4.20	4.0	6.80
Clinical Care				
Treatment of tuberculosis (short course)	1.0	0.60	1.0	6.80
Integrated management of the sick child	14.0	1.60	4.0	0.20
Prenatal and delivery care	4.0	3.80	—	8.80
Family planning	3.0	0.90	1.0	2.20
Treatment of STIs	1.0	0.20	1.0	0.30
Limited care: pain, trauma, infection plus as resources permit	1.0	0.70	1.0	2.10
Sub Total (clinical care)	24.0	7.80	8.0	20.40
Total	32.2	12.00	12.0	27.20

Note: Low income = <$350 GNP per capita; middle income = >$2500 GNP per capita.
Cost per immunized child: $14.60 (0.50/capita); $27.20 ($0.80/capita), respectively, low- and milddle-income countries.
Source: World Bank. World Development Report, 1993.

childhood routine immunization can be expanded in content and strengthened.

In the United States, vaccine costs are high and rising as new effective vaccines are becoming available, but some vaccine prices fall after their initial period of use as manufacturing costs are lessened by improved methods. The measles, mumps, and rubella (MMR) vaccine costs $35 for the recommended two doses if purchased through CDC, but costs $89 if purchased in the private sector in 10-packs of single-dose units of the vaccine (2007). A combined DTaP vaccine costs $13.25 in the public sector 10-pack of single doses, while the same vaccine purchased with hepatitis B and IPV costs $47.25 per dose in the public sector but saves repeated visits and loss of compliance for that reason. The new HPV cervical cancer vaccine costs over $360 for the series of three doses per person. The new vaccine against diarrhea-causing rotavirus costs $190 for the recommended three doses. In addition, vaccine programs must take into account administrative costs and expenses of ordering, storing, inventory control, cold chain, insurance,

wastage, and spoilage. Multiple vaccines in one dose are less costly and less inconvenient for all. Examples include DTap plus Polio and Hib, or MMR (see Chapters 4 and 6).

WHO estimates the cost of all immunization activities in all 117 low- and middle-income countries for the period 2006–2015 to be $75 billion (USD), while low-income countries would need $35 billion (USD). The rate of adoption of presently available and new vaccines will be determined by political decisions in each country, although external aid — such as that of GAVI, an international public-private consortium to promote vaccination — is a resource. UNICEF is concerned about supply problems as well as costs but the key issue relates to political decisions, funding, and capacities of national immunization systems.

The WHO promotes widespread use of basic radiologic units (BRUs) to increase access to low-cost, effective, diagnostic x-rays, especially in rural areas in developing countries. BRUs are hardy, relatively inexpensive pieces of radiologic examination equipment that can be used in harsh field conditions for simple diagnosis of fractures and respiratory infection. The WHO estimates that 80 percent of all diagnostic radiology can be performed adequately by simple, safe, and low-cost equipment, supported by training of local people to operate and maintain the equipment. This is a consensus view of leading radiologists and clinicians helping the WHO to develop model equipment and training material.

In industrialized countries, technological advances in the medical and public health fields have been major contributors to increasing health costs and have led to pressures for greater selectivity in adopting costly innovations without adequate assessment of benefits and costs. Many countries have adopted more cautious policies with regard to financing unlimited expansion of new technology in the field of medical equipment, procedures, or medications. The need for organized assessment of technology is now an essential feature of health management at the international, national, and local levels of service delivery. The major responsibility for technology assessment is at the national levels, even with decentralization of service management.

Appropriate technology in the health field is becoming increasingly complex, laden with economic, legal, and ethical issues. Professional and public opinion demands make this a highly sensitive area of health policy, but responsible management of resources requires decision-making that includes consideration of the effectiveness, costs, and alternatives of any new technology (Box 15.1). Failure to adopt new innovations can result

in obsolescence, while excessive expenditures for hospitals and medical technology prevent a health system from developing more cost-effective preventive approaches, such as improved ambulatory care, or supportive care for the chronically ill.

Technology of unproven value can be a highly emotional and political issue, in criticism of managed care or national regulatory agencies, but spending limited national resources on ultrasound machines, magnetic resonance imaging (MRI) equipment, or inappropriately long hospital stays denies resources needed for other aspects of health care. A society must be able and willing to pay for medical innovation or improving quality of life by medical and public health interventions. Underfunding of a health system can deny these benefits just as misallocation of resources does, and this is a political issue even more than a professional one.

HEALTH TECHNOLOGY ASSESSMENT

Medical and health technology assessment (HTA) is the process of determining the contribution of any form of care to the health of the individual and community. It is a systematic analysis of the anticipated impact of a particular technology in regard to its safety and efficacy as well as its social, political, economic, legal, and ethical consequences. The technology may be a machine, a vaccine, an operation, or a form of organization and management of services. Analysis should include cost-benefit and cost-effectiveness studies (see Chapter 11) as well as clinical outcomes and other performance indicators.

Pressures from medical professionals, manufacturers of new medical equipment, and the public for adoption of new methods can be intense and continuous. Care must be taken that the specialists involved in committees for assessment are not those who may directly or indirectly benefit from the exploitation of technology, and who therefore may have conflicts of interest. Assessment must be multidisciplinary, involving policy analysts, physicians, public health specialists, economists, epidemiologists, sociologists, lawyers, and ethicists. The available information needs to include evidence from clinical trials, critical analysis of the literature, and the economic effect of adopting the technology on allocation of resources.

Medical technology varies in complexity and cost, not only to produce but in utilization. Medical technology that is inexpensive to supply and administer is known as *low technology* or *low-tech*, while *high technology* or *high-tech* refers to costly and complex diagnostic and treatment devices or procedures.

At the low-tech end of the technology scale, ORT (oral rehydration therapy) was developed in the 1960s for oral replacement of fluids and electrolytes lost in diarrheal disease, particularly in children. It was described in *Lancet*

Box 15.1 Health Technology Assessment

Questions that form the basis of technology assessment for a medical innovation include the following:

1. Is it safe and effective for the stated purpose?
2. Is it a new service, or does it replace a less efficient service?
3. Is there a need for it?
4. Where is it in the order of priorities of development of the facility?
5. Does it duplicate a service already available in the community?
6. Does it make medical sense (i.e., does it help in diagnosis and treatment for the patient's benefit)?
7. What are the alternatives?
8. What are the resources needed in terms of supplies, staffing, and upkeep?
9. Can the facility afford it?
10. What could otherwise be done with the resources it requires?

Source: Adapted from Kass, N. 2004. Public health ethics: From foundations and frameworks to justice and global public health. *Journal of Law, Medicine and Ethics*, 32:232–242.

as one of the greatest medical breakthroughs of the twentieth century. The introduction and wide-scale use of ORT for prevention of dehydration from diarrheal diseases throughout the world has saved hundreds of thousands of lives that would otherwise has been lost. In polio eradication, Sabin oral polio vaccine has important advantages over its rival Salk inactivated vaccine, due to its lower cost and ease of administration by less-qualified personnel.

Since the 1990s, advances in endoscopic surgical techniques have greatly improved patient care by reducing trauma, discomfort, and length of hospital stay, and thus have become the surgical approach of choice for many procedures. Since reports of the first 100 operations done in France in 1990, endoscopy spread rapidly to all parts of the world within a short time. It is now recognized by surgeons worldwide as a safer, less traumatic, and more effective approach to traditional invasive surgery. Although the operating time is longer, the patient may be discharged from the hospital within several days and return to work shortly thereafter. In contrast, following traditional abdominal surgery, a patient may require intensive care initially and a recovery period of many weeks.

Endoscopic surgery for cholecystectomy and esophageal, colorectal, hernia repair, renal, orthopedic, and other forms of surgery which previously were done with the patient remaining in hospital for many days are now done on a not-for-admission basis. Not-for-admission surgery has become standard practice in hospitals, extending the range of outpatient surgery and the comfort of patients who can return to their own homes to recuperate and return to regular activities much sooner. Fewer complications arise and patient comfort and economic implications are important. As a result, fewer hospital beds are needed for postoperative care than previously thought necessary, while surgical and ambulatory care facilities may need expansion to accommodate growing elderly populations needing surgical interventions but requiring shorter recovery. This innovation is now accepted as the standard of much of modern surgical care and shows that simple organizational changes can save money and improve patient safety and comfort.

The bacteria *Helicobacter pylori,* was first discovered in 1879 and suspected in 1899 as a cause of peptic ulcer. This was confirmed as the cause of peptic ulcers of the stomach and duodenum in 1979 (Warren and Marshall, Nobel Prize 2005). This led to effective diagnosis and inexpensive treatment of this condition and the elimination of a major component of surgical practice. Surgery for gastrectomies, vagotomies, and other forms of treatment are now virtually gone, resulting in a decreased need for hospital beds even for an aging population. The advent of ambulatory surgical, medical, and mental health care places more of the burden on ambulatory and outpatient services.

The dissonance between high-tech and low-tech procedures may lead to serious consequences in any health system. Choices require well-informed analysis of benefits, costs, alternatives, ethical considerations, and political consequences before distribution of limited health care resources are allocated between hospital-based high-tech medicine versus low-tech primary care.

High-tech procedures are usually applied in hospital settings in the context of other expensive care for seriously ill, often terminal, patients. Computed tomography (CT), invented in the 1960s, quickly proved to be an extremely valuable diagnostic tool. Advances in CT, MRI, and subsequent imaging techniques have proven to be cost-effective and lifesaving, replacing less efficient and dangerous invasive procedures. The CT and MRI scans allow the clinician to come to a rapid diagnosis of many lesions before they could be detected by other invasive and dangerous diagnostic techniques, at stages where the lesions are subject to earlier and more effective interventions. Imaging technology is advancing rapidly and promising inexpensive new systems for long distance transmission of imaging to medical centers may provide an enormous benefit to people living in rural or developing countries.

Technology assessment also examines methods of preventing and managing medical conditions. Treatment protocols or clinical guidelines are based on decision analysis of accumulated weight of evidence. Published clinical studies are assessed in meta-analyses, using statistical methods to combine the results of independent studies, where the studies selected meet predetermined criteria of quality. This provides an overview from pooling of data, but also implies an evaluation of the studies and data used. Clinical guidelines are part of raising standards of care, but also contribute to cost containment. Many countries form professional study groups to carry out meta-analyses on important health policy issues and new technologies.

Technology Assessment in Hospitals

Considering variance among countries, hospitals consume between 40 and 70 percent of total national health expenditures, with pressures for increased staffing and novel medical technology being a continual inflationary factor. Shorter stays and older patients have resulted in a drift toward intensive care, especially for internal medicine patients. Medical innovation is a continuing process with new diagnostic and treatment modalities reaching the market.

Hospitals no longer live in splendid isolation in the medical economy. A state government will have regulatory procedures to rationalize distribution of medical technology. The certificate of need (CON) is a form of technology assessment (TA) used in the United States since the 1960s to assess and regulate the development of hospital services to prevent oversupply and costly duplication of services. It attempts to establish and implement use of rational criteria for diffusion of expensive new technology. Whether this has had a lasting impact on restraining the excesses of high-tech medicine is arguable. This regulatory approach was limited to the hospital setting and failed to stop development of high-tech medical services such as ambulatory for-profit CT and imaging centers.

Many countries have adopted national TA systems to review topics as far-ranging as guidelines for acute cardiac interventions; liver, heart, and lung transplantation; minimal access surgery; and beam and isotope radiotherapy. Other TA guidelines include diagnostic ultrasound, sleep apnea, molecular biology, prostate cancer, magnetic resonance imaging, and new medications for inclusion in national health systems' approved basket of services.

Despite the limitations of this approach, where governments are no longer directly operating health care services, governmental regulation is necessary to prevent inequities in services by excessive development in some geographic areas at the expense of others, or by overexpansion of the institutional sector of health care at the expense of primary care. Regulatory mechanisms are essential in health care planning to restrain excessive and inappropriate use of high-tech services, but need augmentation by fiscal incentives to promote other essential services.

Technology Assessment in Prevention and Health Promotion

Technology assessment of preventive care programs includes evaluation of the methodology itself, along with the costs and measurable benefits, as in reduced burden of disease. The World Bank estimated that measles immunization at 80 percent coverage of the intended target group is more cost-effective per death prevented than 60 percent coverage, by preventing an additional 11,000 deaths in a province in Bangladesh. Ambulatory treatment for tuberculosis costs $1 to $3 per disability-adjusted life year (DALY) gained in several African countries.

Two major epidemics of measles occurred in Canada in 1990–1991, despite high rates of immunization coverage. Following this, a 1993 Delphi conference of experts from 31 countries reached a consensus recommending a two-dose measles immunization policy. As new vaccines enter the field, it is important to evaluate their effectiveness, costs, and the benefits to be derived. The cost of the hepatitis B vaccine initially was over $100 for an immunization schedule of three doses but has come down dramatically to less than $2 for bulk purchases (outside of the United States). The vaccine is now a cost-effective method to prevent liver cancer and long-term effects of chronic hepatitis.

Screening and education for thalassemia in high-prevalence areas have nearly eradicated the clinical disease but not its carrier status in Cyprus, southern Greece, and other countries. Newborn screening and case management for phenylketonuria (PKU), congenital hypothyroidism (CH), Tay-Sachs disease, and many other genetic diseases have been shown to be far less expensive than post facto treatment of severely developmentally delayed and dependent children born with these diseases (see Chapter 6). The success of Papanicolaou smear screening in reducing cancer of the cervix mortality since the 1960s has been dramatic. The discoveries of causation of cancer of the cervix by human papillomavirus (HPV) strains and advent of an effective vaccine since 2006 and that male circumcision can be an effective preventive measure for transmission provide a powerful demonstration of the effectiveness of public health screening and other measures to control this major malignant cause of death of women.

Screening for cervical cancer by Papanicolaou smears is recommended annually for high-risk groups, and every 2 or 3 years for other adult women (Box 15.2). This is vital for many years to come as the new HPV vaccine comes into general use, but its protective effect for individual and herd immunity will not replace the need for ongoing screening for this very common cancer. Routine mammography screening for breast cancer is recommended for women over age 40 every 1–2 years by the U.S. National Cancer Institute and for younger women with high-risk factors (e.g., previous cancer, family history, genetic markers). Cost-effectiveness analysis is now an essential part of decision-making in health policy and priorities. The value of health promotion in reducing exposure to HIV and cigarette smoking has been shown to be very cost-effective despite its low-tech or non-technological methodology, involving primarily group or mass education. Hypertension screening and case management is low-tech but highly effective in preventing strokes and blindness.

Low-tech innovations have had an important impact in reducing death and injury. These include mandatory use of car seat belts (introduced since the 1970s and 1980s in many countries), children's car seats, air bags, and bicycle and motorcycle helmets. Iodization of salt, vitamin A supplementation, and food fortification prevent large numbers of clinical cases of severe retardation, death, and blindness at low cost per child protected. Education for reducing risk factors for the cardiovascular disorders is far less costly than the premature deaths and high medical costs of stroke and congestive heart failure patients. Health education, condom and needle supply, and screening of blood donations are the most important effective community health measures against the spread of HIV. Table 15.3 compares high-tech and low-tech approaches to health problems, which often complement each other.

Technology assessments represent the current consensus derived from reviews of published studies and exchange of views of highly qualified clinicians, epidemiologists, and economists within a context of TA.

Box 15.2 Prevention of Cancer of the Cervix

Cancer of the cervix is the second most common cancer among women worldwide. In 2000, there were over 471,000 new cases diagnosed, and 288,000 deaths from cervical cancer worldwide. Approximately 80 percent of these deaths occurred in developing countries. WHO estimated that the number of prevalent cervical cancer cases diagnosed in the previous 5 years was approximately 1.4 million in 2000 compared with 3.9 million for breast cancer, with over 1 and 1.5 million, respectively, of these occurring in developing countries.

In the United States, cancer of the cervix mortality has been going down steadily since the introduction of Pap smear testing and declined in the period 1996–2004 by 3.7 percent per year. The American Cancer Society estimates that about 11,070 cases of invasive cervical cancer will be diagnosed in the United States and that 3870 women will die from cervical cancer during 2008.

Prevention of cancer of the cervix has until now mainly focused on Pap smears to detect the disease while still in a precancerous (cancer in situ) phase and this procedure has brought rates down dramatically over the latter part of the twentieth century.

The recent development of an effective vaccine against many strains of human papillomavirus (HPV) has brought forward not only a new technology, but debate as to its appropriate use, with a current consensus on vaccination of young girls (aged 11–13) before onset of sexual activity. The high cost of the vaccine precludes its rapid diffusion to most parts of the world and will even limit its use in industrialized countries.

The vaccine could markedly reduce this disease and should, in principle, also be used by adult women, in addition to continuation of routine Pap smear testing.

The role of circumcision among male partners of women has been under dispute for many years but in the past decade, evidence of HPV as the cause of cancer of the cervix and presence of HPV in uncircumcised men has brought this issue back to professional and public debate. Reports from Africa of reduced risk of acquiring HIV among circumcised men have brought new attention to adult male circumcision.

The technological breakthroughs first of the Pap smear in the 1950s, HPV testing in the 1990s, and HPV vaccine in the 2000s should also be seen in the context of prevention by male circumcision. HPV vaccine may in time replace all this, but current measures should continue until that takes place.

Cancer of the cervix remains a major public health problem, despite an excellent screening test and decades of experience in its successful application. The new HPV vaccine is high in cost and this will be an impediment to use where it is most needed in developing and transitional countries in Europe, Africa, and other parts of the world. Because HPV is a sexually transmitted disease, ethical questions are raised about its promotion as a public health measure. Yet the achievement of the scientific and technological basis for reducing the tremendous burden of this disease on mostly young women worldwide, involving ethics, economics, education, and societal norms such as in the practice of male circumcision, is a challenge in the context of a New Public Health.

Sources: Centers for Disease Control. 2007, http://www.cdc.gov/cancer/cervical/ [November 24, 2007]

They may change over time as new data or innovations are reported, and this possibility should be kept in mind in such discussions. Technology assessment mobilizes information and critically analyzes many aspects of medical technology to build a wide community consensus to influence policy decisions. Public opinion, political leadership, administrative practice, as well as the scientific merit of a case are all factors in developing a consensus.

Technology Assessment in National Health Systems

Technology assessment requires an organization within the framework of national regulatory agencies. The U.S. Food and Drug Administration serves this purpose as a statutory body within the U.S. Public Health Service. Sweden, Canada, Australia, the United Kingdom, the Netherlands, Spain, and other countries also have TA advisory or regulatory agencies established by national governments to monitor and examine new technologies as they appear. Sweden has a widely representative national Swedish

Council for Technology Assessment in Health Care, with an advisory role to the national health authorities.

The processes used in traditional systems to regulate food and drugs for efficacy, safety, and cost are more recently being applied to new medical devices and procedures. The unrestricted proliferation of new procedures presents serious dilemmas for national agencies concerned with financing health care and controlling cost increases. Nongovernmental health insurance shares this concern, as does industry which bears much of the cost of health insurance through negotiated, collective bargaining, "voluntary" health insurance in the United States. Most industrialized countries have national health services or national health insurance and are thus vitally interested in health costs and technology assessment. Many industrialized countries maintain technology assessment and cost-control activities. In the United States, the Agency for Healthcare Research and Quality (www.ahrq.gov) maintains oversight and studies related to clinical information, including evidence-based practice, outcomes and effectiveness, comparative clinical effectiveness, risks and benefits, and preventive services.

TABLE 15.3 Examples of High-Tech and Low-Tech Health Problem Solving

Problem	High tech	Low tech
Infectious diseases	Treatment — antibiotics	Vaccination, sanitation, hand washing, infection control in hospitals, health facilities, nursing homes
Breast cancer	Screening — mammography	Nutrition, self-examination, routine medical examination
Colon cancer screening	Colonoscopy	Nutrition, occult blood testing
Acute myocardial infarction, primary, secondary prevention	Coronary angioplasty, stent	Antiplatelet thrombosis treatment (e.g. aspirin, streptokinase intravenously, aspirin, beta blocker), diet, exercise, smoking cessation
Gallstones	Lithotropter Abdominal cholecystectomy	Endoscopic surgical removal
Head injuries	Intensive care	Helmets for bicycle riders, motorcyclists, seat belts front and rear of motor vehicles
Thalassemia	Transfusions, chelating agents; prenatal diagnosis, amniocentesis, chorionic villus biopsy	Screening, education, counseling
Dehydration	Infusions	Oral rehydration
Neural tube defects	Surgery, pregnancy termination	Folic acid fortification of flour and grain products, supplements for women in age of fertility
Liver cirrhosis, liver failure, and cancer	Liver transplant	Hepatitis B vaccine, risk reduction activities among IV drug users, screening blood donors
Cancer of stomach	Surgery, chemotherapy	Dietary change, cure of *H. pylori*-generated gastric ulcers
Cancer of cervix	Pap screening	Human papillomavirus vaccine

In Canada, the Health Protection Branch of the federal Department of Health reviews medical devices and drugs and, with the consent of the provincial governments, now licenses new medical procedures. Concern by governments over cost implications of new procedures led to this practice, and since 1988, a network of government and professional bodies have formed a nonprofit agency for technology assessment. This strengthens the provincial administration of health insurance in resisting professional or political pressures to add untested technology or procedures to the health system as covered benefits. A comparison of rates of procedure performance between provinces shows very high discrepancies, as high as 2:1, in procedures such as coronary artery bypass graft (CABG) or prostatectomy.

Control of acquisition of high-tech equipment by national or state authorities is essential to prevent expenditures on high-cost equipment without adequate assessment. Table 15.4 shows standards adopted by the Israel Ministry of Health as a key to approval control. Some new health technologies, such as *in vitro* fertilization (IVF), have escaped ministry control, and the practice has become completely uncontrolled with anxious clients and willing providers. However, the ministry has established reasonable controls in most areas of major equipment purchases.

DISSEMINATION OF TECHNOLOGY

The rapid spread of medical high-tech equipment has played a substantial role in escalating health costs. Comparison of OECD countries in regard to the number of MRI scanners per million population (Table 15.5) showed Japan and the United States with the highest number with 35.3 and 27.0 per million, respectively, while the median was 6.1. Canada ranked thirteenth among the 20 OECD countries reporting 5.5 MRI scanners per million. MRI availability in Japan increased from 18.6 per million in 1995 to 35 per million population in 2003. Comparing CT scanners showed Japan and Korea had the highest number at 92.6 and 30.9 per million, respectively, while the median was 13.3. Canada was in sixteenth place among the 21 OECD countries reporting with 11.3 per million population. The spread of endoscopic surgery in

TABLE 15.4 Standards for Permitting Acquisition of Costly Medical Equipment, Ministry of Health, Israel, 2006

Technology	Conditions	Ministry permitted rate
Computed tomography (CT)	In all hospitals >300 beds with emergency departments	1 per 125,000 population
Cardiac catheterization unit — (Cath lab)	Hospitals >300 beds with cardiac intensive care >5 beds	1 per 200,000 population
Magnetic resonance imaging (MRI)	Hospitals >400 beds with approved radiology department + CT	1 per 750,000 population
Gamma camera	Standard unit in hospitals >300 beds; MRI (0.5 testa) in all hospitals with neurosurgery	1 per 75,000 population 1 per region
Linear accelerator	Only in ministry-approved centers for radiation therapy with cobalt, CT scanner units	1–4 units in each of 6 regional radiation therapy centers, or 1–3 per million
Lithotripters (extracorporeal) fixed or mobile	Major hospital center	By regions (Jerusalem 1, Central 2, North 1, South 1)
Positron emission tomography (PET)	In hospitals with oncology centers	1 per million
Hyperbaric oxygen chamber	Regional hospitals near sea/lake	By regions (North 2, Central 1, South 2), up to 1 per 70,000 population
Computed tomography, dental	Requires license	No limits

Source: Adapted from Siebzehner, M., Shemer, J. 1995. Regulating medical technology in Israel. In J. Shemer, T. Schersten (eds.). *Technology Assessment in Health Care: From Theory to Practice.* Jerusalem: Gefen Press. Updated to 2006 regulations courtesy of Lotan, Y., Director, Division Licensing Equipment and Hospitals, Ministry of Health, Jerusalem, Israel.

TABLE 15.5 High-Tech Medical Equipment Units — Selected Countries and Years, 1986–2005 (Rate per Million Population)

Country	CTs			MRIs		
Year	1986	1993–1996	2003	1986	1995–1996	2003–2005*
Japan	27.5	69.7	92.6	0.1	18.8	35.3
United States	12.8	26.9	32.0	0.5	16.0	27.0
Sweden	—	13.7	14.2	—	6.8	7.9
Germany	6.9	16.4	13.3	0.7	5.7	6.0
United Kingdom	2.7	6.3	7.0	0.3	3.4	5.0
Canada	—	7.9	11.3	—	1.3	5.5
France	4.7	9.4	8.4	0.5	2.3	2.8

Notes: CTs — computed tomography scanners per million population;
MRIs — magnetic resonance imaging units per million population.
Source: 1993–1996 data from OECD *Health Data 98: A Comparative Analysis of Twenty-Nine Countries.* Paris: Organization of Economic Cooperation and Development, 1998.
2003 data from OECD 2004 and Canadian Institute for Health Information based on OECD data 2005 reported February 8, 2006, http://secure.cihi.ca/cihiweb/dispPage.jsp?cw_page=media_08feb2006_e [accessed May 23, 2008]

the 1990s has been worldwide. Health professionals become almost instantly aware of new developments from the news media as well as professional diffusion of information at conferences, in exchange visits, and in published articles and most dramatically via the Internet.

National policy to foster introduction of appropriate new technology requires a careful program of regulatory and financial incentives and disincentives to encourage or discourage diffusion of new methods of prevention as well as of treatment and community health care. Limitation of new techniques or procedures to selected medical centers allows the passage of time to fully assess the merits and deficiencies of new technology before general diffusion into the health care system. Such limitation, however, is fraught with the danger of depriving the population of benefits of new medical technology, and the possibility of restraint of trade to the economic advantage of selected providers. New technology impacts on insurance and managed care systems are necessarily involved in decision making as to inclusion of new procedures in their service plans.

Publication in professional literature is an accepted method of establishing the scientific merit of a treatment or an intervention. Too rapid diffusion of a medical practice can lead to disillusionment and confusion as to the merits of a particular medical procedure, as happened during the 1960s and 1970s with anticoagulant therapy for acute myocardial infarction and gastric freezing for peptic ulcers. Reviews of the literature should be critical and should assess the scientific merits of published data, as well as the sources of funding. Well-controlled large-scale clinical trials are needed to establish the relative values of alternative therapeutic approaches.

Dissemination of information about new medical innovations in the popular media is almost immediate. Many major newspapers and television networks have well-informed medical writers who have access to electronic mail medical journals as quickly as do specialists in each field. News magazines may carry special articles on new innovations, creating instant demand for them as benefits in a health program. This has both benefits and dangers.

In the United States, health insurers have led the way in developing TA and information synthesis, and in evaluating the costs and benefits of new procedures. The process is affected by public opinion, as well as by court decisions. A landmark decision against an HMO in 1993 awarded $29 million in damages to the family of a terminal breast cancer patient who died following refusal of the HMO to authorize a bone marrow transplant, which was at the time an experimental procedure. Denial of new technology may lead to increases in malpractice suits. In countries with limited financial resources, selection of technological innovations in health care that can benefit patient care or the public health requires a careful balance in order to use limited resources well, and to gain from application of appropriate new health care technology.

Payment systems may be effective in control of technology diffusion. Block budgets for hospitals have been more effective in Canada in restraining proliferation of high-tech equipment than in the United States, but not in reducing hospital bed ratios. This has led to criticism of limited access of Canadians to beneficial medical technology, such as CT, MRI, and advanced cancer therapies. In the United States, universal application of the diagnosis-related group (DRG) payment system for Medicare, Medicaid, and most private insurance had the effect of increasing ambulatory surgery very dramatically, from 16 percent of all surgery in 1980 to 52 percent in 2003 of all surgical procedures in community hospitals (i.e., non-federal short-stay hospitals or 85 percent of all hospitals in the country). Inpatient surgical procedure rates declined from 108.6 in 1980 to 38.0 per 1000 population in 2003.

HMOs and managed care organizations are paid on a per capita basis and have a strong incentive for cost containment. They have developed procedures and medical guidelines for investigation and intervention that seek to reduce unnecessary procedures. At the same time, HMOs are very active in promoting preventive care and non-hospital care insofar as this is compatible with good patient care.

Coronary bypass procedures have increased in frequency, but mainly for white males. In the United States, this procedure is less frequently done in women and African-Americans, because of lesser access to health insurance for African-Americans and possibly because of biases in terms of case assessment criteria in women. Cardiac invasive procedures increased dramatically since the 1980s in most industrialized countries, but with wide variation in their use. The benefits of aggressive invasive management of cardiovascular diseases remain controversial, but many such procedures have proven beneficial in reducing mortality rates and improving quality of life.

Critical analysis of the need for surgery has resulted in lower tonsillectomy and radical mastectomy rates along with increased use of outpatient procedures. Tonsillectomy, a routine procedure until the 1960s, is now performed infrequently since it was found to be of little medical value. Cataract surgery is now largely done on an ambulatory basis. The technology of home care has come to play an important role in early discharge of patients from the hospital, as has the wide use of cancer chemotherapy and radiation therapy on an outpatient basis.

DIFFUSION OF TECHNOLOGY

Innovations in health care through scientific and technological advances are continuing with exciting breakthroughs in effective new treatments and public health interventions coming rapidly and this requires health authorities, practitioners, and the public to maintain constant awareness of the "current state of the art." Diffusion of new technology or adaptations from basic science advances may begin slowly, and then reach a "tipping point," at which time a dramatic change of trend occurs and it becomes the new standard or fashion.

Those with economic interests in the product try to advertise and promote sales, while practitioners are ready to try new methods to help their patients, but those who must pay for services may ask for evidence of effectiveness, safety, added value over present and known methods, and benefit to the length of or quality of life of the individual. This can become a highly charged debate when those responsible for adopting new measures in national health plans must weigh one proposed addition against another, each with its ardent professional, community, or business promoters. The new HPV vaccine approved by the FDA in 2006 for prevention of cancer of the cervix is an example. The HPV vaccine is recommended for pre-teen girls at age 11–12 years and also for females age 13–26 to offset future sexual exposure to HPV-infected boys. It is too costly for most developing countries where cancer of the cervix is at its most lethal levels. The duration of immunity is still not known or whether booster doses will be required. The two manufacturers are naturally interested in increasing their market and market share. Policy people need to consider whether the same money would have greater benefit if used to provide pneumococcal pneumonia and rotavirus vaccine for children in developing countries, which would quickly save hundreds of thousands of lives. It is likely that this wonderful new public health technology (HPV) will be absorbed quickly into public health practice at least in the industrialized countries but there will be a long time lag for its use in developing countries.

QUALITY ASSURANCE

Quality assurance (QA) is an integral part of public health function and involves ensuring the quality of both health practitioners and facilities. It is an approach that measures and evaluates proficiency or quality of services rendered. Hospital accreditation is a long-standing method of QA, providing many generations of health providers in North America with first-hand experience with QA in community hospitals and long-term care facilities, as well as ambulatory and mental health services. Hospital accreditation has contributed to improvement in standards of facilities and patient care throughout Canada and the United States and has provided a working model for replication or adaptation internationally.

Adverse Events and Negligence

Iatrogenic diseases are adverse events that occur as a result of medical management and result in measurable disability.

Negligent adverse events are those events caused by a failure to meet standards of care reasonably expected of the average physician or other provider of care.

Iatrogenic disease is a major cause of morbidity, prolongation of hospitalization, and even death. Hospital-acquired (nosocomial) infections are estimated to occur in 7–10 percent of hospital cases in Britain and the United States. Primarily these are caused by urinary, respiratory tract, and wound infections. Infections increasingly include organisms that are resistant to many antibiotics and are difficult to treat. Infection control in hospitals is therefore an essential part of hospital organization. Because hospitals are increasingly being paid by DRGs, any secondary event prolonging hospital stays may have adverse financial effects on the hospital. In the United States, recent decreases in Medicare reimbursements for nosocomial infections reflect this trend to provide financial incentives to improve hospital infection control. There is, therefore, a strong financial as well as professional interest in reducing hospital-acquired infections. Hospital-acquired infections, anesthesia mishaps, falls, and drug errors are the most common iatrogenic events.

A classic study of 32,000 hospitalizations (out of 2.6 million hospital discharges) in New York State carried out by a Harvard University team, showed that 3.7 percent of hospitalized patients suffered adverse events, or injuries, caused by medical mismanagement which resulted in measurable disability. Of these, 28 percent were due to negligence, so that 1.03 percent of all hospitalizations involved medical negligence leading to measurable injury. Of the total of some 100,000 adverse events in the study group, 57 percent recovered within a month, and 7 percent had severe injury. Some 14 percent or 14,000 persons with adverse events died as a result; 51 percent of these deaths were due to negligence.

A 1999 report of the U.S. National Institute of Medicine estimated that between 44,000 and 98,000 persons die annually in the United States from medical errors occurring in hospitals. Higher rates are seen among the elderly and the poor. Rates were lower in teaching hospitals as compared to community hospitals. About 20 percent of the events were related to drug reactions or dosage errors. Less than 3 percent of those injured brought civil litigation for the negligence. The search for "bad apples" — that is, unethical, criminal, or incompetent health providers — is necessary, but not sufficient to stem the problems created by the health system itself. Prevention requires organized activity. Investigation of adverse events helps to identify methods of prevention and to protect the patient's rights. A program of measures to reduce hospital infection must be based on epidemiologic analysis of recorded events in the search for common causes and preventable factors. Organized surveillance and control require one infection control practitioner per 250 acute care beds, a trained hospital epidemiologist, and routine reporting of wound infections to practicing surgeons (CDC, Hospital Infection Program). Computer-aided medication dispensing, as well as automated and other safety systems are critical elements in minimizing morbidity and mortality resulting from preventable human errors. In response to the high frequency and cost of medical litigation, many states in the United States have enacted legislation to restrict court awards for medical negligence. Proposals for alternatives to the tort system of medical malpractice compensation include arbitration and mediation, an administrative system similar to that used for workers' compensation, and a no-fault system of compensation, such as exists in New Zealand, Sweden, and Finland. In a no-fault system the complainant need not prove negligence on the part of the provider, but only that he or she suffered an adverse event which is compensable at standard rates depending on the degree of disability. In the United States, federal legislation provides compensation for vaccine injuries, and three states have enacted restricted no-fault systems for birth-related neurologic injuries.

In addition, there is greater emphasis on adoption of failsafe mechanisms, such as introducing warning systems in anesthesia machines to alert the anesthetist if oxygen flow in the patient's tubing falls below a safe point. This was tested in Boston hospitals and found to reduce adverse anesthetic events to zero cases over a 3-year period. Vitamin K injection was made mandatory for all newborns in New York State, as was already the case in some other states, when a study showed deaths from hemorrhagic disease of the newborn in cases where vitamin K was not administered.

Inappropriate medical practice patterns are an equal, or even larger, problem for health systems. Comparisons of surgical rates within the United States for coronary bypass procedures, hysterectomies, and cesarean section show wide variation between different areas of the country. The costs of excess surgery are not only economically wasteful, but also involve risks for the patient from the surgery itself or anesthesia mishaps, infection, pain, and discomfort with legal and ethical questions of unwarranted interventions not for the benefit of the patient. Health systems are increasingly required to evaluate and control excess surgical, investigative, or other medical procedures, not only for financial reasons but also for protection against litigation and infringement of patients' rights.

Licensure and Certification

The requirements that society establishes for allowing an individual to practice medicine, and any health profession, are vital to maintaining and improving the quality of care (Chapter 14). These standards require defining the training and experience needed by the individual, examination procedures, and recognition for continued education and maintenance of competence. This requires a statutory base and national bodies operating under a national authority,

separate from the agency operating the health system services. Separation of licensing from operation of the health service is essential in maintaining high professional standards. The licensing authority is accountable to the state and the public. In some cases, this function is delegated to self-regulating professional bodies. In Canada, the licensing of the medical profession and specialty recognition are carried out by medical profession with self-regulation. In the United Kingdom, medical licensing is by a state-appointed board and in the United States by state boards.

Medical schools, postgraduate training programs, and fellowships are all subject to periodic comprehensive assessments. Institutions that fail to meet the standard may have funding or licensure suspended until they have performed adequate remediation.

Health Facility Accreditation

Hospital accreditation in North America is by a voluntary grouping of professional associations, including the Canadian and American Colleges of Physicians and Surgeons, the hospital associations, and the Colleges of Nurses. The Joint Commission, originally operating in both Canada and the United States, carries out regular inspections of hospitals. In Canada, other organizations including the federal Department of Health, provincial ministries of health, Canadian Diabetes Association, Public Health Association, and the Standards Council of Canada participate in the Joint Commission as observers. Initially focusing on acute care hospitals, accreditation has been gradually extended to cover special hospitals, long-term facilities, home care programs, public health departments, and ambulatory care services.

Health facility accreditation is a systematic, multidisciplinary inspection of the physical and organizational structure of the facility or program and the functioning of its component parts. Factors measured include staff qualifications, facilities, organization, record keeping, and continuing education of staff.

The process of accreditation requires a request for accreditation from the board of governors of the hospital or health facility, implying acceptance of the standards of the commission. The accreditation process includes a self-assessment, an on-site survey, and follow-up action for correction of deficits and improvements. The commission is invited to conduct a survey, and resurvey as it sees fit. The hospital pays a fee and commits itself to provide all data requested and to cooperate with the site visit. The commission issues a confidential report, giving the accreditation rating and interim statement of deficiencies, and requests progress reports in correcting deficiencies. It is also empowered to carry out follow-up inspections and resurveys. Table 15.6 lists the areas of a large community or teaching hospital included in the accreditation site visit.

The assessment survey examines the goals and objectives of the organization and its administration, the

TABLE 15.6 Services Evaluated in Accreditation in Large Community and Teaching Hospitals, Canada

1. Ambulatory care
2. Child life services
3. Clinical record services
4. Critical care unit (generic)
5. Discharge planning services
6. Diagnostic services
7. Education services
8. Emergency services
9. Governing body
10. Housekeeping services
11. Human resource services
12. Intensive care, cardiac care, transplant units
13. Laboratory services
14. Laundry and linen services
15. Library services
16. Long-term care/geriatric unit
17. Management services — utilization review, risk management, infection control, health and safety, disaster and emergency planning
18. Material management services
19. Medical equipment services
20. Medical services
21. Neonatal intensive care
22. Nuclear medicine
23. Nursing services
24. Nutrition and food services
25. Obstetrical services
26. Occupational therapy services
27. Operating suite, post-anesthetic recovery unit
28. Palliative care unit
29. Pastoral services
30. Pharmacy services
31. Physical plant and maintenance services
32. Physiotherapy services
33. Psychiatric services
34. Psychology services
35. Rehabilitation services
36. Respiratory therapy services
37. Social work services
38. Speech language and audiology unit
39. Standards for delivery of care by program
40. Volunteer services

Source: Canadian Council on Health Facilities Accreditation.

Each section of the program being accredited is assessed in the following categories:

1. Statement of purposes, goals, and objectives;
2. Organization and administration;
3. Human and physical resources;
4. Orientation, staff development, and continuing education;
5. Patient care;
6. Quality assurance.

These categories are also used in the programs covered by the contracts the Canadian Council on Health Services Accreditation (CCHSA) has with other health and social service agencies.

Hospital accreditation was established in the United Kingdom and Australia in the 1980s and is attracting interest in other countries seeking ways to maintain and promote standards. The procedure for accreditation of hospitals is still voluntary in Canada, but in effect has become universal for hospitals of medium and large size (over 75 beds) and common for smaller hospitals. It is seen as advantageous for the governing board and the community and also for the medical staff in terms of medico-legal protection. In the United States, hospital accreditation has become virtually universal since payment for federally funded health insurance (Medicare and Medicaid) beneficiaries is not allowed for non-accredited hospitals, and many private insurers make this requirement as well. In some states, accreditation is mandatory for all hospitals.

In the past decade, CCHSA's accreditation program has expanded to now cover a diversity of health care and service areas, through contract arrangements with independent non-hospital facilities such as highly specialized programs as well as community health and social service organizations. In 2006, CCHSA introduced standards for child welfare, hospice, palliative and end-of-life care facilities, prison facilities, biomedical laboratories, and supplementary criteria for telehealth. The specialized needs of the ever-changing health and social environment now accommodate specialized needs in a diversity of service areas as an adjunct to the hospital accreditation process. Examples are shown in Box 15.3.

Licensing and regulation of health facilities are a government responsibility, but an independent accreditation authority has advantages. The national authority may fail to monitor its own facilities with the diligence or objectivity needed, and there may be a conflict of interest. Where there is a national system of organization, distinct departmentalization of the operating and certification functions may provide a greater measure of objectivity. Assistance from countries experienced in voluntary accreditation can help to establish accreditation mechanisms and provide technical and professional support to countries wishing to establish such programs.

direction and staffing of the facility, policies, and procedures. Review includes medical staff organization, credentials and review procedures, clinical privileges, selection of department chairpersons and their responsibilities, standing committees, schedule of meetings, bylaws, and the role of the governing board of the hospital. The presence and nature of QA organization, records review procedures, and continuing education are assessed. The quality of clinical records is assessed by examination of charts for the completeness of histories and documentation of the course of the hospital stay including laboratory reports.

Box 15.3 Canadian Council on Health Services Accreditation Program and Expanded CCHSA Accreditation to Other Specialized Programs and Organizations

The Canadian health services accreditation program began in 1917 in conjunction with the American College of Surgeons (ACS) with a hospital standardization program. The first Minimum Standard for Hospitals developed requirements of just one page. In 1918, on-site inspections of hospitals began with 89 of 692 hospitals surveyed meeting the requirements of the Minimum Standard. In 1926, the first Standards Manual was issued.

In 1951, the American College of Physicians, the American Hospital Association, the American Medical Association, and the Canadian Medical Association joined with the ACS to create the Joint Commission on Accreditation of Hospitals (JCAH). It is an independent, not-for-profit organization whose purpose is to provide voluntary accreditation. In 1953, the Canadian Hospital Association (now the Canadian Healthcare Association), the Canadian Medical Association, the Royal College of Physicians and Surgeons, and l'Association des Médecins de Langue Française du Canada, established the Canadian Commission on Hospital Accreditation. The Commission's purpose was to create a Canadian program for hospital accreditation, and in 1958 the Canadian Council on Hospital Accreditation was incorporated.

CCHSA's accreditation program is used by all types of health facilities, from large and complex hospitals, health systems, community health organizations, and residences providing long-term care. CCHSA expanded its scope to include accreditation of a wide range of programs. In 2006, CCHSA introduced new standards on Child Welfare, Hospice Palliative and End-of-Life Care, Biomedical Laboratory Services, Blood Banks, and supplementary criteria for Telehealth. CCHSA's accreditation program covers a diversity of health care and service areas, service programs for brain injury, ambulatory care, assisted reproductive technology — clinical and laboratory services, Canadian Forces — health services, cancer agencies, child welfare organizations, First Nations and Inuit addictions and community health services, the Federal Department of Veterans' Affairs, substance abuse and problem gambling treatment services.

CCHSA accreditation service is on a contract basis with specialized health programs, other federal government departments, for-profit health facilities, and community organizations.

Source: Canadian Council on Health Services Accreditation, 2007, http://www.cchsa.ca/default.aspx [accessed December 28, 2007]

In the current period of transition from central to decentralized management of health services in many countries, health facilities are being transferred from government operation to independent operation as not-for-profit or even for-profit facilities. Present methods of regulation by national or state levels of government will require review as decentralization and privatization take place. Regulation by governmental authorities and nongovernmental professional bodies is mutually complementary in promoting accountability, standards, and quality of services.

Peer Review

A large part of the work of clinical and departmental managers in hospitals or other care settings relates to quality assurance. A major method of improving quality in a health program is through peer review by which the staff organizes systematic review of cases and records, using statistics on performance indicators. In hospitals, this includes review of deaths, maternal mortality and infant mortality cases, surgical rates, complications following surgery, and infection rates. Medical records and computer information systems permit record reviews by diagnosis. These can be utilized to assess other events in hospitals, such as time from admission to surgery, lengths of stay by diagnosis, response to abnormal laboratory findings, and many other indicators of the process of care. Obstetric departments can review the frequency of and criteria for caesarean

section deliveries. Surgical departments review their appendectomy rates to separate pathological findings from normal appendices. Organized peer review has also been called *medical audit* and essentially describes methods of self-policing and education to learn from mistakes and experience and to improve the quality of care.

In 1972, an amendment to the U.S. Social Security Act required hospitals and long-term care facilities to monitor the quality of care given to Medicare and Medicaid patients through professional standards review organizations (PSROs). These were medical audit committees with specified tasks to conduct utilization review, medical care evaluation, and profile analysis of physician or institutional performance compared to accepted standards of the medical community. In 1982, peer review organizations (PROs) were created by federal statutes to replace PSROs. The PROs are non-profit corporations, staffed by physicians and nurses, to review medical necessity, quality, and appropriate level of care under the Medicare and Medicaid programs. The Centers for Medicaid and Medicare Services have an Office of Clinical Standards to conduct surveys, certification, and develop best practices guidelines, in a program for health care quality improvement (HCQIP).

Hospitals have departmental clinical meetings, adverse incident or outcome committees, mortality rounds, and clinical–pathology conferences help to evaluate and learn from difficult cases. The presence of functioning peer review mechanisms indicates that quality is of concern to the professional and administrative network, raising the consumer's confidence in the system.

Maternal mortality committees have been widely used to assess preventable factors in deaths related to maternity and to point out areas of needed improvement in services. Identification of high-risk pregnancies emerged from this process and has become an important part of prenatal care. Infant mortality reviews by professional groups can similarly demonstrate areas of needed improvement in services. Death rounds are held to review cases of death following surgery, or closely following admission, or an "incident" such as inappropriate medication given in error.

The successive waves of peer review initiatives in the United States represent attempts by the federal government to establish mandatory quality of care review by professional peers for facilities providing care to Medicare and Medicaid patients. The concept of requiring standards of care review has probably contributed to greater awareness of accountability of hospital-based practice. Frequent litigation may have contributed more to the sense that the physician is accountable for services and outcomes of care. PROs are a form of quality regulation that represent a commitment by funding agencies to accountability in care systems and to identification of organizational and administrative weaknesses in health care generally and not only in hospitals. The generation of U.S. physicians and health systems managers trained since the 1970s accepts peer review as an integral part of health services. Other countries use this kind of mechanism to maintain and promote quality of care.

Tracer Conditions

Tracer conditions are common medical conditions (or procedures), for which diagnostic criteria are well established and clear, there are effective preventions or treatments, and a lack of treatment can cause significant harm to the patient. Examples of tracer conditions include otitis media, appendectomy, caesarean section, and hysterectomy. These conditions, if evaluated in terms of incidence and actual chart review, can provide useful insights into departmental medical standards. Incident reports by nursing staff and nosocomial infections are an example of the functioning of the tracer condition concept. Incident reports in hospitals are designed to determine the causes of errors and to help develop remedial action and prevention of similar events. Tracer condition studies have become such an accepted part of modern health management that absence of an organized review system could be considered a serious structural flaw in a health service.

Setting Standards

Standards recommended by independent professional organizations or by advisory committees appointed by ministries of health can play important roles in defining standards of care for specified conditions. In addition, organized professional bodies can issue practice guidelines or help governments or health care agencies to develop standards or algorithms for management of specific topics and conditions.

Specifying standards for preventive care, such as for infants and adults, assists local health authorities in planning and evaluating services. The American Academy of Pediatrics has an extensive professional committee structure that publishes periodic guidelines for pediatricians on a wide variety of infant and child topics including nutrition, immunization, prevention of anemia and lead toxicity, child safety, and school health.

The American Public Health Association (APHA) publishes the *Control of Communicable Diseases Manual*, now in its eighteenth edition (2005). It is the authoritative U.S. manual on this topic. The American Academy of Pediatrics *Red Book* on infectious diseases is used across North America by pediatricians in clinical practice. These organizations and their counterparts in obstetrics and many other clinical fields directly relevant to public health continually update practitioners and policy personnel in "the state of the art" or "gold standard," discussed previously. This constitutes a professional self-guidance system in standards. Managed care and other health provider systems also issue guidelines for member practitioners that serve to maintain standards of service.

The wide use of treatment protocols and scoring systems in hospital medicine helps define standards of care in a measurable way. The Apgar score for rating newborn status has been a standard in hospitals worldwide for decades, helping to standardize infant assessment and care. The APACHE system (acute physiology and chronic health evaluation) is a scoring system used widely to assess the chances of survival of patients admitted to intensive care units and to compare outcomes, for example, between teaching hospitals and community hospitals. It is also used in assessing patient outcomes with different modes of treatment. Scoring systems are also used in community health care, as in risk scoring for pregnancy care (see Chapter 6).

Algorithms and Clinical Guidelines

Algorithms are decision trees or a systematic series of decisions based on the outcomes of previous decisions, tests, or findings. Derived from operations research, this approach applied to medicine identifies all available choices (e.g., exposed versus nonexposed) and follow-up decisions based on findings from each previous option substantiated by observation. It is often presented graphically like the branches of a tree, showing the alternatives and subsequent decisions to be made.

A clinical algorithm is a systematic process defining a sequence of alternative, logical steps depending on outcomes of previous ones, incorporating clinical, laboratory, and epidemiologic information, applied to maximize benefits and minimize risks for the patient. It gives the provider a review of the relevant literature and recommended standards of practice on a particular topic for preventive care or case management. These guidelines are usually arrived at by consensus of multidisciplinary working groups taking into account published studies on the topic. The guidelines may suggest that some procedures not be carried out routinely.

Clinical guidelines are meant to establish accepted standards of care and may have important economic implications. *Medical Letter,* published by the Consumers Union, is a long-standing and useful publication that reviews therapeutic issues of everyday medical practice and the relevant studies. It represents a balanced, updated view of medical practice and summaries of current literature, reviewed by respected, experienced, and competent medical authorities. Clinical practice guidelines are produced by hundreds of professional, medical, and governmental agencies in order to standardize and improve medical care.

Clinical guidelines are helpful in clinical practice and in preventive medicine. They are increasingly used in managed care environments to assure standards, quality of care, and cost-effectiveness as well as legal protection. Guidelines for preventive medicine and public health practice are also part of the process of promoting the quality of individual and community health, as discussed in Chapter 11. Annual revision of the infant immunization program, discussed in Chapter 4, is a prime example, as is the set of guidelines for preventive care for adult health maintenance in Table 15.7.

The issue of application of current scientific knowledge for population health is a continuing struggle for recognition of the prime importance of health promotion and preventive care for health of a population. The selection of priorities in use of resources is vital especially in the many developing countries which are in various stages of economic development, or which have abundant income from natural resources such as oil and minerals. Implementation of programs designed to achieve the Millennium Development Goals can help serve this purpose.

Public health standards and clinical practice guidelines are an increasing part of quality improvement. It is important, however, that they are developed as best practices and as little influenced as possible by commercial interests

TABLE 15.7 Adult Health Maintenance Checklist

Procedure	20–39	40–64	65+
Checkup visit	Every 3 years	Every 2 years	Annually
Cholesterol	With checkups	With checkups	With checkups
Fecal occult blood	Age 40–49 if high risk	Annually	Annually
Clinical breast exam (CBE)	Every 1–3 years	Annually*	Annually*
Mammography	Baseline age 35	Age 40–49, every 1–2 years	Over 70, every 2 years
Pelvic exam	Every 1–3 years	Every 1–3 years	Every 1–3 years
Pap smear	Annually if on birth control pill; others every 3 years	Annually if on birth control pill; others every 3 years	Annually if on birth control pill; others every 3 years
Colonoscopy	No	After age 50, every 3–5 years	After age 50, every 3–5 years
Prostate and PSA	No	Annually*	Annually*
Immunizations			
Tetanus-diphtheria	Every 10 years	Every 10 years	Every 10 years
Pneumococcal pneumonia	For high risk	For high risk	Every 6 years
Influenza	For high risk	For high risk	Annually
Skin cancer	Annually*	Annually*	Annually*
Bladder cancer	Annual routine urinalysis	Annual routine urinalysis	Annual routine urinalysis
Lung cancer	Routine exam**	Routine exam**	Routine exam**
Testicular cancer	Routine exam**	Routine exam**	Routine exam**
Oral cancer	Routine exam**	Routine exam**	Routine exam**
Ovarian cancer	Routine exam**	Routine exam**	Routine exam**
Pancreatic cancer	Routine exam**	Routine exam**	Routine exam**
Routine vitamin supplements	Routine**	Routine**	Routine**

*inconclusive, **negative recommendation. The topics are under continuing review, and recommendations are in some cases left to the opinion of the provider as the current cumulative evidence is not affirmative, e.g. Clinical breast exam annually or breast self examination.

Source: U.S. Preventive Services Task Force Ratings: Strength of Recommendations and Quality of Evidence. *Guide to Clinical Preventive Services,* Third Edition: Periodic Updates, 2000–2003.

Agency for Healthcare Research and Quality, Rockville, MD, http://www.ahrq.gov/clinic/3rduspstf/ratings.htm [accessed December 23, 2007]

of drug or vaccine manufacturers. The proliferation of such guidelines by health authorities or professional associations of the United States, United Kingdom, Canada, Australia, and other countries indicates a wide consensus on the importance of such written standards, guidelines, or "best practices" statements. The recommended childhood immunization program put forward annually by the Centers for Disease Control (CDC) in conjunction with the American Academy of Pediatrics and other professional organizations is an example of such best practices and is accepted by health insurers and providers as the gold standard in this field. The concept of promotion of quality in health care and the adoption of current scientific standards are global issues and part of the New Public Health (see Box 15.4).

Box 15.4 Excellence in Science Committee of the Centers for Disease Control (EISC)

The EISC promotes the CDC's scientific infrastructure and facilitates communication and collaboration that enhance scientific areas and activities needed for state-of-the-art conduct of science. EISC serves as a consulting body for science-related issues and makes recommendations to the CDC to foster, support, and protect an environment for the promotion of scientific integrity, quality assurance, and the rapid dissemination of scientific innovations, technology, and information with the ultimate goal of improving public health.

EISC specific functions include:

- Promoting and protecting the scientific infrastructure;
- Providing a forum for information exchange among administration, directors for science, and liaison working members/groups;
- Communicating science-related issues to CDC and related scientists;
- Promoting professional development and training;
- Recognizing and rewarding quality science;
- Acting as an advocate for scientific resources;
- Identifying and disseminating new information; for example, new statistical/epidemiologic techniques or new scientific technologies;
- Developing, revising, and promoting the implementation of crosscutting scientific policies and procedures;
- Serving as a consulting body for science-related issues and making recommendations to the CDC;
- Fostering the development of methods for assessing and monitoring:
  - the environment for quality science and qualitative and quantitative scientific output within CDC and related organizations
  - the impact of CDC science on public health.

Source: Adapted from Centers for Disease Control, Science Coordination and Innovation, at http://www.cdc.gov/od/science/excellence/ [page last modified October 6, 2006, inquiry@cdc.gov, accessed December 28 2007]

The Province of Saskatchewan Health Services Utilization and Research Commission publishes periodic reports presenting consensus positions of panels of medical faculty, clinical specialists in pathology and physical medicine, and public health specialists in nutrition, community health, and epidemiology. Its reports are circulated widely and serve to update medical practitioners, reduce unnecessary testing, promote appropriate use of laboratory and other diagnostic procedures, and provide standards of care for individual patients and community services, such as long-term care facilities and home health agencies.

The Canadian Medical Association issued its *Handbook on Clinical Practice Guidelines* in 2007, based on a systematic review of the literature, interviews of key professionals, consensus conferences, and continuing evaluation of both process and content of such guidelines. The Guideline International Network Fourth International Conference held in Toronto in 2007 involved experts in national and international practice guidelines from 31 countries to share experience and concepts in this ongoing field.

An Institute for Clinical Evaluation (ICES) organization at the University of Toronto was established in 1992 with core funding provided by Ontario's Ministry of Health and Long Term Care, mandated to conduct research that contributes to the effectiveness, quality, equity, and efficiency of health care and health services in Ontario. ICES uses an interdisciplinary research approach to health care, health services, and health policy.

The American College of Cardiology (ACC) provides a framework of evidence-based clinical statements and guidelines developed by leaders in the field of cardiovascular medicine with continuing adoption of new scientific information and experience in many aspects of this field (ACC, 2008).

The Health Care Financing Administration (HCFA) and the National Institutes of Health (NIH) have consensus programs to develop guidelines that are widely disseminated and set standards of practice. In 1977, the NIH issued its first consensus paper on breast screening for cancer, and this has been followed by many other topics each year since. Evidence-based consensus guidelines were issued on the following topics: breast cancer screening for women aged 40–49, interventions to prevent HIV risk behavior, management of hepatitis C, genetic testing for cystic fibrosis, acupuncture, and effective medical treatment for heroin addiction.

Clinical guidelines are increasingly being promoted by professional, governmental, and managed care organizations with the purpose of promoting rational use of health care resources and at the same time promoting standards of care to incorporate good standards of clinical practice. Clinical practice guidelines are now common in the practice of primary care, mental health, and clinical specialties. The University of Southern California's list of clinical guideline Web sites in 2003 (http://medicine.ucsf.edu/resources/guidelines/index.html) provides access to hundreds of Web sites for such practice guidelines.

Clinical guidelines provide practicing doctors, peer review committees, health care managers, managed care companies, governmental bodies, and professional organizations channels to set standards of practice and expectations of care standards. Legal aspects of health care also increasingly recognize the importance of clinical guidelines where committees of appropriate medical professionals convene and set out average or minimum standards of care for defined clinical entities. Thus, peer-reviewed guidelines set an appropriate standard (a silver if not a gold standard) for judging malpractice or adequate practice. Clinical guidelines should be under periodic review and subject to critical discussion and updating using the Cochrane review methods of literature review and analysis. Promotion by advocacy or special interest groups can be constructive, but influence of drug companies can be insidious and reduce the professional objectivity of such reviews and their recommendations, a concern that must be carefully monitored and continuously kept in mind as a potential compromising bias.

The American Academy of Pediatrics' (AAP) policy statements, practice parameters, and model bills have a wide distribution and are published in the Academy's journal, *Pediatrics*. The AAP clinical practice guidelines issued include diagnosis and treatment of urinary tract infection in febrile infants and young children, long-term treatment of the child with simple febrile seizures, management of acute gastroenteritis in young children, management of otitis media with effusion in young children, and others. The policy statements of the AAP cover a wide range of topics from use of bicycle helmets, to 55 mile per hour maximum speed limits, folic acid for the prevention of neural tube defects, to ethics in the care of critically ill infants and children. AAP guidelines are valid for five years only and are reissued or reconfirmed in order to keep up-to-date and to incorporate new or revised knowledge into practice standards.

Empirically derived, peer-reviewed, regularly updated guidelines have become an appropriate standard for practice and for judging malpractice, as well as balancing quality and cost-effectiveness. Clinical guidelines may become restrictive, but they help to reduce practice by whim and unsubstantiated belief to improve the quality of care overall. In large health care organizations they provide a basis for continuing education for staff and advancement of standards of the organization.

The *Community Guide* produced by the CDC provides an excellent source of evidence-based advice for community programs. It serves the needs of public health professionals, health care providers, legislators and policymakers, researchers, community-based organizations, employer-employee groups, and other purchasers of health services. The guide covers a wide range of health issues including alcohol, cancer, diabetes, mental health, motor vehicle safety, nutrition and obesity, oral health, physical activity, pregnancy, sexual behavior, social environment, substance abuse, tobacco, vaccines, violence, and workplace health issues.

The United Kingdom National Health Service established the National Institute for Health and Clinical Excellence (NICE) in 1999 as an independent organization to provide guidelines for public health, health technologies, and clinical practice guidelines for specific conditions. In 2005, the Health Development Agency of the NHS was included in the NICE organization. Guidelines are published and provide a helpful basis for clinical practice and public health as well as other areas in the NHS to update the services provided. Topics for public health include smoking and tobacco control, diet and obesity, exercise and physical activity, sexual and mental health, and alcohol.

ORGANIZATION OF CARE

Administrative and financing systems are essential elements of quality assurance. They can be designed to promote standards of care and to reduce fiscal incentives that foster excess supply and overservicing. The organization of financing health care has important implications for quality, technology, and ethical issues in the New Public Health.

Diagnosis-Related Groups

Diagnosis-related groups (DRGs), discussed extensively in Chapter 11, were developed in the 1960s as an alternative way of paying for hospital care in order to encourage shortened lengths of stay. Experience with payment by days of care (*per diem*) showed that it promoted unnecessary, long, and potentially dangerous use of hospital care, an important factor in escalation of costs in the health system.

The provider hospital is paid by the insurer for a procedure or diagnosis rather than the number of days of stay in hospital. This has led to a large reduction in hospital days of care and a remarkable growth in the number of surgical procedures done on an outpatient basis. Outpatient surgical procedures grew from less than one-fifth to more than half the inpatient cases. Outpatient surgery is safer for the patient and less costly to the insurer.

The DRG system is widely considered to promote quality of care as an active process focusing on quickly addressing the diagnosis and management of the patient with rapid mobilization of treatment and return home. Critics of this system allege that DRGs encourage inappropriate early discharge of patients before optimal patient education and follow-up care have been provided, but long length of hospital stay has not been shown to improve patient outcomes. Critics also suggest that this may promote altering diagnoses to higher-cost units of service. Others think that

DRGs, by reducing length of stay, have turned hospitals into intensive care units with ultra-sick patients. Despite these issues, the trend toward short hospital stays and newer approaches to active treatment seems to be compatible with better care and improved outcomes, according to some measures. The rapid decline in mortality rates from coronary heart disease is thought to be due in large part to the activist treatment approach, with lengths of stay of 1 week or less for acute myocardial infarction compared to 6 weeks on average up to the 1970s.

Managed Care

Managed care systems developed in the United States in response to rapid cost escalation for health care and the successful experience of health maintenance organizations (HMOs). Managed care is based on the concepts of resource management, and quality assurance with rationalized use of technology. The system developed over time with checks and balances to provide comprehensive care at lower cost than traditional fee-for-service systems by discouraging excessive utilization without compromising quality of service. Managed care systems include traditional HMOs and various other organizations which employ physicians or are made up of independent physicians working together who own or contract for hospital services. HMOs, both for-profit and not-for-profit, and managed care itself has been widely criticized as excessively limiting patient access to appropriate care in the interest of cost containment. This is a much debated topic in the United States.

District health systems in the United Kingdom, the Scandinavian countries, and the post-Soviet model of health care incorporate organizational and financial linkage between care systems and funding from tax sources. HMOs, sick benefit funds, and district health systems provide both prepayment and health services. Even in traditional private health insurance systems, the insurer is increasingly taking on the role of regulating reimbursement for medical services in order to contain costs and curb abuses by providers. Clinical indications, utilization review, and organizational and professional standards are now becoming accepted parts of the health insurance milieu.

The competition between hospitals for referrals from managed care plans in the United States has created a market situation in which a high proportion of hospital beds are empty, and in which mergers or closures of hospitals are common. Closures or reductions in hospital bed supply are also occurring in the United Kingdom and in most industrialized countries of Europe.

PERFORMANCE INDICATORS

Performance indicators (PIs) are measures such as morbidity, mortality, functional status, or immunization rates in a community, used to monitor the functioning of a health service. Routinely collected statistics are analyzed to compare performance against objectives, help monitor efficiency and effectiveness, point out problem areas within the service, and plan new health programs. This is based on the use of the concept of management-by-objectives in health administration to promote achievement of national health targets.

The United Kingdom has a strong tradition of mapping diseases as a basis of epidemiologic analysis and has applied this strategy to mapping of PIs to assess health care performance. The U.K. financing system is based on capitation adjusted by standardized mortality rates on the premise that mortality rates standardized and compared to the national average serve as indicators of need. In this way, the approach helps to promote equitable funding among wealthy and poorer regions of the country, and thereby improve services in areas of greater need.

PIs were introduced into the NHS during reforms of the late 1980s, providing a series of outcome or performance measures that are used to adjust payments allocated on a per capita basis to district health authorities (DHA). These authorities can be penalized for low rates of immunization, whereas general practitioners receive incentive payments for full immunization coverage. The result was a rapid improvement in immunization coverage of infants and children compared to rates in the previous decade. Incentive payments in many countries encourage women to go to hospitals for delivery or to attend prenatal care by making social maternity grants conditional on seeking care.

Use of PIs requires development of health information systems with district health profiles to provide ongoing monitoring of health indicators in a district, compared to regional and national rates and targets. Health profiles help to establish and monitor the prevalence of chronic disease and measure the impact of health services. This helps to study the performance of preventive and curative services, such as managing hypertension to reduce the incidence of strokes and related conditions. There are criticisms of PIs alleging a potential for manipulation and abuse of health intervention measures when the financial incentives are used for a specific activity. However, financial incentives are part of the DRG system and have been successfully used in the United Kingdom to improve vaccination coverage and implementation of other preventive health practices by family physicians. In Israel in 2007, payments to hospitals provided a bonus for surgical interventions for hip fracture within 48 hours of the event resulting in a marked rise in early intervention and a reduction in mortality from hip fractures.

CONSUMERISM AND QUALITY

With decentralization and the growth of managed care, health systems must increase their attention to the attitudes of the consumer. Quality is, in part, how the client perceives the system, and how the system meets client needs in an acceptable manner, where privacy, dignity, the right to know, and the right to a defined set of services are protected. However, the rights of the client are not unlimited. A public or private health plan has the duty to manage the basket of services responsibly, which includes limitations such as in access to specialist services.

Patients' rights and consumer protection in health care often (but not always) include the right to select and change a health care provider, as well as the right to receive high-quality care for a designated range of services. The U.K. NHS issued a patient's Charter of Rights during the 1990s, which is perhaps idealistic and may not be actualized in practice, but still outlines an ideal of value both for practical application and for legal rights. The consumer's formal protection includes the right to complain and to seek redress of grievance and compensation for injury suffered from neglect or incompetent care (Table 15.8).

The consumer needs to be informed and conscious of health care costs if efforts to restrain cost increases are to be effective. Public attitudes are vital in terms of self-care, demands on the health service, and limitations to the potential of health care as well as resources for health care. The media and consumer organizations can play important roles in advocacy for health, in raising public consciousness of self-care, and as watchdogs on abuses.

Consumer acceptance is manifested through choice of health plan and practitioner, or by seeking alternative care privately when a service is unacceptable because of quality or style. Erosion of confidence in a public system of care can lead to a two-tier system with the public system serving the poor and a private parallel system serving the middle and wealthy classes. Such a division can seriously undermine a public system unless it is addressed by improving the quality and manner of the service and by establishing supervision and limitations on public and private practice.

The growing inequality caused by the rise of private practice outside of a national health care system is a chronic problem in the United Kingdom's NHS, in Israel's health system, and in many countries developing their health systems through parallel public and private care. The issue is also surfacing in the United States in the transition to managed care with its inherent limitations of choice for people insured through their place of work or covered under the Medicare and Medicaid programs. Extra billing, banned in Canada's national health insurance plan, is a recurring issue with the medical profession.

Consumer knowledge, attitudes, beliefs, and practices are part of the health system, from health promotion to tertiary care. Informed and health-conscious consumers are stronger partners in the health system in achieving improved health than an ill-informed and apathetic public, so that health education and health promotion are fundamental to modern public health. The role of the consumer in health care is unique in that there is a significant information asymmetry between the consumer and provider. Health education programs and wide use of the Internet increase access to health and medical information, but this gap can never completely be eliminated. Patients may use their power as consumers to demand inappropriate care — such as unnecessary surgery or antibiotics when clearly not indicated, because of their preference for intervention and action over watchful waiting. However, there is an equal or perhaps greater danger of provider-induced demand for repeated and possibly unnecessary interventions that may be related to methods of paying the doctor or the hospital. The traditional doctor–patient relationship is still an important factor for the interests of patients and their health. A still effective method of having an individual quit smoking is a brief but stern lecture by the family physician.

THE PUBLIC INTEREST

Population-based interventions are often more effective and less costly ways to reduce morbidity and mortality than individual prevention or treatment services. A population-based preventive program may require behavior change by the individual, such as in mandatory seat belt

---

TABLE 15.8 Patient's Charter of Rights, United Kingdom National Health Service.

1. To receive health care on the basis of clinical need regardless of ability to pay
2. To be registered with a general practitioner (GP)
3. To receive emergency medical care at any time through your GP or the emergency ambulance service, hospital accident and emergency departments
4. To be referred to a consultant acceptable to you when your GP thinks necessary, and to be referred for a second opinion if you and your GP agree this is desirable
5. To be given clear explanations of any treatment proposed, including any risks and alternatives before you decide whether you will agree to the treatment
6. To have access to your health records and to know that those working for the NHS are under a legal duty to keep their contents confidential
7. To choose whether you wish to take part in medical research or medical student training
8. To be given information on local health services including quality standards and maximum waiting times
9. To be guaranteed admission for treatment by a specific date no later than 2 years from the day when your consultant places you on the waiting list
10. To have any complaint about NHS service investigated and to receive a full and prompt reply from the chief executive officer or general manager

Note: The Charter was developed in 1997.
Source: http://www.dh.gov.uk/en/AdvanceSearchResult/index.htm?search-Terms=patient%27s+charter [accessed December 23, 2007]

and motorcycle helmet enforcement or banning smoking in public places. Fortification of flour, milk, and salt with essential micronutrients is a well-established public health measure. There is an element of compulsion in this, with the social gain usually considered to be sufficiently important to outweigh individual rights. The delicate balance between community rights and individual rights is at the heart of many controversies in modern public health and health care, from chlorination of community water supplies to managed care systems for health services. Each issue has to be examined on its merits, especially in terms of what is accepted as good public health practice, based on documented experience, clinical trials, and practice in other countries. The evidence of successful public health measures in improving individual and collective health status is powerful, yet must always be balanced within the context of individual rights and the public interest.

TOTAL QUALITY MANAGEMENT

Total quality management (TQM), discussed in Chapter 12, was also adapted to health care in the 1990s and provides a basis for promoting continuous improvement in health care systems. TQM involves everyone in the system, from all levels of management to production or service personnel and support staff and thus helps raise staff morale because of the shared involvement. Health is provided through multidisciplinary groups which need to approach problems with open and shared scientific inquiry and hypothesis formation, testing, and revision to find operational solutions to problems.

TQM incorporates statistical methods, comparing variations in patterns of service or use of resources. It employs epidemiologic methods to draw conclusions for policy needs. It looks for continuous improvement, encouraging cooperation, and motivation to achieve common goals of service and client satisfaction. Psychological theory helps to foster higher levels of motivation, with early identification and resolution of conflict. Leadership is shared, and there is a basic need for cooperation. Cost and quality are interrelated, as poor quality leads to waste, inefficiency, and dissatisfaction of clients and staff. High-quality, humane, and effective services are especially important in a competitive environment where clients have the right to choose and where costs and efficiency are factors in the well-being and indeed the survival of institutions.

Medical care is increasingly practiced in larger health care programs. To provide technically competent medicine is not by itself sufficient. The patient's rights and sense of personal worth are also of great importance. Financial incentives to redirect health care priorities, such as in reducing hospital length of stay and admissions, may result in the patient or the family feeling they are not receiving the best care. DRGs, HMOs, and other organizational and funding systems meant to increase efficiency of care may have the effect of alienating patients from a health care system. Staff attitudes toward patients are important for client satisfaction. The service must include ready access to a continuum of supportive services, such as home care and counseling, so that the patient and family do not feel abandoned by the system.

A by-product of TQM is continuous quality improvement (CQI) by which institutions wishing to improve quality train and empower the staff to work in teams to assess their own performance and seek solutions to problems in their operational unit. People of different ranks and professions work in a network organization as well as in a traditional hierarchical organization in which rank and seniority provide authority. This community of practice is important for staff morale and a shared sense of responsibility for the patient and the institution.

CQI involves multidisciplinary approaches, not only to review problems, but also to seek better ways of functioning and improving consumer satisfaction. The process includes all those involved in providing care, support services, and administration of a department, hospital, clinic, or community health program. This is not only professional self-policing, but a method to find better ways of meeting needs and using resources. The involvement of all providers improves motivation and promotes a sense of common purpose in the organization.

Applying these principles in a health care setting can take many forms. Selection of topics by TQM/CQI committees in a hospital or other health facility may be based on surveys or interviews with staff, patients, or management. Satisfaction surveys among women following delivery in an obstetrics unit could point out remediable problems. An obstetrics department may be faced with issues related to high or low volume of deliveries, staff training, equipment and supplies, communication among staff and among staff and patients and their families, cleanliness, sterile technique, staff satisfaction, client satisfaction, and many others. The team looking at such a problem should be multidisciplinary, and emphasis should be on client attitudes and satisfaction. Examination of the function of an emergency department (ED) in a hospital would similarly look at many functional and attitudinal aspects of the service including staff attitudes, training needs, waiting times, consultation services, and others. Addressing waiting times, for example, can lead to ways to reduce this substantially, improving both client satisfaction and the efficient management of the ED. Any service is there to serve patients and the community. A service is not primarily for the benefit of the staff, but staff satisfaction and morale are essential for successful service to clientele. CQI can also be applied to assessing and improving compliance with clinical guidelines or evidence. An example is assessing the proportion of diabetics whose HbA1c is measured at least twice annually, who have eye and feet examinations regularly, or the number of diabetics whose blood pressure is managed with an ACE inhibitor.

The European Region of the World Health Organization and the national medical associations in Europe in 1995 agreed that medical associations should take leading roles in programs of CQI to achieve better outcomes of health care in terms of functional ability, patient well-being, consumer satisfaction, and cost-effectiveness. This is in keeping with the European Region's *Health for All* targets: that there should be structures and processes in all member states to ensure continuous improvement in the quality of care and appropriate development and use of health technologies.

The 1990s introduction of general practitioner fund holding for hospital care for patients on the general practitioners' roster in the United Kingdom encourages the hospital to maximize patient satisfaction with the care system. This promotes application of CQI to improving the quality and acceptability of care. Similarly, performance indicators provide regional and district health authorities in the United Kingdom with tools for CQI approaches. The U.K. National Health Service established its National Institute for Health and Clinical Excellence (NICE) as an independent body to promote "national guidance on promoting good health and preventing and treating ill-health." NICE produces guidance in three areas:

- "Public health — guidance on the promotion of good health and the prevention of ill health for those working in the NHS, local authorities and the wider public and voluntary sector
- Health technologies — guidance on the use of new and existing medicines, treatments and procedures within the NHS
- Clinical practice — guidance on the appropriate treatment and care of people with specific diseases and conditions within the NHS"

NICE guidelines are recommended practices with the object of reducing ineffective practices. During 2007, guidelines were issued on topics including asthma, dermatitis, caesarean section, chronic obstructive lung disease, depression (in children and adults), eating disorders, fertility, contraception, multiple sclerosis, post-traumatic stress disorder, and diabetic foot care.

The United States has a number of government and independent organizations dedicated to improving quality in health care systems. The Centers for Disease Control and the Institute of Medicine of the U.S. National Academies of Science play active roles in promoting research quality and methods of CQI in the U.S. health care system. Canada also is very active in this with national and provincial institutes for evaluation of clinical effectiveness and clinical guidelines, as are European countries (Box 15.5).

PUBLIC HEALTH LAW

Law consists of a system of rules, regulations, and orders that govern the behavior of individuals and of society. Law represents the consensus of a society, as enacted by an elected legislature, put into effect by the executive branch of government, and interpreted by the courts as need be from time to time. The legislative and executive branches are separate under the U.S. Constitution, but united in the parliamentary system (see Box 15.6). The authority, responsibility, and power to provide for and protect the public health are basic functions of a sovereign government, which may be delegated to another level of government (higher or lower) or even a nongovernmental agency. The constitution of a sovereign government states explicitly or implicitly that responsibility, but accepted practice and court decisions (i.e., the common law) define the powers of the national, state, or local government to monitor and protect the health of its citizens.

In the United States, national legislation is enacted under the powers of the federal government, namely to regulate interstate commerce and the power to tax and spend for the general welfare. State legislation is enacted under the basic power of the state to protect the health, welfare, and safety of its citizens. Under these federal and state powers, a wide range of health legislation and regulations are enacted affecting public health, labor, and occupational health and safety, environmental controls, public welfare and the financing of health services, agriculture, food, drugs, cosmetics, and medical devices. Public health law relies on a wide range of constitutional, statutory, administrative, and judicial decisions in both civil and criminal actions. Appropriation of funds is a legal act of legislative bodies to achieve objectives directly or indirectly by financial incentives.

Categorical programs may be directed to specific issues such as combating tuberculosis and promoting immunization or for work to combat noncommunicable diseases such as diabetes, or in improving standards of facilities, and in providing health care services. The regulatory, enforcement, policing, and punitive functions of public health laws have evolved over many decades and in many countries lack clear definition. In the United States, efforts are being made to update and reform laws in the public health sector. In 1988, the Institute of Medicine in the United States (the Future of Public Health) called for codification of public health law as essential for the public good, while questioning the soundness of certain U.S. public health laws. More recently, the Model State Emergency Health Powers Act in the United States, the new Quarantine Act in Canada, and the International Health Regulations (2005) have sought to update century-old legislation.

A combination of the regulatory, persuasive, and funding approaches is widely used in public health in control of communicable and noncommunicable disease, in

---

Box 15.5 Organizations to Promote Quality in Health, United States and Canada

1. **NCQA, National Council for Quality Assurance** This non-profit organization founded in 1979 by the managed care industry conducts surveys among managed care plans to evaluate clinical standards, members' rights, and health service performance. It accredits over 550 managed care plans in the United States, and in 2007 published rankings of the "best" health plans. Website http://web.ncqa.org/tabid/577/default.aspx

2. **AHRQ, Agency for Healthcare Research and Quality** This is part of the U.S. Public Health Service. Founded in 1995, it was mandated to develop an evidence-based practice program in 12 centers in the United States and Canada. It conducts systematic reviews of the literature and publishes analyses and findings of these reviews along with clinical guidelines, quality improvement projects, and purchasing decisions for health plans. It funds research on outcomes and cost-effectiveness studies, and disseminates new information and guidelines for medical practice and cost-effectiveness. Website http://www.ahrg.gov/

3. **HCFA, Health Care Financing Administration** HCFA is the federal agency of the Department of Health and Human Services founded in 1977, responsible for administering the Medicare and Medicaid health plans and quality assurance. In the 1990s, the HCFA established requirements for managed care organizations and quality improvement in health care. This requires health plans to provide evidence of improvement in the health care they provide, stressing a move from payment to quality assurance. Website http://www.os.dhhs.gov/about/opdivs/hcfa.html

4. **IHI, Institute for Healthcare Improvement** Founded in 1991, this non-profit organization aims to improve health care in Canada and the United States by fostering collaboration among health care organizations. IHI examines office practices of physicians, educational reform, and promotes interdisciplinary team work in quality improvement. Website http://www.ihi.org/ihi/

5. **NPSF, National Patient Safety Foundation** Sponsored by the American Medical Association as a response to findings of high rates of injury and death from iatrogenic disease in the United States, the NPSF promotes research into human error among health care providers, seeking ways to reduce the frequency and effects of medical error, such as misdiagnosis, medication errors, and mistakes during procedures. Website http://www.npsf.org/au/

6. **JCAHO, Joint Commission on Accreditation of Healthcare Organizations** Originating in 1917 as an activity of the American College of Surgeons, it began accrediting hospitals in 1918. It developed in 1953 as the JCAHO, becoming a national voluntary accreditation organization focusing mainly on hospitals. Its mandate was broadened in 1987 and as of 2007, has accredited more than 15,000 health care organizations. Accreditation is mandatory for Medicare and Medicaid payment. The JCAHO is changing its approach from standards-based assessment every 3 years to one of reviewing performance data quarterly as a continuous surveillance activity for risk reduction. Website http://www.jointcommission.org/

Source: Websites accessed November 24, 2007.

Box 15.6 The Legal Structure of a Federal Country

- Constitution: the supreme law of a country, setting out the divisions of governmental powers including statutory authority, administrative, and taxing powers
- Federal legislature
- State/Provincial legislature
- Local county/municipal/city governments.

In federal countries such as the United States, Canada, Argentina, and Nigeria, there are also state constitutions, legislatures, and court systems, and public agencies. In Unitary governments, there are two levels, the national and local authorities, but the national may appoint regional authorities for various governmental functions. (See Chapter 10.)

---

government is essential for public health to ensure that adequate facilities and access to care are available to all members of the community, especially those in financial need and thus at greater risk for disease.

Medical officers of health (MOH) are limited to issuing orders. Laws may be enacted to fund public health activities whether provided by public health authorities or by acting through official or nonofficial agencies or providers. Public health authorities, namely MOHs, have the legislative power to issue orders to individuals or businesses where there is a threat to the health of the public such as food establishments. Situations which may require enforcement by court proceedings are referred to the justice system. Administrative resources are needed to enforce law, such as the Food and Drug Administration (FDA) and the Environmental Protection Agency (EPA), which come under the aegis of the Department of Health and Human Services. Other departments such as Agriculture, Education, or interdepartmental agencies (e.g., Homeland Security), also are key to public health activities, such as in disaster situations. Other

improving standards of facilities, and in providing health services. The regulatory, enforcement, policing, and punitive functions of public health are important in health promotion and assurance of health care. The taxing power of

intergovernmental activities may require special legislation to empower, finance, and promote their cooperation such as in the case of establishing an authority to manage long-term efforts to clean up a contaminated river or basin, which involves cooperation and coordination of many local authorities.

In the NPH, health protection of individuals and communities may require legal powers to detain a person in order to prevent spread of a reportable communicable disease, to protect a mentally ill patient, or to restrain a violent person. Such powers should be used as a last resort if voluntary compliance and education fail, and where the danger to the community or the individual is sufficient to convince a court of the public need to override the personal liberty of an individual. An example is a 2007 case of a person with multidrug-resistant tuberculosis who was taken into custody on arrival for compulsory treatment after traveling across the Atlantic Ocean on a commercial airline, against specific instructions of his physician, thus endangering fellow passengers. Outbreaks of measles in the United Kingdom (2006–2007) and in Israel via imported cases among ultra-orthodox Jews with transmission among religious people who tend not to immunize their children led to pressure by health authorities to immunize those placed at risk by such contacts at weddings or other large public events. However, these measures are used less, currently, than voluntary isolation/quarantine and placarding homes for reportable infectious diseases such as measles. Powers in extreme cases where refusal to comply with public health measures endangers others are essential. These powers should have been used in recent cases in the United States, the United Kingdom, and Finland where potentially dangerous mentally disturbed individuals first threatened to carry out mass killings and subsequently did so. These powers should have been used more vigorously in the early years of the AIDS epidemic at a time when individual rights took precedence over protection of the population including vulnerable high-risk groups.

Public health has generally evolved with greater reliance on health promotion though voluntary cooperation of a patient than on compulsion. Enabling legislation may permit a local authority to fluoridate its water supply, but enactment of local legislation and funding to implement it may also require a public referendum. In some states in the United States and in Israel, fluoridation of community water supplies is mandatory, which is also part of the health promotion approach to public health.

Appropriation of public funds to promote public health is through approval by the legislature for a specified program. Provision of public funds may take the form of categorical grants for specified services, such as immunization, prenatal care, school health, or specific disease management such as tuberculosis control, cancer control, or AIDS education. Programs may be designed to promote certain types and quality of services, such as the Hill-Burton Act, which provided federal grants for hospital construction in the 1950s to 1970s, conditioning these grants on certain requirements concerning hospital licensure and hospital planning. Such legislation has both a "carrot and stick" effect of attracting lower levels of government to seek such funding but also requiring them to accept the conditions and regulations that accompany the grants. The Canadian federal government's cost sharing of provincial health insurance programs is based on federal criteria requiring public administration, portability between provinces, accessibility without payment, comprehensiveness, and banning extra billing by physicians (Chapter 13).

Public funds are also appropriated in the context of legislated programs in which people are entitled to the services defined in the appropriation legislation, such as in the amendments to the Social Security Act providing Medicare and Medicaid programs, or national health insurance legislation in many countries. These acts and the regulations spell out categories and specified entitlement benefits.

Legislation and court decisions to protect the rights of the individual are part of public health. Public health law covers individual and community life, including the need to protect the individual from potential abuse, in keeping with protection under law, as in the U.S. Bill of Rights. Enforcement of public health law may infringe on individual rights by enforcing sanitation laws and on civil rights by rarely used mandatory treatment of a person with dangerous contagious disease or mental illness. Freedom of religion may come into conflict with other laws in public health where restrictive practices may deny use of publicly supported health facilities, as when a religiously affiliated hospital may refuse an abortion procedure in a case of rape. On the other hand, religious practices or other beliefs may endanger others in the community, such as in refusal to immunize children so that an infectious disease may spread among nonimmunized persons and even affect those who are immunized, as occurs with imported measles cases even when domestic transmission of the disease has been eradicated. General legislative provisions applied to public health forbid misleading or unethical advertising. Legislative provisions may also ban advertising for products, such as tobacco, which are legal but may be harmful to health. These laws affect public health but are provisions in other statutes such as the regulation of business enterprises. Legislation may also make smoking in public places illegal with fines for the offender and the operator of a place such as a bar or pub.

Since the 1973 U.S. Supreme Court decision of *Roe v. Wade*, the law allows women to seek safe and legal abortion. This remains a highly controversial political issue in the United States and several other countries. The potential conflict between community and individual interests and rights is part of the dynamics of public health law and public health practice. The issues involved are complex, highly politicized, and often involve ethical distinctions where "the greatest good for the greatest number" may limit legitimate rights of individuals and vice versa. The legal aspects of public health are vital to its operation and are increasingly complex as they overlap with ethical issues and public debate.

Environmental Health

There is growing concern by the public and by governments over air and water pollution, other noxious and harmful industrial processes, and global warming. Environmental laws affecting the public health include legislation on clean air, clean water, toxic substances, solid waste control, and other noxious substances. Noncompliance with the legislative provisions can result in prosecution in the civil or criminal courts or both.

Infringement of public health laws and regulations may lead to criminal action as an increasingly common method of sanction. While these may not be seen as "truly" criminal and might be treated in the courts as misdemeanors, they may lead to fines or even jail. Such cases are increasingly being addressed seriously in the judicial system.

The CDC, in 1999, defined ten great achievements of public health of the twentieth century. The ten great achievements are identified (Goodman et al., 2006) as control of infectious disease, motor vehicle safety, fluoridation of drinking water, recognition of tobacco use as a health hazard, immunization, decline in deaths from coronary heart disease and stroke, safer and healthier foods, healthier mothers and babies, family planning, and safer workplaces. This was a result of a combination of supportive laws and legal tools at the local, state, and federal levels. These are examples which have similar legislation in the industrialized countries where equal or greater achievements were made in public health over the past century.

Public Health Law Reform

Public health law is scattered through many legislative and administrative documents developed historically. Efforts to codify public health law may contribute to greater understanding and enforceability of the many now separate pieces of legislation (Box 15.7). This will enhance understanding in the legislative, judicial, and administrative branches of government as well as in business, nongovernmental organizations, and the community. Box 15.8 suggests topics for model public health consolidation or compendia for states. The principles of this formulation may also apply to other countries at the national and state/provincial levels.

Box 15.7 Centers for Disease Control: Institute of Public Health Law

The CDC Foundation Institute of Public Health Law as an operating agency complements CDC's Public Health Law Program and promotes research, training, and outreach to strengthen and create networks, study trends, encourage best practices, share lessons learned, and offer practical solutions for putting public health law into action during public health emergencies and crises.

"The Institute will focus on five major areas:

Creating a network of international public health law experts to promote professional collaboration and communication around urgent public health issues and emergency preparedness;

Conducting outreach to businesses in the U.S. and abroad to explore shared interests and concerns about public health law and emergency preparedness;

Addressing community development issues and "smart growth" practices, such as zoning policies, traffic congestion, and environmental concerns;

Developing training tools to educate public health professionals about how law is used in a public health crisis;

Coordinating interdisciplinary research about the growing impact of communications technology on public health law and policies during emergencies."

Source: Centers for Disease Control Foundation, http://www.cdcfoundation.org/programs/iphl/index.aspx [accessed May 23, 2008]

Box 15.8 Model Topics for a State Public Health Act

- Mission and Functions
- Public Health Infrastructure
- Collaboration and Relationships
- Public Health Authorities and Powers
- Public Health Emergencies
- Public Health Information privacy
- Criminal/Civil
- Enforcement

A legislative response to the need to reform core public health powers like surveillance, reporting, epidemiologic investigations, partner notification, testing, screening, quarantine, isolation, vaccination, and nuisance abatement.

Source: Center for the Law and the Public's Health at Georgetown and Johns Hopkins Universities (CDC Collaborating Center) at Website http://www.publichealthlawnet/ [accessed November 2007]

ETHICAL ISSUES IN PUBLIC HEALTH

The field of public health includes a wide range of activities and professional disciplines, ranging from health promotion to disease protection, epidemiology to environmental health, as well as financing and supervision, or provision of clinical care. Each of these disciplines works within systems that face ethical dilemmas, and the public health worker's understanding and motivation within the ethical guidelines of his profession and role are important in his/her training and practice conduct. Ethical frameworks have evolved in part due to bitter experience with ethical failures later recognized and affecting public health standards of practice for future generations (Box 15.9).

Ethics in health are based on the fundamental values and concepts of a society. If the principle of saving a life is valued above all other considerations (i.e., Sanctity of Life or *Pikuah Nefesh*) (see Chapter 1), then all measures available are to be used, irrespective of the condition of the patient or the cost. If sickness and death are seen as acts of God, possibly as punishment for sin, then prevention and treatment may be considered to be interfering with the divine will, and the ethical obligation may be limited to relief of

suffering. Humanism balances these two ethical imperatives: saving of life and relief of suffering. Materialism may see health care as primarily a function to preserve health for economic productivity. The role of society in protecting the health of the population grew during the nineteenth century with the sanitation movement, while medical care became an effective part of public health during the twentieth century. The Lalonde concept prevalent during the 1970s regarded individual behavior as one of the key health determinants (Chapter 2), placing much of the onus of illness and its prevention on the individual. All these points of view are involved in the ethical issues of the New Public Health (see Box 15.10).

Resources for health care are limited even in industrialized countries, and as such, priority setting and judicious allocation of scarce resources is always an issue. Money spent on new technology with only marginal medical advantages is often at the expense of well-tried and proven lower-cost techniques to prevent or treat disease. The potential benefits gained by the patient from more and more interventions are sometimes very limited in terms of length or quality of life. These are difficult issues when the physician's commitment to do all to preserve the life of the patient conflicts with the patient's concept of quality of life and his or her right to decline or terminate heroic measures of intervention. The suffering that a terminal patient may endure during radical treatment and which

---

Box 15.9 Study and Practice of Public Health Ethics

Ethics is a branch of philosophy that deals with distinctions between right and wrong — with the moral consequences of human actions. The ethical principles that arise in epidemiologic practice and research include:

- Informed consent
- Confidentiality
- Respect for human rights
- Scientific integrity

"As a field of study, public health ethics seeks to understand and clarify principles and values which guide public health actions. Principles and values provide a framework for decision making and a means of justifying decisions. Because public health actions are often undertaken by governments and are directed at the population level, the principles and values which guide public health can differ from those which guide actions in biology and clinical medicine (bioethics and medical ethics) which are more patient or individual-centered.

As a field of practice, public health ethics is the application of relevant principles and values to public health decision making. Public health ethics inquiry carries out three core functions,

1) identifying and clarifying the ethical dilemma posed,
2) analyzing it in terms of alternative courses of action and their consequences, and
3) resolving the dilemma by deciding which course of action best incorporates and balances the guiding principles and values (CDC)."

Sources: Last, J. M. (ed.). 2001. A *Dictionary of Epidemiology*, Fourth Edition. Centers for Disease Control, Science Coordination and Innovation. Public Health Ethics, http://www.cdc.gov/od/science/phec/ [accessed May 23, 2008]

Box 15.10 Public Health — Ethical Principles

1. **Harm Principle** The only purpose for which power can be rightfully exercised over any member of a democratic community, against his will, is to prevent harm to others.
2. **Least Restrictive or Coercive Means** Restriction on civil liberties must be legal, legitimate, necessary, and use the least restrictive means available.
3. **Reciprocity Principle** Public health interventions impose burdens on individuals or populations and where feasible, reasonable compensation should be considered.
4. **Transparency Principle** The manner and context in which decisions are made must be clear and accountable.
5. **Precautionary Principle** Decision-makers have a general duty to take preventive action to avoid harm before scientific certainty has been established.
6. **Failure to Act** Public health officials and policymakers have a duty to act and implement preventive health measures demonstrated to be effective, safe, and beneficial to population health. Failure to enforce public health regulations with resulting disease or deaths may constitute negligence on the part of responsible officials with civil or criminal penalties.

Source: Modified from Upshur, R. E. G. 2002. Principles for the justification of public health interventions. *Canadian Journal of Public Health*, 93:101–103.

may prolong life by only hours or days clashes with the physician's ethical obligation to do no harm to the patient. The ethical value of sustaining the life of a terminally ill patient suffering extensively is an increasing medical dilemma. The issue is even more complex when economic values are included in the equation. There is a potential conflict between the economic issues, the role of the physician in preserving life, the physician's obligation to do no harm, the felt needs of the patient and his or her family, and the needs of the community as a whole.

The state represents organized society and has, among its responsibilities, a duty to promote healthful conditions and to provide access to health care and public health services. The conflict between individual rights and community needs is a continuous issue in public health. Application of accepted public health measures for the benefit of some people in society may require applying an intervention to everyone in a community or a nation. The majority thus are subject to a public health activity to protect a minority, without designating which individual's life may be saved. Further, a society may in special cases need to restrict individual liberties to achieve the goal of reducing disease or injury in the population. Raising taxes on alcohol and tobacco products, mandatory speed limits, driving regulations, and seat belt usage laws are examples of public health interventions which interfere with individual liberty but protect those individuals and thereby the community at large, from potential harm.

Some forms of mass medication are accepted forms of public health practice to reduce the risk of disease in the population. Chlorination of community water supplies is a well-established, effective, and safe intervention to protect the public health. Fluoridation of drinking water to prevent tooth decay in children means that other persons are also drinking the same fluoridated water, which is of less direct benefit to them. Fortification of foods with vitamins and minerals is also a cost-effective community health measure with advocates and opponents. The addition of folic acid to food as the most effective way to prevent neural tube defects in newborns is an intervention mandated by the U.S. FDA since 1998.

Confidentiality to assure the right of the individual to privacy involves ethical issues in the use of health information systems. Birth, death, reportable conditions (not all reportable diseases are infectious), and hospitalization data are basic tools of epidemiology and health management. The use of detailed individual data is needed for case-finding and follow-up activities which are vital to good epidemiologic management of diseases, including STIs. However, caution is needed in data use to avoid individual identification that could be used punitively, for example, in denial of access to health insurance for smokers, alcoholics, or AIDS patients because health damage may be attributable to a self-inflicted risk factor. Increasingly, however, reporting is also mandatory for

physical or sexual abuse and criminally linked injuries as essential for protection of individuals at risk or the general public from serious harm.

Individual and Community Rights

The protection of the individual's rights to privacy, and freedom from arbitrary and harmful medical procedures or experiments may come up against the rights of the community to protect itself against harmful health issues. This conflict comes into much of what is done in public health practice which has both an enforcement basis in law and practice as well as a humanitarian and protective aspect based on education, persuasion, and incentives. Society permits its governments to act for the common good, but sets limits which are protected by the courts and administrative appeal mechanisms.

Society has the right to legislate the side of the road on which one is permitted to drive, the speed permitted, and the wearing of seat belts and non-use of cellular phones during driving. Offenders may be punished by significant fines and are subject to strong educational efforts to persuade them to comply. Similarly the community must ensure sanitary conditions and prevent hazards or nuisances from bothering neighbors. Society must act to protect the environment against unlawful poisoning of the atmosphere, the water supply, or the ground. Enforcement is thus a legitimate and necessary activity of the public health network to protect the community from harm and danger to health. Table 15.9 shows a list of topics where individual rights and responsibilities predominate, and a second set of rights that are the prerogative of the community to protect its citizens against public health hazards. Sometimes the issues overlap and sometimes come to legal action where court decisions are needed to adjudicate precedents for the future.

The AIDS epidemic in the 1980s and 1990s raised a host of public health and ethical issues. Management of the AIDS epidemic is in some respects in conflict with the long-established role of society in contacting and quarantining persons suffering from transmissible diseases. It is not acceptable or feasible in modern society to isolate HIV carriers, but failure or delay of public health authorities even in the late 1980s to close public bathhouses in New York and other cities in the United States where exposure to multiple same-sex partners promoted transmission of the infection could be interpreted as negligence. During the 1980s, the politics of AIDS in the United States centered on concerns in the community of men who have sex with men (MSM) that HIV testing would be used in a discriminatory manner. AIDS was initially addressed as a civil liberties issue and not as a public health problem. Screening, reporting, and case contact follow-up were seen as an invasion of privacy and counterproductive by increasing resistance to and

TABLE 15.9 Individual and Community Rights and Responsibility in Health: Ethical/Legal Issues

Ethical/Legal issues	Individual rights and responsibility	Community rights and responsibility
Sanctity of human life	Right to health care; responsibility for self-care and risk reduction	Responsible for providing feasible basket of services, equitable access for all
Individual vs. community rights	Immunization for individual protection	Immunization for herd immunity and community protection; education; community may mandate immunization
Right to health care	All are entitled to needed emergency, preventive, curative care	Community right to care regardless of location, age, sex, ethnicity, medical condition, economic status
Personal responsibility	Individual responsible for health behavior	Community education to health-promoting lifestyles; avoid "blame the victim"
Corporate responsibility	Producer and individual manager accountability to criminal and civil action	Producer, purveyor of health hazard accountable for individual and community damage
Provider responsibility	Professional, ethical care and communication with patient	Access to well-organized health care, accredited to accepted standards
Personal safety	Protection from individual and family violence	Public safety, law enforcement, protection of women, children, and elderly; safety from terrorism
Freedom of choice	Choice of health provider; limitations of gate-keeper function; right to second opinion; right of appeal	Community responsibility to control costs while ensuring individual rights; limitations of self-referrals to specialist
Euthanasia	The individual's right to die; limitation by societal, ethical, and legal standards	Assurance of individual and community interests; prevention of abuse by family or others with conflict of interests
Confidentiality	Right to privacy and security of medical information	Mandatory reporting of specified diseases; data for epidemiologic analysis
Informed consent	Right to know risks vs. benefits and agree or disagree to treatment or participation in experiment	Helsinki committee approval of research; regulate fair practice in right-to-know; Patient's Bill of Rights
Birth control	Right to information and access to birth control and fertility treatment; right to safe termination of pregnancy; woman's control of her body	Political, religious promotion of fertility; alternatives to abortion; protection of women's rights to choose
Resources for health	Universal access, prepayment; individual contribution through the workplace or taxes	Solidarity principle and adequate funding; right to cost containment, limitations on service benefits
Regulation and incentives to promote community health	Social security payments (e.g., for hospital delivery), attendance for prenatal care; care in the community and ambulatory care; supportive community care services	Incentive grants to assist communities for programs of national interest; limit institutional facilities, transfer of resources to primary care
Global health	Human rights and aspirations; economic development, health, education, and jobs	Transfer of health risks; occupational hazards and environmental damage
Rights of minorities	Equality in universal access	Special support for high-needs groups
Prisoners' health	Human rights; prevention of torture, executions, harsh conditions in prisons	Security and human rights
Allocation of resources	Lobbying, advocacy for equity and innovation	Equitable distribution of resources; targeting high-risk groups; cost containment

avoidance of testing. The educational approach was adopted as most feasible and acceptable. The AIDS epidemic and public anxiety about contracting AIDS through casual contact reinforced the need for public education on safe sex. This has been raised as an ethical issue because such education may be construed as condoning teenage and extramarital relations. The issue of HIV screening of pregnant women in general or in high-risk groups took on a new significance with the findings that treatment of the pregnant woman reduces the risk of HIV infection of the newborn, and that breastfeeding may be contraindicated. This issue is arising anew in the context of using human papillomavirus (HPV) vaccine for preteen girls to prevent the sexually transmitted infection, and in the United States this will be mandatory for school entry.

A preeminent ethical issue in public health is that of assuring universal access to services, and/or the provision of services according to need. An important ethical, political, and social issue in the United States in the twenty-first century is how to achieve universal access to health care. The solidarity principle of socially shared responsibility for funding universal access to health care is based on equitable prepayment for health care for all by nationally regulated mechanisms through place of work or general revenues of government. A society may see universal access to health care as a positive value, and at the same time utilize incentives to promote use of services of benefit to the individual such as hospital care, immunization, screening programs, and others. Some services may be arbitrarily excluded from health insurance, such as dental care, although this is to the detriment of children and a financial hardship for many. Strategies for program inclusion are often based on historical precedent rather than cost-effectiveness or evidence. While efforts are being made to include more children in the program, the Medicaid system in the United States defines eligibility at income levels of 185 percent of the poverty line, thus excluding a high percentage of the working poor. This is a topic addressed in other chapters and a topic of continuing political importance in electoral platforms in the United States to address the challenge of the uninsured and poorly insured working poor population. Health is also a political issue in countries with universal health systems where funding may be inadequate or patient dissatisfaction common.

Choices in health policy are often between one "good" and another. Limitations in resources may make this issue even more difficult in the future, with aging populations, increasing population prevalence of physical disabilities, and rapid increases in technology and its associated costs. For example, the United Kingdom's National Health Service at one point refused to provide dialysis to persons over age 65. When computed tomography was first introduced, Medicare in the United States refused to insure this service as an untested medical technique. Due to a lack of facility resources such as incubators and poor prospects for the survivors, the Soviet health system considered newborns as living only if they weighed over 1000 grams and survived more than seven days. Those under 1000 grams who would be considered living by other international definitions would be placed in a freezer to die. At the opposite extreme, many western medical centers use extreme and costly measures to prolong life in terminally ill patients, preserving life temporarily but often with much suffering for the person and at great expense to the public system of financing health care.

In many countries, such as those in the former Soviet system of health care, spending for hospital services, in some cases grossly in excess of need, is accompanied by lack of funds for adding new vaccines for the immunization program for children. In the United States, there is a lack of funds for immunization of poor children but virtually unlimited funding for procedures such as cardiac bypass procedures that are not equitably distributed by need alone. In other countries, delay in updating immunization programs may be due to lack of funding or to delay in professional or governmental acceptance of "new" vaccines.

Closure of hospitals involves difficult decisions and is a source of friction between central health authorities, the medical professions, and local communities. Health reforms in many industrialized countries, such as reducing hospital bed supplies and managed care systems promoting cost containment and reallocation of resources, raise ethical and political issues often based on vested interests such as private insurance systems, hospitals, and private medical practitioners.

Where there is a high level of cumulative evidence from the professional literature and from public health practice in "leading countries" with a strong scientific base and case for action on a public health issue, when does it become bad practice or even unethical public health practice to ignore and fail to implement such an intervention? Such ethical failures occur frequently and widely. For example, is it "unethical" to not fortify grain products with folic acid, and salt with iodine? Should there be a recommended European immunization program; should milk be fortified with vitamin D; should vitamin and mineral supplements be given to women and children; should all newborns be given vitamin K intramuscularly routinely? Other examples include the issues of fluoridation of water supplies and opposition to genetically modified crops or generic drugs in African countries. These issues are continuously debated and the responsibility of the trained public health professional is to review the international literature on a topic and formulate a position based on the cumulative weight of evidence. It is not possible to wait for indisputable evidence because in epidemiology and public health this rarely occurs. This is another reason that guidelines established by respected agencies and professional bodies free from financial obligations to vested interest groups are essential for review of the evidence which continues to accumulate on a continuing basis on many issues thought to have been resolved or which appear *de novo*.

Ethics in Public Health Research

The border between practice and research is not always easy to define in public health, which has as one of its major tasks surveillance of population health. This surveillance is mostly anonymous but relies on individually identifiable data needed for reportable and infectious disease control as well as for causes of death, birth defects, mass screening programs and other special disease registries. It may also be necessary to monitor the effects of chronic disease, for example, to ascertain repeat hospitalizations of patients with congestive heart failure to assess the long-term effects of treatment, and the effects strengthening ambulatory and outreach services to sustain chronic patients at a safe and functional level in their own homes. Hospitalizations, immunizations, preventive care practices (e.g., Pap smears, mammography, and colonoscopies), are all part of the New Public Health. Impact assessment of preventive programs may require special surveys and are important to assess smoking and nutritional status and other measures of health status and risk factors. Every effort must be made to preserve anonymity and privacy for the individual but in some cases where the disease is contagious, case contact is crucial. This can entail identifying people who attended an event or on an airplane where an infected person may have been, so as to take appropriate preventive measures.

The general distinction between research and practice has to do with the intent of the activity. Clinical research uses experimental methods to establish efficacy and safety of new interventions or unproved interventions; many drugs and procedures in common use have never been subjected to randomized controlled trials. In practice, many methods are devised that are held to be effective and safe by expert opinion and documented as such. Researchers comparing HIV or hepatitis B transmission rates among intravenous drug users not using needle exchange programs would be doing unethical research, according to accepted current standards, by giving needles to the experimental group and withholding them from the control group. The scientific justification of an experiment must be made explicit and justifiable. Clinical equivalence is a necessary condition of all clinical and public health research and provision of standard of care treatment to control groups is a minimal requirement for most research ethics boards. Determination of the standard, and whether it should be place, time, and community specific, is an area of ongoing controversy.

A 1996 U.S. Public Health Service study supported by the NIH and WHO compared a short course of zidovudine (AZT) to a placebo given late in pregnancy to HIV-positive women in Thailand, measuring the rate of HIV infection among the newborns. The experiment was terminated when a protest editorial appeared in a prominent medical journal. This study confirmed previous findings that AZT given during late pregnancy and labor reduced maternal-fetal HIV transmission by half. The findings indicate that AZT should be used in developing countries, and the manufacturers agreed to make it available at reduced costs.

Public health may face the challenge of pandemic influenza such as the avian flu with decisions regarding allocation of vaccines, treatment of massive numbers of patients arriving at hospitals in acute respiratory distress with very limited resources to cope, coping with sick or absent staff, and many other issues requiring not only individual life and death situations, but mortality *en masse*. The ethical questions will be replaced by struggles to cope. Preparation for such potential catastrophic events will be a challenge to public health organizations and the health system generally.

Ethics in Patient Care

Ethical issues between the individual patient and health care provider are important in the New Public Health. A doctor is expected to use diligence, care, knowledge, skill, discretion, and caution in keeping with practice standards accepted at the time by responsible medical opinion and to maintain the basic medical imperative to do no harm to the patient. Patients have the right to know his or her condition, available alternatives for treatment, and the risks and benefits involved. They also have a right to seek alternative medical opinions, but this right is not unlimited, as any insurance plan or health service may place restrictions on payment for further opinions and consultation without the agreement of a primary care provider.

Health care has a responsibility beyond that of payment of health service bills and individual care by a physician, in institutions, or through services in the community or the home. The contract for service is becoming less between an individual physician and his or her patient, but increasingly among a health system, its staff, and the client. This places a new onus on the physician to ensure that patients receive the care they require. Conversely, the U.S. provider often faces the dilemma of knowing that a patient may not access needed services because of a lack of adequate health insurance.

Ethics in Public Health

The twentieth century was replete with mass murders, executions, and genocide with nationalistic, ideological, and racist motives perpetrated by fascist, Stalinist, and xenophobic political movements when gaining governmental power by election or by revolution, in some cases using then-common public health terminology and concepts (Box 15.11).

During the early part of the twentieth century, a segment of the social hygiene movement adopted ideas of racial improvement by sterilization of the mentally ill, retarded, and other "undesirable persons." By 1933, when the Nazi sterilization laws were passed, there were about

Box 15.11 Public Health and "the Slippery Slope"

- 1915–1917: Armenian genocide in Ottoman Empire
- 1920s–1930s: Eugenics movement — United States, Sweden
- 1920s–1940: Mass executions, deportations, starvation as policy in fascist and Stalinist regimes
- 1930s–1940s: Mass sterilization of "defectives" in the United States and Sweden
- 1930–1940s: Mass murder of "defectives" in Nazi Germany (750,000)
- 1940s: Quarantining as pretext for ghettos by Nazis
- 1940s: Concentration camps, human experimentation
- 1940s: The Holocaust (6 million Jews and others)
- 1975–1979: Cambodian genocide (1.7 million)
- 1990s: Former Yugoslavian republics and Rwandan genocides
- 2000s: Darfur genocide

20 states in America that already had sterilization laws in effect. American eugenics policies were praised by Hitler, and these ideas were adopted in Nazi Germany leading to execution of half a million "undesirables" under the eugenics concept, and were adapted for mass extermination of the Jews, Gypsies, and others in the Holocaust.

The reappearance of genocide in the late twentieth century in the Balkans and Rwanda and in the twenty-first century in Darfur highlight genocide as a public health concern and its prevention as a public health and international political responsibility. Incitement to genocide is now considered a crime against humanity and was the basis for trials and convictions of leaders of the Rwandan Tutsi tribe, as well as inciters to ethnic violence and the political leaders and perpetrators of mass murders in the former Yugoslav republic.

Human Experimentation

Human experimentation has been a subject of great concern since the Nazi and Imperial Japanese armed forces' experiments on prisoners and concentration camp victims during World War II. The Nuremberg trials set forth standards of professional responsibility to comply with internationally accepted medical behavior (see Table 15.10).

The Helsinki Declaration was first adopted by the World Medical Assembly in 1964, and amended in 1975,

TABLE 15.10 Ethical Issues of Medical Research Derived from the Nuremberg Trials, the Universal Declaration of Human Rights and Declaration of Helsinki

Nuremberg Trials, 1946	The voluntary consent of a human subject is absolutely essential, with the exercise of free power of choice without force, fraud, deceit, duress, or coercion. Experiments should be such as to bear fruitful results, based on prior experimentation and the natural history of the problem under study. They should avoid unnecessary physical and mental suffering. The degree of risk should not exceed the humanitarian importance of the experiment. Persons conducting experiments are responsible for adequate preparations and resources for even the remote possibility of death or injury resulting from the experiment. The human subject should be able to end his participation at any time. The scientist in charge is responsible to terminate the experiment if continuation is likely to result in injury, disability, or death.
Universal Declaration of Human Rights, 1948	Everyone has the right to a standard of living adequate for the health and well-being of himself and of his family, including food, clothing, housing, and medical care and necessary social services.
Declaration of Helsinki, 1964	Research must be in keeping with accepted scientific principles, and should be approved by specially appointed independent committees. Biomedical research should be carried out by scientifically qualified persons, only on topics where potential benefits outweigh the risks, with careful assessment of risks, where the privacy and integrity of the individual is protected, and where the hazards are predictable. Publication must preserve the accuracy of research findings. Each human subject in an experiment should be adequately informed of the aims, methods, anticipated benefits, and hazards of the study. Informed consent should be obtained, and a statement of compliance with this code. Clinical research should allow the doctor to use new diagnostic or therapeutic measures if they offer benefit as compared to current methods. In any study, the patient and the control group should be assured of the best available methods. Refusal to participate should never interfere with the doctor–patient relationship. The well-being of the subject takes precedence over the interests of science or society.

Source: Summarized from the Nuremberg Trials (1948) and World Medical Association, Declaration of Helsinki; Website sources include http://www.wma.net/ http://www.unmc.edu/irb/source_documents/nuremberg.htm; http://www.health.gov.au/nhmrc/ethics/helsinki.htm [accessed November 21, 2007]

Box 15.12 The Tuskegee Experiment

The Tuskegee experiment carried out by the U.S. Public Health Service, between 1932 and 1972, was meant to follow the natural course of syphilis in 399 African-American men in Alabama and 201 uninfected men. The men were not told that they were being used as research subjects. The experiment had been intended to show the need for additional services for those infected with syphilis. However, when penicillin became available, the researchers did not inform or offer the men treatment, even to those who were eligible when drafted into the army in 1942. The experiment was stopped in 1972 as "ethically unjustified" when the media exposed it to public scrutiny.

The case is considered unethical research practice because the best interest of the individuals to receive available care was put aside in the interest of the descriptive study.

In 1997, President Bill Clinton apologized to the survivors and families of the men involved in the experiment on behalf of the U.S. government. The Tuskegee experiment is the source of lingering widespread suspicion in the African-American community to the present time.

Sources: Lombardo, P. A., Dorr, G. M. 2006. Eugenics, medical education and public health: Another perspective on the Tuskegee syphilis experiment. *Bulletin of the History of Medicine*, 80:291–316.
Centers for Disease Control, http://www.cdc.gov/tuskegee/timeline.htm [accessed May 23, 2008]

1983, 1989, and 1996. It delineates standards of medical experimentation and requires informed consent from subjects of medical research. These standards have become an international norm for experiments, with national, state, and hospital Helsinki committees regulating research proposals within their jurisdiction. Funding agencies require standard approval by the appropriate Helsinki committee before considering any proposal, with informed consent on any research project.

The Tuskegee experiment (Box 15.12) was a grave and tragic violation of medical ethics, but in the context of the 1930s was consistent with widespread and institutionalized racism. It provides an important case study which has repercussions until the present time in suspicion of public health endeavors, particularly among the African-American community in the United States.

Sanctity of Life Versus Euthanasia

The ethical imperative to save a life is an important ethical and practical issue in health care. Advocates of physician-assisted suicide (euthanasia) argue for the right of the patient to die with dignity when the illness is terminal and the individual is suffering excessively. This is not a medical decision alone, and is an agonizing issue for

society to address. The Nazi euthanasia program and its human experiments provided the direst of warnings to societies of what may follow when the principle of the sanctity of the individual human life is breached. The issue, however, returned to the public agenda in the 1980s and 1990s as advances in medical science have allowed the prolongation of human life beyond all hope of recovery. Legislation in the Netherlands, United States, and northern Australia has legally sanctioned euthanasia with various safeguards in a variety of circumstances, such as long-term comas or terminal illnesses.

Doctors, patients, relatives, and health care organizations need clear guidelines, orientation, procedures, legal protection, and limitations where failure to take utmost steps to "save" the patient by intubation, resuscitation, or transplantation may cause legal jeopardy. Even though a distinction can be drawn theoretically between permitting and facilitating death, in practice, doctors in intensive care units face such decisions regularly where the line is often blurred. Hospital doctors routinely take extreme measures to prolong the life of hopeless cases. Such decisions should not be considered for economic reasons alone, but in practice, the costs of care of the terminally ill will be a driving force in debate of the issue. Living wills allow a patient to refuse heroic measures such as resuscitation, with "do not resuscitate" standing orders and assignment of power of attorney to family members to make such decisions. Family attitudes are important, but the social issue of redefining the right of a patient to opt for legal termination of life by medical means will be an increasingly important issue in the twenty-first century.

The Imperative to Act or Not Act in Public Health

As in other spheres of medicine and health, in public health the decision whether to intervene on an issue is based on identification and interpretation of the problem, the potential of the intervention to improve the situation, to do no harm, and to convince the public and political levels of the need for such intervention along with the resources to carry it out. This process requires patience and a longer-term time frame than many other fields in health.

Failure to act *is* an action, and when there is convincing evidence of a problem that can be alleviated or prevented entirely by an accepted and demonstrably successful intervention, then the onus is on the public health worker to advocate such action and to implement it as best as possible under the existing conditions. Failure to do so is a breach of "good standards of practice" and could be unethical. Inertia of the public health system in the face of evidence of a demonstrably effective modality such as adoption of state-of-the-art vaccines or fortification of

> **Box 15.13 Principles of Ethical Public Health Practice: American Public Health Association, 2006**
>
> 1. Public health should address principally the fundamental causes of disease and requirements for health, aiming to prevent adverse health outcomes.
> 2. Public health should achieve community health in a way that respects the rights of individuals in the community.
> 3. Public health policies, programs, and priorities should be developed and evaluated through processes that ensure an opportunity for input from community members.
> 4. Public health should advocate and work for the empowerment of disenfranchised community members, aiming to ensure that the basic resources and conditions necessary for health are accessible to all.
> 5. Public health should seek the information needed to implement effective policies and programs that protect and promote health.
> 6. Public health institutions should provide communities with the information they have that is needed for decisions on policies or programs and should obtain the community's consent for their implementation.
>
> 7. Public health institutions should act in a timely manner on the information they have within the resources and the mandate given to them by the public.
> 8. Public health programs and policies should incorporate a variety of approaches that anticipate and respect diverse values, beliefs, and cultures in the community.
> 9. Public health programs and policies should be implemented in a manner that most enhances the physical and social environment.
> 10. Public health institutions should protect the confidentiality of information that can bring harm to an individual or community if made public. Exceptions must be justified on the basis of the high likelihood of significant harm to the individual or others.
> 11. Public health institutions should ensure the professional competence of their employees.
> 12. Public health institutions and their employees should engage in collaborations and affiliations in ways that build the public's trust and the institution's effectiveness.
>
> Source: American Public Health Association, www.apha.org [accessed November 25, 2007]

flour with folic acid to prevent birth defects would come under this categorization and may even constitute neglect and unethical practice. This is not an easy categorization, because there is often disagreement and even opposition to public health interventions, as was the case in opposition to vaccination long after Jenner's crucial discovery of this procedure in the late eighteenth century. It is also true today with opposition to many proven measures such as fluoridation or fortification of basic foods.

The use of ethical and high standards of practice in public health (Box 15.13) requires an ideological commitment to the advancement of health standards and use of best practices of international standards to the maximum extent possible under the local conditions in which the professional is working. This is not an easy commitment as there is often dispute and outright hostility to public health activities in part because of ethical distortions of great magnitude in the past. But this is an optimistic field of activity because of the great achievements it has brought to humankind. Preparation for disasters and unanticipated health emergencies in addition to addressing current issues is a vital part of the New Public health and our ethical and professional commitments.

SUMMARY

Governments have the responsibility to legislate, tax, regulate, and enforce for the public health. This includes protection and proactive services for the weak and the needy, for establishing equity and social solidarity. There

are many ethical considerations concerning the rights of society, the rights and responsibilities of individuals, all playing key roles in the totality of population health. In epidemiology and clinical research, informed consent for research and precaution versus inertia are all vital elements of the New Public Health.

In order to maintain and improve standards of care, health systems need quality assurance and technological assessment as part of their ongoing operation. Poor-quality care is costly in terms of iatrogenic diseases and prolonged or repeated hospitalization, and failure to prevent disease or complications with currently available methods. If innovations such as vaccines against cancer-causing viruses (hepatitis B and human papillomavirus) are not introduced, then cancer of the liver and cervix will continue to cause hundreds of thousands of preventable deaths, long hospital stays, liver transplants, and much unnecessary suffering and premature death. New innovations such as the diagnosis and management of *H. pylori* in peptic ulcers have already reduced surgical interventions. If innovations are not introduced, then longer hospital stays are needed with impacts on patient's suffering, time, and productivity while utilizing expensive health care resources and incurring the risks associated with invasive surgery.

Health care is provided by people, as well as by institutions and technology. Since health care is a knowledge-based service, the quality of care is set by the individuals providing care as much as, if not more than, the technological facilities. Nevertheless, progress on the technological side of medical care is vital to continuing development of the field. New medications, monitoring equipment, laboratory services, and imaging devices have made an enormous contribution to advances in medical care. Appropriate technology is a critical issue for international health, since the most advanced technology may be completely inappropriate in a setting that cannot afford to maintain it or lacks the trained personnel to operate it, or where it comes in place of more vital basic primary care services. Technology assessment needs to be seen in the context of the country and its resources for health care.

Ethical issues in public health are no less demanding than those related to medical research and personal medical care. Yet in public health, the rights of the individual and those of the community are sometimes in conflict, generating controversy which may impede adoption of workable, effective solutions that do not adversely affect individuals but are important for community health, and are used in liberal and progressive countries.

Technology, quality, law, and ethics are closely interrelated in public health. Well-informed and sensitive analysis of all aspects of their development is a part of the New Public Health. The balance between individual and community rights is very sensitive and must be under continuous surveillance. The New Public Health is replete with technological and ethical questions, especially in times of cost restraint, increasing technological potential, the public expectation of universal access to health care, and the assumption that everyone will live a healthy and long life. Health status has always been linked with socioeconomic status, and, despite enormous gains, that remains true even in the most egalitarian countries.

Market mechanisms, such as control of supply of hospital beds and providers, limiting access to referrals, competition, and incentives/disincentives in payment systems for hospital and managed care systems, are all part of health management responsibilities. Management and policy in the context of high legal, ethical, and technological standards, are fundamental in the New Public Health. This is the social responsibility for health for all, using community and personal care modalities as well as health promotion as part of the struggle to achieve the goals of Health for All and quality of individual and societal life.

ELECTRONIC RESOURCES

Agency for Healthcare Research and Quality. http://www.ahrq.gov/ [accessed May 23, 2008]

Agency for Healthcare Research and Quality (AHRQ). http://www.ahrq.gov/ [accessed May 23, 2008]

American Academy of Pediatrics. http://www.aap.org [accessed May 23, 2008]

American College of Cardiology (ACC), 2008. Clinical Practice Support site, [accessed May 24, 2008]

American Public Health Association. The APHA Code of Ethics, www.apha.org [accessed May 23, 2008]

American Public Health Association. Principles of the Ethical Practice of Public Health, http://www.apha.org/NR/rdonlyres/1CED3CEA-287E-4185-9CBD-BD405FC60856/0/ethicsbrochure.pdf [accessed May 23, 2008]

Canadian Cardiovascular Research Outcomes Team. http://www.ccort.ca/ [accessed May 23, 2008]

Canadian Institute for Health Information. http://secure.cihi.ca/cihiweb/dispPage.jsp?cw_page=home_e [accessed May 23, 2008]

Canadian Council of Health Services Accreditation. http://www.cchsa.ca [accessed May 24, 2008]

Canadian Medical Association. Handbook on Clinical Practice Guidelines, http://www.cma.ca/index.cfm/ci_id/54316/la_id/1.htm [accessed May 23, 2008]

Centers for Disease Control. CDC and Public Health Ethics, http://www.cdc.gov/od/science/phec/ [accessed May 24, 2008]

Centers for Disease Control. The *Community Guide*, http://www.thecommunityguide.org/ [accessed May 24, 2008]

Centers for Disease Control. Collaborating Center for the Law and the Public's Health, http://www.publichealthlaw.net/ [accessed May 23, 2008]

Centers for Disease Control Foundation. Institute of Public Health Law, http://www.cdcfoundation.org/programs/iphl/index.aspx [accessed May 24, 2008]

Centers for Disease Control. *MMWR* selected special reports, http://www.cdc.gov/mmwr [accessed May 23, 2008]

Centers for Disease Control. U.S. Public Health Service Syphilis Experiment at Tuskegee, http://www.cdc.gov/tuskegee/timeline.htm [accessed May 24, 2008]

Center for Evidence-Based Medicine (CEBM). http://www.cebm.net/ [accessed May 23, 2008]

GAVI, Saving 10 million more lives through immunization. March 2006, http://www.who.int/immunization/givs/GAVI_Imm_Forum_piece.pdf [accessed July 3, 2008]

Department of Health and Human Services. Health Care Financing Administration, http://www.os.dhhs.gov/about/opdivs/hcfa.html [accessed May 23, 2008]

Department of Health and Human Services. Centers for Medicare and Medicaid Services. Office of Clinical Standards and Quality. 2008. http://www.cms.hhs.gov/CMSLeadership/11_Office_OCSQ.asp [accessed May 24, 2008]

Health Services/Technology Assessment (HSTAT). http://www.text.nlm.nih.gov/

Hospital Infection Program (HIP). Centers for Disease Control. National Nosocomial Infection Surveillance System, http://www.cdc.gov/ncidod/dhqp/nnis_pubs.html [accessed May 23, 2008]

Institute for Health Care Improvement. http://www.ihi.org/ihi/ [accessed May 23, 2008]

International Network of Agencies for Health Technology Assessment (INAHTA). http://www.inahta.org/ [accessed May 23, 2008]

Joint Commission on Accreditation of Health Care Organizations. http://www.jointcommission.org/ [accessed May 23, 2008]

National Cancer Agency. http://www.cancer.gov [accessed May 23, 2008]

National Council for Quality Assurance. http://web.ncqa.org/tabid/577/Default.aspx [accessed May 23, 2008]

National Institutes of Health (NIH) consensus development conference and technology assessment protocols

National Patient Safety Foundation (Georgetown and Johns Hopkins Universities). http://www.npsf.org/au/ [accessed May 23, 2008]

Nuremberg Trials (1948). http://www.unmc.edu/irb/source_documents/nuremberg.htm [accessed May 23, 2008]

Pocket Guide to Clinical Preventive Services, 2007. http://www.ahrq
.gov/clinic/USpstfix.htm#pocket [accessed May 23, 2008]

U.K. National Institute for Clinical Excellence (NICE). http://www.nice
.org.uk/ [accessed May 23, 2008]

U.S. Food and Drug Administration. www.fda.org [accessed May 23, 2008]

U.S. National Library of Medicine (NICHSR). http://www.nlm.nih.gov/
nichsr/ [accessed May 23, 2008]

U.S. National Library of Medicine. National Information Center on Health
Services Research. U.S. Preventive Services Task Force, Guide to Preventive Services, Report of the U.S. Preventive Services Task Force
(USPSTF), http://www.ahrq.gov/clinic/USpstfix.htm [accessed May 23,
2008]

United Health Foundation. http://www.unitedhealthfoundation.org/ahr2007/
index.html [accessed May 23, 2008]

United Kingdom. National Health Service. National Institute for Health
and Clinical Excellence. National Practices Review: Recommendation
Reminders, http://www.nice.org.uk/usingguidance/optimalpracticere-
viewrecommendationreminders/ [accessed May 24, 2008]

U.S. National Library of Medicine. National Institute of Medicine.
National Information Center on Health Services Research and Health
Care Technology (NICHSR). Etext on Health Technology Assessment
(HTA) Information Resources, http://www.nlm.nih.gov/archive//
2060905/nichsr/ehta/ehta.html [accessed May 23, 2008]

World Health Organization. Department of Immuniation, vaccines and
Biologicals. 2005. The Cost of Immunization Programmes in the
Next 10 Years, http://www.path.org/files/Costing_10_years_WHO.
pdf;http://www.who.int/vaccines/ [accessed May 23, 2008]

World Health Organization. 2008. Ethics and Health, http://www.who
.int/ethics/topics/en/ [accessed May 23, 2008]

World Medical Association. http://www.wma.net [accessed May 23, 2008]

World Medical Association. Declaration of Helsinki, http://www.wma
.net/ [accessed May 23, 2008]

World Medical Association, Ethics Unit. Helsinki Declaration, http://
www.wma.net/e/ethicsunit/helsinki.htm [accessed May 23, 2008]

RECOMMENDED READINGS

Annas, G. 1998. Human rights and health — The Universal Declaration
of Human Rights at 50. *New England Journal of Medicine*,
339:1778–1781.

Birne, A.-E., Molina, N. 2005. In the name of public health. *American
Journal of Public Health*, 95:1095–1097.

Centers for Disease Control. 2000. Monitoring hospital-acquired infections to promote patient safety, United States, 1990–1999. *Morbidity
and Mortality Weekly Reports*, 49:149–153.

Coleman, H., Bouësseau, M.-C., Reis, A., Capron, A. M. 2007. How
should ethics be incorporated into public health policy and practice?
Bulletin of the World Health Organization, 85:501–568.

Dannenberg, A. L., Bhatia, R., Cole, B. L., et al. 2006. Growing the field
of Health Impact Assessment: An agenda for research and practice.
American Journal of Public Health, 96:262–270.

Dickens, B. M. 2005. [editorial]. The challenges and opportunities of
ethics. *American Journal of Public Health*, 95:1094.

Fattal-Valevski, A., Kesler, A., Sela, B. A., Nitzan-Kaluski, D., Rotstein,
M., Mesterman, R., Toledano-Alhadef, H., Stolovitch, C., Hoffmann,
C., Globus, O., Eshel, G. 2005. Outbreak of life-threatening thiamine
deficiency in infants in Israel caused by a defective soy-based formula. *Pediatrics*, 115:233–238.

Fielding, J. E., Briss, P. A. 2006. Promoting evidence-based public health
policy: Can we have better evidence and more action? *Health
Affairs*, 25:969–978.

Food and Drug Administration, HHS. 2007. Amendment to the current
good manufacturing practice regulations for finished pharmaceuticals. Direct final rule. *Federal Register*, 72:68064–68070.

Food and Drug Administration, HHS. 2007. Revision of the requirements
for live vaccine processing. Direct final rule. *Federal Register*,
72:59000–59003.

Goodman, R. A., Moulton, A., Matthews, G., Shaw, F. P., Kocher, P.,
Mensah, G., Zaza, S., Besser, R. 2006. Centers for Disease Control.
Law and Public Health at CDC. *Morbidity and Mortality Weekly
Report*, 55(SUP02): 29–33.

Greenberg, M. R. 2007. The diffusion of public health innovations. *American Journal of Public Health*, 96:209–210.

Grodin, M., Annas, G. J. (editorial). 1996. Legacies of Nuremberg: Medical ethics and human rights. *Journal of American Medical Association*, 276:1682–1683.

Gruskin, S., Dickens, B. 2006. Human rights and ethics in public health.
American Journal of Public Health, 96:1903–1905.

Larson, E. L., Quiros, D., Lin, S. X. 2007. Dissemination of the CDC's
Hand Hygiene Guideline and impact on infection rates. *American
Journal of Infection Control*, 35:666–675.

Lehoux, P., Tailliez, S., Denis, J. L., Hivon, M. 2004. Redefining health
technology assessment in Canada: Diversification of products and
contextualization of findings. *International Journal of Technology
Assessment in Health Care*, 20:325–336.

McClellan, M., Kessler, D., for the TECH Investigators. 1999. A global
analysis of technological change in health care: The case of heart
attacks. *Health Affairs*, 18:250–255.

Mariner, W. K. 1997. Public confidence in public health research ethics.
Public Health Reports, 112:33–36.

Martelli, F., Torre, G. L., Ghionno, E. D., Staniscia, T., Neroni, M., Cic-
chetti, A., Bremen, K. V., Ricciardi, W. NI-HTA Collaborative
Group 2007. Health technology assessment agencies: An international overview of organizational aspects. *International Journal of
Technology Assessment in Health Care*, 23(4):414–424.

Murray, A., Lourenco, T., de Verteuil, R., Hernandez, R., Fraser, C.,
McKinley, A., Krukowski, Z., Vale, L., Grant, A. 2006. Clinical
effectiveness and cost-effectiveness of laparoscopic surgery for colorectal cancer: Systematic reviews and economic evaluation. *Health
Technology Assessment*, 10:1–141.

Noorani, H. Z., Husereau, D. R., Boudreau, R., Skidmore, B. 2007. Priority
setting for health technology assessments: A systematic review of
current practical approaches. *International Journal of Technology
Assessment in Health Care*, 23:310–315.

The Nuremberg Code, from *Trials of War Criminals Before the Nuremberg Military Tribunals under Control Council Law*. 2(10):181–
182. Washington, DC: U.S. Government Printing Office, 1949.

Parry, J. M., Kemm, J. R. 2005. Evaluation of Health Impact Assessment
Workshop. Criteria for use in the evaluation of health impact assessments. *Public Health*, 119:1122–1129.

Sterman, J. D. 2006. Learning from evidence in a complex world. *American Journal of Public Health*, 96:505–514.

Terris, M. 2006. Concepts of health promotion: Dualities in public health
theory. *Journal of Public Health Policy*, 13:267–276.

Thieren, M., Mauron, A. 2007. Nuremberg code turns 60. *Bulletin of the
World Health Organization*, 85:573.

United Nations General Assembly. *Universal Declaration of Human
Rights* (A/RES/217), 10 December 1948 at Palais de Chaillot, Paris.

BIBLIOGRAPHY

Agency for Healthcare Research and Quality. 2003. Nominations of topics for evidence-based practice centers (EPCs). *Federal Register*, 68(18), January 28, 2003:4213–4216.

American College of Epidemiology. 2000. Ethics guidelines. *Annals of Epidemiology*, 10:487–497.

American College of Obstetricians and Gynecologists. Committee on Obstetric Practice. ACOG Committee Opinion. Circumcision 2001. *Obstetrics and Gynecology*, 98:707–708.

Balk, E. M., Bonis, P. A. L., Moskowitz, H., et al. 2002. Correlation of quality measures with estimates of treatment effect in meta-analyses of randomized controlled trials. *Journal of the American Medical Association*, 287:2973–2982.

Bayer, R., Fairchild, A. 2004. The genesis of public health ethics. *Bioethics*, 18:473–492.

Bell, J. A., Hyland, S., DePellegrin, T., Upshur, R. E., Bernstein, M., Martin, D. K. 2004. SARS and hospital priority setting: A qualitative case study and evaluation. *BMC Health Service Research*, 4:36.

Botalden, P. B., Stoltz, P. K. 1993. A framework for the continual improvement of health care: Building and applying professional improvement knowledge to test changes in daily work. *Journal on Quality Improvement*, 19:424–452.

Callahan, D., Jennings, B. 2002. Ethics and public health: Forging a strong relationship. *American Journal of Public Health*, 92:169–176.

Castellsague, X., Bosch, F. X., Munoz, N., Meijer, C. J., Shah, K. V., de Sanjose, S., Eluf-Neto, J., Ngelangel, C. A., Chichareon, S., Smith, J. S., Herrero, R., Moreno, V., Franceschi, S. International Agency for Research on Cancer Multicenter Cervical Cancer Study Group. 2002. Male circumcision, penile human papillomavirus infection, and cervical cancer in female partners. *New England Journal of Medicine*, 346:1105–1112.

Childress, J., Faden, R., Gostin, L. O., et al. 2002. Public health ethics: Mapping the terrain. *Journal of Law, Medicine and Ethics*, 30:1073–1105.

Cottam, R 2005. Public health is coercive health. *Lancet*, 366:1592–1594.

Coughlin, S. 2006. Hope, ethics, and public health. *Journal of Epidemiology and Community Health*, 60:826–827.

Coughlin, S. S. 2006. Ethical issues in epidemiologic research and public health practice. *Emerging Themes Epidemiology*, 3:3–16.

Council on Ethical and Judicial Affairs. 1995. Ethical issues in managed care. *Journal of the American Medical Association*, 273:330–335.

Cutler, D. M., McClellan, M. 2001. Is technological change in medicine worth it? *Health Affairs*, 20:11–29.

Dawson, A. 2004. Vaccination and the prevention problem. *Bioethics*, 18:515–530.

Dawson, A. 2006. Commentary on the limits of the law in the protection of public health and the role of public health ethics. *Public Health*, 120:77–80.

Dawson, A., Paul, Y. 2006. Mass public health programs and the obligations of sponsoring and participating organizations. *Journal of Medical Ethics*, 32:580–583.

Deeks, J. J. 2001. Systematic reviews in health care: Systematic reviews of evaluations of diagnostic and screening tests. *British Medical Journal*, 323:157–162.

Deyo, R. A. 2002. Cascade effects of medical technology. *Annual Reviews of Public Health*, 23:23–44.

Forslund, O., Hansson, B. G., Rylander, E., Dillner, J. 2007. Human papillomavirus and Papanicolaou tests to screen for cervical cancer. *New England Journal of Medicine*, 357:1589–1597.

Fox, D., Kramer, M., Standish, M. 2003. From public health to population health: How can law redefine the playing field? *Journal of Law and Medical Ethics*, 31:21–29.

Gardiner, P. 2003. A virtue ethics approach to moral dilemmas in medicine. *Journal of Medical Ethics*, 29:297–302.

Gostin, L. O. 2000. Public health law in a new century. Part II: Public health powers and limits. *Journal of the American Medical Association*, 283:2979–2984.

Gostin, L. O. 2000. Public health law in a new century: Part III: Public health regulation, A systematic evaluation. *Journal of the American Medical Association*, 283:3118–3122.

Gostin, L. O., et al. 2002. The Model State Emergency Health Powers Act. *Journal of the American Medical Association*, 288:622–628.

Grad, F. P. 1990. *The Public Health Law Manual*, 2nd Edition. American Public Health Association. Washington, DC: APHA.

Grill, K., Hanson, S. O. 2005. Epistemic paternalism in public health. *Journal of Medial Ethics*, 31:648–653.

Guttman, N., Salmon, C. 2004. Guilt, fear, stigma, and knowledge gaps: Ethical issues in public health communication interventions. *Bioethics*, 18:531–552.

Harris, R. P., Helfand, M., Woolf, S. H., et al. 2001. Current methods of the U.S. Preventive Services Task Force. A review of the process. *American Journal of Preventive Medicine*, 20:21–35.

Institute of Medicine. 1985. *Assessing Medical Technologies*. Washington, DC: National Academies Press.

Institute of Medicine. 1985. *Clinical Practice Guidelines*. Washington, DC: National Academies Press.

Institute of Medicine. 2007. *Advancing Quality Improvement Research: Challenges and Opportunities. Workshop Summary. Forum on the Science of Health Care Quality Improvement and Implementation. Board of Health Care Services*. Institute of Medicine of the National Academies of Science. Washington, DC: National Academies Press.

Institute of Medicine. 2007. *State of Quality Improvement and Implementation Research: Expert Views, Workshop Summary*. Institute of Medicine of the National Academies of Science. Washington, DC: National Academies Press.

Institute of Medicine, Committee on Quality of Health Care in America. 2001. *Crossing the Quality Chasm: A New Health System for the 21st Century*. Washington, DC: National Academies Press.

Kass, N. 2001. An ethics framework for public health. *American Journal of Public Health*, 91:1776–1782.

Kass, N. 2004. Public health ethics: From foundations and frameworks to justice and global public health. *Journal of Law, Medicine and Ethics*, 32:232–242.

Kohn, L. T., Corrigan, J. M., Donaldson, M. (eds.). Institute of Medicine, Committee on Quality of Health Care in America. 1999. *To Err Is Human: Building a Safer Health System*. Washington, DC: National Academies Press.

Kotalik, J. 2005. Preparing for an influenza pandemic: Ethical issues. *Bioethics*, 19(4):422–431.

Last, J. M. (ed.). 2000. Ethical issues in public health research. *A Dictionary of Epidemiology*. New York: Oxford University Press.

Last, J. M (ed.). 2007. *A Dictionary of Public Health*. New York: Oxford University Press.

Lo, B., Katz, M. H. 2005. Clinical decision making in public health emergencies: Ethical considerations. *Annals of Internal Medicine*, 143:493–498.

McCormack, K., Wake, B., Perez, J., Fraser, C., Cook, J., McIntosh, E., Vale, L., Grant, A. 2005. Laparoscopic surgery for inguinal hernia repair: Systematic review of effectiveness and economic evaluation. *Health Technology Assessment*, 9:1–203, iii–iv.

McDonald, I. G. 2000. Quality assurance and technology assessment: Pieces of a larger puzzle. *Journal of Quality in Clinical Practice*, 20:87–94.

McGlynn, E. A., Asch, S. M., Adams, J., Keesey, J., Hicks, J., DeCristofaro, A., Kerr, E. A. 2003. The quality of health care delivered to adults in the United States. *New England Journal of Medicine*, 26;348(26): 2635–2645.

McKeown, R. E., Weed, D. L. 2002. Ethics in epidemiology and public health I. Technical terms. *Journal of Epidemiology and Community Health*, 56:739–741.

McNeil, B. J. 2001. Shattuck Lecture. Hidden barriers to improvement in the quality of care. *New England Journal of Medicine*, 345:1612–1620.

Mackie, P., Sim, F. 2004. The ethics of public health decision making. *Public Health*, 118:311–312.

Mann, J. M. 1997. Health and human rights: If not now when? *Health and Human Rights*, 2:113–120.

Marer, S., Sutija, M., Rajagopalan, S. 2004. Bioterrorism, bioethics and the emergency physician. *Topics in Emergency Medicine*, 26:44–48.

Mariner, W. K. 1997. Public confidence in public health research ethics: Counterpoint on human subjects research. *Public Health Reports*, 112:33–36.

Martin, D. K., Giacomini, M., Singer, P. 2002. Fairness, accountability for reasonableness and the views of priority setting decision makers. *Health Policy*, 61:279–290.

Martin, R. 2006. The limits of the law in the protection of public health and the role of public health ethics. *Public Health*, 120:71–77.

May, T. 2005. Public communication, risk perception, and the viability of preventive vaccination against communicable diseases. *Bioethics*, 19:407–421.

Mehta, D. K., Ryan, R. S. M., Hogerzeil, H. 2004. *WHO Model Formulary 2004: Based on the 13th Model List of Essential Medicines 2003.* World Health Organization, Geneva.

Miettinen, O. 2005. Idealism and ethics of public health practitioners. *European Journal of Epidemiology*, 20:805–807.

National Information Center on Health Services Research and Health Care Technology.

National Library of Medicine. 2007. National Information Center on Health Services Research & Health Care Technology. *Etext on Health Technology Assessment (HTA) Information Resources*, http://www.nlm.nih.gov/nichsr/ehta [accessed May 23, 2008].

Neumann, P. J., Claxton, K., Weinstein, M. C. 2000. The FDA's regulation of health economic information. *Health Affairs*, 19:129–137.

Nixon, J., Stoykova, B., Glanville, J., Christie, J., Drummond, M., Kleijnen, J. 2000. The U.K. NHS Economic Evaluation Database. Economic issues in evaluations of health technology. *International Journal of Technology Assessment in Health Care*, 16:731–742.

Nixon, S. 2006. Critical public health ethics and Canada's role in global health. *Canadian Journal of Public Health*, 97:32–34.

Olsen, L. A., Aisner, D., McGinnis, J. M. (eds.). 2007. *The Learning Healthcare System: Workshop Summary: IOM Roundtable on Evidence-Based Medicine.* Institute of Medicine of the National Academies of Science. Washington, DC: National Academies Press.

Olson, O., Gøtzsche, P. C. 2001. Cochrane review on screening for breast cancer with mammography. *Lancet*, 358:1340–1342.

Parry, J. M., Kemm, J. R. 2005. Evaluation of Health Impact Assessment Workshop. Criteria for use in the evaluation of health impact assessments. *Public Health*, 119:1122–1129.

Poland, G. A., Tosh, P., Jacobson, R. M. 2005. Requiring influenza vaccinations for healthcare workers: Seven truths we must accept. *Vaccine*, 23:2251–2255.

Public Health Leadership Society. Principles of the Ethical Practice of Public Health Version 2.2, 2002, Adapted from *The Future of Public Health*, Institute of Medicine, 1988, Washington, DC.

Reich, M. R., Hershey, J. H., Hardy, G. E., Childress, J. F., et al. 2003. Workshop on Public Health Law and Ethics I & II: The challenge of public/private partnerships. *Journal of Law and Medical Ethics*, 31:90.

Roberts, M. J., Reich, M. R. 2002. Ethical analysis in public health. *Lancet*, 59:1055–1059.

Roelcke, V., Maio, G. (eds.). 2004. *Twentieth Century Ethics of Human Subjects Research. Historical Perspectives on Values, Practices, and Regulations.* Stuttgart: Franz Steiner Verlag.

Rogers, W. A. 2006. Feminism and public health ethics. *Journal of Medical Ethics*, 32:351–354.

Ruhl, S., Stephens, M., Locke, P. 2005. The role of NGOs in public health law. *Journal of Law, Medicine and Ethics*, 31:76–77.

Saarni, S. I., Gylling, H. A. 2004. Evidence-based medicine guidelines: A solution to rationing or politics disguised as science. *Journal of Medical Ethics*, 30:171–175.

Sassi, F. 2003. Setting priorities for the evaluation of health interventions: When theory does not meet practice. *Health Policy*, 63:141–154.

Schoenbaum, S. C., Holmgren, A. L. 2006. *The State of Health Care Quality: 2006.* Washington, DC: National Committee for Quality Assurance Commonwealth Fund.

Shaffer, N. 1999. Short course zidovudine for perinatal HIV-1 transmission in Bangkok, Thailand: Transmission study group. *Lancet*, 354:156–158.

Shekelle, P. G., Ortiz, E., Rhodes, S., Morton, S. C., Eccles, M. P., Grimshaw, J. M., Woolf, S. H. 2001. Validity of the Agency for Healthcare Research and Quality clinical practice guidelines: How quickly do guidelines become outdated? *Journal of the American Medical Association*, 286:1461–1467.

Sidel, V. 1996. The social responsibilities of health professionals: Lessons from their role in Nazi Germany. *Journal of the American Medical Association*, 276:1679–1681.

Siebzehner, M., Shemer, J. 1995. Regulating medical technology in Isreal. In J. Shemer, T. Schersten (eds.). *Technology Assessment in Health Care: From Theory to Practice.* Jerusalem: Gefen.

Sindal, C. 2002. Does health promotion need a code of ethics? *Health Promotion International*, 17:201–203.

Singer, S. J., Bergthold, L. A. 2001. Prospects for improved decision making about medical necessity. *Health Affairs*, 20:200–206.

Soto, J. 2002. Health economic evaluations using decision analytic modeling. Principles and practices — utilization of a checklist to their development and appraisal. *International Journal of Technology Assessment in Health Care*, 18:94–111.

Speakman, J., Gonzalez-Martin, F., Perez, T. 2003. Quarantine in SARS and other emerging infectious diseases. *Journal of Law and Medical Ethics*, 31:63–64.

Stern, A. M. 2005. Sterilized in the name of public health. *American Journal of Public Health*, 95:1128–1138.

Sterne, J. A., Egger, M., Smith, G. D. 2001. Systematic reviews in health care: Investigating and dealing with publication and other biases in meta-analysis. *British Medical Journal*, 323:101–105.

Stoykova, B., Drummond, M., Barbieri, M., Kleijnen, J. 2003. The lag between effectiveness and cost-effectiveness evidence of new drugs.

Implications for decision-making in health care. *European Journal of Health Economics*, 4:313–318.

Stroup, D. F., Berlin, J. A., Morton, S. C., et al. 2000. Meta-analysis of observational studies in epidemiology. A proposal for reporting. Meta-analysis Of Observational Studies in Epidemiology (MOOSE) Group. *Journal of the American Medical Association*, 283:2008–2012.

Tenke, P., Kovacs, B., Bjerklund-Johansen, T. E., Matsumoto, T., Tambyah, P. A., Naber, K. G. 2007. European and Asian guidelines on management and prevention of catheter-associated urinary tract infections. *International Journal of Antimicrobial Agents*, 31: S68–S78.

Thiel, K. S. 2003. New developments in public health case law. *Journal of Law, Medicine and Ethics*, 31:86–87.

Thomas, J. C., Sage, M., Dillenberg, J., Guillory, V. J. 2002. A code of ethics for public health. *American Journal Public Health*, 92:1057–1060.

United States Preventive Services Task Force. 2008. *Pocket Guide to Clinical Preventive Services 2007. Electronic Selective Services Selector (ePSS),* http://www.ahrq.gov/clinic/USpstfix.htm#pocket [accessed May 24, 2008].

United States Preventive Services Task Force.. 2002. Screening for prostate cancer: Recommendations and rationale. *Annals of Internal Medicine*, 137:915–916.

Upshur, R. E. G. 2002. Principles for the justification of public health interventions. *Canadian Journal of Public Health*, 93:101–103.

Varmus, H., Satcher, D. 1997. Ethical complexities of conducting research in developing countries. *New England Journal of Medicine*, 337:1003–1005.

Verweij, M., Dawson, A. 2004. Ethical principles for collective immunization programs. *Vaccine*, 22:3122–3126.

Weed, D. L., McKeown, R. E. 2003. Science, ethics, and professional public health practice. *Journal of Epidemiology and Community Health*, 57:4–5.

Williams, I., Bryan, S., McIver, S. 2007. How should cost-effectiveness analysis be used in health technology coverage decisions? Evidence from the National Institute for Health and Clinical Excellence approach. *Journal of Health Services Research and Policy*, 12:73–79.

Woods, K. 2002. Health technology assessment for the NHS in England and Wales. *International Journal of Technology Assessment Health Care*, 18:161–165.

World Bank. World Development Report. 1993. *Investing in Health.* New York: Oxford University Press.

World Health Organization. 2006. *Preparing for the Introduction of HPV Vaccine.* Geneva: World Health Organization.

Yoshida, E. M. 1998. Selecting candidates for liver transplantation: A medical ethics perspective on the microallocation of a scarce and rationed resource. *Canadian Journal of Gastroenterology*, 12(3): 209–215.

Zambon, M. C. 2004. Ethics versus evidence in influenza vaccination. *Lancet*, 364:2161–2163.

Zasa, S., Clymer, J., Upmeyer, L., Thacker, S. 2003. Using science-based guidelines to shape public health law. *Journal of Law, Medicine and Ethics*, 31:65–67.

Chapter 16

Globalization of Health

Introduction
The Global Health Situation
Priorities in Global Health
  Poverty–Illness–Population–Environment
  Child Health
  Maternal Health
  Population Growth
  Malnutrition
  The Fight Against HIV/AIDS and Other Communicable
    Diseases
  Chronic Disease
  Disaster Management
  Environment
  Global Partnership for Development
Development and Health
Organization for International Health
The World Health Organization
  The United Nations Children's Fund (UNICEF)
Nongovernmental Organizations
The World Bank
Trends in Global Health
Emerging Infectious Disease Threats
Expanding National Health Capacity
Global Health and the New Public Health
Summary
Electronic Resources
Recommended Readings
Bibliography
Publications and Journals

INTRODUCTION

Globalization of health is the growth of international transfer of diseases including those fostered by lifestyle, and cooperation to address disease threats and health conditions common to many countries in the world. Globalization of health includes rapid movement of large numbers of people, foods, drugs, vaccines, medical education, and technology from place to place. It recognizes that health and

economic development are interlinked, and that social equity in health is essential to achieve the newly reiterated goals of Health for All. In addition, many new conditions, including rapid mass travel, global communication, and entertainment, promote broadening of the effect of communicable diseases and common risk factors leading to epidemics of chronic diseases.

More generally, globalization is a process of international trade liberalization, privatization, domestic deregulation, by which the peoples of the world are becoming part of a single society. The globalization process has intensified since the 1990s with the removal of barriers to international trade and foreign direct investments. At the same time, a combination of economic, technological, environmental, sociocultural, and political forces came into play which increased the commonality of global health concerns and promoted shared attempts to address global warming, poverty, ill health, and inequities. Globalization has a complex influence on health, both negative and positive. The Millennium Development Goals constitute an attempt by the United Nations to address global inequity of poverty, maternal and child health, communicable and other diseases, environmental degradation, and other markers of an increasingly interactive world.

Dr. Margaret Chan, Director-General of the World Health Organization, put it this way in 2007:

First, in matters of health, I believe our world is out of balance, possibly as never before in history. We have never had such a sophisticated arsenal of technologies for treating disease and prolonging life. Yet the gaps in health outcomes keep getting wider. Life expectancy can vary by as much as 40 years between rich and poor countries. This is unacceptable. An estimated 10.5 million children under the age of five die each year. At least 60% of these deaths could have been prevented by just a handful of inexpensive measures. This is not fair. Nor is it fair that more than one million people still die each year from such an easily preventable disease as malaria.

Looking back, we are approaching the 30th anniversary of another historical set of commitments: the Declaration of Alma-Ata.

We are struggling against challenges that have grown enormously in their complexity. The world did not face HIV/AIDS in 1978. Since then, many diseases, including tuberculosis and malaria, have dramatically resurged.

Primary health care is not cheap. It is not a bargain-basement way for governments to fulfill their duty to protect all citizens from risks and dangers to health. This, then, is part of our common humanity, as expressed in the Millennium Declaration. These are our shared traits of compassion, inspiration, aspiration, and great ingenuity. Our common humanity gives us reason to care. It is why we must act with urgency in the face of an emergency. It is also why we have so much to gain, in the name of social justice.[1]

Events in any part of the world can affect the status of health of people in other parts of the globe. There is a dynamic interaction and interdependence so that a global approach is essential to achieve health targets, even locally. Without restating arguments elaborated throughout the text, future generations of public health professionals must be well aware of what is occurring outside of their communities and countries. This means not only learning from the news media of outbreaks of exotic diseases, but also recognizing that the political, social, and economic upheavals that define everyday existence, across the street or halfway around the world, affect us all. Even the most remote communities in the world are not immune to the global impacts of distant military coups, civil wars, natural disasters, economic crises, or epidemics.

The world has become global in political, economic, and health terms. Yet inequities in health are still blatant and call for mutual help as well as self-help. This inequality is not only between countries, but within even developed and much more so in the rapidly developing countries with middle and upper classes living with luxuries and a comfortable lifestyle. Globalization has a complex influence on health. Some components of globalization such as trade liberalization and technology transfer could increase efficiency, welfare, and health. However, under the existing barriers to the international markets, the weaker countries need help to address the stagnation preventing them from improving their market position, thereby increasing wealth and health of the population. Financial aid alone can have destructive effects by creation of dependency and misuse of resources through corruption and promotion of military primacy use of funds, while fundamental poverty and health issues are not a high priority on national agendas.

The tools to prevent and control disease are available and widely successful, yet not well-applied in many countries. In the health sector, there are inappropriate balances of health resource allocation so that important and effective preventive measures are underfunded and often left to international donors to provide. The common interest and social solidarity represented by international health efforts have achieved much but international public health efforts are necessary to combat disease and create a healthy world. Such achievements are shown in global initiatives to eradicate smallpox and greater progress in reducing poliomyelitis from 125 countries in 1988 to only 3 countries in 2007 (WHO, 2007).

Previous generations of public health advocates have made tremendous contributions to understanding disease, how it spreads, and how it affects all forms of life. But mistakes have been made; HIV infection was not detected early enough, or its impact realized, until it had already reached pandemic proportions. In the twenty-first century public health is globalized as the health of all humans is globally linked. Yesterday's local SARS, Ebola, measles, or cholera outbreak 10,000 miles away may manifest itself today in the arrival hall of your local airport. With this very real possibility in mind, the future of public health will require advocacy of international economic, political, and social justice policies, with the necessity of organized common efforts to improve health around the world.

Transportation, colonization, and commerce have been responsible for the dissemination of disease throughout history. With rapid movement of large numbers of people by sailing ships, steamships, rail, and later air, the possibility of disease transmission by travelers has increasingly become a global public health problem. Noncommunicable diseases have also been transferred between populations by adoption of risk factors, such as smoking, the automobile, and western diets heavy in fats, bringing rising waves of mortality from these causes to areas with previously low rates. The impact of economic, demographic, and epidemiologic changes are not uniform, neither within nor between countries. Domestically and internationally, poor health and poverty affect the stability of us all. A popular late twentieth-century slogan in which to "think globally, act locally," expresses the interdependent realities of public health. Another aspect of globalization of health is the increased mobility of health professionals, health consumers, and international organizations which work across boundaries, resulting in a massive flow of trained health workers from developing to developed nations. This aspect of globalization of professional training and mobility threatens poor countries with loss of skilled professionals they so badly need to build their health infrastructure capacity.

Previous chapters addressed demographic and epidemiologic issues with examples from different countries, as well as regional and global trends. In this chapter, major trends and contemporary patterns of health and disease in the world are presented, along with policy issues for improving those patterns. Global trends can be analyzed by grouping countries by geographic regions, by levels of economic development, and political, cultural, or ethnic characteristics. Global health requires both official and unofficial international health organizations

---

[1] From the Opening Address: The contribution of primary health care to the Millennium Development Goals. Dr. Margaret Chan, Director-General of the World Health Organization at the International Conference on Health for Development, Buenos Aires, Argentina, August 16, 2007.

to stimulate and facilitate joint efforts to achieve common goals, such as preventing the transmission of communicable disease or, more generally, promoting *Health for All.*

THE GLOBAL HEALTH SITUATION

Global health status involves a wide diversity of social and economic standards, disease, disability, and mortality throughout the world. Environmental and socioeconomic factors and health interventions all play a role in health status. Differences between and among developed and developing countries in these factors are great, yet there are common concerns and shared interests in health development. Studies of countries classified by geographic region, such as the World Health Organization (WHO) regions, or by economic status, such as the Organization for Economic Cooperation and Development (OECD) countries, European Union (EU), Eastern Europe (CEE), and the former Soviet Union (Commonwealth of Independent States or CIS), help to provide an overall picture of demographic transitions and epidemiologic shifts. Economic groupings of countries are usually measured by gross domestic product (GDP) per capita, a measure of national productivity, which in industrialized countries is more than 20 times greater than that of the developing countries.

In the past several decades , the average life expectancy at birth has increased globally by about 20 years from 46.5 years in 1950–1955 to 67 years in 2002 (World Health Statistics, WHO, 2008). This is a gain of almost 3 months a year in this period, but varied by region. Life expectancy at birth increased globally from 1990 to 2006 from 63 to 67 years. For sub-Saharan Africa, there was no gain (51 and 51 years); for the Americas, the gain was four year (71–75); Euro region four years (70–74); South East Asia six years (from 58–64); and West Pacific five years (from 69–74). The low-income developing countries gained four years life expectancy (from 55–59), lower middle-income countries gained five years (66–74), upper middle-income had a gain of only one year (68–69), and the high-income countries gained five years (75–80).

There are stark differences in average life expectancy at birth. For women in developed countries in 2002, life expectancy was 78 years, and 46 years for men in sub-Saharan Africa. In high-income countries in 2006, life expectancy for women was 82 years and 50 years for men in the Africa region. The differences are due to many factors including quality of sanitation, housing, education, family planning measures, and effective public health measures like immunizations against infectious diseases. Life expectancy has increased in the developed world but also in the low mortality developing countries. Countries like China and India have shown great improvements in the under age 5 mortality in the past 50 years in

contrast to most African countries. The under 5 child mortality rate in India declined from 115 in 1990 to 76 in 2006; in China during the same period the mortality rate for this group declined from 46 to 24 (WHO, 2008).

Life expectancy at birth for the world was 65.8 years (63.9 years for males and 67.8 years for females) in 2007, and is projected to increase to 67.2 years (65.0 years for males and 69.5 years for females) for 2005–2010 (United Nations World Population Prospects, 2006). In 2006, life expectancy at birth ranged from 41 in Switzerland to over 80 in many industrialized countries. In the European region, average life expectancy for men in Eastern Europe in 2005 was 13 years less than for men living elsewhere in Europe; half of this mortality was due to cardiovascular disease and a further 20 percent due to trauma.

Countries now considered developed had in the past disease patterns similar to developing countries today. It should be noted that infant and maternal mortality rates in many developing countries today are similar to those in the United States in the 1920s (Table 16.1). Further, within industrialized countries, there are social, ethnic, or immigrant groups whose current health status is characteristic of developing countries. In many developing countries, the rising middle-class populations show epidemiologic patterns similar to those in developed countries, such as rising rates of heart disease.

The enormous differences in GDP per capita and in birthrates are reflected by differences in almost all health status indicators. Population growth in the developing countries, due to high fertility rates and declining child mortality, is a key factor in poverty and poor health status, and thus is a major health problem.

Trends in demographic and health indicators for countries classified as industrialized, developing, and least developed all show positive changes in health status: fertility and birthrates have declined in developing countries, but also more recently in the least developed countries. Female literacy rates are increasing in the least developed and developing countries. Immunization coverage has improved globally, as have infant, child, and crude mortality rates, so that life expectancy is generally rising. However, there remains a large discrepancy between rich and poor in health status indicators; in the least developed countries, maternal mortality is more than 60 times higher and infant mortality more than 18 times that of the industrialized countries.

Despite the fact that health status indicators have improved for the least developed countries, on the basis of current trends, the United Nations Population Division estimates the world infant mortality rate was 53.9 in 2000–2005 and projects that it will decline to 45.1 by 2015 and to 23.4 by 2030. In 1990, some 13 million children under age 5 died, and in 2006, the number of these deaths declined to 9.7 million, still mostly in developing

TABLE 16.1 Health Status Indicators by Country Development Status, Selected Years, 1970–2006

	Least developed	Developing	CEE/CIS[a]	Industrialized
GNI per capita in $US 2006	438	1967	4264	32,7217
Life expectancy in years at birth				
1970	43	53	66	72
2006	55	66	68	79
Crude birth rate/1000 population				
1970	48	38	20	17
2004	37	23	14	11
Total fertility rate/adult women				
1990	6.8	3.60	12.3	1.7
2006	4.7	2.8	1.7	1.7
% Female adult literacy				
1980	24	46	na[b]	96
2004	45	70	96	na[b]
% Births attended by trained attendant 2000–2006	38	55	95	99
Maternal mortality ratio				
1990	1100	470	85	13
2005	870	450	46	8
Infant mortality rate				
1990	113	70	43	9
2006	90	54	24	5
Under-5 mortality rates[c]				
1990	180	103	63	10
2006	142	79	27	6
% Low birth weight 1999–2006	17	16	7	6
% Immunization 2006				
DPT 3[d]	78	78	96	95
Polio	74	78	97	93
Measles	74	78	97	93

[a]CEE = Countries of Eastern Europe; CIS = Commonwealth of Independent States (including Baltic States), i.e., republics of the former Soviet Union.
[b]na, Not available.
[c]Under-5 mortality rate = deaths from birth to 5 years per 1000 live births.
[d]Diphtheria, pertussis, tetanus vaccine 3 doses.
Source: UNICEF, *State of the World's Children*, 2008: *Child Suvival*. 2008. United Nations Childrens Fund: New York.

countries and largely from preventable or easily treatable diseases. The MDG to cut child mortality in half by the year 2015 would mean that "5 million would die, still largely from preventable diseases" (UNICEF, 2008) (Figure 16.1).

While the gap between developed and developing countries in health is very wide, the adaptation and dissemination of health technology are having profound effects. Each can learn, for good or ill, from the other. A developing country may spend most of its health resources in central teaching hospitals, while primary care is neglected. Adoption of appropriate priorities, including new vaccines and other health technologies, can bring dramatic improvements in

health in developing countries. Conversely, innovations in the health field from developing countries can also be applied in developed countries. For example, oral rehydration therapy and community health workers, which provide care in developing country conditions, can be applied to unmet needs in industrialized countries.

PRIORITIES IN GLOBAL HEALTH

Poverty and disease are interactive. Sick people have reduced or no capacity to perform well economically. Health is not only a development goal, but also a means to promote

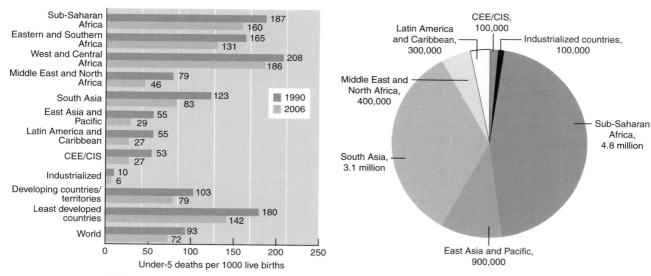

FIGURE 16.1 Declining global mortality for children, 1990–2006, and causes of death, 2006. Note: The global child mortality rate declined by almost one-quarter between 1990 and 2006. Fewer than 10 million children under 5 died in 2006. Source: UNICEF. 2008. *State of the World's Children 2008. Child Survival.* New York: UNICEF.

development. The eight Millennium Development Goals agreed upon by all the countries of the world target the poor from all over the world with a community-based bottom-up approach as well as national targeted programs which must be integrated at the community level (Box 16.1). Three

of the eight Millennium Development Goals directly address health, and all the others are intricately linked to health. Many other global issues in health include climate change, bioterrorism, global epidemiologic surveillance, drug-resistant organisms for tuberculosis, malaria and other infectious diseases,

---

Box 16.1 The World Health Organization, Agenda 2007

1. **Promoting development** During the past decade, health is recognized as a key driver of socioeconomic progress, and more resources than ever are being invested in health. Yet poverty continues to contribute to poor health, and poor health keeps large populations in poverty. Health development is directed by the ethical principle of equity: Access to life-saving or health-promoting interventions should not be denied for economic or social reasons. WHO activities are aimed at health development with priority to health outcomes in poor and disadvantaged countries or vulnerable groups. The Millennium Development Goals addressing poverty reduction, and other programs preventing and treating chronic diseases, and the neglected tropical diseases are cornerstones of the health and development agenda.

2. **Fostering health security** One of the greatest threats to international health security arises from outbreaks of emerging and epidemic-prone diseases. Such outbreaks are occurring in increasing numbers, fueled by such factors as rapid urbanization, environmental mismanagement, the way food is produced and traded, and the way antibiotics are used and misused. Shared vulnerability to health security threats demands joint action to collectively defend against outbreaks. This was strengthened in June 2007 by newly revised International Health Regulations.

3. **Strengthening health systems** For health improvement to help reduce poverty, health services must reach poor and underserved populations. Health systems in many parts of the world are unable to do this, so strengthening of health systems is a high priority. This includes addressing the provision of adequate numbers of appropriately trained staff, sufficient financing, suitable systems for collecting vital statistics, with access to appropriate technology and essential drugs.

4. **Harnessing research, information, and evidence** Evidence provides the foundation for setting priorities, defining strategies, and measuring results. WHO generates authoritative health information, in consultation with leading experts, to set norms and standards, articulate evidence-based policy options, and monitor the evolving global health situation.

5. **Enhancing partnerships** WHO works with the support and collaboration of many partners, including UN agencies and other international organizations, donors, civil society, and the private sector. WHO uses the strategic power of evidence to encourage partners implementing programs within countries to align their activities with best technical guidelines and practices, as well as with the priorities established by member countries.

6. **Improving performance** WHO participates in reforms to improve its efficiency and effectiveness, both at the international level and within countries. WHO aims to ensure that its strongest asset — its staff — works in a motivating and rewarding environment.

Source: WHO, at http://www.who.int/about/agenda/en/index.html [accessed January 3, 2008]

aging, obesity and diabetes, alcohol and drug abuse, violence, prostitution, and human trafficking.

Inadequate levels of health spending by the world's poorest countries are a major factor in continuance of the vicious cycle of poverty and illness.

WHO has in 2006 and 2007 World Health Reports addressed human sources for health and global health security as main themes. The deficiency in trained health workforce in developing counties and movement of health workers from poor to richer nations are matters of global concern. The adoption of the revised International Regulations on Health present a new code for public health based on long traditions derived from the plague and quarantine, cholera and sanitation, and smallpox and immunization. It includes issues related to preparedness for chemical and bioterrorist emergencies, and attempts to provide a basis for localization and control of potentially highly dangerous infectious diseases from spreading uncontrollably.

Poverty–Illness–Population– Environment

The Millennium Development Goals (MDGs) represent an effort by the international community to address the various pressing issues in the world as an global village in which problems of one country cannot be seen in isolation. With 2015 approaching, an MDG monitoring system has been established to help track the progress of individual countries, to learn about new challenges, and to support organizations worldwide working on the MDGs. The first MDG has two targets: to halve, between 1990 and 2015, the proportion of people whose income is less than $1 a day; and to halve, between 1990 and 2015, the proportion of people who suffer from hunger (Box 16.2).

The process evaluation method to track the work done in this regard is being done by the World Bank, UNICEF, WHO, and the Food and Agricultural Organization (FAO), using indicators such as underweight children under the

age of 5 years and proportion of population with an income below $1 (USD) per day.

The past decade has seen a substantial decline in poverty all over the world, largely because of the economic growth in China and India, leading to a spurt in the GDP of the countries in Asia, especially in the Indian subcontinent. However, the situation in sub-Saharan Africa has worsened with the GDP per capita falling by more than 14 percent, the number of people living below the poverty line was increased from 41 percent in 1981 to 46 percent in 2001, and the number of people living in extreme poverty increased by more than 140 million. Despite impressive economic growth, the situation in India and China was no better, as the data in 2001 showed that around 700 million people living below the poverty line (with less than $1 PPP a day) lived in this region. This accounted for slightly less than two-thirds of the total number of poverty-stricken people in the world. The second MDG (target 3) is devoted to equality of boys and girls in completing primary education and is being monitored by UNESCO. The child mortality rate declined from 93 in 1990 to 72 in 2006 (UNICEF, 2008).

The fifth target of the MDG is to reduce the 1990 under-5 mortality rate by two-thirds by 2015. This is primarily a measure of success in immunization and basic child nutrition. The proportion of 1-year-old children immunized is monitored by WHO and UNICEF. The fifth MDG is to reduce by three-quarters the maternal mortality ratio between 1990 and 2015, but this has proven thus far to be resistant to change. The WHO, UNICEF, GAVI, and others have combined efforts to halt or reverse the spread of HIV, malaria, and other major diseases like TB across the globe.

The seventh MDG has three targets which deal with sustainable environment, sanitation, and housing (for at least 100 million slum dwellers). This is a very crucial goal as the great economic growth of many countries will be halted because of environmental degradation. Table 16.2 shows important indicators from developing countries on their track to achieve the MDGs.

The interactions among poverty, population growth, and environmental degradation combine to adversely affect many developing countries and hundreds of millions of people globally. In many developing countries, economic stagnation and political instability compound these issues, causing an inability to address basic human needs and condemning more generations to ill health and early death. Although the effects of low income, lack of basic sanitation, and crowding in rural poverty or the slums of megacities cannot be overcome by health measures alone, the potential for raising the quality of life and survival rates by public health measures is very great. The leading causes of disease and mortality among adults are shown in Table 16.3.

Industrialized countries have overcome many of these problems but must still cope with very substantial pockets of poverty, homelessness, violence, preventable disease,

---

Box 16.2 Millennium Development Goals by 2015

1. Eradicate extreme poverty and hunger
2. Achieve universal primary education
3. Promote gender equality and empower women
4. Reduce child mortality
5. Improve maternal health
6. Combat HIV/AIDS, malaria, and other diseases
7. Ensure environmental sustainability
8. Develop a global partnership for development

Source: United Nations, 2005, http://www.un.org/millenniumgoals/ [accessed November 6, 2007]

TABLE 16.2 Global Progress in Reducing Child Mortality Is Insufficient to Reach MDG 4*
Average Annual Rate of Reduction (AARR) in the Under-5 Mortality Rate (U5MR) Observed for 1990–2006 and
Percent Decline Required During 2007–2015 in Order to Reach MDG 4

| | U5MR No. of deaths per 1000 live births | | AARR | | |
	1990	2006	Observed % 1990–2006	Required % 2007–2015	Progress toward the MDG target
Sub-Saharan Africa	187	160	1.0	10.5	Insufficient progress
Eastern and South Africa	165	131	1.4	9.6	Insufficient progress
West and Central Africa	208	186	0.7	11.0	No progress
Middle East and North Africa	79	46	3.4	6.2	Insufficient progress
South Asia	123	83	2.5	7.8	Insufficient progress
East Asia and Pacific	55	29	4.0	5.1	On track
Latin America and Caribbean	55	27	4.4	4.3	On track
CEE/CIS	53	27	4.2	4.7	On track
Industrialized countries/territories	10	6	3.2	6.6	On track
Developing countries/territories	103	79	1.7	9.3	Insufficient progress
World	93	72	1.6	9.4	Insufficient progress

*Progress toward MDG 4, with countries classified according to the following thresholds:
On track—U5MR is less than 40, or U5MR is 40 or more and the average annual rate of reduction (AARR) in under-5 mortality rate observed from 1990 to 2006 is 4.0 percent or more.
Insufficient progress—U5MR is 40 or more and AARR observed for the 1990–2006 period is between 1.0 percent and 3.9 percent.
No progress—U5MR is 40 or more and AARR observed for 1990–2006 is less than 1.0 percent.
Note: UNICEF estimates based on the work of the Interagency Child Mortality estimation group.
Source: UNICEF. 2008. State of the World's Children 2008. Child Survival. New York, UNICEF.

TABLE 16.3 Leading Causes of Mortality and Disease Burden (DALYs) Among Adults, Worldwide (2002),
Adults Aged 15–59

A. Total global mortality

Rank	Cause	Deaths (000s)
1	HIV/AIDS	2279
2	Ischemic heart disease	1332
3	Tuberculosis	1036
4	Road traffic injuries	814
5	Cerebrovascular disease	783
6	Self-inflicted injuries	672
7	Violence	473
8	Cirrhosis of the liver	382
9	Lower respiratory infections	352
10	Chronic obstructive pulmonary disease	343

(Continued)

TABLE 16.3 Leading Causes of Mortality and Disease Burden (DALYs) Among Adults, Worldwide (2002), Adults Aged 15–59—Cont'd

B. Adults aged 15–59 — disease burden (DALYs) by cause

Rank	Cause	DALYs (000s)
1	HIV/AIDS	68,661
2	Unipolar depressive disorders	57,843
3	Tuberculosis	28,380
4	Road traffic injuries	27,264
5	Ischemic heart disease	26,155
6	Alcohol use disorders	19,567
7	Hearing loss, adult onset	19,486
8	Violence	18,962
9	Cerebrovascular disease	18,749
10	Self-inflicted injuries	18,522

C. Adults aged 60+ — global mortality by cause

Rank	Cause	Deaths (000s)
1	Ischemic heart disease	5825
2	Cerebrovascular disease	4689
3	Chronic obstructive pulmonary disease	2399
4	Lower respiratory infections	1396
5	Trachea, bronchus, lung cancers	928
6	Diabetes mellitus	754
7	Hypertensive heart disease	735
8	Stomach cancer	605
9	Tuberculosis	495
10	Colon and rectum cancers	477

D. Adults aged 60+ disease burden — (DALYs) by cause

Rank	Cause	DALYs (000)
1	Ischemic heart disease	31,481
2	Cerebrovascular disease	29,595
3	Chronic obstructive pulmonary disease	14,380
4	Alzheimer's and other dementias	8569
5	Cataracts	7384
6	Lower respiratory infections	6597
7	Hearing loss, adult onset	6548
8	Trachea, bronchus, lung cancers	5952
9	Diabetes mellitus	5882
10	Vision disorders, age-related and other	4766

Source: Adapted from WHO, *World Health Report*, 2003, Table 1.3.

environmental degradation, and limited access to health care. The southern hemisphere is largely made up of developing countries with massive economic and social needs. The north–south socioeconomic divide is one that will shape global health and politics in the twenty-first century.

During the 1990s, a number of developing countries have entered a phase of rapid economic and industrial growth, combining the advantages of educated low-wage work forces with market economies. Some Asian countries moved ahead rapidly with economic development, creating strong rates of growth. The breakdown of traditional social patterns, traditional rural ways of life and intergenerational family structure to better education, upward mobility, and small family units in southeast Asian economies, will compound these problems.

The recognition that poverty and ill health are interactive led the industrialized nations (G7 plus Russia) to take an important step in mid-1999 to relieve debt and provide aid-related loans to very poor countries by some $118 billion. This will help many countries in sub-Saharan Africa by reducing their debt repayment by one-third to one-half. Despite this important step, most poor countries pay more on debt service than they do on health and education for their people.

Child Health

Almost 19 percent of total deaths in the world (10.5 million) are of children under 5 years of age and almost 98 percent is in the developing world. The figures seem to have improved in many countries since 1970 when the figure was 17 million, but in 14 African countries the present levels of under-5 mortality are higher than they were in 1990. About 35 percent of Africa's children are at a higher risk of death than they were 10 years ago. Nineteen African countries figure in the list of the top 20 countries in the world with highest rates of under-5 mortality. A baby born in an African country, for example, in Sierra Leone, is three and a half times more likely to die before its fifth birthday than a child born in India and more than a hundred times more likely to die than a child from a developed European country or even Singapore (WHO, 2003). The leading causes of under-5 mortality in the poor countries are perinatal conditions, lower respiratory tract infections, diarrheal diseases, and malaria. Sub-Saharan Africa has undergone a severe onslaught by the HIV epidemic, wiping out an estimated 332,000 children in 2002 (WHO, 2003). There is a growing health inequality in children's health all over the world, with a higher rate of death if they are poor and undernourished. The Millennium Development Goal indicator 14 (infant mortality) for 1990 and 2005 by developmental region for infants is shown in Table 16.4.

Most of the deaths in children under the age of 5 are preventable, as can be seen from the list given in Table 16.5.

TABLE 16.4 Millennium Development Goal Indicator 14, Infant Mortality Rate by Region

Year	1990	2005
World	65	52
Developing regions	71	57
Northern Africa	66	30
Sub-Saharan Africa	110	99
Latin America and the Caribbean	43	26
Eastern Asia	37	23
Southern Asia	87	62
Southeastern Asia	53	31
Western Asia	53	45
Oceania	59	47
Commonwealth of Independent States	39	33
Commonwealth of Independent States, Asia	67	60
Commonwealth of Independent States, Europe	21	14
Developed regions	10	5
Transition countries of Southeastern Europe	25	14

Note: Deaths of children before reaching the age of 1 year, per 1000 live births.
Source: Millennium Development Goals Indicators. The Official UN Site for MDG Indicators, http://unstats.un.org/unsd/mdg/Host.aspx?Content=Data/trends [accessed January 17, 2008]

The 10 leading causes listed in the table constitute 86 percent of all causes of under-5 mortality. Five diseases — pneumonia, diarrhea, malaria, measles, and AIDS — account for about half of under-5 deaths. Most of these lives can be saved by expanding coverage of existing interventions, especially among poor families using a bottom-up approach. Many vaccine-preventable diseases are listed in the top causes, and it has been seen (i.e., polio) that a concerted effort by health authorities in promoting immunization can dramatically reduce child mortality. With simple methods like oral rehydration therapy (ORT), mortality related to diarrhea can be reduced within a short time.

Maternal Health

The fifth MDG as seen in the previous section is to reduce the maternal mortality ratio by three-quarters between 1990 and 2015. In developing countries, complications in the antepartum, intrapartum, and postpartum periods are leading causes of death among women. Every year,

TABLE 16.5 Leading Causes of Death in Children in Developing Countries, 2002

Rank	Cause of death	Numbers (000)	% of all deaths
1	Perinatal conditions	2375	23.1
2	Lower respiratory infections	1856	18.1
3	Diarrheal diseases	1566	15.2
4	Malaria	1098	10.7
5	Measles	551	5.4
6	Congenital anomalies	386	3.8
7	HIV/AIDS	370	3.6
8	Pertussis	301	2.9
9	Tetanus	185	1.8
10	Protein-energy malnutrition	138	1.3
	Other causes	1437	14.0
Total		10,263	100%

Source: World Health Report, 2003, Table 1.1. WHO, Geneva.

approximately 9 million women suffer some form of injury from pregnancy or childbirth which has lasting effects on their health. Africa and Asia have the majority of pregnancy-related deaths accounting for 95 percent of the total maternal deaths in the world.

The global maternal mortality ratio in 2004 was 400 per 100,000 live births. There was wide variation among WHO regions and by level of CDP per capita countries grouping. Maternal mortality ratios for the regions were: Africa—900, South East Asia—450, Eastern Mediterranean—420, Western Pacific—82, and the Americas—99 per 100,00 live births. Rates among the countries by level of economic development were: low—650 per 100,000, lower mid-level—180, upper mid-level—91, and high level of development—9 per 100,000 (WHOSIS, 2008). In 2005, 536,000 women died of maternal causes, compared to 576,000 in 1990. Ninety-nine percent of these deaths occurred in developing countries. Just over half of all maternal deaths (270,000) occurred in the sub-Saharan African region, followed by south Asia (188,000). The MMR is declining too slowly to meet the MDG 5, to reduce the number of women who die in pregnancy and childbirth by three-quarters by 2015. The current rate of decline is less than 1 percent while an annual decline of 5.5 percent is needed to achieve the MDG (WHO, UNICEF, UNFPA, World Bank, October 2007). The MMR in CIS countries was 51 per 100,000 live births and 9 in developed countries. The countries with high rates have made only modest progress on this measure since 1990. In these countries, the maternal mortality ratio was 450 deaths per 100,000 live births in developing countries, the highest being in sub-Saharan Africa (920) followed by Southern Asia (540).

Institutionalized delivery or the presence of a skilled attendant decreases the risk of maternal deaths to a great extent. The number of deliveries with a skilled attendant increased significantly between 1990 and 2003, from 41 to 57 percent (UNDESA, 2005).

The health of women in relation to fertility is fundamental to national health standards. Education and improved nutrition for girls and women, better access to modern birth control, spacing of pregnancies, and adequate care in all stages of pregnancy are the vital means to achieve improvement in women's reproductive health. Traditional birth attendants (TBAs) provide care during most deliveries in developing countries. There are no adequate substitutes for good prenatal medical care, but the work of TBAs can be improved by a strict program of licensing, training, and supervision (Chapter 14).

Simple, inexpensive measures can improve outcomes: education about the right to care for safe pregnancy; routine iron and folic acid during pregnancy; prenatal care stations (MCH [maternal and child health]); HIV, STI screening and care; professionally supervised birth centers (in hospitals if possible); high-risk identification and referral systems (see Chapter 5); training, licensing, and supervision of traditional birth attendants; and deployment of well-trained community health workers for preventive health care (Chapter 15) can all make a difference.

The failure to advance in lowering maternal mortality in many countries is largely a failure of political will and initiative. The means to prevent most of these deaths is available knowledge, but applying the national and international will and investment of necessary resources to this task have been wanting. Safe motherhood has taken a back seat to HIV and child health in priorities in many high MMR countries, and stagnation of effort and results has followed. A combination of international and national organization, improved databases, and program entrepreneurship are needed to reduce the staggering toll of maternal mortality, even in countries such as Nigeria with oil wealth, and India with rapid economic development.

Population Growth

Despite the fact that population growth is a religious and political controversy in many societies, falling birthrates are now seen in most parts of the world. Developing countries are increasingly recognizing that high fertility rates hinder economic development, perpetuating poverty, a fundamental cause of ill health. The politics of population have traditionally rested on the assumption that population increase is essential for economic growth and national power. At the micro level in traditional farming societies the assumption is that more children provide greater security for the family. In recent decades the expansion of family planning technology has been accompanied by a gradual shift to the view that unrestrained population growth is a barrier to economic development.

In many poor countries, high rates of population growth perpetuate poverty and ill health for mothers and children. Improved child survival and reduced economic imperatives for more children to work farms or to contribute to family incomes have led most countries to lower overall birthrates. Higher education levels for women have increased knowledge and use of birth control. Religious injunctions against birth control no longer have the power to prevent use, so that birthrates have fallen worldwide, and in many countries to below replacement levels (i.e., negative population growth).

Despite the decline in fertility rates in most regions of the world, the world population has passed the 6 billion mark and is growing at an average annual rate of 1.73 percent (Table 16.6). Asia accounts for almost 60 percent of the world's population, while Europe is less than 10 percent. In many Asian countries, the birthrate has declined precipitously to rates close to those in developed countries. Over the period 1960–1992, total fertility rates fell: in southeast Asia and the Pacific from 5.8 births per woman to 2.6; in Latin America and the Caribbean from 5.8 births per woman to 3.1; and in the Middle East and north African region from 7 births per woman to 5.1. By 2005, total fertility was 5.4 in sub-Saharan Africa, 3.1 in the Middle East and North Africa, 2.5 in Latin America and the Caribbean, and 1.7 in CCE/CIS.

Many countries in sub-Saharan Africa will see their population double within 20 years, although even here there has been what appears to be the beginning of a decline in total fertility rates. Population growth is below replacement level in most industrialized countries and falling in most developing countries as well, but the fertility gap is still high with growth rates of 1.5 percent currently and projected to decline to 0.5 percent by 2030. The State of the World's Children

reports that in 2005 the fertility rate in industrialized countries was 1.6 compared to 4.9 for developing countries and 2.6 for the world (UNICEF, 2007).

Governments have a crucial role in family planning. Distribution of information and promotion of family planning as a national policy and priority must be part of a new emphasis on primary health care. In China in the 1950s, Chairman Mao Tse Tung called birth control a new form of genocide of the developed countries against the developing countries. The legacy of this tragic pronouncement was no less destructive to public health than the pronouncements of religious bodies which still equate birth control with mortal sin. Both had the effect of promoting high fertility rates in those populations that can least afford the health and economic burden of raising large numbers of children. For the past four decades, birth control has been promoted in India and China, the two countries with the world's largest populations, but the momentum of population growth continues and is unlikely to level off in the next 20 years. In addition, China's one-child-per-family policy has reportedly led to female infanticide, forced abortions, and sterilization in a primarily rural society valuing male children.

A demographic transition occurs when the age makeup of the population shifts. As countries move from developing to a developed or industrialized status, population age patterns change. With greater life expectancy and declining birthrates, the ages of the population shift toward older age groups. Developed countries are experiencing a rapid growth of the old-old, more dependent population (i.e., over 75 and over 85 years). These trends are of vital importance to the future of individual countries as they try to sustain or improve economic and social conditions. All countries need a working-age population sufficient to sustain the elderly and the young dependent groups.

High birthrates in developing countries still restrict the potential for economic growth, and the care and nurturing of children. Food supplies have been expanded by improved agriculture, but this may not be able to sustain high rates of population growth. In addition, rising standards of living and aspirations place further demand on natural resources and the environment with great stress on the Earth's fragile ecology.

TABLE 16.6 World Population, in Billions, by Economic Level, Actual and Projected, 1965–2030

Level of development	1965	1980	2000	2010	2020	2030
Low- and middle-income countries	2.6	3.7	5.3	5.8	6.4	7.0
High-income economies	0.7	0.8	0.9	1.0	1.1	1.1
Total world population	3.3	4.5	6.2	6.8	7.5	8.9

Source: World Bank. *World Development Report*, 2003.

Malnutrition

Food production has increased in most parts of the world, but has steadily declined per capita in sub-Saharan Africa, along with GNP per capita. Increased production in other parts of the developing world during the 1960s and 1970s slowed during the 1980s. In developing countries, the capacity to produce food faster than population growth is limited. The developed countries, with one-quarter of the world's population, produce over half of the world's food supply. They dominate food production but have low rates of population growth. Developing countries may purchase this surplus of food, but many lack the hard currency to do so. Gross national product alone cannot measure wealth; it must also be weighed in terms of the capacity to produce food.

Hunger, adaptation, and starvation are difficult to measure. Hunger is a subjective phenomenon; adaptation occurs when persons adjust to lower energy intake; and when energy output exceeds intake, starvation occurs. Starvation may be acute or chronic. Hunger and famine are associated with natural disasters and war, but they also occur chronically in settings where food production cannot keep up with population growth. Although hunger and famine affect all ages and sexes, the most vulnerable groups in the population are infants and children, pregnant women, women as a whole, and the elderly. Men are affected in terms of reduced capacity to work. The Chinese famine of 1959–1961, one of the most tragic disasters of the twentieth century, killed up to 36 million persons.

Estimating the number of persons lacking food is difficult because of limited data. Few countries maintain monitoring systems of national nutrition status because of lack of financial and human resources. The nutritional status of the population, more specifically of children, is usually measured by birth weight, weight-for-age, and height-for-age. Low height for a given age, or stunting, is the most prevalent symptom of protein–energy malnutrition. Approximately 40 percent of all 2-year-olds in developing countries are stunted. The prevalence of stunting may be as high as 65 percent in India, about 40 percent in China and sub-Saharan Africa, and more than 50 percent in the rest of Asia. According to WHO standards, some 780 million persons worldwide are estimated to be energy deficient or in a state of protein–energy malnutrition (PEM). This is not always manifested by hunger, but rather represents long-term inadequate food intake, especially protein and calories for energy needs. Malnutrition may be so widespread among children that parents and health providers assume the children's lethargy and stunting to be normal.

Micronutrient deficiency conditions affect some 2 billion people worldwide, with serious sequelae including premature death, poor health, blindness, growth stunting, mental retardation, learning disabilities, and low work capacity. Iodine, iron, and vitamins A, B, C, and D are commonly deficient in diets in developing countries, adversely affecting the health of the whole population but especially vulnerable subgroups. Iron deficiency is the most common of these, affecting mainly women and children but also men and the elderly. In developing countries, children and women are especially vulnerable because of frequent childbirth and poor diets.

Anemia is, as referred to in Chapter 8, the most common nutritional deficiency in the world. It is primarily a micronutrient dietary deficiency but is often exacerbated by other deficiencies (e.g., vitamin C), concomitant parasitic infection in children, and multiple pregnancies in women. In industrialized countries, anemia of pregnancy affects 18 percent of pregnant women, but 40 percent are affected in China and Latin America and 88 percent in India. Iron deficiency in Russian women exceeds 50 percent, and iron supplementation is not routinely practiced; that is, without hemoglobin testing. Iron deficiency in infancy causes reduced growth (in height) and potential learning capacity in school. In adults, it reduces work potential. Distribution of inexpensive iron (ferrous sulfate) to pregnant and lactating women could largely prevent this onerous health burden, which was estimated by WHO to affect 1.8 billion persons.

Iodine deficiency affects some 1.5 billion persons globally, with severity varying from subclinical deficiency to cretinism and severe retardation. Iodine is deficient in soil and water in many parts of the world, and deficiency conditions at subclinical and clinical levels are widespread. Routine iodizing of salt is widely used to prevent iodine-deficiency disorders and has been recommended by WHO and UNICEF; it was adopted as a major objective of the 1990 World Summit of Children, along with elimination of vitamin A deficiency, but implementation remains problematic.

Vitamin A supplements reduce mortality from measles and prevent xerophthalmia and blindness in children. This has created a major change in public health nutrition needs in developing countries by demonstrating nutritional co-morbidity and the vital importance of nutritional interventions to prevent morbidity and mortality in vulnerable population groups. The extent of iodine and vitamin A deficiency conditions is enormous and entirely within the scope of current technology to prevent at low cost. WHO estimates the cost of eradication of iodine deficiency by iodination of salt at $0.05 per person per year. Elimination of vitamin A deficiency can be achieved by giving vitamin A capsules to children over 6 months of age three times per year at a cost of $0.02 per capsule, by dietary modification to promote vitamin A – rich foods, and/or fortification of basic foods (oil, margarine, milk, or sugar).

Food fortification is now recognized as a major need and cost-effective intervention necessary in all countries, particularly in mid- and lower levels of development. Since the 1990s, fortification of flour with folic acid has been implemented in a number of countries to prevent birth defects (neural tube defects). This has provided a new impetus to promote food fortification and new deficiency conditions of public health importance are reported for vitamin D, vitamin K, vitamin B complex (including B_2), selenium, zinc, and others (see Chapter 8). WHO issued new guidelines for food fortification in 2006, which will have great importance in international aid and development policies for the second decade of the twenty-first century.

In many countries in the African, southeast Asian, and eastern Mediterranean regions, infectious and parasitic diseases occur in association with malnutrition and continue to be major public health problems. The infection–malnutrition co-morbidity causes much of the mortality among infants and children and shortens life expectancy (see Chapter 6).

The Fight Against HIV/AIDS and Other Communicable Diseases

Globalization of the spread of disease is as old as the migration of humans, animals, or disease vectors. The emergence of the AIDS pandemic has affected all countries of the world, regardless of their level of development. It has been estimated that there are around 40 million people in the world currently infected with the HIV virus out of which Africa alone has 25 million. There are more than 10 million children in sub-Saharan Africa who have been orphaned by AIDS. The prevalence rate of HIV infection among adults in sub-Saharan Africa was estimated to be more than 7 percent in 2004. Though the prevalence rate of HIV in countries in Asia, especially in India and China, seems to be low, there is a definite potential danger of epidemics if not controlled soon. The youngest population of the world resides in China and India and worldwide about one-third of those currently living with HIV/AIDS are between 15 and 24 years of age. This figure is rising and with the rising level of substance abuse there is an urgent need for a concerted international action.

As discussed in Chapter 4, the emergence of "new" infectious diseases and the reemergence of well-known but still uncontrolled diseases pose great challenges for public health and clinical care. The problems of these diseases are compounded by the rise of resistant microbial strains. The basic priorities in control of infectious diseases remain the need for universal coverage with childhood immunization; high standards of food and water safety and sanitation; education to reduce the spread of HIV and STIs; improved primary care for prevention, diagnosis, and management of TB and malaria; and provision of antimicrobial therapy.

The HIV epidemic has engulfed the economies of many nations and has been the cause of the rising spread of poverty, reversal of human development, worsening

health inequalities, and crippling government machineries in various parts of the world, thus reducing the provision of essential services. The very obvious health inequalities observed in the world led the WHO to declare a Global Health Emergency to combat HIV/AIDS when it was found that only 5 percent of those in the developing world who require antiretrovirals (ARVs) get them. It will be a pity if one part of the world is oblivious of the health situation in the other not only for moral reasons but for practical ones; as the world is a global village, ill health in one part of the world will definitely affect the other. WHO in 2008 lowered its estimates of HIV-infected persons globally to between 30 and 37 million.

Another very important global problem is malaria which is endemic in many poor countries, especially in the tropical and subtropical regions of Africa, the Americas, and Asia. Although successful treatment and prevention for malaria have been available for a long time, there are still an estimated 300 to 500 million clinical malaria cases, and about 1 million deaths from malaria every year. There are approximately 3 billion people living in malaria-endemic areas, making them vulnerable to the disease.

Tuberculosis is a disease which experts thought could be eradicated in the 1970s, but it has reemerged as one of the major killer diseases in the world, partly because of co-morbidity with HIV. If TB is detected early and fully treated, people with the disease quickly become noninfectious and eventually cured. Most are treated with directly observed therapy (DOT). In 2005, an estimated 8.8 million new cases of TB occurred, including 1.6 million deaths, with some half million HIV infected. TB is a global phenomenon but is mainly restricted to poor countries where it has been increasing particularly in sub-Saharan Africa and south Asia, and is compounded by the additional problem of multidrug-resistant organisms. Multidrug-resistant TB (MDR-TB) and extensively drug-resistant TB (XDR-TB), HIV-associated TB, and weak health systems are major challenges. TB is high on the agenda of WHO, which hopes to reduce the burden of TB, and halve TB deaths and prevalence by 2015, with the help of international donor agencies and national governments.

Important policy and technological advances are needed before TB, HIV, and malaria can be controlled globally, but much can be accomplished with existing technology. Providing ARV therapy and risk-reduction measures at primary care levels along with antimalarial activities are essential to control current epidemics. New vaccines research will hopefully provide more effective preventive measures, but improved use of current methods can markedly reduce the current toll of these diseases.

Adoption of currently available vaccines for children can eliminate deaths from measles, and markedly reduce deaths from diarrheal and respiratory diseases with their high burden of morbidity and mortality. Important new vaccines used in the developed countries are gradually being incorporated in internationally supported vaccination programs, such as *Haemophilus influenzae* b (Hib), but pneumococcal pneumonia, rotavirus, influenza, hepatitis A, and varicella vaccines are not yet widely used in developing countries. Along with immunizations, nutrition counseling, monitoring of child development, and providing of vitamin A supplements, and deworming, and environmental control measures are coming into widening use of DDT-coated bed nets, may reduce child malaria mortality. Breakthroughs in development of new antimicrobial therapies and vaccines for HIV and malaria may occur in the second decade of the twenty-first century. It is hoped that new ways will be found to address the multidrug-resistant organisms.

Improved food technology will be needed to prevent *Salmonella* and *E. coli* from infecting food sources. Medical care will need to improve its methods of control of infectious diseases to avoid emergence of resistant organisms through more restrained use of antibiotics. The achievements toward eradication of important infectious diseases such as smallpox, polio, measles, guinea worm, leprosy, and onchocerciasis provides a basis for cautious optimism even though tempered by realistic appraisal of unsolved and new challenges of infectious diseases.

Chronic Disease

Tobacco is one of the largest causes of preventable deaths globally, killing between one-third and half of those using

it from ischemic heart diseases, stroke, and chronic lung disease, accounting for 5 and 8 million deaths per year (WHO, 2008). Progress in the industrialized countries in reducing all smoking, however, masks a large increase in teenage smoking. The tobacco interests promote smoking in developing and transition countries which are less able to carry out the legal and other issues associated with prevention of smoking. Antismoking legislation has advanced in North America and in the European Union. This is one of the great challenges of public health.

The WHO in its World Health Report 2003 used the term *neglected global epidemics* to emphasize three important and growing threats to the world: cardiovascular diseases (CVDs), tobacco, and motor vehicle accidents (Figure 16.2). Developing countries suffer from the dual burden of communicable diseases and increasing rates of noncommunicable diseases and injury. These conditions are now greater causes of morbidity and mortality than the communicable diseases affecting the poorest countries around the world and the poor in the developed countries. Risk factors for CVDs are indicative of future health status and 5 of the top 10 risks worldwide are specific to noncommunicable diseases, namely, elevated blood pressure, tobacco use, alcohol abuse, increasing fat consumption with cholesterol, and obesity (*World Health Report 2000*). An estimated 32 million deaths are attributable to noncommunicable diseases and around 16.7 million are because of CVD (WHO, 2003).

In developing countries like India, CVDs have become the leading cause of death, responsible for one-third of all

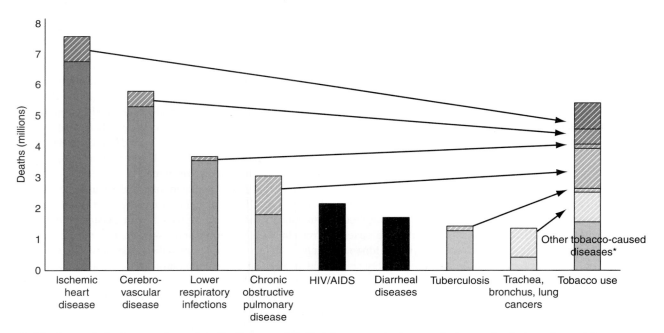

Hatched areas indicate proportions of deaths that are related to tobacoo use and are marked according to the column of the respective cause of death. *Includes mouth and oropharyngeal cancer, stomach cancer, liver cancer, other cancers, cardiovascular diseases other than ischemic heart disease and cerebrovascular disease, diabetes mellitus, as well as digestive diseases.

FIGURE 16.2 The eight leading causes of death worldwide and deaths attributable to tobacco use, 2005.

deaths. Developing countries have double the number of deaths due to CVD in comparison to developed countries. Overall, in developing countries, CVD ranks third in disease burden (after injuries and neuropsychiatric disorders). Even in high-mortality developing countries, CVD morbidity is ranked very high. The major risk factors for CVDs have been identified, even while mortality rates are declining (Chapter 6) and their effects through globalization can be seen all over the world with diets rich in saturated fats, sugar, and salt in vogue everywhere, with increasing sedentary life habits. With the lack of fruits and vegetables in the diet, increase in tobacco use and lack of physical activity, the "global" diet leads to an increase in cases of CVD worldwide.

The epidemiologic transition from a predominance of infectious diseases to the chronic conditions that occurred in the industrialized countries by the mid-twentieth century is also occurring in the developing nations. The cardiovascular diseases, cancer, other degenerative conditions and mental disorders, and trauma are already the major causes of death in many developing countries. Trauma also constitutes a vast public health issue, with serious individual, social, and economic consequences. Worldwide, some 7.2 million deaths occur from ischemic heart disease and another 4.6 million from strokes. Some 2.7 million deaths from injury and poisoning are reported yearly, 2 million of which occur in developing countries, and they result in considerable loss of potentially productive years of life. Motor vehicle accidents (MVAs) rank first in causality, followed by domestic accidents, including falls, burns, poisonings, and drowning, all of which are particularly prevalent among young people and the elderly.

A survey in England predicts that more than 12 million adults and 1 million children will be obese by 2010 if no action is taken. The high and rising prevalence of overweight and obesity in the United States makes this a leading public health problem. From 1980 to 2002, obesity prevalence doubled in adults and overweight prevalence in children and adolescents tripled. Currently, some 119 million, or 64.5 percent, of U.S. adults are either overweight or obese. The prevalence in the United States continues to rise.

Diseases related to smoking, overeating, and unhealthy diet are increasing in developing countries among the middle-class and working-class populations. Rising death rates from coronary heart disease and strokes in the former Soviet countries constitute an enormous contributor to premature death and a burden on underfinanced health systems. As infectious diseases are better controlled and as eating patterns shift in the urban middle and working classes to high meat and fat intake, patterns of cardiovascular disease seen in the industrialized countries are occurring in developing nations. Public health practitioners need to prepare for this epidemiologic transition with interventions such as antismoking campaigns, nutrition education, and other health promotion programs. Similarly, the burden of mental, dental, and other health needs of societies are part of global health planning for developed countries in transition.

Mental health is gaining increasing recognition as a health issue of global proportions affecting hundreds of millions with moderate to severe disability, not only in the industrialized countries, but also in developing countries. As measures of the burden of disease include morbidity as well as mortality (such as in disability-adjusted life years), major depression (unipolar), alcohol dependence, bipolar disorders, and especially schizophrenia become high on the list of causes of disability, particularly in young adults aged 15–24. These require attention in the health system and especially in primary care. Drug abuse and co-morbidity with personality disorders and other mental illnesses are global problems, associated with related issues such as HIV, STIs, tuberculosis, hepatitis B and C, violence, crime, and other destructive behavior with great cost to society.

Disaster Management

Tragic events leading to large-scale loss of property and life created by nature and by man require organized international response to limit the damage, to reduce suffering, and to restore normality. These situations may be natural disasters such as hurricanes, floods, droughts, earthquakes, or volcanic eruptions with severe consequences for public health. They may be larger-scale events created by human initiatives, such as bi-national and civil wars, genocide, and civil strife or repression. Such events can take on enormous proportions as displacement, murder, and other forms of violence disrupt human norms and civil society.

The public health aspects of such events are within the context of restoration of safety, provision of safe water, shelter, food, and sustenance, and efforts to restore civil life. Such events are now brought to the immediate attention of the world's community in television coverage. International action is forthcoming, but often inadequately coordinated with overwhelmed local civil and security authorities. Preparation and organization for such disasters are important elements of global health. The recent tsunami in southeast Asia, with massive loss of life and property, repeated floods, and other disasters, resulted in large-scale displacement of people and public health challenges of the most basic kind. Provision of safety, shelter, safe water and food, disposal of the dead, and many other burdens can overwhelm local resources and require national and international assistance to reduce the scope of the tragedy and to help restore normal life.

Environment

The environment is a global health concern, not only because it affects every country but also because its maintenance requires joint action (Chapter 9). Air pollution caused by industry, power plants, and domestic use of coal is common in urban areas worldwide. The quality of air in the industrialized countries has improved over the 1980s and 1990s, but in many developing countries and the former socialist countries, air quality has deteriorated because of poor quality in power generation and urban congestion. Excessive production of carbon dioxide by use of fossil fuels is contributing to a global warming effect, and chemicals used in industrialized societies cause ecological damage, with potentially serious global consequences. Climate change is an overarching issue affecting health, economics, political developments, and human society generally.

Global Partnership for Development

The eighth goal in the MDGs calls for a partnership between the developed and the developing world, and official development assistance (ODA). The ODA includes measures to ensure debt sustainability on a long-term basis, rule-based, along with predictable and non-discriminatory multilateral trading, and financial systems to address the special needs of the least developed countries. The developed nations undertook to share responsibility for ensuring the global partnership. The UN General Assembly proposed an ODA target of 0.7 percent of the donors' national income but the Organization for Economic Co-operation and Development (OECD) countries have contributed to just around half of the promised amount for many years, which fell to about one-third in the 1990s. World leaders met in 2002 at the International Conference on Financing for Development in Monterrey, Mexico, and established a new framework for a global development partnership (MDG-goal 8, Box 16.2). Since then the signatories started to deliver on commitments made during the conference and aid reached a record high of $79 billion, with countries like Denmark, Netherlands, Norway, and Sweden honoring the initial commitment in the Declaration by the UN General Assembly.

It is very important for developing countries to increase their participation in the global economy, and this depends on reduction in trade barriers imposed by developed countries on imports from developing countries and tariffs on goods that are strategically important to developing economies, such as textiles, clothing, and farm products. There need to be steps taken to write off the debts on very poor countries, especially the economically stagnant African economies. The total debt burden on African economies was estimated to be $206 billion at the end of 2000. Many spend their entire income to service the debts on them, which is many times more than the amount spent on health care.

DEVELOPMENT AND HEALTH

Some 1.3 billion persons in developing countries lack access to clean water; nearly 2 billion lack adequate sanitation. Poverty, low educational and job skills, poor nutrition, an unsanitary environment, and poor housing conditions all contribute to the enormous burden of disease of all kinds in developing countries. Indoor pollution from use of cooking fuels with improper ventilation in developing countries contributes to high rates of acute respiratory disease and deaths in children, as well as to chronic lung disease in the elderly and damage to the fetus in pregnancy.

Health status and economic development are interdependent, and the prevailing social and political philosophies have a vital impact on health, not only in terms of the amount of funds allocated to health, but also in the form of the health care delivery system adopted. Rapid economic development also has its price. Environmental pollution and increases in occupational health hazards occur when new technology is transferred to developing countries. Further degradation also occurs with the tendency of the rural poor to move to cities, where basic sanitation and other infrastructures are often lacking.

Measurement of economic development by GNP alone is misleading. The distribution of wealth in a country is an important variable along with other measures such as school enrollment. The human development index (HDI) includes life expectancy, educational attainment, and measures of income (giving lower weight to income above the poverty level, since extra income for upper income groups is less important to survival). The human development index along with DALYs and QALYs (see Glossary and Chapters 3 and 11) adds an element of the quality of life to the usual economic indices.

Equally important to the amount of money spent on health is how the money is used. Some countries have succeeded in achieving marked improvements in health, while they remain poor as measured by GNP per capita. Some countries have higher ratings in terms of HDI than their ranking by GNP. China, with a GNP per capita of $370, has succeeded in attaining infant mortality and life expectancy rates of mid-level developing countries by bringing primary care to the vast rural population. Sri Lanka, with a per capita GNP of $500, has an infant mortality rate of 15 per 1000, comparable to well-advanced developing countries. The Indian state of Kerala is well above Indian national standards in HDI, even though economically poorer than the national average. On the other hand, some countries with high per capita GNP have lower HDIs; for example, Kuwait and Saudi Arabia have large

GNP per capita but fewer public health achievements than much poorer countries such as Cuba, Costa Rica, and Jamaica. In some countries, this may be due to the large economic gap between the small, very wealthy ruling class and the large, poor population.

ORGANIZATION FOR INTERNATIONAL HEALTH

As seen in Chapter 1, from the decline of the Roman Empire in the fifth century CE, Europe passed into a millennium of scientific repression. Knowledge, including medical knowledge, passed into the hands of the Church, and the Greek and Roman writings that were preserved in the west survived in isolated monasteries of Ireland and Europe, and in Arab civilization, where during the next few hundred years Arabian, Byzantine, and Jewish scholars translated and preserved ancient medical knowledge in Europe. In the ninth century, a medical school was founded in Salerno near Naples, and medical schools spread to cities throughout Europe and the Arab world.

European colonial expansionism, beginning in 1415 with the Portuguese attack on Muslim settlements in nearby north Africa, had extremely important effects on international health. European ships brought smallpox and measles to the natives of the South Pacific and the Americas, decimating their populations. Syphilis, possibly originating from yams, may have been introduced into Europe by sailors returning from the Americas. European adventurers and settlers often suffered severely from the many endemic diseases to which they had scant resistance. Additionally, the slave trade brought communicable diseases from Africa to favorable habitats in the Americas.

Colonialism led to near-eradication of many of the world's native peoples, and changed the character of many populations, most dramatically in North America, Australia, New Zealand, South Africa, and Latin America. Colonial governments introduced western medical organization and practice, including public health and professional education systems, and influenced health with respect to concepts of causality and the treatment of diseases. Widespread education and medical training were important legacies in many developing countries that gained their independence in the mid-twentieth century.

The development of sanitation and later microbiology depended on the scientific and technological underpinning provided by the Industrial Revolution. In the latter half of the nineteenth century, repeated epidemics of cholera in Europe and continuing havoc from other communicable diseases were intense stimuli for researchers to identify the causal agent and means of transmission of almost every major bacterial and parasitic disease. Asiatic cholera arrived in Europe in 1832 and spread throughout the continent in repeated epidemics during the nineteenth century. This led to a convening of the International Sanitation Conference in Paris in 1851, with follow-up meetings held in 1874, 1881, and 1885. These conferences were held more frequently between 1892 and 1903 regarding maritime quarantine and control of international transmission of cholera, yellow fever, and typhus. In the early 1880s, a pioneering step in international public health occurred when, at the request of the International Cholera Commission, Robert Koch led a team to investigate cholera epidemics in Egypt. This resulted in identification of the *Vibrio cholerae* organism and recommendation of preventive procedures.

The Health Organization of the League of Nations (1921–1946), established in Geneva, was an attempt to develop the idea of international collective security for health. As part of its function, the Health Office provided an Epidemic Intelligence Service. The Health Office organized many expert committees on infectious diseases and other public health problems, including the establishment of standards for biologicals, maternal and child health, nutrition, health insurance, and medical education. Malaria, leprosy, and rabies control activities were promoted, as were the establishment of cancer registries and preparation for an international classification of disease; pharmacopoeias were coordinated and standards for housing and nutrition developed. The scope of organized international work was broadened from prevention of international transmission of disease to disease control and improved health conditions for vulnerable groups in the population. The collapse of peace in the late 1930s led to the disbandment of the League of Nations.

During World War II, the United Nations Relief and Rehabilitation Agency was established by the allied powers to assist in resettlement of the millions of displaced persons. This became part of the initiative to establish a new international health organization as part of an international consensus to build a better world after the war, in the context of a stronger, more coordinated United Nations.

THE WORLD HEALTH ORGANIZATION

The World Health Organization (WHO) was founded in 1948 as a United Nations agency in the spirit of cooperation and idealism following World War II. The WHO charter states that one of the fundamental rights of every human being is "the highest attainable standard of health," and the United Nations' Universal Declaration of Human Rights in 1948 stated, "Everyone has the right to a standard of living adequate for the health and well-being of himself and his family."

The World Health Organization has made an enormous contribution to global health. WHO serves as the central, unified, intergovernmental organization representing all countries, covering all fields of health. WHO consists of 193 member states working together and with other organizations toward achievement of the highest possible level of health. It replaced previous organizations, especially the Health Organization of the League of Nations and the Pan American Sanitary Bureau. A Technical Preparatory Commission developed the organization and, in the new optimism of the time, undertook the enormous task of dealing with global health problems. Its direction and coordinating functions are the primary assets of the organization, especially in definition of health goals and in initiating international cooperation to achieve them.

WHO carried this optimistic approach further in the Alma-Ata Conference and Declaration of 1978, in the successful pursuit of smallpox eradication and great success in nearing polio eradication in major reduction of measles morbidity and mortality and in many other fields of disease control. WHO has also been effective in its technical services, epidemiology functions, statistics, standardized nomenclatures for disease and drugs, and publications. WHO has established good working relations with major donors and others in GAVI and in parallel efforts in a wide variety of fields (Box 16.3). Lack of progress in important issues such as women's health and maternal mortality in many large-population countries is primarily due to lack of political commitment within the countries.

The organizational structure of the WHO includes headquarters in Geneva and seven regional offices, including Europe (EURO-Copenhagen), the Eastern Mediterranean (EMRO-Alexandria), Africa (AFRO-Brazzaville), Southeast Asia (SEARO-Delhi), western Pacific (WPRO-Manila), and the western hemisphere (the Pan American Health Organization [PAHO] in Washington, DC). The central headquarters in Geneva has many offices dealing with a diversity of topics (Table 16.7).

The WHO has led in the formulation of a worldwide consensus on a new direction in health policy. It formulated a strategy that incorporated the principles of governments' responsibility for the health of their peoples, the right of people to take part in developing and controlling their health care, and equality in health. It helped to formulate and promote the concept that cooperative activity among different parts of the public and private sectors (intersectoral cooperation) is necessary to advance health causes. The concept of appropriate technology is also a WHO initiative (see Chapter 15).

The problems and limitations of the WHO are important to assess. The organization is part of the UN system and is governed by its membership countries. It cannot avoid being influenced by political conflicts as during the Cold War period and in regional wars, such as in the Middle East, and genocides such as that in Darfur. This politicization can be a detriment to its leadership and

---

Box 16.3 Targets for the Eighth Millennium Development Goal

1. Further develop an open, rule-based, predictable, non-discriminatory trading and financial system (including a commitment to good governance, development, and poverty reduction) both nationally and internationally.
2. Address the special needs of the least developed countries through measures including tariff-and quota-free access for exports, an enhanced program of debt relief, and a cancellation of official bilateral debt, and more generous official development assistance for countries committed to poverty reduction.
3. Address special needs of land-locked countries and small island developing states.
4. Deal comprehensively with the debt problem of developing countries through national and international measures in order to make debt sustainable in the long term.
5. In cooperation with developing countries, develop and implement strategies for decent and productive work for youth.
6. In cooperation with pharmaceutical companies, provide access to affordable essential drugs in developing countries.
7. In cooperation with the private sector, make available the benefits of new technologies, especially information and communication technologies.

---

moral authority. It can also limit contacts of the WHO with the highest quality of professional leadership, impairing its ability to relate to the forefront of medical science, epidemiology, and public health practice. It has also led to inadequate leadership in areas where the response of the WHO to important issues might offend national pride.

The global struggle against tobacco is being led by WHO as a major contributor to the spread of chronic diseases.

Progress in many fields of health as outlined by Box 16.4 has been a major contribution by WHO and shows its important role in global health and the world community. Although hampered by its political nature, it exists as an international body in health representing all countries and dealing with health in a broad definition. The WHO's leadership in the Declaration of Alma-Ata and Health for All 2000 represented an important step forward in international health with its major commitment to primary health care (see Chapter 2).

Tropical disease work on malaria, bilharzia, filariasis, tuberculosis, onchocerciasis, leishmaniasis, schistosomiasis, helminthic diseases, and diarrheal disease control are of particular importance to the developing countries. WHO leadership in eradication of smallpox and virtual eradication of guinea worm disease, onchocerciasis, and poliomyelitis have been outstanding contributions to improved global health. Its initiatives in reducing

TABLE 16.7 National and International Strategies for Coping with Emerging Infectious Diseases

Goals/Topics	Activities	Examples
I. Surveillance	Detect, promptly investigate, and monitor emerging pathogens, the diseases they cause, and the factors influencing their emergence.	Monitoring in sentinel surveillance networks (e.g., blood banks, emergency departments, laboratories, sentinel settings); Population-based surveillance; Increase field investigation of outbreaks; Dissemination of epidemiologic data locally and internationally using electronic media; Internet, websites, PROMED, and others; Rapid laboratory diagnosis; Monitor vector-borne diseases.
II. Applied Research	Integrate laboratory science and epidemiology to optimize public health practices.	Promote reporting by sentinel laboratories and clinical settings; Improve laboratory diagnostic techniques, genotyping, subtyping, and mapping "fingerprinting" (e.g., E. *coli*, cholera, polio, measles, meningitis).
III. Prevention and Control	Enhance communication of public health information about emerging diseases and ensure rapid implementation of preventable strategies.	Wide and immediate dissemination of health information on infectious diseases to health professionals, general public, groups at special risk; Promote health education on prevention of spread of communicable diseases.
IV. Infrastructure	Strengthen local, state, and federal public health infrastructures to support surveillance and implement prevention and control programs.	Improve laboratories, reporting, and training.
V. International cooperation	Strengthen international effort and funding to promote immunization, nutrition interventions, maternal and child health.	WHO, UNICEF, UNDP, WB, FAO, GAVI, Médecins sans Frontières, bilateral aid programs; Rotary International, Gates Foundation, and many others working side by side and in growing cooperation to help national governments achieve MDGs.
VI. International Health Regulations (IHR 2005)	Approved by WHO and in effect in 2007, ratified by most countries.	Legal obligations to report infectious or other public health emergencies (chemical, radiation) of international public health significance. All countries agreed to increase surveillance capacity for emergencies such as SARS or human influenza.

Sources: Modified from Centers for Disease Control, *Addressing Emerging Infectious Disease Threats*. Atlanta, Georgia: U.S. Public Health Service, 1994. WHO International Health Regulations. 2005, http://www.who.int/features/qa/39/en/index.html (accessed January 17, 2008)

nutritional deficiency conditions, in chronic disease control, in defining health personnel needs, and in health services financing have also been important for both developing and the developed countries.

The World Health Organization develops programs of work that guide its activities and its regional offices as well as member states. The WHO has defined 15 objectives and a number of targets for each objective. This involved a global strategy for health, including promotion of food production and distribution, social progress in literacy, poverty reduction, and economic growth. Also included were the following: intersectoral cooperation; development of health care systems with a stress on primary care and improved management skills and efficiency; community involvement;

improved levels of health resources, including financial support by governments and universities involved in training health personnel; research, technology, and cooperation between countries; and environmental sanitation. All were included as areas for action within this program. The WHO policy framework stresses work with member governments, international organizations, banks, nongovernmental organizations, and other organizations related to health, economic, and social development.

Many other UN agencies and other organizations play important roles in international health. These include UNICEF (the United Nations Children's Fund), the United Nations High Commission for Refugees (UNHCR), the United Nations Development Programme (UNDP), the International Labour

---

Box 16.4 Successful Areas of International Health Leadership

Successes and important initiatives of the international health movement led by the WHO and UNICEF include the following:

1. Eradication of smallpox;
2. Massive increase in immunization coverage (EPI);
3. Control and possible eradication of poliomyelitis;
4. Reduced measles incidence (still 400,000 deaths per year in 2006);
5. Massive reduction in incidence of tetanus, diphtheria, pertussis;
6. Tuberculosis control using DOTS and DOTS plus for multi-drug-resistant cases;
7. Improving control of diarrheal disease (CDD) and reduced death rates;
8. Improving control of acute respiratory illness (ARI);
9. Improved control in tropical diseases: onchocerciasis, leprosy, yaws, and potential eradication of dracunculiasis;

10. Leadership in principles of primary health care: influencing national health programs particularly in developing countries;
11. Raising public and political consciousness of health issues;
12. Health for All initiatives;
13. Health targets initiatives;
14. The "Healthy Cities" movement;
15. Health promotion, raising health in national priorities;
16. Increasing awareness of health information needs;
17. Intersectoral cooperation in vaccines and immunization;
18. Health systems reforms;
19. Health human resources issues;
20. Global tobacco control;
21. Millennium Development Goals promotion.
22. International Health Regulations.

Organization (ILO), the Food and Agriculture Organization (FAO), and the International Atomic Energy Commission (IAEC).

The United Nations Children's Fund (UNICEF)

Following World War II, the new UN General Assembly created the United Nations International Children's Emergency Fund (UNICEF, now the United Nations Children's Fund), principally to assist the children of war-torn Europe. The program gradually expanded to include other activities and other areas, particularly in developing countries.

This agency has spent large sums of money, especially on food and supplies, for the promotion of child and maternal health and welfare activities throughout the world. Beyond this, usually through partnership with the WHO, UNICEF has been carrying out large and significant programs of BCG vaccination and yaws and malaria control. The promotion of family planning in developing countries is one of its major activities. UNICEF plays an important leadership role in fostering primary care and community preventive approaches worldwide.

NONGOVERNMENTAL ORGANIZATIONS

Nongovernmental organizations (NGOs) are numerous and carry out specialized activities worldwide. They vary widely in content, funding, ideology, and *modus operandi*. Many provide important support for developing countries, often succeeding where international agencies have failed, precisely because they work outside of the national political framework. This is particularly true in the case of emergencies and areas of conflict.

The earliest NGOs were those of the various church missions and sectarian organizations. Among the many that might be mentioned are the Unitarian Services Committee, the American Friends Services Committee, Catholic Relief Services, the American Jewish Joint Distribution Committee, the International Rotary Club, and the American Bureau for Medical Aid to China. The International Committee of the Red Cross (ICRC), Médicins sans Frontières (MSF), Terres des Hommes, and other European-based NGOs provide direct assistance in developing countries during crises. In 1999, Médecins sans Frontières was awarded the Nobel Peace Prize in recognition of its worldwide health achievements.

Philanthropic foundations have made and continue to make major contributions to international health. Private foundations such as the Ford, Soros, and Rockefeller Foundations carry out important international health work within their own exclusive structures. They are important sources of grants to promote pilot programs and research in health care systems. In addition, they contribute extra governmental funding that can stimulate development of innovative programs later affecting general health services. Since its inception in 1999 through late 2007, the Gates Foundation has donated some $8.5 billion (USD) to the international child vaccination program, aid to small farmers, and other health and education programs, mainly in Africa.

Among the foundations, the Rockefeller Foundation is the best known in the field of international assistance in health. Since its inception in 1913, it has operated in almost every country worldwide. Its significant contributions are many and include support of control programs for malaria and yellow fever, the development of recognized centers of learning in medicine and public health, postgraduate fellowships, and the demonstration of sound methods of organization and operation of health programs.

Despite the many positive aspects of NGOs, they can be a source of distortion in health care services in both developed and developing countries. They tend to focus on one kind of service, are very proud of their independence of government, and can create pressure for services that will place a burden on the system of financing or provision of health care. NGOs or bilateral aid can promote hospital development in places where there is already an oversupply and limited primary care. They can provide a primary service but be unwilling to coordinate with the basic governmental program in immunization, so that no one agency is fully responsible.

Coordination of NGO services into a comprehensive population-based service program may be compromised by political and international sensibilities, which can create chaos in an emergency situation. The balance of services for a population requires inclusion of governmental, NGO, and private services as a coordinated if not integrated whole. This may be impossible with highly independent NGOs, but the state public health authorities are responsible for overseeing the functions of NGOs, no matter how well meaning or charitable the cause.

THE WORLD BANK

The International Bank for Reconstruction and Development (IBRD), also known as the World Bank, is based in Washington, DC. It was established by the industrialized countries following the Bretton Woods Conference toward the end of World War II. It was an important financial institution to facilitate the reconstruction of postwar Europe. It has since become a major source of financing for development projects throughout the world. Traditionally it focused on large-scale infrastructure, industrial, and farming development projects. Its policies in health development focused on promoting market mechanisms and privatization of health care in countries lacking infrastructure. This fostered an inappropriate stress on medical and hospital care when a community health orientation was needed. The World Bank examined the health sector and its importance for economic development.

The 1993 *World Development Report: Investing in Health,* which examined the interaction of health status, health policy, and economic development, stated that, contrary to the views held by many traditional economists, health is essential for economic growth, and not a burden on the economy (Box 16.5). The report advocated a four-pronged approach by governments to improve health in developing and former Soviet countries:

1. Foster an economic environment that will enable households to improve their own health by promoting income gains for the poor and expanding social investment in raising standards of education, especially for girls;

Box 16.5 Health and Economic Development

"Good health ... is a crucial part of well-being, but spending on health can also be justified on purely economic grounds. Improved health contributes to economic growth in four ways: it reduces production losses caused by worker illness; it permits the use of natural resources that have been totally or nearly inaccessible because of disease; it increases enrollment of children in school and makes them better able to learn; and it frees for alternative uses resources that would otherwise have to be spent on treating illness."

Source: World Bank, *World Development Report,* 1993, p. 17.

---

2. Redirect government spending away from specialized care toward low-cost and highly effective activities such as immunization, programs to combat micronutrient deficiencies, and the control and treatment of infectious diseases;

3. Encourage greater diversity and competition in the provision of health services by decentralizing government services, promoting competitive procurement practices; and

4. Foster greater involvement of NGOs and private organizations, and regulate insurance markets.

World Bank assistance to health-related projects is growing steadily. The World Bank and the WHO have worked together on projects such as a special Program for Research and Training in Tropical Diseases (TDR) and the Onchocerciasis Control Project in west Africa. Long-standing World Bank pro-privatization policies and practices, most notably structural adjustment programs, have led to reduced social welfare infrastructure of developing countries in areas such as housing, education, health services, subsidies, and family transfers. While the World Bank recognizes the importance of health to development its advocation of privatization of health services has exacerbated poor health outcomes by reducing access to health services for those unable to pay for care.

The promotion of privatization mirrored the global trend toward more market-oriented economic policies. The World Bank promoted privatization and health insurance to replace direct provision of services by the state. This created problems in many developing countries, where most citizens earn less than $2 a day, and private sector services limited access to care of acceptable quality and affordable prices. World Bank policies in health have also obligated many countries to large repayment loans for programs of questionable value while basic services such as adding more successful vaccines to immunization programs are delayed so that many countries lag behind in adoption of internationally proven vaccines.

TRENDS IN GLOBAL HEALTH

It is now widely understood that the socioeconomic environment is a basic determinant of the state of health of an individual or population, even though the precise nature of intervening variables may not be sufficiently elucidated. The southern hemisphere has witnessed, along with its demographic explosion, the persistence of chronic problems plaguing the education, food, and housing sectors. In addition, more acute situations have emerged over the past few decades in relation to conflict, employment, migration, trade, and degradation of the physical environment. The northern hemisphere has enjoyed a rising level of affluence, with negative aspects that have made a sizable impact on public health: overeating, overdrinking, smoking, pollution, illicit drugs, and motor vehicle accidents. Global changes in the twenty-first century hold the promise of improvements in diagnostic and therapeutic measures, such as targeting drug therapy with nanotechnology, improved diagnostic methods including long-distance imaging methods, and less invasive diagnostic measures. Climate change and the potential for spread of vector-borne diseases, and food supplies and the issues related to genetically modified foods are all issues of huge importance for public health in the coming decades.

Progress in control of vaccine-preventable diseases has been one of the most important public health achievements with eradication of smallpox, closing in on polio eradication, and increasingly effective control of measles since adoption of the two dose policy. The development and increasingly wide use of vaccines with combinations such as measles, mumps, rubella (MMR), hepatitis B, *Haemophilus influenzae* b, rotavirus, varicella, pneumococcal pneumonia and influenza vaccine is one of the vital issues for achieving the Millenium Development Goal of reducing child mortality. Coverage has improved globally, but expanding the content of child and adult immunization has been painfully slow, despite aid from GAVI and other international donors.

WHO reports the maternal mortality ratio in 2005 in developing regions, with 450 maternal deaths per 100,000 live births, is in stark contrast to the average rate of 9 in developed regions and 51 in the countries of the Commonwealth of Independent States (CIS). The small drop in the global maternal mortality ratio reflects mainly the declines that have taken place in countries with relatively low levels of maternal mortality. Countries with the highest initial levels of mortality have made virtually no progress over the past 15 years. While gains are being made in middle-income countries, the annual decline between 1990 and 2005 in sub-Saharan Africa was only 0.1 percent. No region achieved the necessary 5.5 percent annual decline during the same period. Northern Africa, southeastern Asia, Latin America, and the Caribbean experienced relatively faster declines than sub-Saharan Africa (WHO, 2007).

The concept of primary health care as the basis of health system development has been almost universally accepted, yet evidence of public commitment to its implementation is still lacking. Problems reported include poor distribution of resources and inadequate orientation of health workers to primary health care, with continuing emphasis primarily on curative services. The community is often insufficiently aware of the role it should play and is frequently willing to accept competing demands for expensive secondary and tertiary care. Lack of resources to develop preventive services and health promotion is likely to erode the confidence and commitment of health workers and the community to primary health care.

The formulation and analysis of health personnel policy have emerged as growing concerns in the world (Chapter 14). There is a consensus regarding the urgent need to ensure the relevance and quality of human resources to the requirements of the health system, and to avoid imbalances in the production of health professionals, especially with regard to physicians, nurses, and dentists. In most developing countries, health personnel development plans either do not exist or are in the process of being developed.

While there are rising expectations for better health for all, global changes also constitute challenges to continued progress in health. Population growth, aging of the population, increasing incidence of chronic diseases, high expectations of the public for health care, increasing costs and medical technology, economic recession, and limited resources for health have all contributed pressures for health system reforms to maintain universal coverage. During the later part of the twentieth century, many industrialized countries developed health reforms that included reduction in hospital bed supply, financial incentives to promote development of community-based services, and a combination of decentralized management and integration of services in those countries with national health services (e.g., United Kingdom). Control of oversupply and excess utilization of hospital beds is also a feature of reforms for cost containment. In the United States, rapidly rising costs led to the expansion of managed care systems seeking cost-effective health care combined with health promotion to reduce disease prevalence and dependency on treatment services (Chapters 13 and 14).

The relationship between disease and society is such that many of the factors needed to reduce preventable diseases largely lie outside of the biomedical framework of genetics, medical care, public health, and health promotion, but are determined by social preconditions that are in the realm of human rights. This, however, does not absolve governments or the health community from the imperative of applying known measures of prevention and curative services for all as a basic human right.

EMERGING INFECTIOUS DISEASE THREATS

International health began as an activity to prevent the spread of epidemics and communicable diseases. This involved the collection and dissemination of information in a timely fashion, preventive measures such as appropriate immunization campaigns to control the spread of a disease, and subsequent follow-up. Success in eradication of smallpox and increasing control of the vaccine-preventable diseases led to enthusiastic assessments that epidemic diseases were under control. This optimism has been tempered by setbacks in malaria and TB control, along with a host of other emerging and reemerging disease issues. The experience of SARS and the threat of a potential pandemic of avian influenza indicate a new scale of public health threat and the need for global preparation.

Globalization of food and medical products marketing has become an enormous worldwide phenomenon. This has been accompanied by the emergence and dissemination of new infections in human populations, with geographic spread of disease such as HIV, hepatitis C via blood products, and variant Creutzfeldt-Jakob disease (vCJD). Marketing and growth of global demand for beef and animal feed and for human anticoagulant factors for medical treatment have contributed to transmission of HIV, hepatitis C, and vCJD during the decade prior to the detection of diseases in humans. The incubation period of BSE (bovine spongiform encephalopathy), or mad cow disease, is also on the order of years. Consumption of beef was declining in the 1980s in keeping with changing lifestyles. Changes in processing may have permitted prions to cross the species barrier from ruminant to human. Slaughterhouse practices in the United Kingdom in the 1990s were ineffective in deactivating prions, so that vCJD appeared and mad cow disease reappears periodically, requiring large-scale culling of cattle.

Commercial production of blood factors developed in the United States grew between the mid-1970s and the late 1980s from approximately $50 million (USD) in 1975 to $325 million (USD) in 1988. During this period, it is estimated that half of the hemophiliacs in the United States were infected with HIV through this route, and an untold number of hemophiliacs were infected worldwide. In some countries, such as Japan, this was probably the primary route of transmission into the population. By the end of the 1990s, there were 400 commercial centers for plasmapheresis operating in the United States. These centers, which employ paid donors, provided 60 percent of the worldwide requirement for plasma.

By mid-1982 the possible link between AIDS and the blood supply was reported and became widely known and accepted even though AIDS was seen as primarily a sexually transmitted disease of gay men. In the following year the occurrence of cases in hemophiliacs living in geographically dispersed areas led to epidemiologic investigations which identified contaminated blood as the source of infection. The blood factor industry used pooled serum largely from paid donors who were often from groups with high exposure to HIV, such as homosexual men. Self-exclusion was relied on for screening because of fear of potential lawsuits for discrimination by excluding high-risk donors, and the voluntary sector failed to stop blood collection in high-risk areas.

Viral inactivation methods were under development since the early 1970s to reduce hepatitis transmission in blood. However, the industry leaders considered such steps proprietary information, so the work toward successful strategies was not shared across the corporate competitors. In 1984 the major producers had all been licensed to distribute heat-treated products to cut down on the threat of hepatitis and HIV infection, but recall orders were not issued by the Food and Drug Administration as soon as the risk of HIV transmission in blood factors became known in March of 1983, a failure later criticized by the U.S. Institute of Medicine.

By 1985, the ELISA screening tests for HIV and hepatitis C came into use. This, coupled with heat treatment of the factor product, has markedly reduced the risk of HIV transmission in blood. However, global spread of the virus, in part facilitated by the global trade in factor VIII, had already occurred in the 1980s, affecting the majority of Japanese hemophiliacs and AIDS affected people during 1983–1985 by non-heat-treated factor concentrates imported from the United States, according to the World Federation of Hemophilia.

Continued sale of contaminated blood products to hemophiliacs in the mid-1980s led to legal action against four French officials of the national blood bank and the ministry of health. Their trial initially for "poisoning" resulted in convictions and jail terms for several of them for their role in HIV transmission to some 4000 hemophiliacs in France. In Japan in 2000, three former drug company executives were convicted of selling blood products tainted with HIV and given prison terms.

The spectrum of infectious disease in a community evolves rapidly with changing conditions of the environment and society. Population growth, crowding in urban slums, homeless populations, massive migration, and travel contribute to transmission of once-localized diseases internationally. Resistance to available antimicrobial medications is creating a new dilemma for modern medicine and public health. The synergism of infectious diseases such as AIDS with TB or *Cryptosporidium* causes deterioration in the patient and spread by secondary infection to other persons. The "postantibiotic era" is widely discussed as a serious threat to modern public health. Strategies to prevent the loss of some of the important gains of the twentieth century in communicable disease control will require new research and strategies. Among the lessons learned from AIDS are that improvements in early warning systems and attention to new threats are vital tasks of public health.

In the United States, a number of new or resurgent infectious diseases are of increasing public health concern. These include HIV/AIDS, *E. coli* O157:H7 disease, cryptosporidiosis, coccidioidomycosis, multidrug-resistant pneumococcal disease, multidrug-resistant tuberculosis (MDR-TB), vancomycin-resistant enterococci, influenza A Beijing/32/92, hantavirus infections, leishmaniasis in veterans of the 1991 Gulf War, Legionnaire's disease, and Lyme disease.

Newly emerging and reemerging diseases of concern internationally include HIV/AIDS, multidrug-resistant malaria, tuberculosis, cholera, *Shigella dysenteriae*, diphtheria, and *E. coli* O157:H7. Tropical diseases, such as yellow fever and dengue, are reappearing in Asia and Latin America, and Rift Valley fever in Egypt, Saudi Arabia, and Yemen; Lassa fever in west Africa; Ebola virus in Congo (formerly Zaire); Marburg virus via imported monkeys; Oropouche arbovirus and Sabia in Brazil; Junin virus in Argentina; and Machupo virus in Bolivia all are health concerns as they may spread from their natural habitat to other countries before the carrier shows symptoms.

West Nile fever spread from the Middle East to New York City and then spread to other parts of the United States in 1999, with 62 severe cases and 7 deaths, but spread further in 2000 and 2001, followed by a dramatic increase in 2002 with over 4000 human cases in North America and Europe. The disease has become endemic in mosquito populations and the cause of hundreds of deaths annually in the United States alone. The threat of spread of new disease threats was dramatically demonstrated during the SARS episode of 2003 and the threat of avian influenza since 2005 (Chapter 4). In 2007, Chikungunya fever spread from southeast Asia to Italy and southern France, where it has become endemic.

It can no longer be assumed that such diseases will remain in their natural habitats; they can be transmitted all over the world via human or animal carriers and, in appropriate conditions, become serious local or even general public health concerns. The 1995 outbreak of Ebola virus in the former Zaire raised international concern of the very real possibility of widespread transmission of this deadly disease (Box 16.6). The public has been made aware of this kind of situation in graphic detail in the news media, novels, and movies. Again, the threat of avian influenza remains a looming public health disaster of global proportions.

In 1996, a large-scale epidemic of food poisoning in Japan involving *E. coli* O157:H7 (first described in 1982) spread through contaminated school lunches and caused over 3000 illnesses, hundreds of hospitalizations

---

Box 16.6 Ebola Virus

Ebola virus is named after a river in the Democratic of the Congo (formerly Zaire), where it was first recognized in 1976. It is a member of the family of RNA viruses called the *Filoviridae* with identified subtypes including Ebola-Zaire, Ebola-Sudan, and Ebola-Ivory Coast. It is the cause of a deadly hemorrhagic fever, with high mortality. An outbreak of Ebola virus in Zaire in 1995 was of major international concern, because of its 77 percent mortality rate and a fear that it could spread rapidly. This outbreak in Kikwit was limited to 316 cases with 245 deaths, many of whom were hospital workers. The organism has been isolated in specific species of monkeys and can be transmitted to humans via blood and secretions.

An international team mobilized by the WHO and CDC came to the site to assist the Democratic Republic of Congo (DRC) public health staff. Stress on case detection, laboratory confirmation, isolation, and staff protection helped to limit the disease spread. The WHO Collaborating Center on Arboviruses and the Viral Hemorrhagic Fevers reference laboratory at the CDC in Atlanta played an important part in management of this epidemic. Rapid international response and heightened surveillance are part of the global concern for newly emerging infectious diseases, partly resulting from the lessons learned from the slow response to the AIDS epidemic. Outbreaks occurred in South Sudan in 2004, and in the DRC in 2007.

Source: WHO, World Health Report, 1996.
Center for Disease Control, 2005, http://www.cdc.gov/ncidod/dvrd/Spb/mnpages/dispages/ebola/qa.htm; WHO, 2007, http:www.who.int/csr/disease/ebola/en/ and http://www.who.int/mediacentre/factsheets/fs103/en/index.html [accessed July 9, 2008]

---

for bloody diarrhea, many with severe hemolytic uremic syndrome, and 7 deaths. The identification of the source proved to be very difficult. Milder epidemics have occurred in many other countries, including Australia, Canada, the United States, and various European countries, so that there is continued concern for recurrence of this potentially severe form of food poisoning. A website called Promed monitors infectious diseases globally with email reports on virtually a daily basis. Other web-based infectious disease monitoring and information-sharing systems are invaluable public health teaching and service programs that are available via the CDC website.

Food safety monitoring is vital in control of foodborne disease outbreaks in both domestic and international trade. The WHO established the Global Public Health Intelligence Network (GPHIN), a Web-based system that monitors news reports of infectious disease outbreaks around the world; Salm-Surv, a global network linking laboratories tracking the incidence of *Salmonella* and other food-borne diseases; the Global Outbreak Alert and Response Network (GOARN), which provides technical assistance within 24 hours to governments facing potential epidemics; and the International Food Safety Authorities Network (INFOSAN), which enables trans-border collaboration and assistance among food safety officials. These supplement individual national surveillance and diagnostic functions.

In 2005, WHO adopted new International Health Regulations (IHR) which have been adopted by most countries and came into effect in June 2007. The IHR address all diseases and health events that may constitute a public health emergency of international concern, to contain the threat of international spread of diseases such as SARS, or a new human influenza virus. The IHR includes threats of public health emergencies that may spread across borders, such as chemical spills, leaks, and dumping or nuclear meltdowns.

They replace the previous IHR (1969) which addressed only four diseases: cholera, plague, yellow fever, and smallpox, since then eradicated. The old IHR focused on the control at borders and relatively passive notification and control measures. The new IHR provides a legal basis for global disease surveillance, alert, and response. It defines the rights, obligations, and procedures in ensuring international health security without unnecessary interference in international traffic and trade. It requires all member states to strengthen their existing capacity for disease surveillance and response.

International standards of food, plant, and animal safety are addressed in the *Codex Alimentarius* ("Food Law") Commission (World Health Organization and Food and Agriculture Organization of the UN), the International Plant Protection Commission (IPPC), and the Organisation for Animal Health (Office International des Épizooties, or OIE). The WTO works to break down — not erect — trade barriers. While the health and safety measures the WHO and WTO provide are considerable, they are limited by the respective organizations' resources and priorities. Global agricultural products trade in 2002 was $583 billion (USD). Food sciences work both in industry and academic centers to discover new products or processing, quality assurance of specific products, and standards for the production, processing, marketing, and distribution of foods, with improved nutritional value and safety.

International cooperation to identify new infectious or other health threats to prevent global epidemics is an urgent priority for both international agencies and national health systems throughout the world. Control of communicable diseases requires medical, laboratory, and epidemiologic intelligence services of a high order, with rapid means of communication, publication, and coordination, backed up by skilled professional services (see Table 16.7). Examples of international activity in control of infectious disease are numerous (Box 16.7). The crowning achievement in this field was the eradication of smallpox. This great feat may soon be matched by the international eradication of poliomyelitis. Such achievements can only be made with major efforts in international commitment and cooperation, working with international and national agencies, donor organizations, and ultimately with the people themselves.

Box 16.7 GAVI (Global Alliance for Vaccines and Immunization): an Innovative Partnership

GAVI was established in 2000 at the World Economic Forum at Davos as an alliance of different stakeholders from the private and public sectors, with the mission of saving children's lives through the worldwide expansion of mass vaccination programs. GAVI's partners include UN agencies and institutions (UNICEF, WHO, the World Bank), civil society organizations (International Pediatric Association), public health institutes (Johns Hopkins Bloomberg School of Public Health), donor and implementing country governments, the Bill & Melinda Gates Foundation, other private philanthropists, vaccine industry representatives, the financial community, and others whose collective efforts and expertise are enabling great progress to be made in this field.

The GAVI Alliance is a unique, multidimensional partnership of public and private sector resources with a single, shared focus: to improve child health in the poorest countries by extending the reach and quality of immunization coverage within strengthened health services.

These efforts are directed through the financing mechanisms of the GAVI Fund and the work of the Geneva-based GAVI Secretariat, which channels funding, optimizes product availability and market pricing, and coordinates the field support necessary to plan and implement programs in the world's poorest countries.

Funding is time-limited. The intent is to enable countries to develop sustainable programs and progress toward integrating them into national health service budgets. The sustainability of these gains depends on national governments placing financial and political priority on these programs and adopting them into their regular budget or financing systems. More than 40 countries now have multi-year immunization plans in keeping with GAVI objectives.

GAVI's efforts are vital to achieving the Millennium Development Goal on child health, which calls for reducing childhood mortality by two-thirds by 2015. Of the more than 10 million children who die before reaching their fifth birthday every year, 2.5 million die from diseases that could be prevented with currently available or new vaccines. It is estimated that more than 2.3 million early deaths will have been prevented as a result of support by GAVI up to the end of 2006.

Source: GAVI Alliance, http://www.gavialliance.org/about/in_partnership/index.php [accessed January 4, 2008]

Governmental agricultural policies have important impacts on health. Traditionally governmental subsidies on production of meat, dairy products, and sugar have led to their increased consumption. At the same time such subsidies penalize fruit and vegetable production and consumption to the health detriment of populations due to their cost or limited supply. This adversely affects vulnerable groups in the population both within countries and globally. High intakes of fat and sugar with low consumption of fruits and vegetables are associated with obesity, hypertension, stroke, and early coronary heart disease. Domestic and international policies to reduce beef and other saturated fat consumption while increasing fruit and vegetable consumption are vital to a healthy nutrition program. This needs to be coupled with fortification of commonly used foods to address micronutrient deficiencies common in developing and developed countries and as part of larger programs to prevent the rapid spread of cardiovascular disease, obesity, diabetes, and other chronic conditions.

In 2006–2007, high prices for oil led to public demand for alternative energy sources for motor vehicles with increased use of grains and corn for ethanol fuel production, causing sharp rises in price of wheat, corn, and flour-based food products. Use of foods for energy is seen as inefficient and uneconomic, but the search for nonpolluting energy sources will generate major economic growth in new sectors of technology and industry in the decade ahead. These agro-economic and effective energy policies along with poverty reduction, educational opportunities, and global and equitable economic growth are all issues of great public health importance internationally.

Despite progress and optimism, the tragic global toll of death and mental or physical disability continues with preventable infectious or vitamin deficiency conditions numbering in the hundreds of millions. Other issues remain to be dealt with in global health, especially regarding women's health, including family planning, reduction of maternal mortality and morbidity, reduction of violence and abuse against women, and improved education and job opportunities. Child abuse, child labor, female-child murder, and sexual exploitation remain large-scale global health problems. Increasingly, noninfectious conditions affecting young adult and middle-aged males are becoming issues of global health—for example, cardiovascular disease, diabetes mellitus, trauma, and cancer. Care of special needs groups in the population, such as the mentally ill, the disabled, and the elderly (see Chapters 6 and 7), are global health problems that require attention in each country and locality.

International, individual government, and community action are vital to deal with these issues, not only in the poorest countries, but also in developed countries. Known public health measures applied effectively can reduce the burden of these conditions within a very few years. This is a challenge of historical importance and necessity.

EXPANDING NATIONAL HEALTH CAPACITY

The idea of a *cordon sanitaire* to protect a nation's health from invading epidemics is a form of passive defense that has not been effective in major epidemics in the latter part of the twentieth century. A forward defense is now part of the New Public Health. Countries need to reach out to other countries to help improve international public health capacity as their own first line of defense. The tragedy of the late discovery of AIDS and inadequate early response was matched by the equally poor handling of the first phase of the 1991–1996 cholera epidemic in South America.

Building the first line of defense means strengthening the capacity of individual countries to detect, report, and request help in controlling potentially serious disease outbreaks. Help is available from the WHO, the Centers for Disease Control (CDC) in Atlanta, and newly strengthened counterparts in France and the United Kingdom, as well as international organizations such as the International Red Cross, Médecins sans Frontières, and many others. Training in basic epidemiology, sterile techniques, and laboratory services can mean the difference between local containment and widespread infection of hemorrhagic fever viruses, with person-to-person transmission amplified in a hospital setting.

Even the industrialized countries are in need of strengthening of epidemiologic capacity. Few have adequate information systems to collect hospitalization data that can provide vital measures of morbidity and the economics of health services. Few have the training capacity for public health epidemiologists, economists, sociologists, sexologists, psychologists, or anthropologists, let alone entomologists, geneticists, and the many other professionals making up the New Public Health team.

Many industrialized countries, satisfied with universal access to doctors and hospitals and a feeling that infectious diseases are going away under the power of sanitation, vaccination, and antibiotics, allowed their public health infrastructure to decline with poor pay, reward, recognition, and motivation, and lack of training capacity. The 1990s brought a different reality of emerging and reemerging infectious diseases and other plagues such as violence and trauma, drugs, heart disease, cancer, and stroke. Failure to prepare public health professionals and support systems is an invitation to disaster, both epidemiologically and economically. No country can afford such laxity. Training of public health professionals requires graduate schools of public health, which are more essential than the excess of medical schools that already exist in most countries.

The 2006 World Health Report (*Working Together for Health*) focused on worldwide shortages of health personnel, especially in the countries with the most severe health problems. The supply of health workers ranges from 2.3 per 1000 population in Africa and 4.3 in Southeast Asia to 18.9 and 24.8 in Europe and the Americas, respectively.

Issues relate not only to quantity but also quality and support systems for health workers. Migration of educated people in the population tends to drain doctors and nurses from poor countries to wealthy countries, exacerbating shortages and the problems of developing and sustaining standards of care (see Chapter 14).

GLOBAL HEALTH AND THE NEW PUBLIC HEALTH

The New Public Health is concerned with globalization of health in several senses. First, it includes all health activities in any one country, and second, what happens in the rest of the world, including the effects of globalization, is of direct interest to each country, no matter how wealthy, industrialized, or isolated. The lessons of the bubonic plague may seem as remote history to the generation raised on concepts of success of public health in control of communicable diseases, but the lessons of HIV should surely be. John Donne's idea that "no man is an island unto himself" expresses the issue clearly. Global health means identifying and addressing the acute infectious and chronic diseases as early as possible before they spread or amplify by common risk factors.

In the twenty-first century, more developing countries are reaching an epidemiologic transition that took place in the industrialized world in the mid-twentieth century. The resurgence of long known diseases and the emergence of new and sinister infectious disease threats are occurring worldwide. The industrialized countries are again facing serious infectious disease challenges, including those imported from developing countries.

In the 1950s and 1960s, control of infectious diseases looked extremely promising. Vaccines and antibiotics seemed to provide the answer to age-old infectious diseases. But in the 1970s and 1980s new infectious organisms appeared, as well as a frightening increase in resistance of microorganisms to therapeutic agents. Diseases spread from country to country, as did HIV in the 1980s, and cholera in Peru and diphtheria in Russia in the 1990s, and the plague outbreak in India in 1994.

Conversely, the chronic diseases associated with overnutrition and smoking are invading the nonindustrialized countries just when the public health field is gaining momentum in controlling infectious and childhood diseases. Additionally, all countries are facing the strains of health expenditures and the painful process of health reform. The legal, ethical, and technological challenges are increasingly important in managing health care systems (Box 16.8).

All health systems are obliged to face these challenges through the sharing of information, and improved monitoring of the use of resources, as well as seeking effective

---

Box 16.8 The Burden and Costs of Chronic Diseases in Low-Income and Middle-Income Countries

The World Health Organization, Geneva, Switzerland, "estimates the disease burden and loss of economic output associated with chronic diseases — mainly cardiovascular diseases, cancer, chronic respiratory diseases, and diabetes — in 23 selected countries account for around 80 percent of the total burden of chronic disease mortality in developing countries. In these 23 selected low-income and middle-income countries, chronic diseases were responsible for 50 percent of the total disease burden in 2005. For 15 of the selected countries where death registration data are available, the estimated age-standardized death rates for chronic diseases in 2005 were 54 percent higher for men and 86 percent higher for women than those for men and women in high-income countries.

"If nothing is done to reduce the risk of chronic diseases, an estimated $84 billion (USD) of economic production will be lost from heart disease, stroke, and diabetes alone in these 23 countries between 2006 and 2015. Achievement of a global goal for chronic disease prevention and control — an additional 2 percent yearly reduction in chronic disease death rates over the next 10 years — would avert 24 million deaths in these countries, and would save an estimated $8 billion, which is almost 10 percent of the projected loss in national income over the next 10 years."

Source: Abegunde, D. O., Mathers, C. D., Adam, T., Ortegon, M., Strong, K. 2007. The burden and costs of chronic diseases in low-income and middle-income countries. *Lancet*, 370:1929–1938.

---

ways of preventing diseases and managing them to promote early and complete return to function. All industrialized countries are facing serious problems financing health care in its traditional form, and reform is taking place amid aging of the population, increasing technology, and high expectations of health care. Reforms shifting emphasis and resources from hospital to ambulatory and primary care show a strong return to the idea of health promotion by regulation and education.

Some answers to unconquered infectious diseases have come from simple technology, such as the use of oral rehydration therapy to reduce morbidity and mortality from diarrheal diseases. The resurgence of TB and multidrug-resistant organisms has been successfully handled by another simple innovation of directly observed therapy by community health workers to ensure compliance and completion of treatment, especially in high-risk groups. Malaria control, using specially trained community health workers, is another application of inexpensive, simple, appropriate technology.

Simpler technologies are also having a major impact on the chronic diseases. Cardiovascular mortality rates are falling in most industrialized countries as a result of healthier lifestyles and improved treatments. Trauma deaths are falling as a result of both mandatory improvements in car and road safety, occupational safety, and treatment of substance abuse. For chronically ill patients, simpler technology of home care allows them to return to their homes with less lengthy, costly, and risky hospitalizations. Ambulatory care can be provided safely for very many conditions previously requiring hospitalization at greater cost and danger from hospital-acquired infections.

The future of public health and health care will see tremendous change and adoption of new modalities of preventing and managing disease: recombinant vaccines will reduce costs and introduce new vaccines, bringing more infectious diseases under control, including the viral hepatitides and respiratory and diarrheal diseases. Vaccine technology for cancer and genetic disorders is evolving. Congenital disorders will be controlled by education, screening, and appropriate interventions. Dietary change will help in the control of cancer, as will screening and reduced exposure to carcinogens. It is now established that infectious diseases can cause chronic disease, such as *Helicobacter pylori* and peptic ulcer, and cancers such as cancer of stomach, liver, and cervix. There are synergies between micronutrient deficiency conditions with infectious and chronic disease, such as folic acid deficiency and birth defects, and these associations open many new vistas for research, preventive breakthroughs, and applied public health.

Health for All means access to care for everyone. This requires sound management of finances and other resources to provide the needed services efficiently and by reducing the waste and extravagances of unnecessary servicing. It also requires a social and physical environment that enables people to experience healthful, satisfying, and productive lives. To attain these lofty goals, broad partnerships or coalitions of health services and providers working with communities and an increasingly knowledgeable and participating general public must be achieved. This is especially important for compliance with immunization, healthful infant and child nutrition and care, self-care in pregnancy, and healthful adult nutrition. Paternalistic, traditional services of doctors dominating both the health systems and patients are not able to raise the level of patient and community participation needed.

The goal of better health requires a sharing of tasks and resources between the clinical and community levels, and between countries. Assisting countries in developing the staff and infrastructure of epidemiology in infectious and chronic disease is an investment in the frontline of public health protection and self-defense. This is the substance of work by international organizations and bilateral aid. In international partnerships in Europe, the industrialized countries help each other, and this needs to be applied to help promote public health infrastructure in developing countries as well.

SUMMARY

Health for All sounded like a hopeless idealistic dream when first promulgated by the WHO in 1977. Yet the progress since in lowering mortality and birthrates, raising longevity rates, and improving quality of life has been dramatic. *Globalization of health* means that what happens anywhere is the concern of everyone everywhere, as the world learned with plague in the fourteenth century and AIDS in the late twentieth century. At the same time, globalization means new aspects of health for a population, because of the interaction of health care, economics, and the political priority given to health.

In the globalized world of the twenty-first century, public health of one country cannot be considered in isolation. Globalization has bridged countries together, intensified human interactions, and made the international boundaries increasingly irrelevant. *Global health* is a very complex term which is influenced by actions or circumstances in countries other than the one affected directly. Today the determinants of global health include poverty, environmental degradation, climate change, violence, terrorism, illegal drug trafficking, and international or bilateral trade laws. With all its drawbacks, globalization also has its benefits in the transfer of education, science, and technology, helping to provide the benefits of development from developed to developing countries. Many countries are emerging from economic stagnation with rapid development of industry and trade based on domestic and global markets. The middle- and low-income countries are experiencing rapid growth in their middle-class populations and trends in disease prevalence of heart disease, stroke, overnutrition, obesity, diabetes, and growing gaps between rich and poor. The economic burden of these diseases in developing countries justify programs for intervention to prevent them from undermining economic and social development.

Global action means that countries must be committed to health at all levels, including state and local governments as well as voluntary, educational, and many other elements of a society. The potential gain is enormous, and this requires systematic organization, information, with well-defined targets, strategy, and tactics. The WHO Framework Convention on Tobacco Control is the first adopted global public health treaty in history and is meant to help developing countries address the tobacco epidemic promoted by the tobacco industry.

Measuring disease (infectious and chronic), family health, special groups in the population, nutrition, environmental and occupational health, organization of public health, management of health systems, comparing with other national health systems, human resources, technology assessment, quality assurance, law, and ethics, all chapter topics in this book, are the substance of the New Public Health. All together, they are the subjects of day-to-day life in health systems.

Eradication of smallpox will be followed by other successes with great health and economic benefits. Poliomyelitis eradication is progressing with 2007 being the lowest year of polio experience. WHO reports 735 polio cases caused by wild poliovirus reported from 11 countries (to November 2007), compared with 1686 cases from 16 countries for the same period in 2006. The four polio-endemic countries (Afghanistan, India, Nigeria, and Pakistan) account for 88 percent of all cases, with India and Nigeria reporting 53 percent and 31 percent, respectively. The remaining 85 cases were from 7 reinfected countries in Africa. Continuous monitoring and special immunization efforts in still-endemic areas are using type 1 OPV vaccines to reduce this most virulent strain, while also addressing type 2 and type 3 areas as well. Polio eradication is within sight but requires sustained efforts of the donors and national governments, as well as international organizations such as WHO, Rotary International, Gates Foundation, and others.

The experience, skills, and infrastructure of infectious disease control will also bring change in chronic disease control. New acute and chronic disease challenges will emerge; preparation will increase the chances of coping with them before they reach epidemic proportions. Recognizing the rising tide of cardiovascular disease mortality in the former Soviet Union and many developing countries is no less important than recognizing the resurgence of TB or multidrug-resistant infectious diseases.

The conceptual basis of the New Public Health provides an idealized yet practicable model for developing countries. This idea, however, has since grown with many influences, including health promotion and health targets. It has come to involve the actual management of health systems and integration of secondary and tertiary care services of hospitals, and the whole range of programs or services that relate to improving the health of the individual and the society.

The international health community has succeeded, in part, in changing the health agenda of many countries toward prevention, primary care, and health promotion. The development of goals and targets with international sanction helps each country to resist pressure to place most of its health care resources into curative and tertiary services. An international commitment to Health for All has taken on an important meaning in member countries. It has helped national and regional health leadership tackle the difficult task of changing priorities to an emphasis on primary health care and modern public health.

Coalitions of forces are needed to take up the challenges that the health community cannot do alone. Isolation of health from other sectors, or of parts of the health spectrum from each other, lowers the capacity of all to reach common goals. Networks of international agencies, including the WHO, UNICEF, the World Bank, the Food and Agriculture Organization, UNDP, private donor organizations, the private sector, and many others are needed to face the health challenges and tasks. Similarly, at the national, state, and local levels, globalized approaches and networks of organizations can help to define targets and mobilize the resources needed to achieve them.

It is appropriate to end this book with the International Declaration of Health Rights, adopted by schools of public health in the United States as a personal commitment for public health professionals graduating from their training programs (Box 16.9). This statement of personal mission

---

Box 16.9 Graduates' Pledge of Public Health

We, as professionals, do hereby commit ourselves to advocacy and action to promote health rights of all human beings.

The enjoyment of the highest attainable standard of health is one of the fundamental rights of every human being. It is not a privilege reserved for those with power, money or social standing.

Health is more than the absence of disease, but includes prevention of illness, development of individual potential, a positive sense of physical, mental and social well-being.

Health care should be based on dialogue and collaboration among citizens, professionals, communities and policy makers. Health services should emphasize equity, accessibility, community participation, prevention and sustainability.

Health begins with healthy development of the child and a positive family environment.

Health must be sustained by the active role of men and women in health and development.

The role of women, and their rights, must be recognized, respected and promoted.

Health care for the elderly should preserve dignity, respect and concern for quality of life and not merely extend life.

Health requires a sustainable environment with balanced human population growth and preservation of cultural diversity.

Health depends on more than access to health care. It depends on healthy living conditions and the availability to all people of basic essentials: food, water, housing, education, productive employment, protection form pollution and prevention of social alienation.

Health depends on protection from exploitation and discrimination on account of race, religion, political belief, ethnic group, national origin, gender, sexual preference or economic or social status.

Health requires peaceful and equitable development and collaboration of all people.

Source: UCLA School of Public Health, 1998, confirmed 2008.

and values represents the best ideals of public health and is relevant to the many training programs in public health not only in the United States, but also in Europe and other parts of the world.

The New Public Health is a conceptual framework and methodology for implementation of these lofty, but achievable goals. It addresses managing of health systems as well as health promotion and disease prevention so that changes in priorities can be implemented by suitable shifts in resources to meet the health needs of individuals, vulnerable groups, and national and international communities. This means reduced inequities, maximum use of available technologies and statistical, epidemiologic, social, and basic sciences of public health with a renewed global commitment to Health for All.

ELECTRONIC RESOURCES

Bill and Melinda Gates Foundation. http://www.gatesfoundation.org/GlobalHealth/Grants/default.htm?showYear=2007 [accessed May 26, 2008]

Centers for Disease Control. *Morbidity and Mortality Weekly Report,* www.cdc.gov/mmwr [accessed May 26, 2008]

Centers for Disease Control, Innovative information-sharing strategies, *Emerging Infectious Diseases,* http://www.cdc.gov/ncidod/eid/vol+no3/kay.htm

Emerging Infectious Diseases. http://www.cdc.gov/ncidod/eid/ [accessed May 26, 2008]

Copenhagen Consensus Center. http://www.copenhagenconsensus.com/Default.aspx?ID=788 [accessed May 26, 2008]

Copenhagen Consensus original 2004 priorities. http://www.copenhagenconsensus.com/Files/Filer/CC/Press/UK/copenhagen_consensus_result_FINAL.pdf [accessed May 26, 2008]

Disease Control Priorities Project. http://www.dcp2.org/page/main/Home.html (homepage) and http://www.dcp2.org/pubs/DCP (report) [accessed May 26, 2008]

Food and Agriculture Organization. http://www.fao.org/ [accessed May 26, 2008]

Health for All Indicators database. WHO European Region, http://www.euro.who.int/ [accessed May 26, 2008]; may be downloaded to personal computers.

International Infectious Disease Monitoring Network. http://www.promedmail.org [accessed May 26, 2008]

Médecins sans Frontières. http://www.msf.org/ [accessed May 26, 2008]

Millennium Development Goals. http://www.mdgmonitor.org/ [accessed May 26, 2008]

Organization for Economic Cooperation and Development. http://www.oecd.org/ [accessed May 26, 2008]

United Nations Development Programme. Human Development Report, http://hdr.undp.org/en/statistics/ [accessed May 26, 2008]

United Nations High Commission for Refugees. http://www.unhcr.ch/ [accessed May 26, 2008]

United Nations Children's Fund (UNICEF). http://www.unicef.org/ [accessed May 26, 2008]

Weekly Epidemiologic Record (WHO). http://www.who.int/wer/ [accessed May 26, 2008]

World Bank. http://www.worldbank.org/ [accessed May 26, 2008]

World Bank. 2006. Repositioning nutrition as central to development: a strategy for large scale development, http://web.worldbank.org/WBSITE/EXTERNAL/TOPICS/EXTHEALTHNUTRITIONANDPOPULATION/EXTNUTRITION/0,,contentMDK:20146328~menuPK:282580~pagePK:148956~piPK:216618~theSitePK:282575,00.html [accessed May 26, 2008]

World Bank, World Development Report 2003. http://siteresources.worldbank.org/EXTHNPSTATS/Resources/Popprojectiontotal.xls [accessed July 9, 2008]

World Bank, Health, Nutrition, Population Statistics. http://web.worldbank.org/WBSITE/EXTERNAL/TOPICS/EXTHEALTHNUTRITIONANDPOPULATION/EXTDATASTATISTICSHNP/EXTHNPSTATS/0,,contentMDK:21198825~menuPK:3387931~pagePK:64168445~piPK:64168309~theSitePK:3237118,00.html [accessed May 26, 2008]

World Health Organization Global Burden of Disease Project. http://www.who.int/healthinfo/bodproject/en/index.html [accessed May 26, 2008]

World Health Organization TB Control. http://www.who.int/tb/publications/global_report/en/index.html [accessed May 26, 2008]

WHO International Health Regulations. 2005. http://www.who.int/features/qa/39/en/index.html [accessed May 26, 2008]

World Population Prospects. 2006 Revision. Population Division United Nations, NY, http://www.un.org/esa/population/publications/wpp2006/WPP2006_Highlights_rev.pdf [accessed May 26, 2008]

World Health Organization mail and email addresses for headquarters and regional offices:

WHO Headquarters, Avenue Appia 20, 1211, Geneva, Switzerland. http://www.who.int/hac/en/; (postmaster@who.int)

WHO Regional Office for Africa (AFRO), Parirenyatwa Hospital, PO Box BE 773, Harare, Zimbabwe. (regafro@whoafr.org).

WHO Regional Office for the Americas (PAHO), 525 23rd St., N.W., Washington, D.C. 20037, USA. (postmaster@paho.org).

WHO Regional Office for the Eastern Mediterranean (EMRO), PO Box 1517, Alexandria 21511, Egypt. (postmaster@who.sci.eg).

WHO Regional Office for Europe (EURO), 8 Scherfigsvej, DK 2100, Copenhagen, Denmark. (post-master@who.dk).

WHO Regional Office for South East Asia (SEARO), World Health House, Indaprastha Estate, Mahatma Ghandi Road, New Delhi, India 110002. (postmaster@whosea.org)

WHO Regional Office for the Western Pacific (WPRO), PO Box 2932, 1000 Manila, Philippines. (post-master@who.org.ph)

WHO. Chan, M 2007. Director-General of the World Health Organization. The contribution of primary health care to the Millennium Development Goals. Opening address at the International Conference on Health for Development, Buenos Aires August 16, 2007: Argentina. http://www.who.int/dg/speeches/2007/20070816_argentina/en/print.html [accessed May 26, 2008]

WHO Maternal Mortality. http://www.who.int:80/reproductive-health/global_monitoring/ [accessed May 26, 2008]

World Health Organization. Avian influenza, http://www.who.int/csr/disease/avian_influenza/ai_timeline/en/index.html [accessed May 26, 2008]

RECOMMENDED READINGS

Abegunde, D. O., Mathers, C. D., Adam, T., Ortegon, M., Strong, K. 2007. The burden and costs of chronic diseases in low-income and middle-income countries. *Lancet,* 370:1929–1938.

Berlinguer, G. 1992. The interchange of disease and health between the old and new worlds. *American Journal of Public Health*, 82:1407–1413.

Black, R. E., Allen, L. H., Bhutta, Z. A., Caulfield, L. E., de Onis, M., Ezzati, M., Mathers, C., Rivera, J. (for the Maternal and Child Undernutrition Study Group) 2008. Maternal and child undernutrition: Global and regional exposures and health consequences. *Lancet*, 371:9608, 9619.

Centers for Disease Control. 1994. *Addressing Emerging Infectious Disease Threats*. Atlanta, Georgia: U.S. Public Health Service.

Centers for Disease Control. 2007. CDC's 60th Anniversary: Director's Perspective — Centers for Disease Control. Jeffrey P. Koplan, M. D., M.P.H., 1998–2002. *Morbidity and Mortality Weekly Report*, 56:846–850.

Centers for Disease Control. 2007. Update: Chikungunya Fever Diagnosed Among International Travelers — United States, 2006. *Morbidity and Mortality Weekly Report*, 56:276–277.

Centers for Disease Control. 2007. West Nile Virus Update — United States, January 1–November 13, 2007. *Morbidity and Mortality Weekly Report*, 56:1191–1192.

Denny, J., Boelaert, F., Borck, B., Heuer, O. E., Ammon, A., Makela, P. 2007. Zoonotic infections in Europe: Trends and figures — A summary of the EFSA-ECDC annual report. *Eurosurveillance Weekly Report*, 12:12.

Easterly, W. 2006. *The White Man's Burden: Why the West's Efforts to Aid the Rest Have Done So Much Ill and So Little Good*. New York: Penguin Press.

Eisenberg, J. N. S., Scott, J. C., Porco, T. 2007. Integrating disease control strategies: Balancing water sanitation and hygiene interventions to reduce diarrheal disease burden. *American Journal of Public Health*, 97:846–852.

Greenberg, M. R. 2007. Contemporary environmental and occupational health issues: More breadth and depth. *American Journal of Public Health*, 97:395–397.

Hutton, G., Bartram, J. 2008. Global costs of attaining the Millennium Development Goal for water supply and sanitation. *Bulletin of the World Health Organization*, 86:2–3.

Jahan, R. 2007. Securing maternal health through comprehensive health service: Lessons from Bangladesh. *American Journal of Public Health*, 97:1186–1190.

Jamison, D. T. 2006. Investing in health. In Jamison, D. T. (ed.). *Disease Control Priorities in Developing Countries* Second Edition, New York: Oxford University Press. DOI: 10.1596/978-0-821-36179-5/ Chpt-1, 3–36.

Levine, O. S., O'Brien, K. L., Knoll, M., Adegbola, R. A., Black, S., Cherian, T., Dagan, R., Goldblatt, D., Grange, A., Greenwood, B., Hennessy, T., Klugman, K. P., Madhi, S. A., Mulholland, K., Nohynek, H., Santosham, M., Saha, S. K., Scott, J. A., Sow, S., Whitney, C. G., Cutts, F. 2006. Pneumococcal vaccination in developing countries. *Lancet*, 367:1880–1882.

Lopez, A. D., Mathers, C. D., Ezzati, M., Jamison, D. T., Murray, C. J. 2006. Global and regional burden of disease and risk factors, 2001: Systematic analysis of population health data. *Lancet*, 67: 1747–1757.

Lomborg, B. 2007. *Solutions for the World's Biggest Problems: Costs and Benefits*. Cambridge, UK: Cambridge University Press.

Mathers, C. D., Loncar, D. Projections of global mortality and burden of disease from 2002 to 2030. *Public Library of Science on Medicine*, 3: e442 doi:10.1371/journal.pmed.0030442.

Morse, S. S. (editorial). 1992. Global microbial traffic and the interchange of disease. *American Journal of Public Health*, 82:1326–1327.

Shiffman, J. 2007. Generating political priority for maternal mortality reduction in 5 developing countries. *American Journal of Public Health*, 97:796–803.

Sommer, A. 1997. Vitamin A deficiency, child health and survival. *Nutrition*, 13:484–485.

Stephenson, J. 2006. HIV and circumcision. *Journal of the American Medical Association*, 296:759.

Suba, E. J., Murphy, S. K., Donnelly, A. D., Furia, L. M., Huynh, M. L., Raab, S. S. 2006. Systems analysis of real-world obstacles to successful cervical cancer prevention in developing countries. *American Journal of Public Health*, 96:480–487.

UNICEF. 1993 through 2006. *The State of the World's Children*. New York: Oxford University Press.

Wibulpolprasert, S., Tangcharoensathien, V., Kanchanachitra, C. 2008. Three decades of primary health care: Reviewing the past and defining the future. *Bulletin of the World Health Organization*, 86:1–80.

Wolfson, L. J., *et al*. 2008. Estimating the costs of achieving the WHO–UNICEF Global Immunization Vision and Strategy, 2006–2015. *Bulletin of the World Health Organization*, 86:27–39.

World Health Organization. 2007. *Maternal Mortality in 2005*. Estimates developed by WHO, UNICEF, UNFPA, and World Bank. Geneva: World Health Organization.

World Health Organization. 2007. Outbreak and spread of chikungunya. *Weekly Epidemiologic Record*, 82:409.

World Health Organization. 2008. Conclusions and recommendations of the Advisory Committee on Poliomyelitis Eradication, Geneva, 27–28 November 2007. *Weekly Epidemiological Report*, 83:25–36.

World Health Organization. 2008. Meeting of the Immunization Strategic Advisory Group of Experts, November 2007, conclusions and recommendations. *Weekly Epidemiologic Record*, 83:1–16.

BIBLIOGRAPHY

Allen, L., de Benoist, B., Dary, O., Hurrell, R. 2006. *Guidelines on Food Fortification with Micronutrients*. Geneva: World Health Organization.

Basch, P. F. 1990. *Textbook of International Health*. New York: Oxford University Press.

Centers for Disease Control. 1994. *Addressing Emerging Infectious Disease Threats*. Atlanta, Georgia: U.S. Public Health Service.

Cornia, G. A. 2001. Globalization and health: Results and options. *Bulletin of the World Health Organization*, 79:834–841.

Garrett, L. 2007. *The Challenge of Global Health. Foreign Affairs,* January/February 2007, http://www.foreignaffairs.org/20070101faessay86103-p90/laurie-garrett/the-challenge-of-global-health.html [accessed May 26, 2008]

Gostin, L. O. 2007. The "Tobacco Wars"—Global Litigation Strategies. *Journal of the American Medical Association*, 298: 2537–2539.

Hoges, J. R., Kimball, A. M. 2005. The global diet: Trade and novel infections. *Globalization and Health*, 1:4.

Howson, C. P., Kennedy, E. T., and Horowitz, A. (eds.). 1998. *Prevention of Micronutrient Deficiencies: Tools for Policymakers and Public Health Workers*. Washington, DC: Institute of Medicine, National Academy Press.

Institute of Medicine. 1992. *Emerging Infections, Microbial Threats to Health in the United States*. Washington, DC: National Academy Press.

Jamison, D. T., Breman, J. G., Meashan, A. R., Alleyne, G., Cleason, M., Evans, D. B., Jha, P., Mills, A., Musgrove, P. (eds.). 2006. *Priorities in Health*. Washington, DC: The World Bank.

Kimbal, A. M., Arima, Y., Hoges, J. R. 2005. Trade related infections: Farther, faster, quieter. *Globalization and Health*, 1:3.

Lopez, A. D., Mathers, C. D., Ezzati, M., Jamison, D. T., Murray, C. J. L. (eds.). 2006. *Global Burden of Disease and Risk Factors*. Washington, DC: Co-publication of Oxford University Press and The World Bank. The International Bank for Reconstruction and Development. The World Bank.

UNICEF. 2007. *The State of the World's Children 2007: Women and Children — The Double Dividend of Gender Equality*. New York: UNICEF.

UNICEF. 2008. *The State of the World's Children 2007: Child Survival*. New York: UNICEF.

World Bank. 1993. *World Development Report 1993: Investing in Health*. New York: Oxford University Press.

World Bank. 2006. *World Development Report 2006: Equity and Development*. Washington, DC: World Bank.

World Health Organization. 2000. *The World Health Report 2000: Health Systems: Improving Performance*. Geneva: World Health Organization.

World Health Organization. 2001. *The World Health Report 2001: Mental Health: New Understanding, New Hope*. Geneva: World Health Organization.

World Health Organization. 2002. *The World Health Report 2002: Reducing Risks, Promoting Healthy Life*. Geneva: World Health Organization.

World Health Organization. 2003. *The World Health Report 2003: Shaping the Future*. Geneva: World Health Organization.

World Health Organization. 2004. *The World Health Report 2004: Changing History*. Geneva: World Health Organization.

World Health Organization. 2005. *The World Health Report 2005: Make Every Mother and Child Count*. Geneva: World Health Organization.

World Health Organization. 2006. *The World Health Report 2006: Working Together for Health*. Geneva: World Health Organization.

World Health Organization. 2007. *World Health Report 2007: A Safer Future: Global Public Health Security in the 21st Century*. Geneva: World Health Organization.

PUBLICATIONS AND JOURNALS

Bulletin of the Pan American Health Organization: Pan American Health Organization Division of Vaccines & Immunization, 525 Twenty-third Street, N.W., Washington, DC 20037 USA.

Bulletin of the World Health Organization: World Health Organization, 1211 Geneva 27, Switzerland.

Centers for Disease Control. Morbidity and Mortality Weekly Report and many other publications available online and published form via http://www.cdc.mmwr/

Emerging Infectious Diseases: Centers for Disease Control and Prevention. 1600 Clifton Rd NE, #A23, Atlanta, GA, USA, http://www.cdc.gov/ncidod/eid

Health for All Series: World Health Organization, 1211 Geneva 27, Switzerland, 1978–1982.

International Agency for Research on Cancer: IARC Press, International Agency for Research on Cancer, 150 cours Albert Thomas, F-69372 Lyons cedex 08, France.

International Digest of Health Legislation: World Health Organization, Avenue Appia 20, 1211 Geneva 27, Switzerland.

Organization for Economic Cooperation and Development. OECD Health Data 2008: Statistics and Indicators for 30 Countries, http://www.oecd.org/document/30/0,3343,en_2649_34631_12968734_1_1_1_37407,00.html

Population Reports: Population Information Program, the Johns Hopkins School of Public Health, 111 Market Place, Suite 310, Baltimore, MD 21202-4012, USA.

Technical Report Series: World Health Organization, Avenue Appia 20, 1211 Geneva 27, Switzerland.

UNICEF. State of the World's Children annual report, http://www.unicef.org

United Nation Millennium Development Goals publications and reports, http://www.un.org/millenniumgoals/

United States Census Bureau publications at http://www.census.gov/population/www

Weekly Epidemiologic Record: World Health Organization, Avenue Appia 20, 1211 Geneva 27, Switzerland.

World Health Statistics Quarterly: World Health Organization, Avenue Appia 20, 1211 Geneva 27, Switzerland.

World Health Organization. World Health Statistics 2008, http://www.who.int/whosis/whostat/2008/en/index/html

WHO publications and documents are available for free download from the WHO Library database. This includes multiple languages with discounts on orders from developing countries for periodicals, book series, and thematic packages, http://www.who.int/publications/en/

Glossary of Terms

INTRODUCTION

A glossary is provided in order that the students may have an easy reference to understanding basic definitions in public health in all its major aspects. We have attempted to cover the primary concepts in clear and internationally accepted terminology and are grateful to John Last for permission to use his dictionaries of epidemiology and public health as the primary source.

ABBREVIATIONS

ACSD	Accelerated Child Survival and Development
AIDS	Acquired immune deficiency syndrome
APHA	American Public Health Association
ARI	Acute respiratory infection
BCG	Bacillus Calmette-Guérin (Antituberculosis vaccine)
BOD	Burden of disease
BSS	Behavioral Surveillance Survey
CBA	Cost–benefit analysis
CBR	Crude birth rate
CDC	Centers for Disease Control and Prevention
CDD	Control of Diarrhoeal Diseases–WHO
CHW	Community health worker
CIMCI	Community Integrated Management of Childhood Illness
CINDI	Countrywide Integrated Noncommunicable Disease Intervention
COPC	Community-Oriented Primary Care
CVD	Cardiovascular disease
DALYs	Disability-adjusted life years
DHHS	Department of Health and Human Services
DHS	District health system
DOTS	Directly Observed Treatments (Therapy)/ Short Course
DPT	Diphtheria, pertussis, tetanus vaccines combined
DTaP	Diptheria and tetanus toxoid with a cellular pertusis vaccine
EPI	Expanded program of immunization, includes DPT, BCG, measles, polio
EPI Plus	EPI, hepatitis B, yellow fever vaccines, plus vitamin A and iodine supplementation
FAO	Food and Agriculture Organization
FFF	Food supplementation, family spacing, female education
FGM	Female genital mutilation
GAVI	Global Alliance for Vaccines and Immunization
GDP	Gross domestic product
GFATM	Global Fund to Fight AIDS, Tuberculosis, and Malaria
GNI	Gross national income
GNP	Gross national product
GOBI	Growth monitoring, oral rehydration, breastfeeding, immunization
GOBI/ FFF	GOBI plus family planning, food production, female education
HCFA	Health Care Financing Administration
HepB	Hepatitis B vaccine
HFA	Health for All
Hib	*Haemophilus influenzae* type B vaccine
HIS	Health information system
HIV	Human immunodeficiency virus
HMD	Health Manpower Development–WHO
HMO	Health maintenance organization
HRD	Human resource development
HRP	Human resource planning
HRM	Human resource management
IBRD	International Bank for Reconstruction and Development–World Bank
IDA	Iron-deficiency anemia
IDD	Iodine-deficiency diseases
ILO	International Labour Organization
IMNCI	Integrated management of neonatal and childhood illness
IMR	Infant mortality rate
IPV	Inactivated polio vaccine
ITN	Insecticide-treated mosquito net
KABP	Knowledge, attitudes, beliefs, and practices
LDC	Least developed country
MAP	Malaria Action Program–WHO
MCH	Maternal and child health
MDG	Millennium Development Goals

MMR	Maternal mortality rate
MMR	Measles, mumps, rubella combined vaccine
MMWR	*Morbidity and Mortality Weekly Report*
MOH	Ministry of Health, Medical Officer of Health
MONICA	Multinational **MONI**toring of Trends and Determinants in **CA**rdiovascular Disease–WHO
MTEF	Medium-term expenditure framework
NGO	Nongovernmental organization
NHANES	National Health and Nutrition Examination Surveys
NHS	National Health Service, United Kingdom
NID	National immunization days
NIH	National Institutes of Health
OECD	Organization for Economic Cooperation and Development
OPV	Oral polio vaccine
ORT	Oral rehydration therapy
PAB	Protection at birth
PAHO	Pan American Health Organization
PHC	Primary health care
PPP	Purchasing Power Parities
QALY	Quality-adjusted life year
SARS	Severe acute respiratory syndrome
STI	Sexually transmitted infections
TB	Tuberculosis
TBA	Traditional birth attendant
TT	Tetanus toxoid vaccine
U5MR	Under-five mortality rate
UIS	UNESCO Institute for Statistics
UNAIDS	Joint United Nations Programme on HIV/AIDS
UNDP	United Nations Development Programme
UNESCO	United Nations Educational, Scientific and Cultural Organization
UNFPA	United Nations Population Fund
UNHCR	United Nations High Commissioner for Refugees
UNICEF	United Nations Children's Fund
USAID	United States Agency for International Development
VHW	Village (community) health worker
VPDs	Vaccine-prentable diseases
WHO	World Health Organization
YPLL	Years of potential life lost

INTRODUCTION

A glossary is provided in order that the students may have an easy reference to understanding basic definitions in public health in all its major aspects. We have attempted to cover the primary concepts in clear and internationally accepted terminology. The terms listed below are very much based on the outstanding *Dictionary of Epidemiology* (2001) and *Dictionary of Public Health* (2007) both edited by John Last, as well as *Health, United States 2007*, UNICEF, and others listed in the references on page 658.

Abatement: The process of preventing, removing, reducing, or minimizing public health dangers and nuisances, usually supported by regulation or legislation.

Absolute Poverty Level: The income level below which a minimum nutritionally adequate diet plus essential non-food requirements are not affordable (UNICEF).

Accessibility: A measure to assess the quality of a health care system or a public health service; the quality of being available in practice to clients and users of a health system. The potential to reach needed and appropriate health services by members of a population by local means of transport in no more than 1 hour; it is affected by distribution of services, but also by convenience, cost, insurance coverage, and transportation as well as cultural and social factors.

Accountability: The state of being answerable for decisions and actions that can be used as an indicator of how well a health system functions. The care provider, health care organization, or public health system can be held responsible administratively, professionally, morally, and legally.

Accreditation: A systematic, periodic, governmental, or public non-profit (voluntary) external evaluation of health care facilities and its organization by an outside agency. Accreditation is a formal process including self-assessment by the facility or program and an external multidisciplinary professional team site visit. It includes assessment of staff, facilities, programs, and deficiencies according to written standards; it is based on review of organized material for self-assessment and site visits by an external professional team to confer official recognition on a health care facility, service program, or educational institution.

Accuracy: The degree to which a measurement or an estimate based on a measurement represents the true value of that which is being measured, and can be confirmed by repeat testing.

Activities of Daily Living (ADL): Functional capability for activities related to mobility and personal care and the individual's ability to perform daily tasks such as dressing, toileting, feeding, getting out of bed or a chair without assistance; it is used to measure outcomes of interventions for various stages of aging, a disease process, or a chronic or disabling condition.

Acute Illness: A disease or condition that is sudden in onset, a brief, intense or short-term episode of illness, or exposure to a pathogen; sometimes used to indicate severity.

Adaptation: The process by which an organism or person adapts to environmental conditions; the changes that proceed over the course of many generations of an organism.

Addiction: A pathological, physiological, and psychological dependence on a substance or compulsion to act in a particular manner, inability to limit use of a mood-modifying or consciousness-altering substance, or alcohol, or compulsive gambling.

Admission, Hospital: Entry to a hospital inpatient facility for care for a 24-hour period or more, except for newborns.

Adult Literacy Rate: The percentage of persons aged 15 and over who can read and write.

Adverse Event: Any disease or injury (e.g., premature death or unnecessary morbidity), associated with medical management that results in measurable disability.

Adverse Reaction: An undesirable or unwanted consequence or side effect of a preventive, diagnostic, or therapeutic procedure or regimen, not necessarily harmful. In many hospitals and states these are reportable events, and an Adverse Drug Reaction Registry provides a mechanism to detect medically or legally important side effects.

Age-Adjusted Rate: The application of age-specific rates in a population to a standardized age distribution to eliminate differences in observed rates that are the result of age differences in the population.

Age-Specific Rate: The rate at which an event or a condition is observed in a specific group. The numerator and denominator refer to the same age group.

Agent, Infective/Toxic: A pathogenic organism, dietary ingredient, chemical substance, or physical condition of the environment that is necessary for disease to occur if the agent is present in excess, or in some circumstances, if it is absent or deficient (e.g., micronutrient). A microorganism, chemical substance, or form of radiation that is essential but not necessarily sufficient to cause a disease or disorder. A disease may have a single agent as a cause, or may occur as a result of the agent in company with other contributory factors.

Aging of the Population: A demographic term meaning an increase over time in the number and percentage of older persons in the population, lower death rates, and increasing life expectancy, associated with declining birth rates and higher deaths rates in infancy and childhood over a generation or more.

Alcoholism: Physiological or psychological dependence on alcohol in which the person consistently drinks in amounts (a daily intake of 75 grams or more) affecting ability to perform usual work, and in advanced stages with neurologic impairment.

Allocation of Resources: Distribution of resources, including money, personnel, and supplies to different categories of health service, either nationally, regionally, locally, or within an organization providing health care, preferably based on epidemiologic evidence and sometimes other pragmatic considerations.

Alma-Ata: A city in Kazakhstan where the Joint WHO/UNICEF International Conference adopted a Declaration on Primary Health Care in 1978, providing the basis for the WHO program of *Health for All*.

Ambient: Surrounding conditions of the environment, usually referring to air quality or noise.

Ambulatory Care: Medical and paramedical care services provided to persons attending a care setting on a visiting basis, usually in outpatient or community clinics. The care may be primary, episodic, or part of continuing care for an existing condition.

Ambulatory Surgery: Surgical procedures, major or minor with or without general anesthesia, provided to a patient who does not remain in hospital overnight.

Analytic Study: A study designed to examine associations, hypothesized to be causal relationships, to identify or measure risk or health effects of specific exposures; cross-sectional, case–control, and cohort studies.

Anemia: Abnormally low hemoglobin concentration in the blood, below accepted normal values, depending on age and sex, due to loss of blood, rapid breakdown of red cells (hemolysis), or a dietary deficiency of iron or other essential elements of red cells.

***Anopheles* Mosquito:** The mosquito responsible for transmission of malaria parasites from one human host to another, with a life span of 2–3 weeks, in climates above about 24° C.

Anorexia Nervosa: An eating disorder in which the patient suffers from an obsession to lose weight leading to deliberate self-starvation, sometimes resulting in death.

Antenatal (Prenatal) Care: The monitoring methods and procedures for care of the pregnant woman throughout pregnancy, ideally beginning before pregnancy and certainly during the first trimester up through delivery, with follow-up afterward in postnatal care.

Antibody: A protein molecule produced in the body in response to a foreign substance (i.e., an antigen) or acquired by passive transfer. Antibodies bind to a specific antigen that elicits its production, causing its destruction, thereby providing an important mechanism for protection against infectious disease. Antigen-antibody response, along with cellular response, is a vital immunologic defense to protect the individual and populations from specific infectious agents. Concentration can be measured in individuals and therefore in populations.

Antigen: A substance (protein, polysaccharide, glycolipid, tissue transplant, etc.) alien to the body that is capable of inducing a specific immune response and antibody production in a host, (e.g., by invasion of infectious organisms, immunization, ingestion, inhalation, or others). An immune reaction occurring to self-antigens may cause an autoimmune disease.

Antigenicity: The ability of an external agent to produce a systemic or local immunologic reaction in a host.

Antigenic Drift: Evolutionary genetic changes which may occur in viruses during passage from one host to another and are responsible for antigenic changes and immunologic responses of individuals and populations. This necessitates annual reformulation of the influenza vaccine.

Antigenic Shift: A sudden mutation in the molecular structure in the organism, producing completely new subtypes of the organism, which may result in epidemics since hosts previously exposed to other strains may have little or no immunity to the new strains.

Antiretroviral therapy (ART): Use of drugs to treat or retard the onset and progress of AIDS and the transmission of HIV (e.g., from mother to fetus and newborn).

Arthropod borne disease: Diseases (viral, rickettsial, bacterial, or protozoal) transferred to a host by arthropods such as mosquitoes, sandflies, blackflies, ticks, spiders, or other invertebrates.

Apgar Score: A widely accepted composite index used to evaluate neonatal status by assigning numeric scores (0–2) to vital signs and activity level of a newborn performed 1 and 5 minutes after birth for a total score of 0 to 10.

Appropriate Health Technology: Diagnostic, therapeutic facilities and services that are economically sustainable in keeping with the financial ability of a country. They require fewer resources, are easier to maintain, have less impact on the environment, and are compatible with community needs.

Appropriation: Placement of funds in a budget category previously authorized as the final step in a legislative process intended to lead to action in the administrative branch of government.

Arbovirus: A mixed group of animal viruses sharing a common epidemiologic mode of transmission between vertebrate hosts. That is blood-feeding arthropod vectors, such as mosquitoes, ticks, or sandflies, from an infected to a susceptible host. The diseases these viruses cause tend to have distributions dependent on the seasonality of the vector but survive in the vector during periods of hibernation or over-wintering.

Atherosclerosis: Accumulation in the inner lining of arteries of fatty deposits causing fatty deposits, plaque formation, narrowing of the lumen, and blockage of the arteries along with reduced elasticity, resulting in decreased flow of blood in the vessel. It is the underlying pathogenesis of two leading causes of death, stroke, and coronary heart disease.

Attributable Risk: The rate or proportion of a disease, or other outcome, in exposed individuals in a defined population that can be attributed to the exposure to a particular risk factor (i.e., beyond the normal level of the disease in a similar population group not exposed).

Average Length of Stay (ALS): The total number of patient days counting the date of admission but not the date on which patients were discharged during a reporting period, divided by the number of patients discharged from an inpatient care facility.

Avian Influenza: The H5N1 avian influenza is located in domestic and wild fowl, and can be transferred to humans by direct contact; it is a global pandemic threat if person-to-person transmission occurs.

Avoidable (Premature) Mortality: Any death from specific and all causes that occurs a considerable number of years before average life expectancy is achieved; defined as the average years of life expectancy in a low mortality population.

Bacillus Calmette-Guérin (BCG) Vaccine: A weakened or attenuated strain of tubercle bacillus used widely for vaccination to prevent tuberculosis, especially in children in high-incidence areas.

Bacteriology: The branch of science concerned with the study of bacterial microorganisms as part of the science of microbiology, which includes bacteria, viruses, parasites, and other organisms.

Bacteriostasis: Slowing or cessation of multiplication of bacteria, enabling the host's defenses to destroy the invading organism.

Basket of Services: A defined set of public health and clinical interventions included within a national or other health insurance or service program, usually including preventive care, medical and diagnostic services, hospital and home care, possibly prescribed medications, but often excluding mental health services and usually dental care.

Behavioral Risk Factor: A characteristic, behavior, addiction, or fixation associated with increased probability of a specific disease or outcome; does not imply a causal relationship. Common risk factors associated with ways people behave include, for example, reckless driving, insufficient exercise, and eating to excess.

Bed Occupancy Rate: The percentage of hospital beds, cribs, pediatric beds and pediatric bassinets (excluding newborn) staffed and occupied over a specified time period, such as a week, month, or year, divided by the total of beds at the national, regional, or local, hospital or departmental level, to assist management in tracking utilization of facilities. Average daily census is the total annual number of inpatients excluding newborns divided by 365 days. For facilities other than hospitals occupancy rate is the number of residents at the facility on the day of reporting divided by the number of reported beds.

Benefit–Cost Ratio: The ratio of net present value of measurable benefits to costs; used to determine the economic feasibility or success of a program.

Bias: Deviation of results or interferences from the truth, or processes leading to such deviation; may be due to a trend in the collection, analysis, interpretation, publication or review of data that can lead to conclusions that are systematically different from the truth.

Bimodal Distribution: A distribution of two regions of high frequency separated by a region of low frequency of observations; a twin peak distribution curve.

Bioassay: The quantitative evaluation of the potency of a substance by assessing its effects on tissues, cells, live experimental animals, or humans.

Biological Plausibility: An association thought to be causal is strengthened when explainable in terms of current biological or medical knowledge.

Bioterrorism: The intentional release, or threatened release, of disease-producing living organisms or biologically active substances derived from organisms for the purpose of causing death, illness, incapacity, economic damage, or fear.

Births Attended %: Percentage of births attended by physicians, nurses, midwives, trained primary health care workers or trained traditional birth attendants (TBAs).

Birth Certificate: A state-mandated legal document recording details of a live birth. It provides the basis for vital statistics of birth and birth rates, infant and neonatal mortality rates, and other statistical data in a political or administrative jurisdiction.

Birth Interval: The time between completion of one pregnancy and the termination of the next pregnancy.

Birth Weight: The weight of an infant recorded at the time of birth and sometimes on the birth certificate. Those below 2500 grams are *Low Birth Weight (LBW)*; *Very Low Birth Weight (VLBW)* includes those below 1500 grams, and *Ultra-Low Birth Weight (ULBW)* is a birth weight below 1000 grams.

Blind Study: A study in which the observer and/or the subjects are kept unaware of the group the subject is in within a study. When both observer and subject are unaware, the study is called "double-blind," and if the statistical analysts are also unaware then it may be called "triple-blind."

Block Budget: A method of budgeting a health facility, such as a hospital, which is not designated line-by-line in detail, nor by department or other internal breakdown, but is a global sum negotiated with the funding agency, allowing the management freedom to change internal allocations, staff realignment, and reduction.

Block Grant: A financial disbursement of money to an agency or research group without explicit prespecified requirements regarding the manner of operating a health program based on units of services or specific category of service, but directed toward more general program content.

Body Mass Index (BMI): An anthropometric measure is an indicator of fatness and obesity defined as the weight in kilograms divided by the height in meters squared. It correlates closely with body density and skinfold thickness, calculated as follows: BMI = Weight (kg)/(Height) (m) squared. Underweight = <18.5; Normal weight = 15.5–24.9; Overweight = 25–29.9; Obese = BMI of 30 or greater.

Burden of Disease (BOD): Quantification of the impact of a disease in a defined population, as measured in terms of numbers of deaths, as well as the impact of premature death and disability on the population, combined into a single unit of measurement (see DALYs and QALYs and World Development Report).

Capital Investment/Costs: Investments in land, buildings, and equipment that give a tangible and lasting result and will not need to be repeated on an ongoing basis, as distinct from maintenance and operating costs.

Capitation: A method of paying for designated health services, whereby funds are allocated per person registered as a service beneficiary for a specified period of time. Weighted capitation adjusts for factors such as age- and sex-specific mortality rates, representing differences in health care needs of the population.

Carcinogen: An agent (e.g., genetic, chemical, biological, radiological) that can cause cancer.

Carcinogenesis: The process by which cancer is produced; the irreversible changes in a cell's growth-regulatory processes where the potential for unregulated growth is established, usually through genetic damage by a chemical or physical carcinogen.

Cardiovascular Disease (CVD): Diseases of the heart and circulatory system which include coronary heart disease, cerebrovascular accidents, rheumatic heart disease, congenital heart disease, and some others.

Carrier: A person or animal that harbors a specific infectious agent in the absence of discernible clinical disease and serves as a potential source of infection. The carrier state may occur in an individual with an unapparent (subclinical) infection, or during the incubation period, convalescence, or postconvalescence of an individual with clinically recognizable disease. The carrier state may be of short or long duration. A genetic carrier is a person who carries a gene transmissible to subsequent generations where it is expressed as an abnormality.

Case–Control Study: The observational epidemiologic study of persons with the disease or other outcome variable and a suitable control group of persons without the disease. The history of exposure to a suspected risk factor is compared to "cases" and "controls" (persons who resemble the cases in age and sex but do not have the disease or condition of interest). The frequency of exposure is measured in each group, often given as an odds ratio.

Case Fatality Rate (CFR): The proportion of cases of a specified condition which are fatal within a specified time period of follow-up after diagnosis and/or medical intervention. CFR (usually expressed as a percentage) = No. deaths from a disease/number of diagnosed cases of the disease in the same period × 100.

Case for Action: The weight of evidence accumulated from all possible sources of clinical, epidemiologic, societal, and economic aspects of a health or illness condition which justifies an intervention program.

Case Management: The method, often a formal protocol or clinical guideline, for care of persons identified as having cases of specific conditions, such as notifiable communicable diseases, cancer at specified sites, etc., organized to help clients/patients receive the services they need, setting priorities for these services, and helping to integrate services with monitoring of progress.

Categorical Service: A service program funded and developed specifically to address one category of health problem, service, or risk group.

Cause of Death: Causes of death include those diseases, conditions, or injuries that resulted in or contributed to death, and the accident or violence which produced the injuries. The *underlying cause of death* is recorded on the death certificate as the disease or injury that initiated the train of events leading to death, or the circumstances of the accident or violence which produced the fatal injury. These are coded according to the *International Classification of Disease* (currently ICD-10).

Cause-Specific Rate: A rate that specifies events, such as deaths according to their cause.

Census: An official periodic enumeration of a population, recording all persons, in every place of residence, with age or birth date, sex, occupation, national origin, language, marital status, income, relationship to head of household, in addition to information on the dwelling place. It may include additional information, such as educational levels, and health-related data, such as a permanent disability.

Cerebrovascular Disease: Atherosclerosis and other diseases of the cerebral arterial system supplying the brain which can result in cerebrovascular accidents or strokes.

Certificate of Need (CON): A formal procedure in which a health facility wishing to invest in capital equipment, renovation, or expansion must present a detailed plan and justification for the plan to the state or national regulatory authority for consideration, modification, and approval, even when the source of funding is nongovernmental. It may also be the form used by health workers to record reasons why a family or person may require social assistance.

Certification: The completion of an official form for statutory and/or legal purposes as in vital statistics or completion of requirements to practice a profession or function as a health facility with specified rights in providing health services. The state may issue a permit to operate, retaining the right to inspect and cancel the license or require changes in operation of facilities.

Child Mortality Rate: The annual number of deaths of children under age 5 years expressed as a rate per thousand live births. This is often averaged over the previous 5 years in communities where the age of young children may not be known precisely (UNICEF).

Chronic Disease: Chronic refers to a health-related state, lasting a long time; to exposure, prolonged or long-term, often with implication of low intensity; to a condition of 6–12 weeks or longer.

Countrywide Integrated Noncommunicable Disease Intervention (CINDI): A WHO program established to improve community health by prevention and control of risk factors for cardiovascular diseases, cancer, chronic respiratory diseases, accidents, diabetes, and mental disorders.

Circulating Antibodies: Antibodies which are measurable in the blood, in contrast to antibodies within the cells, such as polio antibodies in the cells of the intestinal tract.

Clinical Guidelines: Evidence-based practice standards based on meta-analysis by multidisciplinary professional bodies with comprehensive reviews and rigorous analysis of relevant scientific information. They provide the practitioner and health manager with professionally acceptable criteria for preventive and curative management of specific medical conditions of clinical, epidemiologic, and economic importance, and may be required for accreditation and/or reimbursement.

Cluster Sample: A sample in which each unit selected is a group of persons (all persons in a city block, a school, or a home) rather than an individual.

Coagulation: The process by which bleeding is arrested as a result of physical and biological changes in the composition of circulating blood; or aggregation and molecular changes in suspended protein particles as in clotting of milk or under the influence of bacterial fermentation, or a method of water treatment in which particles suspended in water are adsorbed and settle to the bottom.

Cohort Studies: The analytic method of an epidemiologic study in which subjects of a defined population can be identified who are, have been, or in future may or may not be exposed to factors thought to influence the probability or occurrence of a given disease or outcome. It is the observation of large numbers over a long period of time with comparison of incidence rates in groups that differ in exposure levels. The purpose is to generate reliable incidence or mortality rates in the population subsets.

Coliform Test: A standard bacteriologic examination of drinking water safety, using simple culture methods to test for *Escherichia coli* as an indicator of fecal contamination.

Cold Chain: A system of protection against high environmental temperatures for heat-labile vaccines, sera, and other active biological preparations. Unless the cold chain is preserved, such preparations are inactivated and immunization procedures will be ineffective. It extends from the point of manufacture to the point of use, including all stages of storage and transportation. It is also used in reverse to take specimens to regional laboratories.

Communicable Disease: An illness due to a specific infectious agent or its toxic products which arises through transmission of that agent from an infected person, animal, or inanimate reservoir to a susceptible host, either directly or indirectly through an intermediate plant, animal host, vector, or the environment.

Community-Based Programs: Programs and services serving a population living in a localized area, or a definable population with a common set of norms, values, and organization. Health services in the community usually implies services or programs offered to people in their homes, or in other community settings and through ambulatory care centers.

Community Health Workers (CHWs): Nonprofessional health care providers selected from a designated community and trained to provide general or specific primary care services providing access to care for underserved population groups.

Community Medicine: The study of disease and health in the population of a specified community or group in order to identify the population's needs, the means by which these needs should be met, and to evaluate the extent to which health services effectively meet these needs. It is the practice of medicine concerned with communities and includes the organization and provision of care at the community level.

Community-Oriented Primary Care (COPC): Integrates community medicine with primary health care of individuals at the community or population level. It investigates health care needs and provides systematic, coordinated management of identified health problems by members of a primary health care team comprised of physicians, nurses, and other health professionals.

Community Participation: The involvement of the community in health care issues, through organizational activities to promote health awareness and health programs.

Compliance: The degree to which an individual patient or target group implements health recommendations specifically designated as needed to treat or prevent illness in accordance with recommendations of a health care provider or investigator in a research project. The word *adherence* is used by behavioral scientists to avoid the authoritarian association of compliance.

Confidence Interval (CI): The computed interval with a given probability (e.g., 95 percent) within which the true value of a variable such as a mean, proportion, or rate is contained.

The upper and lower boundaries of the confidence interval are called the *confidence limits*.

Confounding Variable (Confounder): A factor or variable which is not separated from the causal factor being investigated, and is not an intermediate variable, but can potentially cause or contribute to the health outcome of interest. Unless it is possible to separate confounding variables, their effect cannot be distinguished from those of the factor(s) being studied and may invalidate a conclusion of potential causal link.

Contact: A person (or animal) who has been in association with an infected person or animal or contaminated inanimate object or environment that might provide an opportunity to transmit the infective agent. *Direct contact* is a mode of transmission of infection between an infected host and a susceptible host. A *primary contact* is a person who has been in direct contact or associated with a case of the communicable disease; a *secondary contact* has been in contact or associated with a primary contact. A secondary contact is a type of *indirect contact*.

Containment: The process of preventing the spread of transmissible infectious disease by limiting the movements of cases and close contacts by voluntary or mandatory quarantine. Also used in the concept of regional eradication of a communicable disease. It demands a globally coordinated effort to establish local or regional control to prevent transmission and reintroduction of the organism of an infectious disease during an outbreak or epidemic.

Contamination: The presence of toxins, infective, or noxious agents on a body surface, in food, water, soil, air, or inanimate objects. Contamination of a body surface does not imply a carrier state. It is distinct from pollution which is offensive but not infectious. Contamination also refers to a study population that possesses conditions or factors other than those under investigation that may modify the results of the study.

Contraception: Prevention of conception through one of the following methods: hormones (pills, injectables, and subdermal implants), intrauterine devices (IUD), barrier devices (condom, diaphragm), chemical (spermicides), and natural methods (rhythm and withdrawal).

Contraceptive Prevalence Rate: The percentage of women aged 15–49 who are practicing or whose sexual partners are practicing any form of contraception, whether modern or traditional.

Co-Payments: The proportion of a fee that the consumer pays at the time of service, which is thought to deter wasteful and unnecessary utilization of services, a user fee.

Cost: The value of resources invested in a service including costs of buildings, equipment, operating costs (e.g., salaries, food, electricity, and laundry). *Indirect costs* are lost productivity caused by the disease, borne by the individual, the family, society, or the employer. *Intangible costs* are the cost of pain, grief, suffering, loss of leisure time.

Cost–Benefit Analysis (CBA): An analysis in which the economic and social costs of health service/medical care and the benefits of reduced loss of net earnings due to preventing premature death or disability are considered. The general rule for the allocation of funds is that the ratio of the benefit of preventing an additional case to the cost of preventing an additional case should be equal to or greater than 1.

Cost–Benefit Ratio (CBR): An economic analysis in which all costs and benefits are converted into monetary values, and results are usually expressed as dollars (USD) of benefit per dollar expended.

Cost Containment: Reducing the rate of increase in overall health care costs to the rate of increase of an economic indicator of growth, such as GNP per capita.

Cost-Effectiveness Analysis (CEA): A form of economic evaluation where all the costs are expressed in monetary terms, but the consequences are expressed in nonmonetary terms, such as life years gained, cases detected, or cases prevented. The analysis seeks to determine the costs and effectiveness of an activity or to compare similar alternative activities to determine the relative degree to which they will attain the desired objectives or outcomes.

Cost Utility: An economic evaluation in which the outcomes of alternative procedures or programs are expressed in terms of a single "utility-based" unit of measurement. A widely used utility measure is the quality-adjusted life year, a health state on a continuum from full functional status to total dependency on others.

Coverage: A measure of the extent to which services rendered cover the potential need for these services in a community; the extent to which a population is protected against a communicable disease by appropriate vaccination or other preventive measure, (e.g., percent of children vaccinated according to schedule or eligible women having a Pap smear as per the recommended schedule).

Cross-Sectional Studies: Studies that examine the relationship between diseases or other health-related characteristics and other variables of interest in a defined population at a particular time without regard for past conditions. This usually includes disease prevalence and the presence of the characteristic being studied in the diseased as compared to the non-diseased population.

Crude Birth Rate (CBR): The annual number of live births per 1000 population in a calendar year in a defined administrative jurisdiction.

Crude Death Rate: The number of deaths per 1000 population during a specified time period, usually the calendar year.

Days of Care: The total number of patient days accumulated by patients at the time of discharge from short-stay hospitals during a reporting period, including day of admission, but not the day of discharge; usually expressed as rates (i.e., days of care per 1000 population or as age-specific rates).

Death Certificate: A vital record required by law, signed by a licensed physician or by another designated health worker that includes the cause of death, decedent's name, sex, birth date, place of residence and of death. The immediate cause of death recorded on the first line, followed by conditions giving rise to the immediate cause. The underlying cause is recorded last. Also noted is whether there was a medical attendance before death (i.e. at the time or immediately prior, or in the presence of another health professional).

Decentralization of Health Services: Transfer of operational authority and budget for health services, including management of budget and personnel from a central authority at the national or state level to a local health authority, or to independent not-for-profit corporations or trusts.

Demand for Health Services: Willingness, desire, and ability to seek, use, and pay for health services. *Expressed demand* is actual use, while *potential demand* is an expression of need.

Demographic Transition: Long-term trends of declining fertility and mortality to low rates are accompanied changes in the age composition of a population, with increased life expectancy and "aging of the population."

Demography: The study of populations, especially with reference to size and density, fertility, mortality, growth, age distribution, migration, vital statistics, and the interaction of all these with social and economic conditions.

Dependency Ratio: The ratio of dependent persons (infants, children, frail elderly, and the infirm) to the others in the population who support them either directly or through mandatory social security systems. The Dependency Ratio is the proportion of economically inactive compared to the economically active.

Dependent Variable: An outcome the value of which is dependent on the influence of other variable(s) — independent variables — in the relationship under study. In statistics, the dependent variable is the one predicted by a regression equation.

Deregulation: An administrative strategy to reduce government procedures for supervision of an economic or service sector, while retaining overall responsibility for the standards of the outcome.

Developed and Developing Countries: Countries are defined as to their national socioeconomic status, the GDP per capita; people in developed countries have on average sufficient income and access to a broad range of private and publicly provided services; and those in developing countries lack the services and facilities of developed countries.

Development Indicators: Measures which reflect the health, educational, and social progress made in a country, for comparison with other countries.

Diagnosis-Related Group (DRG): Classification of hospital patients according to diagnosis and nature and level of intensity of medical care required, often used by insurance carriers to set reimbursement scales.

Diarrheal Diseases: Passage of frequent loose or watery stools, as a result of virus, bacterial, or parasitic infection.

Dietary Reference Intakes (DRIs): These are guidelines for macro- and micronutrient intake based on caloric, vitamin, and mineral needs for daily nutrition in relation to age, height, sex, and weight, previously called *Recommended Dietary Allowance* and *Recommended Daily Intakes.*

Direct Costs: Costs directly associated with prevention and treatment activities in the health care system for a condition, as opposed to indirect costs, such as loss of productivity or wages from work during illness.

Disability: Any temporary or long-term reduction of a person's capacity to perform usual functions; often the consequence of impairment such as impaired mobility or intellectual impairment.

Disability-Adjusted Life Year (DALY): A measure of the burden of disease and injury in a defined population and the effectiveness of interventions expressed in terms of hypothetical healthy life years that are lost as a result of specified diseases. DALYs are calculated as the present value of the future years of disability-free life that are lost as a result of premature mortality or disability occurring in a particular year.

Discharge, Hospital: The formal release and completion of any continuous period of stay of one night or more in hospital as an inpatient, excluding the period of stay of a well newborn.

Discounting: An adjustment to cost estimates to allow for the fact that future monetary units will have a different value from those at present, usually a smaller value because of the effects of inflation; applies the interest rate equivalent to the rate of inflation to produce estimates of future values at specified future times; used mainly in economics and in clinical decision analysis.

Disease: Any physiological or psychological departure from normal function. *Disease, illness,* and *sickness* are sometimes used interchangeably. Disease is a physiological or psychological dysfunction defined by clinical, pathological, and epidemiologic criteria.

Disinfection/Disinfestation: Any physical or chemical process serving to destroy or remove undesired small animal forms, particularly arthropods or rodents, insects, or microorganisms present on the person, the clothing, or in the environment of an individual, or on domestic animals.

Distributional Effects: The manner in which costs and benefits of a preventive strategy affect different groups of people, in terms of demographics, geographic location, and other descriptive factors.

District Health Authority (DHA): A term describing a health service operated by a local government authority that involves administrative responsibility for a number of services and overall responsibility for the health of the population of the district, with accountability for these to a higher level of government.

DPT Vaccine: Diphtheria, pertussis (whooping cough), and tetanus combined vaccines.

Ecology: The study of the interrelationships among living organisms and their environment. *Human ecology* means the study of human groups as influenced by environmental factors, including social and behavioral factors.

Effectiveness: In epidemiology, effectiveness is a measure of the extent to which a specific intervention, procedure, regimen, or service, used in routine circumstances, does what it is intended to do for a specified population; a measure of the extent to which a health care intervention such as preventive immunization, dietary regimen, or surgical procedure does what it is supposed to do and fulfills it objectives.

Efficacy: The extent to which a specific intervention, procedure, regimen, or service produces a beneficial result under ideal conditions; the benefit or utility to the individual or the population of the service, treatment, regimen, or intervention, ideally measured by a randomized controlled trial.

Efficiency: The extent to which the resources such as money, effort, and time used to provide a specific intervention, procedure, regimen, or service of known efficacy and effectiveness are minimized. A measure of the economy (or cost in resources) with which a procedure of known efficacy and effectiveness is carried out. The process of making the best use of scarce resources; types include technical, productive, allocative, and statistical.

Elimination: Reduction of case transmission of a disease to a predetermined very low level to achieve a situation where the infecting agent cannot sustain itself in the population and is no longer a major public health problem.

Endemic: A disease that is continuously present in a given area or population group.

Enriched Foods: Basic foods to which nutritionally important vitamins and minerals are added to enhance the nutritional value of the food (fortification), or restored to replace those lost in food processing or preparation (restoration).

Enteric Disease: Diseases of the bowel usually due to infection with specific enteric pathogenic organisms including bacteria, viruses, parasites, and helminths; contacted by eating, drinking, or swallowing food or water contaminated with a pathogenic organism or via fecal-oral transmission; usually manifesting with gastro-intestinal symptoms.

Environment: All physical, biological, social, cultural, or other factors external to the organism (e.g., the individual human host) that may impinge on its behavior or well-being. All influences apart from genes that are determinants of health and well-being.

Environmental Protection Agency (EPA): A U.S. federal government agency with responsibility for assessing and managing environmental risks to human health and monitoring standards of all environmental media, including air, water, and land quality.

Epidemic: The occurrence in a community or region of cases of an illness, specific health-related behavior, or other health-related events clearly in excess of normal expectancy for a given period of time. It is relative to the usual frequency of the disease in the same area, among the specified population, at the same season of the year. A single case of a communicable disease long absent from the population, or first invasion by a disease not previously recognized in the area requires immediate reporting and full field investigation and should be considered a threatened epidemic. For example, even one imported case of measles or SARS in an area previously free of cases warrants active investigation and immunization of contacts and other control measures to prevent spread of the disease.

Epidemic Curve: A graphic plotting the course of the epidemic with the distribution of cases by time of onset to help define epidemiologic characteristics of the episode.

Epidemiology: The study of distribution and determinants of health-related states or events in specified populations, and the application of this study to control of health problems. *Study* includes surveillance, observation, hypothesis testing, analytic research, and experiments. *Distribution* refers to analysis by time, place, and classes of persons affected. *Determinants* are all the factors that influence health. The definition has broadened from concern with communicable disease epidemics to take in all phenomena related to health in populations.

Epidemiologic Intelligence Service (EIS): A training and service program of the U.S. Centers for Disease Control and Prevention in which clinicians are trained to carry out epidemiologic investigations as part of their development as public health professionals.

Epidemiologic Transition: Several phases in the interactions of human populations with disease agents resulting in changes in the patterns of predominant diseases in a society (e.g., a decline in the absolute and relative importance of infectious disease as compared to an increase in the importance of chronic disease and injury). New epidemiologic patterns include an increase in significance of newly emerging or recurring infectious diseases and antibiotic resistance.

Eradication: Ending all transmission of infection by extermination of the infectious agent through surveillance and containment (e.g., the eradication of smallpox); WHO defines *eradication* as achievement of a status whereby no further cases of a disease occur anywhere, and continued measures are unnecessary.

Essential Drug List: The *Model List of Essential Drugs* developed by WHO suggests a list of drugs considered essential to treat and prevent the commonly occurring life-threatening diseases. The list is aimed mainly at developing countries.

Etiology: The science of causes, causality; in common usage, cause.

Evaluation: A process that attempts to determine as systematically and objectively as possible the relevance, effectiveness, and impact of health-related activities in the light of their objectives. Evaluation is multidimensional, incorporating description, comparison and analysis of population, infrastructure, input (i.e., resources), process (e.g., utilization), and outcome measures (e.g., morbidity, mortality, and functional status) indicators; costs in relation to benefits.

Evidence-Based Practice: Application of the best available evidence from epidemiologic, sociologic, economic, and other relevant sources including published peer-reviewed articles and reports and best practices as established and tested in advanced public health systems.

Expanded Program on Immunization (EPI): The *Health for All* immunization program (of WHO and UNICEF), including diphtheria, pertussis, tetanus, poliomyelitis, measles, and tuberculosis vaccines for infants and children, conducted especially in developing countries.

Expanded Program on Immunization Plus (EPI Plus): EPI and immunization against hepatitis B, and yellow fever as well as vitamin A and iodine supplementation, expanding to include insecticide impregnated bednets, tetanus toxoid for mother, Hib, and other recent vaccines.

Exposure: Proximity and or contact with a source of a disease agent that leads to transmission of the agent to a new host or harmful effects of the agent. A disease-causing agent such as a toxic substance or radiation level to which a group or individual was causing clinical signs that may be dose-related.

Externalities: Social benefits and costs that are not included in the market price of an economic good. Examples include benefits to others of treating a case of infectious disease, adverse health effects of industrial air pollution not included in the price of the industrial product, and impact on national economy of natural resource depletion not included in the calculation of national income.

Family: A group of two or more persons united by blood, adoptive, marital, or equivalent ties, usually sharing the same dwelling unit. The extended family is multigenerational; the nuclear family, in contrast, is a single-generation family, usually husband-wife-children, but is often headed by a single parent, and sometimes by partners of the same sex. A family is an established entity in law and not dependent on marital status and gender.

Family Physician: A doctor specialist in family medicine, who provides general medical care to individuals and families in the community, with continuity through long-term familiarity with a family and its medical and functional profile. The family physician's activities include curative, preventive, and health promotion services.

Family Violence: Physical, sexual, financial, or emotional abuse among family members. This can include spousal, child, and elder abuse.

Federal: A system of government of a country with elected national, state, or provincial governments, each with legislative and taxing powers as defined in a national constitution and interpreted by supreme court decisions.

Fee-for-Service: Payment of a designated charge per item of health service received, such as an examination, consultation, or diagnostic test.

Feldsher: A mid-level health worker with 2 to 3 years of training who operates independently in rural outposts in Russia.

Fertility Rates: A measure of the fertility of a population. *Total fertility rate* is the average number of children that would be liveborn alive to a woman during her lifetime if she were to go through her child-bearing years conforming to the age-specific fertility rates of a given year. *General fertility rate* is a more refined measure of fertility than the crude birth rate; the number of live births per 1000 women of child-bearing age (15–44) in a year.

Filtration: A stage in the treatment process of community water supplies, following preliminary treatment, for removal of particulate matter in source water by percolation through sand or other porous material before further treatment with disinfection.

Fluoridation: The systematic addition of carefully measured amounts of sodium fluoride to drinking water to a level of 0.7 to 1.2 parts per million (ppm), in order to enable formation and renewal of dental enamel and reduce dental decay in a population.

Food Balance Sheet: A table presenting an overall picture of the pattern of a country's food supply showing types and quantities of food produced, imported, exported, and used, and per capita supplies available for human consumption, in terms of energy and specific nutrients.

Food Security: The situation in which all people, at all times, have access to sufficient, safe, and nutritious food to meet their dietary needs and food preferences to maintain an active healthy lifestyle.

Food Consumption: Food consumption is the actual amount of food consumed in the population surveyed, in contrast to national food balances which relate to supplies of food.

Formerly Socialist Economies (FSE): The republics of the former Soviet Union and the formerly socialist countries of Eastern and Central Europe.

Formula, Baby: Substitutes for maternal milk, mostly based on modified cow's milk, recommended for use during the first year of life after breastfeeding is stopped.

Fortification: See *enriched foods.*

Frequency Distribution: A complete summary of the frequencies of the values or categories of a measurement made on a group of persons. The distribution tells how many or what proportion of the group had each value, or range of values, out of all possible values that the quantitative measure can have.

Friendly Societies: Mutual benefit associations, sick funds, or insurance programs to help cover the cost of medical care, prescribed drugs, and partial support for inpatient hospital care; preceded comprehensive tax-supported health care insurance systems.

Fund Holding: Funding of general practitioners (GPs) on a per capita basis, not only for their services but also for hospital and specialist care for their registered patients, as an experimental program in Britain's National Health Service since 1991.

Gatekeeper: A provider or system that regulates or controls access to a health care service. The role of the general practitioner or other primary care provider to control or regulate the referral of patients to specialist or hospital services, as in the United Kingdom's NHS, or in U.S. managed care plans.

General Hospital: A licensed facility with at least six beds providing diagnostic and therapeutic services for medical conditions staffed by at least one doctor and continuous nursing services. Provides short-term diagnostic and treatment services for patients, with a variety of medical departments, including surgical and nonsurgical.

General Practitioner (GP): A primary care physician trained to provide continuing care for the family or individual. Sometimes used as synonym for *family physician.*

Genocide: Incitement, threats, or acts to destroy, in whole or in part, a national, ethnic, racial, or religious group and calculated to bring about its physical destruction.

Germ Theory: The theory that specific microorganisms cause characteristic infectious disease, confirmed after the introduction of the microscope. This is in contrast to the miasma theory, which attributed disease to influences spread in the air as a result of decaying organic matter.

GINI Coefficient Index: Used mainly in health economics, it is a measure of the degree of inequality and dispersion of income or wealth distribution in a society. A larger GINI coefficient means a larger spread between social class groups.

Global Burden of Disease (GBD): An indicator of the loss of healthy life years in a specified population from disease measured in disability-adjusted life years. Developed by the World Bank and World Health Organization, it is used to measure and compare trends over time, and between countries, and as an outcome measure of effectiveness of specific interventions.

Global Fund: The Global Fund was created to increase resources to fight AIDS, tuberculosis, and malaria, and to direct those resources to areas of greatest need. It is a partnership among governments, civil society, the private sector, and affected communities, and an innovative approach to international health financing.

Global Funding: A method of providing money by a funding agency for health services that is based on a block of funds, instead of a specific amount of money being paid for each program or other unit of service, such as clinic visits, hospital admissions, days of care, or procedures.

Gold Standard: A term for recognized "best practices" or recommended guidelines considered to be desirable and the best available (e.g., GMP or a recommended immunization program).

Good Manufacturing Practice (GMP): Standards of manufacturing of pharmaceuticals, vaccines, biologicals, medical devices, or food products which are mandated by government to ensure high levels of product safety.

Gross Domestic Product (GDP): A measure in U.S. dollars of the amount of goods and services produced by a national economy over a specified period, generally a year, calculated by summing the input of all goods and services that reach the market, are consumed, or are invested; income from investments and possessions or work performed outside the nation are not included.

Gross National Product (GNP): The GDP plus income that accrues from investment and services outside the nation minus income earned within the nation that is sent abroad expressed in current U.S. dollars.

Gross National Product per Capita (GNP per Capita): A measure of the value of the GNP in the country divided by the number of the population, expressed in U.S. dollars.

Hawthorne Effect: The effect (usually positive or beneficial) of being under study upon the persons being studied due to knowing that they are under observation. Knowledge of the study often influences behavior that those being observed believe the observer hopes or expects to see.

Health: A state of complete mental, physical, social, and emotional well-being, not merely the absence of disease (WHO, 1946). It is a sustainable state of equilibrium or harmony between humans and their physical, biological, and social environments that enables them to co-exist indefinitely, with adaptations as needed, and a resource for everyday living.

Health Education: Processes by which individuals and groups of people learn to behave in a manner conducive to the promotion, maintenance, or restoration of health. The aim is to provide information for an understanding of body function, health principles and methods to reduce risk factors, and promote healthy lifestyles.

Health Expenditures: In national accounting, it is money spent for prevention and care of sickness by governments at all levels, nongovernmental organizations, and private expenditures, comprising a substantial amount of gross national expenditure. Includes spending for health care by the type of service delivered (hospital care, physician services, nursing home care); includes sources of funding for those services (public or private health insurance, Medicare, out-of-pocket spending); recurrent expenditures or current operating costs include items that recur year after year, including staff, food, "hotel costs," equipment, and maintenance of buildings; capital expenditures include land, building, and equipment to establish or extend health facilities.

Health Field Theory (Lalonde Report): A report entitled *New Perspectives on the Health of Canadians* developed the concept that heredity, environment, lifestyle, and health care organization (medical care services) are the four principal determinants of health.

Health for All: The social objective of health policy, originally stated in the 1978 Alma-Ata Declaration, as a goal to be achieved by 2000 or as an aspiration that might be realized, by implementing access to essential primary care and the reduction of the burden of disease for all the people of the world, a country, or a region.

Health Insurance: A system of financial protection of a defined population against the costs of care for illness and injury to cover the cost of health services based on regular payments, whether privately, through place of employment, or through payroll deduction or taxation via governmental mechanisms.

Health Maintenance Organization (HMO): A prepaid health care plan that assumes or shares both the financial risks and the delivery risks associated with providing comprehensive health maintenance and treatment services through its own clinics or affiliated services to a voluntarily enrolled population in a particular geographic area in return for a fixed prepaid fee. The HMO may be a nonprofit or a for-profit organization.

Health Promotion: The process of enabling people to increase control over and improve their health. It involves the population as a whole in the context of their everyday lives, rather than focusing on people at risk for specific diseases, and is directed toward action on the determinants or causes of health. It is based on public policies such as provision of bicycle pathways and legislation for smoke-free zones.

Health Targets: Governmental or other organizational sets of measurable health status indicators that a program seeks to achieve by a specified target date.

Health Technology Assessment: Measuring the cost-effectiveness and feasibility of application of existing and new technology methods of providing preventive and curative services.

Health Technology Transfer: The process of ensuring the wide application of scientific discoveries, including best practices and guidelines, methods, procedures, techniques, and equipment that may promote health and socioeconomic development; the process of developing practical applications of new scientific discoveries.

Healthy Communities: An initiative to enhance quality of life by making communities more conducive to healthy living, providing resources and facilities for recreation, easy access to settings for exercise, sport, and physical activity such as walking, swimming, and cycling, and designing dwellings that are amenable to good living with much emphasis on sustainable development.

Healthy Worker Effect: The effect of prior selection of workers with lower overall illness, disability, or death rates than the general population, so that comparison with the general population may be inappropriate.

Herd Immunity: The resistance of a group to invasion and spread of an infectious agent, based on the resistance to infection of a high proportion of individual members of the group. The resistance is because a high proportion of individual members of the group have actively or passively acquired immunity to that disease.

Hierarchy of Needs: An ascending hierarchy of needs, beginning with the basic need for self-preservation in terms of safety, shelter, warmth, water, food, clothing, and medical care before addressing less essential needs of acceptance and self-fulfillment. This has been very influential in promoting attention to worker satisfaction in job performance and in promoting personal behavior contributing to good health.

Home Care Services: Services which are brought into the home for those who need acute, rehabilitative, palliative, or supportive services, including nursing care, meal preparation, personal care, and housekeeping that can be managed effectively in the home as an alternative to or prolongation of care in a hospital.

Hospital: A licensed establishment with at least six beds which offers in-patient accommodation and outpatient care with active medical and nursing care. It provides diagnostic and therapeutic patient services for medical conditions, with an organized physician staff and continual nursing services by registered nurses. Hospitals are classified by duration of stay (short-term or long-term), services provided (general or multiple service department facility, single specialty), ownership (private, voluntary, municipal, religious, or state), and purpose (for-profit or not-for-profit). Also see *general hospital.*

Hospital Bed: A bed, crib, and pediatric bassinet set up, maintained, and staffed for the accommodation and full-time care of inpatients, where continuous medical care is provided.

Hospital Bed Ratio (Acute Care): Hospital beds per 1000 population.

Hospital Day: A hospital inpatient day is the number of adult and pediatric days of care rendered during a 24-hour reporting period, and excludes days of care for newborns.

Hospital Utilization: Use of hospital bed services are measured by admissions or separations per 1000 population and days of care per 1000 population. Average lengths of stay (ALOS) are calculated by total days divided by the number of discharges. Occupancy rates are the average percentage of rated beds being filled by patients over a given period of time.

Host: A person, population group, or other living animal, including birds and arthropods, that harbor(s) a disease agent or an infectious agent under natural conditions. A *transport host* is a carrier in which the organism remains alive but does not develop.

Human Capital: In addition to physical and financial resources, a service (or production) industry depends on trained human resources to manage, operate, and provide the services. Investment in training enhances staff motivation, skills, and satisfaction, as well as productivity.

Human Development Index (HDI): It is a composite index based on measures of life expectancy at birth, educational attainment, and adjusted real GDP per capita income. It is expressed as a value between 0 and 1. HDI was developed by the United Nations Development Programme to give a holistic view of a country's development status, compared to per capita income to rank countries.

Immunity: Resistance acquired by a host as a result of previous exposure to a natural pathogen of foreign substance for the host, developed in response to an antigen; usually characterized by the presence of antibody produced by the host, naturally or as a preventive measure.

Immunization: Protection of susceptible individuals from communicable diseases by administration of a living modified agent (e.g., measles), a suspension of killed organisms (e.g., pertussis), or an inactivated toxin (e.g., tetanus toxoid). Temporary passive immunity can be produced by transfer of maternal antibodies via the placenta or breast milk or by administration of immune globulin in some conditions, but is not long-lasting.

Incidence Rate: The rate at which new health events occur in a population. The numerator is the number of events that occur during a defined period, the denominator is the population at risk of experiencing the event during the same period. Also see *prevalence.*

Incubation Period: The time interval between initial contact with an infectious, toxic, carcinogenic, or teratogenic agent and the appearance of the first sign or symptom of the disease in question. For infectious diseases, each infection has a characteristic incubation period which may range from several hours (e.g., *E. coli*) to some years (e.g., HIV).

Index Case: The initial case in an outbreak or epidemic whose identification is crucial for investigation and control measures.

Indicator: An attribute that can be used to measure and/or record an event, process, or phenomenon; helps to measure changes in health, directly or indirectly, and helps to assess the extent to which the objectives and targets of a program have been met.

Inequality in Health: Differences in access to public health and medical services; in levels of health indicators such as mortality rates and disease incidence rates associated with and often correlated with regional inequalities of socioeconomic levels. All of these may be related to the effects of income, unemployment, housing, educational, environmental, and occupational factors, gender, ethnicity, and behavioral or lifestyle factors such as smoking.

Inequity: An imbalance in resource allocation or access to services between one population group and another, whether the difference is based on location of residence, ethnicity, or socioeconomic group as opposed to others in the population.

Infant Mortality Rate (IMR): A measure of the yearly rate of deaths in children younger than 1 year old. The denominator is the number of live births in the same year. More specifically, this is the probability of dying between birth and exactly 1 year of age.

Infectious Disease: A disease in humans or animals caused by a specific pathogenic agent or its toxic products which may or may not be clinically manifested.

Inpatient Care: Health services that require a sick, injured, or convalescent patient to stay overnight in a health care facility.

International Classification of Disease (ICD): A classification of specific conditions and groups of conditions according to their mode of causation, or body system affected with numbers assigned to each as determined by an internationally representative group of experts who advise the WHO. The complete list is published in periodic revisions. The tenth revision of the manual (ICD-10) was published in 1990. ICD-11 is planned for 2011. The ICD is important to standardize vital statistics, disease registries, and hospital morbidity data for time trends, regional, national, and international comparisons.

Intervention: A specific activity by health professionals to reduce disease risks, treat illness, ameliorate the consequences of disease and disability, or prevent a health problem.

Intersectoral Collaboration: A term used to describe activities involving several components of the body politic that, working together, can enhance health conditions and community health more effectively than when working independently of one another.

Iodine-Deficiency Disorders (IDD): All the effects on human growth and development of disorders or conditions due to inadequate dietary intake of iodine that are the commonest causes of preventable brain damage and mental retardation.

Knowledge, Attitudes, Beliefs, and Practices (KABP): Levels of awareness of information on a topic in health are important for how people feel about it and the beliefs and behaviors or practices associated with it. Attempts to change health status need to take these factors into account if there is to be a chance to succeed.

Latency Period: The delay between exposure to a disease-causing agent and the appearance of clinical signs and symptoms of the disease, mainly in relation to environmental and occupational agents (e.g., asbestos exposure and mesothelioma). In infectious disease, this is the time between exposure of the host to the organism and onset of the disease (e.g., several days for influenza, years for HIV).

Levels of Care: This refers to *primary, secondary,* and *tertiary* health care which are defined, respectively, as ambulatory care in the community, hospital care at the basic level of service, and specialized referral care in a teaching hospital or referral center.

Life Expectancy at Birth: The average number of years a newborn baby can be expected to live if current mortality trends continue. It serves as an indicator of population health status, usually partly dependent on mortality in the first year of life and therefore lower in less wealthy countries because of higher infant and child mortality rates.

Life Expectancy, Age-Specific: The average number of additional years a person at a particular age will live if current mortality rates continue to apply, based on the age-specific death rates for a given year, e.g., at birth, at age 45, at age 65.

Live Birth: The complete expulsion or extraction from its mother of a product of conception, irrespective of the duration of the pregnancy, which after separation breathes independently or shows any other evidence of life such as a heartbeat, umbilical cord pulsation, or definite movement of voluntary muscles, whether or not the umbilical cord has been cut or the placenta is attached.

Local Health Authority: The municipal or county government traditionally responsible for sanitation, including safe drinking water; sewage and garbage collection and disposal; town planning; zoning; control of nuisances, animals, and pests; basic public health services; inspection of businesses, factories, and food products.

Macronutrient: Food components that the body needs in large quantities, including proteins (amino acids), carbohydrates (starch, sugars), and fats.

Malnutrition: A pathological state resulting from relative or absolute deficiency or excess in the diet of one or more essential nutrients (e.g., protein, mineral, or vitamin), as clinically manifest conditions or those detectable only by nutritional histories and surveys, or biochemical or physiological tests.

Malpractice: A form of behavior or negligence of a health professional that deviates far enough from currently accepted standards to constitute misconduct because it is unethical, dishonest, unprofessional, or incompetent, or personal behavior that exploits patient vulnerability for financial or sexual favors.

Managed Care Organizations: Health insurance plans which provide or contract for a comprehensive range of health services to an enrolled population in competition with alternative plans, such as a fee-for-service plan; they are common both as for-profit and not-for-profit organizations in the United States to provide effective care at reduced cost to employers and employees. Includes health maintenance organizations (HMOs) and other models of health care organization.

Management by Objective (MBO): Results-based management setting specified objectives, targets, or aims and adapting by assessing the extent to which the targets are achieved. Measurement or quantifying work output is implicit in the procedures.

Maternal and Child Health (MCH): Health services related to women and fertility, including pregnancy and childbirth, as well as care of infants and children through adolescence with an emphasis on promoting and protecting health.

Maternal Mortality Ratio (MMR): The rate is the annual number of maternal deaths per 100,000 live births. A maternal death is the death of a woman while pregnant or within 42 days of termination of pregnancy from pregnancy-related causes (excluding accident or incidental causes). Where reliable data are lacking this is calculated based on regional patterns every 5 years by WHO/UNICEF/UNFPA.

Maternity Hospital or Home: Separate "lying-in hospitals" for delivery, as opposed to maternity units in general hospitals.

Measles, Mumps, Rubella (MMR) Vaccine: A combined vaccine of live attenuated measles, mumps, and rubella given at 12–15 months of age and with a second dose recommended at age 6 years. It is heat-labile, preserving the cold chain is essential when given in the field in developing countries.

Median: A measure of central tendency; the division of a set of measurements arranged in sequence into two parts, the lower and upper half. The point on the scale that divides the group is called the *median.*

Medical Audit: A health service evaluation procedure in which selected data from patients' charts are summarized in tables displaying such data as average length of stay or duration of an episode of care, the frequency of diagnostic and therapeutic procedures, and outcomes of care arranged by diagnostic category. These are part of quality improvement internal peer review, with analysis and comparison with norms (e.g., caesarian rates).

Medical Technology: Applied medical science and technical developments and not simply the machines used in medicine. The term is used in reference to low-tech, inexpensive interventions (e.g., ORT), as well as high-tech or costly complex procedures or equipment (e.g., MRI).

Medicare: The national health insurance plan for the elderly, disabled, and end-stage renal disease patients, introduced in 1965 in the United States under the Social Security Act of 1935. Medicare is administered directly by the federal government Health Care Financing Administration (HCFA). This term is also used for Canada's national health insurance program.

Medicaid: The state-operated, national health insurance plan for the poor established in 1965 under the U.S. Social Security Act of 1935; it is administered and cost-shared, approximately equally, by federal, state, and local governments.

Mental Disorder: A heterogeneous group of disorders ranging from exaggerated response to stressful events to altered mental state from specific neurologic or genetic abnormalities. Serious mental illnesses include major depression, schizophrenia, bipolar disorder, obsessive-compulsive disorder (OCD), panic disorder, post-traumatic stress disorder (PTSD), and borderline personality disorder, as listed in the ICD-10 (WHO) and the *Diagnostic and Statistical Manual of Mental Disorders* (DSM-IV) of the American Psychiatric Association (APA).

Mental Health: The emotional and social well-being of the individual and his or her psychological resources and functional status. The branch of health care and public health concerned with the prevention and management of dysfunction associated with mental illness.

Mental Disability: Reduced mental, emotional, and/or intellectual capacity renders individuals unable to function fully or normally in society as a result of many possible causes.

Meta-Analysis: A systematic synthesis of the data from separate but similar (comparable) studies of a causal relationship or a therapeutic or preventive regimen, leading to a quantitative summary of the pooled results (often from a set of randomized controlled trials). It applies predetermined criteria of quality and a quantitative component (integration of the numerical information).

Miasma Theory: An eighteenth-century explanation of the origin of epidemics attributing epidemics to ill-defined emanations from rotting matter affecting persons breathing that air and transmitting contagious disease to susceptible persons. Although later displaced by the germ theory of disease transmission, it led to many sanitary reforms that resulted in enormous progress in public health, and has a place in modern public health theory relating social and environmental conditions to health inequalities.

Micronutrient: Essential nutrients (vitamins or minerals) that the body needs in small quantities as part of daily dietary intake.

Micronutrient Deficiency States: Subclinical or clinical deficiency states due to inadequate regular intake of essential minerals and vitamins.

Millennium Development Goals (MDGs): A blueprint established by the United Nations in 2001 setting targets for progress in eight areas: poverty and hunger, primary education, women's equality, child mortality, maternal health, disease, environment, and a global partnership for development with a call for substantial improvement by 2015. These goals are monitored by the UNDP.

Mission Statement: A succinct statement of the purpose for which the organization exists; associated with values, objectives, targets.

Morbidity: Sickness or any departure, subjective or objective, from a state of physiological or psychological well-being.

***Morbidity and Mortality Weekly Report* (MMWR):** A weekly publication of the Centers for Disease Control and Prevention (CDC) in Atlanta, in which epidemiologic data are reported, along with special reports on reportable infectious diseases and noncommunicable diseases of epidemiologic interest. Many other countries have similar reports as a feature of their public health systems.

Mortality: Death — the cessation of life. Despite problems of reporting and diagnostic criteria, mortality statistics are the most powerful of all vital statistical data.

Nanotechnology: The interactions of cellular and molecular components and engineered materials — clusters of atoms, molecules, and molecular fragments at the most elemental level of biology.

National Health and Nutrition Examination Survey (NHANES): First conducted by the National Center for Health Statistics (NCHS) in 1971, it monitors indicators of health and nutrition status of the U.S. population through questionnaires, dietary intake data, biochemical tests (e.g., lead), physical examination (e.g., blood pressure, obesity). NHANES findings are available on CD-ROM and in reports and articles published in peer-reviewed journals.

National Health Expenditures: The total amount of money spent on all health services, supplies, and health-related research and construction during a calendar year, compiled from many sources. This includes estimates of out-of-pocket payments, private health insurance, and government programs, and is broken down by type of service (e.g., hospital care, nursing home care, physician services, dental care, home care, drugs and costs of local, regional, and national public health services, prisons, and the military).

National Health Insurance (NHI): A system for public funding of the costs citizens incur when they require personal medical and hospital care, administered or regulated by the state, with universal coverage, and comprehensive services.

National Health Service (NHS): The health care system of the United Kingdom is a tax-supported national program of comprehensive service free to all at the time of service, including hospital, ambulatory care, and public health. The U.K. NHS model has been adopted in many countries.

National Plan of Action: When a national health strategy has been defined, a national plan of action should be drawn up, including the steps and programs needed to meet objectives and specified targets to be attained.

Natural History of Disease: Both infectious and noninfectious diseases have well-defined stages that characterize their natural history. The course of a disease from onset to resolution may be changed by clinical and public health interventions or other factors such as improved standards of living and nutrition.

Needs: In health economics, needs are the minimum amount of resources required to exhaust an individual or a specified population's capacity to benefit from an intervention. In other contexts need is all things considered necessary or essential but often vaguely defined. Needs may be perceived or professionally defined.

Needs Assessment: A formal systematic procedure for determining the nature and extent of problems experienced by a specified population, especially those at high risk, that affect their health directly or indirectly so that resources can be deployed to manage them efficiently. It may include *demands* for health care, and needs assessments of expert opinion (*normative*), by the individual (*felt*), in comparison with other individuals or groups (*comparative*), and acted upon (*expressed*).

Neonatal Mortality Rate: The number of deaths in infants under 28 days of age per 1000 live births in the same period, usually a year.

Nongovernmental Organization (NGO): An organization in the health field operated by a private non profit organization, separate and independent from government, but may be partially supported by government funds.

Norm: Norms are used to set minimum or maximum levels of facilities for planning or evaluating health care systems. However, norms change with changing technology and epidemiology of health needs (e.g., hospital bed supply and average length of hospital stay, manpower, performance, risk factors or health behavior).

Normal (Gaussian) Distribution or Bell Curve: The continuous frequency distribution of infinite range with the frequency of each value shown on the vertical axis. A normal distribution is a continuous, symmetrical pattern of graphed data observations appearing as a bell-shaped curve with both tails extending to infinity. The arithmetic mean and mode are identical. The shape of the bell is determined by the mean and standard deviation.

Nosocomial Infection: An infection acquired by a patient or staff in a hospital or other health care setting.

Notifiable (Reportable) Diseases/Conditions: Specified infectious diseases due to their public health importance by law must be reported by health care providers to public health officials in the local jurisdiction. Other reportable categories include birth defects; certain industrial, occupational, and environmental diseases; suspected child abuse; patients no longer fit to drive a car; and cancer.

Null Hypothesis: A hypothesis that two or more populations do not differ from one another statistically, and that any observed differences are due to chance alone. The hypothesis is then tested statistically to show the degree to which the observed differences may be greater than could be explained by chance factors. If differences are shown, then inferences can be made regarding a possible causal relationship.

Objectives: The precisely stated ends to which efforts are directed, specifying the population outcomes and variable(s) to be measured.

Observer, Variation: Variation or error due to a failure of the observer to measure or identify a phenomenon accurately. All observations are subject to variation, including between the same observer (*intra-observer*) and between different observers (*inter-observer*).

Occupational Health: The specialized practice of medicine, public health, epidemiology, and other health professions in an occupational setting. It includes promotion of health as well as prevention of occupational diseases and injuries and treatment of work-related injury or illness.

Occupational Safety: Prevention of injury and poisonings from exposure at the workplace and prohibition of the practice of child labor.

Occupational Safety and Health Administration (OSHA): The U.S. federal government agency charged with responsibility for setting standards and enforcing regulations to safeguard the safety and health of workers in the country.

Operating Costs: The costs of operating an enterprise or service, calculated annually, which vary with the volume of the service, in contrast to fixed overhead costs (*recurrent costs*), which are total expenditures relating to non-salary operating items. Capital costs are buildings and equipment whose value extend over a long period of time.

Operations Research: The study by observation and experiment of the workings of a system, such as a health service, with a view to finding ways by which it can be improved with aims of enhancing its quality, efficiency, and reducing costs or shifting resources to more effective services (e.g., ambulatory surgery in place of inpatient surgery).

Oral Rehydration Salts/Therapy (ORS/T): A combination of salts and sugar used to replace debilitating and life-threatening electrolyte loss from diarrheal diseases. The ORT salt solution includes sodium chloride, potassium bicarbonate, and glucose. It includes use of the supplement plus continued feeding.

Organization of Economic Cooperation and Development (OECD): An organization based in Paris which includes many developed or industrialized countries. Its mission is to promote trade and socioeconomic cooperation among its members and internationally; in less developed nations to conduct studies of, and seek solutions for, social and economic problems.

Outbreak: An epidemic limited to a localized cluster increase in the incidence of a disease with more cases than usual for the area, but not at the level of a general epidemic, depending on the disease and the level of control achieved. This does not imply lesser importance epidemiologically, as an outbreak of polio, measles, or Ebola virus can be of very great public health significance.

Outcome Measures: Indicators of all morbidity, mortality, disability, and physiological indicators of health risk factors that are the target of prevention or treatment programs.

Out-of-Pocket Payments: Money spent by the consumer on health care for services not covered under his or her insurance plan or a national health service; or payments, legal or illegal, to health care providers for services rendered to which he or she is entitled.

Outpatient Services: Care and treatment services including all clinic visits, referred visits, observation services, outpatient surgeries, and emergency department visits in a facility for patients who are not admitted overnight and do not occupy a bed.

Output: The professional or institutional health care activities expressed as units of service, such as patient days, outpatient visits, laboratory tests performed, or infants immunized.

Overnutrition: A state of poor nutrition due to caloric intake in excess of body requirements for energy, growth, and maintenance, usually measured by body mass index exceeding 25.

Palliative Care: Palliative care is comprehensive, interdisciplinary care of patients and families facing a terminal illness, focusing primarily on comfort and support. It focuses on quality of life, which includes good symptom control irrespective of setting.

Pan American Health Organization (PAHO): The autonomous regional office of the World Health Organization for the Americas, located in Washington, DC, that deals with the health affairs and disease problems of the Americas and circumpolar problems in the High Arctic.

Parasite: An animal or vegetable organism that lives on or in another organism whence it derives its nourishment. Some, called *obligate* parasites, cannot lead an independent life, while others, *facultative* parasites, can lead either parasitic or independent existence.

Particulate Matter: Particles of solid or liquid matter in the air, including non-toxic materials (soot, dust, dirt), heavy metals (e.g., lead), and toxic materials (e.g., asbestos, suspended sulfates, nitrates).

Pasteurization: The process of heat treating of milk, dairy products, or other perishable foodstuffs to a specific temperature for a given period of time required to kill potentially harmful microorganisms and pathogens that cause spoilage and bad odors and tastes.

Pathogen(ic): An organism, toxin, or other agent capable of causing human, animal, or plant disease.

Pathogenicity: The property of an agent to cause overt disease in an affected host (person or population); for instance, the power of an organism to produce disease. It is measured by the ratio of the number of persons developing clinical illness to the number exposed to the infection. The organism may be an infectious, toxic, carcinogenic, or teratogenic agent or a nutrient deficiency.

Payer: An organization responsible for payment of health care costs.

Payroll Tax: A deduction from the individual worker's pay or of the firm's overall payroll established by the national government for a designated program, such as for health service or health insurance or other social programs such as pensions and unemployment insurance.

Peer Review: The process of review of research proposals, manuscripts submitted for publication, abstracts submitted for presentation at scientific meetings, and judged for scientific and technical merit by other scientists in the same field. It also refers to review of clinical performance when it is a form of medical audit for quality assurance and promotion involving inspection of facilities, qualifications, records or performance of health facilities, departments, and individual health providers, performed by professional colleagues as a regular method of operation of the department or unit to improve quality of care and service.

Per Diem: A method of reimbursement for hospital expenditures, for the days of care received by a hospitalized patient, usually based on average daily costs and not reflective of actual costs of treating an individual case.

Perinatal Mortality Rate: The number of late fetal deaths in utero, with gestation of 28 weeks or more, plus infant deaths within 7 days of birth per 1000 live births and stillbirths; the perinatal mortality ratio is calculated per 1000 live births in developing countries.

Physician: A medical graduate from a recognized and accredited medical school, who is licensed by the state to practice medicine and who is subject to professional standards and discipline by a government-appointed or designated body which supervises medical professional licensure.

Physician Contact: A medical visit or consultation in person or by telephone to a physician for the purpose of examination, diagnosis, treatment, or advice.

Physician-to-Population Ratio: The supply of physicians as measured in relation to the population for comparison purposes; often expressed as population per physician ratio (e.g., 400 persons per physician).

Planning: A continuous process which includes identifying problems and establishing priorities, defining objectives and strategies, monitoring progress or failure to achieve the goals selected, and developing alternative courses of action and resource allocation.

Point Source: The point source of a disease is a single identifiable person, place, and time that is the origin; exposure is brief and essentially simultaneous; the resultant cases all develop within one incubation period of the disease.

Poison Control Center: A national or regional reference center for reporting cases of acute, subacute, and chronic poisoning staffed and available to provide detailed information and advice about emergency treatment, diagnosis, and management of poisonings and acute toxicities of all kinds.

Policy Analysis: A process of analysis and revision of policy based on changing population health needs and technology in order to maintain effective and efficient use of resources.

Pollutant: Any undesirable solid, liquid, or gaseous matter that contaminates people, animals, soil, food, water, clothing, or the atmosphere with a toxic or noxious substance.

Population-at-Risk: A defined population or group with higher than average risk for a specific disease or for all diseases, or the population base for whom a specific service program is developed.

Population-Based Health Services: Services directed toward all members of a community such as in a district health program, or for a specific population subgroup, such as infant and child immunization.

Population Pyramid: A graphic representation of the age and sex composition of a population; the shape varies with the age composition of the population.

Post-Neonatal Mortality Rate: The number of deaths that occur from 28 days to 365 days after birth in a given year per 1000 live births in that year.

Preferred Provider Organization (PPO): A prepaid health insurance plan that designates the physicians, other health professionals, clinics, and hospitals for which it will reimburse charges for services rendered for enrolled members.

Premiums: Payments by the beneficiary or on his behalf for health insurance.

Prepaid Group Practice (PGP): Group medical practice coupled with prepayment with subscriber selection based on voluntary choice; precursor of health maintenance organizations.

Prevalence (Rate): The total number of cases of a disease or health-related events (e.g., motor vehicle accidents) or attributes (e.g., smokers), including new and previous cases, in a given population at a specified point in time (point prevalence), or time period (prevalence rate expressed usually as a rate per 1000 persons during the calendar year.

Prevention: Policies and actions aimed at eradicating, eliminating, or minimizing the impact of disease and disability, or retarding the progress of disease and disability. Primary prevention aims to prevent a disease from occurring at all; secondary prevention aims to delay onset of clinical disease, shorten its duration, and reduce its complications; tertiary prevention aims to promote and assist return to and sustain maximum functional recovery.

Primary Health Care (PHC): As defined by WHO in the Declaration of Alma-Ata, PHC is essential health care made accessible at a cost the country and the community can afford, with methods that are practical, scientifically sound, and socially acceptable. It includes a wide range of health education, preventive care, and treatment of common diseases, proper nutrition, safe water and environmental health, maternal and child health, control of communicable diseases and immunization, and provision of essential drugs. Primary medical care includes first contact care with a family doctor or health care provider.

Private Health Insurance: Health insurance operated by nongovernmental organizations for individual and group subscribers, on a nonprofit or for-profit basis, subject to regulation under state insurance and health law.

Private Practice: A physician or other health care provider working independently or as part of a partnership providing services to patients who agree to pay for the service or on contract to a sick fund or insurance agency to pay for the services provided to a beneficiary.

Private Voluntary Organization (PVO): Nongovernmental organizations with the legal status of a not-for-profit corporation entitled to engage in providing health care, hold assets and properties, and hire staff or contract for services.

Procedure: A surgical or nonsurgical treatment ordered by the attending physician and recorded in the medical record of the patient that may be done in a hospital as an inpatient or that may done safely on an outpatient basis or in other approved settings. Procedures may include medical (e.g., chemotherapy), surgical, or other special procedure (e.g., radiation therapy for cancer).

Professional Standards Review Organizations (PSRO): Peer review groups to monitor services to assure and improve standards of medical practice in medical and hospital practice and ensure services are provided only when medically necessary and in the most economical and medically appropriate settings to restrain cost increases.

Prospective Payment System: Paying for hospital services by diagnosis-related groups (DRGs) is now the standard practice in the United States for Medicare and Medicaid patients following the 1982 and 1983 amendments to the Social Security Act and state laws. DRGs have largely replaced payment for hospital care by length of stay.

Provider: General term for health care facilities and licensed professionals who give care to patients.

Public Administration: Administration of a health program by a governmental agency.

Public Sector: The part of the economy of a country comes within the scope of central, regional, and local government or public corporations, administered and publicly accountable.

Purchasing Power Parities (PPP): These are currency conversion rates to convert to a common currency and equalize the purchasing power of different currencies; this eliminates differences in price levels between countries.

Qualitative Research: Any type of research that includes non-numeric information to explore individual or group characteristics, to augment quantitative data. Examples include clinical or quality of care studies, narrative studies of behavior, ethnography, and organizational or social studies.

Quality-Adjusted Life Years (QALYs): An adjustment of life expectancy that reduces the overall life expectancy by amounts that reflect the existence of chronic conditions causing impairment and/or disability as assessed from health survey data, hospital discharge data, etc. It is a unit of health gain attributable to an intervention typically used in cost-utility analysis.

Quality Promotion: A system of procedures, checks, audits, and corrective actions to ensure that all preventive and therapeutic practices, research, and other technical and reporting activities are conducted in conformity with best practices and standard operating procedure and are aimed to promote the highest achievable quality and level of outcomes in health care.

Quarantine: Legal isolation of a person or animal who is a known contact of a case of contagious disease during the period of its communicability to prevent onward transmission of the disease.

Rapid Assessment Procedure: Methods that can be used to yield results rapidly, usually inexpensively, and as efficiently as possible within available resources to assess health problems and evaluate health programs or to delineate the health impact of a public health emergency such as a disaster or epidemic with unusual features.

Rate: A measure of the frequency with which a specified event occurs in a defined population in a specified time period. This can be presented as a ratio (fraction). It is essential for comparison of experience between populations at different times, different places, or among different classes of persons.

Recurrent Costs: Costs that recur in running an enterprise, such as salaries and supplies, also known as *operating costs*.

Reform: The process of change in health systems organization, financing, and allocation of administrative responsibility and financial allocations.

Regional Hospitals: Hospitals located in urban regional centers which provide primary hospital care to residents of part or all of the city, and secondary care to people from the city and the surrounding communities and health service region.

Regulation: Legally binding rules and procedures for governmental supervision and licensing of food, drugs, cosmetics, medical devices, health facilities, insurance systems, businesses, restaurants, recreation facilities, or anything that might endanger the public interest in health or safety.

Rehabilitation: Programs and services to assist persons with disabilities to achieve levels of function that restore them to the fullest degree of physical, mental, social, vocational, and economic potential of which an individual is capable following disabling illness or injury.

Reimbursement: Methods of paying for doctor or other health services, including salary, fee-for-service, capitation or a mix of them.

Relative Risk (Risk or Odds Ratio): The ratio of risk of disease or death in a population exposed to a specific risk as compared to the nonexposed population (i.e., with the risk ratio).

Replacement Rate: The birth and total fertility rate at which births exceed deaths in a defined population group.

Reservoir of Infection: Any person, animal, plant, soil, substance, or combination of these in which an infectious agent normally lives and multiplies, on which it depends primarily for survival, and where it reproduces itself so that it can be transmitted to a susceptible host. It is the natural habitat of the infectious agent.

Resources: The inputs used to produce and distribute goods and services; these are classified as land and natural resources, labor (people), and capital (goods made to produce other goods); in the health sector, this includes patients' time.

Respite Care: Provides temporary care in a facility for a limited time of chronically ill or disabled persons who normally reside at home and are dependent on family members or friends for continuous or intermittent care in order to give family caregivers a brief rest or respite.

Restructuring: Reorganization of a health service which changes the quantitative balance in resources personnel and program emphasis within an organization or from one sector to others (e.g., from hospital inpatient care to ambulatory and home care).

Risk Approach: The planning of services and interventions based on the highest risk groups in the population, described as "something for all, but more for those in need — in proportion to that need" (Chapter 2).

Risk Factor: An aspect of personal behavior or lifestyle, environmental exposure, or an inborn or inherited characteristic that has been identified to be associated with health-related conditions, particularly increased likelihood of unfavorable health outcomes.

Risk-Taking Behavior: The phenomenon of people engaging in unhealthy or potentially injurious behaviors, despite knowledge of the risks involved or from ignorance. Health workers must avoid blaming people who participate in risk-taking behavior when it has adverse outcomes.

Sample: A selected subset of a population that may be random or nonrandom, representative or nonrepresentative, with the following major categories:

Cluster sample The unit selected is a group of persons rather than an individual (i.e., all persons in a city block);

Grab sample A sample selected by easily employed methods; for example, a simple survey among people who happen by or show up at a service offered, such as a street fair to examine blood pressure, from which no general conclusions can be drawn;

Probability (random) sample All individuals have an equal or known chance of being selected; if a stratified sampling method is used, the rate at which individuals from several subsets are sampled can be varied to produce greater representation from some classes than others;

Simple random sample Each person has an equal chance of being selected out of the entire population with each person assigned a number (e.g., according to a random numbers table), until the needed sample size is achieved;

Stratified random sample The population is divided into distinct subgroups according to some important characteristic (e.g., age or socioeconomic status) and a random sample is selected out of each subgroup. The proportion of the sample in each subgroup is the same as the proportion of the total population in each subgroup;

Systematic sample The sample is selected with a systematic method, such as alphabetical order or birth dates, but this can also lead to sampling bias.

Sampling: The method of selecting a number of subjects from all the subjects in a particular group or "universe."

Sanitation: A set of municipal public health policies and procedures consistent with the Public Health Act and Regulations to ensure safe drinking water, food quality, and hygienic disposal of wastes, including sewage and industrial waste disposal and treatment, garbage collection and disposal, and waste management to maximize recycling, reduction of ambient air pollution, and conservation of the environment.

Screening: The presumptive identification of unrecognized disease, disease precursor, or defect by use of tests, examinations, or other procedures applied rapidly to separate apparently healthy persons who probably have some disease process from persons who do not. *Mass screening* involves a whole population. *Prescriptive screening* aims at early case finding in presumably healthy individuals in specific risk groups (e.g., newborns for birth defects). *Multiphasic screening* involves a variety of tests carried out on the same occasion (e.g., in middle-aged women and men).

Secondary Care Facility: A hospital or other facility that offers services to a community or district that includes medicine, general surgery, obstetrics and gynecology, and pediatric services at a minimum, supported by adequate diagnostic and support facilities to provide high-quality care, and with possibilities for referral to tertiary care centers for more specialized services.

Selective Primary Care: Focusing programs of prevention or treatment of common conditions on those issues selected as being the most common and most health-damaging, with a high benefit-to-cost ratio.

Self-Care: The element of personal responsibility in health and the methods of improving one's own health outside of the professional network of services.

Sensitivity: The proportion of persons truly having a disease in a population screened who are identified as diseased by the screening test. It is a measure of the probability of correctly diagnosing a case or the probability that any given case will be identified by the test. A test with high sensitivity will be positive whenever the condition is present (i.e., 100 percent sensitive), whereas a test of low sensitivity will give false negative results in other cases.

Sentinel Center: Convenient but generally representative points of service, such as in a primary care practice or public health unit, where systematic monitoring of cases presenting, or the population served, provides an index of events of public health importance.

Sentinel Health Event: The first detected case of an unusual condition or unexplainable surge in sickness absences, or a health-related condition or event that can be used to assess the stability or change in the health level of a population, indicated by reported morbidity, mortality, or health service utilization data. It can be used to establish preventive measures.

Sexually Transmitted Infections (STIs): Any infection transmitted through sexual contact.

Sick Funds: Health organizations which insure and provide medical care to their members.

Sin Taxes: Taxes on nonessential, "health-harming" consumer goods, such as tobacco products, tobacco, alcohol, or luxury goods to deter use of these goods increasingly defined as risk factors for disease and injury for the user and for others.

Social Class Classification: Defining persons in a national population by groups according to such characteristics as prior education, occupation, and income. Analysis usually reveals a strong correlation with health-related characteristics and meaningful predictive properties; useful for epidemiologic purposes. The British Registrar-General's Classification of Occupations is used on death certificates and used to study health experiences of social classes even within a universal access health system.

Social Cost: The cost of an activity or disease to society and not merely the agency carrying out the activity.

Social Diagnosis/Policy: Assessment of factors influencing the quality of life, and promotion of changes to raise those standards.

Social Pathology: Structural and functional malfunction of society associated with pervasive problems with adverse consequences for health such as unemployment, alcohol and substance abuse, crime, and violence in a community setting. It is when a segment of the community is alienated and suffers from excess pathology of acute and chronic diseases, or antisocial acts by individuals that are destructive to others or to the individual or community property.

Social Security: Nationally legislated insurance systems with varying levels of comprehensiveness providing public entitlements to the individual for injury or illness related to work, old-age pensions, survivor's benefits, health insurance, pensions for orphans, widows, disabled persons, financed by regular payments of the employee and employer.

Social Support Systems: Family, friends, work, and neighborhood networks that provide an informal social support for most people that help an individual cope with stress and adaptation needs in life. At the community or national level, this refers to legislative and other support systems providing welfare and pensions as well as other mechanisms to prevent people from falling below minimal standards of living.

Social Welfare: This involves programs of governments at all levels often combining publicly administered and funded agencies and a range of private and voluntary services to assist persons in distress with sustenance of basic needs by financial assistance, or entitlement to food purchasing, or special entitlement to health service where these are not a universal right or entitlement.

Standardization: A set of techniques used to adjust sets of data in ways to remove as much as possible the effects of a difference in age or another confounding variable, when comparing two or more populations using either a direct method or an indirect method.

Standardized Mortality (Morbidity) Rate (SMR): The ratio of the number of events (e.g., deaths or diseases) observed in the study group or population to the number that would be expected if the study population had the same specific rates as the standard population, multiplied by 100; usually expressed as a percentage.

Standards of Practice: Practice standards are the professional level practiced by the average physician or other licensed professional in the category of service being rendered and is that which is expected of a practitioner in that field. Professional associations, health insurance, and managed care plans provide leadership in determining practice standards and education of members to reach and maintain those standards.

Staple Food: A food from which a substantial portion of total dietary energy is obtained and which is regularly consumed in a country or community.

Statistical Significance: An observation that is unlikely to have occurred by chance alone; statistical methods allow an estimate to be made of the probability of the observed degree of association between independent and dependant variables under a null hypothesis in a study group and a comparison group. Statistical tests help to indicate the degree of probability that the observed differences occurred by chance alone, or by a possibly causal relationship between the variables.

Strategic Planning: Planning is long-range consideration of predictable health-related contingencies such as changing demographics, etc., and all such factors that can legitimately be described as strategic, as compared to tactical or methodologic. Defining the broad lines of direction, actions, and resources required in an organization to define and achieve its mission and objectives.

Stress: Response to the challenge ("fright-fight-flight") of changes in status quo of an individual causing overt or hidden psychological pressure or stress which may manifest itself in overt psychological symptoms or in physical illness; prolonged or repeated exposure may cause lasting changes.

Stroke: Stroke or cerebrovascular accidents occur when the blood supply to the brain is blocked by a thrombus, blood clot or bleeding in brain tissue, causing brain damage with loss of speech, memory, or other cerebral functions, and paralysis (usually one-sided).

Stunting: Moderate and severe growth retardation in linear growth — below –2 standard deviations from median height-for-age of the standard international reference population.

Subclinical: A condition in which disease is detectable by special tests but does not reveal itself by signs or symptoms. *Preclinical* is disease with no signs or symptoms. Predisposition for disease develops gradually in terms of risk factors or biological changes which precede the onset of clinical symptoms and signs apparent on medical examination.

Substance Abuse: Nontherapeutic use of any licit (prescription drugs / alcohol) or illicit material (drugs) or chemical compound (solvents) with the intention of altering the state of consciousness or emotions, dependancy, with physical, mental, or social harm to the individual or to others, as well as if done repeatedly or habitually, leads to impaired social or occupational functioning.

Supplementation: Providing nutritional supplements of specific vitamins or minerals to prevent deficiency conditions in vulnerable population groups, and to promote optimal health status.

Supply-Induced Demand: Availability of a service increases its utilization.

Surveillance of Disease: Systematic ongoing collection, collation, and analysis of health-related data from many sources and the timely dissemination of information to those who need to know so that action (for prevention and control) can be taken.

Susceptible: A person with insufficient resistance or with associated risk factors to a particular pathogenic agent so that there is real danger of this person contracting the specific disease if or when exposed to the agent.

SWOT Analysis: A rapid review of the internal and external environment of an organization as a basis for the development of strategic planning. It assesses internal factors: strengths, weaknesses, opportunities, and threats.

Targets: Explicitly stated desired outcomes (e.g., reduction of smoking among specified population group).

Taxes: The source of funds for government and are of various kinds including progressive personal income tax, excise tax, resources tax, property tax, sin tax, social security tax, value-added tax, and others.

Terms of Reference: The definition of objectives and the task of a working group or committee established to carry out an investigation, review, or development of a task defined by an organization.

Tertiary Care Facility: A hospital or other facility that offers specialized, highly technical diagnostic and therapeutic health services that are usually only available in major medical centers associated with teaching programs and serving the population of a large region.

Total Fertility Rate: The average number of children who would be born per woman if all women lived to the end of their childbearing years and bore children according to a given set of age-specific fertility rates.

Total Quality Management (TQM): A management approach in a production or service industry to promote a sense that employees at all levels feel part of a common effort to produce the highest quality of goods or services with customer or consumer satisfaction as the measurable end point.

Tracer Conditions: Indicator conditions or health states which are easily diagnosed, reasonably frequent, whose outcomes are believed to be affected by health care, and which may be used to study the quality of care in the health care system (e.g., treatment for acute myocardial events); also called *sentinel health events*.

Traditional Birth Attendant (TBA): A person who delivers babies in the community but whose training has not been formal in a health care setting, and who learned by an apprenticeship or is self-taught by experience and observation.

Transmission: Any mechanism by which an infectious agent is spread from any source or reservoir to a person (host). *Direct transmission* is by contact as in touching, biting, sexual intercourse, or droplet transmission at close quarters such as by sneezing, coughing, or spitting. *Indirect transmission* (i.e., remote from the original source) is by vehicle-borne contact with contaminated inanimate objects such as bedding, handkerchiefs, instruments, water, food, milk, contaminated blood or blood products, or sputum. Vector-borne is via an insect in which the organism may or may not multiply. Airborne is the dissemination of microbial aerosols or particles suspended in air.

Undernutrition: A condition arising from inadequate intake of food and micronutrients. Inadequate energy intake results in reduced body weight as its principal manifestation, but in children results in *stunting* (low height for age), and *wasting* (low weight for height). Inadequate micronutrients result in a variety of deficiency states with individual and community health importance, such as iodine deficiency diseases.

Underweight: *Moderate* — below –2 standard deviations from median weight-for-age of the reference population; *Severe* — below –3 standard deviations from the median weight-for-age of the reference population.

Urban Population: Percentage of the population living in urban areas according to the national definition used in the most recent census.

User Fees: A charge for a medical or other health service at the time of service in addition to the coverage the patient has in a health insurance program, often with deterrent effect for those most in need.

Vaccine: Biological substance acting as an antigen for active immunization by introducing into the body a live, attenuated, or killed (inactivated) infectious organism or its toxin, in order to induce an immune response by the host, so that the host becomes resistant to future infection from that specific organism.

Vaccine-Preventable Disease (VPD): Disease preventable or controllable by currently available vaccines.

Values: Fundamental moral or ethical beliefs of societies, individuals, and organizations that guide their self-identification and actions (e.g., sanctity of human life, human rights, social equity).

Variables: A variable is any quantitative measure of an attribute, phenomenon, or event that can have different values. An *independent variable* is the factor suspected of the causal relationship with the dependent variable or outcome effect being studied. A *confounding variable* is a parallel factor which may cause or hinder the outcome being studied and, unless adjusted for in an epidemiologic study, may limit the potential for defining the relationship of the independent and dependent variables.

Vector: An insect or any living carrier that transports an infectious agent from an infected individual (animal or human), or its wastes, to a susceptible individual or its food or immediate surroundings. The organism so transported may undergo a development cycle during transportation.

Virulence: The degree of pathogenicity of an infectious agent; the disease-evoking power of a microorganism in a given host; may be indicated by case complication or fatality rates and/or its ability to invade and damage tissues of the host.

Vital Record/Statistics: Certificates of births, deaths, marriages, divorces, and immigration/emigration required for legal and demographic purposes compiled by public authorities at the national, regional, and sometimes local level.

Wasting: Moderate and severe — two or three standard deviations below the median for weight-for-height of the reference population (see *undernutrition*).

Waterborne Disease: Communicable diseases and diseases from chemical toxins or microbiologic agents transmitted by contamination of community or noncommunity water supplies.

Wellness: A state of dynamic physical, mental, social, and spiritual well-being that enables a person to achieve full potential and an enjoyable life.

Women of Reproductive Age: All women in the childbearing age group of 15–44 years.

Women's Health: A broad category of illnesses and health conditions associated with being female including those related to reproduction, pregnancy and childbirth, abortion and contraception, management of menopausal problems, and screening for common female cancers. Violence against women, STIs, chronic disease manifestations in women, nutrition, osteoporosis, and many other issues of public health concern come under this grouping.

World Bank: The International Bank for Reconstruction and Development, Washington, DC, USA. This bank was established following World War II by the industrialized countries and has become a key source of long-term loans for development of most of the countries in the United Nations.

***World Development Report* (WDR):** The annual publication of the World Bank relating to international development. The 1993 WDR *Investing in Health* examines the relationship between human health, health policy, and economic development.

World Health Assembly: The WHA is the supreme decision-making body of the WHO. It meets annually, with delegations from all-member countries, to determine policy of the organization and to promote priorities and resource allocation for health by member countries.

World Health Organization (WHO): WHO is a UN–affiliated agency established in 1946 currently with representation from 189 countries. WHO serves as the international forum for cooperation in health. It has six regional offices which carry out regional assessment and cooperative activities to promote health including surveillance and control of health problems affecting the world's people.

Years of Potential Life Lost (YPLL): A measure of the effect of disease and injuries in reducing the life span below national or a hypothetical ideal life expectancy, but does not take disability into account. It highlights the loss to society as a result of early deaths, calculated for specific causes of death by summing the expected years of life remaining to age 65 years for each individual at death for the specified cause. The total years of potential life years lost can be summed for all age groups.

Zoning: Restriction on types of buildings that can be constructed or operated in specific areas of a community, so as to reduce exposure to nuisances or health risks to a residential population.

Zoonoses: Diseases occurring in animals which may be transferred to humans by vertebrate animals under natural conditions.

BIBLIOGRAPHY

Center for Disease Control. Agency for the Toxic Substances and Disease Registry, http://www.atsdr.cdc.gov/glossary.html [accessed July 12, 2008]

Centers for Disease Control and Prevention. National Center for Health Statistics. *Health, United States 2007 with Chartbook on Trends in the Health of Americans.* Hyattsville, MD: U.S. Department of Health and Human Services.

Data Resource Center for Child and Adolescent Health. 2008, http://www.childhealthdata.org/content/Glossary.aspx [accessed July 14, 2008]

European Observatory. Glossary http://www.euro.who.int/observatory/glossary/toppage.

Heymann, D. L. 2004. *Control of Communicable Diseases Manual,* Eighteenth Edition. Washington, DC: American Public Health Association.

Larson, L. 2002. Glossary of Health Care and Health Care Management Terms. University of Washington School of Public Health and Community Medicine. Health Services Library and Information Center, http://www.depts.washington.edu/hsic/resources/glossary.html [accessed July 12, 2008]

Last, J. M. 2001. *A Dictionary of Epidemiology,* Fourth Edition. New York: Oxford University Press.

Last, J. M. 2007. *A Dictionary of Public Health,* New York: Oxford University Press.

Last, J. M. 2007. Public Health Agency of Canada. Glossary of Terms Relevant to Public Health Competencies, http://www.phac-aspc.gc.ca/php-psp/core_competencies_glossary_html [accessed July 12, 2008]

National Center for Health Statistics. Glossary of terms, http://www.cdc.gov/nchs/datawh/nchsdefs/list.htm.

OECD. Glossary of terms, http://www.oecd.org/document/21/0 [accessed May 22, 2008]

PAHO Glossary. http://www.paho.org/Engllish/SHA/glossary.htm

UNICEF. 2008. The State of the World's Children. Child Survival. New York: UNICEF.

United Nations Millennium Goals. http://www.un.org/milleniumgoals/goals.html

United States National Library Association. Medical Dictionary. 2003, http://www.nlm.nih.gov/medicineplus/mplusdictionary.html [accessed July 12, 2008]

World Health Organization. Glossary of globalization, trade and health terms, http://www.who.int/trade/glossary/en/ [accessed May 22, 2008]

World Health Organization. 1998. Health Promotion Glossary, http://www.who.int/hpr/NPH/docs/hp_glossary_en.pdf [accessed July 12, 2008]

World Health Organization. Health impact assessment glossary of terms, http://www.who.int/hia/about/glos/en/ [accessed May 22, 2008]

World Health Organization. List of acronyms, who.int/immunization_monitoring/glossary/en/ [accessed May 22, 2008]

World Health Organization. Regional Office for Europe. European Health for All Database, http://www.euro.who.int/hfadb [accessed May 22, 2008]

World Health Organization. Terminology information system, www.who.int/health-system-performance/docs/glossary.htm [accessed May 22, 2008]

World Health Organization. Terminology information system, http://www.euro.who.int/HEN/Syntheses/hcfunding/20040704_10.

Index

Note: Page Numbers followed by f indicate figures; t, tables; b, boxes.

A

AAMC. *See* Association of American Medical
    Colleges
AAP. *See* American Academy of Pediatrics
Accreditation, 386, 547, 548, 579, 580b
    health facility, 578–580
    of medical educational or training facilities,
        547–548
    of public health facilities, 558
    services evaluated in, 579t
ACER. *See* Average cost effectiveness ratio
Acid rain, 337, 347
ACOG. *See* American College of Obstetrics and
    Gynecology
Acquired immunodeficiency syndrome. *See* AIDS
Acute flaccid paralysis (AFP), 509
Acute Respiratory Diseases (ARD), 50
Acute respiratory infections, 173–174
Adenoviruses, 172
Adequate intake (AI), 300, 300b
Adhocracy, 448
Adolescents, health of, 241–245, 241t, 242b
Adults, health of, 245–246, 246b
    older, 249–254, 251t, 252b, 253b
Advanced practice nurses (APNs), 551
Adverse events, 577–578
Advocacy, 355
    consumerism and, 55–58, 55b
    health promotion and, 457, 476
    professional, 55–57, 56b
    resistance to, 55–57, 56b
Aedes albopictus mosquitos, 335
Affordable Choice Initiative, 477
AFP. *See* Acute flaccid paralysis
Africa. *See also* Nigeria
    CHWs in, 552
    health systems in, 506, 507
    Human Development Indices for, 467, 508t
    human resources in, 538
    immunizations in, 506, 507, 509
African trypanosomiasis (sleeping sickness), 164
Agent Orange, 354, 366, 368
Agricultural and environmental hazards, 346
Agriculture, 2, 3, 7
AI. *See* Adequate intake
AIDS (Acquired immunodeficiency syndrome),
        168–170, 168b, 169f, 593. *See also* HIV/
        AIDS
    control of, 41
    epidemic, 79, 80f
    epidemiology of, 168–170, 168b, 169f
    globalization of, 617–618

Air pollution, 346–349
Albany County Health Department, 389, 391b
Alcohol, 204, 235, 243, 266
Algorithms, 581–584
Alma-Ata (Health for All), 47–49, 48b, 70, 407,
    497, 524
Alternative medicine, 553–554
Alzheimer's disease, 238
AMA. *See* American Medical Association
Amebiasis, 164
American Academy of Pediatrics (AAP),
        229b, 584
    Genetics Committee of, 228
American Board of Preventive Medicine, 545
American College of Obstetrics and Gynecology
        (ACOG), 223, 236
American Medical Association (AMA), 56, 235,
        394, 540, 545
American Pediatric Society, 235
American Public Health Association (APHA), 12,
        47, 56, 349, 364, 394, 581
    guidelines and standards of, 63, 66t
Anorexia nervosa, 312
Anthrax, control/eradication of, 155–156
Antibiotic(s)
    advent of, 15, 23
    for brucellosis, 154–155
    for infectious diseases, 19
Antibodies, 126b
Anticipatory counseling, 238–240, 239t
Antigenic drift, 134
Antigens, 126b
*On the Antiseptic Principle in the Practice of
    Surgery* (Lister), 17
Antisepsis, 17
Antisera, 126b
Anti-Submarine Detection Committee (ASDIC), 445
Antitoxins, 126b
Anxiety, 267
APHA. *See* American Public Health Association
APNs. *See* Advanced practice nurses
Arboviruses, 157–160
ARD. *See* Acute Respiratory Diseases
Arthritis, 205
Arthropod-borne viral diseases, 157–160
Asbestos, 335, 349, 361, 364–365, 365t
Asbestosis, 197
Ascariasis, 164
ASDIC. *See* Anti-Submarine Detection Committee
Asia, 514–520. *See also* China; Japan
    demographic and health indicators in, 515t
    public health education in, 555, 556, 557

ASPH. *See* Association of Schools of Public
    Health
Association of American Medical Colleges
        (AAMC), 537, 540, 547
Association of Schools of Public Health (ASPH),
        555
Asthma, 195–196, 197
Attlee, Clement, 469, 482
Australia, 63
Average cost effectiveness ratio (ACER), 419
Avian influenza, 39, 382

B

Bacille Calmette-Guérin (BCG), 509
Back syndromes, low, 206
Bacon, Francis, 9
BACT. *See* Best available control technology
Bacteriology, 16–17
Baker, George, 364
Bamboo fuels, 349
Barefoot doctors, 516
Basic Assessment Scheme for Intervention Costs
        and Consequences (BASICC), 419
Basic radiologic units (BRU), 569
BASICC. *See* Basic Assessment Scheme for
        Intervention Costs and Consequences
BCG. *See* Bacille Calmette-Guérin
Behavior modification, 450–451
Behavioral syndromes with psychological
        disturbances, 268
Beijing Conference on Women (1995), 248
Bentham, Jeremy, 11
Beriberi, 308–309
Bernard, Claude, 36
Best available control technology (BACT), 348
Beveridge, William, 469, 482, 520
Bias, sources of, 95
Bills of Mortality, 7, 12
Bin Laden, Osama, 354
Binge eating, 312
Biological pollutants, 350
Bioterrorism, 10b
Birth
    control, 221, 221t
    defects, 343, 344, 345t, 351
    disorders, 230–235
        causes of, 230t
        prevention of, 231b
    live, 236
    rates, 75
    vaccinations at, 10
Bismarck, Otto von, 467, 468, 492

Bismarckian national health insurance program, 440, 467, 468, 491, 492–493, 520
Black Death, 6*b*
Black Report, 40, 484
Blair, Tony, 483, 484
Blindness, 207–208, 208*b*
BMI. *See* Body mass index
BOD. *See* Burden of disease
Body mass index (BMI), 298–299, 299*f*, 299*t*
Bologna Declaration, 557, 558, 558*b*
B₁. *See* Thiamine deficiency
Bovine spongiform encephalopathy (BSE), 342
Breslow, Lester, 186, 187*b*
British Poor Law Amendment Act (1834), 11
British Registrar General Classification of Occupation(s), 75*b*
BRU. *See* Basic radiologic units
Brucellosis
    antibiotics for, 154–155
    clinical cases of, 155
    control/eradication of, 154–155
Brundtland, Gro Harlem, 406
BSE. *See* Bovine spongiform encephalopathy
Budgets, factors in, 455
Bulimia, 312
Bunker's Law, 421*b*
Burden of disease (BOD), 88, 90*b*
Bureaucratic pyramidal organizations, 442–443, 443*f*, 446
Burma, 382
Bush, George W.
    on Affordable Choice Initiative, 477
    on prescription drugs, 472
    on SCHIP, 471
Byssinosis, 197, 366

C

Canada, 469, 520
    accreditation of medical schools in, 547, 548
    health expenditures in, 478, 479, 480
    health insurance plans in, 58
    health status in, 480–481
    human resources in, 540, 554
    licensure and supervision in, 548
    life expectancy in, 478, 480, 540
    medical education in, 542, 544
    Medicare in, 478, 479, 480, 481
    mortality rates in, 480
    public health education in, 555
    reform pressures and initiatives in, 479–480, 481, 524
    summary on, 471–472
Canada Health Act, 479, 480
Canada Health and Social Transfers (CHST), 480
Canadian Community Health Survey (2002-2003), 198
Canadian Council on Health Services Accreditation (CCHSA), 579, 580*b*
Cancer(s), 200–204, 201*b*, 201*f*, 202*f*, 203*f*
    of cervix, 573*b*
    epidemiology of, 200–204, 201*b*, 201*f*, 202*f*, 203*f*
    lung, 349, 365, 365*t*
    in Nigeria, 511
    nutrition and, 313–314, 314*t*
    oral, 274
    prevention of, 203–204, 204*t*
    radiation and, 349, 352
CAP. *See* Change Agent Programme

Capitation, 482
Capone's Law, 421*b*
Carbohydrates, 295
Carbon monoxide (CO), 348, 349–350
Cardiovascular diseases (CVDs), 184
    deaths associated with, 186
    epidemiology of, 188–195, 189*t*
    mortality from, 189–192, 190*f*, 191*f*
    nutrition and, 313
    prevention of, 192–195, 193*b*, 193*f*, 195*b*
    risks for, 186, 187*b*
    in Russia, 495, 495*f*
Caribbean. *See* Latin America and the Caribbean
Case mix groups, 485
CBA. *See* Cost-benefit analysis
CCHSA. *See* Canadian Council on Health Services Accreditation
CDC. *See* Centers for Disease Control
CEA. *See* Cost-effective analysis
Census, 7, 74, 75*b*
Centers for Disease Control (CDC), 387, 582
    EISC for, 583*b*
    establishment of, 470
    MMWR, 474
    on occupational health, 356, 360, 362, 364
CEO. *See* Chief executive officer
CEPH. *See* Council on Education for Public Health
Cerebral palsy (CP), 197
Cerebrovascular disease
    hospitalization for, 481, 481*t*
    in Nordic countries, 490, 490*f*, 491*t*
CERES. *See* Coalition for Environmentally Responsible Economics
Certificate of Need (CON), 385, 571
Certification, 578
CF. *See* Cystic fibrosis
CFCs. *See* Chlorofluorocarbons
CH. *See* Congenital hypothyroidism
CHA. *See* Community health assessment
Chadwick, Edwin, 10, 12, 63
Chagas' disease, 164
Chancroid, 166
Change Agent Programme (CAP), 511
Charter of Rights, 586*t*
Chemicals. *See* Hazardous substances; Hazardous wastes; Toxic wastes; Toxicology; Toxins
Chernobyl, 353, 349
Chicken pox, 143
Chief executive officer (CEO), 455–456
Chikungunya, 158, 335
Children, 19–20, 56, 172, 416*t*
    adolescent health, school and, 241–245, 241*t*, 242*b*
    behavioral and emotional disorders during adolescence of, 268–269
    health of, 19–20, 235–237, 613
    immunizations for, 20
    lead and, 343, 344–346, 345*b*, 346*b*, 348, 349, 364
    preschooler, 240–241
CHIMP. *See* Commission for Health Improvement
China, 56, 382, 515–518
    Black Death in, 5
    concept of health in, 2, 6*b*
    11th Five-Year Plan, 518
    health indicators in, 516, 517*t*
    high order medicine in, 515
    infectious diseases, emerging, 517–518

    one child policy in, 516, 517
    reform in, 524, 538
Chlamydia, 167
Chlorination, of water, 338, 339, 340, 341
Chlorofluorocarbons (CFCs), 337
Cholera, 13–14, 13*t*
    acute, 171, 171*b*
    pandemics in India, 13
    re-emergence of, 335
    Snow on, 13–14, 13*t*, 335
    waterborne transmission of, 14, 15*b*, 338, 340
Christianity, 3, 4
Chronic conditions, 213
Chronic disease(s), 618–619
    burden of, 183–184, 183*t*, 631*b*
    causation and risk factors of, 184–187, 184*b*, 185*t*, 186*b*
    introduction to, 181–182
    in Nigeria, 509
    rise of, 182–183, 183*t*
    summary of, 213
Chronic liver disease
    cirrhosis as, 204
    epidemiology of, 204–205
Chronic lung disease (CLD), 195–197
    deaths associated with, 195
    types of, 195–197
Chronic obstructive pulmonary disease (COPD), 183*t*, 196
CHST. *See* Canada Health and Social Transfers
CHW. *See* Community health worker
Cirrhosis, 204
Cities
    health in, 63, 64*b*, 64*t*
    during late medieval period, 5
    during Renaissance, 7
Citrus fruits, 8, 9, 9*b*
Civil societies, 439, 457
Clean Air Act, 348
Clean Water Act (CWA), 338, 347
Climate change, 334, 337. *See also* Global environmental change; Global warming
Clinical guidelines, 581–584
Clinton, Bill, 457
    on national health insurance, 473, 477, 478
    on patients' rights, 477
CMH. *See* Commission on Macroeconomics and Health
CMS. *See* Cooperative medical services
CO. *See* Carbon monoxide
Coal worker's pneumoconiosis, 197
Coalition for Environmentally Responsible Economics (CERES), 353
Code of Hammurabi, 2
Codex Alimentarius, 629
Colombia, 512–514, 552
Commission for Health Improvement (CHIMP), 485
Commission on Macroeconomics and Health (CMH), 406
Communicable disease(s), 2
    classifications of, 124–125
    control of, 122, 122*b*, 130–131, 131*t*, 335
    inequalities in, 174–175, 175*f*
    NPH and, 175
    endemic, 128–130
    epidemic, 128–130
    globalization of, 617–618
    HAI as, 128, 128*b*

immunity from, 125–126, 126b
introduction to, 121–122
nature of, 122–123, 123f
in Nigeria, 509
NPH and control of, 175
prevention of, 130–131, 131t
summary on, 176
surveillance of, 126–128, 127t
transmission modes of, 125, 125t
vaccine-preventable, 37, 131–144, 132t, 133t, 138f, 142b
Community
  health, 78b
    models of promotion, 63
    needs, 259–260, 260b
    participation in, 62, 62f
    social medicine and, 45–46
    summary on, 286
  health and social medicine, 45–46
  medicine, 45
  trials in experimental epidemiology, 99–100
Community Guide, 584
Community health assessment (CHA), 109
Community health worker (CHW), 552–553, 552b
Community participation, in management, 456
Community-oriented primary care (COPC), 46–47, 47b, 104b, 270, 270b
The Communist Manifesto (Marx), 12
A Complete System of Medical Police (Franck), 11
Composting, 341, 342
Comprehensive care, 426–427
CON. See Certificate of need
Condyloma. See Human papilloma virus
Confounders, 91, 91f
Congenital hypothyroidism (CH), 204–205
Consumerism, 55–58, 55b, 585–586
Contagion theory of disease, 7
Continuous Quality Improvement (CQI), 566
  WHO and, 48
Control of Diarrhoeal Diseases, 50
Cooke, W. E., 364
Cooperative medical services (CMS), 516
COPC. See Community-oriented primary care
Cordova Medical Academy, 4
Coronary bypass, 576
Corporate model, 453–454, 454f
Cost-benefit analysis (CBA), 417–419, 418b
Cost-effective analysis (CEA), 416–417, 416t, 417b
Cost-utility analysis, 417–419, 418b
Cotton dust, 366
Council on Education for Public Health (CEPH), 555, 558
CQI. See Continuous Quality Improvement
Crede, Franz, 16b
Creutzfeldt-Jakob disease, 80, 156
Crude rates, 83, 83b
Crusades, 4
Cryptosporidium, 172, 339
Cultural Revolution, China's, 516
CWA. See Clean Water Act
Cybernetics, 444–445, 444f
Cyclone disasters, 382
Cyprus, 193b
Cystic fibrosis (CF), 199–200

D

DALYs. See Disability-adjusted life years
Data, value of, 107b

Death(s). See also Mortality
  associated with chronic lung disease, 195
  associated with CVDs, 186
  due to vaccine-preventable diseases, 144, 144t
  injuries and, 209
  mesothelioma, 94t
  occupational health and, 359, 360, 361t, 365, 367
  U.S. standard certificates of, 86b
Degenerative osteoarthritis, 206
Delaney Clause, food, and FDA, 368
Demand, in health economics, 408–412, 410f
  elasticities of, 413–415
  Grossman's model of, 411–412
Deming, W. Edwards, 449
Demographic transition, 74
Dengue hemorrhagic fevers, 159, 159b, 335
Denmark, 489
Dental care, 274
Department of Health (DH), 485
Developing countries
  health systems in, 466, 506–512
  human resources in, 537
  public health education in, 559
DFLE. See Disability-free life expectancy
DH. See Department of Health
DHAs. See District health authorities
Diabetes mellitus
  complications associated with, 199–200, 199b
  epidemiology of, 197–200, 198f
  ESRD and, 196, 198, 200
  major types of, 197
  prevalence of, 197–200, 198f
  prevention of, 199–200, 199b
Diagnosis-related groups (DRG), 103, 429, 430, 473, 493, 584–585. See also Case mix groups
Diagnostic and Statistical Manual of Mental Disorders (DSM-IV), 260
Diarrheal diseases, 170–173, 170t, 171b
  causes of, 333
  control of, 172–173, 173b
  epidemiology of, 170–173, 170t, 171b
Dietary guidelines, 315
Dietary reference intakes (DRIs), 300, 300b, 301t
Dietary Reference Intakes for Calcium, Phosphorus, Magnesium, Vitamin D and Fluoride, 300
Dietary risk behaviors, 244–245
Diets, healthy, 315
Diphtheria, 135
Diploma in Public Health (DPH), 554
Directly Observed Therapy (DOTS), 518, 552, 553
Disabilities
  intellectual, 197
  learning, 269
    as mental disorder syndromes, 269
  mental, 271–272
  physical, 274–276, 284t
Disability-adjusted life years (DALYs), 88, 90b, 415, 416t, 526
Disability-free life expectancy (DFLE), 88
Disabling condition(s), 205–206, 205f
  arthritis as, 205
  degenerative osteoarthritis as, 206
  gout as, 206
  low back syndromes as, 206
  musculoskeletal disorders as, 205
  osteoporosis as, 205
  RA as, 206

Disasters
  cyclone, 382
  health during, 283–286, 284t
  management, 619
  man-made, war, terrorism and, 354
  natural, 382
  planning/preparedness, 355, 356b, 369
  and preparedness of public health systems, 381–382, 381b, 382b
  in workplace, preventing, 368–369, 368t
Disease(s). See also specific diseases
  arthropod-borne viral, 157–160
  burden of, 88
  cardiovascular, 495f, 495
  causation of, 14
    establishment of, 100b, 101b, 100
  cerebrovascular, 490f, 481, 490, 481t, 491t
  chronic
    burden of, 183–184, 183t, 631b
    causation and risk factors of, 184b, 186b, 184–187, 185t
    introduction to, 181–182
    in Nigeria, 509
    rise of, 182–183, 183t, 183t
    summary of, 213
  church interpretation of, 4
  classifications of, 106, 103t
  communicable, 2
    classifications of, 124–125
    control of, 122b, 122, 130–131, 130, 130–131, 131t
      inequalities in, 175f, 174–175
      NPH and, 175
    endemic, 128–130
    epidemic, 128–130, 129–130
    HAI as, 128b, 128
    immunity from, 126b, 125–126
    introduction to, 121–122
    nature of, 123f, 122–123
    NPH and control of, 175
    prevention of, 130–131, 131t
    summary on, 176
    surveillance of, 126–128, 127t
    transmission modes of, 125, 125t
    tropical, 161b
    vaccine-preventable, 37, 142b, 138f, 131–144, 133–144, 135, 135–136, 136–137, 137–138, 138, 139, 139–141, 140, 140–141, 141, 141–143, 143, 143–144, 144, 132t, 133t
  contagion theory of, 7
  enteric, 14
  environmental impact on health burden of, 337–338, 337t
  food-borne, 342, 352
  globalization and, 467, 477
  health and, 36–38, 37–38
  infectious
    antibiotics for, 19
    chronic manifestations of, 188b, 193b, 195b, 190f, 191f, 192f, 193f, 187–195, 188–195, 189–192, 192–195, 189t
    control/eradication of, 147, 150b, 151b, 153b, 152f, 147–156, 148–149, 149, 149–156, 150–153, 153–154, 154–156, 154–155, 155, 155–156, 156, 150t

Disease(s) *(Continued)*
  epidemiology of, 81*b*
  epidemiology of non, 81*b*
  immunology of, 126*b*
  treatment of, 19
  vaccines for, 19
  lung
    occupational, 196–197, 197
    restrictive, 196
  lung, chronic, 195–197, 195–196, 196,
    196–197, 197
    deaths associated with, 195
    types of, 195–197, 195–196, 196,
      196–197, 197
  microorganisms as causes of, 17*b*, 17
  natural history of, 38*f*, 38–39
  noncommunicable, 181–182
  notification of, 100–101
  NPH, health and, 37*f*, 36–38, 37–38
  occupational, 8
  of overnutrition, 312–314, 312–313, 313,
    313–314, 314*t*
  parasitic, 161*b*, 161–165, 161–162, 162,
    162–163, 163, 163–164, 164,
    164–165, 164
  periodontal, 273
  pinworm, 164
  pneumococcal, 143
  prevention of
    modes, 41*b*, 41–43
    primary, 42
    secondary, 43
    tertiary, 43
  primary prevention of, 42
  reporting, publication and surveillance of,
    107*b*, 108*b*, 107
  reporting systems for, 444
  screening for, 96*b*, 95–96
  secondary prevention of, 43
  social, prevention of, 496
  streptococcal, control/eradication of, 153–154
  surveillance of, 126–128
  tertiary prevention of, 43
  transmission of, 125, 125*t*
  vector-borne, 159*b*, 160*b*, 156–160,
    156–160, 160
  waterborne, 335, 338, 339–340
Disease Control in Developing Countries, 406
Diseases of Infancy and Childhood (Holt), 236
Disinfection, 7
Disorder(s). *See also* Mental disorder syndromes
  birth, 230–235
    causes of, 230*t*
    prevention of, 231*b*
  dissociative, 267, 268
  eating, 311–312
  genetic, 230–235
    causes of, 230*t*
    prevention of, 231*b*
  hearing, 208
  mood, 267
  musculoskeletal, 205
  neurotic, 267
  personality, 268
  of psychological development, 268
  somatoform, 268
  of undernutrition, 300–312, 303*t*, 305*t*,
    307*b*, 309*b*
  visual, 207–208, 208*b*

Dissociative disorders, 267, 268
District health authorities (DHAs), 482–484
Divisional model, 454*f*, 453
Doctors. *See* Physicians
Documentation, 8, 238–240, 239*t*
Doll, Richard, 185
DOTS. *See* Directly Observed Therapy
Douglas, Tommy, 469, 479
Down syndrome, 233
Downsizing, 369, 473, 478, 501, 525
DPH. *See* Diploma in Public Health
Dracunculiasis, 162–163
DRGs. *See* Diagnosis-related groups
DRIs. *See* Dietary reference intakes
Drug(s), 51, 243–244

E

Earthquakes, 382
Eating disorders, 311–312
Eberus Papyrus, 3
Ebola fever, 160
Ebola virus, 628*b*
Echinococcosis, 161–162
Ecology, medical, 46
Economic development, 625*b*
Economics, of health
  basic concepts in, 408, 408*f*
  capital costs in, 421*b*, 430, 431*t*
  CBA of, 417–419, 418*b*
  CEA of, 416–417, 416*t*, 417*b*
  cost measurement of, 415
  demand in, 408–412
    elasticities of, 413–415
    Grossman's model of, 411–412
  introduction to, 405
  investments in, 407–408, 407*b*
  measures, 415–416
  modified market forces in, 432, 432*b*,
    433*b*, 433*t*
  national health systems issues in, 521–523
  needs in, 408–412
    comparative, 410
    demand, 410–411, 410*f*
    expressed, 410
    felt, 409–410
    normative, 409
  NPH and, 432–434
  summary on, 434
  supply in, 408–412
  systems, issues associated with, 405–408, 406*t*,
    407*b*, 407*f*
  utilization in, 408–412
Ectoparasites, 164–165
Education. *See also* Medical education; Medical
  schools
  for public health and health management,
    554–560
  and training, in health professions, 535,
    536, 541
EHEA. *See* European Higher Education Area
Emergency Planning and Community-Right-to-
  Know Act (EPCRA), 354
Empowerment, 451
Encephalitides, 158–159
Endoscopic surgery, 571
End-stage renal disease (ESRD), 198, 200
  diabetes mellitus and, 196, 198, 200
  epidemiology of, 200
Enlightenment, 7–9, 9–10

Enterobiasis. *See* Pinworm disease
Environment, 111, 620
  consciousness of, 334, 335, 458
  impact on health burden of disease,
    337–338, 337*t*
  U.S. government agencies responsible for,
    95, 356*t*
  U.S. milestones regarding, 357*b*
Environmental and geographic epidemiology, 335
Environmental emergencies, 352–354,
  354–355, 356*b*
Environmental health, 333. *See also* Climate
  change; Global environmental change;
  Hazardous wastes; Pollution; Toxins
  agricultural and environmental hazards, 346
  introduction to, 333–334
  issues, 334–335, 335*b*, 369
  organization, 355–356
  radiation, 351–352
  risk assessment, 355
  sewage collection and treatment, 340–343
  summary on, 369
  targets, 335, 336*t*
  water and, 333, 334, 335, 338–340, 339*b*, 340*b*
Environmental impact, 352–355
  emergency events involving hazardous
    substances, 352–354, 356*b*
  emergency events, preventing and managing,
    354–355
  man-made disasters, war and terrorism, 354
Environmental Protection Agency (EPA), 352
  establishment of, 347, 356
  on radon gas, 349
  on toxic waste management, 351
EPA. *See* Environmental Protection Agency
EPCRA. *See* Emergency Planning and
  Community-Right-to-Know Act
EPI. *See* Expanded Programme on Immunization
Epidemic(s)
  AIDS, 79, 80*f*
  communicable disease, 128–130
  HIV, 79, 80*f*
  during late medieval period, 4, 5, 6
  during Renaissance, 6, 7
Epidemiology
  during 1960s-1980s, 37*b*
  of AIDS, 168–170, 168*b*, 169*f*
  applied, 8–9
  of cancer, 200–204, 201*b*, 201*f*, 202*f*, 203*f*
  definitions and methods of, 82–87, 83*b*
  descriptive, 96–97
  of diabetes mellitus, 197–200, 198*f*
  of diarrheal diseases, 170–173, 170*t*, 171*b*
  experimental, 99–100
    community trials in, 99–100
    controlled trials in, 99
    field trials in, 99
  field, 558
  foundations of, 9–10
  geographic and environmental, 335
  goals and methods of, 79*b*
  hazardous wastes and, 342, 343*t*, 369
  health of populations and, 77–80, 78*b*, 93*t*
  health policies and, 81*b*, 82
  of HIV, 168–170, 168*b*, 169*f*
  of infectious diseases, 81*b*
    non, 81*b*
  landmarks in, 81*b*
  measurement in, 89–91, 90*b*, 91*f*

modern, 10
reporting systems for, 101–102, 102b, 102t
of salmonella, 170
social, health of populations and, 80–81,
    81b, 93t
special registries for, 101–102, 102b, 102t
studies, 13–14, 96–100, 97f, 335, 351, 447
    analytical, 97–99
    case-control, 98
    cohort, 98–99
    cross-sectional/prevalence, 98
    ecological, 97–98
    observational, 96–99
    transition into, 23–25, 24b, 36b
Epilepsies/seizures, 207
Escherichia coli, 171
Estimated average requirement (EAR), 300, 300b
Ethanol, 348, 349
Ethics
    human experimentation, 597–598, 598b
    patient care, 596
    public health, 565–566, 596–597
        principles of, 599b
        research, 595–596, 597t
        study/practice of, 592b
    rights, individual/community, issues with,
        593–595, 594t
    sanctity of live v. euthanasia in, 598
    slippery slope in, 597b
    Tuskegee experiment, 598, 598b
Europe. See also specific countries
    Healthy Cities Movement in, 63, 64, 64b, 65
    during late medieval period, 6
    life expectancy in, 493, 494f
    occupational health in, 364, 365
    public health education in, 555, 556, 557,
        558, 559t
    Western, 491–495
European Higher Education Area (EHEA), 557
Euthanasia, 598
Evaluation
    of health of populations, 73–74
    monitoring of nutrition, 320–322, 320t
Evans Postulates, 101b, 184b
Excellence in Science Committee (EISC), 583b
Expanded Programme on Immunization (EPI), 50
Exxon Valdez, 353

F

Failure to thrive, 304
Familial Mediterranean fever, 235
Family
    health
        delivery, labor and, 226–230
        high-risk pregnancies and, 224–226, 225t
        introduction to, 217–218
        maternal health and, 219–223, 220t, 221t,
            222t, 223t
        newborn's care and, 227–229, 228t,
            229b, 231b
        pregnancy care and, 223–226
        puerperium's care and, 229–230
        safe motherhood initiatives and, 227
        summary on, 254
        unit, 218–219
    planning, 188–195, 221t
Family Health Service Authorities (FHSAs),
    482–484
Family practice and specialization, 544–545

Faroe Islands, 15b
Fats, 295–296
FDA. See Food and Drug Administration
Feedback system, 444–445, 444f
Fee-for-service (FFS), 492
Feldshers, 550, 552
Female genital mutilation (FGM), 508–509
Fertility
    concerns regarding, 220
    decline of, 75
    epidemiology of, 219–220
    health of population and, 75, 75b, 76b
    rates, 76b, 220t
    toxic effects on, 343–344
Fetal alcohol syndrome, 235
Fetal mortality, 236–237, 237b, 237t
FETPs. See Field epidemiology training programs
Fever(s)
    Ebola, 160
    familial Mediterranean, 235
    hemorrhagic, 159, 160
        dengue, 159
        other, 160
    Lassa, 160
    yellow, 17, 18b, 159
FFS. See Fee-for-service
FGM. See Female genital mutilation
FHSAs. See Family Health Service Authorities
Field epidemiology training programs (FETPs),
    558
Finland, 63, 65, 489–491
Finlay, Carlos, 17, 18b
Flexnor, Abraham, 542
Flexnor Report, 542, 542b, 543
Fluoridation, 273
Fluoride, 20
Folate deficiency, 309
Food
    basic, 315–317, 316f, 317f
    enrichment, 316–317
    functions of, 294
    nutrition policies and, 317–320, 318b
    quality and safety of, 322–326, 323f, 324f,
        325t, 326b, 326t
Food and Drug Administration (FDA), 474,
    566, 573
    Delaney Clause, food and, 368
Food and Drug Control Act, 470
Food-borne disease, 342, 352
Franck, Johann Peter, 11
Frederick II (Emperor), 4
Friendly Societies, 6
Functional model, 453, 453f
The Future of Public Health (U.S. Institute of
    Medicine), 554

G

Gang behavior, 245
Gastroenteritis
    parasitic, 172
    viral, 171–172
Gates, Bill, 457
Gates Foundation, 57, 406
GAVI Alliance (Global Alliance for Vaccine and
    Immunization), 57, 457
Gays, 276–277
General Motors, 59
General practitioners (GPs), 482–484,
    545, 550

General Social Security System for Health
    (GSSSH), 513
Genetic disorders, 230–235, 230t, 231b
Geographic and environmental epidemiology, 335
Germ theory, miasma theory v., 14, 24b
German measles. See Rubella
Germany, 491–493
    Bismarckian national health insurance
        program, 440, 467, 468, 491, 492–493
    health reform in, 493
Giardia, 339
Giardiasis, 172
Global Alliance for Vaccines and Immunization.
    See GAVI
Global environmental change, 335–338
Global warming, 334, 348. See also Greenhouse
    effect
Globalization
    communicable diseases in, 617–618
    development partnership in, 620
    diseases and, 467, 477
    of health, 605–607
        indicators of, 608t
        situation with, 607–608
        trends in, 626
    health leadership in, 624b
    HIV/AIDS in, 617–618
    malnutrition in, 616–617
    population growth in, 614–615, 615t
    poverty in, 610–613
    public health, 631–632
    vaccinations in, 629b
Glucose-6-phosphate dehydrogenase deficiency
    (G6PD), 235
GMP. See Good Manufacturing Practices
Goldberger, Joseph, 21b
Gonorrhea, 16b, 166, 166t
Good Manufacturing Practices (GMP), 514
"Good worksite practice" principle, 369
Gorgas, William, 17, 18b
Gout, 206
GPs. See General practitioners
Grange, Adenike, 512
Great Britain. See United Kingdom
Great Depression
    in Canada, 479
    in U.K., 469
    in U.S., 468, 470
Great Society programs, 553
Greco-Roman concepts of health, 4
Greece, 3
Greenhouse effect, 336, 338
Greenpeace, 56
Grossman demand model, 411–412
Growth, 296–297
G6PD. See Glucose-6-phosphate dehydrogenase
    deficiency
GSSSH. See General Social Security System for
    Health

H

Habitation, communal, 2
Hadassah, 502
Haemophilus Influenzae Type B (Hib), 141, 458,
    467
HAI. See Health care-associated infections
Hamilton, Alice, 344, 345b
Hawthorne effect, 446
Hazardous substances, 352–354, 356b

Hazardous wastes, 338, 342, 350–351. *See also*
    Toxic wastes
  epidemiology and, 342, 343*t*, 369
  minimata disease, 351
HCFA. *See* Health Care Financing Administration
HDN. *See* Hemorrhagic disease of newborn
Health. *See also* Public health
  of adolescents, 241–245, 241*t*, 242*b*
  of adults, 245–246, 246*b*
    older, 249–254, 251*t*, 252*b*, 253*b*
  in body and mind, 34
  children's, 19–20, 235–237, 613
  Chinese concept of, 2
  in cities, 63, 64*b*, 64*t*
  community, 78*b*
    models of promotion, 63
    needs, 259–260, 260*b*
    participation in, 62, 62*f*
    social medicine and, 45–46
    summary on, 286
  definitions of, 47–49
  development and, 53–54, 620–621
  during disasters, 283–286, 284*t*
  disease and, 36–38
  economic development and, 625*b*
  environmental, 591
  facility accreditation, 578–580
  family
    delivery, labor and, 226–230
    high-risk pregnancies and, 224–226, 225*t*
    introduction to, 217–218
    maternal health and, 219–223, 220*t*,
      221*t*, 222*t*, 223*t*
    newborn's care and, 227–229, 228*t*,
      229*b*, 231*b*
    pregnancy care and, 223–226
    puerperium's care and, 229–230
    safe motherhood initiatives and, 227
    unit, 218–219
  field concept, 58, 58*b*
  globalization of, 605–607
    trends in, 626
  Greco-Roman and Hebraic concepts of, 4
  improved, outcome indicators for, 521
  individual, 78*b*, 107–108, 108*b*
    participation in, 62, 62*f*
  of individuals, 107–108
  information systems, 105, 105*b*, 106, 106*f*
  insurance, 423
    compulsory, 39
    for illnesses, 13
      and injuries, mandatory, 13
    NPH, universal, 393–395, 394*f*
    plans in Canada, 58
    uninsured, public health systems and,
      399–400
  international leadership in, 624*b*
  internationalization of, 23
  maintenance checklist, 582
  maternal, 19–20, 613–614
  of men, 248–249, 249*t*
  municipal boards of, 11
  in municipalities, 63, 64*b*, 64*t*
  needs
    of gays and lesbians, 276–277
    of homeless population, 280–281
    of migrant population, 280
    of native peoples, 277–278
    of prisoners, 278–279, 279*t*

  of refugees, 281–282
  of special groups, 276
  NPH, diseases and, 36–38, 37*f*
  oral, 272–274
  organization for international, 621
  participation in
    community, 62, 62*f*
    individual, 62, 62*f*
  policies, epidemiology and, 81*b*, 82
  political economy and, 52–53
  of population
    assessment of, 108–111, 110*t*, 111–116,
      113*b*, 115*b*
    demography of, 74–76, 75*t*
    epidemiology and, 77–80, 78*b*, 93*t*
    evaluation and measurement, 73–74
    fertility and, 75, 75*b*, 76*b*
    life expectancies and, 76–77, 77*f*
    life expectancy and, 76–77, 77*f*
    sentinel events in, 87–88
    social epidemiology and, 80–81,
      81*b*, 93*t*
    summary on, 116–117, 116*b*
  prevalence rates of, 85
  promotion, 41–42, 42*b*
    advocacy and, 457, 476
    community models of, 63
    human ecology and, 65, 65*t*
    in Israel, 504, 505*t*
    mental health and, 271
    Ottawa Charter for, 62–63, 479
    prevention and, 441
    reform and, 524
    state models of, 63
    in U.K., 484, 486
  self-assessment of, 114
  sentinel events in, 87–88
  services
    interdependence of, 44
    providers, 413*b*
  social classifications of, 87, 87*b*
  society and, 39–41
  state-operated, 13
  statistics, foundations of, 9–10
  systems
    case for reform with, 54
    creation/management of, 26
    district, 428–429
    issues associated with economics of,
      405–408, 406*t*, 407*b*, 407*f*
  targets, 59–60
    international, 61
    of MDGs, 67, 68*b*
      adoption of, 59
    of public health systems, national, 393,
      393*b*, 394*b*
    of Scotland, 61, 61*b*
    U.K., 61, 61*b*
    U.S., 59–60, 60*t*
    WHO, 149
  in towns, 63, 64*t*
  of women, 246–248, 247*t*
Health 21, 504
Health care
  barriers to, 466
  benefits and costs of, 114
  choice in, 523
  competition in, 412–413, 412*b*, 413*b*
  financing, 111–116, 525

  macroeconomic level of, 421–423, 421*b*,
    421*t*, 422*b*, 422*t*, 424*f*, 424*t*
  organizations, 111–116
  outcomes, 112–113, 113*b*
  paradigms, transformation of, 452*t*
  quality of care with, 113–114
  resources, 111–112
  rising costs of, 361
  spending, national, 408
  utilization of services for, 112
Health care expenditures. *See also specific*
    countries
  human resources and, 538
  as percentage of GNP, 441, 461, 478, 521
  as share of GDP, 469, 470*f*, 474, 477, 496,
    521, 522*t*
Health Care Financing Administration (HCFA),
    387, 474
Health care providers, 549–550
Health care-associated infections (HAI), 128, 128*b*
Health field concept, 58, 58*b*, 479, 524
Health for All (HFA)
  in Finland, 491
  WHO and, 441, 445, 504, 506, 524
  World Bank on, 441
Health for All 2000, 407
Health maintenance organizations (HMO), 49,
    112, 425, 445, 472, 473, 585
  in Colombia, 513
  group model, 427*b*
  history of, 427–428, 427*b*
  staff model, 427*b*
Health management, 554–560
Health of Towns Association, 63
Health policies
  epidemiology and, 82
  goals, issues, strategies and tactics for, 524*t*
  human resources management and, 560–561
  and planning, as context, 440–441
Health professions. *See also* Community health
    worker; Feldshers; Medical education;
    Nurses; Physician assistants; Physicians
  education and training in, 535, 536, 539, 541
  new, 550–553
  regulation/licensing of, 539, 544, 548–549
  specialization and family practice, 544–545
  supervision of, 548–549
  types of, in public health and clinical services,
    548, 549*t*
Health promotion, 572–573
Health Promotion and Disease Prevention, 34
Health reform, 451, 456, 468
  in Canada, 479–480, 481, 524
  in China, 524, 538
  in Germany, 493
  in Israel, 503–504, 524
  national health systems and, 523–525
  in Nigeria, 511
  in Russia, 499–500, 501
  in U.K., 482–484, 484–485
Health Reform Foundation of Nigeria (HERFON),
    511
Health status, 369, 480–481
Health system(s). *See also* National health
    systems; *specific countries*
  case for reform with, 54
  creation/management of, 26
  delegation of responsibilities in, 42
  in developed countries, 443–445, 467–469

in developing countries, 466, 506–512
evolution of, 467–469
introduction to, 439–440
national, 440–441
new projects and their evaluation, 57–58,
    459b, 460b
NPH and, 460
strategic management of, 46, 452b
summary on, 461
WHO definition of, 466
Health system organization models, 446, 452–454
basic elements of, 458f
corporate, 453–454, 454f
divisional, 453, 454f
functional, 446, 453f
matrix, 454, 455f
new models for, 458
Health targets, 59–60
international, 61
U.K., 61, 61b
U.S., 59–60, 60t, 363, 363t, 475–476,
    475t, 524
Health, United States, 393
Healthy China 2020, 518
Healthy Israel 2020, 504–505
Healthy People, 393, 393b
Healthy People (Surgeon General), 363, 475, 478
Healthy People 2000, 363, 445, 475, 524
Healthy People 2000: National health Promotion
    and Disease Prevention Objectives, 393
Healthy People 2010, 60, 60t, 221, 444, 476, 477,
    504, 524
Healthy People 2010 Midcourse Review, 393
Healthy Worker Effect, 359
Hearing disorders, 208
Heart disease, impacts on, 441
Hebraic concepts of health, 4
Hebrew Mosaic Law, 3
Helicobacter pylori, 172, 571
history of, 172
treatment for, 188
Helsinki Committees, 115
Helsinki Declaration, 597
Hemorrhagic disease of newborn (HDN),
    184b, 308
Hemorrhagic fevers, 159, 160
Hepatitis
A, 140
B, 140, 204
C, 140–141, 204
D, 141
E, 141
viral, 139–141
Herbicides and pesticides, 346, 366
HERFON. See Health Reform Foundation of
    Nigeria
Herpes simplex, 167
Herpes zoster, 143
Herzberg, Frederick, 448
HFA. See Health for All
HHS. See U.S. Department of Health and Human
    Services
HI. See Hospital Insurance trust fund
Hib. See Haemophilus Influenzae Type B
High order medicine, in China, 515
Hill, Bradford, 185
Hill-Burton Act, 384, 384b, 470
HIV (Human immunodeficiency virus), 168–170,
    168b, 169f

control and treatment of, 457, 459
epidemic, 79, 80f
epidemiology of, 168–170, 168b, 169f
HIV/AIDS
in China, 517, 518
globalization of, 617–618
in Nigeria, 511
HMO. See Health maintenance organizations
Homeless population, health needs of, 280–281
Homicides, 212
Homo erectus, 1
Homo sapiens, 1
Hospital(s), 112
-acquired infections, 15
acute care beds, 488t, 497, 498f, 522, 523, 546
boards, 456
care
    microeconomics of, 426–427
    payments for, 429–430
costs, supply, utilization, 430–431, 431t, 432b,
    432f
discharge information, 103–105, 104b
    system, 10
lateral and vertical integration and, 456
leprosy, 8
Los Angeles county, 398, 398b, 399f
monastery, 4
Muslim, 4
NPH and, 394f, 395–399
    beds, supply of, 396–398
    changing roles of, 398–399, 398b, 399f
    classifications of, 396, 397b
    regulation of, 399
payment systems, 454, 456, 457, 460,
    473, 493
privileges, for physicians, 550
reforms, 15
Roemer's law and, 522
wastes, 350
Hospital Insurance (HI) trust fund, 472
Host-agent-environment
paradigm, 37–38, 37f
triad, 123–124, 123b
Howard, John, 15
HPV. See Human papilloma virus; Human
    papillomavirus
Human behavior, 450–451
Human body, composition of, 294
Human ecology, 65, 65t
Human experimentation, 597–598, 598b
Human immunodeficiency virus. See HIV
Human life, value of, 304–306, 420b
Human papillomavirus (HPV), 36, 74, 167
Human relations management, 445–448
"Human Relations School", 448
Human resources. See also Health professions;
    Medical education
changing the balance in, 554
health policy and management of, 560–561
infrastructure of health system and, 538, 539b
introduction to, 535–536
overview of, 536–538
principle problems in, 536
in Russia, 499, 500t
summary on, 561
Human resources planning, 538–541
issues in, 539, 539b
supply and demand, 539–541
Hurricane Katrina, 382

Hygiene
ancient world and, 2
departments of, 12
factors, 448
sanitary movement and, 11–14
social, 12, 45, 46
Hypertension, secondary prevention of, 193b

I

ICDs. See International Classification of Diseases;
    International Classification of Diseases
ICES. See Institute of Clinical Evaluation
IDDs. See Iodine deficiency disorders
Illnesses
costs of, 424–426
    containment of, 425–426, 425t
    variations in medical practice, 424–425
health insurance for, 13
ILO. See International Labour Organization
Immunity, 10, 125–126, 126b
Immunization(s), 47
in Africa, 506, 507, 509
in Canada, 481
in China, 516
coverage, 133–144
programs, 144–147, 145b, 444, 445, 457
for rabies, 16
social inequities and, 476
in U.K., 457
for women and children, 20
Immunization Plus Days (IPDs), 509
Immunizing agents/processes, 132b
Immunoglobulins, 126b
Immunology
microbiology and, 18–19, 19–20
Pasteur on, 16
Implicit Social Value of Life, 294, 420b
Independent practice associations (IPAs), 473
India, 57
Bhopal chemical accident in, 353, 354
cholera pandemics in, 13
medicine in, 2
Individual Practice Association (IPA), 427b
Indoor pollution, 349–350
biological pollutants, 350
radon gas, 349
sick building syndrome, 350
Industrial humanism, 446
Infant(s)
care of, 50, 237–238, 238b
feeding of, 237–238, 238b
health of, 235–237
mortality, 236–237, 237b, 237t
Infection(s), 7, 126b. See also Health care-
    associated infections
acute respiratory, 173–174
hospital-acquired, 15
nutrition and, 293–294
Rickettsial, 157
Infectious disease(s)
antibiotics for, 19
China's emerging, 517–518
chronic manifestations of, 187–195, 188b, 189t,
    190f, 191f, 192f, 193b, 193f, 195b
control/eradication of, 147–156, 150b, 150t,
    151b, 152f, 153b
emerging threats with, 627–630
epidemiology of, 81b
    non, 81b

Infectious disease(s) *(Continued)*
  globalization and spread of, 467
  immunology of, 126*b*
  national/international strategies for, 623*t*
  treatment of, 19, 557
  vaccines for, 19
Infectious wastes, 350
Influenza, 141–143, 142*b*
  avian, 39, 382
  vaccinations, 142*b*
Informatics, 105*b*, 106, 106*f*
Injuries, 208–213
  deaths and, 209
  motor vehicle, 209, 209*b*, 210–211, 210*t*, 211*f*
Input, 565
Input-process-output model, 443, 443*f*
Inquisition, 5
Insane asylums, 8
Institute of Clinical Evaluation (ICES), 583
Insulin, 198
Insurance, health, 423
  compulsory, 39
  for illnesses, 13
    and injuries, mandatory, 13
  for illnesses and injuries, mandatory, 13
  NPH, universal, 393–395, 394*f*
  plans in Canada, 58
  uninsured, public health systems and,
    399–400
Integration, lateral and vertical, 456
Intellectual disabilities, 197
Internal review boards (IRBs), 115
International Classification of Diseases (ICDs),
  103, 103*t*
International Classification of Diseases (ICDs),
  260, 261*t*
International Labour Organization (ILO), 359
International Task Force for Disease Eradication
  (ITDFE), 149
Intervention, effects of, 114
INVIMA. *See* National Institute for the
  Surveillance of Drugs and Food
Iodine deficiency disorders (IDDs), 56, 310–311
IPA. *See* Individual Practice Association
IPDs. *See* Immunization Plus Days
Iron-deficiency anemia, 310
Islam, 3
Israel, 502–506
  accreditation of medical schools in, 548
  CHWs in, 552
  health promotion in, 504, 505*t*
  health reform in, 503–504, 524
  health resources and expenditures, 503, 503*t*
  health systems origins, 502–503
  Healthy Israel 2020, 504–505
  mental health in, 504
  summary on, 505–506
  water in, 338, 339
Italy, chemical factory explosion in, 353
ITDFE. *See* International Task Force for Disease
  Eradication
Ivan IV (Czar), 7

J

Japan, 519–520
  financing and services, 519–520
  minimata disease in, 351
  NHI in, 519
  summary on, 520

JCAHO. *See* Joint Commission for Accreditation
  of Healthcare Organizations
JCHA. *See* Joint Commission on Hospital
  Accreditation
Jenner, Edward, 9, 10*b*, 56
Jews, 3, 6
Jobs and Growth Tax Relief Reconciliation Act, 472
Joint Commission for Accreditation of Healthcare
  Organizations (JCAHO), 386
Joint Commission on Hospital Accreditation
  (JCHA), 386
Judaism, 3
The Jungle (Sinclair), 347*b*

K

KABP. *See* Knowledge, attitudes, beliefs and
  practices
Kaiser Permanente, 49
Kaiser Permanente health plans, 472, 473
Kazakhstan, TB in, 159*b*
Knowledge, attitudes, beliefs and practices
  (KABP), 109
Knowledge management, 444
Koch, Robert, 14, 16, 17*b*
  on tubercle bacillus, 16, 17, 17*b*
Koch-Henle Postulates, 17, 17*b*, 100*b*
Koop, C. Everett, 387
Kwashiorkor, 304, 305*t*

L

Labor law reforms, 11
Lactation, 314–315
LaLonde Heath Field Concepts, 65
LaLonde, Marc, 58, 65, 65*t*, 479
Land mines, 354
Landfills, 342, 350
Lassa fever, 160
Lateral and vertical integration, 456
Latin America and the Caribbean, 512–514,
  512–514, 552, 552
Laws
  federal country, structure of, 589*b*
  public health, 565–566, 588–591
  rights, individual/community, issues with,
    593–595, 594*t*
Lead, toxic effects of, 343, 344–346, 345*b*, 346*b*,
  348, 349, 364
League of Nations, 359
Learning disabilities, 269
Legge axioms, 359, 359*t*
Legionnaires' disease, 165, 350
Leishmaniasis, 163–164
Leprosy, 5, 6, 8, 165
Lesbians, 276–277
Licensure, 578
Life. *See also specific countries*
  expectancy, 76–77, 77*f*, 535
  Implicit Social Value of, 294, 420*b*
  tables, 10
  value of human, 304–306, 420*b*
Lind, James, 9, 9*b*
London epidemiological society, 10
London "killer fog", 347
Longevity, 75
Los Angeles county hospitals, 398, 398*b*
Love Canal, 350, 351
Lung disease(s), 196–197. *See also* Chronic lung
  disease
Lyme disease, 160, 160*b*

M

Maimonides, Moses, 4
Malaria, 156–157
  cases of, 156–157
  parasites associated with, 156
  prevention/elimination of, 335, 457, 496
Malnutrition, 616–617
Malpractice insurance, 550
Malpractice lawsuits, 477
Managed care, 585. *See also* Health maintenance
  organizations
  organization models, 427–428, 427*b*
  organizations, 445
  programs, 456
  in U.S., 472, 473, 474*t*, 477, 545
Management
  community participation in, 456
  cybernetics and, 444–445, 444*f*
  disaster, 619
  health, education for public health and,
    554–560
  human relations, 445–448
  of human resources, health policy and,
    560–561
  knowledge, 444
  motivation and, 445, 447–448, 461
  national and, responsibilities for occupational
    health, 361–362, 362*t*
  skills for, 454–455, 455*f*, 461
  strategic, of health systems, 46, 452*b*
  target-oriented, 445
  total quality, 449–450, 450*b*
  toxic waste, 351
  traditional/standard, 449, 450*b*
Management by objectives (MBO), 445
Mandatory health insurance (MHI), 496
Manhattan Project, 445
Man-made disasters, war, and terrorism, 354
Marburg disease, 160
March of Dimes, 380, 381*b*
Marine Hospital Service, 387
Market economy theory, 525
Maslow's hierarchy of needs, 446–447, 447*f*
Massachusetts Sanitary Commission, 12
Master of public health (MPH), 555, 558
Maternal and Child Health, 20
Maternal health, 19–20
  family health and, 219–223, 220*t*, 221*t*,
    222*t*, 223*t*
Maternal Mortality Ratio (MMR), 614
Matrix model, 454, 455*f*
MBO. *See* Management by objectives
MDG. *See* Millennium Development Goals
Measles, 6
  acute, 137–138
  eradication of, 449
  on Faroe Islands, 15*b*
  German, 139, 139*b*
Measurement(s)
  economics of health and cost, 415
  in epidemiology, 89–91, 90*b*, 91*f*
  of morbidity, 84–85, 84*b*
    qualitative, 88, 89*t*
  of mortality, 85–87, 86*b*, 110*t*
    qualitative, 88, 89*t*
  potential errors in, 94–95
  in public health, 89–91, 90*b*, 91*f*
Medicaid, 394, 423, 468, 471–472,
  476, 477

Medical care
    microeconomics of, 426–427
    public health and, 58–59
Medical ecology, 46
Medical education, 537
    accreditation of facilities, 547–548
    basic, 541–543
    continuing, 536, 539, 547
    in-service, 547
    international conferences on, 543
    issues for the1990s, 543–544, 543t
    nursing, 545–547
    postgraduate, 543–544, 544b
    preventive medicine, 545
    reform in, 543
    supply and demand and, 539–541, 540t
Medical practice, 382–383, 424–425
Medical schools, 5, 7, 541–543, 547–548, 559
Medicare, 394, 423
    in Canada, 478, 479, 480, 481
    in U.S., 468, 471–472, 477
Medicine
    alternative, 553–554
    causal relationships in, 100b, 101b, 100
    clinical, 8
    community, 45
    Greek, 3
    military, 21–23, 282–283, 283t
        Roman, 3
    Muslim, 4
    preventive, 45
    social, 45
        community health and, 45–46
    traditional, 2, 5
Men, health of, 248–249, 249t
Meningococcal meningitis, 143–144
Mental disabilities, 271–272
Mental disorder syndromes, 265–269
    adolescence/childhood behavioral and
        emotional disorders, 268–269
    anxiety as, 267
    behavioral syndromes with psychological
        disturbances, 268
    disorders of psychological development as, 268
    dissociative disorders as, 267, 268
    learning disabilities as, 269
    mood disorders as, 267
    neurotic disorders as, 267
    organic, 265–266
    personality disorders as, 268
    PSTDs as, 267–268
    schizophrenia as, 266–267
    substance abuse as, 266
    suicide as, 269
Mental health, 260–271, 261t, 262b
    controversies in, 269–270
    epidemiology of, 262b, 263–265, 265b
    health promotion and, 271
    history of methods and treatments for,
        262–263, 263t
    in Israel, 504
    prevention of, 271
Mercury, 351
Mergers, 456
Mesothelioma, deaths from, 94t
Methane gas, 341
Methyl Tertiary Butyl Ether (MTBE), 348–349
MHI. See Mandatory health insurance
Miasma theories, 14, 24b

Microbiology, 18–19, 19–20
Microorganism(s)
    as causes of diseases, 17, 17b
    media characteristics of, 16
    Pasteur on, 16
    vector-borne, 17
Microscopy, 7
Migrant population, health needs of, 280
Millennium Development Goals (MDG), 67–68,
        68b, 406, 444, 559, 610, 610b, 613t
    on environmental health, 333, 337
    health targets of, 67, 68b
        adoption of, 59
    on mortality, 185
    target for 8th, 622b
    United Nations and, 441, 445, 526
Minerals, 296, 298t
Minimata disease, 351
Mintzberg, Henry, 455
MMR. See Maternal Mortality Ratio
MMWR. See Morbidity and Mortality Weekly
    Report
Mood disorders, 267
Morbidity
    burden of occupational mortality and, 359–360
    maternal, 221–223, 222t
    measurement of, 84–85, 84b
        qualitative, 88, 89t
Morbidity and Mortality Weekly Report (MMWR),
    107, 474
Morphology, 16
Morris, Jeremy, 185
Mortality. *See also* Standardized mortality rates
    analysis of, 7
    burden of occupational morbidity and, 359–360
    from CVDs, 189–192, 190f, 191f
    declining, 609f
    fetal, 236–237, 237b, 237t
    global, 611t
    infant, 236–237, 237b, 237t, 611t, 613t, 614t
    living conditions and, 7
    maternal, 221–223, 222t
    MDGs, 185
    measurement of, 85–87, 86b, 110t
        qualitative, 88, 89t
    poverty, illness and, 523
    rates, 86b, 93t
    ratios, 86b
Mortality rates. *See specific countries*
Motherhood initiatives, safe, 227
Motivation, 445, 447–448, 461
The Motivation to Work (Herzberg), 448
Motor vehicle injuries, 209, 209b, 210–211, 210t,
    211f
MPH. See Master of public health
MS. See Multiple sclerosis
MTBE. See Methyl Tertiary Butyl Ether
Multiple sclerosis (MS), 206–207
Mumps, 138, 138f
Municipalities, 11, 63, 64b, 64t
Musculoskeletal disorders, 205
Muslim medicine, 4

N

National Academy of Science, 300
National Center for Health Statistics (NCHS),
    474, 476
National Center for Public Health Informatics
    (NCPHI), 474

National Comorbidity Survey, 263
National drug formularies (NDF), 568
National Economic Empowerment and
    Development Strategy (NEEDS), 507
National Emergency Medical Service, 469, 482
National Environmental Policy Act (NEPA), 352
National health insurance (NHI), 466, 469
    Bismarck on, 440, 467, 468
    in Israel, 503
    in Japan, 519
National Health Insurance Act, 468
National Health Service (NHS), 440, 456, 469,
    482–484
    reforms in, 482–484, 484–485
    structure of, 485f
    "values" of, 484
National health systems, 440–441. *See also*
    specific countries
    Canadian, 574
    comparing, 520–523
    economic issues in, 521–523
    introduction to, 466–467
    Israel, 575t
    key elements of, 466b
    reforming, 523–525
    stakeholders in, 520b, 520
    summary on, 508–509
    technology assessment in, 573–574
    typology of financing and administration of,
        521, 521t
    United Kingdom, 586t
National immunization days (NIDs), 516
National Institute for Clinical Excellence
    (NICE), 485
National Institute for Health and Clinical
    Excellence (NICE), 584
National Institute for the Surveillance of Drugs
    and Food (INVIMA), 514
National Institute of Occupational Safety and
    Health (NIOSH), 360, 362, 364
National Institutes of Health (NIH), 385, 385b, 474
    Alternative Medicine office in, 554
    establishment of, 470
National Mental Health Act (1946), 262
National Traffic and Motor Safety Act (1966), 57
Native peoples, health needs of, 277–278
*Natural and Political Observations Upon the Bills
    of Mortality*, 7
NCDIs. See Non Communicable Diseases and
    Injuries
NCHS. See National Center for Health Statistics
NCPHI. See National Center for Public Health
    Informatics
NDF. See National drug formularies
NEEDS. See National Economic Empowerment
    and Development Strategy
Negligence, 577–578
NEPA. See National Environmental Policy Act
Net present value, 418, 418b
The Netherlands, 494–495
Network organization, 448–449, 449f
Neural tube defects (NTDs), 231–232, 232b,
    309, 309b
Neurologic disorders, 206–207
    Alzheimer's disease as, 206
    epilepsies/seizures as, 207
    MS as, 206–207
    Parkinson's disease as, 206
    TBI and spinal cord injuries as, 207

Neurotic disorders, 267
New Orleans, 382
New Perspectives on the Health of Canadians, 479
New Public Health (NPH), 33
    applications of, 70*b*
    chronic conditions and, 213
    communicable disease control in, 175
    concepts of, 33–35, 35*b*
    diseases, health and, 36–38, 37*f*
    economics and, 432–434
    evolution of, 35–36, 36*b*, 68, 69*t*
    health systems and, 460
    hospitals in, 394*f*, 395–399
        beds, supply of, 396–398
        changing roles of, 398–399, 398*b*, 399*f*
        classifications of, 396, 397*b*
        regulation of, 399
    mission of, 35*b*
    nutrition and, 326–327, 327*f*
    occupation and, 369
    origins and synthesis of, 69*t*
    summary on, 68–70, 70*b*
    universal health coverage and, 393–395, 394*f*
New York State
    agencies, 389, 390*b*, 391*b*
    Department of Health, 386
Newborn(s)
    care of, 227–229, 228*t*, 229*b*, 231*b*
    hemorrhagic disease of, 184*b*, 308
    rhesus hemolytic disease of, 231
NGO. *See* Nongovernmental organizations
NHI. *See* National health insurance
NHS. *See* National Health Service
Niacin deficiency, 308
NICE. *See* National Institute for Clinical
        Excellence; National Institute for Health
        and Clinical Excellence
NIDs. *See* National immunization days
Nigeria, 507–511
    cancer in, 511
    chronic diseases in, 509
    communicable diseases in, 509
    FGM in, 508–509
    health professionals in, 510, 510*t*, 511
    health reform in, 511
    health system in, 509–511
    HIV/AIDS in, 511
    medical schools in, 559
    summary on, 511–512
    violence in, 511
    vital statistics indicators for, 507, 507*t*
Nightingale, Florence, 10, 15, 103
NIH. *See* National Institutes of Health
9/11 terrorist attacks, 382
NIOSH. *See* National Institute of Occupational
        Safety and Health
Non Communicable Diseases and Injuries
        (NCDIs), 502
Nongovernmental organizations (NGO), 624–625
Non-system, 478
Nordic countries, 486–491. *See also* Denmark;
        Finland; Norway; Sweden
    acute care beds in, 488*t*
    health expenditures in, 486, 487, 487*t*, 488*t*
    life expectancy in, 486, 487*f*
    mortality rates in, 487
Normal distribution, 92, 92*f*, 93*b*
Norms, performance indicators and, 457
North Karelia project, 490

Norway, 489
NPH. *See* New Public Health
NTDs. *See* Neural tube defects
Nuclear accidents, 352, 353, 353*b*
Null hypothesis, 90–91
Nurses
    education, 545–547
    practitioners, 550–551
    ratio to population/shortage, 538, 538*t*, 541,
        546, 547*t*
    schools and graduates, 545, 546, 546*t*
Nutrition, 2, 111, 474
    in Canada, 479
    CVDs and, 313
    diseases of over, 312–314, 312–313, 313,
        313–314, 314*t*
    evaluation and monitoring of, 320–322, 320*t*
    in global context, 292–293, 293*t*
    health providers and, 319–320
    infections and, 293–294
    introduction to, 291–292
    issues, 318, 319*t*
    NPH and, 327*f*, 326–327
    policies
        food and, 318*b*, 317–320
        U.S., 318*b*, 317–320
    private sector/NGOs' role in, 319
    proper, 21
    proteins as, 295
    in public health, 21*b*, 20–21
        development of, 292*f*, 292
    summary on, 327–328
    under, disorders of, 307*b*, 309*b*, 300–312,
        303–304, 304, 304–306, 306–308, 308,
        308–309, 309, 310, 310–311, 311,
        311–312, 312, 303*t*, 305*t*
    vitamins as, 296, 297*t*
Nutritional requirement(s)
    carbohydrates as, 295
    fats and oils as, 295–296
    human, 294–296, 295*t*
    in lactation and pregnancies, 314–315
    minerals and trace minerals as, 296, 298*t*

O

Obesity, 312–313
Occupation, 111
Occupational health, 334. *See also* Toxins
    burden of occupational morbidity and
        mortality, 359–360
    in clinical practice, 367–368
    deaths and, 359, 360, 361*t*, 365, 367
    development of, 358–359
    health of workers, 359
    international issues in, 360, 361*b*
    introduction to, 356–358
    national and management responsibilities for,
        361–362, 362*t*
    principal tasks of, 362*b*
    risk assessment, 368
    standards and monitoring for, 362
    summary on, 369–370
    in U.S., 359–360, 361*t*
    workplace violence, 366–367
Occupational Safety and Health Act, 362
Occupational Safety and Health Administration
        (OSHA), 352, 362, 364, 366
ODA. *See* Official development assistance
Odds ratio, 83

OECD. *See* Organization for Economic
        Cooperation and Development;
        Organization of Economic Cooperation
        and Development
Office of Minority Health (OMH), 477
Official development assistance (ODA), 620
Oils, 295–296
Onchocerciasis, 162
Open Society Institute (Soros Foundation),
        459, 557
Operations research, 445
Oral cancer, 274
Oral health, 272–274
Oral rehydration therapy (ORT), 506
    adaptation of, 173
    use of, 49, 66, 170–174, 173*b*
Organization(s). *See also* Health system
        organization models
    bureaucratic pyramidal, 442–443, 443*f*, 446
    CEO of, 455–456
    elements of, 442
    as energy systems, 443–445, 443*f*
    environmental health, 355–356
    formal systems, 442
    managed care, 445
    natural systems, 442
    network, 448–449, 449*f*
    nongovernmental, 624–625
    objectives of, 451
    open systems, 442
Organization for Economic Cooperation and
        Development (OECD), 112, 351, 469
Organization for international health, 621
Organizational theory, 446
ORT. *See* Oral rehydration therapy
OSHA. *See* Occupational Safety and Health
        Administration
Osteomalacia, 306–308, 307*b*
Osteoporosis, 205, 311
Ottawa Charter for Health Promotion, 41, 62–63
Our Seamen (Plimsoll), 55–58, 55*b*
Overnutrition, 312–314, 314*t*
Overweight, 312–313

P

Pakistan, 382
Palestine, 502
Panama Canal Zone, 17, 18*b*
Panum, Peter, 14, 15*b*
Parasites, associated with malaria, 156
Parasitic disease(s), 161–165, 161*b*
Parasitic gastroenteritis, 172
Parkinson's disease, 206
PAs. *See* Physician assistants
Pasteur, Louis, 14, 16
Pasteurization, 16
Patient care, ethics in, 596
Patients' rights
    in U.K., 550
    in U.S., 474, 477
PCGs. *See* Primary care groups
PCTs. *See* Primary care trusts
Peer review, 580–581
Pellagra, 21*b*, 308
PEM. *See* Protein-energy malnutrition
Per diem, 473
Performance indicators, norms and, 457
Performance indicators (PI), 585
Periodontal diseases, 273

Personality disorders, 268
Pertussis, 135
Pesticides and herbicides, 346, 366
PGP. *See* Prepaid group practice
Pharmacies, development of, 7
Phenylketonuria (PKU), 193*b*, 200–204
Philanthropy, 457–458
PHS. *See* Public Health Service
PHWD. *See* Public Health Workforce
    Development
Physical activities, 245
Physical disabilities, 274–276, 284*t*
Physician(s)
    hospital privileges for, 550
    medical, 112
    as monks, 4
    primary care, 544*t*
    ratio to population, 537, 538, 538*t*
    Russian, 7
    services, payment for, 426
    supply and demand and, 539–541
Physician assistants (PAs), 551–552
PI. *See* Performance indicators
Pinworm disease (enterobiasis), 164
PKU. *See* Phenylketonuria
Plague(s), 6, 335
Plimsoll Line, 55*b*
Plimsoll, Samuel, 55–58, 55*b*
Pneumococcal diseases, 143
Point of Service Plan (POS), 427*b*
Poison gas, 354
Poliomyelitis, 18–19, 19*b*, 52
    control/eradication of, 148–149, 445, 509, 516
    growth of, 136–137
Pollution, 334. *See also* Acid rain
    air, 346–349
    indoor, 349–350
    water, 340, 340*b*, 347, 349, 351
Polyclinics, 497
Population(s)
    confounders and, 91, 91*f*
    definition of, 109
    demographic transition of, 43–44
    epidemiologic transition of, 43–44
    growth, 614–615, 615*t*
    health needs
        of homeless, 280–281
        of migrant, 280
        of native, 277–278
        of prisoners, 278–279, 279*t*
        of refugees, 281–282
    health of
        assessment of, 108–111, 110*t*, 111–116,
            113*b*, 115*b*
        demography of, 74–76, 75*t*
        epidemiology and, 77–80, 78*b*, 93*t*
        evaluation and measurement, 73–74
        fertility and, 75, 75*b*, 76*b*
        life expectancy and, 76–77, 77*f*
        sentinel events in, 87–88
        social epidemiology and, 80–81, 81*b*, 93*t*
        summary on, 116–117, 116*b*
    normal distribution of, 92, 92*f*, 93*b*
    null hypothesis and, 90–91
    pyramid, 76, 76*f*
    randomization, 91
        simple, 91
        stratified, 91
        systemic, 91

risk approach to, 50
sampling, 91, 92*b*
    cluster, 91
    multi-step, 91
socioeconomic status of, 109–110, 110*t*
standard reference, 320–322, 321*t*, 322*t*
standardization of rates and, 92–94
    direct method of, 93, 93*t*
    indirect method of, 93–94, 94*b*, 94*t*
variables of, 90
POS. *See* Point of Service Plan
Post-traumatic stress disorder (PTSD), 267–268, 355
Poverty, 523, 610–613
Poverty-population-environment (PPE), 506
PPE. *See* Poverty-population-environment
PPO. *See* Preferred Provider Organization
Preferred Provider Organization (PPO), 427*b*, 473
Pregnancies
    care during, 223–226
    delivery, labor during, 226–230
    high-risk, 224–226, 225*t*
    nutritional requirements in lactation and,
        314–315
    spacing of, 221
    teenage, 221
Prepaid group practice (PGP), 472
Prescription drugs, 472, 477
Prevention
    of cancer, 203–204, 204*t*
    cancer of cervix, 573*b*
    of CVDs, 192–195, 193*b*, 193*f*, 195*b*
    of diabetes mellitus, 199–200, 199*b*
    of diseases
        modes, 41–43, 41*b*
        primary, 42
        secondary, 43
        tertiary, 43
    Hawthorne effect and, 446
    health promotion and, 441
    of hypertension, secondary, 193*b*
    of mental health, 271
    modes of, 41–43, 41*b*
    performance indicators and, 457
    preventive medicine, training in, 545
    of social diseases, 496
    technology assessment in, 572–573
    of violence, 212–213
Primary care
    definition of, 47
    selective, 49–50
Primary care groups (PCGs), 485
Primary care trusts (PCTs), 485–486
Primum inter pares (first among equals), 455
Prisoners, health needs of, 278–279, 279*t*
Privatization, deregulation and, 441
Process, 565
Productivity, Hawthorne effect and, 446
Projects, new, and their evaluation, 57–58
    program evaluation information needs, 459*b*
    project proposal summary, 460*b*
Promatoras (Community Health Workers), 514, 552
Proportion, 83
Protein-energy malnutrition (PEM), 303–304
Proteins, 295
Psychological development, disorders of, 268
PTSD. *See* Post-traumatic stress disorder
Public health
    accreditation of, 558
    action, 10

administration of, 584–585
ancient world and, 2–4
capacity, expanding national, 630–631
case for action in, 50–52, 51*t*
competencies for providing essential services,
    557*b*
concepts of, 33–35, 35*b*
    introduction to, 33
core and cross disciplines in, 556*b*
corruption of, 46
cost-effective interventions in, 569*t*
definition of, 44–47, 45*f*
during early medieval period, 4
education in, 554–560, 559*t*
eighteenth-century reforms in, 8
elements of training, 556*b*
during Enlightenment, 7–9, 9–10
ethics/law in, 565–566, 588–591, 592*b*,
    595–596, 596–597, 597*b*, 599*b*
future of, 67–68, 68*b*
global health as new, 631–632
governments' approach to, 11–14, 566
graduates' pledge of, 633*b*
historical markers, 28–31
history of, 1, 28–31
innovation in, 566–568, 567*t*
integrative approaches to, 66–67, 66*b*
introduction to, 1
during late medieval period, 4–6
measurement in, 89–91, 90*b*, 91*f*
medical care and, 58–59
new projects and their evaluation,
    459, 460
nutrition in, 20–21, 21*b*
    development of, 292, 292*f*
organization of, 584–585
prehistoric societies and, 1–2
quality in, 565
religious beliefs/practices and, 2
during Renaissance, 6–7
slippery slope in, 597*b*
social reforms, sanitary movement and, 11–14
standards, definition of, 65–66, 66*t*
summary on, 26–28
technology in, 566
during twentieth century, 25, 25*b*
Public Health Service (PHS), 107, 387, 393,
    474, 476
Public health system(s)
    accreditation and quality regulation of, 386
    disasters and preparedness of, 381–382,
        381*b*, 382*b*
    functions of, 378–380, 379*b*
        direct/indirect methods of providing or
            assuring services as, 379–380
        regulatory, 379, 380*b*
    government, nation's health and, 376–378
        checks and balances in, 377–378, 378*f*
        federal, unitary states and, 377
        individual and, 378
    incentives and regulations within, 383–386,
        383*b*, 384*b*
    introduction to, 375–376, 376*b*
    local, 389–392, 391*b*
    medical practice and, 382–383
    monitoring, 392–393
    national government, 386–388, 387*b*, 388*f*
    national health targets of, 393, 393*b*, 394*b*
    nongovernmental roles in, 380, 381*b*

Public health system(s) *(Continued)*
  promotion of research and teaching, 385–386,
    385*b*
  state government, 388–389, 389*b*, 390*b*, 391*b*
  summary on, 400–401
  uninsured and, 399–400
Public Health Workforce Development (PHWD),
  557
Public interest, 586–587
Puerperium, care of, 229–230
Pyramidal organizations. *See* Bureaucratic
  pyramidal organizations

Q

QALYs. *See* Quality adjusted life years
QMAS. *See* Quality Management and Analysis
  System
QR. *See* Qualitative research
Qualitative research (QR), 115*b*, 114–116
Quality
  consumerism and, 585–586
  input in, 565
  organizations promoting health, 589*b*
  process in, 565
Quality adjusted life years (QALYs), 88, 90*b*, 113,
  415, 526
Quality assurance (QA), 549, 577–584
Quality control, 566–568
Quality Management and Analysis System
  (QMAS), 486
Quality of care, with health care, 113–114
Quarantine, 6

R

RA. *See* Rheumatoid arthritis
Rabies, 16, 155
Race, 476, 477
Radiation, 351–352
  atomic, 352
  cancer and, 349, 352
  ionizing, 351–352
  low-dose, food and, 352
  non-ionizing, 352
Radioactive wastes, 350
Radiologic waste, 336
Radon gas, 349
Ramazzini, 7, 8, 358, 366*b*
Randomization, 91
  simple, 91
  stratified, 91
  systemic, 91
Rates, 83
RAWP. *See* Regional Allocation Working Party
RDAs. *See* Recommended daily allowances
Recommended daily allowances (RDAs), 299–300
  definitions, 300*b*
  development of, 299
  values, 300
Recycling, 341, 342, 343
Reed, Walter, 17, 18*b*
Refugees, health needs of, 281–282
Regional Allocation Working Party (RAWP), 428
Regional health authorities (RHAs), 480, 482–484
Regression, 104
Rehabilitation, 274–276, 284*t*
Relative risk (RR), 82, 83
Reliability, 94–95
Religious beliefs/practices, 2
Religious ceremonies and burials, 5

Renaissance
  epidemics during, 6, 7
  plagues during, 6
  public health during, 6–7
Report on Health Promotion and Disease
  Prevention (Healthy People), 59
Report on Primary Care, 47
Report on the Sanitary Conditions of the
  Labouring Population of Great Britain,
  10, 12
Research
  of health systems, 51
  medical education and, 541
  methods, 89–90, 90*b*
  public health systems' promotion of teaching
    and, 385–386, 385*b*
  qualitative, 114–116, 115*b*
  quantitative, 114–116, 115*b*
RFV. *See* Rift Valley fever
RHAs. *See* Regional health authorities
Rhesus hemolytic disease of newborn, 231
Rheumatoid arthritis (RA), 206
Rickets, 21, 306–308, 307*b*
Rickettsial infections, 157
Rift Valley fever (RFV), 158–159
Rights, individual/community, 593–595, 594*t*
Risk group, 83
Roe v. Wade, 590
Roemer's law, 421*b*, 522
Roentgen, Wilhelm, 351
Rome, 3
Rotaviruses, 171–172
RR. *See* Relative risk
Rubella (German measles), 139, 139*b*
Rudolph's Pediatrics. See Diseases of Infancy
  and Childhood (Holt)
Russia, 7, 495–502. *See also* Soviet Union
  acute care beds in, 497, 498*f*
  epidemiologic transition in, 497–499
  feldshers in, 550, 552
  future prospects for, 501
  health reform in, 499–500, 501
  HIV prevention in, 459
  human resources in, 499, 500*t*
  life expectancy in, 495, 497, 498*t*, 499*f*
  mortality rates in, 498, 499*f*
  physicians in, 500*t*
  public health education in, 556
  during Renaissance, 7
  summary on, 501–502

S

St. Vincent Declaration, 200
Salmonella, 155, 170
Salt, 20, 49, 56. *See also* Iodine deficiency
  disorders
Sample size, 92*b*
Sampling, 91, 92*b*
  cluster, 91
  multi-step, 91
Sanctity of live, 598
Sanitary movement
  communicable disease control and, 335
  hygiene and, 11–14
  public health and, 11–14
Sanitary-epidemiologic stations (SESs), 497,
  500
Sardinia, 193*b*
SARS. *See* Severe Acute Respiratory Syndrome

SCHIP. *See* State Children's Health Insurance
  Program
Schistosomiasis, 163
Schizophrenia, 266–267
School lunch programs, 476, 478
Science, 407
Scotland, health targets for, 61, 61*b*
Screening
  for diseases, 95–96, 96*b*
  multiphasic, 95
  technology assessment in, 572
Scurvy, 8, 9, 9*b*
SEEDS. *See* State Economic Empowerment and
  Development Strategies
Seizures/epilepsies, 207
Semashko, Nikolai, 496, 520, 556
Semmelweiss, Ignaz, 15, 17
Sensitivity, 96, 96*b*
September 11, 2001 terrorist attack, 354, 355
SESs. *See* Sanitary-epidemiologic stations
Severe Acute Respiratory Syndrome (SARS), 39,
  335, 382, 481, 517
Sewage collection and treatment, 340–343.
  See also Hazardous wastes
Sexual risk behaviors, 244
Sexually transmitted diseases (STIs),
  166–170, 166*t*
  control of, 167–168, 167*t*
  other, 166–167
  vaccines and, 37
SHAs. *See* Strategic health authorities
Shattuck, Lemuel, 10, 12
Sheppard-Towner Act, 470
SHI. *See* Statutory health insurance
Shigella, 170
Shingles, 143
Sick building syndrome, 350
Sick Funds, 468, 492–493, 493
Sickle-cell disease, 200
Sickness insurance, 491
Silica, 365–366
Silicosis, 197, 365–366, 366*b*
Silver nitrate, 16*b*
SIRs. *See* Standardized incidence rates
Sleeping sickness. *See* African trypanosomiasis
Smallpox, 6
  control/eradication of, 147–148
  vaccination for, 9, 10*b*, 56
SMART (Specific, Measurable, Achievable,
  Relevant, and Time-Based), 458
SMI. *See* Supplementary Medical Insurance
Smith, Adam, 7, 11
Smoking
  epidemiology of, 242–243
  Framingham heart study and, 185, 186*b*
  lung cancer and, 349, 365, 365*t*
  passive, 349
  as risk factor, 185
SMRs. *See* Standardized mortality rates
Snow, John, 13–14, 13*t*
  on cholera, 13–14, 13*t*, 335
Social diseases, prevention of, 496
Social entrepreneurship, 457
Social inequities
  in U.K., 484
  in U.S., 476–477, 478
Social marketing, 457
Social Security Act (SSA), 384, 423, 470
  Title XIX of, 471

Title XVII of, 471
Title XXI of, 471
Social security model, of health insurance, 468, 525
Social security (Bismarckian) system, 459, 493
Societies
Cretan and Minoan, 3
eastern, 2
health and, 39–41
London epidemiological, 10
prehistoric and public health, 1–2
Socioeconomic status (SES), 40, 109–110, 110*t*
Somatoform disorders, 268
Soviet Union (U.S.S.R.), 469, 496–497, 541
Spain, 4
Specialization and family practice, 544–545
Specificity, 96, 96*b*
Spinal cord injury, 207
SSA. *See* Social Security Act
Staining, 16
Standardization of rates, populations and, 92–94
direct method of, 93, 93*t*
indirect method of, 93–94, 94*b*, 94*t*
Standardized incidence rates (SIRs), 93, 93*t*, 94, 94*b*
Standardized mortality rates (SMRs), 93, 93*t*, 94, 94*b*, 94*t*, 483
Standards, setting, 581
State Children's Health Insurance Program (SCHIP), 423, 471
State Economic Empowerment and Development Strategies (SEEDS), 507
Statistic(s), 8, 9–10, 74
Statutory health insurance (SHI), 492–493
Stillborn, 236
Stomatology, 272
Strategic health authorities (SHAs), 485
Strengths, weaknesses, opportunities, and threats (SWOT), 458
Streptococcal diseases, control/eradication of, 153–154
Substance abuse, 266
Suicides, 212, 269
Superfund program, 351
Superintendent, 455
Supplementary Medical Insurance (SMI), 472
Supply
and demand, 539–541
in health economics, 408–412
of hospital beds and NPH, 396–398
hospital costs, utilization and, 430–431, 431*t*, 432*b*, 432*f*
Surgeon General, U.S., 474
Healthy People, 363, 475, 478
reports, 474, 475, 478
Surgery, 8
Survey methods, 89–90, 90*b*
Sutton's Law, 421*b*
Sweden, 487–489, 524
SWOT. *See* Strengths, weaknesses, opportunities, and threats
Syphilis, 6, 7, 166

T

Tapeworms, 162
Target-oriented management, 445
Task-oriented working group, 448–449
Tay-Sachs disease, 205
TB. *See* Tuberculosis

TBI. *See* Traumatic brain injury
Technology
acquisition standards for, 575*t*
assessment, 570–574, 570*b*
certificate of need for, 571
national health systems in, 573–574
prevention/health promotion with, 572–573
diffusion of, 576–577
dissemination of, 574–576
payment systems in, 576
publication in, 576
health-related, 566
appropriate, 568–570
Teratogens, 344, 345*t*
Terrorism, man-made disasters, and war, 354
Tetanus, 135–136
Thailand, 382
Thalassemia, 193*b*, 200–204
Thalidomide, 568
Thatcher, Margaret, 440, 482, 484
Theory *X*-Theory *Y*, 447–448, 447*t*
Therapeutics, 8
Thiamine deficiency (B$_1$), 308–309
Three Mile Island, 353
Tipot Halav (mother and child health stations), 502
Tobacco consumption, 196, 618–619. *See also* Smoking
in Nordic countries, 490, 491*t*
Tolerable upper intake level (UL), 300*b*, 300
Total quality environmental management (TQEM), 351
Total quality management (TQM), 449–450, 450*b*, 587–588
Touch therapy, 554
Towns, 63, 64*t*
health in, 63, 64*t*
Toxic Substances Control Act, 349, 365
Toxic wastes, 336, 337, 342, 350–351
management, 351
Toxicology, 343, 344*b*
Toxins, 343–346. *See also specific toxins*
classification of, 344*t*
fertility affected by, 343–344
in workplace and environment, 363–366
TQEM. *See* Total quality environmental management
TQM. *See* Total quality management
Trace minerals, 296, 298*t*
Tracer conditions, 581
Trachoma, 165–166
Traumas, 208–213
Traumatic brain injury (TBI), 207
Treatise on the Scurvy: An Inquiry on the Nature, Causes and Cure of That Disease (Lind), 8, 9, 9*b*
Trichinosis (pork tapeworm), 342
Trichomoniasis, 167
Trypanosomiasis, 164
Tsunamis, 382
Tuberculosis (TB)
in Africa, 509
control/eradication of, 37, 150–153, 150*t*, 151*b*, 152*f*, 153*b*
DOTS and, 518, 552, 553
human resources planning and, 541
in Kazakhstan, 159*b*
numbers and rates of, 152*f*
Tuskegee experiment, 598, 598*b*

Typhoid Fever: Its Nature, Mode of Transmission and Prevention (Budd), 14
Typhus, 6

U

U.S. National Center for Health Statistics (NCHS), 103, 239
UHDIS. *See* Uniform Hospital Discharge Information System
U.K. *See* United Kingdom
U.K. Mental Health Act (1959), 262
U.K. National Health Service, 40, 49, 54, 383, 384
UL. *See* Tolerable upper intake level
U.N. *See* United Nations
U.N. Children's Education Fund (UNICEF), 47, 48*b*, 222
U.N. Millennium Project, 70
UNICF. *See* United Nations International Children's Fund
Uniform Hospital Discharge Information System (UHDIS), 103, 104, 104*b*
United Kingdom (U.K.), 482–486
GPs in, 482–484, 545, 550
Great Depression, 469
health care provider constraint in, 550
health promotion in, 484, 486
health reforms in, 482, 484–485
health targets for, 61, 61*b*
immunizations in, 457
Legge axioms, 359, 359*t*
licensure and supervision in, 548
London "killer fog", 347
mortality rates in, 484, 526
NHS, 440, 456, 469, 482–484, 484–485, 485*f*
public health education in, 554, 555
social inequities in, 484
summary on, 486
United Nations (U.N.), 169
United Nations International Children's Fund (UNICF), 624
United States (U.S.), 469–478
changing health care environment in, 472–474
data on standard certificates of death, 86*b*
environmental milestones in, 357*b*
federal health initiatives, 469–471
government agencies with environmental responsibilities, 95, 356*t*
government of, 469
Great Depression in, 468, 470
health expenditures in, 466, 471, 471*t*, 474, 477
health information in, 474
health targets for, 59–60, 60*t*, 363, 363*t*, 475–476, 475*t*, 524
life expectancy in, 469, 478
managed care in, 472, 473, 474*t*, 477, 545
Medicaid in, 468, 471–472, 476, 477
medical education trends in, 540, 540*t*
Medicare in, 468, 471–472, 477
mortality rates in, 469, 475, 476
nutrition policies, 317–320, 318*b*
occupational health in, 359–360, 361*t*
population of, 469
presidential election, 2008, 477, 478
reform in, 524, 525
social inequities in, 476–477, 478
summary on, 478
uninsured in, dilemma of, 477
Unsafe at Any Speed (Nader), 57

U.S. *See* United States
U.S. Department of Agriculture (USDA), 20
U.S. Department of Health and Human Services
  (HHS), 56, 46, 387, 388*f*, 445, 477
U.S. Institute of Medicine, 47, 334, 554
U.S. Surgeon General, 185, 387
USDA. *See* U.S. Department of Agriculture
U.S.S.R. *See* Soviet Union
Utilization
  in health economics, 408–412
  hospital costs, supply and, 430–431, 431*t*,
    432*b*, 432*f*
  of services for health care, 112

V

VA. *See* Veteran's Administration
Vaccination(s)
  at birth, 10
  global alliance for, 629*b*
  influenza, 142*b*
  for smallpox, 9, 10*b*, 56
Vaccine(s)
  advent of, 16
  development of, 146–147, 458
  for HPV, 80
  for infectious diseases, 19
  new, 51
  -preventable diseases
    communicable, 37, 142*b*, 138*f*, 131–144,
      133–144, 135, 135–136, 136–137,
      137–138, 138, 139, 139–141, 140,
      140–141, 141, 141–143, 143,
      143–144, 144, 132*t*, 133*t*
    deaths due to, 144, 144*t*
  regulation of, 146
  STIs and, 37
Validity, 95, 96*b*
Variables, 90
Varicella, 143
Variolation, 9, 10, 147
Vector-borne diseases, 156–160, 159*b*, 160*b*, 160
Vector-borne microorganisms, 17
Vertical and lateral integration, 456
Veteran's Administration (VA), 470
Vinyl chloride, 366
Violence, 208–213
  domestic, 211–212
  gang behavior and, 245
  in Nigeria, 511

prevention of, 212–213
Viral gastroenteritis, 171–172
Viral hepatitis, 139–141
Visual disorders, 207–208, 208*b*
Vitamin(s)
  B deficiency, 308–309
  B$_{12}$ deficiency, 309
  C, 20
    deficiency, 308
  D, 20
    deficiency, 306–308, 307*b*
  A deficiency, 304–306
  K, 184*b*
    deficiency, 308
  as nutritional requirement, 296, 297*t*
Volunteerism, philanthropy and, 457–458

W

Wagner-Murray-Dingell Bill, 384
War, 354, 366, 368
Water
  Acts regarding, 347
  chlorination of, 338, 339, 340, 341
  contamination/pollution, 340, 340*b*, 347,
    349, 351
  desalination of, 341
  environmental health and, 333, 334, 335,
    338–340, 339*b*, 340*b*
  evaporated, in air, as water source, 341
  lead in, 346
  protected sites of, 5
  protection/management of, 338, 339, 339*b*
Water and Environmental Pollution Act, 484
Water, protected sites of, 5
Waterborne diseases, 335, 338, 339–340
Weekly Epidemiologic Record, 107
Welfare model, 486
West Nile Virus (WNV), 158, 335
WFME. *See* World Federation for Medical
  Education
WHA. *See* World Health Assembly
WNV. *See* West Nile Virus
Women, 20, 246–248, 247*t*
Working group, task-oriented, 448–449
Workplace
  disasters in, preventing, 368–369, 368*t*
  inspections, 367*t*, 368
  toxins in environment and, 363–366
  violence, 366–367

World Bank, 54, 57, 185, 406, 407*b*, 625
  HFA and, 441
  Russia and, 500, 502
World Bank Group, 406
World Development Report: Investing in Health,
  54, 406
World Federation for Medical Education
  (WFME), 543
World Health Assembly (WHA), 48
World Health Organization (WHO), 142, 621–624
  basic radiologic units promoted by, 569
  on Colombia, 514
  Commission on Inequities, 40
  Commission on Social Determinants of
    Health, 40
  CQI and, 48
  definitions, 47–49
  on environment, 334, 335, 336*t*, 337, 338, 345
  European Region for All Database, 106, 106*f*
  Expanded Program of Immunization, 47
  on FGM, 509
  on health system definition, 466
  health targets of, 149
  HFA and, 441, 445, 504, 506, 524
  immunization activities estimates by, 569
  national drug formularies promoted by, 568
  reports, 535, 536, 557
  2007 agenda for, 609*b*
*World Health Organization Commission on Health
  and Environment Report*, 355
World Summit for Children (WSC), 56, 172
World Summit on Sustainable Development, 334
WSC. *See* World Summit for Children

X

X-rays. *See* Radiation

Y

Years of potential life lost (YPLL), 88, 89*t*, 90*b*
Yellow fever, 17, 18*b*, 159
Yersinia pestis, 5
Yin-yang principle, 515
YPLL. *See* Years of potential life lost

Z

Zoonoses
  control/eradication of, 154–156
  other major, 156